Nottingham Trent University
**BRACKENHURST LIBRARY**
www.ntu.ac.uk/llr
Enquiries: 0115 848 2175

22 AUG 2013

42 0044914 9

# Microbial decontamination in the food industry

© Woodhead Publishing Limited, 2012

## Related titles:

*Case studies in novel food processing technologies*
(ISBN 978-1-84569-551-4)
Novel processing technologies can bring benefits in terms of product quality and process efficiency, but the commercialisation of foods produced using newer technologies often involves many unknowns. The diverse selection of case studies presented in this book record the experiences of those who have been involved in the development and commercialisation of foods produced by novel processing technologies. Case studies in Parts I and II focus on foods processed by high pressure, pulsed electric fields and other novel processing techniques. Subsequent chapters explore case studies in food preservation using novel antimicrobials, packaging and storage techniques. The final section contains case studies in established food processing techniques and microbial models.

*Food preservation by pulsed electric fields*
(ISBN 978-1-84569-058-8)
Pulsed electric field (PEF) food processing is a novel, non-thermal preservation method that has the potential to produce foods with excellent sensory and nutritional quality and shelf life. This important book reviews the technology, from research into product safety and technology development to issues associated with its commercial implementation.

*Emerging food packaging technologies*
(ISBN 978-1-84569-809-6)
Packaging technologies remain a key area of interest for the food industry. When they are employed successfully, product safety and shelf life can be greatly improved. Four themes in particular are driving packaging innovation at the present time: development and implementation of active packaging techniques, development and implementation of intelligent packaging techniques, advances in packaging materials and concerns about packaging sustainability. Each part of this book focuses on one of these themes, reviewing key developments in the area.

Details of these books and a complete list of Woodhead's titles can be obtained by:

- visiting our web site at www.woodheadpublishing.com
- contacting Customer Services (e-mail: sales@woodheadpublishing.com; fax: +44 (0) 1223 832819; tel.: +44 (0) 1223 499140 ext. 130; address: Woodhead Publishing Limited, 80 High Street, Sawston, Cambridge CB22 3HJ, UK)
- contacting our US office (e-mail: usmarketing@woodheadpublishing.com; tel.: (215) 928 9112; address: Woodhead Publishing, 1518 Walnut Street, Suite 1100, Philadelphia, PA 19102-3406, USA)

If you would like e-versions of our content, please visit our online platform: www.woodheadpublishingonline.com. Please recommend it to your librarian so that everyone in your institution can benefit from the wealth of content on the site.

© Woodhead Publishing Limited, 2012

Woodhead Publishing Series in Food Science, Technology and Nutrition:
Number 234

# Microbial decontamination in the food industry

## Novel methods and applications

Edited by
Ali Demirci and Michael O. Ngadi

Oxford Cambridge Philadelphia New Delhi

© Woodhead Publishing Limited, 2012

Published by Woodhead Publishing Limited,
80 High Street, Sawston, Cambridge CB22 3HJ, UK
www.woodheadpublishing.com
www.woodheadpublishingonline.com

Woodhead Publishing, 1518 Walnut Street, Suite 1100, Philadelphia, PA 19102-3406, USA

Woodhead Publishing India Private Limited, G-2, Vardaan House, 7/28 Ansari Road, Daryaganj, New Delhi – 110002, India
www.woodheadpublishingindia.com

First published 2012, Woodhead Publishing Limited
© Woodhead Publishing Limited, 2012. Chapter 11 was prepared by US government employees; this chapter is therefore in the public domain and cannot be copyrighted.
The authors have asserted their moral rights.

This book contains information obtained from authentic and highly regarded sources. Reprinted material is quoted with permission, and sources are indicated. Reasonable efforts have been made to publish reliable data and information, but the authors and the publishers cannot assume responsibility for the validity of all materials. Neither the authors nor the publishers, nor anyone else associated with this publication, shall be liable for any loss, damage or liability directly or indirectly caused or alleged to be caused by this book.

Neither this book nor any part may be reproduced or transmitted in any form or by any means, electronic or mechanical, including photocopying, microfilming and recording, or by any information storage or retrieval system, without permission in writing from Woodhead Publishing Limited.

The consent of Woodhead Publishing Limited does not extend to copying for general distribution, for promotion, for creating new works, or for resale. Specific permission must be obtained in writing from Woodhead Publishing Limited for such copying.

Trademark notice: Product or corporate names may be trademarks or registered trademarks, and are used only for identification and explanation, without intent to infringe.

British Library Cataloguing in Publication Data
A catalogue record for this book is available from the British Library.

Library of Congress Control Number: 2012938392

ISBN 978-0-85709-085-0 (print)
ISBN 978-0-85709-575-6 (online)
ISSN 2042-8049 Woodhead Publishing Series in Food Science, Technology and Nutrition (print)
ISSN 2042-8057 Woodhead Publishing Series in Food Science, Technology and Nutrition (online)

The publisher's policy is to use permanent paper from mills that operate a sustainable forestry policy, and which has been manufactured from pulp which is processed using acid-free and elemental chlorine-free practices. Furthermore, the publisher ensures that the text paper and cover board used have met acceptable environmental accreditation standards.

Typeset by Replika Press Pvt Ltd, India
Printed by TJ International Ltd, Padstow, Cornwall, UK

© Woodhead Publishing Limited, 2012

# Contents



© Woodhead Publishing Limited, 2012

© Woodhead Publishing Limited, 2012

© Woodhead Publishing Limited, 2012

© Woodhead Publishing Limited, 2012

© Woodhead Publishing Limited, 2012

© Woodhead Publishing Limited, 2012

© Woodhead Publishing Limited, 2012

© Woodhead Publishing Limited, 2012

© Woodhead Publishing Limited, 2012

# Contributor contact details

(* = main contact)

## Editors

Ali Demirci*
Department of Agricultural and Biological Engineering
The Pennsylvania State University
University Park, PA 16802
USA

E-mail: demirci@psu.edu

Michael O. Ngadi
Department of Bioresource Engineering
McGill University
Ste Anne de Bellevue
Quebec, H9X 3V9
Canada

E-mail: michael.ngadi@mcgill.ca

## Chapter 1

Sally C. C. Foong-Cunningham and Katherine M. J. Swanson*
Ecolab Inc.
655 Lone Oak Drive
Eagan, MN 55121
USA

E-mail: Sally.FoongCunningham@ecolab.com; Katie.Swanson@ecolab.com

Edward L. C. Verkaar
Ecolab Deutschland GmbH
Reisholzer Werftstrasse 38-42
L 34 - Room 060
40589 Düsseldorf
Germany

E-mail: Edward.Verkaar@ecolab.com

© Woodhead Publishing Limited, 2012

**Chapter 2**

Alex Gill
Health Canada
Bureau of Microbial Hazards
Sir F. G. Banting Research Centre
251 Sir Frederick Banting Driveway
P. L. 2204E
Ottawa
Ontario, K1A 0K9
Canada

E-mail: alex.gill@hc-sc.gc.ca

Colin O. Gill*
Agriculture and Agri-Food Canada
Lacombe Research Centre
6000 C & E Trail
Lacombe
Alberta, T4L 1W1
Canada

E-mail: Colin.Gill@agr.gc.ca

**Chapter 3**

Claudio Zweifel* and Roger Stephan
Institute for Food Safety and Hygiene
Vetsuisse Faculty
University of Zurich
Winterthurerstrasse 272
8057 Zurich
Switzerland

E-mail: zweifelc@fsafety.uzh.ch; stephanr@fsafety.uzh.ch

**Chapter 4**

T. Skåra*
Nofima
P.O. Box 8034
NO-4068 Stavanger
Norway

E-mail: Torstein.Skara@nofima.no

and

Katholieke Universiteit Leuven
Department of Chemical Engineering
BioTeC – Chemical and Biochemical Process Technology and Control
Leuven
Belgium

C. Leadley
Campden BRI
Station Road
Chipping Campden GL55 6LD
UK

E-mail: c.leadley@campden.co.uk

J. T. Rosnes
Nofima
P.O. Box 8034
NO-4068 Stavanger
Norway

E-mail: Thomas.Rosnes@nofima.no

**Chapter 5**

Griffiths G. Atungulu
Department of Biological and Agricultural Engineering
University of California – Davis
One Shields Avenue
Davis, CA 95616
USA

E-mail: ggatungulu@ucdavis.edu

© Woodhead Publishing Limited, 2012

Zhongli Pan*
Processed Foods Research Unit
Western Region Research Center
USDA-ARS
800 Buchanan Street
Albany, CA 94710
USA

E-mail: zlpan@ucdavis.edu; Zhongli.pan@ars.usda.gov

and

Department of Biological and Agricultural Engineering
University of California – Davis
One Shields Avenue
Davis, CA 95616
USA

## Chapter 6

Michelle D. Danyluk*
Citrus Research and Education Center
Institute of Food and Agricultural Sciences
University of Florida
700 Experiment Station Road
Lake Alfred, FL 33850
USA

E-mail: mddanyluk@ufl.edu

and

Department of Food Science and Human Nutrition
Institute of Food and Agricultural Sciences
University of Florida
P.O. Box 110370
Gainesville, FL 32611
USA

Mickey E. Parish
FDA/CFSAN Office of Food Safety
5100 Paint Branch Parkway
HFS-300
College Park, MD 20740
USA

E-mail: Michey.Parish@fda.hhs.gov

Renee M. Goodrich-Schneider
Department of Food Science and Human Nutrition
Institute of Food and Agricultural Sciences
University of Florida
P.O. Box 110370
Gainesville, FL 32611
USA

E-mail: goodrich@ufl.edu

Randy W. Worobo
Department of Food Science and Technology
Cornell University
Geneva, NY 14456
USA

E-mail: rww8@cornell.edu

## Chapter 7

M. W. Griffiths and M. Walkling-Ribeiro*
Canadian Research Institute for Food Safety
Department of Food Science
University of Guelph
43 McGilvray Street
Guelph
Ontario, N1G 2W1
Canada

E-mail: mgriffit@uoguelph.ca; markuswr@uoguelph.ca

© Woodhead Publishing Limited, 2012

**Chapter 8**

Graham Purnell* and Christian James
Food Refrigeration and Process Engineering Research Centre (FRPERC)
The Grimsby Institute of Further and Higher Education (GIFHE)
HSI Building
Origin Way, Europarc
Grimsby DN37 9TZ
UK

E-mail: purnellg@grimsby.ac.uk; jamesc@grimsby.ac.uk

**Chapter 9**

Satyanarayan R. S. Dev and G. S. Vijaya Raghavan
McGill University
21111, Lakeshore Road
Ste Anne de Bellevue
Quebec H9X 3V9
Canada

E-mail: satyanarayan.dev@gmail.com; vijaya.raghavan@mcgill.ca

Sohan L. Birla* and Jeyamkondan Subbiah
University of Nebraska
207 LW Chase Hall
Lincoln, NE 68583
USA

E-mail: sbirla2@unl.edu; jsubbiah2@unl.edu

**Chapter 10**

Bin Zhou, Hyoungill Lee and Hao Feng*
Department of Food Science and Human Nutrition
University of Illinois at Urbana-Champaign
1304 West Pennsylvania Ave
Urbana, IL 61801
USA

E-mail: haofeng@illinois.edu

**Chapter 11**

Christopher H. Sommers
Food Safety and Intervention Technologies Research Unit, Eastern Regional Research Center
USDA ARS
600 East Mermaid Lane, Wyndmoor, PA 19038
USA

E-mail: Christopher.Sommers@ars.usda.gov

**Chapter 12**

Nene M. Keklik*
Cumhuriyet University
Faculty of Engineering
Department of Food Engineering
58140 Sivas
Turkey

E-mail: meltemkeklik@gmail.com

© Woodhead Publishing Limited, 2012

Kathiravan Krishnamurthy
National Center for Food Safety and Technology
Illinois Institute of Technology
6502 S Archer Road
Bedford Park, IL 60501
USA

E-mail: kkrishn2@iit.edu

Ali Demirci*
Department of Agricultural and Biological Engineering
The Pennsylvania State University
University Park, PA 16802
USA

E-mail: demirci@psu.edu

## Chapter 13

Hossein Daryaei and V. M. (Bala) Balasubramaniam*
Food Safety Engineering Laboratory
Department of Food Science and Technology
The Ohio State University
2015 Fyffe Road
Columbus, OH 43210
USA

E-mail: balasubramaniam.1@osu.edu

## Chapter 14

Malek Amiali
Ecole Nationale Superieure Agronomique
El Harrach
Algiers
Algeria

E-mail: m.amiali@ina.dz; malek.amiali@mail.mcgill.ca

Michael O. Ngadi*
Department of Bioresource Engineering
McGill University
Ste Anne de Bellevue
Quebec H9X 3V9
Canada

E-mail: michael.ngadi@mcgill.ca

## Chapter 15

Raghupathy Ramaswamy*
Heinz Innovation Center
Heinz North America
Warrendale, PA 15086
USA

E-mail: raghu.ramaswamy@us.hjheinz.com

Kathiravan Krishnamurthy
Institute for Food Safety and Health
Illinois Institute of Technology
6502 S Archer Road
Bedford Park, IL 60501
USA

E-mail: kkrishn2@iit.edu

Soojin Jun
Department of Human Nutrition, Food and Animal Sciences
University of Hawaii
1955 East West Road
Honolulu, HI 96822
USA

E-mail: soojin@hawaii.edu

© Woodhead Publishing Limited, 2012

### Chapter 16

Michael G. Kong
School of Electronic, Electrical and Systems Engineering
Loughborough University
Loughborough LE11 3TU
UK

E-mail: M.G.Kong@lboro.ac.uk

### Chapter 17

Amrish S. Chawla, David R. Kasler, Sudhir K. Sastry and Ahmed E. Yousef*
Department of Food Science and Technology
The Ohio State University
2015 Fyffe Court
Columbus, OH 43210
USA

E-mail: yousef.1@osu.edu

### Chapter 18

Valentina Trinetta* and Mark Morgan
NLSN, Hall of Food Science
Purdue University
745 Agricultural Mall Drive
West Lafayette, IN 47906
USA

E-mail: vtrinett@purdue.edu; mmorgan@purdue.edu

Richard Linton
Department of Food Science and Technology
The Ohio State University
110 Parker Food Science Building
2015 Fyffe Court
Columbus, OH 43210
USA

E-mail: Linton.60@osu.edu

### Chapter 19

Kuan-Chen Cheng
#1, Sec 4 Roosvelt Rd
Institute of Food Science and Technology
National Taiwan University
Taipei, 10617 Taiwan

Satyanarayan R. S. Dev, Katherine L. Bialka and Ali Demirci*
Department of Agricultural and Biological Engineering
The Pennsylvania State University
University Park, PA 16802
USA

E-mail: demirci@psu.edu

### Chapter 20

Alexandra Lianou and Konstantinos P. Koutsoumanis
Department of Food Science and Technology
Laboratory of Food Microbiology and Hygiene
School of Agriculture
Aristotle University of Thessaloniki
Thessaloniki 54124
Greece

E-mail: alianou@agro.auth.gr; kkoutsou@agro.auth.gr

© Woodhead Publishing Limited, 2012

John N. Sofos*
Department of Animal Sciences
Center for Meat Safety and Quality and Food Safety Cluster
Colorado State University
Fort Collins, CO 80523-1171
USA

E-mail: john.sofos@colostate.edu

## Chapter 21

Murat O. Balaban*
Department of Chemical and Materials Engineering
University of Auckland
Auckland
New Zealand

E-mail: m.balaban@auckland.ac.nz

Giovanna Ferrentino and Sara Spilimbergo
Department of Materials Engineering and Industrial Technologies
University of Trento
via Mesiano 77
I-38050 Trento
Italy

E-mail: giovanna.ferrentino@ing.unitn.it; sara.spilimbergo@ing.unitn.it

## Chapter 22

Minal Lalpuria, Ramaswamy Anantheswaran and John Floros*
The Pennsylvania State University
206 Food Science Building
University Park, PA 16802
USA

E-mail: jdf10@psu.edu

## Chapter 23

Hudaa Neetoo, Haiqiang Chen and Dallas G. Hoover*
044 Townsend Hall
Department of Animal and Food Sciences
University of Delaware
Newark, DE 19716
USA

E-mail: hudaa@udel.edu; dgh@udel.edu; haiqiang@udel.edu

© Woodhead Publishing Limited, 2012

# Woodhead Publishing Series in Food Science, Technology and Nutrition

1 **Chilled foods: a comprehensive guide** *Edited by C. Dennis and M. Stringer*
2 **Yoghurt: science and technology** *A. Y. Tamime and R. K. Robinson*
3 **Food processing technology: principles and practice** *P. J. Fellows*
4 **Bender's dictionary of nutrition and food technology Sixth edition** *D. A. Bender*
5 **Determination of veterinary residues in food** *Edited by N. T. Crosby*
6 **Food contaminants: sources and surveillance** *Edited by C. Creaser and R. Purchase*
7 **Nitrates and nitrites in food and water** *Edited by M. J. Hill*
8 **Pesticide chemistry and bioscience: the food-environment challenge** *Edited by G. T. Brooks and T. Roberts*
9 **Pesticides: developments, impacts and controls** *Edited by G. A. Best and A. D. Ruthven*
10 **Dietary fibre: chemical and biological aspects** *Edited by D. A. T. Southgate, K. W. Waldron, I. T. Johnson and G. R. Fenwick*
11 **Vitamins and minerals in health and nutrition** *M. Tolonen*
12 **Technology of biscuits, crackers and cookies Second edition** *D. Manley*
13 **Instrumentation and sensors for the food industry** *Edited by E. Kress-Rogers*
14 **Food and cancer prevention: chemical and biological aspects** *Edited by K. W. Waldron, I. T. Johnson and G. R. Fenwick*
15 **Food colloids: proteins, lipids and polysaccharides** *Edited by E. Dickinson and B. Bergenstahl*

© Woodhead Publishing Limited, 2012

16 **Food emulsions and foams** *Edited by E. Dickinson*
17 **Maillard reactions in chemistry, food and health** *Edited by T. P. Labuza, V. Monnier, J. Baynes and J. O'Brien*
18 **The Maillard reaction in foods and medicine** *Edited by J. O'Brien, H. E. Nursten, M. J. Crabbe and J. M. Ames*
19 **Encapsulation and controlled release** *Edited by D. R. Karsa and R. A. Stephenson*
20 **Flavours and fragrances** *Edited by A. D. Swift*
21 **Feta and related cheeses** *Edited by A. Y. Tamime and R. K. Robinson*
22 **Biochemistry of milk products** *Edited by A. T. Andrews and J. R. Varley*
23 **Physical properties of foods and food processing systems** *M. J. Lewis*
24 **Food irradiation: a reference guide** *V. M. Wilkinson and G. Gould*
25 **Kent's technology of cereals: an introduction for students of food science and agriculture Fourth edition** *N. L. Kent and A. D. Evers*
26 **Biosensors for food analysis** *Edited by A. O. Scott*
27 **Separation processes in the food and biotechnology industries: principles and applications** *Edited by A. S. Grandison and M. J. Lewis*
28 **Handbook of indices of food quality and authenticity** *R. S. Singhal, P. K. Kulkarni and D. V. Rege*
29 **Principles and practices for the safe processing of foods** *D. A. Shapton and N. F. Shapton*
30 **Biscuit, cookie and cracker manufacturing manuals Volume 1: ingredients** *D. Manley*
31 **Biscuit, cookie and cracker manufacturing manuals Volume 2: biscuit doughs** *D. Manley*
32 **Biscuit, cookie and cracker manufacturing manuals Volume 3: biscuit dough piece forming** *D. Manley*
33 **Biscuit, cookie and cracker manufacturing manuals Volume 4: baking and cooling of biscuits** *D. Manley*
34 **Biscuit, cookie and cracker manufacturing manuals Volume 5: secondary processing in biscuit manufacturing** *D. Manley*
35 **Biscuit, cookie and cracker manufacturing manuals Volume 6: biscuit packaging and storage** *D. Manley*
36 **Practical dehydration Second edition** *M. Greensmith*
37 **Lawrie's meat science Sixth edition** *R. A. Lawrie*
38 **Yoghurt: science and technology Second edition** *A. Y. Tamime and R. K. Robinson*
39 **New ingredients in food processing: biochemistry and agriculture** *G. Linden and D. Lorient*
40 **Benders' dictionary of nutrition and food technology Seventh edition** *D. A. Bender and A. E. Bender*

© Woodhead Publishing Limited, 2012

41 **Technology of biscuits, crackers and cookies Third edition** *D. Manley*
42 **Food processing technology: principles and practice Second edition** *P. J. Fellows*
43 **Managing frozen foods** *Edited by C. J. Kennedy*
44 **Handbook of hydrocolloids** *Edited by G. O. Phillips and P. A. Williams*
45 **Food labelling** *Edited by J. R. Blanchfield*
46 **Cereal biotechnology** *Edited by P. C. Morris and J. H. Bryce*
47 **Food intolerance and the food industry** *Edited by T. Dean*
48 **The stability and shelf-life of food** *Edited by D. Kilcast and P. Subramaniam*
49 **Functional foods: concept to product** *Edited by G. R. Gibson and C. M. Williams*
50 **Chilled foods: a comprehensive guide Second edition** *Edited by M. Stringer and C. Dennis*
51 **HACCP in the meat industry** *Edited by M. Brown*
52 **Biscuit, cracker and cookie recipes for the food industry** *D. Manley*
53 **Cereals processing technology** *Edited by G. Owens*
54 **Baking problems solved** *S. P. Cauvain and L. S. Young*
55 **Thermal technologies in food processing** *Edited by P. Richardson*
56 **Frying: improving quality** *Edited by J. B. Rossell*
57 **Food chemical safety Volume 1: contaminants** *Edited by D. Watson*
58 **Making the most of HACCP: learning from others' experience** *Edited by T. Mayes and S. Mortimore*
59 **Food process modelling** *Edited by L. M. M. Tijskens, M. L. A. T. M. Hertog and B. M. Nicolaï*
60 **EU food law: a practical guide** *Edited by K. Goodburn*
61 **Extrusion cooking: technologies and applications** *Edited by R. Guy*
62 **Auditing in the food industry: from safety and quality to environmental and other audits** *Edited by M. Dillon and C. Griffith*
63 **Handbook of herbs and spices Volume 1** *Edited by K. V. Peter*
64 **Food product development: maximising success** *M. Earle, R. Earle and A. Anderson*
65 **Instrumentation and sensors for the food industry Second edition** *Edited by E. Kress-Rogers and C. J. B. Brimelow*
66 **Food chemical safety Volume 2: additives** *Edited by D. Watson*
67 **Fruit and vegetable biotechnology** *Edited by V. Valpuesta*
68 **Foodborne pathogens: hazards, risk analysis and control** *Edited by C. de W. Blackburn and P. J. McClure*
69 **Meat refrigeration** *S. J. James and C. James*

© Woodhead Publishing Limited, 2012

70 **Lockhart and Wiseman's crop husbandry Eighth edition** *H. J. S. Finch, A. M. Samuel and G. P. F. Lane*
71 **Safety and quality issues in fish processing** *Edited by H. A. Bremner*
72 **Minimal processing technologies in the food industries** *Edited by T. Ohlsson and N. Bengtsson*
73 **Fruit and vegetable processing: improving quality** *Edited by W. Jongen*
74 **The nutrition handbook for food processors** *Edited by C. J. K. Henry and C. Chapman*
75 **Colour in food: improving quality** *Edited by D MacDougall*
76 **Meat processing: improving quality** *Edited by J. P. Kerry, J. F. Kerry and D. A. Ledward*
77 **Microbiological risk assessment in food processing** *Edited by M. Brown and M. Stringer*
78 **Performance functional foods** *Edited by D. Watson*
79 **Functional dairy products Volume 1** *Edited by T. Mattila-Sandholm and M. Saarela*
80 **Taints and off-flavours in foods** *Edited by B. Baigrie*
81 **Yeasts in food** *Edited by T. Boekhout and V. Robert*
82 **Phytochemical functional foods** *Edited by I. T. Johnson and G. Williamson*
83 **Novel food packaging techniques** *Edited by R. Ahvenainen*
84 **Detecting pathogens in food** *Edited by T. A. McMeekin*
85 **Natural antimicrobials for the minimal processing of foods** *Edited by S. Roller*
86 **Texture in food Volume 1: semi-solid foods** *Edited by B. M. McKenna*
87 **Dairy processing: improving quality** *Edited by G. Smit*
88 **Hygiene in food processing: principles and practice** *Edited by H. L. M. Lelieveld, M. A. Mostert, B. White and J. Holah*
89 **Rapid and on-line instrumentation for food quality assurance** *Edited by I. Tothill*
90 **Sausage manufacture: principles and practice** *E. Essien*
91 **Environmentally-friendly food processing** *Edited by B. Mattsson and U. Sonesson*
92 **Bread making: improving quality** *Edited by S. P. Cauvain*
93 **Food preservation techniques** *Edited by P. Zeuthen and L. Bøgh-Sørensen*
94 **Food authenticity and traceability** *Edited by M. Lees*
95 **Analytical methods for food additives** *R. Wood, L. Foster, A. Damant and P. Key*
96 **Handbook of herbs and spices Volume 2** *Edited by K. V. Peter*
97 **Texture in food Volume 2: solid foods** *Edited by D. Kilcast*
98 **Proteins in food processing** *Edited by R. Yada*

© Woodhead Publishing Limited, 2012

99 **Detecting foreign bodies in food** *Edited by M. Edwards*
100 **Understanding and measuring the shelf-life of food** *Edited by R. Steele*
101 **Poultry meat processing and quality** *Edited by G. Mead*
102 **Functional foods, ageing and degenerative disease** *Edited by C. Remacle and B. Reusens*
103 **Mycotoxins in food: detection and control** *Edited by N. Magan and M. Olsen*
104 **Improving the thermal processing of foods** *Edited by P. Richardson*
105 **Pesticide, veterinary and other residues in food** *Edited by D. Watson*
106 **Starch in food: structure, functions and applications** *Edited by A.-C. Eliasson*
107 **Functional foods, cardiovascular disease and diabetes** *Edited by A. Arnoldi*
108 **Brewing: science and practice** *D. E. Briggs, P. A. Brookes, R. Stevens and C. A. Boulton*
109 **Using cereal science and technology for the benefit of consumers: proceedings of the 12th International ICC Cereal and Bread Congress, 24–26 May, 2004, Harrogate, UK** *Edited by S. P. Cauvain, L. S. Young and S. Salmon*
110 **Improving the safety of fresh meat** *Edited by J. Sofos*
111 **Understanding pathogen behaviour: virulence, stress response and resistance** *Edited by M. Griffiths*
112 **The microwave processing of foods** *Edited by H. Schubert and M. Regier*
113 **Food safety control in the poultry industry** *Edited by G. Mead*
114 **Improving the safety of fresh fruit and vegetables** *Edited by W. Jongen*
115 **Food, diet and obesity** *Edited by D. Mela*
116 **Handbook of hygiene control in the food industry** *Edited by H. L. M. Lelieveld, M. A. Mostert and J. Holah*
117 **Detecting allergens in food** *Edited by S. Koppelman and S. Hefle*
118 **Improving the fat content of foods** *Edited by C. Williams and J. Buttriss*
119 **Improving traceability in food processing and distribution** *Edited by I. Smith and A. Furness*
120 **Flavour in food** *Edited by A. Voilley and P. Etievant*
121 **The Chorleywood bread process** *S. P. Cauvain and L. S. Young*
122 **Food spoilage microorganisms** *Edited by C. de W. Blackburn*
123 **Emerging foodborne pathogens** *Edited by Y. Motarjemi and M. Adams*
124 **Benders' dictionary of nutrition and food technology Eighth edition** *D. A. Bender*

© Woodhead Publishing Limited, 2012

125 **Optimising sweet taste in foods** *Edited by W. J. Spillane*
126 **Brewing: new technologies** *Edited by C. Bamforth*
127 **Handbook of herbs and spices Volume 3** *Edited by K. V. Peter*
128 **Lawrie's meat science Seventh edition** *R. A. Lawrie in collaboration with D. A. Ledward*
129 **Modifying lipids for use in food** *Edited by F. Gunstone*
130 **Meat products handbook: practical science and technology** *G. Feiner*
131 **Food consumption and disease risk: consumer-pathogen interactions** *Edited by M. Potter*
132 **Acrylamide and other hazardous compounds in heat-treated foods** *Edited by K. Skog and J. Alexander*
133 **Managing allergens in food** *Edited by C. Mills, H. Wichers and K. Hoffman-Sommergruber*
134 **Microbiological analysis of red meat, poultry and eggs** *Edited by G. Mead*
135 **Maximising the value of marine by-products** *Edited by F. Shahidi*
136 **Chemical migration and food contact materials** *Edited by K. Barnes, R. Sinclair and D. Watson*
137 **Understanding consumers of food products** *Edited by L. Frewer and H. van Trijp*
138 **Reducing salt in foods: practical strategies** *Edited by D. Kilcast and F. Angus*
139 **Modelling microorganisms in food** *Edited by S. Brul, S. Van Gerwen and M. Zwietering*
140 **Tamime and Robinson's Yoghurt: science and technology Third edition** *A. Y. Tamime and R. K. Robinson*
141 **Handbook of waste management and co-product recovery in food processing Volume 1** *Edited by K. W. Waldron*
142 **Improving the flavour of cheese** *Edited by B. Weimer*
143 **Novel food ingredients for weight control** *Edited by C. J. K. Henry*
144 **Consumer-led food product development** *Edited by H. MacFie*
145 **Functional dairy products Volume 2** *Edited by M. Saarela*
146 **Modifying flavour in food** *Edited by A. J. Taylor and J. Hort*
147 **Cheese problems solved** *Edited by P. L. H. McSweeney*
148 **Handbook of organic food safety and quality** *Edited by J. Cooper, C. Leifert and U. Niggli*
149 **Understanding and controlling the microstructure of complex foods** *Edited by D. J. McClements*
150 **Novel enzyme technology for food applications** *Edited by R. Rastall*
151 **Food preservation by pulsed electric fields: from research to application** *Edited by H. L. M. Lelieveld and S. W. H. de Haan*
152 **Technology of functional cereal products** *Edited by B. R. Hamaker*

© Woodhead Publishing Limited, 2012

153 **Case studies in food product development** *Edited by M. Earle and R. Earle*
154 **Delivery and controlled release of bioactives in foods and nutraceuticals** *Edited by N. Garti*
155 **Fruit and vegetable flavour: recent advances and future prospects** *Edited by B. Brückner and S. G. Wyllie*
156 **Food fortification and supplementation: technological, safety and regulatory aspects** *Edited by P. Berry Ottaway*
157 **Improving the health-promoting properties of fruit and vegetable products** *Edited by F. A. Tomás-Barberán and M. I. Gil*
158 **Improving seafood products for the consumer** *Edited by T. Børresen*
159 **In-pack processed foods: improving quality** *Edited by P. Richardson*
160 **Handbook of water and energy management in food processing** *Edited by J. Klemeš, R. Smith and J.-K. Kim*
161 **Environmentally compatible food packaging** *Edited by E. Chiellini*
162 **Improving farmed fish quality and safety** *Edited by Ø. Lie*
163 **Carbohydrate-active enzymes** *Edited by K.-H. Park*
164 **Chilled foods: a comprehensive guide Third edition** *Edited by M. Brown*
165 **Food for the ageing population** *Edited by M. M. Raats, C. P. G. M. de Groot and W. A. Van Staveren*
166 **Improving the sensory and nutritional quality of fresh meat** *Edited by J. P. Kerry and D. A. Ledward*
167 **Shellfish safety and quality** *Edited by S. E. Shumway and G. E. Rodrick*
168 **Functional and speciality beverage technology** *Edited by P. Paquin*
169 **Functional foods: principles and technology** *M. Guo*
170 **Endocrine-disrupting chemicals in food** *Edited by I. Shaw*
171 **Meals in science and practice: interdisciplinary research and business applications** *Edited by H. L. Meiselman*
172 **Food constituents and oral health: current status and future prospects** *Edited by M. Wilson*
173 **Handbook of hydrocolloids Second edition** *Edited by G. O. Phillips and P. A. Williams*
174 **Food processing technology: principles and practice Third edition** *P. J. Fellows*
175 **Science and technology of enrobed and filled chocolate, confectionery and bakery products** *Edited by G. Talbot*
176 **Foodborne pathogens: hazards, risk analysis and control Second edition** *Edited by C. de W. Blackburn and P. J. McClure*

© Woodhead Publishing Limited, 2012

177 **Designing functional foods: measuring and controlling food structure breakdown and absorption** *Edited by D. J. McClements and E. A. Decker*
178 **New technologies in aquaculture: improving production efficiency, quality and environmental management** *Edited by G. Burnell and G. Allan*
179 **More baking problems solved** *S. P. Cauvain and L. S. Young*
180 **Soft drink and fruit juice problems solved** *P. Ashurst and R. Hargitt*
181 **Biofilms in the food and beverage industries** *Edited by P. M. Fratamico, B. A. Annous and N. W. Gunther*
182 **Dairy-derived ingredients: food and neutraceutical uses** *Edited by M. Corredig*
183 **Handbook of waste management and co-product recovery in food processing Volume 2** *Edited by K. W. Waldron*
184 **Innovations in food labelling** *Edited by J. Albert*
185 **Delivering performance in food supply chains** *Edited by C. Mena and G. Stevens*
186 **Chemical deterioration and physical instability of food and beverages** *Edited by L. H. Skibsted, J. Risbo and M. L. Andersen*
187 **Managing wine quality Volume 1: viticulture and wine quality** *Edited by A. G. Reynolds*
188 **Improving the safety and quality of milk Volume 1: milk production and processing** *Edited by M. Griffiths*
189 **Improving the safety and quality of milk Volume 2: improving quality in milk products** *Edited by M. Griffiths*
190 **Cereal grains: assessing and managing quality** *Edited by C. Wrigley and I. Batey*
191 **Sensory analysis for food and beverage quality control: a practical guide** *Edited by D. Kilcast*
192 **Managing wine quality Volume 2: oenology and wine quality** *Edited by A. G. Reynolds*
193 **Winemaking problems solved** *Edited by C. E. Butzke*
194 **Environmental assessment and management in the food industry** *Edited by U. Sonesson, J. Berlin and F. Ziegler*
195 **Consumer-driven innovation in food and personal care products** *Edited by S. R. Jaeger and H. MacFie*
196 **Tracing pathogens in the food chain** *Edited by S. Brul, P. M. Fratamico and T. A. McMeekin*
197 **Case studies in novel food processing technologies: innovations in processing, packaging, and predictive modelling** *Edited by C. J. Doona, K. Kustin and F. E. Feeherry*
198 **Freeze-drying of pharmaceutical and food products** *T.-C. Hua, B.-L. Liu and H. Zhang*
199 **Oxidation in foods and beverages and antioxidant applications**

© Woodhead Publishing Limited, 2012

**Volume 1: understanding mechanisms of oxidation and antioxidant activity** *Edited by E. A. Decker, R. J. Elias and D. J. McClements*

200 **Oxidation in foods and beverages and antioxidant applications Volume 2: management in different industry sectors** *Edited by E. A. Decker, R. J. Elias and D. J. McClements*

201 **Protective cultures, antimicrobial metabolites and bacteriophages for food and beverage biopreservation** *Edited by C. Lacroix*

202 **Separation, extraction and concentration processes in the food, beverage and nutraceutical industries** *Edited by S. S. H. Rizvi*

203 **Determining mycotoxins and mycotoxigenic fungi in food and feed** *Edited by S. De Saeger*

204 **Developing children's food products** *Edited by D. Kilcast and F. Angus*

205 **Functional foods: concept to product Second edition** *Edited by M. Saarela*

206 **Postharvest biology and technology of tropical and subtropical fruits Volume 1: Fundamental issues** *Edited by E. M. Yahia*

207 **Postharvest biology and technology of tropical and subtropical fruits Volume 2: Açai to citrus** *Edited by E. M. Yahia*

208 **Postharvest biology and technology of tropical and subtropical fruits Volume 3: Cocona to mango** *Edited by E. M. Yahia*

209 **Postharvest biology and technology of tropical and subtropical fruits Volume 4: Mangosteen to white sapote** *Edited by E. M. Yahia*

210 **Food and beverage stability and shelf life** *Edited by D. Kilcast and P. Subramaniam*

211 **Processed Meats: improving safety, nutrition and quality** *Edited by J. P. Kerry and J. F. Kerry*

212 **Food chain integrity: a holistic approach to food traceability, safety, quality and authenticity** *Edited by J. Hoorfar, K. Jordan, F. Butler and R. Prugger*

213 **Improving the safety and quality of eggs and egg products Volume 1** *Edited by Y. Nys, M. Bain and F. Van Immerseel*

214 **Improving the safety and quality of eggs and egg products Volume 2** *Edited by F. Van Immerseel, Y. Nys and M. Bain*

215 **Animal feed contamination: effects on livestock and food safety** *Edited by J. Fink-Gremmels*

216 **Hygienic design of food factories** *Edited by J. Holah and H. L. M. Lelieveld*

217 **Manley's technology of biscuits, crackers and cookies Fourth edition** *Edited by D. Manley*

218 **Nanotechnology in the food, beverage and nutraceutical industries** *Edited by Q. Huang*

219 **Rice quality: A guide to rice properties and analysis** *K. R. Bhattacharya*

© Woodhead Publishing Limited, 2012

220 **Advances in meat, poultry and seafood packaging** *Edited by J. P. Kerry*
221 **Reducing saturated fats in foods** *Edited by G. Talbot*
222 **Handbook of food proteins** *Edited by G. O. Phillips and P. A. Williams*
223 **Lifetime nutritional influences on cognition, behaviour and psychiatric illness** *Edited by D. Benton*
224 **Food machinery for the production of cereal foods, snack foods and confectionery** *L.-M. Cheng*
225 **Alcoholic beverages: sensory evaluation and consumer research** *Edited by J. Piggott*
226 **Extrusion problems solved: food, pet food and feed** *M. N. Riaz and G. J. Rokey*
227 **Handbook of herbs and spices Volume 1 Second edition** *Edited by K. V. Peter*
228 **Handbook of herbs and spices Volume 2 Second edition** *Edited by K. V. Peter*
229 **Bread making: improving quality Second edition** *Edited by S. P. Cauvain*
230 **Emerging food packaging technologies: principles and practice** *Edited by K. L. Yam and D. S. Lee*
231 **Infectious disease in aquaculture: prevention and control** *Edited by B. Austin*
232 **Diet, immunity and inflammation** *Edited by P. C. Calder and P. Yaqoob*
233 **Natural food additives, ingredients and flavourings** *Edited by D. Baines and R. Seal*
234 **Microbial decontamination in the food industry: novel methods and applications** *Edited by A. Demirci and M. O. Ngadi*
235 **Chemical contaminants and residues in foods** *Edited by D. Schrenk*
236 **Robotics and automation in the food industry: current and future technologies** *Edited by D. G. Caldwell*
237 **Fibre-rich and wholegrain foods: improving quality** *Edited by J. A. Delcour and K. Poutanen*
238 **Computer vision technology in the food and beverage industries** *Edited by D.-W. Sun*
239 **Encapsulation technologies and delivery systems for food ingredients and nutraceuticals** *Edited by N. Garti and D. J. McClements*
240 **Case studies in food safety and authenticity** *Edited by J. Hoorfar*

© Woodhead Publishing Limited, 2012

# Preface

Foodborne illnesses remain a major global public health concern. Every year, 2.2 million people, including 1.9 million children, die due to the consumption of contaminated food and water according to the World Health Organization. More interestingly, outbreaks are increasing in spite of the significant efforts by both government regulatory agencies and the food industry to ensure production of safe products. The increased incidences of pathogens in foods and the resulting disease outbreaks might be attributed to several multidimensional factors including: improved detection methods; changes in eating habits; introduction of new food production methods; better epidemiological data collection; demographic changes; as well as changes in scale and techniques of food processing, handling, and preparation. In particular, the appearance of new and resistant pathogens have placed a huge burden on food processors in their efforts to meet consumers' demand for high quality and safe fresh-like products. Apart from health problems and foodborne illnesses, outbreaks as a result of contamination by pathogens result in huge economic and productivity losses. Therefore, the major challenge of food processing is to develop innovative means of delivering high-quality, shelf-stable and safe products in the most efficient, economical, and sustainable ways.

Today, there is strong and increasing consumer demand for fresh and minimally processed health-focused foods. Food decontamination has traditionally been achieved using chemical and thermal methods. Since Nicolas Appert's introduction of canning that set the stage for modern food processing, the industry has seen the application of several innovative techniques for microbial inactivation in food systems. Food quality and safety may not necessarily be complementary. Indeed, they may be opposing in

© Woodhead Publishing Limited, 2012

some cases. The management and optimization of the relevant parameters to achieve appropriate balance between quality and safety continue to present major challenges in food processing. The use of intense and broad-spectrum thermal treatment is known to be effective for microbial inactivation, but unfortunately it accelerates degradation of important sensory, nutritive, and functional quality attributes of the food products. There is also the aspect of excessive energy use associated with thermal processing. Therefore, more recent advancements have allowed the application of mild processing of food with the promise of safe products without quality degradation.

Some emerging decontamination methods are only suited for certain food products and they have their peculiar challenges with respect to food quality and safety. For instance, fresh fruits and vegetables are known to have high amounts of heat sensitive antioxidants, which can easily be destroyed with intense thermal treatment. In general, thermal processing of a given food matrix is based on time-temperature combinations that are required to inactivate the most heat-resistant pathogens and spoilage organisms in the matrix. Thus, heat penetration characteristics in the product must be assessed in order to evaluate required temperature distributions. The final choice of process depends then on the difficult task of identifying the minimum possible heat treatment that should guarantee freedom from pathogens and toxins and give the desired quality and storage life. The use of chemicals also works well in some cases, but they are largely restricted in their applications. Therefore, researchers are working hard to find effective and economical processing methods for various food products to save lives and also to extend the shelf life of foods.

Alternative non-chemical and non-thermal methods have been proposed and evaluated for the decontamination of foods. These methods include high hydrostatic pressure, pulsed electric field, pulsed UV light, power ultrasound, etc. Although several assessments of these technologies have shown very encouraging results, there may be a need to evaluate their applications and effectiveness for decontamination as well as their effect on food quality.

Therefore, this book has been designed to shed light on developments in food decontamination technologies. We have worked diligently to cover almost every food commodity group as well as recent developments in both conventional and emerging processing technologies. The content of the book is divided into four parts. Part I deals with various food commodities such as fresh produce, meats, seafood, nuts, juices, and dairy products, and provides background in terms of contamination routes, outbreaks as well as proposed processing methods for each commodity. Part II addresses conventional and emerging chemical processing methods, whereas Part III focuses on emerging and non-chemical processing methods. Each chapter in Parts II and III describes principles of the technology, how it works, mode of inactivation, applications on various foods, effects on food quality as well as limitations of the technology, if any. Finally, Part IV includes the current and emerging packaging methods and post-packaging decontamination. Each

© Woodhead Publishing Limited, 2012

chapter is written by internationally renowned authors who are expert in their respective fields. This book would not be possible without their hard work.

We hope that this book will become the go-to reference book for: processors who will need good understanding of the various technologies in order to implement them in their operations; researchers who need to know the current developments in the application of the technologies in food processing as well as need to have new ideas and approaches to improve safety and quality of food that we consume; and regulatory personnel who will need to understand both the advantages and limitations of the technologies. A section on sources of further information has been included in each chapter in order to facilitate further studies in the different technologies.

*Ali Demirci*
*Michael O. Ngadi*

© Woodhead Publishing Limited, 2012

# Part I

# Microbial decontamination of different food products

© Woodhead Publishing Limited, 2012

# 1

# Microbial decontamination of fresh produce

**S. Foong-Cunningham, Ecolab Inc., USA, E. L. C. Verkaar, Ecolab Deutschland GmbH, Germany and K. Swanson, Ecolab Inc., USA**

**Abstract**: Fresh fruit and vegetable consumption has increased dramatically in the past four decades, and fresh produce has been associated with multiple foodborne illness outbreaks involving bacteria, viruses and parasites. Since there is usually no terminal inactivation step prior to consumption, there is great concern over the safety of these products. This chapter discusses fresh produce in the raw or minimally processed state with general information on epidemiology, processing steps, good agricultural practices (GAP) and decontamination strategies. Details of pathogen, produce type and contamination pathways are provided. This chapter also focuses on decontamination methods commonly used in certain regions for fresh produce, recognizing that some antimicrobials may not be approved in some regions. Considerations for comparing the effectiveness of different treatments are addressed, as is the type of application. Chlorine and electrolyzed oxidizing water, chlorine dioxide, acidified sodium chlorite, organic peroxides, hydrogen peroxide, ozone, organic acids, and mild heat treatments are discussed. Novel treatments such as irradiation, essential oils, high hydrostatic pressure and atmospheric plasma inactivation are briefly discussed. Future trends for produce decontamination and differences in approaches among regions of the world are included in this chapter. Understanding the standards related to produce decontamination for different regions of the world is important to determine the proper intervention to use.

**Key words**: fresh produce, decontamination, antimicrobial, chlorine, electrolyzed water, chlorine dioxide, acidified sodium chlorite, organic peroxides, hydrogen peroxide, ozone, organic acids.

## 1.1 Introduction

During the past few decades fresh fruit consumption in the US has increased by 25.8% and fresh vegetable consumption by 32.6% (Barth *et al.*, 2010).

© Woodhead Publishing Limited, 2012

Factors influencing this trend include changes in consumers' dietary habits, shifts in social demographics, more 'outside-home' meals, the growing popularity of salad bars, and recommendations to increase consumption of fruits and vegetables as part of a healthy diet.

The term 'produce' includes the edible components of plants such as leaves, stalks, roots, tubers, bulbs, flowers, fruits, sprouted seeds, and mushrooms. Many fresh produce items are frequently consumed without cooking and in many countries they are considered as raw agricultural commodities that come directly from the field, orchard, or vineyard. Fresh produce, especially vegetables, can be trimmed, rinsed, or washed prior to shipment to the market. Sliced or cut ready-to-eat produce is part of a rapidly growing segment of the food industry in North America, Europe and elsewhere. Once cut, different types of produce are sometimes blended together prior to sale. Much fresh produce is ready to eat in the form of salads and fruit blends.

The link between contaminated fresh produce and outbreaks of foodborne illness is well recognized. Some cultures have a tradition of thorough cooking of produce prior to consumption and others rely on good agricultural and processing practices to minimize the risk of foodborne illness. Travelers' diarrhea is frequently associated with contaminated produce, hence the US Centers for Disease Control and Prevention (CDC) recommends that travelers avoid eating raw fruits and vegetables in regions where travelers' diarrhea is a concern unless they are able to peel the produce (CDC, 2006b). Even in developed countries, fresh produce is associated with outbreaks. From 1998 to 2006, five commodity groups comprised 76% of produce-related US outbreaks; i.e., lettuce/leafy greens, tomatoes, cantaloupe, herbs (basil, parsley), and green onions (Stopforth *et al.*, 2008). In 2007, produce was associated with 24% of outbreaks attributed to a single food commodity in the US, with 14% attributed to leafy green vegetables alone (CDC, 2010c). The Rapid Alert System for Food and Feed (RASFF) system in Europe provides an effective tool to exchange information about measures taken to respond to serious food and feed risks (EU, 2011). Approximately 10–15% of EU border rejections, alerts, and information notifications are associated with fruits, vegetables, herbs, and spices. The microbiological hazards for herbs, spices, fruits, and vegetables in the RASFF system are mainly *Bacillus cereus*, *Campylobacter*, *E. coli*, norovirus, *Salmonella*, and *Shigella* spp.

Leafy vegetables and fruits/nuts are the most common types of produce items associated with outbreaks from a single food, while fish, poultry and beef are each associated with more outbreaks (CDC, 2009b, 2010c). In Europe, sources of foodborne illnesses are also more frequently associated with fresh meat products than with produce items; however, produce safety is still a concern.

The increase in reported foodborne illness associated with consumption of raw produce can, in part, be attributed to improved surveillance systems as well as insufficient hygiene practices, changes in global trade, increased frequency of consuming meals at food service establishments and produce

© Woodhead Publishing Limited, 2012

production, processing, and marketing practices (Park and Beuchat, 1999). The increased per capita consumption of fresh and minimally processed produce, as well as increased imports from regions where good agricultural practices may not be practiced, results in the elevated interest in fresh produce safety.

### 1.1.1 Microorganisms in fresh produce

Fresh vegetables and fruits that have not been exposed to raw human or other animal waste material generally do not contain human pathogens. However, direct contamination can occur through several routes including raw sewage or manure fertilizer; contaminated water for irrigation, cooling or washing; contaminated ice for display or transport; and unhygienic handling. Indirect contamination can also occur during preparation of food where fresh produce is contaminated by other foodstuffs through cross-contamination such as using the same cutting board used for meat preparation and fresh produce preparation. These potential sources of contamination must be controlled for produce that is intended to be consumed raw.

Intact cell structures of a plant act as a protective barrier against contamination of microorganisms. Once this barrier is compromised through aging, wilting, or injury (chopping, bruising, shredding, and juicing), both spoilage and pathogenic microorganisms may be able to grow due to the availability of nutrients and moisture within the plants. Pathogens causing human diseases include bacteria, viruses, and parasites. Common produce/pathogen pairings include *Salmonella* in fruits/nuts and in vine-stalk vegetables, norovirus and Shiga toxin-producing *E. coli* (STEC) in leafy vegetables (CDC, 2009b, 2010c).

Many foodborne illnesses associated with fresh produce consumption are influenced by the type of produce, the physical conditions (e.g. temperature) under which the produce is handled and stored, and the amount consumed. For instance, melons are commonly associated with outbreaks and several types, such as cantaloupe, have intricate, webbed surfaces on the rind. These surfaces make pathogenic bacterial attachment harder to remove and can provide protection against antimicrobials. When the melon is cut, microorganisms from the surface are transferred to the cut fruit surface. Cut melons have been documented to support growth of pathogens (Park and Beuchat, 1999) and therefore once cut should be stored at or below 7°C and displayed for <4 hours if not chilled (ICMSF, 2005).

Availability of fresh-cut tomatoes meets consumer demand for convenience and offers the customer a standardized product. However, cut tomatoes have also been demonstrated to support the growth of *Salmonella* under improper temperature conditions, thus the risk of illness is increased when cut tomatoes are improperly handled, e.g., during production, handling, processing, storage, or distribution (Zhuang *et al.*, 1995; Prakash *et al.*, 2007a).

© Woodhead Publishing Limited, 2012

Outbreaks associated with consumption of sprouts are disproportional to the volume of product consumed. The source of pathogens appears to originate from seeds and not contamination of final products (Nei *et al.*, 2010). While the prevalence of contamination in seeds is low, the temperature and wet conditions required for seed germination and growth are also ideal for rapid growth of pathogens. Therefore, screening suppliers for seeds produced in a manner to minimize the potential for contamination, as well as decontaminating seeds before germination are both important control strategies. The National Advisory Committee on Microbiological Criteria for Foods (NACMCF, 1999) recommended soaking sprout seeds in 20,000 ppm calcium hypochlorite (measured as available chlorine) for 10 to 20 minutes, whereas only 200 ppm is allowed for sanitizing food contact equipment surfaces in the US (EPA, 2011).

### 1.1.2 Good agricultural practices (GAP)

The microbiological quality and safety of fresh produce depend largely on agricultural and harvesting practices. Microorganisms present on fresh produce may be transferred from soil, air, water, and people, and the initial microbiota can remain on the fresh produce during transportation, further processing, and storage. Fresh produce decontamination treatments that are commercially practiced in 2011 are not capable of totally eliminating all microbial contamination. Therefore, application of GAP is essential to grow safe produce for raw and minimally processed produce consumption. General guidance on GAP is available (FDA, 1998; Codex Alimentarius, 2003; ICMSF, 2005, 2011). There is no universal approach to GAP, because water sources, weather patterns, animal activity, and horticultural practices vary considerably in different regions and for different commodities. However, general considerations include use of uncontaminated irrigation water, soil amendments such as composted animal manure, avoidance of water run off from animal production areas, and hygienic practices of field workers (ICMSF, 2005, 2011).

### 1.1.3 Processing steps

Once produce is harvested, further processing, such as sorting, washing, trimming, cutting, slicing, peeling, mixing, packaging, etc., may take place, either in a processing location or just prior to use. All of these steps provide an opportunity for recontamination of the product.

Recommended guidelines to maintain microbial safety in fresh produce are available (Codex Alimentarius, 2003; FDA, 2008). Because growing, harvesting, processing, storage and distribution practices can vary considerably for different commodities and growing regions, commodity specific guidance has been developed for melons, leafy green vegetables, and tomatoes (FDA, 2009a, b, c). Decontamination steps may be applied during processing

© Woodhead Publishing Limited, 2012

through use of antimicrobial treatments in flumes or as sprays. There is also a potential for use of irradiation where this treatment is allowed.

## 1.2 Pathogens of concern and pathways of contamination in fresh produce

Table 1.1 provides examples of produce associated outbreaks involving a variety of pathogens. Pathogens of primary concern for fresh produce are *Salmonella* spp., enterotoxigenic *Escherichia coli*, *Listeria monocytogenes*, *Shigella* spp., *Cryptosporidium* spp., *Cyclospora* spp., hepatitis A virus (HAV) and norovirus (Stopforth *et al.*, 2008; Inatsu *et al.*, 2010). Specific considerations for specific pathogens are addressed subsequently; however, viruses are emerging as the dominant causative agent associated with fruit and vegetable outbreaks. For example, in Europe during 2009, viruses were identified as the causative agent in 65.1% of 43 fruit and vegetable outbreaks with verified causes (EFSA, 2011). The pathway of contamination may vary by agent and this is discussed in the following sections.

### 1.2.1 *Salmonella* spp.

Animals, birds, and insects are natural enteric reservoirs for *Salmonella* and the microorganism has been isolated from a wide variety of produce types (Zhuang *et al.*, 1995; Beuchat, 1996; ICMSF, 2005). *Salmonella* is the most commonly reported bacterial cause of foodborne outbreaks associated with fresh produce in the US (CDC, 2010c) and contamination commonly occurs at the farm, during packing, or during preparation. Examples of produce associated *Salmonella* outbreaks are listed in Table 1.1. Several multi-state outbreaks have been reported in the US including sprouts, cantaloupe, and tomatoes (CDC, 2011c, d, e), indicating contamination in the field, during transport, or during processing.

A contributing factor to some produce outbreaks is holding cut produce at room temperature for extended periods of time. As previously mentioned, *Salmonella* has been reported to grow on cut tomato surfaces stored at 20–30°C while storage at 5°C prevented growth (Beuchat, 1996; Cummings *et al.*, 2001).

### 1.2.2 *Escherichia coli* O157:H7 and other Shiga toxin-producing *E. coli*

A number of produce associated outbreaks have involved enterotoxigenic *E. coli*. A large *E. coli* O104:H4 outbreak associated with fenugreek sprouts occurred in Germany and elsewhere in Europe and involved 4,321 cases and 50 deaths. This enterohaemorrhagic strain of *E. coli* can cause severe cases

© Woodhead Publishing Limited, 2012

**Table 1.1** Fresh produce implicated outbreaks

| Year | Etiology | Produce implicated | Location | No. of cases | References |
|---|---|---|---|---|---|
| 1981 | *Listeria monocytogenes* | cabbage | CA | 41 | Schlech *et al.* (1983) |
| 1986 | *Shigella sonnei* | lettuce | US | 347 | Davis *et al.* (1988) |
| 1988 | *Salmonella* St. Paul | raw bean sprouts | UK, Sweden | 143 | O'Mahony *et al.* (1990) |
| 1988 | Hepatitis A virus | green salads | US | 202 | Rosenblum *et al.* (1990) |
| 1990 | Norovirus | fresh-cut fruit | US | 217 | Herwaldt *et al.* (1994) |
| 1990 | *Salmonella* Chester | cantaloupe | US | 245 | Ries *et al.* (1990) |
| 1991 | *Salmonella* Poona | cantaloupe | US, CAN | 400 | CDC (1991) |
| 1994 | *Shigella flexneri* | scallions | CA | 72 | Cook *et al.* (1995) |
| 1995 | *E. coli* O157:H7 | lettuce | US | 92 | Ackers *et al.* (1998) |
| 1995 | *Salmonella* Stanley | alfalfa sprouts | US, Finland | 50 | Mahon *et al.* (1997) |
| 1996 | *E. coli* O157:H7 | radish sprouts | Japan | ca.10,000 | Watanabe *et al.* (1999) |
| 1996 | *Cyclospora cayetanensis* | raspberries | US, CAN | 850 | CDC (1996) |
| 1997 | Hepatitis A virus | strawberries | US | 153 | CDC (1997) |
| 1998 | Hepatitis A virus | scallions | US | 43 | Dentinger *et al.* (2001) |
| 1999 | *Salmonella* Baildon | fresh tomatoes | US | 86 | Cummings *et al.* (2001) |
| 2003 | Hepatitis A virus | green onions | US | 555 | CDC (2003) |
| 2005–06 | *Salmonella* | tomatoes | US (4 outbreaks) | 459 | CDC (2007) |
| 2006 | Norovirus | raspberries | Sweden | 43 | Hjertqvist *et al.* (2006) |
| 2006 | *Salmonella* Thompson | ruccola (arugula) | EU | 21 | Nygård *et al.* (2008) |
| 2006 | *E. coli* O157:H7 | spinach | NA | 206 | Lynch *et al.* (2009) |
| 2006 | *Salmonella* St. Paul | cantaloupe | Australia | 36 | Munnoch *et al.* (2009) |
| 2007 | *Salmonella* Weltevreden | alfalfa sprouts | EU | 45 | Emberland *et al.* (2007) |
| 2007 | *Salmonella* Seftenberg | fresh basil | EU and NA | 55 | Pezzoli *et al.* (2008) |
| 2008 | *Salmonella* St. Paul | hot peppers/tomatoes | NA | 1442 | CDC (2008) |
| 2008 | Norovirus | lettuce salad and soup | Portugal | 19 | Mesquita and Nascimento (2009) |

© Woodhead Publishing Limited, 2012

| | | | | | |
|---|---|---|---|---|---|
| 2008 | Norovirus | lettuce salad and soup | Portugal | 19 | Mesquita and Nascimento (2009) |
| 2009 | *Salmonella* St. Paul | alfalfa sprouts | US | 235 | CDC (2009a) |
| 2009 | *Cryptosporidium parvum* | raw produce | US | 46 | CDC (2011a) |
| 2010 | *E. coli* O145 | romaine lettuce | US | 33 | CDC (2010a) |
| 2010 | *Salmonella* Newport | alfalfa sprouts | US | 44 | CDC (2010b) |
| 2010 | *Listeria monocytogenes* | chopped celery | US | 10 | FDA (2010) |
| 2010 | Norovirus | lettuce | Denmark | 260 | Ethelberg *et al.* (2010) |
| 2010–11 | *Salmonella* | alfalfa sprout | US | 140 | CDC (2011d) |
| 2011 | Norovirus | frozen raspberries | Finland | 900 | Sarvikivi *et al.* (2011) |
| 2011 | *E. coli* O104:H4 | sprouts | Germany | 4,321 | RKI (2011) |
| 2011 | *Salmonella* Agona | papaya | US | 106 | CDC (2011b) |
| 2011 | *Salmonella* Enteritidis | sprouts | US | 25 | CDC (2011c) |
| 2011 | *Salmonella* Panama | cantaloupe | US | 20 | CDC (2011e) |
| 2011 | *Listeria monocytogenes* | cantaloupe | US | 146 | CDC (2011f) |

EU = Europe
UK = United Kingdom
US = United States
NA = North America
CA = Central America
CAN = Canada

© Woodhead Publishing Limited, 2012

of haemolytic-uremic syndrome (HUS), an illness that can lead to acute kidney failure and cannot be treated with antibiotics. Of the total number of cases, 852 were HUS cases, and of those who died, 32 had developed HUS (RKI, 2011).

*E. coli* O145 was associated with a romaine lettuce outbreak involving 33 cases (CDC, 2010a) and a multi-state outbreak of *E. coli* O157:H7 gastroenteritis was associated with fresh spinach, involving 204 cases and 3 deaths (CDC, 2006a; FDA, 2006). Other outbreaks have involved consumption of mixed fresh salads on an airplane and at a lodge. Cases of hemorrhagic colitis were also reported for an outbreak associated with eating at a salad bar contaminated with *E. coli* O157:H7 where uncooked items were cross-contaminated by raw ground beef and stored at improper temperatures (Beuchat, 1996). Gastroenteritis has also been linked to consumption of contaminated cantaloupe. *E. coli* O157:H7 was reported to grow rapidly in cantaloupe and watermelon cubes stored at 25°C (Beuchat, 1996).

*E. coli* O157:H7 can persist in a number of environments previously thought to be protective against pathogens. It tolerates acidic environments. Modified atmosphere packaging has no effect on the survival or growth of *E. coli* O157:H7 on shredded lettuce or sliced cucumber when stored at 12°C and 21°C (Beuchat, 1996).

### 1.2.3 *Listeria monocytogenes*

*Listeria* is a ubiquitous microorganism commonly found in soil. Only *L. monocytogenes* is considered to be pathogenic. Human listeriosis is linked to consumption of coleslaw made from cabbage grown in fields that were fertilized with manure from sheep that had active cases of meningoencephalitis or circling disease (Schlech *et al.*, 1983). Chopped celery (FDA, 2010), hummus and cut melons (Varma *et al.*, 2007) have also been involved in listeriosis outbreaks and growth occurred in the product during refrigerated storage. *L. monocytogenes* was detected at retail on cabbage, potatoes, cucumbers, tomatoes, bean sprouts, and leafy vegetables and is able to grow on radishes, asparagus, broccoli, cauliflower lettuce, and chicory endive when stored at refrigeration temperatures from 4 to 6.5°C. Controlled atmosphere storage does not affect growth rates of *L. monocytogenes* (Beuchat, 1996).

### 1.2.4 *Shigella*

*Shigella* outbreaks have been associated with consumption of contaminated raw lettuce, green onion and parsley. The primary source of *Shigella* contamination is infected personnel, either working in the field or during preparation. *Shigella sonnei* was reported in sugar peas (snow peas) from Kenya (EU, 2011) and shown to survive on lettuce at 5°C for 3 days (Beuchat, 1996).

© Woodhead Publishing Limited, 2012

### 1.2.5 Viruses

A variety of viral outbreaks have been transmitted via fresh produce and humans which are the reservoir for these viruses. Beuchat (1996) reported that, in 74% of salad associated viral outbreaks, the source of contamination was a food handler. Norovirus and hepatitis A virus are the most common viral food contaminants. Viruses are excreted from infected individuals and transmitted via fecal-oral route. Viruses do not grow on contaminated vegetables and fruits but can survive long enough to cause foodborne illness. Exclusion of symptomatic workers is essential for control and proper handwashing is also important to reduce the spread of disease. Gloves may reduce risk if hands are properly washed prior to donning gloves.

*Norovirus*

Noroviruses are the most common cause of foodborne gastroenteritis in the US (CDC, 2010c). While norovirus transmission occurs more frequently through person-to-person rather than foodborne routes (Dreyfuss, 2009), transmission through food is still a common vector, especially in restaurant settings. Fresh produce that has been implicated in outbreaks includes green salad, celery, and raspberries (Tauxe *et al.*, 1997; Lynch *et al.*, 2009; EFSA, 2011). Five cases of norovirus contaminated products have been identified by the European Rapid Alert System for Food and Feed (RASFF), four of which involved frozen raspberries from Poland and Serbia. Previously, frozen raspberries were reported as the cause of large outbreaks (EU, 2010).

*Hepatitis A virus (HAV)*

Fresh produce outbreaks have been associated with HAV (Table 1.1). An outbreak that demonstrated the role of harvesting workers occurred in Scotland, where raspberry mousse prepared using frozen raspberries was associated with a HAV outbreak. The raspberries were picked by a HAV excreter (Tauxe *et al.*, 1997). Outbreaks have also been associated with green onions (CDC, 2003).

HAV outbreaks associated with processed foods may originate from an infected worker contaminating fresh produce that persists through processing and distribution. An example is a multi-country outbreak involving semi-dried tomatoes. In 2009, 250 people in Australia became ill after consuming semi-dried tomatoes. The European RASFF system identified 49 cases in France and 19 cases in the Netherlands, all epidemiologically linked to semi-dried tomatoes. It appeared that the tomato processing in a plant in Turkey did not use any pasteurizing process (Gallot *et al.*, 2011). During processing, frozen tomatoes were purchased, and oil and herbs were added prior to drying. Sampling detected one positive sample for HAV in Australia. The point of contamination is uncertain but could have been during processing in France, production in Turkey, or in the field during harvesting.

© Woodhead Publishing Limited, 2012

### 1.2.6 Parasites

Protozoan parasites such as *Cryptosporidium parvum*, *Cyclospora cayetanensis*, and *Giardia lamblia* have been associated with raw or minimally processed vegetable outbreaks. In 2009, a *Cryptosporidium parvum* outbreak occurred at a camp setting after consuming ham from a sandwich bar with camp-grown produce and staying with an ill person (CDC, 2011a). Salads with celery, mesclun lettuce, and basil have also been implicated in outbreaks of cryptosporidiosis. *G. lamblia* was associated with a lettuce and onion outbreak in which contaminated water was used to wash vegetables. *Fasciola hepatica* infection was linked to consumption of contaminated watercress. Waterborne parasites are transmitted to water crops (e.g., watercress, cater caltrop, water chestnut, and water bamboo) cultivated in ponds that are subjected to run-off from grazing pastures (Buchanan *et al.*, 1999; ICMSF, 2005).

## 1.3 Current decontamination methods for fresh produce and their limitations

Washing is a primary method to reduce contamination of fresh produce. Washing, with or without additional antimicrobials, may reduce pathogen levels, but cannot completely eliminate pathogens for a variety of reasons. The degree of reduction is influenced by the antimicrobial used, the conditions of application, and the produce type. Pathogens may be protected if they are internalized into the plant tissue or in surface structures such as stomata. Erickson *et al.* (2010) studied the potential for pathogen uptake through roots and demonstrated that while uptake through irrigation water may occur when high inocula are used, persistent contamination through internalization via roots does not appear to be a common situation under GAP. Considerations for various washing treatments are summarized below.

### 1.3.1 Water

Washing produce in water removes sand, soil, and other debris, but may not be reliable enough to completely remove microorganisms. Fresh produce may contain up to $10^4$ to $10^6$ CFU/g. At best, a 1- to 2-log reduction of surface microbiota may be achieved from washing in water, depending on the volume of water used (Beuchat, 1996; ICMSF, 2005). Water can also serve as a vector to spread microbial contamination that may be present at a discrete location (e.g., a bird dropping) throughout an entire batch. Such transfer can be prevented or minimized with the use of effective antimicrobial treatments.

© Woodhead Publishing Limited, 2012

### 1.3.2 Antimicrobial chemicals

Some countries allow the use of antimicrobial chemicals in the wash water to reduce the potential spread of microorganisms while others do not. Therefore, the decision to use such chemicals must be based on local requirements. In the European Union, a regulation on antimicrobiological treatment of fresh produce does not exist, although Regulation 2073/2005 clearly describes the microbiological limits to which certain types of cut and uncut produce and juices need to comply. For example, fresh fruit on retail shelves needs to be free of *Salmonella*. Nevertheless, the EU Commission is implementing a range of guidelines to assist the fresh produce industry. On production of fresh produce, EU Regulation 852/2004 states that, in order to prevent foodstuffs from becoming contaminated, food premises must have adequate supply of potable water (Council Directive 98/83/EC) which can be used as necessary. If non-potable water is used for non-food purposes it must be connected to a separate water supply, labelled as such. This water must not be in contact with potable water or supplies thereof. However, flume water used in fresh produce production is process water; not potable water and as such, local national legislation applies. Use of chemical decontaminants in flume water may be possible after consulting local authorities and some countries (e.g., Spain, The Netherlands) may accept such treatments of process water.

Any antimicrobial chemical must be thoroughly evaluated for potential toxicological issues before approval. A wide range of antimicrobials including ozone, peracids (e.g., peroxyacetic acid), chlorine dioxide, chlorinated trisodium phosphate, hypochlorite and hypochlorous acid (including electrolyzed oxidizing water), and acidified sodium chlorite have been evaluated on their effectiveness for pathogen reduction in fresh produce (Beuchat, 1998; Stopforth *et al.*, 2008). Many studies have been published comparing the effectiveness of antimicrobial treatments. Most of these studies focused on the log reduction that occurs on product surfaces. No standardized protocol to evaluate the effectiveness of antimicrobial treatments exists, thus comparing the relative effectiveness of different antimicrobial treatments through a review of the literature is difficult because the conditions used in inactivation studies can have a significant impact on the results. For example, if the inoculum is not allowed to attach to the surface of the produce tissue prior to treatment, the cells may readily wash off through physical removal usually resulting in higher log reductions than when the inoculum is allowed to attach. It is important to study the addition of organic material to simulate potential use conditions because an increased organic load can reduce the effectiveness of certain antimicrobial treatments such as chlorine – many studies do not evaluate the impact of organic load. Some studies use antimicrobial levels above those that have received regulatory approval. While this is necessary for investigational studies during the development of antimicrobial treatments, consideration should be given to the actual use concentration that may be applied and the potential impact on flavor, shelf life, and toxicological issues. Finally, the contact time for the antimicrobial treatment will influence

© Woodhead Publishing Limited, 2012

results, with longer application times usually resulting in higher kill rates. Incorporation of a neutralizing step is important to reduce the potential for continued residual activity of certain antimicrobial treatments to more accurately determine effectiveness with a specific contact time.

Complete elimination of pathogens is rarely achieved on surfaces. The extent of inactivation can vary considerably depending on the produce type, even when the same antimicrobial treatments are applied. For example, Park and Beuchat (1999) demonstrated different patterns of reduction of *Salmonella* and *E. coli* O157:H7 inoculated on melons and asparagus when a variety of antimicrobial treatments were applied at the levels specified on the product label (Fig. 1.1).

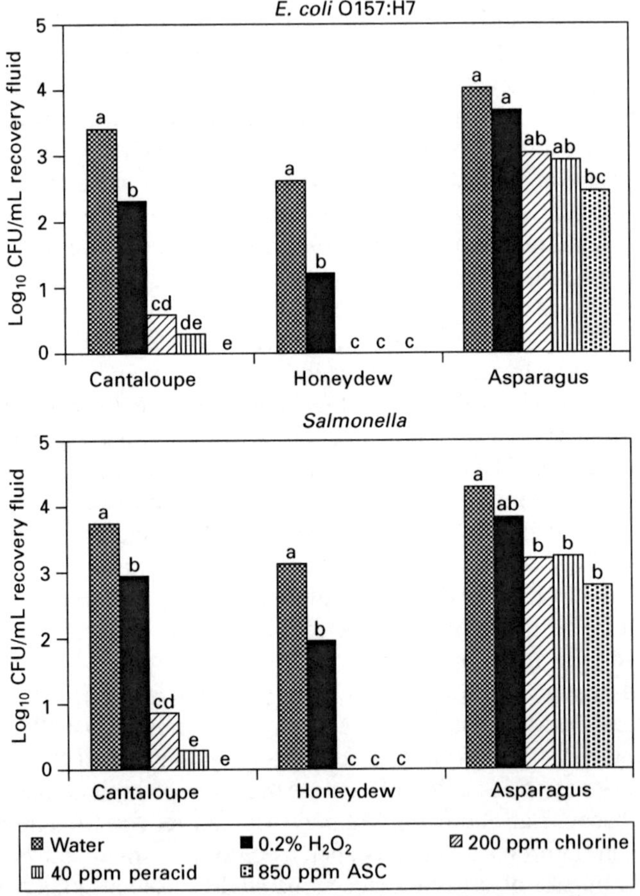


**Fig. 1.1** Effectiveness of water, 0.2% hydrogen peroxide ($H_2O_2$), 200 ppm chlorine, 40 ppm peroxyacetic acid antimicrobial for fruit and vegetable processing water (peracid) and 850 ppm acidified sodium chlorite (ASC) in inactivation of *E. coli* O157:H7 and *Salmonella* on cantaloupe, honeydew and asparagus. Adapted from Park and Beuchat (1999), Center for Food Safety, University of Georgia. Reprinted with permission from *Dairy, Food and Environmental Sanitarian*. Copyright held by the International Association for Food Protection.

© Woodhead Publishing Limited, 2012

The effectiveness of the treatment may be limited or unpredictable on fruit and vegetable surfaces due to the inability to penetrate surface, neutralization of the antimicrobial by soil or other organic material present on the produce, the temperature of application, and the concentration of the active agent (ICMSF, 2005; Pao and Davis, 1999).

Antimicrobial agents are often added to water used in the transportation or washing of fresh produce to reduce the number of microorganisms and to prevent the recycled process water from becoming a vector of cross-contamination. Process water antimicrobials have been demonstrated to be more effective in reducing microorganisms in the water than on the fruit and vegetable surfaces (Foong-Cunningham *et al.*, 2008; Zhang *et al.*, 2009). This may be due to shielding of the microbes from the antimicrobial through internalization or bubble formation on product surfaces. Application of an effective antimicrobial treatment can help to prevent transfer of contamination from the small fraction of product that may be contaminated during growing to the rest of the product, especially in manufacturing settings.

*Chlorine and electrolyzed oxidizing water*

In the US, chlorine is a common wash water treatment for fresh produce at levels from 50 to 200 ppm total chlorine at pH of 6.0 to 7.5, with a contact time of 1 to 2 minutes (FDA, 1998). At this pH range, optimal balance of hypochlorous acid and hypochlorite is available to provide maximum antimicrobial activity. Maintenance of dump tank water quality with antimicrobials is critical to minimize the transfer of microorganisms and it is recommended that levels should be evaluated as part of a hazard analysis and critical control point (HACCP) program (Niemira, 2007; Stopforth *et al.*, 2008).

The effect of chlorine on inactivation of various pathogens on different types of produce is well documented (Tauxe *et al.*, 1997; Buchanan *et al.*, 1999; Cummings *et al.*, 2001; Wachtel and Charkowski, 2002; Prakash *et al.*, 2007b; Inatsu *et al.*, 2010). Chlorine is present in water in three different forms, i.e., elemental chlorine ($Cl_2$), hypochlorous acid (HOCl), which is most bactericidal, and hypochlorite ion ($OCl^-$). Hypochlorite in water produces hypochlorous acid depending on the pH (Cords *et al.*, 2005). Hypochlorite inactivates microorganisms by damaging the respiratory and transport processes of the cell membrane. Hypochlorite is more effective against planktonic microorganisms in water than on vegetable surfaces. Exposure to sunlight or organic matter adversely affects chlorine levels (Erkmen, 2010). Behrsing *et al.* (2000) reported no significant difference in reduction of *E. coli* inoculated onto broccoli florets when exposed to 100 ppm free chlorine at temperatures ranging from 4 to 25°C.

Some advantages of using chlorine include low cost and water solubility. Disadvantages include a low tolerance to the presence of organic material and the initial microbial load, which requires more routine monitoring and dosing to maintain the appropriate level. Maintenance of pH is critical

© Woodhead Publishing Limited, 2012

and also required to ensure that the active antimicrobial species is present. Furthermore, chlorine has been implicated in the formation of carcinogenic organochlorine compounds like chloramines, trihalomethanes (THMs) and semicarbizides, and use as an antimicrobial carcass treatment has been forbidden in the EU (EFSA, 2005a, 2005b; Hoenicke *et al.*, 2004; Stopforth *et al.*, 2008). However, in order to maintain a certain bacteriostatic level (e.g., drinking water regulation), a chlorine level of 0.5 ppm is allowed. As with $ClO_2$, a fixed level is not easy to maintain; with increasing organic load, more chlorine is required, which results in dangerously high levels in the flume water basin. In some regulatory settings, a post-processing fresh water rinse may be required. In addition to safety concerns, high concentrations may be corrosive to processing equipment.

Inatsu *et al.* (2010) recovered 1.3 log of inoculated *E. coli* O157:H7 from tomatoes when sodium hypochlorite (0.1 g/L as free chlorine); 0.5 g/L sodium chlorite, and 1.0 g/L lactic acid; and 0.5 g/L sodium chlorite and 1.0 g/L phytic acid were used compared to 5- to 6-logs recovered when using distilled water alone.

Electrolyzed oxidizing (EO) water is generated by using electricity to split salt into sodium and chlorine which then reacts with water to form hypochlorite (chlorine bleach which is the same as active chlorine), various other chlorine species, and sodium hydroxide. Different on-site generators are available depending on the number of cells in the system; one system combines both the acidic and alkaline streams to form neutral chlorine bleach (NaOCl) for food tissue treatment, while another generator delivers hypochlorous acid for food tissue treatment and sodium hydroxide as a degreaser or general purpose cleaner (multi-stream generator). However, only a small part of the salt is actually electrolyzed, hence the larger part of the initial product remains unchanged and may have adverse effects on corrosion when combined with hypochlorite. Electrolyzed water showed the same effectiveness as chlorine in pathogen reduction on leafy greens (Stopforth *et al.*, 2008). Sharma and Demirci (2003) applied electrolyzed oxidizing water treatment to alfalfa seeds and sprouts, demonstrating greater reduction in inoculated *E. coli* O157:H7 on sprouts than on seeds. Electrolyzed oxidizing water has been evaluated to reduce naturally occurring or inoculated microorganisms on a variety of produce types. For example, neutral electrolyzed water reduced *E. coli* O157:H7, *Salmonella* Enteritidis and *L. monocytogenes* levels inoculated on tomato surfaces (Deza *et al.*, 2003). Acidic electrolyzed water reduced aerobic mesophiles, coliforms, and fungi 0.5 to 2 logs on cucumbers and strawberries depending on treatment time, and reductions were significantly greater than those for ozonated water (Koseki *et al.*, 2004). Conversely, Wang *et al.* (2004) reported that ozone treatment achieved the highest overall quality for cilantro compared to treatments with acidic electrolyzed water.

*Chlorine dioxide gas and acidified sodium chlorite (ASC)*

Chlorine dioxide is a water soluble, aggressive, and unstable gas – even explosive at concentrations higher than 10%. Therefore, chlorine dioxide

© Woodhead Publishing Limited, 2012

is produced with a chemical or electrochemical generator and immediately dissolved in water. The effectiveness of dissolved chlorine dioxide against bacteria is reported on surfaces of lettuce, cabbage, cucumber, and green pepper (Inatsu *et al.*, 2010). Chlorine dioxide gas is more water soluble than chlorine and is a stronger oxidizing agent, with oxidizing capacity of $ClO_2$ approximately 2.5 times higher than hypochlorous acid (Benarde *et al.*, 1965). This oxidation capacity can inactivate bacteria, bacterial spores and viruses depending on the concentration used, temperature, and soil load (Inatsu *et al.*, 2010; Nei *et al.*, 2010). In contrast to active chlorine, chlorine dioxide does not react with organic matter and thereby does not form potential carcinogenic chlorinated organic compounds as easily as chlorine (Nei *et al.*, 2010).

Mixing low concentrations of sodium chlorite with a generally recognized as safe (GRAS) organic acid produces acidified sodium chlorite (ASC, also known as chlorous acid, $HClO_2$). Low levels, 2–3 ppm of chlorine dioxide are also formed within this system in the immediate post-mixing phase. To generate primarily chlorine dioxide, other processes are used. These include combining high concentrations of sodium chlorite and a strong, low pH (often mineral) acid and high concentrations of sodium chlorite and a strong bleach solution. Electrolysis of sodium chlorite can also be used to produce chlorine dioxide. The FDA and USDA have approved ASC as a 'secondary direct food additive permitted in food for human consumption' for poultry, red meat, comminuted meat products, and processed fruits and vegetables for bacterial reduction (Nei *et al.*, 2010). Due to the higher oxidizing potential of its constituents, acidified sodium chlorite is more effective than chlorine. Furthermore, ASC is less affected by organic material, does not form chlorinated organic compounds, does not adversely affect the physical appearance of leafy green leaves (Stopforth *et al.*, 2008), and is relatively pH independent. Studies also reported that ASC is effective in eliminating *E. coli* O157:H7 from shredded carrots (Gonzalez *et al.*, 2004), Chinese cabbage (Inatsu *et al.*, 2005), spinach leaves (Nei *et al.*, 2009), and lettuce leaves (Keskinen *et al.*, 2009). Stopforth *et al.* (2008) studied the effects of acidified sodium chlorite, chlorine and acidic electrolyzed water on *E. coli* O157:H7, *Salmonella*, and *L. monocytogenes* inoculated on leafy greens. Acidified sodium chlorite showed 1 log higher reductions compared to chlorinated water.

*Organic peroxides*

Organic peroxides include peroxyacetic acid (PAA) which is most widely used, peroxyoctanoic acid, cumene peroxide, hydroperoxides, diacyl peroxides and peroxyesters. Peroxyacetic acid provides its antimicrobial activity by disrupting the respiratory, oxidation, and other cellular processes. It reacts with cell organelles and attacks and disrupts DNA (Cords *et al.*, 2005). In contrast to other oxidative antimicrobials such as active chlorine, it does not form carcinogenic compositions such as semicarbazides or THMs.

© Woodhead Publishing Limited, 2012

When validating the effectiveness of antimicrobial treatments, it is critical to understand the impact of organic load. Peroxyacetic acid (peracid) is less affected by organic load and thus may demonstrate more robust results than chlorine for fresh-cut produce (Gonzalez *et al.*, 2004; Foong-Cunningham *et al.*, 2008). Zhang *et al.* (2009) compared different levels of hypochlorite and peracids with and without organic soil load against the transfer of *E. coli* O157:H7 from contaminated lettuce to uncontaminated lettuce through in-process water. With the absence of organic load, peroxyacetic acid and chlorine significantly reduced *E. coli* O157:H7 in processing water by 1.50 to 1.83 log CFU/ml compared to washing with water alone. These antimicrobials also significantly reduced transfer of the bacteria from an inoculated leaf to uninoculated leaves in the processing water by 0.96 to 2.57 log CFU per piece. The organic load in the processing water, however, reduced efficacy of antimicrobial agents in that peroxyacetic acid at 10 and 20 ppm, and chlorine at 30 ppm produced effects similar to those of water alone (Zhang *et al.*, 2009). Baert *et al.* (2009) reported that PAA provided an additional 1 log reduction over tap water for murine norovirus (MNV-1) on shredded iceberg lettuce. No survivors were found in the residual wash water treated with antimicrobial.

*Hydrogen peroxide*

Hydrogen peroxide had minimal antimicrobial effects and was less effective compared to chlorine, acidified sodium chlorite and peracetic acid in inactivating *Salmonella* and *E. coli* O157:H7 on melon cubes but results were not as different for asparagus (see Fig. 1.1; Park and Beuchat, 1999). Sapers and Simmons (1998) suggested exposure of whole cantaloupes to hydrogen peroxide vapor at a concentration of 3 mg/L for 60 minutes to reduce the number of microorganisms and prevent decay during storage at 2°C for up to 4 weeks.

*Ozone*

Ozone is a strong oxidizer and bactericidal at very low concentrations (ppb). The antimicrobial effects of ozone and oxidation of ethylene retard metabolic activity for ripening. Ozone has been used for decades to extend the shelf life of a wide variety of whole fruits including oranges, strawberries, raspberries, grapes, apples, and pears (Xu, 1999). This technology may not be suitable for some produce processing in that it causes adverse sensory effects and physiological injury due to its strong oxidizing effects. While the potential use of ozone in decontaminating wash water during produce processing has been suggested, ozone is highly sensitive to the presence of organic material and very unstable; therefore, it may not be as robust as other treatments in controlling microbial levels in water systems with high organic load.

*Organic acids*

Organic acids have been evaluated for produce decontamination treatments, with the results depending on the conditions used. For example, Akbas and Ölmez (2007) reported a greater reduction of natural flora using citric or

© Woodhead Publishing Limited, 2012

lactic acids over that using chlorine or water alone for cut lettuce. Zhao *et al.* (2009) found that treating lettuce with a combination of 3% levulinic acid plus 1% sodium dodecyl sulfate (SDS) for <20 s reduced both *Salmonella* and *E. coli* O157:H7 populations by >6.7 log CFU/g. Alfalfa seeds treated with 0.5% levulinic acid plus 0.05% SDS for 5 min at 21°C resulted in a >3.0 log inactivation of *E. coli* O157:H7 and *Salmonella* (Zhao *et al.*, 2010). Ortega *et al.* (2011) reported that levulinic acid with SDS is not effective in eliminating parasites from foods.

### 1.3.3 Mild heat treatment

The use of mild heat treatments has been studied for certain fruit products; however, mild heat would not be suitable for sensitive produce tissues such as leafy greens intended to be consumed as a fresh product. Pao and Davis (1999) reported a 5 log reduction of *E. coli* on the surface of inoculated oranges by immersing in water at 80°C for 1 min or 70°C for 2 min. The treatment was less effective at removing microflora from the stem-scar area. Fouladkhah and Avens (2010) reported a 3 log reduction of natural microflora surface contamination when fresh melons were immersed in water at 95°C. Undetectable levels of *E. coli* O157:H7 and *Salmonella* in inoculated mung beans were achieved by immersing mung bean seeds in hot water at 85°C for 40 s, followed by dipping in cold water for 30 s, and soaking in chlorine water (2000 ppm) for 2 h (Bari *et al.*, 2010). Li *et al.* (2002) found that mild heat treatment (dipped in 20 or 50°C water for 90 s) of cut lettuce leaves enhances the growth of *L. monocytogenes* during subsequent storage at 5 or 15°C. Mild heat treatment is used more for fruits because the lower surface to volume ratio makes them more resistant to the adverse effects of the heat treatment compared to leafy greens where wilting can occur.

## 1.4 Novel methods of fresh produce decontamination

Other chapters provide more information on novel methods of decontamination for a variety of products. The following discussion briefly addresses the application of certain novel methods specifically to decontamination of fresh produce.

### 1.4.1 Irradiation

Irradiation of fresh produce is recognized as a potential, safe, and reliable decontamination technology. However, the technology is not widely applied as it is not generally accepted in many regions. The mechanism of microbial inactivation in ionization radiation is damage of nucleic acids by direct or indirect damage due to oxidative radicals originating from the radiolysis of water (Nei *et al.*, 2010). Ionizing radiation penetrates food tissues, which

© Woodhead Publishing Limited, 2012

potentially makes it a good candidate for a terminal control step to target internalized pathogens and to reduce contamination in fresh produce (Niemira, 2007; Lynch *et al.*, 2009). Irradiation is also lethal to spoilage molds of fruits. Use of irradiation is successful in preserving herbs, mushrooms, strawberries, grapes, and other berries. Irradiation is considered to be safe, efficient, environmentally clean and energy-efficient (Nei *et al.*, 2010). However, the effectiveness of irradiation varies by produce type. For example, Niemira (2007) reported differences among reduction of *E. coli* O157:H7 on leafy greens depending on leaf types, with lower efficacy observed for spinach.

### 1.4.2 Non-thermal plasma

An emerging technology is non-thermal plasma, which may have future commercial applications for produce. Neutral particles, electrons, and positively or negatively charged atoms and molecules make up plasma. When a gas goes through plasma, the gas becomes excited, ionized, or dissociated by electron or ion collisions with the background gas. This leads to the formation of active species such as highly energized photons (UV spectra), atomic oxygen, ozone, and free radicals (e.g., hydroxyl, superoxide, and nitrogen oxides). These reactive species have antimicrobial activity through alterations in lipids, proteins, and nucleic acids. These alterations may result in microbial death or injury. A form of cold plasma, one atmosphere uniform glow discharge plasma (OAUGDP), is a novel, atmospheric, radio frequency plasma which can be generated over large areas in large volumes. Gram-positives are more resistant than Gram-negatives and spore formers are more resistant than non-spore formers (Kayes *et al.*, 2007). Exposure time, pH, incubation temperature, and culture age affected survival of *E. coli* O157:H7, *Listeria monocytogenes*, *Staphylococcus aureus*, *Bacillus cereus*, *Salmonella* Enteritidis, *Vibrio parahaemolyticus*, *Yersinia enterocolitica* and *Shigella flexneri* exposed to plasma (Kayes *et al.*, 2007). Critzer *at al.* (2007) treated *E. coli* O157:H7, *Salmonella*, and *L. monocytogenes* inoculated apples, cantaloupe, and lettuce with plasma and found >2 log reduction of *E. coli* O157:H7 in 2 min and >3 log reduction of *Salmonella*, and *L. monocytogenes* in 3 min. Niemira and Sites (2008) found that treating apples with 40 liters/min for 3 min of cold plasma provided reductions ranging from 2.9 to 3.7 log CFU/ml of *Salmonella* Stanley and 3.4–3.6 log CFU/ml of *E. coli* O157:H7.

### 1.4.3 Other novel methods

*Essential oils*

A number of essential oils from spices have antimicrobial properties; however, the high cost of essential oils may prohibit broad application of their use for produce decontamination. The active constituents in thyme oil are thymol, p-cymene and carvacrol. Reductions of viable bacteria were

© Woodhead Publishing Limited, 2012

reported when fresh cut lettuce was washed with basil oil or grape tomatoes with thymol (Erkmen, 2010; Lu and Wu, 2010). On whole tomato surfaces, allyl isothiocyanate in vapor phase (8.3 μl/liter of air) reduced *Salmonella* to the detection limit of <2 log CFU per tomato at 4 and 10°C in 10 days (Obaidat and Frank, 2009).

*Phage*

Leverentz *et al.* (2003) demonstrated reduction of *Salmonella* on fresh-cut vegetables using bacteriophage. Phages are strain-specific, thus successful application outside of the laboratory may require extensive study to optimize delivery and phage strain mix because of the extensive number of *Salmonella* serovars. Phage technology is very specific to the bacterium that it is targeting and will not be useful where there is a broad spectrum of pathogens present on the produce. Furthermore, this technology will not be applicable to decontamination of produce contaminated with human viruses.

*High hydrostatic pressure (HHP)*

High pressure processing may be applied to certain types of fresh produce if textural changes are acceptable. Kingsley *et al.* (2005) demonstrated inactivation of hepatitis A virus in strawberry purée and sliced green onions after 5 min exposures to pressures of 375 MPa with log reductions in plaque forming units (PFU) of 4.32 and 4.75, respectively; however, the technology had adverse effects on sensory qualities.

The technology may have application opportunities for seed decontamination. Pressures of 300 MPa for 15 min at 20°C produced a 4 to 6 log reduction on garden cress seeds inoculated with *S.* Typhimurium, *E. coli* MG1655, *L. innocua*, *S. flexneri* and pressure-resistant *E. coli* LMM1010 (Wuytack *et al.*, 2003). Neetoo *et al.* (2009) concluded that high pressure of 550 MPa for 2 min at 40°C provided a 5 log reduction of *E. coli* O157:H7 and had no adverse effects on germination of alfalfa seeds.

## 1.5 Conclusions and future trends

Produce associated foodborne illness outbreaks continue to be identified, which is likely to stimulate advances in produce decontamination strategies. It is preferable to use GAP to reduce the risk of contamination at the beginning of the supply stream; however, large-scale production and the potential to spread point source contamination during washing and flume conveyance favors use of decontamination treatments of water, at a minimum, to reduce the risk of spreading contamination. Such treatments reduce surface contamination through dilution and to some extent inactivation on the surface of produce, but prevent additional contamination and amplification of pathogens throughout the processing chain. Proper hygiene is crucial, starting with use of properly treated soil amendments, to proper handwashing stations, effective harvest

© Woodhead Publishing Limited, 2012

and transport equipment sanitation, good hygiene in processing facilities and in the kitchen, and measures to prevent cross-contamination.

Ultimately a pasteurization step for fresh produce is desired, and treatment combinations have been studied as a means to achieve this. For example, Sagong *et al.* (2011) demonstrated increased pathogen reduction of 0.8 to 1 log on lettuce when ultrasound was combined with organic acid treatments over that for acids alone. The greatest total log reductions observed were 2.75 for *E. coli* O157:H7, 3.18 for *S.* Typhimurium and 2.87 for *L. monocytogenes*. Similarly, Nei *et al.* (2010) discussed possible options for treating seeds prior to germination including chemical treatments combined with irradiation, high-pressure processing, heat, and soaking in water. Such combinations could be additive, synergistic or antagonistic, and treatment order can play a role in effectiveness (Nei *et al.*, 2010), thus specific combinations require validation under actual use conditions.

Atmospheric packaging has been applied for many years to assist in control of fresh produce ripening and spoilage. Typically these technologies have not been designed as a disinfection method for fresh produce. This may be a potential application in the future using one of the technologies described above or other new technologies. Similarly, in-field decontamination methods have not been the focus of commercial applications or research efforts. Technologies described above may have application for treatment of irrigation water, for example. With concerns related to on-farm contamination, innovative decontamination for specific pathogens may be developed in the future.

Predictive modeling has been applied to evaluate the potential for growth of pathogens in the supply chain, e.g. Tromp *et al.* (2010) modeled the growth of various pathogens on lettuce, while Pan and Schaffner (2010) studied the growth of *Salmonella* on cut tomatoes. Application of models to evaluate pathogen inactivation for different types of produce using different antimicrobials and the influence of time, temperature, concentration, organic load, pH, and other factors would be very useful to establish a robust and useful tool to manage produce wash systems using a science-based approach.

## 1.6 Sources of further information and advice

Enhanced epidemiological surveillance will continue to identify produce associated outbreaks that previously would have gone undetected. This will continue to stimulate investigation of produce decontamination strategies that reduce risk and maintain expected sensory characteristics. Control is needed throughout the food supply chain because of the perishable nature of these ready-to-eat products.

Regulatory guidance related to fresh produce is under development by the US Food and Drug Administration (FDA, 2011), risk assessments are

© Woodhead Publishing Limited, 2012

being conducted by FAO/WHO (2008), and no doubt innovative solutions will be identified in the future.

## 1.7 References


ACKERS M L, MAHON B E, LEAHY E, GOODE B, DAMROW T, HAYES P S, BIBB W F, RICE D H, BARRETT T J, HUTWAGNER L, GRIFFIN P M and SLUTSKER L (1998), 'An outbreak of *Escherichia coli* O157:H7 infections associated with leaf lettuce consumption', *J Infect Dis*, 177, 1588–1593.

AKBAS M Y and ÖLMEZ H (2007), 'Effectiveness of organic acid, ozonated water and chlorine dippings on microbial reduction and storage quality of fresh-cut iceberg lettuce', *J Sci Food Agric*, 87, 2609–2616.

BAERT L, VANDEKINDEREN I, DEVLIEGHERE F, VAN COILLIE E, DEBEVERE J and UYTTENDAELE M (2009), 'Efficacy of sodium hypochlorite and peroxyacetic acid to reduce murine norovirus 1, B40-8, *Listeria monocytogenes*, and *Escherichia coli* O157:H7 on shredded iceberg lettuce and in residual wash water', *J Food Prot*, 72, 1047–1054.

BARI M L, ENOMOTO K, NEI D and KAWAMOTO S (2010), 'Practical evaluation of mung bean seed pasteurization method in Japan', *J Food Prot*, 73, 752–757.

BARTH M, HANKINSON T R, ZHUANG H and BREIDT F (2010), 'Microbiological spoilage of fruits and vegetables', in Sperber W H and Doyle M P, *Compendium of the microbiological spoilage of foods and beverages*, New York, Springer, 135–183.

BEHRSING J, WINKLER S, FRANZ P and PREMIER R (2000), 'Efficacy of chlorine for inactivation of *Escherichia coli* on vegetables', *Postharvest Biol Technol*, 19, 187–192.

BENARDE M A, ISRAEL B M, OLIVIERI V P and GRANSTROM M L (1965), 'Efficiency of chlorine dioxide as a bactericide', *Appl Microbiol*, 13, 776–780.

BEUCHAT L R (1996), 'Pathogenic microorganisms associated with fresh produce', *J Food Prot*, 59, 204–216.

BEUCHAT L R (1998), 'Surface decontamination of fruits and vegetables eaten raw: a review'. World Health Organization, Food Safety Unit WHO/FSF/FOS/98.2. Available from: http://www.who.int/foodsafety/publications/fs_management/en/surface_decon.pdf (accessed 19 April 2011).

BUCHANAN R, EDELSON S and MILLER R (1999), 'Contamination of intact apples after immersion in an aqueous environment containing *Escherichia coli* O157:H7', *J Food Prot*, 62, 444–450.

CDC (US CENTERS FOR DISEASE CONTROL AND PREVENTION) (1991), 'Epidemiologic notes and reports multistate outbreak of *Salmonella* Poona infections – United States and Canada, 1991', *Morbidity Mortality Weekly Rep*, 40, 549–552.

CDC (1996), 'Update: Outbreaks of *Cyclospora cayetanensis* infection – United States and Canada, 1996', *Morbidity Mortality Weekly Rep*, 45, 611–612.

CDC (1997), 'Hepatitis A associated with consumption of frozen strawberries – Michigan, March 1997', *Morbidity Mortality Weekly Rep*, 46, 288, 295.

CDC (2003), 'Hepatitis A outbreak associated with green onions at a restaurant – Monaca, Pennsylvania, 2003', *Morbidity Mortality Weekly Rep*, 52, 1155–1157.

CDC (2006a), 'Ongoing multistate outbreak of *Escherichia coli* serotype O157:H7 infections associated with consumption of fresh spinach – United States, September 2006', *Morbidity Mortality Weekly Rep*, 55(Dispatch), 1–2.

CDC (2006b), 'Travelers' diarrhea'. Available from: http://www.cdc.gov/ncidod/dbmd/diseaseinfo/travelersdiarrhea_g.htm#prevent (accessed 14 April 2011).

CDC (2007), 'Multistate outbreaks of *Salmonella* infections associated with raw tomatoes eaten in restaurants – United States, 2005–2006', *Morbidity Mortality Weekly Rep*, 56, 909–911.


© Woodhead Publishing Limited, 2012

CDC (2008), 'Outbreak of *Salmonella* serotype Saintpaul infections associated with multiple raw produce items – United States, 2008', *Morbidity Mortality Weekly Rep*, 57, 929–934.

CDC (2009a), 'Investigation of an outbreak of *Salmonella* Saintpaul infections linked to raw alfalfa sprouts'. Available from: http://www.cdc.gov/salmonella/saintpaul/alfalfa/index.html (accessed 1 September 2011).

CDC (2009b), 'Surveillance for foodborne disease outbreaks – United States, 2006', *Morbidity Mortality Weekly Rep*, 58, 609–615.

CDC (2010a), 'Investigation update: Multistate outbreak of human *E. coli* O145 infections linked to shredded romaine lettuce from a single processing facility'. Available from: http://www.cdc.gov/ecoli/2010/ecoli_o145/index.html (accessed 1 September 2011).

CDC (2010b), 'Investigation update: Multistate outbreak of human *Salmonella* Newport infections linked to raw alfalfa sprouts'. Available from: http://www.cdc.gov/salmonella/newport/index.html (accessed 1 September 2011).

CDC (2010c), 'Surveillance for foodborne disease outbreaks – United States, 2007', *Morbidity Mortality Weekly Rep*, 59, 973–979.

CDC (2011a), 'Cryptosporidiosis outbreak at a summer camp – North Carolina, 2009', *Morbidity Mortality Weekly Rep*, 60, 918–922.

CDC (2011b), 'Investigation update: Multistate outbreak of human *Salmonella* Agona infections linked to whole, fresh imported papayas'. Available from: http://www.cdc.gov/salmonella/agona-papayas/index.html (accessed 1 September 2011).

CDC (2011c), 'Investigation update: Multistate outbreak of human *Salmonella* Enteritidis infections linked to alfalfa sprouts and spicy sprouts'. Available from: http://www.cdc.gov/salmonella/sprouts-enteritidis0611/070611/index.html (accessed 1 September 2011).

CDC (2011d), 'Investigation update: Multistate outbreak of human *Salmonella* I 4, [5], 12:i:- infections linked to alfalfa sprouts'. Available from: http://www.cdc.gov/salmonella/i4512i-/021011/index.html (accessed 1 September 2011).

CDC (2011e), 'Investigation update: Multistate outbreak of *Salmonella* Panama infections linked to cantaloupe'. Available from: http://www.cdc.gov/salmonella/panama0311/062311/index.html (accessed 1 September 2011).

CDC (2011f), 'Multistate outbreak of Listeriosis linked to whole cantaloupes from Jensen Farms, Colorado'. Available from: http://www.cdc.gov/listeria/outbreaks/cantaloupes-jensen-farms/index.html (accessed 11 April 2012).

CODEX ALIMENTARIUS (2003), Code of Hygienic Practice for Fresh Fruits and Vegetables, CAC/RCP 53-2003, Joint FAO/WHO Food Standards Program, FAO, Rome.

COOK K A, BOYCE T, LANGKOP C, KUO K, SWARTZ M, EWERT D, SOWERS E, WELLS J and TAUXE R (1995), 'Scallions and shigellosis: a multistate outbreak traced to imported green onions', In *Epidemic intelligence service 44th annual conference, March 27–31*, CDC, Atlanta, GA, p. 36.

CORDS B R, BURNETT D L, HILGREN J, FINLEY M and MAGNUSON J (2005), 'Sanitizers: Halogens, surface-active agents, and peroxides', in Davidson P M, Sofos J N and Branen A L, *Antimicrobials in food*, Boca Raton, FL, CRC Press, 507–572.

CRITZER F J, KELLY-WINTENBERG K, SOUTH S L and GOLDEN D A (2007), 'Atmospheric plasma inactivation of foodborne pathogens on fresh produce surfaces', *J Food Prot*, 70, 2290–2296.

CUMMINGS K, BARRETT B, MOHLE-BOETANI J C, BROOKS J T, FARRAR J, HUNT T, FIORE A, KOMATSU K, WERNER S B and SLUTSKER L (2001), 'A multistate outbreak of *Salmonella enterica* serotype Baildon associated with domestic tomatoes', *Emerg Inf Dis*, 7, 1046–1048.

DAVIS H, TAYLOR J P, PERDUE J N, STELMA JR. G N, ROWNTREE III R and GREENE K D (1988), 'A Shigellosis outbreak traced to commercially distributed shredded lettuce', *Am J Epidemiol*, 128, 1312–1321.

DENTINGER C, BOWER W A, NAINAN O V, COTTER S M, MYERS G, DUBUSKY L M, FOWLER S,

© Woodhead Publishing Limited, 2012

SALEHI E D P and BELL B P (2001), 'An outbreak of hepatitis A associated with green onions', *J Infect Dis*, 183, 1273–1276.

DEZA M A, ARAUJO M and GARRIDO M J (2003), 'Inactivation of *Escherichia coli* O157:H7, *Salmonella enteritidis* and *Listeria monocytogenes* on the surface of tomatoes by neutral electrolyzed water', *Letters Appl Microbiol*, 37, 482–487.

DREYFUSS M (2009), 'Is norovirus a foodborne or pandemic pathogen? An analysis of the transmission of norovirus-associated gastroenteritis and the roles of food and food handlers', *Foodborne Pathog Dis*, 6, 1219–1228.

EFSA (EUROPEAN FOOD SAFETY AUTHORITY) (2005a), 'Opinion of the Scientific Panel on food additives, flavourings, processing aids and materials in contact with food on a request from the Commission related to semicarbazide in food. Question number EFSA-2003-235', *EFSA J*, 219, 1–35.

EFSA (2005b), 'Opinion of the Scientific Panel on food additives, flavourings, processing aids and materials in contact with food (AFC) on a request from the Commission related to treatment of poultry carcasses with chlorine dioxide, acidified sodium chloride, trisodium phosphate and peroxyacids. Question number EFSA Q-2005-002', *EFSA J*, 297, 1–27.

EFSA (2011), 'EU summary report on trends and sources of zoonoses and zoonotic agents and food-borne outbreaks 2009', *EFSA J*, 9, 260–301.

EMBERLAND K E, ETHELBERG S, KUUSI M, VOLD L, JENSVOLL L, LINDSTEDT B-A, NYGÅRD K, KJELSØ C, TORPDAHL M, SØRENSEN G, JENSEN T, LUKINMAA S, NISKANEN T and KAPPERUD G (2007), 'Outbreak of *Salmonella* Weltevreden infections in Norway, Denmark and Finland associated with alfalfa sprouts, July–October 2007', *Eurosurveillance*, 12, 389–390.

EPA (2011), 40 CFR 180.940 Tolerance exemptions for active and inert ingredients for use in antimicrobial formulations (food-contact surface sanitizing solutions). Available from: http://www.gpo.gov/fdsys/pkg/CFR-2010-title40-vol23/pdf/CFR-2010-title40-vol23-sec180-940.pdf (accessed 24 July 2011).

ERICKSON M C, WEBB C C, DIAZ-PEREZ J C, PHATAK S C, SILVOY J J, DAVEY L, PAYTON A S, LIAO J, MA L and DOYLE M P (2010), 'Infrequent internalization of *Escherichia coli* O157:H7 into field-grown leafy greens', *J Food Prot*, 73, 500–506.

ERKMEN O (2010), 'Antimicrobial effects of hypochlorite on *Escherichia coli* in water and selected vegetables', *Foodborne Pathog Dis*, 7, 953–958.

ETHELBERG S, LISBY M, BÖTTIGER B, SCHULTZ A C, VILLIF A, JENSEN T, OLSEN K E, SCHEUTZ F, KJELSØ C and MÜLLER L (2010), 'Rapid communications: outbreaks of gastroenteritis linked to lettuce, Denmark, January 2010', *Eurosurveillance*, 15, 1–3.

EU (EUROPEAN UNION) (2010), 'The Rapid Alert System for Food and Feed (RASFF) Annual Report, 2009', Available from: http://ec.europa.eu/food/food/rapidalert/docs/report2009_en.pdf (accessed 2 September 2011).

EU (2011), 'Rapid Alert System for Food and Feed', Available from: http://ec.europa.eu/food/food/rapidalert/index_en.htm (accessed 25 April 2011).

FAO/WHO (2008), Microbial hazards in fresh fruits and vegetable: microbial risk assessment series, pre-publication version. Food and Agriculture Organization and World Health Organization. Available from: http://www.fao.org/ag/agn/agns/files/FFV_2007_Final.pdf. (accessed 2 September 2011).

FDA (US FOOD AND DRUG ADMINISTRATION) (1998), 'Guidance for industry: guide to minimize microbial food safety hazards for fresh fruits and vegetables'. Available from: http://www.fda.gov/Food/GuidanceComplianceRegulatoryInformation/GuidanceDocuments/ProduceandPlanProducts/UCM064574.htm (accessed 18 April 2011).

FDA (2006), 'FDA statement on foodborne *E. coli* O157:H7 outbreak in spinach'. Available from: http://www.fda.gov/NewsEvents/Newsroom/PressAnnouncements/2006/ucm108761.htm (accessed 18 April 2011).

FDA (2008), 'Guidance for industry: Guide to minimize microbial food safety hazards of fresh-cut fruits and vegetables'. Available from: http://www.fda.

© Woodhead Publishing Limited, 2012

gov/Food/GuidanceComplianceRegulatoryInformation/GuidanceDocuments/ProduceandPlanProducts/ucm064458.htm (accessed 25 March 2011).

FDA (2009a), 'Guidance for industry: guide to minimize microbial food safety hazards of melons; draft guidance'. Available from: http://www.fda.gov/Food/GuidanceComplianceRegulatoryInformation/GuidanceDocuments/ProduceandPlanProducts/ucm174171.htm (accessed 15 April 2011).

FDA (2009b), 'Guidance for industry: guide to minimize microbial food safety hazards of leafy greens; draft guidance'. Available from: http://www.fda.gov/Food/GuidanceComplianceRegulatoryInformation/GuidanceDocuments/ProduceandPlanProducts/ucm174200.htm (accessed 15 April 2011).

FDA (2009c), 'Guidance for industry: guide to minimize microbial food safety hazards of tomatoes; draft guidance'. Available from: http://www.fda.gov/Food/GuidanceComplianceRegulatoryInformation/GuidanceDocuments/ProduceandPlanProducts/ucm173902.htm (accessed 15 April 2011).

FDA (2010), 'DSHS orders Sangar Produce to close, recall products'. Available from: http://www.fda.gov/Safety/Recalls/ucm230709.htm (accessed 18 April 2011).

FDA (2011), 'Produce and plant products guidance to the industry'. Available from: http://www.fda.gov/Food/GuidanceComplianceRegulatoryInformation/GuidanceDocuments/ProduceandPlanProducts/default.htm (accessed 2 September 2011).

FOONG-CUNNINGHAM S, WESSINGER A, HINES J, GADBOIS M, HERDT J and SWANSON K (2008), 'Efficacy of antimicrobial agents to reduce transfer of *Escherichia coli* O157:H7 on lettuce pieces', Poster presentation at IAFP Annual Meeting, Columbus, OH.

FOULADKHAH A and AVENS J (2010), 'Effects of combined heat and acetic acid on natural microflora reduction on cantaloupe melons', *J Food Prot*, 73, 981–984.

GALLOT C, GROUT L, ROQUE-AFONSO A-M, COUTURIER E, CARRILLO-SANTISTEVE P, POUEY J, LETORT M-J, HOPPE S, CAPDEPON P, SAINT-MARTIN S, DE VALK H and VAILLANT V (2011), 'Hepatitis A associated with semidried tomatoes, France, 2010', *Emerg Inf Dis*, 17, 566–567.

GONZALEZ R J, LUO Y, RUIZ-CRUZ, S and CEVOY J L M (2004), 'Efficacy of sanitizers to inactivate *Escherichia coli* O157:H7 on fresh-cut carrot shreds under simulated process water conditions', *J Food Prot*, 67, 2375–2380.

HERWALDT B L, LEW J F, MOE C L, LEWIS D C, HUMPHREY C D, MONROE S S, PON E W and GLASS R I (1994), 'Characterization of a variant strain of Norwalk virus from a food-borne outbreak of gastroenteritis on a cruise ship in Hawaii', *J Clin Microbiol*, 32, 861–866.

HJERTQVIST M, JOHANSSON A, SVENSSON N, ÅBOM P E, MAGNUSSON C, OLSSON M, HEDLUND K O and ANDERSSON Y (2006), 'Four outbreaks of norovirus gastroenteritis after consuming raspberries, Sweden, June–August 2006'. Available from: http://www.eurosurveillance.org/ViewArticle.aspx?Articleid=3038 (accessed 28 April 2011).

HOENICKE K, GATERMANN R, HARTIG L, MANDIX M and OTTE S (2004), 'Formation of semicarbazide (SEM) in food by hypochlorite treatment: is SEM a specific marker for nitrofurazone abuse?', *Food Addit Contam*, 21, 526–537.

ICMSF (INTERNATIONAL COMMISSION ON MICROBIOLOGICAL SPECIFICATIONS FOR FOODS) (2005), *Microorganisms in foods 6: microbial ecology of food commodities*, 2nd edn, New York, Kluwer Academic/Plenum Publishers.

ICMSF (2011), *Microorganisms in foods 8: use of data for assessing process control and product acceptance*, New York, Springer.

INATSU Y, BARI M L, KAWASAKI S, ISSHIKI K and KAWAMOTO S (2005), 'Efficacy of acidified sodium chlorite treatments in reducing *Escherichia coli* O157:H7 on Chinese cabbage', *J Food Prot*, 68, 251–255.

INATSU Y, KITAGAWA T, BARI M L, NEI D, JUNEJA V and KAWAMOTO S (2010), 'Effectiveness of acidified sodium chlorite and other sanitizers to control *Escherichia coli* O157:H7 on tomato surfaces', *Foodborne Pathog Dis*, 7, 629–635.

KAYES M M, CRITZER F J, KELLY-WINTENBERG K, ROTH J R, MONTIE T C and GOLDEN D

© Woodhead Publishing Limited, 2012

A (2007), 'Inactivation of foodborne pathogens using one atmosphere', *Foodborne Pathog Dis*, 4, 50–59.

KESKINEN L A, BURKE A and ANNOUS B A (2009), 'Efficacy of chlorine, acidic electrolyzed water and aqueous chlorine dioxide solutions to decontaminate *Escherichia coli* O157:H7 from lettuce leaves', *Int J Food Microbiol*, 132, 134–140.

KINGSLEY D, GUAN D and HOOVER D (2005), 'Pressure inactivation of hepatitis A virus in strawberry purée and sliced green onions', *J Food Prot*, 68, 1748–1751.

KOSEKI S, YOSHIDA K, ISOBE S and ITOH K (2004), 'Efficacy of acidic electrolyzed water for microbial decontamination of cucumbers and strawberries', *J Food Prot*, 67, 1247–1251.

LEVERENTZ B, CONWAY W S, CAMP M J, JANISIEWICZ W J, ABULADZE T, YANG M, SAFTNER R and SULAKVELIDZE A (2003), 'Biocontrol of *Listeria monocytogenes* on fresh-cut produce by treatment with lytic bacteriophages and a bacteriocin', *Appl Environ Microbiol*, 69, 4519–4526.

LI Y, BRACKETT R E, CHEN J and BEUCHAT L R (2002), 'Mild heat treatment of lettuce enhances growth of *Listeria monocytogenes* during subsequent storage at 5°C or 15°C', *J Appl Microbiol*, 92, 269–275.

LU Y and WU C (2010), 'Reduction of *Salmonella enterica* contamination on grape tomatoes by washing with thyme oil, thymol, and carvacrol as compared with chlorine treatment', *J Food Prot*, 73, 2270–2275.

LYNCH M F, TAUXE R V and HEDBERG C W (2009), 'The growing burden of foodborne outbreaks due to contaminated fresh produce: risks and opportunities', *Epidemiol Infect*, 137, 307–315.

MAHON B E, PÖNKÄ A, HALL W N, KOMATSU K, DIETRICH S E, SIITONEN A, CAGE G, HAYES P S, LAMBERT-FAIR M A, BEAN N H, GRIFFIN P M and SLUTSKER L (1997), 'An international outbreak of *Salmonella* infections caused by alfalfa sprouts grown from contaminated seeds', *J Infect Dis*, 175, 876–882.

MESQUITA J R and NASCIMENTO M S J (2009), 'Surveillance and outbreak reports: a foodborne outbreak of norovirus gastroenteritis associated with a Christmas dinner in Porto, Portugal, December 2008', *Eurosurveillance*, 14, 1–3.

MUNNOCH S A, WARD K, SHERIDAN S, FITZSIMMONS G J, SHADBOLT C T, PIISPANEN J P, WANG Q, WARD T J, WORGAN T L M, OXENFORD C, MUSTO J A, MCANULTY J and DURRHEIM D N (2009), 'A multi-state outbreak of *Salmonella* Saintpaul in Australia associated with cantaloupe consumption', *Epidemiol Infect*, 137, 367–374.

NACMCF (1999), 'Microbiological safety evaluations and recommendations on sprout seeds', *Int J Food Microbiol*, 52, 123–153.

NEETOO H, PIZZOLATO T and CHEN H (2009), 'Elimination of *Escherichia coli* O157:H7 from alfalfa seeds through a combination of high hydrostatic pressure and mild heat', *Appl Environ Microbiol*, 75, 1901–1907.

NEI D, CHOI J-W, BARI M L, KAWASAKI S, KAWAMOTO S and INATSU Y (2009), 'Efficacy of chlorine and acidified sodium chlorite on microbial population and quality changes of spinach leaves', *Foodborne Pathog Dis*, 6, 541–546.

NEI D, BARI M L, INATSU Y, KAWASAKI S, TODORIKI S and KAWAMOTO S (2010), 'Combined effect of low-dose irradiation and acidified sodium chlorite washing on *Escherichia coli* O157:H7 inoculated on mung bean seeds', *Foodborne Pathog Dis*, 7, 1217–1223.

NIEMIRA B (2007), 'Relative efficacy of sodium hypochlorite wash versus irradiation to inactivate *Escherichia coli* O157:H7 internalized in leaves of romaine lettuce and baby spinach', *J Food Prot*, 70, 2526–2532.

NIEMIRA B A and SITES J (2008), 'Cold plasma inactivates *Salmonella* Stanley and *Escherichia coli* O157:H7 inoculated on Golden Delicious apples', *J Food Prot*, 71, 1357–1365.

NYGÅRD K, LASSEN J, VOLD L, ANDERSSON Y, FISHER I, LÖFDAHL S, THRELFALL J, LUZZI I, PETERS T, HAMPTON M, TORPDAHL M, KAPPERUD G and AAVITSLAND P (2008), 'Outbreak of *Salmonella* Thompson infections linked to imported rucola lettuce', *Foodborne Pathog Dis*, 5, 165–173.

© Woodhead Publishing Limited, 2012

OBAIDAT M M and FRANK J F (2009), 'Inactivation of *Salmonella* and *Escherichia coli* O157:H7 on sliced and whole tomatoes by allyl isothiocyanate, carvacrol, and cinnamaldehyde in vapor phase', *J Food Prot*, 72, 315–324.

O'MAHONY M, COWDEN J, SMYTH B, LYNCH D, HALL M, ROWE B, TEARE E L, TETTMAR R E, RAMPLING A M, COLES M, GILBERT R J, KINGCOTT E and BARTLETT C L R (1990), 'An outbreak of *Salmonella* Saint-Paul infection associated with bean sprouts', *Epidemiol Infect*, 104, 229–235.

ORTEGA Y, TORRES M and TATUM J (2011), 'Efficacy of levulinic acid-sodium dodecyl sulfate against *Encephalitozoon intestinalis*, *Escherichia coli* O157:H7, and *Cryptosporidium parvum*', *J Food Prot*, 74, 140–144.

PAN W and SCHAFFNER D (2010), 'Modeling the growth of *Salmonella* in cut red round tomatoes as a function of temperature', *J Food Prot*, 73, 1502–1505.

PAO S and DAVIS C L (1999), 'Enhancing microbiological safety of fresh orange juice by fruit immersion in hot water and chemical sanitizers', *J Food Prot*, 62, 756–760.

PARK C and BEUCHAT L (1999), 'Evaluation of sanitizers for killing *Escherichia coli* O157:H7, *Salmonella*, and naturally occurring microorganisms on cantaloupes, honeydew melons, and asparagus', *Dairy Food Environ Sanitation*, 19, 842–847.

PEZZOLI L, ELSON R, LITTLE C L, YIP H, FISHER I, YISHAI R, ANIS E, VALINSKY L, BIGGERSTAFF M, PATEL N, MATHER H, BROWN D J, COIA J E, VAN PELT W, NIELSEN E M, ETHELBERG S, DE PINNA E, HAMPTON M D, PETERS T and THRELFALL J (2008), 'Packed with Salmonella – investigation of an international outbreak of *Salmonella* Senftenberg infection linked to contamination of prepacked basil in 2007', *Foodborne Pathog Dis*, 5, 661–668.

PRAKASH A, CHEN P-C, PILLING R L, JOHNSON N and FOLEY D (2007a), '1% Calcium chloride treatment in combination with gamma irradiation improves microbial and physiochemical properties of diced tomatoes', *Foodborne Pathog Dis*, 4, 89–98.

PRAKASH A, JOHNSON N and FOLEY D (2007b), 'Irradiation D values of *Salmonella* spp. in diced tomatoes dipped in 1% calcium chloride', *Foodborne Pathog Dis*, 4, 84–88.

RIES A A, ZAZA S, LANGKOP C, TAUXE R V and BLAKE P A (1990), 'A multistate outbreak of *Salmonella* chester linked to imported cantaloupe' (Abstract). In: American Society for Microbiology. Program and abstracts of the 30th Interscience Conference on Antimicrobial Agents and Chemotherapy. Washington, DC, American Society for Microbiology, 238.

RKI (ROBERT KOCH INSTITUTE) (2011), 'EHEC/HUS O104:H4 – The outbreak is considered over'. Available from: http://www.rki.de/cln_116/nn_217400/EN/Home/PM__EHEC.html (accessed 17 August 2011).

ROSENBLUM L S, MIRKIN I R, ALLEN D T, SAFFORD S and HADLER S C (1990), 'A multifocal outbreak of hepatitis A traced to commercially distributed lettuce', *Amer J Pub Health*, 80, 1075–1079.

SAGONG H G, LEE S Y, CHANG P S, HEU S, RYU S, CHOJ Y J and KANG D H (2011), 'Combined effect of ultrasound and organic acids to reduce *Escherichia coli* O157:H7, *Salmonella* Typhimurium, and *Listeria monocytogenes* on organic fresh lettuce', *Intl J Food Microbiol*, 145, 287–292.

SAPERS G and SIMMONS G (1998), 'Hydrogen peroxide disinfection of minimally processed fruits and vegetables', *Food Technol*, 52, 48–52.

SARVIKIVI E, ROIVAINEN M, MAUNULA L, NISKANEN T, KORHONEN T, LAPPALAINEN M and KUUSI M (2011), 'Multiple norovirus outbreaks linked to imported frozen raspberries', *Epidemiol Infect*, 22, 1–8.

SCHLECH III W F, LAVIGNE P M, BORTOLUSSI R A, ALLEN A C, HALDANE E V, WORT A J, HIGHTOWER A W, JOHNSON S E, KING S H, NICHOLLS E S and BROOME C V (1983), 'Epidemic Listeriosis – Evidence for transmission by food', *N Engl J Med*, 308, 203–206.

SHARMA R R and DEMIRCI A (2003), 'Treatment of *E. coli* O157:H7 inoculated alfalfa seeds and sprouts with electrolyzed oxidizing water', *Int J Food Microbiol*, 86, 231–237.

STOPFORTH J, MAI T, KOTTAPALLI B and SAMADPOUR M (2008), 'Effect of acidified

© Woodhead Publishing Limited, 2012

sodium chlorite, chlorine, and acidic electrolyzed water on *Escherichia coli* O157:H7, *Salmonella*, and *Listeria monocytogenes* inoculated onto leafy greens', *J Food Prot*, 71, 625–628.

TAUXE R, KRUSE H, HEDBERG C, POTTER M, MADDEN J and WACHSMUTH K (1997), 'Microbial hazards and emerging issues associated with produce – A preliminary report to the National Advisory Committee on Microbiologic Criteria for Foods', *J Food Prot*, 60, 1400–1408.

TROMP S O, RIJGERSBERG H and FRANZ E (2010), 'Quantitative microbial risk assessment for *Escherichia coli* O157:H7, *Salmonella enterica*, and *Listeria monocytogenes* in leafy green vegetables consumed at salad bars, based on modeling supply chain logistics', *J Food Prot*, 73, 1830–1840.

VARMA J K, SAMUEL M C, MARCUS R, HOEKSTRA R M, MEDUS C, SEGLER S, ANDERSON B J, JONES T F, SHIFERAW B, HAUBERT N, MEGGINSON M, MCCARTHY P V, GRAVES L, VAN GILDER T and ANGULO F J (2007), '*Listeria monocytogenes* infection from foods prepared in a commercial establishment: a case-control study of potential sources of sporadic illness in the United States', *Clin Inf Dis*, 44, 521–528.

WACHTEL M R and CHARKOWSKI A O (2002), 'Cross-contamination of lettuce with *Escherichia coli* O157:H7', *J Food Prot*, 65, 465–470.

WANG H, FENG H and LUO YAGUANG (2004), 'Microbial reduction and storage quality of fresh-cut cilantro washed with acidic electrolyzed water and aqueous ozone', *Food Res Int*, 37, 949–956.

WATANABE Y, OZASA K, MERMIN J H, GRIFFIN P M, MASUDA K, IMASHUKU S and SAWADA T (1999), 'Factory outbreak of *Escherichia coli* O157:H7 infection in Japan', *Emerg Inf Dis*, 5, 424–428.

WUYTACK E Y, DIELS A M J, MEERSSEMAN K and MICHIELS C W (2003), 'Decontamination of seeds for seed sprout production by high hydrostatic pressure', *J Food Prot*, 66, 918–923.

XU L (1999) 'Use of ozone to improve the safety of fresh fruits and vegetables', *Food Technol*, 53, 58–62.

ZHANG G, MA L, PHELAN V H and DOYLE M P (2009), 'Efficacy of antimicrobial agents in lettuce leaf processing water for control of *Escherichia coli* O157:H7', *J Food Prot*, 72, 1392–1397.

ZHAO T, ZHAO P and DOYLE M P (2009), 'Inactivation of *Salmonella* and *Escherichia coli* O157:H7 on lettuce and poultry skin by combinations of levulinic acid and sodium dodecyl sulfate', *J Food Prot*, 72, 928–936.

ZHAO T, ZHAO P and DOYLE M P (2010), 'Inactivation of *Escherichia coli* O157:H7 and *Salmonella* Typhimurium DT 104 on alfalfa seeds by levulinic acid and sodium dodecyl sulfate', *J Food Prot*, 73, 2010–2017.

ZHUANG R, BEUCHAT L and ANGULO F (1995), 'Fate of *Salmonella* montevideo on and in raw tomatoes as affected by temperature and treatment with chlorine', *Appl Environ Microbiol*, 61, 2127–2131.

© Woodhead Publishing Limited, 2012

# 2

# Microbial decontamination of raw and ready-to-eat meats

A. Gill, Health Canada, Canada and C. O. Gill, Agriculture and Agri-Food Canada, Canada

**Abstract**: The safety and shelf life of meat products can be increased by the application of appropriate decontamination strategies. In this chapter the technologies for the decontamination of fresh and processed red meats that are suitable for application on a commercial scale are discussed and the effects on the bacterial flora described. The decontamination strategies covered include: thermal processing, freezing and chilling, the removal of contamination by washing or excision, irradiation by ionizing and ultraviolet radiation, high pressure processing and the application of antimicrobial chemicals.

**Key words**: meat; pasteurizing; irradiation; antimicrobials; high pressure processing.

## 2.1 Introduction

Regulations intended to ensure that only wholesome raw and processed meats are offered for sale to consumers have been in place in most developed countries since the early years of the 20th century (Derbyshire, 2006). Initially, meat hygiene regulations were largely concerned with the exclusion from the meat supply of meat from moribund animals, overtly diseased tissue from apparently healthy animals and meat contaminated with filth (Thornton and Gracey, 1974). Regulations regarding the distribution and further processing of raw meats commonly stipulated only general requirements for hygienic handling of product, and the cleanliness of premises and equipment used for meat processing (Blamire, 1984). Improvements in animal husbandry during the latter half of the 20th century greatly reduced the incidences of zoonotic diseases among meat animals. Consequently, contamination of meat

© Woodhead Publishing Limited, 2012

with enteric pathogens, often carried by asymptomatic animals, emerged as the greatest public health risk associated with the consumption of meat (Morris, 2003). The identification of previously unknown pathogens as causes of diseases acquired from meats emphasized the need for enhanced control over the contamination of meats with enteric pathogens (Mor-Mur and Yuste, 2009). Moreover, changes within meat packing and processing industries have resulted in meat being produced from decreasing numbers of plants of increasing size (Broadway, 2002; MacDonald, 2003). Failure of control over hazardous bacterial contamination of product at bigger plants inevitable results in instances of large amounts of compromised product being widely distributed, with corresponding large and widespread risks for consumers.

In commercial practice, enteric pathogens can be largely eliminated from the surfaces of beef carcasses by pasteurising dressed carcasses with hot water or steam. Similar results might be attained by spraying carcasses with relatively high volumes of lactic acid. Such decontaminating treatments could be applied to other types of red meat carcasses with equal effect, provided that the adverse affects of such harsh treatments on carcass appearance are acceptable. Decontaminating treatments that degrade meat appearance generally can not be applied to cuts because the appeal of the meat to consumers would be adversely affected. However, appearance degrading treatments can be applied to manufacturing meat that will be ground or further processed, because discolouration of meat surfaces does not substantially affect the appearance of the final products. Spraying carcasses or cuts with antimicrobial solutions, other than organic acids at relatively high concentration, are in practice largely ineffective for decontaminating fresh meats. The surfaces of ready-to-eat meats that are recontaminated after cooking can be pasteurised by heating packaged products by immersion in hot water or radio frequency, alternatively by the application of steam to the product before sealing of the pack. Surfaces can also be decontaminated by exposure to ultraviolet light. Both surface and internal contaminants of processed meat can be inactivated by high pressure processing or gamma or electron beam irradiation. Fresh meat also can be decontaminated by irradiation, but adverse effects on product appearance generally preclude the use of high pressure processing with fresh meats.

Fresh and processed meats may potentially serve as vehicles for many bacterial pathogens. The sources of pathogens on fresh meats are the hides or intestinal contents of animals, from which bacteria are transferred to meat during carcass dressing, or contamination events during subsequent handling. Certain pathogens are more commonly associated with particular meats and product types than others. A survey of foodborne outbreaks from 1988 to 2007 found that the most common cause of outbreaks involving beef was pathogenic *Escherichia coli* (34.6%), particularly *E. coli* O157:H7, followed by *Salmonella* (27.5%), *Clostridium perfringens* (19.5%), and *Staphylococcus aureus* (5.1%) (Greig and Ravel, 2009). For pork, *Salmonella* accounted for 41.3% of outbreaks, *Staphylococcus aureus* 19.8%, *Clostridium perfringens*

© Woodhead Publishing Limited, 2012

8.2% and *E. coli* for only 1.1% of outbreaks (Greig and Ravel, 2009). Although *C. perfringens* and *S. aureus* have a high association with beef and pork, outbreaks involving these organisms are invariably a consequence of post production contamination and failure to maintain proper temperature control. Thus, the focus of the decontamination of fresh meats is the enteric pathogens, pathogenic *E. coli*, *Salmonella*, *Shigella* and *Yersinia*.

Outbreaks associated with processed meats more commonly involve *S. aureus* and *Listeria monocytogenes* as these pathogens have greater tolerance of the low water activity and pH of these products, which restricts the growth of many other pathogens (Pearson and Tauber, 1996). Although their growth may be restricted on some products, due to their low infectious dose, contamination of processed meats with *E. coli* O157 or *Salmonella* has led to outbreaks (Williams *et al.*, 2000; MacDonald *et al.*, 2004; CDC, 2010).

The concerns about contamination of meats with enteric pathogens were first addressed by enhancement of traditional inspection procedures (Hathaway and McKenzie, 1991). As such an approach signally failed to prevent widespread outbreaks of enteric diseases acquired from meat, regulatory agencies sought improvement of meat safety by requiring the implementation of hazard analysis critical control point (HACCP) systems at meat packing and processing plants (Hulebak and Schlosser, 2002; Miller, 2008).

In the United States, there was early recognition that decontaminating treatments for raw meats would be a desirable, and ultimately, a required element of effective HACCP systems for meat (Sofos, 2008). Regulatory authorities elsewhere have allowed only restricted use of decontaminating treatments for raw meats (Hugas and Tsigarida, 2008). Therefore, much of the reported work on raw meat decontamination has been carried out and is related to practices in North America. Regional restrictions on decontaminating treatments for processed meats have been less pronounced, so treatments for that purpose have been extensively investigated within as well as outside North America.

Processed meats are a highly diverse range of products prepared from either intact primal cuts or chopped or emulsified tissue through a combination of such processes as curing, drying, smoking, fermentation and cooking (Pearson and Tauber, 1996). Traditional processed meats (fermented sausage, ham, bacon, etc.) have their origins in the trial and error development of methods to extend the storage life of meat by inhibiting growth of spoilage organisms. Significant changes to the appearance, flavour, aroma and texture of the product that may occur during processing have become desirable to consumers accustomed to the products. With the development of refrigeration, canning and pasteurization as preservation methods, the role of traditional processes in preservation has been reduced. Producers are now more likely to view processing as a means of adding value, by the creation of high value traditional products or converting less desirable tissues into forms more appealing to consumers.

Ongoing research on decontamination of processed meats is motivated by the economic benefits of extending shelf life and the desire to improve safety.

© Woodhead Publishing Limited, 2012

Additionally, there is perceived need for decontamination processes to maintain the safety and shelf life of products while responding to continuing demands for reduction of the concentrations of traditional preservatives such as nitrates and sodium chloride. However, an important consideration in the application of decontamination technologies to many processed meats, particularly traditional products, is that they must not alter the organoleptic properties which are crucial to the desirability of the product to consumers.

## 2.2 Decontamination of carcasses

### 2.2.1 Microbiological concerns and commercial considerations

The carcasses of most animals used for human food are skinned in the first stages of the carcass dressing process. However, pig carcasses are usually dressed without removal of the skin (Gill *et al.*, 2000a), although pig carcasses are skinned in some processes; and all poultry carcasses are dressed with the skin on, although the carcasses of large birds such as ostrich and emu are skinned (Gill *et al.*, 2000b). The skins of all animals and birds will be contaminated with large numbers of bacteria that include spoilage organisms and pathogens from faecal material. Some bacteria from the skin will inevitably be transferred to the previously sterile meat during carcass skinning operations (Gill *et al.*, 1998b). Pathogenic and other bacteria may also be transferred to edible tissues from the mouth and throat, during operations on the head (Gill and Jones, 1998); from the anus during operations to free the intestine for its removal (Sheridan, 1998); by spillage of gut contents because of rupture of the intestine during evisceration (Gill and Landers, 2004), and/or from equipment in which bacteria persist after routine cleaning (Arnold and Yates, 2009). Thus, meat surfaces exposed during skinning and evisceration as well as any skin retained on dressed carcasses will be contaminated with bacteria.

The principal objective of applying decontaminating treatments to carcasses is to reduce the numbers of enteric pathogens present on the meat. The pathogens of greatest concern are verotoxigenic *E. coli* (VTEC) (also termed Shiga-toxigenic *E. coli*, STEC), particularly VTEC of the O157 serotype on beef (Erickson and Doyle, 2007), *Salmonella* on both pork and beef (O'Brien *et al.*, 2007), and *Campylobacter jejuni* on poultry meats (Sheppard *et al.*, 2009). Reductions in the numbers of spoilage bacteria on carcasses as a result of decontaminating treatments are largely incidental, and have not been widely investigated, although reductions in total aerobic counts have been and are often determined for indication of the efficacies of decontaminating treatments.

Relatively harsh decontaminating treatments, such as pasteurizing with steam or hot water, can result in undesirable changes in the appearance of the treated surfaces. This is now recognized as being of little or no commercial concern if the carcasses are fabricated to primal cuts, with removal of much

© Woodhead Publishing Limited, 2012

or all treated tissues at packing or carcass breaking plants. Such is the case with beef carcasses in most developed countries, where beef is distributed to facilities that prepare retail cuts mostly, if not entirely, as vacuum packaged primal cuts (Gill *et al.*, 2002). The discoloured meat trimmed from carcasses and primal cuts during their fabrication can be used along with other trimmings for the production of ground beef without adverse effect on the product's organoleptic qualities (Gill *et al.*, 2001a). Harsh treatments that discolour meat may, however, be commercially problematic if the product is traded at the wholesale or retail level as split or whole carcasses that must appear attractive to customers.

### 2.2.2 Physical methods for decontamination of carcasses

*Washing*

Carcasses of all types are washed at the end of the carcass dressing process. In addition, carcasses that are skinned are usually washed after the skin is removed and before the carcass is eviscerated. Poultry carcasses are usually washed after plucking and after evisceration as well as at the end of the carcass dressing process. The main purpose of carcass washing is to remove visible contamination from the carcass. Visible contamination can include filth, faecal material, hair, bone dust, tissue fragments, blood, etc. It is generally assumed that removal of visible contamination will improve the microbiological conditions of carcasses. However, there is no necessary relationship between visible and microbiological contamination (Gill, 2004), so the efficacy of washing as a decontaminating treatment cannot be assessed from its effect on carcass appearance.

The effects of washing on numbers of bacteria on naturally contaminated or inoculated meat have been extensively studied, with water being sprayed onto meat surfaces at various temperatures, pressures, volumes, distance from and angle to the surface, etc. (Bacon, 2005). Although such studies have generally indicated that washing can remove substantial numbers of bacteria from carcasses, the findings with commercial carcasses suggest that the effects of washing treatments are limited. With both beef and poultry carcasses subjected to repeated washing, it was found that the first wash could reduce bacterial counts by up to 1 log unit, but subsequent washes had little or no effect (Gill and Landers, 2003b; Gill *et al.*, 2006). With beef carcasses, the initial wash after skinning was also ineffective if the numbers of bacteria deposited on the meat during the skinning operations were relatively few. Such findings indicated that washing is effective for removing particulate matter and associated bacteria from carcasses, but it may only redistribute bacteria from sites of heavy contamination to less contaminated parts of the carcass (Bell, 1997). Washing will not reduce bacterial numbers if contamination during dressing operations is well controlled; and repeated washing will have no microbiological effects, unless the carcass is heavily recontaminated after a first, effective wash.

© Woodhead Publishing Limited, 2012

*Trimming and spot cleaning*

Removal of visible contamination from carcasses by cutting off the affected tissues has for long been required by meat inspection agencies. Hair and filth deposited on parts of skinned carcasses where such contamination is difficult to prevent, such as the hocks of beef carcasses, have been cleaned by vacuuming, and hairs have sometimes been removed by blowing them from the carcass by a stream of air. During the last few years, cleaning visible contamination from commonly contaminated parts of the carcass, such as in the region of opening cuts, by vacuuming while hot water is delivered onto the meat from the vacuum head, has become a usual treatment applied to skinned carcasses. These various treatments are effective for removing visible contamination and bacteria that may be associated with it (Kochevar *et al.*, 1997; Reagan *et al.*, 1996). However, sites that are heavily contaminated with bacteria may not be visibly contaminated, and bacteria on visibly contaminated sites may be few (Gill, 2004). Moreover, the areas treated by trimming or spot cleaning are relatively small. Consequently, trimming or spot cleaning treatments remove bacteria from only some sites where they are present in relatively large numbers; and the effects of the treatments on the overall microbiological conditions of carcasses are trivial (Gill *et al.*, 1996; Gill and Landers, 2003b). The decontaminating effects of trimming or spot cleaning might be greater if they were applied to sites of high probability for contamination, irrespective of whether or not the sites were visibly contaminated (Gill and McGinnis, 1999), but there has been no report of the effects of trimming or spot cleaning treatments used in this way.

*Scalding, singeing and pasteurizing*

Thermal treatments currently applied to carcasses that can inactivate bacteria include, scalding, singeing and pasteurizing of pig carcasses, and cleaning of hide-on beef carcasses and pasteurizing of beef carcass sides. Pig carcasses are scalded by immersion in water at temperatures between 60 and 70°C for between 5 and 10 min (Bolder, 2004). The pH of the water is commonly adjusted by addition of alkaline additives to between 11 and 12 (Gill and Bryant, 1992). The numbers of bacteria on the skin are greatly reduced by such scalding treatments, but the carcasses are recontaminated during mechanical dehairing by faeces forced from carcasses, and bacteria that persist on the dehairing equipment (Gill and Bryant, 1992; Berends *et al.*, 1997). The numbers of bacteria can again be reduced by singeing the carcass to burn off remaining hair, but carcasses can then be recontaminated when carcasses are scraped, brushed and/or flailed (polished) to remove carbonized deposits from the skin. Thus, while scalding can reduce the numbers of aerobes on pig carcasses from >5 to <2 log cfu/cm$^2$ (Spescha *et al.*, 2006), polished carcasses often carry aerobes at numbers of about 3 log cfu/cm$^2$ (Gill *et al.*, 2000a). The numbers of aerobic bacteria on polished carcasses were reduced by >2 log units at plants where the carcasses were subjected to a second singeing after polishing (Delhalle *et al.*, 2008) or were pasteurized

© Woodhead Publishing Limited, 2012

by deluging with recirculated water at a temperature of 85°C for 10 s (Gill *et al.*, 1997). The same pasteurizing treatment was found to similarly reduce the numbers of generic *E. coli* on the carcasses.

Pasteurizing of pork carcass sides has commonly been regarded as commercially impracticable because of the discolouration of cut muscle surfaces as a result of effective treatments (Gill *et al.*, 1998a). Despite that, the continuing concern over *Salmonella* contamination of pork has led to renewed interest in the pasteurization of dressed pork carcasses. Thus, recent studies have shown that the numbers of bacteria on commercial dressed pork carcass sides can be reduced by >2 log units by treatment with water at > 80°C for > 10 s (Alban and Sørenson, 2010; Hamilton *et al.*, 2010).

In recent years, some North American beef packing plants have implemented washing of beef carcasses before they are skinned (Carlson *et al.*, 2008). The one commercial treatment of that sort for which microbiological effects have been reported involved spraying the carcasses with a 1.5% solution of NaOH at a temperature of 65°C for 10 s, then rinsing with water supplemented with hypochlorite delivered at a rate of 900 l/min (Bosilevac *et al.*, 2005). The treatment reduced the numbers of aerobes and Enterobacteriaceae recovered from the carcasses by 2 and 3 log units respectively, but the numbers of those organisms on the carcasses after skinning were reduced by <1 log unit.

Pasteurizing of beef carcass sides at the end of the carcass dressing process has become a usual practice at North American beef packing plants. Sides are generally pasteurized by exposure to steam or hot water at temperatures ≥ 85°C for ≥ 10 s (Gill and Bryant, 2000; Retzlaff *et al.*, 2005), although shorter treatment times are possible with steam at temperatures > 100°C (Gill and Bryant, 1997a). For steam treatments, carcass surfaces must be clean, otherwise contaminated surfaces can be protected by overlying detritus; and must be dry, otherwise the temperature at the meat surface will not be raised sufficiently to rapidly inactivate bacteria. Thus, the surfaces of washed carcass sides are blown dry before they are subjected to steam pasteurizing (Nutsch *et al.*, 1997). For hot water treatments, the water must be applied in a manner that limits cooling by evaporation. Sprayed hot water can cool rapidly because of the large surface area of water in the form of droplets (Davey, 1989). Cooling to temperatures ineffective for pasteurizing surfaces is avoided by delivery of the hot water as sheets through which the carcass side passes (Gill *et al.*, 1999), or as sprays of large drop size delivered from oscillating heads. Effective pasteurizing treatments reduce the numbers of aerobes on beef carcass sides by >1 log unit, and numbers of *E. coli* by >2 log units to numbers that can be <1 cfu/1000 cm$^2$ (Gill and Jones, 2006).

*Chilling and freezing*

Meat hygiene regulations required that carcasses of meat animals be placed in refrigerated facilities immediately after dressing, and that they be cooled to chiller temperatures before they are moved from abattoirs or subject to further processing. Beef carcasses produced in North America are commonly

© Woodhead Publishing Limited, 2012

sprayed with water during the first hours of the cooling process, to prevent the carcasses losing weight because of the evaporation of water from the warm tissues. Otherwise, red meat carcasses are generally cooled without spraying, and drying of carcass surfaces during cooling can reduce the numbers of, particularly, Gram negative bacteria on the meat (Nottingham, 1982; Crowley *et al.*, 2010).

The numbers of bacteria on spray chilled carcasses can also be reduced, if the carcasses are exposed to air at freezing temperatures after spraying is ended (Gill and Bryant, 1997b), or if rapid chilling of the surfaces inactivated mesophilic bacteria that have been injured by pasteurizing of carcasses immediately before they are cooled (Gill and Landers, 2003a). The rates and extents of drying or freezing of carcass surfaces are greatly affected by the speed of refrigerated air flowing over the surfaces. The air speed can vary greatly in different parts of a chiller and at different sites on the same carcass. Consequently, reduction of bacteria numbers on red meat carcasses during a chilling process will not be uniform, while growth of psychotrophic spoilage bacteria on at least some parts of some carcasses can be expected. However, previous reductions in the numbers of mesophilic bacteria, such as *E. coli*, may be maintained.

Bacteria can be inactivated by freezing, but the extent of any inactivation is highly variable with the species and physiological state of the bacteria, rates of freezing and thawing, and the environment in which freezing occurs (Gill, 2006). Generally, Gram-negative bacteria are more susceptible to freezing injury than are Gram-positive bacteria, and bacteria in the log phase of growth are more susceptible than those in the stationary phase; slow freezing and thawing are more injurious than rapid freezing and thawing; and complex media can provide protection against injury from freezing. Most red meat carcasses are not frozen as such. Beef carcasses are generally exposed to air temperatures above, or slightly below zero during chilling, and any freezing of carcass surfaces that occurs is fortuitous. Pig carcasses are commonly blast chilled with air of below –20°C immediately after dressing. Such treatments can reduce the numbers of indicator organisms and *Campylobacter*, although total aerobic counts may be little affected (Gill and Jones, 1992; Nesbakken *et al.*, 2008). Lamb carcasses for export from Australia and New Zealand are frozen (Pham and Willix, 1985), but studies on the microbiological effects of freezing lamb carcasses are lacking.

*UV light and irradiation*

Bacteria on meat surfaces can be destroyed by exposure to UV light, electrons or x-rays generated as beams, or gamma rays from a radioisotope source. A major difficulty with the use of radiation for treatment of items like carcasses that have a complex surface topography is that treatment of all parts of the surface with an adequate but not excessive dose cannot be easily achieved (Kim *et al.*, 2007). In addition, shadowing may prevent any treatment of some indented surfaces, particularly by relatively low energy radiations such as UV light.

© Woodhead Publishing Limited, 2012

Use of UV light and electron beam irradiation to treat the surfaces of portions of pork and beef inoculated with pathogenic bacteria has been reported. Reductions of about 2 log cfu in numbers of *E. coli* and *Salmonella* on pork muscle and skin were achieved with exposure to UV light for several minutes (Wong *et al.*, 1998); and reduction of >4 log cfu in the numbers of such organisms on beef surfaces was achieved with electron beam irradiation (Arthur *et al.*, 2005). Although the reported work was carried out in relation to the possible treatment of carcasses, there has been no report of the use of either of those or other types of radiation treatments for the pilot scale or commercial treatment of red meat carcasses.

## 2.3 Chemical methods for decontamination of carcasses

### 2.3.1 Organic acids

The microbiological effects of treatment of carcasses with various organic acids have been extensively studied (Acuff, 2005). Most studies involved the treatment of whole carcasses, or portions of beef carcasses, with acetic, lactic or citric acids, alone or in mixtures that include commercial preparations of buffered lactic and citric acids, and mixtures of those and inorganic acids (Kalchayanand *et al.*, 2008; Laury *et al.*, 2009). Reported studies have included investigation of treatments of beef carcasses by spraying with solutions of organic acid after skinning but before evisceration (Bosilevac *et al.*, 2006); and of beef carcass sides or quarters at the end of the carcass dressing process (Dormedy *et al.*, 2000) or after cooling to chiller temperatures (Castillo *et al.*, 2001). Studies with poultry carcasses have involved dipping carcasses in or spraying them with solutions of organic acids at various stages of processing (Sakhare *et al.*, 1999) or, most commonly, after dressing and chilling (Del Rio *et al.*, 2007).

Numbers of enteric pathogens inoculated onto poultry carcasses have been reported to be reduced by ≥3 log units as a result of some treatments with organic acids (Cutter, 1999; Fabrizio *et al.*, 2000). Although reductions in numbers of indicator or pathogenic bacteria in natural flora have sometimes been ≥2 log units (Killinger *et al.*, 2010), reductions have mostly been <2 log units, and often <1 log unit (Barboza de Martinez *et al.*, 2002; Sinhamahapatra *et al.*, 2004).

There are a number of reports on the microbiological effects of the routine application of acetic acid or lactic acid solutions to beef carcass sides at the end of the carcass dressing process, or of lactic acid solutions to skinned but uneviscerated beef carcasses. The latter treatments were found to be ineffective, probably at least in part because the carcasses were sprayed with the acid solution immediately after being washed (Bosilevac *et al.*, 2006; Gill and Landers, 2003a), which could result in the acid being diluted to

© Woodhead Publishing Limited, 2012

sublethal concentrations by the water remaining on the carcasses. Routine spraying of warm or chilled beef carcass sides with acetic or lactic acid solutions were also found to be ineffective in some studies (Avens *et al.*, 1996; Bacon *et al.*, 2002; Gill and Landers, 2003b). However, in other instances such treatments reduced the numbers of aerobes and indicator organisms by about 1 log unit (Algino *et al.*, 2007; Dormedy *et al.*, 2000). Even so, at least some current treatments of beef carcasses with lactic acid solutions may be more effective, because of the use of 5% lactic acid instead of 2%, as use of the higher concentration has been permitted in North America in recent years (FSIS, 2009); and because of the separation of the operations of skinned carcass washing and spraying with lactic acid at some beef packing plants.

### 2.3.2 Oxidizing agents

Oxidizing agents investigated for the decontamination of carcasses include chlorine, chlorine dioxide, hypochlorite, chloramines, hydrogen peroxide, ozone, peroxyacetic acid, and the bromine release agent dibromodimethyl hydantoin. Chlorine solutions are prepared with the gas or by electrolysis of weak solutions of sodium chloride, also known as electrolyzed oxidizing (EO) water. Chlorine dioxide solutions are prepared with the gas or generated from commercial preparations containing chlorite (stabilized chlorine dioxide), usually by reaction with citric acid. Ozone is generated on site in air that is bubbled through the treatment water, or it is generated directly in the water by an electrocatalytic process.

The effects of some of these antimicrobial agents sprayed onto carcasses or inoculated portions of beef have been studied. With peroxyacetic acid at the recommended concentration of about 200 ppm, Penney *et al.* (2007) obtained reductions of >3 log units in the numbers of *E. coli* O157:H7 and *Salmonella* inoculated on beef in high numbers. However, other researchers have generally obtained reductions of ≤1 log unit in the numbers of inoculated bacteria or the natural flora on beef by spraying with solutions of peroxyacetic acid or chlorine dioxide (Gill and Badoni, 2004; King *et al.*, 2005), chlorine or hypochlorite (Cutter and Siragusa, 1995; Kenney *et al.*, 1995), or EO water, hydrogen peroxide or ozone (Arthur *et al.*, 2008; Castillio *et al.*, 2003; Reagan *et al.*, 1996). Treatment of carcasses with hydrogen peroxide stabilized with glycerol reduced numbers of the natural flora by about 0.5 log unit (Wagenaar and Snijders, 2004). Reports on the effects of any of these antimicrobial solutions applied routinely to beef carcasses in commercial practice are lacking.

### 2.3.3 Other antimicrobials

The quaternary ammonium compounds, cetylpyridinium chloride and benzalkonium chloride at concentrations of 0.1–1% in waters used for

© Woodhead Publishing Limited, 2012

spraying or dipping samples of beef have been found to be more effective decontaminants than organic acid solutions (Cutter *et al.*, 2000; Riedel *et al.*, 2009; Stopforth *et al.*, 2004). Their possible use with meats is restricted by their acceptability at only very low concentrations in foods.

Treatment of cattle hides with a quaternary ammonium industrial sanitizer reduced the numbers of bacteria recovered from the hide by about 2 log units, but treatment of hides with a food compatible detergent reduced numbers by only 1 log unit (Small *et al.*, 2005).

Various substances not considered in the previous sections, including some organic acids not commonly found in foods, have been tested for their possible use for decontamination of dressed carcasses. Sodium tripolyphosphate (trisodium phosphate), a cleaning agent and component of brines used with meat and other foods for moisture enhancement, generally was found to have effects on bacteria similar to or less than those of 2% solutions of lactic or acetic acids when applied to beef carcasses by dipping or spraying (Del Rio *et al.*, 2007; Dorsa *et al.*, 1997). Use of tripolyphosphate for carcass decontamination is problematic because of concerns over environmental pollution as a result of large amounts of phosphate in waste waters.

In general, use of solutions of 2% levulinic acid (Carpenter *et al.*, 2011), 20% salicylic acid (Hinton and Cason, 2007), 1% saponin (Cutter, 1999), 0.1% or 0.01% sodium hydroxide (Stopforth *et al.*, 2004), or nisin at 500 IU/ml (Barboza de Martinez *et al.*, 2002) were all found to be no more or less effective than treatment with lactic or acetic acid.

### 2.3.4 Effects of hurdle processing on decontamination of carcass

Various studies with inoculated portions of beef or naturally contaminated broiler carcasses have indicated that the effects of decontaminating treatments applied sequentially during and at the end of carcass dressing will be cumulative (Castillo *et al.*, 1999; Koutsoumanis *et al.*, 2004; Phebus *et al.*, 1997; Sakhare *et al.*, 1999). This is apparently not so for beef carcasses undergoing commercial processing, at least when the skinning and eviscerating operations in the process are well controlled. In those circumstances only the pre-evisceration wash and the final pasteurizing treatment of carcass sides were found to be effective (Gill and Landers, 2003b). Cleaning, trimming and treatments with antimicrobial solutions during carcass dressing had little or no effect. Even so, the numbers of bacteria on carcasses produced in such processes can be very small (Gill, 2009).

Broiler carcasses are dressed mechanically, and are washed after singeing of the plucked carcasses, after evisceration and at the end of the dressing process. In addition, the carcasses are repeatedly sprayed with water during operations for the removal of crops, lungs, kidneys and necks. Studies of commercial processes have variously indicated that the repeated washing with chlorinated water reduces the numbers of bacteria on carcasses progressively by small increments (Stopforth *et al.*, 2007), or that substantial reductions

© Woodhead Publishing Limited, 2012

occur only during the spraying incidental to the operations for removal of crops, lungs, etc. (Barbut *et al.*, 2009; Gill *et al.*, 2006). Reductions in bacterial contamination during dressing of broiler carcasses might then be due to the cumulative effects of all or only some of the washing or spraying operations. Those same studies show that, in practice, further reductions in bacterial numbers are obtained by spraying carcasses with solutions of chlorine dioxide or trisodium phosphate before chilling and/or chilling in water containing chlorine dioxide with or without chlorine.

## 2.4 Decontamination of fresh meats

Wholesale trading of red meats in carcass form has become increasingly uncommon. Instead, in developed countries, most red meat carcasses are fabricated to primal cuts and manufacturing meat in facilities at or associated with slaughtering plants. Primal cuts are usually vacuum packaged for distribution to central or retail store cutting facilities where retail cuts are prepared. Primal cuts may be mechanically tenderized or injected with brine before they are vacuum packaged, or before they are fabricated to retail cuts.

Manufacturing meat is collected in bulk containers for distribution to grinding or other fabricating facilities; but some may be vacuum packaged for distribution to retail store facilities, or be coarsely ground before vacuum packaging for distribution. Fabrication of meat to retail cuts is carried out at some slaughtering plants.

Decontaminating treatments used with commercial beef carcasses can effectively eliminate pathogenic bacteria from the meat (Gill, 2009). Despite that, beef cuts and manufacturing beef used for production of ground beef may still be contaminated with enteric pathogens, because carcass decontaminating treatments are not used or are applied ineffectively; or because the decontaminated meat is recontaminated during carcass breaking by bacteria from detritus that persists in conveyors and other equipment in carcass breaking facilities (Gill and McGinnis, 2000; Gill *et al.*, 2001b). Decontaminating treatments are not generally used with pig or sheep carcasses, and decontaminating treatments for poultry carcasses have only limited effects. Therefore, decontamination of all raw meat prepared in bulk or retail forms would be desirable, if only to preclude cross-contamination of foods that are ready to eat by pathogens from raw meats that are prepared in the same household kitchen or commercial facility (Ravishankar *et al.*, 2010; Van Asselt *et al.*, 2008).

In addition, there are specific concerns about the presence of pathogens in some types of raw meat. Pathogens in ground meats in general, and ground beef in particular, are of concern, because grinding distributes bacteria that were only on the surface of intact meat through the mass of ground product. Inadequate cooking of ground meat products such as hamburger patties

© Woodhead Publishing Limited, 2012

can then allow pathogens such as *E. coli* O157:H7 to survive within the prepared food to infect consumers (Slutsker *et al.*, 1998). Bacteria may also be introduced into the tissues within meat cuts by mechanical tenderizing and/or injection of brines (Gill *et al.*, 2005; Laine *et al.*, 2005).

As many consumers prefer underdone beef (McKenna *et al.*, 2004), and some prefer lamb and pork that is underdone (Moeller *et al.*, 2010), non-intact non-comminuted meat cuts may be intentionally cooked to attain temperatures at the centre that are inadequate for killing any pathogens that may be there. Therefore decontamination of the surfaces of pieces of meat that are to be comminuted or pierced for tenderizing or brine injection is desirable.

Decontaminating treatments of cuts or manufacturing beef that compromise the appearance of product that will be offered for retail sale are not commercially acceptable, because consumer decisions on whether or not to purchase individual items of fresh meats are greatly affected by product appearance (Kennedy *et al.*, 2005; Killinger *et al.*, 2004). Treatments that adversely affect the odour, flavour or tenderness of products are also unacceptable.

### 2.4.1 Physical methods for decontamination of fresh meats

Red meats are not usually washed during the fabrication of carcasses and cuts, and trimming is carried out to obtain cuts of the required form and specified fat cover. Pasteurizing treatments cannot be applied to cuts because cut muscle surfaces are inevitably discoloured by heating to effective pasteurizing temperatures. Nevertheless, pasteurizing treatments can be applied to manufacturing meats that are to be ground, because the relatively small amounts of tissue discoloured by heat do not substantially affect the appearance of the ground product (Gill and Badoni, 2002; Gill *et al.*, 2001a). However, pasteurizing of manufacturing meat has not been adopted in commercial practice.

Electron beam radiation, which cannot penetrate deeply into meat is used commercially for treatment of hamburger patties (Satin, 2002), although it could also be used for decontamination of the surfaces of beef cuts (Davis *et al.*, 2004). Gamma irradiation from a cobalt 60 source can be applied commercially to bulk and packaged meat (Satin, 2002). Use of x-ray beams with meat is not commercially attractive, because x-rays are generated by imaging a high energy electron beam on a target of dense material such as tungsten (Van Lancker and Bastiaansen, 2000). The efficiency of energy conversion for the process is relatively low, so the cost of x-ray generation is high in comparison to the costs associated with other forms of irradiation.

Most countries have not as yet approved the use of radiation with fresh meats or allow irradiation of only some types of meat (FSA, 2011), but all fresh meats are allowed to be irradiated with up to 4.5 kGy in the United States (USDA, 1999). Various studies have shown that electron beam or gamma irradiation of meat at levels less than the USDA permitted maximum can be

© Woodhead Publishing Limited, 2012

effective, as treatments with 2–3 kGy have been shown to reduce numbers of inoculated enteric pathogens and other bacteria by between >3 and > 7log units (Kim *et al.*, 2004; Levanduski and Jaczynski, 2008; Min *et al.*, 2007; Sinanoglou *et al.*, 2009). Minimizing the irradiation treatments applied to meat is desirable, because irradiation can adversely affect the colour, odour and flavour of meat (Ahn and Nam, 2004; Ismail *et al.*, 2009). The adverse effects of radiation on meat quality can be reduced by preventing oxidation by the inclusion of antioxidants, freezing of, or exclusion of oxygen from the meat (Brewer, 2004); but only the latter could generally be used for fresh meat treated by irradiation.

Irradiated beef is used in US government school meal programs (CDE, 2010), and irradiated ground beef and hamburger patties from some suppliers have been on retail sale since the beginning of this century (Omaha Steaks, 2009). Even so, the quantities of irradiated meat that are traded may be relatively small. Information on the matter is limited, but the global quantity of irradiated 'meat and seafood' traded in 2005 was estimated to be only 33,000 tonnes (Kume *et al.*, 2009). Use of irradiation with meat continues to be constrained by widespread consumer uncertainty about the wholesomeness of irradiated foods in general (Stefanova *et al.*, 2010).

### 2.4.2 Chemical methods for decontamination of fresh meats with antimicrobials

Treatment of beef trimmings with solutions of peroxyacetic or lactic acids reduced the numbers of pathogens or coliforms in faeces inoculated on the meat by ≤1 log unit (Ellebracht *et al.*, 2005; Kang *et al.*, 2001a). Multiple treatments with 2% lactic acid did not generally increase the decontaminating effects of a single treatment (Kang *et al.*, 2001a). Treatment of 10% fat:90% lean trimmings with solutions of acidified sodium chlorite gave similar reductions in the numbers of bacteria in the natural flora; but the reductions of 50% fat:50% lean trimmings were about 2 log units (Bosilevac *et al.*, 2004). The latter finding suggests that fat surfaces can be more readily decontaminated by antimicrobial solutions than cut muscle surfaces. When multiple treatments with lactic acid were applied to commercial trimmings in combination with multiple hot water and hot air treatments, reductions of about 2 log units in the numbers of bacteria in the natural flora were obtained, but the specific effects of the lactic acid treatments were not identified (Kang *et al.*, 2001b).

Results obtained when inoculated cuts were treated with warm or cold lactic acid solutions at concentrations between 2 and 5%, or acidified sodium chlorite, were similar to those obtained for inoculated trimmings treated with lactic acid (Echeverry *et al.*, 2009; Heller *et al.*, 2007). The reductions in numbers of pathogens were less when cuts inoculated with pathogens and stored in vacuum packs for 14 or 21 days were subjected to the same treatments (Echeverry *et al.*, 2010). In the only reported study of the routine

© Woodhead Publishing Limited, 2012

commercial treatment of beef with a decontaminating solution during carcass breaking processes, the use of a 2% lactic acid solution was found to be ineffective (Bacon *et al.*, 2002).

## 2.5 Decontamination of processed meats

The microbial stability of processed meats prepared by traditional methods is a consequence of the complex interplay between the physicochemical changes occurring as combinations of drying, smoking, curing and fermentation are applied to the raw ingredients (Pearson and Tauber, 1996). Drying functions to control microflora by reducing the water content and water activity of the product below the level required for the growth of undesirable organisms. In curing, the water activity is reduced not by lowering water content but by the addition of solutes, such as sodium chloride, sugars and nitrate to the product. Microbial inhibition occurs as a consequence of low water activity, osmotic stress and the specific inhibitory activities of components of the curing mix. Depending on the specific smoking process, microbial numbers can be reduced by cooking and the stability of the product improved by drying of the product and antimicrobial compounds in the smoke. Meat fermentations commonly utilize lactic acid bacteria to produce lactic and acetic acids by the metabolism of carbohydrates (Lücke, 2000). Inhibition of undesirable flora is primarily a consequence of the reduction in pH, with additional antimicrobial effects due to the production of organic acids, bacteriocins and other metabolic byproducts (Lücke, 2000). Compounds added as spices to the product may have antimicrobial properties (Simpson and Sofos, 2009). Since meat spoilage requires that the numbers of spoilage organisms increase to levels sufficient to cause organoleptic changes in the product, extension of shelf life does not require significant reductions in microflora numbers but only the inhibition of microbial growth. Processing may increase the safety of products by inhibiting the growth or reducing the numbers of specific pathogens, but in traditional production this is often an unintended side effect of delaying spoilage. Treatments applied to processed meats for the specific purpose of decontamination are then relatively few.

### 2.5.1 Physical methods for decontamination of processed meats

*Thermal pasteurization*

The surfaces of cooked, ready-to-eat meat products can be contaminated with pathogenic bacteria during operations such as the removal of moulds or skins, slicing, portioning, etc. before packing. Contamination of products with *Listeria monocytogenes*, which is commonly found in meat processing environments, is of particular concern in this respect (Sofos and Geornaras, 2010). Heating of packaged products for pasteurization, as opposed to cooking, can be achieved by immersion of packs in hot water (Huang and

© Woodhead Publishing Limited, 2012

Sites, 2007), heating packs with steam (Murphy *et al.*, 2003), or heating the product within the sealed pack by radio frequency energy (Orsat *et al.*, 2004; Laycock *et al.*, 2003; Lyng, 2007; Zhang *et al.*, 2004) or microwave energy (Huang and Sites, 2007). Products may also be pasteurized by the direct application of steam to the products in trays before lids are sealed to the trays (Sommers *et al.*, 2009). Ohmic heating of bologna sausages (Piette *et al.*, 2004), pork luncheon meat and white pudding (McKenna *et al.*, 2006) has been investigated but cannot be performed in packaged meats as direct contact of the food with electrodes is required.

The process of heat inactivation of bacteria is the same with all means of heating. Therefore, their efficacies as decontaminating treatments will depend upon the organisms that are targeted, the locations of the organisms on and/or in product units, and the extent to which the product can be heated without adverse effects on the qualities expected by customers. Processors will also likely consider the speed and uniformity of product heating when selecting a pasteurizing treatment for specific products. Thus, flash pasteurization of frankfurters by direct application of steam to the sausages in trays has become a common commercial practice, because most sausage surfaces are exposed to rapid heating by condensing steam (Canadian Meat Council, 2008). Other methods must be used when prolonged heating is required because contaminated surfaces are overlayed by product, as is usual with packs of sliced meats.

*High pressure processing*

High pressure processing (HPP) inactivates microorganisms by applying hydrostatic pressures in the range of 100 MPa to 1000 MPa (1000–10,000 bar) to food products suspended in a liquid medium (Patterson, 2005). HPP treatments of meats are usually applied to products vacuum packed in flexible packs.

Inactivation of vegetative bacterial cells and fungi occurs at pressures greater than 300 MPa, while bacterial spores may resist pressures as high as 1000 MPa (Rendueles *et al.*, 2011). Though research-scale HPP units are capable of reaching 1000 MPa, current industrial units are limited to a maximum of 600 MPa. This effectively limits the commercial application of HPP to the control of vegetative bacteria and fungi. Inactivation of vegetative cells by HPP treatments involves both disruption of the lipid bilayers of the cell membranes and enzyme inactivation, which are the result of changes in protein conformation at increased pressures (Benito *et al.*, 1999; Casadei *et al.*, 2002; Mañas and Mackey, 2004).

HPP treatment can achieve substantial reductions in the numbers of both pathogens and spoilage bacteria on fresh meats. Reductions of 4–8 log for bacteria on ground beef have been reported following 20 min at 200 MPa for *Pseudomonas fluorescens*, 280 MPa for *Citrobacter freundii*, and 400 MPa for *Listeria innocua* (Carlez *et al.*, 1993). Treatment of ground beef at 400 MPa has been reported to reduce *E. coli* O157:H7 by 3–4 log (Morales *et al.*,

© Woodhead Publishing Limited, 2012

2008; Black *et al.*, 2010). Reductions of 4–8 log in *Salmonella* Typhimurium, *E. coli*, and *L. monocytogenes* on chicken breast have been reported for 450 MPa treatment (Kruk *et al.*, 2010). It should be noted when evaluating the effectiveness of HPP treatments that microbial inactivation does not follow the first-order kinetics associated with thermal treatment, except under extreme conditions of heat and pressure for short treatment times (Rendueles *et al.*, 2011). Instead significant shoulders and tails in inactivation curves are observed. Additionally, the response of specific microorganisms is highly variable between species and even strains of the same species (Hauben *et al.*, 1997; Robey *et al.*, 2001; Margosch *et al.*, 2004; Whitney *et al.*, 2007). Due to the absence of reliable kinetic models and standard strains for process validation, careful consideration should be given to the application of HPP to products as the critical kill steps to ensuring product safety.

Significant reductions in pathogens and spoilage organisms on fresh meats can be achieved by HPP, but commercial application has been limited due to the organoleptic changes that occur at bactericidal pressures. Changes in the conformation of meat proteins under pressure result in the denaturation, aggregation and gelation, with consequent impacts on appearance, texture and other organoleptic properties (Sun and Holley 2010; Carlez *et al.*, 1995; Jung *et al.*, 2003; McArdle *et al.*, 2010; Del Olmo *et al.*, 2010; Kruk *et al.*, 2010; Morales *et al.*, 2008; Black *et al.*, 2010). HPP treated ground beef patties have recently been introduced to the American restaurant market by Cargill (2011). Presumably the discolouration that can occur with HPP treatment of fresh meat due to the conversion of myoglobin to metmyoglobin at pressures greater than 400 MPa (Carlez *et al.*, 1995; Jung *et al.*, 2003) is not a problem with this market.

In contrast to fresh meats, HPP treatment of processed meats at pressures as high as 600 MPa has become common, with HPP treated products such as hams and dry and semi-dry sausages becoming widely available in European and North American markets. Processed meats are suitable products for the application of HPP as they are less susceptible than fresh meats to changes in colour and texture. Processed meats containing nitrites do not undergo a significant change in colour, as their bright red appearance is due to the presence of the myoglobin derivative nitrosomyoglobin, which is stable under high pressure (Carlez *et al.*, 1995). The texture of products in which meat is chopped, cooked or coagulated during fermentation is minimally impacted by gelation of proteins under HPP. The stability of the sensory characteristics of a variety of processed meats following treatments up to 600 MPa has been confirmed (Hayman *et al.*, 2004; Morales *et al.*, 2006; Gill and Ramaswamy, 2008; Omer *et al.*, 2010). Treatments in this range substantially reduced the numbers of inoculated *L. monocytogenes* (Garriga *et al.*, 2004, Hayman *et al.*, 2004; Morales *et al.*, 2006), *Salmonella* (Garriga *et al.*, 2004) and *E. coli* O157 (Gill and Ramaswamy, 2008; Omer *et al.*, 2010).

© Woodhead Publishing Limited, 2012

*Irradiation and ultraviolet light*

The application of ionizing radiation for the decontamination of a wide range of processed foods, including meat products, is well established (Ahn and Lee, 2007). Five-log reductions in *L. monocytogenes* on frankfurters, bologna, ham and deli turkey meat could be achieved by gamma ray exposure in the range of 2.45–3.75 kGy (Sommers *et al.*, 2004). In composite foods containing frankfurters or cooked ground beef, 5 log reductions of the pathogens *Salmonella* spp., *S. aureus*, *L. monocytogenes*, *E. coli* O157:H7, and *Y. enterocolitica* required gamma ray doses of 3.05, 2.70, 2.35, 1.80 and 0.75 kGy, respectively (Sommers and Boyd, 2006).

As with fresh meats, the rate of free radical reactions increases with increasing irradiation dosage which brings organoleptic changes in processed meats, including discolouration and oxidation of lipids (Johnson *et al.*, 2000; Cava *et al.*, 2005, 2009). To reduce these undesirable effects, the addition of antioxidants or low oxygen packing for irradiated products has been proposed (Aymerich *et al.*, 2008).

Ultraviolet irradiation was originally developed as a technology for water treatment and has since been investigated primarily for the decontamination of liquid foods such as apple cider (Sastry *et al.*, 2000). The penetration depth of ultraviolet radiation is very low, but it does have potential for decontamination of the surfaces of processed meats. When used for treatment of frankfurters' surface inoculated with *L. monocytogenes*, a dose of 4 $J/cm^2$ was found to reduce the pathogen numbers by 1.9 log with no observable effects on colour and texture (Sommer *et al.*, 2009). However, the organoleptic properties of other products may be more sensitive to ultraviolet light. Wambura and Verghese (2011) reported that the colour, texture and oxidative stability of ham slices was negatively impacted by treatment with pulsed ultraviolet light.

### 2.5.2 Chemical methods for decontamination of processed meats

In this review of the decontamination of processed meats, only physical decontamination methods are discussed. The use of chemical antimicrobials and biopreservatives is a highly complex issue, with the effectiveness and application of many agents resulting from complex interactions between multiple antimicrobial agents and interrelated with fermentative processes. For these reasons we recommend the reader to specialized reviews of the topic (Lücke, 2000; Työppönen *et al.*, 2003; Holley and Patel, 2005; Simpson and Sofos, 2009).

## 2.6 Conclusions and future trends

Although hazardous microbial contamination of red meat carcasses can be minimized by careful attention to carcass dressing operations, it cannot be

© Woodhead Publishing Limited, 2012

prevented entirely. Therefore, the microbial safety of meat can be ensured only with the proper implementation of effective carcass decontamination treatments. The pasteurization of beef carcasses and their treatment with relatively high concentrations of lactic acid have led to at least some beef packing plants routinely producing dressed carcasses that are essentially free of enteric pathogens. There is, in principle, no reason why similar results could not be obtained at all beef packing plants. Moreover, the treatments used for beef carcasses can undoubtedly be applied to carcasses of other species, provided that some adverse effects on carcass appearance are accepted.

Unfortunately, many regulating authorities outside North America continue to ban the use of decontaminating treatments with raw meats, for reasons that are not clearly related to the safety or quality of the product. Moreover, the success of carcass decontaminating treatments is obscured by the plants' adherence to microbiological testing mandated by regulatory authorities, with levels of detection of indicator organisms several orders of magnitude above the numbers actually present on the meat. There has, however, recently been movement towards carcass pasteurizing, albeit with unnecessary restrictions, in some regulatory jurisdictions; while the utility of current regulatory testing to ensure safety is being increasingly questioned. It can then be expected that in the not too distant future decontamination of red meat carcasses by treatments validated in commercial practice will become the norm worldwide. It is to be hoped that this will be accompanied by the adoption of appropriate routine testing for indicator organisms, to ensure the maintenance of microbiological safety.

If carcasses are freed of hazardous contaminants, then it should in principle be possible to maintain meat in that same condition throughout further processing. Recognition of the fact that fresh meats can be contaminated with pathogens during fabrication of carcasses and cuts as well as during carcass dressing appears to be increasing, while it has long been acknowledged that processed meats that are free of pathogens after processing can be cross-contaminated during their preparation for retail sale. Contamination of fresh and processed meats during their fabrication can be wholly prevented by hygienic design of fabricating equipment and scrupulous attention to facility and equipment cleaning, and rigorous maintenance of hygienic work practices. These matters are likely to be given increasing consideration in the future, with consequent reduction in the need for and reliance on decontaminating treatments to ensure product safety.

Even so, decontaminating treatments for raw and processed meats may still be retained or implemented to provide a measure of insurance against unforeseeable loss of control over contamination. For most fresh meats the only effective option may be irradiation, the use of which is likely to continue to be constrained by consumer aversion as well as regulation. For many processed meats, HPP appears to be a practical decontaminating treatment. The use of HPP with processed meats has certainly been increasing in recent years, so its widening application to such products can be expected.

© Woodhead Publishing Limited, 2012

## 2.7 Sources of further information and advice

Further information on the matters discussed in this chapter can be found in the following books.

Doona C J and Feeherry FE (2007), *High pressure processing of foods*, Hoboken, NJ, Wiley.

Holdsworth S D and Simpson R (2007), *Thermal processing of packaged foods*, Cambridge, Woodhead Publishing.

Lelieveld H L M, Mostert M A and Holah J (2005), *Handbook of hygiene control in the food industry*, Cambridge, Woodhead Publishing.

Marriott N G and Gravani R B (2006), '*Principles of food sanitation*, 5th edn, Basel, Birkhäuser.

Sofos J N (2005), *Improving the safety of fresh meat*, Cambridge, Woodhead Publishing.

Sommers C H and Fan X (2006), *Food irradiation research and technology*, Ames, IA, Blackwell Publishing.

Sun D-W (2005), *Thermal food processing*, London, Taylor and Francis.

Information on current regulations and recommendations regarding the application of decontamination technologies can be found at the following websites:

- Canadian Food Inspection Agency, Meat and Poultry Products: http://www.inspection.gc.ca/english/fssa/meavia/meaviae.shtml
- Health Canada, Guidance Documents: http://www.hc-sc.gc.ca/fn-an/legislation/guide-ld/index-eng.php
- European Food Safety Authority, Decontamination of Carcasses http://www.efsa.europa.eu/en/topics/topic/decontamination.htm?wtrl=01
- United Kingdom Food Standards Agency, Meat and Meat Hygiene: http://www.food.gov.uk/foodindustry/meat/
- United States Department of Agriculture Food Safety and Inspection Service, Regulations and Policies: http://www.fsis.usda.gov/Regulations_&_Policies/index.asp

## 2.8 References

ACUFF G R (2005), 'Chemical decontamination strategies for meat', in Sofos J N, *Improving the safety of fresh meat*, Baton Rouge, LA, CRC Press, 350–363.

AHN D U and LEE J (2007), 'Mechanisms and prevention of quality changes in meat by irradiation', in Sommers C H and Fan X, *Food Irradiation Research and Technology*, Ames, IA, Blackwell Publishing, 127–142.

AHN D U and NAM K C (2004), 'Effects of ascorbic acid and antioxidants on color, lipid oxidation and volatiles of irradiated ground beef', *Radiat Phys Chem*, 71, 151–156.

ALBAN L and SØRENSEN L L (2010), 'Hot water decontamination – an effective way of reducing risk of *Salmonella* in pork', *Fleischwirtschaft Int*, 25, 60–64.

ALGINO R J, INGHAM S C and ZHU J (2007), 'Survey of antimicrobial effects of beef carcass

© Woodhead Publishing Limited, 2012

intervention treatments in very small state-inspected slaughter plants', *J Food Sci*, 72, M173–M179.

ARNOLD J W and YATES I E (2009), 'Interventions for control of *Salmonella*: clearance of microbial growth from rubber picker fingers', *Poult Sci*, 88, 1292–1298.

ARTHUR T M, WHEELER T L, SHACKELFORD S D, BOSILEVAC J M, XIANGWU N and KOOHMARAIE M (2005), 'Effects of low-dose, low-penetration electron beam irradiation of chilled beef carcass surface cuts on *Escherichia coli* O157:H7 and meat quality', *J Food Prot*, 68, 666–672.

ARTHUR T M, KALCHAYANAND N, BOSILEVAC J M, BRICHTA-HARHAY D M, SHACKELFORD S D, BONO J L, WHEELER T L and KOOHMARAIE M (2008), 'Comparison of effects of antimicrobial interventions on multidrug-resistant *Salmonella*, susceptible *Salmonella*, and *Escherichia coli* O157:H7', *J Food Prot*, 71, 2177–2181.

AVENS J S, CLAYTON P, JONES D K, BOLIN R, LLOYD W and JANKOW D (1996), 'Acetic acid sprays ineffective on beef carcasses with low bacteria counts', *Lebensm Wiss Technol*, 29, 28–32.

AYMERICH T, PICOUET P A and MONTFORT J M (2008), 'Decontamination technologies for meat products', *Meat Sci*, 78, 114–129.

BACON R T (2005), 'Physical decontamination strategies for meat', in Sofos J N, *Improving the safety of fresh meat*, Cambridge, Woodhead, 318–349.

BACON R T, SOFOS J N, BELK K E and SMITH G C (2002), 'Commercial application of lactic acid to reduce bacterial populations on chilled beef carcasses, subprimal cuts and table surfaces during fabrication', *Dairy, Food Environ Sanit*, 22, 674–682.

BARBOZA DE MARTINEZ Y, FERRER K and MARQUEZ-SALAS E (2002), 'Combined effects of lactic acid and nisin solution in reducing levels of microbiological contamination in red meat carcasses', *J Food Prot*, 65, 1780–1783.

BARBUT S, MOZA L F, NATTRESS F, DILTS B and GILL C O (2009), 'The microbiological conditions of air- or water-chilled carcasses produced at the same poultry packing plant', *J Appl Poult Res*, 18, 501–507.

BELL R G (1997), 'Distribution and sources of microbial contamination on beef carcasses', *J Appl Microbiol*, 82, 292–300.

BENITO A, VENTOURA G, CASADEI M, ROBINSON T and MACKEY B (1999), 'Variation in resistance of natural isolates of *Escherichia coli* O157 to high hydrostatic pressure, mild heat, and other stresses', *Appl Environ Microbiol*, 65, 1564–1569.

BERENDS B R, KNAPEN F V, SNIJDERS J M A and MOSSEL D (1997), 'Identification and quantification of risk factors regarding *Salmonella* spp on pork carcasses', *Int J Food Microbiol*, 36, 199–206.

BLACK E P, HIRNEISEN K A, HOOVER D G and KNIEL K E (2010), 'Fate of *Escherichia coli* O157:H7 in ground beef following high-pressure processing and freezing', *J Appl Microbiol*, 108, 1352–1360.

BLAMIRE R V (1984), 'Meat hygiene: the changing years', *J R Soc Health*, 104, 180–185.

BOLDER D J (2004), 'Slaughter-line operation/pigs', in Jensen W K, Devine C and Dikeman M, *Encylopedia of meat sciences*, Amsterdam, Elsevier, 1243–1249.

BOSILEVAC J M, SHACKELFORD S D, FAHLE R, BIELA T and KOOHMARAIE M (2004), 'Decreased dosage of acidified sodium chlorite reduces microbial contamination and maintains organoleptic qualities of ground beef products', *J Food Prot*, 67, 2248–2254.

BOSILEVAC J M, NOU X, OSBORN M S, ALLEN D M and KOOHMARAIE M (2005), 'Development and evaluation of an on-line hide decontamination procedure for use in a commercial beef processing plant', *J Food Prot*, 68, 265–272.

BOSILEVAC J M, NOU X, BARKOCY-GALLAGHER G A, ARTHUR T M and KOOHMARAIE M (2006), 'Treatments using hot water instead of lactic acid reduce levels of aerobic bacteria and *Enterobacteriaceae* and reduce the prevalence of *Escherichia coli* O157:H7 on preevisceration beef carcasses', *J Food Prot*, 69, 1808–1813.

BREWER S (2004), 'Irradiation effects on meat color – a review', *Meat Sci*, 68, 1–17.

© Woodhead Publishing Limited, 2012

BROADWAY M J (2002), 'The British slaughtering industry: a dying business', *Geography*, 87, 268–280.

CANADIAN MEAT COUNCIL (2008), RapidPak technology flash pasteurization, www.cmc-cvc.com/english/documents/Alkar.pdf (accessed 3 August 2011).

CARGILL (2011), 'Fressure™ fresh ground beef'. Available at: http://www.cargill.com/products/foodservice/beef/brands/fressure-fresh-ground-beef/index.jsp (accessed 28 August 2011).

CARLEZ A, ROSEC J P, RICHARD N and CHEFTEL J-C (1993), 'High pressure inactivation of *Citrobacter freundii*, *Pseudomonas fluorescens* and *Listeria innocua* in inoculated minced beef muscle', *Lebensm Wiss Technol*, 26, 357–363.

CARLEZ A, VECIANA-NOGUES T and CHEFTEL J-C (1995), 'Changes in colour and myoglobin of minced beef meat due to high pressure processing', *Lebensm Wiss Technol*, 28, 528–538.

CARLSON B A, GEORNARAS I, YOHAN Y, SCANGA J A, SOFOS J N, SMITH G C and BELK K E (2008), 'Studies to evaluate chemicals and conditions with low-pressure applications for reducing microbial counts on cattle hides', *J Food Prot*, 71, 1343–1348.

CARPENTER C E, SMITH J V and BROADBENT J R (2011), 'Efficacy of washing meat surfaces with 2% levulinic, acetic, or lactic acid for pathogen decontamination and residual growth inhibition', *Meat Sci*, 88, 256–260.

CASADEI M A, MAÑAS P, NIVEN G, NEEDS E and MACKEY B M (2002), 'Role of membrane fluidity in pressure resistance of *Escherichia coli* NCTC 8164', *Appl Environ Microbiol*, 68, 5965–5972.

CASTILLO A, LUCIA L M, GOODSON K J, SAVELL J W and ACUFF G R (1999), 'Decontamination of beef carcass surface tissue by steam vacuuming alone and combined with hot water and lactic acid sprays', *J Food Prot*, 62, 146–151.

CASTILLO A, LUCIA L M, MERCADO I and ACUFF G R (2001), 'In plant evaluation of a lactic acid treatment for reduction of bacteria on chilled beef carcasses', *J Food Prot*, 64, 738–740.

CASTILLO A, MCKENZIE K S, LUCIA L M and ACUFF G R (2003), 'Ozone treatment for reduction of *Escherichia coli* O157:H7 and *Salmonella* serotype Typhimurium on beef carcass surfaces', *J Food Prot*, 66, 775–779.

CAVA R, TÁRREGA R, RAMIREZ M R, MINGOARRANZ F J and CARRASCO A (2005), 'Effect of irradiation on colour and lipid oxidation of dry-cured hams from free-range reared and intensively reared pigs', *Innovative Food Sci Emerg Technol*, 6, 135–141.

CAVA R, TÁRREGA R, RAMIREZ R and CARRASCO J A (2009), 'Decolouration and lipid oxidation changes of vacuum-packed Iberian dry-cured loin treated with E-beam irradiation (5 kGy and 10 kGy) during refrigerated storage', *Innovative Food Sci Emerg Technol*, 10, 495–499.

CDC (2010), '*Salmonella montevideo* infections associated with salami products made with contaminated imported black and red pepper – United States, July 2009–April 2010', *MMWR Morb Mortal Wkly Rep*, 59, 1647–1650.

CDE (2010), Information Alert – Irradiated beef Management Bulletin 04–402. Available from: http://wwwcdecagov/ls/nu/fd/iausdabeefasp (accessed 29 June 2011).

CROWLEY K M, PRENDERGAST D M, SHERIDAN J and MCDOWELL D A (2010), 'Survival of *Pseudomonas fluorescens* on beef carcass surfaces in a commercial abattoir', *Meat Sci*, 85, 550–554.

CUTTER C N (1999), 'Combination spray washes of saponin with water or acetic acid to reduce aerobic and pathogenic bacteria on lean beef surfaces', *J Food Prot*, 62, 280–283.

CUTTER C N and SIRAGUSA GR (1995), 'Application of chlorine to reduce populations of *Escherichia coli* on beef', *J Food Saf*, 15, 67–75.

CUTTER C N, DORSA W J, HANDIE A, RODRIGUEZ-MORALES S, XIANG Z, BREEN P J and COMPADRE C M (2000), 'Antimicrobial activity of cetylpyridinium chloride washes against pathogenic bacteria on beef surfaces', *J Food Prot*, 63, 593–600.

© Woodhead Publishing Limited, 2012

DAVEY K R (1989), 'Theoretical analysis of two hot water cabinet systems for decontamination of sides of beef', *Int J Food Sci and Tech*, 24, 291–304.

DAVIS K J, SEBRANEK J G, HUFF-LONERGAN E, AHN D U and LONERGAN S M (2004), 'The effects of irradiation on quality of injected fresh pork loins', *Meat Sci*, 67, 395–401.

DEL OLMO A, MORALES P, ÁVILA M, CALZADA J and NUÑEZ M (2010), 'Effect of single-cycle and multiple-cycle high-pressure treatments on the colour and texture of chicken breast fillets', *Innovative Food Sci Emerg Technol*, 11, 411–444.

DEL RIO E, MURIENTE R, PRIETO M, ALONSO-CALLEJA C and CAPITA R (2007), 'Effectiveness of trisodium phosphate, acidified sodium chlorite, citric acid, and peroxyacids against pathogenic bacteria on poultry during refrigerated storage', *J Food Prot*, 70, 2063–2071.

DELHALLE L, SADELEER L D, BOLLAERTS K, FARNIR F, SAEGERMAN C, KORSAK N, DEWULF J, ZUTTER L D and DAUBE G (2008), 'Risk factors for *Salmonella* and hygiene indicators in the 10 largest Belgian pig slaughterhouses', *J Food Prot*, 71, 1320–1329.

DERBYSHIRE J B (2006), 'Early history of the Canadian federal meat inspection service', *Can Vet J*, 47, 542–549.

DORMEDY E S, BRASHEARS M M, CUTTER C N and BURSON D E (2000), 'Validation of acid washes as critical control points in hazard analysis and critical control point systems', *J Food Prot*, 63, 1676–1680.

DORSA W J, CUTTER C N and SIRAGUSA G R (1997), 'Effects of acetic acid, lactic acid and trisodium phosphate on the microflora of refrigerated beef carcass surface tissue inoculated with *Escherichia coli* O157:H7, *Listeria innocua*, and *Clostridium sporogenes*', *J Food Prot*, 60, 619–624.

ECHEVERRY A, BROOKS J C, MILLER M F, COLLINS J A, LONERAGAN G H and BRASHEARS M M (2009), 'Validation of intervention strategies to control *Escherichia coli* O157:H7 and *Salmonella* Typhimurium DT 104 in mechanically tenderized and brine-enhanced beef', *J Food Prot*, 72, 1616–1623.

ECHEVERRY A, BROOKS J C, MILLER M F, COLLINS J A, LONERAGAN G H and BRASHEARS M M (2010), 'Validation of lactic acid bacteria, lactic acid, and acidified sodium chlorite as decontaminating interventions to control *Escherichia coli* O157:H7 and *Salmonella* Typhimurium DT 104 in mechanically tenderized and brine-enhanced (nonintact) beef at the purveyor', *J Food Prot*, 73, 2169–2179.

ELLEBRACHT J W, KING D A, CASTILLO A, LUCIA L M, ACUFF G R, HARRIS K B and SAVELL J W (2005), 'Evaluation of peroxyacetic acid as a potential pre-grinding treatment for control of *Escherichia coli* O157:H7 and *Salmonella* Typhimurium on beef trimmings', *Meat Sci*, 70, 197–203.

ERICKSON M C and DOYLE M P (2007), 'Food as a vehicle for transmission of shiga toxin-producing *Escherichia coli*', *J Food Prot*, 70, 2426–2449.

FABRIZIO K A, SHARMA R R, DEMERCI A and CUTTER C N (2004), 'Comparison of electrolysed oxidizing water with various antimicrobial interventions to reduce *Salmonella* species in poultry', *Poult Sci*, 81, 1598–1605.

FSA (2011), Irradiated Food, UK Food Standards Agency. Available from: http://www.food.gov.uk/safereating/rad_in_food/irradfoodqa/(accessed 28 August 2011).

FSIS (2009), US Department of Agriculture, Food Safety and Inspection Service, Safe and suitable ingredients used in the production of meat and poultry products, FSIS Directive 71201, Amendment 20, Attachment 1.

GARRIGA M, GRÈBOL N, AYMERICH M T, MONFORT J M and HUGAS M (2004), 'Microbial inactivation after high-pressure processing at 600 MPa in commercial meat products over its shelf life', *Innovative Food Sci Emerg Technol*, 5, 451–457.

GILL A O and RAMASWAMY H S (2008), 'The application of high pressure processing to kill *Escherichia coli* O157 in ready to eat meats', *J Food Prot*, 71, 2182–2189.

GILL C O (2004), 'Visible contamination on animals and carcasses and the microbiological condition of meat', *J Food Prot*, 67, 413–419.

© Woodhead Publishing Limited, 2012

GILL C O (2006), 'Microbiology of frozen foods', in Sun D W, *Handbook of frozen food processing and packaging*, Boca Raton, FL, CRC Press, 85–100.

GILL C O (2009), 'Effects on the microbiological condition of product of decontaminating treatments routinely applied to carcasses at beef packing plants', *J Food Prot*, 72, 1790–1801.

GILL C O and BADONI M (2002), 'Microbiological and organoleptic qualities of vacuum-packaged ground beef prepared from pasteurized manufacturing beef', *Int J Food Microbiol*, 74, 111–118.

GILL C O and BADONI M (2004), 'Effects of peroxyacetic acid, acidified sodium chlorite or lactic acid solutions on the microflora of chilled beef carcasses', *Int J Food Microbiol*, 91, 43–50.

GILL C O and BRYANT J (1992), 'The contamination of pork with spoilage bacteria during commercial dressing, chilling and cutting of pig carcasses', *Int J Food Microbiol*, 16, 51–62.

GILL C O and BRYANT J (1997a), 'Decontamination of carcasses by vacuum-hot water cleaning and steam pasteurizing during routine operations at a beef packing plant', *Meat Sci*, 47, 267–276.

GILL C O and BRYANT J (1997b), 'Assessment of the hygienic performance of two beef carcass cooling processes from product temperature history data or enumeration of bacteria on carcass surfaces', *Food Microbiol*, 14, 593–602.

GILL C O and BRYANT J (2000), 'The effects on product of a hot water pasteurizing treatment applied routinely in a commercial beef carcass dressing process', *Food Microbiol*, 17, 495–504.

GILL C O and JONES T (1992), 'Assessment of the hygienic efficiencies of two commercial processes for cooling pig carcass', *Food Microbiol*, 9, 335–343.

GILL C O and JONES T (1998), 'Control of the contamination of pig carcasses by *Escherichia coli* from their mouths', *Int J Food Microbiol*, 44, 43–48.

GILL C O and JONES T (2006), 'Setting control limits for *Escherichia coli* counts in samples collected routinely from pig or beef carcasses', *J Food Prot*, 69, 2837–2842.

GILL C O and LANDERS C (2003a), 'Effects of spray-cooling processes on the microbiological conditions of decontaminated beef carcasses', *J Food Prot*, 66, 1247–1252.

GILL C O and LANDERS C (2003b), 'Microbiological effects of carcass decontaminating treatments at four beef packing plants', *Meat Sci*, 65, 1005–1011.

GILL C O and LANDERS C (2004), 'Microbiological conditions of detained beef carcasses before and after removal of visible contamination', *Meat Sci*, 66, 335–342.

GILL C O and MCGINNIS J C (1999), 'Improvement of the hygienic performance of the hindquarters skinning operations at a beef packing plant', *Int J Food Microbiol*, 51, 123–132.

GILL C O and MCGINNIS J C (2000), 'Contamination of beef trimmings with *Escherichia coli* during a carcass breaking process', *Food Res Int*, 33, 125–130.

GILL C O, BADONI M and JONES T (1996), 'Hygienic effects of trimming and washing operations in a beef-carcass-dressing process', *J Food Prot*, 59, 666–669.

GILL C O, BEDARD D and JONES T (1997), 'The decontaminating performance of a commercial apparatus for pasteurizing polished pig carcasses', *Food Microbiol*, 14, 71–79.

GILL C O, JONES T and BADONI M (1998a), 'The effects of hot water pasteurizing treatments on the microbiological conditions and appearances of pig and sheep carcasses', *Food Res Int*, 31, 273–278.

GILL C O, MCGINNIS J C and BRYANT J (1998b), 'Microbial contamination of meat during the skinning of beef carcass hindquarters at three slaughtering plants', *Int J Food Microbiol*, 42, 175–184.

GILL C O, BRYANT J and BEDARD D (1999), 'The effects of hot water pasteurizing treatments on the appearances and microbiological conditions of beef carcass sides', *Food Microbiology*, 16, 281–289.

GILL C O, DUSSAULT F, HOLLEY R A, HOUDE A, JONES T, RHEAULT N, ROSALES A and

© Woodhead Publishing Limited, 2012

QUESSY S (2000a), 'Evaluation of the hygienic performances of the processes for cleaning, dressing and cooling pig carcasses at eight packing plants', *Int J Food Microbiol*, 58, 65–72.

GILL C O, JONES T, BRYANT J and BRERETON D A (2000b), 'The microbiological conditions of the carcasses of six species after dressing at a small abattoir', *Food Microbiol*, 17, 233–239.

GILL C O, BRYANT J and BADONI M (2001a), 'Effects of hot water pasteurizing treatments on the microbiological condition of manufacturing beef used for hamburger patty manufacture', *Int J Food Microbiol*, 63, 243–256.

GILL C O, MCGINNIS J C and BRYANT J (2001b), 'Contamination of beef chucks with *Escherichia coli* during carcass breaking', *J Food Prot*, 64, 1824–1837.

GILL C O, JONES T, RAHN K, CAMPBELL S, LEBLANC D I, HOLLEY R A and STARK R (2002), 'Temperatures and ages of boxed beef packed and distributed in Canada', *Meat Sci*, 60, 401–410.

GILL C O, MCGINNIS J C, RAHN K, YOUNG D, LEE N and BARBUT S (2005), 'Microbiological condition of beef mechanically tenderized at a packing plant', *Meat Sci*, 69, 811–816.

GILL C O, MOZA L F, BADONI M and BARBUT S (2006), 'The effects on the microbiological condition of product of carcass dressing, cooling, and portioning processes at a poultry packing plant', *Int J Food Microbiol*, 110, 187–193.

GREIG J D and RAVEL A (2009), 'Analysis of foodborne outbreak data reported internationally for source attribution', *Int J Food Microbiol*, 130, 77–87.

HAMILTON D, HOLDS G, LORIMER M, KIERMEIER A, KIDD C, SLADE J and POINTON A (2010), 'Slaughterfloor decontamination of pork carcases with hot water or acidified sodium chlorite – a comparison in two Australian abattoirs', *Zoonoses Public Hlth*, 57(Suppl 1), 16–22.

HATHAWAY S C and MCKENZIE A I (1991), 'Postmortem meat inspection programs; separating science and tradition', *J Food Prot*, 54, 471–475.

HAUBEN K J, BARTLETT D H, SOONTJENS C C, CORNELIS K, WUYTACK E Y and MICHIELS C W (1997), '*Escherichia coli* mutants resistant to inactivation by high hydrostatic pressure', *Appl Environ Microbiol*, 63, 945–950.

HAYMAN M, BAXTER H I, O'RIORDAN P J and STEWART C M (2004), 'Effects of high-pressure processing on the safety, quality and shelf life of ready-to-eat meats', *J Food Prot*, 67, 1709–1718.

HELLER C E, SCANGA J A, SOFOS J N, BELK K E, WARREN-SERNA W, BELLINGER G R, BACON R T, ROSSMAN M L and SMITH G C (2007), 'Decontamination of beef subprimal cuts intended for blade tenderization or moisture enhancement', *J Food Prot*, 70, 1174–1180.

HINTON A JR, and CASON J A (2007), 'Changes in the bacterial flora of skin of processed broiler chickens washed in solutions of salicylic acid', *Int J Poult Sci*, 6, 960–966.

HOLLEY R A and PATEL D (2005), 'Improvement in shelf-life and safety of perishable foods by plant essential oils and smoke antimicrobials – a review', *Food Microbiol*, 22, 273–292.

HUANG L and SITES J (2007), 'Automatic control of a microwave heating process for in-package pasteurization of beef frankfurters', *J Food Eng*, 80, 226–233.

HUGAS M and TSIGARIDA E (2008), 'Pros and cons of carcass decontamination: the role of the European Food Safety Authority', *Meat Sci*, 78, 43–52.

HULEBAK K L and SCHLOSSER W (2002), 'Hazard analysis and critical control point (HACCP) history and conceptual overview', *Risk Anal*, 22, 547–552.

ISMAIL H A, LEE E J, KO K Y, PAIK H D and AHN D U (2009), 'Effect of antioxidant application methods on the color, lipid oxidation, and volatiles of irradiated ground beef', *J Food Sci*, 74, C25–C32.

JOHNSON S C, SEBRANEK J G, OLSON D G and WIEGAND B R (2000), 'Irradiation in contrast to thermal processing of pepperoni for control of pathogens: effects on quality indicators', *J Food Sci*, 65, 1260–1265.

© Woodhead Publishing Limited, 2012

JUNG S, GHOUL M and DE LAMBALLERIE-ANTON M (2003), 'Influence of high pressure on the color and microbial quality of beef meat', *Lebensm Wiss Technol*, 36, 625–631.

KALCHAYANAND N, ARTHUR T M, BOSILEVAC J M, BRICHTA-HARHAY D M, GUERINI M N, WHEELER T L and KOOHMARAIE M (2008), 'Evaluation of various antimicrobial interventions for the reduction of *Escherichia coli* O157:H7 on bovine heads during processing', *J Food Prot*, 71, 621–624.

KANG D H, KOOHMARAIE M, DORSA W J and SIRAGUSA G R (2001a), 'Development of a multiple-step process for the microbial decontamination of beef trim', *J Food Prot*, 64, 63–71.

KANG D H, KOOHMARAIE M and SIRAGUSA G R (2001b), 'Application of multiple antimicrobial interventions for microbial decontamination of commercial beef trim', *J Food Prot*, 64, 168–171.

KENNEDY O B, STEWART-KNOX B J, MITCHELL P C and THURNHAM D I (2005), 'Flesh colour dominates consumer preference for chicken', *Appetite*, 44, 181–186.

KENNEY P B, PRASAI R K, CAMPBELL R E, KASTNER C L and FUNG D Y C (1995), 'Microbiological quality of beef carcasses and vacuum-packaged subprimals: process intervention during slaughter and fabrication', *J Food Prot*, 58, 633–638.

KILLINGER K M, CALKINS C R, UMBERGER W J, FEUZ D M and ESKRIDGE K M (2004), 'Consumer visual preference and value for beef steaks differing in marbling level and color', *J Anim Sci*, 82, 3288–3293.

KILLINGER K M, KANNAN A, BARY A I and COGGER C G (2010), 'Validation of a 2 percent lactic acid antimicrobial rinse for mobile poultry slaughter operations', *J Food Prot*, 73, 2079–2083.

KIM B H, AERA J, LEE S O, MIN J S and MOOHA L (2004), 'Combined effect of electron-beam (beta) irradiation and organic acids on shelf life of pork loins during cold storage', *J Food Prot*, 67, 168–171.

KIM J, MOREIRA R G, HUANG Y and CASTELL-PEREZ M E (2007), '3-D dose distributions for optimum radiation treatment planning of complex foods', *J Food Eng*, 79, 312–321.

KING D A, LUCIA L M, CASTILLO A, ACUFF G R, HARRIS K B and SAVELL J W (2005), 'Evaluation of peroxyacetic acid as a post-chilling intervention for control of *Escherichia coli* O157:H7 and *Salmonella* Typhimurium on beef carcass surfaces', *Meat Sci*, 69, 401–407.

KOCHEVAR S L, SOFOS J N, BOLIN R R, REAGAN J O and SMITH G C (1997), 'Steam vacuuming as a pre-evisceration intervention to decontaminate beef carcasses', *J Food Prot*, 60, 107–113.

KOUTSOUMANIS K P, ASHTON L V, GEORNARAS I, BELK K E, SCANGA J A, KENDALL P A, SMITH G C and SOFOS J N (2004), 'Effect of single or sequential hot water and lactic acid decontamination treatments on the survival and growth of *Listeria monocytogenes* and spoilage microflora during aerobic storage of fresh beef at 4, 10, and 25 degrees C', *J Food Prot*, 67, 2703–2711.

KRUK Z A, YUN H, RUTLEY D L, LEE E J, KIM Y J and JO C (2010), 'The effect of high pressure on microbial population, meat quality and sensory characteristics of chicken breast fillet', *Food Control*, 22, 6–12.

KUME T, FURUTA M, TODORIKI S, UENOYAMA N and KOBAYASHI Y (2009), 'Quantity and economic scale of food irradiation in the world', *Radioisotopes*, 58, 27–35.

LAINE E S, SCHEFTEL J M, BOXRUD D J, VOUGHT K J, DANILA R N, ELFERING K M and SMITH K E (2005), 'Outbreak of *Escherichia coli* O157:H7 infections associated with nonintact blade-tenderized frozen steaks sold by door-to-door vendors', *J Food Prot*, 68, 1198–1202.

LAURY A M, ALVARADO M V, NACE G, ALVARADO C Z, BROOKS J C, ECHEVERRY A and BRASHEARS M M (2009), 'Validation of a lactic acid- and citric acid-based antimicrobial product for the reduction of *Escherichia coli* O157:H7 and *Salmonella* on beef tips and whole chicken carcasses', *J Food Prot*, 72, 2208–2211.

LAYCOCK L, PIYASENA P and MITTAL G S (2003), 'Radio frequency, cooking of ground, comminuted and muscle meat products', *Meat Sci*, 65, 959–965.

© Woodhead Publishing Limited, 2012

LEVANDUSKI L and JACZYNSKI J (2008), 'Increased resistance of *Escherichia coli* O157:H7 to electron beam following repetitive irradiation at sub-lethal doses', *Int J Food Microbiol*, 121, 328–334.

LÜCKE F-K (2000), 'Utilization of microbes to process and preserve meat', *Meat Sci*, 56, 105–115.

LYNG J G (2007), 'Rapid pasteurization of meats using radio frequency or ohmic heating', *New Food*, 3, 58–62.

MACDONALD D M, FYFE M, PACCAGNELLA A M, TRINIDAD A, LOUIE K and PATRICK D (2004), '*Escherichia coli* O157:H7 outbreak linked to salami, British Columbia, Canada, 1999', *Epidemiol Infect*, 132, 283–289.

MACDONALD J M (2003), 'Beef and pork packing industries', *Vet Clin N AM-Food A*, 19, 419–443.

MAÑAS P and MACKEY B M (2004), 'Morphological and physiological changes induced by high hydrostatic pressure in exponential- and stationary-phase cells of *Escherichia coli*: relationship with cell death', *Appl Environ Microbiol*, 70, 1545–1554.

MARGOSCH D, GÄNZLE M G, EHRMAN M A and VOGEL R F (2004), 'Pressure inactivation of Bacillus endospores', *Appl Environ Microbiol*, 70, 7321–7328.

MCARDLE R, MARCOS B, KERRY J P and MULLEN A (2010), 'Monitoring the effects of high pressure processing and temperature on selected beef quality attributes', *Meat Sci*, 86, 629–634.

MCKENNA B M, LYNG J, BRUNTON N and SHIRSAT N (2006), 'Advances in radio frequency and ohmic heating of meats', *J Food Eng*, 77, 215–229.

MCKENNA D R, LORENZEN C L, POLLOK K D, MORGAN W W, MIES W L, HARRIS J J, MURPHY R, MCADAMS M, HALE D S and SAVELL J W (2004), 'Interrelationships of breed type, USDA quality grade, cooking method and degree of doneness on consumer evaluations of beef in Dallas and San Antonio, Texas, USA', *Meat Sci*, 66, 399–406.

MILLER J A (2008), 'The regulation of sanitary food transportation in the United States: a slow journey on a long road', *Food Drug Law J*, 63, 35–73.

MIN J S, LEE S O, JANG A, JO C and LEE M (2007), 'Control of microorganisms and reduction of biogenic amines in chicken breast and thigh by irradiation and organic acids', *Poult Sci*, 86, 2034–2041.

MOELLER S J, MILLER R K, EDWARDS K K, ZERBY H N, LOGAN K E, ALDREDGE T L, STAHL C A, BOGGESS M and BOX-STEFFENSMEIER J M (2010), 'Consumer perceptions of pork eating quality as affected by pork quality attributes and end-point cooked temperature', *Meat Sci*, 84, 14–22.

MORALES P, CALZADA J and NUÑEZ M (2006), 'Effect of high-pressure treatment on the survival *of Listeria monocytogenes* Scott A in sliced vacuum-packaged Iberian and Serrano cured hams', *J Food Prot*, 69, 2539–2543.

MORALES P, CALZADA J, AVILA M and NUÑEZ M (2008), 'Inactivation of *Escherichia coli* O157:H7 in ground beef by single-cycle and multiple-cycle high-pressure treatments', *J Food Prot*, 71, 811–815.

MOR-MUR M and YUSTE J (2009), 'Emerging bacterial pathogens in meat and poultry: an overview', *Food Bioprocess Technol*, 3, 24–35.

MORRIS J G (2003), 'The color of hamburger: slow steps toward the development of a science-based food safety system in the United States', *Trans Am Clin Climatol Assoc*, 114, 191–202.

MURPHY R Y, DRISCOLL K H, ARNOLD M E, MARCY J A and WOLFE R E (2003), 'Lethality of *Listeria monocytogenes* in fully cooked and vacuum packaged chicken leg quarters during steam pasteurization', *J Food Sci*, 68, 2780–2783.

NESBAKKEN T, ECKNER K and ROTTERUD O J (2008), 'The effect of blast chilling on occurrence of human pathogenic *Yersinia enterocolitica* compared to *Campylobacter* spp and numbers of hygienic indicators on pig carcasses', *Int J Food Microbiol*, 123, 130–133.

NOTTINGHAM P M (1982), 'Microbiology of carcass meats', in Brown M H, *Meat microbiology*, London, Applied Science, 13–65.

© Woodhead Publishing Limited, 2012

NUTSCH A L, PHEBUS R K, RIEMANN M J, SCHAFER D E, BOYER J E JR, WILSON R C, LEISING J D and KASTNER C L (1997), 'Evaluation of a steam pasteurization process in a commercial beef processing facility', *J Food Prot*, 60, 485–492.

O'BRIEN S B, LENAHAN M, SWEENEY T and SHERIDAN J J (2007), 'Assessing the hygiene of pig carcasses using whole-body carcass swabs compared with the four-site method in EC Decision 471', *J Food Prot*, 70, 432–439.

OMAHA STEAKS (2009), *Product safety information*. Available from: http://wwwomahasteakscom/servlet/OnlineShopping?Dsp=765 (accessed 25 February 2011).

OMER M K, ALVSEIKE O, HOLCK A, AXELSSON L, PRIETO M, SKJERVE E and HEIR E (2010), 'Application of high pressure processing to reduce verotoxigenic *E coli* in two types of dry-fermented sausage', *Meat Sci*, 86, 1005–1009.

ORSAT V, BAI L, RAGHAVAN G S V and SMITH J P (2004), 'Radio-frequency heating of ham to enhance shelf-life in vacuum packaging', *J Food Process Eng*, 27, 267–283.

PATTERSON M F (2005), 'Microbiology of pressure-treated foods', *J Appl Microbiol*, 98, 1400–1409.

PEARSON A M and TAUBER F W (1996), *Processed Meats*, 3rd edn, Gaithersburg, MD, Aspen Publishers.

PENNEY N, BIGWOOD T, BAREA H, PULFORD D, LEROUX G, COOK R, JARVIS G and BRIGHTWELL G (2007), 'Efficacy of a peroxyacetic acid formulation as an antimicrobial intervention to reduce levels of inoculated *Escherichia coli* O157:H7 on external carcass surfaces of hot-boned beef and veal', *J Food Prot*, 70, 200–203.

PHAM Q T and WILLIX J (1985), 'Weight loss from lamb carcasses in frozen storage: influence of environmental factors', *Int J Refrig*, 8, 231–235.

PHEBUS R K, NUTSCH A L, SCHAFER D E, WILSON R C, RIEMANN M J, LEISING J D, KASTNER C L, WOLF J R and RAM K P (1997), 'Comparison of steam pasteurization and other methods for reduction of pathogens on surfaces of freshly slaughtered beef', *J Food Prot*, 60, 476–484.

PIETTE G, BUTEAU M L, DE HALLEUX D, CHIU L, RAYMOND Y, RAMASWAMY H S and DOSTIE M (2004), 'Ohmic cooking of processed meats and its effects on product quality', *J Food Sci*, 69, FEP71–FEP78.

RAVISHANKAR S, ZHU L and JARONI D (2010), 'Assessing the cross contamination and transfer rates of *Salmonella enterica* from chicken to lettuce under different food-handling scenarios', *Food Microbiol*, 27, 791–794.

REAGAN J O, ACUFF G R, BUEGE D R, BUYCK M J, DICKSON J S, KASTNER C L, MARSDEN J L, MORGAN J B, NICKELSON R, SMITH G C and SOFOS J N (1996), 'Trimming and washing of beef carcasses as a method of improving the microbiological quality of meat', *J Food Prot*, 59, 751–756.

RENDUELES E, OMER M K, ALVSEIKE O, ALONSO-CALLEJA C, CAPITA R and PRIETO M (2011), 'Microbiological food safety assessment of high hydrostatic pressure processing: a review', *LWT – Food Sci Technol*, 44, 1251–1260.

RETZLAFF D, PHEBUS R, KASTNER C and MARSDEN J (2005), 'Establishment of minimum operational parameters for a high-volume static chamber steam pasteurization system (SPS 400-SC) for beef carcasses to support HACCP programs', *Foodborne Pathog Dis*, 2, 146–151.

RIEDEL C T, BRØNDSTED L, ROSENQUIST H, HAXGART S N and CHRISTENSEN B B (2009), 'Chemical decontamination of *Campylobacter jejun*i on chicken skin and meat', *J Food Prot*, 72, 1173–1180.

ROBEY M, BENITO A, HUTSON R H, PASCUAL C, PARK S F and MACKEY B M (2001), 'Variation in resistance to high hydrostatic pressure and rpoS heterogeneity in natural isolates of *Escherichia coli* O157:H7', *Appl Environ Microbiol*, 67, 4901–4907.

SAKHARE P Z, SACHINDRA N M, YASHODA K P and NARASIMHA RAO D (1999), 'Efficacy of intermittent decontamination treatments during processing in reducing the microbial load on broiler chicken carcass', *Food Control*, 10, 189–194.

© Woodhead Publishing Limited, 2012

SASTRY S K, DATTA A K and WOROBO R W (2000), 'Ultraviolet light', *J Food Sci*, 65 (Suppl.), 90–92.

SATIN M (2002), 'Use of irradiation for microbial decontamination of meat: situation and perspectives', *Meat Sci*, 62, 277–283.

SHEPPARD S K, DALLAS J F, STRACHAN N J C, MACRAE M, MCCARTHY N D, WILSON D J, GORMLEY F J, FALUSH D, OGDEN I D, MAIDEN M C J and FORBES K J (2009), 'Campylobacter genotyping to determine the source of human infection', *Clin Infect Dis*, 48, 1072–1078.

SHERIDAN J J (1998), 'Sources of contamination during slaughter and measures for control', *J Food Saf*, 18, 321–339.

SIMPSON C A and SOFOS J N (2009), 'Antimicrobial ingredients', in Tarté R, *Ingredients in meat products: properties, functionality and applications*, New York, Springer, 301–378.

SINANOGLOU V J, KONTELES S, BATRINOU A, MANTIS F and SFLOMOS K (2009), 'Effects of gamma radiation on microbiological status, fatty acid composition, and color of vacuum-packaged cold-stored fresh pork meat', *J Food Prot*, 72, 556–563.

SINHAMAHAPATRA M, BISWAS S, DAS A K and BHATTACHARYYA D (2004), 'Comparative study of different surface decontaminants on chicken quality', *Brit Poult Sci*, 45, 624–630.

SLUTSKER L, RIES A A, MALONEY K, WELLS J G, GREEN K D and GRIFFIN P M (1998), 'A nationwide case-control study of *Escherichia coli* O157:H7 infection in the United States', *J Infectious Diseases*, 177, 962–966.

SMALL A, WELLS-BURR B and BUNCIC S (2005), 'An evaluation of selected methods for the decontamination of cattle hides prior to skinning', *Meat Sci*, 69, 263–268.

SOFOS J N (2008), 'Challenges to meat safety in the 21st century', *Meat Sci*, 78, 3–13.

SOFOS J N and GEORNARAS I (2010), 'Overview of current meat hygiene and safety risks and summary of recent studies on biofilms, and control of *Escherichia coli* O157:H7 in nonintact, and *Listeria monocytogenes* in ready-to-eat meat products', *Meat Sci*, 86, 2–14.

SOMMERS C H and BOYD G (2006), 'Variations in the radiation sensitivity of foodborne pathogens associated with complex ready-to-eat food products', *Radiat Phys Chem*, 75, 773–778.

SOMMERS C H, FAN X, NIEMIRA B and RAJKOWSKI K (2004), 'Irradiation of ready-to-eat foods at USDA'S Eastern Regional Research Center – 2003 update', *Radiat Phys Chem*, 71, 511–514.

SOMMERS C H, COOKE P H, FAN X and SITES J E (2009), 'Ultraviolet light (254 nm) inactivation of *Listeria monocytogenes* on frankfurters that contain potassium lactate and sodium diacetate', *J Food Sci*, 74, M114–M119.

SPESCHA C, STEPHAN R and ZWEIFEL C (2006), 'Microbiological contamination of pig carcasses at different stages of slaughter in two European Union-approved abattoirs', *J Food Prot*, 69, 2568–2575.

STEFANOVA R, VASILEV N V and SPASSOV S L (2010), 'Irradiation of foods, current legislation framework, and detection of irradiated foods', *Food Anal Method*, 3, 225–252.

STOPFORTH J D, YOON Y, BELK K E, SCANGA J A, KENDALL P A, SMITH G C and SOFOS J N (2004), 'Effect of simulated spray chilling with chemical solutions on acid-habituated and non-acid-habituated *Escherichia coli* O157:H7 cells attached to beef carcass tissue', *J Food Prot*, 67, 2099–2106.

STOPFORTH J D, O'CONNOR R, LOPES M, KOTTAPALLI B, HILL W E and SAMADPOUR M (2007), 'Validation of individual and multiple sequential interventions for reduction of microbial populations during processing of poultry carcasses and parts', *J Food Prot*, 70, 1393–1401.

SUN X D and HOLLEY R A (2010), 'High hydrostatic pressure effects on the texture of meat and meat products', *J Food Sci*, 75, R17–R23.

THORNTON H and GRACEY J F (1974), *Textbook of meat hygiene*, London, Baillier Tindall.

© Woodhead Publishing Limited, 2012

TYÖPPÖNEN S, PETÄJÄ E and MATTILA-SANDHOLM T (2003), 'Bioprotectives and probiotics for dry sausages', *Int J Food Microbiol*, 83, 233–244.

USDA (1999), Irradiation of meat food products. United States Department of Agriculture, Final Rule 9CFR81 and Docket No 97–076 C.

VAN ASSELT E D V, JONG A E I D, JONGE R D and NAUTA M J (2008), 'Cross-contamination in the kitchen: estimation of transfer rates for cutting boards, hands and knives', *J Appl Microbiol*, 105, 1392–1401.

VAN LANCKER M and BASTIAANSEN L (2000), 'Electron beam sterilization: trends and developments', *Med Dev Technol*, 11, 18–21.

WAGENAAR C L and SNIJDERS J M A (2004), 'Decontamination of broilers with hydrogen peroxide stabilized with glycerol during processing', *Int J Food Microbiol*, 91, 205–208.

WAMBURA P and VERGHESE M (2011), 'Effect of pulsed ultraviolet light on quality of sliced ham', *LWT – Food Sci Technol*, 44, 2173–2179.

WHITNEY B M, WILLIAMS R C, EIFERT J and MARCY J (2007), 'High pressure resistance variation of *Escherichia coli* O157:H7 strains and *Salmonella* serovars in tryptic soy broth, distilled water, and fruit juice', *J Food Prot*, 70, 2078–2083.

WILLIAMS R C, ISAACS S, DECOU M L, RICHARDSON E A, BUFFETT M C, SLINGER R W, BRODSKY M H, CIEBIN B W, ELLIS A, HOCKIN J and THE *E. COLI* O157:H7 WORKING GROUP (2000), 'Illness outbreak associated with *Escherichia coli* O157:H7 in Genoa salami', *CMAJ Can Med Assoc J*, 162, 1409–1413.

WONG E, LINTON R H and GERRARD D E (1998), 'Reduction of *Escherichia coli* and *Salmonella senftenberg* on pork skin and pork muscle using ultraviolet light', *Food Microbiology*, 15, 415–423.

ZHANG L, LYNG J G and BRUNTON N P (2004), 'Effect of radio frequency cooking on the texture, colour and sensory properties of a large diameter comminuted meat product', *Meat Sci*, 68, 257–268.

© Woodhead Publishing Limited, 2012

# 3

# Microbial decontamination of poultry carcasses

**C. Zweifel and R. Stephan, University of Zurich, Switzerland**

**Abstract**: The present chapter on the decontamination of poultry carcasses first discusses contamination sources and routes of poultry carcasses as well as major bacterial pathogens of concern, in particular *Campylobacter* and *Salmonella*. The main part of the chapter then addresses the antibacterial activity of physical, chemical, or biological intervention treatments applied to poultry carcasses. To appraise the efficacy, *Campylobacter* and *Salmonella* reductions after interventions are compared. Basic advantages and disadvantages as well as the present and future application of various methods and substances for the decontamination of poultry carcasses are also described.

**Key words**: poultry carcasses, decontamination, *Campylobacter*, *Salmonella*.

## 3.1 Introduction

Foodborne diseases are widespread, thus affecting lives, businesses, and economies worldwide. They have a major health impact in industrialized countries and remain responsible for high levels of morbidity and mortality in the general population but particularly for at-risk groups such as infants, young children, pregnant women, elderly, or immunocompromised people. The Centers for Disease Control and Prevention (CDC) recently published updated estimates of foodborne diseases in the United States (Scallan

---

Adapted from *Food Control* 21, M. Loretz, R. Stephan and C. Zweifel, 'Antimicrobial activity of decontamination treatments for poultry carcasses: a literature survey', 791–804, copyright (2010), with permission from Elsevier.

© Woodhead Publishing Limited, 2012

*et al.*, 2011a,b). Illnesses caused by 24 major pathogens and illnesses caused by unspecified agents were thereby considered (Scallan *et al.*, 2011a,b). If the two estimates were combined, the totals were 47.8 million foodborne illnesses, 127,839 hospitalizations, and 3,037 deaths per year in the US. In many countries, *Campylobacter*, *Salmonella* and Shiga toxin-producing *Escherichia coli* (STEC) are among the most important bacterial foodborne pathogens (EFSA/ECDC, 2011; Scallan *et al.*, 2011a).

With regard to meat production, healthy food animals were recognized in recent years as carriers of pathogens causing human illness. To counter this threat, the focus is currently on preventive systems in accordance with the hazard analysis and critical control point (HACCP) principles instead of only final product testing. Intervention strategies in the production of poultry can be applied on-farm or at slaughter and processing. In terms of pre-harvest control options, reduction of environmental exposures by on-farm bio-security measures (e.g. effective cleaning and disinfection, minimizing the use of invasive practices such as thinning, restricting personnel access, securing premises from wild birds and mammals), an increase in poultry's host resistance to reduce pathogen carriage in the gut (e.g. competitive exclusion, vaccination), and the use of antimicrobial alternatives for reduction/elimination of pathogens from colonized chicken (e.g. bacteriophage treatment) have been proposed (see Section 3.9). However, the effectiveness of some measures is controversially discussed, many interventions are still under development, and regulatory issues remain to be solved for certain methods. On the other hand, interventions at slaughter basically comprise measures aimed at preventing faecal contamination (slaughter hygiene measures, logistic or scheduled slaughter) and decontamination treatments aimed at reducing the bacterial load on carcasses (see Section 3.9). Decontamination treatments including chemicals are widely used in the US and Canada for poultry carcasses and such interventions have also attracted renewed attention in Europe.

Hence, the potential role and impact of the various decontamination treatments need to be accurately assessed. The present chapter on the decontamination of poultry carcasses first briefly discusses the contamination sources and routes of poultry carcasses as well as the major pathogens of concern, in particular *Campylobacter* and *Salmonella*. The main part of this chapter then addresses the antibacterial activity of intervention treatments applied for poultry carcasses. The focus is thereby on reductions obtained for *Campylobacter* and *Salmonella*.

## 3.2 Contamination of poultry carcasses and major pathogens of concern

### 3.2.1 Contamination sources and routes

The modern slaughter of poultry is a complex, rapid, and highly automated process. Despite all advancements in slaughter technologies, significant

© Woodhead Publishing Limited, 2012

bacterial contamination of carcasses might occur during slaughter and there are considerable opportunities for the spread of bacteria (Jackson *et al.*, 2001). Upon arrival at the slaughter plant, chickens have substantial numbers of bacteria associated with them. These bacteria are found both on the surface of feathers and in the alimentary tract (Berrang *et al.*, 2000a). Although some bacteria are resident on feathers, much of the external contamination results from fecal contamination. Of importance in this context are contaminations during the transport to the slaughter facility, leakage of fecal content from the cloaca, and hygienically critical stages of the slaughter process, such as defeathering or evisceration. Equipment like the defeathering machine can thereby contribute to the spread of bacteria on carcasses. In addition, carcasses might be directly or indirectly cross-contaminated at multiple sites throughout transport, slaughter, and processing. Cross-contamination is a complex process and sources include transport crates, various equipments, knives, workers' hands, and transfer belts. The concept of bacterial cross-contamination and the impact of slaughterline contamination have been intensively studied (see Section 3.9).

The extent of carcass contamination and the relevance of sources are dependent on the carriage of pathogens by incoming animals, the slaughter technology, the cleaning and disinfection regime, and the process hygiene including behavior in terms of personnel hygiene. In this context, it must be emphasized that carcasses might be microbiologically contaminated despite the absence of visible debris (Gill, 2004). Transferred bacteria may affect meat quality and cause spoilage (such as *Pseudomonas*, *Moraxella*, *Lactobacillus*, or *Brochothrix thermosphacta*) or act as human pathogens (such as *Campylobacter*, *Salmonella*, or STEC). Amongst the pathogens of concern, a good deal originates from the enteric flora of healthy animals (Nørrung and Buncic, 2008). This is also the case with chickens that constitute an important reservoir for *Campylobacter*, for example. By contamination and cross-contamination such pathogens might be introduced into the food chain, spread to processing facilities and consumers, and thereby pose a threat for the contamination of other foods at consumer level and for foodborne diseases.

### 3.2.2 Major bacterial pathogens of concern

Several pathogenic bacteria have been associated with raw poultry, including *Campylobacter* spp., *Salmonella enterica* subsp. *enterica*, *Arcobacter* spp., *Clostridium difficile*, *Clostridium perfringens*, *Listeria monocytogenes*, and *Staphylococcus aureus*. The main pathogens associated with human illness resulting from consumption of poultry belong to the genera *Salmonella* and *Campylobacter*, which are briefly discussed below.

*Campylobacter* and *Salmonella* are leading causes of acute bacterial gastroenteritis in the developed world. Within the genus *Campylobacter*, *C. jejuni* and *C. coli* are basically the most common species associated with human

© Woodhead Publishing Limited, 2012

disease, whereas *S.* Enteritidis and *S.* Typhimurium are the predominating *Salmonella* serovars in this context. In the European Union (EU), 198,252 confirmed human cases of campylobacteriosis (45.6 cases per 100,000 population) and 108,614 cases of salmonellosis (23.7 cases per 100,000 population) were reported in the year 2009 (EFSA/ECDC, 2011). In many countries, the incidence of campylobacteriosis exceeds that of *Salmonella* infections. Illness is primarily characterized by diarrhoea, in the case of campylobacteriosis often accompanied by abdominal pain. Most infections are self-limiting, lasting a few days. With regard to campylobacteriosis, post-infection complications such as reactive arthritis and neurological disorders including Guillain-Barré syndrome might occur. *Campylobacter* and *Salmonella* infections frequently result from foodborne pathways, with some cases being associated with environmental exposure. Using a population genetic approach, Wilson *et al.* (2008) found that the majority of sporadic campylobacteriosis cases were attributed to animals farmed for poultry and meat.

The intestinal tract of a large number of mammals and birds can be colonized by *Campylobacter* and *Salmonella.* Chickens thereby constitute an important reservoir. Amongst broiler flocks in the EU examined in 2009, 20.5% (range between member states: 0–78.4%) and 5.0% (range between member states: 0–32.4%) tested positive for *Campylobacter* and *Salmonella*, respectively (EFSA/ECDC, 2011). Enteric pathogens can be introduced to chickens through a variety of sources. Due to implementation of successful control programs throughout the production chain in many countries, the *Salmonella* situation has been improved in recent years. But measures that substantially reduced *Salmonella* in poultry have been largely ineffective against *Campylobacter*. The epidemiology of *Campylobacter* is complex and still not entirely understood. Although horizontal transmission is believed to be the common way for poultry flock colonization, there is a degree of dispute over which are the most important sources and transmission routes. Thus, contaminated water, domestic and wild animals, personnel working in the chicken house, the external environment, persistence in the chicken house, transport crates, or airborne transmission have been implicated (see Section 3.9). Identification of major transmission routes enables bio-security measures to be targeted towards the areas posing the greatest risk.

When chickens harbor pathogenic bacteria, these might be transferred to the carcasses and in this way be introduced into the food chain. Many studies have investigated the contamination of poultry carcasses with *Campylobacter* and *Salmonella* during slaughter (see Section 3.9). A baseline survey examining broiler carcasses from different EU member states showed that *Campylobacter* and *Salmonella* were detected in 75.8% (range between member states: 4.9–100%) and 15.7% (range between member states: 0–85.6%), respectively (EFSA, 2010a). Moreover, amongst fresh broiler meat samples examined in 2009, 31.0% tested positive for *Campylobacter* and 5.4% for *Salmonella* (EFSA/ECDC, 2011). Consumption and handling of

© Woodhead Publishing Limited, 2012

poultry meat is considered an important risk factor for human disease (Guo *et al.*, 2011; Humphrey *et al.*, 2007; Kimura *et al.*, 2004). Such contaminated poultry also constitutes a source for cross-contamination of other foods in the kitchen. Based on epidemiological data, outbreak investigations, case control studies, or molecular typing, data poultry meat has been implicated as major source in particular for campylobacteriosis (EFSA, 2010b; Friedman *et al.*, 2004; Müllner *et al.*, 2010; Sheppard *et al.*, 2009; Stafford *et al.*, 2008; Wingstrand *et al.*, 2006). Efforts are therefore attempted to reduce the number of colonized flocks being delivered for slaughter and to prevent or eliminate contamination of carcasses during slaughter. Rosenquist *et al.* (2003) estimated that the burden of *Campylobacter* infections associated with the consumption of poultry would be reduced 30-fold by introducing a 2 log reduction of the number of *Campylobacter* on poultry carcasses. These findings underline the potential impact of control measures at the slaughterhouse level.

## 3.3 Antibacterial activity of decontamination treatments for poultry carcasses

With regard to interventions at slaughter, hygienic measures aimed at preventing fecal contamination of carcasses must be differentiated from decontamination treatments aimed at reducing the bacterial level on carcasses. With regard to preventing carcass contamination, improvements of slaughter hygiene measures, prevention/minimizing of fecal contamination, logistic or scheduled slaughter, or hygienic design of equipment have been proposed. On the other hand, decontamination treatments comprise a wide variety of methods and substances. This chapter is focusing on treatments used for the decontamination of poultry carcasses. Interventions applied at slaughter basically comprise physical, chemical, and biological treatments applied alone or in combination. Various demands have thereby to be met: improvement of food safety, no changes in organoleptic and nutritional attributes, no residues on foods, convenient to apply, cheap, and no objections from consumers or legislators.

In Europe, carcass interventions with substances other than potable water are not categorically banned for use on foods of animal origin (Regulation (EC) No 853/2004), but approval is tied to strict prescriptions and can only be authorized after the European Food Safety Authority (EFSA) has provided a risk assessment (Hugas and Tsigarida, 2008). For a long time, interventions were discouraged because such treatments were perceived to be means of concealing poor hygiene practices. On the other hand, decontamination treatments for poultry carcasses including chemicals are widely used in various other countries. Substances that may be used in the production of poultry in the US are listed by the Food Safety and Inspection Service

© Woodhead Publishing Limited, 2012

(FSIS, 2011). Oyarzabal (2005) reviewed antimicrobials applied in poultry processing from a US perspective. In particular, the topic of chlorine-washed poultry carcasses and related trade restrictions are controversially discussed between the US and the EU. On the other hand, decontamination of poultry carcasses with chemicals has recently been reconsidered in Europe. The European Commission issued a proposal (COM, 2008) in which the use of chlorine dioxide, acidified sodium chlorite, trisodium phosphate, and peroxyacids has been proposed for the removal of surface contamination from poultry carcasses and framing conditions have been defined. No consensus has yet been reached among the involved competent authorities. The EU Council rejected the mentioned proposal (Council Decision 121/2009) and also the EFSA panel on biological hazards (BIOHAZ) was cautious in its appraisal.

Sections 3.4–3.7 address the antibacterial activity of interventions used for poultry carcasses, whereby the focus is on *Campylobacter* and *Salmonella*. For this purpose, the literature published since 1990 was reviewed. These sections are partly based on a review previously published by Loretz *et al.* (2010) (copyright clearance: http://s100.copyright.com/CustomerAdmin/PLF.jsp?lID=2010070_1278310891747). To appraise the antibacterial activity, *Campylobacter* and *Salmonella* reductions after interventions were compared, whereas investigations mainly addressing growth inhibition or processed meat products were not considered. Selected resulting reductions are summarized in Tables 3.1 to 3.6. Most studies addressed the efficacy of physical and chemical treatments.

## 3.4 Physical decontamination treatments for poultry carcasses

### 3.4.1 Hot water

Washing with water is routinely used and has been shown to be effective in removing visible contaminants such as soil, feathers, and other debris. On the other hand, water spraying might spread bacteria on carcass surfaces. With regard to the slaughter of poultry, spraying, immersion, and immersion chilling are basically distinguished. In several studies, the use of hot water was investigated. Mechanisms are thereby twofold: first the water facilitates removal of dirt and second there is a lethal effect caused by the heat. Probably due to the absent heat inactivation, cold and warm water tends to yield, in general, lower reductions. In the following section, mainly studies based on industrial data (commercial conditions) or pilot-scale data (experiments using industrial equipment in non-industrial settings) are discussed.

Based on the evaluated studies, *Campylobacter* and *Salmonella* on poultry carcasses and parts were reduced by 0.1–2.8 orders of magnitude (log reductions) after hot water treatment (Table 3.1). Reductions of *Campylobacter*

© Woodhead Publishing Limited, 2012

**Table 3.1** Reductions of *Campylobacter* and *Salmonella* by hot water and steam on the surface of poultry carcasses and parts

| Agent/ microorganism | Reduction ($log_{10}$ CFU) | Treated material | Application[h] | Contamination | Temp (°C) | Exposure time (min) | References |
|---|---|---|---|---|---|---|---|
| **Hot water** | | | | | | | |
| *Campylobacter* spp. | 2.1–2.8/ml[a] | Carcass | SP | Artificial | 21–54 | 0.1 | Northcutt *et al.* (2005) |
| | 1.1–1.6/ml[b] | Carcass | IM/SC | Natural | 70/12–15 | 0.7/0.2 | Purnell *et al.* (2004) |
| | >1.0/ml[b] | Carcass | IM/SC | Natural | 75/12–15 | 0.5/0.2 | Purnell *et al.* (2004) |
| | 0.9–1.1/ml[c] | Carcass | IM | Natural | 75–80 | 0.3–0.5 | Purnell *et al.* (2004) |
| | <0.5/ml[d] | Carcass | IM | Natural | 60 | 0.5 | Berrang *et al.* (2000b) |
| | 0.1–0.4/ml[d] | Carcass | SP | Natural | 70–73 | 0.3 | Berrang *et al.* (2000b) |
| *Campylobacter jejuni* | 2.4–2.5/carcass[d] | Carcass | SP/IC | Artificial | 55–60/1 | 0.2/50 | Li *et al.* (2002) |
| | 1.8–1.9/g | Leg | IM | Artificial | 80–85 | 0.3 | Whyte *et al.* (2003) |
| | 1.7/$cm^2$[e] | Carcass | IM | Artificial | 75 | 0.5 | Corry *et al.* (2007) |
| | 1.3–1.4/carcass[d] | Carcass | SP | Artificial | 55–60 | 0.2–0.3 | Li *et al.* (2002) |
| | 1.3/$cm^2$[e] | Carcass | IM | Artificial | 80 | 0.3 | Corry *et al.* (2007) |
| | 1.0/$cm^2$[e] | Carcass | IM | Artificial | 70 | 0.7 | Corry *et al.* (2007) |
| | 0.9/g | Leg | IM | Artificial | 75 | 0.3 | Whyte *et al.* (2003) |
| | 0.8–1.1/g | Leg | IM | Artificial | 75–85 | 0.2 | Whyte *et al.* (2003) |
| | 0.6–1.3/g | Leg | IM | Natural | 70–85 | 0.3 | Whyte *et al.* (2003) |
| | 0.3–0.4/g | Leg | IM | Natural | 80–85 | 0.2 | Whyte *et al.* (2003) |
| | 0.6–1.4/$cm^2$ | Carcass | IM | Artificial | 55–65 | 0.3 | Li *et al.* (2002) |
| *Salmonella* spp. | 0.7–1.1/ml[a] | Carcass | SP | Artificial | 21–54 | 0.1 | Northcutt *et al.* (2005) |
| **Steam** | | | | | | | |
| *Campylobacter* spp. | 0.5–1.3/g[f] | Carcass | Steam | Natural | 90 | 0.2–0.4 | Whyte *et al.* (2003) |
| *Campylobacter jejuni* | 1.8–3.3/$cm^2$[g] | Carcass | Steam | Artificial | 100 | 0.2–0.3 | James *et al.* (2007) |
| *Salmonella* Typhimurium | 6.2/$cm^2$ | Breast | Steam | Artificial | 100 | 1 | McCann *et al.* (2006) |

[a] Treatment using a spray cabinet (pilot-scale).
[b] Treatment in poultry slaughter pilot plant, on-line processing unit.
[c] Treatment in poultry slaughter pilot plant, initial trials.
[d] Treatment in poultry slaughter plant.
[e] Treatment using a pilot hot water immersion vessel.
[f] Treatment using a steam pasteurization unit of a beef abattoir (pilot-scale).
[g] Treatment using a steam cabinet (pilot-scale).
[h] IM, immersion; SP, spraying; IC, immersion chilling; SC, spray chilling.

© Woodhead Publishing Limited, 2012

were mainly in the range from 1.0 to 1.7 logs. Purnell *et al.* (2004) reported that immersion in water at 65–70°C did not reduce *Campylobacter* on naturally contaminated carcasses, but increasing the temperature yielded reductions of about 1.0 log CFU/ml. After these initial trials, Purnell *et al.* (2004) also investigated the prototype of an on-line processing unit, which combined immersion with a chill spray cabinet. Using an immersion or spray rescald treatment, Berrang *et al.* (2000b) found only low reductions of *Campylobacter* (<0.5 log CFU/ml). The other surveys were based mainly on inoculated carcasses and parts (Table 3.1). On inoculated chicken legs, Whyte *et al.* (2003) reported an increase in reductions of about 1.0 log CFU/g by increasing the water temperature from 75 to 85°C, but temperatures of >75°C caused skin damage resulting in discoloration and deterioration in the appearance. In comparison with hot water treatments, cold and warm water often yielded 0.2–1.4 log reductions of *Campylobacter* and *Salmonella* (Del Río *et al.*, 2007; Fabrizio *et al.*, 2002; Huezo *et al.*, 2007; Hwang and Beuchat, 1995; Kim *et al.*, 2005; Li *et al.*, 2002; Oyarzabal *et al.*, 2004; Russel and Axtell, 2005; Whyte *et al.*, 2001). Moreover, Wang *et al.* (1997) evaluated the influence of water temperature (10–60°C) and pressure (414–1034 kPa) for reducing *S.* Typhimurium. Highest reductions (1.6 log CFU/38.5 $cm^2$) were obtained by spraying with water at 35°C and a pressure of 621 kPa. The authors hypothesized that the use of higher pressures might force bacteria into the skin instead of washing them off.

Direct comparison of spraying, immersion, and immersion chilling is hampered by varying framing conditions (Table 3.1). Li *et al.* (2002) found similar reductions of *C. jejuni* at the upper level for immersion and spraying, but immersion results depended strongly on the water temperature. Moreover, some studies evaluated the efficacy of combined water treatments. Immersion in hot water followed by spray chilling thereby tended to yield slightly higher reductions of *Campylobacter* than immersion alone (Purnell *et al.*, 2004). Similarly, the combination of water spraying and immersion chilling yielded higher reductions of *C. jejuni* than spraying alone (Li *et al.*, 2002). With regard to practical application, it must be considered that spraying generally requires less liquid than the other methods and does not need installation of immersion tanks, which could be a source of cross-contamination.

### 3.4.2 Steam

An alternative to hot water is the application of steam (Table 3.1). But more data, especially obtained under commercial conditions are required. On naturally contaminated poultry carcasses, Whyte *et al.* (2003) reported reductions of *Campylobacter* by up to 1.3 log CFU/g. The higher reductions (1.8–3.3 log CFU/ $cm^2$) in the study of James *et al.* (2007) might be correlated with inoculation. However, the steam treatments mentioned resulted in deterioration in the appearance of carcasses in both studies. Adverse effects must primarily be expected when great heat in combination with long

© Woodhead Publishing Limited, 2012

exposure times are used. Besides, under laboratory conditions, a 1 min steam treatment reduced *S.* Typhimurium by 6.2 log CFU/cm$^2$ (McCann *et al.*, 2006). Highest inactivation rates were thereby reported during the first seconds of treatment.

### 3.4.3 Irradiation

Irradiation of foods (often referred to as 'cold pasteurization') is based primarily on the use of three kinds of ionizing radiation: gamma rays (from the radionuclides cobalt-60 or much less frequently cesium-137), machine generated X-rays at or below an energy level of 5 MeV, or accelerated electrons at or below an energy level of 10 MeV (EFSA, 2011a; Farkas, 1998; Farkas and Mohácsi-Farkas, 2011). The antimicrobial activity of ionizing radiation is due to direct damage of DNA and the effect of generated free radicals. The efficacy of irradiation depends on target organisms, the reduction required, the type of food, the presence of oxygen, or the content of water (EFSA, 2011a). Moreover, irradiation of frozen products requires a greater radiation dose than irradiation of fresh food. Recently there has been increased interest in food irradiation. In this context, EFSA has published updated scientific opinions on the efficacy and microbiological safety of food irradiation and on its chemical safety (EFSA, 2011a,b).

For poultry, most data on the antibacterial efficacy of irradiation originate from studies performed under laboratory conditions and from studies investigating (processed) meat products (O'Bryan *et al.*, 2008). Further research in terms of practical application for poultry carcasses is therefore required. In a study investigating inoculated poultry carcasses irradiated at 2–7 kGy (Nassar *et al.*, 1997), the proportion of *Salmonella* (*S.* Virchow positive poultry carcasses was clearly reduced at 4–6 kGy and *S.* Virchow was completely eliminated at 7 kGy. One of the few other surveys showed that levels as low as 1 kGy would reduce *Salmonella* on chicken breasts and legs (Heath *et al.*, 1990). Santos *et al.* (2003) investigated the effect of different doses of gamma irradiation on chicken legs and they recommended a dose of 3.8 kGy to assure safety with regard to the presence of *Salmonella.* In addition, Patterson (1995) investigated the sensitivity of *Campylobacter* spp. to irradiation in poultry meat and results indicated that *Campylobacter* were more radiation sensitive than *Salmonella.*

### 3.4.4 Ultrasound

Only limited and mainly experimental data exist on the application of ultrasound to solid foods (Piyasena *et al.*, 2003). Sams and Feria (1991) hypothesized that the irregular skin surface of poultry carcasses might provide protection for bacteria and reduce the efficacy. However, Boysen and Rosenquist (2009) showed that steam treatment combined with ultrasound at a poultry slaughter plant reduced *Campylobacter* by >2.5 log CFU per carcass.

© Woodhead Publishing Limited, 2012

The use of ultrasound coupled with heat therefore seems to be a promising alternative. Moreover, Haughton *et al.* (2011b) investigated the efficacy of high and low intensity ultrasound (20 kW/L and 20 W/L, respectively) under laboratory conditions. Following sonication or thermosonication in the high intensity unit, inoculated *Campylobacter* were no longer detected on chicken skin (reductions >3.8 log CFU/g), whereas reductions in the low intensity unit were between 0.6 and 0.8 log CFU/g. In another study, sonication in peptone (20 kHz, 30 min) reduced *S.* Typhimurium on chicken breast skin by 1.0–1.5 logs (Lillard, 1993).

### 3.4.5 Ultraviolet light

Ultraviolet light (UV) is commonly used in the food industry for decontamination of packaging material or water (Bintsis *et al.*, 2000). For use on poultry carcasses, the restricted penetration depth (skin folds, hair follicles), the potential impact on fat oxidation, and the lack of data obtained under commercial conditions must be considered. In a recent study, Chun *et al.* (2010) investigated the effect of UV-C on the inactivation of pathogens inoculated on chicken breasts. *C. jejuni* and *S.* Typhimurium were thereby reduced by 1.2–1.3 log CFU/g after treatment at 5 kJ/m$^2$. Haughton *et al.* (2011a) investigated the efficacy of UV-C treatment on inoculated skinless chicken breast and chicken skin. On skinless chicken breast, *C. jejuni* was reduced by 0.8 log CFU/g and *S.* Enteritidis by 1.3 log CFU/g, whereas on chicken skin reductions of 0.5 and 0.7 log CFU/g were obtained for *C. jejuni* and *S.* Enteritidis, respectively (Haughton *et al.*, 2011a). Other studies investigating the efficacy of UV-C on inoculated samples reported reductions of *S.* Typhimurium mainly below 1.0 log (Kim *et al.*, 2002) or of *C. jejuni* by 0.4 log (carcasses) and 0.8 log (chicken skin samples) (Isohanni and Lyhs, 2009). The study by Isohanni and Lyhs (2009) discouraged the use of UV-C as the primary decontamination method for poultry. Various other studies, on the other hand, showed that UV-C radiation scarcely affects the sensory quality (Chun *et al.*, 2010; Haughton *et al.*, 2011a; Isohanni and Lyhs, 2009; Wallner-Pendleton *et al.*, 1994).

Of growing interest is also the use of pulsed UV for the decontamination of poultry (Keklik *et al.*, 2010a,b; Paskeviciute *et al.*, 2011). Pulsed UV light is a novel technology, which is designated to inactivate microorganisms on surfaces in short times under non-thermal conditions and it is considered an alternative decontamination technique to irradiation. Under varying framing conditions, Keklik *et al.* (2010b) reported reductions of *S.* Typhimurium by 1.2–2.4 log CFU/cm$^2$ after pulsed UV light treatment of unpackaged chicken breast samples. No significant changes in lipid peroxidation or color parameters were found after mild or moderate treatments (Keklik *et al.*, 2010b). In another study, high-power pulsed light treatment reduced viability of *S.* Typhimurium inoculated skinless chicken breast by 2.0 log CFU/ml and changes in lipid peroxidation or sensory characteristics were

© Woodhead Publishing Limited, 2012

not significant when short exposures were used (Paskeviciute *et al.*, 2011). The results of these studies indicate that pulsed UV light has potential to be used for decontamination of poultry.

### 3.4.6 Air chilling

The antibacterial activity of air chilling on red meat carcasses is based mainly on the surface desiccation achieved by high air velocity. A comparable effect is commonly not achieved and not desired for poultry carcasses for quality reasons. Yet some studies evaluated the antibacterial potential of forced air chilling. In the study by Huezo *et al.* (2007), performed in a poultry slaughter pilot plant, air chilling (150 min, –1.1°C, air speed: 3.5 m/s) reduced inoculated *C. jejuni* and *Salmonella* by 1.4 and 1.0 log CFU/ml, respectively. Reductions were thereby 0.4 log higher than those obtained after immersion chilling. Similarly, James *et al.* (2007) reported reductions of inoculated *C. jejuni* by up to 1.8 log CFU/$cm^2$ using an impingement air chilling system. Enhanced reductions were obtained by combining rapid air chilling with previous steam or hot water treatment. However, at a Danish slaughter plant, mean reductions of *Campylobacter* only accounted for about 0.4 log CFU per carcass after air chilling or crust freezing (Boysen and Rosenquist, 2009). The practicability and efficacy of air chilling in the slaughter of poultry as well as potential adverse effects on carcass appearance need to be further evaluated.

### 3.4.7 Freezing

With regard to *Campylobacter*, freezing is of interest and seems to constitute an effective decontamination method. Many of the published studies originate from European countries, especially from Scandinavia (Birk *et al.*, 2006; Boysen and Rosenquist, 2009, El-Shibiny *et al.*, 2009; Georgsson *et al.*, 2006; Rosenquist *et al.*, 2006; Sandberg *et al.*, 2005). In these studies, reductions of *Campylobacter* ranged mainly from 1.3 to 2.2 orders of magnitude (log reductions) and the majority of results were obtained from naturally contaminated poultry carcasses. In a study from a Danish slaughter plant, freezing was more effective in reducing *Campylobacter* than air chilling, crust freezing, or steam–ultrasound (Boysen and Rosenquist, 2009). US studies based on inoculated carcasses also proved the efficacy of freezing in reducing *Campylobacter* on poultry (Bhaduri and Cottrell, 2004; Solow *et al.*, 2003; Zhao *et al.*, 2003). For example, Bhaduri and Cottrell (2004) showed that frozen storage at –20°C, alone and with pre-refrigeration, reduced counts of *C. jejuni* by 1.4–3.4 log CFU/g on chicken skin over a two-week period. Moreover, the results obtained by Zhao *et al.* (2003) indicated that freezing conditions, including temperature and holding time, greatly influenced the rate of inactivation of *C. jejuni* on poultry.

© Woodhead Publishing Limited, 2012

### 3.4.8 Summary of physical treatments

Amongst the physical methods used for the decontamination of poultry carcasses, water-based treatments predominate, especially with regard to industrial or pilot-scale data. Bacterial reductions were dependent on framing conditions, including application modes (spraying, immersion, immersion chilling), temperatures, pressures, exposure times, or contamination levels (natural, artificial).

The bactericidal effect of hot water and steam is due to the combination of removal and heat inactivation. Critical for the second effect is the temperature actually achieved on carcass surfaces. Although hot water and steam were quite effective in reducing *Campylobacter* and *Salmonella* on poultry carcasses, the additional investments, costs, and potential adverse effects on the appearance and quality of carcasses must be considered. In this regard, the use of hot water followed by ultrasound or air chilling might constitute interesting options, but further investigations under practical conditions are required. Similarly, the use of pulsed UV treatment deserves attention. With special regard to *Campylobacter*, freezing must also be mentioned. Furthermore, the application of irradiation at adequate dosages is effective to reduce bacteria on poultry carcasses, but costs for the infrastructure, the appropriate application during processing, and the limited acceptance of this technology by consumers are limiting factors.

## 3.5 Chemical decontamination treatments for poultry carcasses

Chemical compounds used for the decontamination of poultry carcasses comprise a wide variety of substances and combinations. Reductions of *Campylobacter* and *Salmonella* discussed in the following are generally those compared to 'no treatment'. But on an industrial scale, it is difficult to appraise whether reductions obtained are mainly due to physical removal by water or by the antibacterial activity of chemicals.

### 3.5.1 Organic acids

Organic acids have considerable potential for acceptance by the industry because they are quite inexpensive and generally recognized as safe. However, discoloration of skin may occur at high concentrations or low pH. Based on the evaluated studies, most data reporting reductions of *Campylobacter* and *Salmonella* originate from laboratory surveys.

Two studies used pilot-scale equipment to investigate the antibacterial efficacy of lactic acid spraying (Li *et al.*, 1997; Yang *et al.*, 1998). Inoculated *S.* Typhimurium was thereby reduced by 0.6–1.8 log CFU/carcass (Table 3.2). With the exception of the study by Killinger *et al.* (2010) reporting

© Woodhead Publishing Limited, 2012

**Table 3.2** Reductions of *Campylobacter* and *Salmonella* by organic acid treatments on the surface of poultry carcasses and parts

| Agent/ microorganism | Reduction ($log_{10}$ CFU) | Treated material | Application[c] | Concentration | Contamination | Temp (°C) | Exposure time (min) | References |
|---|---|---|---|---|---|---|---|---|
| **Acetic acid** | | | | | | | | |
| *Campylobacter jejuni* | 1.2–1.4/g | Wing | IM | 2% | Artificial | 4 | 0.3–0.8 | Zhao and Doyle (2006) |
| *Salmonella* spp. | 1.1/cm$^2$ | Breast | SP | 2% | Artificial | 55 | 0.3 | Carpenter *et al.* (2011) |
| *Salmonella* Hadar | 1.8–2.0/10 cm$^2$ | Breast | SP | 2.5% | Artificial | 55 | 0.5 | Jiménez *et al.* (2007) |
| | 1.2–1.8/10 cm$^2$ | Breast | SP | 1–2.5% | Artificial | 25 | 0.2–0.4 | Jiménez *et al.* (2007) |
| *Salmonella* Typhimurium | 1.4/ml | Carcass | IC | 20 ppm | Artificial | 4 | 45 | Fabrizio *et al.* (2002) |
| | 0.8/ml | Carcass | SP | 20 ppm | Artificial | NA[d] | 0.3 | Fabrizio *et al.* (2002) |
| **Citric acid** | | | | | | | | |
| *Campylobacter jejuni* | 1.2/g | Carcass | IM or SP | 15% | Artificial | 30 | 0.3–0.5 | Ellerbroek *et al.* (2007) |
| | 0.8–0.9/g | Carcass | SP | 10% | Artificial | 10–30 | 0.3–0.5 | Ellerbroek *et al.* (2007) |
| | 0.6–0.8/g | Carcass | IM | 10% | Artificial | 10–30 | 0.3–0.5 | Ellerbroek *et al.* (2007) |
| *Salmonella* Enteritidis | 0.2/g | Leg | IM | 2% | Artificial | 18 | 15 | Del Río *et al.* (2007) |
| **Lactic acid** | | | | | | | | |
| *Campylobacter jejuni* | 1.7/ml | Back | IM | 2.5% | Artificial | 20 | 1 | Riedel *et al.* (2009) |
| | 1.5/g | Carcass | IM | 15% | Artificial | 30 | 0.3–0.5 | Ellerbroek *et al.* (2007) |
| | 0.9/g | Carcass | SP | 15% | Artificial | 30 | 0.3–0.5 | Ellerbroek *et al.* (2007) |
| | 0.3/g | Carcass | IM | 10% | Artificial | 10 | 0.3–0.5 | Ellerbroek *et al.* (2007) |
| | 0.2/g | Carcass | SP | 10% | Artificial | 10 | 0.3–0.5 | Ellerbroek *et al.* (2007) |
| *Salmonella* spp. | 5.4/wing | Wing | IM | 2% | Artificial | NA | 3 | Killinger *et al.* (2010) |
| | 2.0/cm$^2$ | Breast | IM | 1% | Artificial | 25 | 30 | Hwang and Beuchat (1995) |
| | 1.3/cm$^2$ | Breast | SP | 2% | Artificial | 55 | 0.3 | Carpenter *et al.* (2011) |
| *Salmonella* Enteritidis | 0.8–1.7/g | Breast | IM | 0.5–2% | Artificial | 25 | 10–30 | Anang *et al.* (2007) |
| *Salmonella* Typhimurium | 2.2/ml | Breast | SP | 1–2% | Artificial | 20 | 0.5 | Xiong *et al.* (1998) |
| | 1.8/carcass[a] | Carcass | SP | 2% | Artificial | 35 | 0.3 | Yang *et al.* (1998) |
| | 0.6–0.7/carcass[b] | Carcass | SP | 1% | Artificial | 22 | 0.5 | Li *et al.* (1997) |

[a] Treatment using a modified inside–outside birdwasher for spraying (pilot-scale).
[b] Treatment using a spraying test chamber (pilot-scale).
[c] IM, immersion; SP, spraying; IC, immersion chilling.
[d] NA, not available.

© Woodhead Publishing Limited, 2012

strikingly high reductions, lactic acid treatment reduced *Campylobacter* and *Salmonella* in laboratory studies by 0.2–1.7 and 0.6–2.2 orders of magnitude (log reductions), respectively (Table 3.2). In another study, Carpenter *et al.* (2011) compared the efficacy of 2% acetic, lactic, or levulinic acid for reducing *Salmonella*. Spraying with these acids yielded similar reductions ranging from 1.1 to 1.3 log CFU $cm^2$ (and from 0.7 to 0.9 log CFU $cm^2$ when compared to water washed samples).

Only a few laboratory studies investigated the efficacy of acetic or citric acid against *Campylobacter* or *Salmonella* on poultry carcasses and parts. Acetic acid treatment yielded reductions between 0.8 and 2.0 logs (Table 3.2). Jiménez *et al.* (2007) investigated the effect of single and double spraying, different concentrations, exposure times, and application temperatures. Moreover, Dickens and Whittemore (1994, 1995) observed distinct reductions of the prevalence of *Salmonella* when using varying framing conditions. Citric acid treatment yielded reductions between 0.2 and 1.2 log CFU/g (Table 3.2). In the study by Del Río *et al.* (2007), reductions of *S.* Enteritidis were comparable to those obtained by a water dip control. By the use of the proprietary product Chicxide® (a buffered blend of lactic and citric acid), Laury *et al.* (2009) reported reductions of *Salmonella* inoculated on poultry carcasses by 1.3 and 2.3 log CFU/ml after spraying and immersion, respectively.

### 3.5.2 Chlorine, hypochlorite, sodium hypochlorite, and sodium chlorite

Chlorine-containing compounds are widely used for pre- and post-chill spraying/washing of poultry carcasses or in carcass chillers, especially in the US and Canada. An overview on benefits and risks of different compounds is provided in the report of a FAO/WHO expert meeting (FAO/WHO, 2008). Chlorine, whether in the form of a gas or a solid (e.g. sodium hypochlorite), dissolves in water to form hypochlorous acid and hypochlorite ions. The efficacy is greatest in the acid form, but hypochlorite is also an effective, but slower-acting biocide. On the other hand, chlorine-containing compounds have the potential to interact with organic matter.

Chlorine treatment of inoculated poultry carcasses and parts reduced *Campylobacter* (*Campylobacter* spp., *C. jejuni*) and *Salmonella* (*Salmonella* spp., *S.* Typhimurium) by 0.2–3.0 and 0.1–2.4 logs, respectively (Table 3.3). In the following, mainly studies based on industrial data (commercial conditions) or pilot-scale data (experiments using industrial equipment in non-industrial settings) are discussed.

In industrial studies, chorine-containing carcass washers reduced *Campylobacter* by 0.3–0.7 log CFU/ml on naturally contaminated carcasses (Bashor *et al.*, 2004; Oyarzabal *et al.* 2004). But in another study, chlorinated carcass washes before evisceration did not affect post-chill counts (Berrang *et al.*, 2007). Pilot-scale studies reported reductions of *Campylobacter* by up to

© Woodhead Publishing Limited, 2012

**Table 3.3** Reductions of *Campylobacter* and *Salmonella* by chlorine, hypochlorite, sodium hypochlorite, and sodium chlorite on the surface of poultry carcasses and parts

| Microorganism | Reduction ($\log_{10}$ CFU) | Treated material | Application[e] | Concentration | Contamination | Temp (°C) | Exposure time (min) | References |
|---|---|---|---|---|---|---|---|---|
| *Campylobacter* spp. | 2.5–2.6/ml[a] | Carcass | SP | 55 ppm | Artificial | 21–54 | 0.1 | Northcutt *et al.* (2005) |
| *Campylobacter jejuni* | 2.6–3.0/g | Wing | IM | 50 ppm | Artificial | 4 or 23 | 10 or 30 | Park *et al.* (2002) |
| | 2.3/g[b] | Carcass | SP/IM | 73 ppm | Artificial | NA[f]/3 | 0.03/40 | Kim *et al.* (2005) |
| | 2.1/g[b] | Carcass | IM | 73 ppm | Artificial | 4 | 40 | Kim *et al.* (2005) |
| | 1.9–2.5/carcass[c] | Carcass | SP/IC | 50 ppm | Artificial | 20–60/NA | 0.2/50 | Li *et al.* (2002) |
| | 1.8–2.2/carcass[c] | Carcass | SP | 50 ppm | Artificial | 55–60 | 0.2 | Li *et al.* (2002) |
| | 1.6/ml[a] | Carcass | SP | 50 ppm | Artificial | NA | 0.1–0.3 | Northcutt *et al.* (2007) |
| | 1.3–1.7/$cm^2$ | Carcass | IM | 50 ppm | Artificial | 60–65 | 0.3 | Li *et al.* (2002) |
| | 1.2/carcass[c] | Carcass | SP | 50 ppm | Artificial | 20 | 0.2 | Li *et al.* (2002) |
| | 0.8–1.4/g | Carcass | IM | 0.2–0.4% | Artificial | 30 | 0.3–0.5 | Ellerbroek *et al.* (2007) |
| | 0.5–0.7/$cm^2$ | Carcass | IM | 50 ppm | Artificial | 20–55 | 0.3 | Li *et al.* (2002) |
| | 0.2–0.8 | Carcass | SP | 0.2–0.4% | Artificial | 30 | 0.3–0.5 | Ellerbroek *et al.* (2007) |
| *Salmonella* spp. | 2.4/ml[a] | Carcass | SP | 50 ppm | Artificial | NA | 0.1–0.3 | Northcutt *et al.* (2007) |
| | 1.2/ml[d] | Carcass | IC | 50 ppm | Artificial | 5 | 60 | Russel and Axtel (2005) |
| | 0.9–1.1/ml[a] | Carcass | SP | 55 ppm | Artificial | 21–54 | 0.1 | Northcutt *et al.* (2005) |
| | 0.1/wing | Wing | IM | >50ppm | Artificial | NA | 3 | Killinger *et al.* (2010) |
| *Salmonella* Typhimurium | 1.4–2.1/carcass[c] | Carcass | SP/IC | 50 ppm | Artificial | 20/4 | 0.3/45 | Yang *et al.* (1999) |
| | 0.8/ml | Carcass | SP | 2% | Artificial | NA | 0.3 | Fabrizio *et al.* (2002) |
| | 0.5–1.5/carcass[c] | Carcass | SP | 50 ppm | Artificial | 20 | 0.3 | Yang *et al.* (1999) |
| | 0.1–0.3/$cm^2$ | Leg | IM | 1% | Artificial | 23 | 10 | Li *et al.* (1994) |

[a] Treatment using a spray cabinet (pilot-scale).
[b] Simulated industrial processing (pilot-scale).
[c] Treatment using an inside–outside birdwasher for spraying (pilot-scale).
[d] Treatment using pilot-scale poultry chillers.
[e] IM, immersion; SP, spraying; IC, immersion chilling.
[f] NA, not available.

© Woodhead Publishing Limited, 2012

2.6 log CFU/ml after spray washing with chlorine (Table 3.3). But reductions obtained by Northcutt *et al.* (2005) were comparable to those obtained by water spraying alone. Dependent on the application temperature, Li *et al.* (2002) reported reductions after immersion and spraying ranging from 0.5 to 1.7 and from 1.2 to 2.2 log CFU/cm$^2$, respectively. In the laboratory study of Park *et al.* (2002), immersion in chlorine solution yielded high reductions of *C. jejuni* (2.6–3.0 log CFU/g). Compared with a water control, reductions were enhanced by 1.5–1.8 log CFU/g, but contact times were substantially longer than those employed at commercial premises.

With regard to *Salmonella*, the pilot-scale studies of Northcutt *et al.* (2005) and Yang *et al.* (1999) reported reductions of 0.5–1.5 log CFU/ml after chlorine treatment (Table 3.3), but results were mainly comparable to those obtained by water spraying alone. In another study by Northcutt *et al.* (2007), *Salmonella* were reduced by 2.4 log CFU/ml after spraying with hypochlorite. Stopforth *et al.* (2007) evaluated the effect of sequential interventions in the industrial slaughter process and most of the chlorine-containing carcass wash steps reduced the prevalence of *Salmonella*. Under laboratory conditions, the comparatively low reductions found by Killinger *et al.* (2010) and Li *et al.* (1994) were striking (Table 3.3).

In poultry carcass chillers, maintenance of residual free chlorine concentration is difficult due to interaction with organic matter. To maintain the activity of chlorine, continual dosing in the chiller water is required. On the other hand, chiller water can be a source of cross-contamination when chlorine is absent. In the industrial study by Oyarzabal *et al.* (2004), *Campylobacter* were reduced by 1.1–1.3 log CFU/ml. Moreover, Berrang *et al.* (2007) showed that the use of chlorine in the chill tank reduced *Campylobacter* on postchill carcasses by 2.3 log CFU/ml (compared to 1.8 log CFU/ml when no chlorine was added). But in other studies, immersion chilling was not effective in reducing *Campylobacter* or *Salmonella* (Bashor *et al.*, 2004; Stopforth *et al.*, 2007; Yang *et al.*, 2001). On pilot-scale, combining hypochlorite spraying with immersion chilling in hypochlorite solution reduced inoculated *S.* Typhimurium by 1.4–2.1 log CFU/carcass (Yang *et al.*, 1999). In another study, the use of a pilot-scale poultry chiller yielded reductions of *Salmonella* by 1.2 log CFU/ml, but the addition of chlorine had no additional effect (Russel and Axtell, 2005). Furthermore, Mead *et al.* (1995) examined the effect of process water chlorination at several stages in poultry slaughter, including the chiller water. Considering the whole process, numbers of *Campylobacter* were significantly reduced.

### 3.5.3 Chlorine dioxide

Chlorine dioxide ($ClO_2$) inactivates microorganisms by altering nutrient transport and disrupting protein synthesis after penetrating into cells, but mechanisms are not entirely understood. The effect of food composition on the efficacy of chlorine dioxide has recently been reviewed and the

© Woodhead Publishing Limited, 2012

antibacterial activity seems to be less counteracted by organic matter than that of chlorine (Vandekinderen *et al.*, 2009). Only few studies examined the effect of chlorine dioxide on pathogenic bacteria in poultry. In a recent survey, aqueous chlorine dioxide reduced *C. jejuni* inoculated on chicken breast and legs by 1.0–1.2 log CFU/g (Hong *et al.*, 2007). In commercial processing plants, reprocessing with chlorine dioxide yielded *Campylobacter* reductions that were about 0.6 log CFU/ml higher than after reprocessing without chemicals (Berrang *et al.*, 2007). Moreover, Stopforth *et al.* (2007) found only slight reductions in the prevalence of *Salmonella* after spraying of carcasses with chlorine dioxide.

### 3.5.4 Acidified sodium chlorite

The antimicrobial activity of acidified sodium chlorite (ASC, $NaClO_2$) is derived from chlorous acid and chlorine dioxide, which inactivate microorganisms by damage of cellular membranes and oxidation of cellular constituents. In the following, mainly studies based on industrial data are discussed. The absence of controls for water treatment often makes it difficult to draw conclusions regarding the effect of including ASC in wash water.

Industrial studies by Kere Kemp *et al.* (2001) and Oyarzabal *et al.* (2004) reported reductions of *Campylobacter* by 0.9–2.6 log CFU/ml after ASC treatment (Table 3.4). Reductions reported by Kere Kemp *et al.* (2001) were obtained on visibly contaminated carcasses during continuous online processing, whereas Oyarzabal *et al.* (2004) investigated the efficacy of an ASC dip applied after chilling. Moreover, Bashor *et al.* (2004) reported that addition of ASC to carcass washers increased *Campylobacter* reductions by 1.3 log above that seen with chlorine spraying. Reprocessing with the proprietary product Sanova® yielded reductions that were about 0.6 log CFU/ml higher than after reprocessing without chemicals (Berrang *et al.*, 2007). The prevalence of *Campylobacter* or *Salmonella* was also significantly reduced by ASC spraying or dipping of naturally contaminated carcasses (Sexton *et al.*, 2007; Stopforth *et al.*, 2007). Under laboratory conditions, ASC treatment reduced inoculated *C. jejuni* and *S.* Enteritidis by 1.5 and 2.1 orders of magnitude (log reductions), respectively (Table 3.4). Water controls thereby achieved 0.2–0.3 log CFU/g reductions (Arritt *et al.*, 2002; Del Río *et al.*, 2007). Compared with immersion in water, Özdemir *et al.* (2006) reported reductions of *C. jejuni* on inoculated breast skin samples by 1.6–1.9 log CFU/g after immersion in ASC. By the use of the proprietary product Sanova®, *C. jejuni* and *Salmonella* were reduced on inoculated legs by 1.6 and 1.1 log CFU/g, respectively (Mehyar *et al.*, 2005).

### 3.5.5 Monochloramine

Monochloramine ($NH_2Cl$) belongs to the N-chloramines and is generated by reaction between ammonia and chlorine. Monochloramine constitutes a

© Woodhead Publishing Limited, 2012

**Table 3.4** Reductions of *Campylobacter* and *Salmonella* by acidified sodium chlorite and cetylpyridinium chloride on the surface of poultry carcasses and parts

| Agent/ microorganism | Reduction ($log_{10}$ CFU) | Treated material | Application[d] | Concentration | Contamination | Temp (°C) | Exposure time (min) | References |
|---|---|---|---|---|---|---|---|---|
| **Acidified sodium chlorite** | | | | | | | | |
| *Campylobacter* spp. | 2.6/ml[a] | Carcass | SP | 1100 ppm | Natural | 14–18 | 0.3 | Kere Kemp *et al.* (2001) |
| | 0.9–1.2/ml[a] | Carcass | IM | 600–800 ppm | Natural | NA[e] | 0.3 | Oyarzabal *et al.* (2004) |
| *Campylobacter jejuni* | 1.5/breast | Breast | SP | 0.1% | Artificial | 21 | 0.1 | Arritt *et al.* (2002) |
| *Salmonella* Enteritidis | 2.1/g | Leg | IM | 1200 ppm | Artificial | 18 | 15 | Del Río *et al.* (2007) |
| **Cetylpyridinium chloride** | | | | | | | | |
| *Campylobacter jejuni* | >4.2/ml | Back | IM | 0.5% | Artificial | 20 | 1 | Riedel *et al.* (2009) |
| | 2.9/breast | Breast | SP | 0.5% | Artificial | 21 | 0.1 | Arritt *et al.* (2002) |
| | 1.4/breast | Breast | SP | 0.1% | Artificial | 21 | 0.1 | Arritt *et al.* (2002) |
| *Salmonella* Typhimurium | 4.9/leg | Leg | IM | 4 mg/ml | Artificial | 25 | 3 | Breen *et al.* (1997) |
| | 2.0/carcass[b] | Carcass | SP | 0.5% | Artificial | 35 | 0.3 | Yang *et al.* (1998) |
| | 1.5–2.5/38.5 $cm^2$ | Breast | SP | 0.1% | Artificial | 10–60 | 0.5 | Wang *et al.* (1997) |
| | 1.5–1.9/ml | Breast | SP | 0.1–0.5% | Artificial | 20 | 0.5 | Xiong *et al.* (1998) |
| | 1.4/leg | Leg | IM | 4 mg/ml | Artificial | 25 | 1 | Breen *et al.* (1997) |
| | 1.2–1.6/carcass[c] | Carcass | SP | 0.1% | Artificial | 22 | 1.5 | Li *et al.* (1997) |
| | 1.0–1.6/$cm^2$ | Breast | IM | 0.1% | Artificial | NA | 1–3 | Kim and Slavik (1996) |
| | 0.9–1.7/$cm^2$ | Breast | SP | 0.1% | Artificial | 15 or 50 | 1 | Kim and Slavik (1996) |
| | 0.6–0.8/carcass[c] | Carcass | SP | 0.1% | Artificial | 22 | 0.5 | Li *et al.* (1997) |
| | 0.5–1.1/leg | Leg | IM | 2 mg/ml | Artificial | 25 | 1–3 | Breen *et al.* (1997) |
| | 0.5–0.6/leg | Leg | IM | 1 mg/ml | Artificial | 25 | 1–3 | Breen *et al.* (1997) |

[a] Treatment in poultry slaughter plant under commercial conditions.
[b] Treatment using an inside–outside birdwasher for spraying (pilot-scale).
[c] Treatment using a spraying test chamber (pilot-scale).
[d] IM, immersion; SP, spraying.
[e] NA, not available.

© Woodhead Publishing Limited, 2012

chlorine species that is tasteless, odorless, stable, highly soluble, persistent in water, biocidal, and unlike free chlorine does not react readily with organic material. Using a pilot-scale poultry chiller, Russel and Axtell (2005) reported reductions of inoculated *Salmonella* by 2.0 log CFU/ml. Monochloramine thereby exerted stronger antibacterial activity than sodium hypochlorite.

### 3.5.6 Cetylpyridinium chloride

Cetylpyridinium chloride (CPC) belongs to the group of quaternary ammonium compounds. CPC is a cationic surfactant, which has a neutral pH and the chloride portion is not functional. Its antibacterial activity results from interaction with acidic groups at the surface or within bacteria to form weakly ionized compounds that inhibit bacterial metabolism.

On inoculated poultry carcasses and parts, CPC treatment reduced *C. jejuni* and *S.* Typhimurium by 0.5–4.9 logs (Table 3.4). Li *et al.* (1997) and Yang *et al.* (1998) used pilot-scale spraying equipment and reductions of *S.* Typhimurium ranged from 0.6 to 2.0 log CFU/carcass. Different exposure times and application pressures explain the range of results reported by Li *et al.* (1997). The other results presented in Table 3.4 originate from laboratory surveys. In the study by Riedel *et al.* (2009), CPC along with benzalkonium chloride was the most effective of 11 chemicals tested. Different studies showed that the bactericidal efficacy was dependent on concentration, exposure time, or temperature (Arritt *et al.*, 2002; Kim and Slavik, 1996; Xiong *et al.*, 1998). In the study by Wang *et al.* (1997), application pressure (414–1034 kPa) played only a minor role at 10°C, but significantly influenced the efficacy at 60°C. Furthermore, by the use of the proprietary product Cecure®, *C. jejuni* and *Salmonella* inoculated on poultry legs were reduced by 1.4 log CFU/g (Mehyar *et al.*, 2005).

### 3.5.7 Phosphate-based compounds

The use of trisodium phosphate (TSP, $Na_3PO_4$) for the decontamination of poultry is well documented (Capita *et al.*, 2002). Important factors are the high pH and the ionic strength causing bacterial cell autolysis, but mechanisms are not entirely understood. The majority of the evaluated studies were performed under laboratory conditions. Li *et al.* (1997) and Yang *et al.* (1998) used pilot-scale spraying equipment, but inoculated carcasses were investigated (Table 3.5). Great ranges of results were found in the study by Li *et al.* (1997). Increasing the TSP concentration from 5 to 10% distinctly enhanced the reductions. Different exposure times and application pressures also influenced the results. In commercial processing plants, reprocessing with TSP yielded *Campylobacter* reductions on naturally contaminated carcasses that were about 1.2 log CFU/ml higher than after reprocessing without chemicals (Berrang *et al.*, 2007). Bashor *et al.* (2004) reported that addition of ASC to carcass washers increased *Campylobacter* reductions by

© Woodhead Publishing Limited, 2012

**Table 3.5** Reductions of *Campylobacter* and *Salmonella* by trisodium phosphate on the surface of poultry carcasses and parts

| Microorganism | Reduction ($log_{10}$ CFU) | Treated material | Application[c] | Concentration | Contamination | Temp (°C) | Exposure time (min) | References |
|---|---|---|---|---|---|---|---|---|
| *Campylobacter* spp. | 1.7/g | Carcass | IM | 10% | Natural | 20 | 0.3 | Whyte *et al.* (2001) |
| | 1.5/carcass | Carcass | IM | 10% | Natural | 50 | 0.3 | Slavik *et al.* (1994) |
| | 0.2/carcass | Carcass | IM | 10% | Natural | 10 | 0.3 | Slavik *et al.* (1994) |
| *Campylobacter jejuni* | 1.9/g | Carcass | IM | 10% | Artificial | NA[d] | 1 | Mehyar *et al.* (2005) |
| | 1.7/ml | Back | IM | 10% | Artificial | 20 | 1 | Riedel *et al.* (2009) |
| | 1.6/breast | Breast | SP | 10% | Artificial | 21 | 0.1 | Arritt *et al.* (2002) |
| *Salmonella* spp. | 1.7/ $cm^2$ | Breast | IM | 1% | Artificial | 25 | 25 | Hwang and Beuchat (1995) |
| | 1.6/g | Carcass | IM | 10% | Artificial | NA | 1 | Mehyar *et al.* (2005) |
| | 1.4/g | Carcass | IM | 10% | Natural | 20 | 15 | Whyte *et al.* (2001) |
| *Salmonella* Enteritidis | 1.9/g | Leg | IM | 12% | Artificial | 18 | 15 | del Río *et al.* (2007) |
| *Salmonella* Typhimurium | 3.7–3.8/carcass[b] | Carcass | SP | 10% | Artificial | 22 | 1.5 | Li *et al.* (1997) |
| | 2.3/ml | Leg | IM | 210 mM | Artificial | 37 | 10 | Mullerat *et al.* (1994) |
| | >2.2/$cm^2$ | Leg | IM | 10% | Artificial | 10 | 0.3 | Kim and Slavik (1994) |
| | 2.1–2.2/ml | Breast | SP | 5–10% | Artificial | 20 | 0.5 | Xiong *et al.* (1998) |
| | 1.8/carcass[a] | Carcass | SP | 10% | Artificial | 35 | 0.3 | Yang *et al.* (1998) |
| | 1.6–1.8/carcass[b] | Carcass | SP | 10% | Artificial | 22 | 0.5 | Li *et al.* (1997) |
| | 1.5–2.3/38.5 $cm^2$ | Breast | SP | 10% | Artificial | 10–60 | 0.5 | Wang *et al.* (1997) |
| | 1.5–1.6/carcass[b] | Carcass | | 5% | Artificial | 22 | 1.5 | Li *et al.* (1997) |
| | 1.4/ml | Carcass | IC | 10 ppm | Artificial | 4 | 45 | Fabrizio *et al.* (2002) |
| | 0.9/ml | Carcass | SP | 10 ppm | Artificial | NA | 0.3 | Fabrizio *et al.* (2002) |
| | 0.8/carcass[b] | Carcass | SP | 5% | Artificial | 22 | 0.5 | Li *et al.* (1997) |
| | 0.6–0.9/$cm^2$ | Leg | IM | 1% | Artificial | 23 | 10 | Li *et al.* (1994) |

[a] Treatment using a modified inside–outside birdwasher for spraying (pilot-scale).
[b] Treatment using a spraying test chamber (pilot-scale).
[c] IM, immersion; SP, spraying; IC, immersion chilling.
[d] NA, not available.

© Woodhead Publishing Limited, 2012

1 log (log CFU/ml) above that seen with chlorine spraying. Other surveys evaluating the efficacy of TSP for reducing *Campylobacter* and *Salmonella* under industrial conditions were performed by Coppen *et al.* (1998) and Salvat *et al.* (1997).

Laboratory studies mainly investigated immersion treatments and yielded reductions of *Campylobacter* and *Salmonella* by 0.2–1.9 and 0.6–3.8 orders of magnitude (log reductions), respectively (Table 3.5). Arritt *et al.* (2002) and Del Río *et al.* (2007) found reductions by >1.5 log CFU/g, whereas water controls achieved 0.2–0.3 log CFU/g reductions. Compared with immersion in water, Özdemir *et al.* (2006) reported reductions of *C. jejuni* inoculated on breast skin samples by 1.7–2.4 log CFU/g after immersion in TSP. Increasing application temperatures and/or pressures influenced the reductions obtained (Slavik *et al.*, 1994; Wang *et al.*, 1997). Furthermore, immersion (25°C, 30 min) in sodium acid pyrophosphate, monosodium phosphate, sodium hexametaphosphate, and sodium tripolyphosphate reduced *Salmonella* inoculated on breast skin by 0.8–1.1 log CFU/cm$^2$ (Hwang and Beuchat, 1995).

An important factor to consider when evaluating the bactericidal efficacy of TSP and other phosphate applications is the high pH values resulting in carcass rinses, which can interfere with the recovery of bacteria (Bourassa *et al.*, 2004, 2005; Lillard, 1994a; Rathgeber and Waldroup, 1995). Lillard (1994a) initially reported reductions of *S.* Typhimurium by about 2 logs, but when residual TSP was washed off and pH was neutralized, *S.* Typhimurium was recovered even at low inoculation levels. Similarly, *Salmonella* recovery was reduced in the study of Bourassa *et al.* (2004), but no difference in prevalence was found between control and TSP-treated carcasses after pH adjustment (Bourassa *et al.*, 2005).

### 3.5.8 Electrolyzed oxidizing water and ozonated water

Electrolyzed water (EW) is gaining popularity in the food industry for use on food and processing surfaces (Hricova *et al.*, 2008). By electrolysis a dilute sodium chloride solution dissociates into electrolyzed oxidizing water (EO water) and electrolyzed reducing water (ER water). EO water thereby exerts strong antimicrobial properties against a variety of microorganisms. EO water has proved to be environmentally friendly, showed nonselective antimicrobial properties, and did not negatively affect the organoleptic properties of various foods. On the other hand, the adverse impact of organic matter on its efficacy must be considered. With regard to *Campylobacter*, Gellynck *et al.* (2008) reported that treatment of poultry carcasses with EO water showed the best cost–benefit ratio.

Treatment of inoculated poultry carcasses and parts with EO water reduced *C. jejuni* by 2.0–3.2 logs (Kim *et al.*, 2005; Northcutt *et al.*, 2007; Park *et al.*, 2002) and *Salmonella* (*Salmonella* spp., *S.* Typhimurium) by 0.6–2.7 logs (Fabrizio *et al.*, 2002; Northcutt *et al.*, 2007; Yang *et al.*, 1999). Three

© Woodhead Publishing Limited, 2012

of these studies were performed on pilot-scale. Kim *et al.* (2005) evaluated the effect of immersion in EO water and combinations of spraying and immersion under simulated industrial processing. They reported reductions for *C. jejuni* in the range from 2.0 to 2.3 log CFU/g. The use of a spray cabinet in the study by Northcutt *et al.* (2007) also yielded considerable reductions for *C. jejuni* (1.9 log CFU/ml) and *Salmonella* (2.7 log CFU/ml). Using an inside–outside birdwasher, Yang *et al.* (1999) reported reductions of *S.* Typhimurium by 0.9–1.8 log CFU/carcass and by 1.4–1.9 log CFU/carcass after combining spraying with immersion chilling in EW. Under laboratory conditions, immersion chilling with acidic EW reduced *C. jejuni* inoculated on chicken wings by 2.8–3.2 log CFU/g (Park *et al.*, 2002). Comparatively low reductions of *S.* Typhimurium were reported by Fabrizio *et al.* (2002) after immersion chilling or spraying (0.6–0.8 log CFU/ml). However, by combination of ER water spraying followed by EO water immersion, significant reductions (2.1 log CFU/ml) were obtained on poultry carcasses (Fabrizio *et al.*, 2002).

Furthermore, since the recent commercial development of portable ozone generators, the use of ozone has become more practical for the food industry. Ozone is characterized by a strong oxidizing effect. Because of the instability of ozone in gaseous and aqueous states, it cannot be generated and stored for later application and it must be generated on site as needed. Owing to its high oxidation potential, ozone reacts with a large number of compounds including organic substrates. The application of ozone in the food industry for enhancing the microbiological safety and quality of foods has been evaluated in the comprehensive reviews of Kim and co-workers (Kim *et al.*, 1999, 2003). For poultry, ozone has been tested for disinfecting hatchery, hatching eggs, poultry chiller water, poultry carcasses, and contaminated eggs (Kim *et al.*, 1999; Vadhanasin *et al.*, 2004; Waldroup *et al.*, 1993). However, in an experimental study (Fabrizio *et al.*, 2002), *S.* Typhimurium inoculated on poultry carcasses was reduced after spraying (0.3 min) and immersion chilling (45 min) with ozonated water by only 0.6–0.7 log CFU/ml.

### 3.5.9 Other chemical treatments

Occasionally, the antibacterial activity of other chemicals such as peroxides, peroxyacids, sulfate-based compounds, or sodium hydroxide (NaOH) have been evaluated, but mainly under laboratory conditions. Selected examples are briefly mentioned below. Immersion treatment with 0.05% NaOH reduced *Salmonella* by about 0.8 log CFU/cm$^2$ on chicken breast skin (Hwang and Beuchat, 1995), whereas *C. jejuni* was reduced in another study (0.1 N NaOH) by 3.5–3.7 log CFU/g on chicken wings (Zhao and Doyle, 2006). Moreover, spraying of carcasses with 5% sodium bisulfate (35°C, 0.5 min) reduced *S.* Typhimurium by 1.7 log CFU/carcass (Yang *et al.*, 1998) and immersion of leg skin in 1% sodium carbonate (23°C, 10 min) by 0.8–1.0 log CFU/cm$^2$ (Li *et al.*, 1994). Moreover, Zhao and Doyle (2006) showed

© Woodhead Publishing Limited, 2012

that the Safe$_2$O-Poultry Wash® (based on acidic calcium sulfate) was highly effective in reducing *C. jejuni*. In another survey, the Safe$_2$O-Poultry Wash® reduced *C. jejuni* and *Salmonella* on chicken legs by 1.2–1.7 log CFU/g (Mehyar *et al.*, 2005).

Furthermore, as there is increasing interest in 'natural foods' by consumers, plants gain importance as a source of antimicrobials. Valtierra-Rodríguez *et al.* (2010) showed that some extracts from fruits and their mixtures reduced the viability of *Campylobacter* on chicken skin by >4.0 log. Moreover, grapefruit seed extract yielded reductions of *S.* Typhimurium (0.5 min spray) and *C. jejuni* (1 min dip) by 1.6–1.8 and 3.1 log CFU/ml, respectively (Riedel *et al.*, 2009; Xiong *et al.*, 1998).

### 3.5.10 Summary of chemical treatments

Chemical compounds comprise a variety of substances, but organic acids, chlorine-based treatments, or TSP were most frequently used. However, varying framing conditions hampered comparisons and many studies investigated inoculated samples under laboratory conditions. For appraisal of the applicability in poultry processing, it must also be considered that the activity of some chemicals is counteracted by organic matter, concentrated substances might constitute a health hazard or ecological menace, some agents show corrosive properties, or their stability is limited in solution.

Based on the evaluated studies, compounds such as organic acids, chlorine, ASC, CPC, or TSP yielded promising result for reducing *Campylobacter* and *Salmonella* on poultry carcasses, in particular when compared with untreated controls. ASC thereby yielded considerable reductions also on naturally contaminated carcasses. In addition, some chemicals (e.g. organic acids) show some residual effect, but discoloration of carcasses might occur when high acid concentrations or low pH are used. When using TSP, pH neutralization is required for accurate appraisal of the bactericidal activity. Some other chemicals such as monochloramine require further investigations to appraise their eligibility under practical conditions. Another promising alternative is the use of EO water. However, as is also the case with some other compounds, the adverse impact of organic matter must be considered.

## 3.6 Combinations of chemical and physical or of chemical decontamination treatments for poultry carcasses

Different combinations of chemical and physical or of chemical decontamination treatments were tested for reducing *Campylobacter* and *Salmonella* on poultry carcasses and parts. With regard to practical application, the location in the slaughter process and the effect of multiple sequential interventions must also be considered. Selected combinations are shown in Table 3.6. Though

© Woodhead Publishing Limited, 2012

**Table 3.6** Reductions of *Campylobacter* and *Salmonella* by selected combinations of chemical and physical treatments or of chemical treatments on the surface of poultry carcasses and parts

| Microorganism | Reduction ($\log_{10}$ CFU) | Combination | Application[c] | References |
|---|---|---|---|---|
| **Physical–chemical combinations** | | | | |
| *Salmonella* Typhimurium | 2.4–3.9/ml | Ultrasound + chlorine | IM | Lillard (1993) |
| | 1.9/cm$^2$ | Hot water + trisodium phosphate | IM | Rodriguez de Ledesma *et al.* (1996) |
| | 1.8/cm$^2$ | Hot water + sodium carbonate | IM | Rodriguez de Ledesma *et al.* (1996) |
| | 1.6–1.9/cm$^2$ | Electricity + trisodium phosphate | IM | Li *et al.* (1994) |
| | 1.0/cm$^2$ | Electricity + sodium carbonate | IM | Li *et al.* (1994) |
| | 0.9–1.0/cm$^2$ | Electricity + sodium chlorite | IM | Li *et al.* (1994) |
| **Chemical–chemical combinations** | | | | |
| *Campylobacter* spp. | 2.7/ml[a] | Tripotassium phosphate + lauric acid | R | Hinton and Ingram (2005) |
| | 1.4/ml[a] | Tripotassium phosphate + myristic acid | R | Hinton and Ingram (2005) |
| *Campylobacter jejuni* | 2.4–2.5/g | Acidified sodium chlorite + trisodium phosphate | IM | Özdemir *et al.* (2006) |
| | 1.1–1.4/g | Trisodium phosphate + acidified sodium chlorite | IM | Özdemir *et al.* (2006) |
| *Salmonella* Enteritidis | 2.6–4.0/g[b] | Levulinic acid + sodium dodecyl sulfate (SDS) | IM | Zhao *et al.* (2009) |
| *Salmonella* Typhimurium | 3.0/ml | Salmide® + trisodium phosphate | IM | Mullerat *et al.* (1994) |
| | 2.0/ml | Acetic acid + chlorine | SP/IM | Fabrizio *et al.* (2002) |
| | 2.0/ml | Trisodium phosphate + chlorine | SP/IM | Fabrizio *et al.* (2002) |
| | 1.7–2.7/ml | Salmide® + EDTA | IM | Mullerat *et al.* (1994) |
| | 1.2–1.7/ml | Salmide® + sodium lauryl sulfate | IM | Mullerat *et al.* (1994) |

[a] Highest reduction obtained.
[b] Reductions up to 7.0 log CFU/g after increasing concentrations of levulinic acid and SDS to 3 and 2%, respectively.
[c] IM, immersion; SP, spraying; R, rinsing.

© Woodhead Publishing Limited, 2012

strongly influenced by framing conditions, some combinations enhanced the reductions compared to the single treatments. For example, ultrasound in combination with chlorine yielded reductions of *S.* Typhimurium inoculated on chicken skin by 2.4–3.9 logs and was thus more effective than ultrasound or chlorine alone (Lillard, 1993, 1994b). With regard to chemical combinations, Özdemir *et al.* (2006) reported that TSP followed by ASC treatment was more effective than TSP followed by ASC in reducing *C. jejuni* (Table 3.6). The TSP–ASC combination yielded even lower reductions than the TSP or ASC treatment alone. In addition, Fabrizio *et al.* (2002) reported that spraying with acetic acid or TSP followed by immersion in chlorinated water yielded reductions of *S.* Typhimurium that were about one order of magnitude (log CFU/ml) higher than those obtained by spraying alone. Furthermore, the antibacterial activity of the sodium chlorite-based oxy-halogen disinfectant Salmide® was enhanced by combination with sodium lauryl sulfate, EDTA, or TSP (Mullerat *et al.*, 1994).

## 3.7 Biological decontamination treatments for poultry carcasses

Biological interventions such as bacteriophages show some promise as decontamination treatments. Bacteriophages are increasingly used in the food industry and they are considered as highly host specific (Greer, 2005; Hudson *et al.*, 2005). Yet their practical use on foods is still impaired by factors such as the guarantee of a sufficient threshold level or the challenge of potential resistance development. For poultry carcasses, studies on the efficacy of bacteriophages and bacteriocins are so far very limited. Higgins *et al.* (2005) investigated poultry and turkey carcasses originating from *Salmonella*-positive commercial flocks. Following phage treatment, a significant reduction in the proportion of *Salmonella*-positive carcasses was reported. Two other experimental surveys investigated the efficacy of bacteriophages on inoculated chicken skin. Dependent on phage titer (3.0–7.0 log PFU) and amount of inoculum, Atterbury *et al.* (2003) reported reductions of *C. jejuni* by up to 1.2 log CFU/2 $cm^2$. In the other study, *C. jejuni* and *S.* Enteritidis were reduced by up to 2.0 log CFU/$cm^2$ and Goode *et al.* (2003) also reported that phages may protect carcasses against cross-contamination. Overall, further investigations are required for appraisal of the practicability of phage treatment under commercial conditions.

## 3.8 Conclusions and future trends

A comprehensive strategy for foodborne pathogen control has to be based on an integrated approach involving measures at pre-harvest, harvest, processing,

© Woodhead Publishing Limited, 2012

storage, distribution, preparation, and consumption (Sofos, 2008). For on-farm control, the applicability of novel approaches such as vaccination, bacteriocins, or bacteriophages in addition to bio-security measures must be evaluated. On the other hand, the retail, food service, and consumer level must not be neglected, and adequate education thereby plays an important role. But hygienic practices and habits of consumers are hard to change, and associated costs to reach consumers might be large (Gellynck *et al.*, 2008).

During poultry production, despite all efforts targeted on the maintenance of good hygiene practices, complete prevention of microbial carcass contamination can hardly be guaranteed. This is of special concern as healthy food-producing animals can harbor important human pathogens such as *Campylobacter* or *Salmonella*. Antimicrobial intervention technologies are therefore gaining interest in order to reduce bacterial contamination through implementation of decontamination treatments. Based on the evaluated studies, which investigated the antibacterial activity of decontamination treatments for poultry carcasses, some interventions effectively reduced *Campylobacter* and *Salmonella*. Bacterial reductions obtained by different physical, chemical, or biological methods applied alone or in combination and their main advantages and disadvantages are elucidated in Sections 3.4–3.7 of this chapter. However, decontamination treatments for carcasses must always be seen as part of an integral food safety system. Such treatments cannot compensate for poor hygiene practices or replace good manufacturing and slaughter hygiene practices along with risk-based preventive measures. In fact, adherence to strict hygiene practices provides the foundation upon which intervention technologies are most effective.

For appraisal of the applicability of antibacterial interventions in poultry slaughter and processing, many different factors are of importance. Apart from microbial safety, the application of decontamination treatments for carcasses must also consider other issues such as product quality (e.g. discoloration or defects in organoleptic properties), consumer acceptance, associated costs, environmental and health occupational matters, and the practical application during slaughter or processing. Selection and adaptation of decontamination steps have thereby to be customized to plant- and process-specific circumstances. Moreover, accurate evaluation of the overall effects of decontamination treatments is difficult for many applications because there is a lack of data on industrial-scale processes and extrapolation of experimental results to commercial practices is restricted. Surveys examining inoculated carcasses tend to overestimate the efficacy of a disinfectant, partly due to inefficient attachment of bacteria to the surface. Thus, more data evaluating disinfection effects of serial or sequential control strategies under commercial conditions (multiple hurdle approach) are required.

Consequently there is a need for more commercial data on the antibacterial activity of decontamination methods for poultry carcasses (e.g. bacteriophages, thermosonication). For better comparability of results, more standardized protocols for surveys evaluating microbial reductions should be developed. On

© Woodhead Publishing Limited, 2012

the other hand, a variety of new or novel technologies have been evaluated for treatment of various food items (Aymerich *et al.*, 2008; Rajkovic *et al.*, 2010; Sofos, 2008). Such technologies include irradiation, high hydrostatic pressure, electroporation with pulsed electric fields, ultrasonic waves, oscillating magnetic fields, cell lysis with bacteriophages or enzymes, and various combinations of such treatments or processes such as manothermosonication involving ultrasonic radiation, pressure and heat, or irradiation and heat (Sofos, 2008). Although some of these new technologies will not be applicable to poultry carcasses and some have become controversial, the development and practical application of new processes and technologies have to be carefully monitored.

## 3.9 Sources of further information and advice

Selected sources are mentioned in the following, but this list is not exhaustive. Basic information on foodborne pathogens can be found, for example, on the homepage of the CDC (http://www.cdc.gov) or the EFSA (http://www.efsa.europe.eu). The Foodborne Diseases Active Surveillance Network (FoodNet) and the updated US estimates of foodborne diseases (Scallan *et al.*, 2011a,b) deserve special mention. For the situation in Europe, the annual zoonoses report constitutes a good source of information (EFSA/ECDC, 2011) and a baseline survey on the prevalence of *Campylobacter* and *Salmonella* on poultry carcasses was recently published (EFSA, 2010a). Data on the (cross-)contamination of poultry carcasses are available for *Campylobacter*, for example in the studies from Allen *et al.* (2007), Ellerbroek *et al.* (2010), Normand *et al.* (2008), Peyrat *et al.* (2008), or Rasschaert *et al.* (2006), and for *Salmonella*, for example in the studies from Corry *et al.* (2002), Heyndrickx *et al.* (2002), Nde *et al.* (2007), Olsen *et al.* (2003), or Rasschaert *et al.* (2008).

A general overview on the topic of *Campylobacter* provides the review 'Campylobacters as zoonotic pathogens: a food production perspective' (Humphrey *et al.*, 2007). With regard to poultry flock colonization, the reviews of Newell and Fearnley (2003) and Sahin *et al.* (2002) provide overviews, while selected recent results are available in the studies from Bull *et al.*, (2006), Hastings *et al.* (2010), McDowell *et al.* (2008), Ridley *et al.* (2011), or Zweifel *et al.* (2008). Moreover, a recent opinion of the EFSA BIOHAZ panel addressed the risk posed by broiler meat to human campylobacteriosis (EFSA, 2010b). With special regard to risk management and assessment, the homepage of the CARMA project deserves to be mentioned (http://www.rivm.nl/carma/index_eng.html).

Basic information on pre-harvest and harvest control measures can be found in a recent opinion of the EFSA BIOHAZ panel on control options for *Campylobacter* in broiler meat production (EFSA, 2011c) or the 'proceedings of an international meeting on *Campylobacter* reduction in chicken' (FSA,

© Woodhead Publishing Limited, 2012

2010). Further information is also available in the reviews by Doyle and Erickson (2006), Havelaar *et al.* (2007), Hermans *et al.* (2011), Lin (2009), or Wagenaar *et al.* (2006). A detailed overview on benefits and risks of chlorine-containing disinfectants provides the report of a FAO/WHO expert meeting (FAO/WHO, 2008). This report also includes information on the chemistry, formation of by-products, unintended consequences, or risk–benefit assessments.

## 3.10 References

ALLEN V M, BULL S A, CORRY J E, DOMINGUE G, JØRGENSEN F, FROST J A, WHYTE R, GONZALEZ A, ELVISS N and HUMPHREY T J (2007), '*Campylobacter* spp. contamination of chicken carcasses during processing in relation to flock colonisation', *Int J Food Microbiol*, 113, 54–61.

ANANG D M, RUSUL G, BAKAR J and LING F H (2007), 'Effects of lactic acid and lauricidin on the survival of *Listeria monocytogenes*, *Salmonella* Enteritidis and *Escherichia coli* O157:H7 in chicken breast stored at 4°C', *Food Control*, 18, 961–969.

ARRITT F M, EIFERT J D, PIERSON M D and SUMNER S S (2002), 'Efficacy of antimicrobials against *Campylobacter jejuni* on chicken breast skin', *J Appl Poultry Res*, 11, 358–366.

ATTERBURY R J, CONNERTON P L, DODD C E, REES C E and CONNERTON I F (2003), 'Application of host-specific bacteriophages to the surface of chicken skin leads to a reduction in recovery of *Campylobacter jejuni*', *Appl Environ Microbiol*, 69, 6302–6306.

AYMERICH T, PICOUET P A and MONFORT J M (2008), 'Decontamination technologies for meat products', *Meat Sci*, 78, 114–129.

BASHOR M P, CURTIS P A, KEENER K M, SHELDON B W, KATHARIOU S and OSBORNE J A (2004), 'Effects of carcass washers on *Campylobacter* contamination in large broiler processing plants', *Poult Sci*, 83, 1232–1239.

BERRANG M E, DICKENS J A and MUSGROVE M T (2000a), '*Campylobacter* recovery from external and internal organs of commercial broiler carcasses prior to scalding', *Poult Sci*, 79, 286–290.

BERRANG M E, BUHR R J and CASON J A (2000b), 'Effects of hot water application after defeathering on levels of *Campylobacter*, coliform bacteria and *Escherichia coli* on broiler carcasses', *Poult Sci*, 79, 1689–1693.

BERRANG M E, BAILEY J S, ALTEKRUSE S F, PATEL B, SHAW W K JR, MEINERSMANN R J and FEDORKA-CRAY P J (2007), 'Prevalence and numbers of *Campylobacter* on broiler carcasses collected at rehang and postchill in 20 US processing plants', *J Food Prot*, 70, 1556–1560.

BHADURI S and COTTRELL B (2004), 'Survival of cold-stressed *Campylobacter jejuni* on ground chicken and chicken skin during frozen storage', *Appl Environ Microbiol*, 70, 7103–7109.

BINTSIS T, LITOPOULOU-TZANETAKI E and ROBINSON R K (2000), 'Existing and potential applications of ultraviolet light in the food industry – a critical review', *J Sci Food Agric*, 80, 637–645.

BIRK T, ROSENQUIST H, BRØNDSTED L, INGMER H, BYSTED A and CHRISTENSEN B B (2006), 'A comparative study of two food model systems to test the survival of *Campylobacter jejuni* at –18°C', *J Food Prot*, 69, 2635–2639.

BOURASSA D V, FLETCHER D L, BUHR R J, BERRANG M E and CASON J A (2004), 'Recovery of salmonellae from trisodium phosphate-treated commercially processed broiler carcasses after chilling and after seven-day storage', *Poult Sci*, 83, 2079–2082.

© Woodhead Publishing Limited, 2012

BOURASSA D V, FLETCHER D L, BUHR R J, CASON J A and BERRANG M E (2005), 'Recovery of salmonellae following pH adjusted pre-enrichment of broiler carcasses treated with trisodium phosphate', *Poult Sci*, 84, 475–478.

BOYSEN L and ROSENQUIST H (2009), 'Reduction of thermotolerant *Campylobacter* species on broiler carcasses following physical decontamination at slaughter', *J Food Prot*, 72, 497–502.

BREEN P J, SALARI H and COMPADRE C M (1997), 'Elimination of *Salmonella* contamination from poultry tissues by cetylpyridinium chloride solutions', *J Food Prot*, 60, 1019–1021.

BULL S A, ALLEN V M, DOMINGUE G, JØRGENSEN F, FROST J A, URE R, WHYTE R, TINKER D, CORRY J E, GILLARD-KING J and HUMPHREY T J (2006), 'Sources of *Campylobacter* spp. colonizing housed broiler flocks during rearing', *Appl Environ Microbiol*, 72, 645–652.

CAPITA R, ALONSO-CALLEJA C, GARCÍA-FERNÁNDEZ M C and MORENO B (2002), 'Trisodium phosphate (TSP) treatment for decontamination of poultry', *Food Sci Technol Int*, 8, 11–24.

CARPENTER C E, SMITH J V and BROADBENT J R (2011), 'Efficacy of washing meat surfaces with 2% levulinic, acetic, or lactic acid for pathogen decontamination and residual growth inhibition', *Meat Sci*, 88, 256–260.

CHUN H H, KIM J Y, LEE B D, YU D J and SONG K B (2010), 'Effect of UV-C irradiation on the inactivation of inoculated pathogens and quality of chicken breasts during storage', *Food Control*, 21, 276–280.

COM, Commission of the European Communities (2008), *Proposal for a council regulation implementing Regulation (EC) No 853/2004 of the European Parliament and of the Council as regards the use of antimicrobial substances to remove surface contamination from poultry carcasses*: 430 final, 29 October 2008, Brussels.

COPPEN P, FENNER S and SALVAT G (1998), 'Antimicrobial efficacy of AvGard® carcase wash under industrial processing conditions', *Br Poult Sci*, 39, 229–234.

CORRY J E, ALLEN V M, HUDSON W R, BRESLIN M F and DAVIES R H (2002), 'Sources of *Salmonella* on broiler carcasses during transportation and processing: modes of contamination and methods of control', *J Appl Microbiol*, 92, 424–432.

CORRY J E, JAMES S J, PURNELL G, BARBEDO-PINTO C S, CHOCHOIS Y, HOWELL M and JAMES C (2007), 'Surface pasteurization of chicken carcasses using hot water', *J Food Eng*, 79, 913–919.

COUNCIL DECISION 121/2009, 'Decision of 18 December 2008 rejecting the proposal from the Commission for a council regulation implementing Regulation (EC) No 853/2004 of the European Parliament and of the Council as regards the use of antimicrobial substances to remove surface contamination from poultry carcasses', *Off J Eur Union*, L42, 13–15.

DEL RÍO E, MURIENTE R, PRIETO M, ALONSO-CALLEJA C and CAPITA R (2007), 'Effectiveness of trisodium phosphate, acidified sodium chlorite, citric acid and peroxyacids against pathogenic bacteria on poultry during refrigerated storage', *J Food Prot*, 70, 2063–2071.

DICKENS J A and WHITTEMORE A D (1994), 'The effect of acetic acid and air injection on appearance, moisture pick-up, microbiological quality, and *Salmonella* incidence on processed poultry carcasses', *Poult Sci*, 73, 582–586.

DICKENS J A and WHITTEMORE A D (1995), 'The effects of extended chilling times with acetic acid on the temperature and microbiological quality of processed poultry carcasses', *Poult Sci*, 74, 1044–1048.

DOYLE M P and ERICKSON M C (2006), 'Reducing the carriage of foodborne pathogens in livestock and poultry', *Poult Sci*, 85, 960–973.

EFSA, EUROPEAN FOOD SAFETY AUTHORITY (2010a), 'Analysis of the baseline survey on the prevalence of *Campylobacter* in broiler batches and of *Campylobacter* and *Salmonella* on broiler carcasses in the EU, 2008, Part A: *Campylobacter* and *Salmonella* prevalence estimates', *EFSA Journal*, 8(3), 1503.

© Woodhead Publishing Limited, 2012

EFSA, EUROPEAN FOOD SAFETY AUTHORITY, PANEL ON BIOLOGICAL HAZARDS (BIOHAZ) (2010b), 'Scientific opinion on quantification of the risk posed by broiler meat to human campylobacteriosis in the EU', *EFSA Journal*, 8(1), 1437.

EFSA, EUROPEAN FOOD SAFETY AUTHORITY, PANEL ON BIOLOGICAL HAZARDS (BIOHAZ) (2011a), 'Scientific opinion on the efficacy and microbiological safety of irradiation of food', *EFSA Journal*, 9(4), 2103.

EFSA, EUROPEAN FOOD SAFETY AUTHORITY, PANEL ON FOOD CONTACT MATERIALS, ENZYMES, FLAVOURINGS and PROCESSING AIDS (CEF) (2011b), 'Scientific opinion on the chemical safety of irradiation of food', *EFSA Journal*, 9(4), 1930.

EFSA, EUROPEAN FOOD SAFETY AUTHORITY, PANEL ON BIOLOGICAL HAZARDS (BIOHAZ) (2011c), 'Scientific opinion on *Campylobacter* in broiler meat production: control options and performance objectives and/or targets at different stages of the food chain', *EFSA Journal*, 9(4), 2105.

EL-SHIBINY A, CONNERTON P and CONNERTON I (2009), 'Survival at refrigeration and freezing temperatures of *Campylobacter coli* and *Campylobacter jejuni* on chicken skin applied as axenic and mixed inoculums', *Int J Food Microbiol*, 131, 197–202.

EFSA/ECDC, EUROPEAN FOOD SAFETY AUTHORITY/EUROPEAN CENTRE FOR DISEASE PREVENTION AND CONTROL (2011), 'The European Union summary report on trends and sources of zoonoses, zoonotic agents and food-borne outbreaks in 2009', *EFSA Journal*, 9(3), 2090.

ELLERBROEK L, LIENAU J-A, ALTER T and SCHLICHTING D (2007), 'Effectiveness of different chemical decontamination methods on the *Campylobacter* load of poultry carcasses', *Fleischwirtsch*, 87(4), 224–227.

ELLERBROEK L, LIENAU J-A and KLEIN G (2010), '*Campylobacter* spp. in broiler flocks at farm level and the potential for cross-contamination during slaughter', *Zoonoses Public Health*, 57, e81–e88.

FABRIZIO K A, SHARMA R R, DEMIRCI A and CUTTER C N (2002), 'Comparison of electrolyzed oxidizing water with various antimicrobial interventions to reduce *Salmonella* species on poultry', *Poult Sci*, 81, 1598–1605.

FAO/WHO (2008), *Benefits and risks of the use of chlorine-containing disinfectants in food production and food processing*: report of a Joint FAO/WHO Expert Meeting, 27–30 May 2008, Ann Arbor, MI. Available from: http://www.who.int (accessed 15 April 2011).

FARKAS J (1998), 'Irradiation as a method for decontaminating food', *Int J Food Microbiol*, 44, 189–204.

FARKAS J and MOHÁCSI-FARKAS C (2011), 'History and future of food irradiation', *Trends Food Sci Technol*, 22, 121–126.

FRIEDMAN C R, HOEKSTRA R M, SAMUEL M, MARCUS R, BENDER J, SHIFERAW B, REDDY S, AHUJA S D, HELFRICK D L, HARDNETT F, CARTER M, ANDERSON B and TAUXE R V (2004), 'Risk factors for sporadic *Campylobacter* infection in the United States: a case-control study in FoodNet sites', *Clin Infect Dis*, 38, S285–S296.

FSA, FOOD STANDARD AGENCY (2010), *Proceedings of an international meeting on Campylobacter reduction in chicken*, 30–31 March 2010, London. Available from: http://www.food.gov.uk (accessed 15 April 2011).

FSIS, FOOD SAFETY and INSPECTION SERVICE (2011), *Directive 7120.1/2011, Safe and suitable ingredients used in the production of meat, poultry, and egg products*: revision 6, dated April 05 2011. Available from: http://www.fsis.usda.gov (accessed 15 April 2011).

GELLYNCK X, MESSENS W, HALET D, GRIJSPEERDT K, HARTNETT E and VIAENE J (2008), 'Economics of reducing *Campylobacter* at different levels within the Belgian poultry meat chain', *J Food Prot*, 71, 479–485.

GEORGSSON F, ÞORKELSSON À E, GEIRSDÓTTIR M, REIERSEN J and STERN N J (2006), 'The influence of freezing and duration of storage on *Campylobacter* and indicator bacteria in broiler carcasses', *Food Microbiol*, 23, 677–683.

© Woodhead Publishing Limited, 2012

GILL C O (2004), 'Visible contamination on animals and carcasses and the microbiological condition of meat', *J Food Prot*, 67, 413–419.

GOODE D, ALLEN V M and BARROW P A (2003), 'Reduction of experimental *Salmonella* and *Campylobacter* contamination of chicken skin by application of lytic bacteriophages', *Appl Environ Microbiol*, 69, 5032–5036.

GREER G G (2005), 'Bacteriophage control of foodborne bacteria', *J Food Prot*, 68, 1102–1111.

GUO C, HOEKSTRA R M, SCHROEDER C M, PIRES S M, ONG K L, HARTNETT E, NAUGLE A, HARMAN J, BENNETT P, CIESLAK P, SCALLAN E, ROSE B, HOLT K G, KISSLER B, MBANDI E, ROODSARI R, ANGULO F J and COLE D (2011), 'Application of Bayesian techniques to model the burden of human salmonellosis attributable to US food commodities at the point of processing: adaptation of a Danish model', *Foodborne Pathog Dis*, 8, 509–516.

HASTINGS R, COLLES F M, MCCARTHY N D, MAIDEN M C and SHEPPARD S K (2010), '*Campylobacter* genotypes from poultry transportation crates indicate a source of contamination and transmission', *J Appl Microbiol*, 110, 266–276.

HAUGHTON P N, LYNG J G, CRONIN D A, MORGAN D J, FANNING S and WHYTE P (2011a), 'Efficacy of UV light treatment for the microbiological decontamination of chicken, associated packaging, and contact surfaces', *J Food Prot*, 74, 565–572.

HAUGHTON P N, LYNG J G, MORGAN D J, CRONIN D A, NOCI F, FANNING S and WHYTE P (2011b), 'An evaluation of the potential of high-intensity ultrasound for improving the microbial safety of poultry', *Food Bioprocess Technol*, doi: 10.1007/s11947–010–0372-y.

HAVELAAR A H, MANGEN M J, DE KOEIJER A A, BOGAARDT M J, EVERS E G, JACOBS-REITSMA W F, VAN PELT W, WAGENAAR J A, DE WIT G A, VAN DER ZEE H and NAUTA M J (2007), 'Effectiveness and efficiency of controlling *Campylobacter* on broiler chicken meat', *Risk Anal*, 27, 831–844.

HEATH J L, OWENS S L and TESCH S (1990), 'Effects of high-energy electron irradiation of chicken meat on *Salmonella* and aerobic plate count', *Poult Sci*, 69, 150–156.

HERMANS D, VAN DEUN K, MESSENS W, MARTEL A, VAN IMMERSEEL F, HAESEBROUCK F, RASSCHAERT G, HEYNDRICKX M and PASMANS F (2011), '*Campylobacter* control in poultry by current intervention measures ineffective: urgent need for intensified fundamental research', *Vet Microbiol*, 152, 219–228.

HEYNDRICKX M, VANDEKERCHOVE D, HERMAN L, ROLLIER I, GRIJSPEERDT K and DE ZUTTER L (2002), 'Routes for *Salmonella* contamination of poultry meat: epidemiological study from hatchery to slaughterhouse', *Epidemiol Infect*, 129, 253–265.

HIGGINS J P, HIGGINS S E, GUENTHER K L, HUFF W, DONOGHUE A M, DONOGHUE D J and HARGIS B M (2005), 'Use of a specific bacteriophage treatment to reduce *Salmonella* in poultry products', *Poult Sci*, 84, 1141–1145.

HINTON A JR and INGRAM K D (2005), 'Microbicidal activity of tripotassium phosphate and fatty acids towards spoilage and pathogenic bacteria associated with poultry', *J Food Prot*, 68, 1462–1466.

HONG, Y-H, KU G-J, KIM M-K and SONG K B (2007), 'Inactivation of *Listeria monocytogenes* and *Campylobacter jejuni* in chicken by aqueous chlorine dioxide treatment', *J Food Sci Nutr*, 12, 279–283.

HRICOVA D, STEPHAN R and ZWEIFEL C (2008), 'Electrolyzed water and its application in the food industry', *J Food Prot*, 71, 1934–1947.

HUDSON J A, BILLINGTON C, CAREY-SMITH G and GREENING G (2005), 'Bacteriophages as biocontrol agents in food', *J Food Prot*, 68, 426–437.

HUEZO R, NORTHCUTT J K, SMITH D P, FLETCHER D L and INGRAM K D (2007), 'Effect of dry air or immersion chilling on recovery of bacteria from broiler carcasses', *J Food Prot*, 70, 1829–1834.

HUGAS M and TSIGARIDA E (2008), 'Pros and cons of carcass decontamination: the role of the European Food Safety Authority', *Meat Sci*, 78, 43–52.

© Woodhead Publishing Limited, 2012

HUMPHREY T, O'BRIEN S and MADSEN M (2007), 'Campylobacters as zoonotic pathogens: a food production perspective', *Int J Food Microbiol*, 117, 237–257.

HWANG C A and BEUCHAT L R (1995), 'Efficacy of selected chemicals for killing pathogenic and spoilage microorganisms on chicken skin', *J Food Prot*, 58, 19–23.

ISOHANNI P M and LYHS U (2009), 'Use of ultraviolet irradiation to reduce *Campylobacter jejuni* on broiler meat', *Poult Sci*, 88, 661–668.

JACKSON T C, MARSHALL D L, ACUFF G R and DICKSON J S (2001), 'Meat, poultry and seafood', in Doyle M P, Beuchat L R and Montville T J, *Food Microbiology: Fundamentals and Frontiers*, Washington, DC, ASM Press, 91–109.

JAMES C, JAMES S J, HANNAY N, PURNELL G, BARBEDO-PINTO C, YAMAN H, ARAUJO M, GONZÀLEZ M L, CALVO J, HOWELL M and CORRY J E (2007), 'Decontamination of poultry carcasses using steam or hot water in combination with rapid cooling, chilling or freezing of carcass surfaces', *Int J Food Microbiol*, 114, 195–203.

JIMÉNEZ S M, CALIUSCO M F, TIBURZI M C, SALSI M S and PIROVANI M E (2007), 'Predictive models for reduction of *Salmonella* Hadar on chicken skin during single and double sequential spraying treatments with acetic acid', *J Appl Microbiol*, 103, 528–535.

KEKLIK N M, DEMIRCI A and BOCK R G (2010a). 'Decontamination of whole chicken carcasses by using a pilot-scale pulsed UV-light system', ASABE Paper No. 1008677. American Society of Agricultural Engineers, St. Joseph, MI.

KEKLIK N M, DEMIRCI A and PURI V M (2010b), 'Decontamination of unpackaged and vacuum packaged boneless chicken breast with pulsed UV-light', *Poult Sci*, 89, 570–581.

KERE KEMP G, ALDRICH M L, GUERRA M L and SCHNEIDER K R (2001), 'Continuous online processing of fecal- and ingesta-contaminated poultry carcasses using an acidified sodium chlorite antimicrobial intervention', *J Food Prot*, 64, 807–812.

KILLINGER K M, KANNAN A, BARY A I and COGGER C G (2010), 'Validation of a 2 percent lactic acid antimicrobial rinse for mobile poultry slaughter operations', *J Food Prot*, 73, 2079–2083.

KIM C, HUNG Y-C and RUSSEL S M (2005), 'Efficacy of electrolyzed water in the prevention and removal of fecal material attachment and its microbicidal effectiveness during simulated industrial poultry processing', *Poult Sci*, 84, 1778–1784.

KIM J-G, YOUSEF A E and DAVE S (1999), 'Application of ozone for enhancing the microbiological safety and quality of foods: a review', *J Food Prot*, 62, 1071–1087.

KIM J-G, YOUSEF A E and KHADRE M A (2003), 'Ozone and its current and future application in the food industry', *Adv Food Nutr Res*, 45, 167–218.

KIM J-W and SLAVIK M F (1994), 'Removal of *Salmonella* Typhimurium attached to chicken skin by rinsing with trisodium phosphate solution: scanning electron microscopic examination', *J Food Safety*, 14, 77–84.

KIM J-W and SLAVIK M F (1996), 'Cetylpyridinium chloride (CPC) treatment on poultry skin to reduce attached *Salmonella*', *J Food Prot*, 59, 322–326.

KIM T, SILVA J L and CHEN T C (2002), 'Effects of UV irradiation on selected pathogens in peptone water and on stainless steel and chicken meat', *J Food Prot*, 65, 1142–1145.

KIMURA A C, REDDY V, MARCUS R, CIESLAK P R, MOHLE-BOETANI J C, KASSENBORG H D, SEGLER S D, HARDNETT F P, BARRETT T and SWERDLOW D L; Emerging Infections Program FoodNet Working Group (2004), 'Chicken consumption is a newly identified risk factor for sporadic *Salmonella enterica* serotype Enteritidis infections in the United States: a case-control study in FoodNet sites', *Clin Infect Dis*, 38 Suppl 3, S244–S252.

LAURY A M, ALVARADO M V, NACE G, ALVARADO C Z, BROOKS J C, ECHEVERRY A and BRASHEARS M M (2009), 'Validation of a lactic acid- and citric acid-based antimicrobial product for the reduction of *Escherichia coli* O157:H7 and *Salmonella* on beef tips and whole chicken carcasses', *J Food Prot*, 72, 2208–2211.

LI Y, KIM J-W, SLAVIK M F, GRIFFIS C L, WALKER J T and WANG H (1994), '*Salmonella*

© Woodhead Publishing Limited, 2012

Typhimurium attached to chicken skin reduced using electrical stimulation and inorganic salts', *J Food Sci*, 59, 23–25.

LI Y, SLAVIK M F, WALKER J T and XIONG H (1997), 'Pre-chill spray of chicken carcasses to reduce *Salmonella* Typhimurium', *J Food Sci*, 62, 605–607.

LI Y, YANG H and SWEM B L (2002), 'Effect of high-temperature inside-outside spray on survival of *Campylobacter jejuni* attached to prechill chicken carcasses', *Poult Sci*, 81, 1371–1377.

LILLARD H S (1993), 'Bactericidal effect of chlorine on attached salmonellae with and without sonification', *J Food Prot*, 56, 716–719.

LILLARD H S (1994a), 'Effect of trisodium phosphate on salmonellae attached to chicken skin', *J Food Prot*, 57, 465–469.

LILLARD H S (1994b), 'Decontamination of poultry skin by sonication', *Food Technol*, 48, 72–73.

LIN J (2009), 'Novel approaches for *Campylobacter* control in poultry', *Foodborne Pathog Dis*, 6, 755–765.

LORETZ M, STEPHAN R and ZWEIFEL C (2010), 'Antimicrobial activity of decontamination treatments for poultry carcasses: a literature survey', *Food Control*, 21, 791–804.

MCCANN M S, SHERIDAN J J, MCDOWELL D A and BLAIR I S (2006), 'Effects of steam pasteurization on *Salmonella* Typhimurium DT104 and *Escherichia coli* O157:H7 surface inoculated onto beef, pork and chicken', *J Food Eng*, 76, 32–40.

MCDOWELL S W, MENZIES F D, MCBRIDE S H, OZA A N, MCKENNA J P, GORDON A W and NEILL S D (2008), '*Campylobacter* spp. in conventional broiler flocks in Northern Ireland: epidemiology and risk factors', *Prev Vet Med*, 84, 261–276.

MEAD G C, HUDSON W R and HINTON M H (1995), 'Effect of changes in processing to improve hygiene control on contamination of poultry carcasses with campylobacter', *Epidemiol Infect*, 115, 495–500.

MEHYAR G, BLANK G, HAN J H, HYDAMAKA A and HOLLEY R A (2005), 'Effectiveness of trisodium phosphate, lactic acid and commercial antimicrobials against pathogenic bacteria on chicken skin', *Food Prot Trends*, 25, 351–362.

MULLERAT J, KLAPES A and SHELDON B W (1994), 'Efficacy of Salmide®, a sodium chlorite-based oxy-halogen disinfectant, to inactivate bacterial pathogens and extend shelf-life of broiler carcasses', *J Food Prot*, 57, 596–603.

MÜLLNER P, COLLINS-EMERSON J M, MIDWINTER A C, CARTER P, SPENCER S E, VAN DER LOGT P, HATHAWAY S and FRENCH N P (2010), 'Molecular epidemiology of *Campylobacter jejuni* in a geographically isolated country with a uniquely structured poultry industry', *Appl Environ Microbiol*, 76, 2145–2154.

NASSAR T J, AL-MASHHADI A S, FAWAL A K and SHALHAT A F (1997), 'Decontamination of chicken carcasses artificially contaminated with *Salmonella*', *Rev Sci Tech Off Int Epiz*, 16, 891–897.

NDE C W, MCEVOY J M, SHERWOOD J S and LOGUE C M (2007), 'Cross contamination of turkey carcasses by *Salmonella* species during defeathering', *Poult Sci*, 86, 162–167.

NEWELL D G and FEARNLEY C (2003), 'Sources of *Campylobacter* colonization in broiler chickens', *Appl Environ Microbiol*, 69, 4343–4351.

NORMAND V, BOULIANNE M and QUESSY S (2008), 'Evidence of cross-contamination by *Campylobacter* spp. of broiler carcasses using genetic characterization of isolates', *Can J Vet Res*, 72, 396–402.

NØRRUNG B and BUNCIC S (2008), 'Microbial safety of meat in the European Union', *Meat Sci*, 78, 14–24.

NORTHCUTT J K, SMITH D P, MUSGROVE M T, INGRAM K D and HINTON A JR (2005), 'Microbiological impact of spray washing broiler carcasses using different chlorine concentrations and water temperatures', *Poult Sci*, 84, 1648–1652.

NORTHCUTT J, SMITH D, INGRAM K D, HINTON A JR and MUSGROVE M (2007), 'Recovery of bacteria from broiler carcasses after spray washing with acidified electrolyzed water or sodium hypochlorite solutions', *Poult Sci*, 86, 2239–2244.

© Woodhead Publishing Limited, 2012

O'BRYAN C A, CRANDALL P G, RICKE S C and OLSON D G (2008), 'Impact of irradiation on the safety and quality of poultry and meat products: a review', *Crit Rev Food Sci Nutr*, 48, 442–457.

OLSEN J E, BROWN D J, MADSEN M and BISGAARD M (2003), 'Cross-contamination with *Salmonella* on a broiler slaughterhouse line demonstrated by use of epidemiological markers', *J Appl Microbiol*, 94, 826–835.

OYARZABAL O A (2005), 'Reduction of *Campylobacter* spp. by commercial antimicrobials applied during the processing of broiler chickens: a review from the United States perspective', *J Food Prot*, 68, 1752–1760.

OYARZABAL O A, HAWK C, BILGILI S F, WARF C C and KERE KEMP G (2004), 'Effects of postchill application of acidified sodium chlorite to control *Campylobacter* spp. and *Escherichia coli* on commercial broiler carcasses', *J Food Prot*, 67, 2288–2291.

ÖZDEMIR H, GÜCÜKOGLU A and KOLUMAN A (2006), 'Acidified sodium chlorite, trisodium phosphate and populations of *Campylobacter jejuni* on chicken breast skin', *J Food Process Pres*, 30, 608–615.

PARK H, HUNG Y-C and BRACKETT R E (2002), 'Antimicrobial effect of electrolyzed water for inactivating *Campylobacter jejuni* during poultry washing', *Int J Food Microbiol*, 72, 77–83.

PASKEVICIUTE E, BUCHOVEC I and LUKSIENE Z (2011), 'High-power pulsed light for decontamination of chicken from food pathogens: a study on antimicrobial efficiency and organoleptic properties', *J Food Safety*, 31, 61–68.

PATTERSON M F (1995), 'Sensitivity of *Campylobacter* spp. to irradiation in poultry meat', *Lett Appl Microbiol*, 20, 338–340.

PEYRAT M B, SOUMET C, MARIS P and SANDERS P (2008), 'Recovery of *Campylobacter jejuni* from surfaces of poultry slaughterhouses after cleaning and disinfection procedures: analysis of a potential source of carcass contamination', *Int J Food Microbiol*, 124, 188–194.

PIYASENA P, MOHAREB E and MCKELLAR R C (2003), 'Inactivation of microbes using ultrasound: a review', *Int J Food Microbiol*, 87, 207–216.

PURNELL G, MATTIK K and HUMPHREY T (2004), 'The use of "hot wash" treatments to reduce the number of pathogenic and spoilage bacteria on raw retail poultry', *J Food Eng*, 62, 29–36.

RAJKOVIC A, SMIGIC N and DEVLIEGHERE F (2010), 'Contemporary strategies in combating microbial contamination in food chain', *Int J Food Microbiol*, 141, S29–S42.

RASSCHAERT G, HOUF K, VAN HENDE J and DE ZUTTER L (2006), '*Campylobacter* contamination during poultry slaughter in Belgium', *J Food Prot*, 69, 27–33.

RASSCHAERT G, HOUF K, GODARD C, WILDEMAUWE C, PASTUSZCZAK-FRAK M and DE ZUTTER L (2008), 'Contamination of carcasses with *Salmonella* during poultry slaughter', *J Food Prot*, 71, 146–152.

RATHGEBER B M and WALDROUP A L (1995), 'Antibacterial activity of a sodium acid pyrophosphate product in chiller water against selected bacteria on broiler carcasses', *J Food Prot*, 58, 530–534.

Regulation (EC) No 853/2004 of the European Parliament and of the Council of 29 April 2004 laying down specific hygiene rules for food of animal origin, *Off J Eur Union*, L139, 55–205.

RIDLEY A M, MORRIS V K, CAWTHRAW S A, ELLIS-IVERSEN J, HARRIS J A, KENNEDY E M, NEWELL D G and ALLEN V M (2011), 'Longitudinal molecular epidemiological study of thermophilic campylobacters on one conventional broiler chicken farm', *Appl Environ Microbiol*, 77, 98–107.

RIEDEL C T, BRØNDSTED L, ROSENQUIST H, HAXGART S N and CHRISTENSEN B B (2009), 'Chemical decontamination of *Campylobacter jejuni* on chicken skin and meat', *J Food Prot*, 72, 1173–1180.

RODRIGUEZ DE LEDESMA A M, RIEMANN H P and FARVER T B (1996), 'Short-time treatment with alkali and/or hot water to remove common pathogenic and spoilage bacteria from chicken wing skin', *J Food Prot*, 59, 746–750.

© Woodhead Publishing Limited, 2012

ROSENQUIST H, NIELSEN N L, SOMMER H M, NØRRUNG B and CHRISTENSEN B B (2003), 'Quantitative risk assessment of human campylobacteriosis associated with thermophilic *Campylobacter* species in chickens', *Int J Food Microbiol*, 108, 226–232.

ROSENQUIST H, SOMMER H M, NIELSEN N L and CHRISTENSEN B B (2006), 'The effect of slaughter operations on the contamination of chicken carcasses with thermotolerant *Campylobacter*', *Int J Food Microbiol*, 108, 226–232.

RUSSEL S M and AXTELL S P (2005), 'Monochloramine versus sodium hypochlorite as antimicrobial agents for reducing populations of bacteria on broiler chicken carcasses', *J Food Prot*, 68, 758–763.

SAHIN O, MORISHITA T Y and ZHANG Q (2002), '*Campylobacter* colonization in poultry: sources of infection and modes of transmission', *Anim Health Res Rev*, 3, 95–105.

SALVAT G, COPPEN P, ALLO J C, FENNER S, LAISNEY M J, TOQUIN M T, HUMBERT F and COLIN P (1997), 'Effects of AvGard$^{TM}$ treatment on the microbiological flora of poultry carcasses', *Br Poult Sci*, 38, 489–498.

SAMS A R and FERIA R (1991), 'Microbial effects of ultrasonication of broiler drumstick skin', *J Food Sci*, 56, 247–248.

SANDBERG M, HOFSHAGEN M, ØSTENSVIK Ø, SKJERVE E and INNOCENT G (2005), 'Survival of *Campylobacter* on frozen broiler carcasses as a function of time', *J Food Prot*, 68, 1600–1605.

SANTOS A F, VIZEU D M, DESTRO M T, FRANCO B D and LANDGRAF M (2003), 'Determination of gamma radiation doses to reduce *Salmonella* spp in chicken meat', *Ciênc Tecnol Aliment*, 23, 200–205.

SCALLAN E, HOEKSTRA R M, ANGULO F J, TAUXE R V, WIDDOWSON M-A, ROY S L, JONES J L and GRIFFIN P M (2011a), 'Foodborne illness acquired in the United States – major pathogens', *Emerg Infect Dis*, 17, 7–15.

SCALLAN E, GRIFFIN P M, ANGULO F J, TAUXE R V and HOEKSTRA R M (2011b), 'Foodborne illness acquired in the United States – unspecified agents', *Emerg Infect Dis*, 17, 16–22.

SEXTON M, RAVEN G, HOLD G, POINTON A, KIERMEIER A and SUMNER J (2007), 'Effect of acidified sodium chlorite treatment on chicken carcasses processed in South Australia', *Int J Food Microbiol*, 115, 252–255.

SHEPPARD S K, DALLAS J F, STRACHAN N J, MACRAE M, MCCARTHY N D, WILSON D J, GORMLEY F J, FALUSH D, OGDEN I D, MAIDEN M C and FORBES K J (2009), '*Campylobacter* genotyping to determine the source of human infection', *Clin Infect Dis*, 48, 1072–1078.

SLAVIK M F, KIM J-W, PHARR M D, RABEN D P, TSAI S and LOBSINGER C M (1994), 'Effect of trisodium phosphate on *Campylobacter* attached to post-chill chicken carcasses', *J Food Prot*, 57, 324–326.

SOFOS J N (2008), 'Challenges to meat safety in the 21st century', *Meat Sci*, 78, 3–13.

SOLOW B T, CLOAK O M and FRATAMICO P M (2003), 'Effect of temperature on viability of *Campylobacter jejuni* and *Campylobacter coli* on raw chicken or pork skin', *J Food Prot*, 66, 2023–2031.

STAFFORD R J, SCHLUTER P J, WILSON A J, KIRK M D, HALL G and UNICOMB L; OzFoodNet Working Group (2008), 'Population-attributable risk estimates for risk factors associated with *Campylobacter* infection, Australia', *Emerg Infect Dis*, 14, 895–901.

STOPFORTH J D, O'CONNOR R, LOPES M, KOTTAPALLI B, HILL W E and SAMADPOUR M (2007), 'Validation of individual and multiple sequential interventions for reduction of microbial populations during processing of poultry carcasses and parts', *J Food Prot*, 70, 1393–1401.

VADHANASIN S, BANGTRAKULNONTH A and CHIDKRAU T (2004), 'Critical control points for monitoring salmonellae reduction in Thai commercial frozen broiler processing', *J Food Prot*, 67, 1480–1483.

VALTIERRA-RODRÍGUEZ D, HEREDIA N L, GARCÍA S and SÁNCHEZ E (2010), 'Reduction of *Campylobacter jejuni* and *Campylobacter coli* in poultry skin by fruit extracts', *J Food Prot*, 73, 477–482.

© Woodhead Publishing Limited, 2012

VANDEKINDEREN I, DEVLIEGHERE F, VAN CAMP J, KERKAERT B, CUCU T, RAGAERT P, DE BRUYNE J and DE MEULENAER B (2009), 'Effects of food composition on the inactivation of foodborne microorganisms by chlorine dioxide', *Int J Food Microbiol*, 131, 138–144.

WAGENAAR J A, MEVIUS D J and HAVELAAR A H (2006), '*Campylobacter* in primary animal production and control strategies to reduce the burden of human campylobacteriosis', *Rev Sci Tech Off Int Epiz*, 25, 581–594.

WALDROUP A L, HIERHOLZER R E, FORSYTHE R H and MILLER M J (1993), 'Recycling of poultry chill water using ozone', *J Appl Poult Res*, 2, 330–336.

WALLNER-PENDLETON E A, SUMNER S S, FRONING G W and STETSON L E (1994), 'The use of ultraviolet radiation to reduce *Salmonella* and psychrotrophic bacterial contamination on poultry carcasses', *Poult Sci*, 73, 1327–1333.

WANG W-C, LI Y, SLAVIK M F and XIONG H (1997), 'Trisodium phosphate and cetylpyridinium chloride spraying on chicken skin to reduce attached *Salmonella* Typhimurium', *J Food Prot*, 60, 992–994.

WHYTE P, COLLINS J D, MCGILL K, MONAHAN C and O'MAHONY H (2001), 'Quantitative investigation of the effects of chemical decontamination procedures on the microbiological status of broiler carcasses during processing', *J Food Prot*, 64, 179–183.

WHYTE P, MCGILL K and COLLINS J D (2003), 'An assessment of steam pasteurization and hot water immersion treatments for the microbiological decontamination of broiler carcass', *Food Microbiol*, 20, 111–117.

WILSON D J, GABRIEL E, LEATHERBARROW A J, CHEESBROUGH J, GEE S, BOLTON E, FOX A, FEARNHEAD P, HART C A and DIGGLE P J (2008), 'Tracing the source of campylobacteriosis', *PLoS Genet*, 4(9), e1000203.

WINGSTRAND A, NEIMANN J, ENGBERG J, NIELSEN E M, GERNER-SMIDT P, WEGENER H C and MØLBAK K (2006), 'Fresh chicken as main risk factor for campylobacteriosis, Denmark', *Emerg Infect Dis*, 12, 280–285.

XIONG H, LI Y, SLAVIK M F and WALKER J T (1998), 'Spraying chicken skin with selected chemicals to reduce attached *Salmonella* Typhimurium', *J Food Prot*, 61, 272–275.

YANG H, LI Y and JOHNSON M G (2001), 'Survival and death of *Salmonella* Typhimurium and *Campylobacter jejuni* in processing water and on chicken skin during poultry scalding and chilling', *J Food Prot*, 64, 770–776.

YANG Z, LI Y and SLAVIK M F (1998), 'Use of antimicrobial spray applied with an inside-outside birdwasher to reduce bacterial contamination on prechilled chicken carcasses', *J Food Prot*, 61, 829–832.

YANG Z, LI Y and SLAVIK M F (1999), 'Antibacterial efficacy of electrochemically activated solution for poultry spraying and chilling', *J Food Sci*, 64, 469–472.

ZHAO T and DOYLE M P (2006), 'Reduction of *Campylobacter jejuni* on chicken wings by chemical treatments', *J Food Prot*, 69, 762–767.

ZHAO T, EZEIKE G O, DOYLE M P, HUNG Y-C and HOWELL R S (2003), 'Reduction of *Campylobacter jejuni* on poultry by low-temperature treatment', *J Food Prot*, 66, 652–655.

ZHAO T, ZHAO P and DOYLE M P (2009), 'Inactivation of *Salmonella* and *Escherichia coli* O157:H7 on lettuce and poultry skin by combinations of levulinic acid and sodium dodecyl sulfate', *J Food Prot*, 72, 928–936.

ZWEIFEL C, SCHEU K D, KEEL M, RENGGLI F and STEPHAN R (2008), 'Occurrence and genotypes of *Campylobacter* in broiler flocks, other farm animals, and the environment during several rearing periods on selected poultry farms', *Int J Food Microbiol*, 125, 182–187.

© Woodhead Publishing Limited, 2012

# 4

# Microbial decontamination of seafood

**T. Skåra and J. T. Rosnes, Nofima, Norway and C. Leadley, Campden BRI, UK**

**Abstract**: The sensory properties of many seafoods change more rapidly than those of most other food products, when exposed to a high thermal load. Hence the surface, being the main habitat of bacteria in most fresh and newly processed seafoods, is an area of interest for efficient bacterial reduction with minimal thermal load. Recent findings in the area of surface decontamination of particular relevance for seafood are summarised, either as reported results for seafood, or as principles applicable to seafoods. In this chapter, the microbiological issues are described first, followed by the technologies and applications. However, the chapter separates the traditional physical/chemical methods from the emerging novel technologies, of which the latter typically lack published data for seafood, but are included to give some foresights into future developments.

**Key words**: surface decontamination, seafood, fish, steam, dipping, acid, chlorine.

## 4.1 Introduction

The ability of microorganisms to adhere firmly to surfaces of processing equipment, as well as to muscle and skin, is well documented and represents an ongoing concern for the food industry (Cortesi *et al.*, 2009). This chapter aims to summarise the findings in the area of surface decontamination with particular relevance for seafood. Sections 4.2 and 4.3 deal with the microbiological issues. Sections 4.4 and 4.5 describe technologies and applications separating the traditional physical/chemical methods from the emerging novel technologies. The latter typically lack published data for seafood, but are included to give some foresights into future developments.

© Woodhead Publishing Limited, 2012

Consumers are generally advised to increase their consumption of seafood due to the health benefits of both lean and fatty fish. It has long been well established that fish is an important source for vital nutrients, and marine omega fatty acids can reduce the risk of both heart disease and mental diseases (Hibbeln *et al.*, 2007; Ruxton *et al.*, 2004). Changing lifestyles have had a deep impact on food consumption habits in most developed countries. An increasing demand for quality, variety and service characteristics has triggered the development of many new products designed to meet the need created by the fact that more people are eating their meals outside their homes (Kennedy and Wall, 2007), as well as new and more efficient production systems that are being introduced in food service. In addition, there is a huge increase in retail ready meals with a shelf life of one to four weeks. Meeting the three consumer mega-trends of convenience, health and sensory enjoyment are vital for a product's success in today's marketplace (Datamonitor, 2006). For seafood products, the demand for convenience is a trend that needs to be addressed by research and development. The combination of convenience and sensory quality, however, must never compromise food safety.

Fishery products are of great importance for human nutrition worldwide. Aquaculture production in 2000 was 45.7 million metric tons (mmt) or 32.2% of the total world fisheries landings of 141.8 mmt (FAO, 2003). During the past 20 years, there has been an intensive global increase in fish farming. Food safety issues associated with aquaculture products differ from region to region and from habitat to habitat and vary according to the method of production, management practices and environmental conditions (Reilly, 1998). Risks to consumer health from fishery products from the unpolluted marine environment are low. Potential food safety hazards of aquaculture products vary according to the culture system and may include foodborne trematode infections, foodborne disease associated with pathogenic bacteria and viruses, veterinary drug residues, and contamination by agrochemicals and toxic metals. During storage, indigenous spoilage bacteria will outgrow the indigenous pathogenic bacteria, and normally fish will spoil before becoming toxic due to the presence of greater amounts of pathogens. However, exceptions exist; for example, *Listeria monocytogenes* may grow to concentrations harmful for vulnerable groups (pregnant, immuno-compromised, elderly) without any warning from spoilage. Pathogenic species of bacteria can be introduced into coastal regions and aquaculture ponds by animal manure and human waste and are usually found in fish and crustaceans after catch at fairly low levels. Hence seafood may carry a variety of microorganisms from both aquatic and terrestrial sources, and in addition to spoilage microorganisms, may contain various potential pathogens, which are public health hazards. A microorganism-specific food source profile for foodborne outbreaks, reported internationally between 1988 and 2007, showed that *Clostridium botulinum*, *Vibrio* spp. and *L. monocytogenes* constituted the highest proportion (%) in seafoods (Greig and Ravel, 2009) and a thermal process should target a safe destruction of these pathogens.

© Woodhead Publishing Limited, 2012

It is often difficult to maintain the quality of seafood and seafood products because of the considerable distance between consumers and harvesting areas, which provides opportunities for microbial growth and recontamination. Ready-to-eat (RTE) products are of particular concern since they are consumed without further treatment or processing that can reduce the bacterial numbers. All appropriate processes that can reduce bacterial numbers in such products, including decontamination, are therefore of prime importance in order to reduce the health hazard.

## 4.2 Organisms of concern: pathogens that may contaminate fish surfaces

Bacteria on fish from temperate waters are generally psychrotrophic and able to spoil fish at a high rate at refrigerated temperatures. This is a major difference, compared to bacteria found on warm blooded animals, which are often mesophilic and adapted to higher temperatures. On fish skin surfaces, microorganisms grow in the scale pockets and subsequently invade the flesh by moving between the muscle fibres. Deskinned fillets are subjected to a direct attachment to the muscle cells. The microflora on temperate water fish and shellfish is dominated by psychrotrophic or psychrophilic, Gram-negative, rod-shaped bacteria belonging to the genera *Pseudomonas*, *Moraxella*, *Alkaligens*, *Acinetobacter*, *Shewanella*, *Flavobacterium* and members from *Vibrionaceae* and *Aeromonadaceae* (Gram and Huss, 2000). The bacterial flora on farmed fish from temperate waters is similar to the microflora of wild fish (Spanggaard *et al.*, 1993). Gram-negative dominate, but Gram-positive bacteria appear in varying concentrations, e.g. *Bacillus*, *Clostridium*, *Lactobacillus* and coryneforms. Several pathogens are indigenous in the environments for fish and seafood and may therefore be candidates to contaminate fish products during handling and processing. Among the toxin-producing bacteria are psychrotrophic non-proteolytic *C. botulinum* types B, E and F, and psychrotolerant histamine producing bacteria (photobacteria). The infective microorganisms are *Listeria monocytogenes*, *Vibrio cholerae*, *Vibrio parahaemolyticus*, *Vibrio vulnificus*, *Aeromonas hydrophila* and *Plesiomonas shigelloides*. Non-indigenous bacteria are *Staphylococcus aureus*, *C. botulinum* proteolytic types A and B, mesophilic histamine producing bacteria (*Morganella morganii*) and *Salmonella* spp., *Shigella* spp. and *Escherichia coli* (Nilsson and Gram, 2002). Pathogenic toxin-producing *Bacillus cereus* are not associated with raw fish materials, but may be a risk factor from ingredients in mixed or minced fish products.

Seafood borne pathogenic bacteria may conveniently be divided into two groups, namely indigenous and non-indigenous organisms as shown in Table 4.1. Bacteria commonly found in the environment, like psychrotropic *Clostridium botulinum* type E and *Listeria monocytogenes* may easily

© Woodhead Publishing Limited, 2012

**Table 4.1** Indigenous and non-indigenous organisms of concern in fish handling and processing

| | | Mode of action | | | | |
|---|---|---|---|---|---|---|
| | Microorganism | Infection | Pre-formed toxin | Heat stability | Estimated minimum infectious dose | Reference |
| Indigenous | *Clostridium botulinum* type E | | + | Low | 0.1–1 μg toxin lethal dose | Feldhusen, 2000 |
| | *Vibrio* sp | + | | | High | |
| | *V. cholera* (serovar O1 and O139) | | | | $10^3$ to $10^8$–$10^9$ CFU/g | Feldhusen, 2000 |
| | *V. parahaemolyticus* | | | | $10^5$–$10^6$ CFU/g | Feldhusen, 2000 |
| | Other vibrios[a] | | | | – | |
| | *Aeromonas hydrophila* | + | | | Not known | |
| | *Pleisomonas shigelloidis* | + | | | Not known | |
| | *Listeria monocytogenes* | + | | | $>10^2$ CFU/g | Feldhusen, 2000 |
| Non-indigenous | *Salmonella* sp. | + | | | $10^2$ CFU/g depend on species | Feldhusen, 2000 |
| | *Shigella* | + | | | $10^1$–$10^2$ | Feldhusen, 2000 |
| | *E. coli* | + | | | $10^1$–$10^3$ [b] | ACMSF, 1995 |
| | *Staphylococcus aureus* | | + | High | Toxin level $10^5$–$10^6$ CFU/g<br>0.14–0.19 μg/kg bodyweight | Feldhusen, 2000 |
| | *Bacillus cereus* | | + | High (emetic) | $10^6$–$10^9$ CFU/g | Feldhusen, 2000 |

[a]Other vibrios are: *V. vulnificus*, *V. hollisae*, *V. mimicus*, *V. Fluvialis*.

[b]For verotoxin-producing strain O157:H7.

© Woodhead Publishing Limited, 2012

contaminate fish and shellfish. *L. monocytogenes* has been considered as a leading cause of death amongst the foodborne bacterial pathogens (Paoli *et al.*, 2005). The main origin of *L. monocytogenes* in catfish fillets was found to be the food contact surfaces of the processing line (Chen *et al.*, 2010). The prevalence of *L. monocytogenes* in RTE food varies with the product type and the stage in the production-to-consumption chain at which it is monitored. High incidences have been found in lightly preserved fish products (e.g. cold smoked and low salted) (Beaufort *et al.*, 2007; Lappi *et al.*, 2004). Cooked RTE products which may be consumed without further heating are obviously of concern, but as the levels of *L. monocytogenes* associated with contamination of these products are typically low, the risks are minimal if multiplication does not or cannot occur during storage, distribution and preparation. Epidemiologic data indicate that foods involved in listeriosis outbreaks are those in which the organism has multiplied and in general have contained levels significantly higher than 100 CFU/g (Buchanan *et al.*, 1997; ICMSF, 2002). Based on risk assessment in RTE foods, the Codex Alimentarius (2002) recommended that the maximum contamination level for *L. monocytogenes* in food at consumption should be less than 100 CFU/g.

Most microbiological research effort has been applied to the organisms indigenous to seafood rather than to those cross-contaminated during processing. This may have been a reasonable approach but with the globalisation of the fish industry, the increasing proportion of the total catch in international trade, and the current change in technological practices, also non-indigenous microorganisms may severely affect the safety of seafood products. The trend towards the consumption of more lightly processed fish with fewer preservatives increases the risk of food poisoning. Some cases of listeriosis with fish as the most likely source have been reported (Ericsson *et al.*, 1997; Miettinen *et al.*, 1999; Tham *et al.*, 2000). The Scientific Panel on Biological Hazards (BIOHAZ) of the European Food Safety Authority (EFSA) warned that *Listeria* was on the rise, after a general decline in the 1990s. Surveys carried out by EFSA have revealed associations with food packaging type, preparation practices such as the use of slicing machines for meat products, storage temperatures, the stage of sampling with respect to shelf life, the lack of an effective HACCP system, and lack of education and training of food handlers. Samples from investigations conducted in 23 EU member states showed that the legal safety criterion of 100 *Listeria* CFU/g was most often observed in RTE fishery products (EFSA, 2007).

## 4.3 Pathways of contamination

### 4.3.1 Bacterial contamination of fish

Post mortem, the immune system of fish collapses and bacteria residing on the surface start to degrade the sterile muscle tissue. The initial quality loss in fish is primarily caused by autolytic changes and is not related to

© Woodhead Publishing Limited, 2012

microbial activity. The degradation of nucleotides (ATP-related compounds) is in this respect of particular importance. Fish muscle provides an excellent substrate for the growth of most heterotrophic bacteria. Raw fish materials are initially contaminated with a wide variety of microorganisms, but it has been shown that only a limited number of bacteria actually invade the muscle during storage on ice, with no difference in the invasive pattern of specific spoilage bacteria and non-spoilage bacteria.

The bacterial flora on newly caught fish depends on the surrounding environment, rather than the species itself (Shewan, 1977). Fish caught in clean, temperate waters carry lower levels of microorganisms than fish caught in warm waters. The muscle tissue and internal organs of freshly caught, healthy finfish are normally sterile, but bacteria are found on the skin, chitinous shell, gills, as well as in their intestinal tract. On live fish in temperate waters, bacteria are present in the mucus layer of the skin ($10^2$–$10^7$ CFU/cm$^2$) (Liston, 1980), the gills ($10^3$–$10^7$ CFU/g) and the guts ($10^7$–$10^9$ CFU/g) (Shewan, 1962).

The factors that influence microbial contamination and growth include fish species and size, method of catch, on-board handling, fishing vessel sanitation, processing, and storage conditions (Chen and Chai, 1982; Ward and Baj, 1988). Fish are subjected to rapid microbial contamination and growth if handling and storage are inadequate. It is estimated that about 10% of the total world catch is lost due to bacterial spoilage (James, 1986).

After capture or slaughter and death, finfish are normally stored in crushed ice or in chilled brines, giving rise to changes in the microflora. The fraction of the microflora which will ultimately grow on the products will be determined by the intrinsic and extrinsic growth parameters. Such factors may include the poikilotherm nature of the fish and its aquatic environment, a high post mortem pH in the flesh (usually $> 6.0$), the presence of large amounts of non-protein-nitrogen (NPN) and the presence of trimethylamine (TMAO) as a part of the NPN factor (Gram and Huss, 1996).

### 4.3.2 Bacterial cell attachment on surfaces

In the phenomenon of bacterial adhesion to surfaces, the physicochemical properties and substrates or surface topography play important roles. The potential for attachment and development of microorganisms on different surfaces has been, and continues to be, extensively studied. Adhesion of pathogenic bacteria to food contact surfaces, and biofilm formation has attained the greatest focus. Biofilm formation is not an actual case for regular fish processing, although the first steps of attachments and colonisation may be similar (Dickson and Koohmaraie, 1989).

Attachment of bacteria is influenced by cell surface charge (Fletcher and Loeb, 1979), hydrophobicity (Dahlback *et al.*, 1981), and structures, including extracellular polysaccharides (Fletcher and Floodgate, 1973) and flagella (Notermans and Kamplelmacher, 1974). The role of surface structures has

© Woodhead Publishing Limited, 2012

been debated, since non-fimbriated (Meadows, 1971) and non-flagellated (Lillard, 1986) cells have been reported to attach at rates similar to those of cells which possess these structures. However, other reports indicate that motile bacteria attach to surfaces more rapidly than non-motile strains (Butler *et al.*, 1979; Farber and Idziak, 1984). The actual role of flagella in attachment is probably dependent on the specific strain of bacterium as well as growth conditions. The cell wall of bacteria cells is negatively charged (Corpe, 1970). The magnitude of this charge varies between strains, but the cell surface is generally considered a significant factor in bacterial attachment to surfaces. Many studies suggest that microbial cell surface charge and hydrophobicity play an important role in the initial steps of microbial adhesion (Briandet *et al.*, 1999; Chavant *et al.*, 2002; Gilbert *et al.*, 1991; Palmer *et al.*, 2007).

During chilled storage the bacteria will move between the muscle fibres. Murray and Shewan (1979) found that very limited numbers of bacteria invaded the flesh during iced storage, and bacteria have been detected by microscopy in the flesh when the numbers of organisms on the skin surface increased above $10^6$ CFU/cm$^2$. Because, only limited numbers of organisms on the skin surface actually invade the flesh and microbial growth takes place mainly at the surface, spoilage is probably to a large extent a consequence of bacterial enzymes diffusing into the flesh and nutrients diffusing to the outside. Bacteria on fish caught in temperate waters enter the exponential growth phase almost immediately post mortem. This is probably because the microflora is already adapted to the chill temperatures. During ice storage, the aerobic count increases with a doubling time of approximately 24 hours and will, after 2–3 weeks, reach numbers of $10^8$–$10^9$ CFU/g flesh or/cm$^2$ of skin. The bacteria on fish caught in tropical waters often pass through a lag phase up to 2 weeks if the fish is stored on ice, thereafter exponential growth begins and the numbers reach a level similar to that on temperate water fish when spoiled. The bacterial level on tropical fish has been found to be similar to those in temperate fish species at the time of spoilage (Acuff *et al.*, 1984; Gram *et al.*, 1990; Lannelongue *et al.*, 1982). Foong and Dickson (2004) found *Listeria monocytogenes* to rapidly attach to surfaces of RTE food (meat surface), some serotypes better than others. The time at which the organism comes in contact with the surface until irreversible attachment occurs is important and strong attachment to the meats by the cells occurred within 5 min. The initial numbers of contaminants were directly proportional to the numbers of organisms attached.

## 4.4 Current methods of seafood decontamination

Fish muscle interior is sterile immediately after slaughter. Hence the initial post process bacterial load is exterior, and decontamination may be targeted

© Woodhead Publishing Limited, 2012

as such. When production involves high temperature processes such as baking or frying of non-packaged products, a surface heating step may be used immediately before the packaging step to prevent any recontamination on the surface of the product.

The following section aims to summarise the methods that have been investigated for seafood products. Some technologies are regarded as potentially suitable, but when no published data are available, only general effects are described. Furthermore, the section aims to focus on surface decontamination. In some instances, dipping treatments can be used to add components to a product, and the active components will penetrate into the product. As explained for a number of components, the growth inhibiting effect during the subsequent storage can be an important asset of the treatment. The current methods of decontamination can be divided into three main categories: chemical, thermal and combinations of two or more of these.

### 4.4.1 Chemical methods

A number of chemicals show a bactericidal effect. This can be based on effects on physiological cellular processes, or disruption of membranes or other cellular constituents (Loretz *et al.*, 2010). An overview of chemical agents used for seafood is shown in Table 4.2.

*Organic acids*

The effect of organic acids on the surface of muscle food is an initial pH drop with a corresponding reduction in bacterial numbers. Moreover, the potential inhibitory effect could also contribute to shelf life extension. Different organisms have demonstrated different rankings for the inhibitory effect of organic acids (Matsuda *et al.*, 1994). The inhibitory effect of 17 organic acids commonly found in food systems has been modelled as a function of physical and chemical properties (Hsiao and Siebert, 1999), which could be helpful if targeting of specific organisms is an objective. Seafoods typically have very delicate flavours, and are thus very susceptible to taints and off-flavours. This must be taken into account when organic acid and treatment conditions are selected. Bal'a and Marshall (1998) investigated the effects of dipping catfish fillets inoculated with *Listeria monocytogenes* in 2% solutions of acetic, citric, hydrochloric, lactic, malic or tartaric acid at 4°C. Their findings show that acid dipping reduced surface pH and *L. monocytogenes*, coliform and aerobic microbial loads (Table 4.2). Also, the acid treatment had a bleaching effect on the fillets, but other than colour, no sensory data were recorded. Shirazinejad *et al.* (2010) tested lactic acid as a decontaminant for shrimp, dipping for 10–30 min in 1.5 and 3% lactic acid at room temperature. They found significant reduction in *Vibrio cholerae*, *V. parahaemolyticus*, *Salmonella* Enteritidis and *E. coli* O157:H7. The treatment time affected the sensory scores, and 10 min was the treatment time that demonstrated only slight sensory effects, while reducing the pathogens

© Woodhead Publishing Limited, 2012

**Table 4.2** Antibacterial activity of chemical decontamination treatments for seafood

| Agent/microorganism | Reduction (log CFU) (days until analysed) | Species | Sampling point | Concentration | Contamination | $T$ [°C] | $t$ [min] | Reference |
|---|---|---|---|---|---|---|---|---|
| **Organic acids** | | | | | | | | |
| **Acetic acid** | | | | | | | | |
| Areobic | ≈ 3/g (5) | Catfish | Fillets | 2% | Artificial | 4 | 10 | Bal'a and Marshall, 1998 |
| Coliform | ≈ 3.5/g (5) | Catfish | Fillets | 2% | Artificial | 4 | 10 | Bal'a and Marshall, 1998 |
| *Listeria monocytogenes* | ≈ 2/g (5) | Catfish | Fillets | 2% | Artificial | 4 | 10 | Bal'a and Marshall, 1998 |
| **Citric acid** | | | | | | | | |
| Areobic | ≈ 2/g (5) | Catfish | Fillets | 2% | Artificial | 4 | 10 | Bal'a and Marshall, 1998 |
| Coliform | ≈ 2.2/g (5) | Catfish | Fillets | 2% | Artificial | 4 | 10 | Bal'a and Marshall, 1998 |
| *Listeria monocytogenes* | ≈ 2/g (5) | Catfish | Fillets | 2% | Artificial | 4 | 10 | Bal'a and Marshall, 1998 |
| **Lactic acid** | | | | | | | | |
| Areobic | ≈ 3/g (5) | Catfish | Fillets | 2% | Artificial | 4 | 10 | Bal'a and Marshall, 1998 |
| Coliform | ≈ 2.2/g (5) | Catfish | Fillets | 2% | Artificial | 4 | 10 | Bal'a and Marshall, 1998 |
| *Escherichia coli* O157:H7 | 2.3/g | Shrimp | After catch | 3% | Artificial | 20 | 10 | Shirazinejad *et al.*, 2010 |
| *Listeria monocytogenes* | ≈ 2/g (5) | Catfish | Fillets | 2% | Artificial | 4 | 10 | Bal'a and Marshall, 1998 |
| *Salmonella* Enteritidis | 2.3/g | Shrimp | After catch | 3% | Artificial | 20 | 10 | Shirazinejad *et al.*, 2010 |
| *Vibrio cholerae* | 1.9/g | Shrimp | After catch | 3% | Artificial | 20 | 10 | Shirazinejad *et al.*, 2010 |
| *Vibrio parahemolyticae* | 1.2/g | Shrimp | After catch | 3% | Artificial | 20 | 10 | Shirazinejad *et al.*, 2010 |
| **Malic acid** | | | | | | | | |
| Areobic | ≈ 2/g (5) | Catfish | Fillets | 2% | Artificial | 4 | 10 | Bal'a and Marshall, 1998 |
| Coliform | ≈ 2.2/g (5) | Catfish | Fillets | 2% | Artificial | 4 | 10 | Bal'a and Marshall, 1998 |
| *Listeria monocytogenes* | ≈ 2/g (5) | Catfish | Fillets | 2% | Artificial | 4 | 10 | Bal'a and Marshall, 1998 |

© Woodhead Publishing Limited, 2012

| | | | | | | | | |
|---|---|---|---|---|---|---|---|---|
| **Tartaric acid** | | | | | | | | |
| Areobic | ≈ 2/g (5) | Catfish | Fillets | 2% | Artificial | 4 | 10 | Bal'a and Marshall, 1998 |
| Coliform | ≈ 2.2/g (5) | Catfish | Fillets | 2% | Artificial | 4 | 10 | Bal'a and Marshall, 1998 |
| *Listeria monocytogenes* | ≈ 2/g (5) | Catfish | Fillets | 2% | Artificial | 4 | 10 | Bal'a and Marshall, 1998 |
| **Chlorine and Chlorine dioxide** | | | | | | | | |
| **Chlorine** | | | | | | | | |
| Aerobic | $0/cm^2$ (6-12) | Mullet | Gutted | 1 mg/mL | Natural | | | Kosak and Toledo, 1981 |
| | 1.4/g | Shrimp | Tail | 40 mg/L | Artificial | 15.6 | 2 | Andrews *et al.*, 2002 |
| | 0.9/g | Crawfish | Tail | 40 mg/L | Artificial | 15.6 | 2 | Andrews *et al.*, 2002 |
| Psychrotrophic | 1.2/g | Shrimp | Tail | 40 mg/L | Artificial | 15.6 | 2 | Andrews *et al.*, 2002 |
| | 0.9/g | Crawfish | Tail | 40 mg/L | Artificial | 15.6 | 2 | Andrews *et al.*, 2002 |
| **Chlorine dioxide** | | | | | | | | |
| Aerobic | 0.5/g | Salmon | Fillet | 200 ppm | Natural | 0 | 5 | Kim *et al.*, 1999b |
| | 0.6/g | Salmon | Whole (skin) | 200 ppm | Natural | 0 | 5 | Kim *et al.*, 1999b |
| | 2/g | Salmon | Whole (muscle) | 200 ppm | Natural | 0 | 5 | Kim *et al.*, 1999b |
| | 0.5/g | Red grouper | Fillet | 200 ppm | Natural | 0 | 5 | Kim *et al.*, 1999b |
| | 1.2/g | Red grouper | Whole (skin) | 200 ppm | Natural | 0 | 5 | Kim *et al.*, 1999b |
| | 0.5/g | Scallops | Whole | 200 ppm | Natural | 0 | 5 | Kim *et al.*, 1999b |
| | 0.9/g | Shrimp | Headless with shell | 200 ppm | Natural | 0 | 5 | Kim *et al.*, 1999b |
| Aerobic | 3.7/g | Shrimp | Tail | 40 mg/L | Artificial | 15.6 | 2 | Andrews *et al.*, 2002 |
| | 3.6/g | Crawfish | Tail | 40 mg/L | Artificial | 15.6 | 2 | Andrews *et al.*, 2002 |

*(Continued)*

© Woodhead Publishing Limited, 2012

**Table 4.2** Continued

| Agent/microorganism | Reduction (log CFU) (days until analysed) | Species | Sampling point | Concentration | Contamination | $T$ [°C] | $t$ [min] | Reference |
|---|---|---|---|---|---|---|---|---|
| Psychrotrophic | 3.3/g | Shrimp | Tail | 40 mg/L | Artificial | 15.6 | 2 | Andrews *et al.*, 2002 |
| | 4.6/g | Crawfish | Tail | 40 mg/L | Artificial | 15.6 | 2 | Andrews *et al.*, 2002 |
| **Electrolysed oxidising water** | | | | | | | | |
| Aerobic | 2.8/cm$^2$ | Carp (skin) | After slaughter | 41 ppm* | Artificial | 25 | 15 | Mahmoud *et al.*, 2004 |
| | 2.0/g | Carp | Fillets | 41 ppm* | Artificial | 25 | 15 | Mahmoud *et al.*, 2004 |
| *Enterobacter aerogenes* | 1.3/cm$^2$ | Salmon (skin) | Skin only | 100 ppm* | Artificial | | 120 | Phuvasate and Su, 2010 |
| | 2.4/cm$^2$ | Tuna (skin) | Skin only | 100 ppm* | Artificial | ≈ 23 | 1440 | Phuvasate and Su, 2010 |
| *Escherichia coli* | 0.7/cm$^2$ | Tilapia (skin) | After slaughter | 120 ppm* | Artificial | 23 | 1 | Huang *et al.*, 2006 |
| *Escherichia coli* O157:H7 | 1.1/g | Salmon | Fillets | 76–90 ppm* | Artificial | 35 | 64 | Ozer and Demirci, 2006a |
| *Listeria monocytogenes* | 0.4/g | Salmon | Fillets | 76–90 ppm* | Artificial | 35 | 64 | Ozer and Demirci, 2006a |
| *Morganella morganii* | 2.2/cm$^2$ | Salmon (skin) | Skin only | 100 ppm* | Artificial | ≈ 23 | 120 | Phuvasate and Su, 2010 |
| | 3.5/cm$^2$ | Tuna (skin) | Skin only | 100 ppm* | Artificial | 0 | 1440 | Phuvasate and Su, 2010 |
| *Vibrio parahemolyticae* | 2.6/cm$^2$ | Tilapia (skin) | After slaughter | 120 ppm* | Artificial | 23 | 5 | Huang *et al.*, 2006 |
| **Ozone and Peroxide** | | | | | | | | |
| Aerobic | 3/g | Shrimp | After peeling | 1–3 ppm | Artificial | 10 | 0.33–1 | Chawla *et al.*, 2007 |
| *Pseudomonas fluorescens* | 3/g | Shrimp | After peeling | 1–3 ppm | Artificial | 10 | 0.33–1 | Chawla *et al.*, 2007 Mahapatra *et al.*, 2005 |

*Active chlorine concentration, SLE = shelf life extension (immediate reduction not relevant).

© Woodhead Publishing Limited, 2012

investigated by approximately 1.5 log units (Table 4.2). Tumbling with lactic acid/lactates has been shown to extend the shelf life of catfish (Kim *et al.*, 1995; Williams *et al.*, 1995), but treatments typically last 15–20 min and cannot be regarded as surface treatment.

*Chlorine-based treatments*

Chlorine has a long history in the fishing industry and continues to be one of the most widely used disinfectants in seafood processing industries (Andrews *et al.*, 2002). The efficiency of chlorine is very dependent on the amount of free available chlorine (HOCl) in the wash water, and also on pH, temperature, treatment duration and organic matter (Beuchat, 1998). Concerns regarding formation of potentially carcinogenic by-products like chloramines and trihalomethanes, has led to restrictions in use and even banning for some applications in a number of European countries (Beltran *et al.*, 2005).

Kosak and Toledo (1981) investigated the effect of using chlorine dips as a method of surface decontamination and hence prolonging shelf life of mullet. Fillets were dipped in 1000 µg/mL free chlorine solution for 3.5 min, before being wrapped or vacuum packaged, with ice, tap water or water with free chlorine. Quantitative figures are not given, but the authors report that immediately after treatment the colony forming units (CFU) dropped to very low levels. Moreover the results show that the treatment combined with subsequent packaging in water may extend shelf life (12 d) more than packaging in vacuum (6 d) or ice (4 d) and stored at –2°C.

Chlorine dioxide ($ClO_2$) is a neutral compound of chlorine which disinfects by oxidation. It does not chlorinate, but functions as a highly selective oxidant due to its unique, one-electron transfer mechanism where it is reduced to chlorite ($ClO_2^-$) (Gomez-Lopez *et al.*, 2009). Kim *et al.* (1999a) discussed the potential of using chlorine dioxide as an oxidative surface decontaminant. Formation of mutagenic compounds after treatment of salmon and red grouper with chlorine dioxide was tested using the Ames Salmonella/microsome assay, but no mutagenic activity was detected. Kim *et al.* (1999b) investigated the effect of using chlorine dioxide as a surface decontaminant. The log reductions in fish fillets and scallop were approximately 0.5. For shrimp with shell the reduction was 0.9 log and whole fish showed even larger reductions. A 1.2 log reduction was seen in Red grouper skin and a 2 log reduction was seen in whole salmon muscle. Overall the shelf life of the treated samples was increased compared to the control samples; however, it was commented that a rusty discolouration was seen in the Red grouper samples treated at 100 and 200 ppm. For the scallops there was a significantly lower microbial count even after 7 days.

Andrews *et al.* (2002) compared the decontamination effect of chlorine dioxide with the traditional method of a chlorine wash on shrimps and crawfish tails. The study used shrimps sourced from the sea and pond raised crawfish. The shellfish were dipped in aqueous chlorine (hypochlorite) (10–40

© Woodhead Publishing Limited, 2012

mg/L) or chlorine dioxide solutions (10–40 mg/L) for 2 min, with dipping in water as control. Aerobic plate counts showed that the chlorine dioxide samples showed a 4 log reduction whereas the chlorine wash (40 ppm) resulted in a 1–2 log reduction in the microbial counts compared to the control (Table 4.2).

*Electrolysed oxidising water*

Electrolysed oxidising (EO) water is produced by passing diluted salt water through an electrical current between an anode and a cathode separated by a membrane (electrolysis) (Huang *et al.*, 2008). This creates two types of water: acidic (pH 2.3–2.7) with high oxidation-reduction potential (ORP), dissolved oxygen and variable amounts of free chlorine, and alkaline (pH 10.0–11.5), with high dissolved hydrogen and low ORP. Acidic EO water is not corrosive to skin, mucous membrane or organic material. It has been shown to be capable of killing pathogenic microorganisms, but the efficacy varies between microbial species. Huang *et al.* (2006) studied the effect of using EO water to reduce the amount of *Escherichia coli* and *Vibrio parahaemolyticus* from the skin surface of tilapia. *E. coli* was reduced by 0.7 log after 1 min, but subsequent treatment time did not seem to increase the reduction, whereas *V. paraheamolyticus* was reduced by 1.5 log reduction after 5 min and by 2.6 log after exposure for 10 min. Ozer and Demirci (2006a) investigated the effectiveness of EO acidic treatment followed by EO alkaline treatment at different temperatures, to decontaminate the surface of salmon fillets. Acidic EO water treatment at 22°C (2 min) caused a 0.4 log reduction, and at 35°C (64 min) a 1.1 log reduction in *L. monocytogenes* Scott A. A similar effect was seen for *E. coli* O157:H7.

*Ozone and peroxide*

Ozone or its decomposition products, e.g. hydroxyl radical, can rapidly inactivate microorganisms by reacting with intracellular enzymes, nucleic material or the cell envelope (Khadre *et al.*, 2001). Peroxide is considered to inactivate cells by passing through cell walls/membranes and reacting with internal cellular components, or severely damage microbial structure causing the release of intracellular components (Finnegan *et al.*, 2010). (Manousaridis *et al.*, 2005) investigated the effect of ozonation via aqueous solution of shucked vacuum packed mussels. After treatment for 90 min there was a 1.1 log reduction for *Pseudomonas* spp., and a 2.5 log reduction for sulfide-producing bacteria. This allowed an increase in shelf life of 3 days with no sensory changes detected.

Gelman *et al.* (2005) treated live tilapia fish with ozone for 1 h prior to storage, and the ozone dosage used in the treatment was approximately 6 ppm. They found that this treatment prolonged the shelf life of the dead fish by 12 days at 0°C, and by 3 days at 5°C. Crapo *et al.* (2004) evaluated the application of ozonated water to treat fish fillets and roe and found that it had no significant effect on the microbial load, most probably due to the presence

© Woodhead Publishing Limited, 2012

of organic material which greatly reduced the efficiency of the ozone. It was also observed that samples treated with ozone had an accelerated rate of rancidity. Mahapatra *et al.* (2005) reviewed the use of ozone in a multitude of foods, and stated that 1.4 mL/L ozone causes a reduction of *E. coli* and *Salmonella* Typhimurium on shrimp. The effect of gaseous ozone on the most commonly found bacteria on fish was investigated by da Silva *et al.* (1998). When grown on agar, even very low levels ($<0.27 \times 10^{-3}$ g $L^{-1}$) of ozone were required to cause a 1.5–2.5 log reduction in newly inoculated plates after 15 min. Studies of similar treatments with plates that were grown for 4 h, showed that 30 min treatment was required for most strains to achieve a significant (approximately 2 log) reduction, and that no effect was seen on plates that were grown for 10 h.

*Phosphate-based treatments*
Polyphosphates can be used in muscle foods to improve water binding and increase firmness. They may enhance the preservative effect by (1) acting as metal ion chelators, (2) acting as pH buffers, (3) interacting with proteins to promote hydration and water binding capacity, and thus (4) preventing lipid oxidation and microbial growth (Ellinger, 1972). Trisodium phosphate is used for food decontamination and significant reductions have been shown for *Salmonella* (Kanellos and Burriel, 2005). Masniyom *et al.* (2005) studied the combination effect of three phosphate compounds and modified atmosphere on quality and shelf life of seabass slices, and found that, although there was little effect on microbial growth, sodium pyrophosphate was most effective in reducing the amount of exudates and maintaining the sensory properties.

*Naturally occurring antimicrobial compounds*
With the recent scepticism towards synthetic additives and food processing agents occurring on product labels, an increased interest has been observed in naturally occurring antimicrobial compounds. These substances may exhibit antimicrobial properties in foods in which they occur naturally, or they may be used as additives to other foods requiring preservation. The compounds, which are sometimes produced as a result of a bacteria host defence mechanism, can be extracted from living tissues (Roller and Board, 2003).

Essential oils are aromatic oily liquids obtained from plant material (flowers, buds, seeds, leaves, twigs, herbs, fruits and roots). They are among the most widely used naturally occurring antimicrobial agents (Burt, 2004) and regarded as natural alternatives to chemical preservatives. Hence the effect is measured as extension of shelf life, rather than as immediate reduction after treatment (Harpaz *et al.*, 2003). Most available studies include one or more preserving agents or factors, in addition to essential oils (Frangos *et al.*, 2010; Mejlholm and Dalgaard, 2002; Pyrgotou *et al.*, 2010). In fish, as in meat products, a high fat content appears to reduce the effectiveness of essential oils. For example, oregano oil at 0.5 μl $g^{-1}$ is more effective against

© Woodhead Publishing Limited, 2012

the spoilage organism *Photobacterium phosphoreum* on modified atmosphere packaged cod fillets than on salmon (Mejlholm and Dalgaard, 2002).

### 4.4.2 Thermal methods

The thermal load required for pasteurisation will affect most fish products negatively with respect to sensory quality. At certain stages of the production chain, however, reducing the bacterial load through targeting the surface may be an option. The amount of heat needed to kill 100 bacteria per $cm^2$, is 15 million times less than that required to cook the surface to a depth equal to the length of a bacteria (Morgan *et al.*, 1996). Based on this, surface pasteurisation appears to be a sound approach. However, there are a number of complicating elements and most of them are related to the target area, namely the surface.

As opposed to liquid media, often used to generate models, a structured food product will immobilise bacteria and thus lead to colony growth (Malakar *et al.*, 2003). The surface also interacts with the environment, and key parameters like water activity may change quickly, and contribute to increasing the thermal resistance of bacteria. Finally there is the issue of channels and microchannels that may harbour reservoirs which a rapid surface treatment will not inactivate. The results of available data from thermal decontamination of seafoods are summarized in Table 4.3.

*Hot water*

Hot water washing can inactivate bacteria under the product surface and may thus be more efficient than chemical washes. Since water provides excellent heat transfer to the product, a uniform temperature profile on the product surface can be established in a very short time (Couey, 1989). For unprocessed products, hot water is a potential method of reducing pathogenic bacteria on surfaces.

Vaz *et al.* (1994) investigated the use of hot water on the surface mircoflora of horse mackerel (*Trachurus trachurus*). The study showed that after treatment of whole fish in water at 60°C for 20 s there was no significant decrease in total viable counts (TVC) compared to the controls (Table 4.3). After storage for 12 days, it was found that the growth of surface microflora was delayed by four days compared to the control sample.

*Steam*

A much improved heat transfer is achieved using steam condensing on the surface. The size of a water molecule (approx. $2 \times 10^{-4}$ μm) and the free path length of water vapour molecules, compared to the size bacteria (e.g. *Salmonella*; 4 μm long and 0.7 μm thick) should facilitate the access of steam to all crevices or pores where bacteria might enter (Morgan *et al.*, 1996). At atmospheric pressure, the steam temperature is 100°C, and the surface treatment temperature will typically be <100°C due to steam being

© Woodhead Publishing Limited, 2012

**Table 4.3** Antibacterial activity of thermal decontamination treatments for seafood

| Agent/ microorganism | Reduction (log CFU) (days until analysed) | Species | Sampling point | *T* (°C) | Contamination | Exposure time (min) | Reference |
|---|---|---|---|---|---|---|---|
| **Hot water** | | | | | | | |
| | 0 | Horse mackerel | Whole fish | 60 | | 0.33 | Vaz *et al.*, 1994 |
| **Steam** | | | | | | | |
| Aerobic | 2.3/cm$^2$ (4) | Catfish | After slaughter | | Natural | 0.5–2 | Bal'a *et al.*, 1999 |
| | 1.6/g (4) | Catfish | After slaughter | | Natural | 0.25–2 | Bal'a *et al.*, 2000 |
| Coliform | 3.5/cm$^2$ (4) | Catfish | After slaughter | | Natural | 0.5–2 | Bal'a *et al.*, 1999 |
| | 1.7 log CFU/g (4) | Catfish | After slaughter | | Natural | 0.25–2 | Bal'a *et al.*, 2000 |
| *Listeria monocytogenes* | 4/fish | Salmon | After slaughter | | Artificial | 0.12 | Bremer *et al.*, 2002 |
| *Listeria innocua* | 2/mL (4) | Catfish | After slaughter | 143 | Artificial | 0.007 | Kozempel *et al.*, 2001 |
| Psychrotrophic | 3.4/cm$^2$ (4) | Catfish | After slaughter | | Natural | 0.5–2 | Bal'a *et al.*, 1999 |
| | 1.9/g (4) | Catfish | After slaughter | | Natural | 0.25–2 | Bal'a *et al.*, 2000 |

© Woodhead Publishing Limited, 2012

mixed with air on its way from the outlet to the food product surface. When steam is directed towards food surfaces, however, the air surrounding the product compresses a very thin film of air against the food surface which insulates the food surface against direct contact with the steam (Annous and Kozempel, 2006). Huang (2005) suggested that a layer of condensate could act as an insulation blanket which together with bacteria were filling pores and preventing the free access of steam, thus explaining the increased required steam processing time. The United States Department of Agriculture (USDA) has developed a process that seeks to overcome these phenomena, through exposing the solid food surface to vacuum, then steam and then vacuum again. The typical process conditions are 0.1 sec for vacuum and steam, and 0.5 sec for the final vacuum. Steam is saturated at 138–143°C, and two or three cycles are preferred (Kozempel *et al.*, 2003). When applied to catfish samples inoculated with *L. innocua*, a 3 log reduction was achieved using this process (Kozempel *et al.*, 2001). Bremer *et al.* (2002) reported a 4 log reduction of *L. monocytogenes* when subjected to steam in their steam treatment system, but their use of selective medium (PALCAM) to enumerate survivors could lead to an overestimation of the lethal effect.

### 4.4.3 Hurdle processes

Further improvements may be achieved with technological solutions; process/product adaptations and combinations with growth inhibiting agents may constitute a useful approach when designing a surface decontamination process. A device for steam surface pasteurisation, described by Sommers *et al.* (2008), was very efficient in removing *Listeria innocua* from frankfurters with added lactate and diacetate. This approach is probably also applicable to hot smoked seafood products, to ensure elimination of pathogens prior to the final packaging step. Other, new combinations of agents/processes that eliminate organisms from the surface and agents that either reduce the thermal resistance of target organisms, or inhibit growth of potentially surviving organisms are likely to be developed.

## 4.5 Novel methods of seafood decontamination

There is significant interest from the seafood sector for novel processing methods that could achieve a significant surface decontamination effect. Frequently, such emerging technologies are non-thermal in nature because there is a driver to minimise undesirable quality changes brought about by thermal methods. Technologies that are potentially relevant to the sector are briefly described in the following section along with, where possible, examples of pertinent research studies. Some of the technologies (such as pulsed light) are near market, but many novel preservation technologies remain very much in the research arena; in the case of some of the technologies discussed (such

© Woodhead Publishing Limited, 2012

as cold plasma), there is little or no published data specifically related to seafood products.

### 4.5.1 Ultrasound

Power ultrasound (typically in the range of 20–100 kHz) has numerous non-preservation applications such as cutting, cleaning, emulsifying, de-gassing and foam breaking (Mason, 1998; Patist and Bates, 2008). It is not thought to be currently in use for surface pasteurisation in food production but there is evidence at laboratory scale demonstrating general anti-microbial effects (Miles *et al.*, 1995; Ordoñez *et al.*, 1984; Patist and Bates, 2008; Piyasena *et al.*, 2003; Raso *et al.*, 1989a, 1998b). For the inactivation of microorganisms, it is usually suggested that ultrasound is combined with moderate heating and, in some cases, slightly elevated pressure; processes known respectively as thermosonication and manothermosonication (Bermudez-Aguirre and Barbosa-Canovas, 2008; Hurst *et al.*, 1995; Sala *et al.*, 1995). Greater understanding is required relating to process scale-up, energy requirements, inactivation mechanisms and food quality effects before power ultrasound can realistically be considered for preservation applications. It could, however, prove very useful as a tool for augmenting conventional heat processes and is already being used for this purpose in some food processing facilities (personal communication with companies using the technology). Similarly, it also has potential as a means of enhancing surface cleaning of seafoods. In a US patent, Robinson (1996) claimed that ultrasound could be used as a means of accelerating depuration and enhancing the detachment and subsequent precipitation of microorganisms from shellfish. In another patent application (WO2007100261 A1), Egebjerg (2007) claimed that ultrasound could improve the uniformity of application of flavourings on to seafood surfaces and enhance the penetration of these flavourings into the tissue of the product. Given that ultrasound is known to enhance mass transfer, it is also conceivable that ultrasound could be used to enhance the transfer of antimicrobial compounds into the surface of seafood products.

### 4.5.2 Pulsed UV light

Pulsed light is a surface preservation method in which a material is subjected to very short pulses, typically of the order of 300 micro-seconds, of broad-spectrum white light (180–1100 nm) produced by capacitor discharge to xenon-filled lamps. Flashes are controlled and focused by reflectors that are bespoke designed for each application. Each flash produces a large amount of energy, power output being around 1 kW per square centimetre of treated surface (Claranor, 2011). There are at least 10 commercial units in operation in Europe but none is thought to be in use for surface decontamination of foods. Many of the current applications in Europe are for the surface decontamination of packaging such as drinks bottle-caps.

© Woodhead Publishing Limited, 2012

On smooth surfaces such as stainless steel, vegetative bacterial cells can be reduced from $10^6$ CFU/$cm^2$ to the limits of detection after only 1 pulse of 300 microsecond duration (Shaw *et al.*, 2009). The level of microbial inactivation that can be achieved on food surfaces is, however, more limited, most likely due to surface roughness effects as is seen with continuous UV processing. Shaw *et al.* (2009) observed a 1–3 log reduction in total viable counts on food surfaces such as fish, meat, cheese and fruit, and similar results have been reported by other groups (Dunn *et al.*, 1995; Ozer and Demirci, 2006b).

Ozer and Demirci (2006b) observed a 1 log reduction of *Listeria monocytogenes* after pulsed light treatment of salmon fillets. Shaw (2008) explored the effects of pulsed light treatment for the inactivation of naturally occurring *Pseudomonads*, total viable counts and surface inoculated *Listeria innocua* on a range of seafood products including whelks, cod and smoked salmon. Minimal antimicrobial effects were observed in all but smoked salmon. However, for this latter product, *Pseudomonads* were reduced by up to 3.6 log and *Listeria innocua* was reduced by 1.8 log cycles. These levels of inactivation are commercially significant and work is ongoing with industry to further explore this application.

### 4.5.3 Cold plasmas

Cold plasmas have been proposed as a potential method for the non-thermal surface decontamination of foods. The plasma state can be considered as the fourth state of matter, created when sufficient energy is applied to a gas. Plasmas occur naturally (e.g. lightning and the *aurora borealis*) and man-made plasmas have found widespread use, everything from candles to the production of plasma TVs and plasma coating of jet turbine blades (Baylis, 2011). The use of cold plasmas for the inactivation of surface microbial contamination is still very much in its infancy, but data are available to suggest it has promise. For example, Wang *et al.* (2003) demonstrated a 4 log reduction of *Listeria monocytogenes* on stainless steel when treated with a plasma. Both Creaney *et al.* (2007) and Niemira and Sites (2008) indicate that very few cold plasma *food* studies exist. The latter demonstrated that a reduction of 3.7 log CFU/ml was achievable for *Salmonella* Stanley when spot inoculated onto the surface of Golden Delicious apples and then treated with a cold plasma. Ragni *et al.* (2010) achieved up to a 4.5 log reduction of *Salmonella* Enteritidis and *Salmonella* Typhimurium on the surface of eggshells using cold plasma in a 65% relative humidity environment. These studies suggest that meaningful levels of inactivation could be achieved using cold plasma for the surface decontamination of seafoods, but the technology remains experimental, the mechanisms of inactivation are not yet fully understood (Clark, 2011), and data on microbial inactivation on seafoods are not currently available.

© Woodhead Publishing Limited, 2012

## 4.6 Regulatory issues surrounding decontamination of seafood

Regulatory aspects are dynamic, with respect to both time and region. They will not be covered here, with the exception of the EU, where recent EC regulation (EC, 2004b) provides the legal basis for the use of substances other than potable water to remove surface contamination from foods of animal origin intended for human consumption. This has led to an increasing interest in the development and commercial application of decontamination procedures (Cortesi *et al.*, 2009). As a general rule, however, no substance other than potable water (EC, 1998) is to be used to remove surface contamination from products of animal origin, which includes seafood. The European Commission considers that the use of hot potable water to remove surface contamination on carcases may be allowed provided that the hot water application does not result in any irreversible discolouration of the meat; and that all the relevant requirements of EC Regulations (EC, 2004a, 2004b, 2004c) for the production of fresh meat are respected.

With respect to technologies used, there will also be legal obstacles to overcome because foods preserved using some of these technologies may be considered novel as defined by the EU (EC, 1997) and will therefore require pre-market approval by EU member states. A novel food application can be a complex, time-consuming and expensive process. In the first instance, it is suggested that processors interested in applying ultrasound to their products should discuss the matter at an early stage with the relevant competent authority in the EU member state in which the product will be first marketed. This will give the user an opportunity to get an initial opinion and will help to ensure that any subsequent novel food application will proceed as smoothly as possible.

Decontamination can constitute a useful element in further reducing the number of pathogens, provided that an integrated control strategy is applied throughout the entire food chain, including hygienic measures applied at primary production, during transport and in the slaughter and processing plant. When steam or hot potable water decontamination is used, it shall be considered as a potential critical control point in the HACCP-based approach.

## 4.7 Conclusions and future trends

The surfaces of most fresh and newly processed seafoods are the main habitat of bacteria. Hence these food surfaces are often the target for efficient bacterial reduction with minimal thermal load.

Regulatory aspects may limit the use of chemical agents, but it seems that combinations of added growth inhibiting agents combined with surface heat treatments (Sommers *et al.*, 2008, 2009) may be an interesting approach.

© Woodhead Publishing Limited, 2012

Also, the combination of steam and ultrasound has been tested on broilers (Boysen and Rosenquist, 2009) and pork (Morild *et al.*, 2011), with significant log reductions in the range between 1.1 and 3.6 logs. The technology is patented and commercially available from SonoSteam (2011), who claim that the ultrasound causes an enhanced steam efficacy. Although this enhanced efficacy has not been documented in available literature, it may well be that this technology is also suitable for seafood.

The emerging possibilities given by visible (VIS)/near-infrared (NIR) spectroscopy for detecting bacterial growth (Al Qadiri *et al.*, 2008a, 2008b) and fish freshness (Sivertsen *et al.*, 2011; Sone *et al.*, 2011) may provide opportunities in combination with surface treatment, mainly through ensuring that raw material input does not exceed specifications. VIS/NIR spectroscopy can also be used to determine end-point processing temperature (Uddin *et al.*, 2002a, 2002b), which can also be a valuable asset for verification of processing conditions.

As non-thermal approaches offer the best potential for surface inactivation of microorganisms with minimal changes to the sensory properties of the product, use of emerging technologies seems to be of particular interest. Power ultrasound is used industrially for non-preservation applications such as continuous flow extraction, viscosity modification and the enhancement of heat and mass transfer. It is also well established as a cleaning technique. Ultrasonic baths are in widespread use for applications such as cleaning of pipework, and continuous flow ultrasonic systems are in use for continuous cleaning of materials such as wires and cables before further processing (Hielscher – Ultrasound Technology, 2011). Hence the technology is well developed for industrial processing and bath type technologies could be applied to seafood products. This approach is certainly worthy of investigation, although the potential exists that ultrasound could damage delicate fish products as a result of cavitation.

Pulsed light is now commercially available for high throughput production. For example, the French pulsed light equipment supplier Claranor (2011) cites commercial plants that are decontaminating packaging at a rate of 40,000 units per hour (Riedel, 2007). Current pulsed light systems are, however, not designed specifically for food decontamination; food processing systems would need to be bespoke designed for each individual application with the associated costs. There are also technical and legislative barriers that still need to be overcome in order to use pulsed light for seafood decontamination. For example, ideally, pulsed light would be applied to a packaged seafood product in order to minimise the risk of recontamination post-process. This requires packaging that is UV-C transmissive (when most packaging is designed to have moderate to good UV barrier properties) and there are practical difficulties in ensuring that all surfaces of the product are exposed to the pulsed light treatment – particularly if there is a requirement to use printed packaging for cosmetic/marketing reasons. The novel foods regulatory issues outlined above are also pertinent to pulsed light processing.

© Woodhead Publishing Limited, 2012

As cold plasma technology improves, we can expect to see greater interest in this fledgling technique. Commercial cold plasma systems for food decontamination are not yet thought to be available. The Dutch company OMVE (2011) supplies laboratory-scale plasma demonstrator units, designed to enable users to carry out microbial inactivation kinetic studies using well-defined and controlled plasma conditions. The unit produces a plasma jet from a nitrogen source at temperature as low as 40°C. It may be several years before industrial scale cold plasma systems are available for surface decontamination of seafoods. In the meantime research work is likely to focus on fully elucidating the mechanisms for microbial inactivation and demonstrating efficacy on a range of substrates. Much of the work to date has focused on non-food surfaces, but there is a need to generate more data on inactivation effects on food surfaces. In parallel with this work on inactivation mechanisms, there is a need to generate data on the effects of cold plasmas on food quality.

## 4.8 Sources of further information and advice

The Institute of Food Technology (IFT) has a very active 'Non-Thermal Processing Division' which produces regular newsletters for its members that highlight developments in new preservation technologies. The Division also serves as a powerful networking tool for technologists interested in this field. A large EU project, NovelQ, completed in February 2011, was dedicated to identifying and removing barriers to the commercial adoption of emerging preservation technologies, i.e. cold plasma. The project website (NovelQ 2011) contains useful information on the project and its outputs and also Campden BRI (2011) offers wide-ranging information on various emerging technologies.

## 4.9 References

ACMSF 1995, *Report on vero cytotoxin-producing Escherichia coli*, HMSO, London.

ACUFF G, IZAT AL and FINNE G (1984), 'Microbial-flora of pond-reared tilapia (*Tilapia aurea*) held on ice', *Journal of Food Protection*, 47, 778–780.

AL QADIRI HM, AL ALAMI NI, LIN M, AL HOLY M, CAVINATO AG and RASCO BA (2008a), 'Studying of the bacterial growth phases using Fourier transform infrared spectroscopy and multivariate analysis', *Journal of Rapid Methods and Automation in Microbiology*, 16, 73–89.

AL QADIRI HM, LIN M, AL HOLY MA, CAVINATO AG and RASCO BA (2008b), 'Detection of sublethal thermal injury in *Salmonella enterica* serotype Typhimurium and *Listeria monocytogenes* using Fourier transform infrared (FT-IR) spectroscopy (4000 to 600 $cm^{-1}$', *Journal of Food Science*, 73, M54–M61.

ANDREWS LS, KEYS AM, MARTIN RL, GRODNER R and PARK DL (2002), 'Chlorine dioxide wash of shrimp and crawfish: an alternative to aqueous chlorine', *Food Microbiology*, 19, 261–267.

© Woodhead Publishing Limited, 2012

ANNOUS BA and KOZEMPEL MF (2006), 'Surface pasteurization with hot water and steam', in Sapers GM, Gorny JR and Yousef AE, *Microbiology of fruits and vegetables*, Boca Raton, FL, Taylor and Francis, 479–496.

BAL'A MF and MARSHALL DL (1998), 'Organic acid dipping of catfish fillets: effect on color, microbial load, and *Listeria monocytogenes*', *Journal of Food Protection*, 61, 1470–1474.

BAL'A MF, PODOLAK R and MARSHALL DL (1999), 'Steam treatment reduces the skin microflora population on deheaded and eviscerated whole catfish', *Food Microbiology*, 16, 495–501.

BAL'A MF, PODOLAK R and MARSHALL DL (2000), 'Microbial and color quality of fillets obtained from steam-pasteurized deheaded and eviscerated whole catfish', *Food Microbiology*, 17, 625–631.

BAYLIS D (2011), Cold atmospheric plasmas: probing inactivation mechanisms. Presentation at Campden BRI 'Manufacturing Technologies' Panel, 2 February 2011.

BEAUFORT A, RUDELLE S, GNANOU-BESSE N, TOQUIN MT, KEROUANTON A, BERGIS H, SALVAT G and CORNU M (2007), 'Prevalence and growth of *Listeria monocytogenes* in naturally contaminated cold-smoked salmon', *Letters in Applied Microbiology*, 44, 406–411.

BELTRAN D, SELMA MV, TUDELA JA and GIL MI (2005), 'Effect of different sanitizers on microbial and sensory quality of fresh-cut potato strips stored under modified atmosphere or vacuum packaging', *Postharvest Biology and Technology*, 37, 37–46.

BERMUDEZ-AGUIRRE D and BARBOSA-CANOVAS GV (2008), 'Study of butter fat content in milk on the inactivation of *Listeria innocua* ATCC 51742 by thermo-sonication', *Innovative Food Science and Emerging Technologies*, 9, 176–185.

BEUCHAT LR (1998), *Surface decontamination of fruits and vegetables eaten raw: a review*. WHO/FSF/FOS/98.2.

BOYSEN L and ROSENQUIST H (2009), 'Reduction of thermotolerant *Campylobacter* species on broiler carcasses following physical decontamination at slaughter', *Journal of Food Protection*, 72, 497–502.

BREMER PJ, MONK I, OSBORNE CM, HILLS S and BUTLER R (2002), 'Development of a steam treatment to eliminate *Listeria monocytogenes* from king salmon (*Oncorhynchus tshawytscha*)', *Journal of Food Science*, 67, 2282–2287.

BRIANDET R, MEYLHEUC T, MAHER C and BELLON-FONTAINE MN (1999), '*Listeria monocytogenes* Scott A: cell surface charge, hydrophobicity, and electron donor and acceptor characteristics under different environmental growth conditions', *Applied and Environmental Microbiology*, 65, 5328–5333.

BUCHANAN RL, DAMERT WG, WHITING RC and VAN SCHOTHORST M (1997), 'Use of epidemiologic and food survey data to estimate a purposefully conservative dose-response relationship for *Listeria monocytogenes* levels and incidence of listeriosis', *Journal of Food Protection*, 60, 918–922.

BURT S (2004), 'Essential oils: their antibacterial properties and potential applications in foods – a review', *International Journal of Food Microbiology*, 94, 223–253.

BUTLER JL, STEWART JC, VANDERZANT C, CARPENTER ZL and SMITH GC (1979), 'Attachment of microorganisms to pork skin and surfaces of beef and lamb carcasses', *Journal of Food Protection*, 42, 401–406.

CAMPDEN BRI (2011), New and innovative technologies. Available from: http://www.campden.co.uk/new-technologies.htm (accessed 16 May 2011).

CHAVANT P, MARTINIE B, MEYLHEUC T, BELLON-FONTAINE MN and HEBRAUD M (2002), '*Listeria monocytogenes* LO28: surface physicochemical properties and ability to form biofilms at different temperatures and growth phases', *Applied and Environmental Microbiology*, 68, 728–737.

CHAWLA A, BELL JW and JANES ME (2007), 'Optimization of ozonated water treatment of wild-caught and mechanically peeled shrimp meat', *Journal of Aquatic Food Product Technology*, 16, 41–56.

© Woodhead Publishing Limited, 2012

CHEN BY, PYLA R, KIM TJ, SILVA JL and JUNG YS (2010), 'Incidence and persistence of *Listeria monocytogenes* in the catfish processing environment and fresh fillets', *Journal of Food Protection*, 73, 1641–1650.

CHEN HC and CHAI TJ (1982), 'Microflora of drainage from ice in fishing vessel fish-holds', *Applied and Environmental Microbiology*, 43, 1360–1365.

CLARANOR (2011), Pulsed light by Claranor – how does it work? Available from: http://claranor.com/pulsed-light-sterilization (accessed 11 May 2011).

CLARK JP (2011), 'Novel nonthermal technologies', *Food Technology*, 65, 80–81, 84.

CODEX ALIMENTARIUS (2002), *Report on the thirty-fourth session of the Codex Committee on food hygiene*, FAO/WHO, Rome, Italy, ALINORM 03/13.

CORPE WA (1970), 'Attachment of marine bacteria to solid surfaces', in Manly RS, *Adhesion in biological systems*, New York, Academic Press.

CORTESI ML, PANEBIANCO A, GIUFFRIDA A and ANASTASIO A (2009), 'Innovations in seafood preservation and storage', *Veterinary Research Communications*, 33, S15–S23.

COUEY HM (1989), 'Heat-treatment for control of postharvest diseases and insect pests of fruits', *HortScience*, 24, 198–202.

CRAPO C, HIMELBLOOM B, VITT S and PEDERSEN L (2004), 'Ozone efficacy as a bactericide in seafood processing', *Journal of Aquatic Food Product Technology*, 13, 111–123.

CREANEY C, GREEN A, SHAW H, and LEADLEY CE (2007), *New Technologies Bulletin 34*. Published by Campden BRI, Station Road, Chipping Campden GL55 6LD, UK.

DA SILVA M, GIBBS PA and KIRBY RM (1998), 'Sensorial and microbial effects of gaseous ozone on fresh scad', *Journal of Applied Microbiology*, 84, 802–810.

DAHLBACK B, HERMANSSON M, KJELLEBERG S and NORKRANS B (1981), 'The hydrophobicity of bacteria – an important factor in their initial adhesion at the air-water interface', *Archives of Microbiology*, 128, 267–270.

DATAMONITOR (2006), Changing cooking behaviors and attitudes: beyond convenience. Available from: http://www.marketresearch.com/product/display.asp?productid=1424626&xs=r (accessed 15 April 2011).

DICKSON JS and KOOHMARAIE M (1989), 'Cell-surface charge characteristics and their relationship to bacterial attachment to meat surfaces', *Applied and Environmental Microbiology*, 55, 832–836.

DUNN J, OTT T and CLARK W (1995), 'Pulsed-light treatment of food and packaging', *Food Technology*, 49, 95–98.

EC (1997), Regulation (EC) No 258/97 of the European Parliament and of the Council of 27 January 1997 concerning novel foods and novel food ingredients.

EC (1998), Council Directive 98/83/EC of 3 November 1998 on the quality of water intended for human consumption.

EC (2004a), Regulation (EC) No 852/2004 of the European Parliament and of the Council of 29 April 2004 on the hygiene of foodstuffs.

EC (2004b), Regulation (EC) No 853/2004 of the European Parliament and of the Council of 29 April 2004 laying down specific hygiene rules for food of animal origin.

EC (2004c), Regulation (EC) No 854/2004 of the European Parliament and of the Council of 29 April 2004 laying down specific rules for the organisation of official controls on products of animal origin intended for human consumption.

EFSA (2007), 'The Community Summary Report on Trends and Sources of Zoonoses, Zoonotic Agents, Antimicrobial Resistance and Foodborne Outbreaks in the European Union in 2006', *The EFSA Journal*, 130. Available from: www.efsa.europe.eu/en/efsajournal/pub/130r.htm (accessed 31 January 2012).

EGEBJERG J (2007), *A method of processing a raw product in the form of fish fillet and shellfish*. International patent WO 2007/100261 A1.

ELLINGER RH (1972), 'Phosphates in food processing', in Furia T, *Handbook of food additives*, Boca Raton, FL, CRC Press, 617–780.

© Woodhead Publishing Limited, 2012

ERICSSON H, EKLOW A, DANIELSSON-THAM ML, LONCAREVIC S, MENTZING LO, PERSSON I, UNNERSTAD H and THAM W (1997), 'An outbreak of listeriosis suspected to have been caused by rainbow trout', *Journal of Clinical Microbiology* 35, 2904–2907.
FAO (2003), Review of the state of world aquaculture, Rome, Available from: ftp://ftp.fao.org/docrep/fao/005/y4490e/y4490e00.pdf (accessed 16 May 2011).
FARBER JM and IDZIAK ES (1984), 'Attachment of psychrotrophic meat spoilage bacteria to muscle surfaces', *Journal of Food Protection*, 47, 92–95.
FELDHUSEN F (2000), 'The role of seafood in bacterial foodborne diseases', *Microbes and Infection*, 2, 1651–1660.
FINNEGAN M, LINLEY E, DENYER SP, MCDONNELL G, SIMONS C and MAILLARD JY (2010), 'Mode of action of hydrogen peroxide and other oxidizing agents: differences between liquid and gas forms', *Journal of Antimicrobial Chemotherapy*, 65, 2108–2115.
FLETCHER M and FLOODGATE GD (1973), 'Electron-microscopic demonstration of an acidic polysaccharide involved in adhesion of a marine bacterium to solid surfaces', *Journal of General Microbiology*, 74, 325–334.
FLETCHER M and LOEB GI (1979), 'Influence of substratum characteristics on the attachment of a marine pseudomonad to solid surfaces', *Journal of General Microbiology*, 37, 85–89.
FOONG SCC and DICKSON JS (2004), 'Attachment of *Listeria monocytogenes* on ready-to-eat meats', *Journal of Food Protection*, 67, 456–462.
FRANGOS L, PYRGOTOU N, GIATRAKOU V, NTZIMANI A and SAVVAIDIS IN (2010), 'Combined effects of salting, oregano oil and vacuum-packaging on the shelf-life of refrigerated trout fillets', *Food Microbiology*, 27, 115–121.
GELMAN A, SACHS O, KHANIN Y, DRABKIN V and GLATMAN L (2005), 'Effect of ozone pretreatment on fish storage life at low temperatures', *Journal of Food Protection*, 68, 778–784.
GILBERT P, EVANS DJ, EVANS E, DUGUID IG and BROWN MRW (1991), 'Surface characteristics and adhesion of *Escherichia coli* and *Staphylococcus epidermidis*', *Journal of Applied Bacteriology*, 71, 72–77.
GOMEZ-LOPEZ VM, RAJKOVIC A, RAGAERT P, SMIGIC N and DEVLIEGHERE F (2009), 'Chlorine dioxide for minimally processed produce preservation: a review', *Trends in Food Science and Technology*, 20, 17–26.
GRAM L and HUSS HH (1996), 'Microbiological spoilage of fish and fish products', *International Journal of Food Microbiology*, 33, 121–137.
GRAM L and HUSS HH (2000), 'Fresh and processed fish and shellfish', in Lund BM, Baird-Parker T, and Gould G, *The microbial safety and quality of foods*, Gaithensberg, MD, Aspen Publishers, 472–506.
GRAM L, WEDELL-NEERGAARD C and HUSS HH (1990), 'The bacteriology of fresh and spoiling Lake Victorian Nile perch (*Lates niloticus*)', *International Journal Food Microbiology*, 10, 303–316.
GREIG JD and RAVEL A (2009), 'Analysis of foodborne outbreak data reported internationally for source attribution', *International Journal Food Microbiology*, 130, 77–87.
HARPAZ S, GLATMAN L, DRABKIN V and GELMAN A (2003), 'Effects of herbal essential oils used to extend the shelf life of freshwater-reared Asian sea bass fish (*Lates calcarifer*)', *Journal of Food Protection*, 66, 410–417.
HIBBELN JR, DAVIS JM, STEER C, EMMETT P, ROGERS I, WILLIAMS C and GOLDING J (2007), 'Maternal seafood consumption in pregnancy and neurodevelopmental outcomes in childhood (ALSPAC study): an observational cohort study', *The Lancet*, 369, 578–585.
HIELSCHER – ULTRASOUND TECHNOLOGY (2011), Ultrasonic Cleaning of Wire, Rod and Strip, Available from: http://www.hielscher.com/ultrasonics/wire_01.htm (accessed 11 May 2011).
HSIAO CP and SIEBERT KJ (1999), 'Modeling the inhibitory effects of organic acids on bacteria', *International Journal of Food Microbiology*, 47, 189–201.

© Woodhead Publishing Limited, 2012

HUANG L (2005), 'Dynamic measurement and mathematical modeling of the temperature history on hot dog surfaces during vacuum-steam-vacuum process', *Journal of Food Engineering*, 71, 109–118.

HUANG YR, HSIEH HS, LIN SY, LIN SJ, HUNG YC and HWANG DF (2006), 'Application of electrolyzed oxidizing water on the reduction of bacterial contamination for seafood', *Food Control*, 17, 987–993.

HUANG YR, HUNG YC, HSU SY, HUANG YW and HWANG DF (2008), 'Application of electrolyzed water in the food industry', *Food Control*, 19, 329–345.

HURST RM, BETTS GD and EARNSHAW RG (1995), *The antimicrobial effect of power ultrasound*, Chipping Campden, Campden & Chorleywood Food Research Association.

ICMSF (2002), 'Micro-organisms in foods,' *Microbiological testing in food safety management*, New York, Kluwer Academic, 285–309.

JAMES DG (1986), *The prospects of fish for the undernourished food and nutrition*, FAO, Rome, Italy, 12.

KANELLOS TS and BURRIEL AR (2005), 'The *in vitro* bactericidal effects of the food decontaminants lactic acid and trisodium phoshate', *Food Microbiology*, 22, 591–594.

KENNEDY J and WALL P (2007), 'Food safety challenges', in Storrs M, Devoluy M-C, and Cruveiller P, *Safety handbook: microbiological challenges*, France, Bio-Mérieux Education, 8–19.

KHADRE MA, YOUSEF AE and KIM J-G (2001), 'Microbiological aspects of ozone applications in food: a review', *Journal of Food Science*, 66, 1242–1252.

KIM CR, HEARNSBERGER JO and EUN JB (1995), 'Gram-negative bacteria in refrigerated catfish fillets treated with lactic culture and lactic acid', *Journal of Food Protection*, 58, 639–643.

KIM JM, HUANG TS, MARSHALL MR and WEI CI (1999a), 'Chlorine dioxide treatment of seafoods to reduce bacterial loads', *Journal of Food Science*, 64, 1089–1093.

KIM JM, MARSHALL MR, WEN XD, OTWELL WS and CHENG IW (1999b), 'Determination of chlorate and chlorite and mutagenicity of seafood treated with aqueous chlorine dioxide', *Journal of Agricultural and Food Chemistry*, 47, 3586–3591.

KOSAK PH and TOLEDO RT (1981), 'Effects of microbiological decontamination on the storage stability of fresh fish', *Journal of Food Science*, 46, 1012–1014.

KOZEMPEL MF, MARSHALL DL, RADEWONUK ER, SCULLEN OJ, GOLDBERG N and BAL'a MFA (2001), 'A rapid surface intervention process to kill *Listeria innocua* on catfish using cycles of vacuum and steam', *Journal of Food Science*, 66, 1012–1016.

KOZEMPEL M, GOLDBERG N and CRAIG JC (2003), 'The vacuum/steam/vacuum process', *Food Technology*, 57, 30–33.

LANNELONGUE M, HANNA MO, FINNE G, NICKELSON R and VANDERZANT C (1982), 'Storage characteristics of finfish fillets (*Archosargus probatocephalus*) packaged in modified gas atmospheres containing carbon dioxide', *Journal of Food Protection*, 45, 440–444.

LAPPI VR, THIMOTHE J, NIGHTINGALE KK, GALL K, SCOTT VN and WIEDMANN M (2004), 'Longitudinal studies on *Listeria* in smoked fish plants: impact of intervention strategies on contamination patterns', *Journal of Food Protection*, 67, 2500–2514.

LILLARD HS (1986), 'Distribution of attached *Salmonella typhimurium* cells between poultry skin and a surface-film following water immersion', *Journal of Food Protection*, 49, 449–454.

LISTON J (1980), 'Microbiology in fisheries science', in Connell JJ, *Advances in fish science and technology*, Farnham, Fishing News Books, 138–157.

LORETZ M, STEPHAN R and ZWEIFEL C (2010), 'Antimicrobial activity of decontamination treatments for poultry carcasses: a literature survey', *Food Control*, 21, 791–804.

MAHAPATRA AK, MUTHUKUMARAPPAN K and JULSON JL (2005), 'Applications of ozone, bacteriocins and irradiation in food processing: a review', *Critical Reviews in Food Science and Nutrition*, 45, 447–461.

© Woodhead Publishing Limited, 2012

MAHMOUD BSM, YAMAZAKI K, MIYASHITA K, IL-SHIK S, DONG-SUK C and SUZUKI T (2004), 'Decontamination effect of electrolysed NaCl solutions on carp', *Letters in Applied Microbiology*, 39, 169–173.

MALAKAR PK, BARKER GC, ZWIETERING MH and VAN'T RIET K (2003), 'Relevance of microbial interactions to predictive microbiology', *International Journal of Food Microbiology*, 84, 263–272.

MANOUSARIDIS G, NERANTZAKI A, PALEOLOGOS EK, TSIOTSIAS A, SAVVAIDIS IN and KONTOMINAS MG (2005), 'Effect of ozone on microbial, chemical and sensory attributes of shucked mussels', *Food Microbiology*, 22, 1–9.

MASNIYOM P, BENJAKUL S and VISESSANGUAN W (2005), 'Combination effect of phosphate and modified atmosphere on quality and shelf-life extension of refrigerated seabass slices', *LWT – Food Science and Technology*, 38, 745–756.

MASON TJ (1998), 'Power ultrasound in food processing', in Povey MJW and Mason TJ, *The way forward in ultrasound in food processing*, London, Blackie Academic and Professional, 105–126.

MATSUDA T, YANO T, MARUYAMA A and KUMAGAI H (1994), 'Antimicrobial activities of organic acids determined by minimum inhibitory concentrations at different pH ranged from 4.0 to 7.0', *Journal of the Japanese Society for Food Science and Technology – Nippon Shokuhin Kagaku Kogaku Kaishi*, 41, 687–702.

MEADOWS PS (1971), 'Attachment of bacteria to solid surfaces', *Archiv für Mikrobiologie*, 75, 374–381.

MEJLHOLM O and DALGAARD P (2002), 'Antimicrobial effect of essential oils on the seafood spoilage micro-organism *Photobacterium phosphoreum* in liquid media and fish products', *Letters in Applied Microbiology*, 34, 27–31.

MIETTINEN MK, SIITONEN A, HEISKANEN P, HAAJANEN H, BJORKROTH KJ and KORKEALA HJ (1999), 'Molecular epidemiology of an outbreak of febrile gastroenteritis caused by *Listeria monocytogenes* in cold-smoked rainbow trout', *Journal of Clinical Microbiology* 37, 2358–2360.

MILES CA, MORLEY MJ, HUDSON WR and MACKEY BM (1995), 'Principles of separating microorganisms from suspensions using ultrasound', *Journal of Applied Bacteriology*, 78, 47–54.

MORGAN AI, GOLDBERG N, RADEWONUK ER and SCULLEN OJ (1996), 'Surface pasteurization of raw poultry meat by steam', *Lebensmittel-Wissenschaft und Technologie*, 29, 447–451.

MORILD RK, CHRISTIANSEN P, SØRENSEN AH, NONBOE U and AABO S (2011), 'Inactivation of pathogens on pork by steam-ultrasound treatment', *Journal of Food Protection*, 74, 769–775.

MURRAY CK and SHEWAN JM (1979), 'The microbial spoilage of fish with special reference to the role of psychrotrophs', in Russel AD and Fuller R, *Cold tolerant microbes in spoilage and the environment*, London, Academic Press, 117–136.

NIEMIRA BA and SITES J (2008), 'Cold plasma inactivates *Salmonella* Stanley and *Escherichia coli* O157:H7 inoculated on Golden Delicious apples', *Journal of Food Protection*, 71, 1357–1365.

NILSSON L and GRAM L (2002), 'Improving the control of pathogens in fish products', in Bremner HA, *Safety and quality issues in fish processing*, Cambridge, Woodhead Publishing, 54–84.

NOTERMANS S and KAMPLELMACHER EH (1974), 'Attachment of some bacterial strains to skin of broiler chickens', *British Poultry Science*, 15, 573–585.

NOVELQ (2011), Cold plasma processing. Available from: http://www.novelq.org/download/Short_Description_Plasma.pdf (accessed 16 May 2011).

OMVE (2011), Cold plasma demonstrator. Available from: http://www.omve.com/technologies/?omvetechnology_categorie=CP+Cold+Plasma+Technologies (accessed 11 May 2011).

© Woodhead Publishing Limited, 2012

ORDOÑEZ JA, SANZ B, HERNANDEZ PE and LOPEZ-LORENZO P (1984), 'Effect of combined ultrasonic and heat treatment (thermoultrasonication) on the survival of a strain of *Staphylococcus aureus*', *Journal of Dairy Research*, 54, 61–67.

OZER NP and DEMIRCI A (2006a), 'Electrolyzed oxidizing water treatment for decontamination of raw salmon inoculated with *Escherichia coli* O157:H7 and *Listeria monocytogenes* Scott A and response surface modeling', *Journal of Food Engineering*, 72, 234–241.

OZER NP and DEMIRCI A (2006b), 'Inactivation of *Escherichia coli* O157:H7 and *Listeria monocytogenes* inoculated on raw salmon fillets by pulsed UV-light treatment', *International Journal of Food Science and Technology*, 41, 354–360.

PALMER J, FLINT S and BROOKS J (2007), 'Bacterial cell attachment, the beginning of a biofilm', *Journal of Industrial Microbiology and Biotechnology*, 34, 577–588.

PAOLI GC, BHUNIA AK and BAYLES DO (2005), '*Listeria monocytogenes*', in Fratamico PM, Bhunia AK and Smith JL, *Foodborne pathogens*, Wymondham, Caister Academic Press, 295–324.

PATIST A and BATES D (2008), 'Ultrasonic innovations in the food industry: from the laboratory to commercial production', *Innovative Food Science and Emerging Technologies*, 9, 147–154.

PHUVASATE S and SU YC (2010), 'Effects of electrolyzed oxidizing water and ice treatments on reducing histamine-producing bacteria on fish skin and food contact surface', *Food Control*, 21, 286–291.

PIYASENA P, MOHAREB E and MCKELLAR RC (2003), 'Inactivation of microbes using ultrasound: a review', *International Journal of Food Microbiology*, 87, 207–216.

PYRGOTOU N, GIATRAKOU V, NTZIMANI A and SAVVAIDIS IN (2010), 'Quality assessment of salted, modified atmosphere packaged rainbow trout under treatment with oregano essential oil', *Journal of Food Science*, 75, M406–M411.

RAGNI L, BERDARDINELLI A, VANNINI L, MONTANARI C, SIRRI F, GUERZONI E and GUANIERI A (2010), 'Non-thermal atmospheric gas plasma device for surface decontamination of shell eggs', *Journal of Food Engineering*, 100, 125–132.

RASO J, PAGAN R, CONDON S and SALA FJ (1998a), 'Influence of temperature and pressure on the lethality of ultrasound', *Applied and Environmental Microbiology*, 64, 465–471.

RASO J, PALOP A, PAGAN R and CONDON S (1998b), 'Inactivation of *Bacillus subtilis* spores by combining ultrasonic waves under pressure and mild heat treatment', *Journal of Applied Microbiology*, 85, 849–854.

REILLY PJA (1998), 'Emerging food safety issues and the seafood sector', in 26th Session of the Asia Fisheries Commission, 24–30 September 1998, Beijing, China.

RIEDEL C (2007), Pulsed light treatment for food and packaging. Presentation at NovelQ seminar, 9–10 July 2007, held at Campden BRI, Chipping Campden, Gloucestershire.

ROBINSON WL (1996), *Method for reducing contamination of shellfish*, US patent 5482726, US Harvest Technologies Corporation.

ROLLER S and BOARD RG (2003), 'Naturally occuring antimicrobial systems', in Russell NJ and Gould GW, *Food Preservatives*, Berlin, Springer, 262–283.

RUXTON CHS, REED SC, SIMPSON MJA and MILLINGTON KJ (2004), 'The health benefits of omega-3 polyunsaturated fatty acids: a review of the evidence', *Journal of Human Nutrition and Dietetics*, 17, 449–459.

SALA FJ, BURGOS J, CONDON S, LOPEZ P and RASO J (1995), 'Effect of heat and ultrasound on microorganisms and enzymes', in Gould GW, *New methods of food preservation*, London, Blackie Academic & Professional, 176–205.

SHAW H (2008), Report on pulsed light processing of seafood. Available from: www.sin.seafish.org (accessed 6 May 2011).

SHAW H, LEADLEY CE and GREEN A (2009), *Pulsed light for surface decontamination*, Campden BRI, Chipping Campden, R&D report number 281.

© Woodhead Publishing Limited, 2012

SHEWAN JM (1962), 'The bacteriology of fresh and spoiling fish and some related chemical changes', in Hawthorn J and Leitch JM, *Recent advances in food science*, London, Butterworth, 167–193.

SHEWAN JM (1977), 'The bacteriology of fresh and spoiling fish and the biochemical changes induced by bacterial action', in *Proceedings of the Conference on Handling, Processing and Marketing of Tropical Fish*, London, Tropical Products Institute.

SHIRAZINEJAD A, ISMAIL N and BHAT R (2010), 'Lactic acid as a potential decontaminant of selected foodborne pathogenic bacteria in shrimp (*Penaeus merguiensis* de Man)', *Foodborne Pathogens and Disease*, 7, 1531–1536.

SIVERTSEN AH, KIMIYA T and HEIA K (2011), 'Automatic freshness assessment of cod (*Gadus morhua*) fillets by VIS/NIR spectroscopy', *Journal of Food Engineering*, 103, 317–323.

SOMMERS CH, GEVEKE DJ and FAN X (2008), 'Inactivation of *Listeria innocua* on frankfurters that contain potassium lactate and sodium diacetate by flash pasteurization', *Journal of Food Science*, 73, M72–M74.

SOMMERS CH, COOKE PH, FAN X and SITES JE (2009), 'Ultraviolet light (254 nm) inactivation of *Listeria monocytogenes* on frankfurters that contain potassium lactate and sodium diacetate', *Journal of Food Science*, 74, M114–M119.

SONE I, OLSEN RL, DAHL R and HEIA K (2011), 'Visible/near-infrared spectroscopy detects autolytic changes during storage of Atlantic Salmon (*Salmo salar* L.)', *Journal of Food Science*, 76, S203–S209.

SONOSTEAM (2011), The SonoSteam® Technology. Available from: http://www.simflex.dk/cms/site.aspx?p=8561 (accessed 16 May 2011).

SPANGGAARD B, JORGENSEN F, GRAM L and HUSS HH (1993), 'Antibiotic-resistance in bacteria isolated from 3 fresh-water fish farms and an unpolluted stream in Denmark', *Aquaculture*, 115, 195–207.

THAM W, ERICSSON H, LONCAREVIC S, UNNERSTAD H and DANIELSSON-THAM ML (2000), 'Lessons from an outbreak of listeriosis related to vacuum-packed gravad and cold-smoked fish', *International Journal of Food Microbiology*, 62, 173–175.

UDDIN M, ISHIZAKI S, ISHIDA M and TANAKA M (2002a), 'Assessing end-point temperature of heated fish and shellfish meats', *Fisheries Science*, 68, 768–775.

UDDIN M, ISHIZAKI S, OKAZAKI E and TANAKA M (2002b), 'Near-infrared reflectance spectroscopy for determining end-point temperature of heated fish and shellfish meats', *Journal of the Science of Food and Agriculture*, 82, 286–292.

VAZ PP, CAPELL C and KIRBY R (1994), 'Low-level heat-treatment to extend shelf-life of fresh fish', *International Journal of Food Science and Technology*, 29, 405–413.

WANG Y, SOMERS EB, MANOLACHE S, DENES FS and WONG ACL (2003), 'Cold plasma synthesis of poly(ethylene glycol)-like layers on stainless-steel surfaces to reduce attachment and biofilm formation by *Listeria monocytogenes*', *Journal of Food Science*, 68, 2772–2779.

WARD DR and BAJ NJ (1988), 'Factors affecting microbiological quality of seafoods', *Food Technology*, 42, 85–89.

WILLIAMS SK, RODRICK GE and WEST RL (1995), 'Sodium lactate affects shelf life and consumer acceptance of fresh catfish (*Ictalurus nebulosus*, marmoratus) fillets under simulated retail conditions', *Journal of Food Science*, 60, 636–639.

© Woodhead Publishing Limited, 2012

# 5

# Microbial decontamination of nuts and spices

**G. G. Atungulu, University of California Davis, USA and Z. Pan, USDA-ARS, USA**

**Abstract**: The social and economic impacts of outbreaks of foodborne illness and food recalls have become important food safety concerns. These outbreaks are usually due to the consumption of microbiologically contaminated foods, including nuts and spices. Regulatory and public health agencies, industry and research organizations have undertaken initiatives targeting the microbial safety of these food commodities. In order to ensure that these commodities are safe, the responsibility for implementing scientifically validated risk-reduction practices must be shared among all individuals involved in the farm-to-fork continuum. This chapter focuses on the aspects related to microorganisms of concern, routes of microbial contamination, risks of foodborne illness, and available decontamination technologies for nuts and spices. Emerging decontamination technologies, strategic issues for achieving microbiological safety, and research and development priorities related to nuts and spices are also discussed.

**Key words**: nuts and spices, food safety, decontamination, contamination pathways, microorganisms of concern, pathogens.

## 5.1 Introduction

Nuts and spices form a significant part of the human diet and increasing attention is being paid to the microbiological safety of these commodities. Technically, nuts are defined as a one-seeded, hard, fruit developed from many carpels which unite to form a compound ovary. However, the term nut is also applied indiscriminately to seeds, fruits, and tubers with hard coverings, as well as to single oily or starchy kernels. The Food and Drug Administration Compliance Policy Guideline (FDA-CPG Sec. 525.750) recognizes the term spice to refer to aromatic vegetable substances in whole, broken, or ground

© Woodhead Publishing Limited, 2012

form, whose significant function in food is seasoning rather than nutrition and from which no portion of any volatile oil or other flavoring principle has been removed.

In recent decades, the global consumption of edible nuts, kernels and spices has demonstrated an upward trend. In the US, consumption of tree nuts increased by 45% between the mid-1990s and mid-2000s and the trend continues to grow. The increasing consumer tendency towards healthy snacks such as ready-shelled nuts has almost certainly contributed to the upswing in per capita nut consumption (CBI, 2008). The impact of promotional programs advertising the nutritional value of nuts (e.g. the beneficial levels of vitamin E and omega fatty acids) should also be taken into account. At the same time, the increasing popularity of highly spiced cuisines, as well as consumer demand for flavorful foods which are low in sodium and fat, has sparked an increase in the use of spices. Between 1970 and 2005, the United States overall per capita consumption of spices doubled, increasing from 1.6 to 3.3 lbs per year (Kaefer and Milner, 2008). The demand for spices has propelled many countries to have an economic interest in the international spice trade (FAOSTAT, 2005).

Although nuts and spices have continued to be popular food commodities, possible contamination with certain pathogenic microorganisms is a major safety concern for both the industry and consumers. The US Food and Drug Administration (FDA) has continued to note an increased number of recalls of nut and spice products due to microbial contamination (CDC, 2004, 2009; McKee, 1995). In March 2011, there was a recall of hazelnuts in Canada as a result of an *Escherichia coli* outbreak in the US (http://www.fda.gov/NewsEvents/Newsroom/PressAnnouncements/ucm245902.htm and http://active.inspection.gc.ca/eng/corp/recarapp_dbe.asp). Severe outbreaks and subsequent recalls have raised awareness of the increased potential of nuts and spices to act as vehicles for foodborne illness (CDC, 2004, 2009; Isaacs *et al.*, 2005; Kirk *et al.*, 2004; Ledet-Muller *et al.*, 2007). Concern for maintaining quality of exported and imported nuts and spices has prompted an upsurge in monitoring and research related to these food commodities.

Good agricultural practices (GAP) and hygienic practices in the production of tree nuts have been made available as tools to contain outbreaks of nut-associated foodborne illnesses, as embodied in the International Code of Hygienic Practice (CAC, 1972). Effective decontamination technologies are available for processed and/or packed nuts and products. In recent years, not only regulatory and public health agencies but also industry and research organizations have undertaken initiatives targeting microbial safety of nuts and spices. New regulations based on the concept of Hazard Analysis and Critical Control Points (HACCP) have been developed and established. The US Department of Agriculture (USDA) and the FDA have promulgated rules and guidelines concerning the handling, growth and subsequent processing of these food commodities in an effort to contain foodborne illness. At the same time, scientific research into better control of the pathogens and

© Woodhead Publishing Limited, 2012

microbes in question has been emphasized. As a result, a variety of effective decontamination methods have been put in place for nuts and spices. These have attained increased importance and evaluation due to both the frequent use of spices in ready-to-eat foods and the fact that nuts often receive minimal processing at the plant and may receive no further processing before the end use. Identification of pathogenic microorganisms in these food commodities and understanding their survival and growth in the entire farm-to-fork continuum, and providing effective inactivation means are of extreme interest to the food industry and regulatory agencies concerned with consumer protection.

## 5.2 Microorganisms of concern in nuts and spices and related outbreaks

### 5.2.1 Bacteria

Until recently, *Salmonella* and *Escherichia coli* O157:H7 stand out as the most problematic pathogenic microorganisms to the nut and spice industry. Other less harmful bacteria such as *Pseudomonas*, *Xanthomonas*, and *Clostridium* spp. as well as *Penicillium*, *Fusarium*, and *Eurotium* spp. are also occasionally found in contaminated nuts.

*Salmonella enterica* is known widely for causing outbreaks of salmonellosis (Sobel *et al.*, 2001). A variety of *Salmonella* serotypes including Enteritidis, Typhimurium, Anatum, Virchow, Hadar, Paratyphi B, Saphra, Javiana and Agona have been associated with outbreaks of foodborne illnesses (Peters *et al.*, 2003). Typhimurium and Enteritidis serotypes have been implicated as the most likely cause of outbreaks related to nut and nut products.

The first recognized outbreak of salmonellosis associated with tree nuts was identified in 2001 (Isaacs *et al.*, 2005). Consumption of raw almonds contaminated with *Salmonella enterica* serotype Enteritidis (SE) was linked to the outbreak of SE infections, mostly in Canada. In 2004, a second *Salmonella* outbreak transpired in Oregon that was also linked to raw almonds, wherein the *Salmonella* strain was very similar to that identified in 2001. Since 1996, three outbreaks involving store-brand peanut butter and peanut butter products contaminated with *Salmonella enterica* serovars have been recorded (Killalea *et al.*, 1996; Shohat *et al.*, 1996; Ng *et al.*, 1996; Scheil *et al.*, 1998). Contamination of roasted peanut in store-brand peanut butters with *S.* Mbandaka resulted in an outbreak of salmonellosis in South Australia in 1996 (Scheil *et al.*, 1998). In 2006 and 2007 water leakage and dormant *S.* Tennessee located in a peanut processing plant in Sylvester, GA, led to contamination of processed peanut butter, which made 628 individuals ill (CDC, 2007). The largest recall in the US (2008 to 2009) involved an *S.* Typhimurium outbreak which implicated peanut butter products. An FDA investigation is still ongoing regarding the presence of this pathogen in

© Woodhead Publishing Limited, 2012

foods containing the peanut-based ingredients. This contamination may be responsible for illnesses in more than 700 people with more than 150 hospitalizations and nine recorded deaths (CDC, 2009; FDA, 2009) (http://www.fda.gov/NewsEvents/Newsroom/PressAnnouncements/ucm245902.htm; http://active.inspection.gc.ca/eng/corp/recarapp_dbe.asp). In April 2009, a far-reaching recall of pistachios resulted from the recovery of multiple strains of *Salmonella* from pistachios in different forms (raw, roasted in shell, roasted shelled). A number of human illnesses with matching strains were reported in the time frame of the outbreak (http://outbreakdatabase.com/details/setton-pistachio-of-terra-bella-inc.-2009/?vehicle=nuts). A rare strain of *Salmonella* Enteritidis was announced in 2004 in contaminated almonds, which resulted in an outbreak with illnesses traced back to as early as 2002 (http://www.salmonellalitigation.com/salmonella_caseupdates/view/paramount_farms_salmonella_outbreak_litigation/). Ultimately illnesses linked to the almonds were found throughout the US and Canada. Environmental sampling conducted as part of this effort to backtrack did not reveal the outbreak strain of *Salmonella*, but did reveal several other *Salmonella* serotypes. As a result of these outbreaks, there has been considerable concern on the part of the nut industry and regulators. In 2006, new proposed rules were published by the USDA outlining a mandatory program to reduce the potential of *Salmonella* to contaminate almonds (USDA, 2006). The regulation requires that handlers subject their almonds to a process that achieves a minimum of 4 log reduction of *Salmonella* prior to shipment. These rules became effective in March 2007, and mandatory compliance began in September 2007.

Enterohaemorrhagic *Escherichia coli* (EHEC) is particularly associated with hemorrhagic colitis, hemolytic uremic syndrome (HUS), and thrombotic thrombocytopenic purpura (TTP). Several serotypes of EHEC are known. The most common and problematic one to the nut industry in the United States, Canada, Great Britain, and parts of Europe, is *E. coli* O157:H7. In April 2011, Canadian health officials in Quebec announced 13 illnesses, including one death, from *E. coli* O157:H7 associated with consumption of shelled walnuts. The illnesses ranged across three provinces, nine in Quebec, two in Ontario, and two in New Brunswick (http://www.marlerblog.com/case-news/e-coli-tainted-walnuts-from-us-sicken-13-with-1-death/). Eight *E. coli* O157:H7 infections in three US states between late 2010 and early 2011 were linked to hazelnuts from California. Illnesses were reported in Michigan, Minnesota, and Wisconsin (http://www.marlerblog.com/case-news/defranco-hazelnut-e-coli-o157h7-outbreak-hits-eight-in-three-states/).

A survey of recent food poisoning outbreaks involving spices shows that a number of vehicles were involved (Vij *et al.*, 2006). Between 1970 and 2003, the FDA monitored 21 recalls involving 12 spice types contaminated with bacterial pathogens. In all but one instance, the recalled spices contained *Salmonella* spp. *S. oranienburg*. It has been implicated in contaminated black peppercorns, which resulted in an outbreak in Norway (of 126 cases), while both black and white peppercorns contaminated with *S. weltevreden* caused

© Woodhead Publishing Limited, 2012

outbreaks of salmonellosis in Newfoundland in Canada (Robillard, 1983). An outbreak of cyclosporiasis involving over 300 cases occurred in 1997 in which fresh basil was implicated as the cause of the infection (CDC, 1997). An outbreak resulting from *Shigella sonnei* contamination of uncooked parsley occurred in 1998 (Campbell *et al.*, 2001). In 1999, an outbreak of *S.* Thompson was observed on cilantro (Campbell *et al.*, 2001). Typically, *S.* Thompson is not usually found in food poisoning incidences. Other bacteria commonly found to contaminate spices include *Clostridium perfringens* and *Bacillus cereus*.

Typically, spices are produced in different regions. Depending on origin, microflora type and load on specific spices may vary. Bacterial microflora and their loads on popular spices grown in various regions have been studied by many researchers. Antai (1988) studied alligator pepper, red pepper, black pepper, thyme, and curry powder, which were collected from the main markets in Nigeria. Total microbial counts ranged from $1.8 \times 10^4$ to $1.1 \times 10^8$ CFU/g. Most of the spices were reported to have high loads of *B. cereus*. Significant numbers of *B. polymyxa*, *B. subtilis* and *B. coagulans* were also detected. Baxter and Holzapfel (1982) evaluated the microbial load of 36 spices in South Africa. Source, manufacturing process, age, and type of condiment were found to affect microbial contamination. Contamination varied from several hundred to several million per gram with the highest levels of contamination ($>10^6$ CFU/g) found in black pepper, coriander, pimento, paprika, mace, and white pepper. The presence of *B. cereus* was confirmed in cardamom, marjoram, onion powder, paprika, white pepper, and pimento. Other microorganisms, including *E. coli* isolated from rosemary, *Salmonella* from paprika, fecal streptococci isolated from white and black peppers, pimento, and paprika, were also detected.

The microbial profiles of cumin seeds and chilli powder were investigated by Bhat *et al.* (1987) in Bombay, India. Samples were evaluated for aerobic plate count, *E. coli*, *B. cereus*, *Staphylococcus aureus*, anaerobic sporeformers *Salmonella*, *Shigella*, *Vibrio*, and enterococci. Aerobic plate counts ranging from $10^6$ to $10^8$ CFU/g were detected for chilli powder and $10^4$ to $10^8$ CFU/g for cumin seeds. Although no *Salmonella*, *Shigella*, or *VibHo* were detected in any of the samples, *E. coli* and *B. cereus* were detected in chilli powder. The microbiology of 150 samples of 54 different spices, spice mixtures, and herbs from three retail suppliers was studied by de Boer *et al.* (1985) in the Netherlands. Curry, paprika, white and black peppers, ginger, basil, mixed spices, and curcuma were reported to have total aerobic counts of $>10^7$ CFU/g, while thyme, dill, coriander, basil, and chervil had total aerobic counts of $>10^6$ CFU/g. Total spore counts were $>10^7$ CFU/g in paprika, white and black peppers, ginger, curcuma, and caraway. Enterobacteriaceae were detected in oregano, tarragon, parsley, basil, and chervil at $>10^5$ CFU/g. *Bacillus cereus* population was found to be $>10^4$ CFU/g for white pepper, ginger, and mixed spices. The microflora of black and red pepper was studied by Christensen *et al.* (1967). A composite sample prepared from six of the

© Woodhead Publishing Limited, 2012

black pepper samples was cultured for bacterial identification. *E. coli*, *E. freundii*, *Serratia* sp., *Klebsiella* sp., *Bacillus* sp., *Staphylococcus* sp., and *Streptococcus* sp. were identified from the composite sample (Christensen *et al.*, 1967).

### 5.2.2 Fungi

Fungi are predominant spice and nut contaminants and are regarded as commensal residents on the plants and can survive during drying and storage processes. Molds of the genus *Aspergillus* frequently decay kernels of nuts such as pistachios (Mojitahedi *et al.*, 1979), almonds (Phillips *et al.*, 1976; Purcell *et al.*, 1980), chestnuts (Wells *et al.*, 1975), and pecans (Wells and Payne, 1976). Many *Aspergillus* species infect and decay nuts before harvest (Mojitahedi *et al.*, 1979). The presence of *Aspergillus* on the nuts can be a serious problem, because they produce toxins harmful to humans and animals (Pier and Richard, 1992). The most important of these toxins is aflatoxin, produced by *A. flavus* and *A. Parasiticus* during their growth on nuts. Over 18 aflatoxins belonging to the mycotoxin group exist, out of which aflatoxins $B_1$ and $G_1$ are the two most common toxins (Wilson and Hayes, 1973). Aflatoxin $B_1$ is the most potent hepatocarcinogen known. Aflatoxins (AFTs) have been found in pistachios nuts (Sommer *et al.*, 1986), almonds (Fuller *et al.*, 1977; Schade *et al.*, 1975) and walnuts (Fuller *et al.*, 1977). Another mycotoxin of concern is ochratoxin, produced by *A. ochraceus* K. Wilh and closely related to *Aspergillus* species (Pier and Richard, 1992).

The Food and Drug Administration (1996) set a maximum guidance level limit of 20 ng/g (20 ppb) for tree nuts (shells included) intended for human consumption within the US, and the European Community has recently imposed even more restrictive tolerance levels of 2 ng/g for aflatoxin $B_1$ and 4 ng/g total AFTs (CEC, 1998). Codex Alimentarius Commission in its 31st session agreed on maximum tolerable levels (MTL) of 10 and 15 ng/g for total aflatoxin in almonds, hazelnuts and pistachios 'ready-to-eat' and for further processing, respectively (FAO and WHO, 2008).

Like nuts, fungi-related mycotoxins relevant to spice contamination include aflatoxin, sterigmatocystin, citrinin, zearalenone, and T-2 toxin (Saxena *et al.*, 1988). Among these mycotoxins, aflatoxins (AFTs) pose a significant threat to human health. Results from compliance programs in the US in 1986 reported that 19% of spices imported from 19 countries contained measurable levels of AFTs (Wood, 1989). Flannigan and Llewellyn (1986) and Llewellyn *et al.* (1988) provide exhaustive triennial reviews touching on fungal and mycotoxin contamination of spices for the period 1986–1992. These reviews, in addition to other previous reports, denote numerous genera of fungi and the presence of mycotoxins on spices (Seenappa and Kempton, 1980; Misra, 1987; Tabata *et al.*, 1987; Wood, 1989).

Other studies have identified and mapped fungal and mycotoxin prevalence in popular spices from different regions. El-Rahman (1987) examined selected

© Woodhead Publishing Limited, 2012

spices for the presence of *A. flavus* and aflatoxin $B_1$ ($AFTB_1$). White pepper, red pepper, black pepper, capsicum, and cumin were evaluated. There was great variation in the mold counts which ranged from $2 \times 10^2$ to $3.4 \times 10^6$ spores/g. Black pepper and white pepper were shown to have the highest levels of spore contamination. Black pepper and capsicum are of particular interest because of their wide use in the US and elsewhere. Ath-Har *et al.* (1988) examined the mycoflora of the following Indian spices: *Piper nigrum* (black pepper), *Coriandrum sativum* (coriander), *Capsicum frutescens* (red pepper, paprika), *Cuminum cyminum* (cumin), *Foeniculum vulgare* (fennel), *Trigonella foenum-graecum* (fenugreek) and *Brassica nigra* (black mustard). The most frequent fungi encountered in these spices were *A. flavus*, *A. niger*, *A. nidulans*, *A. sydowii*, *A. ochraceus*, *Penicillium* and *Rhizopus* spp. The authors reported that 16 of the isolates were capable of producing $AFTB_1$ and that the distribution of contamination followed the sequence: pepper = mustard > coriander > cumin > chilli > fenugreek > fennel. Misra (1987) isolated AFTs from coriander seeds and reported $AFTB_1$, aflatoxin $B_2$ ($AFTB_2$), and aflatoxin $G_1$ ($AFTG_1$) at levels of 75, 21, and 16 mg/g, respectively. The reported level for $AFTB_1$ alone was nearly four times the FDA maximum allowable concentration for total AFTs. Abdel-Hafez *et al.* (1989) determined the thermophilic and thermotolerant fungi in caraway, cumin, fennel, anise, and coriander seeds from the Egyptian crops of the spices in 1986. The main fungal contaminant was reported to be *A. fumigatus*, which was present in all but one anise and one coriander seed sample. *Emericellan idulans* and *Rhizomucor pusillus* were also identified from the samples. Fully thermophilic fungi isolated from the seeds included *Malbranchea pulchella*, *Humicola grisea* subsp, *thermoidea*, *Myceliophtora thermophila*, and *Talaromyces dupontii*.

### 5.2.3 Others

In addition to bacteria and fungi, nuts and spices are also susceptible to contamination by protozoa and viruses (Herwaldt *et al.*, 1994). Viruses transmitted by foods or water fall into three groups including, hepatovirus, enterovirus, and norovirus. Of these, the hepatovirus and norovirus appear to be of greatest concern with nuts (Susanne, 2009).

## 5.3 Contamination pathways and persistence of microorganisms in nuts and spices

### 5.3.1 Pathways of microbial contamination of nuts and spices

Ordinarily, foodborne pathogens are not considered to be part of the normal epiphytic populations on food commodities that provide an environment hostile to their growth and survival (Brandl, 2006). In fact, none of the usual human

© Woodhead Publishing Limited, 2012

pathogens causing foodborne illnesses such as EHEC, *Salmonella*, *Shigella* species, *Cryptosporidium*, or *L. monocytogenes* is considered endogenous microflora of nuts, but all may occur as contaminants. In the case of spices, microbes commensally residing on plants may survive drying and storage and contaminate spices. Primary sources of pathogenic bacteria including *Salmonella* and *E. coli* are intestinal tracts of animals (e.g., cattle, goats, sheep, and deer) and humans.

Depending on the pre- and post-harvest processes, nuts and spices could have different critical points in the production line at which contamination is the highest (Podolak *et al.*, 2010). Therefore, emphasis on good agricultural management and practices from farm to table are vital features in ensuring safety of these commodities. Water, animal manure, and municipal biosolid wastes should be used with care as they pose a huge potential pathway for contamination. Worker hygiene, sanitation practices as well as adherence to applicable laws and regulations are critical measures for prevention of contamination rather than correction following contamination. In most cases, the conditions and primary pathways for *Salmonella* contamination of nuts and spices include the following:

- field fertilized with untreated manure or sewage as a soil amendment
- fields irrigated with water contaminated with animal waste
- untreated surface water (ponds, rivers) with runoff from livestock operations
- wildlife grazing on or near fields
- leaks in roof on which birds congregate
- rodent and insect activity, especially if facility is near livestock operations
- forklift and transport equipment exposed to mud, water or contaminated soil outdoors brought into sheds and warehouses without prior cleaning and disinfection
- cattle, livestock or wildlife grazing in orchards or farms especially near trees
- fertilizing soil in orchards or farms with untreated fecal waste
- in the case of nuts, shells of many tree nuts such as pecans are porous and can become contaminated by contact with soil
- exposure to contaminated water from rain or processing
- insufficient hygiene, cross-contamination, processing or storage in inadequate locations, contaminated equipment, and contamination by personnel.

For most nuts, the current harvesting operation easily allows for nuts to contact with the orchard soil. For instance, almonds are harvested after kernels have reached maturation and the hulls have begun to split and dry (Reil *et al.*, 1996). The nuts are shaken from the trees and left to dry on the orchard floor for 1–2 weeks. Almonds kernels in hulls and shells are then gathered from the orchard floor by mechanical sweeping, which unintentionally mixes the

© Woodhead Publishing Limited, 2012

almonds with the top layers of soil. Uesugi and Harris (2006) demonstrated that almond hull and shell slurries provide an excellent medium for the growth of *S.* Enteritidis PT30. When almonds drop prematurely before pre-harvest irrigation is terminated, rainfall occurs during harvest or almonds remain in the orchard following harvest, the resulting conditions favor the growth of *S.* Enteritidis PT30.

Maintaining livestock in nut orchards has been associated with the risk of infield contamination by *E. coli*, especially for nuts (Anon., 1970). Literature records indicate that there were about six times as many *E. coli*-contaminated pecan collected from grazed orchards as from ungrazed ones (Marcus and Amling, 1973). In a grazed orchard, cracks induced by water absorption or mechanical harvesting could provide an easy entry point for *E. coli* into nut kernels. If subsequent cleaning processes employed by growers and shellers do not include removal of cracked but otherwise intact nuts, hazards of contaminating shelling equipment which result in cross-contamination of nuts will definitely manifest (Kokal and Thorpe, 1969).

Contamination of nuts and spices with mycotoxins may start before harvesting, and may not be solely from improper shipping, processing, and storage. In the case of nuts, natural splitting of the shell of nuts, for instance of pistachio, in the orchard prior to harvesting exposes the kernel to molds and insects. These nuts, called early splits, have relatively high levels of aflatoxins (Sommer *et al.*, 1986). Moreover, the Navel Orangeworm also commonly infests nuts with ruptured hulls, like early splits, and has been associated with very high levels of aflatoxins (Rice, 1978). In pistachio orchards, besides early splits, there are instances when bird damage also causes hull rapture exposing nut kernel to molds. Some published reports indicated that aflatoxin contamination of pistachio nuts is also correlated with rainfall after ripening and before harvest time (Danesh *et al.* 1979). In such cases, combined interventions which include farming, harvesting, processing, storage and shipping, and possibly decontamination procedures might be the route to curb aflatoxin contamination. In the case of peanuts, toxigenic strains of *A. flavus* present in the soil may invade the nut, which allows AFTs to be produced during the pre- and post-maturity stages in the field, during post-harvest drying, or in storage (Mehan, 1988). In contrast to other nuts with the potential to be contaminated by AFTs, especially peanuts, walnuts are well protected against fungal infection by the hull, shell, and the seed coat tissue surrounding the kernel. Nevertheless, all these barriers can be penetrated by insect attack, providing an entrance for fungal spores.

### 5.3.2 Tracebacks, monitoring and surveillance of foodborne illness

*Regulation and protocols of traceback, monitoring and surveillance in the United States*

In 1997, the US launched a Food Safety Initiative with the goal of improving food safety and reducing the incidence of foodborne illness to the greatest

© Woodhead Publishing Limited, 2012

extent feasible. The initiative, which affects the nut and spice industry, establishes that while individual industries have the primary responsibility for the safety of the food they produce and distribute, federal, state, and local governments' roles are to verify that these industries are carrying out their responsibility and to initiate appropriate regulatory action if necessary. The initiative seeks to improve coordination, communication, and information exchange among federal, state, and local government agencies, and enhance collaboration between the public and private sectors. The initiative further seeks to improve responses to outbreaks of illness caused by contamination from microorganisms through better coordination and communication during traceback investigations.

The FDA is mandated with the responsibility of accomplishing traceback investigation related to foodborne disease outbreaks and performing surveillance and monitoring to ensure food safety. Traceback investigation is a method typically used to determine the source and distribution of the implicated product associated with an outbreak and to identify potential points where contamination could have occurred. Monitoring involves the performance and analysis of routine measurements, aimed at detecting changes in the environment or health status of populations while surveillance is the ongoing systematic collection, collation, analysis, and interpretation of data, followed by the dissemination of information to all those involved so that directed actions may be taken (Wong *et al.*, 2004).

Most of the identified foodborne disease outbreaks in the US including those implicating nuts and spices are reported by consumers who suspect an association between the food they have eaten and an illness they are suffering. These individuals report such information to their local health departments. Linking sporadic cases or clusters of reportable diseases such as salmonellosis helps to identify other foodborne illness outbreaks. These include sporadic case surveillance of cases reported by clinical laboratories and physicians at the state and local level, and through FoodNet and PulseNet at the national level. The Foodborne Diseases Active Surveillance Network (FoodNet) is the principal foodborne disease component of the Centers for Disease Control and Prevention (CDC) Emerging Infections Program (EIP). FoodNet is a collaborative project of the CDC, nine state sites (California, Colorado, Connecticut, Georgia, New York, Maryland, Minnesota, Oregon and Tennessee), the USDA and the FDA. The project consists of active surveillance for foodborne diseases and related epidemiological studies designed to help public health officials better understand the epidemiology of foodborne diseases in the US. PulseNet is a collaborative project between the CDC, FDA, USDA, and state health departments and uses a national computer network to confirm outbreaks of foodborne illness and to link cases/clusters occurring in multiple states. Public health laboratories across the country perform DNA 'fingerprinting' on bacteria that may be foodborne and use the system to exchange findings when outbreaks of foodborne disease

© Woodhead Publishing Limited, 2012

occur. The network permits rapid comparison of these 'fingerprint' patterns through an electronic database at the CDC.

The FDA becomes involved in an outbreak investigation when surveillance identifies disease clusters or outbreaks and an FDA-regulated product is implicated. The FDA's roles in foodborne outbreak investigations include:

- assisting in investigation and coordination in multistate outbreaks
- reviewing epidemiological, laboratory, and environmental data with the CDC and state/local agencies
- providing investigational and laboratory assistance, as needed
- conducting tracebacks of implicated foods and removal from the market
- monitoring recalls
- taking other appropriate regulatory actions
- identifying how the food became contaminated at its source (such as the farm, in the case of produce)
- evaluating data from investigation findings to identify trends and make recommendations to prevent similar problems.

Foodborne disease investigations have three components: epidemiological, laboratory, and environmental. Epidemiological investigations verify a diagnosis through case interviews and laboratory confirmation; identify the range of onset of symptoms, provide case definitions, conduct epidemiology studies, and determine statistical associations between eating various foods and becoming ill. The laboratory component includes analysis of clinical samples, food samples (if leftovers are available), and environmental samples. The environmental component focuses on food preparation methods and the potential for temperature abuse or cross-contamination and the location of preparation. The environmental component also identifies possible modes of contamination at the food's source. Should the environmental investigation determine that the contamination most likely did not occur at the point of food preparation, then a traceback investigation may be initiated.

*The traceback process*

Public health agencies conduct traceback activities to determine the source and distribution of the implicated product associated with the outbreak and to subsequently identify potential points where contamination could have occurred. This action helps prevent additional illnesses by providing a foundation for recalls of contaminated foods remaining in the market place and identifying hazardous practices or violations.

Traceback studies have been used as vital tools by food safety bodies to provide insight into steps necessary to mitigate future outbreaks of foodborne illnesses. For instance, traceback investigation associated with the 2000–2001 outbreak of salmonellosis from the consumption of raw almonds detected *S.* Enteritidis PT30 in drag swabs of almond orchard floors over a large geographic area one year after the harvest of the outbreak almonds (Isaacs

© Woodhead Publishing Limited, 2012

*et al.*, 2005). The bacterial strain was not isolated from animal or bird feces (i.e. from rodents, migratory waterfowl, or crows) collected from the farms and surrounding areas according to California Department of Health Services (CDHS, 2002). The original source of this strain remains unknown. However, from these traceback studies, important guidelines were issued by the Almond Board of California and the FDA on pasteurization procedures of the almond nut.

### 5.3.3 Factors affecting microbial survival and persistence in nuts and spices

The survival of microorganisms in foods is of prime concern from both food safety and economic perspectives. Techniques for processing and the storage stability of foods are influenced by the microflora present. Microbial behavior differs under different conditions of temperature, pH, water activity, and/or environment.

On the farms or orchards where spices and nuts are grown, soil type has been identified as an important variable in pathogen survival and persistence with clay soils supporting greater survival of enteric pathogens than sandy soils (Roper and Marshall, 1978; Natvig *et al.*, 2002). Production practices that disturb the upper soil layers and generate large volumes of dust have the potential to mix the surface and subsurface layers of soil and spread existing microorganisms throughout the environment (Uesugi and Harris, 2006). *S.* Enteritidis PT30 is capable of extended survival, at least 180 days, in typical almond orchard soils even in the absence of growth substrates. Soil moisture level and temperature, planting density, tree size, and irrigation management also may vary and depend on location within the orchard or farm. A better understanding of the influence of these factors on the persistence of *S.* Enteritidis PT30 is important for developing specific good agricultural practices.

A combination of suitable nutrient, moisture, and temperature conditions typically regulates survival of *Salmonella* for months to years. *Salmonella* heat resistance increases with decreased moisture content/water activity. Such adaptation effects may pose health risks as nuts and spices could be subjected to long storage periods before distribution and consumption.

Typically, *E. coli* O157:H7 can survive in soil and water for months and has exceptional tolerance to acidic conditions and thus can persist in acidic environments for exceptionally long periods of time. *E. coli* field contamination poses a huge health risk to consumers because the pathogen could persist through the processing stages of nuts and spices. It has been shown that some *E. coli* on walnut shells remain through the tempering process and subsequently contaminate the nutmeats (Meyer and Vaughn, 1969).

The prevalence and persistence of fungi and subsequent toxin production in spices depend on factors such as drought, humidity, temperature, insect infestation, and handling (McKee, 1995). Spices are often the products

© Woodhead Publishing Limited, 2012

of tropical climates which frequently provide extreme ranges of rainfall, temperature, and humidity. These growing conditions, harvesting and processing methods, storage conditions, and post-harvest treatments should be carefully controlled to prevent potential food spoilage and foodborne illnesses due to contaminated spices.

The wide variation in microbial profiles and survival on spices could also be attributed to their chemical composition, antimicrobial factors, and environmental conditions among other factors.

## 5.4 Decontamination of nuts

Two general decontamination strategies are available for inactivation of pathogenic and non-pathogenic microorganism of most nuts: chemical, heat, radiation, and non-thermal treatments of the commodity in order to kill the microorganism, decompose or otherwise change the toxic compounds to chemically innocuous species, and physical removal of the contaminated portion of the product.

In this section, microbial and pathogen reduction options available for most nuts such as almonds, brazil nuts, cashews, hazelnuts, macadamia nuts, peanuts, pecans, pistachios, pumpkin seeds, sunflower seeds, and walnuts including dry heat, moist heat, oil roasting, sanitizers, gas treatments, high pressure, irradiation, and physical sorting techniques are discussed.

### 5.4.1 Thermal methods

Thermal treatments do not pose significant health risks and, as a result, are appealing to consumers. The effectiveness of the process depends both on the temperature of exposure and the time required at this temperature to accomplish the desired rate of destruction. The extent of the pasteurization treatment required is determined by the heat resistance of the most heat-resistant microorganism or pathogen in the food. For example, for almond processors, the identified pathogen is *S.* Enteritidis PT30. The almond industry has determined that a certain temperature and time period are required to destroy this pathogen (ABC, 2004). A 5-log reduction, or a 100,000 count reduction, is the goal of any treatment of *S.* Enteritidis PT30. Thermal treatments are commonly applied in the form of hot water/liquid, high-temperature forced air, or steam (vapor heat). Popular methods applied in the nut industry are described in the following subsections.

*Oil roasting*

Pasteurization of almonds by roasting in oil has been approved by the FDA. However, almonds pasteurized by this method have the tendency to lose fresh-like properties (color, taste, and texture). It has been shown by Du

© Woodhead Publishing Limited, 2012

*et al.* (2010) that dipping the almonds in hot oil (127°C) for 1 min provides 5-log reduction of *S.* Enteritidis PT30.

*Steam pasteurization*

Steam pasteurization involves heating a product to a specific temperature for a specific period of time. Hence, its most common method is called high temperature short time (HTST) treatment. The FDA has approved steam processing as an acceptable means of pasteurizing almonds. The key to steam pasteurizing almonds is in the condensation power of the steam. During pasteurization, the steam is heated to supersaturation and then allowed to condense on the cool surface of the product. One system combines steam with fluidization, then drying and cooling via a shaking mechanism. The almonds are fed into the steam pasteurizer, where superheated steam is released in a specially designed pressure area. The steam then 'fluidizes' the nuts with the assistance of the shaking mechanism via a specially designed distribution plate. It is with this condensation and consequent hot moisture that the pathogen *S.* Enteritidis PT30 is inactivated and the 5-log reduction occurs. The shaking and fluidization method ensures that all surfaces are exposed and treated equally. The mean free path of the steam molecule at 140°C is 0.4 μm which is about half the diameter of the smallest cavity capable of containing *Salmonella*. Hence steam can quickly reach all organisms on the surface of foodstuffs and therefore incidences of the revival of the bacteria hidden in crevices are eliminated (Bari *et al.*, 2009).

The processes of steam heating for pasteurization and sterilization are widely used and accepted worldwide. In particular, organic almonds are designated by the FDA to be pasteurized by steam, which meets the USDA Organic Program's national standards. Although the FDA-approved steam pasteurization can effectively pasteurize raw almonds, it is energy intensive and purported to increase nut moisture content which reduces nut flavor quality and structural integrity, and may need additional drying steps (Perren, 2008). Chang *et al.* (2010) observed that 25 s exposure of contaminated almond to steam at 143 kPa provided 5-log reduction of S. Enteritidis. The authors also noted that prolonged exposure to steam compromised product quality by increasing nut moisture content, which loosens nut skin thereby forming visible wrinkles.

Using *Enterococcus faecium* as a surrogate for *S.* Enteritidis PT30, Perren (2008) introduced the idea of controlled condensation of steam which was reported to achieve 5-log reductions in 5 min without causing any change in appearance or color of almond nuts.

*Hot water pasteurization*

Bari *et al.* (2009) dipped raw almonds into hot water at 88°C for 20 s followed by infrared (IR) drying for 70 s, which reduced *Salmonella* to undetectable limits without significantly affecting the overall product quality. However, after 24 h, bacteria were detected in five out of eight samples. Hot water may

© Woodhead Publishing Limited, 2012

fail to reach all the contaminated surfaces that are large enough to contain bacteria because of the high surface tension of aqueous fluids (Morgan *et al.*, 1996).

*Dry heat or hot air pasteurization*

Dry heat can be used to pasteurize nuts but higher temperature and longer processing time are needed for the same killing effect as other methods such as steam. The use of hot air at 70°C to pasteurize product contaminated with *S.* Enteritidis to achieving a 6-log reduction has been reported (James *et al.*, 2002). The efficacy of hot air treatment, however, is dependent on relative humidity (RH). It has been documented that condensation and evaporation of moisture on the almond surface alters the thermal resistance of bacteria on the product surface, which significantly accelerates the pasteurization process (Jeong *et al.*, 2009).

### 5.4.2 Chemical methods

Bacterial spores typically have the highest resistance to chemicals followed by fungal spores, then non-enveloped viruses, then fungi, then vegetative bacteria, and lastly enveloped viruses in respective order (Ransom, 2006). The following chemical methods are commonly used to decontaminate nuts in the industry.

*Propylene oxide*

The FDA has approved propylene oxide (PPO) processes as acceptable forms of pasteurization for almonds. However, PPO pasteurized nuts can only be sold in markets in the US, while the European Union, Canada, Mexico, and most other countries restrict the usage of PPO for treating foodstuffs for human consumption. The US Environmental Protection Agency (EPA) requires that exposure time to PPO should not be more than 4 h and the residue on the product should not be more than 300 ppm. In pasteurization of *Salmonella* Typhosa and other Gram negative bacteria, it has been shown that as the RH of the air within the chamber containing the propylene oxide vapor increased, the *D*-values of the bacteria decreased significantly (Himmelfarb *et al.*, 1962). For example, the authors showed that at 1% RH, the *D*-value for *S.* Typhosa was 42.2 min whereas at 65% RH it was 4.2 min.

*Chlorine dioxide*

The pecan industry at present employs chlorine and/or PPO to disinfect inshell pecans, halves, and pieces. Chlorine dioxide ($ClO_2$) has been evaluated as an alternative to PPO and found to reduce *Salmonella* population, but resulted in bleaching of the nut kernels. Wihodo *et al.* (2005) treated the raw almonds with 5–10 mg/L $ClO_2$ gas for 10–30 min at 22°C and 80–90% RH under atmospheric pressure or vacuum conditions (20–80 kPa) and found that the efficacy of $ClO_2$ gas increased with increasing concentrations

© Woodhead Publishing Limited, 2012

and treatment time. Under atmospheric pressure, a treatment with 10 mg/L $ClO_2$ gas for 30 min provided 5.29-log reductions on almonds, but with a bleaching effect on almond surface which is not desirable and attractive to customers. However, under vacuum conditions, application of 10 mg/L $ClO_2$ for 10 min achieved 4.5-log reductions without significantly changing the color of the almonds.

*Organic acid sprays and rinses*

Organic acid sprays and rinses are also used by the food industry to reduce or eliminate surface contamination on nuts (Ravishankar and Juneja, 2003). Pao *et al.* (2006) reported that using acid sprays such as acetic, citric or peroxyacetic acid or acidified sodium chlorite was effective in reducing *Salmonella*. The authors showed that a single spray (20 min holding at 24°C) of 10% acetic acid, 10% citric acid, acidified sodium chlorite, 500 ppm peroxyacetic acid, or a mixture of hydrochloric, phosphoric, and citric acids on shelled almonds provided 0.38–1.42-log CFU/g reductions for *Salmonella enterica* population. The authors also reported that spraying the above-mentioned solutions multiple times and extending the holding times enhanced the *Salmonella* reduction. To meet the 4-log pasteurization requirements, three spray applications of the solutions followed by at least 3 days storage are needed.

*Ammoniation*

The use of ammonia to remove aflatoxins (AFTs) from peanuts, among other food products, has been widely studied. The effectiveness of ammoniation in the destruction of AFTs has been shown to be dependent on pressure, temperature, reaction time, particle size of the commodity, and the type of ammonia (e.g. anhydrous gas or aqueous solution) used for treatment. A review by Park *et al.* (1988) thoroughly discusses aflatoxin decontamination by ammoniation. Depending on the conditions and the foodstuff, foods containing very high aflatoxin levels (>1000 $\mu g\ kg^{-1}$) can be decontaminated by as much as 99% by using ammoniation.

### 5.4.3 Irradiation

Decontamination of food by ionizing radiation is a safe, efficient, environmentally clean, and energy efficient process which involves exposure of food to a carefully controlled amount of energy in the form of high-speed particles or rays. Examples of irradiation treatments employed in the food industry may involve gamma rays, energetic electrons, and X-rays. Irradiation is particularly valuable as an end product decontamination procedure that offers various advantages for disinfestations and pasteurization of nuts. Radiation treatment at doses of 2–7 kGy – depending on condition of irradiation and the food – can effectively eliminate potentially pathogenic non-spore-forming bacteria including both long-time recognized pathogens such as *Salmonella*

© Woodhead Publishing Limited, 2012

and *Staphylococcus aureus* as well as emerging or 'new' pathogens such as *Campylobacter*, *Listeria monocytogenes* or *Escherichia coli* O157:H7 from suspected food products without altering the general characteristics of the food and raising the food temperature.

Most pathogens are sensitive to irradiation. *Salmonella*, *E. coli* O157:H7 and *Shigella* are among the microorganisms that have a low tolerance for irradiation. Sensitivity of microbial cells to irradiation is a function of moisture content or water activity of the host product. Susceptibility to irradiation is greater in high-moisture environments. In low-moisture conditions such as in nuts, water molecules produce fewer radicals and thus have less indirect effects on the DNA of microbial cells. Ionizing radiation applied to raw almonds at 5 kGy yielded a 4-log decrease in *S.* Enteritidis PT30 levels, but resulted in significant loss of kernel quality (Prakash *et al.*, 2010). Sanchez-Bela *et al.* (2008) similarly reported the lowering of sensory scores of raw almonds with respect to sweetness and color when electron beam irradiation was used for decontamination.

Low irradiation doses can be used for disinfestations of arthropod pests according to the current FDA phytosanitary regulations (21 CFR 179.26). A 1 kGy treatment has been proven to have the potential to eliminate most insects, prevent emergence of adults, or induce sterility. Johnson and Vail (1987) have conducted studies on Nonpareli almonds and Hartley walnuts, which were treated with irradiation doses from 0.144 to 0.921 kGy to observe the effect on the Indian meal moth, *Plodia interpunctella*. Irradiation effects were the same for both Nonpareli almonds and Hartley walnuts. Up to 0.269 kGy treatment gave no significant reduction in adult emergence. Irradiation treatments between 0.594 and 0.607 kGy resulted in significantly reduced adult emergence. At 0.822 kGy, more reduction was observed in adult emergence. In a separate study, Johnson and Vail (1987) also irradiated larvae of Indian meal moth and Navel Orangeworm at 0.337 to 0.497 kGy. The authors indicated that although radiation-induced mortality was delayed, damage to product quality was significantly reduced and overall appearance of the product was improved.

Irradiation can affect molds, but typically is not effective in treatment aimed at destroying aflatoxins. Wilson (1990) observed that English walnuts irradiated with doses up to 20 kGy had a greater decrease in mold compared to propylene oxide-treated nuts. No significant difference was observed for yeast counts. Irradiation of walnut with doses of 0.1 to 1 kGy is reported to have given a significant though not complete reduction in number of kernels infested with *Aspergillus* sp., and did not affect other fungi (Changa *et al.*, 1988). Chiou *et al.* (1990) inoculated peanuts with *A. parasiticus* NRRL 2999 and exposed them to radiation of up to 15 kGy. Doses of 5 kGy and above significantly reduced the growth of *A. parasiticus* and naturally occurring mold, although complete elimination was not achieved at any dose level. Aziz and Moussa (2004) reported that irradiation at 4.0 kGy reduced mold growth in groundnut seeds, no growth was observed at 5 kGy, and irradiation

© Woodhead Publishing Limited, 2012

at 6.0 kGy detoxified aflatoxin $B_1$ by 74.3–76.7%. In general, mold growth and toxin formation can be inhibited by irradiation at 3–6 kGy. However, toxin inactivation requires much higher levels of treatment, especially in dry matter. A dose of 50–60 kGy may be required to eliminate the effect of aflatoxin in contaminated peanut meal.

A synergistic effect is achieved when irradiation and mild heat are used in combination to decontaminate nuts. In the case of fungi, heat treatment preceding irradiation usually results in a greater antimicrobial effect of the combination process compared with heating after irradiation. The usefulness of mild heat treatment prior to low-dose irradiation has been demonstrated to achieve inactivation of toxigenic molds on nuts and other food products (Farkas, 1990). With bacterial spores, pre-irradiation followed by heating proved to be synergistic and the combination may be utilized for preservation of foods where heat resistance of bacterial spores is a critical factor (Farkas *et al.*, 1973). In groundnuts, a combination of heat plus irradiation (65°C for 35 min and 0.5 kGy) was found to inactivate toxigenic fungi like *A. flavus* (Padwal-Desai, 1974). The irradiated nuts were shelf-stable for several months when vacuum sealing or packaging under nitrogen was employed. Encouraging results have also been obtained for cashew nuts and raisins by using a combination of treatments.

In general, there is significant recent interest and controversy concerning irradiation (Farkas and Mohacsi-Farkas, 2011). In spite of pioneering past R&D activities in Europe and North America, the utilization of the process is growing faster and increasingly, mainly for sanitary purposes, in fast-developing countries in the South-East Asian region and some Latin American countries. In the EU, food irradiation has been regulated since 1999 by a General Directive, but its implementing directive, the Community List of EU approved irradiated foods, contains only a single class of items: 'dried aromatic herbs, spices and vegetable seasonings'. The key to changing the sluggishness of implementation of the manifold potential uses of food irradiation technology is a better appreciation of its potential role in controlling foodborne diseases and spoilage, as well as the willingness to pay for processing for food safety.

### 5.4.4 Oven and microwave roasting

Pluyer *et al.* (1987) used both oven and microwave roasting to destroy aflatoxins in contaminated peanuts. Oven roasting (30 min at 150°C) and microwave roasting (0.7 kW for 8.5 min) destroyed 48–61% of aflatoxin $B_1$ and 32–40% of aflatoxin $G_1$ in peanuts that initially contained 300–1200 $\mu g\ kg^{-1}$ aflatoxin $B_1$ and 700–1800 $\mu g\ kg^{-1}$ aflatoxin $G_1$. Basaran and Akhan (2010) studied microwave treatment of hazelnuts for the control of aflatoxin producing *A. parasiticus*. In their study, the effects of microwave treatment on hazelnuts artificially contaminated with aflatoxigenic fungi were evaluated qualitatively and quantitatively. A significant 3-log reduction

© Woodhead Publishing Limited, 2012

in *A. parasiticus* contamination was observed after 120 s treatment; similar changes in quality were observed during the storage of microwave treated and non-treated hazelnuts.

### 5.4.5 Physical screening methods

Physical removal of contaminated portions of the foodstuff is highly attractive, because it can meet the following criteria, which are vital for decontamination: no production of any toxic or carcinogenic compounds, retention of food nutritive value and palatability, and it does not significantly alter the physical properties of the foodstuff. However, this method is not always effective, since the toxins can diffuse throughout the material and are not associated exclusively with damaged, discolored or malformed parts. Also the target microorganism may not be destroyed, removed, or inactivated. In addition, the method of physical removal can be labor or equipment intensive and may not always be economically feasible. Nevertheless, some research results have demonstrated the successful removal of toxins by physical methods. Blanching of the shelled nuts to remove the nut testa, followed by both electronic color sorting and hand-picking, can substantially decrease overall aflatoxin contamination (Read, 1989). For example, in one study, electronic sorting and hand-picking resulted in 2.8% of the nuts being removed. The pickings (consisting of discolored or malformed nuts) contained 1528 μg $kg^{-1}$ of AFTs, while the remaining 97.2% of the nuts contained no detectable AFTs.

A technique known as belt screening can significantly decrease the aflatoxin content of unshelled peanuts (Cole and Dorner, 1989). In belt screening, the unshelled nuts are passed over a set of closely spaced belts. Sound, full-sized nuts remain on the belts and are called 'overs'; broken, immature, or undersized nuts pass through the spaces in the belts and are called 'thrus'. For peanuts containing 11.32 μg $kg^{-1}$ aflatoxin, belt screening resulted in aflatoxin levels of 9.63 μg $kg^{-1}$ in the 'overs' and 48.94 μg $kg^{-1}$ in the 'thrus'. In addition to demonstrating that belt screening can decrease aflatoxin levels in the 'overs', the study demonstrated that the AFTs were concentrated in the broken, immature, or otherwise smaller nuts that fell through the belt screen.

Henderson *et al.* (1989) have patented a screening method for reducing the aflatoxin contents of peanuts and other seeds. Nuts containing most of the toxins tend to be damaged and are less dense than the relatively toxin-free nuts. When placed in a flotation medium (water worked well for peanuts), the undamaged nuts sank. The average aflatoxin content of the nuts was 10 μg $kg^{-1}$; 90% sank in water. The 'sinkers', after color sorting, contained 8 μg $kg^{-1}$ aflatoxins. The 10% of the nuts that floated contained an average of 49 μg $kg^{-1}$ AFTs.

© Woodhead Publishing Limited, 2012

### 5.4.6 Emerging methods

*Infrared pasteurization*

Infrared (IR) radiation is energy in the form of an electromagnetic wave, which is more efficient in heat transfer than the conventional convection and conduction forms. The efficient heat transfer can provide a high heating rate and reduce the heating time to reach the required product temperature. IR heating has shown great promise as a potential and effective non-chemical technology for pasteurization of nuts such as almonds, and has the capacity to preserve high product quality.

Pan and Atungulu (2010a, 2010b) reported on the efficacy of IR heating to pasteurize almonds including (1) measures of IR pasteurization effectiveness and product quality under different combinations of IR heating temperature, holding temperature, and time; and (2) optimized processing conditions of IR pasteurization methods for commercial implementation by the almond industry. Based on bacterial reduction and preservation of sensory quality, any of the following three processing conditions is recommended to almond processors for pasteurization achieving the best results: IR heating 120°C, holding at 90°C for 5 min; IR heating 110°C, holding at 90°C for 10 min; and IR heating 100°C, holding at 90°C for 10 min. Any of these treatment conditions provides over 5.5-log reductions of *Pediococcus* (the surrogate for *S.* Enteritidis), which is above the industrially required minimum 4-log bacterial reduction.

Other researchers have evaluated combined IR treatment with other methods to inactivate pathogens from nuts. Bari *et al.* (2009) injured the bacterial cells by dipping almonds in ozonated water or mild or strong acidic electrolyzed water for 10 s followed by holding at room temperature for 3 min and benefited from a catalytic IR heater for 70 s to dry the wetted almonds. The authors found that combination treatments with acidic electrolyzed water followed by IR drying reduced *S.* Enteritidis by 3-log.

IR heating has also been studied and applied to improve the safety and processing efficiency for dry-roasting almonds. Key findings realized in case studies developed at the USDA-ARS Western Region Research Center and the University of California, Davis include:

- the appropriate IR heating conditions to achieve the desired product temperatures with minimum heating/roasting time
- the pasteurization efficacy of IR compared to sequential IR radiation and hot air roasting (SIRHA) and to traditional hot air roasting
- the quality of the almond kernels so produced, and
- recommendations about the technology for scaling up for commercial applications that are clearly detailed by Pan and Atungulu (2010a, 2010b) and Yang *et al.* (2010).

A sample of the significant research findings is reported in Table 5.1, which shows the reductions in *Pediococcus* population size on medium roasted almonds under different conditions. It was observed that SIRHA

© Woodhead Publishing Limited, 2012

**Table 5.1** Reductions in *Pediococcus* (the surrogate for *Salmonella enterica* serovar Enteritidis) population size on medium roasted almonds under different conditions

| Treatment temperature (°C) | 130 | 140 | 150 |
|---|---|---|---|
| Hot air treatment | 3.58a AB* | 4.62a B | 5.39a B |
| Infrared treatment | 2.94a B | 3.21a C | 4.12b B |
| Sequential IR and hot air treatment | 4.1a A | 5.82b A | 6.96c A |

*The same letters in lower case in the same row mean no significant difference at $p \leq 0.05$; the same letters in upper case in the same column mean no significant difference at $p \leq 0.05$.

roasting produces medium roasted almonds with 4.10-, 5.82- and 6.96-log bacterial reductions at roasting temperatures of 130, 140 and 150°C and roasting times of 21, 11 and 5 min, respectively. IR heating alone produced medium roasted almonds with a 4.12-log bacterial reduction at 150°C for 4 min, compared to 13 min with hot air at 150°C. Hot air roasting at 140 and 150°C resulted in 4.62- and 5.39-log bacterial reductions, which required 18 and 13 min of roasting, respectively, and is much longer than the IR or SIRHA roasting.

SIRHA roasting is a substantially faster method for producing pasteurized roasted almonds with tremendous potential to reduce costs associated with the longer roasting times of the current hot air method. The roasting using IR alone is recommended only for pasteurization that targets 4-log bacterial reduction.

*Combined superheated steam and infrared heat pasteurization*

Combined superheated steam (SHS) and IR heat treatments have recently been investigated for inactivation of *Salmonella* on raw almonds (Bari *et al.*, 2010). The combination has various advantages over other heating system, including high heat transfer rate due to condensation and radiation, accelerated drying rate, and an oxygen-free environment. It was observed that condensation was followed by evaporation of moisture on the materials (Konishi *et al.*, 2004; Iyota *et al.*, 2005).

A novel oven system of SHS has been developed at the National Food Research Institute in Japan. The system uses SHS containing micro droplets of hot water for wide application including pasteurization of nuts. Bari *et al.* (2010) studied the SHS treatments followed by IR heat treatment to inactivate *Salmonella* populations on raw almond and found that SHS treatment for 70 s followed by IR treatment for 70 s was able to reduce *Salmonella* population by 5.73-log CFU/g, and no survivors were found in the enrichment medium. The quality parameters, along with the microbiological parameter, suggest that SHS treatment is effective in eliminating the population of pathogens on almonds without significantly affecting their overall quality (Bari *et al.*, 2010).

© Woodhead Publishing Limited, 2012

*Other non-thermal pasteurization methods*

In non-thermal processing, the key objective is to ensure that food temperatures are kept low enough so as to keep at minimum or initiate no degradation and changes in vitamins, nutrients, and flavors during decontamination of foods (Barbosa-Canovas *et al.*, 1998). Examples of non-thermal pasteurization techniques, which have shown great potential at research level for applications involving nut pasteurization, are described below.

High hydrostatic pressure

Efficacy of high hydrostatic pressure treatment to eliminate *S.* Enteritidis PT30 on raw almonds was investigated by Goodridge *et al.* (2006). The authors documented that pressurizing the raw almonds at 413.68 MPa (60,000 psi) at 50°C provided 1-log reduction in 9.78 min, while increasing the pressure level further resulted in an oily almond surface which may protect the bacteria. The authors suggested that soaking the inoculated almonds in sterile water to adjust the water activity and then applying pressure followed by air drying could lead to greater decrease of *S.* Enteritidis population. The rationale was proved by Willford *et al.* (2008), who pressurized SE PT30 at 414 MPa for 6 min followed by air drying at either ambient temperatures or at 115°C for 25 min and achieved 4.03- and 6.70-log reductions, respectively. The efficiency of this technology especially for dry materials on an industrial scale is yet to be elucidated.

Non-thermal plasma

Plasma is one of the newest technologies in food decontamination. It has emerged as a novel, non-thermal, non-contact, and water-free antimicrobial intervention that can eliminate foodborne pathogens such as *E. coli* O157:H7, *Salmonella*, *L. monocytogenes* and other microorganisms of concern on produce and nuts (Niemira, 2010). The antimicrobial efficacy of plasma is related to the specific technology used, the power level used to generate the plasma, the gas mixture used in the plasma emitter and the intensity and length of exposure (Marsilia *et al.*, 2002).

Application of non-thermal plasma (NTP) to inactivate *E. coli* on almonds was proposed by Deng *et al.* (2007). During NTP generation in a chamber consisting of air and almonds, strong oxidizers such as oxygen atoms, ozone, OH-radicals, and N-radicals are produced. Moreover, discharging the electrons directly onto raw almonds might also cause inactivation of pathogens. The authors showed that when the potential differences between the electrodes of NTP setup were 16 and 25 kV, 1.0 and 4.12-log *E. coli* population could be inactivated, respectively, in 30 s. The authors also reported that increasing the application frequency at a constant voltage increased the inactivation rate. Deng *et al.* (2007) reported nearly 5-log reduction of *E. coli* counts on almonds using non-thermal plasma. The reduction was achieved after 30 s treatment at 30 kV and 2000 Hz. The inactivation effects varied with the

© Woodhead Publishing Limited, 2012

treatment time, power voltage and frequency and varieties and grades of almonds.

Although still in its infant stage of development, scientific studies have demonstrated that low pressure cold plasma (LPCP) using air gases and sulfur hexafluoride ($SF_6$) possesses anti-fungal efficacy against *A. parasiticus* on artificially contaminated hazelnuts, peanuts, and pistachio nuts (Basaran *et al.*, 2008). In the case of hazelnuts, *D*-values close to 1.1 min and greater than 4.2 min are achieved with application of $SF_6$ and air gas plasma, respectively. In addition, a reduction of aflatoxin ($AFB_1$, $AFB_2$, $AFG_1$, and $AFG_2$) contamination on nut surfaces by nearly up to half of initial levels has been reported (Basaran *et al.*, 2008). Park *et al.* (2007) reported that the mycotoxins, $AFB_1$, $AFB_2$, $AFG_1$, and $AFG_2$ were completely removed after 5 s of microwave-induced argon plasma at an atmospheric pressure.

### Ozone

Ozonation is a relatively new method for food processing. High reactivity, penetrability, and spontaneous decomposition to a nontoxic product (i.e., $O_2$) make ozone a viable disinfectant for ensuring the microbiological safety of food products.

The effectiveness of ozone for the inactivation of *E. coli* and *Bacillus cereus* in kernels, shelled and ground pistachios has been reported by Akbas and Ozdemir (2006a). In their study, substantial reductions in *E. coli* and *B. cereus* populations in kernels, shelled, and ground pistachios were observed at the end of a 360 min ozonation. Ozone concentration $<1.0$ ppm for ground pistachios and ozone concentration of 1.0 ppm for kernels and shelled pistachios were successfully used to reduce initial counts of *E. coli* and *B. cereus* without affecting any physicochemical properties of pistachios such as pH, free fatty acids and peroxide values, color and fatty acid compositions.

Ozonation has been used to control fungal growth and mycotoxin contamination in nuts (Kim *et al.*, 1999). Maeba *et al.* (1988) studied the reaction of aflatoxins $B_1$, $B_2$, $G_1$, and $G_2$ with ozone and determined that aflatoxins $B_1$ and $G_1$ rapidly degraded, while aflatoxins $B_2$ and $G_2$ reacted more slowly. Akbas and Ozdemir (2006b) demonstrated that ozonation can be used to degrade aflatoxins in pistachio nuts. Aflatoxin degradation was more substantial in pistachio kernels than ground pistachios, possibly owing to the localized flow and poor contact. Aflatoxin B1 was more sensitive to ozonation than the other types of aflatoxins ($B_2$, $G_1$, and $G_2$) inoculated into pistachios. Substantial reductions in aflatoxin $B_1$ and total aflatoxin levels were achieved when pistachio kernels and ground pistachios were ozonated at 9.0 mg $L^{-1}$ ozone concentration for 420 min. Ozone treatments did not change pH, color, moisture content, free fatty acid values, or organoleptic properties of pistachio kernels. However, ozone treatments affected sweetness, flavor, appearance, and overall palatability of ground pistachios.

© Woodhead Publishing Limited, 2012

## 5.5 Decontamination of spices

### 5.5.1 Ethylene oxide

Ethylene oxide has been proven to significantly reduce microbial population of bacteria and molds in spices (Leistritz, 1997; Toofanian and Stegeman, 1988). However, the processed spices are characterized with aroma and color alterations and volatile compounds are lost because of the low pressure that is necessary to remove the sanitizing agent (Farkas and Andrassy, 1988; Vajdi and Pereira, 1973). Ethylene oxide is extremely toxic and is a known carcinogen and mutagen (EPA, 2006; Steenland *et al.*, 2004; Fowles *et al.*, 2001, Scudamore and Heuser, 1971; Tateo and Bononi, 2006). Due to safety and environmental concerns, the use of ethylene oxide for any food fumigation has been banned in the European Union. The use of ethylene oxide for treatment of ground spice is also banned in the United States (Loaharanu, 1994; Fowels *et al.*, 2001).

### 5.5.2 Irradiation

Irradiation, including the application of gamma rays, electron beam, and X-rays, has proven to be a more effective, environmentally clean and energy efficient spice decontamination technique than fumigation (Govindarajan, 1985; Farkas, 1988; Kiss and Farkas, 1988; Cho *et al.*, 1986; Farkas and Andrassy, 1988; Byun *et al.*, 1996; IAEA, 1988). Studies have shown that radiation doses of 3–10 kGy produced equal or better microbicidal effects than fumigation, and reduced microbial cell counts to a satisfactory level of less than $10^4$ colony forming units per gram (Nieto-Sandoval *et al.*, 2000). Bolander *et al.* (1995) irradiated spices in sausage. Mexican sausages containing treated spices had significantly lower yeast, mold and spore counts than control sausages. The effects of gamma irradiation on microbial load, total aflatoxins and phyto constituent content of *Trigonella foenum-graecum* were studied by Prakash *et al.* (2009). Gamma irradiation at a dose of 2.5 kGy resulted in a 2-log reduction of the total aerobic microbial count. A complete sterilization was, however, observed at 10 kGy. The total aflatoxin level decreased gradually with increase in gamma irradiation dose as compared to its unirradiated counterparts. The results suggest that gamma irradiation at a dose of 5.0 kGy was very effective for microbial decontamination because it did not adversely affect the active components of *T. foenum-graecum*. Al-Bachir (2007) irradiated seeds of anise (*Pimpinella anisum*) at doses of 0, 5, 10, 15, and 20 kGy in a $^{60}Co$ package irradiator. The results indicated that a 10 kGy dose of gamma irradiation was effective for microbial decontamination of aniseeds. Jalili *et al.* (2010) evaluated the effect of gamma radiation on reduction of mycotoxins, i.e. ochratoxin A (OTA) and aflatoxins $B_1$, $B_2$, $G_1$, and $G_2$ ($AFB_1$, $AFB_2$, $AFG_1$ and $AFG_2$) in black pepper. At 60 kGy, reductions of 52%, 43%, 24%, 40%, and 36% for OTA, $AFB_1$, $AFB_2$, $AFG_1$, and $AFG_2$, respectively, were attained. Only less than 30% reduction of mycotoxins was achievable at a dose of 30 kGy. Generally, a maximum overall average

© Woodhead Publishing Limited, 2012

absorbed dose of 30 kGy is acceptable for the decontamination of dried aromatic herbs, spices, and vegetable seasonings (Farkas, 2006).

Although ionizing radiation of spices has been officially approved, consumer acceptance of irradiated foods has generally been poor. Additionally, some studies also indicated slight modifications in sensorial and antioxidative properties of irradiated spices (Farag *et al.*, 1995; Nieto-Sandoval *et al.*, 2000; Suhaj *et al.*, 2006; Topuz and Ozdemir, 2003). Moreover, irradiation of spices, which are usually pre-packaged to prevent secondary contamination, may cause the formation of low-molecular-weight volatile or non-volatile radiolysis products emanating from the packaging material (Goulas *et al.*, 2004). It could be potentially harmful and may migrate into the food and impair food flavor (Chytiri *et al.*, 2005; Krzymien *et al.*, 2001).

### 5.5.3 Steam

Steam treatment of spices has gained considerably wide consumer acceptance, especially because ethylene oxide has been phased out and irradiated products still face consumer skepticism regarding their quality. In steam treatments of spices, high temperature steam is usually applied to whole spices before grinding. The decontamination technique is characterized by moisture condensation on the surface of the particles, which must be removed after treatment to prevent unwanted mold growth. Thus, the application of this method is difficult, particularly when applied to ground spices. An effective decontamination is associated with discoloration and reduction of volatile oil contents (Almela *et al.*, 2002; Modlich and Weber, 1993; Schneider, 1993; Rico *et al.*, 2010).

As an alternative approach for quality retention in steam treatments, a modification of vacuum-steam-vacuum (VSV) procedures has been tested for decontamination of natural inoculated whole black pepper. In the modified procedure of VSV, an abrupt evacuation of the treatment chamber takes place, which leads to an intensive re-evaporation of the superficial condensate layer. In this way, the microorganisms on the surface are removed from the product. Each steam temperature, with its corresponding pressure, leads to a specific evacuation velocity after the steam treatment. This evacuation velocity significantly influences the lifting force which has to overcome any adhesion force between the microorganisms and the product surface. Simultaneously, microorganisms are thermally inactivated. The procedure combines mechanical and thermal decontamination effects, while applying only low thermal stress to the product. Using this modified procedure, Lilie *et al.* (2007) decontaminated pepper with steam treatment of 10 s at 125°C or 20 s at 120°C and achieved a 3-log reduction of aerobic mesophile bacteria.

### 5.5.4 Emerging methods

Research has continued to seek other potential techniques that could be used to produce high quality and microbiologically safe spices. Some of

© Woodhead Publishing Limited, 2012

the techniques currently being studied include IR radiation, high hydrostatic pressure treatment, heat sterilization, microwave radiation and ultraviolet radiation techniques among others. Of these techniques, heat sterilization, microwave and ultraviolet radiation techniques have been applied, but with limited success so far. Heat sterilization has been unsuccessful because of the heat sensitivity of the spice aroma and flavor components. On the other hand, microwave treatments have also proved ineffective because of the low moisture content of spices, while ultraviolet radiation lacks the necessary penetration capacity for decontamination application to spices (Weber, 1980; Kiss and Farkas, 1988).

Based on recent studies, IR radiation for spice decontamination seems to hold great promise if the processing conditions can be optimized. Staack *et al.* (2008) investigated IR decontamination of paprika powder of different water activities and inoculated with *B. cereus* spores. Unsatisfactory *B. cereus* spore reduction with IR was noted for paprika powder at low water activity (0.5 and 0.8). At water activity of 0.96, *B. cereus* spore reduction was greater than 6 log CFU/g. Process parameters such as IR power, wavelength, and water activity should be carefully selected to significantly reduce microbial numbers and avoid changes in product quality. Therefore, it can be suggested that microbial safety would be ensured by first conducting IR decontamination of spices while at high water activity, and then drying the products to safe storage moisture.

Recent studies have indicated that high hydrostatic pressures (HHP) ranging from 100 to 1000 MPa can be effective in producing microbiologically safe and shelf-stable spice products (Guerrero-Beltran *et al.*, 2005; Manas and Pagan, 2005). Reports, however, indicated that spice samples with water activities below 0.66 showed no reduction in their microbial count when treated with HHP (Butz *et al.*, 1994). Windyga *et al.* (2008) has reported on microbiological decontamination of coriander and caraway with a 30 min treatment of HHP at 800 and 1000 MPa, which was applied in an elevated temperature range of 60–121°C in helium atmosphere. The authors observed 2-log reduction of aerobic mesophilic bacteria and total elimination of coliform, yeast, and molds. The authors concluded that high pressure treatment under helium atmosphere, and combination of proper high pressure and time improved the microbiological quality of spices. However, the strong dependence of inactivation of microorganisms on water activity is still a significant hindrance to the use of HHP as a sanitation method in spice production.

## 5.6 The limitations of technologies and the challenges to adoption of technologies for decontamination of nuts and spices

The new decontamination technologies are faced with various challenges: to be attractive in terms of consumers' psychological perception, adoption

© Woodhead Publishing Limited, 2012

cost, legislative factors, retention of consumers' preferred product quality and sensory characteristics, environmental sustainability, and availability, as well as level of information on the technology understandable to stakeholders, among other factors. These factors limit full-scale realization of some benefits embedded within some of the decontamination technologies. Due to lack of public knowledge on wholesomeness of treated food, sensory changes frequently constitute or dictate a dose limitation. Proper information about the safety and benefits of the new decontamination technologies for foods could increase the consumers' level of acceptance of products. It is also important that the World Health Organization (WHO) and several respected nonprofit organizations such as the American Medical Association, the American Dietetic Association, the Institute of Food Technologists, and the Council for Agricultural Science and Technology in the US have a positive attitude towards scientifically validated decontamination technologies, while maintaining that the benefits should never be considered as an excuse for poor quality or for poor handling and storage conditions, or as a substitute for good manufacturing and hygienic practices. Greater efforts need to be made to develop standard equipment based on novel technologies and to educate the diverse users of the potential benefits. Economic factors restricting the development and commercialization of the technologies deny new avenues for delivering safe and value-added foods desirable to consumers. If commercial equipment manufacturing is not addressed, some of the technology will remain interesting for research purposes only, but with limited application to the commercial food process.

In general, most technologies do not stand as a panacea, and a hurdle approach involving a combination of decontamination techniques optimized for producing safe and high quality nuts and spices is ideal.

## 5.7 Strategic issues and research and development priorities

The nutritional significance of nuts and spices and their prevalence in the daily diet of many different cultures emphasize the importance of ensuring that these food commodities are safe from pathogenic contamination. The social and economic impacts of outbreaks experienced so far in connection with contaminated nuts, spices or their products are far-reaching. It is imperative that lessons are learned from past outbreaks, and it is vital that attention is paid to the microbiological safety of foods from production to consumption. A farm-to-fork continuum, which comprises primary production, processing, distribution/retail and consumers, must be used to emphasize the importance of food safety practices to all handlers at all stages. Ultimately, the safety of nuts and spices is a shared responsibility, relying on the implementation of scientifically validated risk-reduction practices by all individuals involved in the farm-to-fork continuum.

© Woodhead Publishing Limited, 2012

A number of effective and relatively safe methods have been devised to decrease the risk of contamination of nuts and spices by pathogenic microorganisms. However, even in the ideal case in which the toxin is completely removed, destroyed, or inactivated, none of the reaction products exhibit toxicity and the treatment does not decrease nutritive value, the decontamination procedure invariably adds to the cost of the commodity. Thus, it remains of prime importance that appropriate good agricultural and management practices be followed to prevent contamination whenever and wherever possible. Adherence to post-harvest good manufacturing practice (GMP), food safety laws and industry initiated HACCP strategies at processing plants should lead to the decline of foodborne illnesses.

At commercial and research level, the challenge is still to produce safe and high quality processed nuts, spices, and their products. It is vital to make research effort geared to establishing new technologies and their process requirements, as well as efficiency and equivalency evaluation of these new methods to conventional sterilization and pasteurization. Because certain consumers remain skeptical of some new technologies, such as irradiation, food safety regulatory bodies need to obtain scientific evidence from these treatments that clearly delineates the boundaries of safety.

It is a well understood fact that the general microflora on most nuts and spices is very diverse. Some microorganisms can overcome one or more 'hurdles' or methods of decontamination, but may not be able to jump over all the hurdles used together. A hurdle approach, involving a combination of decontamination techniques optimized for producing safe and high quality nuts and spices, needs to be explored. The approach could provide an opportunity to overcome the limitations associated with individual treatments in terms of nutritional and qualitative degradation. Utilizing a hurdle approach to take advantage of the merits of some popular technologies such as steam, hot water treatment, pulsed electric field, pulsed light, infrared processing, ultraviolet radiation, microwave processing, ohmic/inductive heating, high pressure processing, nonthermal plasma, oscillating magnetic field and ultrasound is likely to result in better quality and microbiologically safe products. At present, little work has been devoted to the investigation of hurdle approaches for decontaminating nuts and spices.

Once a technology is fully developed, there is still a need for it to be adopted readily in industry, in order that the benefits can be realized. Primarily, there are three reasons why the adoption of improved technologies by end-users can be a challenge. The first is simply that they are not aware of the technologies' existence or the benefits that they would provide. The end-users may also have misconceptions about the costs and benefits of the technologies. The second reason is that the technologies are not available, or not available at the times when they would be needed. The third reason is that the technologies are not always profitable, given the complex sets of decisions that end-users are making about how to allocate their resources across agricultural and non-agricultural activities. Institutional factors, such

© Woodhead Publishing Limited, 2012

as the policy environment, affect the availability of inputs and markets for credit and outputs and thus the profitability of a technology. Therefore, in developing the technologies, all the elements that will promote their rapid adoption must be equally delineated, strategically planned and considered alongside technological advancement.

## 5.8 Sources of further information and advice

Notable suppliers of food safety information include both the private sector and governmental agencies. The literature identifies five primary sources of food safety information: (a) experiential or family, (b) government agencies, (c) professional associations, (d) formal education, and (e) media.

Within the US government, there are six main agencies that regulate food production/processing and they are responsible for a large portion of the food safety standards and the available safety information. These agencies are the US Department of Agriculture (USDA), the Food and Drug Administration (FDA), the US Public Health Service (USPHS), Centers for Disease Control and Prevention (CDC), the Environmental Protection Agency (EPA), and the National Marine Fisheries Service (NMFS). The USDA is responsible for the grading and inspection of a variety of agricultural products shipped across state boundaries. In addition, the USDA and the Food Safety Inspection Service (FSIS) research, compile, and provide data for the public. Many brochures, pamphlets, and fact sheets are prepared by the USDA for consumers. The UDSA also sponsors a number of educational websites offering consumers information on a variety of topics. The FDA also plays a large role in food safety education. The Model Food Code is written by the FDA and details the government's recommendations for food service regulations. The FDA is also involved with the interpretation of food safety standards and random food safety inspections to food service operations. The CDC is a multi-functional federal agency whose primary goal is to detect and assess threats to public health. It investigates foodborne illness outbreaks, studies the causes and control of diseases, publishes data in the *Morbidity and Mortality Weekly Report (MMWR)*, and provides education regarding proper food sanitation practices. Water quality regulation, including the use of pesticides and sanitizers, and handling of wastes are the responsibility of the EPA.

Many professional associations conduct and report food safety research. The International Association for Food Protection (IAFP) is a group of industry professionals that provides its members with food safety information via two of its journal publications (*Journal of Food Protection* and *Dairy, Food and Environmental Sanitation)*. The Institute of Food Technologists (IFT) is an organization of food science and technology professionals whose primary goal is to be a source of scientific and professional-based food science and technology information, including two journals, the *Journal of Food Science*

© Woodhead Publishing Limited, 2012

and *Food Technology*. Food companies and other related organizations in the private sector also provide safety-related information. In addition to industry and professional associations, many universities extensively research and publish food safety information, often disseminated through the Cooperative Extension Service.

## 5.9 References


ABC (Almond Board of California) (2004), Almonds from Bloom to Market. Available from: http://www.almondboard.com/Resources/content.cfm?ItemNumber=637&snItemNumber=461.

ABDEL-HAFEZ AII, MOHARRAM AMM, ABDELMALLEK AY (1989), Thermophilic and thermotolerant fungi associated with seeds of five members of the *Umbelliferae* from Egypt, *Cryptogamie Mycologie*, 8, 315–320.

AKBAS MY, OZDEMIR M (2006a), Effectiveness of ozone for inactivation of *Escherichia coli* and *Bacillus cereus* in pistachios. *International Journal of Food Science and Technology*, 41, 513–519.

AKBAS MY, OZDEMIR M (2006b), Effect of different ozone treatments on aflatoxin degradation and physicochemical properties of pistachios, *Journal of the Science of Food and Agriculture*, 86, 2099–2104.

AL-BACHIR M (2007), Effect of gamma irradiation on microbial load and sensory characteristics of aniseed (*Pimpinella anisum*), *Bioresource Technology*, 98(10), 1871–1876.

ALMELA L, NIETO-SANDOVAL JM, FERNÁNDEZ-LÓPEZ JA (2002), Microbial inactivation of paprika by a high-temperature short-*X* time treatment – influence on color properties, *Journal of Agricultural and Food Chemistry*, 50, 1435–1440.

ANON (1970), Clean isn't good enough. *Diamond Walnut News*. Diamond Walnut Growers, Inc., Stockton, CA, August, p. 19.

ANTAI SP (1988), Study of the *Bacillus* fora of Nigerian spices, *International Journal of Food Microbiology*, 6, 259–261.

ATH-HAR MA, PRAKASH HS, SHETTY HS (1988), Mycoflora of Indian spices with special reference to aflatoxin producing isolates *of Aspergillus flavus*, *Indian Journal of Microbiology*, 28, 125–127.

AZIZ NH, MOUSSA LAA (2004), Reduction of fungi and mycotoxins formation in seeds by gamma-radiation, *Journal of Food Safety*, 24, 109–127.

BARBOSA-CANOVAS GV, POTHAKAMURY UR, PALOU E, SWANSON BG (1998), *Nonthermal Preservation of Foods*, Marcel Dekker, New York.

BARI L, NEI D, SOTOME I, NISHINA I, ISOBE S, KAWAMOTO S (2009), Effectiveness of sanitizers, dry heat, hot water, and gas catalytic infrared heat treatments to inactivate *Salmonella* on almonds, *Foodborne Pathogens and Disease*, 6(8), 953–958.

BARI L, NEI D, SOTOME I, NISHINA I, HAYAKAWA F, ISOBE S, KAWAMOTO S (2010), *Foodborne Pathogens and Disease*, 7(7), 845–850.

BASARAN P, AKHAN U (2010), Microwave irradiation of hazelnuts for the control of aflatoxin producing *Aspergillus parasiticus*, *Innovative Food Science and Emerging Technologies*, 11(1), 113–117.

BASARAN P, BASARAN-AKGUL N, OKSUZ L (2008), Elimination of *Aspergillus parasiticus* from nut surface with low pressure cold plasma (LPCP) treatment, *Food Microbiology*, 25(4), 626–632.

BAXTER R, HOLZAPFEL WH (1982), A microbial investigation of selected spices, herbs and additives in South Africa, *Journal of Food Science*, 47, 570–578.


© Woodhead Publishing Limited, 2012

BHAT R, GEETA H, KULKARNI PR (1987), Microbial profile of cumin seeds and chilli powder sold in retail shops in the city of Bombay, *Journal of Food Protection*, 50, 418–419.

BOLANDER CR, TOMA RB, DAVIS RM, MEDORA NP (1995), Irradiated versus fumigated spices in sausage, *International Journal of Food Sciences and Nutrition*, 46, 319–325.

BRANDL M (2006), Fitness of human enteric pathogens on plants and implications for food safety, *Annual Review of Phytopathology*, 44, 367–392.

BUTZ P, HEINISCH O, TAUSCHER B (1994), Hydrostatic high pressure applied to food sterilization III: application to spices and spice mixtures, *High Pressure Research*, 12, 239–243.

BYUN MW, YOOK HS, KWON JH, KIM JO (1996), Improvement of hygiene quality and long-term storage of dried red pepper by gamma-irradiation, *Korean Journal of Food Science and Technology*, 28, 482–489.

CAC (CODEX ALIMENTARIUS COMMISSION) (1972), Codex Alimentarius Commission Recommended international code of hygienic practice for tree nuts, CAC/RCP 6–1972. Available from: www.codexalimentarius.net/download/standards/266/CXP_006e.pdf

CAMPBELL JV, MOHLE-BOETANI J, REPORTER R, ABBOTT S, FARRAR J, BRANDL M, MANDRELL R, WERNER SB (2001), An outbreak of *Salmonella* serotype Thompson associated with fresh cilantro, *Journal of Infectious Disease*, 183, 984–987.

CBI (CENTER FOR PROMOTION OF IMPORTS FROM DEVELOPING COUNTRIES) (2008), The Preserved Fruit and Vegetables Market in the EU, June 2008. Available from: www.cbi.eu (accessed 12 June 2009).

CDC (CENTERS FOR DISEASE CONTROL and PREVENTION) (1997), Update: Outbreaks of cyclosporiasis – United States and Canada, 1997, *Morbidity and Mortality Weekly Report*, 46, 521–523.

CDC (CENTERS FOR DISEASE CONTROL and PREVENTION) (2004), Outbreak of *Salmonella* serotype Enteritidis infections associated with raw almonds – United States and Canada, 2003–2004, *Morbidity and Mortality Weekly Report*, 53, 484–487.

CDC (CENTERS FOR DISEASE CONTROL and PREVENTION) (2007), Multistate outbreak of *Salmonella* serotype Tennessee infections associated with peanut butter – United States, 2006–2007, *Morbidity and Mortality Weekly Report*, 56(21), 521–524.

CDC (CENTERS FOR DISEASE CONTROL and PREVENTION) (2009), Multistate outbreak of *Salmonella* infections associated with peanut butter and peanut butter containing products, United States, 2008–2009, *Morbidity and Mortality Weekly Report*, 1–6 January 29/58 (Early Release).

CDHS (CALIFORNIA DEPARTMENT OF HEALTH SERVICES) (2002), Environmental Investigation of *Salmonella* Enteritidis, Phage Type 30 Outbreak associated with Consumption of Raw Almonds, Sacramento California Department of Health Services, Food and Drug Branch. Available from: http://www.dhs.ca.gov/ps/fdb/local/PDF/2001AlmondReportWebVersion.pdf

CEC (COMMISSION OF THE EUROPEAN COMMUNITY) (1998), Commission Directive 98/53/EC of 16 July 1998 laying down the sampling methods and the methods of analysis for the official control of the levels of certain contaminants in foodstuffs. *Official Journal of the European Community Legislation*, L201, 93–101.

CHANG SS, HAN AR, REYES-DE-CORCUERA JI, POWERS JR, KANG DH (2010), Evaluation of steam pasteurization in controlling *Salmonella* serotype Enteritidis on raw almond surfaces, *Letters in Applied Microbiology*, 50(4), 393–398.

CHANGA T, MATTA F, SILVA J (1988), Pecan kernel quality as affected by antioxidant and irradiation, *Proceedings of the Annual Convention – Southern Pecan Growers Association*, 81, 59–63.

CHIOU R, LIN CM, SHYU S (1990), Property characterization of peanut kernels subjected to gamma irradiation and its effect on the outgrowth and aflatoxin production by *Aspergillus parasiticus*, *Journal of Food Science*, 56, 210–217.

© Woodhead Publishing Limited, 2012

CHO HO, BYUN MW, KWON JH, LEE JW, YANG JS (1986), Comparison of ethylene oxide (EO) and irradiation treatment on the sterilization of spices, *Korean Journal of Food Science and Technology*, 18, 283–287.

CHRISTENSEN CM, FANSE HA, NELSON GH, BATES F, MIROCHA CJ (1967), Microflora of black and red pepper, *Applied Microbiology*, 15, 622–626.

CHYTIRI S, GOULAS AE, BADEKA A, RIGANAKOS KA, KONTOMINAS MG (2005), Volatile and non-volatile radiolysis products in irradiated multilayer coextruded food-packaging films containing a buried layer of recycled low-density polyethylene, *Food Additives and Contaminants*, 22, 1264–1273.

COLE R, DORNER L (1989), In A Rice, G Hargelt, B Paulk, R Wheeler (eds), *Peanut Quality Enhancement Project Report*, National Peanut Council, VA, p. 13.

DANESH D, MOJTAHEDI H, BARNETT R, CAMPBELL A (1979), Correlation between climatic data and aflatoxin contamination of Iranian pistachio nuts, *Phytopathology*, 69, 715–716.

DE BOER E, SPIEGELENBERG WM, JANSSEN FW (1985), Microbiology of spices and herbs, *Antonie van Leeuwenhoek*, 51, 435–438.

DENG S, RUAN R, MOK CK, HUANG G, LIN X, CHEN P (2007), Inactivation of *Escherichia coli* on almonds using nonthermal plasma, *Journal of Food Science*, 72, M62–M66.

DU WX, ABD SJ, MCCARTHY KL, HARRIS LJ (2010), Reduction of *Salmonella* on inoculated almonds exposed to hot oil, *Journal of Food Protection*, 73(7), 1238–1246.

EL-RAHMAN HA (1987), Mycological studies on some selected spices with special reference to aflatoxin producing *Aspergillus flavus* species, *Assuit Veterinary Medical Journal*, 19, 92–100.

EPA (2006), Technology transfer networks. Available from: http://www.epa.gov/ttn/atw/hlthef/ethylene.html

FAO and WHO (2008), *Codex Alimentarius Commission 31st session*, June 30–July 4, Geneva, Switzerland, p. 104.

FAOSTAT (2005), FAO Statistical Database. Available from: http://www.fao.org

FARAG S A, AZIZ N H, ATTIA A (1995), Effect of irradiation on the microbiological status and flavouring materials of selected spices, *Zeitschrift für Lebensmittel Unterschung und Forschung*, 201, 283–298.

FARKAS J (1988), Irradiation of Dry Food Ingredients, CRC Press, Boca Raton, FL, p. 153.

FARKAS J (1990), Combination of irradiation with mild heat treatment, *Food Control*, 1(4), 223–229.

FARKAS J (2006), Irradiation for better food, *Trends in Food Science and Technology*, 17, 148–152.

FARKAS J, ANDRASSY E (1988), Comparative analysis of spices decontaminated by ethylene oxide or gamma radiation, *Acta Aliment*, 17(1), 77–94.

FARKAS J, MOHACSI-FARKAS C (2011), History and future of irradiation, *Trends in Food Science and Technology*, 22(2–3), 121–126.

FARKAS J, INCZE K, ZUKAL E (1973), Effect of nitrite on the microbiological stability of canned Vienna sausages preserved by mild heat treatment or combinations of heat and irradiation, *Acfa Aliment*, 2, 361–376.

FDA (FOOD and DRUG ADMINISTRATION) (1996), Compliance policy guides manual. 1996. Sec. 555.400, 268; Sec. 570.500, 299, FDA, Washington, DC.

FDA Compliance Policy Guidelines. CPG Sec.525.750 Spices. Available from: http://www.fda.gov/ICECI/ComplianceManuals/CompliancePolicyGuidanceManual/ucm074468.htm

FDA (FOOD and DRUG ADMINISTRATION) (2009), Peanut butter and other peanut containing products recall list. Washington, DC: Food and Drug Administration. Available from: http://www.accessdata.fda.gov/scripts/peanutbutterrecall/index.cfm

FDA (FOOD and DRUG ADMINISTRATION) (2011), *E. coli* O157:H7 cases linked to hazelnuts, FDA Press release. Available from: http://www.fda.gov/NewsEvents/Newsroom/PressAnnouncements/ucm245902.htm

© Woodhead Publishing Limited, 2012

FLANNIGAN B, LLEWELLYN GC (1986), The microbiology and mycotoxicology of spices: a review. In: S Barry, DR Houghton, GC Llewellyn, CE O'Rear (eds), *Biodeterioration* 6, Commonwealth Agriculture Bureau, International Mycological Institute, London, pp. 273–279.

FOWLES J, MITCHELL J, MCGRATH H (2001), Assessment of cancer risk from ethylene oxide residues in spices imported to New Zealand, *Food and Chemical Toxicology*, 39, 1055–1062.

FULLER G, SPOONCER WW, KING AD, SCHADE J, MACKEY B (1977), Survey of aflatoxins in California tree nuts, *Journal of the American Oil Chemists Society*, 54, A231–234A.

GATTO A (2003), The Salm-gene project – a European collaboration for DNA fingerprinting for food-related salmonellosis. *Eurosurveillance*, 8, 46–50.

GOODRIDGE LD, WILLFORD J, KALCHAYANAND N (2006), Destruction of *Salmonella* Enteriditis inoculated onto raw almonds by high hydrostatic pressure, *Food Research International*, 39, 408–412.

GOULAS AE, RIGANAKOS KA, KONOTMINAS MG (2004), Effect of ionizing radiation on physicochemical and mechanical properties of commercial monolayer and multilayer semirigid plastics packaging materials, *Radiation Physics and Chemistry*, 69, 411–417.

GOVINDARAJAN VS (1985), Capsicum-production, technology, chemistry, and quality. Part 1: history, botany, cultivation, and primary processing. *CRC Critical Review Food Science Nutrition*, 22(2), 109–175.

GUERRERO-BELTRAN J, BARBOSA-CANOVAS G, SWANSON B (2005), High hydrostatic pressure processing of fruit and vegetable products, *Food Reviews International*, 21, 411–425.

HENDERSON J, KREULZER S, SCHMIDT A, SMITH C, HAGEN W (1989), Flotation Separation of Aflatoxin Contaminated Grain or Nuts, US Patent 4 795 651.

HERWALDT BL, LEW JF, MOE CL, LEWIS DC, HUMPHREY CD, MONROE SS, PON EW, GLASS RI (1994), Characterization of a variant strain of Norwalk virus from a food-borne outbreak of gastroenteritis on a cruise ship in Hawaii, *Journal of Clinical Microbiology*, 32, 861–866.

HIMMELFARB P, EL-BISI HM, READ RB, LITSKY W (1962), Effect of relative humidity on the bactericidal activity of propylene oxide vapor, *Applied Microbiology*, 10, 431–435.

IAEA (International Atomic Energy Agency) (1988), Supplement to *Food Irradiation Newsletter*, 12(1), International Atomic Energy Agency, Vienna, 1–15.

ISAACS S, ARAMINI J, CIEBIN B, FARRAR JA, AHMED R, MIDDLETON D, CHANDRAN AU, HARRIS, LJ, HOWES M, CHAN E, PICHETTE AS, CAMPBELL K, GUPTA A, LIOR LY, PEARCE M, CLARK C, RODGERS F, JAMIESON F, BROPHY I, ELLIS A (2005), *Salmonella* Enteritidis PT30 outbreak investigation group, An international outbreak of salmonellosis associated with raw almonds contaminated with a rare phage type of *Salmonella* Enteritidis, *Journal of Food Protection*, 68, 191–198.

IYOTA H, KONISHI Y, INOUE T, YOSHIDA K, NISHIMURA N, NOMURA T (2005), Popping of amaranth seeds in hot air and superheated steam, *Drying Technology*, 23, 1273–1278.

JALILI M, JINAP S, NORANIZAN A (2010), Effect of gamma radiation on reduction of mycotoxins in black pepper, *Food Control*, 21(10), 1388–1393.

JAMES C, LECHEVALIER V, KETTERINGHAM L (2002), Surface pasteurization of shell eggs, *Journal of Food Engineering*, 53, 193–197.

JEONG S, MARKS BP, ORTA-RAMIREZ A (2009), Thermal inactivation kinetics for *Salmonella* Enteritidis PT30 on almonds subjected to moist-air convection heating, *Journal of Food Protection*, 72(8), 1602–1609.

JOHNSON JA, VAIL PV (1987), Adult emergence and sterility of Indian meal moths (Lepidoptera: Pyralidae) irradiated as pupae in dried fruits and nuts, *Journal of Economic Entomology*, 80, 497–501.

© Woodhead Publishing Limited, 2012

KAEFER CM, MILNER JA (2009), The role of herbs and spices in cancer prevention, *Journal of Nutrition Biochemistry*, 19(6), 347–361.

KILLALEA D, WARD LR, ROBERTS D, DE LOUVOIS J, SUFI F, STUART JM, WALL PG, SUSMAN M, SCHWIEGER M, SANDERSON PJ, FISHER IS, MEAD PS, GILL ON, BARTLETT CL, ROWE B (1996), International epidemiological and microbiological study of outbreak of *Salmonella* agona infection from a ready to eat savoury snack – I: England and Wales and the United States, *British Medical Journal*, 313, 1105–1107.

KIM JG, YOUSEF AE, DAVE SA (1999), Application of ozone for enhancing the microbiological safety and quality of foods: a review, *Journal of Food Protection*, 62, 1071–1087.

KIRK DM, LITTLE CL, LEM M, FYFE M, TAN A, THRELFALL J, PACCAGNELLA A, LIGHTFOOT D, GENOBILE D, LI H, MCINTYRE L, CRAWFORD C, WARD L, BROWN DJ, SURMAN S, FISHER IST (2004), An outbreak due to peanuts in their shell caused by *Salmonella enterica* serotypes Stanley and Newport-sharing molecular information to solve international outbreaks, *Epidemiology and Infection*, 132, 571–577.

KISS I, FARKAS J (1988), Irradiation as a method for decontaminating spices, *Food Review International*, 4, 77–92.

KOKAL D, THORPE DW (1969), Occurrence of *Escherichia coli* in almonds of nonpareil variety, *Food Technology*, 23, 227–232.

KONISHI Y, IYOTA H, YOSHIDA K, MURITANI J, INOUE T, NISHIMURA N, NOMURA T (2004), Effect of moisture content on the expansion volume of popped amaranth seeds by hot air and superheated steam using a fluidized bed system, *Bioscience Biotechnology Biochemistry*, 68, 2186–2189.

KRZYMIEN MIE, CARLSSON DJ, DESCHENES L, MERCIER M (2001), Analyses of volatile transformation products from additives in gamma-irradiated polyethylene packaging, *Food Additives and Contaminants*, 18, 739–749.

LEDET-MULLER L, HJERTQVIST M, PAYNE L, PETTERSSON H, OLSSON A, PLYM F L, ANDERSSON Y (2007), Cluster of *Salmonella enteritidis* in Sweden 2005-2006 – suspected source: almonds, *European Surveillance*, 12, E9-10.

LEISTRITZ W (1997), Methods of bacterial reduction in spices. In: SJ Risch, CT Ho (eds), *Spices: Flavor Chemistry and Antioxidant Properties*, ACS Symposium Series, Vol. 660, American Chemical Society, Washington, DC, pp. 7–10.

LILIE M, HEIN S, WILHELM P, MÜLLER U (2007), Decontamination of spices by combining mechanical and thermal effects – an alternative approach for quality retention, *International Journal of Food Science and Technology*, 42, 190–193.

LLEWELLYN GC, O'DELL VL, FLANNIGAN B (1988), Spices and mycotoxins: a review of biodeterioration and health implications. In DR Houghton, RN Smith, HOW Eggings (eds), *Biodeterioration 7*, Elsevier Applied Science, London, pp. 634–641.

LOAHARANU P (1994), Status and prospects of food irradiation, *Food Technology*, 52, 124–131.

MAEBA H, TAKAMOTO Y, KAMIMURA M, MIURA T (1988), Destruction and detoxification of aflatoxins with ozone, *Journal of Food Science*, 53, 667–669.

MANAS P, PAGAN R (2005), Microbial inactivation by new technologies of food preservation, *Journal of Applied Microbiology*, 98, 1387–1399.

MARCUS AK, AMLING HJ (1973), *Escherichia coli* field contamination of pecan nuts, *Applied Microbiology*, 26(3), 279–281.

MARSILIA L, ESPIE S, ANDERSON JG, MACGREGOR, SJ (2002), Plasma inactivation of food-related microorganisms in liquids, *Radiation Physics and Chemistry*, 65, 507–513.

MCKEE LH (1995), Microbial contamination of spices and herbs: a review, *Lebensmittel-Wissenschaft und -Technologie*, 28, 1–11.

MEHAN VK (1988), In: PS Reddy (ed.), *Groundnut*, Indian Council of Agricultural Research, New Delhi, India, pp. 526–541.

MEYER MT, VAUGHN RH (1969), Incidence of *Escherichia coli* in black walnut meats, *Applied Microbiology*, 18, 925–931.

MISRA N (1987), Mycotoxins in spices III, Investigation on the natural occurrence of

© Woodhead Publishing Limited, 2012

aflatoxins in *Coriandrum sativum* L, *Journal of Food Science and Technology, India*, 24, 324–325.

MODLICH G, WEBER H (1993), Comparison of methods for sterilization of spices: microbiological and sensory aspects, *Fleischwirtschaft*, 73, 337–343.

MOJITAHEDI H, RABIE CJ, LUBBEN A, STYEN M, DANESH D (1979), Toxic *Aspergilli* from pistachio nuts, *Mycopathologia*, 67, 123–127.

MORGAN AI, GOLDBERG N, RADEWONUK ER, SCULLEN OJ (1996), Surface pasteurization of raw poultry meat by steam, *LWT-Food Science and Technology*, 29, 447–451.

NATVIG EE, INGHAM SC, INGHAM BH, COOPERBAND LR, ROPER MM (2002), *Salmonella enterica* serovar Typhimurium and *Escherichia coli* contamination of root and leaf vegetables grown in soils with incorporated bovine manure, *Applied Environmental Microbiology*, 68, 2737–2744.

NG S, ROUCH G, DEDMAN R, HARRIES B, BOYDEN A, MCLENNAN L, BEATON S, TAN A, HEATON S, LIGHTFOOT D, VULCANIS M, HOGG G, SCHEIL W, CAMERON S, KIRK M, FELDHEIM J, HOLLAND R, MURRAY C, ROSE N, ECKERT P (1996), Human salmonellosis and peanut butter, *Communicable Diseases Intelligence*, 20, 326.

NIEMIRA B (2010), Short pulse cold plasma treatment reduces *Salmonella* and *E. coli* O157:H7 on almonds [abstract], Institute of Food Technologies Annual Meeting, Chicago, IL, p. 1.

NIETO-SANDOVAL JM, ALMELA L, FERNÁNDEZ-LÓPEZ JA, MUNOZ JA (2000), Effect of electron beam irradiation on color and microbial bioburden of red paprika, *Journal of Food Protection*, 63, 633–637.

PADWAL-DESAI SR (1974), Studies on control of foodborne moulds by gamma-irradiation, PhD Thesis, Bhabha Atomic Research Centre, Bombay, University of Bombay, India.

PAN Z, ATUNGULU GG (eds) (2010a), *Infrared Heating for Food and Agricultural Processing*, CRC Press, Boca Raton, FL.

PAN Z, ATUNGULU GG (2010b), The potential novel infrared food processing technologies – case studies of those developed at the USDA-ARS Western Region Research Center and the University of California at Davis. In: CJ Doona, K Kustin, and FE Feeherry (eds), *Case Studies in Processing, Packaging, and Predictive Modeling*, Cambridge, Woodhead Publishing.

PAO S, KALANTARI A, HUANG G (2006), Utilizing acidic sprays for eliminating *Salmonella enterica* on raw almonds, *JFS Food Microbiology and Safety*, 71(1), M14–M19.

PARK BJ, TAKATORI K, SUGITA-KONISHI K, KIM IH, LEE MH, HAN DW, CHUNG KH, HYUN SO, PARK JC (2007), Degradation of mycotoxins using microwave-induced argon plasma at atmospheric pressure, *Surface Coating Technology*, 201, 5733–5737.

PARK D, LEE L, PRICE R, POHLAND A (1988), Review of the decontamination of aflatoxins by ammoniation: current status and regulation, *Journal of the Association of Official Analytical Chemists*, 71, 685–703.

PERREN R (2008), Almond pasteurization-controlling moisture and quality, *The Manufacturing Confectioner*, 88(4), 77–84.

PETERS TM, MAGUIRE C, THRELFALL EJ, FISHER IST, GILL N, GATTO AJ (2003), The Salm-gene project – a European collaboration for DNA fingerprinting for food-related salmonellosis, *Eurosurveillance*, 8, 46–50.

PHILLIPS DJ, UOTA M, MONTICELLI D, CURTIS C (1976), Colonization of almond by *Aspergillus flavus*, *Journal of the American Society for Horticultural Science*, 101, 19–23.

PIER AC, RICHARD JL (1992), Mycoses and mycotoxicoses of animal caused by *Aspergillus*. In: JW Bennett, MA Klich (eds), *Aspergillus: Biology and Industrial Applications*, Butterworth-Heinnemann, Boston, MA, pp. 233–248.

PLUYER H, AHMED E, WEI C (1987), Destruction of aflatoxins on peanuts by oven- and microwave-roasting, *Journal of Food Protection*, 50, 504–508.

PODOLAK R, ENACHE E, STONE W, BLACK GD, ELLIOTT HP (2010), Sources and risk factors for contamination, survival, persistence, and heat resistance of *Salmonella* in low-moisture foods, *Journal of Food Protection*, 73(10), 1919–1936.

© Woodhead Publishing Limited, 2012

PRAKASH CG, VIVEK B, MISHRA V, SAXENA RK, SURENDRA S (2009), Effect of gamma irradiation on microbial load, aflatoxins and phytochemicals present in *Trigonella foenum-graecum*, *World Journal of Microbiology and Biotechnology*, 25(12), 2267–2271.

PRAKASH A, LIM FT, DUONG C, CAPORASO F, FOLEY D (2010), The effects of ionizing irradiation on *Salmonella* inoculated on almonds and changes in sensory properties, *Radiation Physics and Chemistry*, 79, 502–506.

PURCELL SL, PHILIP DJ, MACKEY BE (1980), Distribution of *Aspergillus flavus* and other fungi in several almond-growing areas of California, *Phytopathology*, 70, 926–929.

RANSOM G (2006), Supplement – Requisite scientific parameters for establishing the equivalence of alternative methods of pasteurization, *Journal of Food Protection*, 69(5), 1190–1216.

RAVISHANKAR S, JUNEJA VK (2003), Adaptation or resistance responses of microorganisms to stresses in the food processing environment. In: AE Yousef, VK Juneja (eds), *Microbial Stress Adaptation and Food Safety*, CRC Press, Boca Raton, FL, pp. 105–158.

READ M (1989), Removal of aflatoxin contamination from the Australian groundnut crop. In D McDonald, V Mehan (eds), *Aflatoxin Contamination of the Groundnut: Proceedings of the International Workshop*, pp. 133–140.

REIL W, LABAVITCH JM, HOLMBERG D (1996), Harvesting. In WC Micke (ed.), *Almond Production Manual*, Division of Agriculture and Natural Resources, University of California, Oakland, CA, pp. 260–264.

RICE RE (1978), Navel Orangeworm (*Lepidoptera-ptraziuae*) – pest of pistachio nuts in California, *Journal of Economic Entomology*, 71, 822–824.

RICO WC, KIM G, AHN J, KIM H, FURUTA M, KWON J (2010), The comparative effect of steaming and irradiation on the physicochemical and microbiological properties of dried red pepper (*Capsicum annum* L.), *Food Chemistry*, 119(3), 1012–1016.

ROBILLARD P (1983), *Salmonella* in spices and chocolate. *Canada Disease Weekly Report*, 9, 35.

ROPER MM, MARSHALL KC (1978), Effects of a clay mineral on microbial predation and parasitism of *Escherichia coli*, *Microbial Ecology*, 4, 279–289.

SANCHEZ-BELA P, EGEA I, ROMOJARO F, MARTINEZ-MADRID MC (2008), Sensorial and chemical quality of electron beam irradiated almonds (*Prunus amygdalus*), *LWT-Food Science and Technology*, 41, 442–449.

SAXENA J, PANDEY N, PANDEY SK, PANDEY H, MEHROTRA BS (1988), Mycotoxin producing potentials of the fungi isolated from some common spices, *Proceedings of the National Academy of Science, India, Section B (Biol. Sci.)*, 58, 427–430.

SCHADE JE, MCGREEVY K, KING AD, MACKEY B, FULLER G (1975), Incidence of aflatoxin in California almonds, *Applied Microbiology*, 29, 48–53.

SCHEIL W, CAMERON S, DALTON C, MURRAY C, WILSON D (1998), A South Australian *Salmonella* Mbandaka outbreak investigation using a database to select controls, *Australia New Zealand Journal of Public Health*, 22, 536–539.

SCHNEIDER B (1993), Steam sterilization of spices, *Fleischwirtschaft*, 73, 646–648.

SCUDAMORE KA, HEUSER SG (1971), Ethylene oxide and its persistent reaction products in wheat flour and other commodities: residues from fumigation or sterilization, and effects of processing, *Pesticide Science*, 2, 80–91.

SEENAPPA M, KEMPTON AG (1980), *Aspergillus* growth and aflatoxin production on black pepper, *Mycopathologia*, 70, 135–137.

SHOHAT T, GREEN MS, MEROM D, GILL ON, REISFELD A, MATAS A, BLAU D, GAL N, SLATER PE (1996), International epidemiological and microbiological study of outbreak of *Salmonella* agona infection from a ready to eat savoury snack – II: Israel, *British Medical Journal*, 313, 1107–1109.

SOBEL J, SWERDLOW DL, PARSONNET J (2001), Is anything safe to eat? *Current Clinical Topics in Infectious Disease*, 21, 114–134.

SOMMER NF, BUCHANAN JR, FORTLAGE RJ (1986), Relation of early splitting and tattering of pistachio nuts to aflatoxin in the orchard, *Phytopathology*, 76, 692–694.

© Woodhead Publishing Limited, 2012

STAACK N, AHRNÉ L, BORCH E, KNORR D (2008), Effect of infrared heating on quality and microbial decontamination in paprika powder, *Journal of Food Engineering*, 86(1), 17–24.

STEENLAND K, STAYNER L, DEDDENS J (2004), Mortality analyses in a cohort of 18235 ethylene oxide exposed workers: follow-up extended from 1987 to 1998, *Occupational and Environmental Medicine*, 61, 2–7.

SUHAJ M, RACOVA J, POLOVKA M, BREZOVA V (2006), Effect of γ-irradiation on antioxidant activity of black pepper (*Piper nigrum* L.), *Food Chemistry*, 97, 696–704.

SUSANNE EK (2009), Tree fruit and nuts: outbreaks, contamination sources, prevention and remediation. In: MS Gerald, BS Ethan, RM Karl (eds), *The Produce Contamination Problem: Causes and Solutions*, Academic Press, New York, pp. 249–260.

TABATA S, KAMIMURA H, TAMURA Y, YASUDA K, USHIYAMA H, HASHIMOTO H, NISHIJIMA M, NISHIMA T (1987), Aflatoxin contamination in foods and foodstuffs, *Journal of the Food Hygiene Society of Japan*, 28, 395–401.

TATEO F, BONONI M (2006), Determination of ethylene chlorohydrin as marker of spice fumigation with ethylene oxide, *Journal of Food Composition and Analysis*, 19, 83–87.

TOOFANIAN F, STEGEMAN H (1988), Comparative effect of ethylene oxide and gamma irradiation on the chemical sensory and microbial quality of ginger, cinnamon, fennel and fenugreek, *Acta Alimentaria*, 17, 271–281.

TOPUZ A, OZDEMIR F (2003), Influences of γ-irradiation and storage on the carotenoids of sun dried and dehydrated paprika, *Journal of Agricultural and Food Chemistry*, 51, 4972–4977.

UESUGI AR, HARRIS LJ (2006),Growth of *Salmonella* Enteritidis Phage Type 30 in almond hull and shell slurries and survival in drying almond hulls, *Journal of Food Protection*, 69, 712–718.

UESUGI AR, DANYLUK MD, MANDRELL RE, HARRIS LJ (2007), Isolation of *Salmonella* Enteritidis Phage Type 30 from a single almond orchard over a 5-year period, *Journal of Food Protection*, 70(8), 1784–1789.

US DEPARTMENT OF AGRICULTURE (2006), 7 CFR Part 981: Almonds grown in California, Outgoing quality control requirements and request or approval of new information collection. Federal Register 71, 70683–70692.

VAJDI M, PEREIRA RR (1973), Comparative effects of ethylene oxide, gamma irradiation and microwave treatments on selected spices, *Journal of Food Science*, 38, 893–895.

VIJ V, AILES E, WOLYNIAK C, ANGULO FJ, KLONTZ KC (2006), Recalls of spices due to bacterial contamination monitored by the US Food and Drug Administration: the predominance of salmonellae, *Journal of Food Protection*, 69, 233–237.

WEBER FE (1980), Controlling microorganisms in spices, *Cereal Foods World*, 25, 319–321.

WELLS JM, PAYNE JA (1976), Toxigenic species of *Penicillium*, *Fusarium* and *Aspergillus* from weevil-damaged pecans, *Canadian Journal of Microbiology*, 22, 281–285.

WELLS JM, COLE R J and KIRKSEY JW (1975), Emodin, a toxic metabolite of *Aspergillus wentii* isolated from weevil-damaged chestnuts, *Applied Environmental Microbiology*, 30(1), 26–28.

WIHODO M, HAN Y, SELBY TL, LORCHEIM P, CZARNESKI M, HUANG G, LINTON RH (2005), Decontamination of raw almonds using chlorine dioxide gas (abstract), Institute of Food Technologist Annual Meeting, New Orleans, LA, July 15–20.

WILLFORD J, MENDONCA A, GOODRIDGE LD (2008), Water pressure effectively reduces *Salmonella enterica* Serovar Enteritidis on the surface of raw almonds, *Journal of Food Protection*, 71(4), 825–829.

WILSON BJ, HAYES W (1973), Microbial toxins. In *Toxicants Occurring Naturally in Foods*, National Academy of Sciences, Washington, DC, pp. 372–423.

WILSON GY (1990), The effectiveness of gamma processing as compared to propylene oxide for controlling mold on English walnut (*Juglans regia*), Chapman University, Orange, CA. Available from: Food Science and Nutrition Department, Orange, CA.

© Woodhead Publishing Limited, 2012

WINDYGA B, FONBERG-BROCZEK M, SCIEZYŃSKA H, SKAPSKA S, GÓRECKA K, GROCHOWSKA A, MORAWSKI A, SZCZEPEK J, KARłOWSKI K, POROWSKI S (2008), High pressure processing of spices in atmosphere of helium for decrease of microbiological contamination, *Rocz Panstw Zakl Hig*, 59(4), 437–443.

WONG LO FO, ANDERSEN JK, NORRUNG B, WEGENER HC (2004), Food contamination monitoring and food-borne disease surveillance at national level. In *Second FAO/WHO Global Forum of Food Safety Regulators – Proceedings of the Forum in Bangkok, Thailand, 12–14 October 2004*, FAO Corporate Document Repository, Agriculture and Consumer Protection.

WOOD GE (1989), Aflatoxins in domestic and imported foods and feeds, *Journal of the Association of Official Analytical Chemists*, 72, 543–548.

YANG J, BINGOL G, PAN Z, BRANDL MT, MCHUGH TH, WANG H (2010), Infrared heating for improved safety and processing efficiency of dry-roasted almonds, *Journal of Food Engineering*, 101, 273–280.

© Woodhead Publishing Limited, 2012

# 6

# Microbial decontamination of juices

**M. D. Danyluk, University of Florida, USA, M. E. Parish, FDA/ CFSAN Office of Food Safety, USA, R. M. Goodrich-Schneider, University of Florida, USA and R. W. Worobo, Cornell University, USA**

**Abstract**: The microbiological concerns of juice decontamination historically involved prevention of spoilage. A number of *Salmonella*, *Escherichia coli* O157:H7, and *Cryptosporidium parvum* outbreaks associated with raw (unpasteurized) juices in the 1990s led to the introduction of regulation requiring 100% juice products be produced under a Hazard Analysis Critical Control Point (HACCP) program by the US Food and Drug Administration. This regulation requires that juice processors obtain at least a 5-log reduction of the 'pertinent microorganism' as a decontamination step. As a result of juice HACCP, most decontamination methods are measured to the 5-log reduction of the pertinent pathogen performance standard this regulation requires. In this chapter, outbreaks associated with juices will be summarized and we will discuss means by which microorganisms are eliminated in juices through both classical processing and novel methods for juice decontamination, what future trends in juice decontamination may be and provide further sources of information.

**Key words**: apple, orange, *Salmonella*, *E. coli*, *Cryptosporidium*, HACCP.

## 6.1 Introduction

Most juices processed are made with acidic or acidified fruits and are sold as frozen concentrated, pasteurized and refrigerated, or shelf stable. Fruit juices may be squeezed directly from the fruit (i.e., citrus) or be prepared from macerated or crushed materials (i.e., grapes, berries, apples). In the case of many vegetable juices (i.e., carrot), they may be pulped. Juices

---

The opinions and conclusions expressed in this manuscript are solely the views of the authors and do not necessarily reflect those of the Food and Drug Administration.

© Woodhead Publishing Limited, 2012

may be highly clarified, or they may contain considerable amounts of suspended solids. Even though most fruit juices have a pH below 4.6, the acidity is dependent on the raw material and can vary greatly. Recently, a number of drinks and beverages containing less than 100% juice have gained popularity, and come under a number of names, including 'ades', 'nectars', and 'cocktails'. Pulp and other derivatives of juice can also be used as ingredients in the production of other beverages. In this chapter we will focus on 100% juice products.

Historically, the microbiological concerns of juice processors and packagers involved prevention of spoilage in intermediate and final products, and control of microorganisms in the processing and distribution systems. Stratford *et al.* (2000) described the processing of fruit juices in detail and provides an overview of spoilage issues. More recently, the juice industry has focused on microbiological food safety. Vojdani *et al.* (2008) reviewed outbreaks of juice-associated human illness from 1995 to 2005.

As a consequence of outbreaks of *Salmonella*, *Escherichia coli* O157:H7, and *Cryptosporidium parvum* in the 1990s associated with raw juices processed at commercial facilities, the US Food and Drug Administration (US FDA) introduced regulation (21 Code of Federal Regulations 120; US FDA, 2001) mandating that 100% juice be produced under a HACCP plan having supporting Good Manufacturing Practices (GMPs) and Sanitation Standard Operating Procedures (SSOPs). This regulation further requires that juice processors produce at least a 5-log reduction of the 'pertinent microorganism' which is defined as 'the most resistant microorganism of public health significance that is likely to occur in the juice.' Determination of the public health microorganism that is most pertinent for any particular juice may be based upon disease outbreak data as well as any other appropriate information available. Currently, *Salmonella* is generally accepted as the pertinent pathogen in citrus juices, whereas *E. coli* O157:H7 as well as *C. parvum* need to be taken into consideration for apple juice (US FDA, 2001). The juice HACCP regulation applies to domestic and imported 100% juice and has implications for juice producers in countries that export juice into the United States.

While not the focus of this chapter, it is important to note that growth of fungi in fruit used in juice production or in the fruit juice itself may result in production of mycotoxins, which also pose a public health risk. Specifically, patulin is a widespread mycotoxin produced by several species of *Penicillium*, as well as a few species of *Aspergillus* and *Byssochlamys* (Davis and Diener, 1987). This mycotoxin has been found in commercial apple juice in concentrations as high as 1 mg/l (1 ppm) (Doores, 1983; Scott *et al.*, 1972). Patulin production has also been reported in juices of grape, blueberry, red raspberry, and boysenberry (Beuchat, 1986). The US FDA has drafted guidance for apple juice processors establishing 50 μg/kg (50 ppb) patulin in apple juice as a critical action level under juice HACCP.

As a result of juice HACCP, most decontamination methods are measured

© Woodhead Publishing Limited, 2012

to the 5-log reduction of the pertinent pathogen performance standard this regulation requires. Here, we will discuss means by which microorganisms are eliminated in juices through both classical processing and novel methods for juice decontamination, what future trends in juice decontamination may be and provide further sources of information.

## 6.2 Pathogens of concern and potential for contamination

### 6.2.1 Outbreaks associated with juices and potential routes of contamination

Despite early documented evidence that specific pathogens are capable of survival in fruit juices (Douglas, 1930; Forgacs and Tanner, 1943; Goverd *et al.*, 1979; Mitscherlich and Marth, 1984; Paquet, 1923), it was widely accepted through the mid-1990s that most low pH, high acid juices were of minimal concern for foodborne illness, because they did not support the growth of pathogenic microorganisms. Although there is a long history of juice-related outbreaks, they have been infrequent and, until 1995, were generally associated with small commercial processors or restaurant or home-prepared products (Table 6.1) (Parish, 1997). Most outbreaks with raw juices have been associated with enteric pathogens.

Raw apple juice outbreaks have been associated with *Salmonella*, Shiga toxin-producing *E. coli*, and *Cryptosporidium parvum* (Harris *et al.*, 2003). Although *E. coli* O157:H7 was not described as a human pathogen until 1982, it is probable that a 1980 outbreak of diarrhea and hemolytic uremic syndrome (HUS) from raw apple cider was caused by this organism (Steele *et al.*, 1982; Zhao *et al.*, 1993). The use of animal manure as fertilizer and collection of dropped apples was indicated as a possible source of the *S.* Typhimurium in a 1974 outbreak associated with apple cider (CDC, 1975). Manure is suspected to be the primary source of contamination in a *Cryptosporidium* outbreak where apples were shaken from the trees and gathered from the ground after cattle were allowed to graze on the grass beneath the trees (Miller and Kaspar, 1994). Although dropped apples were not used in another *C. parvum* outbreak linked to raw apple cider, a dairy farm was located nearby and *E. coli* was detected in well water samples from the farm (CDC, 1997). In one outbreak, implicated apples were mostly from juice production dates late in the harvest season, and were of minimal quality by company standards. Extensive grading was necessary to remove unacceptable fruit before milling and pressing (Cody *et al.*, 1999).

Citrus juices have been associated with outbreaks from a number of pathogenic viruses and bacteria (Parish, 1997; Vojdani *et al.*, 2008). In some cases an asymptomatic human handler (shedding pathogens in their feces without showing signs of illness) has contaminated the juice with hepatitis A or *S.* Typhi as it was being reconstituted or prepared at food service. In one outbreak, the source of contamination was thought to be the water used

© Woodhead Publishing Limited, 2012

**Table 6.1** Outbreaks associated with juices, 1922–2010

| Type | Product | Pathogen | Year | Location | Venue | Cases (deaths) | Reference |
|---|---|---|---|---|---|---|---|
| Apple | Unpasteurized | *S.* Typhi | 1922 | France | NR[a] | 23(0) | Paquet, 1923 |
| | Unpasteurized | *S.* Typhimurium | 1974 | USA (NJ) | Farm, small retail outlets | 296 (0) | CDC, 1975 |
| | Unpasteurized | *E. coli* O157:H7 (suspected) | 1980 | Canada (ON) | Local market | 14 (1) | Steele *et al.*, 1982 |
| | Unpasteurized | *E. coli* O157:H7 | 1991 | USA (MA) | Small cider mill | 23 (0) | Besser *et al.*, 1993 |
| | Unpasteurized | *Cryptosporidium* | 1993 | USA (ME) | School | 213 (0) | Millard *et al.*, 1994 |
| | Unpasteurized | *C. parvum* | 1996 | USA (NY) | Small cider mill | 31 (0) | CDC, 1997 |
| | Unpasteurized | *E. coli* O157:H7 | 1996 | USA (CT) | Small cider mill | 14 (0) | CDC, 1997 |
| | Unpasteurized | *E.coli* O157:H7 | 1996 | USA (WA) | Small cider mill | 6 (0) | US FDA, 2011 |
| | Unpasteurized | *E. coli* O157:H7 | 1996 | Canada (BC), USA (CA, CO, WA) | Retail | 70 (1) | CDC, 1996; Cody *et al.*, 1999 |
| | Unpasteurized | *E. coli* O157:H7 | 1997 | USA (IN) | Farm | 6 | INS DOH, 1997 |
| | Unpasteurized | *E. coli* O157:H7 | 1998 | Canada (ON) | Farm/home | 14 (0) | Tamblyn *et al.*, 1999 |
| | Unpasteurized | *E. coli* O157:H7 | 1999 | USA (OK) | NR | 25 | CDC, 2011 |
| | Unpasteurized | *C. parvum* | 2003 | USA (OH) | Farm/retail | 144 | Vojdani *et al.*, 2008 |
| | Unpasteurized | *E. coli* O111 and *C. parvum* | 2004 | USA (NY) | Farm/home | 212 | Vojdani *et al.*, 2008 |
| | Unpasteurized | *E. coli* O157:H7 | 2005 | Canada (ON) | NR | 4 (?) | LSDEPC, 2005 |
| | Unpasteurized | *E. coli* O157:H7 | 2007 | USA (MA) | NR | 9 | CDC, 2011 |
| | Unpasteurized | *E. coli* O157:H7 | 2008 | USA (IA) | Fair | 7 | CDC, 2011 |
| | Unpasteurized | *E. coli* O157:H7 | 2010 | USA (MD) | Retail | 7 | US FDA, 2010 |
| Carrot | Homemade | *C. botulinum* | 1993 | USA (WA) | Home | 1 (0) | Buzby and Crutchfield, 1999 |

© Woodhead Publishing Limited, 2012

| | | | | | | | |
|---|---|---|---|---|---|---|---|
| | Pasteurized | *C. botulinum* | 2006 | USA | Retail | 4 | CDC, 2006 |
| Coconut | Milk | *Vibrio cholerae* | 1991 | USA (MD) | Home/picnic | 4 | CDC, 1991; Taylor *et al.*, 1993 |
| Mamey | Frozen Puree | *S.* Typhi | 1999 | USA | NR | 19 | Katz *et al.*, 2002 |
| | Frozen Pulp | *S.* Typhi | 2010 | USA | Retail | 9 | CDC, 2010 |
| Mixed Fruit | Unspecified | *Shigella sonnei* | 2002 | Canada, USA, UK, British West Indies | Resort | 78 | CDC, 2011 |
| | Acai, banana, strawberry, sugar cane | Hepatitis A | 2007 | USA (FL) | Food service | 3 | CDC, 2011 |
| Orange | Unpasteurized | Enterotoxigenic *E. coli* | 1992 | India | Roadside vendor | 6 (0) | Singh *et al.*, 1996 |
| | Unpasteurized | *Salmonella* serovars Gaminera, Hartford and Rubislaw | 1995 | USA (FL) | Retail | 63 (0) | CDC, 1995; Cook *et al.*, 1998; Parish, 1998 |
| | Unpasteurized | *Shigella flexneri* | 1995 | South Africa | Restaurant | 14 (NR) | Thurston *et al.*, 1998 |
| | Unpasteurized | Virus suspected | 1996 | USA | Food service | 2 | Parish, 2000 |
| | Unpasteurized | *S.* Muenchen | 1999 | Canada and USA | Restaurant | 423 (1) | CDC, 1999 |
| | Unpasteurized | *S.* Anatum | 1999 | USA (FL) | Roadside stand | 6 (0) | Krause *et al.*, 2001 |
| | Unpasteurized | *S.* Typhimurium | 1999 | Australia | Retail | 405 (0) | National Center for Disease Control, 1999 |
| | Unpasteurized | *S.* Enteritidis | 2000 | USA (6 states) | Retail and food service | 88 | Butler, 2000 |

*(Continued)*

© Woodhead Publishing Limited, 2012

**Table 6.1** Continued

| Type | Product | Pathogen | Year | Location | Venue | Cases (deaths) | Reference |
|---|---|---|---|---|---|---|---|
| | Unpasteurized | *Salmonella* serovars Typhimurium and Saintpaul | 2005 | USA (14 states) | Retail and food service | 157 | Jain *et al.*, 2009 |
| | Reconstituted | *S.* Typhi | 1944 | USA (OH) | Hotel | 18 (1) | Duncan *et al.*, 1946 |
| | Reconstituted | Hepatitis A | 1962 | USA (MO) | Hospital | 24 | Eisenstein *et al.*, 1963 |
| | Reconstituted | Unknown | 1965 | USA (CA) | Football game | 563 | Tabershaw *et al.*, 1967 |
| | Reconstituted | *S.* Typhi | 1989 | USA (NY) | Hotel | 69 | Birkhead *et al.*, 1993 |
| | NR | Norwalk-like virus | 1991 | Australia | Airline | 3,053 | Lester *et al.*, 1991 |
| Watermelon | Homemade | *Salmonella* spp | 1993 | USA (FL) | Home | 18 (0) | US FDA, 1998 |

[a] NR – Not reported.

© Woodhead Publishing Limited, 2012

to dilute a juice concentrate. In another one, the water source was thought to contaminate food service workers hands with *Shigella flexneri* that subsequently contaminated the juice as they squeezed the oranges. In contrast to outbreaks related to human handling in food-service settings, *Salmonella* has been exclusively involved in outbreaks associated with single-strength raw citrus juices prepared in commercial processing facilities (Parish, 1997; Vojdani *et al.*, 2008). In some cases the cause of the outbreak was thought to be poor sanitation practices in the juicing facility (Parish, 1998). In one outbreak, salmonellae were isolated from various samples including unopened bottles of orange juice, unwashed fruit surfaces, and amphibians found in close proximity to the processing facility (Parish, 1998).

While strict production controls have made outbreaks rare in commercially processed juices, cases of botulism have been reported. The vast majority of foods causing botulism are low acid (> pH 4.6); however, there have been occasional outbreaks associated with typically acidic foods including at least four outbreaks attributed to home-canned tomato juice (1935, 1965, 1969, 1974) and one linked to huckleberry juice in 1953 (Odlaug and Pflug, 1978). Later, six cases of botulism were linked to refrigerated carrot juice in the US and Canada in September and October 2006. The implicated products, all produced at a single firm, were pasteurized, but were not heated to a temperature that would eliminate spores of proteolytic (the most heat resistant type) *Clostridium botulinum*. Subsequent testing of leftover carrot juice recovered from the home of one of the affected persons found botulinum toxin in the juice (Sheth *et al.*, 2008).

### 6.2.2 Pathogen isolation from juices

There have been very few surveys of retail juices for the presence of pathogens in part due to the very low probability of finding pathogens in these products (Harris *et al.*, 2003). Coliforms isolated from citrus products include *Enterobacter* spp., *Serratia* spp., *Klebsiella* spp., *Citrobacter* spp. and *E. coli* (Weihe, 1986). Coliform levels in unpasteurized apple cider ranged from < 1.0 to 4.6 CFU/ml (59 samples tested; Silk *et al.*, 1997) and from < 3 to 4.0 MPN/ml (169 samples tested; US FDA, 1999). However, there is little relationship between 'total coliforms' and pathogens in raw juices, because coliforms are commonly found on fruit and vegetable surfaces and are not recommended as indicators of food safety for these products (Kornacki and Johnson, 2001).

### 6.2.3 Fate of pathogens in juice

Acidic juices were not historically considered to support the growth or survival of foodborne pathogens. A report detailing a 1991 raw apple juice outbreak indicated that *E. coli* O157:H7 was capable of surviving 20 days at refrigerated temperature when inoculated into apple cider (Besser *et al.*,

© Woodhead Publishing Limited, 2012

1993). Miller and Kaspar (1994) determined that two strains of *E. coli* O157:H7 were able to survive for at least 21 days at 4°C, Zhao *et al.* (1993) reported from 10 to 31 days survival at 8°C. However, at 25°C survival of only 2–3 days is reported.

The acidity (pH 2.8–4.0) of orange juice prevents growth of *Salmonella* spp. and influences its survival (Parish *et al.*, 1997). Storage for approximately 15 (pH 3.5, 0°C) to 60 (pH 4.4, 0°C) or 68 days (pH 4.4, 4°C) was required for an approximately 6-log reduction in populations of *S.* Gaminara inoculated into pasteurized orange juice. Lag times before initial cell populations began to decline were correlated with pH and ranged from < 1 day at pH 3.5 to 27 days at pH 4.4.

Similarly, *Listeria monocytogenes* was capable of growing in orange serum adjusted to pH 4.8 (4°C) or 5.0 (30°C) with NaOH (Parish and Higgins, 1989). When the pH was adjusted to below pH 4.8 (4°C) or 4.6 (30°C), populations declined at progressively increasing rates as the pH declined. At pH 3.6, populations declined by approximately 6 log CFU/ml in 25 (4°C) to 5 days (30°C). Growth of *L. monocytogenes* in orange juice, at a pH of as low as 2.6 when cells are acid adapted, has also been reported (Caggia *et al.*, 2009). Populations of *L. monocytogenes* inoculated into commercially processed tomato juice held at 5°C remained constant over a 12-day storage period (Beuchat and Bracket, 1991). In contrast, carrot juice is anti-listerial, and this activity is influenced by pH, NaCl content, temperature, and time (Beuchat *et al.*, 1994).

## 6.3 Current methods of juice decontamination

Juice processing has been carried out on a large scale for almost a century, and there are several decontamination methods that have proven effective with respect to both quality retention and microbial decontamination. The microbiology of juice products, including foodborne illness associated with such products and need for and types of processing treatments to prevent spoilage and foodborne illness, has been extensively discussed (McLellan and Padilla-Zakour, 2005; Rutledge, 1996; Vojdani *et al.*, 2008; Worobo and Splittstoesser, 2005). This section summarizes common types of processing used to stabilize juice products.

### 6.3.1 Intrinsic factors

Intrinsic factors are those that are inherent to the juice itself or those that become part of the juice. Historically, the relatively low pH of fruit juices was thought to contribute to some inherent level of protection against pathogenic organisms. Preservatives are added to processed foods to protect against undesirable chemical and microbiological changes in the product over its shelf life. Antimicrobials, as the name suggests, control against undesirable

© Woodhead Publishing Limited, 2012

microorganisms and are part of that larger group of food additives known as preservatives. In most developed countries, these additives are highly regulated and require labeling if added to foods. For that reason, and also due to the effectiveness and ubiquity of thermal processing systems developed for juice, the use of preservatives in 100% fruit and vegetable juices in the US market is not extensive. Much of the current juice preservative research focuses on methods to achieve 5-log reductions in specific pathogen populations, either with the preservative itself or in conjunction with other technologies such as UV processing, thermal processing and others (the 'hurdle' concept), largely in response to the US Food and Drug Administration (FDA) Juice HACCP Regulation (US FDA, 2001).

*Juice acidity*

The pH of many fruit juices ranges from up to 5.0–5.5 for juices such as some apple and pear, to below pH = 2.0 for lemon juice. Vegetable juices such as carrot, beet and celery have somewhat higher pHs (pH = 6 and above) with concomitantly lower levels of titratable acidity. Spoilage of unpasteurized high and low pH juices is well documented, and contamination post-processing can become evident, particularly if the juice is not maintained below 5°C (41°F) or less. Many microorganisms are associated with juice spoilage, and are generally those that have some degree of acid resistance. These include such organisms as lactic acid bacteria (LAB) in the genera *Lactobacillus* and *Leuconostoc*; acetic acid bacteria including *Gluconobacter* and *Acetobacter* spp.; yeasts, molds, and spore-forming organisms such as *Alicyclobacillus* spp. that are capable of causing spoilage in high-acid shelf stable juices (Worobo and Splittstoesser, 2005).

Foodborne illness outbreaks have been associated with unpasteurized fruit juices, indicating that juice pH itself is not sufficient to ensure microbial stability and food safety. This had been recognized by early researchers (Goverd *et al.*, 1979; Murdock and Hatcher, 1975) and discussed by Parish (1997). A series of juice-related outbreaks occurring in the mid-1990s highlighted the food safety risk of selling and distributing fresh (unprocessed) juice products and led to the development of food safety regulations pertaining to juice. Parish (1998) described the isolation of several *Salmonella* serovars from orange juice and processing environments that were associated with a 1995 salmonellosis outbreak. In addition to plant sanitation issues, it was noted that the juice was extracted from very mature oranges with somewhat higher pH levels than average for orange juice, a risk factor that had not generally been considered for citrus juices. In an inoculated study with orange juice adjusted to different pHs and held under refrigeration, Parish *et al.* (1997) demonstrated death rates of *Salmonella* spp. were inversely correlated with juice pH. Numerous outbreaks related to *E. coli* O157:H7 in apple juice or cider (unpasteurized) were reported in the mid-1990s (CDC, 1996, 1997). Subsequent research demonstrated the survival of *E. coli* O157:H7 in unpasteurized apple juice, as well as providing a possible

© Woodhead Publishing Limited, 2012

mechanism of contamination during apple processing (Fisher and Golden, 1998a; 1998b).

*Antimicrobial preservatives*

Acidification

Acidification is recognized as a means of controlling the growth of undesirable microorganisms, including pathogens. Fermentation, a form of naturally-occurring acidification has long been used for food preservation, as has acidification by direct addition of organic and other appropriate acids (Brown and Booth, 1991). Acidulants such as citric acid and malic acid are commonly used in juice beverages and fruit products for both pH adjustment and flavor purposes (Somogyi, 2005). While acidification is rarely used as the sole control mechanism for pathogenic organisms in fruit juices, it has been recommended as a control step for producing pasteurized, chilled lower-acid juices such as carrot that can be contaminated with *C. botulinum* spores that survive pasteurization and can subsequently outgrow if the consumer package is subjected to temperature abuse (US FDA, 2007). This guidance was issued in response to the outbreak of botulism linked to refrigerated carrot juice that occurred in 2006.

Sulfites

Sulfites, including sodium sulfite, sodium metabisulfite, and sulfur dioxide are used as preservatives in processed fruit products including dried fruits, juices, and wine. They are both effective antioxidants that slow browning reactions and enhance ascorbic acid retention in certain fruit products, as well as inhibitors of selected microorganisms which makes them useful in the winemaking industry. They are used in the juice and juice beverage industry to inhibit malolactic bacteria, acetic acid bacteria, and spoilage yeasts and molds (Raybaudi-Massilia *et al.*, 2009). Sulfilte compounds demonstrate optimal antimicrobial effects below pH = 4 (Davidson and Taylor, 2007), which favors their use in high acid food systems such as fruit juices.

Sulfur dioxide is fungicidal at low concentrations and is used extensively in winemaking (before fermentation) to inhibit fungal spoilage organisms such as *Penicillium*, *Mucor*, and *Aspergillus* which can grow on freshly harvested grapes and contaminate the pressed juice. Residual sulfites are effective post-fermentation to control ethanol-tolerant spoilage yeasts, including *Brettanomyces bruxellensis* which can be responsible for a serious spoilage defect known as 'mousy aftertaste' (Hutkins, 2006).

Fisher and Golden (1998b) investigated treatment of apple juice (unfermented cider, pH = 3.4) with 0.0046% sodium bisulfite as a control for *E. coli* O157:H7, the pertinent pathogen in apple cider. That study demonstrated greater than 5-log reduction of CFU/mL in *E. coli* O157:H7 after 18 days (storage temperature of 4 and 10°C) and after 3 days (25°C) after sodium bisulfite addition to the cider.

© Woodhead Publishing Limited, 2012

Novel antimicrobials

Novel antimicrobial agents (those in development but not widely commercialized) have limited utility in the juice industry but may be useful for the control of pathogenic and spoilage organisms in fresh cut fruits and vegetables and freshly pressed juices (Raybaudi-Massilia, 2009). Natural plant antimicrobial compounds under investigation for fruit and vegetable juice systems include vanillin to control *Listeria* spp. in orange and apple juice (Ferrante *et al.*, 2007) and cinnamon powder to control *L. monocytogenes* and *E. coli* O157:H7 in pasteurized and unpasteurized apple juice (Ceylan *et al.*, 2004; Yuste and Fung, 2002). An obvious limitation to these treatments is the sensory effect on the final product, which is an inherent characteristic of these generally aromatic compounds.

Nisin is an antimicrobial peptide of bacterial origin, and is effective against Gram-positive vegetative organisms such as *Alicyclobacillus* spp., *Bacillus cereus*, and *C. botulinum*, and is sporostatic against the spores of these organisms (Davidson and Zivanovic, 2003). While commonly used in some foods, nisin has not found widespread use in juices. Nisin has been investigated alone and in combination with other preservation techniques in juice systems by a variety of different researchers. Liang *et al.* (2002) were unable to demonstrate significant reductions in *S.* Typhimurium (a Gram-negative organism) in orange juice with nisin treatments of 0.1 μg/mL as compared to control juice, but were able to demonstrate higher levels of microbial inactivation with nisin treatments combined with pulsed electric field (PEF) treatments. Wu *et al.* (2005) have found that nisin in combination with heat (51°C) PEF treatments reduced the natural bacteria population in intentionally spoiled grape juice by up to 6.2 log CFU/mL. The use of alternative processes for juice preservation will be further discussed later in this chapter.

### 6.3.2 Extrinsic factors

Extrinsic factors encompass specific processing operations used to produce a safe juice with sufficient shelf life for distribution and purchasing by the consumer. These include refrigeration, freezing, and thermal processing, which itself encompasses treatments ranging from lightly processed to aseptic sterilization (Farkas, 2007). Although the focus here is on processing to achieve the reduction of spoilage organisms and the destruction of pathogenic microbes, enzymatic inactivation is often a major objective of such processes. Refrigeration and freezing are processing operations that can obviously affect the microbiological character and shelf life of juice products. The most efficient, common, and highly validated process for juice decontamination, however, is thermal processing/pasteurization, which will be discussed in this section.

© Woodhead Publishing Limited, 2012

*Refrigerated storage*

For juices that are not further processed (i.e., fresh juice that is sold in its raw, unpasteurized state), refrigeration (generally defined as holding at or below 4°C) is the key hurdle technology used to achieve adequate shelf life. Refrigeration of product prior to processing and during intermediate steps is considered a 'best practice' in industry, and the literature has described the beneficial impact of decreased temperature on juice microbial shelf stability (Fellers, 1988; Root and Barrett, 2005; Worobo and Splittstoesser, 2005). It is important to note that refrigeration cannot be considered a decontamination method per se, as microbial populations can increase, albeit slowly, at refrigeration temperatures.

Some pathogens such as *Salmonella* spp. and *E. coli* O157:H7 survive storage under refrigerated conditions, and in some systems better under refrigerated conditions than at warmer holding temperatures. Goverd *et al.* (1979) observed *Salmonella* spp. survive at least 30 days at both 4 and 22°C when inoculated into apple cider (unpasteurized apple juice) of pH 4.0 and discussed the public health implication of relying on refrigeration and acidity to prevent foodborne illness. Parish *et al.* (1997) concluded that *Salmonella* spp. could survive sufficient time to cause illness from contaminated orange juice held under refrigerated conditions (0° and 4°C).

*Frozen storage*

Freezing has long been utilized for the long-term storage of juice concentrates and to a lesser degree, single-strength juice. This preservation method has been particularly important in the development of the global orange juice concentrate market. Much of the research has examined the effect of frozen storage on quality preservation, particularly with respect to nutrient and flavor quality, as well as prevention of color degradation due to chemical and enzymatic reactions and microbial degradation due to spoilage (Uljas and Ingham, 1998; Murdock and Hatcher, 1978; Murdock and Brokaw, 1965). For almost all microorganism/food systems, there is no microbial growth below –8°C (Farkas, 2007).

*Thermal processing*

Thermal processing of juices encompasses a wide range of treatments, from relatively mild heat treatments resulting in products that must be refrigerated, to products which have received sufficient treatment to render them shelf-stable. Unlike the dairy industry, where the term 'pasteurization' is well-defined, the term pasteurization as used by the juice industry describes heat processes used to: control pertinent pathogens by a 5-log reduction process and some spoilage organisms ('flash pasteurization' or 'lightly pasteurized', product requires refrigeration); control pathogens and most spoilage organisms (products require refrigeration); and control pathogens and spoilage organisms (shelf-stable products). Temperatures used for thermal processing of juices range from 65°C for lemon/lime juice to as high as 99°C for orange or grapefruit

© Woodhead Publishing Limited, 2012

juices for 6–30 s depending on product. Pasteurization conditions for most citrus juices are designed to inactivate the enzyme pectinmethylesterase and will easily destroy almost all spoilage organisms. The thermal sensitivity of important spoilage microbes in fruit juices varies depending upon the type of cell (vegetative, bacterial spore, fungal spore, yeast), pH, solids content, and water activity of the juice. The most common form of pasteurization uses plate-and-frame heat exchangers (for grape or clear apple juice) and shell-and-tube heat exchangers for more turbid and pulpy juices. Regardless of type, modern heat exchangers are energy efficient and have been optimized to produce high-quality juices.

Juice pasteurization has been identified by the US FDA Hazards and Control Guide (associated with the Juice HACCP Regulation) as the primary means of ensuring a 5-log reduction in the pertinent pathogen (US FDA, 2004), with specific time–temperature regimes recommended. For example, Mazzotta (2001) has provided validated information for pathogen reduction regimes for apple, orange, and other juices.

## 6.4 Novel methods of juice decontamination

Non-thermal processing alternatives for juice and beverages have been investigated as a means to retain flavor and minimize nutritional losses while still achieving a minimum 5-log reduction performance standard as prescribed by the FDA Juice HACCP Regulation 21 CFR 120 (US FDA, 2001). These non-thermal processing methods investigated for juice application include high pressure processing (HPP), ultraviolet light (UV), pulsed light (PL), pulsed electric field (PEF), supercritical carbon dioxide (SCD), and ozone (O). In addition, numerous chemical additives (dimethyl dicarbonate, benzoate, hydrogen peroxide, nisin, lysozyme, and spice/herb extracts) have been investigated as potential preservatives or in combination treatments to enhance the safety or shelf life extension of juices (Basaran-Akgul *et al.*, 2009; Elgayyar *et al.*, 2001; Kniel *et al.*, 2003; Liang *et al.*, 2002; Ukuku *et al.*, 2009; Williams *et al.*, 2005).

As with thermal processing, non-thermal processing must achieve a minimum 5-log reduction with the most resistant pertinent pathogen for the type of juice. Numerous studies of non-thermal processing technologies have evaluated non-pathogenic bacteria in juice.

### 6.4.1 High pressure processing

High pressure processing (HPP), also referred to as ultra high pressure (UHP) or high hydrostatic pressure processing (HHP) relies on the generation of high pressures ranging from 100 to 1000 MPa that results in the disruption of cell integrity and subsequent death. HPP has been shown to be effective in

© Woodhead Publishing Limited, 2012

achieving a greater than 5-log reduction with *Salmonella* spp. in orange juice (600 MPa, 5 s, at 20°C) and *E. coli* O157:H7 in apple juice (Bull *et al.*, 2005; Whitney *et al.*, 2007). It was shown that HPP (55 MPa, 30 s) was capable of achieving a 3.4-log reduction of *C. parvum* oocysts, whereas 60 s at 550 MPa resulted in a 99.995% reduction of *C. parvum* inactivation (Slifko *et al.*, 2000). Due to the low acidity of vegetable juices, *C. botulinum* is the most resistant pertinent pathogen of public health concern. The pressure resistance of *C. botulinum* has been shown to be much greater than vegetative cells for non-juice foods, requiring elevated temperatures (and pressures (600 MPa, 220 s, 121°C) to inactivate *C. botulinum* spores (Juliano *et al.*, 2009).

### 6.4.2 Ultraviolet light

The germicidal properties of ultraviolet light are primarily due to the formation of pyrimidine dimers between adjacent nucleotides which interferes with DNA replication and if sufficient UV exposure occurs, results in cell death (Bintsis *et al.*, 2000). UV has been shown to be effective in achieving a greater than 5-log reduction with 14.2 mJ of exposure (254 nm) for multiple strains of *E. coli* O157:H7 as well as *C. parum* oocysts in apple cider (Basaran *et al.*, 2004; Hanes *et al.*, 2002). The log reduction achieved with 14.2 mJ of UV exposure is not influenced by apple variety or pH, and does not alter the chemical composition or the organoleptic quality of the juice (Basaran *et al.*, 2004; Quintero-Ramos *et al.*, 2004). The effectiveness of UV for citrus juices is significantly reduced to the absorption of UV by vitamin C and pulp issues (Basaran *et al.*, 2004; Quintero-Ramos *et al.*, 2004). The UV resistance of bacterial spores and fungi is greater than vegetative bacteria or oocysts. As a result, the shelf life of UV treated juices is diminished compared to thermally processed juices (Tandon *et al.*, 2003).

### 6.4.3 Pulsed light

Pulsed light treatment consists of short intense pulses of light in the UV to near infrared range. Recently, the germicidal properties of pulsed light have been primarily due to the UV portion of the light spectra emitted (Woodling and Moraru, 2007). Earlier studies have reported additional non-UV cellular effects such as cell membrane distortion, cell shape, enlarged vacuoles, and protein elution. The additional cellular damage may be explained by the thermal effects that have been reported with pulsed light (Krishnamurthy *et al.*, 2010; Ozer and Demirci, 2006).

Pulsed light has been used to achieve the minimum 5-log reduction with *E. coli* O157:H7 in apple cider (Sauer and Moraru, 2009), but additional juices have not been tested to evaluate the effectiveness of pulsed light to inactivate pathogens.

© Woodhead Publishing Limited, 2012

### 6.4.4 Pulsed electric field

The mode of microbial inactivation of PEF is believed to be the disruption of cellular membranes due to the external pulsed electric field that is applied. The cell membrane disruption causes an efflux of intracellular components and, if sufficient damage is caused and regeneration of the membrane is prevented, cell death occurs (Wouters and Smelt, 1997; Wouters *et al.*, 2001). PEF germicidal efficacy is enhanced with mild heating (45–55°C) and heating of the juice may occur if the electric field strength or treatment time is excessive. Researchers have reported a greater than 5-log reduction for *E. coli* O157:H7 in apple juice (80 kV, 30 pulses, 42°C) and for *S.* Typhimurium in orange juice (90 kV/cm, 50 pulses, 55°C), but the lethality due to the thermal treatment was not solely evaluated so the reported log reduction may be higher than reported for the *Salmonella* study (Ju *et al.*, 2001; Liang *et al.*, 2002). The inactivation of *C. parvum* oocysts in apple cider using PEF has not been reported, and is needed for the application of PEF as a recognized 5-log treatment for apple juice not from concentrate.

### 6.4.5 Supercritical carbon dioxide

Carbon dioxide when pressurized to greater than 7.4 MPa and heated to above 31°C exists as supercritical carbon dioxide (Tomasula, 2003). In this state, supercritical carbon dioxide has properties that are between liquid and gas that are believed to allow for penetration and subsequent disruption of bacterial cell membranes. The disruption of the cell membrane integrity is believed to result in cellular efflux and if sufficient damage occurs, cell death results (Shimoda *et al.* 1998). Limited published research has been reported for supercritical carbon dioxide inactivation of pathogens for juices. Yuk *et al.* (2010) reported that supercritical carbon dioxide at 7.6 MPa and 34–42°C, in apple cider, resulted in a greater than 5-log reduction of *E. coli* K12, a non-pathogenic bacterial strain. Rasanayagam (2004) reported that orange juice inoculated with *Salmonella* spp., *E. coli* O157:H7, and *L. monocytogenes*, subjected to 7.9 MPa at 40°C for 4 min, achieved a greater than 6-log reduction, but a full publication of this report has not been made. *Bacillus* spp. spores have been reported to exhibit higher temperature and pressure requirements (30 MPa, 45–50°C) compared to vegetative cells but this study did not involve juice as the medium. No published research has been conducted on the effectiveness of supercritical carbon dioxide with *C. parvum* in juices.

### 6.4.6 Ozone

Ozone is the highly reactive species of oxygen that is formed by several methods that include ultraviolet light, coronal discharge, cold plasma, and electrolytic ozone generation. Ozone quickly decays and must be produced on site for use. Ozone has been investigated as a potential treatment method

© Woodhead Publishing Limited, 2012

for orange and apple juice (Williams *et al.*, 2005). It was reported that at least 5-log reductions of *E. coli* O157:H7 and *Salmonella* spp. in orange and apple juices were achieved following addition of dimethyl dicarbonate (500 ppm) or hydrogen peroxide (300 or 600 ppm), with 0.9g/h ozone and storage at 4°C for 24 h. There are no published reports for ozone inactivation of *C. parvum* in any juice, but it is believed that *C. parvum* has higher ozone requirements for inactivation since ozonated apple cider was responsible for a 2003 *C. parvum* outbreak that occurred in Ohio (Blackburn *et al.*, 2006). Additional research has been reported that 75–78 μg/mL of ozone resulted in a greater than 5-log reduction of non-pathogenic *E. coli* inactivation in orange juice (Patil *et al.*, 2009). As with other thermal and non-thermal processing treatments, bacterial spores are more resistant to ozone than vegetative cells of the same species.

### 6.4.7 Conclusion

The commercialization of non-thermal processing technologies has progressed in the past twenty years; initiated by the desire for improved organoleptic qualities and retention of nutritional qualities compared to that of thermal processed juice. However, additional research is needed to fully evaluate the various technologies to meet minimum regulatory safety requirements. Numerous non-thermal processing methods have not been evaluated for their effectiveness against pertinent pathogens for the various types of juice, or have been evaluated only on commercial scale processing equipment with appropriate and validated non-pathogenic surrogate organisms. In addition, further research is needed to address the minimum shelf life requirements that are expected by the juice and beverage industries.

## 6.5 Future trends

The only constant in the world is change. During the past half century, consumer demands for convenient high quality products at reasonable costs have led to newer and faster juice processing and packaging technologies. With each new product and packaging innovation, the potential for the emergence or re-emergence of microbiological problems has existed. For example, in the 1950s after commercialization of juice evaporators for production of concentrated orange juice, colonization of evaporators and process lines by lactic acid bacteria (LAB) yielded concentrates with off-flavors and other quality problems (Hays and Riester, 1952). This required additional advances in processing and decontamination technology, such as development of the thermally accelerated short time evaporator, in order to control spoilage and produce consistently high quality juice concentrates.

There are other examples where marketing and/or technological advances led to microbiological issues. Technologies for production of shelf-stable fruit

© Woodhead Publishing Limited, 2012

juices led to the realization by industry that heat resistant molds (HRM) can grow in these products and must be controlled with appropriate processing, packaging, sanitation, and ingredient sourcing (Oliver and Rendle, 1934). The advent of oxygen-barrier gable top cartons in the 1990s allowed for the extension of refrigerated juice shelf life from approximately 35 days to 70 days leading to growth of psychrotrophic mold propagules contained within the carton paperboard fibers to cause mold spoilage (Narciso and Parish, 2000). Additionally, growth of the raw, unpasteurized juice markets in the 1990s was a reminder to some segments of the juice industry that certain pathogens, such as *Salmonella*, *E. coli* O157:H7 and *C. parvum*, will survive the acidic conditions of refrigerated juices for a period of weeks thereby allowing disease outbreaks to occur (Parish, 1997; Vojdani *et al.*, 2008).

The above examples show us that even subtle changes in a product or process can lead to new microbiological spoilage and safety issues. This is important to consider as we explore new technologies and new types of products to meet consumer demands. For the past two decades, there has been an increase in research efforts on processing technologies that will reduce thermal treatments thereby enhancing flavor quality of juices. Research on process technologies such as high hydrostatic pressure (Bayindirli *et al.*, 2006; Rendueles *et al.*, 2011), pulsed electric field (Morales-de la Penã *et al.*, 2010; Mosqueda-Melgar *et al.*, 2008), UV treatment (Basaran *et al.*, 2004; Hanes *et al.*, 2002), high pressure homogenization (Maresca *et al.*, 2011; Suárez-Jacobo *et al.*, 2010), sonication (Adekunte *et al.*, 2010), irradiation (Song *et al.*, 2007), radio frequency electric field (Ukuku and Geveke, 2010) and super critical dense phase $CO_2$ (Mantoan and Spilimbergo, 2011; Xu *et al.*, 2011) has been conducted with varying degrees of success. Some well-established processes, such as pasteurization and aseptics, are widely utilized commercially and capable of producing wholesome and safe juices and beverages. Other newer processes remain experimental or are applied in a limited commercial fashion to satisfy niche markets. While the potential for newer processes to be more widely adopted remains a possibility, further research that addresses the mechanism of microbial inactivation and the application of the technology for control of specific organisms may be needed. Additionally, new antimicrobial packaging concepts, some based on nanoparticles (Emamifar *et al.*, 2011), may offer alternative methods for controlling microbial growth in final products.

Marketing trends suggest that the 100% juice market is mature with consumption per capita relatively unchanged in recent years. Future growth may focus on niche marketing to specific population groups as well as growth in consumption of juice-containing beverages, juice flavored waters, teas, and other beverages. Development of health conscious products is leading to incorporation of new ingredients, such as vitamins, dietary supplements, probiotics, and new non-nutrient sweeteners into traditional juice-containing drinks and beverages. These products also may contain exotic juices (e.g., açaí and pomegranate) or juices from common fruit usually eaten whole or

© Woodhead Publishing Limited, 2012

cut (e.g., berries and melons). Concerns about higher pH values for several of these newer juice types require that processes be validated to control pathogens as well as spoilage agents.

Incorporation of new ingredients often results in a need to make slight changes to process parameters or to merge new technologies into existing process lines due to functionality issues related to the new ingredients. As mentioned previously, experience teaches us that such changes may lead to unforeseen or re-emerging microbiological challenges. For example, research on the use of chitosan as a preservative in fruit juice suggests that this compound may enhance survival of *E. coli* O157 in refrigerated apple juice while inhibiting growth of spoilage yeasts (Kiskó *et al.*, 2005). Therefore the use of chitosan may require additional scrutiny to ensure control of vegetative pathogens. Another case in point is the 2006 outbreak of botulism from thermally abused carrot juice (Sheth *et al.*, 2008). Juices and beverages of pH > 4.5 should not depend solely on refrigeration to inhibit growth of *C. botulinum*. Additional process hurdles such as acidification or use of appropriate preservatives would be needed in such products to inhibit growth of *C. botulinum*.

Of particular concern in recent years is spoilage of juices and beverages by the alicyclobacilli (Steyn *et al.*, 2011). The genus *Alicyclobacillus* contains species of thermoacidophilic sporeforming bacteria capable of growing in low pH shelf-stable foods at ambient temperatures. While substantial research has been conducted in recent years on control methods, this organism will remain an issue the industry and academia must address in the future (Walker and Phillips, 2008). Basic research is needed to understand the mechanisms of both spore outgrowth and spore/cell inactivation in juices so that process technologies can be developed that will prevent, reduce or eliminate the ability of this organism to cause product spoilage.

In summary, future trends toward new products with unique ingredients and adoption of new process technologies will require additional process validations to ensure that spoilage and public health microorganisms of concern are adequately addressed. Emphasis on environmental monitoring to ensure control of microorganisms in the processing environment will also provide opportunities for control of microorganisms in the final product.

## 6.6 Sources of further information

Additional information regarding the decontamination of juices and beverages can be found in a number of sources depending upon your specific needs. Books, scientific journals, professional societies and trade associations are common resources for information on juices, juice processing, and public health and spoilage microorganisms. There are numerous refereed journals that serve as primary sources of scientific information related to juice processing and microbiology. Selected journals of note include the *International Journal*

© Woodhead Publishing Limited, 2012

*of Food Microbiology*, *Journal of Food Science*, *Journal of Food Protection*, *Journal of Food Engineering*, *Journal of Applied Microbiology*, *Food Research International*, and *Food Science & Technology (LWT)*.

Numerous books addressing juice processing and microbiology have been published for several decades and cover a variety of topics related to juice processing, including microbial inactivation and decontamination issues (Barrett *et al.*, 2004; Bates *et al.*, 2001; Braddock, 1999; Foster and Vasavada, 2003; Hui *et al.*, 2006; Kimball, 1999; Nagy *et al.*, 1993). Specific information on methods used for microbiological analysis of juices can be found in the *Compendium of Methods for the Microbiological Examination of Foods*, 4th edn (Downes and Ito, 2001).

In the US, a particularly effective source of information is the State Cooperative Extension System. Extension agents and specialists affiliated with the land-grant universities are located on campus, at research centers and in county offices within the state. A list of local Cooperative Extension offices is available at http://www.nifa.usda.gov/Extension/index.html.

Professional societies and trade associations often provide information to their members and the general public. Professional societies with sections that actively address juice processing microbiology include the International Association for Food Protection (http://foodprotection.org), International Society of Beverage Technologists (http://bevtech.org) and the Institute of Food Technologists (http://www.ift.org), among others. A few of the trade associations and other organizations that provide information on juice processing include the Juice Products Association (http://juiceproducts.org), Grocery Manufacturers Association (http://www.gmaonline.org), International Fruit-Juice Union (http://www.ifu-fruitjuice.com), Florida Citrus Processors Association (http://fcplanet.org), and Apple Processors Association (http://www.appleprocessors.org).

## 6.7 References

ADEKUNTE A, TIWARI B, SCANNELL A, CULLEN P, O'DONNELL C (2010) Modelling of yeast inactivation in sonicated tomato juice. *International Journal of Food Microbiology* 137:116–120.

BARRETT D, SOMOGYI L, RAMASWAMY H (2004) *Processing Fruits, Science and Technology*, 2nd edn CRC Press, Boca Raton, FL.

BASARAN N, QUINTERO-RAMOS A, MOAKE M, CHUREY J, WOROBO R (2004) Influence of apple cultivars on inactivation of different strains of *Escherichia coli* O157:H7 in apple cider by UV irradiation. *Applied and Environmental Microbiology* 70:6061–6065.

BASARAN-AKGUL N, CHUREY JJ, BASARAN P, WOROBO RW (2009) Inactivation of different strains of *Escherichia coli* O157:H7 in various apple ciders treated with dimethyl dicarbonate (DMDC) and sulfer dioxide ($SO_2$) as an alternative method. *Food Microbiology* 26:8–15.

BATES R, MORRIS J, CRANDALL P (2001) Principles and Practices of Small- and Medium-Scale Fruit Juice Processing. *FAO Agricultural Services Bulletin 146*. Food and Agricultural Organization, Rome.

© Woodhead Publishing Limited, 2012

BAYINDIRLI A, ALPAS H, BOZO LU F, HIZAL M (2006) Efficiency of high pressure treatment on inactivation of pathogenic microorganisms and enzymes in apple, orange, apricot and sour cherry juices. *Food Control* 17:52–58.

BESSER RE, LETT SM, WEBER JT, DOYLE MP, BARRETT TJ, WELLS JG, GRIFFIN PM (1993) An outbreak of diarrhea and hemolytic uremic syndrome from *Escherichia coli* O157:H7 in fresh-pressed apple cider. *Journal of the American Medical Association* 269:2217–2220.

BEUCHAT LR (1986) Mold spoilage of fruit products. In: Matthews RF (ed) *Proceedings of the 26th Annual Short Course for the Food Industry: Citrus Processing New Developments and Adaptations*, University of Florida and Institute of Food Technologists Florida Section, pp. 39–53.

BEUCHAT LR, BRACKETT RE (1991) Behavior of *Listeria monocytogenes* inoculated into raw tomatoes and processed tomato products. *Applied and Environmental Microbiology* 57:1367–1371.

BEUCHAT LR, BRACKETT RE, DOYLE MP (1994) Lethality of carrot juice to *Listeria monocytogenes* as affected by pH, sodium chloride, and temperature. *Journal of Food Protection* 57:470–474.

BINTSIS T, LITOPOULOU-TZANETAKI E, ROBINSON RK (2000) Existing and potential applications of ultraviolet light in the food industry – a critical review. *Journal of the Science of Food and Agriculture* 80:637–645.

BIRKHEAD GS, MORSE DL, LEVINE WC, FUDALA JK, KONDRACKI SF, CHANG HG, SHAYDGANI M, NOVICK L, BLAKE PA (1993) Typhoid fever at a resort hotel in New York: a large outbreak with an unusual vehicle. *Journal of Infectious Diseases* 167:1228–1232.

BLACKBURN BG, MAZUREK JM, HLAVSA M, PARK J, TILLAPAW M, PARRISH M, SALEHI E, FRANKS W, KOCH E, SMITH F, XIAO L, ARROWOOD M, HILL V, D A SILVA A, JOHNSTON S, JONES JL (2006) Cryptosporidiosis associated with ozonated apple cider. *Emerging Infectious Diseases* 12(4):684–686.

BRADDOCK R (1999) *Handbook of Citrus By-Products and Processing Technology*. John Wiley & Sons, New York.

BROWN MH, BOOTH IR (1991) Acidulants and low pH. In: Russell NJ, Gould GW (eds) *Food Preservatives*. Blackie, Glasgow, pp. 22–43.

BULL MK, SZABO EA, COLE MB, STEWART CM (2005) Toward validation of process criteria for high-pressure processing of orange juice with predictive models. *Journal of Food Protection* 68(5):949–954.

BUTLER ME (2000) Salmonella outbreak leads to juice recall in Western states. *Food Chemical News*, April 24.

BUZBY JC, CRUTCHFIELD SR (1999) New Juice Regulations Underway. *Food Review*. Available from: http://www.ers.usda.gov/publications/foodreview/may1999/frmay99f.pdf (accessed Oct 5, 2011).

CAGGIA C, SCIFO GO, RESTUCCIA C, RANDAZZO CL (2009) Growth of acid-adapted *Listeria monocytogenes* in orange juice and in minimally processed orange slices. *Food Control* 20:59–66.

CENTERS FOR DISEASE CONTROL (1975) *Salmonella typhimurium* outbreak traced to a commercial apple cider – New Jersey. *Morbidity and Mortality Weekly Report* 24:87–88.

CENTERS FOR DISEASE CONTROL (1991) Cholera associated with imported frozen coconut milk – Maryland. *Morbidity and Mortality Weekly Report* 40:844–845.

CENTERS FOR DISEASE CONTROL (1995) Outbreak of *Salmonella* Hartford infections among travellers to Orlando, Florida, EPI-AID Trip Report 95–62.

CENTERS FOR DISEASE CONTROL (1996) Outbreak of *Escherichia coli* O157:H7 infections associated with drinking unpasteurized commercial apple juice – British Columbia, California, Colorado, and Washington, October 1996. *Morbidity and Mortality Weekly Report* 45:975.

CENTERS FOR DISEASE CONTROL (1997) Outbreaks of *Escherichia coli* O157:H7 infection and cryptosporidiosis associated with drinking unpasteurized apple cider – Connecticut and New York, October 1996. *Morbidity and Mortality Weekly Report* 46:4–8.

© Woodhead Publishing Limited, 2012

CENTERS FOR DISEASE CONTROL (1999) Outbreak of Salmonella serotype Muenchen infections associated with unpasteurized orange juice – United States and Canada, June 1999. *Morbidity and Mortality Weekly Report* 48:582–585.

CENTERS FOR DISEASE CONTROL (2006) Botulism associated with commercial carrot juice – Georgia and Florida. *Morbidity and Mortality Weekly Report* 55:1098–1099.

CENTERS FOR DISEASE CONTROL (2010) Investigation Update: Multistate outbreak of human Typhoid Fever infections associated with frozen mamey pulp. Available from: http://www.cdc.gov/salmonella/typhoidfever/index.html (accessed Oct 4, 2011).

CENTERS FOR DISEASE CONTROL (2011) Foodborne Outbreak Online Database (FOOD). Available from: http://wwwn.cdc.gov/foodborneoutbreaks/Default.aspx (accessed Oct 4, 2011).

CEYLAN E, FUNG DYC, SABAH JR (2004) Antimicrobial activity and synergistic effect of cinnamon with sodium benzoate or potassium sorbate in controlling *E. coli* O157:H7 in apple juice. *Journal of Food Science* 69:102–106.

CODY SH, GLYNN K, FARRAR JA, CAIRNS KL, GRIFFIN PM, KOBAYASHI J, FYFE M, HOFFMAN R, KING AS, LEWIS JH, SWAMINATHAN B, BRYANT RG, VUGIA DJ (1999) An outbreak of *Escherichia coli* O157:H7 infection from unpasteurized commercial apple juice. *Annals of Internal Medicine* 130:202–209.

COOK KA, DOBBS TE, HLADY WG, WELLS JG, BARRETT TJ, PUHR ND, LANCETTE GA, BODAGER DW, TOTH WL, GENESE CA, HIGHSMITH AK, PILOT KE, FINELLI L, SWERDLOW DL (1998) Outbreak of *Salmonella* serotype Hartford infections associated with unpasteurized orange juice. *Journal of the American Medical Association* 280:1504–1509.

DAVIDSON PM, TAYLOR TM (2007) Chemical preservatives and natural antimicrobial compounds. In: Doyle MP, Beuchat LR (eds) *Food Microbiology Fundamentals and Frontiers*. ASM Press, Washington, DC, pp. 713–746.

DAVIDSON PM, ZIVANOVIC S (2003) The use of natural antimicrobials. In: Zeuth P, Bogh-Sorensen L (eds) *Food Preservation Techniques*. Woodhead Publishing, Cambridge.

DAVIS ND, DIENER UL (1987) Mycotoxins. In: Beuchat LR (ed) *Food and Beverage Mycology*, 2nd edn. AVI/Van Nostrand Reinhold, New York, pp. 517–70.

DOORES S (1983) The microbiology of apples and apple products. *CRC Critial Reviews in Food Science and Nutrition*. 19:133–149.

DOUGLAS M (1930) Some principles regulating the life and death of pathogenic intestinal bacteria in artificial media and in fruit juices. *Lancet* 2:789–791.

DOWNES F, ITO K (2001) *Compendium of Methods for the Microbiological Examination of Foods*, 4th edn. American Public Health Association, Washington, DC.

DUNCAN TG, COULL JA, MILLER ER, BANCROFT H (1946) Outbreak of typhoid fever with orange juice as the vehicle illustrating the value of immunization. *American Journal of Public Health* 36:34–36.

EISENSTEIN AB, AACH RD, JACOBSON W, GOLDMAN A (1963) An epidemic of infectious hepatitis in a general hospital. *Journal of the American Medical Association* 185:171–174.

ELGAYYAR M, DRAUGHON FA, GOLDEN DA, MOUNT JR (2001) Antimicrobial activity of essential oils from plants against selected pathogenic and saprophytic organisms. *Journal of Food Protection* 64:1019–1024.

EMAMIFAR A, KADIVAR M, SHAHEDI M, SOLEIMANIAN-ZAD S (2011) Effect of nanocomposite packaging containing Ag and ZnO on inactivation of *Lactobacillus plantarum* in orange juice. *Food Control* 22:408–413.

FARKAS J (2007) Physical methods of food preservation. In: Doyle MP, Beuchat LR (eds) *Food Microbiology Fundamentals and Frontiers*. ASM Press, Washington, DC, pp. 685–712.

FELLERS PJ (1988) Shelf-life and quality of freshly squeezed, unpasteurized, polyethylene-bottled citrus juice. *Journal of Food Science* 53:1699–1702.

FERRANTE S, GUERRERO S, ALZAMORA MS (2007) Combined use of ultrasound and natural antimicrobials to inactivate *Listeria monocytogenes* in orange juice. *Journal of Food Science* 70:1850–1856.

© Woodhead Publishing Limited, 2012

FISHER TL, GOLDEN DA (1998a) Fate of *Escherichia coli* O157:H7 in ground apples used in cider production. *Journal of Food Protection* 61:1372–1374.

FISHER TL, GOLDEN DA (1998b) Survival of *Escherichia coli* O157:H7 in apple cider as affected by dimethyl dicarbonate, sodium bisulfite, and sodium benzoate. *Journal of Food Science* 63:1–3.

FORGACS J, TANNER FW (1943) Longevity of pathogenic bacteria in apple juice. *Fruit Products Journal and American Vinegar Industry* 22:295–304.

FOSTER T, VASAVADA P (2003) *Beverage Quality and Safety*. CRC Press, Boca Raton, FL.

GOVERD KA, BEECH FW, HOBBS RP, SHANNON R (1979) The occurrence and survival of coliforms and salmonellas in apple juice and cider. *Journal of Applied Microbiology* 46(3):521–530.

HANES DE, ORLANDI PA, BURR DH, MILIOTIS MD, BIER JW, JACKSON JG, ARROWOOD MJ, CHUREY JJ, WOROBO RW (2002) Inactivation of *Cryptosporidium parvum* oocysts in fresh apple cider using ultraviolet irradiation. *Applied and Environmental Microbiology* 68(8):4168–4172.

HARRIS LJ, FARBER JN, BEUCHAT LR, PARISH ME, SUSLOW TV, GARRETT EH, BUSTA FF (2003) Outbreaks associated with fresh produce: incidence, growth and survival of pathogens in fresh and fresh-cut produce. *Comprehensive Reviews in Food Science and Food Safety* 2S:78–141.

HAYS G, RIESTER D (1952) The control of 'off-odor' spoilage in frozen concentrated orange juice. *Food Technology* 6:386–389.

HUI Y, BARTA J, PILAR CANO M, GUSEK T, SIDHU J, SINHA N (2006). *Handbook of Fruits and Fruit Processing*. Blackwell Publishing, Oxford.

HUTKINS RW (2006) *Microbiology and Technology of Fermented Foods*, Blackwell Publishing, Ames, IA, pp. 349–395.

INDIANA STATE DEPARTMENT OF HEALTH (INS DOH) (1997) Summary of Special Disease Outbreak Investigations – 1997, Appendix E. Available from: http://www.in.gov/isdh/21186.htm (accessed Oct 5, 2011).

JAIN S, BIDOL SA, AUSTIN JL, BERL E, ELSON F, WILLIAMS ML, DEASSY III, M, MOLL ME, REA V, VOJDANI JD, YU PA, HOEKSTRA RM, BRADEN CR, LYNCH MF (2009) Multistate outbreak of *Salmonella* Typhimurium and Saintpaul infections associated with unpasteurized orange juice – United States, 2005. *Clinical Infectious Diseases* 48:1065–1071.

JU J, MITTAL GS, GRIFFITHS MW (2001) Reduction in levels of *Escherichia coli* O157:H7 in apple cider by pulsed electric fields. *Journal of Food Protection* 64(7):964–969.

JULIANO P, KNOERZER K, FRYER PJ, VERSTEEG C (2009) *C. botulinum* inactivation kinetics implemented in a computational model of a high-pressure sterilization process. *Biotechnology Progress*, 25(1):163–175.

KATZ DJ, CRUZ MA, TREPKA MJ, SAREZ JA, FIORELLA PD, HAMMOND RM (2002) An outbreak of Typhoid Fever in Florida associated with an imported frozen fruit. *Journal of Infectious Diseases* 186:234–239.

KIMBALL D (1999) *Citrus Processing: A Complete Guide*, 2nd edn. Aspen Publishers, Gaithersburg, MD.

KISKÓ G, SHARP R, ROLLER S (2005) Chitosan inactivates spoilage yeasts but enhances survival of *Escherichia coli* O157:H7 in apple juice. *Journal of Applied Microbiology* 98:872–880.

KNIEL KE, SUMNER SS, LINDSAY DS, HACKNEY CR, PIERSON MD, ZAJAC AM, GOLDEN DA, FAYER R (2003) Effect of organic acids and hydrogen peroxide on *Cryptosporidium parvum* viability in fruit juices. *Journal of Food Protection* 66:1650–1657.

KORNACKI JL, JOHNSON JL (2001) *Enterobacteriaceae*, coliforms and *Escherichia coli* as quality and safety indicators. In: Downes FP, Ito K (eds) *Compendium of Methods for the Microbiological Examination of Foods*, 4th edn. American Public Health Association, Washington, DC.

KRAUSE G, TERZAGIAN R, HAMMOND R (2001) Outbreak of *Salmonella* serotype Anatum

© Woodhead Publishing Limited, 2012

infection associated with unpasteurized orange juice. *Southern Medical Journal* 94:1168–1172.

KRISHNAMURTHY K, TEWARI JC, IRUDAYARAJ J, DEMIRCI A (2010) Microscopic and spectroscopic evaluation of inactivation of *Staphylococcus aureus* by pulsed UV light and infrared heating. *Food and Bioprocess Technology* 3:93–104.

LABORATORY SURVEILLANCE DATA FOR ENTERIC PATHOGENS IN CANADA (LSDEPC). Annual summary 2005. Available from: http://www.nml-lnm.gc.ca/NESP-PNSME/index-eng.htm (accessed Oct 4, 2005).

LESTER R, STEWART T, CARNIE J, NG S, TAYLOR R (1991) Air travel-associated gastroenteritis outbreak, August 1991. *Communicable Disease Intelligence* 15:292–293.

LIANG Z, MITTAL GS, GRIFFITHS MW (2002) Inactivation of *Salmonella* Typhimurium in orange juice containing antimicrobial agents by pulsed electric field. *Journal of Food Protection* 65(7):1081–1087.

MANTOAN D, SPILIMBERGO S (2011) Mathematical modeling of yeast inactivation of freshly squeezed apple juice under high-pressure carbon dioxide. *Critical Reviews in Food Science and Nutrition* 51:91–97.

MARESCA P, DONSÌ F, FERRARI G (2011) Application of a multi-pass high-pressure homogenization treatment for the pasteurization of fruit juices. *Journal of Food Engineering* 104:364–372.

MAZZOTTA AS (2001) Thermal inactivation of stationary-phase and acid-adapted *Escherichia coli* O157:H7, *Salmonella*, and *Listeria monocytogenes* in fruit juices. *Journal of Food Protection* 64(3):315–320.

MCLELLAN M, PADILLA-ZAKOUR O (2005) In: Barrett DM, Somogyi L, Ramaswamy H (eds) *Processing Fruits*, 2nd edn. CRC Press, Boca Raton, FL.

MILLARD PS, GENSHEIMER KF, ADDISS DG, SOSIN DM, BECKETT GA, HOUCK-JANKOSKI A, HUDSON A (1994) An outbreak of cryptosporidiosis from fresh-pressed apple cider. *Journal of the American Medical Association* 272:1592–1596.

MILLER LG, KASPAR CW (1994) *Escherichia coli* O157:H7 acid tolerance and survival in apple cider. *Journal of Food Protection* 57:460–464.

MITSCHERLICH E, MARTH EH (1984) *Microbial Survival in the Environment*. Springer Verlag, New York.

MORALES-DE LA PENÃ M, SALVIA-TRUJILLO L, ROJAS-GRAÜ M, MARTÍN-BELLOSO O (2010) Impact of high intensity pulsed electric field on antioxidant properties and quality parameters of a fruit juice–soymilk beverage in chilled storage. *LWT-Food Science and Technology* 43:872–881.

MOSQUEDA-MELGAR J, ELEZ-MARTÍNEZ P, RAYBAUDI-MASSILIA R, MARTÍN-BELLOSO O (2008) Effects of pulsed electric fields on pathogenic microorganisms of major concern in fluid foods: a review. *Critical Review in Food Science and Nutrition* 48:747–759.

MURDOCK DI, BROKAW CH (1965) Spoilage developing in 6 oz. cans of frozen concentrated orange juice during storage at room temperature and 40°F with particular reference to microflora present. *Food Technology* 19:234–238.

MURDOCK DI, HATCHER WS (1975) Growth of microorganisms in chilled orange juice. *Journal of Milk and Food Technology* 38:393–396.

MURDOCK DI, HATCHER WS (1978) Effect of temperature on survival of yeast in 45°and 65° Brix orange concentrate. *Journal of Food Protection* 41:689–691.

NAGY S, CHEN C, SHAW P (1993) *Fruit Juice Processing Technology*, AgScience Inc., Auburndale, FL.

NARCISO J, PARISH M (2000) Relationship of molds in paperboard packaging to food spoilage. *Dairy, Food and Environmental Sanitation* 20:944–951.

NATIONAL CENTRE FOR DISEASE CONTROL/COMMUNICABLE DISEASES NETWORK AUSTRALIA NEW ZEALAND, AUSTRALIAN DEPARTMENT OF HEALTH and AGED CARE (1999) Salmonellosis outbreak, South Australia. *Communicable Diseases Intelligence* 23:73.

ODLAUG TE, PFLUG IJ (1978) *Clostridium botulinum* and acid foods. *Journal of Food Protection* 41:566–573.

© Woodhead Publishing Limited, 2012

OLIVER M, RENDLE T (1934) A new problem in fruit preservation. Studies on *Byssochlamys fulva* and its effect on the tissues of processed fruit. *Journal of the Society of Chemical Industry* 53:166–172.

OZER NP, DEMIRCI A (2006) Inactivation of *Escherichia coli* O157:H7 and *Listeria monocytogenes* inoculated on raw salmon fillets by pulsed UV-light treatment. *International Journal of Food Science and Technology* 41:354–360.

PAQUET P (1923) Épidémie de fièvre typhoide: Determinée par la consummation de petit citre. *Revue d'Hygiene et de Medcine Sociale* 45:165–169.

PARISH ME (1997) Public health and non-pasteurized fruit juices. *Critical Review in Microbiology* 23:109–119.

PARISH ME (1998) Coliforms, *Escherichia coli* and *Salmonella* serovars associated with a citrus-processing facility implicated in a salmonellosis outbreak. *Journal of Food Protection* 61:280–284.

PARISH M (2000) Relevancy of *Salmonella* and pathogenic *E. coli* to fruit juices. Proceedings IFU-Workshop 'Microbiology', *Fruit Processing* 10:246–250.

PARISH ME, HIGGINS DP (1989) Extinction of *Listeria monocytogenes* in single-strength orange juice: comparison of methods for detection in mixed populations. *Journal of Food Safety* 9:267–277.

PARISH ME, NARCISO JA, FRIEDRICH LM (1997) Survival of *Salmonella*e in orange juice. *Journal of Food Safety* 17(1997):273–281.

PATIL S, BOURKE P, FRIAS JM, TIWARI BK, CULLEN PJ (2009) Inactivation of *Escherichia coli* in orange juice using ozone. *Innovative Food Science and Emerging Technologies* 10:551–557.

QUINTERO-RAMOS A, CHUREY J, HARTMAN P, BARNARD J, WOROBO RW (2004) Modeling of *Escherichia coli*: O157:H7 inactivation by UV irradiation at different pH values in apple cider. *Journal of Food Protection* 67(6):1153–1156.

RASANAYAGM V (2004) Processing of orange juice using supercritical carbon dioxide. Abstract 49H-23. 2004 IFT Annual Meeting, 12–16 July, Las Vegas, NV.

RAYBAUDI-MASSILIA RM, MOSQUEDA-MELGAR J, SOLIVA-FORTUNY R, MARTIN-BELLOSO O (2009) Control of pathogenic and spoilage microorganisms in fresh-cut fruits and fruit juices by traditional and alternative natural antimicrobials. *Comprehensive Review of Food Science and Food Safety* 8:157–180.

RENDUELES E, OMER M, ALVSEIKE O, ALONSO-CALLEJA C, CAPITA R, PRIETO M (2011) Microbiological food safety assessment of high hydrostatic pressure processing: a review. *LWT-Food Science and Technology* 44:1251–1260.

ROOT WH and BARRETT DM (2005) Apples and apple processing. In: Barrett DM, Somogyi L and Ramaswamy H (eds) *Processing Fruits*, 2nd edn. CRC Press, Boca Raton, FL, pp. 455–480.

RUTLEDGE P (1996) Production of non-fermented fruit products. In: Arthey D, Ashust PR (eds) *Fruit Processing*. Chapman and Hall, Glasgow.

SAUER A, MORARU CI (2009) Inactivation of *Escherichia coli* ATCC 25922 and *Escherichia coli* O157:H7 in apple juice and apple cider, using pulsed light treatment. *Journal of Food Protection* 72:937–944.

SCOTT PM, MILES WF, TOFT P, DUBE JG (1972) Occurrence of patulin in apple juice. *Journal of Agriculture and Food Chemistry* 20:450–451.

SHETH AN, WIERSMA P, ATRUBIN D, DUBEY V, ZINK D, SKINNER G, DOERR F, JULIAO P, GONZALEZ G, BURNETT C, DRENZEK C, SHULER C, AUSTIN J, ELLIS A, MASLANKA S, SOBEL J (2008) International outbreak of severe botulism with prolonged toxemia caused by commercial carrot juice. *Clinical Infectious Diseases* 47(10):1245–1251.

SHIMODA M, YAMAMOTO Y, COCUNUBO-CASTELLANOS J, TONOIKE H, KAWANO T, ISHIKAWA H, OSAJIMA Y (1998) Antimicrobial effects of pressured carbon dioxide in a continuous flow system. *Journal of Food Science* 63(4):709–712.

SILK TM, RYSER ET, DONNELLY CW (1997) Comparison of methods for determining coliform and *Escherichia coli* levels in apple cider. *Journal of Food Protection* 60:1302–1305.

© Woodhead Publishing Limited, 2012

SINGH BR, KULSHRESHTHA SB, KAPOOR KN (1996) An orange juiceborne outbreak due to enterotoxigenic *Eshcerichia coli*. *Journal of Food Science and Technology-India* 34:504–506.

SLIFKO TR, RAGHUBEER E, ROSE JB (2000) Effect of high hydrostatic pressure on *Cryptosporidium parvum* infectivity. *Journal of Food Protection* 63(9):1262–1267.

SOMOGYI LP (2005) Direct food additives in fruit processing. In: Barrett DM, Somogyi L and Ramaswamy H (eds) *Processing Fruits*, 2nd edn. CRC Press, Boca Raton, FL, pp. 285–338.

SONG H-P, BYUN M-W, JO C, LEE C-H, KIM K-S, KIM D-H (2007) Effects of gamma irradiation on the microbiological, nutritional and sensory properties of fresh vegetable juice. *Food Control* 18:5–10.

STEELE BT, MURPHY N, RANCE CP (1982) An outbreak of hemolytic uremic syndrome associated with ingestion of fresh apple juice. *Journal of Pediatrics* 101:963–966.

STEYN C, CAMERON M, WITTHUHN R (2011) Occurrence of *Alicyclobacillus* in the fruit processing environment – a review. *International Journal of Food Microbiology* 147:1–11.

STRATFORD M, HOFMAN PD, COLE MB (2000) Fresh and processed fruits. In: Lund BM, Baird-Parker TC, Gould GW (eds) *The Microbiological Safety and Quality of Food, Volume I*. Aspen Publishers, MD.

SUÁREZ-JACOBO Á, GERVILLA R, GUAMIS B, ROIG-SAGUÉS AX, SALDO J (2010) Effect of UHPH on indigenous microbiota of apple juice: a preliminary study of microbial shelf-life. *International Journal of Food Microbiology*. 136:261–267.

TABERSHAW IR, SCHMELZER LL, BRUHN HB (1967) Gastroenteritis from an orange juice preparation. *Archives of Environmental Health* 15:72–77.

TAMBLYN S, DE GROSBOIS J, TAYLOR D, STRATTON J (1999) An outbreak of *Escherichia coli* O157:H7 infection associated with unpasteurized non-commercial, custom-pressed apple cider – Ontario, 1998. *Canada Communicable Disease Report* 25:113–117; discussion 117–120.

TANDON K, WOROBO RW, CHUREY JJ, PADILLA-ZAKOUR OI (2003) Storage quality of pasteurized and UV treated apple cider. *Journal of Food Processing and Packaging* 27:21–35.

TAYLOR JL, TUTTLE J, PRAMUKUL T, O'BRIEN K, BARRETT TJ, JOLBAITO B, LIM YL, VUGIA DJ, MORRIS JR. JG, TAUXE RV, DWYER DM (1993) An outbreak of cholera in Maryland associated with imported commercial frozen fresh coconut milk. *Journal of Infectious Diseases* 167:1330–1335.

THURSTON H, STUART J, MCDONNELL B, NICHOLAS S, CHEASTY T (1998) Fresh orange juice implicated in an outbreak of *Shigella flexneri* among visitors to a South African game reserve. *Journal of Infectious Diseases* 36:350.

TOMASULA PM (2003) Supercritical fluid extraction of foods. In: Heldman D (ed) *Encyclopedia of Agricultural, Food, and Biological Engineering*. Marcel Dekker, New York.

UKUKU D, GEVEKE D (2010) A combined treatment of UV-light and radio frequency electric field for the inactivation of *Escherichia coli* K-12 in apple juice. *International Journal of Food Microbiology* 138:50–55.

UKUKU DO, ZHANG HQ, HUANG L (2009) Growth parameters of *Escherichia coli* O157:H7, *Salmonella* and *Listeria monocytogenes* and aerobic mesophilic bacteria of apple cider amended with nisin-EDTA. *Foodborne Pathogens and Disease* 6(4):487–494.

ULJAS HE, INGHAM SC (1998) Survival of *Escherichia coli* O157:H7 in synthetic gastric fluid after cold and acid habituation in apple juice or trypticase soy broth acidified with hydrochloric acid or organic acids. *Journal of Food Protection* 61:939–947.

US FDA (1998) Federal Register Proposed Rules – 63 FR 20449 April 24, 1998 – HACCP; Procedures for the safe and sanitary processing and importing of juice; food labeling: warning notice statements; labeling of juice products. Available from: http://www.fda.gov/Food/FoodSafety/HazardAnalysisCriticalControlPointsHACCP/JuiceHACCP/ucm082031.htm (accessed Oct 4, 2011).

© Woodhead Publishing Limited, 2012

US FDA (1999) Report of 1997 inspections of fresh, unpasteurized apple cider manufacturers – summary of results – January 1999. Available from: http://www.fda.gov/Food/FoodSafety/HazardAnalysisCriticalControlPointsHACCP/JuiceHACCP/ucm085512.htm.

US FDA (2001) Hazard analysis and critical control point (HACCP); procedures for the safe and sanitary processing and importing of juices; final rule. Federal Register: January 19, 2001.66, 6137–6202.

US FDA (2004) Guidance for Industry: Juice HACCP Hazards and Controls Guidance First Edition; Final Guidance. Available from: http://www.fda.gov/Food/GuidanceComplianceRegulatoryInformation/GuidanceDocuments/Juice/ucm072557.htm

US FDA (2007) Guidance for Industry: Refrigerated Carrot Juice and Other Refrigerated Low-Acid Juices. Available from: http://www.fda.gov/Food/GuidanceComplianceRegulatoryInformation/GuidanceDocuments/Juice/ucm072481.htm (accessed Feb 27, 2011).

US FDA (2010) DHMH Issues Consumer Alert Regarding Recall of Baugher's Apple Cider. Available from: http://www.fda.gov/Safety/Recalls/ucm232878.htm (accessed Oct 4, 2011).

US FDA (2011) Outbreaks associated with fresh and fresh-cut produce. Incidence, growth, and survival of pathogens in fresh and fresh-cut produce. Available from: http://www.fda.gov/Food/ScienceResearch/ResearchAreas/SafePracticesforFoodProcesses/ucm091270.htm (accessed Oct 4, 2011).

VOJDANI J, BEUCHAT L, TAUXE R (2008) Juice-associated outbreaks of human illness in the United States, 1995 through 2005. *Journal of Food Protection* 71(2):356–364.

WALKER M, PHILLIPS C (2008) The effect of preservatives on *Alicyclobacillus acidoterrestris* and *Propionibacterium cyclohexanicum* in fruit juice. *Food Control* 19:974–981.

WEIHE JL (1986) Citrus and beverage microbiology. In: Matthews RF (ed) *Proceedings of the 26th Annual Short Course for the Food Industry: Citrus Processing New Developments and Adaptations*, University of Florida and Institute of Food Technologists Florida Section, pp. 29–37.

WHITNEY BM, WILLIAMS RC, EIFERT J, MARCY J (2007) High-pressure resistance variation of *Escherichia coli* O157:H7 strains and *Salmonella* serovars in tryptic soy broth, distilled water, and fruit juice. *Journal of Food Protection*; 70(9):2078–2083.

WILLIAMS RC, SUMNER SS, GOLDEN DA (2005) Inactivation of *Escherichia coli* O157:H7 and *Salmonella* in apple cider and orange juice treated with combinations of ozone, dimethyl dicarbonate, and hydrogen peroxide. *Journal of Food Science* 70(4):197–201.

WOODLING SE, MORARU CI (2007) Effect of spectral range in surface inactivation of *Listeria innocua* using broad-spectrum pulsed light. *Journal of Food Protection* 70: 909–916.

WOROBO RW, SPLITTSTOESSER DF (2005) Microbiology of fruit products. In: Barrett DM, Somogyi L and Ramaswamy H (eds) *Processing Fruits*, 2nd edn. CRC Press, Boca Raton, FL, pp. 261–284.

WOUTERS PC, SMELT JPPM (1997) Inactivation of microorganisms with pulsed electric fields: potential for food preservation. *Food Biotechnology* 11(3):193–229.

WOUTERS PC, BOS AP, UECKERT J (2001) Membrane permeabilization in relation to inactivation kinetics of *Lactobacillus* species due to pulsed electric fields. *Applied and Environmental Microbiology* 67(7):3092–3101.

WU Y, MITTAL GS, GRIFFITHS MW (2005) Effect of pulsed electric field on the inactivation of microorganisms in grape juices with and without antimicrobials. *Biosystems Engineering* 90:1–7.

XU Z, ZHANG L, WANG Y, BI X, BUCKOW R, LIAO X (2011) Effects of high pressure $CO_2$ treatments on microflora, enzymes and some quality attributes of apple juice. *Journal of Food Engineering* 104:577–584.

© Woodhead Publishing Limited, 2012

YUK H, GEVEKE DJ, ZHANG HQ (2010) Efficacy of supercritical carbon dioxide for nonthermal inactivation of *Escherichia coli* K12 in apple cider. *International Journal of Food Microbiology* 138:91–99.
YUSTE J, FUNG DYC (2002) Inactivation of *Listeria monocytogenes* Scott A 49594 in apple juice supplemented with cinnamon. *Journal of Food Protection* 65:1663–1666.
ZHAO T, DOYLE MP, BESSER RE (1993) Fate of enterohemorrhagic *Escherichia coli* O157:H7 in apple cider with and without preservatives. *Applied and Environmental Microbiology* 59:2526–2530.

© Woodhead Publishing Limited, 2012

# 7

# Microbial decontamination of milk and dairy products

**M. W. Griffiths and M. Walkling-Ribeiro, University of Guelph, Canada**

**Abstract**: Despite advances in dairy production and processing methods, outbreaks of illness associated with milk and milk products continue to occur. At the same time there has been a resurgence of interest by consumers in organic, 'slow' and 'raw' foods. To offset issues related to food safety and to comply with the demand for these fresh foods, there is an urgent need to develop novel processing concepts that provide effective public health protection with minimal treatments to retain nutrient and flavour characteristics of the product. In the first part of this chapter all milk and dairy pathogens of relevance are discussed with particular focus on their relative risk ranking. The second part of the chapter reviews the features and potential of conventional and most promising emerging preservation techniques for pathogen mitigation in milk and dairy products, highlighting the advantages of combined decontamination strategies and indicating hurdle processing as the most efficacious approach.

**Key words**: milk and dairy pathogens, conventional disinfection methods, emerging decontamination techniques, hurdle technology, minimal processing.

## 7.1 Introduction

A growing number of farm and food industry mergers domestically and worldwide in the last few decades has resulted in larger farming facilities, food processing capacities and plants. A potential risk and common disadvantage of upscaling, frequently overlooked by businesses, is the build-up of new interdependencies resulting from the accumulation of more operations under the same roof for the sake of increased efficiency, productivity and ultimately profit. With higher production volumes, simultaneous improvements of hygiene

© Woodhead Publishing Limited, 2012

practice and supervision and an upgrade of decontamination techniques to the same order of magnitude are necessary to maintain the previous level of safety and, indeed, additional safety procedures may be necessary to ensure that product safety and quality is improved. However, due to the substantial potential for cost cutting offered by the increase in scale of operations and the lack of willingness to maintain or enhance previous standards of food safety, the opposite is often the reality. The latter is a fundamental factor that contributes to the likelihood of pathogen contamination during milk production and processing from the raw material to the distribution stage.

Typical sources for pathogen contamination of milk and dairy products include inadequate sanitation of the environment (e.g., animals, equipment, storage facilities) for the processing stages, improper application of milk and dairy processing conditions (e.g., temperature, time) and equipment (e.g., milking installation, pasteurization units), cross-contamination and disruption of the cold chain. In addition to traditional principles of pre-emptive preservation (e.g., hygienic processing and storage, aseptic packaging), the introduction of prevention and management systems to achieve food safety such as Good Farming Practice (GFP), Good Manufacturing Practice (GMP), Good Laboratory Practice (GLP), Hazard Analysis and Critical Control Point (HACCP), International Organisation of Standardisation 9001 certification and total quality management has enhanced the quality and safety of milk and dairy products considerably. However, outbreaks related to infection with pathogens from milk and dairy products still occur periodically on a global scale.

According to the World Health Organisation (WHO, 2002) the main reasons for the emergence of foodborne pathogens are the increase in international travel and trade (e.g., globalisation of the food supply; exposure of travellers, refugees and immigrants to unfamiliar foodborne hazards while staying abroad), adaptation of microorganisms (e.g., inadvertent introduction of pathogens into new geographic areas; changes in microorganisms), alterations in the food production system (e.g., large-scale production and processing) and human demographic and behavioural factors (e.g., changes in life style and change in the human population such as aging). Most of the pathogens that have emerged in food over the last decades (Tauxe, 2002) are of great relevance for the safety of milk and dairy products as their nutrient density enables the propagation and subsequent transmission of many microorganisms, thereby, underlining the epidemiological significance of these complex food systems. In a summary of foodborne outbreaks and illnesses in the United States between 1998 and 2007, of all Food and Drug Administration (FDA) regulated food categories dairy products were recorded with the fourth highest number (6.2% share of the FDA-regulated food total) of outbreaks and illnesses of which 35.2% were associated with milk, 31.4% were documented for cheese, 15.4% were apportionable to ice cream and 18.0% were accounted for by other dairy commodities (CSPI, 2009).

© Woodhead Publishing Limited, 2012

## 7.2 Important pathogens and pathways of contamination in milk and dairy products

Pathogenic microorganisms implicated in milk and dairy foodborne outbreaks on four different continents over the last three decades (NSW Food Authority, 2009; Food Protection Services, 2011a; Sharp, 1987; Safe Food International, 2011), the associated sources of infection and their percentages in the resulting illnesses are summarized in Table 7.1. The data suggest that bacteria of the genera *Salmonella* and *Campylobacter* account for the highest rate of infectivity, but also a noticeable amount of diseases resulting from *Listeria* and *Escherichia coli* contamination of milk and dairy products were determined on a fairly regular basis within the last 30 years. The importance of these four pathogens is also reflected in the consistently high output of research studies that focus on their presence and impact in milk and dairy products and the fact that campylobacteriosis, salmonellosis, verotoxigenic *Escherichia coli* infections and listeriosis have been listed among the top five zoonoses with reported incidences in humans (European Food Safety Authority, 2007). Nonetheless, there are a number of pathogenic microbiota that merit subsequent characterisation as these microorganisms represent serious food safety hazards and may also evolve into a greater public health concern in the future.

### 7.2.1 Bacterial pathogens

Most milk and dairy borne outbreaks originate from bacterial pathogens (Food Protection Services, 2011b). Examples of their genus or strain-specific, intrinsic characteristics, that render them adaptive and survivable with regard to different compositions of milk and dairy products as well as changing environmental conditions, are development of resistance to antibiotics,

**Table 7.1** Overview of pathogens most commonly associated with outbreaks in milk and dairy products in the last three decades in Asia, Australia, Europe and North America

| Microorganisms | Products | Incidences [%][a] |
|---|---|---|
| Asia 2007–2010 (adapted from Safe Food International, 2011) | | |
| *Brucella* | Raw goats' milk | 4.5 |
| *Escherichia coli* | Milk | 95.5 |
| Australia 1995–2008 (adapted from NSW Food Authority, 2009) | | |
| *Campylobacter* | Dairy products and foods with dairy ingredients | 15.3 |
| *Clostridium* | Dairy products and foods with dairy ingredients | 4.8 |
| *Cryptosporidium* | Dairy products and foods with dairy ingredients | 1.4 |

© Woodhead Publishing Limited, 2012

| Unidentified | Dairy products and foods with dairy ingredients | 15.4 |
|---|---|---|
| *Norovirus* | Dairy products and foods with dairy ingredients | 22.1 |
| *Salmonella* | Dairy products and foods with dairy ingredients | 40.6 |
| *Staphylococcus aureus* | Dairy products and foods with dairy ingredients | 0.4 |
| Europe 2007–2010 (adapted from Safe Food International, 2011) | | |
| *Clostridium* | Cheese | 0.2 |
| *Listeria monocytogenes* | Curd cheese | 33.9 |
| *Salmonella* | Infant formula milk | 3.7 |
| *Shigella* | Dairy products, milk, sour cream | 60.4 |
| *Staphylococcus* | Unpasteurised soft cheese | 1.8 |
| Europe and North America 1980–1985 (adapted from Sharp, 1987) | | |
| *Brucella* | Cheese, goats' and sheep's milk products, raw cows' milk | 0.3 |
| *Bacillus* | Cheese | 0.2 |
| *Campylobacter* | Goats' and sheep's milk products, raw cows' milk, heat-treated cows' milk | 12.3 |
| *Corynebacterium* | Goats' and sheep's milk products, raw cows' milk | ≤ 0.1 |
| *Escherichia coli* | Cheese | 1.3 |
| *Listeria monocytogenes* | Cheese | 0.6 |
| *Salmonella* | Cheese, goats' and sheep's milk products, raw cows' milk, heat-treated cows' milk | 81.1 |
| *Staphylococcus* | Cheese, goats' and sheep's milk products, heat-treated cows' milk | 3.4 |
| *Streptococcus* | Cheese, raw cows' milk | 0.1 |
| *Yersinia* | Heat-treated cows' milk | 0.6 |
| North America 2000–2010 (adapted from Food Protection Services, 2011a) | | |
| *Campylobacter* | Raw milk, unpasteurised fresh cheese, unpasteurised cheese curds | 66.9 |
| *Cryptosporidium* | Raw milk | 0.5 |
| *Escherichia coli* | Raw milk, raw goats' milk, raw milk cheese | 15.0 |
| *Listeria monocytogenes* | Raw milk, raw milk cheese, unpasteurised soft cheese | 2.6 |
| *Salmonella* | Raw milk, raw milk fresh cheese and products | 14.8 |
| *Yersinia* | Raw milk | 0.2 |

[a]The proportion between the illness cases of outbreaks associated with the listed, specific pathogens and the overall number of illness cases from dairy and milk borne outbreaks in the respective regions is reflected by the incidence percentages.

© Woodhead Publishing Limited, 2012

formation of endospores and biofilm production (Griffiths, 2004). Pathogenic bacteria that exhibit several of these defence mechanisms and cause disease at low infectious dosages in milk and dairy foods and over short incubation periods are regarded as particularly hazardous.

#### Aeromonas *species*

*Aeromonas* spp. cause gastroenteritis and wound infections and they are Gram-negative, facultative anaerobic rods that occur naturally in fresh and brackish water. Species of this bacterium are also common in raw milk and other milk products, in particular, *Aeromonas hydrophilia*, *A. caviae*, *A. sobria*, *A. veronii*, *A. salminicida* (Papageorgiou *et al.*, 2006; Martins *et al.*, 2002, Melas *et al.*, 1998).

#### Bacillus *species*

Due to their endospore-forming capability and their production of heat stable toxins, both with the chance of surviving conventional thermal processing of milk and dairy products, *Bacillus* spp. can be considered as one of the most challenging microorganisms. Species that are potentially pathogenic to humans include *Bacillus cereus* (Hassan *et al.*, 2010), *Bacillus anthracis* (Perdue *et al.*, 2003), *Bacillus weihenstephanensis*, *Bacillus thuringiensis*, *Bacillus mycoides*, *Bacillus pseudomycoides* (Bednarczyk and Daczkowska-Kozon 2008), *Bacillus amyloliquefaciens* (Mikkola *et al.*, 2004), *Bacillus pumilus* (Suominen *et al.*, 2001) and *Bacillus licheniformis* (Salkinoja-Salonen *et al.*, 1999), whereas the latter two species have been reported to cause mastitis in cows (Nieminen *et al.*, 2007).

#### Brucella *species*

Infection with this zoonotic pathogen, of which *Brucella abortus* and *Brucella melitensis* have been identified as particularly problematic (Rezaei *et al.*, 2010), causes brucellosis that leads to arthralgia, fever and hepatomegaly most frequently (Buzgan *et al.*, 2010). The majority of *Brucella* outbreaks have been due to ingestion of raw milk or dairy products and, in spite of eradication of this pathogen in some countries (Pappas *et al.*, 2006; Black, 2004), brucellosis is still considered the most common zoonotic disease globally with approximately half a million humans affected each year (Buzgan *et al.*, 2010).

#### Campylobacter jejuni

Campylobacteriosis outbreaks in raw milk have mainly been associated with *Campylobacter jejeuni* (Heuvelink *et al.*, 2009, Unicomb *et al.*, 2009) with typical symptoms appearing within five days. These include nausea, vomiting, diarrhea, bloody diarrhea and abdominal pain (Peterson, 2003) which can last up to two weeks. At present *C. jejeuni* is regarded as a major threat in developing countries and it has been related to numerous foodborne bacterial diseases there (Moore *et al.*, 2005). About 1 person in

© Woodhead Publishing Limited, 2012

every 1000 that contract campylobacteriosis goes on to develop Guillain-Barré syndrome, a severe autoimmune disease that leads to death in 2–12% of sufferers, depending on age.

Cronobacter sakazakii

Increased interest in *Cronobacter sakazakii* has emerged in recent years following contamination of reconstituted infant formula and powdered milk (Craven *et al.*, 2010). Initially designated as *Enterobacter sakazakii*, the organism can elicit medical conditions, namely, meningitis, or necrotising enterocolitis in neonates, particularly those with a low birth weight (Friedemann, 2009). Other food products, including cheese and milk, have also been reported as possible contamination sources of *C. sakazakii* (Beuchat *et al.*, 2009; Kandhai *et al.*, 2010) but so far cases of illness have been related exclusively to the intake of contaminated milk reconstituted from powdered infant formula.

Clostridium *species*

Similar to bacilli, also vegetative clostridia cells come with the attribute of transformation to endospores, whenever the surrounding environmental conditions are unfavourable for the growth of the bacterium. Many clostridial species such as *Clostridium botulinum*, *Clostridium difficile*, *Clostridium perfringens*, *Clostridium tetani* and *Clostridium sordellii* also produce toxins that affect humans. In a study by Julien *et al.* (2008), approximately 75% of all samples from sources associated with raw milk production in farms in Quebec, Canada contained *Clostridium* spp. indicating their importance. Although clostridia typically found in a milk or dairy food production environment are not pathogens, the natural transfer of neurotoxin genes from *C. botulinum* to other clostridia is of great concern (Lindström *et al.*, 2010) as contamination of silage and gastrointestinal tracts in cattle with this pathogen have been reported in the past, thus, representing a possible safety risk for the dairy chain.

Corynebacterium *species*

Besides *Corynebacterium diphtheriae* and *Corynebacterium pseudodiphtheriticum*, nondiphterial pathogenic species include *Corynebacterium amycolatum*, *Corynebacterium pseudotuberculosis*, *Corynebacterium urealyticum* and *Corynebacterium jeikium* (Denis and Irlinger, 2008; Baird *et al.*, 2005), which can cause, for example, diphtheria, pneumonitis, pharyngitis and endocarditis in humans and may also cause farm animal diseases. Cases resulting from consumption of raw milk infested with *C. diphtheriae*, *C. pseudotuberculosis* and *C. ulcerans* have been reported (Goldberger *et al.*, 1981; Hart, 1984) but it was shown for *C. diphtheriae* (Wilson and Tanner, 1945) that their survival in fermented milk products is rather unlikely. Because of vaccination programs the incidence of milk borne diphtheria cases in the developed world is extremely low.

© Woodhead Publishing Limited, 2012

*Verotoxigenic* Escherichia coli

The most prominent and pathogenic members of *Escherichia coli* are the enterohemorrhagic O157 strains that are responsible for the majority of outbreaks associated with consumption of *E. coli* contaminated milk and other dairy products (Guh *et al.*, 2010; Keene *et al.*, 1997; Honish *et al.*, 2005; Rangel *et al.*, 2005). Other verotoxigenic *E. coli*, for instance, *E. coli* O145 and *E. coli* O26, are also a safety concern and have been identified as the source of milk and dairy product outbreaks occurring in the recent past (De Schrijver *et al.*, 2008; Lorusso *et al.*, 2009). In the case of *E. coli* O157:H7, few cells are necessary to cause an infection that will result in medical conditions that vary depending on the type of *E. coli* encountered; examples are bloody diarrhea, peritonitis, inflammatory bowel diseases, Crohn's disease, ulcerative colitis and fever.

Listeria monocytogenes

Another foodborne pathogen that has frequently been linked to outbreaks associated with contaminated milk and dairy products (Koch *et al.*, 2010; Kells and Gilmour, 2004; Cagri-Mehmetoglu *et al.*, 2011; Lundén *et al.*, 2004) is *Listeria monocytogenes.* As a psychrotroph bacterium, the latter is capable of surviving and propagating during long storage periods in cold environments making it an omnipresent pathogenic contaminant in all stages of milk and dairy production, notably problematic when occuring post pasteurisation. Listeriosis follows from invasive infection with *L. monocytogenes* and its symptoms comprise septicaemia, meningitis, encephalitis, corneal ulcer, pneumonia and spontaneous abortion or stillbirth in pregnant women, while non-invasive exposure to the pathogen brings about febrile gastroenteritis. In general, a trend towards an increase in listeriosis cases was observed during the last decade, for example, in England and Wales (Gillespie *et al.*, 2009), Denmark (Kvistholm Jensen *et al.*, 2010) and Germany (Koch and Stark, 2006), indicating the overall growing significance of *L. monocytogenes.* Moreover, there is also an increase in the number of elderly contracting listeriosis.

Mycobacterium *species*

The likelihood of raw milk contamination with *Mycobacterium avium* subspecies *paratuberculosis* (MAP) was deemed significant in a review by Eltholth *et al.* (2009). The organism has been attributed to Crohn's disease in humans (Hermon-Taylor and Bull, 2002) as well as Johne's disease in cattle and sheep (Collins, 1997). However, the link with human disease is far from being established (Griffiths, 2009). Analyses of pasteurised milk and cheese samples were not found to be completely free of these bacteria indicating limited efficacy of heat pasteurisation, which may be related to their initial concentration, but again there is some controversy surrounding the heat resistance of the bacterium. *Mycobacterium bovis* is another pathogenic representative causing human tuberculosis (TB) and cases of the latter have

© Woodhead Publishing Limited, 2012

been reported in the UK (Rowe and Donaghy, 2008) as a result of growing popularity and consumption of artisan cheeses made from unpasteurised *M. bovis* contaminated milk. Both *M. bovis* and the closely related *Mycobacterium tuberculosis* were identified in Egyptian milk samples (Guindi *et al.*, 1980). In addition, there has been a number of outbreaks of TB resulting from consumption of cheese made from raw milk contaminated with *M. bovis* in the UK (Evans *et al.*, 2007).

Pseudomonas aeruginosa

This opportunistic and psychrotrophic pathogen was shown to be less sensitive to antibacterial effects emanating from fermented dairy products than other foodborne pathogens (Yesillik *et al.*, 2011). Its low susceptibility to antibiotics, metabolic diversity, quorum sensing and capability of biofilm formation (Rybtke *et al.*, 2011) are indicative of the microorganism's adaptability and make it a key human pathogen that can inflict diarrhea, pneumonia, necrosis and septic shock among other things on infected individuals. *P. aeruginosa* is also a pathogen commonly implicated in outbreaks of clinical mastitis in sheep and goats (Leitner and Krifucks, 2007) or cattle (Daly *et al.*, 1999) as it is often ubiquitous in milk farm environments (Kapur *et al.*, 1986), although it is primarily associated with contamination through water. However, human infections are rare and invariably involve infants.

Salmonella

Outbreaks due to contamination of milk and dairy products with *Salmonella* (Dominguez *et al.*, 2009; Lanzas *et al.*, 2010; Van Duynhoven *et al.*, 2009) are relatively frequent and, therefore, increased attention is paid to this group of bacterial pathogens from institutional or corporate networks as well as organisations and governments that monitor food safety worldwide. Immuno-compromised and elderly are more vulnerable to *Salmonella* infections as a lower infectious dose may lead to typhoid, paratyphoid fever, or salmonellosis as a result of food poisoning, and a high number of fatalities are caused every year, especially in regions where poor hygienic conditions are prevalent. Concerns have arisen due to the emergence of multidrug resistant *Salmonella*, such as *S.* Typhimurium DT104, which has caused infections related to consumption of raw milk cheese.

Shigella *species*

Foodborne gastrointestinal illness can be induced by all four different *Shigella* groups which are *Shigella flexneri*, *Shigella sonnei*, *Shigella boydii* and *Shigella dysenteriae*, typically at a relatively low infectious dose between 10 and 500 cells (DuPont *et al.*, 1989). Contamination of fresh pasteurised milk cheese (García-Fulgueiras *et al.*, 2001), milk-based sauce (Castell Monsalve *et al.*, 2008), unpasteurised milk curds (Zagrebneviene *et al.*, 2005) and milk (Tucker *et al.*, 1954) have been reported as sources of shigellosis outbreaks.

© Woodhead Publishing Limited, 2012

### Staphylococcus aureus

Both exo- and enterotoxins can be produced by *Staphylococcus aureus*, bringing about toxic shock syndrome and gastroenteritis, respectively. While vegetative *S. aureus* cells are usually inactivated during milk processing, the thermostable enterotoxins may survive heat pasteurisation conditions and retain their activity and, consequently, cause food poisoning in products such as cream filling (Anunciacao *et al.*, 1995), pasteurised chocolate (Evenson *et al.*, 1988), regular milk (Schmid *et al.*, 2009), goats' milk (Akineden *et al.*, 2008), sheep milk cheese (Bone *et al.*, 1989) and milk powder (Asao *et al.*, 2003). Because a cell concentration of above $10^5$ CFU/ml is necessary for the bacterial pathogen to elicit skin diseases, septic arthritis, endocarditis, or pneumonia, a staphylococcal infection because of ingestion of sufficient amounts of bacteria through milk and dairy products could be regarded as unlikely. As *S. aureus* is a well-known cause of mastitis (Virgin *et al.*, 2009) in addition to thorough hygiene practice early monitoring of the cattle at the farm level is important to prevent critical contamination that could lead to substantial formation of staphylococcal enterotoxins in the milk.

### Streptococcus *species*

Infections with *Streptococcus* spp. are usually related to mastitis in dairy herds (Chaffer *et al.*, 2005; Pisoni *et al.*, 2009; Zadoks *et al.*, 2001) and by comparison are rarely found in humans (Poulin and Boivin, 2009) either from consumption of unpasteurised milk (Edwards *et al.*, 1988) or dairy products (Bordes-Benítez *et al.*, 2006). Typical symptoms that can arise in humans affected by streptococcal transmission are, for instance, meningitis, endocarditis, arthritis, pharyngitis, erysipelas and necrotising fasciitis, depending on haemolysis (i.e. alpha haemolytic, beta haemolytic or non-haemolytic) and on the group (pneumococci or Viridians, A to G and D, respectively) of bacteria.

### Yersinia enterocolitica

Cases of the zoonotic disease yersiniosis as a result of contact with the pathogen *Yersinia entertocolitica* have been observed in cattle and other farm animals (Brewer and Corbel, 1983) as well as in humans after intake of pasteurised milk (Ackers *et al.*, 2000), raw milk (Schiemann, 1987) and fermented milk (Okwori *et al.*, 2009). The psychrotrophic pathogen was also isolated from commercial and non-commercial ice cream samples (Barbini de Pederiva *et al.*, 2000). In addition to diarrhea and fever, more serious medical conditions include septicaemia, faecal infection and immunosuppression and, because beta-lactamase production of *Y. enterocolitica* is responsible for the resistance of the bacterium to some antibiotics, the selection of the latter for treatment of infections is limited.

© Woodhead Publishing Limited, 2012

### 7.2.2 Pathogenic protozoon, rickettsia and viruses

Coxiella burnetii

Several cases of Q fever in humans, as a result of drinking milk from *Coxiella burnetti* infected dairy herds, have been reported (Bosnjak *et al.*, 2009; Kim *et al.*, 2005) and due to its heat resistance and high probability of survival following pasteurisation treatment, the pathogen is considered to be one of the most problematic and challenging microorganisms for milk and dairy product manufacturers (Cerf and Condron, 2006). Symptoms of Q fever are many and can vary from flu-like conditions, headache, sudden fever, muscle and joint pains, loss of appetite, respiratory problems, chills, nausea and diarrhea, to very severe medical conditions such as pneumonia, hepatitis and endocarditis-like inflammation. At present no effective control mechanisms for *C. burnetti* have been established by the milk and dairy processing industry (Loftis *et al.*, 2010). Due to its high transmittability, for example there is a 50% probability of infection by inhalation ($ID_{50}$), it requires containment level 3 biosafety facilities for scientific research. In addition, it is resistant to standard disinfectants as well as thermal, osmotic pressure and ultraviolet irradiation stresses, systematic vaccination (Q-vax) and antibiotics (erythromycin-rifampin combination). In a 2007 study by the Animal and Plant Health Inspection Service (APHIS, 2011) in the US, about 77% of bulk tank milk from dairy herds, representing about 80% of the dairy operation in the country, were shown to be positive for *C. burnetii* by polymerase chain reaction assay. Thus, a search for alternative decontamination methods appears essential to avoid the scenario of a possible epidemic in the future that could have grave consequences.

Cryptosporidium parvum

Pathogenic protozoans of the genus *Cryptosporidium* are commonly associated with infection of dairy herds (Lassen *et al.*, 2009; Robinson *et al.*, 2006; Esteban and Anderson, 1995) but the resulting disease, known as cryptosporidiosis, can also affect the health of humans ingesting milk and dairy products contaminated with the parasites (Fretz *et al.*, 2003; Laberge and Griffiths, 1996), especially, if the persons are immuno-compromised. Good hygienic practices for feeding and housing in dairy farms were suggested as an effective means to reduce the occurrence of *Cryptosporidium parvum* (Trotz-Williams *et al.*, 2008), thus, limiting possible health hazards like diarrhea, nausea, dehydration, pancreatitis and cholangitis due to milk contamination.

Entamoeba histolytica

Several diagnostic findings of another diarrhea-causing parasitic protozoon, *Entamoeba histolytica*, were also reported in raw milk samples (Rai *et al.*, 2008) with a higher likelihood of occurrence in developing countries with poor sanitation than in industrialised nations (WHO, 2008). In addition to being a virus carrier and modulating human viruses, the cyst-producing *E.*

© Woodhead Publishing Limited, 2012

*histolytica* can destroy tissue and inflict weight loss, fatigue, abdominal pain, amoebic dysentery and lytic necrosis.

Giardia lamblia

Parasitic infection of humans with *Giardia lamblia* is normally followed by giardisis, provoking diarrhea, flatulence, epigastric pain, nausea, weight loss and temporary lactase deficiency among other symptoms and its occurence in raw milk was reported to be more likely than for other parasitic organisms (Rai *et al.*, 2008). Moreover, this zoonotic protozoan has been categorised as a pathogen of great importance for dairy cattle and dairy producers (O'Handley *et al.*, 2000), thereby being of public health concern.

*Hepatitis virus*

While acute viral hepatitis is exhibited by hepatitis A, B, C, D or E, yellow fever and infection with adenoviruses, both causing chronic viral hepatitis, arise from hepatitis B, partially in conjunction with hepatitis C and D. Recovery of adenovirus 12 from milk, ice cream mix, chocolate milk and heavy cream was achieved by Sullivan and Read (1968). An increase in the fat content of milk products was shown to increase protection of hepatitis A virus against heat treatments (Bidawid *et al.*, 2000) and lactoferrin content of milk has been suggested to affect hepatitis C virus accounting for a lower transmission rate compared to hepatitis B virus (Yi *et al.*, 1997). Symptoms of acute hepatitis range from loss of appetite and muscle and joint aches to jaundice and liver failure at worst, whereas the chronic disease is manifested by liver cirrhosis, swelling, fluid accumulation as well as bleeding, confusion, coma and kidney dysfunction at an advanced stage.

*Norovirus*

The Norwalk-like virus, or norovirus, is considered as the foremost non-bacterial initiator of gastroenteritis infections worldwide and evidence of its involvement in milk and dairy outbreaks has also been provided (NSW Food Authority, 2009), although for the most part the viral pathogen has been linked to consumption of other foods products (e.g., shellfish, sandwiches, water). As notification of norovirus infection is not required by law one can presume that the reported and published cases vastly underestimate the true burden of disease (Mead *et al.*, 2000). However, outbreaks related to gastroenteritis give a vague indication of the pathogen's spread and frequency of occurrence. Perhaps of particular concern is the recent finding that human norovirus strains can be found in cattle and isolated from faeces (Mattison *et al.*, 2007).

*Rotavirus*

Reduction in milk production of dairy cows as a consequence of rotavirus group C infections had been reported (Mawatari *et al.*, 2004; Kiyohito *et al.*, 2006) and it was suggested that cattle act as the host of the diarrhea-causing

© Woodhead Publishing Limited, 2012

virus. In addition to causing severe diarrhea in infants, the virus is also the source of stomach flu in humans, with group A being most commonly associated with outbreaks. Rotaviruses still present a major epidemic challenge in developing countries (Simpson *et al.*, 2007) and have been listed as a cause of enteric infections associated with raw milk consumption (NSW Food Authority, 2009).

Toxoplasma gondii
Milk and dairy products were examined as a potential source of the parasitic protozoan *Toxoplasma gondii* (Hiramoto *et al.*, 2001) and reports of cases of toxoplasmosis in children after ingestion of goats' milk are mentioned in the literature (Dubey and Jones, 2008; Nichols, 2000). The zoonosis is a public health concern in both developed and developing countries (Inpankaew *et al.*, 2010) and, in particular, immuno-compromised persons are affected by symptoms that can include encephalitis or miscarriage and, furthermore, behavioural changes such as hallucinations and schizophrenia have also been linked to toxoplasmosis.

## 7.3 Decontamination methods for milk and dairy products

### 7.3.1 Biological antimicrobials

The simplest approach to biopreservation is most likely the enrichment of the milk and dairy foods with an adequately large number of lactic acid bacteria that not only create an unfavourable environment for most pathogens (Zalan *et al.*, 2010; Buriti *et al.*, 2007; Caridi, 2002) but also may benefit the consumer's health if the lactic acid bacteria have probiotic properties and are consumed in large enough quantities on a regular basis. Another aspect is that this form of natural preservation method applies predominantly to fermented foods and, in addition, the presence of prebiotics is recommendable to maximise the activity and effect of the probiotics. An alternative, less product-specific approach is the use of tetracycline antibiotics (e.g., tetracyclines and oxytetracyclines) as well as aminoglycosides (e.g., gentamicin, kanamycin, streptomycin and neomycin), which are classic antibiotics originating from live microorganisms. An advantage of aminoglycosides and partially also tetracyclines is their observed stability against heat pasteurisation and sterilisation (Zorraquino *et al.*, 2009; Hassani *et al.*, 2008). However, drawbacks are the gradual development of resistance to these natural antibiotics (Rodriguez-Alonso *et al.*, 2009; Nam *et al.*, 2011) and the restriction to dosages defined by the food legislation (Loomans *et al.* 2003). The addition of antibiotics to foods has to be discouraged because of the possible role of this practice in the emergence of "super bugs" that are resistant to a broad spectrum of these agents.

A powerful decontamination method that was long forgotten and for which interest re-emerged in recent years is the application of viruses naturally occurring in the habitat of bacterial pathogens. These bacterial

© Woodhead Publishing Limited, 2012

viruses or bacteriophages are very effective for the inactivation of their host organisms and closely related strains and at the same time these viruses are generally recognised as safe (GRAS), an important prerequisite for broad consumer acceptance. Mitigation of pathogenic bacteria by phage in pathogen contaminated milk and dairy environments has scarcely been investigated to date, although Modi *et al.* (2001) showed that they could be used to inactivate *Salmonella* in Cheddar cheese made from raw milk. However, most of the research on bacteriophage has investigated their use in combination with other decontamination techniques to enhance the overall antimicrobial effect (Martinez *et al.*, 2008; Garcia *et al.*, 2010). The high specificity of the phage and possible development of resistance to one type of phage by the bacteria suggests that the application of cocktails consisting of several bacteriophages may be the preferred mode of use. As high concentration of phages could be required to effectively inactivate pathogens, in some cases the number of different phages for incorporation in a cocktail might be limited by the phage titres and therefore, a combination with other decontamination methods seems advisable.

Bacteriocins are produced by bacteria with lantibiotics (class I bacteriocins), which are formed by lactic acid bacteria, being the most widely used. Although some of the dairy research on lantibiotics focuses on lacticin (McAuliffe *et al.*, 1999; Martinez-Cuesta *et al.*, 2010) the lantibiotic representative that still attracts the highest interest is nisin, which has been studied thoroughly for isolation and preservation in numerous milk and dairy products, for instance, bovine milk (Da Silva Malheiros *et al.*, 2010; Boussouel *et al.*, 2000), fermented bovine milk (Mitra *et al.*, 2009), goats' milk (Cocolin *et al.*, 2007), raw ewe's milk (Bravo *et al.*, 2009), milk curd (Arques *et al.*. 2008), yogurt and cottage cheese (Aslim *et al.*, 2004), raw milk and soft cheese (Ortolani *et al.*, 2010) and ricotta-type cheeses (Davies *et al.*, 1997).

Epidermin is another antimicrobial polypeptide secreted by staphylococci and known for its application as an anticancer agent. Both nisin and epidermin are long flexible molecules that belong to type A lantibiotics, which bind electro-statically to phospolipids of microbial cell walls and cause inactivation by pore formation (Breukink *et al.*, 1997), but while nisin is typically effective against Gram-positive organisms and may exhibit universal antimicrobial activity when used in combination with chelating agents, epidermin is particularly effective against staphylococci and streptococci similar to colicins and subtilins that target coliforms and bacilli, respectively.

By contrast, the globular molecules of type B lanthionine-containing peptides such as cinnamycin, duramycin and mersacidin act as antimicrobials by repressing the biosynthesis of petidoglycans (Sahl and Bierbaum, 1998). Enterocins, lactococcins and pediocins obtained from respective strains of *Enterococcus faecium* and *Pediococcus acidilactici* are examples of small heat-stable proteins (class II bacteriocins) that have drawn increased attention for milk and dairy decontamination (Achemchem *et al.*, 2006; Ananou *et al.*, 2010; Garcia *et al.*, 2004; Laukova *et al.*, 1999; Munoz *et al.*, 2004;

© Woodhead Publishing Limited, 2012

Sivakumar *et al.*, 2010; Somkuti and Steinberg, 2010) as they possess higher efficacy than lantibiotics (Rodriguez *et al.*, 2001). Due to the thermal processing encountered during milk production, research on larger heat-labile bacteriocins (class III bacteriocins) such as lactacin B (Tabasco *et al.*, 2009) and helveticin (Thompson *et al.*, 1996) is comparatively difficult to find in the literature. Relatively few studies are also available on complex bactericidal compounds that are composed of a protein and one or more chemical groups (class IV bacteriocins (Klaenhammer, 1993)), although findings for the application of plantaricin indicated good potential for the inactivation of Gram-positive bacteria and some *Enterobacteriaceae* (Hernandez *et al.*, 2005). With the introduction of better hygiene procedures in dairy farming and milk processing coupled with the prolongation of the cold chain, the chance of survival for native lactic acid bacteria has been reduced and therefore a sufficient amount of natively occurring bacteriocins, which could sufficiently inhibit pathogens in milk and dairy products, is probably not achieved. For this reason the screening for new bacteriocin-producing bacteria in raw milk is important (Mirhosseini *et al.*, 2010) as the addition of bacteriocins to milk and dairy products offers a relatively inexpensive and natural method of biopreservation and their application has GRAS status. However, the use of bacteriocins for food decontamination is limited in large part due to the presence of proteases in foods that inactivate these antimicrobial peptides (Phillips *et al.*, 1983). Although a mixture of different bacteriocins may enhance the treatment effectiveness, a combination with other conventional or novel preservation techniques is needed.

### 7.3.2 Chemical decontamination methods

*Synthetic and semi-synthetic preservatives*

Based on the reaction of sulfonyl chloride with ammonia or amine, sulfonamides, such as sulfadiazine, sulfamerazine, sulfamethazine sulfachloropyridazine, sulfathiazole and sulfamethoxazole, are formed that function as synthetic preservatives in milk (Rodriguez *et al.*, 2010; Yucel and Citak, 2003). However, uncontrolled addition and overconsumption of these drug additives will promote allergies and, thus, mixtures of sulfonamides combining multiple antibiotics or other methods of preservation to counteract the development of resistance will result in more efficient decontamination of milk and dairy products. Additional synthetic antibiotics that could be added to milk and dairy products include beta-lactams with saturated five membered (e.g., penicillins), unsaturated five membered (e.g., meropenem) and six membered ring structures (e.g., cephalosporins) or not fused to any ring (e.g. aztreonam); semi-synthetic tetracyclines for a broad-spectrum approach (e.g. methacycline, doxycycline); chemotherapeutic quinolones and even more potent oxazolidones that represent the most sophisticated group of synthetic antibiotics available, which will be considered for decontamination of extremely virulent and decontamination-resistant pathogens (Adrian

© Woodhead Publishing Limited, 2012

*et al.*, 2009). It is noteworthy that, with the exception of an epidemic scenario, the addition of the latter two groups of antibiotics to commercial foods is very questionable due to prospective side effects. As stated above, the use of antibiotics to control pathogens in foods is to be strongly discouraged and, in many cases, is illegal.

*Chemically preserving additives*

Milk and dairy products are among foods that are supplemented with antioxidants either directly, as in the case of the oxygen scavenging vitamin C to milk for which a bacteriostatic effect on pathogens was suggested (Dojchinova *et al.*, 1990), or indirectly, as exemplified by the addition of berry fruits and cocoa powder to diverse dairy foods (e.g. yogurt, ice cream, cake cream fillings and milk drinks), thus fortifying the products with polyphenols and procyanidins, respectively. Furthermore, raising the concentration of sugars in the food to high levels will lower the water activity in the food and, hence, decrease the risk of contamination but is not exactly regarded as healthy from a nutritional perspective. Common chemical preservatives added to some dairy products are also organic acids that reduce the pH, for instance, propionic acid and sorbic acid, both used for the preservation of cream-based cake and cheese. Also certain fatty acids such as monolaurin, used as an emulsifier, have antimicrobial properties and their use in cottage cheese has been proposed (Bautista *et al.*, 1993).

### 7.3.3 Physical decontamination methods

*Heat pasteurisation and sterilisation*

The principal treatment methods still applied to milk and dairy products are heat processing techniques, which achieve preservation by means of thermal degradation of the microbial cell membrane and cell constituents. Depending on the treatment time and the heat load applied thermal processing methods can be distinguished into:

- ultra-high temperature (UHT) treatment sterilising the liquid product at 135–150°C for 2–20 s (Hinrichs and Kessler, 1995).
- high-temperature short-time (HTST) pasteurisation (at 72°C or above for 15 s or longer; Richter *et al.*, 1992),
- low-temperature long-time (LTLT) pasteurisation (e.g. at 63°C for 30 min; FSANZ, 2003).

While batch LTLT pasteurisation conducted in containers under steady state conditions has limited industrial applications, the more common HTST or UHT treatments are carried out continuously, mainly in plate or tubular heat exchangers. Conventional thermal processing is still the most widespread and accepted type of preservation treatment for foods, because it features most of the factors of an 'ideal' food preservation technique such as prolongation of the shelf life and microbial safety, preservation of quality characteristics (i.e.

© Woodhead Publishing Limited, 2012

organoleptical and nutritional), avoidance of residues, low processing costs and simple handling. In addition, it attracts few objections from consumers and legislators (Raso and Barbosa-Cánovas, 2003). However, it may produce milk and dairy foods of reduced quality due to the level of heat necessary for the inactivation of thermotolerant enzymes and the presence of heat resistant spoilage or pathogenic microorganisms.

Some research groups have suggested that certain heat pasteurisation conditions may not suffice for the inactivation of certain milk and dairy borne pathogens (Bidawid *et al.*, 2000; Van Brandt *et al.*, 2011; Novak *et al.*, 2005) in contrast to others (Osaili *et al.*, 2009; Van Der Veen *et al.*, 2009; Rademaker *et al.*, 2007; Sullivan *et al.*, 1971) reporting that pathogen control is adequately ensured in milk and dairy products with present thermal pasteurisation processes. In some cases the difference in the findings could be explained by the nature of the investigated products, the specificity of the microorganisms used and the treatment conditions used as all of these parameters may account for a more or less pronounced microbial resistance to the heat pasteurisation applied. Another plausible explanation for the experimental deviations could be the level of expressed virulence that affects the thermal stress adaptation and determines the heat resistivity of the microorganisms subjected to heat pasteurisation experiments. Although phosphatase and peroxidase tests have been widely applied to assess the effectiveness of conventional heat pasteurisation treatments, indicated by negative and positive enzymatic responses, respectively, the accuracy of these analyses could be affected by souring of milk (Kiermeier and Kayser, 1960), enzyme reactivation in high fat content products (Varnam and Sutherland, 2001) and higher heat loads (Ulberth, 2003) rendering these test methods possibly unreliable for the verification of heat pasteurisation and sterilisation efficacies.

*Microfiltration-assisted heat pasteurisation and sterilisation*

Extended shelf life (ESL) milk and dairy products have steadily grown in consumer popularity and are a result of enhancing the antimicrobial process of conventional heat pasteurisation (usually consisting of a heat pasteurised milk portion that is blended and standardised with a cream portion treated at a slightly lower level than UHT or in the low UHT temperature range to reach the designated fat content) further with an additional membrane microfiltration (MF) step in combined processing systems (Tuchenhagen Dairy Systems, 2006, Tetra Pak Cheese and Powder Systems, 2010). The milk obtained with these systems offers better shelf stability and also improved sensory perception of milk freshness and taste. Microfiltration works particularly effectively for the removal of bacterial endospores (Gésan-Guiziou, 2010; Guerra *et al.*, 1997; Elwell and Barbano, 2006), but it also enhances removal of somatic cells and vegetative bacteria from milk based on the applied membrane pore size, the latter typically being 1.4 μm for milk processing applications.

© Woodhead Publishing Limited, 2012

*Aseptic and active packaging*

Another achievement of modern food processing technology is aseptic and active packaging contributing primarily to longer shelf life (Henyon, 1999), but also making sure that pathogen invasion of milk and dairy products is avoided. Aseptic packaging is achieved by either eliminating any possible sources of contamination between the, typically heat, processed food and the sterile package of the end product or by indirect treatment of the food in its final shipment container. Furthermore, packaging materials or conditions that actively prevent contamination with spoilage or pathogen microorganisms, e.g., antimicrobial and bioactive agents immobilized on surfaces or in layers of the package (Anany, 2010, Granda-Restrepo *et al.*, 2009), generation and maintenance of a modified atmosphere that suppresses microbial growth (Stoermer and Welle, 2006) and sensors and intelligent barrier self-regulation depending on the environmental conditions (Yam *et al.*, 2005), play a vital role for successful pathogen prevention and mitigation in milk and dairy foods (Kamei *et al.*, 1991). This is of particular importance in developing countries because availability of state-of-the-art packaging technology is often scarce and its application in the hygienically more challenging surroundings of the developing world would most likely make a greater difference from a food safety point of view than in industrialised nations. Nevertheless, the key focus of aseptic and active packaging in the industry is on the retention of a high nutrient and organoleptical product quality and extension of the storage stability of ESL milk as well as conventionally heat-treated milk and dairy commodities.

## 7.4 Novel techniques for the decontamination of milk and dairy products

While several novel techniques for food processing have been engineered in recent years, not all of them have been regarded as suitable and applied to milk and dairy products. A selection of emerging food technologies that were introduced to milk and dairy processing to a greater or lesser extent is presented in Table 7.2, also indicating their mode of operation and mechanism of decontamination. The significance of these technologies is elucidated in more detail in subsequent sections.

### 7.4.1 Emerging processing technologies

*Microfiltration*

Incorporation of microfiltration (MF) in the milk manufacturing process has led to an improvement in the microbial and organoleptical quality of milk following the principle of liquid separation by membranes with a pore size in the range of 0.1–10 μm (Brans *et al.*, 2004; Kaufmann and Kulozik,

© Woodhead Publishing Limited, 2012

**Table 7.2** Emerging food processing technologies differentiated by their operation modes, decontamination mechanisms, and milk and dairy product group specific research

| Processing technology | Operation mode | Decontamination mechanism | Applied research |
|---|---|---|---|
| Bactofugation | Batch | Centrifugal separation | Milk |
| High hydrostatic pressure | Batch | Cell permeability, shape, and constituents' modifications | Milk, butter products, cream products, cheese, powdered products, curdled milk products |
| Gamma irradiation | Batch | Inhibited DNA transcription and replication | Milk, butter products, cheese, powdered products, ice cream, whey |
| High pressure homogenisation | Continuous | Pressure drop, cavitation, shear forces, turbulence, collision | Milk |
| Microfiltration | Continuous | Cross-flow membrane separation | Milk |
| Ohmic heating | Continuous | Thermoelectric degradation | Milk, cream products, curdled milk products |
| Pulsed electric fields | Continuous | Electroporation | Milk, cream products, curdled milk products |
| Ultraviolet irradiation | Batch/ Continuous | Inhibited DNA transcription and replication | Milk, cheese whey |
| Pulsed ultraviolet irradiation | Batch/ Continuous | Inhibited DNA transcription and replication | Milk, milk foam |
| Pulsed high intensity light | Batch/ Continuous | Thermal radiation degradation, inhibited DNA transcription and replication | Milk |
| Sonication | Batch/ Continuous | Cavitation | Milk, curdled milk products |

2006; Daufin *et al.*, 2001), which generally enables high flow rates, improved yield, reduced processing costs and higher quality compared to conventional heat treatments. Nonetheless, MF is combined with HTST for commercial scale production of milk due to current regulatory food safety requirements (Lewis, 2010). The main application for membrane filtration processes (e.g. MF, ultrafiltration, nanofiltration and reverse osmosis) is water desalination (Wagner, 2001) and dialysis (Moresi and Lo Presti, 2003). However, MF has been used as a standard unit operation for food processes such as clarification (Cisse *et al.*, 2005) or concentration (Vaillant *et al.*, 2005), and in recent years a particular interest in microfiltration as a food preservation method has been established (Beolchini *et al.*, 2004).

MF is a fluid separation technology based on membrane modules with pore diameters in the low μm range through which a liquid is pumped. The pore diameter, material and configuration of the MF module as well

© Woodhead Publishing Limited, 2012

as the trans-membrane pressure affect the retentate and permeate fractions. In the food industry MF is mainly applied by dairy processors (Cheryan, 1998; Fritsch and Moraru, 2008; Nielsen, 2000), but it is also employed by producers of fermented and alcoholic beverages, fruit juices and the sugar manufacturing industry (Moresi and Lo Presti, 2003). MF offers high throughput, improved yield, reduced processing costs and higher product quality in comparison to thermal treatments (Girard and Fukumoto, 1999, Youn *et al.*, 2004). Nonetheless, a polarised surface membrane, blocking of membrane surface pores, fouling of support materials causing a drop in performance of the permeation flow as well as cleanability and membrane resistance to solvents are possible bottlenecks for the widespread adoption of this 'cold' preservation technology (Girard and Fukumoto, 1999).

*Bactofugation*

Growing interest in bactofugation (BF) as a milk and dairy decontamination step in addition to conventional heat processing has manifested as indicated by the successful integration in milk and dairy processing plants (Joppen, 2004). After separation from cream causing the majority of bacteria and spores to remain in the skimmed milk portion, the latter is centrifuged at accelerations of up to 10,000 × *g* and temperatures between 55 and 60°C for treatment volumes of up to 40,000 l/h (Stack and Sillen, 1998). During BF the centripetal force drives higher density bacteria and spores to the outside of the centrifuge where the contaminated milk is collected for subsequent UHT treatment and mixing with the purified skim milk and heat-treated cream fractions for standardisation. In a direct comparison of processing methods for the inactivation of coliforms in cheddar cheese milk with BP, hydrogen peroxide and mild heat treatments (Kosikowski and Fox, 1968), the centrifugation method achieved an average reduction of 95.3%, thus proving to be more efficient than the other techniques.

Investment costs for bactofugation are relatively high, processing capacities limited due to the batch mode setup and competition is fierce with other non-thermal processing technologies, that may exhibit higher effectiveness in reducing bacteria or spores in milk and dairy products such as, for instance, MF, pulsed electric fields (PEF), pulsed ultraviolet light (PUV) and high hydrostatic pressure (HHP) (Joppen, 2004; Rysstad and Kolstad, 2006).

*High hydrostatic pressure*

Because HHP is generally acknowledged as a universal food pasteurisation technology, to date a considerable amount of research has been conducted on the preservation of milk (Bozoglu *et al.*, 2004; Phua and Davey, 2007; Mussa *et al.*, 1998; De Lamo-Castellvi *et al.*, 2005b) and dairy products (Pina-Perez *et al.*, 2009b; Moschopoulou *et al.*, 2010; O'Reilly *et al.*, 2000; Daryaei *et al.*, 2010). More than 4 log cycles of *E. coli* and *L. innocua* were inactivated by pulsed HHP treatment of whole milk using holding times of 20 min (Buzrul *et al.*, 2009) indicating good decontamination potential for

© Woodhead Publishing Limited, 2012

milk but in a very time-consuming and, hence, uneconomic manner that may not meet the time efficiency criteria desired by milk producers. In addition to the requirement for long milk processing times, pathogen-specific variability in resistance to ambient temperature HHP treatment was observed in a study by Chen (2007) that compared several pathogens with regard to their survival rates after HHP processing at a pressure level of 600 MPa, achieving a 5 log cycle reduction after 0.5 min for *L. monocytogenes* and up to 10 min for *S. aureus* and *E. coli*; times for strains of *Salmonella* and *Vibrio parahaemolyticus* treated at 600 and 300 MPa, respectively, fell in between these values. Moreover, volatility in critical pressure levels up to 500 MPa was detected for some pathogenic bacteria HHP-treated in milk (Chen *et al.*, 2006), identifying *S. flexneri*, *E. coli* O157:H7 and *S. aureus* as the most pressure resistant pathogens. When processing temperature levels (4, 21 and 45°C) during HHP treatment of the latter two bacteria in UHT milk were compared (Guan *et al.*, 2006), higher temperature levels led to higher inactivation of the pathogens.

A minimum pressure level of 500 MPa or above was also recommended by Schlesser and Parisi (2009), based on requiring a minimum 5 log cycle inactivation at a relatively short processing time of 30 s for *Yersinia pseudotuberculosis* and *Francisella tularensis* in UHT skim milk, which broadly coincided with the pressure level needed for effective inactivation of *E. coli* in skim milk by other researchers (Linton *et al.*, 2001). However, the latter research group applied substantially longer treatment times. High decontamination potential was reported when exposing *Listeria monocytogenes* to HHP obtaining reductions of more than 6 log cycles in UHT whole milk (Amina *et al.*, 2010), exhibiting a maximum sensitivity to HHP processing in the mid exponential growth phase (Hayman *et al.*, 2007) and increased pressure resistance with exposure to a heat shock treatment prior to HHP (Hayman *et al.*, 2008). *Campylobacter* species were inactivated effectively in HHP-processed UHT milk (Martinez-Rodriguez and Mackey 2005, Solomon and Hoover 2004), which appeared to challenge the HHP processing most among tested foods but was evaluated as no threat under the present HHP conditions applied in food production (425–580 MPa for 5–7 min). By contrast, the findings of Guan *et al.* (2005) and Erkmen (2009) suggested that *Salmonella* Typhimurium is more pressure-resistant in UHT whole milk and raw milk, respectively, necessitating long HHP processing times and the same seems to apply to *E. sakazakii* in reconstituted infant formula milk due to wide variation in HHP resistances among investigated strains (Gonzalez *et al.*, 2006).

With regard to cheese products the survival of *L. monocytogenes* following HHP has been studied most thoroughly, even though published work on the HHP decontamination of other major pathogens of importance such as *E. coli* O157:H7 (Buzrul, 2009; Shao *et al.*, 2007), *S. aureus* (Lopez-Pedemonte *et al.*, 2007b), *Y. enterocolitica* (De Lamo-Castellvi *et al.*, 2005a) and *Salmonella* Enteritidis and *S.* Typhimurium (De Lamo-Castellvi *et al.*, 2007)

© Woodhead Publishing Limited, 2012

is available in the literature. A 5 log cycle reduction of *L. monocytogenes* was obtained in HHP-treated Gorgonzola cheese rinds at 700 MPa for 15 min (Carminati *et al*., 2004), pasteurised and raw milk white cheese following 600 MPa for 10 min (Evrendilek *et al*., 2008), curd cheese using 500 MPa for 10 min (Lopez-Pedemonte *et al*., 2007a) as well as inactivation of 4.5 log cycles of the pathogen in Camembert-type cheese after HHP exposure at 500 MPa for 10 min was reported (Linton *et al*., 2008).

As the high decontamination efficacy against various bacteria comes at the expense of relatively long batch-type treatments, HHP seems economically unfavourable for milk processing compared to continuous non-thermal technologies such as MF and PEF; instead it appears particularly well-suited as a preservation method for semi-solid and solid dairy foods that could justify the high initial equipment cost and longer treatment times.

*Gamma irradiation*

Suitability of gamma irradiation (GI) for ice cream processing at an irradiation dose of 3 kGy was reported by Pietranera *et al*. (2003) after obtaining effective decontamination of microorganisms and favourable product quality at the same time. The use of GI for dairy products at low temperature was investigated by Hashisaka *et al*. (1990) suggesting complete sterilisation of ice cream and frozen yogurt at 40 kGy, but not for mozzarella and cheddar cheeses, which showed survival of *Staphylococcus*, *Streptococcus*, *Lactobacillus* and *Bacillus* bacteria. Dosages of GI between 43 and 50 kGy ensured sufficient reduction of *B. cereus* in the cheeses, also identifying cheeses as the microbiologically most demanding group of dairy products for GI treatment. However, such high dosages of ionising radiation are not recommended for processing as they will substantially deteriorate the quality of products and, therefore, irradiation of up to 10 kGy is used in practice to provide acceptable product quality.

Bougle and Stahl (1994) reported eradication of $10^4$ CFU of *L. monocytogenes* per g raw milk Camembert at a much lower dose of 2.6 kGy. According to Kim *et al*. (2010) in GI-treated sliced and pizza cheeses, *S. aureus* was eliminated at 3 kGy and the lethal irradiation dose for *L. monocytogenes* was 5 kGy. In a study by Konteles *et al*. (2009), the latter pathogen was reduced below detection limit in Feta cheese when exposed to GI dosages of 2.5 and 4.7 kGy, but the higher dose seemed to adversely affect the sensory quality of the cheese. By contrast, no effect on the sensory quality of soft whey cheese (Tsiotsias *et al*., 2002) was detected after GI was applied with 4 kGy that led to a reduction of 3 log cycles of *L. monocytogenes*. In addition, a limited efficacy of GI against psychrotrophs during cheddar cheese ripening was observed by Seisa *et al*. (2004).

The research findings indicate that ionising irradiation has great potential for the decontamination of milk and dairy foods but over-processing may deteriorate the product quality considerably and thus, careful optimisation of the technology is necessary and adds to the high equipment cost as well

© Woodhead Publishing Limited, 2012

as the problematic transport and final disposal of the radioactive material that is generated during the operation of GI facilities.

*High pressure homogenisation*

The established field of high pressure homogenisation (HPH) research is decontamination of milk (Picart *et al.*, 2006; Hayes *et al.*, 2005) and dairy products (Kheadr *et al.*, 2002; Lanciotti *et al.* 2006), but the efficacy of HPH for mitigation of pathogens in juices (Briñez *et al.*, 2007, Campos and Cristianini, 2007; Velazquez-Estrada *et al.*, 2011) and buffer solution (Diels *et al.*, 2005) has also been investigated. The principle of operation of HPH differs from that of microfluidisation as the latter is based on high pressure treatment caused by two high speed liquid jet streams colliding in a reaction chamber, while HPH takes advantage of the same mode of operation as regular homogenisation but at considerably higher pressures of up to 4000 bar. Hence, it has also been termed ultra-high pressure homogenisation. However, it has been suggested that this terminology should be used once pressures above 1000 bar are reached (Pereda *et al.*, 2007). Microbial inactivation is achieved by physical forces, for instance, pressure drops, cavitation, shearing, turbulence or collision (Moroni *et al.*, 2002) and is dependent on the pressure applied, the inlet temperature and number of product passages through the HPH unit (Pereda *et al.*, 2007). In addition, it was suggested that the higher fat content of milk and dairy products exposed to HPH is converted into better microbial inactivation of *L. monocytogenes* (Roig-Sagues *et al.*, 2009). Overall, the research interest in this continuous type decontamination method has remained relatively stable, but HPH still occupies a niche compared to the more prevalent HHP technology and against this background it remains questionable whether or not HPH will achieve industrial application. At present it cannot be determined with certainty if this novel technology will prevail over a longer period of time, also because research progress is limited and equipment and maintenance cost in combination with the achievable product quality and safety may become decisive factors for the commercial implementation of HPH (Hayes *et al.*, 2005).

*Ohmic heating*

Despite the fact that milk and dairy foods are multiphase systems which could be regarded as more suitable for the application of ohmic heating than for conventional heat pasteurisation and sterilisation as the processing time-scale and quality of food is limited by heat conduction to and from particle centres (Davies *et al.*, 1999), only few studies are available (Mainville *et al.*, 2001; Sun *et al.*, 2008) that looked into this technology for possible decontamination of milk and dairy foods. Due to differences in electrical conductivities that affect the heating uniformity of the individual food components (Tulsiyan *et al.*, 2008), where low conductivity regions will divert local electric fields, thus heating less quickly than high conductivity regions, inhomogeneous treatment could occur during ohmic heating. Therefore, uniform and stable

© Woodhead Publishing Limited, 2012

dispersion of the milk and dairy products is an important prerequisite prior to exposure in the ohmic heating system, designed to permit relatively homogeneous processing, to enable successful preservation.

*Pulsed electric fields*

With pulsed electric fields (PEF) being a non-thermal liquid foods processing technology, most of the research has centred on preservation of milk and little on other dairy products such as, for example, yogurt (Yeom *et al.* 2004), cream (Manas *et al.*, 2001) and whey (Gallo *et al.*, 2007). Moreover, PEF was suggested as an alternative pasteurisation technique for cheese making (Yu *et al.*, 2009), reducing *S.* Enteritidis in whole milk by up to 5 log cycles and, when combined with mild heating, also affecting coagulation properties to a lesser extent than thermal pasteurisation which was reflected in better curd firmness and shorter rennet coagulation time of the non-thermally treated product. Although the observed effects vary, a majority of research groups that looked into the reduction of the microbial load by PEF in different milk and dairy foods suggested that decontamination is often significantly affected by more or less complex composition of the products (Walkling-Ribeiro *et al.*, 2009; Martin *et al.*, 1997; Gilliland and Speck, 1967; Rodríguez-González *et al.*, 2011b; Otunola *et al.*, 2008), while others observed that differences in composition between their investigated milk and dairy foods did not affect microbial inactivation significantly (Picart *et al.*, 2002; Reina *et al.*, 1998). Manas *et al.* (2001) reported the reduction of almost 2 log cycles of *E. coli* in cream by PEF using an electric field strength of 33 kV/cm for a treatment time just below 100 μs and Evrendilek and Zhang (2005) indicated equivalent reduction of PEF-treated *E. coli* O157:H7 in skim milk using 24 kV/cm for 141 μs. By contrast, in PEF-treated skim milk at 25 kV/cm for 45 μs, a decrease in *E. coli* bacteria of more than 2 log cycles was reported (Martin *et al.*, 1997), in UHT skim milk more than 4 log cycles of *E. coli* cells were inactivated by PEF at 22.4 kV/cm for 46 μs (Grahl and Maerkl, 1996), in pasteurised fat-free milk inoculated with *E. coli* prior to PEF treatment at 41 kV/cm for 158 μs, more than 5.5 log cycles did not survive the processing (Dutreux *et al.*, 2000) and in simulated milk ultra-filtrate (SMUF) reduction of up to 9 log cycles of *E. coli* was suggested (Zhang *et al.*, 1995) after PEF exposure at 70 kV/cm for 160 μs.

In addition to the possible impact of product characteristics and considerable variations in the PEF resistance of *E. coli* strains, the inoculation with pure culture instead of the enrichment with indigenous pathogens could account for higher resistivity as proposed by Otunola *et al.* (2008) and the occurrence of cross-protective effects that render these microorganisms less prone to PEF (Rodríguez-González *et al.*, 2011a) will have to be considered.

The highest PEF reduction for *Pseudomonas fluorescens* in UHT skim milk was suggested by Grahl and Maerkl (1996), who achieved more than 4 log cycles when PEF was applied at 22 kV/cm for 300 μs, whereas, in a study by Fernandez-Molina *et al.* (2006), PEF at 39 kV/cm led to a *Pseudomonas*

© Woodhead Publishing Limited, 2012

*fluorescens* reduction of approximately 2 log cycles in skim milk which is in agreement with the amount of pseudomonads obtained also in skim milk by Craven *et al.* (2008) following PEF at 29 kV/cm and 40°C and achieving more than 1 extra log cycle of inactivation when the temperature was increased to 55°C. Similarly, *Salmonella* Enteritidis proved to be relatively resistant to pre-heated (42°C) PEF treatment at 45–47 kV/cm for 500 μs without and with additional heat treatment (62°C, 38 s) limiting the reduction of the pathogen to slightly more than 1 and 2 log cycles, respectively (Floury *et al.*, 2006a, 2006b). This coincides with the inactivation reported for *Salmonella* Dublin in skim milk by other researchers (Sensoy *et al.*, 1997) who used PEF at 25 kV/cm for 127 μs. Inactivation of vegetative *Bacillus cereus* cells up to 3 log cycles after PEF treatment at 35 kV/cm for 188 μs was suggested by Michalac *et al.* (2003), but endospores of the pathogen in UHT skim milk were not distinctly affected by PEF (Grahl and Maerkl, 1996). *S. aureus* decontamination of 3 log cycles obtained by Evrendilek *et al.* (2004) in PEF-processed skim milk with 35 kV/cm for 450 μs appeared less efficient than with PEF treatment times extended to 1200 μs, which was applied by a different research group (Sobrino-Lopez *et al.*, 2009) who achieved up to 4.5 log cycles of pathogen reduction in milk.

Decontamination of *L. monocytogenes* with PEF at 30 kV/cm for 600 μs in whole milk led to more than a 4 log cycle reduction (Reina *et al.*, 1998) indicating also that higher processing temperature levels significantly enhanced the pathogen reduction. Moreover, many studies looked into the PEF resistance of *Listeria innocua*, used as non-pathogenic surrogate for *L. monocytogenes*. Examples for PEF application to *L. innocua* are studies by Dutreux *et al.* (2000), in which nearly 4 log cycles of the bacteria in fat-free milk were inactivated at 41 kV/cm for 158 μs and by Picart *et al.* (2002), who achieved approximately 2.5 log cycles in UHT whole milk at 29 kV/cm for 350 μs. Other pathogens that have been tested for their susceptibility to PEF were *C. sakazakii* in infant formula milk (Pina-Perez *et al.*, 2009a) and *M. paratuberculosis* in tyndallised milk (Rowan *et al.*, 2001), showing maximum reductions of more than 2 log cycles at 15 kV/cm for 3000 μs and almost 6 log cycles at 30 kV/cm for 1250 μs with a processing temperature of 50°C, respectively.

Allowing continuous and gentle decontamination of milk and dairy products makes PEF particularly interesting for milk and dairy processing, although it may not have the same required efficacy as a stand-alone pasteurisation technology and lacks the capability to function as a sterilisation method. Unlike other emerging preservation technologies, multiple processing parameters have to be aligned, rendering the optimisation of PEF processing equipment and operating conditions more complicated than for other emerging technologies. This often makes PEF research a time-consuming challenge for food engineers and scientists. However, substantial enhancement of PEF effectiveness can be achieved in combination with heat and other preservation technologies, as discussed in the subsequent section of this chapter

© Woodhead Publishing Limited, 2012

and, therefore, its integration in an efficacious non-thermal hurdle technology for decontamination of milk and dairy products that meets growing consumer quality demands for minimally processed foods appears advisable.

*Sonication*

Several studies are available that have looked into decontamination of milk with high intensity ultrasound (Chouliara *et al.*, 2010; Bermudez-Aguirre *et al.*, 2009; Villamiel and De Jong, 2000; Zenker *et al.*, 2003). Application of ultrasonic waves (UW) alone, also termed (ultra-)sonication, was suggested to have limited effect on the survival of microorganisms in liquid foods (Piyasena *et al.*, 2003). However, combined processing by means of moderate levels of heat and sonication, thermosonication (TS), pressure and sonication, manosonication (MS) and sonication with both pressure and heat stress add-ons, manothermosonication (MTS), have produced better microbial inactivation.

In a study by Cameron *et al.* (2009) it was reported that *Pseudomonas fluorescens* and *E. coli* were eradicated after 6 and 10 min of UW treatment, respectively and 99% of *L. monocytogenes* were inactivated following 10 min of milk exposure to UW. By comparison, D'Amico *et al.* (2006) and Earnshaw *et al.* (1995) suggested more than 5 and 6 decimal reductions in *L. monocytogenes* contaminated UHT milk after TS treatment, respectively, which differs from the findings obtained by Pagan *et al.* (1999) for MS-treated *L. monocytogenes* in skim milk, achieving 2 decimal reductions. In addition, Bermudez-Aguirre and Barbosa-Canovas (2008) observed that an increase in the butter fat content in milk rendered TS processing less effective for the decontamination of *L. innocua*. In reconstituted infant formula milk higher inactivation of *C. sakazakii* under TS than with regular UW processing conditions was reported by Adekunte *et al.* (2010). Moreover, *C. sakazakii* was exposed to MS showing a higher resistance than *Y. enterocolitica*, but it appeared to be more prone to the treatment than *S. enterica* serovar Enteritidis, *L. monocytogenes* and *E. faecium* in phosphate buffer and the efficacy of MS was decreased further when *C. sakazakii* was processed in rehydrated powdered milk. In spite of the considerable decontamination effect exhibited by TS, MS and MTS, detrimental effects caused by sonication on labile food components and physical food properties could occur and limit the application of these processes for milk and dairy processing (Chemat *et al.*, 2004).

*Ultraviolet light*

At present three liquid food processing technologies have been studied that utilise ultraviolet light for the decontamination of milk: conventional (non-intermittent) ultraviolet light (UV), pulsed ultraviolet light (PUV) and pulsed high intensity light (PHIL). While PHIL uses a broadband white light including a 25% fraction from ultraviolet light, thus also bringing about thermal degradation due to heating, UV (100%) and PUV (<50%) focus mostly on the most effective ultraviolet fraction of the light spectrum.

© Woodhead Publishing Limited, 2012

In sterilised whole milk, a maximum inactivation of one log cycle of *E. coli* and *L. innocua* bacteria by PHIL was reported by Palgan *et al.* (2011) whereas Choi *et al.* (2010) obtained a 2.5 log cycle reduction of *L. monocytogenes* in infant powdered milk using the same processing technology. Despite the discrepancy in results between the two PHIL studies, which could be explained by the difference in the product transparencies, equipment and microorganisms used, the findings indicate limited efficacy of PHIL, as surface but not in-depth decontamination of milk takes place in a batch mode setup, thus not suggesting suitability for a continuous stand-alone processing approach at present.

By contrast, PUV research by Krishnamurthy *et al.* (2007) on *S. aureus* suggested more than 7 log cycles inactivation in raw milk and the same researchers (Krishnamurthy *et al.*, 2008a) achieved a reduction of *S. aureus* cells of up to 8.5 log cycles in a follow-up PUV study, also inactivating more than 6.5 log cycles of the bacteria in milk foam. For the decontamination of *M. avium* subspecies *paratuberculosis* in whole and skim milk, maximum reductions of 2.6 and 2.8 log cycles, respectively, were reported by Altic *et al.* (2007), which is slightly higher than the maximum inactivation obtained in another UV study (Donaghy *et al.*, 2009) using *M. avium* subspecies *paratuberculosis* suspended in UHT whole milk, reducing nearly up to 2.0 log cycles.

Although the decontamination reached with UV seems to be lower than that of PUV, it is noteworthy that mycobacteria have generally been proposed to be more resistant to UV than the majority of other bacteria groups. The latter is supported by a study conducted on UV-treated goats' milk inoculated with *L. monocytogenes* (Matak *et al.*, 2005) inactivating more than 5.5 log cycles of the pathogen. In addition, research on cheese whey (Mahmoud and Ghaly, 2004; Singh and Ghaly, 2006) also suggested that it can be successfully decontaminated if fouling on the UV lamps is sufficiently circumvented. Combining UV treatment with infrared radiation heating (Giraffa and Carini, 1984), with the latter being a relatively fast and energy efficient thermal treatment method, led to 2.0 log reductions of yeasts and moulds and vegetative bacteria with subsequent heating up to 75°C, sporeformers remained unaffected, while an almost 8.5 log inactivation of infrared-heated *S. aureus* in milk was reported by Krishnamurthy *et al.* (2008b).

Overall, it can be concluded that UV and PUV could represent promising technologies for the decontamination of milk and dairy products, provided that the nutritional and sensory product quality following irradiation treatments remains acceptable.

### 7.4.2 Minimal processing with hurdle strategies

Most hurdle strategies that have been studied for milk and dairy foods are either HHP-based (Capellas *et al.*, 2000; Morgan *et al.*, 2000; Koseki

© Woodhead Publishing Limited, 2012

*et al.*, 2008; Van Opstal *et al.*, 2004) or PEF-based (Fernandez-Molina *et al.*, 2005c; Sobrino-Lopez and Martin-Belloso, 2008; Walkling-Ribeiro *et al.*, 2011). Inoculation of high pressure processed raw milk cheese at 500 MPa for 5 min with bacteriocin-producing lactic acid bacteria was found to lower the *S. aureus* and *E. coli* O157:H7 populations during cheese ripening by up to 4 log cycles (Arques *et al.*, 2005) and more than 6 log cycles (Rodriguez *et al.*, 2005), respectively. Combining (500 IU/ml) nisin with HHP (500 MPa for 10 min) was reported to reduce *Bacillus subtilis* spores effectively with 3.5 log cycles added to the HHP inactivation while *B. cereus* was less sensitive with an inactivation approximately 1 log cycle below that obtained for *B. subtilis* (Black *et al.*, 2008) indicating the very high decontamination potential of this hurdle technology, yet not accomplishing complete milk sterilisation.

Effective HHP/nisin processing was confirmed by other researchers, who also added heat (Gao and Ju, 2008) for 6 log cycles of *C. botulinum* spore inactivation by HHP (545 MPa for 13.3 min, 51°C) and nisin (129 IU/ml) in UHT milk and who applied a mixture of 400 μg/ml lysozyme and 400 IU/ml nisin with HHP (600 MPa, 15min) for an *E. coli* reduction of more than 5.5 log cycles in skim milk and up to more than 2.5 log cycles in whole milk (Garcia-Graells *et al.*, 1999) that was further increased to more than 8 log cycles in skim milk when additional heating (50°C) was applied at a slightly lower pressure level (550 MPa). By comparison, addition of 224 U/ml lysozymes to skim milk containing *S.* Typhimurium, *E. coli* O157:H7, *Y. enterocolitica* and *S. flexneri* after HHP treatments (225, 350, 235 and 275 MPa, respectively, for 15 min) increased the pathogen reductions by respective 1.5, 0.5, almost 2 and slightly above 2 log cycles (Nakimbugwe *et al.*, 2006). Combined treatments of HHP with heat have also attracted great interest by researchers, for instance, for the inactivation of *B. cereus* (Ju *et al.*, 2008) and *Geobacillus stearothermophilus* (Gao *et al.*, 2006) spores in milk buffer achieving reductions of 6 log cycles at 71°C with 540 MPa for 16.8 min and at 86°C with 625 MPa for 14 min, respectively and for *M. avium* subspecies *paratuberculosis* reduction in milk subjected to HHP at 600 MPa for 5 min with a subsequent mild thermal pasteurisation (72°C, 15 s) step decimating more than 6.6 log cycles of the pathogen (Donaghy *et al.*, 2007).

While HHP-based hurdle strategies for milk and dairy product decontamination have focused on the addition of antimicrobials or heat, a larger selection of hurdles that could be applied in conjunction with PEF was examined for milk processing. The majority of non-thermal or sub-pasteurisation level hurdle treatments involving PEF also include mild (40–50°C) or moderate (50–65°C) heating that is either due to PEF dissipation of heat into the treated food or an additional processing stage preceding or following the high voltage pulse treatment. However, most of these studies have focused on investigating the inactivation of groups of native microbiota by a combination of PEF and heat and the resulting microbiological shelf

© Woodhead Publishing Limited, 2012

stability of milk and dairy products non-thermally (Walkling-Ribeiro *et al.*, 2009; Fernandez-Molina *et al.*, 2005c; Sepulveda *et al.*, 2005) as well as in conjunction with a more severe treatment at HTST pasteurisation level (Evrendilek *et al.*, 2001; Fernandez-Molina *et al.*, 2005b) and not on decontaminating specific pathogens in milk and dairy products. Nearly 4 log cycles of *E. coli* were inactivated when an orange juice-milk beverage was processed with PEF and heat at 15 kV/cm for 700 μs and 55°C (Rivas *et al.*, 2006), while very effective reduction of 8 log cycles of *E. coli* and *P. fluorescens* as well as 3 log cycles of *G. stearothermophilus* in UHT whole milk was reported by Shin *et al.* (2007) following combined PEF and heat processing at 60 kV/cm for 210 μs and 50°C. In addition, Shamsi *et al.* (2008) suggested a 6 log cycle inactivation of pseudomonads and enteric bacteria in PEF (35 kV/cm, 20 μs) and heat (60°C) treated raw skim milk and Guerrero-Beltran *et al.* (2010) referred to a 5.5 log cycle reduction of *L. innocua* in whole milk after exposure to PEF (40 kV/cm, 38 μs) and heat (67°C).

In a recent non-thermal approach, PEF was combined with cross-flow MF achieving a reduction of between 4 and 5 log cycles of native microbiota in skim milk that was comparable to HTST treatment (75°C, 24 s) when MF preceded PEF treatment (Rodríguez-González *et al.*, 2011b; Walkling-Ribeiro *et al.*, 2011), while a reversed PEF/MF processing sequence, that was also applied in the study by Walkling-Ribeiro *et al.* (2011), eliminated up to 7 log cycles of the microorganisms which was significantly higher than that of the heat pasteurization. The same researchers (Walkling-Ribeiro *et al.*, 2011) suggested that PEF possibly altered the steric configuration of the microorganisms by electroporation or induced interactions between microorganisms or microorganisms and milk components that led to agglomeration prior to MF processing, thus enhancing the latter filtration process substantially.

Hurdle technologies consisting of PEF and antimicrobial agents have also been researched relatively well. In most of these studies PEF was combined with nisin as reported by Calderon-Miranda *et al.* (1999) inactivating more than 3.5 log cycles of *L. innocua* in skim milk using 50 kV/cm for 64 μs and 100 IU/ml nisin, by Sobrino-Lopez and Martin-Belloso (2006) indicating a 6 log cycle reduction of *S. aureus* in skim milk at 35 kV/cm for 2400 μs and 20 ppm nisin and by Terebiznik *et al.* (2000) suggesting the reduction of 4 log cycles of *E. coli* in SMUF pulsed at 11 kV/cm with 500 UI/ml of nisin. The effect of nisin and lysozyme in combination with PEF was studied by Smith *et al.* (2002) achieving a 7 log cycle reduction of native microbiota in raw skim milk using 80 kV/cm for 100 μs at 52°C with both 38 and 1638 IU/ml of respective nisin and lysozyme and by Sobrino-Lopez and Martin-Belloso (2008) indicating a *S. aureus* reduction greater than 6 log cycles in skim milk using 35 kV/cm for 1200 μs with both 1 and 300 IU/ml of respective nisin and lysozyme. Combination of PEF, nisin and enterocin was also investigated by Sobrino-Lopez *et al.* (2009) reporting

© Woodhead Publishing Limited, 2012

*S. aureus* inactivation of more than 6 log cycles, when PEF at 35 kV/cm for 800 μs was combined with 28 AU/ml enterocin AS-48 and 20 IU/ml nisin in a hurdle technology. Organic acids were also tested as a different combination of antimicrobials with PEF in skim milk (Fernandez-Molina *et al.*, 2005a) reducing almost 2 log cycles of *P. fluorescens* at 38 kV/cm for 60 μs but obtaining no increase in the antimicrobial effect with the addition of either acetic or propionic acid.

Another novel hurdle approach looked into the combined processing with PEF and heat-assisted ultrasound of UHT skim milk contaminated with *L. innocua* (Noci *et al.*, 2009), reporting a more than 6 log cycle inactivation of the microorganism thermosonicated at 85 W/cm$^2$ for 80 s with an outlet temperature of 55°C and pulsed at 40 kV/cm for 50 μs and a close to 7 log cycle reduction by pre-heating the milk at 55°C for 60 s prior to TS (85W/cm$^2$, 80 s) and subsequent PEF (50 kV/cm, 38 μs) treatment. These findings indicated that TS/PEF achieved *L. innocua* inactivation in UHT skim milk comparable to HTST treatment (72°C, 26 s; 7 log cycles) and a synergism between TS and PEF occurred when milk was not pre-heated prior to TS. Synergistic treatment effects have also been observed in some of the aforementioned studies that analysed HHP-based and PEF-based hurdle technologies.

## 7.5 Conclusions and future trends

The emergence of new food processing technologies has significantly improved the armamentarium that can be used to decontaminate the gradually growing variety of pathogens adaptive to milk and dairy foods. While it is still essential from a food safety perspective to identify and study well-known and new pathogens in a milk and dairy environment and assess their resistances and susceptibility with the microbiological and molecular biological analysis techniques available, the rapid and thorough investigation and documentation of foodborne pathogen outbreaks on a global scale is also vital to prioritise and highlight imminent hazards, so that both research and industry can develop necessary practices that counteract these health threats.

Although first steps have been taken to establish networks and exchange of information among researchers, to date a lack of coordination between government agencies of different countries or continental regions is visible. The latter makes itself felt as pathogen-related incidents are viewed in an isolated manner, predominantly of national interest, although containment of foodborne outbreaks or epidemics across borders has often failed. A more concerted approach similar to that in the case of natural disasters when countries offer their help by sending their specialists to the affected site, without eyeing their own domestic interests, could be desirable.

© Woodhead Publishing Limited, 2012

In recent years a growing number of food scientists and engineers have examined minimal processing strategies based on the combination of the decontamination methods currently available and, due to the obvious advantages of hurdle technologies, an increase of research in this area will be very likely and necessary to meet future food safety challenges. The technological and economic feasibility when combining different hurdles for a new processing approach often depends on product and technology, so that some products are not suitable for certain techniques, for example, limited surface decontamination by UV, PUV and PHIL treatments of opaque food media and techniques that affect the microorganisms with the same mode of action, for instance, HHP, HPH and MTS, will increase the costs significantly without adding to the overall efficacy of the treatment. Hence, for decontamination of liquid milk and dairy products, a continuous processing approach, that combines moderate heat, microfiltration, pulsed electric fields, thermosonication, manosonication, manothermosonication, bacteriocins and bacteriophages in hurdle technologies, appears to be most promising and to a limited extent ultraviolet light and pulsed ultraviolet light may also be suitable in case sufficient product transparency or treatment of thin product food layers are provided.

With regard to pathogen mitigation in semi-solid and solid dairy foods, a batch mode hurdle approach combining moderate heat, high hydrostatic pressure, high pressure homogenisation, pulsed electric fields, bactofugation, gamma irradiation, bacteriocins and bacteriophages could be a potent alternative to conventional HTST pasteurisation or sterilisation by UHT. The availability of more studies on HHP-based and PEF-based hurdle technologies than for other emerging technologies could also be a factor that could benefit the introduction of these technologies in the milk and dairy sector. However, a change in thinking towards these alternative preservation methods will have to take place for legislative authorities, milk and dairy farmers and the processing industry and consumers alike. The introduction of microfiltration in addition to HTST pasteurisation for the production of ESL milk is a good example that increasing consideration of novel hurdle technologies for their commercial application in the milk and dairy sector and faster processing times for the approval of emerging technologies by legislators will be a great step forward to commercialisation of such techniques and subsequent prevention of foodborne outbreaks. In addition, financial support by farmers and manufacturers for research on hurdle strategies that produce more effective decontamination of milk and dairy products with higher quality will also be of great importance for progress in this field. With the support and commitment of lawmakers, milk and dairy farming and producing industry and food researchers to these new methods for decontamination, consumer acceptance is likely to follow and this could play an important role in preventing the occurrence of more pathogen outbreaks associated with milk and dairy foods in the future.

© Woodhead Publishing Limited, 2012

## 7.6 Sources of further information and advice

- Code of hygienic practice for milk and milk products http://www.codexalimentarius.net/download/standards/10087/CXP_057e.pdf
- Resource center for the US dairy industry on food safety http://www.nmpf.org/washington_watch/standardsandsafety
- World ranking of food safety performances http://www.schoolofpublicpolicy.sk.ca/_documents/_publications_reports/food_safety_final.pdf

## 7.7 References and further reading

ACHEMCHEM F, ABRINI J, MARTINEZ-BUENO M, VALDIVIA E and MAQUEDA M (2006), 'Control of *Listeria monocytogenes* in goat's milk and goat's jben by the bacteriocinogenic *Enterococcus faecium* F58 strain', *J Food Protect*, 69, 2370–2376.

ACKERS M L, SCHOENFELD S, MARKMAN S J, SMITH G, NICHOLSON M A, DEWITT W, CAMERON D N, GRIFFIN P M and SLUTSKER L (2000), 'An outbreak of *Yersinia enterocolitica* O:8 infections associated with pasteurised milk', *J Infect Dis*, 181, 1834–1837.

ADEKUNTE A, VALDRAMIDIS V P, TIWARI B K, SLONE N, CULLEN P J, O'DONNELL C P and SCANNELL A (2010), 'Resistance of *Cronobacter sakazakii* in reconstituted powdered infant formula during ultrasound at controlled temperatures: a quantitative approach on microbial responses', *Int J Food Microbiol*, 142, 53–59.

ADRIAN J, PASCHE S, VOIRIN G, ADRIAN J, PINACHO D G, FONT H, SANCHEZ-BAEZA F, PILAR-MARCO M, DISERENS J M and GRANIER B (2009), 'Wavelength-interrogated optical biosensor for multi-analyte screening of sulfonamide, fluoroquinolone, beta-lactam and tetracycline antibiotics in milk', *Trends Anal Chem*, 28, 769–777.

AKINEDEN Ö, HASSAN A A, SCHNEIDER E and USLEBER E (2008), 'Enterotoxigenic properties of *Staphylococcus aureus* isolated from goats' milk cheese', *Int J Food Microbiol*, 124, 211–216.

ALTIC L C, ROWE M T and GRANT I R (2007), 'UV light inactivation of *Mycobacterium avium* subsp. *paratuberculosis* in milk as assessed by FASTPlaqueTB phage assay and culture', *Appl Environ Microbiol*, 73, 3728–3733.

AMINA M, PANAGOU E Z, KODOGIANNIS V S and NYCHAS G J E (2010), 'Wavelet neural networks for modelling high pressure inactivation kinetics of *Listeria monocytogenes* in UHT whole milk', *Chemometr Intell Lab Syst*, 103, 170–183.

ANANOU S, MUNOZ A, MARTINEZ-BUENO M, GONZALEZ-TELLO P, GALVEZ A, MAQUEDA M and VALDIVIA E (2010), 'Evaluation of an enterocin AS-48 enriched bioactive powder obtained by spray drying', *Food Microbiol*, 27, 58–63.

ANANY H (2010), Biocontrol of foodborne bacterial pathogens using immobilized bacteriophages, PhD Thesis, University of Guelph, Ontario, Canada.

ANUNCIACAO L L, LINARDI W R, DO CARMO L S and BERGDOLL M S (1995), 'Production of staphylococcal enterotoxin A in cream-filled cake', *Int J Food Microbiol*, 26, 259–263.

APHIS (2011), 'Prevalence of *Coxiella burnetii* in bulk-tank milk on US dairy operations, 2007', Fort Collins, CO, USA, March 2011 Veterinary Services Technical Brief of the Animal and Plant Health Inspection Service, United States Department of Agriculture.

ARQUES J L, RODRIGUEZ E, GAYA P, MEDINA M, GUAMIS B and NUNEZ M (2005), 'Inactivation of *Staphylococcus aureus* in raw milk cheese by combinations of high-pressure treatments and bacteriocin-producing lactic acid bacteria', *J Appl Microbiol*, 98, 254–260.

© Woodhead Publishing Limited, 2012

ARQUES J L, RODRIGUEZ E, NUNEZ M and MEDINA M (2008), 'Antimicrobial activity of nisin, reuterin and the lactoperoxidase system on *Listeria monocytogenes* and *Staphylococcus aureus* in cuajada, a semisolid dairy product manufactured in Spain', *J Dairy Sci*, 91, 70–75.

ASAO T, KUMEDA Y, KAWAI T, SHIBATA T, ODA H, HARUKI K, NAKAZAWA H and KOZAKI S (2003), 'An extensive outbreak of staphylococcal food poisoning due to low-fat milk in Japan: estimation of enterotoxin A in the incriminated milk and powdered skim milk', *Epidemiol Infect*, 130, 33–40.

ASLIM B, YUCEL N and BEYATLI Y (2004), 'Effect of a bacteriocin-like substance (BLS) produced by *Streptococcus thermophilus* strains on *Listeria* spp. strains', *J Food Process Pres*, 28, 241–250.

BAIRD G, FONTAINE M and DONACHIE W (2005), 'The effects of pasteurization on the survival of *Corynebacterium pseudotuberculosis* in milk', *Res Vet Sci*, 78, 18–19.

BARBINI DE PEDERIVA N B and STEFANINI DE GUZMÁN A M (2000), 'Isolation and survival of *Yersinia enterocolitica* in ice cream at different pH values, stored at –18°C', *Braz J Microbiol*, 31, 174–177.

BAUTISTA D A, DURISIN M D, RAZAVI-ROHANI S M, HILL A R and GRIFFITHS M W (1993), 'Extending the shelf-life of cottage cheese using monolaurin', *Food Res Int*, 26, 203–208.

BEDNARCZYK A and DACZKOWSKA-KOZON E G (2008), 'Pathogenic features of bacteria from the *Bacillus cereus* group', *Postepy Mikrobiologii*, 47, 51–63.

BEOLCHINI F, VEGLIO F and BARBA D (2004), 'Microfiltration of bovine and ovine milk for the reduction of microbial content in a tubular membrane: a preliminary investigation', *Desalination*, 161, 251–258.

BERMUDEZ-AGUIRRE D and BARBOSA-CANOVAS G V (2008), 'Study of butter fat content in milk on the inactivation of *Listeria innocua* ATCC 51742 by thermo-sonication', *Innovat Food Sci Emerg Tech*, 9, 176–185.

BERMUDEZ-AGUIRRE D, CORRADINI M G, MAWSON R and BARBOSA-CANOVAS G V (2009), 'Modeling the inactivation of *Listeria innocua* in raw whole milk treated under thermo-sonication', *Innovat Food Sci Emerg Tech*, 10, 172–178.

BEUCHAT L R, KIM H, GURTLER J B, LIN L C, RYU J H and RICHARDS G M (2009), '*Cronobacter sakazakii* in foods and factors affecting its survival, growth and inactivation', *Int J Food Microbiol*, 136, 204–213.

BIDAWID S, FARBER J M, SATTAR S A and HAYWARD S (2000), 'Heat inactivation of Hepatitis A virus in dairy foods', *J Food Protect*, 63, 522–528.

BLACK E P, LINTON M, MCCALL R D, CURRAN W, FITZGERALD G F, KELLY A L and PATTERSON M F (2008), 'The combined effects of high pressure and nisin on germination and inactivation of *Bacillus* spores in milk', *J Appl Microbiol*, 105, 78–87.

BLACK T F (2004), 'Brucellosis', in Cohen J and Powderly W G, *Infectious diseases*, St. Louis, MO, Mosby, 1665–1667.

BONE F J, BOGIE D and MORGAN-JONES S C (1989), 'Staphylococcal food poisoning from sheep milk cheese', *Epidemiol Infect*, 103, 449–458.

BORDES-BENÍTEZ A, SÁNCHEZ-OÑORO M, SUÁREZ-BORDÓN P, GARCÍA-ROJAS A J, SAÉZ-NIETO J A, GONZÁLEZ-GARCÍA A, ÁLAMO-ANTÚNEZ I, SÁNCHEZ-MAROTO A and BOLAÑOS-RIVERO M (2006), 'Outbreak of *Streptococcus equi* subsp. *zooepidemicus* infections on the island of Gran Canaria associated with the consumption of inadequately pasteurised cheese', *Eur J Clin Microbiol Infect Dis*, 25, 242–246.

BOSNJAK E, HVASS A M S W, VILLUMSEN S and NIELSEN H (2009), 'Emerging evidence for Q fever in humans in Denmark: role of contact with dairy cattle', *Clin Microbiol Infect*, 16, 1285–1288.

BOUGLE D L and STAHL V (1994), 'Survival of *Listeria monocytogenes* after irradiation treatment of Camembert cheeses made from raw milk', *J Food Protect*, 57, 811–813.

© Woodhead Publishing Limited, 2012

BOUSSOUEL N, MATHIEU F, REVOL-JUNELLES A M and MILLIERE J B (2000), 'Effects of combinations of lactoperoxidase system and nisin on the behaviour of *Listeria monocytogenes* ATCC 15313 in skim milk', *Int J Food Microbiol*, 61, 169–175.

BOZOGLU F, ALPAS H and KALETUNC G (2004), 'Injury recovery of foodborne pathogens in high hydrostatic pressure treated milk during storage', *FEMS Immunol Med Microbiol*, 40, 243–247.

BRANS G, SCHROËN C G P H, VAN DER SMAN R G M and BOOM R M (2004), 'Membrane fractionation of milk: state of the art and challenges', *J Membr Sci*, 243, 263–272.

BRAVO D, RODRIGUEZ E and MEDINA M (2009), 'Nisin and lacticin 481 coproduction by *Lactococcus lactis* strains isolated from raw ewes' milk', *J Dairy Sci*, 92, 4805–4811.

BREUKINK E, VAN KRAAIJ C, DEMEL R A, SIEZEN R J, KUIPERS O P and DE KRUIJFF B (1997), 'The C-terminal region of nisin is responsible for the initial interaction of nisin with the target membrane', *Biochemistry*, 36, 6968–6976.

BREWER R A and CORBEL M J (1983), 'Characterisation of *Yersinia enterocolitica* strains isolated from cattle, sheep and pigs in the United Kingdom', *J Hyg*, 90, 425–433.

BRIÑEZ W J, ROIG-SAGUÉS A X, HERRERO M M H and LÓPEZ B G (2007), 'Inactivation of *Staphylococcus* spp. strains in whole milk and orange juice using ultra high pressure homogenization at inlet temperatures of 6 and 20°C', *Food Contr*, 18, 1282–1288.

BURITI F C A, CARDARELLI H R and SAAD S M I (2007), 'Biopreservation by *Lactobacillus paracasei* coculture with *Streptococcus thermophilus* in potentially probiotic and synbiotic fresh cream cheeses', *J Food Protect*, 70, 228–235.

BUZGAN T, KARAHOCAGIL M K, IRMAK H, BARAN AI, KARSEN H, EVIRGEN O and AKDENIZ H (2010), 'Clinical manifestations and complications in 1028 cases of brucellosis: a retrospective evaluation and review of the literature', *Int J Infect Dis*, 14, E469–E478.

BUZRUL S (2009), 'Modeling and predicting inactivation of *Escherichia coli* under isobaric and dynamic high hydrostatic pressure', *Innovat Food Sci Emerg Tech*, 10, 391–395.

BUZRUL S, ALPAS H, LARGETEAU A and DEMAZEAU G (2009), 'Efficiency of pulse pressure treatment for inactivation of *Escherichia coli* and *Listeria innocua* in whole milk', *Eur Food Res Tech*, 229, 127–131.

CAGRI-MEHMETOGLU A, YALDIRAK G, BODUR T, SIMSEK M, BOZKIR H and EREN N M (2011), 'Incidence of *Listeria monocytogenes* and *Escherichia coli* O157:H7 in two Kasar cheese processing environments', *Food Contr*, 22, 762–766.

CALDERON-MIRANDA M L, BARBOSA-CANOVAS G V and SWANSON B G (1999), 'Inactivation of *Listeria innocua* in skim milk by pulsed electric fields and nisin', *Int J Food Microbiol*, 51, 19–30.

CAMERON M, MCMASTER L D and BRITZ T J (2009), 'Impact of ultrasound on dairy spoilage microbes and milk components', *Dairy Sci Tech*, 89, 83–98.

CAMPOS F P and CRISTIANINI M (2007), 'Inactivation of *Saccharomyces cerevisiae* and *Lactobacillus plantarum* in orange juice using ultra high-pressure homogenisation', *Innovat Food Sci Emerg Tech*, 8, 226–229.

CAPELLAS M, MOR-MUR M, GERVILLA R, YUSTE J and GUAMIS B (2000), 'Effect of high pressure combined with mild heat or nisin on inoculated bacteria and mesophiles of goat's milk fresh cheese', *Food Microbiol*, 17, 633–641.

CARIDI A (2002), 'Selection of *Escherichia coli*-inhibiting strains of *Lactobacillus paracasei* subsp. *paracasei*', *J Ind Microbiol Biotechnol*, 29, 303–308.

CARMINATI D, GATTI M, BONVINI B, NEVIANI E and MUCCHETTI G (2004), 'High-pressure processing of Gorgonzola cheese: influence on *Listeria monocytogenes* inactivation and on sensory characteristics', *J Food Protect*, 67, 1671–1675.

CASTELL MONSALVE J, GUTIÉRREZ ÁVILA G, RODOLFO SAAVEDRA R and SANTOS AZORÍN A (2008), 'Shigellosis outbreak with 146 cases related to a fair', *Gac Sanit*, 22, 35–39.

© Woodhead Publishing Limited, 2012

CERF O. and CONDRON R (2006), '*Coxiella burnetii* and milk pasteurization: an early application of the precautionary principle?', *Epidemiol Infect*, 134, 946–951.

CHAFFER M, FRIEDMAN S, SARAN A and YOUNIS A (2005), 'An outbreak of *Streptococcus canis* mastitis in a dairy herd in Israel', *New Zeal Vet J*, 53, 261–264.

CHEMAT F, GRONDIN I, SHUM CHEONG SING A and SMADJA J (2004), 'Deterioration of edible oils during food processing by ultrasound', *Ultrason Sonochem*, 11, 13–15.

CHEN H (2007), 'Use of linear, Weibull and log-logistic functions to model pressure inactivation of seven foodborne pathogens in milk', *Food Microbiol*, 24, 197–204.

CHEN H, GUAN D and HOOVER D G (2006), 'Sensitivities of foodborne pathogens to pressure changes', *J Food Protect*, 69, 130–136.

CHERYAN M (1998), *Ultrafiltration and microfiltration handbook*, Lancaster, PA, Technomic Publ Co.

CHOI M S, CHEIGH C I, JEONG E A, SHIN J K and CHUNG M S (2010), 'Nonthermal sterilization of *Listeria monocytogenes* in infant foods by intense pulsed-light treatment', *J Food Eng*, 97, 504–509.

CHOULIARA E, GEORGOGIANNI K G, KANELLOPOULOU N and KONTOMINAS M G (2010), 'Effect of ultrasonication on microbiological, chemical and sensory properties of raw, thermized and pasteurised milk', *Int Dairy J*, 20, 307–313.

CISSE M, VAILLANT F, PEREZ A, DORNIER M and REYNES M (2005), 'The quality of orange juice processed by coupling crossflow microfiltration and osmotic evaporation', *Int J Food Sci Tech*, 40, 105–116.

COCOLIN L, FOSCHINO R, COMI G and FORTINA M G (2007), 'Description of the bacteriocins produced by two strains of *Enterococcus faecium* isolated from Italian goat milk', *Food Microbiol*, 24, 752–758.

COLLINS M T (1997), '*Mycobacterium paratuberculosis*: a potential food-borne pathogen?', *J Dairy Sci*, 80, 3445–3448.

CRAVEN H M, SWIERGON P, NG S, MIDGELY J, VERSTEEG C, COVENTRY M J and WAN J (2008), 'Evaluation of pulsed electric field and minimal heat treatments for inactivation of pseudomonads and enhancement of milk shelf-life', *Innovat Food Sci Emerg Tech*, 9, 211–216.

CRAVEN H M, MCAULEY C M, DUFFY L L and FEGAN N (2010), 'Distribution, prevalence and persistence of *Cronobacter* (*Enterobacter sakazakii*) in the nonprocessing and processing environments of five milk powder factories', *J Appl Microbiol*, 109, 1044–1052.

CSPI (2009), Outbreak Alert! Analyzing foodborne outbreaks 1998 to 2007 – closing the gaps in our federal food safety net, Washington, DC, USA, Center for the Science in the Public Interest.

DA SILVA MALHEIROS P, DAROIT D J, DA SILVEIRA N P and BRANDELLI A (2010), 'Effect of nanovesicle-encapsulated nisin on growth of *Listeria monocytogenes* in milk', *Food Microbiol*, 27, 175–178.

DALY M, POWER E, BJÖRKROTH J, SHEEHAN P, O'CONNELL A, COLGAN M, KORKEALA H and FANNING S (1999), 'Molecular analysis of *Pseudomonas aeruginosa*: epidemiological investigation of mastitis outbreaks in Irish dairy herds', *Appl Environ Microbiol*, 65, 2723–2729.

D'AMICO D J, SILK T M, JUNRU W and MINGRUO G (2006), 'Inactivation of microorganisms in milk and apple cider treated with ultrasound', *J Food Protect*, 69, 556–563.

DARYAEI H, COVENTRY J, VERSTEEG C and SHERKAT F (2010), 'Combined pH and high hydrostatic pressure effects on *Lactococcus* starter cultures and *Candida* spoilage yeasts in a fermented milk test system during cold storage', *Food Microbiol*, 27, 1051–1056.

DAUFIN G, ESCUDIER J P, CARRÈRE H, BÉROT S, FILLAUDEAU L and DECLOUX M (2001), 'Recent and emerging applications of membrane processes in the food and dairy industry', *Food Bioprod Process*, 79, 89–102.

© Woodhead Publishing Limited, 2012

DAVIES E A, BEVIS H E and DELVES-BROUGHTON J (1997), 'The use of the bacteriocin, nisin, as a preservative in Ricotta-type cheeses to control the food-borne pathogen *Listeria monocytogenes*', *Lett Applied Microbiol*, 24, 343–346.

DAVIES L J, KEMP M R and FRYER P J (1999), 'The geometry of shadows: effects of inhomogeneities in electrical field processing', *J Food Eng*, 40, 245–258.

DE LAMO-CASTELLVI S, CAPELLAS M, LOPEZ-PEDEMONTE T, HERNANDEZ-HERRERO M M, GUAMIS B and ROIG-SAGUES A X (2005a), 'Behavior of *Yersinia enterocolitica* strains inoculated in model cheese treated with high hydrostatic pressure', *J Food Protect*, 68, 528–533.

DE LAMO-CASTELLVI S, ROIG-SAGUES A X, CAPELLAS M, HERNANDEZ-HERRERO M and GUAMIS B (2005b), 'Survival and growth of *Yersinia enterocolitica* strains inoculated in skimmed milk treated with high hydrostatic pressure', *Int J Food Microbiol*, 102, 337–342.

DE LAMO-CASTELLVI S, ROIG-SAGUES A X, LOPEZ-PEDEMONTE T, HERNANDEZ-HERRERO M M, GUAMIS B and CAPELLAS M (2007), 'Response of two *Salmonella enterica* strains inoculated in model cheese treated with high hydrostatic pressure', *J Dairy Sci*, 90, 99–109.

DE SCHRIJVER K, BUVENS G, POSSÉ B, VAN DEN BRANDEN D, OOSTERLYNCK O, DE ZUTTER L, EILERS K, PIÉRARD D, DIERICK K, VAN DAMME-LOMBAERTS R, LAUWERS C and JACOBS R (2008), 'Outbreak of verocytotoxin-producing *E. coli* O145 and O26 infections associated with the consumption of ice cream produced at a farm, Belgium, 2007', *Euro Surveill*, 13, 61–64.

DENIS C and IRLINGER F (2008), 'Safety assessment of dairy microorganisms: aerobic coryneform bacteria isolated from the surface of smear-ripened cheeses', *Int J Food Microbiol*, 126, 311–315.

DIELS A M J, DE TAEYE J and MICHIELS C W (2005), 'Sensitisation of *Escherichia coli* to antibacterial peptides and enzymes by high pressure homogenisation', *Int J Food Microbiol*, 105, 165–175.

DOJCHINOVA A, SLAVKOVA L and TRIFONOVA L (1990), 'Study on the effect of ascorbic acid and retinol on the antimicrobial activity of maternal milk', *Problemy Pediatrii*, 33, 81–92.

DOMINGUEZ M, JOURDAN-DA SILVA N, VAILLANT V, PIHIER N, KERMIN C, WEILL F X, DELMAS G, KEROUANTON A, BRISABOIS A and DE VALK H (2009), 'Outbreak of *Salmonella enterica* serotype Montevideo infections in France linked to consumption of cheese made from raw milk', *Foodborne Pathog Dis*, 6, 121–128.

DONAGHY J A, LINTON M, PATTERSON M F and ROWE M T (2007), 'Effect of high pressure and pasteurisation on *Mycobacterium avium* ssp. *paratuberculosis* in milk', *Lett Appl Microbiol*, 45, 154–159.

DONAGHY J, KEYSER M, JOHNSTON J, CILLIERS F P, GOUWS P A and ROWE M T (2009), 'Inactivation of *Mycobacterium avium* ssp. *paratuberculosis* in milk by UV treatment', *Lett Appl Microbiol*, 49, 217–221.

DUBEY J P and JONES J L (2008), '*Toxoplasma gondii* infection in humans and animals in the United States', *Int J Parasitol*, 38, 1257–1278.

DUPONT H L, LEVINE M M, HORNICK R B and FORMAL S B (1989), 'Inoculum size in shigellosis and implications for expected mode of transmission', *J Infect Dis*, 159, 1126–1128.

DUTREUX N, NOTERMANS S, WIJTZES T, GONGORA-NIETO M M, BARBOSA-CANOVAS G V and SWANSON B G (2000), 'Pulsed electric fields inactivation of attached and free-living *Escherichia coli* and *Listeria innocua* under several conditions', *Int J Food Microbiol*, 54, 91–98.

EARNSHAW R G, APPLEYARD J and HURST R M (1995), 'Understanding physical inactivation processes: combined preservation opportunities using heat, ultrasound and pressure', *Int J Food Microbiol*, 28, 197–219.

EDWARDS A T, ROULSON M and IRONSIDE M J (1988), 'A milk-borne outbreak of serious

© Woodhead Publishing Limited, 2012

infection due to *Streptococcus zooepidemicus* (Lancefield Group C)', *Epidemiol Infect*, 101, 43–51.

ELTHOLTH M M, MARSH V R, VAN WINDEN S and GUITIAN F J (2009), 'Contamination of food products with *Mycobacterium avium paratuberculosis*: a systematic review', *J Appl Microbiol*, 107, 1061–1071.

ELWELL M W and BARBANO D M (2006), 'Use of microfiltration to improve fluid milk quality', *J Dairy Sci*, 89, E20–E30.

ERKMEN O (2009), 'High hydrostatic pressure inactivation kinetics of *Salmonella typhimurium*', *High Pres Res*, 29, 129–140.

ESTEBAN E and ANDERSON B C (1995), '*Cryptosporidium muris*: prevalence, persistency and detrimental effect on milk production in drylot dairy', *J Dairy Sci*, 78, 1068–1072.

EUROPEAN FOOD SAFETY AUTHORITY (2007), 'The community summary report on trends and sources of zoonoses, zoonotic agents, antimicrobial resistance and foodborne outbreaks in the European Union in 2006', *EFSA J*, 130, 1–352.

EVANS J T, SMITH E G, BANERJEE A, SMITH R M M, DALE J, INNES J A, HUNT D, TWEDDELL A, WOOD A ANDERSON C, HEWINSON R G, SMITH N H, HAWKEY P M and SONNENBERG P (2007), 'Cluster of human tuberculosis caused by *Mycobacterium bovis*: evidence for person-to-person transmission in the UK', *Lancet*, 369, 1270–1276.

EVENSON M L, HINDS M W, BERNSTEIN R S and BERGDOLL M S (1988), 'Estimation of human dose of staphylococcal enterotoxin A from a large outbreak of staphylococcal food poisoning involving chocolate milk', *Int J Food Microbiol*, 7, 311–316.

EVRENDILEK G A and ZHANG Q H (2005), 'Effects of pulse polarity and pulse delaying time on pulsed electric fields-induced pasteurization of *E. coli* O157:H7', *J Food Eng*, 68, 271–276.

EVRENDILEK G A, DANTZER W R, STREAKER C B, RATANATRIWONG P and ZHANG Q H (2001), 'Shelf-life evaluations of liquid foods treated by pilot plant pulsed electric field system', *J Food Process Pres*, 25, 283–297.

EVRENDILEK G A, ZHANG Q H and RICHTER E R (2004), 'Application of pulsed electric fields to skim milk inoculated with *Staphylococcus aureus*', *Biosystems Eng*, 87, 137–144.

EVRENDILEK G A, KOCA N, HARPER J W and BALASUBRAMANIAN V M (2008), 'High-pressure processing of Turkish white cheese for microbial inactivation', *J Food Protect*, 71, 102–108.

FERNANDEZ-MOLINA J J, ALTUNAKAR B, BERMUDEZ-AGUIRRE D, SWANSON B G and BARBOSA-CANOVAS G V (2005a), 'Inactivation of *Pseudomonas fluorescens* in skim milk by combinations of pulsed electric fields and organic acids', *J Food Protect*, 68, 1232–1235.

FERNANDEZ-MOLINA J J, BARBOSA-CANOVAS G V and SWANSON B G (2005b), 'Skim milk processing by combining pulsed electric fields and thermal treatments', *J Food Process Pres*, 29, 291–306.

FERNANDEZ-MOLINA J J, FERNANDEZ-GUTIERREZ S A, ALTUNAKAR B, BERMUDEZ-AGUIRRE D, SWANSON, B G and BARBOSA-CANOVAS G V (2005c), 'The combined effect of pulsed electric fields and conventional heating on the microbial quality and shelf life of skim milk', *J Food Process Pres*, 29, 390–406.

FERNANDEZ-MOLINA J J, BERMUDEZ-AGUIRRE D, ALTUNAKAR B, SWANSON B G and BARBOSA-CANOVAS G V (2006), 'Inactivation of *Listeria innocua* and *Pseudomonas fluorescens* by pulsed electric fields in skim milk: energy requirements', *J Food Process Eng*, 29, 561–573.

FLOURY J, GROSSET N, LECONTE N, PASCO M, MADEC M N and JEANTET R (2006a), 'Continuous raw skim milk processing by pulsed electric field at non-lethal temperature: effect on microbial inactivation and functional properties', *Lait*, 86, 43–57.

FLOURY J, GROSSET N, LESNE E and JEANTET R (2006b), 'Continuous processing of skim milk by a combination of pulsed electric fields and conventional heat treatments: does a synergetic effect on microbial inactivation exist?', *Lait*, 86, 203–211.

© Woodhead Publishing Limited, 2012

FOOD PROTECTION SERVICES (2011a), Summary of foodborne illness outbreaks in North America associated with the consumption of raw milk and raw milk dairy products (2000–2010). Available from: http://www.bccdc.ca/NR/rdonlyres/628544F1-0533-48E8-9C96-8391952BEF96/0/RawMilkOutbreakTable2000_2010.pdf (accessed 25 May 2011).

FOOD PROTECTION SERVICES (2011b), Raw milk contaminants and pathogens. Available from: http://www.bccdc.ca/NR/rdonlyres/3A2C5D87-5615-4B4B-9279-F28BE80E7764/0/RawMilkPathogens.pdf (accessed 25 May 2011).

FRETZ R, SVOBODA P, RYAN U M, THOMPSON R C A, TANNER M and BAUMGARTNER A (2003), 'Genotyping of *Cryptosporidium* spp. isolated from human stool samples in Switzerland', *Epidemiol and Infect*, 131, 663–667.

FRIEDEMANN M (2009), 'Epidemiology of invasive neonatal *Cronobacter* (*Enterobacter sakazakii*) infections', *Eur J Clin Microbiol Infect Dis*, 28, 1297–1304.

FRITSCH J A and MORARU C I (2008), 'Development and optimization of a carbon dioxide-aided cold microfiltration process for the physical removal of microorganisms and somatic cells from skim milk', *J Dairy Sci*, 91, 3744–3760.

FSANZ (2003), Food standards code vol. 2 – standard 1.6.2., Canberra, Australia and Wellington, New Zealand, Governments of Australia and of New Zealand, Food Standards Australia New Zealand.

GALLO L I, PILOSOF A M R and JAGUS R J (2007), 'Effect of the sequence of nisin and pulsed electric fields treatments and mechanisms involved in the inactivation of *Listeria innocua* in whey', *J Food Eng*, 79, 188–193.

GAO Y L and JU X R (2008), 'Exploiting the combined effects of high pressure and moderate heat with nisin on inactivation of *Clostridium botulinum* spores', *J Microbiol Meth*, 72, 20–28.

GAO Y L, JU X R and JIANG H H (2006), 'Analysis of reduction of *Geobacillus stearothermophilus* spores treated with high hydrostatic pressure and mild heat in milk buffer', *J Biotechnol*, 125, 351–360.

GARCIA M T, CANAMERO M M, LUCAS R, OMAR N B, PULIDO R P and GALVEZ A (2004), 'Inhibition of *Listeria monocytogenes* by enterocin EJ97 produced by *Enterococcus faecalis* EJ97', *Int J Food Microbiol*, 90, 161–170.

GARCIA P, MARTINEZ B, RODRIGUEZ L and RODRIGUEZ A (2010), 'Synergy between the phage endolysin LysH5 and nisin to kill *Staphylococcus aureus* in pasteurised milk', *Int J Food Microbiol*, 141, 151–155.

GARCÍA-FULGUEIRAS A, SÁNCHEZ S, GUILLÉN J J, MARSILLA B, ALADUEÑA A and NAVARRO C (2001), 'A large outbreak of *Shigella sonnei* gastroenteritis associated with consumption of fresh pasteurised milk cheese', *Eur J Epidemiol*, 17, 533–538.

GARCIA-GRAELLS C, MASSCHALCK B and MICHIELS C W (1999), 'Inactivation of *Escherichia coli* in milk by high-hydrostatic-pressure treatment in combination with antimicrobial peptides', *J Food Protect*, 62, 1248–1254.

GÉSAN-GUIZIOU G (2010), 'Removal of bacteria, spores and somatic cells from milk by centrifugation and microfiltration techniques', in Griffiths M W, *Improving the safety and quality of milk, volume 1: milk production and processing*, Cambridge, Woodhead Publishing, 349–372.

GILLESPIE I A, MCLAUCHLIN J, LITTLE C L, PENMAN C, MOOK P, GRANT K and O'BRIEN S J (2009), 'Disease presentation in relation to infection foci for non-pregnancy-associated human listeriosis in England and Wales, 2001 to 2007', *J Clin Microbiol*, 47, 3301–3307.

GILLILAND S E and SPECK M L (1967), 'Inactivation of microorganisms by electrohydraulic shock', *Appl Microbiol*, 15, 1031–1037.

GIRAFFA G and CARINI S (1984), 'Ultraviolet and infrared utilization for the microbiological safety of milk', *Microbiologie – Aliments – Nutrition*, 2, 91–93.

GIRARD B and FUKUMOTO L R (1999), 'Apple juice clarification using microfiltration and ultrafiltration polymeric membranes', *LWT – Food Sci Tech*, 32, 290–298.

© Woodhead Publishing Limited, 2012

GOLDBERGER A C, LIPSKY B A and PLORDE J J (1981), 'Suppurative granulomatous lymphadentis caused by *Corynebacterium ovis (paratuberculosis)*', *Am J Clin Pathol*, 76, 486–490.

GONZALEZ S, FLICK G J, ARRITT F M, HOLLIMAN D and MEADOWS B (2006), 'Effect of high-pressure processing on strains of *Enterobacter sakazakii*', *J Food Protect*, 69, 935–937.

GRAHL T and MAERKL H (1996), 'Killing of microorganisms by pulsed electric fields', *Appl Microbiol Biotechnol*, 45, 148–157.

GRANDA-RESTREPO D, PERALTA E, TRONCOSO-ROJAS R and SOTO-VALDEZ H (2009), 'Release of antioxidants from co-extruded active packaging developed for whole milk powder', *Int Dairy J*, 19, 481–488.

GRIFFITHS M W (2004), 'Milk and unfermented milk products', in Lund B M, Baird-Parker A C and Gould G W, *The microbiological safety and quality of food, volume 1*, Gaithersburg, MD, Aspen Publishers, 507–534.

GRIFFITHS M W (2009), '*Mycobacterium paratuberculosis*', in Blackburn C D W and McClure P J, *Foodborne pathogens: hazards, risk analysis and control*, Cambridge, Woodhead Publishing, 1060–1118.

GUAN D, CHEN H and HOOVER D G (2005), 'Inactivation of *Salmonella* Typhimurium DT 104 in UHT whole milk by high hydrostatic pressure', *Int J Food Microbiol*, 104, 145–153.

GUAN D, CHEN H, TING E Y and HOOVER D G (2006), 'Inactivation of *Staphylococcus aureus* and *Escherichia coli* O157:H7 under isothermal-endpoint pressure conditions', *J Food Eng*, 77, 620–627.

GUERRA A, JONSSON G, RASMUSSEN A, NIELSEN E W and EDELSTEN D (1997), 'Low cross-flow velocity microfiltration of skim milk for removal of bacterial spores', *Int Dairy J*, 7, 849–861.

GUERRERO-BELTRAN J A, SEPULVEDA D R, GONGORA-NIETO M M, SWANSON B and BARBOSA-CANOVAS G V (2010), 'Milk thermization by pulsed electric fields (PEF) and electrically induced heat', *J Food Eng*, 100, 56–60.

GUH A, PHAN Q, NELSON R, PURVIANCE K, MILARDO E, KINNEY S, MSHAR P, KASACEK W and CARTTER M (2010), 'Outbreak of *Escherichia coli* O157 associated with raw milk, Connecticut, 2008', *Clin Infect Dis*, 51, 1411–1417.

GUINDI S M, AHMED O L, AWAD W M and EL-SABAN M S (1980), 'Incidence of bovine and human tubercle bacilli in milk and milk products', *Agr Res Rev*, 58, 75–84.

HART R J C (1984), '*Corynebacterium ulcerans* in humans and cattle in North Devon', *J Hyg, Cambridge*, 92, 161–164.

HASHISAKA A E, MATCHES J R, BATTERS Y, HUNGATE F P and DONG F M (1990), 'Effects of gamma irradiation at –78°C on microbial populations in dairy products', *J Food Sci*, 55, 1284–1289.

HASSAN G M, AL-ASHMAWY M A M, MESHREF A M S and AFIFY S I (2010), 'Studies on enterotoxigenic *Bacillus cereus* in raw milk and some dairy products', *J Food Saf*, 30, 569–583.

HASSANI M, LAZARO R, PEREZ C, CONDON S and PAGAN R (2008), 'Thermostability of oxytetracycline, tetracycline and doxycycline at ultrahigh temperatures', *J Agr Food Chem*, 56, 2676–2680.

HAYES M G, FOX P F and KELLY A L (2005), 'Potential applications of high pressure homogenisation in processing of liquid milk', *J Dairy Res*, 72, 25–33.

HAYMAN M M, ANANTHESWARAN R C and KNABEL S J (2007), 'The effects of growth temperature and growth phase on the inactivation of *Listeria monocytogenes* in whole milk subject to high pressure processing', *Int J Food Microbiol*, 115, 220–226.

HAYMAN M M, ANANTHESWARAN R C and KNABEL S J (2008), 'Heat shock induces barotolerance in *Listeria monocytogenes*', *J Food Protect*, 71, 426–430.

HENYON D K (1999), 'Extended shelf-life milks in North America: a perspective', *Int J Dairy Tech*, 52, 95–101.

© Woodhead Publishing Limited, 2012

HERMON-TAYLOR J and BULL T (2002), 'Crohn's disease caused by *Mycobacterium avium* subspecies *paratuberculosis*: a public health tragedy whose resolution is long overdue', *J Med Microbiol*, 51, 3–6.

HERNANDEZ D, CARDELL E and ZARATE V (2005), 'Antimicrobial activity of lactic acid bacteria isolated from Tenerife cheese: initial characterisation of plantaricin TF711, a bacteriocin-like substance produced by *Lactobacillus plantarum* TF711', *J Applied Microbiol*, 99, 77–84.

HEUVELINK A E, VAN HEERWAARDEN C, ZWARTKRUIS-NAHUIS A, TILBURG J J H C, BOS M H, HEILMANN F G C, HOFHUIS A, HOEKSTRA T and DE BOER E (2009), 'Two outbreaks of campylobacteriosis associated with the consumption of raw cows' milk', *Int J Food Microbiol*, 134, 70–74.

HINRICHS J and KESSLER H G (1995), 'Thermal processes of milk – processes and equipment' in Fox, P F, *Heat-induced changes in milk*, Brussels, International Dairy Federation, 9–21.

HIRAMOTO R M, MAYRBAURL-BORGES M, GALISTEO JR A J, MEIRELES L R, MACRE M S and ANDRADE JR H F (2001), 'Infectivity of cysts of the ME-49 *Toxoplasma gondii* strain in bovine milk and homemade cheese', *Rev Saude Publica*, 35, 113–118.

HONISH L, PREDY G, HISLOP N, CHUI L, KOWALEWSKA-GROCHOWSKA K, TROTTIER L, KREPLIN C and ZAZULAK I (2005), 'An outbreak of *E. coli* O157:H7 hemorrhagic colitis associated with unpasteurised gouda cheese', *Rev Canad Sante Publique*, 96, 182–184.

INPANKAEW T, PINYOPANUWUT N, CHIMNOI W, KENGRADOMKIT C, SUNUNTA C, ZHANG G, NISHIKAWA Y, IGARASHI I, XUAN X and JITTAPALAPONG S (2010), 'Serodiagnosis of *Toxoplasma gondii* infection in dairy cows in Thailand', *Transbound Emerg Dis*, 57, 42–45.

JOPPEN L (2004), 'Quest for longer shelf lives', *Food Eng Ingredients*, 29, 44–45.

JU X R, GAO Y L, YAO M L and QIAN Y (2008), 'Response of *Bacillus cereus* spores to high hydrostatic pressure and moderate heat', *LWT – Food Sci Tech*, 41, 2104–2112.

JULIEN M C, DION P, LAFRENIÈRE C, ANTOUN H and DROUIN P (2008), 'Sources of clostridia in raw milk on farms', *Appl Environ Microbiol*, 74, 6348–6357.

KAMEI T, SATO J, NAKAI Y, NATSUME A and NODA K (1991), 'Microbiological quality of aseptic packaging and the effect of pin-holes on sterility of aseptic products', *Packag Tech Sci*, 4, 185–193.

KANDHAI M C, HEUVELINK A E, REIJ M W, BEUMER R R, DIJK R, VAN TILBURG J J H C, VAN SCHOTHORST M and GORRIS L G M (2010), 'A study into the occurrence of *Cronobacter* spp. in the Netherlands between 2001 and 2005', *Food Contr*, 21, 1127–1136.

KAPUR M P, SHARMA A and SINGH R P (1986), 'Occurrence and aeruginocine typing of *Pseudomonas aeruginosa* in buffaloes and their environment', *Comp Immunol Microbiol Infect Dis*, 9, 89–93.

KAUFMANN V and KULOZIK U (2006), 'Kombination von Mikrofiltration und thermischen Verfahren zur Haltbarkeitsverlängerung von Lebensmitteln', *Chem Ing Tech*, 78, 1647–1654.

KEENE W E, HEDBERG K, HERRIOTT D E, HANCOCK D D, MCKAY R W, BARRETT T J and FLEMING D W (1997), 'A prolonged outbreak of *Escherichia coli* O157:H7 infections caused by commercially distributed raw milk', *J Infect Dis*, 176, 815–818.

KELLS J and GILMOUR A (2004), 'Incidence of *Listeria monocytogenes* in two milk processing environments and assessment of *Listeria monocytogenes* blood agar for isolation', *Int J Food Microbiol*, 91, 167–174.

KHEADR E E, VACHON J F, PAQUIN P and FLISS I (2002), 'Effect of dynamic high pressure on microbiological, rheological and microstructural quality of cheddar cheese', *Int Dairy J*, 12, 435–446.

KIERMEIER F and KAYSER C (1960), 'Zur Kenntnis der Lactoseperoxydase. IV. Inaktivierung der Lactoseperoxydase durch Mikroorganismen', *Zeitschrift für Lebensmitteluntersuchung und -forschung A*, 113, 203–213.

© Woodhead Publishing Limited, 2012

KIM H J, HAM J S, LEE J W, KIM K, HA S D and JO C (2010), 'Effects of gamma and electron beam irradiation on the survival of pathogens inoculated into sliced and pizza cheeses', *Radiat Phys Chem*, 79, 731–734.

KIM S G, KIM E H, LAFFERTY C J and DUBOVI E (2005), '*Coxiella burnetii* in bulk tank milk samples, United States', *Emerg Infect Dis*, 11, 619–621.

KIYOHITO K, KAZUNORI N, SEIICHI S, RIE M, YOSHIE T, YUKI A and HIROSHI T (2006), 'Outbreak of symptomatic diarrhea, milk drop and inappetence in lactating cows infected with bovine group B rotavirus', *J Japan Vet Med Assoc*, 59, 254–258.

KLAENHAMMER T R (1993), 'Genetics of bacteriocins produced by lactic acid bacteria', *FEMS Microbiol Rev*, 12, 38–85.

KOCH J and STARK K (2006), 'Significant increase of listeriosis in Germany – epidemiological patterns 2001–2005', *Euro Surveill*, 11, 85–88.

KOCH J, DWORAK R, PRAGER R, BECKER B, BROCKMANN S, WICKE A, WICHMANN-SCHAUER H, HOF H, WERBER D and STARK K (2010), 'Large listeriosis outbreak linked to cheese made from pasteurised milk, Germany, 2006–2007', *Foodborne Pathog Dis*, 7, 1581–1584.

KONTELES S, SINANOGLOU V J, BATRINOU A and SFLOMOS K (2009), 'Effects of gamma-irradiation on *Listeria monocytogenes* population, colour, texture and sensory properties of Feta cheese during cold storage', *Food Microbiol*, 26, 157–165.

KOSEKI S, MIZUNO Y and YAMAMOTO K (2008), 'Use of mild-heat treatment following high-pressure processing to prevent recovery of pressure-injured *Listeria monocytogenes* in milk', *Food Microbiol*, 25, 288–293.

KOSIKOWSKI F V and FOX P F (1968), 'Low heat, hydrogen peroxide and bactofugation treatments of milk to control coliforms in Cheddar cheese', *J Dairy Sci*, 51, 1018–1022.

KRISHNAMURTHY K, DEMIRCI A and IRUDAYARAJ J M (2007), 'Inactivation of *Staphylococcus aureus* in milk using flow-through pulsed UV-light treatment system', *J Food Sci*, 72, M233–M239.

KRISHNAMURTHY K, DEMIRCI A and IRUDAYARAJ J (2008a), 'Inactivation of *Staphylococcus aureus* in milk and milk foam by pulsed UV-light treatment and surface response modeling', *Trans ASABE*, 51, 2083–2090.

KRISHNAMURTHY K, JUN S, IRUDAYARAJ J and DEMIRCI A (2008b), 'Efficacy of infrared heat treatment for inactivation of *Staphylococcus aureus* in milk', *J Food Process Eng*, 31, 798–816.

KVISTHOLM JENSEN A, ETHELBERG S, SMITH B, MØLLER NIELSEN E, LARSSON J, MØLBAK K, CHRISTENSEN J J and KEMP M (2010), 'Substantial increase in listeriosis, Denmark 2009', *Euro Surveill*, 15, 12–15.

LABERGE I and GRIFFITHS M W (1996), 'Prevalence, detection and control of *Cryptosporidium parvum* in food', *Int J Food Microbiol*, 32, 1–26.

LANCIOTTI R, VANNINI L, PATRIGNANI F, IUCCI L, VALLICELLI M, NDAGIJIMANA M and GUERZONI M E (2006), 'Effect of high pressure homogenisation of milk on chesse yield and microbiology, lipolysis and proteolysis during ripening of Caciotta cheese', *J Dairy Res*, 73, 216–226.

LANZAS C, WARNICK L D, JAMES K L, WRIGHT E M, WIEDMANN M and GRÖHN Y T (2010), 'Transmission dynamics of a multidrug-resistant *Salmonella* Typhimurium outbreak in a dairy farm', *Foodborne Pathog Dis*, 7, 467–474.

LASSEN B, VILTROP A, RAAPERI K and JÄRVIS T (2009), '*Eimeria* and *Cryptosporidium* in Estonian dairy farms in regard to age, species and diarrhoea', *Vet Parasitol*, 166, 212–219.

LAUKOVA A, CZIKKOVA S, DOBRANSKY T and BURDOVA O (1999), 'Inhibition of *Listeria monocytogenes* and *Staphylococcus aureus* by enterocin CCM 4231 in milk products', *Food Microbiol*, 16, 93–99.

LEITNER G and KRIFUCKS O (2007), '*Pseudomonas aeruginosa* mastitis outbreaks in sheep and goat flocks: antibody production and vaccination in a mouse model', *Vet Immunol Immunopathol*, 119, 198–203.

© Woodhead Publishing Limited, 2012

LEWIS M (2010), 'Improving pasteurised and extended shelf-life milk', in Griffiths M W, *Improving the safety and quality of milk, volume 1: milk production and processing*, Cambridge, Woodhead Publishing, 277–301.

LINDSTRÖM M, MYLLYKOSKI J, SIVELÄ S and KORKEALA H (2010), '*Clostridium botulinum* in cattle and dairy products', *Crit Rev Food Sci Nutr*, 50, 281–304.

LINTON M, MCCLEMENTS J M J and PATTERSON M F (2001), 'Inactivation of pathogenic *Escherichia coli* in skimmed milk using high hydrostatic pressure', *Innovat Food Sci Emerg Tech*, 2, 99–104.

LINTON M, MACKLE A B, UPADHYAY V K, KELLY A L and PATTERSON M F (2008), 'The fate of *Listeria monocytogenes* during the manufacture of Camembert-type cheese: a comparison between raw milk and milk treated with high hydrostatic pressure', *Innovat Food Sci Emerg Tech*, 9, 423–428.

LOFTIS A D, PRIESTLEY R A and MASSUNG R F (2010), 'Detection of *Coxiella burnetii* in commercially available raw milk from the United States', *Foodborne Pathog Dis*, 7, 1453–1455.

LOOMANS E E M G, VAN WILTENBURG J, KOETS M and VAN AMERONGEN A (2003), 'Neamin as an immunogen for the development of a generic ELISA detecting gentamicin, kanamycin and neomycin in milk', *J Agr Food Chem*, 51, 587–593.

LOPEZ-PEDEMONTE T, ROIG-SAGUES A, DE LAMO S, HERNANDEZ-HERRERO M and GUAMIS B (2007a), 'Reduction of counts of *Listeria monocytogenes* in cheese by means of high hydrostatic pressure', *Food Microbiol*, 24, 59–66.

LOPEZ-PEDEMONTE T, ROIG-SAGUES A X, DE LAMO S, GERVILLA R and GUAMIS B (2007b), 'High hydrostatic pressure treatment applied to model cheeses made from cow's milk inoculated with *Staphylococcus aureus*', *Food Contr*, 18, 441–447.

LORUSSO V, DAMBROSIO A, QUAGLIA N C, PARISI A, LA SALANDRA G, LUCIFORA G, MULA G, VIRGILO S, CAROSIELLI L, RELLA A, DARIO M and NORMANNO G (2009), 'Verocytotoxin-producing *Escherichia coli* O26 in raw water buffalo (*Bubalus bubalis*) milk products in Italy', *J Food Protect*, 72, 1705–1708.

LUNDÉN J, TOLVANEN R and KORKEALA H (2004), 'Human listeriosis outbreaks linked to dairy products in Europe', *J Dairy Sci*, 87, E6–E11.

MAHMOUD N S and GHALY A E (2004), 'On-line sterilization of cheese whey using ultraviolet radiation', *Biotechnol Progr*, 20, 550–560.

MAINVILLE I, MONTPETIT D, DURAND N and FARNWORTH E R (2001), 'Deactivating the bacteria and yeast in kefir using heat treatment, irradiation and high pressure', *Int Dairy J*, 11, 45–49.

MANAS P, BARSOTTI L and CHEFTEL J C (2001), 'Microbial inactivation by pulsed electric fields in a batch treatment chamber: effects of some electrical parameters and food constituents', *Innovat Food Sci Emerg Tech*, 2, 239–249.

MARTIN O, QIN B L, CHANG F J, BARBOSA-CANOVAS G V and SWANSON B G (1997), 'Inactivation of *Escherichia coli* in skim milk by high intensity pulsed electric fields', *J Food Process Eng*, 20, 317–336.

MARTINEZ B, OBESO J M, RODRIGUEZ A and GARCIA P (2008), 'Nisin-bacteriophage crossresistance in *Staphylococcus aureus*', *Int J Food Microbiol*, 122, 253–258.

MARTINEZ-CUESTA M C, BENGOECHEA J, BUSTOS I, RODRIGUEZ B, REQUENA T and PELAEZ C (2010), 'Control of late blowing in cheese by adding lacticin 3147-producing *Lactococcus lactis* IFPL 3593 to the starter', *Int Dairy J*, 20, 18–24.

MARTINEZ-RODRIGUEZ A and MACKEY B M (2005), 'Factors affecting the pressure resistance of some *Campylobacter* species', *Lett Appl Microbiol*, 41, 321–326.

MARTINS L M, MARQUEZ R F and YANO T (2002), 'Incidence of toxic *Aeromonas* isolated from food and human infection', *FEMS Immunol Med Microbiol*, 32, 237–242.

MATAK K E, CHUREY J J, WOROBO R W, SUMNER S S, HOVINGH E, HACKNEY C R and PIERSON M D (2005), 'Efficacy of UV light for the reduction of *Listeria monocytogenes* in goat's milk', *J Food Protect*, 68, 2212–2216.

MATTISON K, SHUKLA A, COOK A, POLLARI F, FRIENDSHIP R, KELTON D, BIDAWID S and

© Woodhead Publishing Limited, 2012

FARBER J M (2007), 'Human noroviruses in swine and cattle', *Emerg Infect Dis*, 13, 1184–1188.

MAWATARI T, TANEICHI A, KAWAGOE T, HOSOKAWA M, TOGASHI K and TSUNEMITSU H (2004), 'Detection of a bovine group C rotavirus from adult cows with diarrhea and reduced milk production', *J Vet Med Sci*, 66, 887–890.

MCAULIFFE O, HILL C and ROSS R P (1999), 'Inhibition of *Listeria monocytogenes* in cottage cheese manufactured with a lacticin 3147-producing starter culture', *J Appl Microbiol*, 86, 251–256.

MEAD P S, SLUTSKER L, DIETZ V, MCCAIG L F, BRESEE J S, SHAPIRO C, GRIFFIN P M and TAUXE R V (2000), 'Food-related illness and death in the United States', *J Environ Health*, 62, 9–18.

MELAS D S, PAPAGEORGIOU D K and MANTIS A I (1998), 'Enumeration and confirmation of *Aeromonas hydrophila*, *Aeromonas caviae* and *Aeromonas sobria* isolated from raw milk and other milk products in Northern Greece', *J Food Protect*, 62, 463–466.

MICHALAC S, ALVAREZ V, JI T and ZHANG Q H (2003), 'Inactivation of selected microorganisms and properties of pulsed electric field processed milk', *J Food Process Pres*, 27, 137–151.

MIKKOLA R ANDERSSON M A, GRIGORIEV P, TEPLOVA V V, SARIS N E, RAINEY F A and SALKINOJA-SALONEN M S (2004), '*Bacillus amyloliquefaciens* strains isolated from moisture-damaged buildings produced surfactin and a substance toxic to mammalian cells', *Arch Microbiol*, 181, 314–323.

MIRHOSSEINI M, NAHVI I, EMTIAZI G and TAVASSOLI M (2010), 'Characterisation of anti-*Listeria monocytogenes* bacteriocins from *Enterococcus faecium* strains isolated from dairy products', *Int J Dairy Tech*, 63, 55–61.

MITRA S, CHANDRA MUKHOPADHYAY B and RANJAN BISWAS S (2009), 'A foodgrade preparation of nisin from diluted milk fermented with *Lactococcus lactis* W8', *J Food Protect*, 72, 2615–2617.

MODI R, HIRVI Y, HILL A and GRIFFITHS M W (2001), 'Effect of phage on survival of *Salmonella* Enteritidis during manufacture and storage of cheddar cheese made from raw and pasteurised milk', *J Food Protect*, 64, 927–933.

MOORE J E, CORCORAN D, DOOLEY J S G, FANNING S, LUCEY B, MATSUDA M, MCDOWELL D A, MEGRAUD F, MILLAR B C, O'MAHONY R, O'RIORDAN L, O'ROURKE M, RAO J R, ROONEY P J, SAILS A and WHYTE P (2005), 'Campylobacter', *Vet Res*, 36, 351–382.

MORESI M and LO PRESTI S (2003), 'Present and potential applications of membrane processing in the food industry', *Ital J Food Sci*, 15, 3–33.

MORGAN S M, ROSS R P, BERESFORD T and HILL C (2000), 'Combination of hydrostatic pressure and lacticin 3147 causes increased killing of *Staphylococcus* and *Listeria*', *J Appl Microbiol*, 88, 414–420.

MORONI O, JEAN J, AUTRENT J and FLISS I (2002), 'Inactivation of bacteriophages in liquid media using dynamic high pressure', *Int Dairy J*, 12, 907–913.

MOSCHOPOULOU E, ANISA T, KATSAROS G, TAOUKIS P and MOATSOU G (2010), 'Application of high-pressure treatment on ovine brined cheese: effect on composition and microflora throughout ripening', *Innovat Food Sci Emerg Tech*, 11, 543–550.

MUNOZ A, MAQUEDA M, GALVEZ A, MARTINEZ-BUENO M, RODRIGUEZ A and VALDIVIA E (2004), 'Biocontrol of psychrotrophic enterotoxigenic *Bacillus cereus* in a nonfat hard cheese by an enterococcal strain-producing enterocin AS-48', *J Food Protect*, 67, 1517–1521.

MUSSA D M, RAMASWAMY H S and SMITH J P (1998), 'High pressure (HP) destruction kinetics of *Listeria monocytogenes* Scott A in raw milk', *Food Res Int*, 31, 343–350.

NAKIMBUGWE D, MASSCHALCK B, ANIM G and MICHIELS C W (2006), 'Inactivation of gram-negative bacteria in milk and banana juice by hen egg white and lambda lysozyme under high hydrostatic pressure', *Int J Food Microbiol*, 112, 19–25.

NAM H M, LEE A L, JUNG S C, KIM M N, JANG G C, WEE S H and LIM S K (2011), 'Antimicrobial susceptibility of *Staphylococcus aureus* and characterisation of methicillin-resistant

© Woodhead Publishing Limited, 2012

*Staphylococcus aureus* isolated from bovine mastitis in Korea', *Foodborne Pathog Dis*, 8, 231–238.

NICHOLS G L (2000), 'Food-borne protozoa', *Br Med Bull*, 56, 209–235.

NIELSEN W K (2000), Membrane filtration and related molecular separation technologies, Silkeborg, Denmark, APV Systems.

NIEMINEN T, RINTALUOMA N ANDERSSON M, TAIMISTO A M, ALI-VEHMAS T, SEPPÄLÄ A, PRIHA O and SALKINOJA-SALONEN M (2007), 'Toxinogenic *Bacillus pumilus* and *Bacillus licheniformis* from mastitic milk', *Vet Microbiol*, 124, 329–339.

NOCI F, WALKLING-RIBEIRO M, CRONIN D A, MORGAN D J and LYNG J G (2009), 'Effect of thermosonication, pulsed electric field and their combination on inactivation of *Listeria innocua* in milk', *Int Dairy J*, 19, 30–35.

NOVAK J S, CALL J, TOMASULA P and LUCHANSKY J B (2005), 'An assessment of pasteurization treatment of water, media and milk with respect to *Bacillus* spores', *J Food Protect*, 68, 751–757.

NSW FOOD AUTHORITY (2009), Food safety risk assessment of New South Wales food safety schemes, Newington, Australia, New South Wales Food Authority, NSW/FA/FI039/0903 edn.

O'HANDLEY R M, OLSON M E, FRASER D, ADAMS P and THOMPSON R C A (2000), 'Prevalence and genotypic characterisation of *Giardia* in dairy calves from Western Australia and Western Canada', *Vet Parasitol*, 90, 193–200.

OKWORI A E J, MARTÍNEZ P O, FREDRIKSSON-AHOMAA M, AGINA S E and KORKEALA H (2009), 'Pathogenic *Yersinia enterocolitica* 2/O:9 and *Yersinia pseudotuberculosis* 1/O:1 strains isolated from human and non-human sources in the plateau state of Nigeria', *Food Microbiol*, 26, 872–875.

O'REILLY C E, O'CONNOR P M, KELLY A L, BERESFORD T P and MURPHY P M (2000), 'Use of hydrostatic pressure for inactivation of microbial contaminants in cheese', *Appl Environ Microbiol*, 66, 4890–4896.

ORTOLANI M B T, MORAES P M, PERIN L M, VICOSA G N, CARVALHO K G, SILVA JR A and NERO L A (2010), 'Molecular identification of naturally occurring bacteriocinogenic and bacteriocinogenic-like lactic acid bacteria in raw milk and soft cheese', *J Dairy Sci*, 93, 2880–2886.

OSAILI T M, SHAKER R R, AL-HADDAQ M S, AL-NABULSI A A and HOLLEY R A (2009), 'Heat resistance of *Cronobacter* species (*Enterobacter sakazakii*) in milk and special feeding formula', *J Appl Microbiol*, 107, 928–935.

OTUNOLA A, EL-HAG A, JAYARAM S and ANDERSON W A (2008), 'Effectiveness of pulsed electric fields in controlling microbial growth in milk', *Int J Food Eng*, 4, 1–15.

PAGAN R, MANAS P, ALVAREZ I and CONDON S (1999), 'Resistance of *Listeria monocytogenes* to ultrasonic waves under pressure at sublethal (manosonication) and lethal (manothermosonication) temperatures', *Food Microbiol*, 16, 139–148.

PALGAN I, CAMINITI I M, MUNOZ A, NOCI F, WHYTE P, MORGAN D J, CRONIN D A and LYNG J G (2011), 'Effectiveness of High Intensity Light Pulses (HILP) treatments for the control of *Escherichia coli* and *Listeria innocua* in apple juice, orange juice and milk', *Food Microbiol*, 28, 14–20.

PAPAGEORGIOU D K, MELAS D S, ABRAHIM A and ANGELIDIS A S (2006), 'Growth of *Aeromonas hydrophila* in the whey cheeses myzithra, anthotyros and manouri during storage at 4 and 12°C', *J Food Protect*, 69, 308–314.

PAPPAS G, PAPADIMITRIOU P, AKRITIDIS N, CHRISTOU L and TSIANOS E V (2006), 'The new global map of human brucellosis', *Lancet Infect Dis*, 6, 91–99.

PERDUE M L, KARNS J, HIGGINS J and VAN KESSEL J A (2003), 'Detection and fate of *Bacillus anthracis* (sterne) vegetative cells and spores added to bulk tank milk', *J Food Protect*, 66, 2349–2354.

PEREDA J, FERRAGUT V, QUEVEDO J M, GUAMIS B and TRUJILLO A J (2007), 'Effects of ultra-high pressure homogenisation on microbial and physicochemical shelf life of milk', *J Dairy Sci*, 90, 1081–1093.

© Woodhead Publishing Limited, 2012

PETERSON M C (2003), '*Campylobacter jejeuni* Enteritis associated with consumption of raw milk', *J Environ Health*, 65, 20–21.

PHILLIPS J D, GRIFFITHS M W and MUIR D D (1983), 'Effect of nisin on the shelf-life of pasteurised double cream', *J Soc Dairy Tech*, 36, 17–21.

PHUA S T G and DAVEY K R (2007), 'Predictive modelling of high pressure (<= 700 MPa)-cold pasteurisation (<= 25 degrees C) of *Escherichia coli*, *Yersinia enterocolitica* and *Listeria monocytogenes* in three liquid foods', *Chem Eng Process*, 46, 458–464.

PICART L, DUMAY E and CHEFTEL J C (2002), 'Inactivation of *Listeria innocua* in dairy fluids by pulsed electric fields: influence of electric parameters and food composition', *Innovat Food Sci Emerg Tech*, 3, 357–369.

PICART L, THIEBAUD M, RENE M, GUIRAUD J P, CHEFTEL J C and DUMAY E (2006), 'Effects of high pressure homogenisation of raw bovine milk on alkaline phosphatase and microbial inactivation. A comparison with continuous short-time thermal treatments', *J Dairy Res*, 73, 454–463.

PIETRANERA M S A, NARVAIZ P, HORAK C and KAIRIYAMA E (2003), 'Irradiated icecreams for immunosuppressed patients', *Radiat Phys Chem*, 66, 357–365.

PINA-PEREZ M C, RODRIGO D and MARTINEZ-LOPEZ A (2009a), 'Sub-lethal damage in *Cronobacter sakazakii* subsp. *sakazakii* cells after different pulsed electric field treatments in infant formula milk', *Food Contr*, 20, 1145–1150.

PINA-PEREZ M C, SILVA-ANGULO A B, MUGUERZA-MARQUINEZ B, ALIAGA D R and MARTINEZ-LOPEZ A (2009b), 'Synergistic effect of high hydrostatic pressure and natural antimicrobials on inactivation kinetics of *Bacillus cereus* in a liquid whole egg and skim milk mixed beverage', *Foodborne Pathog Dis*, 6, 649–656.

PISONI G, ZADOKS R N, VIMERCATI C, LOCATELLI C, ZANONI M G and MORONI P (2009), 'Epidemiological investigation of *Streptococcus equi* subspecies *zooepidemicus* involved in clinical mastitis in dairy goats', *J Dairy Sci*, 92, 943–951.

PIYASENA P, MOHAREB E and MCKELLAR R C (2003), 'Inactivation of microbes using ultrasound: a review', *Int J Food Microbiol*, 87, 207–216.

POULIN M F and BOIVIN G (2009), 'A case of disseminated infection caused by *Streptococcus equi* subspecies *zooepidemicus*', *Can J Infect Dis Med Microbiol*, 20, 59–61.

RADEMAKER J L W, VISSERS M M M and GIFFEL M C T (2007), 'Effective heat inactivation of *Mycobacterium avium* subsp *paratuberculosis* in raw milk contaminated with naturally infected feces', *Appl Environ Microbiol*, 73, 4185–4190.

RAI A K, CHAKRAVORTY R and PAUL J (2008), 'Detection of *Giardia*, *Entamoeba* and *Cryptosporidium* in unprocessed food items from Northern India', *World J Microbiol Biotechnol*, 24, 2879–2887.

RANGEL J M, SPARLING P H, CROWE C, GRIFFIN P M and SWERDLOW D L (2005), 'Epidemiology of *Escherichia coli* O157:H7 outbreaks, United States, 1982–2002', *Emerg Infect Dis*, 11, 603–609.

RASO J and BARBOSA-CÁNOVAS G V (2003), 'Nonthermal preservation of foods using combined processing techniques', *Crit Rev Food Sci Nutr*, 43, 265–285.

REINA L D, JIN Z T, ZHANG Q H and YOUSEF A E (1998), 'Inactivation of *Listeria monocytogenes* in milk by pulsed electric field', *J Food Protect*, 61, 1203–1206.

REZAEI M, MOHEBALI S H, ABADI Y K, SURI E, ZARE A, MALAMIR S H, RASULI E S and MAADI H (2010), 'Investigation on the seroprevalence and pollution severity to *Brucella abortus* and *Brucella melitensis* bacteria in cows and sheep living in the villager region of Toyserkan City, Hamdean, Iran', *J Anim Vet Adv*, 9, 2870–2872.

RICHTER R, LEDFORD R and MURPHY S (1992), 'Milk and milk products', in Vanderzant C and Splittstoesser D F, *Compendium of methods for the microbial examination of foods*, Washington, DC, USA, American Public Health Association, 837–856.

RIVAS A, SAMPEDRO F, RODRIGO D, MARTINEZ A and RODRIGO M (2006), 'Nature of the inactivation of *Escherichia coli* suspended in an orange juice and milk beverage', *Eur Food Res Tech*, 223, 541–545.

ROBINSON G, THOMAS A L, DANIEL R G, HADFIELD S J, ELWIN K and CHALMERS R M (2006),

© Woodhead Publishing Limited, 2012

'Sample prevalence and molecular characterisation of *Cryptosporidium andersoni* within a dairy herd in the United Kingdom', *Vet Parasitol*, 142, 163–167.

RODRIGUEZ E, ARQUES J L, GAYA P, NUNEZ M and MEDINA M (2001), 'Control of *Listeria monocytogenes* by bacteriocins and monitoring of bacteriocin-producing lactic acid bacteria by colony hybridization in semi-hard raw milk cheese', *J Dairy Res*, 68, 131–137.

RODRIGUEZ E, ARQUES J L, NUNEZ M, GAYA P and MEDINA M (2005), 'Combined effect of high-pressure treatments and bacteriocin-producing lactic acid bacteria on inactivation of *Escherichia coli* O157:H7 in raw-milk cheese', *Appl Environ Microbiol*, 71, 3399–3404.

RODRIGUEZ N, ORTIZ M C, SARABIA L A and HERRERO A (2010), 'A multivariate multianalyte screening method for sulfonamides in milk based on front-face fluorescence spectroscopy', *Anal Chim Acta*, 657, 136–146.

RODRIGUEZ-ALONSO P, FERNANDEZ-OTERO C, CENTENO J A and GARABAL J I (2009), 'Antibiotic resistance in lactic acid bacteria and micrococcaceae/staphylococcaceae isolates from artisanal raw milk cheeses and potential implications on cheese making', *J Food Sci*, 74, M284–M293.

RODRÍGUEZ-GONZÁLEZ O, WALKLING-RIBEIRO M, JAYARAM S and GRIFFITHS M W (2011a), 'Cross-protective effects of temperature, pH, osmotic and starvation stresses in *Escherichia coli* O157:H7 subjected to pulsed electric fields in milk', *Int Dairy J*, 21, 953–962.

RODRÍGUEZ-GONZÁLEZ O, WALKLING-RIBEIRO M, JAYARAM S and GRIFFITHS M W (2011b), 'Factors affecting the inactivation of the natural microbiota of milk processed by pulsed electric fields and cross-flow microfiltration', *J Dairy Res*, 78, 270–278.

RODWELL T C, KAPASI A J, MOORE M, MILIAN-SUAZO F, HARRIS B, GUERRERO L P, MOSER K, STRATHDEE S A and GARFEIN R S (2010), 'Tracing the origins of *Mycobacterium bovis* tuberculosis in humans in the USA to cattle in Mexico using spoligotyping', *Int J Infect Dis*, 14, E129–E135.

ROIG-SAGUES A X, VELAZQUEZ R M, MONTEALEGRE-AGRAMONT P, LOPEZ-PEDEMONTE T J, BRINEZ-ZAMBRANO W J, GUAMIS-LOPEZ B and HERNANDEZ-HERRERO M M (2009), 'Fat content increases the lethality of ultra-high-pressure homogenization on *Listeria monocytogenes* in milk', *J Dairy Sci*, 92, 5394–5402.

ROWAN N J, MACGREGOR S J ANDERSON J G, CAMERON D and FARISH O (2001), 'Inactivation of *Mycobacterium paratuberculosis* by pulsed electric fields', *Appl Environ Microbiol*, 67, 2833–2836.

ROWE M T and DONAGHY J (2008), '*Mycobacterium bovis*: the importance of milk and dairy products as a cause of human tuberculosis in the UK. A review of taxonomy and culture methods, with particular reference to artisanal cheeses', *Int J Dairy Tech*, 61, 317–326.

RYBTKE M T, JENSEN P O, HOIBY N, GIVSKOV M, TOLKER-NIELSEN T and BJARNSHOLT T (2011), 'The implication of *Pseudomonas aeruginosa* biofilms in infections', *Inflamm Allergy Drug Targets*, 10, 141–157.

RYSSTAD G and KOLSTAD J (2006), 'Extended shelf-life milk – advances in technology', *Int J Dairy Tech*, 59, 85–96.

SAFE FOOD INTERNATIONAL (2011), Food/Water borne illness outbreaks. Available from: http://regionalnews.safefoodinternational.org/page/Food%2FWater+Borne+Illness+Outbreaks (accessed 25 May 2011).

SAHL H G and BIERBAUM G (1998), 'Lantibiotics: biosynthesis and biological activities of uniquely modified peptides from gram-positive bacteria', *Annu Rev Microbiol*, 52, 41–79.

SALKINOJA-SALONEN M S, VUORIO R ANDERSSON M A, KÄMPFER P ANDERSSON M C, HONKANEN-BUZALSKI T and SCOGING A C (1999), 'Toxigenic strains of *Bacillus licheniformis* related to food poisoning', *Appl Environ Microbiol*, 65, 4637–4645.

© Woodhead Publishing Limited, 2012

SCHIEMANN D A (1987), '*Yersinia enterocolitica* in milk and dairy products', *J Dairy Sci*, 70, 383–391.

SCHLESSER J E and PARISI B (2009), 'Inactivation of *Yersinia pseudotuberculosis* 197 and *Francisella tularensis* LVS in beverages by high pressure processing', *J Food Protect*, 72, 165–168.

SCHMID D, FRETZ R, WINTER P, MANN M, HÖGER G, STÖGER A, RUPPITSCH W, LADSTÄTTER J, MAYER N, DE MARTIN A and ALLERBERGER F (2009), 'Outbreak of staphylococcal food intoxication after consumption of pasteurised milk products, June 2007, Austria', *Wien Klin Wochenschr*, 121, 125–131.

SEISA D, OSTHOFF G, HUGO C, HUGO A, BOTHMA C and VAN DER MERWE J (2004), 'The effect of low-dose gamma irradiation and temperature on the microbiological and chemical changes during ripening of Cheddar cheese', *Radiat Phys Chem*, 69, 419–431.

SENSOY I, ZHANG Q H and SASTRY S K (1997), 'Inactivation kinetics of *Salmonella* Dublin by pulsed electric field', *J Food Process Eng*, 20, 367–381.

SEPULVEDA D R, GONGORA-NIETO M M, GUERRERO J A and BARBOSA-CANOVAS G V (2005), 'Production of extended-shelf life milk by processing pasteurised milk with pulsed electric fields', *J Food Eng*, 67, 81–86.

SHAMSI K, VERSTEEG C, SHERKAT F and WAN J (2008), 'Alkaline phosphatase and microbial inactivation by pulsed electric field in bovine milk', *Innovat Food Sci Emerg Tech*, 9, 217–223.

SHAO Y, RAMASWAMY H S and ZHU S (2007), 'High-pressure destruction kinetics of spoilage and pathogenic bacteria in raw milk cheese', *J Food Process Eng*, 30, 357–374.

SHARP J C M (1987), 'Infections associated with milk and dairy products in Europe and North America, 1980–85', *Bull World Health Organ*, 65, 397–406.

SHIN J K, JUNG K J, PYUN Y R and CHUNG M S (2007), 'Application of pulsed electric fields with square wave pulse to milk inoculated with *E. coli*, *P. fluorescens* and *B. stearothermophilus*', *Food Sci Biotechnol*, 16, 1082–1084.

SIMPSON E, WITTET S, BONILLA J, GAMAZINA K, COOLEY L and WINKLER J L (2007), 'Use of formative research in developing a knowledge translation approach to rotavirus vaccine introduction in developing countries', *BMC Publ Health*, 7, 1–11.

SINGH J P and GHALY A E (2006), 'Reduced fouling and enhanced microbial inactivation during online sterilization of cheese whey using UV coil reactors in series', *Bioproc Biosystems Eng*, 29, 269–281.

SIVAKUMAR N, RAJAMANI and SAIF A (2010), 'Partial characterisation of bacteriocins produced by *Lactobacillus acidophilus* and *Pediococcus acidilactici*', *Braz Arch Biol Tech*, 53, 1177–1184.

SMITH K, MITTAL G S and GRIFFITHS M W (2002), 'Pasteurization of milk using pulsed electrical field and antimicrobials', *J Food Sci*, 67, 2304–2308.

SOBRINO-LOPEZ A and MARTIN-BELLOSO O (2006), 'Enhancing inactivation of *Staphylococcus aureus* in skim milk by combining high-intensity pulsed electric fields and nisin', *J Food Protect*, 69, 345–353.

SOBRINO-LOPEZ A and MARTIN-BELLOSO O (2008), 'Enhancing the lethal effect of high-intensity pulsed electric field in milk by antimicrobial compounds as combined hurdles', *J Dairy Sci*, 91, 1759–1768.

SOBRINO-LOPEZ A, VIEDMA-MARTINEZ P, ABRIOUEL H, VALDIVIA E, GALVEZ A and MARTIN-BELLOSO O (2009), 'The effect of adding antimicrobial peptides to milk inoculated with *Staphylococcus aureus* and processed by high-intensity pulsed-electric field', *J Dairy Sci*, 92, 2514–2523.

SOLOMON E B and HOOVER D G (2004), 'Inactivation of *Campylobacter jejuni* by high hydrostatic pressure', *Lett Appl Microbiol*, 38, 505–509.

SOMKUTI G A and STEINBERG D H (2010), 'Pediocin production in milk by *Pediococcus acidilactici* in co-culture with *Streptococcus thermophilus* and *Lactobacillus delbrueckii* subsp. *Bulgaricus*', *J Ind Microbiol Biotechnol*, 37, 65–69.

© Woodhead Publishing Limited, 2012

STACK A and SILLEN G (1998), 'Bactofugation of liquid milks', *Nutr Food Sci*, 5, 280–282.

STOERMER A and WELLE F (2006), 'Packaging of dairy products – developments in line with legal compliance', *Verpackungs-Rundschau*, 57, 63–66.

SULLIVAN R and READ R B (1968), 'Method for recovery of viruses from milk and milk products', *J Dairy Sci*, 51, 1748–1751.

SULLIVAN R, TIERNEY J T, LARKIN E P, READ JR R B and PEELER J T (1971), 'Thermal resistance of certain oncogenic viruses suspended in milk and milk products', *Appl Microbiol*, 22, 315–320.

SUN H, KAWAMURA S, HIMOTO J, ITOH K, WADA T and KIMURA T (2008), 'Effects of ohmic heating on microbial counts and denaturation of proteins in milk', *Food Sci Tech Res*, 14, 117–123.

SUOMINEN I ANDERSSON M A ANDERSSON M C, HALLAKSELA A M, KÄMPFER P, RAINEY F A and SALKINOJA-SALONEN M (2001), 'Toxic *Bacillus pumilus* from indoor air, recycled paper pulp, Norway spruce, food poisoning outbreaks and clinical samples', *Syst Appl Microbiol*, 24, 267–276.

TABASCO R, GARCIA-CAYUELA T, PELAEZ C and REQUENA T (2009), '*Lactobacillus acidophilus* La-5 increases lactacin B production when it senses live target bacteria', *Int J Food Microbiol*, 132, 109–116.

TAUXE R V (2002), 'Emerging foodborne pathogens', *Int J Food Microbiol*, 78, 31–42.

TEREBIZNIK M R, JAGUS R J, CERRUTTI P, DE HUERGO M S and PILOSOF A M R (2000), 'Combined effect of nisin and pulsed electric fields on the inactivation of *Escherichia coli*', *J Food Protect*, 63, 741–746.

TETRA PAK CHEESE and POWDER SYSTEMS (2010), Tetra Alcross Bactocatch – bacteria and spore removal with a ceramic membrane solution, Lund, Sweden, Tetra Pak Group.

THOMPSON J K, COLLINS M A and MERCER W D (1996), 'Characterisation of a proteinaceous antimicrobial produced by *Lactobacillus helveticus* CNRZ450', *J Appl Bacteriol*, 80, 338–348.

TROTZ-WILLIAMS L A, MARTIN S W, LESLIE K E, DUFFIELD T, NYDAM D V and PEREGRINE A S (2008), 'Association between management practices and within-herd prevalence of *Cryptosporidium parvum* shedding on dairy farms in Southern Ontario', *Prev Vet Med*, 83, 11–23.

TSIOTSIAS A, SAVVAIDIS I, VASSILA A, KONTOMINAS M and KOTZEKIDOU P (2002), 'Control of *Listeria monocytogenes* by low-dose irradiation in combination with refrigeration in the soft whey cheese Anthotyros', *Food Microbiol*, 19, 117–126.

TUCHENHAGEN DAIRY SYSTEMS (2006), Extended shelf life milk – solutions with indirect heating applications, Büchen, Germany, GEA Liquid Processing Division.

TUCKER C B, FULKERSON G C and NEUDECKER R M (1954), 'A milkborne outbreak of shigellosis in Madison County, Tenn.', *Publ Health Rep*, 69, 432–436.

TULSIYAN P, SARANG S and SASTRY S K (2008), 'Electrical conductivity of multicomponent systems during ohmic heating', *Int J Food Prop*, 11, 233–241.

ULBERTH F (2003), 'Testing the authenticity of milk and milk products', in Smit G, *Dairy processing: improving quality*, Cambridge, Woodhead Publishing, 208–228.

UNICOMB L E, FULLERTON K E, KIRK M D and STAFFORD R J (2009), 'Outbreaks of campylobacteriosis in Australia, 2001 to 2006', *Foodborne Pathog Dis*, 6, 1241–1250.

VAILLANT F, CISSE M, CHAVERRI M, PEREZ A, DORNIER M, VIQUEZ F and DHUIQUE-MAYER C (2005), 'Clarification and concentration of melon juice using membrane processes', *Innovat Food Sci Emerg Tech*, 6, 213–220.

VAN BRANDT L, VAN DER PLANCKEN I, DE BLOCK J, VLAEMYNCK G, VAN COILLIE E, HERMAN L and HENDRICKX M (2011), 'Adequacy of current pasteurization standards to inactivate *Mycobacterium paratuberculosis* in milk and phosphate buffer', *Int Dairy J*, 21, 295–304.

© Woodhead Publishing Limited, 2012

VAN DER VEEN S, WAGENDORP A, ABEE T and WELLS-BENNIK M H J (2009), 'Diversity assessment of heat resistance of *Listeria monocytogenes* strains in a continuous-flow heating system', *J Food Protect*, 72, 999–1004.

VAN DUYNHOVEN Y T H P, ISKEN L D, BORGEN K, BESSELSE M, SOETHOUDT K, HAITSMA O, MULDER B, NOTERMANS D W, DE JONGE R, KOCK P, VAN PELT W and STENVERS O (2009), 'A prolonged outbreak of *Salmonella* Typhimurium infection related to an uncommon vehicle: hard cheese made from raw milk', *Epidemiol Infect*, 137, 1548–1557.

VAN OPSTAL I, BAGAMBOULA C F, VANMUYSEN S C M, WUYTACK E Y and MICHIELS C W (2004), 'Inactivation of *Bacillus cereus* spores in milk by mild pressure and heat treatments', *Int J Food Microbiol*, 92, 227–234.

VARNAM A and SUTHERLAND J P (2001), *Milk and milk products: technology, chemistry and microbiology*, Gaithersburg, MD, Aspen Publishers.

VELAZQUEZ-ESTRADA R M, HERNANDEZ-HERRERO M M, LOPEZ-PEDEMONTE T J, BRINEZ-ZAMBRANO W J, GUAMIS-LOPEZ B and ROIG-SAGUES A X (2011), 'Inactivation of *Listeria monocytogenes* and *Salmonella enterica* serovar Senftenberg 775W inoculated into fruit juice by means of ultra high pressure homogenisation', *Food Contr*, 22, 313–317.

VILLAMIEL M and DE JONG P (2000), 'Inactivation of *Pseudomonas fluorescens* and *Streptococcus thermophilus* in trypticase soy broth and total bacteria in milk by continuous-flow ultrasonic treatment and conventional heating', *J Food Eng*, 45, 171–179.

VIRGIN J E, VAN SLYKE T M, LOMBARD J E and ZADOKS R N (2009), 'Short communication: methicillin-resistant *Staphylococcus aureus* detection in US bulk tank milk', *J Dairy Sci*, 92, 4988–4991.

WAGNER J (2001), *Membrane filtration handbook: practical tips and hints*, Minnetonka, MN, USA, Osmonics, Inc.

WALKLING-RIBEIRO M, NOCI F, CRONIN D A, LYNG J G and MORGAN D J (2009), 'Antimicrobial effect and shelf-life extension by combined thermal and pulsed electric field treatment of milk', *J Appl Microbiol*, 106, 241–248.

WALKLING-RIBEIRO M, RODRÍGUEZ-GONZÁLEZ O, JAYARAM S and GRIFFITHS M W (2011), 'Microbial inactivation and shelf life comparison of "cold" hurdle processing with pulsed electric fields and microfiltration and conventional thermal pasteurisation in skim milk', *Int J Food Microbiol*, 144, 379–386.

WHO (2002), Factsheet No. 124, Available from: http://www.who.int/mediacentre/factsheets/fs124/en/ (accessed 25 May 2011).

WHO (2008), Foodborne disease outbreaks: guidelines for investigation and control, Geneva, Switzerland, WHO Press.

WILSON F L and TANNER F W (1945), 'Behaviour of pathogenic bacteria in fermented milks', *J Food Sci*, 10, 122–134.

YAM K L, TAKHISTOV P T and MILTZ J (2005), 'Intelligent packaging: concepts and applications', *J Food Sci*, 70, R1–R10.

YEOM H W, EVRENDELIK G A, JIN Z T and ZHANG Q H (2004), 'Processing of yogurt-based products with pulsed electric fields: microbial, sensory and physical evaluations', *J Food Process Pres*, 28, 161–178.

YESILLIK S, YILDIRIM N, DIKICI A, YILDIZ A and YESILLIK S (2011), 'Antibacterial effects of some fermented commercial and homemade dairy products and 0.9% lactic acid against selected foodborne pathogens', *Asian J Anim Vet Adv*, 6, 189–195.

YI M, KANEKO S, YU D Y and MURAKAMI S (1997), 'Hepatitis C virus envelope proteins bind lactoferrin', *J Virol*, 71, 5997–6002.

YOUN K S, HONG J H, BAE D H, KIM S J and KIM S D (2004), 'Effective clarifying process of reconstituted apple juice using membrane filtration with filter-aid pretreatment', *J Membr Sci*, 228, 179–186.

YU L J, NGADI M and RAGHAVAN G S V (2009), 'Effect of temperature and pulsed electric field on rennet coagulation properties of milk', *J Food Eng*, 95, 115–118.

© Woodhead Publishing Limited, 2012

YUCEL N and CITAK S (2003), 'The occurrence, hemolytic activity and antibiotic susceptibility of motile *Aeromonas* spp. isolated from meat and milk samples in Turkey', *J Food Saf*, 23, 189–200.

ZADOKS R N, ALLORE H G, BARKEMA H W, SAMPIMON O C, GRÖHN Y T and SCHUKKEN Y H (2001), 'Analysis of an outbreak of *Streptococcus uberis* mastitis', *J Dairy Sci*, 84, 590–599.

ZAGREBNEVIENE G, JASULAITIENE V, MORKUNAS B, TARBUNAS S and LADYGAITE J (2005), '*Shigella sonnei* outbreak due to consumption of unpasteurised milk curds in Vilnius, Lithuania, 2004', *Euro Surveill*, 10, Available from: http://www.eurosurveillance.org/ViewArticle.aspx?ArticleId=2848 (accessed 19 May 2011).

ZALAN Z, HUDACEK J, STETINA J, CHUMCHALOVA J and HALASZ A (2010), 'Production of organic acids by *Lactobacillus strains* in three different media', *Eur Food Res Tech*, 230, 395–404.

ZENKER M, HEINZ V and KNORR D (2003), 'Application of ultrasound-assisted thermal processing for preservation and quality retention of liquid foods', *J Food Protect*, 66, 1642–1649.

ZHANG Q H, QIN B L, BARBOSA-CANOVAS G V and SWANSON B G (1995), 'Inactivation of *E. coli* for food pasteurization by high-strength pulsed electric fields', *J Food Process Pres*, 19, 103–118.

ZORRAQUINO M A, ALTHAUS R L, ROCA M and MOLINA M P (2009), 'Effect of heat treatments on aminoglycosides in milk', *J Food Protect*, 72, 1338–1341.

© Woodhead Publishing Limited, 2012

# Part II

# Current and emerging non-chemical decontamination methods

© Woodhead Publishing Limited, 2012

# 8

# Advances in food surface pasteurisation by thermal methods

**G. Purnell and C. James, The Grimsby Institute (GIFHE), UK**

**Abstract**: The surfaces of naturally derived meats, fish, fruit, vegetables, etc. (and poorly produced manufactured foods) can be contaminated to a varying degree by both pathogenic and spoilage bacteria. There is no terminal step (such as cooking) to eliminate pathogenic organisms from many raw products, such as meats, until they reach the consumer, and in the case of many fruits and vegetables no terminal step at all before ingestion. The introduction of efficient surface pasteurisation measures to reduce such contamination would reduce any potential or additional food safety risk for human health. To be successful such measures must significantly reduce microbial levels without changing the nature of the food, i.e., a 'fresh', 'raw' food must remain as such. This chapter discusses first the principles of thermal surface pasteurisation treatments and then the details of how such treatments can and have been applied for treating a variety of foods.

**Key words**: surface pasteurisation, steam, hot water, dry heat, hot air, infra-red.

## 8.1 Introduction

The surfaces of naturally derived meats, fish, fruit, vegetables, etc. (and poorly produced manufactured foods) can be contaminated to a varying degree by both pathogenic and spoilage bacteria. Obtaining microbially clean raw food materials is difficult. There is no terminal step (such as cooking) to eliminate pathogenic organisms from many raw products, such as meats, until they reach the consumer, and in the case of many fruits and vegetables no terminal step at all before ingestion. The introduction of efficient pasteurisation measures to reduce such contamination would reduce any potential or additional food safety risk for human health. However, conventional pasteurisation of such

© Woodhead Publishing Limited, 2012

products results in a 'cooked' product. Since microbial contamination on such raw materials is generally on the surface, there has been much interest in the development of thermal pasteurisation treatments that only target these microorganisms on the surface of the food without denaturing the surface of the food. To be successful, such measures must significantly reduce microbial levels without changing the nature of the food, i.e. a 'fresh', 'raw' food must remain as such.

## 8.2 The principles of thermal surface pasteurisation

### 8.2.1 Effect of thermal surface pasteurisation on microorganisms

The principles and effect of thermal preservation treatments such as pasteurisation and sterilisation are well established and discussed in many widely available publications (Stumbo, 1973; Jay *et al.*, 2005; ECFF, 2006). While many thermal surface pasteurisation treatments have been applied to foods, many of which we will discuss in detail in this chapter, few authors have specifically addressed the effect of thermal surface pasteurisation treatments on microorganisms.

Morgan *et al.* (1996a) have presented a specific theoretical basis to thermal surface pasteurisation. This is based on the principle that heat kills microorganisms mainly by inactivating the most sensitive, vital enzymes. They claim that, for meats, 8.38 to 50.28 kJ $g^{-1}$ $mol^{-1}$ is typically required to inactivate these enzymes whilst the heat for irreversible cooking of muscle is substantially higher, in the range 209.5–419 kJ $g^{-1}$ $mol^{-1}$. Thus, if heat energy can be supplied at the correct levels to deactivate the enzymes, whilst not denaturing the product, the foodstuff can be surface pasteurised without damaging the quality. Additionally, they claim that only microgram amounts of enzyme need to be inactivated, compared to the grams of muscle required for denaturation during cooking to be detectable. Thus 'for a square centimetre of surface contaminated with 100 bacteria, 15 million times as much heat is needed to cook the surface to a depth equal to the length of a bacterium, compared to the heat needed to kill all the bacteria' (Morgan *et al.*, 1996a). Since microorganisms are present only on the surface of many foods, even assuming that heating rates are the same for both microorganism and food, in theory since any microorganism will be on the food surface the microorganism should die before there is significant penetration of heat into the food. In fact, the food will take longer, because it requires conductive heat transfer through the food.

Conventional pasteurisation temperature/time treatments are typically designed based on established death kinetics relationships using *D*- and *z*-values. A *D*-value is the time required for a 1 log reduction in microbial numbers at a set temperature. The *z*-value is the temperature rise required to increase death by 1 log. However, the majority of these data relating heat treatments to thermal death kinetics have been obtained by carrying out

© Woodhead Publishing Limited, 2012

*in vitro* tests on small samples of microorganisms in growth medium or food slurries. A valuable insight into the factors influencing microbial death can result from the data gathered in such investigations. However, studies have shown that such data cannot be applied to surface pasteurisation treatments, since the conditions on the surface of a food are very different from that within the food (James and Evans, 2006). In addition, such data are usually generated at much lower temperatures than are used in rapid surface pasteurisation treatments and cannot be extrapolated. It has been recommended that decimal reduction time curves should not be extrapolated more than 5.56°C higher than those actually tested (Thomas *et al.*, 1966).

At present, there are little readily available data on the thermal death kinetics of microorganisms on the surface of foods in rapid heating systems. Data have been produced for chicken skin heated in water (Notermans and Kampelmacher, 1975) and other authors have produced individual data sets during decontamination studies (James *et al.*, 1998; Evans and Brown, 1999; Göksoy *et al.*, 2001). However, the time-temperature treatments were not well enough defined, or comprehensive enough for accurate predictive modelling. In addition Notermans and Kampelmacher's (1975) work on a 'real' food showed that, contrary to the assumptions made in using standard $D$- and $z$-value microbial death models, heat destruction of microorganisms attached to the skin of poultry did not occur at a logarithmic rate at immersion temperatures of 51°C. Other studies have also revealed that the rate of destruction of microorganisms in dry surface heating systems can be orders of magnitude less than those using steam or water, even if the rates of heating were very similar (James and Evans, 2006). A series of microbiological death kinetic data related to surface pasteurisation treatments generated by the EU 'BUGDEATH' project has been published in a number of papers (Lewis *et al.*, 2006; McCann *et al.*, 2006; Gaze *et al.*, 2006; Valdramidis *et al.*, 2006); however, this data remains to be validated under commercial conditions. This project showed that microbiological resistance to thermal surface pasteurisation treatments is product specific and that rapid changes in water activity at the surface of the food during treatment provide a degree of protection to microorganisms, accounting for greater survival than would be predicted using conventionally derived $D$ and $z$ data.

### 8.2.2 Effect of thermal surface pasteurisation on food quality

Avoiding cooking of the foodstuff is the primary issue in determining acceptable thermal surface pasteurisation conditions. While data are available on the death kinetics of pathogenic bacteria, there is little information on the relationship between surface temperature/process time and the appearance and colour of food surfaces. The importance of any surface changes caused by such treatment may depend on whether the food will receive any further processing that may remove such affected areas before sale.

© Woodhead Publishing Limited, 2012

Numerous studies have shown that while heat-treated meat may show an immediate appearance of heat damage, it may recover during subsequent chilled storage, particularly if rapidly cooled immediately after treatment. The appearance of meat carcass surface tissue immediately after exposure to treatments of 80°C or above may appear bleached, gray or 'cooked' to an approximate depth of 0.5 mm (Smith, 1992). Smith and Graham (1978) found that the cooked appearance of the surface of beef, lamb and mutton carcasses immersed in hot water (80°C for 10 s) disappeared almost completely within a few hours of cold storage. Similar observations were made for lamb carcasses after hot water immersion at 90°C for 8 s (C. James *et al.*, 2000) and after hot spray wash at 82°C for 8 s (Hauge *et al.*, 2011). Smith and Davey (1990) reported that the use of a novel hot water cabinet to apply water of 83.5°C for 20 s over the entire surface of sides of beef caused a cooked/bleached appearance, but that the sides regained normal appearance during chilling. Exposure to temperatures above 85°C for more than 20 s may result in permanent damage or surface bloom (Davey and Smith, 1989). A slight visual discoloration observed by Barkate *et al.* (1993) on the carcass surface immediately after treatment at 95°C for 10 s was absent after 24 h of refrigerated storage and the normal carcass colour was restored. Gill and Badoni (1997) immersed portions of post- and pre-rigor pork and beef in 75 or 85°C water for 5, 10, 15 or 20 s and concluded that pasteurising treatments cannot be applied to meat without some degradation of the appearances of cut muscle surfaces. Gill *et al.* (1999) concluded that treatment of beef carcass sides with water of 85°C for 10 s should substantially reduce bacterial counts without unacceptable damage to the appearance of the product. According to Castillo *et al.* (2002), treatment of carcass tissue with water of 80°C or higher did not result in permanent discoloration of the tissue surface.

In general, potential concerns associated with surface tissue discoloration of meats have been discounted by most studies as being transitory under the practical conditions of treatment application. However, although this may not be true for all species, initial surface discolorations, caused by treatments based on high temperature, are usually unnoticed after a few hours of chilling. Lamb/mutton and beef have been reported to recover; however, detailed work on pre- and post-rigor pork and beef by Gill and Badoni (1997) showed that, while pre-rigor beef will recover to a certain extent from heat damage, pre- and post-rigor pork will not. Poultry also appear to be relatively sensitive to temperatures over 60–65°C. Morrison and Fleet (1985) reported that carcasses immersed in water at 60°C for 10 min possessed acceptable surface appearance. Dawson *et al.* (1963) found that water at temperatures higher than 60°C caused partial cooking of poultry and Cox *et al.* (1974a) showed that immersion of carcasses at 71.7°C greatly impaired appearance. The data of Morgan *et al.* (1996a, 1996b) show that it is possible to expose small pieces of chicken breast (10 mm x 10 mm, 50 mm long) to steam at 100°C for 1 second, without cooking. These workers also provide a clear table of cooking times at higher temperatures. Data for lower

© Woodhead Publishing Limited, 2012

temperatures are less clear. With whole poultry carcasses the main effect of hot water treatments is to tighten the skin of the carcass making it difficult to sometimes truss the carcasses after chilling without the skin tearing (Purnell *et al.*, 2004). Purnell *et al.* (2004) found that while an immersion treatment at 75°C for 30 s significantly reduced APCs (total aerobic plate count) and counts of Enterobacteriaceae and *Campylobacter*, the skin tended to tear during trussing. However, treating carcasses at 70°C for 40 s, followed by a 12–15°C spray-chill treatment for 13 s, did not detrimentally affect the skin. In subsequent trials using an 80°C for 20 s immersion treatment, with no spray-chill treatment produced a generally acceptable carcass (Corry *et al.*, 2007).

In general, heat treatments for fruits and vegetables are at lower temperatures than those employed for treating meats. This is principally because they are aimed at controlling insect pests and fungal spoilage rather than thermotolerant foodborne pathogens. Typically, a temperature below 55°C is applied for a number of minutes (Scheerlinck *et al.*, 2004). For strawberries, Garcia *et al.* (1995) used a treatment of 15 min at 44–48°C. However, studies have demonstrated that high temperature treatments can be applied on a wide variety of fruits and vegetables. For cantaloupe, grapefruits, papaya and beets, Kozempel *et al.* (2002) used high pressure steam at 138–143°C for 0.1–0.2 s with no thermal damage.

Few published studies have assessed the effect of thermal surface pasteurisation treatments on the subsequent eating quality of the treated food. Gill *et al.* (2001) found that the flavour of cooked patties manufactured with meat treated with water of 85°C for 60 s was not distinguished from the normal commercial product, and concluded that pasteurising manufacturing beef with water of 85°C for 45 s could be a practicable treatment for enhancing the microbiological safety of frozen hamburger patties.

### 8.2.3 Effect of thermal surface pasteurisation on shelf-life

There is conflicting evidence on the ability of thermal surface pasteurisation treatments to extend shelf-life. From a commercial view point, it is the possibility of extending shelf-life that has the most appeal to much of the food industry. However, since many of the published studies on thermal surface pasteurisation treatments are funded by public bodies with an interest solely in safety, shelf-life or the effect of such treatments on spoilage microorganisms is often not addressed, particularly studies concerning meat and poultry.

Studies on the effectiveness of moderate heat treatment (60°C for 20 s) in extending the shelf-life of fresh fish by Vaz-Pires *et al.* (1994) showed that a warm water immersion treatment was more effective on bacteria in broth cultures than on the surface microflora of fish. Extension of the lag phase of bacteria was not found to extend shelf-life in all cases. Results showed that there is the possibility of more rapid growth subsequent to an extended lag phase. They concluded that more rapid growth might occur due to the

© Woodhead Publishing Limited, 2012

loss of competitive microflora, thus reducing or eliminating the advantage gained from the delayed onset of growth. Anderson *et al.* (1979) reported higher rates of growth on meat samples treated with water (cold and hot), steam and hypochlorite than on control samples. They concluded that this was probably due to the drier surfaces of the controls.

Thermal pasteurisation of produce is primarily performed to combat plant pathogens that cause spoilage and reduce shelf-life. In addition to the antimicrobial action of the heating, there is also some evidence to suggest that the natural waxes in the skins are redistributed to seal fissures and pores of the skin, thus limiting ingress of pathogens and fungal spores (Fallik, 2004). Schirra and d'Hallewin (1997) determined that a 2 min hot water immersion of mandarins at 50–54°C caused a clear redistribution of the epicuticular wax layer and a significant reduction in cuticular cracks, thus improving physical barriers to pathogen penetration.

## 8.3 Wet heat pasteurisation

### 8.3.1 Sub-atmospheric steam

The temperature at which water boils is a function of pressure. At atmospheric pressure, steam will be created initially at 100°C. At pressures below atmospheric (sub-atmospheric), the generation temperature will be lower than 100°C, while at pressures higher than atmospheric it will be above 100°C. Generation at temperatures other than 100°C does not substantially reduce the heat capacity of the steam. Treatment temperatures below 100°C are less likely to cause damage to the surface of the food stuff, but will require longer treatment times than treatments at 100°C and above.

Thermal pasteurisation techniques involving steam fall into three categories based on the pressure (and hence temperature) of the steam.

- Sub-atmospheric steam treatments take place at less than 1 bar pressure and utilise temperatures below 100°C. These are typically batch processing methods because of the need to isolate the treated products from the atmosphere to pull the necessary vacuum.
- Atmospheric steam treatments take place at approximately 1 bar pressure and use temperatures close to 100°C. These methods are relatively simple to engineer, but are restricted to using only the treatment temperature defined by the local atmospheric pressure conditions.
- Pressurised steam treatments are similar to sub-atmospheric treatments in that processing is typically batchwise in a pressure vessel. However, because of the increased steam pressure, temperature in excess of 100°C can be used.

Steam at 100°C has a substantially higher heat capacity than the same amount of water at that temperature. When steam condenses onto the surface of a

© Woodhead Publishing Limited, 2012

(cooler) product, the latent heat of condensation is released onto the surface of the food as the steam turns from the higher energy gaseous state to the lower energy liquid state. This can rapidly raise the surface temperature with little sensible heat change in the heating medium.

Another very attractive feature of heating with condensing steam is its ability to penetrate cavities and condense on any cold surface. Water vapour molecules (typically $2 \times 10^{-4}$ μm in diameter) are much smaller than bacteria, for example, *Salmonella* cells are 0.7 μm thick and 4 μm long (Morgan *et al.*, 1996b). Therefore, steam is capable of reaching any bacteria that occur in cavities. Although the velocity of steam is reduced by cavities of diameter less than the mean free path of the gas density, this does not restrict steam in reaching bacteria. In 140°C saturated steam, the mean free path of the steam molecule is 0.4 μm, half the diameter of the smallest cavity capable of containing a *Salmonella* cell (Morgan *et al.*, 1996b).

To prevent cooking, the steam must condense on the surface rapidly, and re-evaporate equally rapidly (Morgan *et al.*, 1996a). Gases move by either flow or diffusion. Flow is rapid, motivated by a pressure gradient. Diffusion is much slower and motivated by a concentration gradient of the gas through other gases. During pressurised steam treatment, air and any other non-condensable gas present are concentrated by the inrush of condensing steam that forms a layer around the product surface. This prevents steam flow, slowing condensation as the steam diffuses through the layer. Non-condensable gases can come from three sources: gases around the product, when enclosed in a chamber; gases entering with the treatment steam; and gases that have been desorbed by heat from the surface of the food being treated.

*Sub-atmospheric steam pasteurisation of meat*

Utilising the condensation of steam at sub-atmospheric pressure has been shown to be an effective method of decontaminating chicken drumsticks (Klose and Bayne, 1970). Reductions in APCs of up to 5.6 log CFU $ml^{-1}$ of rinse were achieved for drumsticks treated at 76°C for periods of up to 16 min, and the shelf-life at 3°C increased from 6 to 19 days. The work was later extended to surface pasteurisation of whole poultry carcasses (Klose *et al.*, 1971). Reductions of 3–3.7 log CFU $ml^{-1}$ were achieved, using steam at temperatures not exceeding 75°C for 4 min. The orientation of the poultry carcass within the chamber and the method of introducing the steam were found to have significant effects on microbial counts. Toughening of the meat was observed, when steam treatment was applied to a chicken carcass within two hours of slaughter, but not if the steam was applied 22 h after slaughter. This was possibly due to hot-shortening of the muscles (a phenomenon in pre-rigor muscle, similar to cold-shortening, in which the contractile elements of the muscle fibres shorten, which results in an appreciable reduction in the tenderness of meat).

© Woodhead Publishing Limited, 2012

Similar reductions in microbial counts were reported by Evans (1999). Steam at 75 or 85°C for 40 s was required to reduce levels of *S.* Enteritidis inoculated onto chicken portions by 3–4 $\log_{10}$ CFU $cm^{-2}$. However, some degree of cooking was apparent for all the meats investigated. On chicken, the outer layer of muscle was slightly cooked, although the skin was barely affected. Similar conditions were also found to produce similar reductions on pork skin, beef and pepper samples (S. James *et al.*, 2000).

*Sub-atmospheric steam pasteurisation of produce*

Potty (2011) reports a commercially available sub-atmospheric surface pasteurisation system primarily designed for dry produce such as peanuts, herbs, spices and grains. By controlling temperature and pressure, the degree of condensation heating is monitored to manipulate surface moisture on the produce thus maintaining quality. This is achieved in a four-step batch process involving: product preheating, equilibration to the condensation dew-point, pasteurisation and removal (or restoration) of surface moisture. Reductions in the *Salmonella* surrogate species *Enterococcus faecium* of up to 5 log are reported (Potty, 2011).

### 8.3.2 Atmospheric steam

*Atmospheric steam pasteurisation of meat*

A successful steam process, in terms of industrial application, has been that developed in the United States by FMC Frigoscandia; the Steam Pasteurisation System (SPS) for red meat (Fig. 8.1). Studies on this commercially available system for treating red-meat carcasses have been conducted and published by Kansas State University (Nutsch *et al.*, 1997; Phebus *et al.*, 1997). Significant reductions, in the order of 3.5 log units for specific bacteria, have been reported. The full commercial system (SPS 400 Steam Pasteurisation System) consists of a three-stage cabinet. Washed carcasses pass through an air-drying stage to remove residual water from the carcass, before an enclosed steam-treatment stage that may be followed by spray-cooling.

A modified version of this system has been evaluated in a UK red-meat abattoir. In this work a 90°C for 10 s treatment was favoured, producing reductions in inoculated Enterobacteriaceae of up to 3 log CFU $cm^{-2}$ on treated beef carcasses (Eveleigh, 2000). However, a noticeable colour change to treated carcasses was apparent, resulting in this treatment not being adopted by the UK red-meat industry. The same equipment was evaluated by Whyte *et al.* (2003) on poultry carcasses. Conditions similar to those used to treat beef carcasses commercially in the United States (90°C for 12 s) were found to produce statistically insignificant reductions in APCs, Enterobacteriaceae and thermophilic campylobacters. Increasing the treatment time to 24 s decreased cell population by 0.75, 0.69 and 1.3 log CFU $g^{-1}$, respectively. However, visible damage to the outer skin tissue was found.

© Woodhead Publishing Limited, 2012

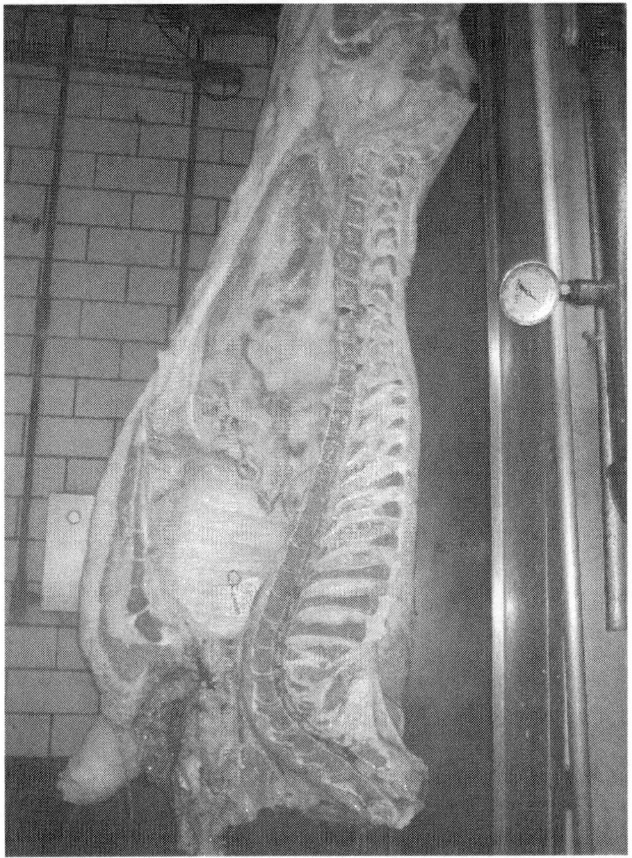

**Fig. 8.1** Beef carcass immediately after treatment in an atmospheric steam cabinet in commercial operation.

Studies at the University of Bristol (Corry *et al.*, 2007), using bespoke pilot equipment (Fig. 8.2), have compared both atmospheric steam and hot water treatments for use on poultry carcasses under both laboratory and commercial conditions. In experimental studies, whole chicken carcasses, inoculated with *Campylobacter jejuni* and *Escherichia coli* K12, were treated with steam at atmospheric pressure for up to 20 s in a pilot-scale cabinet (and with hot water in a pilot immersion system for 20–30 s at 75 and 80°C). In steam, numbers of *C. jejuni* were reduced by ~1.8 log CFU $cm^{-2}$ in 10 s and 3.3 log CFU $cm^{-2}$ in 20 s. Corresponding reductions in numbers of *E. coli* K12 were 1.7 and 2.8 log CFU $cm^{-2}$. However, the 20 s treatments caused the skin to shrink and change colour. The optimum treatment for maximum effect on *C. jejuni* and *E. coli*, least skin shrinkage and change of colour was concluded to be 10–12 s. Trials in a commercial poultry plant using naturally contaminated carcasses compared treatments for 10 s in steam with 20 s in hot water at 80°C. The appearance of the treated

© Woodhead Publishing Limited, 2012

**Fig. 8.2** Batch pilot-scale steam pasteurisation unit used for scientific evaluation of steam treatment of poultry carcasses.

carcasses was assessed visually at intervals until the end of shelf-life, and checks made for pseudomonas, Enterobacteriaceae and campylobacters on breast skin. Initial levels of *Campylobacter* spp. were low (~1 log CFU $cm^{-2}$) and variable, but reductions (similar for steam and hot water) of about 2 log cycles were obtained for the other two groups. Numbers of campylobacters were reduced, but not eliminated. Visual assessment indicated that the hot water treatment caused less change in appearance than the steam treatment. Carcasses produced using either treatment could be used for production of 'skin-off' portions. It was considered that changes to appearance of skin-on carcasses or portions would be acceptable to many consumers.

Most studies on the utilisation of atmospheric steam have aimed to develop processes that produce a substantial reduction in microbial numbers, but do not result in substantial cooking of the product. Avens *et al.* (2002) determined the treatment conditions necessary for a total thermal destruction of microbes (end count of <10 CFU $cm^{-2}$), irrespective of damage. For atmospheric steam (96–98°C), a 3 min treatment was required to reduce natural contamination of retail carcasses from $10^4$ to <6 CFU $cm^{-2}$. This treatment substantially cooked all the samples.

*Atmospheric steam pasteurisation of seafood*

Atmospheric steam treatments have been shown to have potential for treating fish, specifically catfish (Balá *et al.*, 1999) and salmon (Bremer *et al.*, 2002).

© Woodhead Publishing Limited, 2012

Atmospheric steam pasteurisation of fish has been demonstrated by Balá *et al.* (1999) to have potential in reducing skin microflora on deheaded and eviscerated whole catfish. A food steamer was used and samples treated for 30, 60, 90 and 120 s in conditions ranging between 90 and 98°C. Results indicated that as steam treatment duration increased, so did microbial reductions. Initial reductions in total, coliform and psychrotrophic counts caused by the steam treatments were maintained during storage. Steam treatment for 120 s reduced total, coliform and psychrotrophic counts (compared with untreated controls) by 2.3, 3.5 and 3.4 $\log_{10}$ CFU $cm^{-2}$ after 4 days of storage at 4°C. The effect of such treatment on sensory characteristics was not discussed; however, it was noted that exposed muscle surfaces were discoloured and that the skin developed a white film after heating.

Bremer *et al.* (2002) describe trials on steam pasteurisation of gutted and defined whole salmon using bespoke treatment equipment. The fish was hung by a jaw hook and a reservoir of steam initially at approximately 8 bar was released onto and into the carcass. Fish were inoculated with *L. monocytogenes* at 4–5 log CFU/fish, and after an 8 s treatment, no *L. monocytogenes* could be recovered. In-plant trials conducted with naturally occurring organisms showed longer treatments were necessary, but nevertheless, 7 of 10 fish treated for 8 s and all 10 fish treated for 11 s showed no recoverable *L. monocytogenes*.

*Atmospheric steam pasteurisation of produce*

Atmospheric steam treatments before storage have been successfully used to control microorganisms on carrots (Afek *et al.*, 1999; Gan-Mor *et al.*, 2011), potato (Afek *et al.*, 1999), sugarcane (Singh *et al.*, 1987), sweet potato (Afek and Orenstein, 2003) and even lettuce (Martin-Diana *et al.*, 2007). Steaming of produce has also been shown to perform functional operations such as improving the red colour of litchi peels (Kaiser *et al.*, 1995), raising the levels of ascorbic acid and vitamins in broccoli (Petersen, 1993), and to change evaporation levels in mandarin oranges (Lin *et al.*, 1992). It is also used to aid peeling of a wide range of agricultural produce.

A 3 s atmospheric steam treatment (product temperature ~90°C) before packaging reduced the proportion of decayed carrots from 23% to 2% after 60 days storage at 0.5°C followed by 7 days at 20°C (Afek *et al.*, 1999). When carrots were inoculated with spoilage fungi, percentages of decay, after similar periods of storage and shelf-life, were 5% for steam-treated carrot and 65% for the non-treated controls.

Gan-Mor *et al.* (2011) assessed steam jets for the reduction of the plant pathogen *Sclerotinia sclerotiorum*. Carrots were rotated on a roller conveyer as they passed below steam jets combined with electric steam drying elements and heat reflectors. The aim of the rotation was to give steam coverage onto all surfaces and this was monitored by thermal imaging equipment. After water cooling (4°C, 10 min), carrots were subjected to 0, 2, 3 and 4 s steam treatments attaining mean surface temperatures of 57°C, and maximal

© Woodhead Publishing Limited, 2012

temperatures of 65°C for 0.3 s. Carrots not water-cooled before steam pasteurisation were not of a marketable quality, whereas those water-cooled did not possess significantly different crispness, sweetness, bitterness and overall eating quality scores to non-pasteurised carrots. A 3 s treatment gave the best overall compromise treatment with almost no colour change, a slight reduction in sprouting and a 60% reduction in soft rot after storage over non-pasteurised products. Similar benefits were seen for sweet potato storage (Afek and Orenstein, 2003) where a 3 s atmospheric steam treatment reduced the proportion of decay in cured sweet potatoes from 32% to 3% after 5 months of storage and in non-cured sweet potatoes from 86% to 14%.

Whilst many of the produce items treated with steam have been relatively robust, steam pasteurisation of more delicate items such as lettuce has also been assessed (Martin-Diana *et al.*, 2007). Pre-cooled (4°C) lettuces were subjected to a 10 s blanching steam spray at about 100°C, then water-cooled at ~25°C for 1 min with agitation, and finally dried for 5 min using an automatic salad spinner. These were compared to a conventional chorine wash. Steam-treated lettuces showed improved browning and textural properties over the standard chlorine rinse techniques. Both treatments showed similar microbial profiles, sensory panel scores for fresh appearance and acceptability (although outer wrapped leaves were removed after treatments). However, both groups of samples were deemed fit for consumption at the end of a 10-day storage period. The steam treatment was found to have a negative effect in nutrient levels.

### 8.3.3 Pressurised steam

*Pressurised steam pasteurisation of meat*

Morgan *et al.* (1996a,b) have developed a device that permits surface treatment of meat at very high temperatures, without cooking. This system utilises very rapid cycles (for milliseconds) of heating and cooling, using steam under pressure and vacuum-cooling. Meat samples are placed in a rotating chamber. As it rotates, the meat is exposed to three other chambers that provide, respectively, a vacuum, steam and final vacuum. This procedure allows temperatures of up to 145°C to be used. Experimental trials demonstrated that treatment at 145°C for 25 ms produced a 4 log reduction in *L. innocua*, added artificially to raw chicken meat.

The concept has been further developed by Kozempel *et al.* (2000, 2001, 2003a, b) as the vacuum/steam/vacuum (VSV) process. The initial prototype system, developed to treat whole carcasses (Kozempel *et al.*, 2000), was found to have little significant effect on APCs, when assessed using a whole-carcass rinse. Therefore, it was concluded that the abdominal cavity was not receiving a sufficient treatment. Consequently, the pilot system was modified to pass steam directly into the cavity (Kozempel *et al.*, 2001). With this modification, the optimum process conditions (initial vacuum 0.1 s, final vacuum 0.5 s, vacuum absolute pressure of 4.1–7.1 kPa, steam time 0.1 s,

© Woodhead Publishing Limited, 2012

steam temperature 138°C) were shown to reduce counts of *L. innocua* by 0.7–0.8 log CFU $ml^{-1}$. The pilot system has been scaled up to a mobile unit (Kozempel *et al.*, 2003a). Field studies of this system achieved a 1.4 log reduction for *E. coli* and a 1.2 log reduction for campylobacter on naturally contaminated carcasses; however, there was extensive mechanical damage caused by the introduction of steam into the cavity. (Similar problems were encountered in studies on a similar system at the University of Bristol; Purnell *et al.*, 2004). In further trials, the mechanical damage was eliminated and a 1.1–1.5 log reduction was obtained for *E. coli* K12 (added artificially), with a total process time of 1.1 s, although this period did not include the time taken to introduce and withdraw the product.

*Pressurised steam pasteurisation of produce*
Although specifically designed to treat broiler carcasses, the vacuum/steam/vacuum (VSV) system described by Kozempel *et al.* (2000, 2001, 2003a, b) was also evaluated on fruits and vegetables (cantaloupe, grapefruits, papaya and beets). The system was shown to effectively reduce counts of inoculated *L. innocua* and APCs by 3–5 log CFU $ml^{-1}$ of rinse, but successful application of the process was product specific (Kozempel *et al.*, 2002). To prevent thermal damage to the produce, process times had to be restricted to less than 2 s, with optimal treatments being steam temperature of 138–143°C for 0.1–0.2 s. Other produce, namely, bananas, cauliflower, peppers, and broccoli, could not be treated using this process without damage (Kozempel *et al.*, 2002).

In general, high pressure steam pasteurisation systems are batch processes, although continuous systems are possible. A continuous steam treatment tunnel producing superheated steam above 120°C has been developed (Fig. 8.3) and is in continuous use with a herb processor in the UK (Bulut *et al.*, 2006). The product temperature quickly rises to 100°C in about 5 s upon entering the heat treatment chamber and the temperature reaches a maximum of 106°C within 9 s before it leaves the heat treatment chamber. The mean numbers of total viable microorganisms are typically 4–7 log CFU $g^{-1}$ on the raw product and heat treatment consistently reduces this to below 3 log CFU $g^{-1}$ for all the herbs. With final bacterial levels of this magnitude, the herbs can be used as fresh garnishes on high-risk dishes and as ingredients in microbiologically sensitive situations. As the continuous steam treatment unit inactivates enzymes, it also helps to prevent the darkening of the herbs and deterioration of the flavour and aroma. Additionally, as the enzyme inactivation occurs efficiently in a very short time, the unit could be used for blanching of herbs or other leafy vegetables before freezing to improve the quality of the frozen products and to prolong the shelf-life.

### 8.3.4 Hot water

Hot water may be applied by spraying (Fig. 8.4) at high or low pressures, deluging with cascading sheets of hot water, or by immersion (dipping)

© Woodhead Publishing Limited, 2012

**Fig. 8.3** Continuous pressurised steam pasteurisation unit used for commercial treatment of fresh herbs.

**Fig. 8.4** Hot water spraying of turkey carcasses.

of the product. The temperature of the water at the surface of the treated product and the method of applying the water are the two most important factors in removing microbes.

© Woodhead Publishing Limited, 2012

According to Sofos and Smith (1998), each procedure has advantages and disadvantages. Immersion may be more applicable to smaller products (poultry carcasses, fruits and vegetables), while spraying at high pressures may remove visible soil but not achieve the desired high temperatures and may generate aerosols. Low pressure rinsing yields higher tissue temperatures, while deluging with hot water achieves higher temperatures throughout irregularly shaped products.

Some studies, however, have shown that a deluge method of application, where the food is passed under a waterfall, offers a more effective method of coverage (Davey and Smith, 1989).

*Spray applications for hot water*

Extensive studies have been carried out on meat carcass spray treatments by Bailey (1971), Kelly *et al.* (1981, 1982), Anderson *et al.* (1987), and Smith and Davey (Davey, 1988; Davey and Smith, 1989). Together, these studies provide essential information on the parameters that affect the effectiveness of spray treatments. Physical parameters include spray pressure and flow rate, and nozzle type, configuration and the angle of spray. In addition, other variables such as tissue type or temperature of treatment also affect the efficacy of decontamination procedures. Many of these have been discussed in Pordesimo *et al.*'s (2002) review of spray treatment; however, this review concentrates mainly on relatively recent US-based studies. Naturally raising the temperature of the water increases the reduction in microbial counts. However, there are practical problems in using hot water, for instance, a spray-jet water rapidly loses heat by evaporation. Studies have shown that the maximum impact temperature on the carcass from a spray placed 30 cm away and supplied with water at 90°C is approximately 63°C (Bailey, 1971). This problem has led to the development of various cabinet systems (Graham *et al.*, 1978).

Hot water spray pasteurisation of meat

Extensive trials to design an automated beef carcass spray treatment unit were conducted by Anderson and co-workers at the United States Department of Agriculture Agricultural Research Service (USDA-ARS) and the Department of Food Science and Nutrition at the University of Missouri (Anderson *et al.*, 1984). From their work, a commercial in-line spray treatment/sanitising system, the Carcass Acquired Pathogen Elimination Reduction (CAPER) system was developed. Design specifications of a commercial CAPER system unit were described by Anderson *et al.* (1984); the chamber included both an initial treatment unit and a final sanitising unit using water sprays incorporating a sanitising agent.

Graham *et al.* (1978) evaluated a fully enclosed spray cabinet system for sheep carcasses capable of handling over 300 sheep carcasses per hour with mean reductions of 2.5 log units achieved for both total aerobic and coliform counts. Treated carcasses had a slightly cooked appearance, but

© Woodhead Publishing Limited, 2012

reverted to normal after overnight chilling at 1–4°C. There were no significant differences in weight losses between treated and control carcasses. Powell and Cain (1987) adapted this system to a beef slaughter line. Surface impact temperatures were approximately 7°C below that of supply temperatures at the maximum throughput studied (135 carcasses/h). At slower rates the temperature difference was smaller. Carcass size also had an effect on surface temperature.

The effectiveness of a commercial hot water cabinet for treating pig carcasses was evaluated by Gill *et al.* (1998). Treatments used water at 85°C for 10 s. APC, coliform and *E. coli* counts on pig carcasses were reduced by >1, >2 and >2 log, respectively. There was little overall change in the appearance of carcasses. However, cut muscle surfaces were substantially degraded.

Since 2000, hot water decontamination of pig carcasses from MRDT104 ((R-type ACSSuT) *S.* Typhimurium DT104) infected herds has been mandatory in Denmark (Aabo, 2007). Carcasses are subjected to showering with 80°C hot water for 15 s. Aabo reported that in 2007 only one Danish abattoir was equipped with a hot water decontamination system and that this abattoir dealt with all the animals from infected pig herds in Denmark (accounting for approximately 1% of Danish pig carcasses).

Hauge *et al.* (2011) have recently evaluated a commercial hot water spray system for the surface pasteurisation of lamb carcasses. The 8 s treatment sprayed 134 l of water at 82°C ±1°C over every three carcasses in an enclosed cabinet. Levels of naturally occurring *E. coli*, *Enterobacteriaceae*, *B. cereus* and APC were all significantly reduced by the treatment in comparison to untreated carcasses. *E. coli* was isolated from 66% of control carcasses and from 26% of pasteurised carcasses. *E. coli* CFU reduction was 99.5%, corresponding to a reduction of 1.85 log CFU per carcass. After 24 h storage, the reduction in *E. coli* was increased to 2.02 log, and after five days *E. coli* could not be isolated from the pasteurised carcasses. As with other studies some initial discolouration of the pasterised meat was seen, but this dissipated after 24 h of chilled storage.

Relatively few studies have been published on hot water spray treatment of poultry. Thomson *et al.* (1974) spray treated chicken carcasses with water at 20 or 30 psi at 21.1, 34.4, 60.0, 65.6 and 71.1°C. The numbers of surviving microbes decreased as the temperature of the water increased. Water at 65.6°C or higher was required to reduce the mean bacterial count by about 1 log unit. There were no significant effects of water pressure. Berrang *et al.* (2000) investigated spray at 73°C (on the carcass) for 20 s, 30 min after defeathering, and spray at 71°C (on the carcass) for 20 s immediately after defeathering. Overall, neither of the treatments produced significant reductions in *Campylobacter*, coliform or *E. coli* counts. Li *et al.* (2002) did find a significant effect of using hot water in an inside-outside bird washer. Water spray treatments of 55 and 60°C for 12 s reduced inoculated numbers of *C. jejuni* by at least 0.78 log units per carcass compared to a 20°C treatment.

© Woodhead Publishing Limited, 2012

The skin colour of the chicken carcasses was not found to be significantly affected by treatment temperatures of below 60°C.

Hot water spray pasteurisation of produce

Fallik (2004) describes a hot water spray system for cleaning of fruit that combines water sprays with a roller conveyor comprising rotary brushes that move the produce through sprays whilst also providing a mechanical brushing to remove physical surface detritus and other contaminants. An initial ~10 s cold rinse stage removes gross soiling, produce then takes 10–25 s to pass through the spray tunnel with water temperatures of 48–63°C dependent on produce type and cultivar. The final stage is forced air drying for up to 2 min. Reductions in APCs of 3–4 log cycles are reported compared to spray washed and brushed fruit (Fallik *et al.*, 2000; Porat *et al.*, 2000; Smilanick *et al.*, 2003).

*Deluge application*

An alternative approach to spraying is to use a deluge, or waterfall, method of hot water distribution. The advantages of the deluge method over spray stem largely from the reduced surface area of the continuous vertical water wall for reduced heat and evaporative losses and the absence of spray nozzles, which can be subject to fouling (Davey, 1989).

Such a system was developed in laboratory evaluations by Davey and Smith (Davey, 1989, 1990; Davey and Smith, 1989). This system was compared directly with the spray cabinet used by Powell and Cain (1987) and studies showed that greater reductions could be obtained with the deluge system. The deluge system was also shown to be the more cost effective system (by a factor of three), in terms of both achieving bacterial reductions and lower capital costs. A predictive model for processing conditions was developed based on the pilot data (Davey, 1989, 1990). Predictions indicated that, for a production rate of 135 sides/h, a 1 and 2 log reduction in bacteria could be achieved at the site of minimum treatment (the neck); with meat surface temperatures of around 65°C and 80°C, respectively; at added costs of less than 0.016% and 0.03%, respectively, of the value of the meat. It was reported in 1995 (Anon, 1995) that in response to the USDA's proposal to introduce a mandatory antimicrobial treatment before chilling carcasses, the Australian Meat Research Council had funded Australian Meat Technology to develop the deluge system commercially. A cabinet was reported to have been undergoing on-line trials in a Queensland export abattoir and that evaluations during full commercial operations indicated that water could be recycled 'while still retaining potability, and satisfactory organoleptic and microbiological quality' for both the water and the meat. Operating at 79°C (surface temperature of about 76°C), and a contact time of about 17 s, average reductions of 2 log units in *E. coli* numbers were quoted.

A system similar to the cabinet developed by Davey and Smith was used by Gill *et al.* (Gill and Badoni,1997; Gill *et al.*, 1998) for treating

© Woodhead Publishing Limited, 2012

pork carcasses before evisceration but after singeing and polishing. Since dehairing and polishing distributes microorganisms over the whole surface of the carcass, the unit was designed to deliver sheets of water from both above and below. Overall, results were very similar to those of Davey and Smith. A fully commercial version of the apparatus was constructed and evaluated under commercial conditions over three months and further trials (Gill and Badoni, 1997). Carcasses were processed at rates between 600 and 800 carcasses $h^{-1}$ with each carcass being subjected to hot water at 85°C for 15 s. An assessment of the dressing process showed that much of the subsequent *E. coli* contamination occurred during the operation for opening the throat and the floor of the mouth (Gill and Badoni, 1997). Further work by Gill *et al.* (1998) indicated that pasteurisation (85°C for 30 s) should be delayed until after this operation. This reduced counts by around 2 log in comparison with untreated carcasses. However, the treatment bleached the exposed muscle along the edges of the cut lines.

*Immersion applications*

Hot water immersion is a simple pasteurisation process that has a number of advantages which include relative ease of use, short treatment times, reliable monitoring of product and water temperatures, and some washing effect that reduces physical surface detritus and contaminants (Fallik, 2004). This simplicity gives a substantially lower equipment capital cost than for other forms of thermal surface pasteurisation.

Hot water immersion pasteurisation of meat

In general, immersion systems have been more readily applied to poultry carcasses than red meat carcasses, although since pig carcasses are regularly scalded in tanks systems the technology certainly exists for treating red meat carcasses, although immersing a beef side may not be practical.

Smith and Graham (1978) found immersion in hot water to be a very effective method of reducing bacteria on the surface of meat. Immersing whole sheep carcasses in water at 80°C for 10 s reduced bacterial numbers by more than 1 log CFU $cm^{-2}$. Although the carcasses had a 'cooked' appearance immediately after treatment, normal surface colour returned almost completely during storage overnight in a chiller. In trials by C. James *et al.* (2000), immersion in water at 90°C for 8 s and treatment with condensing steam at 100°C produced very similar reductions. Neither system was optimised, but it was suggested that agitating the water would have probably improved the immersion treatment. The authors concluded that inherently the use of atmospheric steam was the more attractive option for commercial use, due to its simplicity. They considered that commercial application of the immersion system presented a number of problems in terms of filtering, recirculating the water, and final disposal of the water. The buoyancy of the lamb carcasses also posed a particular practical problem.

Decontaminating poultry meat by using immersion methods has been

© Woodhead Publishing Limited, 2012

studied by a number of researchers, with some degree of success (Dawson *et al.*, 1963; Pickett and Miller, 1966; Avens and Miller, 1972; Cox *et al.*, 1974a,b; Teotia and Miller, 1972; Thomson *et al.*, 1974, 1979; Morrison and Fleet, 1985; de Ledesma *et al.*, 1996; Berrang *et al.*, 2000; Li *et al.*, 2002). As would be expected, it appears from the literature that treatments are more effective at higher temperatures and with longer immersion or exposure times. Up to 2 log reductions have been obtained in some of these studies (Pickett and Miller, 1966; Cox *et al.*, 1974a; Morrison and Fleet, 1985). Although substantial reductions may be achieved with water temperatures exceeding 60–65°C, such treatments can result in a partially cooked product, with browning of the flesh and tightening of the skin of the carcass.

Studies by a number of workers have shown that immersion of poultry carcasses for pasteurisation by heat treatment causes swelling of skin and tissues, and absorption of bacteria into deep tissues, making them even more difficult to remove in subsequent washing (Notermans and Kampelmacher, 1974; McMeekin and Thomas, 1978; Thomas and McMeekin, 1982; Lillard, 1989).

Purnell *et al.* (2004) developed and evaluated an experimental in-line processing unit for poultry carcasses, using hot water immersion. Treatment at 75°C for 30 s significantly reduced APCs and counts of Enterobacteriaceae and campylobacters, but the skin tended to tear during trussing. However, treating carcasses at 70°C for 40 s, followed by a 12–15°C spray-chill treatment for 13 s, did not detrimentally affect the skin. Microbial counts remained significantly lower than the controls for 8 days under typical chill-storage conditions.

Further work compared both atmospheric steam and hot water treatments, under laboratory and commercial conditions, using bespoke pilot-scale equipment (James *et al.*, 2007; Corry *et al.*, 2007). In experimental studies, whole chicken carcasses, inoculated with *C. jejuni* and *Escherichia coli* K12, were treated with hot water in a pilot immersion system for 20–30 s at 75 and 80°C. A reduction of 1.3 log CFU $cm^{-2}$ in counts of *E. coli* was achieved using a 20 s, 80°C treatment. A 1.66 log CFU $cm^{-2}$ reduction in *C. jejuni* was achieved by a 30 s, 75°C treatment. Trials in a commercial poultry plant with naturally contaminated carcasses used a 20 s immersion in hot water at 80°C. The appearance of the treated carcasses was assessed visually at intervals until the end of shelf-life, and checks made for pseudomonads, Enterobacteriaceae and campylobacters on breast skin. Initial levels of *Campylobacter* spp. were low (~1 $log_{10}$ CFU $cm^{-2}$) and variable. Numbers of campylobacters were reduced, but not eliminated. Reductions of about 2 log cycles were obtained in aerobic counts. Visual assessment indicated that the hot water treatment caused less change in appearance than a steam treatment. It was concluded that carcasses treated with either hot water or steam could easily be used for the production of 'skin-off' portions. It was considered that changes to appearance of skin-on carcasses or portions would be acceptable to many consumers.

© Woodhead Publishing Limited, 2012

Hot water immersion pasteurisation of seafood

Liew *et al.* (1998) evaluated the effects of hot water (~99°C) immersion for 10 or 30 s on the survival of intrinsic flora (*Vibrio* spp.) on cockles. Microbial levels were reduced from 5.7 log CFU $g^{-1}$ to 3.2 log and <1 $log_{10}$ CFU $g^{-1}$, respectively. Cockles inoculated with *V. cholerae* at an initial level of 6 log CFU $g^{-1}$ were also reduced to <1 log CFU $g^{-1}$ after 30 s of heat-treatment (Liew *et al.*, 1998).

Hot water immersion pasteurisation of produce

Hot water immersion has been successfully used on many fruits, primarily for the removal of plant pathogens and spoilage organisms, however, the physiological responses for each fruit type vary dependent on the cultivars, season, growing location, time of harvesting and previous treatment (Fallik, 2004).

Produce evaluated for hot water immersion pasteurisation treatments has included apple, cherry, grapefruit, lemon, mango, melon, papaya, pear and tomato (Parish *et al.*, 2003). Whilst surface changes in colour and textural quality are reported, hot water immersion is often used where the outer skins or rinds of the produce are removed in later processing. A reduction in *E. coli* on the surface of oranges of up to 5 log CFU $cm^{-2}$ is reported by Pao and Davis (1999) for a 70°C, 2 min or 80°C, 1 min hot water immersion treatment. However, the effects were variable over the surface with a lesser reduction at the stem end of the fruit.

Many authors have established that microorganisms can penetrate fruit skins, including pathogens. This migration can be exacerbated with immersion in warm water and pathogens have been shown to penetrate into tomatoes (Bartz, 1982; Bartz and Showalter,1981; Showalter, 1979; Zhuang *et al.*, 1995), apples (Buchanan *et al.*, 1999; Burnett *et al.*, 2000) and mangoes (Anon, 2000). However, these temperatures (15–22°C) and times (~10 min) are unlikely to occur in thermal pasteurisation equipment operating correctly. This penetration has been shown to be greater if the organisms can enter through a cut surface (Takeuchi and Frank, 2000).

Warm water washes in combination with chlorination were investigated by Delaquis *et al.* (1999) for antimicrobial effect on shredded lettuce. Three prime parameters of wash temperature, duration and pH were varied for wash water containing 100 ppm chlorine. Total aerobic populations, psychrotrophs, *Pseudomonas* spp., yeasts and moulds, and lactic acid bacteria were enumerated, and freshness of appearance after 7 days storage was assessed. For temperatures in excess of 45°C, reductions between two and three log cycles were obtained, giving a substantial improvement over the antimicrobial effect of cold chlorinated water, which rarely produces reductions greater than 1 log CFU $g^{-1}$. This effect may only be partially due to the thermal aspects, as increased temperatures are known to improve antimicrobial efficacy of chlorine solutions (Weber and Levine, 1944). Acidifications of the water between pH 8 and pH 6 did not significantly

© Woodhead Publishing Limited, 2012

affect the results. Assessment of 'freshness' showed quality degradation for non-treated and extended duration treatments with a peak freshness for shredded lettuce washed for 3 min at 47°C in water containing 100 ppm total chlorine. Development of the spoilage microflora was retarded by several days under these conditions, with enzymatic browning being delayed until the end of the storage period.

Whilst extension of storage life has been the main driver for hot water treatment of produce, some work with pathogens has taken place with Fleischman *et al.* (2001) reporting reductions of more than 5 log of *Escherichia coli* O157:H7 for apples immersed for 15 s in hot water at 80 and 95°C.

Hot water immersion pasteurisation of eggs

While the washing of eggs is carried out as standard in North America the practice is not permitted at present in the EU for Class 'A' eggs (the class most commonly found at retail). The general reasoning for this is that it is 'considered preferable to produce a clean, quality egg in the first place' (MAFF, 2000). However, microbes, including pathogens, can be present in high numbers even on visibly clean eggs. A number of studies have been published on the use of low temperature (<60°C) water bath pasteurisation processes to fully pasteurise whole shell eggs and a commercial system has been developed (Stadelman *et al.*, 1996; Hou *et al.*, 1996; Schuman *et al.*, 1997). However, by nature this process takes a long time and can result in some detrimental changes to quality and properties of the egg contents. A few studies (Gast, 1993; Himathongkham *et al.*, 1999; James *et al.*, 2002) have investigated the applicability of hot water treatments for surface pasteurising shell eggs, with some success. Himathongkham *et al.* (1999) reported that dipping eggs for three seconds in boiling water resulted in complete destruction of *Salmonella* Enteritidis in shells and membranes, but sometimes caused eggs to crack, while James *et al.* (2002) showed that temperatures sufficient to achieve significant reductions in bacterial numbers can be attained on the outside of an egg without raising interior temperatures to those that would cause coagulation of the egg contents. The water immersion trials showed that at 95°C eggs could be subjected to the conditions for as long as 10 s. No problems of egg cracking were encountered under any of the treatments they used.

## 8.4 Dry heat pasteurisation

While many techniques have involved the application of wet heat (in the form of steam or hot water) in order to decontaminate foods, few have utilised dry heat.

© Woodhead Publishing Limited, 2012

### 8.4.1 Hot air

The use of rapid desiccation by means of dry heat from a forced-air heater has been studied by Cutter *et al.* (1997). This laboratory study on beef carcasses examined combinations of desiccation at 400 or 300°C, before inoculation of carcasses with bovine faeces, with subsequent washing in water at 35°C. Combined treatments were found to be more effective than washing alone. The authors postulated that the desiccation process affected the meat surface in such a way that the degree of bacterial attachment was reduced.

A limited study (Corry *et al.*, 2003) at the University of Bristol used an experimental, pilot-scale system to evaluate controlled, hot-air heating of food surfaces (James and James, 2003). Results showed large reductions in counts of added *C. jejuni* and *E. coli* from chicken skin subjected to 15 min treatments with high velocity (15 m $s^{-1}$) warm air at 10, 40, 50 and 60°C. Reductions of 1–2 log units at 20 and 40°C indicated that there was some form of non-thermal drying effect. However, these data have not been fully validated.

A hot air treatment has been evaluated for treating eggs (Pasquali *et al.*, 2010). Two 8 s shots of air at 600°C with an interval of 30 s of cold air has been demonstrated to reduce inoculated *S.* Enteritidis numbers by up to 1.9 log. No significant changes to any of the quality traits tested were recorded.

### 8.4.2 Infra-red

Infra-red (IR) radiation lies between microwave and visible light in the electromagnetic spectrum. IR radiation causes food molecules to vibrate so quickly that they generate heat, which is transferred to adjacent molecules. Since IR needs no secondary conduit or contact to transfer heat energy, there is far less loss of energy compared to heating by convection or conduction. This makes IR radiation more efficient for delivering high energy quickly to food. Unlike microwaves, IR has little penetration, making it particularly suitable for rapidly heating surfaces. IR technology is used throughout the food industry for cooking and drying many foods.

Huang (2004) achieved mean reductions of inoculated *L. monocytogenes* of 3.5, 4.3 and 4.5 log cycles for turkey frankfurters with treatments raising surface temperatures to 70, 75 and 80°C respectively. No noticeable physical damage to treated products was apparent. Although surface browning occurred during treatment, this dissipated after a few hours chilled storage with no significant changes to $L^*$, $a^*$, $b^*$ colour attributes.

IR heating of almonds has been shown by Bingol *et al.* (2011) to reduce inoculated *Pediococcus* by up to 5 log CFU $g^{-1}$ without altering the organoleptic or colour characteristics. Initial experiments determined 120°C to be the maximum temperature before irreversible surface change (roasting) occurred. Batches of inoculated almonds were then heated to 100, 110 and 120°C with a catalytic IR heater and then held at temperatures 30°C lower than the peak temperature for periods of up to 1 h. Holding at 90°C for

© Woodhead Publishing Limited, 2012

10–15 min significantly ($p > 0.01$) reduced the *Pediococcus* population by more than 5 log CFU $g^{-1}$, and holding at 80°C for >22 minutes provided a >4 $\log_{10}$ CFU $g^{-1}$ reduction. Holding at 70°C for as long as 60 min did not provide satisfactory reductions.

Sequential treatments of 30 s IR heating and 30 s ultraviolet (UV) irradiation have been shown to inactivate *Rhodotorula mucilaginosa* cells on fig fruits with little unfavourable effects on surface colour, hardness score, and respiration of fruits during subsequent storage (Hamanaka *et al.*, 2011). In contrast, single treatments with IR heating or UV irradiation had little effect. The authors hypothesised that the DNA damage caused by UV irradiation and the inhibition of its repair might be enhanced by the thermal energy of IR heating to a sublethal level, since the temperature monitored during IR heating was considerably lower than the lethal level of *R. mucilaginosa* cells.

## 8.5 Selecting the right method

The shapes of many foods are not ideal for thermal surface pasteurisation treatments. Thermal surface pasteurisation treatments rely on physical contact and uniform coverage of the food surface. This is often difficult, since the surfaces of many foods are irregular. Particular foods such as whole poultry carcasses pose specific problems, such as within the abdominal cavity that is reached through a relatively small opening. Also, the outer surfaces of many foods have crevices and folds. These areas are very difficult to treat and tend to provide protection for attached microorganisms. They slow down the penetration of both liquid and gaseous treatments. As well as protecting microorganisms, these areas often acquire physical debris and do not drain well. Accumulation of water lying in these areas can have a detrimental effect on the visual quality of the food and cause difficulties in controlling the treatment contact time.

There is much evidence that the time after harvest or slaughter at which the food is treated greatly affects the efficacy of the thermal pasteurisation treatment. The longer microorganisms reside on the food surface, the more difficult removal or inactivation becomes, because of their ability to attach to tissue. However, organisms differ in their ability to attach to different surfaces and in the time they require to become fully attached. The formation of biofilms may increase microbial resistance to destructive treatments.

There is rarely any distinction in the literature between 'methods' (the method of applying a treatment) and 'treatments' (the treatment that is applied). This often clouds the practical issues; too much emphasis is often placed on the treatment rather than the method of application. A thermal surface pasteurisation process is not simply a matter of applying steam, or spraying the product with hot water. For example, many factors affect the efficiency of aqueous spray systems. In automated spray cabinets, the position

© Woodhead Publishing Limited, 2012

and number of the sprays, the shape of each spray and spray pressure, all have a significant effect on the treatment, irrespective of the nature of the fluid being pumped through the sprays. Many studies have indicated that the method of application is often more important than the treatment itself.

The design of product support and transfer systems are seldom mentioned in the literature. In most thermal surface pasteurisation systems (except for perhaps water immersion), the product is held by clamps or similar as it is presented to the heating medium. The areas where a support system touches the product will experience less heating effect than the fully exposed surfaces.

## 8.6 The limitations of technologies and the challenges to adoption of surface pasteurisation by thermal methods

Heat treatments are very reliant on the method of application. To prevent cooking of the product, such treatments have to provide a uniform heating of all surfaces for a short period. This is not particularly difficult to achieve on a laboratory scale by spraying or dipping small samples, using hot water, for example. Similarly, laboratory studies using steam have shown that, if very high temperatures are applied for very short times, followed by rapid cooling of the surface, large reductions in microbial contamination can be achieved, without affecting the appearance of the meat. However, successfully applying such techniques in processing plants at industrial production rates presents many engineering challenges. The greatest challenge to the adoption of surface thermal pasteurisation treatments is the difficulty in applying enough heat to all surfaces of the food for sufficient time to cause a significant reduction in microbial numbers without cooking the surface of the food.

## 8.7 Conclusions and future trends

There is great interest in 'minimal' or 'invisible' food preservation techniques that will improve shelf-life and food safety whilst preserving the 'fresh' appearance/quality of the food in question. Thermal surface pasteurisation is one of many such techniques for decontaminating food. As a physical treatment it offers advantages over chemical and biological (bacteriophages, chitosan) based decontamination alternatives, as it leaves no residues and is relatively non-specific. In comparison with non-thermal physical technologies most thermal treatments are cheaper and more readily adapted to current processing line operations.

A key challenge of thermal surface pasteurisation is how to achieve the highest log reduction of microbial populations on the food without adversely affecting the sensory quality of the food. The treatments that maximise microbial reduction, without significant quality changes, need to be identified, along with

© Woodhead Publishing Limited, 2012

the development of engineering expertise to produce the required conditions consistently over the surface of the food being treated, at industrial rates of throughput. While some effective commercial systems are now available, there is much evidence that these systems still require further development to achieve full efficiency. Of the thermal methods, steam or hot water are the most common methods applied although infra-red and air heating can also be applied.

Another key challenge, and opportunity, is regulatory. The adoption of a 'zero tolerance' requirement in the USA for *E. coli* has effectively forced their red meat industry to introduce various decontamination interventions. In response to US policy, many countries that export to the US market have also adopted such systems. In the EU, however, there has been a fear that the use of such treatments would lead to the 'disproportionate reliance on the decontamination step and consequent reduction of the process hygiene' (Hugas and Tsigarida, 2008). Current EU Regulations (EC) No. 852/2004, (EC) No. 853/2004 and (EC) No. 854/2004 only allow the use of potable water on foods of animal origin (red meat, poultry and fish). While a debate continues in the EU on permitting chemicals (EFSA, 2010a), it has been recognised that there is no restriction on physical treatments (with the exception of ionising and UV radiation ((EC) No. 854/2004) such as steam and hot water, and recent EFSA opinions support the use of such treatments (EFSA, 2010b, 2011). This may lead to a wider adoption of such treatments throughout the EU in the future.

Finally, there is the economic hurdle to cross. Introducing an additional processing step to any process will always have a cost implication. It is debatable whether consumers are willing to pay more for safer food, when they believe, logically, that the food they have been buying is already safe. While improving food safety is, rightly, an important driver to governmental food safety authorities, there may not be a direct cost benefit to food processors. However, processes that eliminate pathogens may also produce a substantial reduction in the numbers of spoilage organisms and hence an extension of product storage life. This would help the economics of food production, allowing longer production runs, delivery to more distant markets and reduced waste. Reduced spoilage rates will permit the use of (typically slower) more carbon neutral global transportation such as surface shipping in preference to airfreight. Despite the obvious advantages to the industry as a whole and the consumer, such thermal decontamination systems will only be implemented if they are robust, reliable and have low capital and running costs.

## 8.8 Sources of further information and advice

Good general reviews of contamination and decontamination issues have been published in recent years by Corry and Mead (1996), Bolder (1997),

© Woodhead Publishing Limited, 2012

Dorsa (1997), Sofos *et al.* (1999) and Bjerklie (2000). An overview of contamination and decontamination issues involving poultry meat has been produced by Smulders (1999). Recent reviews of hot water systems for thermal decontamination of fruits and produce are provided by Parish *et al.* (2003) and Fallik (2004), and meat by Aymerich (2008). Useful books include those edited by Gould (1995), Smulders (1987) and Ellerbroek (1999).

## 8.9 References

AABO S (2007), 'Control of *Salmonella* in pork by decontamination', *New Food*, 2, 54–55.

AFEK U and ORENSTEIN J (2003), 'Decreased sweetpotato decay during storage by steam treatments', *Crop Protection*, 22, 321–324.

AFEK U, ORENSTEIN J and NURIEL E (1999), 'Steam treatment to prevent carrot decay during storage', *Crop Protection*, 18, 639–642.

ANDERSON M E, MARSHALL R T, STRINGER W C and NAUMANN H D (1979), 'Microbial growth on plate beef during extended storage after washing and sanitising', *Journal of Food Protection*, 42(5), 389–392.

ANDERSON M E, NAUMANN H D and COOK N K (1984), 'Design specifications of a red meat carcass washing and sanitising unit', presented at the 1984 Winter Meeting of the American Society of Agricultural Engineers, Paper No. 84-6546.

ANDERSON M E, HUFF H E, NAUMANN H D, MARSHALL R T, DAMARE J M, PRATT M and JOHNSTON R (1987), 'Evaluation of an automated beef carcass washing and sanitising system under production conditions', *Journal of Food Protection*, 50(7), 562–566.

ANON. (1995), 'Australian hot water pasteurisation (i.e. decontamination) cabinet being commercialised', *Connect*, 1/2, May, p. 4, Publication of Australian Meat Technology Pty Ltd.

ANON. (2000), 'DBMD traces Salmonella outbreak to mangoes', *CDC/NCID Focus* 9(4), 1–2. Available from: <http://www.cdc.gov/ncidod/focus/index.htm> (accessed 6 September 2001).

AVENS J S and MILLER B F (1972), 'Pasteurisation of turkey carcasses', *Poultry Science*, 51, 1781.

AVENS J S, ALBRIGHT S N, MORTON A S, PREWITT B E, KENDALL P A and SOFOS J N (2002), 'Destruction of microorganisms on chicken carcasses by steam and boiling water immersion', *Food Control*, 13, 445–450.

AYMERICH T, PICOUET P A and MONFORT J M (2008), 'Decontamination technologies for meat products', *Meat Science*, 78, 114–129

BAILEY C (1971), 'Spray washing of lamb carcasses', *Proceedings of the 17th European. Meeting of Meat Research Workers*, Bristol, Paper B16, 175–181.

BALÁ M F A, PODOLAK R and MARSHALL D L (1999), 'Steam treatment reduces the skin microflora population on deheaded and eviscerated whole catfish', *Food Microbiology*, 16, 495–501.

BARKATE M L, ACUFF G R, LUCIA L M and HALE D S (1993), 'Hot water decontamination of beef carcasses for reduction of initial bacterial numbers', *Meat Science*, 35, 397–401.

BARTZ J A (1982), 'Infiltration of tomatoes immersed at different temperatures to different depths in suspensions of *Erwinia carotovora* subsp. Carotovora', *Plant Disease*, 66(4), 302–306.

BARTZ J A and SHOWALTER R K (1981), 'Infiltration of tomatoes by aqueous bacterial suspensions', *Phytopathology*, 71(5), 515–518.

© Woodhead Publishing Limited, 2012

BERRANG M E, DICKENS J A and MUSGROVE M T (2000), 'Effects of hot water application after defeathering on the levels of *Campylobacter*, coliform bacteria, and *Escherichia coli* on broiler carcasses', *Poultry Science*, 79, 1689–1693.

BINGOL G, YANG J, BRANDL M T, PAN Z, WANG H and MCHUGH T H (2011), 'Infrared pasteurization of raw almonds', *Journal of Food Engineering*, 104, 387–393.

BJERKLIE S (2000), 'Intervention overview', *Meat Processing: North American Edition*, June, 94, 96–97.

BOLDER N M (1997), 'Decontamination of meat and poultry carcasses', *Trends in Food Science and Technology*, 8, 221–227.

BREMER P J, MONK I, OSBOURNE C M, HILLS S and BUTLER R (2002), 'Development of a steam treatment to eliminate *Listeria monocytogenes* from king salmon (*Oncorhynchus tshawytscha*)', *Journal of Food Science*, 67(6), 2282–2287.

BUCHANAN R L, EDELSON S G, MILLER R L and SAPERS G M (1999), 'Contamination of intact apples after immersion in an aqueous environment containing *Escherichia coli* O157:H7', *Journal of Food Protection*, 62(5), 444–450.

BULUT S, PURNELL G, JAMES C, TAYLOR G and HERBERT R (2006), 'Continuous steam sterilisation of fresh herbs', *Food Factory of the Future 3*, Gothenburg, Sweden, 7–9 June.

BURNETT S L, CHEN J and BEUCHAT L R (2000), 'Attachment of *Escherichia coli* O157:H7 to the surfaces and internal structures of apples as detected by confocal scanning laser microscopy', *Applied and Environmental Microbiology*, 66(11), 4679–4687.

CASTILLO A, HARDIN M D, ACUFF G R and DICKSON J S (2002), 'Reduction of microbial contaminants on carcasses', in Juneja V K and Sofos J N, *Control of Foodborne Microorganisms*, New York, Marcel Dekker, 351–381.

CORRY J E L and MEAD G C (1996), *Microbial Control in the Meat Industry: 3. Decontamination of Meat*, Concerted Action CT94–1456, Bristol, University of Bristol Press.

CORRY J E, JAMES C, O'NEILL D, YAMAN H, KENDALL A and HOWELL M (2003), 'Physical methods, readily adapted to existing commercial processing plants, for reducing numbers of campylobacters, on raw poultry', Proceedings of the 12th International Workshop on Campylobacter, Helicobacter and Related Organisms, Aarhus, Denmark, *International Journal of Medical Microbiology*, 293, 32.

CORRY J E L, JAMES S J, PURNELL G, PINTO C S, CHOCHOIS Y, HOWELL M and JAMES C (2007), 'Surface pasteurisation of chicken carcasses using hot water', *Journal of Food Engineering*, 79, 913–919.

COX N A, MERCURI A J, THOMSON J E and GREGORY D W (1974a), 'Quality of broiler carcasses as affected by hot water treatments', *Poultry Science*, 53, 1566–1571.

COX N A, MERCURI A J, JUVEN B J, THOMSON J E and CHOW V (1974b), 'Evaluation of succinic acid and heat to improve the microbiological quality of poultry meat', *Journal of Food Science*, 39, 985–987.

CUTTER C N, DORSA W J and SIRAGUSA G R (1997), 'Rapid desiccation with heat in combination with water washing for reducing bacteria beef carcass surfaces', *Food Microbiology*, 14, 493–503.

DAVEY K R (1988), 'A novel cabinet for the hot water decontamination of sides of beef', *Proceedings of the 34th International Congress of Meat Science and Technology*, Brisbane, Australia, Part B, 642–645.

DAVEY K R (1989), 'Theoretical analysis of two hot water cabinet systems for decontamination of beef sides', *International Journal of Food Science and Technology*, 24, 291–304.

DAVEY K R (1990), 'A model for the hot water decontamination of sides of beef in a novel cabinet based on laboratory data', *International Journal of Food Science and Technology*, 25, 88–97.

DAVEY K R and SMITH M G (1989), 'A laboratory evaluation of a novel hot water cabinet for the decontamination of sides of beef', *International Journal of Food Science and Technology*, 24, 305–316.

© Woodhead Publishing Limited, 2012

DAWSON L E, MALLMANN W L, BIGBEE D G, WALKER R and ZABIK M E (1963), 'Influence of surface pasteurisation and chlortetracycline on bacterial incidence on fryers', *Food Technology*, 17, 100–104.

DE LEDESMA A M R, RIEMANN H P and FARVER T B (1996), 'Short-time treatment with alkali and/or hot water to remove common pathogenic and spoilage bacteria from chicken wing skin', *Journal of Food Protection*, 59, 746–750.

DELAQUIS P J, STEWART S, TOIVONEN P M A and MOYLS A L (1999), 'Effect of warm, chlorinated water on the microbial flora of shredded iceberg lettuce', *Food Research International*, 32, 7–14.

DORSA W J (1997), 'New and established carcass decontamination procedures commonly used in the beef-processing industry', *Journal of Food Protection*, 60, 1146–1151.

EFSA PANEL ON BIOLOGICAL HAZARDS (BIOHAZ) (2010a), 'Guidance on Revision of the joint AFC/BIOHAZ guidance document on the submission of data for the evaluation of the safety and efficacy of substances for the removal of microbial surface contamination of foods of animal origin intended for human consumption', *EFSA Journal*, 8(4), 1544.

EFSA PANEL ON BIOLOGICAL HAZARDS (BIOHAZ) (2010b), 'Scientific Opinion on the safety and efficacy of using recycled hot water as a decontamination technique for meat carcasses', *EFSA Journal*, 8(9), 1827.

EFSA PANEL ON BIOLOGICAL HAZARDS (BIOHAZ) (2011), 'Scientific Opinion on *Campylobacter* in broiler meat production: control options and performance objectives and/or targets at different stages of the food chain', *EFSA Journal*, 9(4), 2105.

ELLERBROEK L (ed.) (1999), *COST Action 97 – Pathogenic Micro-organisms in Poultry and Eggs. 8. New Technology for Safe and Shelf-stable Products*, Luxembourg, Office for Official Publications of the European Communities.

EUROPEAN CHILLED FOOD FEDERATION (ECFF) (2006), *Recommendations for the Production of Prepackaged Chilled Food*, European Chilled Food Federation.

EVANS J A (1999), 'Novel decontamination treatments for meat', in Ellerbroek L, *COST Action 97 – Pathogenic Micro-organisms in Poultry and Eggs. 8. New Technology for Safe and Shelf-stable Products*, Luxembourg, Office for Official Publications of the European Communities, 14–21.

EVANS J A and BROWN T (1999), *Decontamination of meat and meat products using combinations of condensation at sub-atmospheric pressure and organic acids*, Final report for EU FAIR project ERBFAIR-CT-1027.

EVELEIGH K (2000), 'Carcass decontamination', *Meat and Poultry 2000 – Seminar 1*, Campden & Chorleywood Food Research Association, 17 July 2000.

FALLIK E (2004), 'Prestorage hot water treatments (immersion, rinsing and brushing)', *Postharvest Biology and Technology*, 32, 125–134.

FALLIK E, AHARONI Y, COPEL A, RODOV R, TUVIA-ALKALAI S, HOREV B, YEKUTIELI O, WISEBLUM A and REGEV R (2000), 'A short hot water rinse reduces postharvest losses of "Galia" melon', *Plant Pathology*, 49, 333–338.

FLEISCHMAN G J, BATOR C, MERKER R and KELLER S E (2001), 'Hot water immersion to eliminate *Escherichia coli* O157:H7 on the surface of whole apples: thermal effects and efficacy', *Journal of Food Protection*, 64, 451–455.

GAN-MOR S, REGEV R, LEVI A and ESHEL D (2011), 'Adapted thermal imaging for the development of postharvest precision steam-disinfection technology for carrots', *Postharvest Biology and Technology*, 59, 265–271.

GARCIA J M, AGUILERA C and ALBI M A (1995), 'Postharvest heat treatment on Spanish strawberry (Fragaria x ananassa cv. Tulda)', *Journal of Agricultural Food Chemistry*, 43, 1489–1492.

GAST R K (1993), 'Immersion in boiling water to disinfect egg shells before culturing egg contents for *Salmonella enteritidis*', *Journal of Food Protection*, 56(6), 533–535.

GAZE J E, BOYD A R and SHAW H L (2006), 'Heat inactivation of *Listeria monocytogenes* Scott A on potato surfaces', *Journal of Food Engineering*, 76, 27–31.

© Woodhead Publishing Limited, 2012

GILL C O and BADONI M (1997), 'The effects of hot water pasteurizing treatments on the appearances of pork and beef', *Meat Science*, 46(1), 77–87.

GILL C O, JONES T and BADONI M (1998), 'The effects of hot water pasteurizing treatments on the microbiological conditions and appearances of pig and sheep carcasses', *Food Research International*, 31:4, 273–278.

GILL C O, BRYANT J and BEDARD D (1999), 'The effects of hot water pasteurizing treatments on the appearances and microbiological conditions of beef carcass sides', *Food Microbiology*, 16, 281–289.

GILL C O, BRYANT J and BADONI M (2001), 'Effects of hot water pasteurizing treatments on the microbiological condition of manufacturing beef used for hamburger patty manufacture', *International Journal of Food Microbiology*, 63, 243–256.

GÖKSOY E O, JAMES C, CORRY J E L and JAMES S J (2001), 'The effect of hot water immersions on the appearance and microbiological quality of skin-on chicken breast pieces', *International Journal of Food Science and Technology*, 36(1), 61–69.

GOULD G W (ed.) (1995), *New Methods of Food Preservation*, London, Blackie Academic and Professional, Chapman and Hall.

GRAHAM A, CAIN B P and EUSTACE I J (1978), 'An enclosed hot water spray cabinet for improved hygiene of carcass meat', CSIRO Meat Research Report No. 11/78.

HAMANAKA D, NORIMURA N, BABA N, MANO K, KAKIUCHI M, TANAKA F and UCHINO T (2011), 'Surface decontamination of fig fruit by combination of infrared radiation heating with ultraviolet irradiation', *Food Control*, 22, 375–380.

HAUGE S J, WAHLGREN M, RØTTERUD O-J and NESBAKKEN T (2011), 'Hot water surface pasteurisation of lamb carcasses: microbial effects and cost-benefit considerations', *International Journal of Food Microbiology*, 146, 69–75.

HIMATHONGKHAM S, RIEMANN H and ERNST R (1999), 'Efficacy of disinfection of shell eggs externally contaminated with *Salmonella* Enteritidis, implications for egg testing', *International Journal of Food Microbiology*, 49, 161–167.

HOU H, SINGH R K, MURIANA P M and STADELMAN W J (1996), 'Pasteurisation of intact shell eggs', *Food Microbiology*, 13, 93–101.

HUANG L (2004), 'Infrared surface pasteurization of Turkey frankfurters', *Innovative Food Science and Emerging Technologies*, 5(3), 345–351.

HUGAS M and TSIGARIDA E (2008), 'Pros and cons of carcass decontamination: the role of the European Food Safety Authority', *Meat Science*, 78, 43–52.

JAMES C, THORNTON J A, KETTERINGHAM L and JAMES S J (2000), 'Effect of steam condensation, hot water or chlorinated hot water immersion on bacterial numbers and quality of lamb carcasses', *Journal of Food Engineering*, 43, 219–225.

JAMES C, LECHEVALIER V and KETTERINGHAM L (2002), 'Surface pasteurisation of shell eggs', *Journal of Food Engineering*, 53, 193–197.

JAMES C, JAMES S J, HANNAY N, PURNELL G, PINTO C S, YAMAN H, ARAUJO M, GONZALEZ M L, CALVO J, HOWELL M and CORRY J E L (2007), 'Decontamination of poultry carcasses using steam or hot water in combination with rapid cooling, chilling or freezing of carcass surfaces', *International Journal of Food Microbiology*, 114, 195–203.

JAMES S J and EVANS J A (2006), 'Predicting the reduction in microbes on the surface of foods during surface pasteurisation – the "BUGDEATH" project', *Journal of Food Engineering*, 76, 1–6.

JAMES S J and JAMES C (2003), 'Thermal decontamination of food', *Proceedings of the Institute of Refrigeration*, 100, 1–12.

JAMES S J, BROWN T, EVANS J A, JAMES C, KETTERINGHAM L and SCHOFIELD I (1998), 'Decontamination of meat, meat products and other foods using steam condensation and organic acids', *Proceedings of 3rd Karlsruhe Nutrition Symposium 'European Research towards Safer and Better Food'*, Karlsruhe, Germany, 175–185.

JAMES S J, BROWN T, CORRY J, EVANS J A, JAMES C, KETTERINGHAM L and PURNELL G (2000), 'Surface pasteurisation of food using steam condensation', *IChemE Food & Drink*

© Woodhead Publishing Limited, 2012

*2000. Processing solutions for innovative products*, London, Institution of Chemical Engineers, UK, 22–25.

JAY J M, LOESSNER M J and GOLDEN D A (2005), *Modern Food Microbiology*, New York, Springer.

KAISER C, LEVIN J and WOLSTENHOLME B N (1995), 'Vapour, heat and low pH dips improve litchi (*Litchi chinensis Sonn.*) pericarp colour retention', *Journal of South African Society of Horticultural Science*, 95(1), 7–10.

KELLY C A, DEMPSTER J F and MCLOUGHLIN A J (1981), 'The effect of temperature, pressure and chlorine concentration of spray washing water on numbers of bacteria on lamb carcasses', *Journal of Applied Bacteriology*, 51, 415–424.

KELLY C A, LYNCH J F and MCLOUGHLIN A J (1982), 'The effect of spray washing on the development of bacterial numbers and storage life of lamb carcasses', *Journal of Applied Bacteriology*, 53, 335–341.

KLOSE A A and BAYNE H G (1970), 'Experimental approaches to poultry meat surface pasteurisation by condensing vapours', *Poultry Science*, 49, 504–511.

KLOSE A A, KAUFMAN V F, BAYNE H G and POOL M F (1971), 'Pasteurisation of poultry meat by steam under reduced pressure', *Poultry Science*, 50, 1156–1160.

KOZEMPEL M, GOLDBERG N, RADEWONUK E R and SCULLEN O J (2000), 'Commercial testing and optimization studies of the surface pasteurization process of chicken', *Journal of Food Process Engineering*, 23, 387–402.

KOZEMPEL M, GOLDBERG N, RADEWONUK E R and SCULLEN O J (2001), 'Modification of the VSV surface pasteurizer to treat the visceral cavity and surfaces of chicken carcasses', *Journal of Food Science*, 66, 954–959.

KOZEMPEL M, RADEWONUK E R, SCULLEN O J and GOLDBERG N (2002), 'Application of the vacuum/steam/vacuum surface intervention process to reduce bacteria on the surface of fruits and vegetables', *Innovative Food Science and Emerging Technologies*, 3, 63–72.

KOZEMPEL M, GOLDBERG N and CRAIG J C (2003a), 'The vacuum/steam/vacuum process', *Food Technology*, 57, 30–33.

KOZEMPEL M, GOLDBERG N, DICKENS J A, INGRAM K D and GRAIG J C (2003b), 'Scale-up and field test of the vacuum/steam/vacuum surface intervention process for poultry', *Journal of Food Process Engineering*, 26, 447–468.

LEWIS R J, BALDWIN A, O'NEILL T, ALLOUSH H A, NELSON S M, DOWMAN T and SALISBURY V (2006), 'Use of *Salmonella enterica* serovar Typhimurium DT104 expressing lux genes to assess, in real time and *in situ*, heat inactivation and recovery on a range of contaminated food surfaces', *Journal of Food Engineering*, 76, 41–48.

LI Y, YANG H and SWEM B L (2002), 'Effect of high-temperature inside-outside spray on survival of *Campylobacter jejuni* attached to prechill chicken carcasses', *Poultry Science*, 81, 1371–1377.

LIEW W S, LEISNER J J, RUSUL G, RADU S and RASSIP A (1998), 'Survival of *Vibrio* spp. including inoculated *V. cholerae* 0139 during heat-treatment of cockles (*Anadara granosa*)', *International Journal of Food Microbiology*, 42, 167–173.

LILLARD H S (1989), 'Factors affecting the persistence of *Salmonella* during the processing of poultry', *Journal of Food Protection*, 52, 829–832.

LIN L Y, YU T H, WU C M and LEE M H (1992), 'Changes of volatile compounds of mandarin peels during steam cooking', *Journal of Chinese Agricultural Chemical Society*, 30, 110–118.

MAFF (MINISTRY OF AGRICULTURE, FISHERIES and FOOD) (2000), 'Eggs and poultrymeat – frequently asked questions', available at: http://www.maff.gov.uk/foodrin/poultry/epfaq.htm (accessed December 2000).

MARTIN-DIANA A B, RICO D, BARRY-RYAN C, FRIAS J M, HENEHAN G T M and BARAT J M (2007), 'Efficacy of steamer jet-injection as alternative to chlorine in fresh-cut lettuce', *Postharvest Biology and Technology*, 45, 97–107.

© Woodhead Publishing Limited, 2012

MCCANN M S, SHERIDAN J J, MCDOWELL D A and BLAIR I S (2006), 'Effects of steam pasteurisation on *Salmonella* Typhimurium DT104 and *Escherichia coli* O157:H7 surface inoculated onto beef, pork and chicken', *Journal of Food Engineering*, 76, 32–40.

MCMEEKIN T A and THOMAS C T (1978), 'Retention of bacteria on chicken skin after immersion in bacterial suspension', *Journal of Applied Bacteriology*, 46, 383–387.

MORGAN A I, RADEWONUK E R and SCULLEN O J (1996a), 'Ultra high temperature, ultra short time surface pasteurisation of meat', *Journal of Food Science*, 61, 1216–1218.

MORGAN A I, GOLDBERG N, RADEWONUK E R and SCULLEN O J (1996b), 'Surface pasteurization of raw poultry meat by steam', *Lebensmittel-Wissenschaft und -Technologie*, 29, 447–451.

MORRISON G J and FLEET G H (1985), 'Reduction of *Salmonella* on chicken carcasses by immersion treatments', *Journal of Food Protection*, 48, 939–943.

NOTERMANS S and KAMPELMACHER E H (1974), 'Attachment of some bacterial strains to the skin of broiler chickens', *British Poultry Science*, 15, 573–585.

NOTERMANS S and KAMPELMACHER E H (1975), 'Heat destruction of some bacterial strains attached to broiler skin', *British Poultry Science*, 16, 351–361.

NUTSCH A L, PHEBUS R K, RIEMANN M J, SCHAFER D E, BOYER J E, WILSON R C, LEISING J D and KASTNER C L (1997), 'Evaluation of a steam pasteurisation process in a commercial beef processing facility', *Journal of Food Protection*, 60, 485–492.

PAO S and DAVIS C L (1999), 'Enhancing microbiological safety of fresh orange juice by fruit immersion in hot water and chemical sanitizers', *Journal of Food Protection*, 62(7), 756–760.

PARISH M E, BEAUCHAT L R, SUSLOW T V, HARRIS L J, GARRETT E H, FARBER J N and BUSTA F F (2003), 'Methods to reduce/eliminate pathogens from fresh and fresh-cut produce', *Comprehensive Reviews in Food Science and Food Safety*, 2, 161–173.

PASQUALI F, FABBRI A, CEVOLI C, MANFREDA G and FRANCHINI A (2010), 'Hot air treatment for surface decontamination of table eggs', *Food Control*, 21, 431–435.

PETERSEN M A (1993), 'Infuence of sous vide processing, steaming and boiling on vitamin retention and sensory quality in broccoli florets', *Zeit Lebensmittel Untersuchung Forschung*, 197, 375–380.

PHEBUS R K, NUTSCH A L, SCHAFER D E, WILSON R C, RIEMANN M J, LEISING J D, KASTNER C L, WOLF J R and PRASAI R K (1997), 'Comparison of steam pasteurisation and other methods for reduction of pathogens on freshly slaughtered beef surfaces', *Journal of Food Protection*, 60, 476–484.

PICKETT L D and MILLER B F (1966), 'The effect of high temperature immersion on the microbial population of whole turkey carcasses', *Poultry Science*, 45, 1116.

PORAT R, DAUS A, WEISS B, COHEN L, FALLIK E and DROBY S (2000), 'Reduction of postharvest decay in organic citrus fruit by a short hot water brushing treatment', *Postharvest Biology and Technology*, 18, 151–157.

PORDESIMO L O, WILKERSON E G, WOMAC A R and CUTTER C N (2002), 'Process engineering variables in the spray washing of meat and produce', *Journal of Food Protection*, 65(1), 222–237.

POTTY V H (2011), '"New" emerging technology for pasteurisation for better food safety', available from: http://foodtechupdates.blogspot.com (accessed March 2011).

POWELL V H and CAIN B P (1987), 'A hot water decontamination system for beef sides', *CSIRO Food Research Quarterly*, 47(4), 79–84.

PURNELL G, MATTICK K and HUMPHREY T (2004), 'The use of "hot wash" treatments to reduce the number of pathogenic and spoilage bacteria on raw retail poultry', *Journal of Food Engineering*, 62, 29–36.

SCHEERLINCK N, MARQUENIE D, JANCSÓK P T, VERBOVEN P, MOLES C G, BANGA J R and NICOLAÏ B (2004), 'A model-based approach to develop periodic thermal treatments for surface decontamination of strawberries', *Postharvest Biology and Technology*, 34, 39–52.

© Woodhead Publishing Limited, 2012

SCHIRRA M and D'HALLEWIN G (1997), 'Storage performance of Fortune mandarins following hot water dips', *Postharvest Biology and Technology*, 10, 229–237.

SCHUMAN J D, SHELDON B W, VANDEPOPULIERE J M and BALL JR H R (1997), 'Immersion heat treatments for inactivation of *Salmonella enteritidis* with intact eggs', *Journal of Applied Microbiology*, 83, 438–444.

SHOWALTER R K (1979), 'Postharvest water intake by tomatoes', *Horticultural Science*, 14(2), 125.

SINGH R P, AGNIHOTRI V P, RAYCHAUDHURI S P and VERMA J P (1987), 'Thermotherapy of sugarcane for disease control', *Review of Tropical Plant Pathology*, 4, 305–329.

SMILANICK J L, SORENSON D, MANSOUR M, AIEYABEI J and PLAZA P (2003), 'Impact of a brief postharvest hot water drench treatment on decay, fruit appearance, and microbe populations of California lemons and oranges', *Horticultural Technology*, 13, 333–338.

SMITH M G (1992), 'Destruction of bacteria on fresh meat by hot water', *Epidemiology and Infection*, 109, 491–496.

SMITH M G and DAVEY K R (1990), 'Destruction of *Escherichia coli* on sides of beef by a hot water decontamination process', *Food Australia*, 42(4), 195–198.

SMITH M G and GRAHAM A (1978), 'Destruction of *Escherichia coli* and salmonellae on mutton carcasses by treatment with hot water', *Meat Science*, 2, 119–128.

SMULDERS F J M (ed.) (1987), *Elimination of Pathogenic Organisms from Meat and Poultry*, Amsterdam, Elsevier.

SMULDERS F J M (1999), 'Contamination and decontamination of foods of animal origin, with special reference to poultry meat', in Ellerbroek L, *COST Action 97 – Pathogenic Micro-organisms in Poultry and Eggs. 8. New Technology for Safe and Shelf-stable Products*, Luxembourg, Office for Official Publications of the European Communities, 4–13.

SOFOS J N and SMITH G C (1998), 'Non-acid meat decontamination technologies: model studies and commercial applications', *International Journal of Food Microbiology*, 44, 171–188.

SOFOS J N, BELK K E and SMITH G C (1999), 'Processes to reduce contamination with pathogenic microorganisms in meat', *Proceedings, 45th International Conference of Meat Science and Technology, Yokohama, Japan*, 45, 596–605.

STADELMAN W J, SINGH R K, MURIANA P M and HOU H (1996), 'Pasteurisation of eggs in the shell', *Poultry Science*, 75, 1122–1125.

STUMBO C R (1973), *Thermobacteriology in Food Processing*, 2nd edn, New York, Academic Press.

TAKEUCHI K and FRANK J F (2000), 'Penetration of *Escherichia coli* O157:H7 into lettuce tissues as affected by inoculum size and temperature and the effect of chlorine treatment on cell viability', *Journal of Food Protection*, 63, 434–440.

TEOTIA J S and MILLER B F (1972), 'Destruction of *Salmonella* on turkey carcass skin with chemicals and hot water', *Poultry Science*, 51, 1878.

THOMAS C J and MCMEEKIN T A (1982), 'Effect of water immersion on the microtopography of the skin of chicken carcasses', *Journal of the Science of Food and Agriculture*, 33, 549–554.

THOMAS C T, WHITE J C and LONGREE K (1966), 'Thermal resistance of Salmonellae and Staphylococci in foods', *Applied Microbiology*, 14, 815–820.

THOMSON J E, COX N A, WHITEHEAD W K and MERCURI A J (1974), 'Effect of hot spray washing on broiler carcass quality', *Poultry Science*, 53, 946–952.

THOMSON J E, BAILEY J S and COX N A (1979), 'Phosphate and heat treatments to control *Salmonella* and reduce spoilage and rancidity on broiler carcasses', *Poultry Science*, 58, 139–143.

VALDRAMIDIS V P, GEERAERD A H, GAZE J E, KONDJOYAN A, BOYD A R, SHAW H L and VAN IMPE J F (2006), 'Quantitative description of *Listeria monocytogenes* inactivation kinetics with temperature and water activity as the influencing factors; model prediction

© Woodhead Publishing Limited, 2012

and methodological validation on dynamic data', *Journal of Food Engineering*, 76, 79–88.

VAZ-PIRES P, CAPELL C and KIRBY R (1994), 'Low-level heat treatment to extend shelf-life of fresh fish', *International Journal of Food Science and Technology*, 29, 405–413.

WEBER S J and LEVINE W W (1944), 'Factors affecting germicidal efficiency of chlorine and chloramine', *American Journal of Public Health*, 32, 719–722.

WHYTE P, MCGILL K and COLLINS J D (2003), 'An assessment of steam pasteurization and hot water immersion treatments for the microbiological decontamination of broiler carcasses', *Food Microbiology*, 20, 111–117.

ZHUANG R-Y, BEUCHAT L R and ANGULO F J (1995), 'Fate of *Salmonella montevideo* on and in raw tomatoes as affected by temperature', *Letters in Applied Microbiology*, 22, 97–100.

© Woodhead Publishing Limited, 2012

# 9

# Microbial decontamination of food by microwave (MW) and radio frequency (RF)

**S. R. S. Dev, McGill University, Canada, S. L. Birla, University of Nebraska, USA, G. S. V. Raghavan, McGill University, Canada and J. Subbiah, University of Nebraska, USA**

**Abstract**: Thermal destruction of microorganisms has been proven to be the most effective way of decontaminating food to ensure food safety. Electromagnetic radiation in the microwave and radio frequency ranges can penetrate food materials and generate heat from within the products. By taking advantage of this property, microbiological decontamination of food materials can be accomplished. In order to exploit the benefits of this technology, a deeper understanding of the interaction of these non-ionizing electromagnetic radiations, with various food matrices, is required. Furthermore, due to the organic nature of the microorganisms along with their high moisture content within the cell constituents, there are studies showing enhanced thermal destruction kinetics in certain food materials during the application of these radiations. Despite these advantages, this technology is not very popular for food decontamination, primarily due to its relatively high capital cost and extreme non-uniformity in solid food applications. This chapter will focus on the principles of dielectric heating, application of this technology for food decontamination and their respective advantages and disadvantages.

**Key words**: MW, radio frequency, decontamination, dielectric heating.

## 9.1 Introduction

Historically, heat treatments have been the most common means of food decontamination, either through sterilization or pasteurization. These processes involve a reduction in the number of viable pathogens so that the food products are unlikely to cause disease. For liquid food materials conventional heating techniques, primarily employing the conduction and convection modes of heat transfer, are proven to be economical and effective. For solid food

© Woodhead Publishing Limited, 2012

materials, when a product has high moisture content (50–95% wet basis), the above mentioned conventional heating methods and modern techniques like infrared heating are often the most economical.

Conventional heating techniques can impart undesirable changes in product quality, as it takes longer to achieve the desired thermal treatment needed to inactivate pathogens deeper within the food. This issue is more prominent for solid food products which have intermediate (20–50% wet basis) and low (below 20% wet basis) moisture content. This is because conventional heating is extremely non-uniform when applied to solid food materials and in-package heating of liquid food materials (usually only the surface is heated and the heat must conduct to the interior). It is only possible to produce a uniform temperature distribution if the heat is applied very slowly. However, this increase in process time is undesirable in the context of food quality and energy efficiency.

There is increasing demand for convenience foods which involve in-package processing. However, most modern packaging materials have poor thermal properties, thus rendering in-package thermal processing using conventional heating techniques ineffective. In-package processing enhances food safety, as the food material will not be subjected to further handling after the thermal decontamination has been achieved. Therefore, there is a need for a volumetric heating technique that can penetrate materials with poor thermal properties and effectively heat the food to the required processing temperature. One of the main advantages of dielectric heating is that the heat is generated in the interior of the sample, thus avoiding a delay in heat transmission to the interior caused by low thermal conductivity (Orsat *et al.*, 2005).

Conventional irradiation for food decontamination employs ionizing electromagnetic radiation for microbial destruction; subsequently lysis of the cells occurs due to ionization of the organelles. In contrast, the application of electro-technology, which involves using electromagnetic radiation in the microwave (MW) and radio frequency (RF) ranges, results in thermal destruction of microorganisms in food without altering the chemical composition of the food to a significant extent. Any irreversible change in the physiological state of the food material can be attributed to the heat generated in the food and not to the electric or magnetic fields within the food material (Datta *et al.*, 2005). This technique has several unique advantages compared to conventional thermal treatments.

Dielectric heating can deliver a high temperature in a very short time, minimizing nutritional and sensorial losses to the food products. The application of dielectric heating in food processing dates back to the 1940s. Early efforts attempted to apply RF energy to cook various processed meat products, to heat bread, and to dehydrate and blanch vegetables. However, the work resulted in very few commercial installations, primarily because of the high overall operating costs associated with using RF energy in the 1940s. By the 1960s, studies into the applications of RF energy in food processing focused mainly on thawing of frozen products, which resulted

© Woodhead Publishing Limited, 2012

in several commercial production lines. The next generation of commercial applications of RF energy in the food industry was post-bake drying of cookies and snack foods, starting in the late 1980s.

Commercial dielectric heating systems for the purpose of food pasteurization or sterilization were not widely used until more recently. There are a number of reasons for this, such as the inability to ensure sterilization of the entire package, lack of suitable packaging materials, and unfavorable economics when compared with conventional thermal processing technologies. By the 1990s, due to increased food safety and quality requirements, great attention was given to the use of electromagnetic (EM) energy for pasteurization and sterilization of foodstuffs to provide high quality and safe food products.

Many explorations into different food processing techniques failed due to inadequate knowledge of the interaction of electromagnetic waves with the food matrix. The following sections discuss the fundamental aspects of dielectric heating.

## 9.2 Properties of microwave (MW) and radio frequency (RF) electromagnetic waves

RF and MW are short waves of electromagnetic energy that travel at the speed of light. They have all the basic properties of any electromagnetic radiation. They have excellent penetrating power, which is inversely proportional to their frequency (Pozar, 2005). Figure 9.1 shows the position of RF and MW on the electromagnetic spectrum.

The propagation of one complete cycle in the waveform of any electromagnetic radiation is shown in Fig. 9.2. Because RF and MW are electromagnetic radiation, they create an alternating electric field and an alternating magnetic field perpendicular to each other. This property is exploited in the thermal applications of RF and MW.

Although there is a wide range of frequencies in the electromagnetic spectrum which are classified as RF and MW frequencies, the Federal Communications Commission (FCC) in the United States reserves three RFs and two MW frequencies for industrial, scientific, and medical (ISM) applications. Table 9.1 lists frequencies designated for ISM applications, along with the tolerance percentage of the frequency and the corresponding free space wavelength of the RF and MW. Each individual frequency has its own distinct advantages for different categories of materials and scope for the scale of processing. For example, 2450 MHz is most commonly used in domestic MW ovens where materials and smaller package dimensions do not limit their application. Another frequency, 915 MHz, is suitable for industrial-scale processing of foods (Tang *et al.*, 2008). Radio frequency band 13.56 and 27.12 MHz have both been successfully used in commercial scale dehydration of textiles, paper, and some food products.

© Woodhead Publishing Limited, 2012

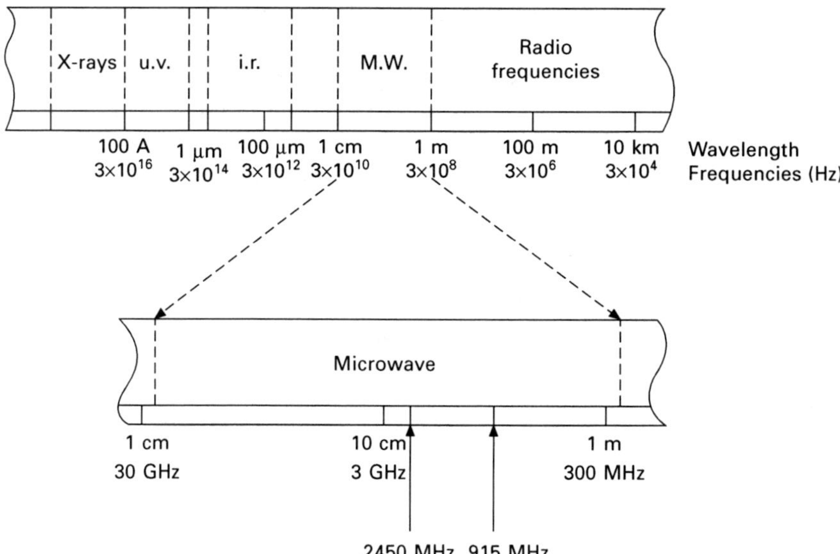


**Fig. 9.1** Locations of RF and MW on the electromagnetic spectrum.

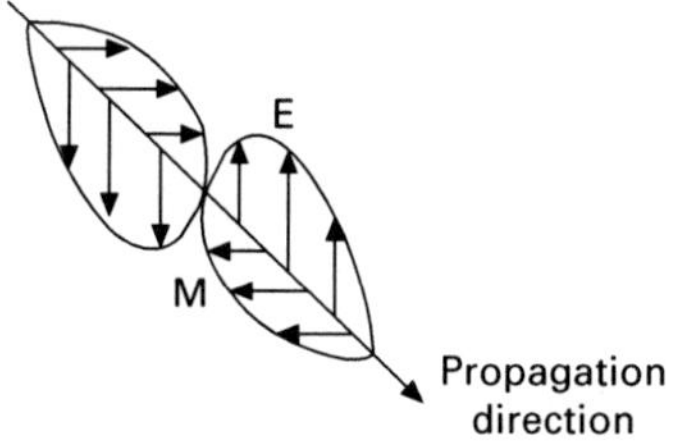


**Fig. 9.2** Electromagnetic wave propagation (E – electric field, M – magnetic field).

**Table 9.1** FCC allocated frequency bands designated for ISM applications

| Frequency spectrum | Center frequency (MHz) | Frequency tolerance (± %) | Free space wavelength (m) |
|---|---|---|---|
| Radio | 13.56 | 0.052 | 22.1 |
| | 27.12 | 0.6 | 11.1 |
| | 40.68 | 0.005 | 7.4 |
| MW | 915 | 1.4 | 0.328 |
| | 2450 | 2.0 | 0.122 |

Source: FCC, 1988.

## 9.3 Dielectric heating

The phenomenon of the dielectric heating of foods was discovered accidentally. In the late 1940s, a candy bar in the shirt pocket of an engineer softened

© Woodhead Publishing Limited, 2012

considerably when the engineer stood in front of a MW transmitter (Murray, 1958). It didn't take long for this phenomenon to be capitalized upon and in just a few years MW ovens began to appear. Although dielectric heating has been successfully applied at the industrial level in other fields, in food processing it has met with limited success.

Water, fat, and sugar molecules in foods absorb MW energy in a process called dielectric heating. Many molecules (such as those of water) are electric dipoles, meaning that they have a positive charge at one end and a negative charge at the other. Because of this, they rotate as they try to align themselves with the alternating electric field induced by the MW beam. This molecular movement creates heat as the rotating molecules hit other molecules and put them into motion. Figure 9.3 shows the dipolar nature of a water molecule with its polar energy field.

Microwave heating is most efficient on liquid water, but much less efficient on fats and sugars (which have less molecular dipole movement), and frozen water (where the molecules are not free to rotate). MW heating sometimes occurs due to rotational resonance of water molecules, which happens only at much higher frequencies, in the tens of Gigahertz.

The electromagnetic energy is absorbed into the outer layers of food in a manner somewhat similar to heat from other methods. The electromagnetic field penetrates deeper into substances that are relatively dry on the surface, as in the case of many common foods, and thus often deposits heat more deeply than other methods of heating. Also, the amount of heat lost from the surface of the food to the surrounding area is much higher than from its interior. This gives the appearance that MWs heat the food from the inside out, although they actually heat up almost every part of the food simultaneously. Depending on moisture content, the depth of the initial heat deposition may be several centimetres or more, in contrast to convection heating, which deposits heat shallowly onto the food surface (Dev *et al.*, 2008a).

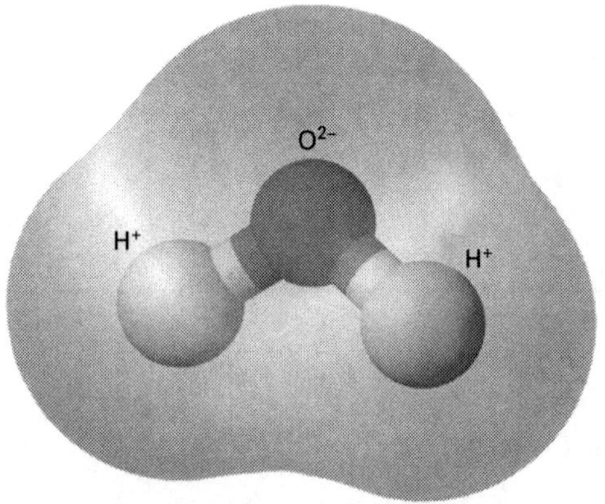


**Fig. 9.3** A dipolar water molecule with its polar energy field.

© Woodhead Publishing Limited, 2012

### 9.3.1 Principles of dielectric heating

As RF and MW are both electromagnetic radiation similar to visible light, they follow all the basic laws of physics like reflection, refraction, interference, diffraction, and polarization. The energy level of microwaves is usually measured in terms of power density (W $g^{-1}$), which corresponds to the dipolar rotational energy level of polar molecules. Therefore the interaction of MW energy with matter is through the dielectric rotation of the molecules. The intermolecular friction between the fast rotating molecules causes a rapid volumetric heating. This is a unique characteristic that differentiates dielectric heating from conventional heating methods. Only those molecules that can couple with the electromagnetic field can be heated using this technique.

The classical mechanisms of orientation and ionic polarizations are sufficient to explain dielectric heating of materials in the RF and MW range of frequencies. However, additional factors are needed to fully understand the heat generation mechanism within the materials (Roland and Kopinke, 2009). In the MW range, dipole orientation is the dominant mechanism for heat generation (Fig. 9.4), whereas in RF frequency range, ionic conduction is the dominant mechanism of heat generation (Fig. 9.5). Both dipole orientation and ionic conduction result in capacitive loss of electric energy inside the food material. These losses result in volumetric heat generation in materials,

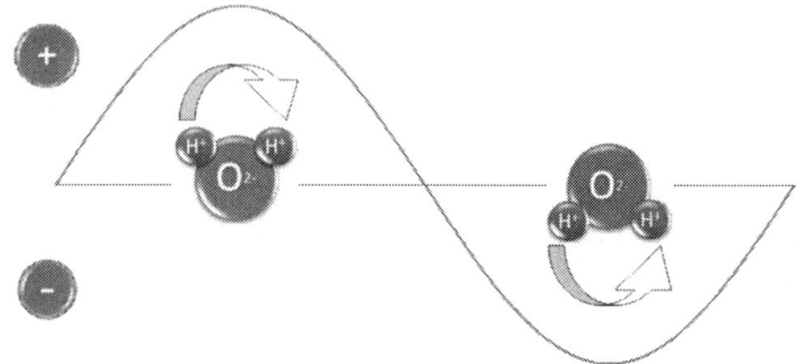


**Fig. 9.4** Dipole orientation in a MW environment.

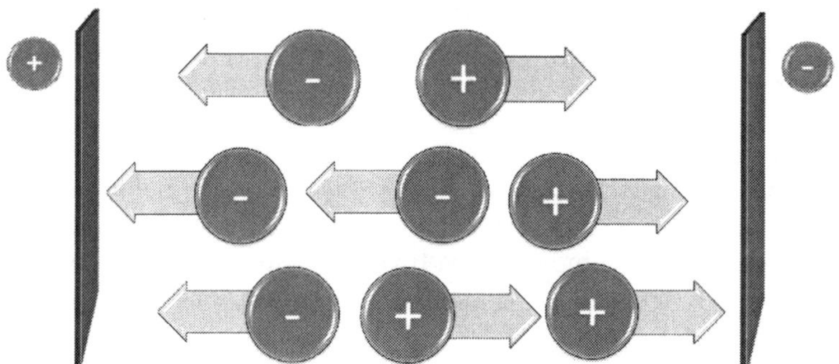


**Fig. 9.5** Ionic conduction in a parallel plate RF applicator.

© Woodhead Publishing Limited, 2012

which is a function of the electromagnetic frequency ($f$, Hz), the dielectric properties and the electric field strength ($E$, V/m).

Electrically, the complex relative permittivity ($\varepsilon^*$) is used to describe the interaction of MW and matter. The complex relative permittivity ($\varepsilon^*$) can be expressed as:

$$\varepsilon^* = \varepsilon' - j\varepsilon'' \qquad [9.1]$$

where $\varepsilon'$ is the dielectric constant, $\varepsilon''$ is the loss factor and $j$ is $\sqrt{-1}$. Dielectric properties of a material consist of two parts: real and imaginary. The 'real part' is known as the dielectric constant ($\varepsilon'$), which describes the capability of molecules to be polarized by the electric field and is directly related to the capacity of the material to hold electromagnetic energy. The 'imaginary part' is known as the dielectric loss factor ($\varepsilon''$), which indicates the efficiency of molecules to convert a fraction of stored electromagnetic energy into heat (Mingos and Baghurst, 1991). These properties are a function of the moisture content and temperature. Knowledge of dielectric properties is very important in deciding what frequency should be used for a better heating rate and uniformity.

Electromagnetic power dissipation is a strong function of electric field strength and materials' dielectric properties as shown by:

$$P_v = 2\pi f \varepsilon_0 \varepsilon'' |E|^2 \qquad [9.2]$$

where $P_v$ is the power absorbed per unit volume (W m$^{-3}$), $f$ is the frequency (Hz), $\varepsilon_0$ is the absolute permittivity of a vacuum (F m$^{-1}$), and $|E|$ is the absolute value of the electric field strength inside the load (V m$^{-1}$). As it can be seen from Eq. [9.2] the power dissipated in a certain volume is proportional to the loss factor of the matter. High moisture food materials (moisture content > 80%), usually have both their dielectric constant and dielectric loss factor close to that of pure water, and therefore have heating characteristics similar to that of pure water in an electromagnetic environment.

From Eq. [9.2], it can be inferred that the heating rate will increase as the loss factor increases. However, this is not true, as the electric field value is a function of the dielectric constant. In most cases within the MW frequencies, an increase in the dielectric loss factor can be associated with a decrease in the dielectric constant, resulting in a net decrease in the heating efficiency of the material in an electromagnetic environment. Figure 9.6 shows this trend in egg yolk for a frequency sweep of 200 MHz to 10 GHz. Ionic solutions or salty food products do not heat quickly when whole, but do benefit from surface heating in MW frequencies. This phenomenon can be observed whilst heating a salty water solution in a dielectric heating system, resulting in the container surface becoming very hot. Figure 9.7 shows that by increasing the loss factor at a fixed value for the dielectric constant, total power absorption in the material continues to increase until it reaches a specific value, after which it begins to decline.

According to Eq. [9.2], power dissipation is directly proportional to

© Woodhead Publishing Limited, 2012

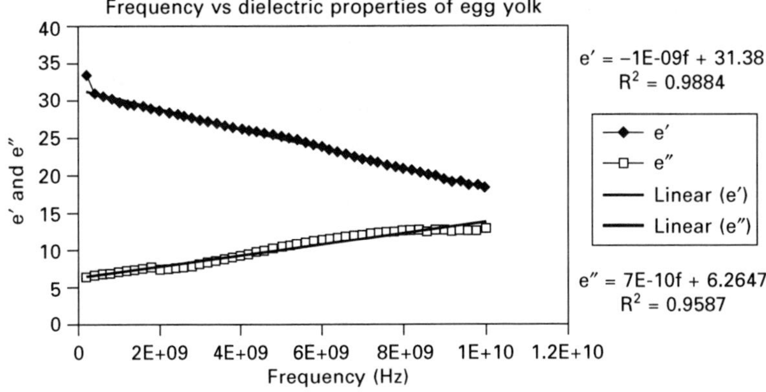


**Fig. 9.6** Change in dielectric properties of egg yolk with frequency (laboratory measured at McGill University).

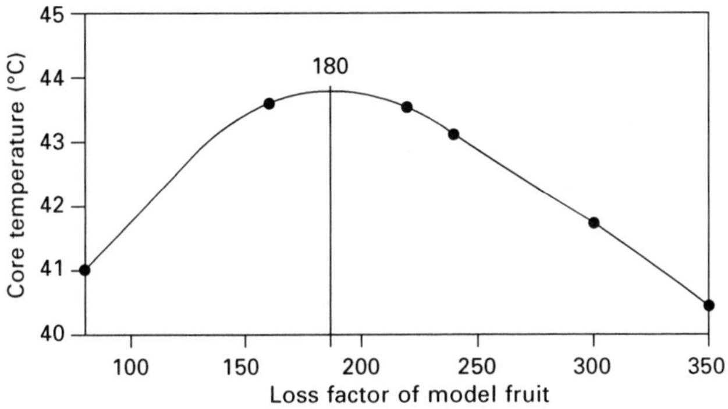


**Fig. 9.7** Simulated effects of the dielectric loss factor of the model fruit (80 mm) on the core temperature of a fruit immersed in water after RF heating for 7 min in a 195 mm electrode gap (laboratory simulated at University of Nebraska).

frequency and therefore MW should hypothetically heat a material faster than RF using similar material properties and electric field conditions. To match the power dissipation in a RF system with MW, the RF electric field needs to be increased by 10 times. The RF electric field value can come close to the air dielectric breakdown threshold (30 kV/cm) and initiate a condition of arcing in the applicator space. Under reduced air pressure, the air dielectric breakdown threshold decreases. This is one of the reasons why it is not advisable to operate RF equipment under vacuum conditions. Arcing should be taken into design consideration, especially for flammable materials.

### 9.3.2 Volumetric but non-uniform heating

The dielectric heating process suffers from two distinct forms of non-uniformity. The first one is the fundamental 'standing wave' effect. This is

© Woodhead Publishing Limited, 2012

a repeated pattern of field intensity variation within an applicator, which, for the most part, follows a half-sine pattern. In a distance of one quarter of the operating wavelength, the field intensity can change from a maximum to zero, or in one-tenth of a wavelength the intensity can change by 60%. Uniformity considerations can also play a role in the choice of frequency. The lower the frequency, the larger will be the uniform volume. This explains why RF heating is more uniform than MW.

The second type of non-uniformity arises from limited penetration of MW in food products. Penetration depth is defined as a distance from the surface of the material at which incident power is reduced to $1/e$ (about 36.79%) of the original value at the surface of the material. The penetration depth is the function of the frequency and dielectric properties of the material as shown in Eq. [9.3] (Meda *et al.*, 2005):

$$D_p = \frac{1}{2\pi f}\sqrt{\frac{2}{\mu_0\varepsilon_0\varepsilon'}\left(\sqrt{\left(1+\left(\frac{\varepsilon''}{\varepsilon'}\right)^2\right)}-1\right)} \approx \frac{\lambda_0\sqrt{\varepsilon'}}{2\pi\varepsilon''} \quad [9.3]$$

where $f$ is the frequency (Hz), $\varepsilon'$ is the dielectric constant, $\varepsilon''$ is the dielectric loss factor, $\varepsilon_0$ is the absolute permittivity of free space (F $m^{-1}$), $\mu_0$ is the permeability of free space (W $m^{-1}$) and $\lambda_0$ is the wavelength in free space. Figure 9.8 gives a graphical representation of the concept of penetration depth.

The electric field attenuates faster within the bulk of high dielectric loss materials. This is particularly troublesome for large-scale processes. Both types of non-uniformity described above are frequency dependent and become less severe with the increase in frequency. Mode mixers and turntables (mechanical methods) in home MW ovens have been designed to reduce the non-uniformity effects that stem from inherent standing wave patterns in the applicator cavity. Figure 9.9 shows the simulated temperature distribution inside an egg heated inside a domestic MW oven at 2450 MHz. The problem of the penetration depth is tackled by either switching over to a lower frequency or feeding the MW simultaneously from the top and bottom of the applicator. In both MW and RF systems, edge heating is always present due to fringe field effects. Figure 9.10 shows simulated RF power density distribution at 27.12 MHz in different object shapes. To some extent, continuous processing minimizes fringe field effects as material enters at one end and exits at the other.

## 9.4 Radio frequency (RF) and microwave (MW) interactions with food constituents

As the interaction of RF and MW with the substance it is exposed to takes place at the molecular level, the thermal and non-thermal effects produced

© Woodhead Publishing Limited, 2012

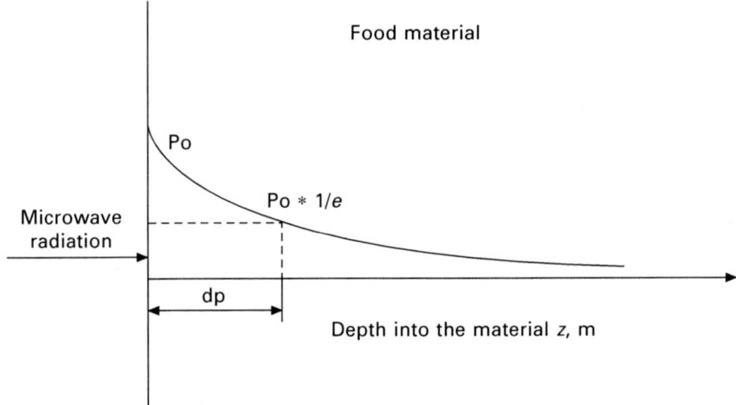


**Fig. 9.8** Graphical representation of the concept of penetration depth.

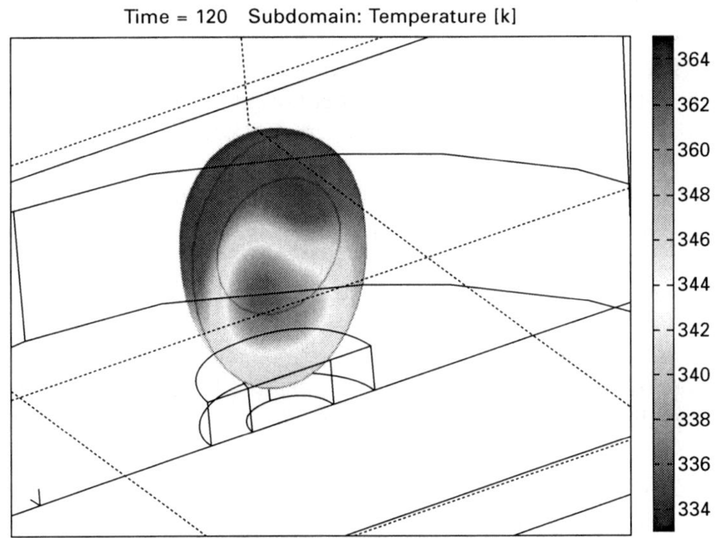


**Fig. 9.9** Simulated temperature profile of shell egg heated in a domestic MW oven at 2450 MHz using a power density of 2 W $g^{-1}$ for 120 s (Dev *et al.*, 2011).

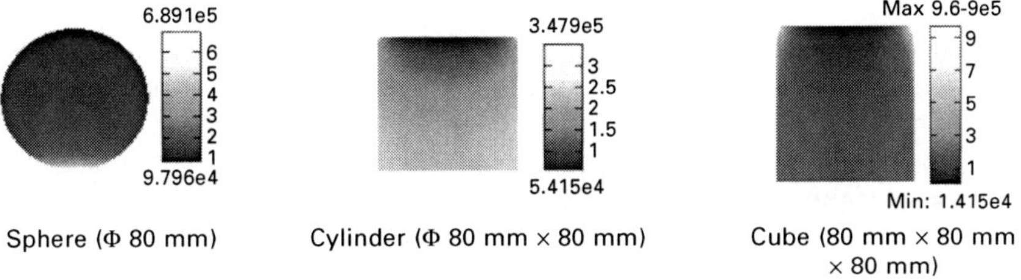


**Fig. 9.10** Simulated RF power density (W $m^{-3}$) distribution in sphere, cylinder and cube made of gellan gel (84–*j*86) placed in 300 mm electrodes in RF system operating at 27.12 MHz (laboratory simulated at University of Nebraska).

© Woodhead Publishing Limited, 2012

are considerably dependent on the molecular structure and intermolecular organization of the material. As food materials contain a high concentration of organic matter, the interaction of food constituents with RF and MW can be categorized based on their biochemical classification as carbohydrates, lipids, and proteins.

### 9.4.1 Carbohydrates

In general, carbohydrates are hydrophilic and are often present binding to moisture in food matrices. This decreases both the dielectric constant and dielectric loss factor of the food material, rendering it less reactive to the electromagnetic radiation. Nonetheless, carbohydrates have hydroxyl (–OH) groups associated with them giving them dipoles that are active in an electromagnetic environment (Roebuck *et al.*, 1972). Moreover, carbohydrates undergo a variety of chemical changes when exposed to different temperatures, resulting in highly variable dielectric properties and corresponding highly variable heating rates, affecting uniformity of heating (Datta *et al.*, 2005). Refractive index and penetration depth are key parameters in the design of MW thermal applicators, especially when heating uniformity is the major focus of the design. Carbohydrates exhibit either dextro-rotatory (d) or levo-rotatory (l) optical activity. Therefore, the refractive index and penetration depth of the electromagnetic radiation are highly variable in carbohydrates depending on the ratio of the 'd' and 'l' components present.

### 9.4.2 Lipids

Lipids may be broadly defined as a group of naturally occurring hydrophobic or amphiphilic small molecules that originate entirely or in part from two distinct types of biochemical subunits or 'building blocks': ketoacyl and isoprene groups. This includes fats, oils, waxes, cholesterol, sterols, fat-soluble vitamins (such as vitamins A, D, E, and K), monoglycerides, diglycerides, phospholipids, and others. The major biological functions of lipids are energy storage, acting as structural components of cell membranes, providing thermal insulation to the body and participating as important signaling molecules (Lira, 1996).

The hydrophobic portions of lipids do not interact significantly with electromagnetic radiation; however, the ionizable carboxyl groups of fatty acids show limited interaction. The extent of saturation of the fatty acids has no effect on the dielectric response of the lipid molecules (Mudgett and Westphal, 1989). Therefore, the dielectric properties of fats and oils are very low. The dielectric constant ($\varepsilon'$) of most lipids is between 2.5 and 3 and their dielectric loss factor ($\varepsilon''$) is in the range of 0.1–0.2. The effect of fat on the dielectric properties of food systems is mainly due to their diluting effect on the system. An increase in fat content reduces the free water content

© Woodhead Publishing Limited, 2012

in the system, which in turn reduces the dielectric properties on the whole (Datta *et al.*, 2005).

Loss factors of lipids at ambient temperature are greater in more liquid forms, such as corn oil and cottonseed oil, when compared to lard and tallow fats which are solids at that temperature. The effect of changing temperatures on the dielectric properties of lipids is not significant (Pace *et al.*, 1968). In many cases in their naturally occurring form, lipids exist as an emulsion in water (or at least the fat-containing cells are surrounded by lots of free water). Under these conditions lipids appear to heat up faster during dielectric heating, but this is because the specific heat capacity of lipids is much lower than that of water and hence they require less MW energy to heat up (Dev *et al.*, 2008a). Also, energy is absorbed from the surrounding water molecules, which efficiently convert electromagnetic energy into heat energy.

Based on the above cited dielectric properties, the refractive index of most lipids at a MW frequency of 2450 MHz is around 1.45–1.75 with an angle of refraction ranging from 35 to 45° for normal incidental MW. The penetration depth of MW for most lipids is in the range of 15–30 cm in any given direction of propagation (Duck, 1990).

### 9.4.3 Proteins

Proteins are relatively MW-inert and do not interact significantly with them. Some portions of proteins are water soluble, whereas some portions are largely insoluble (Datta *et al.*, 2005). Nonetheless, proteins mainly interact with MW on three levels conforming to their primary, secondary, and tertiary structures.

Proteins have ionizable surface regions that can bind with water or salts to give rise to zeta potential and double-layer effects associated with free surface charges (Mudgett and Westphal, 1989). These have a small but significant effect on the dielectric behavior of the proteins at MW frequencies. The solvated or hydrated form of protein, protein hydrolysates and polypeptides, are much more reactive to MW than native proteins (Datta *et al.*, 2005). The dielectric properties of proteins depend on their side chains, which can be non-polar (alanine, glycine, leucine, isoleucine, methionine, phenylalanine, and valine), or polar (in decreasing order: thyrosine, tryptophan, serine, threonine, proline, lysine, arginine, aspartic acid, aspergine, glutamic acid, glutamine, cysteine, and histidine) (Shukla and Anantheswaran, 2001).

Free amino acids are more dielectrically reactive (Pethig, 1984); they contribute to an increase in the dielectric loss factor. Since protein dipole moments are a function of their amino acids, which have their unique pH and dielectric properties, the MW reactivity of cereal, legumes, milk, meat, and fish proteins are all expected to be different. The bound water on the proteins also affects their dielectric properties (Shukla and Anantheswaran, 2001).

© Woodhead Publishing Limited, 2012

The dielectric activity of proteins can be attributed to four main aspects. These are, in decreasing order of significance (Datta *et al.*, 2005):

1. charge effects of the ionization of carboxyl, sulfhydryls and amines
2. hydrogen and ion binding as affected by pH
3. net charges on dissolved proteins
4. relaxation and conductive effects.

Such activities are important for hydrolyzed proteins and free amino acids. Since most proteins are consumed in a cooked form, it is important to determine the dielectric properties during denaturation of proteins in order to understand the MW heating of these foods (Shukla and Anantheswaran, 2001).

The refractive index and penetration depth of MW at 2450 MHz for proteins are highly variable, and depend on the type of protein and its primary, secondary and tertiary structures, along with the above mentioned factors. For example, using 2450 MHz MW, the refractive index of ovalbumin, a major protein in egg white, is around 1.5 (Barer and Tkaczyk, 1954) with an angle of refraction of roughly 40° for normally incident MW.

## 9.5 Dielectric system design and components

Special design of dielectric heating systems to address the complex heat distribution problem is possible if the food is fairly uniform in shape and composition. If the added quality to the food is tangible to the point where the added expense of MW processing can be passed along to the consumer, then an industrial process becomes more viable.

### 9.5.1 RF system

The dielectric properties of food materials change continuously with temperature. This is an important issue in the design considerations for any dielectric heating system, as this would impact the process time and heating uniformity. In RF systems, this issue becomes more important, as the food material becomes a part of the electrical circuit and the change in dielectric properties results in a change in the impedance of the system. This will have a direct impact on the energy efficiency of the system. Newer 50 $\Omega$ technology offers the promise of improved energy efficiency due to the ability of the system to tune to the optimum Debye resonance frequencies of a heated medium with the help of a matching box (Zhao *et al.*, 2000).

A typical RF heating system consists of a RF generator, an adaptor and an applicator as shown in Fig. 9.11. The applicator is the part of the system in which material is kept for heating. In general, it is made of two metal plates separated by a defined distance. The top electrode is fed with a high frequency voltage, generated by an oscillator tube. The material kept between the gap

© Woodhead Publishing Limited, 2012

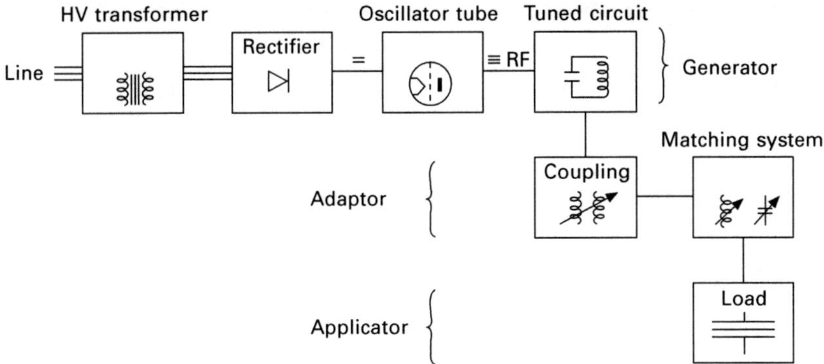


**Fig. 9.11** RF heating system components (Zhao *et al.*, 2000).

forms a capacitor. In some small RF units, an adapter called a matching network is inserted to provide better process control. The RF generator is connected to the applicator by a 50 Ω coaxial cable. The matching network automatically changes the capacitance and inductance to facilitate maximum possible power coupling. In a free oscillator system, varying the electrode gap is the only way to enable process control.

Designing a continuous process applicator for a lower frequency band is quite simple as it does not require sophisticated doors/interlocks to contain radiation. There are lots of continuous RF processing systems in use in the textiles, baking, and paper industries. Figure 9.12 depicts a commercial scale continuous RF applicator operating at 40 MHz.

### 9.5.2 MW system

A typical MW system consists of a magnetron, a waveguide and an applicator cavity. Figure 9.13 shows the schematic of an industrial MW system. The magnetron generates MW and the waveguide carries the waves to the applicator cavity. The material is kept in the applicator cavity. Since the invention of the MW oven, the 2450 MHz frequency has been used for cooking and reheating at a domestic level. Before this, the 2450 MHz frequency had been explored for the industrial scale processing of foods. Many applications and systems failed to live up to expectations, due to limited penetration depth and the inherent non-uniformity of heating caused by standing wave patterns in the multimode cavity, as in a domestic oven.

Use of multimode batch MW systems can result in uneven heating of certain products, depending on their dielectric and thermo-physical properties, as well as the system design. Seeing the problems, researchers started exploring lower frequencies, such as 915 MHz, to mitigate some of the problems that arise due to penetration depth. Edge and corner heating was still a cause of concern. One group of researchers at Washington State University (WSU) started exploring water as a surrounding medium to alleviate the problem of

© Woodhead Publishing Limited, 2012

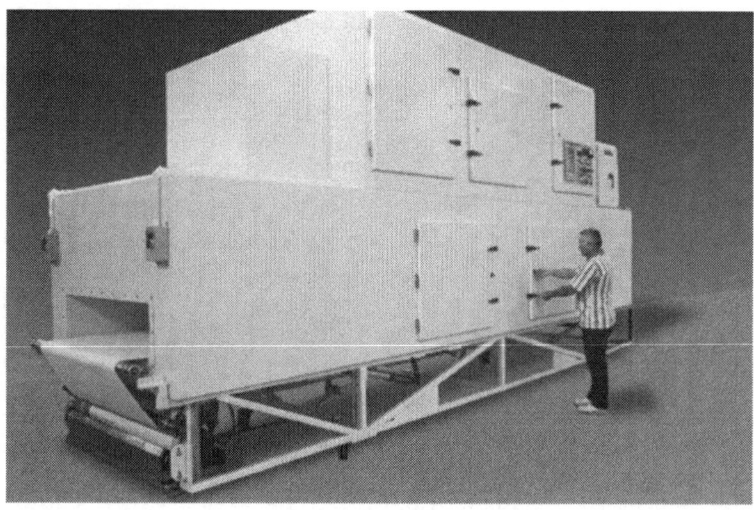

**Fig. 9.12** A commercial scale RF applicator operating at 40 MHz (courtesy: Radio Frequency Co., Millis, MA).

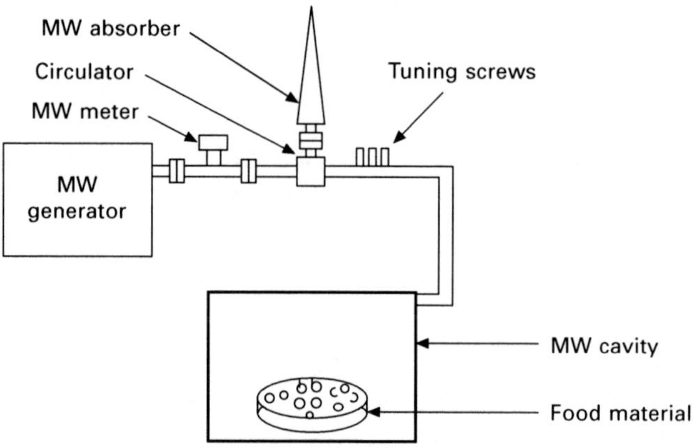


**Fig. 9.13** Components of an industrial MW system (adapted from Dev *et al.*, 2008b).

edge heating (Fig. 9.14). Another group at North Carolina State University started exploring circular focused MW energy systems for MW heating of pumpable foods (Fig. 9.15). Emerging technologies, such as continuous flow focused MW systems in resonant cavities, can alleviate uneven heating by applying an electric field with appropriate distribution (i.e., maximum at the center of the tube where velocity is high and minimum at the edges where velocity is low) to the liquid flowing through an applicator tube in the MW cavity.

In the past, many continuous MW heating systems have been designed and installed in actual food production facilities (Datta and Anantheswaran,

© Woodhead Publishing Limited, 2012

**Fig. 9.14** Continuous MW sterilization system at WSU.

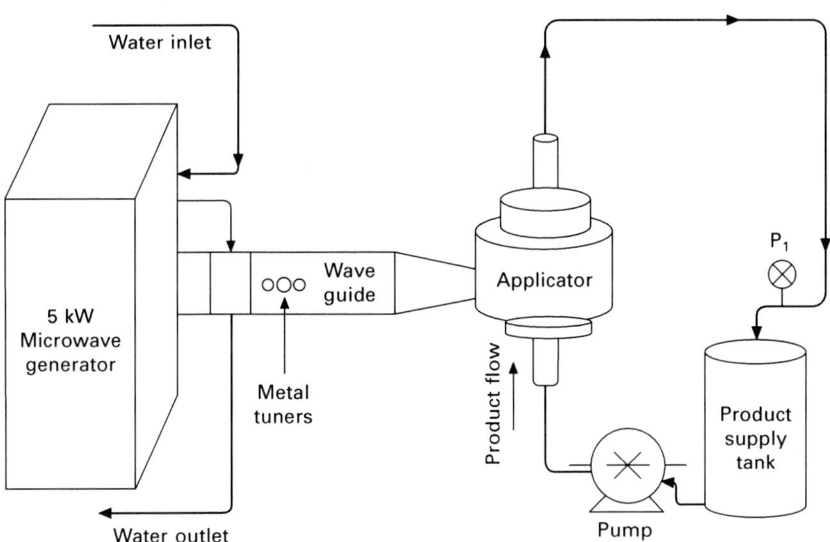


**Fig. 9.15** Schematic of the patented cylindrical MW heating system for pumpable foods (Drozd, 2010).

2001). There are a few continuous MW systems in use in Europe to sterilize ready-to-eat meals (Ramaswamy and Tang, 2008). Such systems could not meet the requirements for food safety, due to their non-uniform heating characteristics. Over the years many efforts have been made to resolve the constant nagging problem of uneven heating of food items using MW-based processes.

© Woodhead Publishing Limited, 2012

### 9.5.3 Mode of microbial inactivation

The mechanism of inactivation of microorganisms in volumetric heating processes is mainly due to thermal effects. Electromagnetic energy in RF and MW inactivates microbes via conventional thermal mechanisms, including thermal irreversible denaturation of enzymes, proteins, and nucleic acids. Most of the biochemical reactions happening in a living cell of microorganisms are catalyzed by enzymes. Enzymes are proteins that denature in their primary, secondary and tertiary structures when subjected to high temperatures. This results in loss of their active sites, thus impairing several essential biochemical activities that are essential for the survival of the microorganism. Nucleic acids, such as DNA (deoxyribonucleic acid) are essential for replication and vegetative multiplication of the microorganisms. They also denature at high temperatures, as the double-stranded DNA becomes single-stranded, due to the breakage of hydrogen bonds between bases forming the strands.

There are a few studies which have claimed to observe a non-thermal effect at 20–40 kHz frequency (Geveke and Brunkhorst, 2008). Shamis *et al.* (2008) showed that at sub-lethal temperatures, repeated exposure using high frequency MW radiation (>18 GHz) was significantly more effective in decontaminating bacteria in raw meats, compared to a single exposure. They concluded that non-thermal inactivation of pathogenic bacteria in raw meats could be achieved in defined conditions using high frequency MW radiation.

An IFT-FDA task force report has extensively elaborated the microbial mechanism and reported that no pathogen is identified as uniquely resistant to RF and MW processing methods (IFT-FDA, 2000). The microbial destruction using dielectric heating in a closed container/package and compared to an open container varies considerably. This is because steam generation in a closed container/package during treatment adds to the lethality of the treatment. If an open container is exposed to dielectric heating, the *D-z* values should be determined in the system which allows for venting of the steam during the treatment. For this purpose calculating the *D-z* values for dielectric heating processes, Chung *et al.* (2008) designed short-time come-up test cells in which both liquid and semi-solid materials with known concentrations of a microorganism of interest can be filled easily and heated using a RF or MW heating system.

## 9.6 Decontamination of foods by radio frequency (RF) and microwave (MW)

The interest in using dielectric heating to pasteurize and sterilize foodstuffs has increased as a result of the technology's capability to achieve rapid and uniform heating. Zhao *et al.* (2000) and Piyasena *et al.* (2003) extensively reviewed various applications of RF heating which had been explored up until

© Woodhead Publishing Limited, 2012

1999. More work has appeared in the literature showing several additional applications for RF heating in food processing which was summarized by Marra *et al.* (2009). Muhamad *et al.* (2010) reviewed some applications of electromagnetic waves in the processing of agriculture crops and listed work to be done in order to industrialize the process. Table 9.2 lists the effectiveness of dielectric heating processes on the food decontamination for different foodborne pathogens.

### 9.6.1 RF pasteurization and sterilization

Commercial radio frequency heating systems for the purpose of food pasteurization or sterilization are not known to be in use, although they have been researched for a number of years. Houben et al (1991) described RF pasteurization of moving sausage emulsions in tubes (inner diameter of 50 mm) made of different materials. The emulsion was pumped through a 50 mm pipe surrounded by RF electrodes. They reported a large difference in temperature of emulsion at the core and near the wall of the pipe. The material near the wall was heating faster than the material at the core. Other studies on heating performance of 1% CMC solution and water continuously flowing through a vertical pipe enveloped with RF electrodes (Zhong *et al.*, 2003, 2004) also corroborated those results.

Although the effectiveness of using radio waves to kill destructive insects in agricultural products has been known for over 70 years, the technique has rarely been applied on a commercial scale. Hallman and Sharp (1994) summarized two decades of research on the potential uses of RF/MW heating in fresh fruit disinfestation. They concluded that damage to the fruit was the major concern in commercial applications of this technology. RF treatment may hold potential for grains, pulses, nuts, dried fruits, cured tobacco and other similar commodities because of their relatively higher thermal tolerance when compared to fresh fruits. Hallman and Sharp (1994) suggested a need

**Table 9.2** Effectiveness of dielectric heating processes on food decontamination

| Target | Product | log reduction (CFU/g) | Process |
|---|---|---|---|
| *Enterococcus Streptococcus* | Vacuum packaged ham | >4.0 | 27.12 MHz; 600 W; 600 s |
| *Salmonella* Enteritidis | Fresh chicken thighs | 6.4 | 2450 MHz; 800 W; 95 s |
| *E. coli* O157:H7 | Chicken portions | 6.0 | 2450 MHz; 650 W; 35 s |
| Total microflora | Meat balls | 2 | 2450 MHz; 800 W; 300 s |
| *L. monocytogenes* | Packaged beef frankfurters | 5.64 | 2450 MHz; 550 W; 360 s |
| *Salmonella* Enteritidis | In-shell eggs | 7.5 | 2450 MHz; 350 W; 120 s |

Source: Aymerich *et al.*, 2008 and Dev *et al.*, 2010.

© Woodhead Publishing Limited, 2012

for research that deals with non-uniform heating of fresh fruit, including a combination of RF heating with other quarantine methods to overcome problems associated with individual methods. The bottom line is that RF heat treatment can be effectively applied if the problem of uneven heating is resolved. To resolve these problems, Birla *et al.* (2004) undertook long-term research projects to understand the problems and suggest ways and means to overcome them.

Awuah *et al.* (2002) studied the continuous RF heating of starch solutions. They reported that system and product parameters greatly influence temperature change across the applicator tube. A US patent describes the development of a radio frequency system for continuous pasteurization of liquid eggs (Ball *et al.*, 2002). In 1997, Proctor and Schwartz Inc., a dryer manufacturer, and Strayfield-Fastran, a UK-based radio frequency dryer manufacturer, developed a 25 kW Magnatube RF pasteurization system. The system was designed and developed to demonstrate the successful cooking and sterilization of pumpable food products using RF energy. The product was pumped at the base of the 4″ Teflon tube which had a pair of ribbon electrodes wrapped around it. The system was successfully tested with various food products, namely meat loaf and cooked rice. Developers were excited about the prospect of commercial production of the unit. However, due to lack of motivation for a capital intensive switchover, the RF-based continuous process was not attractive to the food industry at that time.

Researchers at McGill University have demonstrated that RF at 27.12 MHz can be successfully used for pasteurization of in-shell eggs (Dev *et al.*, 2011). Researchers at the University of Nebraska have demonstrated that continuous decontamination treatment for fresh fruits is feasible by flowing fresh fruits through inclined Teflon. Teflon coating on the electrode eliminated the fouling of the electrode surface by the protein from the soybean milk. Wang *et al.* (2003) showed improved uniformity of heating in-package sterilization of foods in large packages using radio frequency at 27.12 MHz, although enhanced edge heating was an issue. They were able to overcome some of the edge heating by surrounding the large food package with deionized water.

Casals *et al.* (2010) reported that radio frequency heating can be used for controlling brown rot in peaches and nectarines. Uemura *et al.* (2010) used (RF-FH) for inactivation of *B. subtilis* spores in soybean milk. A 4 log reduction of *B. subtilis* spores was realized in the soybean milk by RF-FH at up to 115°C for 0.4 s. Comparative studies revealed that the tofu made by RF-FH processing had higher gel strength than the tofu made by conventional heating.

For building a RF-based continuous process, one major limitation is its adaptability to different package geometry, as change in package geometry changes the RF coupling power. This requires a moving electrode to facilitate the use of the same RF system for a variety of package geometric configurations.

© Woodhead Publishing Limited, 2012

### 9.6.2 MW pasteurization and sterilization

MW sterilization has a major potential advantage over retorting because the heat-up time of MW heating processes can be very short (Buffler, 1993). Ohlsson (1987) demonstrated that a high-temperature short-time MW process (128°C and 3 min cooking time) produced products superior to those from canning (120°C retort temperature and 45 min processing time) and retorting foil pouches (125°C and 13 min cooking time). Earlier studies by Stenstrom (1974) and O'Meara *et al.* (1977) also showed that the MW process produced better products than conventional sterilization processes. Several commercial MW sterilization systems have been reported in the literature (Harlfinger, 1992; Schlegel, 1992). However, commercial applications of MW sterilization processes are predominantly found only in Belgium (TOPS Foods, Belgium) and Japan (Otsuka Chemical Co., Osaka, Japan). All of these systems work successfully for the processing of liquid foods only.

The reasons for the slow adaptation of MW sterilization processes include non-uniform heating and a lack of reliable methods to validate commercial thermal processes for food safety. Large temperature variations in MW heating are due to excessive heating at the corner or edge of the foodstuff, as a result of localized concentrations of the MW field. The means of providing a more uniform heat need to be investigated. The majority of work reported to date on MW sterilization has been at the MW frequency of 2450 MHz (Tang *et al.*, 2008). Potential advantages of using 915 MHz rather than 2450 MHz, are the deeper penetration depth in foodstuffs and, possibly, more uniform field distribution over a confined surface area of packaged foods.

As mentioned above, researchers at Washington State University have developed a continuous MW sterilization system, using 915 MHz frequency, for semisolid and solid food products. During the sterilization process, vacuum-packaged food is immersed in a water solution with a selected salt concentration (based on a dielectric property measurement) in a pressurized vessel (Tang *et al.*, 2008). MW assisted sterilization (MAS) of homogeneous and heterogeneous products has led to several opportunities for embracing continuous MW technology. The pilot-scale MAS has demonstrated improvements in quality and safety of homogeneous products. There are still some challenges that need to be addressed, such as heating uniformity and adaptability to different types of packages.

Another research team led by Drozd (2010) developed a continuous MW system for pumpable liquid foods. A patented focused cylindrical applicator MW system, operating at 915 MHz, was developed and tested (Fig. 9.15). A few large industrial-scale systems, designed and developed by Industrial MW System, LLC, are in use for sterilization of vegetable purées.

In-shell eggs build up pressure and explode when heated in a MW cavity. Dev *et al.* (2010) successfully designed and built a laboratory scale MW in-shell eggs pasteurizer using MW at 2450 MHz. Figure 9.16 portrays a schematic of the system built by the authors. The challenge of heating egg yolk to a higher temperature of 61.1°C, whilst maintaining the temperature of

© Woodhead Publishing Limited, 2012

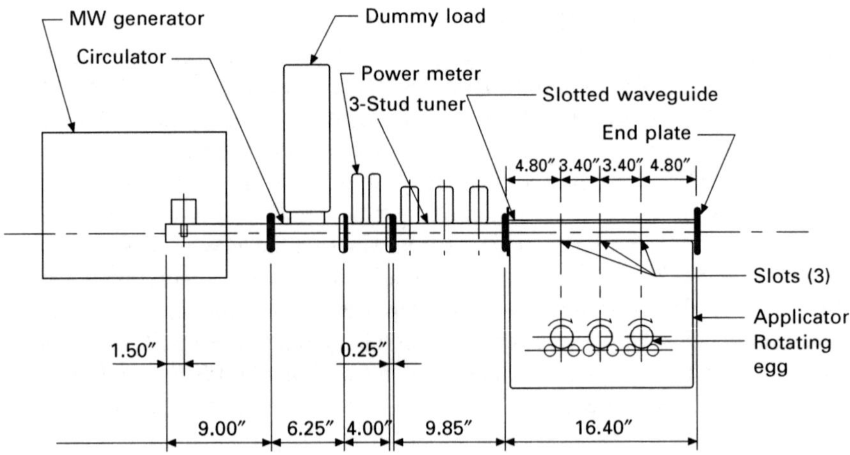


**Fig. 9.16** Schematic of the custom-built MW pasteurization setup.

the egg white at 57.5°C, was achieved by a specifically designed waveguide with non-linear slots.

Yaghmaee and Durance (2007) demonstrated that vacuum MW drying and a combination of MW processing (2450 MHz, 1.8 kW) at atmospheric pressure is effective in decontamination of freshly grated carrots and parsley leaves from naturally occurring microorganisms. When powdered black pepper with varying moisture contents were subjected to intermittent or continuous MW treatments (2450 MHz, 450 W for 150 s), Aydin and Bostan (2006) reported a 90% reduction in the microbial load, with 18% volatile compound losses.

Legnani *et al.* (2001) studied microbiological safety of MW treated (100°C for 15 s) black pepper, red chili, oregano, rosemary, and sage. MW heating was more effective on the molds and fecal indicators, which, after treatment, were within the limits set by the International Commission on Microbiological Specifications (*E. coli* $< 10^3$ cfu/g).

MW irradiation has been applied to control aflatoxin producing *Aspergillus parasiticus* in hazelnuts without affecting taste and odor of in-shell hazelnuts treated for 120 s (Basaran and Akhan, 2010). Application has also been successfully demonstrated in other nuts, including walnut and almond.

### 9.6.3 Effect of RF and MW on quality of processed food

Dielectric heating is used to reduce the exposure time of the food product, so that high-temperature short-time treatments for materials with poor thermal conductivity can be achieved. Rapid cooling techniques must be considered after a high-temperature short-time treatment, as slow cooling of food materials could be detrimental to post-treatment quality.

Wang *et al.* (2006) reported that rapid RF heating of apple was possible, but

© Woodhead Publishing Limited, 2012

the slow cooling of apples resulted in undesirable post-treatment fruit quality. Due to rapid heating and volumetric MW/radio frequency heating, the loss of nutrients is always less than with conventional thermally processed food. Due to rapid heating, protein unfolding and denaturation can be minimized resulting in a better quality end product. Therefore, post-treatment quality monitoring is an important consideration in the design of RF/MW based thermal processes. Little conclusive evidence exists for any real flavor differences between many conventional and MW-heated foods.

Van Roon *et al.* (1994) compared the viscoelastic properties of RF pasteurization of meat dough with conventional pasteurization. The RF-heated products had both higher storage and loss moduli (were more firm). They fractured at higher stress values and were considered to be more firm in the sensory evaluation. The microstructure of dielectrically heated versus conventionally heated samples displayed a more open structure of the protein matrix, with larger irregularly shaped fat particles that were surrounded by relatively thin and compact protein bridges.

Short MW pasteurization (35 s at 652 W) of Granny Smith apple purée reduced microorganisms, but could not inactivate the enzymes present in the product and prevent vitamin C degradation. On the other hand, the treatment did not affect the stability, viscosity, and titratable acidity during the storage period (Picouet *et al.*, 2009). The enzyme inactivation may require a longer holding time at an elevated temperature.

Tang *et al.* (2002) compared MW sterilized Kraft® macaroni and cheese with that of the retorted product and found that the macaroni and cheese sterilized with MW retained the original structure whereas the canned type became 'lumpy and gooey' (Fig. 9.17). Similar research into peas also indicated a better color and texture retention when sterilized with MW when compared to conventionally canned products.

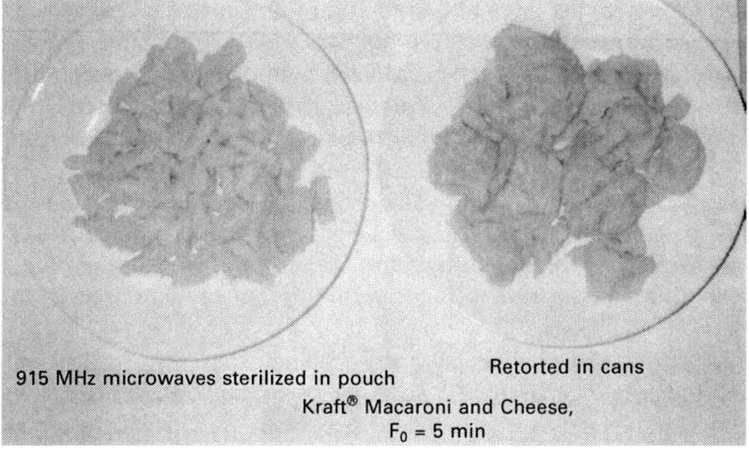


**Fig. 9.17** Microwave and retort sterilized macaroni and cheese (Tang *et al.*, 2002).

© Woodhead Publishing Limited, 2012

### 9.6.4 Process economics

There is currently no thorough investigation available in the literature into the process economics of employing dielectric heating for food processing. Many concerns have been raised about the high overall operational cost of using dielectric heating in food processing, which has hindered the commercialization of this technology.

Dielectric heating has the potential to improve the quality of food products, and this improvement must be taken into account when undertaking an economic study. Sometimes dielectric heating alone is not sufficient to meet process requirements. Dielectric heating should be used sparingly when other heating methods are failing to meet the requirements of quality and safety. For example, for in-shell pasteurization of eggs, where the process demands a higher core temperature, dielectric heating provides a viable solution (Dev *et al.*, 2008a). Birla *et al.* (2005) demonstrated that a combination of hot water treatment with RF heating can help to overcome the problems encountered when these treatments are used alone for disinfection of fruits.

## 9.7 Conclusions and future trends

Dielectric heating tends to create hot and cold spots. In order to find a solution to this problem, complete understanding of geometry, composition, dielectric properties, and packaging is required. Sometimes hot and cold spots cannot be eliminated from a product completely. The future of MW heating in food processing applications is promising, but successful exploration of MW heating applications relies on a thorough understanding of the interaction between MW and foods, and on the ability to predict and provide a desired heating pattern in foods for specific applications. These facts together with the possibility of offering continuous systems are seen as advantages in the food processing industry, although the issue of non-uniformity remains unresolved.

Because the dielectric properties of foods and food components determine their interaction with EM energy, accurate temperature dependent dielectric property data are required in order to design processes using computer simulation models. This is imperative and inevitable for successful implementation of this technology on an industrial scale for food decontamination. There is still a research gap in designing efficient applicators for dielectric heating in order to meet requirements. With the advent of powerful simulation packages and the availability of increasingly fast computers, it will be possible to use computer simulation to help improve the design of MW and RF heating systems and processes.

Partnership between industry and academia has been instrumental in developing these technologies. Such continual support/interest is vital for developing more applications and processes. With significant industry interest, undergoing research at McGill University on optimization and scale-up of the MW assisted in-shell egg pasteurization technique can be commercialized.

© Woodhead Publishing Limited, 2012

## 9.8 References

AWUAH, G. B., RAMASWAMY, H. S. & PIYASENA, P. 2002. Radio frequency (RF) heating of starch solutions under continuous flow conditions: effect of system and product parameters on temperature change across the applicator tube. *Journal of Food Process Engineering*, 25(3), 201–223.

AYDIN, A. & BOSTAN, K. 2006. Microbial decontamination of powdered black pepper (*Piper nigrum* L.) by using MW. *Journal of Food Science and Technology-Mysore*, 43(6), 575–578.

AYMERICH, T., PICOUET, P. A. & MONFORT, J. M. 2008. Decontamination technologies for meat products. *Meat Science*, 78(1–2), 114–129.

BALL, H. R., HAMID-SAMIMI, M. & SWARTZEL, K. R. 2002. Method for the pasteurization of egg products using radio waves. US Patent 6406727.

BARER, R. & TKACZYK, S. 1954. Refractive index of concentrated protein solutions. *Nature*, 173, 821–822.

BASARAN, P. & AKHAN, Ü. 2010. MW irradiation of hazelnuts for the control of aflatoxin producing *Aspergillus parasiticus*. *Innovative Food Science & Emerging Technologies*, 11(1), 113–117.

BIRLA, S. L., WANG, S., TANG, J. & HALLMAN, G. 2004. Improving heating uniformity of fresh fruit in radio frequency treatments for pest control. *Postharvest Biology and Technology*, 33(2), 205–217.

BIRLA, S. L., WANG, S., TANG, J., FELLMAN, J. K., MATTISON, D. S., & LURIE, S. 2005. Quality of oranges as influenced by potential radio frequency heat treatments against Mediterranean fruit flies. *Postharvest Biology and Technology*, 38(1), 66–79.

BUFFLER, C. 1993. *Microwave Cooking and Processing: Engineering Fundamentals for the Food Scientist*. New York: Springer.

CASALS, C., VINAS, I., LANDL, A., PICOUET, P., TORRES, R. & USALL, J. 2010. Application of radio frequency heating to control brown rot on peaches and nectarines. *Postharvest Biology and Technology*, 58(3), 218–224.

CHUNG, H. J., BIRLA, S. L. & TANG, J. 2008. Performance evaluation of aluminum test cell designed for determining the heat resistance of bacterial spores in foods. *LWT-Food Science and Technology*, 41(8), 1351–1359.

DATTA, A. K. & ANANTHESWARAN, R. C. 2001. *Handbook of Microwave Technology for Food Applications*. New York: Marcel Dekker.

DATTA, A., SUMNU, G. & RAGHAVAN, G. S. V. 2005. Dielectric properties of foods. In *Engineering Properties of Foods*, edited by M. A. Rao and A. Datta. Boca Raton, FL: Taylor & Francis.

DEV, S. R. S., RAGHAVAN, G. S. V. & GARIÉPY, Y. 2008a. Dielectric properties of egg components and microwave heating for in-shell pasteurization of eggs. *Journal of Food Engineering*, 86, 207–214.

DEV, S. R. S., PADMINI, T., ADEDEJI, A., GARIÉPY, Y. & RAGHAVAN, G. S. V. 2008b. A comparative study on the effect of chemical, MW, and pulsed electric pretreatments on convective drying and quality of raisins. *Drying Technology*, 26(10), 1238–1243.

DEV, S. R. S, ORSAT, V., GARIÉPY, Y. & RAGHAVAN, G. S. V. 2010. Quality assessment of MW pasteurized in-shell eggs. *XVIIth World Congress of the International Commission of Agricultural and Biosystems Engineering (CIGR) Conference Proceedings*, Québec City, Canada, June 13–17, 2010.

DEV, S. R. S., KANNAN, S., GARIEPY, Y., ORSAT, V. & RAGHAVAN, G. S. V. 2011. Simulation, experimental validation and process optimization of radio frequency heating for pasteurization of in-shell eggs. Presented at the 2011 ASABE Annual International Meeting, Louisville, KY, August 7–10, 2011.

DROZD, E. 2010. Multi-Stage Cylindrical Waveguide Applicator Systems. US Patent 2010/0012650 A1.

DUCK, F. A. 1990. Physical Properties of Tissue: A Comprehensive Reference Book, New York: Academic Press.

© Woodhead Publishing Limited, 2012

FCC 1988. Federal Communication Commission Standard. Available from: www.fcc.gov.
GEVEKE, D. J. & BRUNKHORST, C. 2008. Radio frequency electric fields inactivation of *Escherichia coli* in apple cider. *Journal of Food Engineering*, 85(2), 215–221.
HALLMAN, G. J. & SHARP, J. L. 1994. Radio frequency heat treatments. In *Quarantine Treatments for Pests of Food Plants*. San Francisco, CA: Westview Press, pp. 165–170.
HARLFINGER, L. 1992. MW sterilization. *Food Technology*, 46(12), 57–61.
HOUBEN, J. H., VAN ROON, P. S. & KROL, B. 1991. Radio-frequency pasteurization of sausage emulsions as a continuous process. *Journal of Microwave Power and Electromagnetic Energy*, 26(4), 202–205.
IFT-FDA 2000. Kinetics of Microbial Inactivation for Alternative Food Processing Technologies. Institute of Food Technologists and FDA. Available from: http://www.fda.gov/Food/ScienceResearch/ResearchAreas/SafePracticesforFoodProcesses/ucm100158.htm (accessed 26 November 2011).
LEGNANI, P. P., LEONI, E., RIGHI, F. & ZARABINI, L. A. 2001. Effect of microwave heating and gamma irradiation on microbiological quality of spices and herbs. *Italian Journal of Food Science*, 13(3), 337–345.
LIRA, C.T. 1996. Thermodynamics of supercritical fluids with respect to lipid-containing systems. In *Supercritical Fluid Technology in Oil and Lipid Chemistry*, edited by J.W. King, G. R. List. Urbana, IL: The American Oil Chemists Society, pp. 1–19.
MARRA, F., ZHANG, L. & LYNG, J. G. 2009. Radio frequency treatment of foods: review of recent advances. *Journal of Food Engineering*, 91(4), 497–508.
MEDA, V., ORSAT, V. & RAGHAVAN, G. S. V. 2005. MW heating and dielectric properties of foods. In *The Microwave Processing of Foods*, edited by H. Schubert & M. Regier. Cambridge, Woodhead Publishing.
MINGOS, D. M. P & BAGHURST, D. R. 1991. Application of MW dielectric heating effects to synthetic problems in chemistry. *Chemical Society Reviews*, 20, 1–47.
MUDGETT, R. E. & WESTPHAL, W. B. 1989. Dielectric behavior of an aqueous cation exchanger. *Journal of Microwave Power*, 24, 33–37.
MUHAMAD, I. I. B., SOLTANI, M. A., ENAYATI, A., ALIAKBARIAN, H., AMERI, H. & MOGHAVVEMI, M. 2010. Application of electromagnetic waves as a solution to some agricultural problems. *Egyptian Journal of Biological Pest Control*, 20(1), 79–84.
MURRAY, D. 1958. Percy Spencer and His Itch to Know. *Readers Digest*, August.
O'MEARA, J. P., FARKAS, D. F. & WADSWORTH, C. K. 1977. Flexible pouch sterilization using combined MW-hot water hold simulator. Contact No. (PN)DRXNM 77–120, US Army Natick Research and Development Laboratories, Natick, MA 01760.
OHLSSON, T. 1987. Sterilization of foods by MW. Presented at Int. Sem. New Trends in Aseptic Processing and Packaging of Foodstuffs, Munich, October 22–23.
ORSAT, V., RAGHAVAN, V. & MEDA, V. 2005. MW technology for food processing: an overview. In *The Microwave Processing of Foods*. edited by H. Schubert & M. Regier. Cambridge: Woodhead Publishing, pp. 106–118.
PACE, W. E., WESTPHAL, W. B. & GOLDBLITH, S. A. 1968. Dielectric properties of commercial cooking oils. *Journal of Food Science*, 33, 30–36.
PETHIG, R. 1984. Dielectric properties of biological materials: biophysical and medical applications. IEEE Transactions on Electrical Insulation, EI-19, 5, 453–473.
PICOUET, P. A., LANDL, A., ABADIAS, M., CASTELLARI, M. & VINAS, I. 2009. Minimal processing of a Granny Smith apple purée by MW heating. *Innovative Food Science & Emerging Technologies*, 10(4), 545–550.
PIYASENA, P., DUSSAULT, C., KOUTCHMA, T., RAMASWAMY, H. S. & AWUAH, G. B. 2003. Radio frequency heating of foods: principles, applications and related properties – a review. *Critical Reviews in Food Science and Nutrition*, 43(6), 587–606.
POZAR, D. M. 2005. *Microwave Engineering*, 3rd edn. New York: John Wiley & Sons.
RAMASWAMY, H. & TANG, J. 2008. Microwave and radio frequency heating. *Food Science and Technology International*, 14(5), 423–427.

© Woodhead Publishing Limited, 2012

ROEBUCK, B. D., GOLDBLITH, S. A. & WESTPHAL, W. B. 1972. Dielectric properties of carbohydrate-water mixtures at MW frequencies. *Journal of Food Science*, 37(2), 199–204.

ROLAND, U. & KOPINKE, F. D. 2009. The role of water in dielectric heating with radio waves. *Chemical Engineering & Technology*, 32(5), 754–762.

SCHLEGEL, W. 1992. Commercial pasteurization and sterilization of food products using MW technology. *Food Technology*, 46(12), 62–63.

SHAMIS, Y., TAUBE, A., SHRAMKOV, Y., MITIK-DINEVA, N., VU, B. & IVANOVA, E. P. 2008. Development of a microwave treatment technique for bacterial decontamination of raw meat. *International Journal of Food Engineering*, 4(3), Article 8.

SHUKLA, T. P. & ANANTHESWARAN, R. C. 2001. Ingredient interactions and product development for MW heating. In: *Handbook of Microwave Technology for Food Application*, edited by A. K. Datta & R. K. Anantheswaran. New York: Marcel Dekker.

STENSTROM, L. A. 1974. Heating of Products in Electromagnetic Field. US Patent 3809845.

TANG, J., HAO, F. & LAU, M. 2002. Microwave heating in food processing in *Advances in Bioprocessing Engineering*, edited by X. H. Yang & J. Tang. Singapore: World Scientific Publishing.

TANG, Z., MIKHAYLENKO, G., LIU, F., MAH, J.-H., PANDIT, R., YOUNCE, F. & TANG, J. 2008. Microwave sterilization of sliced beef in gravy in 7-oz trays. *Journal of Food Engineering*, 89(4), 375–383.

UEMURA, K., TAKAHASHI, C. & KOBAYASHI, I. 2010. Inactivation of *Bacillus subtilis* spores in soybean milk by radio-frequency flash heating. *Journal of Food Engineering*, 100(4), 622–626.

VAN ROON, P. S., HOUBEN, J. H., KOOLMEES, P. A. & VANVLIET, T. 1994. Mechanical and microstructural characteristics of meat doughs, either heated by a continuous process in a radiofrequency field or conventionally in a waterbath. *Meat Science*, 38(1), 103–116.

WANG, S., BIRLA, S. L., HANSEN, J. D. & TANG, J. 2006. Postharvest treatment to control codling moth in fresh apples using water assisted radio frequency heating. *Postharvest Biology and Technology*, 40(1), 89–96.

WANG, Y., WIG, T. D., TANG, J. & HALLBERG, L. M. 2003. Sterilization of foodstuffs using radio frequency heating. *Journal of Food Science*, 68(2), 539–544.

YAGHMAEE, P. & DURANCE, T. 2007. Efficacy of vacuum microwave drying in microbial decontamination of dried vegetables. *Drying Technology*, 25(4–6), 1099–1104.

ZHAO, Y. Y., FLUGSTAD, B., KOLBE, E., PARK, J. W. & WELLS, J. H. 2000. Using capacitive (radio frequency) dielectric heating in food processing and preservation – a review. *Journal of Food Process Engineering*, 23(1), 25–55.

ZHONG, Q., SANDEEP, K. P. & SWARTZEL, K. R. 2003. Continuous flow radio frequency heating of water and carboxymethylcellulose solutions. *Journal of Food Science*, 68(1), 217–223.

ZHONG, Q. X., SANDEEP, K. P. & SWARTZEL, K. R. 2004. Continuous flow radio frequency heating of particulate foods. *Innovative Food Science & Emerging Technologies*, 5, 475–483.

© Woodhead Publishing Limited, 2012

# 10

# Microbial decontamination of food by power ultrasound

**B. Zhou, H. Lee and H. Feng, University of Illinois at Urbana-Champaign, USA**

**Abstract**: Power ultrasound has shown great promise as a new food processing and food preservation method, increasing the microbial inactivation rate while providing a high-quality end product. It is regarded as a non-thermal processing method, mainly due to the fact that its mode of action is based on cavitation-generated physical and chemical effects, as opposed to traditional thermal processes where heat is the major lethal factor in microbial destruction. Currently, research and development activities on the topic of power ultrasound as a food decontamination method focus on liquid food processing and product surface decontamination, with some studies looking into the effect of ultrasonication on product quality attributes. In this chapter, an effort is made to provide a brief introduction to the basic concepts of power ultrasound decontamination. Three modes of inactivation are introduced, combining ultrasound with other lethal factors for treatment of liquid food. The recent development in ultrasound-assisted surface decontamination of fresh produce is presented, and the advantages and limitations of power ultrasound are summarized.

**Key words:** power ultrasound, inactivation, surface decontamination, liquid food, fresh produce, cavitation.

## 10.1 Introduction

Consumer demand for fresh, safe, and nutritious foods continues to drive the food industry to pursue new and minimal processing and preservation technologies. Power ultrasound is one such technology that could be used to provide improved food safety and quality for consumers. 'Power ultrasound' refers to sound waves in the frequencies ranging from 20 kHz to around 1 MHz, with a sound intensity of between 10 and 1000 W/cm$^2$ (Mason and

© Woodhead Publishing Limited, 2012

Lorimer, 2002). It has found applications in a number of food processing unit operations such as homogenization, cutting, defoaming, drying, bio-component separation and extraction of bioactive component(s) in foods and plants, among others.

The bactericidal effect of power ultrasound has long been under investigation by researchers (Harvey and Loomis, 1929). In the early years of this research, microbial inactivation by ultrasound was conducted mainly at room or low temperatures with a relatively low acoustic power level, meaning that it took a long time to achieve the low level of microbial survival needed for liquid food pasteurization. Therefore, ultrasound was not intensively investigated by the food industry as an alternative to thermal pasteurization of liquid foods. In recent years, due to new developments in ultrasound technology as well as our increased understanding of the interactions between acoustic energy and food systems, there has been increased interest in examining the use of ultrasound as an alternative food decontamination tool. Combining sonication with other hurdles, such as pH, chemical, mild heat, and low pressure, has been reported to enhance the efficacy of an ultrasound treatment (Raso *et al.*, 1998a; Álvarez *et al.*, 2006; Lee *et al.*, 2009b). An additive or synergistic effect has been observed for microbial inactivation due to a thermal sonication or manothermosonication (MTS) treatment compared to an ultrasound alone treatment (Raso *et al.*, 1998b; Pagán *et al.*, 1999c; Lee *et al.*, 2009a). Food industry concerns about the application of ultrasound as a food decontamination method include the quality of the foods treated with ultrasound, and the potential for scale-up and economic issues. This chapter will summarize the principles of ultrasound technology as a food decontamination tool, our understanding of the mechanisms of how ultrasound works to inactivate microorganisms, and information about the effect of power ultrasound on food quality. Comments on potential future trends and sources of further information and advice will also be provided.

## 10.2 Principles and technology

Ultrasonic waves are stress waves that transmit from one mass to an adjacent mass through direct contact, making it different from light or electromagnetic radiation which can travel through free space. Ultrasonic waves are also elastic waves since its propagation in a medium is determined by the elastic properties of the medium (Ensminger, 1973). Ultrasound applications can be classified into two broad categories: high intensity and low intensity applications. The sound intensities are normally greater than 10 W/cm$^2$ for high intensity applications, while those for low intensity applications are not greater than 1 W/cm$^2$ (Feng and Yang, 2005). Since sound intensity is inversely proportional to sound frequency, high intensity sound waves are generated by relatively low frequency ultrasound generators (20 kHz to around 1 MHz) while those with low intensity are produced by high frequency

© Woodhead Publishing Limited, 2012

transducers. Similar to the seismic P-waves caused by an earthquake, the sound waves propagate in a fluid in the form of longitudinal waves made up of compressions and rarefactions. An important phenomenon takes place in the liquid when a longitudinal wave passes through it, which is the generation of cavitation. Theoretically, cavitation bubbles can only be produced in a liquid when the difference between the negative sound pressure in the rarefaction period of the wave and the local hydrostatic pressure is greater than the tensile strength between the water molecules (Fig. 10.1). As the theoretical tensile strength of pure water (27.7 MPa) is much higher than the pressure differential that can be produced by a sound wave, pure water does not cavitate (Briggs, 1950). It has been reported that regular tap water contains ample nuclei, which provide weak points to facilitate acoustic cavitation (Kentish and Ashokkumar, 2011).

There are two types of cavitation bubbles for power ultrasound applications, namely transient and stable cavitation. Stable cavitation means that the bubbles oscillate around their equilibrium position over dozens or even more refraction/compression cycles. For transient cavitation, the bubbles grow over a few

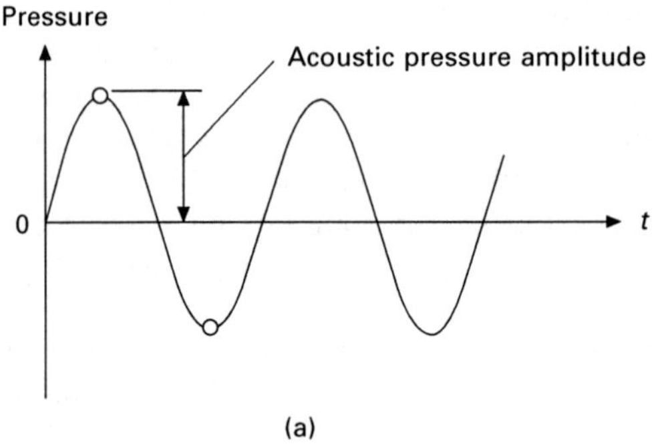


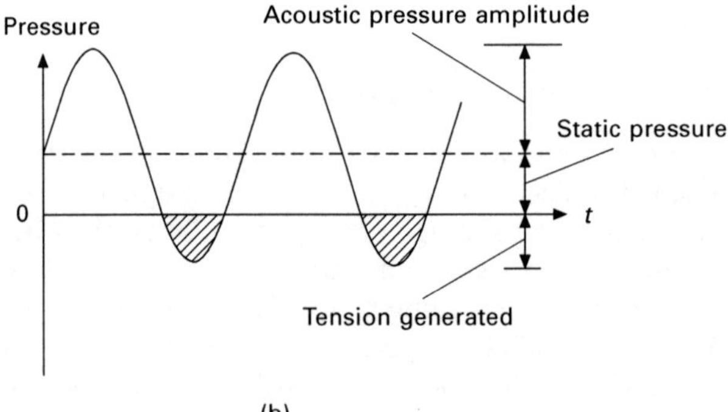


**Fig. 10.1** Cavitation generation by tensile stress in a liquid. (a) Longitudinal ultrasound wave propagation in a liquid without hydrostatic pressure and (b) longitudinal ultrasound wave propagation in a liquid with hydrostatic pressure showing the tension produced in the shadowed areas.

© Woodhead Publishing Limited, 2012

acoustic cycles to rapidly increase their initial size and finally collapse or implode violently. The size, lifespan, and fate of the cavitation bubbles depend on ultrasound frequency, intensity (acoustic pressure), solvent, bubbled gas, and external parameters, such as temperature and pressure. The implosion of transient cavitation bubbles results in formation of shock waves, generation of localized high temperature and pressure, as well as dissociation of water molecules to generate free radicals (Feng and Yang, 2005). At a solid and liquid interface, the asymmetric implosion of transient cavitation bubbles can generate water jets impinging toward the solid surface. For stable cavitation bubbles produced at a lower sound intensity (1–3 W/cm$^2$), the oscillation of the bubbles in the acoustic field stirs the liquid vigorously inducing fluid flow at high velocity, thus forming microstreaming. These physical and chemical effects generated by the two types of cavitation bubbles are the basis for power ultrasound decontamination applications.

Current power ultrasound generation technologies include piezoelectric and magnetostrictive transducers. Piezoelectricity was first discovered in 1880 by Pierre and Jacques Curie. Later, the converse effect of piezoelectricity was used for the generation of mechanical vibration by the application of an alternating electrical field. Nowadays, the widely used transducers are usually piezoelectric transducers made of ceramic elements. On the other hand, a magnetostrictive transducer utilizes the magnetostrictive property of a material to convert the energy in a magnetic field into mechanical energy. The magnetic field, created by a coil of wire wrapped around the magnetostrictive material, causes the magnetostrictive material to contract or elongate, thus producing a sound wave (Mason and Lorimer, 2002). Most ultrasound applications are conducted in a container filled with water in two types of configurations, probes and ultrasonic tanks (Fig. 10.2). The probe type treatment chambers normally have a high sound intensity (W/cm$^2$)

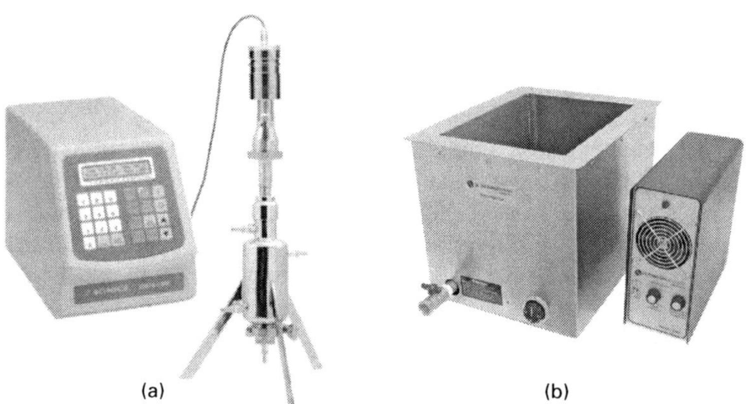


**Fig. 10.2** Commonly used ultrasound processing systems: (a) probe system and (b) tank system.

© Woodhead Publishing Limited, 2012

and high acoustic power density (W/cm$^3$), while that for an ultrasonic tank is at lower energy levels. Since most microbial inactivation tests require a relatively high power density, the use of probe systems has dominated the treatment of liquid foods for the purpose of pasteurization. The ultrasonic tanks find applications mainly in surface decontamination or relatively lower power density usage (Feng and Yang, 2011).

Since most ultrasound applications are conducted in a liquid, a good understanding of the factors influencing the cavitation formation and intensity is critical. The effectiveness of a power ultrasound treatment is influenced by a number of factors, including ultrasound frequency, power level, the size and shape of the ultrasonic bath, the depth, volume, temperature, nature of the liquid, and treatment time (Jeng *et al.*, 1990; Zhou *et al.*, 2009). In general, the intensity of cavitation in liquids decreases with the increase in ultrasonic frequency, and increases with the intensity of ultrasound. Hydrostatic pressure of the liquid and the acoustic pressure amplitude affect the ultrasound intensity proportionally, and an increase in the pressure leads to a more rapid and violent collapse of the cavitation bubbles (Mason and Lorimer, 2002). A high ultrasonic intensity is critical for a liquid with high viscosity and surface tension to generate a cavitational bubble. The existence of monatomic gases (He, Ar, Ne) is preferred to diatomics ($N_2$, air, $O_2$). Employing gases with increased solubility will reduce both the threshold intensity and the intensity of cavitation. A degassing step is necessary in practical ultrasonic cleaning applications (Awad, 2011).

## 10.3 Mode of inactivation by power ultrasound

The lethal effect of ultrasonic waves on microorganisms was first reported in the 1920s (Harvey and Loomis, 1929). It is currently agreed by most researchers that the inactivation of microorganisms by ultrasonic waves is attributed to the physical and chemical effects generated by acoustic cavitation. Transient cavitation bubbles are voids or vapor filled bubbles, generated when sound intensity in a liquid is in excess of 10 W/cm$^2$ with a lifetime of a few acoustic cycles (Kentish and Ashokkumar, 2011). The implosion of tiny bubbles can cause shock wave formation, which might lead to pore formation on cell surface and mechanical disruption of the cell boundaries. The sudden collapse of the cavitating bubbles produces a powerful inrush of liquid at a liquid–solid interface with a maximum speed of up to 156 m/s (Plesset and Chapman, 1971). The impingement of a high speed water jet onto a microbial cell surface may bring about cell wall damage, resulting in cell death. The generation of extremely high temperature (5000°C) and pressure (2000 atm) by transient cavitation may cause dissociation of water molecules into hydroxyl radicals (Koda *et al.*, 2009), which are known to have a bactericidal effect. The reactive radicals can damage microbial cells

© Woodhead Publishing Limited, 2012

by oxidative compounds but studies have concluded that the inactivation due to sonolysis is minimal (Pagán *et al.*, 1999a). Stable cavitation refers to bubbles having a lifespan of up to hundreds of cycles and due to the high speed oscillation of the bubbles in the acoustic field, high velocity microstreaming can be produced. The microstreaming generates mechanical shear forces on the cell membrane, which can also disrupt microorganisms (Gómez-López *et al.*, 2010).

When ultrasound is combined with heat and pressure, a synergistic effect has been reported on *Streptococcus faecium*. This synergistic effect was attributed to the disruption of the bacterial spore cortex, which resulted in protoplast rehydration and loss of heat resistance (Raso *et al.*, 1998a). In the case of sonication assisted by elevated pressure, the increase in inactivation rate was probably due to an increase in bubble implosion intensity, as postulated by Pagán *et al.* (1999a). It needs to be pointed out that for certain microorganisms, such as *Listeria monocytogenes*, *Salmonella* Enteritidis, and *Aeromonas hydrophila*, only additive effects were observed (Pagán *et al.*, 1999b). It was also found that the effect of thermosonication (TS) was dramatically diminished when the temperature approached 100°C for *Bacillus subtilis* (Ordóñez *et al.*, 1987). Ugarte-Romero *et al.* (2007) observed that there exists an upper temperature limit for thermosonication inactivation of bacteria and, when the treatment temperature is above the threshold; no additional killing can be achieved compared to a treatment using only heat at the same temperature. This has been attributed to the cushioning effect when vapor-filled bubbles imploded at a relatively high temperature. Feng *et al.* (2011) proposed that the non-equilibrium thermodynamic behavior of a bacterial system is responsible for the observed abnormality. For ultrasound and pressure combined treatment at sub-lethal temperatures (manosonication), Raso *et al.* (1998a) reported a pressure threshold for inactivation of *B. subtilis*; when the pressure was above 500 kPa, no increase in inactivation rate was observed when pressure was applied.

Many researchers have studied the interaction of ultrasound with microbial cells based on surface topography and cell ultrastructure observations using scanning electron microscopy (SEM) and transmission electron microscopy (TEM). Based on SEM images, Ugarte-Romero *et al.* (2006) reported that *Escherichia coli* cells were ruptured and disintegrated after ultrasonication. The cell walls were deformed and shrunk, cytoplasmic membrane was perforated, and intracellular material was discharged from the cells. For the observations with TEM, Balasundaram and Harrison (2006) reported that ultrasonication retracted the cytoplasmic membrane (CM) of *E. coli* cell from the outer membrane at a large scale, and the yeast cell wall was injured locally together with released periplasmic constitutes. In another study using TEM, Cameron *et al.* (2008) also observed extensive damage of ultrasound to bacterial cells, including inward folding and vesicle formation in *E. coli*. The observation with SEM by Gera and Doores (2011) showed that, after ultrasound treatment, some *E. coli* cells infolded, cell membrane or cell

© Woodhead Publishing Limited, 2012

wall was ruptured in some places, and the cell walls were roughened and discontinued. Some *E. coli* cells were even torn into pieces and cell debris distributed throughout the field. *L. monocytogenes* cells were also damaged physically by ultrasound waves, and the cells were broken into halves (Gera and Doores, 2011). For treatment with the combination of low pressure, heat, and ultrasound, a high inactivation rate has been reported. This is confirmed by SEM observations where extensive and irreversible cell damage can be observed, as shown in Fig. 10.3. These included cell segments, ruptured cells, intracellular constituent release, and physical breakage/damage on cell membranes (Lee *et al.*, 2009a). All the observations seem to confirm that the inactivation of microorganism by ultrasound is due to physical damage of the cell that is hard for the cells to repair.

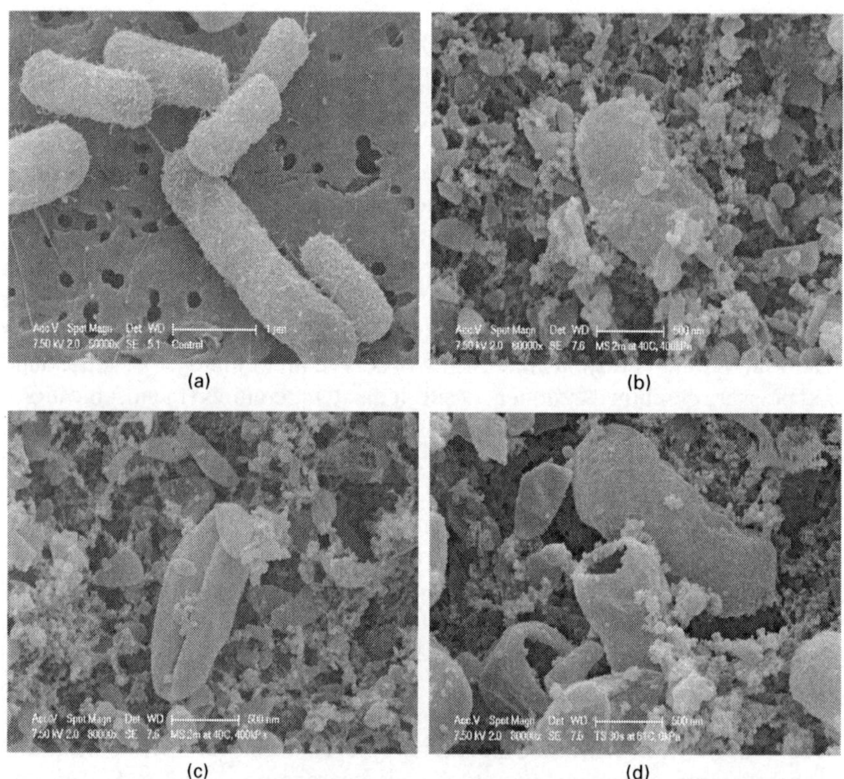


**Fig. 10.3** *E. coli* K12 cells observed with environmental scanning electron microscopy (ESEM): (a) control (50,000 magnification), (b)/(c) manosonication at 40°C and 500 kPa for 2 min (80,000 magnification), (d) thermosonication at 61°C and 100 kPa for 0.5 min (80,000 magnification) (Lee *et al.*, 2009a).

© Woodhead Publishing Limited, 2012

## 10.4 Applications in food decontamination by power ultrasound

### 10.4.1 Surface decontamination

A few reports about how ultrasound has improved microbial inactivation for cells attached to a surface have been published. Mott *et al.* (1998) investigated the application of axially propagated ultrasound (APU) at frequencies ranging from 20 to 350 kHz applied to one end of water-filled glass tubes for the removal of mineralized *Proteus mirabilis* biofilm. The results showed that three 30 s pulses from the 20 kHz transducer removed 87.5% of the biofilm. Berrang *et al.* (2008) reported that quaternary ammonium- and chlorine-based chemicals lowered numbers of planktonic cells to less than 100 CFU/ml. Approximately 6.0 log CFU/$cm^2$ *L. monocytogenes* in biofilms was detected in the inner wall surface of model polyvinyl chloride drain pipes, and a 30 s ultrasound alone treatment did not lower *L. monocytogenes* population. However, the addition of 30 s ultrasonication improved the performance of both quaternary ammonium- and chlorine-based chemicals by 1.29 and 1.14 log CFU/$cm^2$, respectively (Berrang *et al.*, 2008).

Recently, increasing attention has been paid to the application of ultrasound in surface decontamination of fresh produce. As reported by Seymour *et al.* (2002), the combination of ultrasound with chlorinated water enhanced the removal of *Salmonella* Typhimurium attached to iceberg lettuce by 1 log CFU/g compared to chlorine wash alone. Ajlouni *et al.* (2006) demonstrated that washing Romaine lettuce in various sanitizers at different concentrations when ultrasound (40 kHz) was applied reduced the microbiological counts by 1–2.5 log CFU/g immediately after washing, but ultrasonication of Romaine lettuce did not show significant improvement in bactericidal effects ($p > 0.05$) on the total or psychrophilic counts, nor did it have significant effects ($p > 0.05$) on the total or the psychrophilic microorganisms during storage at 10°C. The total plate count in Romaine lettuce reached 9.74 log CFU/g after ultrasonication (2 min at 50°C) in chlorinated water (100 mg/L) after 6 days of storage at 10°C. Moreover, the extension of treatment time (40 kHz) to 20 min did not improve the bactericidal effect of ultrasonication. Huang *et al.* (2006) reported that the combined treatment of $ClO_2$ and 170 kHz ultrasonication resulted in 2.26–2.97 log reductions in *Salmonella enterica* and 1.36–2.26 log reductions in *E. coli* O157:H7 on inoculated lettuce. When using a $ClO_2$ and ultrasonication combined wash to treat *S. enterica* and *E. coli* O157:H7 inoculated apples, the bacterial reductions were 3.12–4.25 and 2.24–3.87 log, respectively. A similar result was obtained by Huang *et al.* (2006) where one more log cycle was reduced for *S. enterica* and *E. coli* O157:H7 on apples by an ultrasound and chlorine dioxide combined treatment, whereas no obvious increase in log reduction was observed for *E. coli* O157:H7 inoculated on lettuce. It has also been demonstrated that ultrasound in combination with 1% calcium hydroxide enhanced the decontamination efficacy on alfalfa seeds inoculated with *S. enterica* and *E. coli* O157:H7 (Scouten and Beuchat, 2002).

© Woodhead Publishing Limited, 2012

Previous reports have shown contradictory results about the efficacy of ultrasound-assisted fresh produce treatments (Ajlouni *et al.*, 2006; Huang *et al.*, 2006). The lack of effectiveness in some studies might be caused by the ultrasonication system and/or the operational procedures used. There are a few key factors important to any ultrasound applications to a produce wash. For instance, dissolved gas in a washing solution is known to decrease the cavitation activity in a cleaning operation (Awad, 2011), and, therefore, degassing is essential for any ultrasonic cleaning applications. More importantly, the acoustic field distribution in an ultrasonic treatment chamber or tank is not uniform, mainly due to a standing wave formation. The non-uniform ultrasound field distribution and hence the non-uniform cavitation will result in variations in microbial inactivation activities at different locations in a washing tank. As a result, samples that have received a good dose of ultrasonication and thus have a low microbial count would be easily cross-contaminated by neighboring samples that have not received a sufficient ultrasound treatment due to a non-uniform sound field distribution, or by blockage of ultrasound propagation in the wash liquid by other samples. A good understanding of the underlying principles of power ultrasound, as well as a good design in wash system and operation procedure, is thus a prerequisite for ultrasound decontamination applications.

### 10.4.2 Liquid foods

Ultrasound techniques have been explored as a non-thermal processing alternative to thermal processing methods for use in liquid food products, such as juice and milk, mostly for the purpose of pasteurization. When log-linear inactivation kinetics applies, traditional kinetic parameters, such as *D*-values and *z*-values have been used to report the microbial inactivation with ultrasound. The treatments with ultrasound in liquid food processing can be classified into three groups: sonication at sub-lethal temperature, thermosonication (TS) at lethal temperature, and manothermosonication (MTS) where pressure at 200–500 kPa is added to TS.

*Sonication*

For ultrasound treatments at sub-lethal temperatures, the inactivation is caused by ultrasound alone. Cameron *et al.* (2008) treated *E. coli* by ultrasound (20 kHz) at sub-lethal temperatures for 10 min, and reported in a *D*-value of 2.0 and 2.0 min, and microbial count reduction of 3.88 and 4.42 log CFU/g in saline buffer and milk, respectively. In a sonication test at sub-lethal temperatures, the *D*-values of *E. coli* strains in Tryptic Soy Broth (TSB) were 15.26, 3.05, and 2.75 min, while that in a model orange juice 6.56, 6.14, and 5.4 min, for amplitudes of 0.4, 7.5, and 37.5 μm, respectively (Patil *et al.*, 2009). Ugarte-Romero *et al.* (2006) observed a non-linear inactivation of *E. coli* K12 in apple cider by sonication (20 kHz, 0.46 W/ml) at 40, 45, 50, 55, and 60°C, mostly at sub-lethal temperatures.

© Woodhead Publishing Limited, 2012

The time required to achieve a 5 log reduction of *E. coli* K12 in apple cider was 3.6, 17.7, and 4.2 min by sonication at 60°C, sonication at 40°C, and heat at 60°C, respectively. In sterile water, more than 5 log reduction for *E. coli* was achieved by ultrasound (24 kHz) treatment at room temperature in 180 s, with a device which had a narrow gap between the tip of the probe to the treatment chamber bottom, termed a squeeze-film device by the authors (Furuta *et al.*, 2004). Stanley *et al.* (2004) investigated the effect of salt on the inactivation of *E. coli* O157:H7 by sonication at sub-lethal temperatures. They found that sonication (20 kHz, 49.2 W/cm$^2$) at 40°C with 5% $AlCl_3$ was the most effective to inactivate *E. coli* O157:H7. Two different sonication (20 kHz) modes, batch and continuous, were tested to inactivate *E. coli* O157:H7 in milk by D'Amico *et al.* (2006). They reported that 4.5 and 18 min were required to achieve a 5 log reduction of *E. coli* O157:H7 by batch and continuous flow modes at 57°C, respectively. In a pilot-scale continuous flow water disinfection treatment at different flow rates (120, 240, and 1000 L/h) with a tubular ultrasound emitting design, a 2.5 log reduction for *E. coli* LMG 2092 T was achieved by circulation for 180 min (Hulsmans *et al.*, 2010).

The *D*-values for *L. monocytogenes* inactivation by thermal treatments were 1.6–117 min while those by sonication (20 kHz, 457 mW/ml) were 1.0–12.3 min at 20–60°C in apple cider (Baumann *et al.*, 2005). *L. monocytogenes* cells in UHT milk were treated by continuous flow sonication (20 kHz) at 57°C, where 5 log reduction of *L. monocytogenes* was achieved in 18 min (D'Amico *et al.*, 2006). The *D*-values for *L. monocytogenes* by sonication (20 kHz, 26°C) in milk were 5.1 and 4.9 min for an initial inoculum load of $10^4$ and $10^6$ CFU/ml, respectively (Cameron *et al.*, 2009). At sub-lethal temperatures, the *D*-values for *L. monocytogenes* Scott A by sonication (22.3 kHz) in a saline solution decreased from 31.5 to 7.3 min as the acoustic energy density increased from 0.49 to 1.43 W/ml (Ugarte-Romero *et al.*, 2007). At above 50°C, sonication increased the inactivation of *L. monocytogenes* Scott A, while at 65°C there was no significant difference between the heat treatment and the TS treatment (Ugarte-Romero *et al.*, 2007).

The *D*-values for *Saccharomyces cerevisiae* inactivation by ultrasound (20 kHz) at 24–26°C were 2.8 and 4.3 min in saline buffer and milk (Cameron *et al.*, 2008). Sonication (20 kHz, 45°C) achieved a 5 log reduction of yeast in tomato juice in 7.5 min at 61.0 μm (Adekunte *et al.*, 2010). López-Malo *et al.* (1999) treated *S. cerevisiae* in water by sonication (20 kHz) at 45–55°C and reported *D*-values of 22.3 to 0.8 min. *S. cerevisiae* was treated by sonication (20 kHz, 71–110 μm) at 35, 45, and 55°C in buffer at pH 3 and 5 and the *D*-values ranged from 0.6 to 31 min (Guerrero *et al.*, 2001). The *D*-values for *S. cerevisiae* were reduced with an increase in ultrasound amplitude at temperatures of up to 45°C. At 55°C, no additional inactivation with sonication was reported (Guerrero *et al.*, 2001). A 1 and 3 log reduction of *S. typhimurium* in whole liquid egg and skim milk were achieved by indirect sonication in an ultrasonic cleaning bath in a 30 min treatment (Wrigley and Llorca, 1992).

© Woodhead Publishing Limited, 2012

*Thermo-sonication*

In TS treatments, heat at lethal temperature is added to enhance the inactivation effect of an ultrasound treatment. Zenker *et al.* (2003) reported that the *D*-values for *E. coli* K12 DH 5 $\alpha$ (60°C, 20 kHz)/thermal (60°C) treatments were 23.1/84.6 s in phosphate buffer (pH 7.0), 23.9/84.3 s in carrot juice (pH 5.9), and 23.0/77.0 s in UHT milk (pH 6.7). Salleh-Mack and Roberts (2007) reported an ultrasound treatment (24 kHz) at 60°C (final temperature) to inactivate *E. coli* ATCC 25922, with a 6.29 log reduction achieved in 3 min in a juice simulation, while at 30°C (final temperature), a 5.54 log reduction was obtained in 10 min. The presence of soluble solids (up to 12 g/ml) in a juice simulation caused a lower reduction of *E. coli* ATCC 25922 cells by up to 1.14 log (Salleh-Mack and Roberts, 2007). In the inactivation tests of Bermúdez-Aguirre *et al.* (2009), TS (24 kHz) at 63°C for 10 min achieved a 5 log reduction of *L. innocua* in raw whole milk while the thermal treatment achieved 0.69 and 5.3 log reductions for treatment of 10 and 30 min, respectively.

*Manothermosonication*

To further increase the efficacy and achieve a 5 log reduction in a timeframe comparable to that used in traditional thermal pasteurization, ultrasound treatment has been combined with heat (lethal temperature) and low pressure (200–500 kPa) to achieve the so-called MTS, the most effective sonication method currently available. Using MTS, Lee *et al.* (2009a) reported that the time used to achieve a 5 log reduction for *E. coli* K12 in a pH 7 buffer at 61°C was 0.5 min. No colonies were detected at pH 3 after a 0.25 min MTS treatment (Lee *et al.*, 2009b). In apple cider, a 5 log reduction of *E. coli* K12 was achieved in 1.4 min by MTS, while that for TS was 3.8 min (Lee *et al.*, 2010).

In manosonication (MS) (20 kHz, 40°C), the $D_{ms}$-values for *L. monocytogenes* inactivation were 4.3, 1.5, and 1.0 min at 0, 200, and 400 kPa, respectively (Pagán *et al.*, 1999a,b). A six-fold reduction in the $D_{ms}$-value was observed when increasing the ultrasound amplitude from 62 to 150 μm (Pagán *et al.*, 1999b). MTS treatment (over 50°C) exhibited a significantly increased inactivation of *L. monocytogenes*, with an additive effect reported over that of heat treatment and MS (Pagán *et al.*, 1999a,b). The *D*-values for *S. enteritidis* inactivation in a citrate-phosphate buffer (pH 7) by heat treatment at 62°C and MS (20 kHz, 117 μm, 40°C) were 0.024 and 0.86 min, respectively (Pagán *et al.*, 1999b). The *D*-values of by *S. enteritidis* ATCC 13076, *S. typhimurium* ATCC 13311 and *S. senftenberg* ATCC 43845 inactivation with MS (20 kHz, 200 kPa, 40°C) were 0.73, 0.78, and 0.84 min in buffer (pH 7.0), and 0.76, 0.84, and 1.4 min for liquid whole egg, respectively (Mañas *et al.*, 2000). Álvarez *et al.* (2003, 2006) reported that *D*-values for *S. senftenberg* 775W and *S. enterica* Serovar treated by MS (20 kHz, 175 kPa) were lower than those by heat treatment at different water activity. At a lower water activity, a synergistic inactivation effect of heat and MS was observed (Álvarez *et al.*, 2003, 2006).

© Woodhead Publishing Limited, 2012

### 10.4.3 Other applications

Besides ultrasound-assisted liquid food processing for the purpose of pasteurization, as well as the use of ultrasound in surface decontamination, there are a number of other applications where power ultrasound is applied to reduce the number of harmful microorganisms.

In the wine industry, there are multiple unit operations where power ultrasound treatment can be used to provide a beneficial effect. First, a sonication treatment can reduce the loads of spoilage organisms before the primary fermentation. The organisms that may exist in crushed grapes include yeasts such as *Kloeckera*, *Hanseniaspora*, *Matschnikowia*, *Candida*, *Pichia*, and *Zygosaccharomyces*, and some Gram-positive lactic acid and Gram-negative acetic acid bacteria (Jiranek *et al.*, 2008). These organisms will have a negative effect on the establishment of inoculated organisms, as well as on the final product composition and sensory properties. A treatment with ultrasound is an attractive option as, in addition to its bactericidal effect, ultrasound is known to play a role in the extraction of colors and flavor compounds from the grape into the liquid. The control of spoilage organisms with ultrasonication during fermentation is another promising application. It can be achieved using a flow-through system or by inserting sonotrodes into a fermentor.

The sanitation of oak barrels used for maturation of wine with ultrasound is especially attractive. Many difficulties are reported in current industry practices where steam, hot water, or chemicals are used to clean the barrels and to inactivate spoilage-causing organisms, such as *Dekkera* spp. and *Brettanomyces* spp. For example, *Dekkera* spp. and *Brettanomyces* spp. can penetrate into the wood to a depth of 8–10 mm (Clack, 2008), which makes common sanitation methods ineffective. In addition, tartrates in the wine would deposit to the barrel walls and form a layer that is hard to remove. Methods based on sonication have been tested by Yap *et al.* (2007a, b) to remove the tartrates and inactivate *Dekkera* spp. and *Brettanomyces bruxellensis*. Yap *et al.* (2007b) reported that the ultrasound treatment not only removed the tartrate, but also killed the spoilage-causing organisms located deep in the pores of the wooden barrels.

After fermentation, wine aging using ultrasound as an aid has also been explored over the years. An advantage of using ultrasonication to treat the wine is that imparted flavor and mouthful complexity of the wine can be observed due to the autolysis of the yeast caused by ultrasound and the subsequent release of enzymes into the wine (Iland and Gago, 2002). It is claimed that it is easy to apply the ultrasound technique in current winemaking practices without additional changes of the production lines (Yap *et al.*, 2007a).

Most microbial inactivation tests are conducted in a liquid medium. However, airborne ultrasound where the ultrasonic waves propagate through air to reach the product to be processed has also been exploited. Hoover *et al.* (2002) developed a non-contact ultrasound transducer for the treatment of envelopes contaminated with spores of *Bacillus thuringiensis*,

© Woodhead Publishing Limited, 2012

which was used as a surrogate for *Bacillus anthracis*, and reported an over 3 log reduction of dried spore samples in 30 s when the frequency was 93 kHz. They estimated that the acoustic pressure produced by the device was 10 MPa in ambient air. The frequency range of the device was 50 kHz to 5 MHz and the distance of the transducer to the sample to be treated was a 5 mm ambient air column (Bhardwaj *et al.*, 2004).

## 10.5 Effects of power ultrasound on food quality

The mechanical and chemical effects generated by ultrasound can contribute to changes in the food components and quality of final products. The effect of ultrasound on two liquid food products, specifically juice and milk will be discussed below.

### 10.5.1 Juices

Apple ciders were treated by heat and different sonication techniques, such as thermosonication (TS), manosonication (MS), and manothermosonication (MTS) to achieve a 5 log reduction of *E. coli* K12 population (Ugarte-Romero *et al.*, 2006; Lee *et al.*, 2010). Titratable acidity, °Brix, and pH did not show any differences among all samples. It was also reported that there was no significant effect of sonication on these quality attributes in tomato and red grape juice (Adekunte *et al.*, 2010; Tiwari *et al.*, 2010). The turbidity values of sonicated apple cider were significantly lower than those of heat-treated and not-treated (control) samples, which indicates lower cloudiness of sonicated samples.

Orange juice treated with a continuous or batch ultrasound system showed an increase of lightness with treatment time (Zenker *et al.*, 2003; Gómez-López *et al.*, 2010). However, there was no significant difference in colors by sensory analysis (Gómez-López *et al.*, 2010). An increase in lightness was also observed in sonicated red grape juice (Tiwari *et al.*, 2010). The authors postulated that partial precipitation of suspended, insoluble particles in the juice probably contributed to the increase in lightness (Zenker *et al.*, 2003). Browning was observed in sonicated orange juice in an open system treatment where the samples were exposed to air. The ultrasound treatment of glucose in an aqueous phase could yield glucosyl radical and polymers in the presence of oxygen, which could contribute to the formation of browning pigments due to the absence of amine groups in the juice (Portenlänger and Heusinger, 1994; Vercet *et al.*, 2001). In a nitrogen-protected MTS system, the browning index of MTS-treated orange juice was significantly lower than that treated by a commercial thermal method, but was significantly higher than that of raw juice (Lee *et al.*, 2005). In a report by Valero *et al.* (2007), orange juice was sonicated by both batch and continuous systems. An increase in brown pigments in ultrasound-treated orange juice was observed

© Woodhead Publishing Limited, 2012

only in the continuous system. Valero *et al.* (2007) attributed this result to a greater exposure of the orange juice to oxygen in the continuous system. Valdramidis *et al.* (2010) reported that sonication temperature, time, and amplitude were significant parameters for browning of orange juice.

It was reported that the ascorbic acid retention in sonicated orange juice was the same or higher than an untreated or thermally processed juice after storage, even though the initial ascorbic content in sonicated juice was lower than that in a counterpart (Zenker *et al.*, 2003; Lee *et al.*, 2005; Tiwari *et al.*, 2009a; Gómez-López *et al.*, 2010). Decrease in ascorbic acid was observed in tomato juice right after sonication (Adekunte *et al.*, 2010a). Improved retentions of anthocyanin and ascorbic acid by sonication were also reported in red grape juice and blackberry juice (Tiwari *et al.*, 2009b, 2010). The lower degradation of ascorbic acid in sonicated orange juice was attributed to degassing of juice by ultrasound (Zenker *et al.*, 2003). Dissolved gases, including oxygen can act as nuclei to form bubbles, which could float to the surface and be removed from the juice. This degassing effect could lower the dissolved oxygen level in the juice and hence reduce oxidative degradation of ascorbic acid during storage. In another study by Feng (2005), an improvement in ascorbic acid retention of MTS-treated juice was observed in three out of five storage tests. The dissolved oxygen levels in MTS- and thermal-pasteurized orange juice were about the same during storage at 4°C and oxygen levels became negligible after 50 days. There might be factors other than dissolved oxygen that contribute to degradation reactions in juice. It has been found that during MTS treatment, due to strong cavitation activities, some metal ions, including iron, manganese, and nickel, were released from the metal container wall. These metal ions may function as catalysts to speed up some degradation reactions. More studies are needed to better understand the degradation reactions in ultrasound treatment of juice products.

### 10.5.2 Milk

In ultrasound treatment of milk, Chouliara *et al.* (2010) reported that the odor and taste scores of ultrasonicated milks decreased as ultrasonication treatment time increased. Lipid oxidation was higher in ultrasonicated milk samples than in the untreated sample. The distinctive volatile compounds in ultrasonicated milk were the compounds through lipid oxidation induced by radicals, including pentanal, hexanal, and heptanal, while the compounds by pyrolysis were 2,2,4-trimethyl pentane, 1,3-butadiene, and 1-buten-3-yne (Riener *et al.*, 2009a; Chouliara *et al.*, 2010).

It was reported that thermosonication (TS) where temperatures were high enough to kill microorganisms did not significantly affect fat content, protein content, and lightness, but did cause decreases in pH, yellowish color, solid-non-fat and density, and increases in acidity, greenish color and freezing point (Bermúdez-Aguirre and Barbosa-Cánovas, 2008). Brown pigments in a

© Woodhead Publishing Limited, 2012

MTS-treated model milk system increased with treatment time compared to heat-treated milk, which was probably caused by the Maillard reaction (Vercet *et al.*, 2001). The disintegration of milk fat globule membrane (MFGM) was observed by TS (Bermúdez-Aguirre *et al.*, 2008). TS contributed to smaller fat globules than thermal and raw milks.

For lightness and fat content, different results were reported, including an increase in lightness and fat content (Bermúdez-Aguirre *et al.*, 2008). The authors ascribed this phenomenon to the milk samples used. The milk used in the study by Bermúdez-Aguirre *et al.* (2008) was raw milk, where TS contributed to homogenizing (higher lightness) and releasing lipid (increase in fat content). In contrast, the milk was already homogenized in another study (Bermúdez-Aguirre and Barbosa-Cánovas, 2008) and thus TS did not exhibit the effects of homogenizing and releasing lipid.

An increase in fat content in ultrasonicated raw milk was reported by Cameron *et al.* (2009). They also noticed no detrimental effect of ultrasonication on protein, casein, and lactose content. In the study, milk pasteurized at ultra-high temperature (UHT) was also ultrasonicated and resulted in an increase in fat content. However, information about homogenizing was not available in the study. In terms of homogenizing capacity of ultrasound, pasteurized milks were homogenized by a conventional homogenizer with pressure (200 bar, 55°C) and TS (20 kHz, 55°C, 10 min). The TS exhibited higher homogenizing efficiency than conventional homogenizer (Ertugay *et al.*, 2004). Villamiel and De Jong (2000) reported reduced fat globule sizes in milk after sonication at 20 kHz and 8 W/ml. Good homogenization was also observed in ultrasound-treated milk at 20 kHz and 3 W/ml (Wu *et al.*, 2000). On the other hand, ultrasound treatment can denature whey protein in milk (Villamiel and De Jong, 2000).

In dairy product processing, ultrasound was tested to improve rheological properties of final products, such as soft cheese and yogurt. The cheese prepared with TS-treated milk showed a whiter color, better microstructure and texture than the counterpart made with thermo-processed raw milk mainly due to homogenized casein, fat, and whey by ultrasonication and their enhanced water-holding capacity (Bermúdez-Aguirre *et al.*, 2010). Similarly, MTS-treated milk was processed to make yogurt (Vercet *et al.*, 2002; Riener *et al.*, 2009b). The MTS or TS yogurt resulted in better rheological properties compared to control samples made with untreated milk. Two times higher water-holding capacity and honeycomb network structure were observed in the TS yogurt. The proposed mechanisms to improve gelling property and firmness of MTS or TS yogurt were modified fat globules and increased number of fat globules to enhance interactions with themselves and casein micelles, and denatured serum protein (Vercet *et al.*, 2002; Riener *et al.*, 2009b). Zisu *et al.* (2010) demonstrated an ultrasonic process in industrial scale operation used to reduce the viscosity of aqueous dairy ingredients and improve the heat stability. The MTS treatment showed no effect on the concentration of thiamin and riboflavin in milk (Vercet *et al.*, 2001).

© Woodhead Publishing Limited, 2012

### 10.5.3 Fresh produce

Fresh produce surface decontamination with ultrasound has been a topic of research for a few years. Most of the previous studies have focused on ways to increase the efficacy of inactivation by combining ultrasound with selected chemical sanitizers (Seymour *et al.*, 2002; Huang *et al.*, 2006; Zhou *et al.*, 2009). Not much research has been conducted to examine the effect of ultrasound on produce quality. Ajlouni *et al.* (2006) reported that a 20 min ultrasonication caused significant ($p < 0.05$) damage to the quality of Romaine lettuce tissues. An ultrasonic tank was used in the work of Ajlouni *et al.* (2006), but the acoustic energy level used was not documented. Not enough information was provided to examine the cause of damage to the lettuce. Nevertheless, a 20 min treatment is much longer than that used in the produce industry. In a two-stage industrial produce washing line, normally 30 s to 1 min wash times are used. In addition, it is known that the ultrasound field distribution in a tank is not uniform if no special measures are taken in the design of the tank. The standing wave that may present in the tank will introduce nodes and anti-nodes and thus areas with localized acoustic energy. The produce leaves exposed to a non-uniform ultrasound treatment will have a high likelihood of being damaged. In a recent work by Feng *et al.* (2009), produce samples (spinach, loose leaf lettuce, lollo rosso, and Romaine lettuce) were placed in a pilot-scale ultrasonic channel and treated for up to 16 min with an acoustic power density of 81 W/L. No visual damage was observed for the samples treated for 8 min after 14 days of storage at 4°C. It is expected that, with a uniform acoustic distribution in the treatment chamber and a carefully controlled acoustic energy level, an ultrasound treatment can deliver acceptable quality while providing an enhanced sanitation of fresh produce.

### 10.5.4 Limitations and challenges to adoption of power ultrasound technology

It should be noted that the application of power ultrasound in food decontamination is mainly a surface treatment for solid foods. The effect of ultrasound on the inactivation of bacteria inside a solid food has not been investigated. Ultrasound will partially transmit into a solid medium, but its pressure amplitude is expected to experience a significant decay. In addition, since the mode of action for ultrasonic inactivation is cavitation which is generated in a liquid, the low moisture content in a solid food will make it less possible to generate cavitation activities inside the solid product. It is also true that ultrasound would not be effective for inactivation of a food product in a package. In liquid food processing, the combined treatment of low pressure, mild temperature and ultrasound (MTS) can effectively reduce the treatment time used to achieve a 5 log reduction in the population of pathogenic organisms, but tests for the purpose of sterilization have not been conducted.

© Woodhead Publishing Limited, 2012

## 10.6 Conclusions and future trends

Ultrasound-assisted inactivation of pathogenic organisms in liquid foods and on food surfaces is a promising technology. It utilizes the cavitation-generated micro-events, such as high shear, micro-streaming, water jets, shock waves, and free radicals to destroy bacteria. The mechanical disruption introduced by cavitation activities is mostly physical damage, especially for treatment with pressure (MS), which is hard for the cells to repair. The percentage of damaged cells may change with the ultrasound treatment mode used, i.e., depending on whether it is a sonication, TS, MS, or MTS treatment, as well as the food matrix and organism type. An advantage of ultrasound treatment over other non-thermal processing technologies is that it results in less or no sub-lethal cells.

Compared to thermal processing, and even to other non-thermal food processing technologies such as high pressure processing (HPP) and pulsed electric field (PEF), the investigation into ultrasound application in food decontamination is less than sufficient in both the number of published works and the microorganisms and food products that have been treated by ultrasound. It is understood that an ultrasound treatment that takes a few minutes to achieve the 5 log reduction required by the US Food and Drug Administration (FDA) to pasteurize a food is not practical when considering both the throughput and the impact on the food quality. Future developments will thus still hinge on whether an ultrasonic inactivation of harmful microorganisms can be achieved in a time frame comparable to the thermal processing counterparts. A solution to the problem is to employ a high intensity short time (HIST) ultrasonic treatment. However, an accompanying problem with a HIST treatment is the potential pitting and corrosion of the metallic ultrasound probes caused by strong cavitation activities near the surface of the probes. The metal powders released from the probes into the liquid will introduce metal ions into the food and may have a negative impact on food quality. Consequently, a non-contact sono-reactor design will be welcomed. Currently, some non-contact designs have been proposed and tested (Dion, 2011). It is expected that more non-contact designs will be envisioned and developed in the future. The key to success is to achieve a uniform acoustic field distribution in the treatment chamber by mitigating the effect of standing waves. The application of variable frequency ultrasound generation technique will be another area for the future.

## 10.7 Sources of further information and advice

A few comprehensive compilations of the literature on the application of ultrasound are available. An early publication by Povey and Mason (1998) summarized the work on the applications of both power ultrasound and high frequency ultrasound in a number of processing unit operations and food

© Woodhead Publishing Limited, 2012

quality inspection applications. There are also a number of books mainly focusing on sonochemistry aspects of power ultrasound usages (Mason, 1999; Mason and Lorimer, 2002). The most comprehensive and up-to-date summary of the topic is published in 2011 by Feng *et al.* Interested readers can refer to the above-mentioned books to get an in-depth knowledge about ultrasound and its applications in food decontamination.

## 10.8 References


ADEKUNTE A O, TIWARI B K, CULLEN P J, SCANNELL A G M and O'DONNELL C P (2010), 'Effect of sonication on colour, ascorbic acid and yeast inactivation in tomato juice', *Food Chem*, 122, 500–507.

AJLOUNI S, SIBRANI H, PREMIER R and TOMKINS B (2006), 'Ultrasonication and fresh produce (Cos lettuce) preservation', *J Food Sci*, 71, M62–M68.

ÁLVAREZ I, MAÑAS P, SALA F J and CONDÓN S (2003), 'Inactivation of *Salmonella enteritidis* by ultrasonic waves under pressure at different water activities', *Appl Envir Microbiol*, 69, 668–672.

ÁLVAREZ I, MAÑAS P, VITRO R and CONDÓN S (2006), 'Inactivation of *Salmonella senftenberg* 775W by ultrasonic waves under pressure at different water activity', *Int J Food Microbiol*, 108, 218–225.

AWAD S B (2011), 'High power ultrasound in surface cleaning', in Feng H, Weiss J and Barbosa-Cánovas G V, *Ultrasound technologies for food and bioprocessing*, New York, Springer.

BALASUNDARAM B and HARRISON S T L (2006), 'Disruption of Brewers' yeast by hydrodynamic cavitation: process variables and their influence on selective release', *Biotech Bioeng*, 94, 303–311.

BAUMANN A, MARTIN S E and FENG H (2005), 'Power ultrasound treatment of *Listeria monocytogenes* in apple cider', *J Food Prot*, 68, 2333–2340.

BERMÚDEZ-AGUIRRE D and BARBOSA-CÁNOVAS G V (2008), 'Study of butter fat content in milk on the inactivation of *Listeria innocua* ATCC 51742 by thermo-sonication', *Innov Food Sci Emerg Technol*, 9, 176–185.

BERMÚDEZ-AGUIRRE D, MAWSON R and BARBOSA-CÁNOVAS G V (2008), 'Microstructure of fat globules in whole milk after thermosonication treatment', *J Food Sci*, 73, E325–E332.

BERMÚDEZ-AGUIRRE D, CORRADINI M G, MAWSON R and BARBOSA-CÁNOVAS G V (2009), 'Modeling the inactivation of *Listeria innocua* in raw whole milk treated under thermo-sonication', *Innov Food Sci Emerg Technol*, 10, 172–178.

BERMÚDEZ-AGUIRRE D and BARBOSA-CÁNOVAS G V (2010), 'Processing of soft hispanic cheese ("Queso Fresco") using thermo-sonicated milk: a study of physicochemical characteristics and storage life', *J Food Sci*, 75, S548–S558.

BERRANG M E, FRANK J F and MEINERSMANN R J (2008), 'Effect of chemical sanitizers with and without ultrasonication on *Listeria monocytogenes* as a biofilm within polyvinyl chloride drain pipes', *J Food Prot*, 71, 66–69.

BHARDWAJ M C, HOOVER K and OSTIGUY N (2004), 'Gas contact ultrasound germicide and therapeutic treaqtment', US Patent 2004/0028552 A1.

BRIGGS L J (1950), 'Limiting negative pressure of water', *J Appl Phys*, 21, 721–722.

CAMERON M, MCMASTER L D and BRITZ T J (2008), 'Electron microscopic analysis of dairy microbes inactivated by ultrasound', *Ultrason Sonochem*, 15, 960–964.

CAMERON M, MCMASTER L D and BRITZ T J (2009), 'Impact of ultrasound on dairy spoilage microbes and milk components', *Dairy Sci & Tech*, 89, 83–98.


© Woodhead Publishing Limited, 2012

CHOULIARA E, GEORGOGIANNI K G, KANELLOPOULOU N and KONTOMINAS M G (2010), 'Effect of ultrasonication on microbiological, chemical and sensory properties of raw, thermized and pasteurized milk', *Int Dairy J*, 20, 307–313.

CLACK P (2008), 'An update on ultrasonics', *Food Technol*, July, 75–77.

CONDÓN S, MAÑAS P and CEBRIÁN G (2011), 'Manothermosonication for microbial inactivation', in Feng H, Weiss J and Barbosa-Cánovas G V, *Ultrasound technologies for food and bioprocessing*, New York, Springer.

D'AMICO D J, SILK T M, WU J and GUO M (2006), 'Inactivation of microorganisms in milk and apple cider treated with ultrasound', *J Food Prot*, 69, 556–563.

DION J-L (2011), 'Contamination-free sonoreactor for the food industry', in Feng H, Weiss J and Barbosa-Cánovas G V, *Ultrasound technologies for food and bioprocessing*, New York, Springer.

ENSMINGER D (1973), *Ultrasonics – The Low- and High-Intensity Applications*, New York, Marcel Dekker.

ERTUGAY M F, ŞENGÜL M and ŞENGÜL M (2004), 'Effect of ultrasound treatment on milk homogenisation and particle size distribution of fat', *Turkish J Vet Animal Sci*, 28, 303–308.

FENG H (2005), 'Manothermosonication for dual-inactivation of thermoresistant pectin-methyl-esterase and acid tolerant foodborne pathogens in orange juice', CAPPS Project Final Report.

FENG H and YANG W (2005), 'Power ultrasound', in Hui Y H, *Handbook of food science, technology, and engineering*, Boca Raton, FL, CRC Press.

FENG H and YANG W (2011), 'Ultrasonic processing', in Zhang H Q, Barbosa G V, Balasubramaniam V M, Dunne C P, Farkas D F and Yuan J T C, *Nonthermal Processing Technologies for Food*. Chichester, Wiley-Blackwell.

FENG H, PEARLSTEIN A and ZHOU B (2009), 'Continuous-flow bacterial disinfection of fruits, vegetables, fresh-cut produce and leafy greens using high-intensity ultrasound', US Patent 61/245,382.

FENG H, BARBOSA-CÁNOVAS G V and WEISS J (2011), *Ultrasound technologies for food and bioprocessing*, New York, Springer.

FURUTA M, YAMAGUCHI M, TSUKAMOTO T, YIM B, STAVARACHE C E, HASIBA K and MAEDA Y (2004), 'Inactivation of *Escherichia coli* by ultrasonic irradiation', *Ultrasonics Sonochem*, 11, 57–60.

GERA N and DOORES S (2011), 'Kinetics and mechanism of bacterial inactivation by ultrasound waves and sonoprotective effect of milk components', *J Food Sci*, 76, M111–M119.

GÓMEZ-LÓPEZ V M, ORSOLANI L, MARTÍNEZ-YÉPEZ A and TAPIA M S (2010), 'Microbiological and sensory quality of sonicated calcium-added orange juice', *LWT – Food Sci Technol*, 43, 808–813.

GUERRERO S, LÓPEZ-MALO A and ALZAMORA S M (2001), 'Effect of ultrasound on the survival of *Saccharomyces cerevisiae*: influence of temperature, pH and amplitude', *Innov Food Sci Emerg Technol*, 2, 31–39.

HARVEY E N and LOOMIS A L (1929), 'The destruction of luminous bacteria by high frequency sound waves', *J Bacteriol*, 17, 373–376.

HOOVER K, BHARDWAJ M, OSTIGUY N and THOMPSON O (2002), 'Destruction of bacterial spores by phenomenally high efficiency non-contact ultrasonic transducers', *Mat Res Innovat*, 6, 291–295.

HUANG T, XU C, WALKER K, WEST P, ZHANG S and WEESE J (2006), 'Decontamination efficacy of combined chlorine dioxide with ultrasonication on apples and lettuce', *J Food Sci*, 71, M134–M139.

HULSMANS A, JORIS K, LAMBERT N, REDIERS H, DECLERCK P, DELAEDT Y, OLLEVIER F and LIERS S (2010), 'Evaluation of process parameters of ultrasonic treatment of bacterial suspensions in a pilot scale water disinfection system', *Ultrason Sonochem*, 17, 1004–1009.

© Woodhead Publishing Limited, 2012

ILAND P G and GAGO P (2002), *Australian wine: styles and tastes*, Campbell Town, Australia, Patrick Iland.

JENG D K, LIN L I and HARVEY L V (1990), 'Importance of ultrasonication conditions in recovery of microbial contamination from material surfaces', *J Appl Bacteriol*, 68, 479–484.

JIRANEK V, GRBIN P, YAP A, BARNES M and BATES D (2008), 'High intensity ultrasonics as a novel tool offering new opportunities for managing wine microbiology', *Biotechnol Lett*, 30, 1–6.

KENTISH S and ASHOKKUMAR M (2011), 'The physical and chemical effects of ultrasound', in Feng H, Weiss J and Barbosa-Cánovas G V, *Ultrasound technologies for food and bioprocessing*, New York, Springer.

KODA S, MIYAMOTO M, TOMA M, MATSUOKA T and MAEBAYASHI M (2009), 'Inactivation of *Escherichia coli* and *Streptococcus* mutans by ultrasound at 500 kHz', *Ultrason Sonochem*, 16, 655–659

LEE H, ZHOU B, LIANG W, FENG H and MARTIN S E (2009a), 'Inactivation of *Escherichia coli* with sonication, manosonication, thermosonication, and manothermosonication: microbial responses and kinetics modeling', *J Food Eng*, 93, 354–364.

LEE H, ZHOU B, LIANG W, FENG H and MARTIN S E (2009b), 'Effect of pH on inactivation of *Escherichia coli* K12 by sonication, manosonication, thermosonication, and manothermosonication', *J Food Sci*, 74, E191–E199.

LEE H, FENG H and MARTIN S E (2010), 'Inactivation of *Escherichia coli* K12 with manosonication, thermosonication, and manothermosonication in apple cider', Institute of Food Technologists Annual Meeting, Chicago, IL.

LEE J W, FENG H and KUSHAD M M (2005), 'Effect of manothermosonication on quality of orange juice', AIChE 2005 Annual Meeting, Cincinnati, OH.

LÓPEZ-MALO A, GUERRERO S and ALZAMORA S M (1999), '*Saccharomyces cerevisiae* thermal inactivation kinetics combined with ultrasound', *J Food Prot*, 62, 1215–1217.

MAÑAS P, PAGÁN R, RASO J, SALA F J and CODÓN S (2000), 'Inactivation of *Salmonella enteritidis*, *Salmonella typhimurium*, and *Salmonella senftenberg* by ultrasonic waves under pressure', *J Food Prot*, 63, 451–456.

MASON T J (1999), *Sonochemistry*, Oxford, Oxford University Press.

MASON T J and LORIMER J P (2002), *Applied sonochemistry. The uses of power ultrasound in chemistry and processing*, Weinheim, Wiley VCH.

MOTT I E C, STICKLER D J, COAKLEY W T and BOTT T R (1998), 'The removal of bacterial biofilm from water-filled tubes using axially propagated ultrasound', *J Appl Microbiol*, 84, 509–514.

ORDÓÑEZ J A, AGUILERA M A, GARCIA M L and SANZ B (1987), 'Effect of combined ultrasonic and heat treatment (thermoultrasonication) on the survival of a strain of *Staphylococcus aureus*', *J Dairy Res*, 54, 61–67.

PAGÁN R, MAÑAS P, ALVAREZ I and CONDÓN S (1999a), 'Resistance of *Listeria monocytogenes* to ultrasonic waves under pressure at sublethal (manosonication) and lethal (manothermosonication) temperatures', *Food Microbiol*, 16, 139–148.

PAGÁN R, MAÑAS P, RASO J and CONDÓN S (1999b), 'Bacterial resistance to ultrasonic waves under pressure (manosonication) and lethal (manothermosonication) temperatures', *Appl Environ Microbiol*, 65, 297–300.

PAGÁN R, MAÑAS P, PALOP A and SALA F J (1999c), 'Resistance of heat-shocked cells of *Listeria monocytogenes* to manosonication and to manothermosonication', *Lett Appl Microbiol* 28, 71–75.

PATIL S, BOURKE P, KELLY B, FRÍAS J M and CULLEN P J (2009), 'The effects of acid adaptation on *Escherichia coli* inactivation using power ultrasound', *Innov Food Sci Emerg Technol*, 10, 486–490.

PLESSET M S and CHAPMAN R B (1971), 'Collapse of an initially spherical vapour cavity in the neighbourhood of a solid boundary', *J Fluid Mech*, 72, 283–290.

© Woodhead Publishing Limited, 2012

PORTENLÄNGER G and HEUSINGER H (1994), 'Polymer formation from aqueous solutions of $\alpha$-D-glucose by ultrasound and $\gamma$-rays', *Ultrason Sonochem*, 1, 125–129.

POVEY M J W and MASON T J (1998), *Ultrasound in food processing*, New York, Springer.

RASO J, PAGÁN R, CONDÓN S and SALA F J (1998a), 'Influence of temperature and pressure on the lethality of ultrasound', *Appl Environ Microbiol*, 64, 465–471.

RASO J, PALOP A, PAGÁN R and CONDÓN S (1998b), 'Inactivation of *Bacillus subtilis* spores by combining ultrasonic waves under pressure and mild heat treatment', *J Appl Microbiol*, 85, 849–854.

RIENER J, NOCI F, CRONIN D A, MORGAN DJ and LYNG J G (2009a), 'Characterisation of volatile compounds generated in milk by high intensity ultrasound', *Int Dairy J*, 19, 269–272.

RIENERA J, NOCI F, CRONIN D A, MORGAN DJ and LYNG J G (2009b), 'The effect of thermosonication of milk on selected physicochemical and microstructural properties of yoghurt gels during fermentation', *Food Chem*, 114, 905–911.

SALLEH-MACK S Z and ROBERTS J S (2007), 'Ultrasound pasteurization: the effects of temperature, soluble solids, organic acids and pH on the inactivation of *Escherichia coli* ATCC 25922', *Ultrason Sonochem*, 14, 323–329.

SCOUTEN A J and BEUCHAT L R (2002), 'Combined effects of chemical, heat and ultrasound treatments to kill *Salmonella* and *Escherichia coli* O157:H7 on alfalfa seeds', *J Appl Microbiol*, 92, 668–674.

SEYMOUR I J, BURFOOT D, SMITH R L, COX L A and LOCKWOOD A (2002), 'Ultrasound decontamination of minimally processed fruits and vegetables', *Int J Food Sci Tech*, 37, 547–557.

STANLEY K D, GOLDEN D A, WILLIAMS R C and WEISS J (2004), 'Inactivation of *Escherichia coli* O157:H7 by high-intensity ultrasonication in the presence of salts', *Foodborne Pathogens Disease*, 1, 267–280.

TIWARI B K, O'DONNELL C P, MUTHUKUMARAPPAN K and CULLEN P J (2009a), 'Ascorbic acid degradation kinetics of sonicated orange juice during storage and comparison with thermally pasteurised juice', *LWT – Food Sci Technol*, 42, 700–704.

TIWARI B K, O'DONNELL C P and CULLEN P J (2009b), 'Effect of sonication on retention of anthocyanins in blackberry juice', *J Food Eng*, 93, 166–171.

TIWARI B K, PATRAS A, BRUNTON, N, CULLEN P J and O'DONNELL C P (2010), 'Effect of ultrasound processing on anthocyanins and color of red grape juice', *Ultrason Sonochem*, 17, 598–604.

UGARTE-ROMERO E, FENG H, MARTIN S E, CADWALLADER K R and ROBINSON S J (2006), 'Inactivation of *Escherichia coli* with power ultrasound in apple cider', *J Food Sci*, 71, E102–E108.

UGARTE-ROMERO E, FENG H and MARTIN S E (2007), 'Inactivation of *Shigella* and *Listeria monocytogenes* with high-intensity ultrasound at sub-lethal and lethal temperatures', *J Food Sci*, 72, M103–M107.

VALDRAMIDIS V P, CULLEN P J, TIWARI B K and O'DONNELL C P (2010), 'Quantitative modelling approaches for ascorbic acid degradation and non-enzymatic browning of orange juice during ultrasound processing', *J Food Eng*, 96, 449–454.

VALERO M, RECROSIO N, SAURA D, MUÑOZ N, MARTÍ N and LIZAMA V (2007), 'Effects of ultrasonic treatments in orange juice processing', *J Food Eng*, 80, 509–516.

VERCET A, BURGOS J and LÓPEZ-BUESA P (2001), 'Manothermosonication of foods and food-resembling system: effect of nutrient content and nonenzymatic browning', *J Agr Food Chem*, 49, 483–489.

VERCET A, ORIA R, MARQUINA P, CRELIER S and LOPEZ-BUESA P (2002), 'Rheological properties of yogurt made with milk submitted to manothermosonication', *J Agr Food Chem*, 50, 6165–6171.

VILLAMIEL M and DE JONG P (2000), 'Influence of high-intensity ultrasound and heat treatment in continuous flow on fat, proteins, and native enzyme of milk', *J Agr Food Chem*, 48, 472–478.

© Woodhead Publishing Limited, 2012

WRIGLEY D M and LLORCA N G (1992), 'Decrease of *Salmonella typhimurium* in skim milk and egg by heat and ultrasonic wave treatment', *J Food Prot*, 55, 678–680.

WU H, HULBERT G J and MOUNT J R (2000), 'Effects of ultrasound on milk homogenization and fermentation with yogurt starter', *Innov Food Sci Emerg Technol*, 1, 211–218.

YAP A, JIRANEK V, GRBIN, BARNES M and BATES D (2007a), 'The application of high power ultrasonics to enhance wine-making processes and wine quality', *Aust NZ Wine Indust J*, 22, 44–48.

YAP A, JIRANEK V, GRBIN, BARNES M and BATES D (2007b), 'Studies on the application of high power ultrasonics for barrel and plank cleaning and disinfection', *Aust NZ Wine Indust J*, 22, 96–104.

ZENKER M, HEINZ V and KNORR D (2003), 'Application of ultrasound-assisted thermal processing for preservative and quality retention of liquid foods', *J Food Prot*, 66, 1642–1649.

ZHOU B, FENG H and LUO Y (2009), 'Ultrasound enhanced sanitizer efficacy in reduction of *Escherichia coli* O157:H7 population on spinach leaves', *J Food Sci*, 74, M308–M313.

ZISU B, BHASKARACHARYA R, KENTISH S and ASHOKKUMAR M (2010), 'Ultrasonic processing of dairy systems in large scale reactors', *Ultrason Sonochem*, 17, 1075–1081.

© Woodhead Publishing Limited, 2012

# 11

# Microbial decontamination of food by irradiation

**C. H. Sommers, USDA-ARS, USA**

**Abstract**: The radiation (ionizing) processing industry is large and encompasses commercial uses such as medical therapeutics, anticancer therapy and medical diagnostics as well as sterilization of pharmaceutical, medical devices, and medical supplies. It is used for everything from polymerization and processing of plastics and polymers, quality control of industrial components to detect cracks and other structural defects, to giving gemstones that 'special sparkle'. Food irradiation, one small part of a very large radiation processing industry, is the process of exposing food to a field of ionizing radiation to inhibit sprouting and delay the ripening of fruits and vegetables, destroy microorganisms, parasites, viruses, or insects that might be present in the food. This chapter reviews the principles of the irradiation process and its advantages, its modes of application, and the theories about its mechanisms of inactivation of various organisms.

**Key words**: irradiation, radiation, ionizing, inactivation, decontamination, dosimetry.

## 11.1 Introduction

Even though every bit of food consumers eat, and every liter of water they drink, contain naturally occurring radioisotopes, the process of food irradiation does not induce additional radioactivity in their food (Diehl, 1995). It is an

---

Mention of trade names or commercial products in this publication is solely for the purpose of providing specific information and does not imply recommendation or endorsement by the US Department of Agriculture.

inherently safe process and is used in over 40 countries around the globe. The amount of food irradiated worldwide on an annual basis is estimated to exceed 500,000 metric tonnes; however, this amount is almost certainly a gross underestimate (Kume *et al.*, 2009). The increasing use of food irradiation as a processing technology, which is substantial, is regularly reviewed and updated by organizations such as the Minnesota Beef Council as part of 'Food Irradiation Update' (http://www.mnbeef.org/foodirradiationupdate.aspx) or the International Atomic Energy Agency (http://www.iaea.org).

The idea of using ionizing radiation to improve the microbiological safety of foods is very old and has been wonderfully reviewed by a number of authors including Satin (1993), Diehl (1995), and Josephson (1983). Patents describing the use of ionizing radiation were issued in the United States and Great Britain in 1905 (Appleby and Banks, 1905; Lieber 1905). One of the single most important uses of ionizing radiation, to kill and sterilize insects, was described in 1916 (Runner, 1916), and the use of irradiation to inactivate parasites was patented in 1921 (Schwartz, 1921). The use of ionizing radiation to extend the shelf life of beef for the US army received extensive attention from scientists at the Massachusetts Institute of Technology during World War II, well before Eisenhower's Atoms for Peace Program. Thus, food irradiation predated nuclear power and nuclear weapons by decades, and it is rather unfortunate that this life-saving technology has somehow become associated with atomic energy and weapons.

In reality food irradiation, along with the entire radiation processing industry, is a 'green technology' as it utilizes no chemicals and releases no pollution. It is a technology which is consistent with the concept of sustainable agriculture in a world with an increasing population and a limited amount of arable land and oceans with declining catches. The objections to the use of irradiation technology, unfortunately, are typically based on issues such as globalization, domestic and international trade, and opposition to the nuclear power industry. It is a technology that can be used to prevent the spread of invasive insects and/or kill bacteria in foods ranging from leafy greens to food for astronauts. The story of food irradiation is one of how science has triumphed over fear of the word 'radiation'.

## 11.2 Types, sources, and units of ionizing radiation

Ionizing radiation refers to forms of radiant energy that possess sufficient energy to create negatively and positively charged ions in the target substrate. The three forms of ionizing radiation approved for treatment of foods are gamma radiation, beta radiation (electron-beam), and x-rays (WHO, 1994).

Gamma rays are produced through the natural decay of isotopes (either $^{60}Co$ or $^{137}Cs$). Most, but not all, gamma irradiators use $^{60}Co$ as the source material. The advantage of gamma radiation is the high penetrability of the photons which are produced, allowing irradiation of high bulk density and volume materials. The other advantage is that, because commercial gamma irradiators are always producing photons, they may be run using back-up generators in the case of regional or national emergencies when electrical power grids are inoperable. $^{60}$Co releases photons having energies of 1.33 MeV and 1.17 MeV, while $^{137}$Cs emits photons at an energy of 0.66 MeV. The approximate penetration of $^{60}$Co and $^{137}$Cs photons in a layer of water sufficient to reduce the signal by 90% is 37, 35, and 25 cm for 1.33, 1.77, and 0.66 MeV photons, respectively. Because of the 5.27 year half-life of $^{60}$Co, approximately 12% of the source material must be replaced annually to maintain the original dose-rate of the facility (WHO, 1994).

Electron-beam radiation is a machine source for production of ionizing radiation. Typically, electrons are produced at a maximum energy of 10 MeV. Unlike gamma radiation, which is highly penetrating, even electrons produced at relatively high energies have limited penetration into foodstuffs. For instance, double-sided irradiation of high density food products such as meat are limited to approximately 10 cm thickness. The advantage of electron-beam radiation is that the machine can be turned on and off as needed; there is no isotope source which needs to be replenished.

Machine generated x-rays with a maximum energy of 5 MeV (7 MeV in the US) may be used to treat foodstuffs (WHO, 1994). X-rays offer the advantage of penetration of gamma rays without the use of isotopes. To produce x-rays, an electron-beam is bounced off a tantalum or gold target. However, most of the energy needed to produce x-rays is lost in the form of heat, making the process more expensive than gamma or electron-beam irradiation.

Each source of ionizing radiation available for treatment of foodstuffs and other materials as part of the overall radiation processing industry is inherently safe. Issues such as worker safety are universal across the radiation processing industry, and are not specific to food irradiation. Information safety of radiation processing industry workers is available through the International Atomic Energy Agency (IAEA, 2011http://www. iaea.org).

The basic unit of absorbed dose is the Gray (Gy), where 1 Gy is equal to 1 joule/kilogram or 1 watt-second/kilogram (Cleland, 2006). Irradiation is considered a nonthermal processing technology because of its relatively low thermal capacity in J/g/°C, where a radiation dose of 1 kGy will result in a temperature rise of only 0.24 °C/kg in water-rich foods such as meat, poultry, fish, and produce (Cleland, 2006).

The radiation processing industry is very different from the nuclear power industry and is relatively safe. Accidents resulting in injury or death are exceedingly rare, are often associated with the unauthorized disposal of radiation source material associated with the medical therapeutic and diagnostics industry. Case studies and after-action reports pertaining to accidental release or exposure to radiation and radioactive materials are routinely published by the IAEA (http://www.IAEA.org). Radiation worker safety in facilities irradiating foods is no different from that in the rest of the radiation processing industry. Appropriate shielding, safety interlocks, and personal dosimeters are required at irradiation facilities to prevent accidental exposure of workers, and to monitor exposure of workers to radiation fields. Unfortunately, there are no uniform standards across countries regarding organ-specific, total body, and pregnant radiation industry workers. Fortunately, the IAEA has published easily accessible radiation worker safety guidelines (IAEA, 1999).

## 11.3 Regulations for food irradiation

Irradiation of one or more food commodities is currently allowed in over 40 countries. As these regulations constantly change, current allowances for individual nations should be checked via their regulatory agencies. Current regulatory allowances for irradiation of foods may be obtained from the IAEA (http://www.iaea.org). In the European Union current allowances are listed in the recent review published by the European Food Safety Agency (EFSA, 2011: http://www.efsa.europa.eu/it/efsajournal/doc/s1930.pdf). In the United States the list of approvals is listed in the Code of Federal Regulations (21 CFR 179: http://ecfr.gpoaccess.gov/cgi/t/text/text-idx?c=ecfr&tpl=%2Findex.tpl).

### 11.3.1 Labeling of irradiated foods

Labeling laws vary from country to country. The Codex Alimentarius, which represents the global standard under World Trade Organization (WTO) agreements, has published general guidelines in CODEX STAN-1 (2005) Labeling of prepackaged food. It should be noted, however, that member nations are free to determine how this is accomplished. Some nations use the Radura symbol (Fig. 11.1) to identify foods as irradiated, and a number of countries use the Radura symbol, or a variation of it, that differs from the Codex symbol.

Primary food products must be labeled 'irradiated', as are any derived directly from an irradiated raw material. Some entities have attempted to use the terms 'cold pasteurization' and 'electronic pasteurization' as substitutes for the term irradiated. However, other technologies such as high pressure processing could also be considered 'cold pasteurization', and technologies such as radio-frequency or pulsed electric fields could be considered 'electronic'. Therefore those terms would not be exclusive to, or appropriate for, the radiation processing industry. Older, and now rarely used, obsolete terms include radurization, radappertization, and radicidation (WHO, 1994). The label which is most relevant to modern nonthermal processing technologies is the general term 'pasteurized' as defined by the National Advisory Committee on Microbiological Criteria for Food (NACMCF), which calls for a 5 $\log_{10}$

**Fig. 11.1** The Radura symbol.

reduction of common foodborne pathogens on a product in order to obtain that designation (NACMCF, 2006).

### 11.3.2 Detection of irradiated foods

While irradiation of one or more foodstuffs is allowed in over 40 countries, many still do not allow irradiation of foods or limit the dose that can be used to treat specific food items. Therefore, a variety of technologies have been developed to detect irradiated foods under programs supervised by the IAEA and Community Bureau of Reference (McMurray *et al.*, 1996; Raffi *et al.*, 1993). These detection technologies include detection of free radicals trapped in irradiated foods by electron spin resonance, thermoluminescence, detection of radiolytic products from lipids, and detection of disrupted nucleic acids (EFSA, 2011; Marchioni, 2006).

### 11.3.3 Dosimetry

When materials are irradiated, the distribution of the absorbed dose throughout the target product is non-homogeneous due to the nature of the imparted energy. While irradiation is a nonthermal process, the closest analogy to this non-homogeneous nature is heat transfer kinetics for thermal processing, where the heating is uneven due to the nature of energy transfer, the product mass, and the product dimensions. Therefore, like measuring a temperature profile during thermal processing, the absorbed radiation dose should be measured during radiation processing. When a food, or virtually any material, is irradiated, the absorbed dose within a specific product configuration should be enough for the intended purpose (e.g. inactivation of *Escherichia coli*

O157:H7 in ground beef or leafy greens) but not high enough that product quality is affected (e.g. off-flavors or aromas in beef, or wilting and browning of leafy greens).

Many times the minimum and maximum radiation doses intended for a specific purpose are set by a nation's regulatory agency. Other times they are set by the customer of the radiation processing facility. It is the responsibility of the radiation processing facility, as the service provider, to ensure that equipment is appropriately installed, calibrated and operating within appropriate process and operational qualifications. It is the responsibility of the food product owner to identify the minimum and maximum radiation absorbed doses the product should receive, as well as other requirements, such as the temperature that should be maintained (refrigerated or frozen) during irradiation. The radiation processing facility then irradiates the product to the specifications provided by the food product owner and provides the appropriate documentation and certificates of analysis.

In order to assure product is exposed to radiation doses within set limits (maximum absorbed dose ($D_{max}$) and minimum absorbed dose ($D_{min}$)), and maintain appropriate quality control and assurance, the radiation dose absorbed by a given product during a production run is routinely monitored. Standards for calibration of radiation sources (irradiators), dosimetry systems, determination of uncertainty during radiation processing, dose mapping, etc., are published by ASTM International (www.ASTM.org) which includes both ISO/ASTM Standards and a number of helpful 'Guides' for irradiation of specific foodstuffs such as dry herbs and spices, meats, produce, seafood, etc. In-depth reference books and conceptual texts describing radiation dosimetry practices include those published by McLaughlin *et al.* (1989).

## 11.4 Toxicological safety of irradiated foods

Ionizing radiation is probably the most studied nonthermal processing technology in terms of its toxicological safety in the history of food science because of the psychological impact of the word 'irradiation'. In contrast to irradiation, there is very little peer-reviewed scientific literature that demonstrates the toxicological safety of other nonthermal processing technologies. The WHO considers foods irradiated up to 10 kGy to be toxicologically safe, and a thorough review of the toxicological safety and nutritional adequacy of irradiated foods has been published by that organization (WHO, 1994), which includes evaluation of nutritional quality such as vitamin content. In 1999, that opinion was revised to call all irradiated foods safe for human consumption. The European Food Safety Agency (EFSA, 2011) and the US Food and Drug Administration (FDA, 2005) have thoroughly reviewed the available literature on short-term genotoxicity, animal feeding studies, and teratology studies of irradiated foods. Unlike other nonthermal food processing technologies, there is a large literature base which indicates that irradiated foods are safe for human consumption.

As noted by the EFSA (2011), the vast majority of chemicals which are produced by the irradiation process such as aldehydes, ketones, peroxides, lipid oxidation products, furan, etc., are also produced by thermal processing. One of the more recent issues regarding the toxicological safety of irradiated foods is the presence of radiolytic products known as 2-alkylcyclobutanones (2-ACBs) which are formed by the radiolysis of fats, and are present in ppb and ppm levels in irradiated foods which contain fat (Sommers, 2006b). The 2-ACBs were originally thought to be unique to irradiated food; however, 2-ACBs were detected in non-irradiated cashew nuts. This follows a much earlier report of 2-methylcyclobutanone being naturally present in the tissues of the plant *Hevea brasiliensis* (Nishimura *et al.*, 1977), which received little attention at the time. Given that gamma rays, x-rays, beta particles, and lipids are produced naturally by solar radiation and naturally occurring isotopes, it is difficult to believe that 2-ACBs do not exist, albeit in infinitesimal quantities, in nature.

While the 2-ACBs are non-mutagenic in bacterial and human cells (Burnouf *et al.*, 2002; Sommers, 2006b, Sommers and Mackay, 2005; Sommers and Schiestl, 2004; Delincée *et al.*, 1999; Delincée and Pool-Zobel, 1998), they have been found to induce oxidative stress, cell membrane damage, the formation of DNA strand breaks, micronuclei, and chromosomal rearrangements in human and animal cells *in vitro* (Sommers, 2006b; Knoll *et al.*, 2006; Delincée *et al.*, 1999; Delincée and Pool-Zobel, 1998). Later, it was found that the parent fatty acids of the 2-ACBs induce the same effects in human and animal cells lines (Ji *et al.*, 2005; Beeharry *et al.*, 2003; Uloth *et al.*, 2003; Udilova *et al.*, 2003; Nogueirra *et al.*, 2005). Thus, there is little difference in the biological activity of 2-ACBs versus those of their parent fatty acids (Sommers, 2007). While 2-ACBs do not induce tumors in animals, it is possible they have tumor promotion activity similar to excess fatty intake observed in humans for those substances (Bartsch *et al.*, 2003; Coquhoun and Guri, 1998). Therefore, 2-ACBs are not considered to be toxicologically important when considered in the context of the biological activities of their parent fatty acids, their detection in non-artificially irradiated foods, and the long-term feeding and teratology studies that have been conducted in animals.

One of the most recent controversies pertaining to food irradiation was the recent death of cats in Australia following the radiation sterilization of cat food at doses of >50 kGy, which was most likely due to destruction of micronutrients as a result of the irradiation process. Because of this, radiation sterilization of cat foods has been discontinued in Australia. Because irradiated foods destined for human consumption are a relatively small fraction of the diet, and receive radiation doses which do not produce significant impact on vitamin and nutrient content (EFSA, 2011; FDA, 2005; WHO 1994), this issue is not relevant to low dose irradiation of foods.

### 11.4.1 Mode of biological inactivation by ionizing radiation

Ionizing radiation kills, sterilizes, or renders insects and microorganisms unable to reproduce by damaging nucleic acid, specifically DNA or RNA (Ward, 1991). It induces DNA strand breaks, transition and transversion mutations, frameshift mutations, and deletion mutations (Glickman *et al.*, 1980; Raha and Hutchinson, 1991; Sargentini and Smith, 1994; Wijker *et al.*, 1996). Ionizing radiation damages DNA by two mechanisms, direct action against DNA by photon-induced breakage of the DNA phosphodiester backbone or indirect damage to the DNA by the radiaolysis of water, primarily caused by hydroxyl radicals. Indirect damage accounts for approximately 70% of the DNA damage caused by ionizing radiation.

The increased radiation resistance of microorganisms at subfreezing temperatures has been attributed to the lower $a_w$ of meat at subfreezing temperatures and to decreased hydroxyl radical mobility following the radiolyis of water when in the frozen state (Taub *et al.*, 1979; Bruns and Maxcy, 1979). Note the increasing radiation resistance of *Yersinia enterocolitica* at subfreezing temperatures shown in Fig. 11.2.

In addition to damage to chromosomes and nucleic acids, ionizing radiation can also disrupt proteins and cellular membranes. Naidu *et al.* (1998) found the lipopolysaccharide (LPS) cell membrane component isolated from

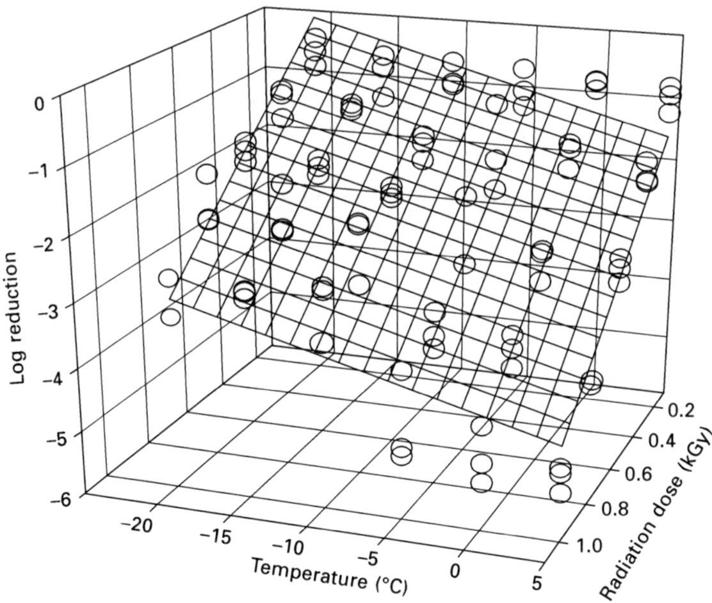


**Fig. 11.2** A 3D mesh plot of log reduction data for *Y. enterocolitica* as a function of radiation dose and product temperature. The predictive equation followed a parabolic fit ($R^2 = 0.85$) where $\log_{10}$ reduction = –0.666 – (2.955 × Dose) – (0.069 × Temp.) – (0.484 × Dose$^2$) – (0.001 × Temp.$^2$). Note the increased radiation resistance of the pathogen at –20°C as opposed to 5°C (reproduced with permission).

*Salmonella* Typhimurium was depolymerized following exposure to ionizing radiation. The irradiated LPS and lipid A were less toxic and mitogenic in animal studies. Ionizing radiation can also disrupt cellular membrane-DNA complexes required for DNA repair and chromosome/plasmid partitioning (Khare *et al.*, 1982; Watkins, 1980).

Another important concept involving the mechanism by which ionizing radiation inactivates microorganisms is the sensitivity of foodborne pathogens to heat following irradiation. Such damage increases the sensitivity of microorganisms to heat through thermalability of the bacterial DNA (Alvarez *et al.*, 2006; Sommers and Schiestl, 2004; Kim and Thayer, 1996). The risk associated with consumption of undercooked foods becomes more serious when consumer preferences and cooking habits (especially meats, e.g medium rare) are considered. In the US current guidelines for the cooking of beef call for reaching an internal temperature of 160°F (71°C) in order to inactivate pathogenic bacteria such as *E. coli* O157:H7 (USDA FSIS, 2004). However, research by US FDA and USDA FSIS showed that only 60% of households have a meat thermometer, and only 6% of consumers report using one on a regular basis (Cates, 2002). It has been shown that less than 5% of participants used a thermometer to determine doneness of meat. Almost 45% of the study participants reported not knowing the recommended cooking temperature for ground beef. In addition to the lack of consumer knowledge is the preference of many consumers for beef cooked to a medium rare temperature. Because ionizing radiation induces DNA strand breakage and DNA damage is maintained by foodborne pathogens during refrigerated storage, the heat sensitivity of some pathogens can be maintained for as long as 3 weeks post-irradiation (Sommers and Fan, 2011) (Fig. 11.3). Additional research is needed to quantify foodborne pathogen heat sensitivity in refrigerated and frozen foods.

### 11.4.2 Induced radiation resistance, antibiotic resistance, and virulence loss

Increased resistance of microorganisms to ionizing radiation following repeated exposure has been observed in a number of studies. Parisi and Antoine (1974) observed increasing radiation resistance of *Bacillus pumilus* following repeated exposure to sublethal doses of ionizing radiation. Erdman *et al.* (1961) demonstrated a similar phenomenon using *E. coli*, *Streptococcus faecalis*, *Clostridium botulinum*, *Staphylococcus aureus*, and *Salmonella*. While it is theoretically possible to create highly virulent radiation resistant 'superbugs', no such occurrence has ever been documented in the last fifty years. In contrast, research indicates that ionizing radiation decreases foodborne pathogen virulence through induced loss of large MDa 'virulence plasmids' common to foodborne pathogens and mutation of chromosomal virulence factors (Sommers and Schiestl, 2004).

Another significant issue has been the emergence of antibiotic resistant

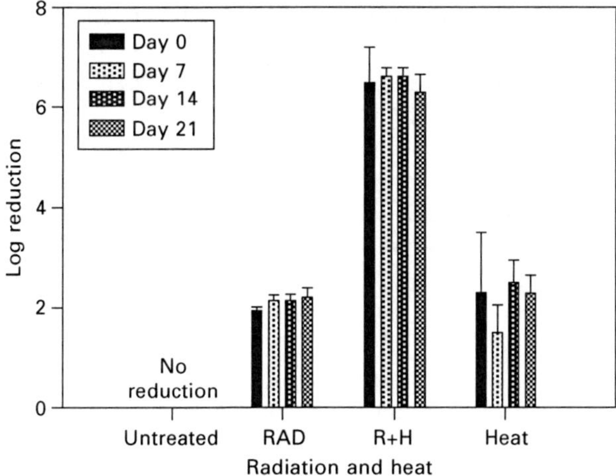


**Fig. 11.3** Inactivation of *Salmonella* Senftenberg in 125 g extra-lean ground beef patties after Week 0 irradiation (1.5 kGy), followed by either cooking (internal temperature 65.5°C: medium rare), or irradiation and cooking (R+H) during a 3 week refrigerated storage period. Untreated = inoculated controls ($n$ = 3). Heat sensitivity is maintained for 3 weeks at refrigeration temperature post-irradiation (from Sommers and Fan, 2011 with permission).

human pathogens such as multi-drug resistant *Salmonella*, *S. aureus* MRSA, and other bacteria containing the NDM-1 antibiotic resistance markers through the gene transfer process and the misuse of antibiotics. Most recently, in a foodborne illness outbreak associated with consumption of contaminated sprouts, the pathogenic *E. coli* isolate was found to be resistant to multiple antibiotics. Thayer *et al.* (1992) demonstrated there was no difference in the radiation resistance of streptomycin resistant and streptomycin sensitive *S.* Typhimurium inoculated onto chicken wings. Later, Niemira *et al.* (2006) demonstrated that resistance to the antibiotics ampicillin, chloramphenicol, and gentamycin had no effect of the radiation resistance of *Salmonella*. In a separate study, Niemira (2005) and Niemira and Lonczynski (2006) demonstrated that nalidixic acid resistant *Salmonella* were more sensitive to ionizing radiation then their antibiotic resistant parents, most likely due to the fact that nalidixic acid resistance is conferred by mutation of genes required for DNA repair and replication. These studies indicate that ionizing radiation is an effective technology for inactivating antibiotic resistant foodborne pathogens.

## 11.5 Microbial inactivation

### 11.5.1 $D_{10}$ value

Microorganisms vary in their resistance to irradiation. The relative resistances of different species can be compared through the use of $D_{10}$ values, where the $D_{10}$ value is the dose required (in kGy) to achieve a 1-log reduction (or 90%) in viable numbers. $D_{10}$ is typically defined as the reciprocal of the slope for the log reduction of the microorganism (Diehl, 1995). Typically, inactivation of vegetative foodborne pathogens ($D_{10}$ values) follows first order kinetics under set conditions (Fig. 11.2). However, the response to irradiation may sometimes be nonlinear. An initial shoulder may be observed when plotting $\log_{10}$ numbers of survivors against dose. This should be considered when determining the radiation dose needed to obtain a 5-log reduction of foodborne pathogens (NACMCF, 2006).

### 11.5.2 Vegetative foodborne pathogens

The most common vegetative (non-spore formers) microorganisms associated with foodborne illness include *Campylobacter*, pathogenic *Escherichia coli*, *Listeria monocytogenes*, *Salmonella* spp., *Shigellae*, *Vibrio* spp., and *Yersinia enterocolitica*. Historical data indicate that the food matrix, water activity ($a_w$), temperature, modified atmosphere, the presence of GRAS antimicrobials, and the growth and physiological state of the microorganism can all affect the radiation resistance of foodborne pathogens (EFSA, 2011). Regardless of these issues, the order of radiation resistances of foodborne pathogens suspended in refrigerated food products follows a hierarchy with *Salmonella* spp. > *Listeria monocytogenes* ≥ *S. aureus*, *Escherichia coli* O157:H7 > (*Campylobacter*, *Y. enterocolitica*, *Vibrio* spp.), with *Salmonella* spp. typically being the most resistant. The radiation resistance of *Y. enterocolitica*, *Campylobacter* spp., *Vibrio* spp., and Shigella spp. are typically below 0.25 kGy on most food products and are easily controlled in foods by ionizing radiation (FDA, 2005; Patterson, 1995). The radiation resistances of common foodborne pathogens (*Escherichia coli* O157:H7, *L. monocytogenes*, *Salmonella* spp., *S. aureus*) are listed in Table 11.1 and can vary widely depending on the food product matrix and product temperature during irradiation.

### 11.5.3 Endospore formers

Endospores are more resistant to ionizing radiation of the same species, possibly due to reduced availability of water in the spores for hydrolysis of water and also the condensed nature of the bacterial chromosome within the spore. *Clostridium botulinum* spores have relatively high $D_{10}$ values that range from 1.29 to 3.34 kGy (Anellis and Koch, 1962). *C. sporogenes* spores have $D_{10}$ values of 6.3, 7.8, and 10.1 kGy when suspended in beef, pork, and chicken fat, respectively (Shamsuzzaman and Lucht, 1993). *C. perfringens* vegetative cells have reported $D_{10}$ values ranging from 0.34 to 0.83 kGy when inoculated into a variety of food matrices (Kamat *et al.*, 1989), while *C. perfringens* spores inoculated into clam meat had $D_{10}$ values of 2.7 kGy (Harewood *et al.*, 1994).

**Table 11.1** Radiation resistance ($D_{10}$ value, kGy) of foodborne pathogens in various food matrices

| | Refrigerated meat and poultry[a] | Frozen (–20°C) meat and poultry[b] | Refrigerated processed meats[c] | Refrigerated vending foods | Frozen (–20°C) seafood[d] | Frozen (–20°C) vegetables[e] | Refrigerated green vegetables | Seeds for sprouting[f] | Sprouts |
|---|---|---|---|---|---|---|---|---|---|
| *E.coli* O157:H7 | 0.29–0.32 | 0.98 | 0.23–0.33 | 0.32–0.39 | 0.32–0.36 | NR | 0.30–0.45 | 0.56–1.43 | 0.30–0.34 |
| *Salmonella* spp. | 0.51–0.71 | 1.18 | 0.65–0.74 | 0.59–0.63 | 0.47–0.70 | NR | 0.38–0.47 | 0.74–1.24 | 0.46–0.54 |
| *S. aureus* | 0.40–0.46 | 0.87 | 0.43–0.62 | 0.48–0.57 | 0.53–0.56 | NR | 0.28–0.42 | NR | NR |
| *L. monocytogenes* | 0.40–0.47 | 1.21 | 0.48–0.67 | 0.43–0.53 | 0.48–0.66 | 0.77–0.92 | 0.19–0.24 | NR | 0.20–0.22 |

NR = Not reported.
Sources:
[a]Thayer *et al.*, 1995.
[b]Thayer and Boyd, 1995, 2001.
[c]Sommers and Mackay (in press).
[d]Black and Jaczynski, 2008.
[e]Niemira *et al.*, 2002.
[f]Bari *et al.*, 2009; Rajkowski *et al.*, 2003; Thayer *et al.*, 2003.

As with *Clostridium* species, the $D_{10}$ values of *Bacillus* spores were significantly higher than their vegetative cells. *B. cereus* spores suspended in phosphate buffer were reported to have $D_{10}$ values of 2.5 to 4.0 kGy, while the microorganism's vegetative cells had $D_{10}$ values of 0.3–0.65 kGy (Kamat *et al.*, 1989). When *B. cereus* vegetative cells and spores were suspended in mechanically deboned chicken meat, $D_{10}$ values were 0.19 kGy for log phase vegetative cells, 0.45 kGy for stationary phase vegetative cells, and 2.67 kGy for the spores. Spores of *B. anthracis* suspended in non-fat dry milk applied to postal envelopes had a $D_{10}$ of 3.32 kGy, while spores on envelopes had a $D_{10}$ of 1.53 kGy (Helfinstine *et al.*, 2005; Niebuhr and Dickson, 2003).

### 11.5.4 Parasites

Parasites commonly found in foods include *Trichenella spiralis*, *Taenia solium*, *Taenia saginata*, *Toxiplasma godi*, and cryptosporidium. Parasite larvae and cysts are relatively sensitive to ionizing radiation. Brake *et al.* (1985) found that 0.15–0.30 kGy prevented the maturation of *T. spiralis* and production of parasite progeny. Verster *et al.* (1977) were able to eradicate *T. solium* and *T. saginata*, beef and pork tapeworms, from carcasses using radiation doses of 0.2–0.6 kGy with no effect on carcass quality. Tolgay (1972) found that a radiation dose of 0.4 kGy was sufficient to prevent maturation of *T. saginata* cysts in beef but that 1.0 kGy was needed for elimination of the adult parasite. Dubey *et al.* (1996, 1998) investigated the use of ionizing radiation to eliminate sporulated and unsporulated *T. gondi*. Unsporulated *T. gondi* oocytes irradiated to doses of 0.4–0.8 kGy were able to undergo sproulation but were not able to infect mice. Sporulated oocytes irradiated to 0.4 kGy were infective, but unable to reproduce. Unlike the relatively low doses of ionizing radiation needed to control other parasites, the radiation dose needed to inactivate the protozoan parasite *Cryptosporidium parvum* was >10 kGy (Lee *et al.*, 2009; Jenkins *et al.*, 2004).

### 11.5.5 Irradiation for phytosanitary purposes

One of the most significant uses of ionizing radiation is its use as a phytosanitary treatment for the control and sterilization of insect pests, and it is an effective replacement for the toxicant methyl bromide that is currently used in many countries as a fumigant for insect control, or for hot water treatments which could adversely affect product quality. Other than irradiation of dry herbs and spices, use of irradiation as a phytosanitary treatment may be the largest application and market for the foreseeable future. As with microorganisms, sterilization of insect larvae is typically due to disruption of DNA in rapidly dividing tissues. The tissues most susceptible to irradiation are the eggs and the reproductive tissues (Koval 1994; Ducoff 1972).

A quarantine pest is a plant pest of potential economic importance to an area that is not yet present there, or present but not widely distributed and being officially controlled. Quarantine or phytosanitary treatments eliminate, sterilize, or kill regulatory pests in exported commodities to prevent their introduction and establishment into new areas. Treatments for quarantine pests are commonly referred to as 'Probit 9'. A response at the Probit 9 level results in a 99.9968% response, and to achieve Probit 9 mortality at the 95% confidence level, 93,613 insects must be tested with no survivors. Maximum radiation doses of 1 kGy applied to produce are sufficient to meet Probit 9 requirements for most quarantine pests.

### 11.5.6 Viruses

While it is effective for the inactivation of vegetative bacteria, ionizing radiation is only partially effective against viruses. Although reduced in titer, polio virus could be recovered from hydrated and dehydrated meat following exposure to radiation doses of 6 kGy (Heidelbaugh and Giron, 1969). Sullivan *et al.* (1971) determined the $D_{10}$ values for 30 viruses of public health significance, which ranged from 3.8–5.0 kGy when suspended in Eagles Minimal Essential Medium. Pirtle *et al.* (1997) investigated the combined effects of ionizing radiation (4.4 and 5.27 kGy) and heat on the destruction of two RNA and DNA viruses suspended in either saline or ground pork. The authors concluded that ionizing radiation did not increase the heat sensitivity to heat inactivation and that the method has no practical application toward the goal of virus removal. Foot and Mouth Disease virus suspended in bovine tissues can be eliminated at radiation doses of 15 of 25 kGy in combination with heat treatment (Lasta *et al.*, 1992). However, these radiation doses are well above the 4.5 or 5.0 kGy doses allowed for treatment of refrigerated or frozen meat in the US. Bidawid *et al.* (2000) found that radiation doses of 2.72 and 2.92 were required to inactivate hepatitis A virus in lettuce and strawberries, respectively, which is well above the radiation threshold for maintaining the quality of those products. Feng *et al.* (2011) found that a gamma radiation dose of 5.6 kGy inactivated only 1.7–2.4 log of murine norovirus-1 and vesicular stomatitis virus, which are surrogates for human norovirus, on fresh produce.

### 11.5.7 Transmissible encephalopathy (TSE) and bovine spongiform encephalopathy (BSE)

While many infectious agents are living organisms, or viruses, a key topic in food safety has been the emergence and incidence of transmissible encephalopathies (TSEs), primarily bovine spongiform encephalopathy (BSE) and its human counterpart Creutzfeld-Jakob Disease (CJD). The infectious agent responsible for TSE has been termed the 'prion', which Prusiner (1982) defined as 'a proteinaceous infectious particle – a small infectious agent consisting largely or solely of protein' (Schreuder, 1994). Alper (1966) first described attempts to inactivate prions isolated from the brains of infected mice by irradiating them and then reinjecting them into healthy mice. Alper (1966) calculated the radiation dose needed to provide a 1-log reduction of prion to be approximately 43 kGy. Ionizing radiation is simply not effective for inactivation of prions.

## 11.6 Consumer acceptance of irradiated foods

For many years the consumer acceptance of irradiated foods has been hampered by the inappropriate association of the technology with nuclear power and nuclear weapons. Consumer acceptance is a slippery thing to grab onto. Who is the consumer, and what role do consumer interest groups play in the acceptance of irradiated foods by the public? This is an issue that was discussed in detail by Morton Satin in '*Food Irradiation, a Guide Book*' (Satin, 1993), as many consumer interest groups have very little in common with the actual consumer. The objections to food irradiation often have to do with issues such as opposition to any nuclear technology, opposition to globalization of trade, and objections to industrial production of agricultural commodities. Persons opposed to food irradiation based on those concepts, as part of their lifestyle or personal philosophy, will almost certainly never accept or purchase irradiated foods.

Objections to food irradiation include:

- it will mask food spoilage;
- it will impact food nutritional quality;
- it will impair food flavor and aroma;
- it will not inactivate all viruses and toxins;
- it will be used as a clean-up process and discourage good manufacturing practices; and
- irradiated foods are potentially harmful to humans.

Of course, these arguments are true for virtually all food process intervention technologies, and are not specific to food irradiation. As noted by Satin (1993), those same objections were made for pasteurization of milk.

As discussed by Eustice and Bruhn (2006), consumer acceptance of irradiated foods improves when factual educational materials are provided to counter negative information, and consumer acceptance tends to increase with increasing educational level of the consumer. Of particular importance are endorsements of the process by organizations such as the American Medical Association and American Dietetic Association. Despite some opposition to food irradiation as a process, consumer acceptance and availability of irradiated foods is slowly increasing as commodities such as irradiated tropical fruits are appearing in the marketplace as noted in monthly news releases from organizations such as the Minnesota Beef Council (http://www.mnbeef.org/foodirradiationupdate.aspx).

## 11.7 Limitations and challenges of irradiation technology

Perhaps the single largest barrier to more widespread acceptance and adoption of irradiated foods is cost. The USDA Economic Research Service estimated the cost of food irradiation in large capacity processing plants to be less than US$0.11 per pound. However, those estimates do not include the cost of transportation to and from radiation processing facilities due to increased fuel costs in recent years. Later cost estimates of irradiated ground beef indicated US$0.13–0.30 per pound more than non-irradiated ground beef (Roos 2003; Burros 2003; Brown 2002; Melgares 2002). In more recent testimony before the United States Congress, the cost of irradiated ground beef was US$0.3–0.4 over that of non-irradiated ground beef. That differential has almost certainly increased as fuel and energy costs have increased since 2011. While there has been a considerable increase in the sales and availability of irradiated produce (e.g. mangoes, papaya, and other exotic tropical fruits) for phytosanitary purposes, much of that produce is shipped by air freight, which is costly. Research which would allow irradiation and ground/sea transport of unripened product sealed containers is desperately needed to reduce the cost of these irradiated commodities.

## 11.8 Conclusion and future trends

Food irradiation is probably the only nonthermal processing technology which can be used to inactivate parasites, pathogenic microorganisms, and insects on the interior of a wide variety of food product, in the high volumes required in today's food processing environment. It is a green technology which is consistent with the practice of sustainable agriculture. While there is an innate fear of the word 'irradiation', which creates a negative consumer perception, the global market for irradiated foods is expanding, especially in the use of irradiation as a phytosanitary treatment to prevent the spread of invasive insects between continents. Perhaps the largest area of emerging research in the field of food irradiation is the harmonization of doses for phytosanitary and pathogen reduction and the ability to irradiate produce in the unripened state for compatibility with the extended times required for sea and ground shipment as opposed to air freight.

## 11.9 Sources of further information and advice

- International Atomic Energy Agency (http://www.iaea.org).
- In the European Union current allowances are listed in the recent review published by the European Food Safety Agency (http://www.efsa.europa.eu/it/efsajournal/doc/s1930.pdf).

- In the United States the list of approvals is listed in the Code of Federal Regulations (21CFR179: http://ecfr.gpoaccess.gov/cgi/t/text/text-idx?c=ecfr&tpl=%2Findex.tpl).
- Minnesota Beef Council as part of 'Food Irradiation Update' (http://www.mnbeef.org/foodirradiationupdate.aspx).

## 11.10 References and further reading

ALPER T. 1966. The exceptionally small size of the scrapie agent. *Biochem Biophys Res Comm* 22: 278–284.

ALVAREZ I, NIEMIRA B A, FAN X, SOMMERS C H. 2006. Inactivation of *Salmonella* serovars in liquid whole egg by heat following irradiation treatments. *J Food Prot* 69: 2066–2074.

ANELLIS A, KOCH R. 1962. Comparative resistance of strains of *Clostridium botulinum* to gamma rays. *Appl Environ Microbiol* 10: 326–330.

APPLEBY J, BANKS A. 1905. British patent, No. 1609.

BARI M, NAZUKA E, SABINA Y, TODORIKI S, ISSHIKI K. 2002. Chemical and irradiation treatments for killing *Escherichia coli* O157:H7 on alfalfa, radish and mung bean seeds. *J Food Prot* 66: 767–774.

BARI M, AL-HAQ M, KAWASAKI T, NAKAUMA M, TODORIKI S, KAWAMOTO S, ISSHIKI K. 2004. Irradiation to kill *Escherichia coli* O157:H7 and *Salmonella* on ready-to-eat radish and mung bean sprouts. *J Food Prot* 67: 2263–2268.

BARI M, NAKAUMA M, TODORIKI S, JUNEJA V, ISSHIKI K, KAWOMOTO S. 2005. Effectiveness of irradiation treatments in killing *Listeria monocytogenes* on fresh vegetables at refrigerated temperature. *J Food Prot* 68: 318–323.

BARI M, NEI D, ENOMOTO K, TODORIKI S, KAWAMOTO S. 2009. Combination treatment for killing *Escherichia coli* O157:H7 on alfalfa, radish, broccoli and mung bean seeds. *J Food Prot* 72: 631–636.

BARTSCH H, NAIR J, OWEN R W. 2003. Dietary polyunsaturated fatty acids and cancers of the breast and colorectum: emerging evidence for their role as risk modifiers. *Carcinogenesis* 20: 2209–2218.

BEEHARRY N, LOWE J, HERNANDEZ A R, CHAMBERS J A, FUCASSI F, CRAGG P J, GREEN M, GREEN I. 2003. Linoleic acid and antioxidants protect against DNA damage and apoptosis induced by palmitic acid. *Mutat Res* 530: 27–33.

BIDAWID S, FARBER J, SATTAR S. 2000. Inactivation of hepatitis A virus (HAV) in fruits and vegetable by gamma irradiation. *Int J Food Microbiol* 57: 91–97.

BLACK J, JACZYNSKI J. 2008. Effect of water activity on the inactivation kinetics of *Escherichia coli* O157:H7 by electron beam in ground beef, chicken breast meat, and trout fillets. *Int J Food Sci Technol* 43: 579–586.

BRAKE R, MURRELL K, RAY E, THOMAS J, MUGGENBURG B, SIVINSKI J. 1985. Destruction of *Trichenella spiralis* by low dose irradiation of pork. *J Food Safety* 7: 127–143.

BROWN J L. 2002. Food Irradiation 3: Labeling and Cost of Irradiated Foods. The Pennsylvania State University, College of Agricultural Sciences, Agricultural Research and Cooperative Extension.

BRUNS M W, MAXCY R B. 1979. Effect of irradiation temperature and drying on survival of highly radiation resistant bacteria in complex menstrua. *J Food Sci* 44: 1743–1746.

BURNOUF D, DELINCÉE H, HARTWIG A, MARCHIONI E, MIESCH M, RAUL F, WERNER D. 2002. Etude toxicologique transfrontalière destinée à évaluer le risque encouru lors de la consommation d'aliments gras ionisés. Toxikologische Untersuchung zur Risikobewertung beim Verzehr von bestrahlten fetthaltigen Lebensmitteln. Eine französisch-deutsche

Studie im Grenzraum Oberrhein. Rapport final/Schlussbericht INTERREG II. Projet/Projekt No. 3.171. Marchioni E, Delincée H; eds. Bundesforschungsanstalt für Ernährung, Karlsruhe.

BURROS M. 2003. Schools in no hurry to buy irradiated beef. *New York Times*, October 8 Accessed December 2006 at: http://query.nytimes.com/gst/fullpage.html?sec=health&res=9B05E1DA113CF93BA35753C1A9659C8B63.

CATES S. 2002. Reported Safe Handling Practices: Cooking (FDA/FSIS Food Safety Survey-2001). Changes in Consumer Knowledge, Behavior, and Confidence Since 1996 PR/HACCP Final Rule. Thinking Globally-Working Locally: A Conference on Food Safety Education, Orlando, FL, September 18, 2002.

CLELAND M. 2006. Advances in gamma ray, electron beam, and X-ray technologies for food irradiation. In: *Food Irradiation Research and Technology* (Sommers C, Fan X, eds). IFT Press-Blackwell Publishing, Ames, IA, pp. 11–36.

CODEX ALIMENTARIUS. 2005. Labeling of Prepackaged Foods (CODEX STAN-1).

COQUHOUN A, GURI R. 1998. Effects of saturated and polyunsaturated fatty acids on human tumor cell proliferation. Gen Pharmacol 30(2): 191–194.

DELINCÉE H, POOL-ZOBEL B L. 1998. Genotoxic properties of 2-dodecylcyclobutanone, a compound formed on irradiation of food containing fat. *Radiat Phys Chem* 52: 39–42.

DELINCÉE H, POOL-ZOBEL BL, RECHKEMMER G. 1999. Genotoxizität von 2-Dodecylcylobutanon. In: *Lebensmittelbestrahlung* (Knörr M, Ehlermann D A E, Delincée H, eds), 5. Deutsche Tagung, Karlsruhe, Berichte der Bundesforschungsanstalt für Ernährung, 11–12 Nov. 1998, *BFE-R–99-01*, pp. 262–269.

DIEHL J. 1995. *Safety of Irradiated Foods*. Marcel Dekker, New York.

DUBEY J, JENKINS M, THAYER D. 1996. Irradiation killing of *Toxoplasma gondii* oocysts. *J Eukaryot Microbiol* 45: 123S.

DUBEY J, THAYER D, SPEER C, SHEN S. 1998. Effect of gamma irradiation on unsporulated and sporulated *Toxoplasma gondii* oocysts. *Int J Parasitol* 28: 369-374.

DUCOFF H. 1972. Causes of cell death in irradiated adult insects. *Biol Rev* 47: 211–240.

EFSA. 2011. Statement summarising the conclusions and recommendations from the opinions on the safety of irradiation of food adopted by the BIOHAZ and CEF Panels. *EFSA Journal* 9(4): 2107.

EUSTICE R, BRUHN C. 2006. Consumer acceptance and marketing of irradiated foods. In: *Food Irradiation Research and Technology* (Sommers C, Fan X, eds). IFT Press-Blackwell Publishing, Ames, IA, pp. 63–84.

ERDMAN I, THATCHER F, MACQUEEN K. 1961. Studies on the irradiation of microorganisms in relation to food preservation. II. Irradiation resistant mutants. *Can J Microbiol* 7: 207–215.

FDA. 2005. Irradiation in the Production, Processing, and handling of Food. 21 CFR Part 179 [Docket No. 1999F-4372]. *Federal Register* 70(157): 48057–48073.

FENG K, DIVERS E, MA Y, LI J. 2011. Inactivation of a human norovirus surrogate, human norovirus virus-like particles, and vesicular stomatitis virus by gamma irradiation. *Appl Environ Microbiol* 77: 3507–3517.

FRENZEN P, MAJCHROWICZ A, BUZBY J, IMHOFF B, FOODNET WORKING GROUP. 2000. Consumer acceptance of irradiated meat and poultry products. Issues in Food Safety Economics. Agriculture Information Bulletin 757/August 2000: 1–8.

GADGIL P, SMITH J. 2004. Mutagenicity and acute toxicity evaluation of 2-dodecylcyclobutanone. *J Food Sci* 69: 713-716.

GADGIL P, SMITH, J. 2006. Metabolism by rats of 2-dodecylcyclobutanone, a radiolytic compound present in irradiated beef. *J Agric Food Chem* 54: 4896–4990.

GLICKMAN B, RIETVELD K, AARON C. 1980. X-Ray induced mutational spectrum in the *lacI* gene of *Escherichia coli*. *Mutat Res* 69: 1–12.

HAREWOOD P, RIPPLEY S, MONTESALVO M. 1994. Effect of gamma-irradiation on shelf life and bacterial and viral loads in hard-shelled clams. *Appl Environ Microbiol* 60: 2666–2670.

HEALTH CANADA. 2003. Evaluation of the significance of 2-dodecylcyclobutanone and other alkylcyclobutanones. Accessed September 2006 at: http://www.hc-sc.gc.ca/food-aliment/fpi-ipa/e-cyclobutanone.html

HEIDELBAUGH N, GIRON D. 1969. Effect of processing on recovery of polio virus from inoculated foods. *J Food Sci* 34: 239–241.

HELFINSTINE S, VARGAS-ABURTO C, URIBE R, WOOLVERTON C. 2005. Inactivation of Bacillus endospores in envelopes by electron beam irradiation. *Appl Environ Microbiol* 71: 7029–7032.

IAEA. 1999. Occupational Radiation Protection, IAEA Safety Standard Series. Safety Guide RS-G-1.1. International Atomic Energy Agency, Vienna.

JENKINS M, HIGGINS J, KNIEL K, TROUT J, FAYER R. 2004. Protection of calves against cryptosporidiosis by oral inoculation with gamma-irradiated *Cryptosporidium parvum* oocysts. *J Parasitol* 90: 1178–1180.

JI J, ZHANG L, WANG P, MU Y, ZHU X, WU Y, YU H, ZHANG B, CHEN S, SUN X. 2005. Saturated free fatty acid, palmitic acid, induces apoptosis in fetal hepatocytes in culture. *Exp Toxicol Pathol* 56: 369–376.

JOSEPHSON E. 1983. An historical review of food irradiation. *J Food Safety* 5: 161–189.

KAMAT A, NERKAR D, NAIR P. 1989. *Bacillus cereus* in some Indian foods: incidence and antibiotic, heat, and radiation resistance. *J Food Safety* 10: 31–41.

KATO T, ODA Y, GLICKMAN B. 1985. Randomness of base substitution mutations induced in the *lacI* gene of *Escherichia coli* by ionizing radiation. *Radiat Res* 101: 402–406.

KHARE S, JAYAKUMAR A, TREVVEDI A, KESEVAN P, PRESAD R. 1982. Radiation effects on membranes. *Radiat Res* 90, 233–243.

KIM A, THAYER D. 1996. Mechanism by which gamma irradiation increases the sensitivity of *Salmonella typhimurium* ATCC14028 to heat. *Appl Environ Microbiol* 62: 1759–1763.

KNOLL N, WEISE A, CLAUSSEN U, SENDT W, MARIAN B, GLEI M, POOL-ZOBEL B L. 2006. 2-Dodecylcyclobutanone, a radiolytic product of palmitic acid, is genotoxic in primary human colon cells and in cells from preneoplastic lesions. *Mutat Res* 594: 10–19.

KOMOLPRASERT V, MOREHOUSE K. 2004. *Irradiation of Food and Packaging*. ACS Symposium Series 875. American Chemical Society, Washington, DC.

KOVAL, T. 1994. Intrinsic stress resistance of cultured lepidopteran cells. In: *Insect Cell Biotechnology* (Marmarorosh K, McIntosh A, eds). CRC Press, Boca Raton, FL, pp. 157–185.

KUME T, FURATA, TODORIKI S, UENOMOYA N, KOBAYASHI Y. 2009. Status of food irradiation worldwide. *Radiat Phys Chem* 78: 222–226.

LASTA J, BLACKWELL J, SADIR A, GALLINGER M, MARCOVECCIO F, ZAMORANO M, LUDDENAND B, RODRIQUEZ R. 1992. Combined treatments of heat, irradiation, and pH effects on infectivity of foot and mouth disease virus in bovine tissues. *J Food Sci* 49: 665–667.

LEE S-U, JOUNG M, NAM T, WOO-YOON PARK W-Y, YU J-R. 2009. Quantitative evaluation of infectivity change of *Cryptosporidium parvum* after gamma irradiation. *Korean J Parasitol* 47: 7–11.

LIEBER H. 1905. US Patent 788480.

MARCHIONI E. 2006. Detection of irradiated foods. In: *Food Irradiation Research and Technology* (Sommers C, Fan X, eds) IFT-Blackwell Publishing, Ames, IA, pp. 85–104.

MCLAUGHLIN W, BOYD A, CHADWICK K, MCDONALD J, MILLER A. 1989. *Dosimetry for Radiation Processing*. Taylor and Francis, New York.

MCMURRAY C, STEWART M, GRAY R, PEARCE J. 1996. *Detection Methods for Irradiated Foods – Current Status*. Royal Society of Chemistry, Cambridge.

MELGARES K. 2002. K-State Food Scientist: Consumers Finding Safety in Irradiated Foods. Kansas State University Research and Extension Service News, November 7. Accessed December 2006 at: http://www.oznet.ksu.edu/news/sty/2002/irradiated_foods110702.htm

NACMCF (NATIONAL ADVISORY COMMITTEE ON MICROBIOLOGICAL CRITERIA FOR FOOD). 2006. Requisite scientific parameters for establishing the equivalence for alternative methods of pasteurization. *J Food Prot* 69: 1190–1216.

NAIDU M, CHANDER R, NAIR P M. 1998. Effect of gamma irradiation on chemical and biological properties of lipopolysaccharide from *Salmonella typhimurium. Indian J. Exp. Biol.* 36: 588–592.

NIEBUHR S, DICKSON J. 2003. Destruction of *Bacillus anthracis* strain Sterne 34F2 spores in postal envelopes by exposure to electron beam irradiation. *Lett Appl Microbiol* 37(1): 17–20.

NIEMIRA B. 2005. Nalidixic acid resistance increases sensitivity of *escherichia coli* O157:H7 to ionizing radiation in solution and on green leaf lettuce. *J Food Sci* 70: M121–M124.

NIEMIRA B, LONCZYNSKI K. 2006. Nalidixic acid resistance influences sensitivity ionizing radiation among *Salmonella isolates. J Food Prot* 69: 1587–1593.

NIEMIRA B, FAN X, SOMMERS C. 2002. Irradiation temperature influences product quality factors of frozen vegetables and radiation sensitivity of inoculated *Listeria monocytogenes. J Food Prot* 65: 1406–1410.

NIEMIRA B, LONCZYNSKI K, SOMMERS C. 2006. Radiation sensitivity of *Salmonella* isolates relative to resistance to ampicillin, chloramphenicol or gentamicin. *J Rad Phys Chem* 75: 1080–1086.

NISHIMURA H, PHILIP R P, CALVIN M. 1977. Lipids of *Hevea brasiliensis* and *Euphorbia coerulescens. Phytochemistry* 16: 1048–1049.

NOGUEIRRA DE SOUSA ANDRADE L, MATINS DE LIMA T, CURI R, LAURO CASTRUCCI A M. 2005. Toxicity of fatty acids on murine and human melanoma cell lines. *Toxicology In Vitro* 19: 553–560.

PATTERSON M. 1995. Sensitivity of *Campylobacter* spp. to irradiation in poultry meat. *Lett Appl Microbiol* 20: 338–340.

PARISI A, ANTOINE A. 1974. Increased radiation resistance of vegetative *Bacillus pumilus. Appl Environ Microbiol* 28: 41–46.

PIRTLE E, PROSCHOLDT T, BERAN G. 1997. Trial of heat inactivation of selected viruses following irradiation. *J Food Prot* 60: 426–429.

POSCHL M, NOLLET L. 2007. *Radionuclide Concentrations in Food and the Environment.* Taylor and Francis, New York.

PRUSINER S B. 1982. Novel proteinaceous infectious particles cause scrapie. *Science* 216: 136–144.

RAFFI J, DELINCEE H, MARCHIONI E, HASSELMAN C, SJOBERG A, LEONARDI M, KENT M, BOGL K, SCHREIBER G, STEVENSON H, MEIER W. 1993. Concerted action of the Community Bureau of Reference on the methods of identification of Irradiated foods. EUR 15261 EN, Brussels.

RAHA M, HUTCHINSON F. 1991. Deletions induced by gamma rays in the genome of *Escherichia coli. J Mol Biol* 220: 193–198.

RAJKOWSKI K, BOYD G, THAYER D. 2003. Irradiation *D*-values for *Escherichia coli* O157:H7 and *Salmonella* sp. on inoculated broccoli seeds and effects of irradiation on broccoli sprout keeping quality and seed viability. *J Food Prot* 66: 760–766.

ROOS M. 2003. USDA to offer irradiated beef to schools next January. *CIDRAP News*, May, 30. Accessed December 30, 2006, at: http://www.cidrap.umn.edu/cidrap/content/fs/irradiation/news/may3003irradiation.html

RUNNER G. 1916. Effect of roentgen rays on the tobacco or cigarette beetle and results with a new form of roentgen tube. *J Agric Res* 6: 383.

SARGENTINI N, SMITH K. 1994. DNA sequence analysis of gamma-radiation (anoxic)-induced and spontaneous *lacI*$^{d}$ mutations in *Escherichia coli* K-12. *Mutat Res* 309: 147–163.

SATIN M. 1993. Technomic Publishing, Lancaster PA.

SCHREUDER B. 1994. BSE agent hypotheses. *Livestock Prod Sci* 38: 23–33.

SCHWARTZ, B. 1921. Effect of x-rays on thrichinae. *J Ag Res* 20: 845.

SHAMSUZZMAN K, LUCHT L. 1993. Resistance of *Clostridium sporogenes* to gamma radiation and heat in various nonaqeous suspension media. *J Food Prot* 56: 10–12.

SMITH J, PILLAI S. 2004. Irradiation and food safety. *Food Technology* 58(11): 48–55.

SOMMERS C. 2003. 2-Dodecylcyclobutanone does not induce mutations in the *Escherichia coli* tryptophan reverse mutation assay. *J Agric Food Chem* 51: 6367–6370.

SOMMERS C. 2006a. Recent advances in food irradiation. In: *Advances in Microbial Food Safety*. (Cherry J, Juneja V, eds). American Chemical Society, Washington, DC.

SOMMERS C. 2006b. Induction of micronuclei in human TK6 lymphoblasts by the unique radiolytic product 2-dodecylcyclobutanone. *J Food Sci* 71: C281–C284.

SOMMERS C. 2007. Induction of micronuclei by palmitic acid and its unique radiolytic product 2-dodecylcyclobutanone. In: *Proceedings of the United States-Japan (UJNR), 36th Annual Meeting*, October 21–25, 2007, Tsukuba, Ibaraki, Japan, pp. 59–62.

SOMMERS C, FAN, X. 2011. Irradiation of ground beef and fresh produce. In: *Nonthermal Processing Technologies for Food* (Zhang H *et al.* eds). IFT, Blackwell Publishing, Ames, IA, pp. 236–246.

SOMMERS C, MACKAY W. 2005. 2-Dodeclycyclobutanone does not induce formation of 5-fluoruracil resistant mutants or increase expression of DNA damage inducible genes in *Escherichia coli. J Food Sci* 70: 254–256.

SOMMERS C, MACKAY W. (In Press). Irradiation of processed meats to improve microbial safety. In: *Food Irradiation Research and Technology*, 2nd edn (Fan X, Sommers C, eds). IFT-Blackwell Publishing, Ames, IA.

SOMMERS C, SCHIESTL R. 2004. 2-Dodecylcyclobutanone does not induce mutations in the *Salmonella* mutagenicity test or intrachromosomal recombination in *Saccharomyces cerevisiae. J Food Prot* 67: 1293–1298.

SOMMERS C, NEIMIRA B, TUNICK M, BOYD G. 2002. Effect of temperature on the radiation resistance of virulent *Yersinia enterocolitica. Meat Sci* 61: 323–328.

SULLIVAN R, FASSOLITIS A, LARKIN E, REED R, PEELER J. 1971. Inactivation of thirty viruses by gamma radiation. *Appl Environ Microbiol* 22: 61–65.

TAUB I, HALLIDAY J, SEVILLA M D. 1979. Chemical reactions in proteins irradiated at subfreezing temperatures. *Adv Chem Serol* 180, 109–140.

THAYER D, BOYD G. 1995. Radiation sensitivity of *Listeria monocytogenes* on beef as affected by temperature. *J Food Sci* 60: 237–240.

THAYER D, BOYD G. 2001. Effect of irradiation temperature on inactivation of *E. coli* O157:H7 and *Staphylococcus aureus. J Food Prot* 64: 1624–1626.

THAYER D, SONGPRASERTCHAI S, BOYD G. 1991. Effects of heat and ionizing on *Salmonella typhimurium* in mechanically deboned chicken meat. *J Food Prot* 54: 718–734.

THAYER D, DICKERSON C, RAO D, BOYD G, CHAWA C. 1992. Destruction of *Salmonella typhimurium* on chicken wings by gamma radiation. *J Food Sci* 57: 586–589.

THAYER D, BOYD G, LAKRITZ L, HAMPSON J. 1995. Variations in radiation sensitivity of foodborne pathogens associated with the suspending meat. *J Food Sci* 60: 63–67.

THAYER D, RAJKOWSKI K, BOYD G, COOKE P, SOROKA D. 2003. Inactivation of *Escherichia coli* O157:H7 and *Salmonella* by gamma irradiation of alfalfa seed intended for production of food sprouts. *J Food Prot* 66:175–181.

TOLGAY Z. 1972. Investigations in invasion capacity and destruction of *Cysticercus bovis* in beef treated by ionizing radiation (gamma-rays from C0-60). *Tuerk Veteriners Hekimleri* 42: 13–19.

UDILOVA N, JUREK D, MARIAN B, GILLE L, SCHULTE-HERMANN R, NOHL H. 2003. Induction of lipid peroxidation in biomembranes by dietary oil components. *Food Chem Toxicol* 41: 1481–1489.

ULOTH J, CASIANO C, DE LEON M. 2003. Palmitic and stearic fatty acids induce caspase-dependent and independent cell death in nerve growth factor differentiated PC12 cells. *J Neurochem* 84(4): 655–668.

USDA FSIS. 2004. FSIS issues alert on the importance of cooking and handling ground beef. Food Safety and Inspection Service, United States Department of Agriculture, Washington, DC. Available online at: http://www.fsis.usda.gov/OA/news/2004/alert012904.htm

VERSTER A, PLESSIS T, VAN DEN HEEVER L. 1977. The eradication of tapeworms in pork and beef carcasses by irradiation. In: *Irradiation Processing*, Elsevier, London, pp. 769–993.

WARD J. 1991. Mechanisms of radiation action on DNA in model systems – their relevance to cellular DNA. In: *The Early Effects of Radiation on DNA* (Felden E, O'Neill P, eds). Springer-Verlag, New York, pp. 1–16.

WATKINS D. 1980. Stimulation of DNA synthesis in bacterial DNA-membrane complexes after low doses of ionizing radiation. *Int J Radiat Biol Relat Chem Med* 38: 247–256.

WHO. 1994. Safety and Nutritional Adequacy of Irradiated Food. World Health Organization, Geneva, pp. 81–107.

WIJKER C, LAFLEUR M, VAN STEEG H, MOHN G, RETEL J. 1996. Gamma-radiation-induced mutation spectrum in the episomal lacI gene of *Escherichia coli* under oxic conditions. *Mutat Res* 349: 229–239.

# 12

# Microbial decontamination of food by ultraviolet (UV) and pulsed UV light

**N. M. Keklik, Cumhuriyet University, Turkey, K. Krishnamurthy, Illinois Institute of Technology, USA and A. Demirci, The Pennsylvania State University, USA**

**Abstract**: The application of ultraviolet (UV) and pulsed UV light to foods has gained attention in recent years as a potential alternative to chemical and thermal disinfection methods. These techniques can be used on packaged foods, and so allow post-processing decontamination of food products. This chapter first reviews the principles and technology behind UV and pulsed UV light, and discusses critical processing factors. It then describes the mechanisms of microbial inactivation and provides an overview of the effects of pulsed UV light on food quality. Research challenges and future needs relating to these technologies are also addressed.

**Key words**: UV light, pulsed UV light, food decontamination.

## 12.1 Introduction to food decontamination by ultraviolet (UV) and pulsed UV light

Thermal and chemical inactivation techniques have traditionally been used for decontamination. However, thermal methods are not suitable for temperature sensitive products and can deteriorate food quality. Chemical methods raise concerns such as toxicity and undesirable residuals. Alternative methods have therefore been developed and evaluated for the decontamination of foods, most of which aim to avoid heat and chemicals.

Among these methods, ultraviolet light (UV) has gained attention in recent years due to its low cost and ease of use. Ultraviolet light is generated by mercury lamps, and consists of electromagnetic radiation in the wavelength range from 100 to 400 nm. The UV light spectrum is divided into four regions: UV-A (315–400 nm), UV-B (280–315 nm), UV-C (200–280 nm),

© Woodhead Publishing Limited, 2012

and vacuum-UV (100–200 nm). UV-C is highly germicidal (Acra *et al.*, 1990) causing photochemical changes to microbial DNA, which eventually inactivates the microorganism (Miller *et al.*, 1999). UV light is a non-ionizing, non-chemical, and non-thermal method, which is environmentally friendly, easy to handle, and cost-efficient. Many studies have demonstrated the effectiveness of UV light on pathogenic microorganisms in water and on foods including bacteria, fungi, viruses, and protozoans. UV disinfection has been practiced since the 1900s (Wright and Cairns, 1998), and the most common applications of this technology include treatment of drinking water and wastewater.

Pulsed UV light is a recent innovative technology utilizing UV light which, like conventional UV light, is non-chemical and non-ionizing, and is considered to be non-thermal when the treatment time is short. Pulsed UV light is generated by inert gas (e.g. xenon) flashlamps, which produce intense light pulses consisting of electromagnetic radiation in the wavelength range from 100 to 1100 nm, which includes the UV, visible, and infrared spectra (Krishnamurthy *et al.*, 2007). Pulsed UV light has many advantages over continuous UV light in terms of microbial inactivation efficacy, penetration ability, and lamp safety (Xenon, 2006).

Both UV and pulsed UV light technologies depend on the transmittance and absorbance of light by the target food. These techniques can be advantageous for the post-processing decontamination of food products due to their suitability for foods packaged with UV-permeable materials. Food attributes and light characteristics play major roles in achieving maximum efficacy. The critical process factors include, but are not limited to, the light transmissibility of the product, wavelength range, light intensity, and the treatment time or number of pulses (for pulsed UV).

This chapter covers the basic knowledge and current trends concerning the use of UV and pulsed UV light technologies in decontaminating foods. The principles and microbial inactivation mechanisms of these technologies are explained, and their effects on food quality are discussed. The possible challenges and needs in future research are also considered.

## 12.2 Fundamentals of ultraviolet (UV) and pulsed UV light

### 12.2.1 UV light

Interactions between light and matter mainly occur as photochemical reactions, which are initiated when photons, or light particles, are absorbed by atoms or molecules. The electrons of atoms enter an excited state after absorbing the photon energy, which promotes the formation of new products with different chemical structures. There are two critical conditions for atoms/molecules to undergo a photochemical reaction: photons must have adequate energy to break or form chemical bonds, and photon energy must be absorbed (Blatchley and Peel, 2001). A basic mathematical expression developed by

© Woodhead Publishing Limited, 2012

German physicist Max Planck is used to define photon energy:

$$E = h \cdot v \qquad [12.1]$$

where $E$ is photon energy (kJ/Einstein) or (eV), $h$ is the Planck constant $\cong 6.63 \times 10^{-34}$ J·s $\cong 4.14 \times 10^{-15}$eV·s, and $v$ is frequency ($s^{-1}$), which is given by:

$$v = {}^{c}/_{\lambda} \qquad [12.2]$$

where $c$ is speed of light $\cong 3.00 \times 10^8$ m/s, and $\lambda$ is wavelength (m).

As can be seen in Eq. [12.1], the photon energy is proportional to the frequency of vibration, implying that electromagnetic radiation at shorter wavelengths carries higher photon energy.

Ultraviolet light is a portion of the electromagnetic spectrum in the wavelength range of 100–400 nm, which constitutes about 10% of total solar radiation (El-Boury *et al.*, 2007; Melquiades *et al.*, 2008). UV-A (315–400 nm), UV-B (280–315 nm), UV-C (200–280 nm), and vacuum-UV (100–200 nm) correspond to photon energies of 3.94–3.10, 4.43–3.94, 12.40–4.43, and 12.40–124 eV, respectively (Krishnamurthy *et al.*, 2008). Photon energies in the UV region of the electromagnetic spectrum are generally sufficient to initiate photochemical reactions in biomolecules. These reactions can produce either favorable or adverse effects on living organisms. For example, UV light enhances the synthesis of vitamin D, but it can also cause skin damage. The DNA structure of a microbial cell can be altered by UV light through the formation of thymine dimers, which can lead to inactivation of the microorganism.

The ability of sunlight to inactivate bacterial cells has been known since English scientists Downes and Blunt demonstrated the inhibition of bacterial growth in solutions exposed to sunlight in 1877 (Reed, 2010). In 1892, Marshall Ward demonstrated that the bactericidal effect of sunlight is caused mainly by the UV portion of the spectrum (Bischof, 1994). The ability of UV light to inactivate microorganisms has since drawn the attention of researchers looking to use UV light as a tool to eliminate pathogens on surfaces, in equipment, and in food. Different technologies have been developed, especially in recent years, to produce UV light with enhanced germicidal effects.

### 12.2.2 Pulsed UV light

Pulsed UV light, also known as pulsed light, is an emerging UV technology capable of inactivating microorganisms on surfaces, in water, and in air. Pulsed UV light is generated by inert gas (e.g. xenon) flashlamps, which have been used for disinfection since the late 1970s in Japan (Wekhof, 2000). A typical flashlamp produces intense light pulses consisting of electromagnetic radiation in the wavelength range from 100 to 1100 nm, which includes the UV, visible, and infrared spectra (Krishnamurthy *et al.*, 2007). Since more

© Woodhead Publishing Limited, 2012

than 50% of the spectrum is in the UV range, it is called pulsed UV light. The UV portion below 400 nm has particularly good germicidal properties. Flashes are delivered at a rate of about 1–20 pulses per second, and the duration of each pulse is as short as a few hundred microseconds (Dunn, 1996). Pulsed UV light is able to kill vegetative bacteria, bacterial and fungal spores, viruses, and protozoan oocysts.

Pulsed UV light is generated when the electrical energy stored in a high energy density electrical storage capacitor is released in high peak power pulses (Dunn, 1996). These pulses are then used to excite an inert gas, usually xenon, which releases short, high-intensity light pulses as it returns to its initial energy level. The high intensity of these pulses creates a unique germicidal effect that is not observed with low-intensity continuous-wave ultraviolet light at the same energy level.

## 12.3 Ultraviolet (UV) light technology

The thermal and chemical treatments of foods are commonly used methods of killing pathogens. However, alternative decontamination techniques have been of major interest to researchers due to inefficiencies, deteriorative effects, or safety issues associated with heat and chemical treatments. Ultraviolet light is effective in inactivating a broad range of microorganisms; however, its ability to penetrate food depends on the color and transparency of the product. The limited UV penetrability and its dependence on food properties need to be considered in the design of UV processing systems for specific foods. Improvements and developments in these techniques may offer a new direction to food processors, and make the production of cost-efficient, safe, and wholesome foods possible.

Radiation in vacuum-UV and UV-C regions exhibits germicidal properties by forming lethal dimers on the DNA of microbial cells (Krishnamurthy *et al.*, 2008). Although the photon energies at wavelengths in the UV portion of the spectrum are high enough to yield photochemical reactions in living matter, UV light is a non-ionizing radiation, unlike gamma and x-rays, i.e. it does not cause ionization of molecules.

When light strikes the surface of an object, its energy is partly reflected and partly absorbed, and, depending on the transparency of the object, may be transmitted through it (Acra *et al.*, 1990). The behavior of light on a food depends on both the optical characteristics of the food and the properties of the light including the duration, intensity, and spectrum. The absorbed portion of the light is the part that can interact with biomolecules, including those of the microbial cells. The maximum absorption of UV light by the nucleic acid occurs at a wavelength of about 260 nm, at which UV light causes the most damage to microbial DNA (Aguiar *et al.*, 1996).

Ultraviolet light was initially used for the disinfection of drinking water. Due to the low efficiency of early UV lamps, the production of UV equipment

© Woodhead Publishing Limited, 2012

slowed down until about 1970, when more durable and reliable UV lamps started to be produced (Solsona and Mendez, 2003). The undesirable byproducts of chemical treatments, particularly those involving chlorine, have led many water suppliers to adopt UV technology for the disinfection of drinking water. UV technology is now used in many other areas, such as disinfection of foods and packaging materials, pharmaceuticals, and wastewater, and equipment in facilities such as food processing plants, hospitals, laboratories, and restaurants.

Ultraviolet light is typically produced by low and medium pressure mercury arc lamps, which consist of a mixture of mercury and inert argon vapor confined in a UV-transmittant silica or quartz tube. The tube contains electrodes at both ends, which are usually made of tungsten with a mixture of alkaline earth metals, and which aid in arc formation (Wright and Cairns, 1998). When high voltage is applied across the electrodes, the electric arc excites the mercury vapor, which emits UV light while returning to a lower energy level. The argon helps to start up the lamp, extend electrode life, and decrease thermal losses. Low pressure mercury arc lamps produce monochromatic radiation at about 254 nm, while medium pressure mercury arc lamps yield a polychromatic radiation in the UV and visible light range (Bohrerova *et al.*, 2008).

For the safe processing and treatment of foods using UV radiation, the FDA stipulates the use of low pressure mercury lamps that emit 90% of the light at a wavelength of 253.7 nm (FDA, 2010). The limitations set by the FDA for the use of UV light to process or treat foods are given in Table 12.1. The critical processing factors include, but are not limited to, the chemical composition of food, uniformity of the radiation, UV food transmissivity and thickness, and UV wavelength range (FDA, 2000). A schematic diagram of a flow-through UV treatment system is given in Fig. 12.1. Many factors need to be taken into account in order to determine the cost-efficiency of a UV disinfection system. UV dosages and flow rates are the main factors affecting the costs for low and medium pressure UV lamp systems, though many other elements affect the cost-efficiency of UV systems based on their objectives and applications (Malley, 2002). The cost of power, periodical replacement of lamps and sleeves, simple routine maintenance, fouling, and cleaning are included in the operation and maintenance costs. Since UV treatment does not involve the use of chemicals, the costs of transport, storage, and handling of chemicals are eliminated (Wright and Cairns, 1998).

## 12.4 Pulsed ultraviolet (UV) light technology

Pulsed UV light is considered to be four to six times more efficient in inactivating microorganisms than continuous UV light (Fine and Gervais, 2004; Krishnamurthy *et al.*, 2010). However, the penetration ability of pulsed UV light is still limited, and thus it is only suitable for surface disinfection

© Woodhead Publishing Limited, 2012

**Table 12.1** Limitations for the use of ultraviolet radiation for the processing and treatment of foods

| Use | Limitations |
|---|---|
| Surface microorganism control for food and food products | Without ozone production<br>High fat content food irradiated in vacuum or in an inert atmosphere<br>Intensity of radiation, 1 W (of 253.7 nm radiation) per 5–10 $ft^2$ |
| Sterilization of potable water used in food production | Without ozone production<br>Coefficient of absorption, 0.19 per cm or less<br>Flow rate, 100 gal/h per watt of 253.7 nm radiation<br>Water depth, 1 cm or less<br>Lamp operating temperature, 36–46°C |
| Reduction of human pathogens and other microorganisms in juice products | Turbulent flow through tubes with a minimum Reynolds number of 2200 |

Source: FDA, 2010.

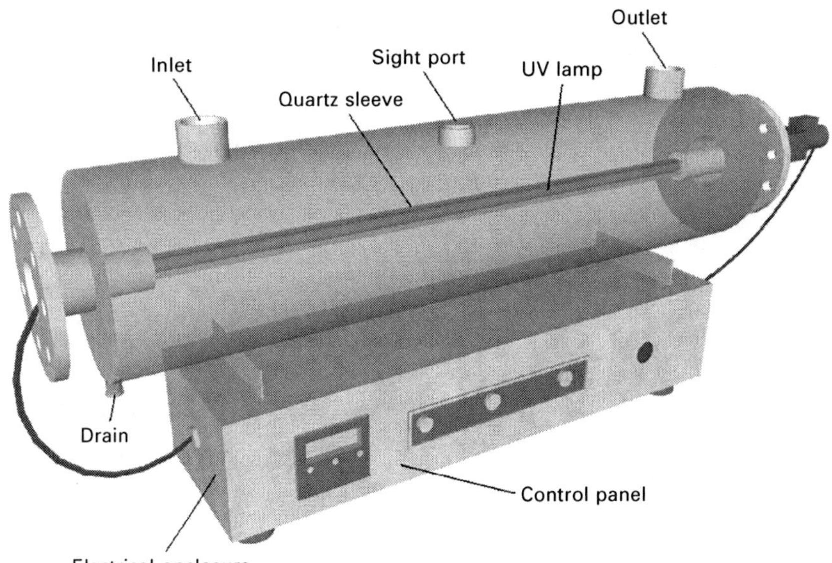


**Fig. 12.1** Schematic diagram of a flow-through UV disinfection system.

of solid foods or transparent liquid foods. It can also provide post-processing decontamination for packaged foods providing that the packaging material is UV-transmittant and UV-resistant. Pulsed light is an FDA-approved method

© Woodhead Publishing Limited, 2012

(Federal Register, 1999), and the conditions for its use for the treatment of foods are given in Table 12.2 (FDA, 2010).

The germicidal effects of pulsed UV light increase with higher energy absorption by the food. It is therefore necessary to adjust the energy dose delivered to the food in order to optimize inactivation. The total dose of light energy ($D$) (J/cm$^2$) is calculated as (Luksiene *et al.*, 2007):

$$D = E_p * t * f \quad [12.3]$$

where, $E_p$ is energy of one pulse (J/cm$^2$), $t$ is treatment time (s), and $f$ is pulse frequency (Hz).

Critical processing factors include the wavelength and intensity of light; duration and number of pulses; and the type, thickness, transparency, and color of packaging material and food (FDA, 2000). Examples of static and flow-through pulsed UV light systems are shown in Figs 12.2 and 12.3. In continuous processes, the food product can be passed under a pulsed UV lamp on a conveyor or in a quartz tube (for liquids), moving or flowing at a speed that allows sufficient time to expose the product to the desired number of pulses. More lamps placed around the food product can eliminate the shadow effect, which occurs when some parts of the food surfaces do not receive sufficient UV light because of their curvatures.

**Table 12.2** Conditions for the use of pulsed UV light for the treatment of foods

| Use | Conditions |
|---|---|
| Radiation sources | Xenon flashlamps designed to emit broadband radiation consisting of wavelength range of 200–1100 nm |
| Pulse duration | No longer than 2 milliseconds (ms) |
| Treatment | Surface microorganism control |
| Total cumulative dose | Shall not exceed 12.0 Joules/square centimeter (J/cm$^2$) |

Source: FDA, 2010.

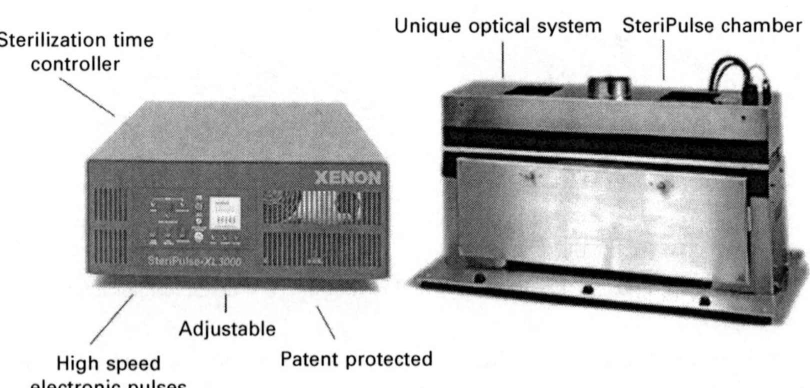


**Fig. 12.2** Static pulsed UV light system. Source: Xenon Corp., Wilmington, MA.

© Woodhead Publishing Limited, 2012

**Fig. 12.3** Flow-through pulsed UV light system. Source: Xenon Corp., Wilmington, MA.

## 12.5 Mechanisms of microbial inactivation in foods by ultraviolet (UV) and pulsed UV light

The germicidal effects of UV light occur as a result of photochemical changes induced when UV light is absorbed by living cells. The peak absorption of UV light by nucleic acids, particularly DNA, in the UV-C range makes them the targets of the photochemical action of UV light. UV light directly causes photo-dimerization between adjacent pyrimidine bases and photo-hydration of cytosine (Miller *et al.*, 1999). Furthermore, UV light indirectly plays a role in the generation of reactive oxygen species, which also react with DNA and proteins (Miller *et al.*, 1999). The replication of microbial DNA is prevented by these lethal photo-dimerizations. As a result, cellular functions are inhibited, and eventually the cell becomes inactivated. Pulsed UV light inactivates microorganisms not only by photochemical reactions, but also by other mechanisms associated with the use of high-intensity pulses. A microbial cell becomes inactive once UV induced photochemical reactions have taken place, but the cell structure remains intact. However, pulsed UV light also damages the cell structure.

Krishnamurthy *et al.* (2010) identified the damage caused by pulsed UV light treatment to *Staphylococcus aureus* using Fourier-transformed infrared spectroscopy (FTIR) and transmission electron microscopy (TEM). A 5 s treatment with pulsed UV light resulted in severe damage to *S. aureus* cells. Cell wall damage, cytoplasmic membrane shrinkage, and internal cellular structure collapse were observed. Consequently, cellular contents leaked from the cytoplasm leading to cell death. This study also confirmed that some of the microbial cells were inactivated without any structural damage, indicating the photochemical effect of pulsed UV light.

The structural damage to microbial cells caused by pulsed UV light has

© Woodhead Publishing Limited, 2012

been explained by several researchers. Thermal stress is one possible cause of cell rupture, which occurs due to the difference in the absorption of pulsed UV light by a microorganism and its surrounding medium (Wekhof, 2000; Fine and Gervais, 2004). This difference in heating rate results in the vaporization of water in the bacterial cell, leading to cell rupture caused primarily by the UV components of the light pulse. Another effect was suggested in which microbial cell damage is caused by high-energy pulses (Krishnamurthy *et al.*, 2010). Overall, the effects of pulsed UV light on microbial cells can be classified as photochemical (e.g. thymine dimer formation in microbial DNA), photothermal (localized heating of bacteria), and photophysical (constant disturbance caused by the high-energy pulses) effects (Krishnamurthy *et al.*, 2010).

### 12.5.1 Contribution of wavelengths to microbial inactivation

As previously mentioned, a target molecule needs to absorb light in order for a photochemical reaction to occur. However, not all molecules absorb the same amount of light at the same wavelength. It is therefore important to match the absorption spectra of a target molecule with the emission spectra of the lamp in order to optimize the photolytic action (Wekhof, 2000). Wavelengths known to cause undesirable changes in foods can be filtered through glass or liquid filters (FDA, 2000). The UV region of the electromagnetic spectrum is mostly responsible for the microbial inactivation caused by pulsed light. The lamp spectrum that includes more UV light, with a higher peak power, is considered to be better for decontamination and disinfection (Wekhof, 2000). In fact, no killing effect was observed when wavelengths below 300 nm were blocked (Takeshita *et al.*, 2002). Since maximum absorption by microbial DNA occurs at about 260 nm, most of the damage will be caused by UV-C light (200–280 nm). However, the absorption of UV by water limits the lethal effect below 230 nm (Koutchma *et al.*, 2009b). While DNA absorbance is highest in the UV-C range, it falls by more than three orders of magnitude in the UV-B range, and is almost negligible in the UV-A range (Miller *et al.*, 1999).

Wekhof (2000) reported that half of the flash disinfection is caused by the UV-C region alone, though UV-B and UV-A can also induce changes in microorganisms. In fact, Wekhof *et al.* (2001) demonstrated that the light consisting of only UV-B and UV-A regions could also lead to the overheating and disintegration of a microorganism. Consequently, besides the germicidal action of UV-C, all the absorbed incident UV photons in the light pulses cause rupture and disintegration of the microorganisms through overheating. However, the photophysical effect of pulsed UV light on microorganisms, mentioned by Krishnamurthy *et al.* (2010), appears to be caused by the high-intensity light pulses, which may be attributed to the entire emission spectrum of the pulsed UV lamp. Bohrerova *et al.* (2008) also demonstrated that the portion of pulsed UV light greater than 400 nm was able to inactivate

© Woodhead Publishing Limited, 2012

phages. Elmnasser *et al.* (2007) also suggested that the visible and infrared portions contribute to microbial inactivation by pulsed UV light.

### 12.5.2 Energy levels and wavelength spectra of various lamps

The selection of a UV source for a disinfection system is critical, since the efficacy and cost-efficiency of the process is closely related to the type of UV lamp used in the system. There are many types of UV lamps, which provide either monochromatic or polychromatic emission, and operate in continuous or pulsed mode. The choice of UV lamp is specific to its application. In other words, no one type of UV lamp is best for all applications. Among the factors affecting this choice are the wavelength spectrum, efficiency, operating temperature, arc length, and the lamp lifetime. Below are summarized the different types of UV sources.

*Low pressure mercury lamps*

Low pressure (LP) mercury lamps are monochromatic with 85–90% of the emission at 253.7 nm, yielding relatively low light irradiance (Linden, 2004). LP mercury lamps have operating temperatures of 30–50°C with mercury vapor pressure varying between 0.1 and 10 Pa. The germicidal efficiency of an LP lamp at 200–300 nm is about 35–40%. There are also low pressure high output (LP-HO) mercury lamps, which provide higher irradiance compared to LP mercury lamps. LP-HO mercury lamps operate at higher temperatures (60–100°C), and their efficiency is slightly lower than LP lamps (30–35%) (Linden, 2004). An LP mercury lamp has a UV intensity of about 0.01 W/cm$^2$, and a lifetime of 18–24 months (Koutchma *et al.*, 2009a).

*Medium pressure mercury lamps*

Medium pressure (MP) mercury lamps emit polychromatic light in the wavelength range of 185–600 nm (Linden, 2004). The mercury vapor pressure varies between 50 and 300 kPa and their range of operating temperatures is 600–900°C. Although MP mercury lamps have higher UV intensity (~12 W/cm$^2$) (Koutchma *et al.*, 2009a), they are germicidally less efficient compared to LP and LP-HO mercury lamps (15–20%) (Linden, 2004). The lamp lifetime is about half a month. The spectrum of light emitted is mainly determined by the gas mixture and temperature. Medium pressure mercury arc lamps operate at higher temperatures due to higher current density, thus yielding more continuum light (McDonald *et al.*, 2000). UV light produced by either low or medium mercury arc lamps is termed continuous wave (CW) or, more briefly, continuous UV light.

*Excimer lamps*

Excimer lamps are another source of monochromatic radiation, in which an electric potential is applied to a mixture of rare gases across a dielectric barrier (Blatchley and Peel, 2001). The characteristic wavelength emitted

© Woodhead Publishing Limited, 2012

by an excimer lamp is determined by the gas mixture used in the system, and thus is 'tunable'. Excimer lamps can operate at much lower surface temperatures than other types of UV sources, which may be advantageous in terms of fouling behavior. An excimer lamp has an electrical efficiency ranging from 10 to 35%, and a lifetime of six months (Blatchley and Peel, 2001).

*Low-pressure amalgam lamps*

Low-pressure amalgam (LPA) lamps have recently been developed as an alternative to mercury lamps. An LPA lamp has a UV spectrum of 185–254 nm with 35% efficiency at 254 nm (Koutchma *et al.*, 2009a). The lamp has a wall temperature of 90–120°C. UV intensity of the lamp is not affected by temperature fluctuations. Heat generation is negligible, and the transmission losses of quartz glass associated with LP mercury lamps are not observed with LPA lamps. High efficiency, low operating costs, and long lifetime are among the other advantages of LPA lamps.

*Microwave UV lamps*

There are also microwave UV lamps, which use microwave energy to excite mercury atoms without electrodes. Microwave UV lamps operate at similar temperatures and pressures to LP mercury lamps. However, unlike LP and MP mercury lamps, they warm up quickly and their lifetime is about three times that of electrode lamps (Koutchma *et al.*, 2009a).

*Pulsed UV lamps*

Pulsed UV lamps, which contain xenon, argon, krypton or other inert gases or mixtures, are characterized by the emission of extremely intense flashes in a broadband spectrum (Linden, 2004). Two types of lamp produce pulsed light: the flashlamp and the surface discharge (SD) lamp (Bohrerova *et al.*, 2008). In a flashlamp, the pulses are produced by a rare gas between two electrodes confined in a small envelope. A typical flashlamp can generate a broadband wavelength spectrum ranging from 100 to 1100 nm. The UV efficiency and intensity of a flashlamp is about 9% and 600 W/cm$^2$, respectively (Koutchma *et al.*, 2009a). A pulsed xenon lamp has a lifetime of about one month. SD lamps have a higher UV intensity (30,000 W/cm$^2$) and efficiency (17%) than flashlamps. In an SD lamp, plasma is produced by a high power electrical discharge along the surface of the dielectric substrate (fused silica tube) inside an envelope containing xenon gas (Schaefer *et al.*, 2007). Unlike flashlamps, the plasma formation and evolution do not involve the outer envelope containing xenon gas. However, the large diameter of the envelope in SD lamps may contribute to a longer lamp lifetime (Schaefer *et al.*, 2007).

© Woodhead Publishing Limited, 2012

## 12.6 Applications of ultraviolet (UV) and pulsed UV light for food decontamination

UV light is commonly used for disinfecting water. Thousands of installations in Europe and the US utilize UV light technology to disinfect drinking water, either alone or in combination with chlorine (Wright and Cairns, 1998). UV light is also utilized in the food industry for disinfecting equipment and other food-contact surfaces, packaging materials such as bottles, caps, and films, and in post-harvest storage of fruits and vegetables to increase their shelf-life by inactivating spoilage fungi (Begum *et al.*, 2009). However, this technology has not been widely adopted for the processing of foods. The lack of commercial-scale UV light systems for treating solid foods limits utilization of this technology in the food industry. Many researchers, however, continue to evaluate the effectiveness of UV light on the microbial content of foods. Pulsed UV light technology offers more efficient, safer, and faster decontamination than non-pulsed UV light, but its ability to penetrate foods is still an issue. The irregular and complex surface properties and the opaqueness of foods limit the use of this technology. Therefore, the 'surface decontamination' of foods and food-contact surfaces such as the equipment, conveyors, and packaging materials using pulsed UV light seems more feasible at the moment. The effectiveness of pulsed UV light in reducing microorganisms in foods has been demonstrated by various researchers. Some selected studies involving the application of UV and pulsed UV light technologies are summarized below. It should be noted that the wavelength used for UV treatments in these studies is 254 nm unless indicated otherwise.

### 12.6.1 Solid foods

*Decontamination by UV light treatment*

Isohanni and Lyhs (2009) studied the effect of UV light on *Campylobacter jejuni* on broiler meat. The UV treatments for 14–18 s at doses of 32.4–32.9 mW s/cm$^2$ yielded the maximum $\log_{10}$ reductions of 0.7, 0.8, and 0.4 on broiler meat, skin, and carcass, respectively. In another study (Sommers *et al.*, 2009b), frankfurters containing sodium diacetate and potassium lactate were treated with UV light at 4.0 J/cm$^2$, followed by flash (steam) pasteurization (3 s steam at 121°C), which resulted in about 3.89 $\log_{10}$ reduction in *Listeria innocua*. No growth of *L. innocua* was observed after the treated frankfurters were stored for 8 weeks at 8°C. This study showed that the combination of UV light with other intervention techniques and antimicrobials could be an effective and cost-efficient method of controlling *Listeria* on frankfurters.

UV light is known to significantly reduce bacteria and molds on shell eggs (Kuo *et al.*, 1997). In the study carried out by Coufal *et al.* (2003), a UV treatment cabinet for eggs was constructed, in which a conveyor system carried the eggs through the cabinet. A 4 min UV treatment at 4–14 mW cm$^2$

© Woodhead Publishing Limited, 2012

yielded about 1.3, 4, and 4–5 $\log_{10}$ reductions in aerobic plate counts (APC), *S.* Typhimurium, and *E. coli* on hatching eggs, respectively.

In a study performed by Fonseca and Rushing (2006), aerobic plate counts on fresh-cut watermelons were reduced by >1 $\log_{10}$ after 3 min UV treatment at 4.1 kJ/m$^2$. In contrast, the treatment of watermelons in polypropylene packages yielded <1 $\log_{10}$ reduction. Fino and Kniel (2008) reported that UV light treatment (40–240 mW s/cm$^2$) of strawberries, green onions, and lettuce yielded 1.9–2.6, 2.5–5.6, and 4.5–4.6 $\log_{10}$ reductions in the infectivity of the viruses (feline calicivirus, hepatitis A virus, and Aichi virus), respectively. Begum *et al.* (2009) demonstrated that UV light was effective in reducing counts of food spoilage fungi including *Aspergillus flavus*, *Aspergillus niger*, *Penicillium corylophilum* and *Eurotium rubrum*. The efficacy of UV against fungal spores was dependent on the method of UV exposure and genera. In the study by Stevens *et al.* (1997), UV treatment of peach, tangerine, tomato, and sweet potato reduced the incidence of brown rot (*Monilinia fructicola*), green mold (*Penicillium digitatum*), and Rhizopus soft rot (*Rhizopus stolonifer*).

*Decontamination by pulsed UV light treatment*

In a study performed by Dunn (1996), *Salmonella* Enteritidis could be reduced by as much as 8 $\log_{10}$ following treatment of shell egg surfaces with 8 light pulses at a total dose of 4 J/cm$^2$. In another study (Keklik *et al.*, 2010b) on shell eggs, complete inactivation of *S.* Enteritidis was achieved at a pulsed UV dose of 23.6 J/cm$^2$. Sharma and Demirci (2003) exposed alfalfa seeds inoculated with *E. coli* O157:H7 to pulsed UV light. In a seed layer 1.02 nm thick, a reduction of about 4.80 $\log_{10}$ CFU/g was achieved during a 30 s treatment. In a study carried out by Ozer and Demirci (2006) on salmon fillets, *E. coli* O157:H7 and *Listeria monocytogenes* were reduced by about 1 $\log_{10}$ cfu/g, following a 60 s treatment at 8 cm from the quartz window in a pulsed UV light system generating an amount of 5.6 J/cm$^2$/pulse energy on the lamp surface. In Bialka and Demirci's study of 2008, a 60 s treatment with pulsed UV light at doses varying between 25.7 and 72 J/cm$^2$ reduced *Escherichia coli* O157:H7 and *Salmonella* spp. by 3.9 and 3.4 $\log_{10}$ cfu/g on raspberries, and by 2.1 and 2.8 $\log_{10}$ cfu/g on strawberries, respectively.

Jun *et al.* (2003) predicted that *Aspergillus niger* spores in corn meal could be reduced by a maximum of 4.93 $\log_{10}$ following a 100 s treatment at a distance of 3 cm from the pulsed UV strobe. The authors recommended modifying the pulsed UV light system so as to minimize heat generation, which otherwise limits the optimization of fungal disinfection by pulsed UV light. Lagunas-Solar *et al.* (2006) investigated the potential of pulsed UV light technology as an alternative to the use of contact pesticides for the surface disinfection of fresh fruits. More than 5 $\log_{10}$ reductions in plant (fungal) pathogens on fresh fruits including apples, kiwi, lemon, nectarines, oranges, peaches, pears, raspberries, and grapes were obtained in less than 10 s. The minimal UV dose for complete fungal inactivation ranged from 100 to 1900

© Woodhead Publishing Limited, 2012

$mJ/cm^2$ on fruit surfaces for each microorganism, among which *A. niger* required the highest energy dose. The authors related this to the presence of UV-absorbent melanin-type pigments on the cells of *A. niger*, which provided a surface shield for the microorganism. Therefore, microorganisms with such surface pigmentation seem to be potentially more resistant to pulsed UV light, requiring higher energy doses to promote the photolytic destruction of the outer pigment layer before the photons reach the cell cytoplasm and nucleus. Surface crevices and injuries on fruits create another shadowing effect by providing shelter for microorganisms in sub-surface areas. The authors suggested that this shadowing could be eliminated if the surfaces of fruits moved randomly on a conveyor are uniformly exposed to a diffuse, multidirectional, and intense UV beam.

Keklik *et al.* (2009) studied the effectiveness of pulsed UV light on *L. monocytogenes* in unpackaged and vacuum-packaged chicken frankfurters. Polypropylene film with a UV transmittance of ~75% at 260 nm was used as the packaging material. After 60 s treatments at 8 cm from the quartz window, about 1.6 and 1.5 $log_{10}$ $cfu/cm^2$ reductions were obtained for unpackaged and vacuum-packaged frankfurters at energy doses of 55.9 and 48.4 $J/cm^2$, respectively. This study demonstrated the effectiveness of pulsed UV light on packaged foods, implying that this technology has potential as a post-processing decontamination technique. About 2 $log_{10}$ reductions of *S.* Typhimurium were obtained on both unpackaged and vacuum-packaged chicken breast after only 15 and 30 s treatments at 5 cm distance from the quartz window (Keklik *et al.*, 2010c). In another study, Keklik *et al.* (2010a) designed, built, and evaluated a pilot-scale pulsed UV light system for the decontamination of whole chicken carcasses, which were moved by a linear rail between two sets of pulsed UV light systems on each side. Accordingly, the $log_{10}$ reductions of *E. coli* K12 on carcasses ranged from 0.87 to 1.43 cfu/ml rinse solution after 30 s and 180 s treatments (at about 0.25 J/pulse/$cm^2$ at 5 cm distance from the quartz window), which correspond to the conveyor velocities of 78 and 13 cm/min, respectively.

### 12.6.2 Liquid foods

*Decontamination by UV light treatment*

Hanes *et al.* (2002) reported >5 $log_{10}$ reduction in *Cryptosporidium parvum* oocyst viability in apple cider after a UV treatment for 1.2–1.9 s at a dose of 14.32 $mJ/cm^2$. Koutchma *et al.* (2006) studied the inactivation of *E. coli* K12 in apple cider using flow-through UV reactors named 'CiderSure', 'Aquionics', and 'UltraDynamics', which respectively involve thin-film laminar flow, turbulent flow, and an annular single-lamp. The UV decimal reduction doses for *E. coli* K12 in apple cider were 18.8–25.1 $mJ/cm^2$ in 'CiderSure', 90–150 $mJ/cm^2$ in 'Aquionics', and 20.4 $mJ/cm^2$ in 'UltraDynamics'. In a study performed by Hakguder (2009), more than 5 $log_{10}$ reduction of *E. coli* K12 was obtained in white grape juice at a UV dose of 75.04 $mJ/cm^2$, while

© Woodhead Publishing Limited, 2012

only 1.76 $\log_{10}$ reduction of naturally grown microorganisms was obtained in fresh squeezed orange juice at a UV dose of 144.36 $mJ/cm^2$.

Matak *et al.* (2005) obtained >5 $\log_{10}$ reduction in *L. monocytogenes* in goat's milk after the milk was passed 12 times through the UV light system, corresponding to a cumulative exposure time of ~18 s and a cumulative UV dose of 15.8 $mJ/cm^2$. The inoculated milk flowed in thin films into the system, where eight UV lamps situated along the same axis in the quartz tube provided a uniform UV exposure throughout the milk. In a study carried out by Unluturk *et al.* (2010), a 20 min exposure of liquid egg white to UV light at 1.314 $mW/cm^{-2}$ yielded 0.896, 1.403, and 0.960 $\log_{10}$ reductions of *E. coli* K12, *E. coli* O157:H7, and *L. innocua*.

*Decontamination by pulsed UV light treatment*

Krishnamurthy *et al.* (2007) investigated the inactivation of *S. aureus* in milk using a flow-through pulsed UV light system. The milk containing *S. aureus* was pumped into the system through a quartz tube. A V-groove reflector setup was used both to adjust the distance between the quartz tube and the light source and to reflect the light back into the quartz window to increase the absorption of energy by the microbial cells in milk. Complete inactivation of *S. aureus* was observed after two treatments: (1) single pass at 20 ml/min flow rate at 8 cm distance from quartz window (at 0.98 $W/cm^2$), and (2) two passes at 20 ml/min at 11 cm sample distance (0.80 $W/cm^2$).

In a study performed by Sauer and Moraru (2009), the effects of pulsed UV light on *E. coli* strains in apple juice and apple cider were investigated. The treatments were performed in static and turbulent modes. Turbulence was created using an orbital shaker placed in the pulsed UV light unit. Low turbulence and high turbulence were obtained at velocities of 500 and 3000 rpm, respectively. The treatment at high turbulence yielded inactivation levels of 5.76 and 7.15 $\log_{10}$ for *E. coli* in apple cider and apple juice, below the energy dose of 12 $J/cm^2$, thus satisfying the requirement of a minimum 5 $\log_{10}$ reduction in pathogen level in juices by the US Food and Drug Administration (FDA, 2001). The inactivation levels were partly attributed to the aid of turbulence in reducing the shadowing effect of the particulates present in the non-clear liquid substrates. The authors also suggested that the use of controlled turbulence in a flow-through system would be useful in the potential commercial application of pulsed light for processing liquid foods.

## 12.7 Effects of ultraviolet (UV) and pulsed UV light on food quality

### 12.7.1 Effects of UV light

Commercial UV disinfection systems are expected to guarantee sufficient microbial inactivation and prolonged shelf life for the foods being processed.

© Woodhead Publishing Limited, 2012

However, such applications can only be safe if these techniques do not lead to undesirable quality changes in the products. The treatment of foods with either continuous or pulsed UV light does not typically involve addition of chemicals to the food being treated, which is one of the main advantages of these technologies over chemical-based disinfection methods. However, the potential of UV light to promote photochemical reactions in biomolecules connotes that UV disinfection may cause formation or changes of certain molecules in target foods.

UV light at 254 nm has a photon energy of 112.8 kcal/Einstein (one Einstein is defined as one mole of photons) which, if absorbed, could possibly affect O-H, C-C, C-H, C-N, H-N, and S-S bonds (Koutchma *et al.*, 2009b). Purine and pyrimidine bases on the nucleic acid strands, about 10% of the proteins (only those containing aromatic amino acids and amino acids with disulfide bonds), and ascorbic acid are among the strong absorbers of UV light at 254 nm. Nutrients regarded as 'light sensitive' include vitamin A, riboflavin (vitamin B2), cyanobalamin (vitamin B12), tocopherols (vitamin E), vitamin D, vitamin K, carotenes, folic acid, tryptophan, solid fats, phospholipids and unsaturated fatty acid residues in oils (Koutchma *et al.*, 2009b). Certain natural pigments and artificial colorants are also light sensitive. Carbohydrates are not considered to be light sensitive in the absence of certain photosensitizers. Light accelerates oxidative reactions in fats and oils, and tocopherol can act as a protector against photooxidation (Koutchma *et al.*, 2009b). Factors affecting the photosensitivity of a compound include the wavelength of the light absorbed, the chemical structure of the compound, and the presence of a photosensitizer.

UV treatment of drinking water is not associated with carcinogenic or mutagenic by-products, or any other negative effects on water quality (Wright and Cairns, 1998). UV light if applied improperly or excessively may lead to sensory defects and/or oxidation in milk (Reinemann *et al.*, 2006). Lyon *et al.* (2007) observed slight changes in meat color after the exposure of raw broiler breast fillets to UV light at 1000 $\mu W/cm^2$ for 5 min, both on the treatment day and at the end of 7-day storage. No significant change was observed in egg-shell conductance or hatchability of the treated eggs following the treatment of broiler hatching eggs with UV light for 3 min at up to 14 $mW/cm^2$ (Coufal *et al.*, 2003). The treatment of fresh-cut watermelons with UV light at up to 6.9 $kJ/m^2$ did not significantly alter juice leakage, flesh darkening, visual quality or color values at the end of 7-day storage at 3°C (Fonseca and Rushing, 2006). The exposure of wheat surfaces to UV light at 97 $W/m^2$ for up to 15 h did not influence the germination rate or the viscosity (Hidaka and Kubota, 2006). The UV treatment of frankfurters with doses up to 4 $J/cm^2$ did not change the color and texture of the frankfurters significantly (Sommers *et al.*, 2009a). In addition, bacterial mutagenicity tests showed no mutagenic potential in the frankfurters exposed to UV doses of up to 16 $J/cm^2$.

A study by Tran and Farid (2004) revealed that exposure of orange juice

© Woodhead Publishing Limited, 2012

to UV light at 100 mJ/cm$^2$ caused about a 17% degradation of vitamin C, which is comparable with levels commonly occurring in thermal sterilization. UV treatment at 73.8 mJ/cm$^2$ did not significantly inactivate the pectin methylesterase, which is an enzyme responsible for reducing cloudiness in juice. UV treatment alone might not be sufficient to prevent this quality defect in orange juice. The color and pH were also unaffected by the UV treatments.

UV light could also lead to favorable changes in foods. A UV treatment (3.7 kJ/m$^2$) together with storage at 16°C and 95% humidity significantly delayed senescence in tomato fruits by at least one week. Consequently, color and lycopene development and chlorophyll loss slowed during the 35-day storage period (Maharaj *et al.*, 2010). Fruits treated with UV light also contained significantly higher amounts of carotenoids than the untreated ones which, the authors suggested, act as an antioxidative phytochemical defense against free radical damage from photosensitized reactions. One negative effect of UV light on tomato fruits was abnormal browning, which was observed only after treatment at a higher dose of 24.4 kJ/m$^2$.

UV-C light stimulates the formation of phenylalanine ammonia-lyase, which induces the production of phenolic compounds such as phytoalexins in fruits and vegetables. These compounds play a role in the reduction of microbial decay by increasing the resistance of fruits and vegetables to microorganisms (Stevens *et al.*, 1999; Guerrero-Beltran and Barbosa-Canovas, 2004). This effect of stimulating a beneficial plant response is termed 'hormesis', and UV light as the stimulatory agent (below lethal doses) is called a 'hormetin'; something that has 'hormetic' effects on plants (Stevens *et al.*, 1999). An increasing accumulation of the phytoalexins, scoparone and scopoletin, in the flavedo tissue of oranges subjected to higher UV irradiation doses was reported by D'hallewin *et al.* (1999). The activity of phenylalanine ammonia-lyase and two other plant defense enzymes, chitinase and beta-1,3-glucanase, were induced in peach fruit by UV-C treatment (El Ghaouth *et al.*, 2003).

The study by Erkan *et al.* (2008) demonstrated that UV-C treatment of strawberries resulted in higher antioxidant capacity and enzyme activity, and less decay compared to the control fruit. UV irradiation has also been reported to enhance the accumulation of anthocyanin in sweet cherries (Arakawa, 1993), red apples (Dong *et al.*, 1995), and strawberries (Baka *et al.*, 1999). UV light (254 nm) had a hormetic effect on fresh-cut cantaloupe melon, increasing the activity of ascorbate peroxidase, which is a defense enzyme protecting the plant cells from damage by oxidative stress (Lamikanra *et al.*, 2005). The reduction of rancidity and improvement of firmness retention in the stored fruit were attributed to decreased lipase activity during storage with other induced anti-senescence defense responses. It has been reported by Cantos *et al.* (2002) that UV-C irradiation can induce the formation of stilbenes, including resveratrol, which are health-beneficial nonflavonoid phenolics present in grapes.

© Woodhead Publishing Limited, 2012

### 12.7.2 Effects of pulsed UV light

During pulsed UV light treatments, the action of polychromatic irradiation involving the UV, visible, and infrared portions of the electromagnetic spectrum causes photodegradation and other chemical reactions in foods. Quality changes occurring in pulsed UV-treated foods cannot be attributed only to UV, but to the entire spectrum of the flashlamp. High intensity light may also lead to the formation of germicidal chemical species such as hydroxyl radicals, ozone, or hydrogen peroxide (Malley, 2002). The most predominant difference between pulsed UV light and continuous UV light, in terms of the effect on food quality, is the generation of heat by pulsed UV lamps due to the infrared portion of the spectrum. Infrared light causes heat accumulation in the system over longer treatment times, increasing the temperature, and thus magnifying heat-related quality changes in foods.

Sharma and Demirci (2003) demonstrated that treating 1.02 mm thick alfalfa seeds for more than 30 s and 1.92 mm thick alfalfa seeds for more than 75 s with pulsed UV light (5.6 $J/cm^2$/pulse at the lamp surface) reduced the germination capability of the seeds due to excessive heating. On the other hand, treating alfalfa seeds of constant thickness (6.25 mm) at the distances of 3 and 5 cm from the lamp reduced the germination ability of the seeds after 30 and 60 s, respectively. This study demonstrated that the heating effect of pulsed UV light on the viability of the alfalfa seeds increased with lower seed layer thickness, longer treatment times, and shorter distance from the lamp.

Fine and Gervais (2004) reported that the thermal effect of pulsed light dominated the UV effect on colored food powders (black pepper and wheat flour). Significant changes in the visual appearance and flavor of the food powders were observed before complete microbial inactivation was reached at 58 $J/cm^2$. The color of black pepper showed more sensitivity to pulsed light compared to that of wheat flour, which is attributed to the absorption of more light energy by darker products. During the treatment of milk in a flow-through pulsed UV light system, the milk closer to the quartz window of the lamp had a higher temperature build-up, which could enhance fouling on the surface and thus decrease the amount of energy absorption (Krishnamurthy *et al.*, 2007). Blueberries treated with pulsed UV light at 8 cm distance from the quartz window for 30 and 60 s did not change in sensory quality or color (Bialka and Demirci, 2007). Pulsed UV light treatment of portabella and sliced white whole mushrooms for less than 1 s yielded quantities of vitamin D greater than 100% daily value (Williams, 2008).

In a study performed by Hierro *et al.* (2009), scanning electron microscopy of an egg shell revealed that damaged cuticles can create shadow zones and pores where bacteria can hide from pulsed light. Thus, the authors pointed out the importance of an intact cuticle in efficacious treatment of shell eggs using pulsed light. Keklik *et al.* (2010b) reported that the complete inactivation in *Salmonella* Enteritidis on eggshells was observed after 20 min pulsed UV

© Woodhead Publishing Limited, 2012

treatment at 23.6 J/cm$^2$, which did not significantly change the albumen height, eggshell strength, or cuticle presence. The treatment of unpackaged and vacuum-packaged chicken breast and frankfurters with pulsed UV light at closer distances (from the lamp) for longer times appeared to cause higher amounts of lipid peroxidation in meat and lower elastic modulus of the packaging material (polypropylene film), while producing changes in meat color and other mechanical properties of the packaging material (Keklik *et al.*, 2009, 2010c).

## 12.8 Limitations and challenges

The biggest issue associated with UV and pulsed UV light appears to be its limited ability to penetrate products. Although the penetration ability of pulsed UV light is better than that of UV light due to the high peak power delivered (Xenon, 2006), it is still not sufficient to reach all parts of foods. In fact, Bialka *et al.* (2008) reported that pulsed UV light can penetrate opaque materials by up to 10 mm, but with decreasing energy levels. The applications of UV and pulsed UV light technologies are therefore more or less limited to surface decontamination of foods.

The mercury lamps used in UV light treatments are another source of difficulty. A leakage of mercury from the lamp and its sleeve would pose major health and environmental hazards. However, pulsed UV lamps do not carry this risk, because they use inert gases such as xenon.

Complex surface properties of foods bring another challenge: microorganisms located in pores and crevices of a food surface can be shaded from light, and thus remain unaffected (Lagunas-Solar *et al.*, 2006). Hence, surface decontamination using UV or pulsed UV light is more suitable for foods with smooth surfaces such as fresh whole fruits, vegetables, hard cheeses, and smooth-surface meat slices (Oms-Oliu *et al.*, 2010). In liquid foods, the shadowing effect is caused by the particulates in turbid or non-clear liquids (Sauer and Moraru, 2009). Consequently, uniform exposure of foods to UV or pulsed UV light remains a challenge.

In the case of pulsed UV light treatment, heat is accumulated over longer treatment times and at shorter distances from the lamp, due to the infrared portion of light (Bialka, 2006). The resulting uncontrolled heat buildup not only affects microbial inactivation as a confounding variable, but also leads to food quality defects associated with temperature increase. In the case of UV light, the nucleic acid repair mechanisms of microorganisms may result in reversible inactivation (also known as reactivation) due to the delivery of low UV flux magnitudes.

© Woodhead Publishing Limited, 2012

## 12.9 Future trends

Despite having the potential to inactivate microorganisms on foods, the treatment of liquid or solid foods in a commercial system using UV or pulsed UV light is challenging. This is mostly due to the complex nature of foods, which prevents the UV light from reaching all parts of the product. In order to improve the efficacy of a UV or pulsed UV light system, the effects of treatment variables on the efficacy of the system need to be determined. In any UV or pulsed UV system, the treatment parameters must relate to certain properties of the food, light, and the system in which the food is treated.

Recent developments of lamps with different spectra, pulse rates, and energies provide options for different applications and the control of ozone generation (Xenon, 2008). The design of reflectors to increase absorbance and uniform exposure is also necessary (Fig. 12.4). Furthermore, researchers continue in their attempts to design systems that could eliminate microorganisms in/on foods more effectively using UV or pulsed UV light.

The establishment of databases of microbial inactivation curves and the classification of pathogens in terms of their resistances to UV and pulsed UV light (Demirci and Panico, 2008), the development of mathematical models describing the microbial inactivation in/on foods, and the toxicological risk analyses of foods treated with these techniques (Sommers *et al.*, 2009a) would all help to determine and control the microbiological and/or quality hazard risks during food processing and storage. Furthermore, a full understanding

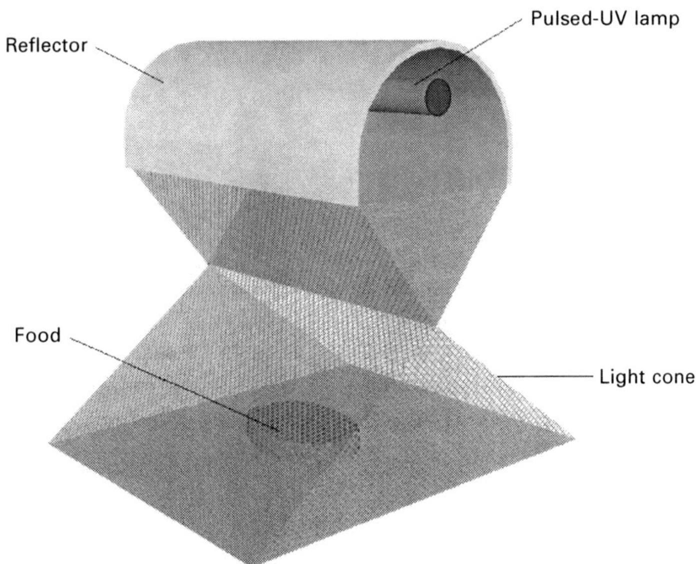


**Fig. 12.4** A light reflector cone.

© Woodhead Publishing Limited, 2012

of the mechanisms of microbial inactivation and hormetic effects of UV and pulsed UV light would provide many clues on how to increase the efficiency and benefits of these technologies for food applications.

## 12.10 Sources of further information and advice

UV and pulsed UV light technologies have the potential to decontaminate foods. While destroying the microbial cells in/on a target food, they do not appear to affect the food's quality significantly unless extreme treatments are performed. Furthermore, they can even initiate beneficial photochemical reactions, which, in turn, can increase the activities of desirable enzymes, vitamins, and antioxidants in certain foods. It is very important to precisely assess the effects of UV or pulsed UV light treatment on microbial content and the physical/chemical quality of the food and packaging material before adopting a commercial application. The efficacy of UV or pulsed UV light treatment is dependent on the characteristics of the light, food, and the system in which the food is treated. Since manipulating food so as to increase its photosensitivity is not desirable, the properties of the light and the system parameters need to be optimized for each type of food. A successful UV or pulsed UV light system designed for commercial applications in food processing must ensure the following:

- Pathogen levels are reduced below the limits set by federal regulations.
- Shelf life is prolonged by eliminating the spoilage microorganisms.
- Food quality is protected/improved.
- Operation, maintenance, and cleaning of the system is safe and practical.
- Processing is fast and cost-efficient.

Several websites providing information and/or advice on the various aspects of UV and/or pulsed UV light technologies are given below:

- NASA (National Aeronautics and Space Administration): Electomagnetic spectrum: http://missionscience.nasa.gov/ems/index.html
- FDA (US Food and Drug Administration): UV light: http://www.fda.gov/Food/ScienceResearch/ResearchAreas/SafePracticesforFoodProcesses/ucm103137.htm
- FDA (US Food and Drug Administration): Pulsed light: http://www.fda.gov/Food/ScienceResearch/ResearchAreas/SafePracticesforFoodProcesses/ucm103058.htm

© Woodhead Publishing Limited, 2012

## 12.11 References

ACRA A, JURDI M, MU'ALLEM H, KARAHAGOPIAN Y and RAFFOUL Z (1990), *Water Disinfection by Solar Radiation – Assessment and Application*, Ottowa, IDRC.

AGUIAR A, BAINBRIDGE D, BOND M, GUPTA C, MATTHEWS K, PLANTE J, RANTANEN E, TURNER C, YANG D, ZHANG W and DREW D (1996), 'Modeling UV-damage to *E. coli* bacteria, Mathematical modeling for instructors', 1422–7, University of Minnesota, Minneapolis, MN. Available from: http://www.ima.umn.edu/preprints/Sept96/1422g.pdf (accessed 20 February 2011).

ARAKAWA O (1993), 'Effect of ultraviolet light on anthocyanin synthesis in light-colored sweet cherry, cv. satonishiki', *J Japan Soc Hort Sci*, 62(3), 543–546.

BAKA M, MERCIER J, CORCRUFF R, CASTAIGNE F and ARUL J (1999), 'Photochemical treatment to improve storability of fresh strawberries', *J Food Sci*, 64, 1068–1072.

BEGUM M, HOCKING A D and MISKELLY D (2009), 'Inactivation of food spoilage fungi by ultra violet (UVC) irradiation', *Int J Food Microbiol*, 129, 74–77.

BIALKA K L (2006), 'Decontamination of Berries with Ozone and Pulsed UV-light', PhD Thesis, University Park, PA: The Pennsylvania State University, Department of Agricultural and Biological Engineering.

BIALKA K L and DEMIRCI A (2007), 'Decontamination of *Escherichia coli* O157:H7 and *Salmonella enterica* on blueberries using ozone and pulsed UV-light', *J Food Sci*, 72(9), M391–M396.

BIALKA K L and DEMIRCI A (2008), 'Efficacy of pulsed UV-light for the decontamination of *Escherichia coli* O157:H7 and *Salmonella* spp. on raspberries and strawberries', *J Food Sci*, 73(5), M201–M207.

BIALKA K L, DEMIRCI A, WALKER P N and PURI V M (2008), 'Pulsed UV-light penetration of characterization and the inactivation of *Escherichia coli* K12 in solid model systems', *Trans ASABE*, 51(1), 195–204.

BISCHOF M (1994), 'History of bioelectromagnetism', in Ho M W, Popp F A and Warnke U, *Bioelectrodynamics and Biocommunication*. Singapore: World Scientific Publishing.

BLATCHLEY III E R and PEEL M M (2001), 'Disinfection by ultraviolet irradiation', in Block S S, *Disinfection, Sterilization, and Preservation*, 5th edn. Philadelphia, PA: Lippincott Williams & Wilkins, 823–851.

BOHREROVA Z, SHEMER H, LANTIS R, IMPELLITTERI C A and LINDEN K G (2008), 'Comparative disinfection efficiency of pulsed and continuous-wave UV irradiation technologies', *Water Research*, 42, 2975–2982.

CANTOS E, ESPIN J C and TOMAS-BARBERAN F A (2002), 'Postharvest stilbene-enrichment of red and white table grape varieties using UV-C irradiation pulses', *J Agric Food Chem*, 50, 6322–6329.

COUFAL C D, CHAVEZ C, KNAPE K D and CAREY J B (2003), 'Evaluation of a method of ultraviolet light sanitation of broiler hatching eggs', *Poultry Sci*, 82, 754–759.

D'HALLEWIN G, SCHIRRA M, MANUEDDU E, PIGA A and BEN-YEHOSHUA S (1999), 'Scoparone and scopoletin accumulation and ultraviolet-C induced resistance to postharvest decay in oranges as influenced by harvest date', *J Am Soc Hortic Sci*, 124(6), 702–707.

DEMIRCI A and PANICO L (2008), 'Pulsed ultraviolet light', *Food Sci Technol Int*,14(5), 443–446.

DONG Y H, MITRA D, KOOTSTRA A, LISTER C and LANCASTER J (1995), 'Postharvest stimulation of skin color in Royal-gala apple', *J Am Soc Hortic Sci*, 120, 95–100.

DUNN J (1996), 'Pulsed light and pulsed electric field for foods and eggs', *Poultry Sci*, 75, 1133–1136.

EL-BOURY S, COUTEAU C, BOULANDE L, PAPARIS E and COIFFARD L J M (2007), 'Effect of the combination of organic and inorganic filters on the Sun Protection Factor (SPF) determined by *in vitro* method', *Int J Pharm*, 340, 1–5.

© Woodhead Publishing Limited, 2012

EL GHAOUTH A, WILSON C L and CALLAHAN A M (2003), 'Induction of chitinase, beta-1,3-glucanase, and phenylalanine ammonia lyase in peach fruit by UV-C treatment', *Phytopathology*, 93, 349–355.

ELMNASSER N, GUILLOU S, LEROI F, ORANGE N, BAKHROUF A and FEDERIGHI M (2007), 'Pulsed-light system as a novel food decontamination technology: a review', *Can J Microbiol*, 53, 813–821.

ERKAN M, WANG S Y and WANG C Y (2008), 'Effect of UV treatment on antioxidant capacity, antioxidant enzyme activity and decay in strawberry fruit', *Postharv Biol Technol*, 48, 163–171.

FDA (2000), Kinetics of Microbial Inactivation for Alternative Food Processing Technologies, A report of the Institute of Food Technologists for the Food and Drug Administration of the US Department of Health and Human Services. Available from: http://www.fda.gov/Food/ScienceResearch/ResearchAreas/SafePracticesforFoodProcesses/ucm100158.htm (accessed 17 February 2011).

FDA (2001), 'Hazard Analysis and Critical Control Point (HAACP); Procedures for the Safe and Sanitary Processing and Importing of Juice', *Fed Regist*, 66(13), 6137–6202.

FDA (2010), Title 21 – Food and drugs, Chapter I – Food and Drug Administration, Department of Health and Human Services, Subchapter B – Food for human consumption (continued), Part 179 – Irradiation in the production, processing and handling of food. Available from: http://www.accessdata.fda.gov/scripts/cdrh/cfdocs/cfcfr/CFRSearch.cfm?CFRPart=179&showFR=1 (accessed 25 February 2011).

FEDERAL REGISTER (1999), 'Pulsed light treatment of food', *Fed Regist*, 66, 338829–338830.

FINE F and GERVAIS P (2004), 'Efficiency of pulsed UV light for microbial decontamination of food powders', *J Food Prot*, 67(4), 787–792.

FINO V R and KNIEL K E (2008), 'UV light inactivation of hepatitis A virus, Aichi virus, and feline calicivirus on strawberries, green onions, and lettuce', *J Food Prot*, 71(5), 908–913.

FONSECA J M and RUSHING J W (2006), 'Effect of ultraviolet-C light on quality and microbial population of fresh-cut watermelon', *Postharv Biol Technol*, 40, 256–261.

GUERRERO-BELTRAN J A and BARBOSA-CANOVAS G V (2004), 'Review: advantages and limitations on processing foods by UV light', *Food Sci Tech Int*, 10(3), 137–147.

HAKGUDER B (2009), 'UV Disinfection of Some of the Fruit Juices, Decontamination of Berries with Ozone and Pulsed UV-light', MSc Thesis, Izmir, Turkey: İzmir Institute of Technology, Department of Food Engineering.

HANES D E, WOROBO R W, ORLANDI P A, BURR D H, MILIOTIS M D, ROBL M G, BIER J W, ARROWOOD M J, CHUREY J J and JACKSON G J (2002), 'Inactivation of *Cryptosporidium parvum* oocysts in fresh apple cider by UV irradiation', *Appl Environ Microbiol*, 68(8), 4168–4172.

HIDAKA Y and KUBOTA K (2006), 'Study on the sterilization of grain surface using UV radiation – development and evaluation of UV irradiation equipment', *Jpn Agr Res Q*, 40(2), 157–161.

HIERRO E, MANZANO S, ORDONEZ J A, DE LA HOZ L and FERNANDEZ M (2009), 'Inactivation of *Salmonella enterica* serovar Enteritidis on shell eggs by pulsed light technology', *Int J Food Microbiol*, 135, 125–130.

ISOHANNI P M I and LYHS U (2009), 'Use of ultraviolet irradiation to reduce *Campylobacter jejuni* on broiler meat', *Poultry Sci*, 88, 661–668.

JUN S, IRUDAYARAJ J M, DEMIRCI A and GEISER D (2003),'Pulsed UV-light treatment of corn meal for inactivation of *Aspergillus niger* spores', *Int J Food Sci Technol*, 38(8), 883–888.

KEKLIK N M, DEMIRCI A and PURI V M (2009), 'Inactivation of *Listeria monocytogenes* on unpackaged and vacuum-packaged chicken frankfurters using pulsed UV-light', *J Food Sci*, 74(8), M431–M439.

KEKLIK N M, DEMIRCI A and BOCK R G (2010a), 'Decontamination of whole chicken

© Woodhead Publishing Limited, 2012

carcasses by using a pilot-scale pulsed UV-light system', *ASABE (American Society of Agricultural and Biological Engineers) Meeting*, Pittsburgh, PA, 20–23 June 2010, Paper No. 1008677.

KEKLIK N M, DEMIRCI A, PATTERSON P H and PURI V M (2010b), 'Pulsed UV light inactivation of *Salmonella* Enteritidis on egg shells and its effects on egg quality', *J Food Prot*, 73(8): 1408–1415.

KEKLIK N M, DEMIRCI A and PURI V M (2010c), 'Decontamination of unpackaged and vacuum-packaged chicken breast with pulsed UV-light', *Poultry Sci*, 89(3), 570–581.

KOUTCHMA T, PARISI B and UNLUTURK S (2006), 'Evaluation of UV dose in flow-through reactors for fresh apple juice and cider', *Chem Eng Commun*, 193(6), 715–728.

KOUTCHMA T N, FORNEY L J and MORARU C I (2009a), 'Sources of UV Light', in *Ultraviolet Light in Food Technology – Principles and Applications*, Boca Raton, FL: CRC Press, 33–47.

KOUTCHMA T N, FORNEY L J and MORARU C I (2009b), 'UV processing effects on quality of foods', in *Ultraviolet Light in Food Technology – Principles and Applications*, Boca Raton, FL: CRC Press, 103–124.

KRISHNAMURTHY K, DEMIRCI A and IRUDAYARAJ J M (2007), 'Inactivation of *Staphylococcus aureus* in milk using flow-through pulsed UV-light treatment system', *J Food Sci*, 72(7), M233–M239.

KRISHNAMURTHY K, IRUDAYARAJ J, DEMIRCI A and YANG W (2008), 'UV pasteurization of food materials', in Jun S and Irudayaraj J M, *Food Processing Operations Modeling: Design and Analysis*, 2nd edn, Boca Raton, FL: CRC Press, 281–299.

KRISHNAMURTHY K, TEWARI J C, IRUDAYARAJ J and DEMIRCI A (2010), 'Microscopic and spectroscopic evaluation of inactivation of *Staphylococcus aureus* by pulsed UV light and infrared heating', *Food Bioprocess Technol*, 3(1), 93–104.

KUO F L, CAREY J B and RICKE S C (1997), 'UV irradiation of shell eggs: effect of populations of aerobes, molds, and inoculated *Salmonella* Typhimurium', *J Food Prot*, 60(6), 639–643.

LAGUNAS-SOLAR M C, PINA C, MACDONALD J D and BOLKAN L (2006), 'Development of pulsed UV light processes for surface fungal disinfection of fresh fruits', *J Food Prot*, 69(2), 376–384.

LAMIKANRA O, KUENEMAN D, UKUKU D and BETT-GARBER K L (2005), 'Effect of processing under ultraviolet light on the shelf life of fresh-cut cantaloupe melon', *J Food Sci*, 70(9), C534–C539.

LINDEN K (2004), 'Introduction', in *Disinfection Efficiency and Dose Measurement of Polychromatic UV Light*, London: IWA Publishing, 1–5.

LUKSIENE Z, GUDELIS V, BUCHOVEC I and RAUDELIUNIENE J (2007), 'Advanced high-power pulsed light device to decontaminate food from pathogens: effects on *Salmonella* Typhimurium viability *in vitro*', *J App Microbiol*, 103, 1545–1552.

LYON S A, FLETCHER D L and BERRANG M E (2007), 'Germicidal ultraviolet light to lower numbers of *Listeria monocytogenes* on broiler breast fillets', *Poultry Sci*, 86(5), 964–967.

MAHARAJ R, ARUL J and NADEAU P (2010), 'UV-C irradiation of tomato and its effects on color and pigments', *Adv Environ Biol*, 4(2), 308–315.

MALLEY J P (2002), 'Ultraviolet disinfection', in Lingireddy S, *Control of Microorganisms in Drinking Water*, Reston, VA: American Society of Civil Engineers, 213–235.

MATAK K E, CHUREY J J, WOROBO R W, SUMNER S S, HOVINGH E, HACKNEY C R and PIERSON M D (2005), 'Efficacy of UV light for the reduction of *Listeria monocytogenes* in goat's milk', *J Food Prot*, 68(10), 2212–2216.

MCDONALD K F, CURRY R D, CLEVENGER T E, UNKLESBAY K, EISENSTARK A, GOLDEN J and MORGAN R D (2000), 'A comparison of pulsed and continuous ultraviolet light sources for the decontamination of surfaces', *IEEE Trans Plasma Sci*, 28(5), 1581–1587.

MELQUIADES F L, FERREIRA D D, APPOLONI C R, LOPES F, LONNI A G, OLIVEIRA F M and DUARTE J C (2008), 'Titanium dioxide determination in sunscreen by energy dispersive X-ray fluorescence methodology', *Analytica Chimica Acta*, 613, 135–143.

© Woodhead Publishing Limited, 2012

MILLER R V, JEFFREY W, MITCHELL J D and ELASRI M (1999), 'Bacterial responses to ultraviolet light', *ASM News*, 65, 535–541.

OMS-OLIU G, MARTIN-BELLOSO O and SOLIVA-FORTUNY R (2010), 'Pulsed light treatments for food preservation. A review', *Food Bioprocess Technol*, 3, 13–23.

OZER N and DEMIRCI A (2006), 'Inactivation of *Escherichia coli* O157:H7 and *Listeria monocytogenes* inoculated on raw salmon fillets by pulsed UV-light treatment', *Int J Food Sci Technol*, 41, 354–360.

REED N G (2010), 'The history of ultraviolet germicidal irradiation for air disinfection', *Public Health Rep*, 125, 15–27.

REINEMANN D J, GOUWS P, CILLIERS T, HOUCK K and BISHOP J R (2006), 'New methods for UV treatment of milk for improved food safety and product quality', in *ASABE (American Society of Agricultural and Biological Engineers) Meeting*, Portland, OR, 9–12 July 2006, Paper No. 066088.

SAUER A and MORARU C I (2009), 'Inactivation of *Escherichia coli* ATCC 25922 and *Escherichia coli* O157:H7 in apple juice and apple cider, using pulsed light treatment', *J Food Prot*, 72(5), 937–944.

SCHAEFER R, GRAPPERHAUS M, SCHAEFER I and LINDEN K (2007), 'Pulsed UV lamp performance and comparison with UV mercury lamps', *J Environ Eng Sci*, 6, 303–310.

SHARMA R R and DEMIRCI A (2003), 'Inactivation of *Escherichia coli* O157:H7 on inoculated alfalfa seeds with pulsed ultraviolet light and response surface modeling', *J Food Sci*, 68(4), 1448–1453.

SOLSONA F and MENDEZ J P (2003), 'Ultraviolet radiation', in *Water Disinfection.* Lima, CEPIS, 71–84. Available from: http://www.bvsde.paho.org/bvsacg/fulltext/desinfeccioneng/chapter4.pdf (accessed 20 February 2011).

SOMMERS C H, COOKE P H, FAN X and SITES J E (2009a), 'Ultraviolet light (254 nm) inactivation of *Listeria monocytogenes* on frankfurters that contain potassium lactate and sodium diacetate', *J Food Sci*, 74(3), M114–M119.

SOMMERS C H, GEVEKE D J, PULSFUS S and LEMMENES B (2009b), 'Inactivation of *Listeria innocua* on frankfurters by ultraviolet light and flash pasteurization', *J Food Sci*, 74(3), M138–M141.

STEVENS C, KHAN V A, LU J Y, WILSON C L, PUSEY P L, IGWEGBE E C K, KABWE K, MAFOLO Y, LIU J, CHALUTZ E and DROBY S (1997), 'Integration of ultraviolet (UV-C) light with yeast treatment for control of postharvest storage rots of fruits and vegetables', *Biol Control*, 10(2), 98–103.

STEVENS C, KHAN V A, LU J Y, WILSON C L, CHALUTZ E, DROBY S, KABWE M K, HAUNG Z, ADEYEYE O, PUSEY L P and TANG A Y A (1999), 'Induced resistance of sweet potato to Fusarium root rot by UV-C hormesis', *Crop Prot*, 18, 463–470.

TAKESHITA K, YAMANAKA H, SAMESHIMA T, FUKUNAGA S, ISOBE S, ARIHARA K and ITOH M (2002), 'Sterilization effect of pulsed light on various microorganisms', *J Antibact Antifung Agents*, 30(5), 277–284.

TRAN M T T and FARID M (2004), 'Ultraviolet treatment of orange juice', *Innovat Food Sci Emerg Tech*, 5, 495–502.

UESUGI A R and MORARU C I (2009), 'Reduction of *Listeria* on ready-to-eat sausages after exposure to a combination of pulsed light and nisin', *J Food Prot*, 72(2), 347–353.

UNLUTURK S, ATILGAN M R, BAYSAL A H and UNLUTURK M S (2010), 'Modeling inactivation kinetics of liquid egg white exposed to UV-C irradiation', *Int J Food Microbiol*, 142, 341–347.

WEKHOF A (2000), 'Disinfection with flash lamps', *PDA J Pharm Sci Technol*, 54(3), 264–276.

WEKHOF A, TROMPETER F-J and FRANKEN O (2001), 'Pulsed UV Disintegration (PUVD): a new sterilisation mechanism for packaging and broad medical-hospital applications', in The First International Conference on Ultraviolet Technologies, Washington, DC, USA,14–16 June 2001.

© Woodhead Publishing Limited, 2012

WILLIAMS R (2008), Vitamin D formation from post-harvest pulsed light treatment of mushrooms, Xenon Corporation, Wilmington, MA. Available from: http://www.xenoncorp.com/PDFs/Xenon%20Corp%20White%20Paper%20Mushroom%20Vitamin%20D%20Enhancement-a.pdf (accessed 22 February 2011).

WRIGHT H B and CAIRNS W L (1998), 'Ultraviolet light', in Regional symposium on water quality: effective disinfection, Lima, CEPIS/OPS, 1–26. Available from: http://www.cepis.org.pe/bvsacg/i/fulltext/symposium/ponen10.pdf (accessed 11 August 2010).

XENON (2006), Pulsed UV treatment for sanitation and sterilization, Pulsed UV technology, Xenon Corporation, Wilmington, MA. Available from: http://www.xenoncorp.com/Literature/PDF/BrochureSteri.pdf (accessed 25 February 2011).

XENON (2008), RC-800 Series, UV Curing Systems, Xenon Corporation, Wilmington, MA. Available from: http://uv-curing.xenoncorp.com/Literature/PDF/RC-800%20Series%20Brochure.pdf (accessed 22 February 2011).

© Woodhead Publishing Limited, 2012

# 13

# Microbial decontamination of food by high pressure processing

**H. Daryaei and V. M. Balasubramaniam, The Ohio State University, USA**

**Abstract**: High pressure processing (HPP) is an alternative method of food preservation capable of inactivating pathogenic and spoilage microorganisms while maintaining the desirable quality attributes of the product, including nutritional and sensory properties. Depending on the process temperature, HPP can be used for either pasteurization or sterilization of foods. This chapter discusses the basic principles and applications of high pressure processing in food industry with emphasis on microbial inactivation under different process conditions. Process and product related factors influencing the microbial efficacy of pressure treatment will be reviewed.

**Key words**: high pressure processing, food preservation, microbial inactivation, pasteurization, sterilization.

## 13.1 Introduction

There has been ongoing interest in recent years in using alternative minimal food preservation technologies including high pressure processing (HPP), pulsed electric field (PEF) processing, irradiation, ozone treatment, and ultraviolet light. Many of these technologies have the promising potential to overcome the limitations associated with traditional thermal processing methods. They can retain natural freshness and have minimal effect on product quality and sensory attributes. Amongst different non-thermal processes, HPP has attracted wide industrial interest, because it can inactivate microorganisms in foods without adversely affecting quality attributes. The recent availability of commercial-scale HPP equipment from a number of manufacturers, as well as the increasing consumer demand for safe, minimally processed,

© Woodhead Publishing Limited, 2012

preservative-free foods with an extended shelf life has also stimulated interests in using HPP in the food industry.

One of the unique advantages of pressure treatment is that the pressure, at the levels used in the food industry, mainly acts on non-covalent bonds, such as hydrogen, ionic and hydrophobic bonds and has only a limited effect on covalent bonds within biological matter (Mozhaev *et al.*, 1994). Consequently, many of the food components responsible for sensory and nutritional properties of foods, such as flavor components, vitamins, and other small molecules are unaffected or only marginally influenced by high pressure, while the structure and functionality of large molecules such as proteins, enzymes, polysaccharides and nucleic acids may be altered (Balci and Wilbey, 1999).

Although the history of HPP of foods dates back more than a century (Hite, 1899), comprehensive studies in this field have been undertaken only in recent years. Earlier studies have shown that by subjecting foods to high pressures in the range of 300–400 MPa, vegetative cells of microorganisms and certain enzymes can be inactivated at ambient temperature without degradation of flavor and nutrients. However, bacterial spores can only be killed by high pressures (600–700 MPa) in combination with heat (>70°C) (Smelt, 1998). Pulsed or oscillating pressurization in conjunction with heat is claimed to be more effective in spore inactivation (Furukawa *et al.*, 2000; Meyer *et al.*, 2000).

Almost eight decades after Hite's attempt to preserve foods with high pressure, the first commercial high pressure processed foods including jams, jellies and sauces were released to the market in Japan (Thakur and Nelson, 1998). Today, a wide range of value-added pressure-treated foods including seafood, processed meats, and vegetable and fruit preparations are available to consumers. HPP has the potential of providing microbiologically safe products with superior sensory quality, nutritional properties, and extended shelf life. As a novel processing technology, the relatively high cost of the equipment acted as a barrier for widespread industry use especially for commodity type products. Nonetheless, several pressure-treated value-added products that can justify modest increase in the cost have been successfully commercialized.

Different aspects of high pressure processing including microbial inactivation and food safety have been reviewed by Farkas and Hoover (2000), Mañas and Pagán (2005), Patterson (2005), Rastogi *et al.* (2007) and Zhang *et al.* (2011). This chapter provides an overview of microbial efficacy of HPP technology. The review will also cover basic principles associated with the technology, including processing and product parameters influencing microbial efficacy, mechanisms of microbial inactivation, and potential injury and recovery of microorganisms during extended storage.

© Woodhead Publishing Limited, 2012

## 13.2 The high pressure processing (HPP) system

### 13.2.1 Definition and basic principles of high pressure processing

Hydrostatic pressure is usually defined as 'isostatic' or 'isobaric' pressure transferred by water. Processing foods with high hydrostatic pressure (or just high pressure) is an emerging non-thermal processing technique for microbial inactivation at refrigeration, ambient, or moderate heating temperatures. HPP is capable of inactivating foodborne pathogens, spoilage bacteria, yeasts, and molds with minimal degradation in texture, color, flavor, and nutritional quality of foods, as compared to conventional food preservation technologies (Knorr, 1993; Cheftel, 1995; Velazquez *et al.*, 2002). In this technology, foods usually in their final flexible package are subjected to a high level of uniform hydrostatic pressure within the range of 150–700 MPa for a period lasting from a few seconds up to several minutes. The choice of processing pressure is dependent on the product to be treated, the processing equipment available and the desired results. The temperature of food increases during pressurization as a result of compression and drops back close to its initial value after decompression (Ting *et al.*, 2002). The basic principles governing the application of high pressure processing in foods are summarized below.

*LeChatelier's principle*

LeChatelier's principle states that the application of pressure shifts the system equilibrium toward the state occupying the smallest volume and, therefore, any phenomenon (phase transition, change in molecular configuration, chemical reaction, etc.) accompanied by a decrease in volume will be enhanced by an increase in pressure and vice versa (Farkas and Hoover, 2000).

*Microscopic ordering principle*

According to the principle of microscopic ordering, an increase in pressure, at a constant temperature, increases the degree of ordering of molecules of a given substance. Therefore, pressure and temperature exert antagonistic forces on molecular structure and chemical reactions (Balny and Masson, 1993).

*Isostatic principle*

According to the isostatic principle, pressure is transmitted quasi-instantaneously and uniformly throughout the sample volume independently of the size and the geometry of the product (Heremans, 1982; Cheftel, 1995; Tauscher, 1995; Torres and Velazquez, 2005). In other words, all parts of foods experience similar pressure intensity during the treatment. Thus, at macroscopic level, products containing high moisture content in general are not distorted or damaged. However, at molecular level, pressure treatment may induce structural changes.

© Woodhead Publishing Limited, 2012

### 13.2.2 Equipment and typical operation

The main components of a high pressure processing system are shown in Fig. 13.1 and are described as follows (Ting, 2011):

(a) *Pressure vessel*: Typically pressure vessels used in the food industry utilize a cylindrical structure. Pressure vessels can be constructed in three different ways. The pressure chamber can be built using monoblock (single forged) chambers. Monolithic chambers are simple to build and less expensive, but they cannot withstand pressure levels above 400 MPa. Multiwall chambers can be built from a shrink fitting series of concentric cylinders. Thus, they can withstand higher pressures and are safer than single wall vessels. The wire wound approach involves winding pre-stressed stainless steel onto a thin wall core of the pressure vessel, and thus can be utilized for pressure vessels with larger diameter and can be operated at higher pressures. Depending on the target process conditions, the pressure vessels may have external jackets for temperature control.

(b) *Top and bottom closures to contain the pressure*: Two end closures contain the product and pressure-transmitting fluid within the pressure vessel during processing. For food process operation, rapid opening and closing of the closure is desired. It is worth noting that the pressure vessel along with the associated closure and yoke are commonly installed with either vertical or horizontal orientation.

(c) *Yoke*: At elevated pressures, often a secondary structure is desired to contain end closures. This secondary external frame, called a yoke, is made of high tensile strength steel or wire wound frame.

(d) *Pressure intensifier and pump for generating the pressure*: Larger industrial scale pressure vessels utilize an external pumping intensifier

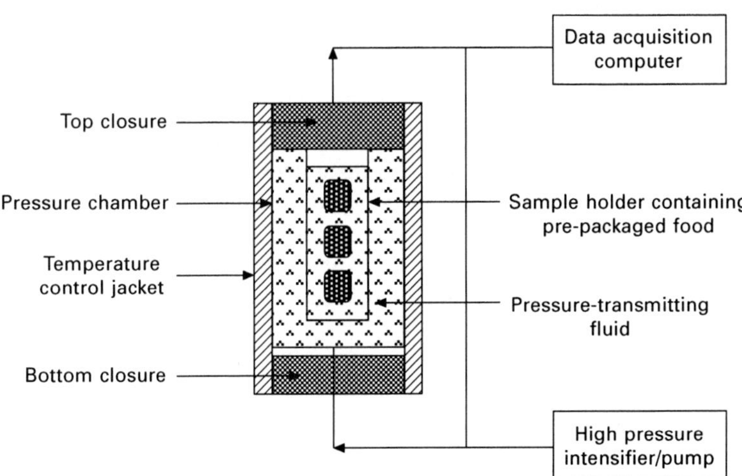


**Fig. 13.1** Schematic diagram of a high pressure processing system.

© Woodhead Publishing Limited, 2012

system to deliver target pressure with minimal pressure come-up time (about 1–2 min). They utilize a motor to run a lower pressure pump to compress the hydraulic fluid that is used in a larger intensifier. Water is the most commonly used pressure-transmitting medium in industrial scale pressure vessels. The larger intensifier compresses the water and sends it to the pressure vessel.

(e) *A material handling system for loading and unloading the test sample:* Typically during batch processing, pre-packaged samples are loaded into a cylindrical shaped canister. The canister containing the sample may be pre-heated or chilled prior to loading into the pressure vessel.

(f) *Process control system for monitoring and recording various process variables*: The control system will help the processors to monitor and document relevant process data (such as pressure, temperature, etc.). The sensors need to be periodically calibrated using established protocols (Balasubramaniam *et al.*, 2004).

In the United States, the operation and safety of high pressure food processing vessels are governed by city and state regulations for boiler inspection and operation (Ting, 2011). The pressure vessels should carry a stamp indicating that they have been designed, constructed, tested, and certified to operate under a target pressure-temperature range. Many states adopted ASME's Section VIII Division 3 'Alternative rules for the construction of pressure vessels' (ASME, 2011). Industrial food processing plants also need to consider relevant Occupational Safety and Health Administration (OSHA) regulations. In Europe, the safety of the high pressure equipment is governed by the Pressure Equipment Directive (PED 97/23/EC) (Ting, 2011).

Water is the most common medium for pressure transmission due to its availability, non-toxicity, and low cost; however, compounds such as castor oil, silicone oil, corn oil, sodium benzoate, ethanol and glycol blends are also used for lubrication and anticorrosion purposes (Mertens and Deplace, 1993; Needs, *et al.*, 2000; Mainville *et al.*, 2001; Walsh-O'Grady *et al.*, 2001; Johnston *et al.*, 2002; Balasubramanian and Balasubramaniam, 2003). Various pressure-transmitting fluids may have different compression heating values depending on their thermal and physical properties (such as specific heat, viscosity, and compressibility). Such differences can affect the magnitude of heat transfer among the fluid, food product, and the environment, and subsequently influence microbial inactivation and quality changes under pressure (Balasubramanian and Balasubramaniam, 2003). The heat of compression values of water and silicone oil are about 3 and 20°C per 100 MPa, respectively.

Industrial high pressure treatment of foods is currently conducted in batch or semi-continuous systems with pressure vessel capacities of 35–600 L. The vessel can be oriented either horizontally, vertically or in a tilted position (Fig. 13.2). Available space and constraints in the food processing facility dictate the configuration of the pressure vessel. In the batch high pressure system, foods are pre-packaged and loaded into the vessel, the lid

© Woodhead Publishing Limited, 2012

(a)

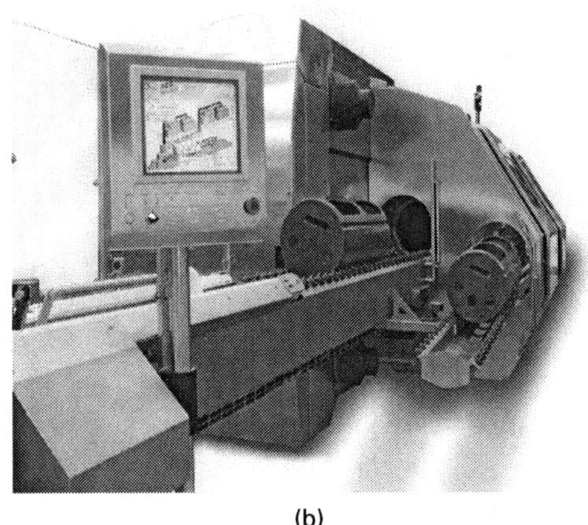

(b)

**Fig. 13.2** Two commercial high pressure processing systems: (a) a vertical 215L-600 MPa system by Avure Technologies; (b) a horizontal 420L-600 MPa system by NC Hyperbaric.

is closed and pressure-transmitting fluid (e.g. water) is pumped into the vessel as the system is vented. Once the desired pressure is reached, the pressure is maintained without further need for energy input. The pressure is released after the desired treatment time has elapsed, the food packages are then unloaded and the system is reloaded with new product either by an operator or automatically (Ting and Marshall, 2002). The total time required for pressurization, holding and depressurization is referred to as a 'cycle'

© Woodhead Publishing Limited, 2012

which is an important technological factor in determining the throughput of the system.

For pumpable products (e.g. juices), a semi-continuous system containing three or more pressure vessels can be used. Unlike batch systems, the product does not require pre-packaging prior to pressure treatment. The liquid product can be filled into the pressure chamber. The pressure vessels in a semi-continuous system can be connected such that, while one vessel discharges the product, the second vessel is being compressed, and the third vessel is being loaded. In this way a continuous output is obtained. After treatment, the material can be pumped to an aseptic filling line, similar to that used for ultra-high-temperature (UHT) processed liquids to be packaged in glass bottles or gable cartons (Ting and Marshall, 2002). A semi-continuous HPP system with a capacity of processing 600 L of liquid per hour is commercially used in Japan for processing grapefruit juice at a maximum pressure of 400 MPa (Palou *et al.*, 2002).

### 13.2.3 Typical pressure-temperature history

When applying HPP to process foods, the pressure-temperature history of the samples during treatment should be carefully monitored. Figure 13.3 shows a typical pressure-temperature history curve for a batch HPP treatment (Farkas and Hoover, 2000; Balasubramaniam *et al.*, 2004).

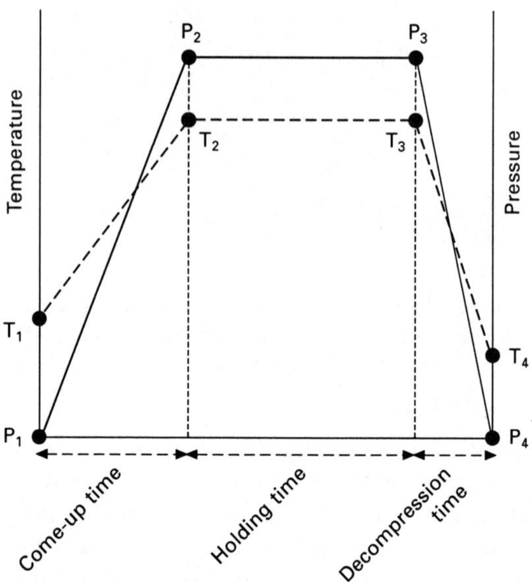


**Fig. 13.3** Typical pressure-temperature response of a food material undergoing high pressure processing.

© Woodhead Publishing Limited, 2012

*Pressure come-up time*

The time required to increase the pressure of the sample from atmospheric pressure (0.1 MPa) to the target process pressure is commonly referred to as pressure come-up time. The pressure come-up time varies depending on the target process pressure. It is also dependent on the volume of the pressure vessel and the horsepower of the pump intensifier employed (Nguyen and Balasubramaniam, 2011). Typical commercial scale high pressure equipment may have come-up time of about 2 min.

*Pressure holding time*

The pressure holding time starts immediately after come-up time and usually lasts from a few seconds up to several minutes depending on the preservation goal. The pressure holding time normally does not include the pressure come-up time or depressurization time (Farkas and Hoover 2000; Balasubramaniam *et al.*, 2004). The product is compressed about 15–20% at 600 MPa during pressurization and regains its original volume upon decompression.

*Decompression time*

The time required to decrease the pressure of a food sample from target process pressure to near atmospheric pressure is referred to as the decompression time. A typical decompression time is less than 20–30 s.

## 13.3 Compression heating of pressure-transmitting fluids and food materials

The increase in the temperature of food material as a result of pressurization is known as 'heat of compression'. This temperature increase is induced by compressive work against intermolecular forces and can be calculated using Eq. [13.1] (Denys *et al.*, 2000):

$$\frac{dT}{dP} = \frac{T\alpha_P}{\rho C_P} \qquad [13.1]$$

where $T$, $P$, $\rho$, $C_P$ and $\alpha_P$ are temperature (K), pressure (Pa), density (kg/m$^3$), specific heat capacity (J/kg K) of the food substance at a constant pressure, and thermal expansion coefficient (K$^{-1}$), respectively.

It has been shown that the application of HPP increases the temperature of the water and high moisture content foods by approximately 3°C per 100 MPa at room temperature (25°C) or 9°C per 100 MPa if the food contains a significant amount of fat (Rasanayagam *et al.*, 2003). Table 13.1 presents the heat of compression values of selected substances treated at 25°C. Unlike water where its heat of compression value is dependent on the initial temperature, for fats and oils the values are not influenced by the initial sample temperature.

© Woodhead Publishing Limited, 2012

**Table 13.1** Heat of compression values of selected substances treated at 25°C

| Medium | Temperature increase (°C per 100 MPa) |
|---|---|
| Water | 3.0 |
| Fruit juice, tomato salsa, skim or 2% fat milk | 2.6–3.0 |
| Tofu | 3.1 |
| Egg albumin | 3.0 |
| Mashed potato | 3.0 |
| Yogurt | 3.1 |
| Honey | 3.2 |
| Salmon | 3.2 |
| Chicken fat | 4.5 |
| Water/glycol (50:50) | 3.7–4.8 |
| Beef fat | 6.3 |
| Olive oil | 6.3–8.7 |
| Soybean oil | 6.2–9.1 |

Source: Otero *et al.*, 2000; Ting *et al.*, 2002; De Heij *et al.*, 2003; Rasanayagam *et al.*, 2003; Balasubramaniam *et al.*, 2004; Patazca *et al.*, 2007.

## 13.4 Microbial inactivation by high pressure processing (HPP)

Microbial inactivation by HPP cannot only be influenced by the process parameters, but also by the characteristics of the microorganisms as well as the substrate in which they are treated. Important factors affecting microbial efficacy of a high pressure process are briefly discussed here.

### 13.4.1 Process parameters

Pressure level, process temperature, and treatment time are the main process parameters influencing the microbial efficacy of high pressure processing (Smelt *et al.*, 2002). Vegetative microorganisms usually show maximum resistance to pressure treatment at room temperature (~25°C). However, pressure treatment either at chilled process temperatures (<20°C) or combined with moderate heat (~45°C) is more effective in killing vegetative microorganisms (Balasubramaniam *et al.*, 2008). Patterson *et al.* (1995) reported a 2-log reduction of *Escherichia coli* O157:H7 NCTC 12079 in UHT milk treated at 600 MPa for 30 min at 20°C. Patterson and Kilpatrick (1998) found an 8-log reduction of the same strain by treatment at 200 MPa for 15 min at 60°C or 700 MPa for 15 min at 40°C. Enhanced microbial efficacy of HPP by heat is usually attributed to a higher degree of damage on proteins (Sonoike *et al.*, 1992). Enhanced microbiological efficacy of the pressure treatment at chilled process temperatures (<20°C) was attributed to reduced fluidity of the cell membrane (Casadei *et al.*, 2002). Bacterial spores are highly resistant to pressure treatment at room temperature and their inactivation is only

© Woodhead Publishing Limited, 2012

possible at moderate or high process temperatures. In summary, depending on the intensity of pressure-thermal treatment, high pressure processing can be used as an alternative pasteurization or sterilization method (see Section 13.5 for more details).

### 13.4.2 Microbial characteristics

It is generally agreed that vegetative cells are more sensitive to high pressure than bacterial spores; however, the sensitivity of cells may vary within the genus, species, and strains of microorganisms (Smelt, 1998; Balci and Wilbey, 1999; Pagán *et al.*, 1999; Patterson, 1999). The more complex structure of the cell membrane in Gram-negative bacteria makes them more susceptible to physiological changes caused by high pressure than Gram-positive bacteria (Cheftel, 1992; Smelt, 1998). Certain strains of *E. coli* O157:H7, *Salmonella* spp., and *Staphylococcus aureus* are known among the bacteria showing extreme resistance to HPP (Patterson *et al.*, 1995; Erkmen and Karatas, 1997; Patterson and Kilpatrick, 1998; Linton *et al.*, 2001).

Stationary phase cells of bacteria are more resistant to pressure than the exponential phase cells due to their more robust cytoplasmic membrane that can better tolerate pressure treatment (Mackey *et al.*, 1995; McClements *et al.*, 2001; Mañas and Mackey, 2004). Rod-shaped bacteria are more sensitive to high pressure than the spherical-shaped bacteria (Ludwig and Schreck, 1997); however, Casal and Gomez (1999) found that lactococci inoculated into 10% (w/v) reconstituted skim milk (RSM) were more sensitive to high pressure treatment (100–350 MPa) than lactobacilli.

Most of the yeasts and molds are very sensitive to high pressure treatment (Patterson and Loaharanu, 2000). The pressure sensitivity of yeasts as an important cause of spoilage in acidic foods makes HPP treatment a viable alternative for extending the shelf life of such products as fruit-based and cultured dairy products.

Very few studies have investigated the inactivation of viruses by HPP. While some viruses including poliovirus (PV) are highly pressure-resistant, others such as feline calicivirus (FCV; a Norwalk virus surrogate), have been found to be very pressure-sensitive (Wilkinson *et al.*, 2001; Kingsley *et al.*, 2002, 2004; Grove *et al.*, 2008).

### 13.4.3 Influence of substrate composition, pH, and water activity

Microbial inactivation by HPP can be influenced by the composition of the substrate, product pH, and water activity. The potential protective effect of certain food constituents (e.g. proteins, carbohydrates, lipids, and minerals) on the inactivation microorganisms by HPP (usually known as 'baroprotective effect') has been reported by a number of researchers, including a strong baroprotective effect of milk constituents on *Listeria monocytogenes* 4a KUEN 136 (Dogan and Erkmen, 2004) and *Listeria innocua* 4202 (Black

© Woodhead Publishing Limited, 2012

*et al.*, 2007a). The baroprotective effect of solutes such as sugars and salts at high concentration on vegetative microorganisms is attributed to the reduced water activity ($a_w$) that may cause cell shrinkage and thickening of the cell membrane, thus reducing the cell size and membrane permeability and fluidity (Knorr, 1993, 1994; Oxen and Knorr, 1993; Palou *et al.*, 1997; Takahashi *et al.*, 1993; Molina-Höppner *et al.*, 2004). The baroprotective effect of reduced water activity has been reported for various vegetative microorganisms, including the yeast *Rhodotorula rubra* (Oxen and Knorr, 1993), *E. coli* and *Saccharomyces cerevisiae* (Takahashi *et al.*, 1993). Reducing water activity has also been reported to increase the resistance of bacterial spores to HPP by increasing the osmotic dehydration of the spore protoplast (Cheftel, 1995; Raso *et al.*, 1998).

Kingsley *et al.* (2002) found an increase in the pressure resistance of hepatitis A virus when treated in sea water as compared to the treatment in tissue culture medium which resulted in an undetectable level of inactivation (≥ 6-log reduction) at 450 MPa (21°C, 5 min). This was attributed to the protective effect of the salts.

A reduction in the pH of suspending media as a result of the pressure-induced transient pH shift or addition of acids usually leads to a greater microbial inactivation by HPP, similar to that for thermal processing. Application of HPP may cause transient pH shift towards acidic values by enhancing ionic dissociation of water and several weak acids (Cheftel, 1995; Earnshaw *et al.*, 1995; Smelt, 1998; Smelt *et al.*, 2002). The pH probably shifts back close to its initial value upon depressurization. The reduction in pH may also result in the prevention of the outgrowth of microorganisms during storage post-pressurization due to the inability of sub-lethally injured cells to recover. Enhanced HPP inactivation by reducing pH has been reported for foodborne vegetative pathogens including *L. monocytogenes* (Mackey *et al.*, 1995; Stewart *et al.*, 1997; Alpas *et al.*, 2000), *S. aureus*, *E. coli*, *Salmonella* Enteritidis and *Salmonella* Typhimurium (Alpas *et al.*, 2000). In contrast, Ogawa *et al.* (1990) reported that the pressure-induced inactivation of yeasts and molds inoculated in Satsuma mandarin juice, was not affected by either the juice pH (2.5–4.5) or the type of organic acids (citric, tartaric, lactic, or acetic acid) used for acidification of juice before treatment at 100–600 MPa. Acidic pH may also enhance bacterial spore inactivation by high pressure treatment. Greater inactivation of *Bacillus coagulans* and *Clostridium sporogenes* spores at pH 4.0 than at pH 7.0 by high pressure treatment at moderate process temperatures has been reported by Roberts and Hoover (1996) and Stewart *et al.* (2000), respectively.

## 13.5 Food pasteurization and sterilization effects

The term 'pasteurization' has recently been redefined by the National Advisory Committee on Microbiological Criteria for Foods (NACMCF) for products

© Woodhead Publishing Limited, 2012

preserved by different lethal agents such as pressure and electric fields. According to NACMCF, pasteurization can refer to any process, treatment, or combination thereof that is applied to food to reduce the most resistant microorganism(s) of public health significance to a level that is not likely to present a public health risk under normal conditions of distribution and storage (NACMCF, 2006).

Generally speaking, 'high-pressure pasteurization' involves an application of pressure in the range of 300–600 MPa at process temperatures <45°C (Lau and Turek, 2007). This effectively inactivates a variety of vegetative pathogenic and spoilage bacteria, yeasts and molds, and viruses. Several pressure-pasteurized products have been successfully commercialized worldwide including fruit and vegetable juices (in France, Japan, USA, Portugal, UK, Italy), acidified avocado purée (guacamole) and salsa dips (USA), sliced cured cooked and raw-cooked hams, tapas and other processed meats and poultry products (Spain, USA), seafood including oysters, clams, mussels, scallops, shrimps, crabs, and squid (USA, Japan), beverage blends (USA), rice cakes (Japan), jams, jellies, tropical fruits, fruit sauces and fruit desserts (Japan, USA), yogurt (Japan) and smoothies (UK) (Torres and Velazquez, 2005; Rastogi *et al.*, 2007; Sáiz *et al.*, 2008).

The microorganism of choice to validate the microbial efficacy of the pressure pasteurization varies depending on the type of food to be processed. *Salmonella*, *E.coli* O157:H7, *L. monocytogenes* and *Vibrio parahaemolyticus* are the common pathogenic bacteria of public health concern in different categories of foods such as eggs, fruit juices, meat, and seafood. Processors should provide adequate pressure treatment to ensure the elimination of these target pathogenic microorganisms to a specified log reduction (normally a 5-log reduction) (NACMCF, 2006, 2010). Often processors may have difficulty in verifying the stated reduction of target pathogens in industrial scale systems. Under these circumstances, food processors can use surrogate (or) biological indicator organisms. These are organisms that have relatively high and stable pressure-thermal resistance characteristics in foods that can be documented. For example, *L. innocua* and non-pathogenic *E. coli* K12 have been proposed as surrogate bacteria for *L. monocytogenes* and *E. coli* O157:H7, respectively (Farkas and Hoover, 2000; NACMCF, 2010).

Pressure-assisted thermal processing (PATP), also known as pressure-assisted thermal sterilization (PATS), has emerged as one of the viable alternatives for producing sterile shelf-stable low-acid foods. Various bacterial spores and other thermophilic microorganisms are successfully inactivated by the simultaneous application of pressure and heat. During PATP, the food material is subjected to a combination of elevated pressure (500–700 MPa) and temperature (90–121°C) for a short duration to a pre-heated food product (Margosch *et al.*, 2004a; Rajan *et al.*, 2006; Ahn *et al.*, 2007). The unique advantages of this technology include a rapid increase in the temperature of treated food samples and the expansion cooling of products

© Woodhead Publishing Limited, 2012

upon depressurization (Ting *et al.*, 2002; Rajan *et al.*, 2006). Consequently, PATP can better preserve product attributes including color, flavor, texture, and nutritional values in comparison to conventional thermal sterilization. Zhu *et al.* (2008) demonstrated that *Clostridium sporogenes* spores in ground beef could be inactivated within a shorter time or at lower temperature than in conventional thermal processing by using high pressure in the range 700–900 MPa at elevated temperatures (80–100°C). *Bacillus amyloliquefaciens* has been identified as one of the pressure-thermal resistant non-pathogenic indicator organisms (Rajan *et al.*, 2006; NACMCF, 2010). More research is needed to compare combined pressure-heat resistance of *Clostridium botulinum* spores against that of non-pathogenic spores such as *B. amyloliquefaciens* spores. Recently, the FDA issued a no objection certificate to an industrial petition for sterilization of a low-acid product using PATP (IFT News, 2009); however, no commercial PATP-treated shelf-stable low-acid food is known to be currently available in the market.

### 13.5.1 Microbial efficacy of pulsed pressure treatment

Pulsed pressure treatment (also referred to as 'oscillatory pressure treatment') is the application of consecutive, short pressure treatments interrupted by brief decompressions. Pulsed pressure treatment has been shown to substantially enhance the inactivation of pathogens such as *L. monocytogenes* (Vachon *et al.*, 2002), *E. coli* (Garcia-Graells *et al.*, 1999; Vachon *et al.*, 2002) and *S.* Enteritidis (Vachon *et al.*, 2002) by increasing the sensitivity of cells to pressure, offering a promising alternative to the cold pasteurization of milk and possibly other low-acid liquid foods (Vachon *et al.*, 2002).

Application of pulsed pressurization may enhance the lethality of combined pressure-thermal treatment against bacterial spores (Hayakawa *et al.*, 1994; Rajan *et al.*, 2006; Ahn and Balasubramaniam, 2007; Ratphitagsanti *et al.*, 2009). Spore inactivation during pressure pulsing is influenced by the target pressure, process temperature, pressure holding time during each pulse, number of pulses, and the time interval between pulses (Ratphitagsanti *et al.*, 2009). Inactivation to undetectable level (≥ 6-log reduction) of *Geobacillus stearothermophilus* spores has been reported by Hayakawa *et al.* (1994) by using six pulses (5 min each) at 600 MPa and 70°C. Margosch *et al.* (2004b) also reported undetectable levels of inactivation of *Bacillus subtilis* and *Bacillus licheniformis* spores by applying the first pulse at 800 MPa and 70°C for 2 min following by the second pulse at 800 MPa and 70°C or 0.1 MPa and 70°C. The increased cost associated with possible equipment wear and tear should be considered if pulsed pressure treatment is to be used for enhancing spore inactivation (Ratphitagsanti *et al.*, 2009; Nguyen and Balasubramaniam, 2011).

© Woodhead Publishing Limited, 2012

### 13.5.2 Mechanisms of microbial inactivation in foods by high pressure

Various studies have been conducted to underline the mechanisms of inactivation of microorganisms by high pressure processing. Identifying the inactivation mechanisms is critical in order to design and develop more efficient high pressure equipment and define conditions for effective inactivation of microorganisms in food products. Several mechanisms have been proposed for the pressure-induced inactivation of vegetative cells such as damage to the sub-cellular structures including cell membrane, nucleoid, ribosomes, and enzymes, as well as morphological, biochemical, and genetic alterations (Landau, 1967; Hoover *et al.*, 1989; Osumi *et al.*, 1992; Gross *et al.*, 1993; Cheftel, 1995; Kobori *et al.*, 1995; Smelt, 1998; Balci and Wilbey, 1999). Damage to the cell membranes and membrane-bound enzymes (ATPases) that control transport phenomena involved in nutrient uptake and waste disposal is generally accepted to be the primary mechanism (Morita, 1975; Earnshaw, 1992; Wouters *et al.*, 1998; Ulmer *et al.*, 2000; Gänzle and Vogel, 2001; Chilton *et al.*, 2001; McClements *et al.*, 2001; Casadei *et al.*, 2002; Torres and Velazquez, 2005). However, inactivation (denaturation) of key enzymes, including those involved in the DNA replication and transcription, as well as the disruption of ribosomes may also play a role (Hoover *et al.*, 1989; Linton and Patterson, 2000). Depending on the process temperature, spores may exhibit different inactivation mechanisms by pressure treatment. At lower process temperatures (<90°C), pressure induces spores to germinate, and the germinated spores are subsequently inactivated by pressure (Ananta *et al.*, 2001; Paidhungat *et al.*, 2002; Black *et al.*, 2007b, 2007c). At higher process temperatures (>90°C), a direct spore inactivation mechanism that bypasses germination is more likely (Ananta *et al.*, 2001). Ahn and Balasubramaniam (2007) and Ratphitagsanti *et al.* (2010) observed no significant germination of *B. amyloliquefaciens* TMW 2.479 Fad 82 spores treated under PATP conditions (700 MPa, 105°C). This was in agreement with an earlier study by Ananta *et al.* (2001), suggesting that elevated pressure-temperature conditions may more likely cause direct inactivation or partial spore injury than spore germination.

Microbial efficacy of HPP may be enhanced by decreasing pH, increasing temperature, or adding antimicrobial compounds such as bacteriocins (e.g. nisin and lacticin), sorbic or benzoic acids, ethanol, spice extracts, lysozyme or sucrose esters (Popper and Knorr, 1990; Papineau *et al.*, 1991; Kalchayanand *et al.*, 1998; ter Steeg *et al.*, 1999; Capellas *et al.*, 2000; Morgan *et al.*, 2000; Masschalck *et al.*, 2001a, 2001b; Karatzas *et al.*, 2001; Black *et al.*, 2005; de Lamo-Castellvi *et al.*, 2010). Table 13.2 shows the combined effects of HPP and selected antimicrobial compounds on microorganisms.

© Woodhead Publishing Limited, 2012

**Table 13.2** Combined effects of high pressure processing and antimicrobial compounds on microorganisms

| Compounds and maximum concentration tested | Process conditions | Medium | Effect | Reference |
|---|---|---|---|---|
| 1 IU/mL* nisin in plating medium | 400 MPa/25–70°C/up to 30 min at different pH levels | Citrate phosphate buffer | Synergistic inhibitory effect of nisin with other parameters (pressure, heat and pH) on the outgrowth of *Bacillus coagulans* spores:<br>• No notable reduction at 25°C<br>• 2–4 log reduction at 45°C (pH-dependent)<br>• 4–5 log reduction at 70°C (pH-dependent)<br>• 6 log reduction at 400 MPa and pH 4.0 at 70°C for 30 min when plated in NA containing 0.8 IU/mL nisin | Roberts and Hoover (1996) |
| 2000 μg/mL limonene | 177 MPa/25°C/1 h | Distilled water/ sodium chloride solution containing Tween 80 | 3 log reduction *Saccharomyces cerevisiae* | Adegoke *et al.* (1997) |
| 0.5% monolaurin | 392 MPa/45°C/10 min | Milk | 3 log reduction of *B. subtilis* spores at a concentration as low as 0.001% | Shearer *et al.* (2000) |
| 500 μg/mL lactoferrin (an iron binding milk protein) | 155–400 MPa/20°C/15 min | 10 mM phosphate buffer (pH 7.0) | Increased pressure inactivation of *Shigella sonnei*, *Pseudomonas fluorescens* and *Staphylococcus aureus* by at least 1 log | Masschalck *et al.* (2001a) |
| 500 μg/mL Na-EDTA + 100 μg/mL lysozyme | 345 MPa/60°C/5 min + storage at 4 or 12°C for 84 d or at 25°C for 7 d | Roast beef | ≤3.23 log reduction of *Clostridium laramie*, *Clostridium sporogenes*, *clostridium perfringens* and *Clostridium tertium* spores; extended shelf life of roast beef up to 7 days at 25°C | Kalchayanand *et al.* (2003) |
| Pediocin + nisin (3:7) at a final concentration of 5000 activity unit per mL (AU/mL) | 345 MPa/60°C/5 min + storage at 4 or 12°C for 84 d or at 25°C for 7 d | Roast beef | ≤2.91 log reduction of *C. laramie*, *C. sporogenes*, *C. perfringens* and *C. tertium*; controlled growth of spores and extended shelf life up to 84 days (temperature-dependent) | Kalchayanand *et al.* (2003) |
| 333 IU/ml* nisin | 300–700 MPa/30–70°C/7.5–17.5 min | UHT milk | ≥6 log reduction of *Clostridium botulinum* spores | Gao and Ju (2008) |

* 1 g of commercial nisin contains $10^6$ IU.

© Woodhead Publishing Limited, 2012

### 13.5.3 Pressure-induced injury and subsequent recovery of microorganisms

High pressure processing of foods may not always result in complete microbial inactivation, but may injure a proportion of the microbial population, depending on the process conditions (pressure level, holding time, etc.) and the substrate in which the microorganisms are treated (Patterson *et al.*, 1995). The injured cells may recover after treatment under optimal storage conditions in a suitable substrate (Mackey, 2000; McClements *et al.*, 2001). The injured cells may not be detected using selective conditions for the enumeration of survivors and this may cause an overestimation of the HPP lethality (Cheftel, 1995; Mañas and Pagán, 2005). The use of non-selective microbiological media along with the incubation at several temperatures between 4 and 30°C is recommended to allow detection of all viable organisms of concern post-treatment (Balasubramaniam and Farkas, 2008). Garcia-Risco *et al.* (1998) found a recovery of sub-lethally injured psychrotrophic bacteria (an increase of 1.7 logs) in pressure-treated milk (400 MPa, 25°C, 30 min) within 45 days of storage at 7°C. Wick *et al.* (2004) detected the recovery of pressure-treated (200–800 MPa, 25°C, 5 min) starter and non-starter lactic acid bacteria in 1-month and 4-month-old cheddar cheeses during subsequent ripening at 10°C. For example, the viable number of *L. lactis* in 1-month-old cheese that was initially reduced by >4 logs by treatment at 400 MPa, increased to the same level as the control cheese within 126 days of ripening.

Daryaei *et al.* (2010) studied the HPP-induced inactivation (≤20°C, 5 min) of wild-type spoilage yeasts of *Candida* species in a fermented milk test system at different pH levels (4.30, 5.20, and 6.50) and their subsequent recovery during refrigerated storage. Treatment of milk at 600 MPa decreased the viable counts of both *Candida zeylanoides* and *Candida lipolytica* to below the detection limit of 1 CFU $mL^{-1}$ (≥5.0 log reduction) with no subsequent outgrowth during storage. However, recovery of *C. lipolytica* was observed in milk processed at 300 MPa at all pH levels from week 3 onwards, reaching $10^6$–$10^7$ CFU $mL^{-1}$ within 8 weeks.

The recovery of pressure-injured cells may be inhibited at acidic pH levels (Stewart *et al.*, 1997). Application of HPP combined with other hurdles such as mild heat or bacteriocins (e.g. nisin or lacticin) has been reported to prevent the recovery of injured microorganisms and improve the product microbial stability (Cheftel, 1995; Chilton *et al.*, 2001; Black *et al.*, 2005). Some compounds such as sodium chloride added to the growth media are known to selectively inhibit the recovery of pressure-injured cells (Patterson *et al.*, 1995).

### 13.5.4 Kinetics of high pressure microbial inactivation

Researchers utilize kinetic models to mathematically describe experimentally determined microbial inactivation data as a function of different process parameters (pressure, temperature, and time) (Heldman and Newsome, 2003;

© Woodhead Publishing Limited, 2012

NACMCF, 2010). These models are essentially empirical curve fitting of the experimental data. Since these models are empirical in nature, extrapolation of models beyond process conditions under which they were developed may lead to erroneous results (Heldman and Newsome, 2003). The kinetic models can be used by the food processors and regulators to investigate the microbial safety of the treated products under different scenarios without much experimentation.

While first-order kinetic equations are commonly employed for estimating kinetic parameters of thermally processed foods, both linear and nonlinear models have been proposed to estimate high pressure processing kinetic inactivation parameters. Some of the commonly utilized nonlinear models may include: the Weibull model (Peleg and Cole, 1998; Chen and Hoover, 2003; Mañas and Pagán, 2005), the log-logistic model (Chen and Hoover, 2003), and the modified Gompertz model (Patterson and Kilpatrick, 1998; Chen and Hoover, 2003) (Table 13.3). Researchers also proposed the use of the quasi-chemical modeling concept for the inactivation of microorganisms during high pressure processing (Feeherry *et al.*, 2006).

Nonlinear response of bacteria to HPP could be due to the presence of resistant subpopulations after treatment. Also, multi-target microbial inactivation and phenomena such as pressure injury and cell clumping during treatment can result in a nonlinear survival curve for various vegetative cells and spores (Peleg and Cole, 1998; Heldman and Newsome, 2003; Mañas and Pagán, 2005). The occurrence of shoulders and tails has been reported for microbial inactivation by HPP (Benito *et al.*, 1999; Tay *et al.*, 2003). The 'tailing' phenomenon may be caused by the inherent variation in resistance of the population, the effects of experimental conditions, or cell-age distribution. Unlike thermal processing, during pressure come-up time certain levels of bacteria can be inactivated and this reduction should be taken into consideration (Rajan *et al.*, 2006).

### 13.5.5 Effects of high pressure processing on food quality

High pressure processing is capable of enhancing safety and prolonging shelf life of foods while maintaining consumer-desired freshness and sensory and nutritional quality attributes. Application of HPP up to 600 MPa at refrigeration, ambient, or moderate heating temperatures allows better preservation of organoleptic properties of foods compared to conventional preservation technologies (Knorr, 1993; Cheftel, 1995; Velazquez *et al.*, 2002). Application of high pressure at elevated temperatures (PATP) has also been reported to reduce quality degradation of food compared to thermal processing (Nguyen *et al.*, 2007). Small molecules such as amino acids, flavor compounds, vitamins, and pigments are not largely affected by high pressure, while changes in the structure of larger molecules are expected (Balci and Wilbey, 1999; Rastogi *et al.*, 2007). The food product to be processed by high pressure must contain water and not have entrapped air. If the food materials contain internal air

© Woodhead Publishing Limited, 2012

**Table 13.3** Kinetics models commonly used to describe the microbial inactivation by high pressure processing

| Model | Equation* | Description |
|---|---|---|
| First-order kinetics | $\log \frac{N_t}{N_0} = -\frac{t}{D}$ | $t$ = treatment time<br>$D$ = decimal reduction time<br>(time required to reduce microbial population by 1.0 log) |
| Weibull model | $\log \frac{N_t}{N_0} = -bt^n$ | $t$ = treatment time<br>$b$ = scale factor<br>$n$ = shape factor |
| Log-logistic model | $\log \frac{N_t}{N_0} = \frac{A}{1 + e^{4\sigma(\tau - \log t)/A}} - \frac{A}{1 + e^{4\sigma(\tau + 6)/A}}$ | $t$ = treatment time<br>$A$ = upper asymptote – lower asymptote<br>$\sigma$ = maximum rate of inactivation<br>$\tau = \log_{10}$ time to the maximum rate of inactivation |
| Modified Gompertz model | $\log \frac{N_t}{N_0} = Ce^{-e^{BM}} - Ce^{-e^{-B(t-M)}}$ | $M$ = time at which the absolute death rate is maximum<br>$B$ = relative death rate at $M$<br>$C$ = upper asymptote – lower asymptote |

* $N_0$ and $N_t$ are the numbers of viable cells in the medium (CFU $mL^{-1}$ or CFU $g^{-1}$) before and after being exposed to a lethal treatment for a specific time $t$.

© Woodhead Publishing Limited, 2012

pockets such as marshmallows, strawberries, and leafy vegetables, the air will escape from the product after pressure treatment, due to the difference in material compressibility, thus the product will be crushed (Nguyen and Balasubramaniam, 2011). Also, HPP of foods lacking sufficient moisture to maintain a water activity above 0.98 will not result in an effective microbial destruction (Nguyen and Balasubramaniam, 2011).

Depending on the process conditions, HPP may result in conformational changes of large molecules, such as starch gelatinization (Douzals *et al.*, 1996, 1998; Bauer and Knorr, 2005; Buckow *et al.*, 2009) and protein denaturation (Messens *et al.*, 1997; Angsupanich and Ledward, 1998; Scollard *et al.*, 2000; Anema, 2008). Gelatinization of 16% (w/w) wheat starch by pressure starts at 300 MPa and starch is completely gelatinized at 600–700 MPa at 25°C as indicated by differential scanning calorimetry (DSC) (Douzals *et al.*, 1996, 1998). It is possible to use HPP to create novel food products with unique structures, textures, or tastes (Balasubramaniam and Farkas, 2008).

High pressure application can effectively reduce the activity of many food quality-related enzymes and thus ensure high quality of pressure-treated products during storage. A number of researchers have reported inactivation of pectinmethylesterase (PME), polyphenoloxidase (PPO) and polygalacturonase (PG) and peroxidase in fruits and fruit products, and proteases (e.g. plasmin) in milk by HPP (see Table 13.4 for more details). These enzymes can adversely affect the quality of the products. HPP has also been reported to improve the extractability of compounds such as flavonols, carotenoids, and lycopene from vegetable products (de Ancos *et al.*, 2000, 2002; Sánchez-Moreno *et al.*, 2004; Hsu *et al.*, 2008; Gupta *et al.*, 2010). Table 13.4 shows some of the major quality changes of different food products as a result of high pressure treatment.

## 13.6 Applications of food decontamination by high pressure

A variety of value-added pressure-pasteurized products have been commercially introduced over the last 15 years. According to some industry estimates, by 2008 about 125 industrial HPP machines were in production for food processing applications (Tonello, 2011). The following sections summarize the commercial applications in various segments of the food industry.

### 13.6.1 HPP of fruit jams

Fruit jams including strawberry, apple, and kiwi jams are examples of the first wave of pressure-pasteurized products (Tonello, 2011). They are normally processed at pressures around 400 MPa for up to 5 min at room temperature. The product is refrigerated after processing to prevent browning and flavor changes caused by enzymatic activities. The refrigerated shelf

© Woodhead Publishing Limited, 2012

**Table 13.4** Selected examples of quality changes in foods by combined pressure-heat treatment

| Product | Process conditions | Quality changes | Reference |
|---|---|---|---|
| Orange juice | 800 MPa/25°C/1 min | Substantial inactivation of pectin methylesterase (PME); good cloud stability, and less loss of ascorbic acid over a period of more than 2 months at 4°C or 37°C | Nienaber and Shellhammer (2001) |
| Orange juice | 350 MPa/30°C/5 min | Higher extraction of carotenoid content as compared to the untreated sample | de Ancos *et al.* (2002) |
| Orange juice | 600 MPa/20°C/1 min | Partial inactivation of PME; no changes in color, browning index, viscosity, concentration, titratable acidity (TA), and levels of alcohol-insoluble acids, ascorbic acid, and β-carotene | Bull *et al.* (2004) |
| Orange juice | 500 MPa/35°C/5 min | Less loss of ascorbic acid and higher retention of flavor, antioxidant capacity, shelf life, sensory scores, and viscosity values as compared to conventionally pasteurized juice (80°C/30 s) | Polydera *et al.* (2003, 2004, 2005) |
| Orange juice | 700 MPa/55°C/2 min | Complete inactivation of PME | Sampedro *et al.* (2008) |
| Precut mango | 300 or 600 MPa/1 min | Slight reduction of fresh flavor and increase of off-flavor and sweetness, and very little change of color, texture, and other sensory attributes during storage at 3°C | Boynton *et al.* (2002) |
| Guava purée | 600 MPa/25°C/15 min | Partial inactivation of enzymes; no change in color, cloudiness, ascorbic acid content, flavor distribution, and viscosity during storage at 4°C up to 40 d; increased levels of methanol, ethanol and 2-ethylfuran during storage | Gow and Hsin (1996, 1999) |

*(Continued)*

© Woodhead Publishing Limited, 2012

**Table 13.4** Continued

| Product | Process conditions | Quality changes | Reference |
|---|---|---|---|
| White peach | 400 MPa/20°C/10 min | Enzymatic formation of benzaldehyde, $C_6$ aldehydes, and alcohols by disruption of fruit tissues | Sumitani *et al.* (1994) |
| Diced apple in pineapple juice | 600 MPa/ 22°C/1–5 min | Significant reduction of the out-of-pack browning during air exposure | Perera *et al.* (2009) |
| Strawberry purée | 500 or 800 MPa/20°C/20 min | Inactivation of PPO and peroxidase; no major changes in aroma profiles up to 500 MPa, and significant changes in the aroma profiles at higher pressure (800 MPa) due to the synthesis of new compounds such as 3,4-dimethoxy-2-methyl-furan and γ-lactone | Lambert *et al.* (1999) |
| Lychee | 600 MPa/60°C/20 min | Less loss of visual quality in both fresh and syrup-processed lychee compared to thermal processing; extensive inactivation of peroxidase and PPO in fresh lychee | Phunchaisri and Apichartsrangkoon (2005) |
| Persimmon purée | ≤400 MPa/25°C/15 min | Increased amount of extractable carotenoids and vitamin A | de Ancos *et al.* (2000) |
| Carrot | 500–700 MPa/95–105°C/≤15 min | Less quality degradation (in terms of texture, color, and carotene content) as compared to thermally processed carrots | Nguyen *et al.* (2007) |
| Onion | 100 and 400 MPa/5°C/5 min | Improved extraction of flavonols (quercetin and quercetin glucosides) | Roldán-Marín *et al.* (2009) |
| Tomato purée | 400 MPa/25°C/15 min | Higher redness, carotenoids, and vitamin C in HPP-treated purée than in low-temperature (70°C, 30 s) or high-temperature (90°C, 1 min) pasteurized samples | Sánchez-Moreno *et al.* (2006) |

© Woodhead Publishing Limited, 2012

| | | | |
|---|---|---|---|
| Tomato juice | 500 MPa/25°C/10 min | Inactivation of pectolytic enzymes, improvement in extractable carotenoids and lycopene contents, and retention of vitamin C compared with fresh juice | Hsu *et al.* (2008) |
| Tomato | 200–500 MPa/4 and 25°C/10 min | Retained color, extractable total carotenoids, lycopene content, and antioxidant activity. Lowest residual activities of PME and PG at lower pressure range (200–400 MPa) and higher at higher pressure (500 MPa) | Kuo (2008) |
| Tomato purée | 600 MPa20°C/15 min | Retained more than 90% of ascorbic acid in tomato purée as compared to thermal processing | Patras *et al.* (2009) |
| Tomato juice | 600 MPa/100°C/10 min or 700 MPa/45°C/10 min | Significant improvement of lycopene extractability as compared to thermal processing (100°C/35 min) | Gupta *et al.* (2010) |
| Raw milk | 600–800 MPa/20°C/8 min | 40% loss of alkaline phosphatase (AlP) activity at 600 MPa and complete inactivation at 800 MPa | Rademacher *et al.* (1998) |
| Milk | 400 MPa/25–60°C/15–30 min | Inactivation of plasmin at temperatures above 25°C (86.5% inactivation at 60°C); improved organoleptic properties of milk at 40–60°C | Garcia-Risco *et al.* (1998, 2000). |
| Mató (a fresh cheese made from goat's milk) | 500 MPa/10 or 25°C/5–30 min | Only minor changes in water-holding capacity, texture, color and microstructure of cheese, as compared with untreated one; more whey loss with a higher total nitrogen content in pressurized cheese than in untreated one | Capellas *et al.* (2001) |

*(Continued)*

© Woodhead Publishing Limited, 2012

**Table 13.4** Continued

| Product | Process conditions | Quality changes | Reference |
|---|---|---|---|
| Fresh lactic curd cheese | 300 and 600 MPa/≤22°C/5 min | No adverse effect of pressure on textural and sensory attributes; substantial decrease in proteolysis during storage (4°C) as a result of treatment at 600 MPa | Daryaei *et al.* (2006) |
| Lean meat | 100–500 MPa/5 min at refrigeration temperature | Increased total activity of calpain in pressurized muscle resulting in meat tenderization | Homma *et al.* (1995) |
| Ready-to-eat meat | 600 MPa/20°C/3 min | Very high consumer acceptability and sensory quality of the product | Hayman *et al.* (2004) |

life of pressure-treated jams is around 30 days. They have superior sensory quality compared with those prepared in a conventional manner (Ludikhuyze and Hendrickx, 2001).

### 13.6.2 HPP of fruit juices

High pressure-treated acidic fruit juices (e.g. orange juice) are commercially available in the United States, Mexico, Canada, Europe, and Japan. They are treated at ≥400 MPa for a few minutes (generally under 10 min) at ≤20°C that can not only significantly reduce the number of yeasts and molds, but also the pathogens, such as *E. coli* O157:H7 (Linton *et al.*, 1999; Ramaswamy *et al.*, 2003; Tonello, 2011). The shelf life of pressurized fruit juices can be extended up to 30 days.

### 13.6.3 HPP of vegetable products

Non-thermal processing of freshly prepared vegetable products to enhance safety is usually problematic because of their high pH and possibility of survival and growth of pathogenic spore-forming organisms. Guacamole (acidified avocado purée) is one of the most successful HPP-treated vegetable products in the United States. Guacamole is usually treated at pressures around 500–600 MPa for 2 min or longer which can cause a 5-log reduction in numbers of pathogens (Patterson, 2005; Tonello, 2011). The shelf life of the product can be extended from 7 to 30 days at refrigeration temperature (Patterson, 2005).

© Woodhead Publishing Limited, 2012

### 13.6.4 HPP of meat products

High pressure treatment of meat and poultry products (e.g. ready-to-eat ham, chicken and turkey products) is usually used as a final preservation step after packaging to reduce the risk of post-processing contamination from pathogens such as *L. monocytogenes* and *E. coli* (Patterson *et al.*, 1995; Patterson, 2005). High pressure-processed sliced cooked hams are available in the United States and Spain which are treated in flexible pouches at 500 MPa for a few minutes. This helps in preservation of the sensory properties and shelf life extension to 60 days under chilled storage (Patterson, 2005). Combination of high pressure with heat is capable of significantly reducing bacterial spore population in meat products. Table 13.5 shows selected examples of microbial inactivation in various meat products.

### 13.6.5 HPP of oysters and other seafood

HPP has proven to be a useful technology for processing seafood including shellfish and oysters. *Vibrio* species associated with seafood products are relatively sensitive to HPP (Styles *et al.*, 1991; Cook, 2003; Koo *et al.*, 2006).

**Table 13.5** Microbial inactivation by high pressure processing in meat products

| Medium | Process conditions | Microbial inactivation | Reference |
|---|---|---|---|
| Strained chicken baby food | 340 MPa/23°C/10 min | 3.0 log reduction of *Salmonella* Senftenberg 775W (heat-resistant), and 2.0 log reduction of *Salmonella* Typhimurium ATCC7136 (heat-sensitive) | Metrick *et al.* (1989) |
| Minced beef | 200 MPa/20°C/20 min | 5.0 log reduction of *Pseudomonas fluorescens* | Carlez *et al.* (1993) |
| Minced meat | 280 MPa/20°C/20 min | 5.0 log reduction of *Citrobacter freundii* | Carlez *et al.* (1993) |
| Poultry meat | 375 MPa/20°C/15 min | 2.0 log reduction of *Listeria monocytogenes* | Patterson *et al.* (1995) |
| Poultry meat | 600 MPa/20°C/15 min | 3.0 log reduction of *E. coli* O157:H7 and *S. aureus* | Patterson *et al.* (1995) |
| Chicken breast | 680 MPa/80°C/20 min | 2.0 log reduction of *Clostridium sporogenes* spores | Crawford *et al.* (1996) |
| Ham | 500 MPa/25°C/5 min | 4.0 log reduction of *Lactobacillus viridescens* | Park *et al.* (2001) |
| Cooked ham | 500 MPa/40°C/10 min | 4.0 log reduction of *Lactobacillus sakei*, and 1.29 log reduction of *Staphylococcus carnosus* | Hugas *et al.* (2002) |
| Chicken purée | 375 MPa/25°C/10 min | 6.0 log reduction of *Campylobacter jejuni* | Solomon and Hoover (2004) |

© Woodhead Publishing Limited, 2012

Oysters are treated at 250–350 MPa for 1–3 min at ambient temperature to eliminate *Vibrio* spp. from the product with little or no effect on sensory quality (He *et al.*, 2002; Cook, 2003). A 7-log reduction of *V. parahaemolyticus* has been reported by Calik *et al.* (2002) by processing oysters at 345 MPa and 22°C for 1.5 min. Cook (2003) also reported a 5-log reduction of *V. parahaemolyticus* O3:K6 (the most resistant strains amongst 10 strains studied) by a 300 MPa treatment at 10°C for 3 min. Furthermore, HPP can extend the limited refrigerated shelf life of oysters while maintaining the overall quality of this product (He *et al.*, 2002). An additional and important benefit of HPP for oysters is the mechanical shucking effect by denaturing and releasing the adductor muscle from the shell (He *et al.*, 2002; Cruz-Romero *et al.*, 2004).

### 13.6.6 HPP of milk and dairy products

Milk is an excellent medium for the growth of many microorganisms that negatively affect its safety and quality. Low-temperature long-time (LTLT) (63°C, 30 min) and high-temperature short-time (HTST) (72°C, 15 s) pasteurizations, and ultra-high-temperature (UHT) (~140°C for a few seconds) treatment are the main methods in combination with hygienic milk harvesting for the elimination of pathogens in milk (Pearce, 2004; Kelly *et al.*, 2006). The application of HPP for milk preservation dates back over a century when Hite (1899) found a 5–6 log reduction in the total microbial load of milk treated at *ca.* 680 MPa for 10 min at room temperature. The combined effect of pressure and temperature (67–71°C) further increased milk shelf life (Hite, 1899). Although many studies documented inactivation of different microorganisms in milk and dairy products (Table 13.6), more

**Table 13.6** Selected examples of microbial inactivation by high pressure processing in milk and dairy products

| Medium | Process conditions | Microbial inactivation | Reference |
|---|---|---|---|
| UHT milk | 600 MPa/20°C/15 min | ≤2.0 log reduction of *Escherichia coli* O157:H7 and *Staphylococcus aureus* | Patterson *et al.* (1995) |
| UHT milk | 375 MPa/20°C/15 min | <1 log reduction of *Listeria monocytogenes* | Patterson *et al.* (1995) |
| Milk | 400 MPa/50°C/15 min | 5.0 log reduction of *E. coli* O157:H7 | Patterson *et al.* (1995); Patterson and Kilpatrick (1998) |
| Fresh lactic curd cheese | 300–600 MPa/≤22°C/5 min | ≤7.0 log reduction of *Lactococcus*; prevention of spoilage yeast and mold outgrowth for 6–8 weeks | Daryaei *et al.* (2006, 2008) |

© Woodhead Publishing Limited, 2012

information is required about the risks associated with pressure resistance of the traditional milk pathogens, i.e. *Mycobacterium tuberculosis* and *Coxiella burnetti* (the most heat-resistant organism of public health significance) (Cerf and Condron, 2006).

## 13.7 Limitations and challenges to adoption of high pressure processing (HPP) technology

Currently, pressure treatment is commercially viable for preserving a variety of value-added food products. Due to its batch nature and limited throughput, the technology may not be commercially viable for commodity products such as milk which may require high throughput. In comparison to thermally-treated products, pressure-treated products may cost 3–20 cents per pound more to manufacture. More efforts are needed to introduce reliable, cost-effective pressure vessels with increased throughput (Tonello, 2011).

## 13.8 Conclusions and future trends

High pressure processing is an effective alternative food preservation method capable of ensuring microbial safety and extending shelf life of a number of perishable products without the use of preservatives or additives. An important advantage of using HPP is that it usually has minimal or no impact on color, flavor, texture, nutritive, and functional qualities of food. HPP-treated foods are comparable to thermally pasteurized products in many respects. It is important to document and validate critical process parameters that can affect pressure inactivation of microorganisms, including initial temperature, the pressure level, pressure holding time, and process temperature. Moreover, it is vital to understand the mechanisms of microbial death by HPP in order to design and develop more efficient equipment and define conditions for effective inactivation of microorganisms. In addition, more studies are needed to understand potential formation of toxic compounds during combined pressure-heat treatment.

The composition of the medium in which microorganisms are exposed to pressure treatment can influence the level of microbial inactivation. More effective inactivation of bacterial contaminants can be achieved by combination of HPP with heat or by applying antimicrobial compounds. This may provide a method to inactivate microorganisms at lower pressure levels, thus reducing processing time and cost.

© Woodhead Publishing Limited, 2012

## 13.9 Sources of further information and advice

- Institute of Food Technologists (IFT) Non-thermal Processing Division: http://www.ift.org/divisions/nonthermal/
- The European High Pressure Research Group (EHPRG): http://www.ehprg.org/
- *High Pressure Research* (An International Journal Published in Association with AIRAPT and EHPRG): http://www.tandf.co.uk/journals/GHPR
- High Pressure Processing Laboratory at The Ohio State University: http://grad.fst.ohio-state.edu/hpp/
- High Pressure Processing Laboratory at Virginia Polytechnic Institute and State University: http://www.hpp.vt.edu/
- Center for Nonthermal Processing of Food at Washington State University: http://www.bsyse.wsu.edu/cnpf/

## 13.10 References

ADEGOKE G O, IWAHASHI H and KOMATSU Y (1997), 'Inhibition of *Saccharomyces cerevisiae* by combination of hydrostatic pressure and monoterpenes', *J Food Sci*, 62(2), 404–405.

AHN J and BALASUBRAMANIAM V M (2007), 'Physiological responses of *Bacillus amyloliquefaciens* spores to high pressure', *J Microbiol Biotechnol*, 17(3), 524–529.

AHN J, BALASUBRAMANIAM V M and YOUSEF A E (2007), 'Inactivation kinetics of selected aerobic and anaerobic bacterial spores by pressure-assisted thermal processing', *Int J Food Microbiol*, 113, 321–329.

ALPAS H, KALCHAYANAND N, BOZOGLU F and RAY B (2000), 'Interactions of high hydrostatic pressure pressurization temperature and pH on death and injury of pressure-resistant and pressure-sensitive strains of foodborne pathogens', *Int J Food Microbiol*, 60, 33–42.

AMERICAN SOCIETY OF MECHANICAL ENGINEERS (2011), '2010 Boiler and pressure vessel codes with addenda'. Available from: http://files.asme.org/Catalog/Codes/PrintBook/30202.pdf (accessed 28 October 2011).

ANANTA E, HEINZ V, SCHLÜTER O and KNORR D (2001), 'Kinetic studies on high pressure inactivation of *Bacillus stearothermophilus* spores suspended in food matrices', *Innov Food Sci Emerg Technol*, 2, 261–272.

ANEMA S G (2008), 'Effect of milk solids concentration on whey protein denaturation, particle size changes and solubilization of casein in high-pressure-treated skim milk', *Int Dairy J*, 18, 228–235.

ANGSUPANICH K and LEDWARD D A (1998), 'High pressure treatment effects on cod (*Gadus morhua*) muscle', *Food Chem*, 63, 39–50.

ASME (American Society of Mechanical Engineers) (2011), An International Code – 2010 ASME Boiler & Pressure Vessel Code Section VIII Rules for Construction of Pressure Vessels – Division 3: Alternative Rules for Construction of High Pressure Vessels, New York, ASME.

BALASUBRAMANIAM V M and FARKAS D (2008), 'High pressure food processing', *Food Sci Technol Int*, 14(5), 413–418.

BALASUBRAMANIAM V M, FARKAS D and TUREK E J (2008), 'Preserving foods through high-pressure processing', *Food Technol*, 62(11), 32–38.

© Woodhead Publishing Limited, 2012

BALASUBRAMANIAM V M, TING E Y, STEWART C M and ROBBINS J A (2004), 'Recommended laboratory practices for conducting high-pressure microbial inactivation experiments', *Innov Food Sci Emerg Technol*, 5, 299–306.

BALASUBRAMANIAN S and BALASUBRAMANIAM V M (2003), 'Compression heating influence of pressure transmitting fluids on bacteria inactivation during high pressure processing', *Food Res Int*, 36, 661–668.

BALCI A T and WILBEY R A (1999), 'High pressure processing of milk – the first 100 years in the development of a new technology', *Int J Dairy Technol*, 52, 149–155.

BALNY C and MASSON P (1993), 'Effects of high pressure on proteins', *Food Rev Int*, 9(4), 611–628.

BAUER B A and KNORR D (2005), 'The impact of pressure, temperature and treatment time on starches: pressure-induced starch gelatinisation as pressure time temperature indicator for high hydrostatic pressure processing', *J Food Eng*, 68(3), 329–334.

BENITO A, VENTOURA G, CASADEI M, ROBINSON T and MACKEY B (1999), 'Variation in resistance of natural isolates of *Escherichia coli* O157 to high hydrostatic pressure, mild heat and other stresses', *Appl Environ Microbiol*, 65, 1564–1569.

BLACK E P, KELLY A L and FITZGERALD G F (2005), 'The combined effect of high pressure and nisin on inactivation of microorganisms in milk', *Innov Food Sci Emerg Technol*, 6, 286–292.

BLACK E P, HUPPERTZ T, FITZGERALD G F and KELLY A L (2007a), 'Baroprotection of vegetative bacteria by milk constituents: a study of *Listeria innocua*', *Int Dairy J*, 17(2), 104–110.

BLACK E P, SETLOW P, HOCKING A D, STEWART C M, KELLY A L and HOOVER D G (2007b), 'Response of spores to high-pressure processing', *Compr Rev Food Sci Food Safety*, 6(4), 103–119.

BLACK E P, WEI J, ATLURI S, CORTEZZO D E, KOZIOL-DUBE K, HOOVER D G and SETLOW P (2007c), 'Analysis of factors influencing the rate of germination of spores of *Bacillus subtilis* by very high pressure', *J Appl Microbiol*, 102, 65–76.

BOYNTON B B, SIMS C A, SARGENT S, BALABAN M O and MARSHALL M R (2002), 'Quality and stability of precut mangos and carambolas subjected to high-pressure processing', *J Food Sci*, 67, 409–415.

BUCKOW R, JANKOWIAK L, KNORR D and VERSTEEG C (2009), 'Pressure-temperature phase diagrams of maize starches with different amylose contents', *J Agr Food Chem*, 57, 11510–11516.

BULL M K, ZERDIN K, HOWE E, GOICOECHE G, PARAMANANDHAN P, STOCKMAN R, SELLAHEWA J, SZABO E A, JOHNSON R L and STEWART C M (2004), 'The effect of high pressure processing on the microbial physical and chemical properties of Valencia and navel orange juice', *Innov Food Sci Emerg Technol*, 5, 135–149.

CALIK H, MORRISSEY M T, RENO P W and AN H (2002), 'Effect of high-pressure processing on *Vibrio parahaemolyticus* strains in pure culture and Pacific oysters', *J Food Sci*, 67(4), 1506–1510.

CAPELLAS M, MOR-MUR M, GERVILLA R, YUSTE J and GUAMIS B (2000), 'Effect of high pressure combined with mild heat or nisin on inoculated bacteria and mesophiles of goat's milk fresh cheese', *Food Microbiol*, 17, 633–641.

CAPELLAS M, MOR-MUR M, SENDRA E and GUAMIS B (2001), 'Effect of high-pressure processing on physico-chemical characteristics of fresh goats' milk cheese (*Mató*)', *Int Dairy J*, 11, 165–173.

CARLEZ A, ROSEC J, RICHARD N and CHEFTEL J (1993), 'High pressure inactivation of *Citrobacter freundii*, *Pseudomonas fluorecens* and *Listeria innocua* in inoculated minced beef muscle', *Lebensm Wiss Technol*, 26, 357–363.

CASADEI M A, MAÑAS P, NIVEN G, NEEDS E and MACKEY B M (2002), 'Role of membrane fluidity in pressure resistance of *Escherichia coli* NCTC 8164', *Appl Environ Microbiol*, 68(12), 5965–5972.

CASAL V and GOMEZ R (1999), 'Effect of high pressure on the viability and enzymatic

© Woodhead Publishing Limited, 2012

activity of mesophilic lactic acid bacteria isolated from caprine cheese', *J Dairy Sci*, 82, 1092–1098.

CERF O and CONDRON R (2006), '*Coxiella burnetii* and milk pasteurization: an early application of the precautionary principle?', *Epidemiol Infect*, 134, 946–951.

CHEFTEL J (1992), 'Effects of high hydrostatic pressure on food constituents: an overview', *High Pressure Biotechnol*, 224, 195–209.

CHEFTEL J C (1995), 'Review: high-pressure microbial inactivation and food preservation', *Food Sci Technol Int*, 1(2/3), 75–90.

CHEN H and HOOVER D G (2003), 'Modeling the combined effect of high hydrostatic pressure and mild heat on the inactivation kinetics of *Listeria monocytogenes* Scott A in whole milk', *Innov Food Sci Emerg Technol*, 4, 25–34.

CHILTON P, ISAACS N S, MAÑAS P and MACKEY B M (2001), 'Biosynthetic requirements for the repair of membrane damage in pressure-treated *Escherichia coli*', *Int J Food Microbiol*, 71, 101–104.

COOK D (2003), 'Sensitivity of *Vibrio* species in phosphate buffered saline and in oysters to high-pressure processing', *J Food Protect*, 66, 2276–2282.

CRAWFORD Y J, MURANO E A, OLSEN DG and SHENOY K (1996), 'Use of high hydrostatic pressure and irradiation to eliminate *Clostridium sporogenes* in chicken breast', *J Food Protect*, 59, 711–715.

CRUZ-ROMERO M, SMIDDY M, HILL C, KERRY J P and KELLY A L (2004), 'Effects of high pressure treatment on physicochemical characteristics of fresh oysters (*Crassostrea gigas*)', *Innov Food Sci Emerg Technol*, 5, 161–169.

DARYAEI H, COVENTRY M J, VERSTEEG C and SHERKAT F (2006), 'Effects of high-pressure treatment on shelf life and quality of fresh lactic curd cheese', *Aust J Dairy Technol*, 61(2), 186–188.

DARYAEI H, COVENTRY M J, VERSTEEG C and SHERKAT F (2008), 'Effect of high pressure treatment on starter bacteria and spoilage yeasts in fresh lactic curd cheese of bovine milk', *Innov Food Sci Emerg Technol*, 9(2), 201–205.

DARYAEI H, COVENTRY M J, VERSTEEG C and SHERKAT F (2010), 'Combined pH and high hydrostatic pressure effects on *Lactococcus* starter cultures and *Candida* spoilage yeasts in a fermented milk test system during cold storage', *Food Microbiol*, 27, 1051–1056.

DE ANCOS B, GONZALEZ E and CANO M P (2000), 'Effect of high-pressure treatment on the carotenoid composition and the radical scavenging activity of persimmon fruit purées', *J Agr Food Chem*, 48, 3542–3548.

DE ANCOS B, SGROPPO S, PLAZA L and CANO M P (2002), 'Possible nutritional and health-related value promotion in orange juice preserved by high-pressure treatment', *J Sci Food Agr*, 82, 790–796.

DE HEIJ W B C, VAN SCHEPDAEL L J M M, MOEZELAAR R, HOOGLAND H, MATSER A M and VAN DEN BERG R W (2003), 'High pressure sterilization: maximizing the benefits of adiabatic heating', *Food Technol*, 57(3), 37–41.

DE LAMO-CASTELLVI S, RATPHITAGSANTI W, BALASUBRAMANIAM V M and YOUSEF A E (2010), 'Inactivation of *Bacillus amyloliquefaciens* spores by a combination of sucrose laurate and pressure-assisted thermal processing', *J Food Protect*, 73(11), 2043–2052.

DENYS S, VAN LOEY A M and HENDRICKX M E (2000), 'A modeling approach for evaluating process uniformity during batch high hydrostatic pressure processing: combination of a numerical heat transfer model and enzyme inactivation kinetics', *Innov Food Sci Emerg Technol*, 1, 5–19.

DOGAN C and ERKMEN O (2004), 'High pressure inactivation kinetics of *Listeria monocytogenes* inactivation in broth milk and peach and orange juices', *J Food Eng*, 62, 47–52.

DOUZALS J P, MARECHAL P A, COQUILLE J C and GERVAIS P (1996), 'Microscopic study of starch gelatinization under high hydrostatic pressure', *J Agr Food Chem*, 44, 1403–1408.

© Woodhead Publishing Limited, 2012

DOUZALS J P, PERRIER CORNET J M, GERVAIS P and COQUILLE J € (1998), 'High-pressure gelatinization of wheat starch and properties of pressure-induced gels', *J Agr Food Chem*, 46, 4824–4829.

EARNSHAW R G (1992), 'High pressure as cell sensitiser: new opportunities to increase the efficacy of preservation processes', in Balny C, Hayashi R, Heremans K and Masson P, *High Pressure and Biotechnology*, Paris, Coloque INSERM/John Libbey Eurotext Ltd, 261–267.

EARNSHAW R G, APPLEYARD J and HURST R M (1995), 'Understanding physical inactivation processes: combined preservation opportunities using heat ultrasound and pressure', *Int J Food Microbiol*, 28(2), 197–219.

ERKMEN O and KARATAS S (1997), 'Effect of high hydrostatic pressure on *Staphylococcus aureus* in milk', *J Food Eng*, 33, 257–262.

FARKAS D F and HOOVER D G (2000), 'High pressure processing', in *Kinetics of microbial inactivation for alternative food processing technologies*, *J Food Sci* Suppl, 65(8), 47–64.

FEEHERRY F E, DOONA C J and ROSS E W (2006), 'The quasi-chemical kinetics model for the inactivation of microbial pathogens using high pressure processing', *Acta Horticulture* 674, 245–251.

FURUKAWA S, NAKAHARA A and HAYAKAWA I (2000), 'Effect of reciprocal pressurisation on germination and killing of bacterial spores', *Int J Food Sci Technol*, 35, 529–532.

GÄNZLE M G and VOGEL R F (2001), 'On-line fluorescence determination of pressure mediated outer membrane damage in *Escherichia coli*', *Syst Appl Microbiol*, 24, 477–485.

GAO Y-L and JU X-R (2008), 'Exploiting the combined effects of high pressure and moderate heat with nisin on inactivation of *Clostridium botulinum* spores', *J Microbiol Meth*, 72, 20–28.

GARCIA-GRAELLS C, MASSCHALCK B and MICHIELS C W (1999), 'Inactivation of *Escherichia coli* in milk by high-hydrostatic-pressure treatment in combination with antimicrobial peptides', *J Food Protect*, 62(11), 1248–1254.

GARCIA-RISCO M R, CORTES E, CARRASCOSA A V and LÓPEZ-FANDIÑO R (1998), 'Microbiological and chemical changes in high-pressure-treated milk during refrigerated storage', *J Food Protect*, 61(6), 735–737.

GARCIA-RISCO M R, OLANO A, RAMOS M and LÓPEZ-FANDIÑO R (2000), 'Micellar changes induced by high pressure. Influence in the proteolytic activity and organoleptic properties of milk', *J Dairy Sci*, 83(10), 2184–2189.

GOW C Y and HSIN T L (1996), 'Comparison of high pressure treatment and thermal pasteurization effects on the quality and shelf life of guava purée', *Int J Food Sci Technol*, 31, 205–213.

GOW C Y and HSIN T L (1999), 'Changes in volatile flavor components of guava juice with high-pressure treatment and heat processing and during storage', *J Agr Food Chem*, 47, 2082–2087.

GROSS M, LEHLE K, JAENICKE R and NIERHAUS K H (1993), 'Pressure-induced dissociation of ribosomes and elongation cycle intermediates: stabilizing conditions and identification of the most sensitive functional state', *Eur J Biochem*, 218, 463–468.

GROVE S F, FORSYTH S, WAN J, COVENTRY J, COLE M, STEWART C M, LEWIS T, ROSS T and LEE A (2008), 'Inactivation of hepatitis A virus, poliovirus and a norovirus surrogate by high pressure processing', *Innov Food Sci Emerg Technol*, 9(2), 206–210.

GUPTA R, BALASUBRAMANIAM V M, SCHWARTZ S J and FRANCIS D M (2010), 'Storage stability of lycopene in tomato juice subjected to combined pressure-heat treatments', *J Agr Food Chem*, 58(14), 8305–8313.

HAYAKAWA I, KANNO T, YOSHIYAMA K and FUJIO Y (1994), 'Oscillatory compared with continuous high pressure sterilization on *Bacillus stearothermophilus* spores', *J Food Sci*, 59(1), 164–167.

HAYMAN M M, BAXTER I, O'RIORDAN P J and STEWART C M (2004), 'Effects of high-pressure

© Woodhead Publishing Limited, 2012

processing on the safety quality and shelf life of ready-to-eat meats', *J Food Protect*, 67, 1709–1718.

HE H, ADAMS R M, FARKAS D F and MORRISSEY M T (2002), 'Use of high-pressure processing for oyster shucking and shelf-life extension', *J Food Sci*, 67(2), 640–645.

HELDMAN D R and NEWSOME R L (2003), 'Kinetic models for microbial survival during processing', *Food Technol*, 57(8), 40–46.

HEREMANS K (1982), 'High pressure effects on proteins and other biomolecules', *Ann Rev Biophys Bioeng*, 11, 1–21.

HITE B H (1899), 'The effect of pressure in the preservation of milk', *W VA Univ AES Bull*, 58, 15–35.

HOMMA N, IKEUCHI Y and SUZUKI A (1995), 'Effects of high pressure treatment on the proteolytic enzymes in meat', *Meat Sci*, 38, 219–228.

HOOVER D G, METRICK C, PAPINEAU A M, FARKAS D F and KNORR D (1989), 'Biological effects of high hydrostatic pressure on food microorganisms', *Food Technol*, 43(3), 99–107.

HSU K C, TAN F J and CHI H Y (2008), 'Evaluation of microbial inactivation and physicochemical properties of pressurized tomato juice during refrigerated storage', *Lebensm Wiss Technol*, 41, 367–375.

HUGAS M, GARRIGA M and MONFORT J M (2002), 'New mild technologies in meat processing: high pressure as a model technology', *Meat Sci*, 62, 359–371.

IFT NEWS (2009), 'Institute of Food Technologists announces 2009 Innovation Award winners'. Available from: http://wwwiftorg/newsroom/news-releases/(2009),/june/07/(2009),-innovation-award-winnersaspx (accessed 25 July 2011).

JOHNSTON D E, RUTHERFORD J A and MCCREEDY R W (2002), 'Ethanol stability and chymosin-induced coagulation behaviour of high pressure treated milk', *Milchwissenschaft*, 57(7), 363–366.

KALCHAYANAND N, SIKES A, DUNNE C P and RAY B (1998), 'Interaction of hydrostatic pressure time and temperature of pressurization and pediocin AcH on inactivation of foodborne bacteria', *J Food Protect*, 61(4), 425–431.

KALCHAYANAND N, DUNNE C P, SIKES A and RAY B (2003), 'Inactivation of bacterial endospores by combined action of hydrostatic pressure and bacteriocins in roast beef', *J Food Safety*, 23, 219–233.

KARATZAS A K, KETS E P W, SMID E J and BENNIK M H J (2001), 'The combined action of carvacrol and high hydrostatic pressure on *Listeria monocytogenes* Scott A', *J Appl Microbiol*, 90, 463–469.

KELLY A L, DATTA N and DEETH H C (2006), 'Thermal processing of dairy products', in Sun D-W, *Thermal food processing: new technologies and quality issues*, Boca Raton, FL, CRC Press, 266–298.

KINGSLEY D H, HOOVER D G, PAPAFRAGKOU E and RICHARDS G P (2002), 'Inactivation of hepatitis A virus and a calicivirus by high hydrostatic pressure', *J Food Protect*, 65, 1605–1609.

KINGSLEY D H, CHEN H and HOOVER D G (2004), 'Inactivation of selected picornaviruses by high hydrostatic pressure', *Virus Res*, 102(2), 221–224.

KNORR D (1993), 'Effects of high-hydrostatic-pressure processes on food safety and quality', *Food Technol*, 47(6), 156–161.

KNORR D (1994), 'Hydrostatic pressure treatment of food: microbiology', in Gould G W, *New methods of food preservation*, London, Blackie Academic and Professional, 159–175.

KOBORI H, SATO M, TAMEIKE A, HAMADA K, SHIMADA S and OSUMI M (1995), 'Ultrastructural effects of pressure stress to the nucleus in *Saccharomyces cerevisiae*: a study by immunoelectron microscopy using frozen thin sections', *FEMS Microbiol Lett*, 132, 253–258.

KOO J, JAHNCKE M L, RENO P W, HU X and MALLIKARJUNAN P (2006), 'Inactivation of *Vibro parahaemolyticus* and *Vibrio vulnificus* in phosphate-buffered saline and in inoculated whole oysters by high-pressure processing', *J Food Protect*, 69(3), 596–601.

© Woodhead Publishing Limited, 2012

KUO C H (2008), 'Evaluation of processing qualities of tomato juice induced by thermal and pressure processing', *Lebensm Wiss Technol*, 41, 450–459.

LAMBERT Y, DEMAZEAU G, LARGETEAU A and BOUVIER J M (1999), 'Changes in aromatic volatile composition of strawberry after high pressure treatment', *Food Chem*, 67, 7–16.

LANDAU J (1967), 'Induction transcription and translation in *Escherichia coli*: a hydrostatic pressure study', *Biochim Biophys Acta*, 149, 506–512.

LAU M H and TUREK E J (2007), 'Determination of quality differences in low-acid foods sterilized by high pressure versus retorting', in Doona C and Feeherry F, *High Pressure Processing of Foods*, Ames, IA, Blackwell Publishing, 195–217.

LINTON M and PATTERSON M F (2000), 'High pressure processing of foods for microbiological safety and quality: a short review', *Acta Microbiol Immunol Hung*, 47(2/3), 175–182.

LINTON M, MCCLEMENTS J M J and PATTERSON M F (1999), 'Survival of *Escherichia coli* O157:H7 during storage in pressure-treated orange juice', *J Food Protect*, 62(9), 1038–1040.

LINTON M, MCCLEMENTS J M J and PATTERSON M F (2001), 'Inactivation of pathogenic *Escherichia coli* in skimmed milk using high hydrostatic pressure', *Innov Food Sci Emerg Technol*, 2, 99–104.

LUDIKHUYZE L and HENDRICKX M E G (2001), 'Effects of high pressure on chemical reactions related to food quality', in Hendrickx M E G and Knorr D, *Ultra high pressure treatments of foods*, New York, Kluwer Academic/Plenum Publishers, 167–188.

LUDWIG H and SCHRECK C (1997), 'The inactivation of vegetative bacteria by pressure', in Heremans K, *High pressure research in the biosciences and biotechnology*, Leuven, Leuven University Press, 221–224.

MACKEY B M (2000), 'Injured bacteria', in Lund B M, Baird-Parker T C and Gould G W, *The microbiological safety and quality of foods*, Gaithersburg, MD, Aspen Publishers, 315–341.

MACKEY B M, FORESTIÈRE K and ISAACS N (1995), 'Factors affecting the pressure resistance of *Listeria monocytogenes* to high hydrostatic pressure', *Food Biotechnol*, 9, 1–11.

MAINVILLE I, MONTPETIT D, DURAND N and FARNWORTH E R (2001), 'Deactivating the bacteria and yeast in kefir using heat treatment irradiation and high pressure', *Int Dairy J*, 11, 45–49.

MAÑAS P and MACKEY B M (2004), 'Morphological and physiological changes induced by high hydrostatic pressure in exponential- and stationary-phase cells of *Escherichia coli*: relationship with cell death', *Appl Environ Microbiol*, 70(3), 1545–1554.

MAÑAS P and PAGÁN R (2005), 'Microbial inactivation by new technologies of food preservation', *J Appl Microbiol*, 98, 1387–1399.

MARGOSCH D, EHRMANN M A, GÄNZLE M G and VOGEL R F (2004a), 'Comparison of pressure and heat resistance of *Clostridium botulinum* and other endospores in mashed carrots', *J Food Protect*, 67(11), 2530–2537.

MARGOSCH D, GÄNZLE M G, EHRMANN M A and VOGEL R F (2004b), 'Pressure inactivation of *Bacillus* endospores', *Appl Environ Microbiol*, 70(12), 7321–7328.

MASSCHALCK B, VAN HOUDT R and MICHIELS C W (2001a), 'High pressure increases bactericidal activity and spectrum of lactoferrin, lactoferricin and nisin', *Int J Food Microbiol*, 64, 325–332.

MASSCHALCK B, VAN HOUDT R, VAN HAVRE E G R and MICHIELS C W (2001b), 'Inactivation of Gram-negative bacteria by lysozyme, denatured lysozyme and lysosyme derived peptides under high hydrostatic pressure', *Appl Environ Microbiol*, 67(1), 339–344.

MCCLEMENTS J M J, PATTERSON M F and LINTON M (2001), 'The effect of growth stage and growth temperature on high hydrostatic pressure inactivation of some psychrotrophic bacteria in milk', *J Food Protect*, 64(4), 514–522.

MERTENS B and DEPLACE G (1993), 'Engineering aspects of high-pressure technology in the food industry', *Food Technol*, 47(6), 164–169.

© Woodhead Publishing Limited, 2012

MESSENS W, VAN CAMP J and HUYGHEBARET A (1997), 'The use of high pressure to modify the functionality of food proteins', *Trends Food Sci Technol*, 8, 107–112.

METRICK C, HOOVER D G and FARKAS D F (1989), 'Effects of high hydrostatic pressure on heat-resistant and heat-sensitive strains of *salmonella*', *J Food Sci*, 54(6), 1547–1549.

MEYER R, COOPER K L, KNORR D and LELIEVELD H L M (2000), 'High pressure sterilization of foods', *Food Technol*, 54 (11), 67–72.

MOLINA-HÖPPNER A, DOSTER W, VOGEL R F and GÄNZLE M G (2004), 'Protective effect of sucrose and sodium chloride for *Lactococcus lactis* during sublethal and lethal high pressure treatments', *Appl Environ Microbiol*, 70(4), 2013–2020.

MORGAN S M, ROSS R P, BERESFORD T and HILL C (2000), 'Combination of hydrostatic pressure and lacticin 3147 causes increased killing of *Staphylococcus* and *Listeria*', *J Appl Microbiol*, 88, 414–420.

MORITA R Y (1975), 'Psychrophilic bacteria', *Bacteriol Rev*, 39, 144–167.

MOZHAEV V V, HEREMANS K, FRANK J, MASSON P and BALNY C (1994), 'Exploiting the effects of high hydrostatic pressure in biotechnological applications', *Trends Biotechnol*, 12, 493–501.

NATIONAL ADVISORY COMMITTEE ON MICROBIOLOGICAL CRITERIA FOR FOODS (NACMCF) (2006), 'Requisite scientific parameters for establishing the equivalence of alternative methods of pasteurization', *J Food Protect*, 69(5), 1190–1216.

NATIONAL ADVISORY COMMITTEE ON MICROBIOLOGICAL CRITERIA FOR FOODS (NACMCF) (2010), 'Parameters for determining inoculated pack/challenge study protocols', *J Food Protect*, 73(1), 140–202.

NEEDS E C, CAPELLAS M, BLAND A P, MANOJ P, MACDOUGAL D and PAUL G (2000), 'Comparison of heat and pressure treatments of skim milk fortified with whey protein concentrate for set yogurt preparation: effects on milk proteins and gel structure', *J Dairy Res*, 67(3), 329–348.

NGUYEN L T and BALASUBRAMANIAM V M (2011), 'Fundamentals of food processing using high pressure', in Zhang H Q, Barbosa-Cánovas G V, Balasubramaniam V M, Dunne C P, Farkas D F and Yuan J T C, *Nonthermal processing technologies for food*, Chicago, IL, Blackwell Publishing, 3–19.

NGUYEN L T, RASTOGI N K and BALASUBRAMANIAM V M (2007), 'Evaluation of instrumental quality of pressure-assisted thermally processed carrots', *J Food Sci*, 72, E264–E270.

NIENABER U and SHELLHAMMER T H (2001), 'High pressure processing of orange juice: combination treatments and a shelf life study', *J Food Sci*, 66, 332–336.

OGAWA H, FUKUHISA K, KUBO Y and FUKUMOTO H (1990), 'Pressure inactivation of yeasts molds and pectinesterase in Satsuma mandarin juice: effects of juice concentration pH and organic acids and comparison with heat sanitation', *Agr Biol Chem*, 54(5), 1219–1225.

OSUMI M, YAMADA N, SATO M, KOBORI H, SHIMADA S and HAYASHI R (1992), 'Pressure effects on yeast cells ultrastructure: change in the ultrastructure and cytoskeleton of the dimorphic yeast *Candida tropicalis*', in Balny C, Hayashi R, Heremans K and Masson P, *High pressure and biotechnology*, Paris, Colloque INSERM/J Libby Eurotext Ltd, 9–18.

OTERO L, MOLINA-GARCÍA A D and SANZ P D (2000), 'Thermal effect in foods during quasi-adiabatic pressure treatments', *Innov Food Sci Emerg Technol*, 1, 119–126.

OXEN P and KNORR D (1993), 'Baroprotective effects of high solute concentrations against inactivation of *Rhodotorula rubra*', *Lebensm Wiss Technol*, 26, 220–223.

PAGÁN R, MAÑAS P, RASO J and CONDÓN S (1999), 'Bacterial resistance to ultrasonic waves under pressure at non lethal (manosonication) and lethal (manothermosonication) temperatures', *Appl Environ Microbiol*, 65, 297–300.

PAIDHUNGAT M, SETLOW B, DANIELS W B, HOOVER D, PAPAFRAGKOU E and SETLOW P (2002), 'Mechanisms of induction of germination of *Bacillus subtilis* spores by high pressure', *Appl Environ Microbiol*, 68, 3172–3175.

© Woodhead Publishing Limited, 2012

PALOU E, LÓPEZ-MALO A, BARBOSA-CÁNOVAS G V, WELTI-CHANES J and SWANSON B G (1997), 'Combined effect of high hydrostatic pressure and water activity on *Zygosaccharomyces bailii* inhibition', *Lett Appl Microbiol*, 24, 417–420.

PALOU E, LÓPEZ-MALO A and WELTI-CHANES J (2002), 'Innovative fruit preservation using high pressure', in Welti-Chanes J, Barbosa-Cánovas G V and Aguilera J M, *Engineering and food for the 21st century*, Boca Raton, FL, CRC Press, 715–726.

PAPINEAU A M, HOOVER D G, KNORR D and FARKAS D J (1991), 'Antimicrobial effect of water-soluble chitosans with high hydrostatic pressure', *Food Biotechnol*, 5 (1), 45–57.

PARK S W, SOHN K H, SHIN J H and LEE H J (2001), 'High hydrostatic pressure inactivation of *Lactobacillus viridescens* and its effects on ultrastructure of cells', *Int J Food Sci Technol*, 36, 775–781.

PATAZCA E, KOUTCHMA T and BALASUBRAMANIAM V M (2007), 'Quasi-adiabatic temperature increase during high pressure processing of selected foods', *J Food Eng*, 80, 199–205.

PATRAS A, BRUNTON N P, PIEVE S D, BUTLER F and DOWNEY G (2009), 'Effect of thermal and high pressure processing on antioxidant activity and instrumental color of tomato and carrot purées', *Innov Food Sci Emerg Technol*, 10, 16–22.

PATTERSON M (1999), 'High-pressure treatment of foods', in Robertson R K, Batt A and Patel P D, *The encyclopedia of food microbiology*, London, Academic Press, 1059–1065.

PATTERSON M F (2005), 'Microbiology of pressure-treated foods', *J Appl Microbiol*, 98, 1400–1409.

PATTERSON M F and KILPATRICK D J (1998), 'The combined effect of high hydrostatic pressure and mild heat on inactivation of pathogens in milk and poultry', *J Food Protect*, 61(4), 432–436.

PATTERSON M and LOAHARANU P (2000), 'Irradiation', in Lund B M, Baird-Parker T C and Gould G W, *The microbiological safety and quality of foods*, Gaithersburg, MD, Aspen Publishers, 65–100.

PATTERSON M F, QUINN M, SIMPSON R and GILMOUR A (1995), 'Sensitivity of vegetative pathogens to high hydrostatic pressure treatment in phosphate-buffered saline and foods', *J Food Protect*, 58(5), 524–529.

PEARCE L E (2004), 'Survey of data on heat resistance of dairy pathogens', *Bull Int Dairy Fed*, 392, 37–41.

PELEG M and COLE M B (1998), 'Reinterpretation of microbial survival curves', *Crit Rev Food Sci*, 38(5), 353–380.

PERERA N, GAMAGE T V, WAKELING L, GAMLATH G G S and VERSTEEG C (2009), 'Colour and texture of apples high pressure processed in pineapple juice', *Innov Food Sci Emerg Technol*, 11(1), 39–46.

PHUNCHAISRI C and APICHARTSRANGKOON A (2005), 'Effects of ultra-high pressure on biochemical and physical modification of lychee (*Litchi chinensis* Sonn)', *Food Chem*, 93, 57–64.

POLYDERA A C, STOFOROS N G and TAOUKIS P S (2003), 'Comparative shelf life study and vitamin C loss kinetics in pasteurized and high pressure processed reconstituted orange juice', *J Food Eng*, 60, 21–29.

POLYDERA A C, STOFOROS N G and TAOUKIS P S (2004), 'The effect of storage on the antioxidant activity of reconstituted orange juice which had been pasteurized by high pressure or heat', *Int J Food Sci Technol*, 39, 783–791.

POLYDERA A C, STOFOROS N G and TAOUKIS P S (2005), 'Effect of high hydrostatic pressure treatment on post processing antioxidant activity of fresh navel orange juice', *Food Chem*, 91, 495–503.

POPPER L and KNORR D (1990), 'Applications of high-pressure homogenization for food preservation', *Food Technol*, 44(7), 84–89.

© Woodhead Publishing Limited, 2012

RADEMACHER B, PFEIFFER B and KESSLER H G (1998), 'Inactivation of microorganisms and enzymes in pressure-treated raw milk', in Isaacs NS, *High pressure food science, bioscience and chemistry*, Cambridge, The Royal Society of Chemistry, 145–151.

RAJAN S, AHN J, BALASUBRAMANIAM V M and YOUSEF A E (2006), 'Combined pressure-thermal inactivation kinetics of *Bacillus amyloliquefaciens* spores in egg patty mince', *J Food Protect*, 69(4), 853–860.

RAMASWAMY H S, RIAHI E and IDZIAK E (2003), 'High-pressure destruction kinetics of *E coli* (29055) in apple juice', *J Food Sci*, 68, 1750–1756.

RASANAYAGAM V, BALASUBRAMANIAM V M, TING E, SIZER C E, BUSH C and ANDERSON C (2003), 'Compression heating of selected fatty food materials during high-pressure processing', *J Food Sci*, 68(1), 254–259.

RASO J, GÓNGORA-NIETO M M, BARBOSA-CÁNOVAS G V and SWANSON B G (1998), 'Influence of several environmental factors on the initiation of germination and inactivation of *Bacillus cereus* by high hydrostatic pressure', *Int J Food Microbiol*, 44(1–2), 125–132.

RASTOGI N K, RAGHAVARAO K S M S, BALASUBRAMANIAM V M, NIRANJAN K and KNORR D (2007), 'Opportunities and challenges in high pressure processing of foods', *Crit Rev Food Sci Nutr*, 47 (1), 69–112.

RATPHITAGSANTI W, AHN J, BALASUBRAMANIAM V M and YOUSEF A E (2009), 'Influence of pressurization rate and pressure pulsing on the inactivation of bacterial spores during pressure-assisted thermal processing', *J Food Protect*, 72(4), 775–782.

RATPHITAGSANTI W, DE LAMO-CASTELLVI S, BALASUBRAMANIAM V M and YOUSEF A E (2010), 'Efficacy of pressure-assisted thermal processing in combination with organic acids against *Bacillus amyloliquefaciens* spores suspended in deionized water and carrot purée', *J Food Sci*, 75(1), M46–M52.

ROBERTS C M and HOOVER D G (1996), 'Sensitivity of *Bacillus coagulans* spores to combinations of high hydrostatic pressure, heat, acidity and nisin', *J Appl Bacteriol*, 81, 363–368.

ROLDÁN-MARÍN E, SANCHEZ M C, LLORIA R, ANCOS B and CANO M P (2009), 'Onion high-pressure processing: flavonol content and antioxidant activity', *Lebensm Wiss Technol*, 42, 835–841.

SÁIZ A H, MINGO S T, BALDA F P and SAMSON C T (2008), 'Advances in design for successful commercial high pressure food processing', *Food Aust*, 60(4), 154–156.

SAMPEDRO F, RODRIGO D and HENDRICKX M (2008), 'Inactivation kinetics of pectin methyl esterase under combined thermal high pressure treatment in an orange juice milk beverage', *J Food Eng*, 86, 133–139.

SÁNCHEZ-MORENO C, PLAZA L, DE ANCOS B and CANO M P (2004), 'Effect of combined treatments of high-pressure and natural additives on carotenoid extractability and antioxidant activity of tomato purée (*Lycopersicum esculentum* Mill.)', *Eur Food Res Technol*, 219, 151–160.

SÁNCHEZ-MORENO C, PLAZA L, DE ANCOS B and CANO M P (2006), 'Impact of high-pressure and traditional thermal processing of tomato purée on carotenoids, vitamin C and antioxidant activity', *J Sci Food Agr*, 86, 171–179.

SCOLLARD P G, BERESFORD T P, NEEDS E C, MURPHY P M and KELLY A L (2000), 'Plasmin activity, beta lactoglobulin denaturation and proteolysis in high pressure treated milk', *Int Dairy J*, 10, 835–841.

SHEARER A E H, DUNNE C P, SIKES A and HOOVER D G (2000), 'Bacterial spore inhibition and inactivation in foods by pressure chemical preservatives and mild heat', *J Food Protect*, 63(11), 1503–1510.

SMELT J P P M (1998), 'Recent advances in the microbiology of high pressure processing', *Trends Food Sci Technol*, 9, 152–158.

SMELT J P P M, HELLEMONS J C, WOUTERS P C and VAN GERWEN S J C (2002), 'Physiological and mathematical aspects in setting criteria for decontamination of foods by physical means', *Int J Food Microbiol*, 78, 57–77.

© Woodhead Publishing Limited, 2012

SOLOMON E B and HOOVER D G (2004), 'Inactivation of *Campylobacter jejuni* by high hydrostatic pressure', *Lett Appl Microbiol*, 38, 505–509.

SONOIKE K, SETOYAMA T, KUMA Y and KOBAYASHI S (1992), 'Effect of pressure and temperature on the death rate of *Lactobacillus casei* and *Escherichia coli*', in Balny C, Hayashi R, Heremans K and Masson P, *High pressure and biotechnology*, Paris, Colloque INSERM/John Libbey Eurotext Ltd, 297–301.

STEWART C M, JEWETT F F, DUNNE C P and HOOVER D G (1997), 'Effect of concurrent high hydrostatic pressure acidity and heat on the injury and destruction of *Listeria monocytogenes*', *J Food Safety*, 17, 23–26.

STEWART C M, DUNNE C P, SIKES A and HOOVER D G (2000), 'Sensitivity of spores of *Bacillus subtilis* and *Clostridium sporogenes* PA 3679 to combinations of high hydrostatic pressure and other processing parameters', *Innov Food Sci Emerg Technol*, 1(1), 49–56.

STYLES M F, HOOVER D G and FARKAS D F (1991), 'Response of *Listeria monocytogenes* and *Vibrio parahaemolyticus* to high hydrostatic pressure', *J Food Sci*, 56(5), 1404–1407.

SUMITANI H, SUEKANE S, NAKATANI A and TATSUKA K (1994), 'Changes in composition of volatile compounds in high pressure treated peach', *J Agr Food Chem*, 42, 785–790.

TAKAHASHI K, ISHII H and ISHIKAWA H (1993), 'Sterilization of bacteria and yeast by hydrostatic pressurization at low temperature: effect of temperature, pH and the concentration of proteins, carbohydrates and lipids', in Hayashi R, *High pressure bioscience and food science*, Kyoto, San-Ei Pub Co, 244–249.

TAUSCHER B (1995), 'Pasteurization of food by hydrostatic high pressure: chemical aspects', *Lebensm Unters Forch*, 28, 3–13.

TAY A, SHELLHAMMER T H, YOUSEF A E and CHISM G W (2003), 'Pressure death and tailing behavior of *Listeria monocytogenes* strains having different barotolerances', *J Food Protect*, 66(11), 2057–2061.

TER STEEG P F, HELLEMONS J C and KOK A E (1999), 'Synergistic actions of nisin sublethal ultrahigh pressure and reduced temperature on bacteria and yeast', *Appl Environ Microbiol*, 65(9), 4148–4154.

THAKUR B R and NELSON P E (1998), 'High pressure processing and preservation of food', *Food Rev Int*, 14(4), 427–447.

TING E (2011), 'High-pressure processing equipment fundamentals', in Zhang H Q, Barbosa-Cánovas G V, Balasubramaniam V M, Dunne C P, Farkas D F and Yuan J T C, *Nonthermal processing technologies for food*, Chicago, IL, Blackwell Publishing, 20–27.

TING E Y and MARSHALL R G (2002), 'Production issues related to UHP food', in Welti-Chanes J, Barbosa-Cánovas G V and Aguilera J M, *Engineering and food for the 21st century*, Boca Raton, FL, CRC Press, 727–738.

TING E, BALASUBRAMANIAM V M and RAGHUBEER E (2002), 'Determining thermal effects in high-pressure processing', *Food Technol*, 56(2), 31–35.

TONELLO C (2011), 'Case studies on high-pressure processing of foods', in Zhang H Q, Barbosa-Cánovas G V, Balasubramaniam V M, Dunne C P, Farkas D F and Yuan J T C, *Nonthermal processing technologies for food*, Chicago, IL, Blackwell Publishing, 36–50.

TORRES J A and VELAZQUEZ G (2005), 'Commercial opportunities and research challenges in the high pressure processing of foods', *J Food Eng*, 67, 95–112.

ULMER HM, GÄNZLE M G and VOGEL R F (2000), 'Effects of high pressure on survival and metabolic activity of *Lactobacillus plantarum* TMW1460', *Appl Environ Microbiol*, 66(9), 3966–3973.

VACHON J F, KHEADR E E, GIASSON J, PAQUIN P and FLISS I (2002), 'Inactivation of foodborne pathogens in milk using dynamic high pressure', *J Food Protect*, 65(2), 345–352.

© Woodhead Publishing Limited, 2012

VELAZQUEZ G, GANDHI K and TORRES J A (2002), 'Hydrostatic pressure processing: a review', *Biotam*, 12(2): 71–78.

WALSH-O'GRADY C D, O'KENNEDY B T, FITZGERALD R J and LANE C N (2001), 'A rheological study of acid-set "simulated yogurt milk" gels prepared from heat- or pressure-treated milk proteins', *Lait*, 81, 637–650.

WICK C, NIENABER U, ANGGRAENI O, SHELLHAMMER T H and COURTNEY P D (2004), 'Texture, proteolysis and viable lactic acid bacteria in commercial Cheddar cheeses treated with high pressure', *J Dairy Res*, 71, 107–115.

WILKINSON N, KURDZIEL A S, LANGTON S, NEEDS E and COOK N (2001), 'Resistance of poliovirus to inactivation by high hydrostatic pressures', *Innov Food Sci Emerg Technol*, 2, 95–98.

WOUTERS P C, GLAASKER E and SMELT J P P M (1998), 'Effects of high pressure on inactivation kinetics and events related to proton efflux in *Lactobacillus plantarum*', *Appl Environ Microbiol*, 64(2), 509–514.

ZHANG H Q, BARBOSA-CÁNOVAS G V, BALASUBRAMANIAM V M, DUNNE C P, FARKAS D F and YUAN J T C (2011), *Nonthermal processing technologies for food*, Chicago, IL, Blackwell Publishing.

ZHU S, NAIM F, MARCOTTE M, RAMASWAMY H and SHAO Y (2008), 'High pressure destruction kinetics of *Clostridium sporogenes* spores in ground beef at elevated temperatures', *Int J Food Microbiol*, 126, 86–92.

© Woodhead Publishing Limited, 2012

# 14

# Microbial decontamination of food by pulsed electric fields (PEFs)

**M. Amiali, Ecole Nationale Superieure Agronomique, Algeria and M. O. Ngadi, McGill University, Canada**

**Abstract**: This chapter presents the current state of the art in the application of pulsed electric field (PEF) technology. The changes in the features of PEF physicochemical parameters (pulse waveform, treatment time and pulse duration, electric field strength, temperature treatment, energy input, pH of the medium, electrical conductivity and ionic strength) and characteristics of microorganisms to be inactivated (shape of the cells, growth stage, etc.) can impact the effectiveness of the PEF process. Several recent studies have been carried out on the inactivation of pathogenic bacteria (such as *E. coli* O157:H7 and *Salmonella* Enteritidis) inoculated in milk, egg and juice products using PEF treatment, with the treated products shown to be safe. This emerging technique can be used as an integral part of non-thermal pasteurization, or can be used in combination with a hurdle method, depending on the characteristics of the liquid food.

**Key words**: pulsed electric field, decontamination, pathogenic bacteria, emerging technology, food safety, liquid food products.

## 14.1 Introduction

Food safety and quality are areas of major concern for consumers. The agri-food industries have traditionally used thermal methods to process liquid and semi-solid foods in order to inactivate pathogenic and spoilage microorganisms. These methods consist of subjecting foods to temperatures of between 60° and 140°C for different lengths of time, ranging from a few seconds to several minutes. However, processing is limited by the considerable losses of flavour and nutritional value observed in the treated products. Therefore, the need to respond to the increasing demand from consumers for

© Woodhead Publishing Limited, 2012

high quality, minimally processed products with fresh-like characteristics has led to the development of innovative and emerging technologies for new food processing methods. Research efforts are currently focused on the development of non-thermal technologies such as high hydrostatic pressures, intense light pulses, oscillating magnetic fields, and pulsed electric fields as alternative or complementary processes to conventional thermal pasteurization of liquid foods. The application of these technologies offers interesting potential for treating food while still maintaining its sensory and nutritional qualities.

Processing by pulsed electric fields (PEFs) has been shown to inactivate microorganisms with minimum losses of flavour and food quality, since it can be applied at lower temperatures, thereby leading to better organoleptic and functional properties in the treated foods. In addition, the low temperatures used in the PEF process allow the treatment to be energy efficient, resulting in lower environmental impact. This technology is based on the application of pulses of high voltage (typically 20–80 kV/cm) to the food product, which is placed between two electrodes in a treatment chamber. Large electric field intensities are achieved in the process by storing a large amount of energy in the capacitor bank and discharging the energy in the form of high voltage pulses (Zhang *et al.*, 1995a).

PEF is effective in the inactivation of most vegetative bacterial cells. The efficacy of the process depends on several factors such as electric field intensity, pulse shape and polarity, treatment time and frequency, temperature, type and concentration of microorganisms, and the PEF operation mode (Ngadi *et al.*, 2009; Amiali *et al.*, 2010). The temperature must be maintained below 30–40°C by providing a cooling system to cool the electrodes of the treatment cell, in order to eliminate thermal effects. Exposing bacterial cells to a critical electric field (>10 kV/cm) for a few microseconds leads to an irreversible electrical breakdown and changes the structure of the cell membrane, thereby resulting in a radical increase in permeability. Based on the dielectric rupture theory, the external electric field induces an electric potential difference (known as the transmembrane potential) across the cell membrane (Zimmermann, 1986). Electroporation or pore formation occurs when the transmembrane potential reaches a threshold or a critical value. The threshold transmembrane potential depends on the specific microorganisms as well as the medium or liquid food in which the microorganisms are suspended. Different bacteria have different sensitivities to electric field treatment: Gram-negative bacteria are more sensitive to electric fields than Gram-positive bacteria and yeasts, but spores are more resistant than vegetative bacteria (Barbosa-Cánovas *et al.*, 1999). The rate of inactivation of microorganisms also depends on product factors including pH and ionic strength, electrical conductivity, and composition of the liquid food (Ngadi *et al.*, 2009; Amiali *et al.*, 2010).

© Woodhead Publishing Limited, 2012

## 14.2 Pulsed electric field (PEF) technology

Electric field pulses have been widely used by gene technologists to induce pores within the cell plasma to facilitate a transmembrane exchange of materials. This process is known as electroporation (Zimmermann, 1986). When a cell is exposed to pulses of high voltage electric fields, the lipid bilayer and proteins of the cell membranes are temporarily destabilized and perforation ensues. When the electric field is removed, the pores reseal. Between the poration and resealing, it is possible to induce external DNA or RNA to enter the cell (Prasanna and Panda, 1997). However, if the intensity of the electric field is too strong, the microorganism will not be able to repair itself and will start leaking small compounds or even undergo lysis (Ngadi *et al.*, 2009).

In the PEF process, food is exposed to short electrical pulses of high electric field intensity, which is usually between 20 and 80 kV/cm (FDA, 2000). The process consists of the application pulses of short duration (1 to 100 μs) and high electric field intensity (20 to 80 kV/cm) to the food product, which is placed between two electrodes (FDA, 2000). This technology presents several advantages over conventional heat pasteurization: better retention of flavour, colour and nutritional properties, improved protein functionality, increased shelf life and reduced pathogen levels (Barbosa-Cánovas *et al.*, 1998).

The apparatus for the PEF process typically consists of a number of components including a high voltage power source, an energy storage capacitor bank, a switch, a treatment chamber(s), voltage, current and temperature probes, a pump to conduct food through the treatment chamber(s) (when a continuous system is employed), cooling devices, and a computer or control panel to control operations (Fig. 14.1). In general, the DC power supply used to charge the capacitor bank is obtained from amplified and rectified alternative current (AC). A switch is used to instantaneously discharge energy from the capacitor storage bank across the food held in the treatment chamber (Barbosa-Cánovas *et al.*, 1999). A cooling device is used to maintain the

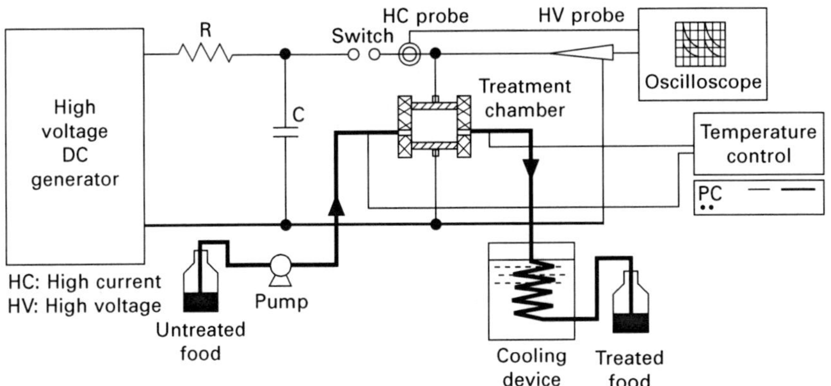


**Fig. 14.1** Major components of continuous pulsed electric field systems.

© Woodhead Publishing Limited, 2012

desired temperature of the food. The voltage and current can be measured using an oscilloscope. Food may be processed in a static manner or may be pumped through a chamber on a continuous basis. The static chamber is suitable for preliminary laboratory-scale studies, while for pilot-plant or industrial scale operations, a continuous chamber is desirable. After treatment, the food is filled into individual consumer packages or bulk storage containers using aseptic packaging equipment.

PEF treatment can be carried out using different types of pulse waveform as well as treatment chambers to effectively pasteurize and/or inactivate pathogenic and spoilage microorganisms.

### 14.2.1 Types of pulse waveforms

The waveform used in PEF treatment is determined by the equipment design. The pulse waveforms include exponential decay, square, oscillatory, biopolar exponential and square, instant charge reversal, and instant charge reversal square pulses (Table 14.1 and Fig. 14.2). Of these, the most commonly used are the exponentially decaying and square waveforms, with the former easier to generate than the latter. Generation of an exponentially decaying voltage wave simply requires a DC power supply to charge a bank of capacitors connected in series with a charging resistor (Barbosa-Cánovas *et al.*, 2000). Upon activation of a trigger signal, the charge in the capacitors flows through the food held in the treatment chamber. An exponential decay voltage wave is a unidirectional voltage rising rapidly to a maximum value and then slowly decaying to zero. Therefore, food is subjected to the peak voltage for only a short period of time. Hence, exponential decay pulses have a long tail with a low electric field, during which excess heat is generated in the food without an antimicrobial effect (Zhang *et al.*, 1995a).

In order to generate a square waveform, a pulse forming network (PFN) consisting of an array of capacitors and inductors is required. Moreover, in order to ensure that the square waveforms produced in this way are as efficient as possible, it is necessary to match the resistance of the food to be treated with the impedance of the PFN. Thus, this method requires that the resistance of the food be accurately determined. Square pulses have a longer peak voltage duration than exponential pulses, and are thus not only more lethal but also more energy efficient (Zhang *et al.*, 1995a; Evrendilek and Zhang, 2005; Yu, 2009; Amiali *et al.*, 2010).

The least efficient type of waveform is the oscillatory decay pulse: this prevents the cell from being continuously exposed to the high intensity electric field for an extended period, meaning that irreversible breakdown of the cell membrane over a large area does not occur (Jeyamkondan *et al.*, 1999).

Bipolar pulses are more lethal than any monopolar pulses (square or exponential decay) because the PEF causes movement of charged molecules in the cell membrane of microorganisms, and a reversal in the orientation or

© Woodhead Publishing Limited, 2012

**Table 14.1** Different waveforms

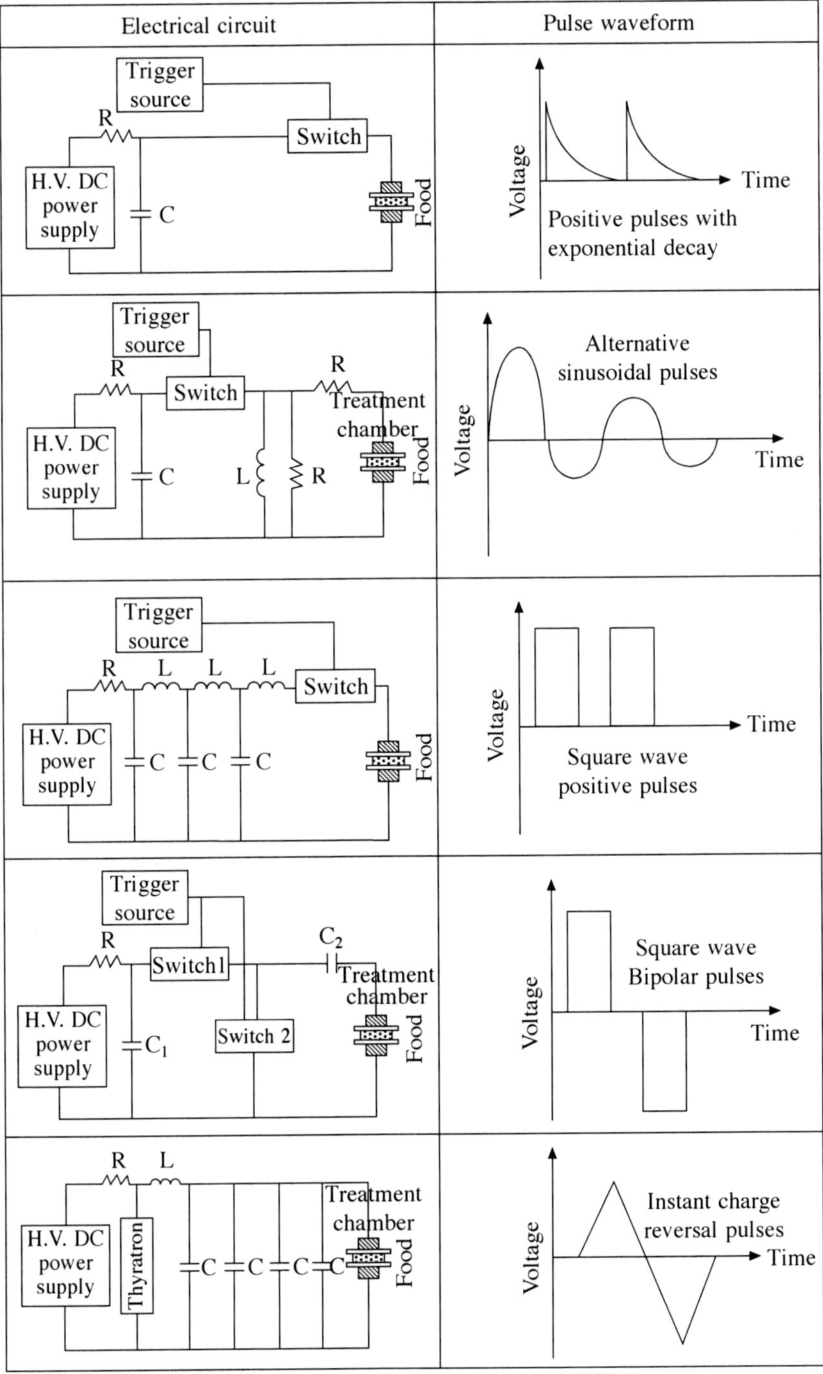


Source: Barbosa-Cánovas *et al.*, 1999.

© Woodhead Publishing Limited, 2012

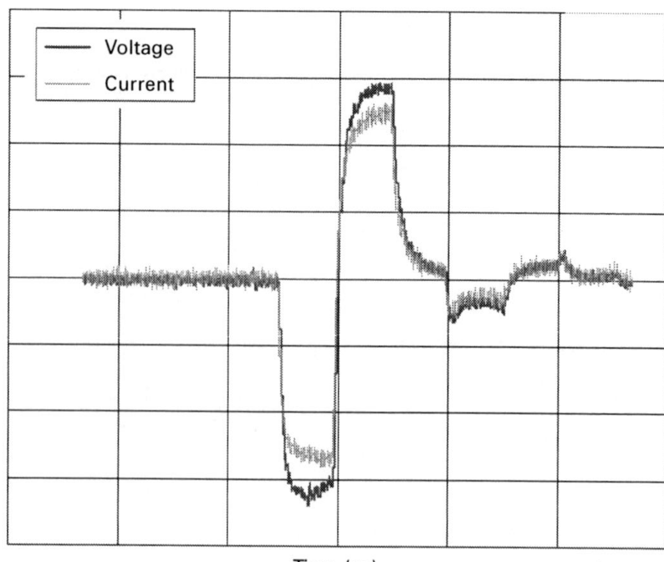


**Fig. 14.2** Instant reversal bi-phase waveform used by food engineering group (McGill University).

polarity of the electric field causes a corresponding change in the direction of charged molecules (Barbosa-Cánovas *et al.*, 1999). The alternating changes in the movement of charged molecules in bipolar pulses cause stresses in the cell membrane and enhances cell lysis. However, the change in polarity of the pulses is obtained by alternating with periods of relaxation, thus preventing the bacterial cells from being continuously subjected to a high electric field. To overcome this problem, Ho *et al.* (1995) proposed instant reversal pulses, whereby the charge is partially positive at first and partially negative immediately thereafter. The inactivation effect of an instant reversal pulse is believed to be due to a significant alternating stress on the microbial cell, causing structural fatigue. This increased rate of activity of instant-charge reversal pulses compared to other pulse types can save between 17 and 20% of total energy and equipment costs.

Amiali *et al.* (2006, 2007) have used an instant reversal square (Fig. 14.2), and after only 210 ms of treatment, up to 5 log reductions were observed of *E. coli* O157:H7 and *Salmonella* Enteritidis inoculated in liquid egg products.

### 14.2.2 Treatment chambers

The design of the treatment chamber is one of the key factors in the development of the pulsed electric field process for non-thermal pasteurization (Alkhafaji and Farid, 2007). The treatment chamber is an important and complex component of a PEF processing system. The major function of the treatment chamber

© Woodhead Publishing Limited, 2012

is to transfer the electric fields to the food. To obtain consistently high field intensity, it is necessary to minimize local field enhancements, since such regions increase the probability of electric breakdown inside the treatment chamber. An ideal chamber design ensures that all microorganisms are exposed to the same electric field intensity, number of pulses and temperature (Qin *et al.*, 1998). The treatment chamber essentially consists of two electrodes held in position by insulating material that also forms an enclosure containing the food (Qin *et al.*, 1996). Several designs of static and continuous chambers have been suggested (Barbosa-Cánovas *et al.*, 1999). The first PEF systems were static and designed to treat small volumes of product. Many have parallel electrodes of stainless steel or carbon, embedded in different kinds of materials, with spacers made of plexiglass, polysulfone, polyvinyl or nylon. The construction also requires filling and withdrawing ports and sometimes cooling jackets on the electrodes. By using different spacers, it is possible to adjust the distance between electrodes. Each treatment chamber has its own electric field distribution; it is consequently difficult to compare the differing results reported in the literature. The electric field across the two electrodes must be uniform, so the gap must be smaller than the dimension of the electrodes. The final design of the treatment chamber should be carried out in conjunction with mathematical modelling of the product's conductivity dependence on temperature (Qin *et al.*, 1995a; Zhang *et al.*, 1995a).

Continuous treatment chambers are used mostly for pilot-plant or industrial scale operations to pasteurize large volumes of liquid food. Various designs of treatment chambers have been reported in the literature, differing in chamber size, capacity and shape of the electrodes. A number of these continuous-flow treatment chambers were modified from similar static treatment chambers. It has been reported that PEF treatments were more effective when continuous treatment chambers were used, due to the greater treatment uniformity achieved in the continuous systems (Martín *et al.*, 1997; Huang and Wang, 2009).

Co-field and coaxial treatment chambers are well known and are the most widely used in PEF pasteurization processing. In the co-field design, liquid food is exposed in parallel to electrical field (EF) intensity, whereas in the coaxial chamber design, the EF intensity is perpendicular to the food (Min *et al.*, 2007; Amiali *et al.*, 2010). The main advantage of co-field cell treatment is the ability to connect a series of chambers (up to eight chambers) in parallel with electric field intensity of a commercial-scale PEF system.

The efficiency of the PEF technology is dependent upon correct design of the treatment chamber (Narsetti *et al.*, 2006; Huang and Wang, 2009), as a number of factors affect the microbial inactivation. These factors can be classified into: treatment parameters, namely electric field intensity, treatment time and temperature, pulse duration and shape; product parameters, namely electrical conductivity, density, viscosity, water activity and pH; and microbial characteristics, such as bacterial cell size and shape, growth stage of the microorganisms, and whether the bacteria is Gram-positive or Gram-

© Woodhead Publishing Limited, 2012

negative. Of these, it is the treatment parameters that can be modified through proper design of the treatment chamber (Knorr *et al.*, 1994; Hülsheger and Niemann, 1980; Huang and Wang, 2009; Ngadi *et al.*, 2009; Amiali *et al.*, 2010). It is generally the case that microbial inactivation by PEF increases when any of these factors is increased.

## 14.3 Critical factors affecting food decontamination by pulsed electric fields (PEFs)

### 14.3.1 Electric field ($E$) intensity

Electric field intensity is defined as electric potential difference ($V$) between two given points in space divided by the distance ($d$).

$$E = \frac{V}{d} \qquad [14.1]$$

Electric field strength and treatment time are the most important treatment parameters affecting the performance of microbial inactivation by PEF (Castro *et al.*, 1993). In order to achieve a lethal effect the EF must be uniform and evenly distributed through the treatment chamber. In addition, an EF of 20 kV/cm or more is usually sufficient to reduce the viability of pathogenic microorganisms from 3 to 6 log cycles (Amiali *et al.*, 2010; Yu, 2009). The fact that microbial inactivation increases with increasing EF intensity can be attributed to the high energy applied to cell suspension in a food system (Alkhafaji and Farid, 2007). Therefore, the higher the field strength at constant energy levels, the higher the lethality of the treatment, according to studies on the effect of PEF on fluid food products, which show that the killing power increases more with electric field strength than with pulse duration (Schoenbach *et al.*, 2000; Huang and Wang, 2009). Some of the studies reported in the literature have used a maximum electric field strength of 30 kV/cm at atmospheric pressure (Ngadi *et al.*, 2009; Amiali *et al.*, 2010). However, other researchers have reported that EF greater than 30 kV/cm and up to 100 kV/cm could be applied to food using a continuous treatment system (Perni *et al.*, 2007).

Bacterial spores appear to resist electric field pulse conditions that would inactivate vegetative cells. *Bacillus cereus* spores are resistant to an electric field of 30 kV/cm (Barbosa-Cánovas *et al.*, 1998). Only a 1.3 log cycle reduction was obtained after subjecting *Bacillus subtilis* spores to 30 kV/cm at a period of 0.5–3 ms (Su *et al.*, 1996). Others reported that spores of *B. cereus* showed high $E$ resistance with a less than 0.5 log cycle reduction when treatment at 25 kV/cm was applied (Cserhalmi *et al.*, 2002). However, the study by Marquez *et al.* (1997) showed a more than 5 log cycle reduction for *B. cereus* spores suspended in 0.15% NaCl solution when treated with an $E$ of 50 kV/cm. This discrepancy in the reported data may be explained

© Woodhead Publishing Limited, 2012

by the different treatment conditions used as well as the media in which the spores were suspended (Rajkovic *et al.*, 2010). In general, $E$ does not induce spore germination but, provided such germination is induced by other factors, the resulting vegetative cells become sensitive to electric pulses (Barsotti *et al.*, 1999b).

Khadre and Yousef (2002) discussed the effect of PEF on viruses, and reported that rotavirus showed high resistance to PEF treatment at maximum $E$ of 29 kV/cm, pulse duration of 3 μs, and total treatment time of 145.6 μs. PEF appears to be less effective against the protein capsids found in viruses than against the lipid-rich membranes in bacteria (Rajkovic *et al.*, 2010).

Several empirical models have been proposed to describe the relationship between inactivation rate and $E$. The first and best known model is that proposed by Hülsheger *et al.* (1981):

$$\ln(S) = -b_E \cdot (E - E_c) \qquad [14.2]$$

where ($S = N/N_0$) is the survival ratio, $E$ is the electric field intensity at given time, $E_c$ is the critical electric field obtained by extrapolating the value of $E$ for a survival ratio of one unit and $b_E$ is the regression coefficient (kV/cm).

Peleg (1995) proposed a sigmoidal model

$$S_P = \frac{100}{(1 + e^{(E - E_{c'})/a})} \qquad [14.3]$$

where $S_p$ is the percentage of surviving microorganisms, $E_{c'}$ is the critical electric field strength when the survival level is 50% (i.e. the inflection point for $S_p$) and $a$ is the parameter indicating the steepness of the survival curve around $E_c$.

$E_{c'}$ and $a$ are described by a single exponential decay model:

$$E_c = E_\infty e^{(-k_1 n)} \qquad [14.4]$$

$$a = a_0 e^{(-k_2 n)} \qquad [14.5]$$

where $E_\infty$ is a constant (kV/cm), $a_0$ is a constant (kV/cm) and $k_1$ ($\mu s^{-1}$) and $k_2$ ($\mu s^{-1}$) are kinetics constants.

At $E >> E_c$, Eq. [14.3] reduces to the following:

$$S_p = \frac{100}{1 + e^{(E/a)}} \qquad [14.6]$$

Peleg (1995) tested the model using the published data of Castro *et al.* (1993) and Ho *et al.* (1995) and found the model to be a very good fit.

### 14.3.2 Treatment time and pulse duration

As the PEF treatment time increases, the inactivation rate of microorganisms also showed a substantial increase and then gently and gradually levelled

© Woodhead Publishing Limited, 2012

out over time (Jayaram *et al.*, 1992). In some cases, tailing in the survival curves of PEF inactivation has been observed (Jayaram *et al.*, 1992; Amiali *et al.*, 2007, 2010). Higher microbial cell death was observed during the first pulses; this gradually decreased after 20 pulses of treatment (Liu *et al.*, 1997; Alkhafaji and Farid, 2007; Amiali *et al.*, 2007; Yu, 2009).

The proper treatment time is derived from the multiplication of the number of pulses and their duration. Optimum processing conditions should be established to obtain the highest inactivation rate with lowest heating effect, because long pulse duration may result in an undesirable rise in the temperature of the product.

Different models relating the survival ratio of microorganisms to treatment time have been proposed. Based on Bigelow's first kinetics model (Bigelow, 1921), PEF processes have been modelled to describe the inactivation of microorganisms and enzymes. The microbial inactivation model results in the following equation:

$$\frac{\mathrm{d}N}{\mathrm{d}t} = -kN \qquad [14.7]$$

where $N$ is the microorganism population, and $t$ is the processing time at a constant rate ($k$) depending on its size. Integration of this expression yields:

$$\log \frac{N}{N_0} = \log S(t) = -\frac{t}{D} = S(t) = e^{-kt} \qquad [14.8]$$

where $N_0$ is the initial number of microorganisms. Equation [14.8] can be rearranged as follows:

$$\log \frac{N}{N_0} = \log S(t) = -\frac{t}{D} = S(t) = e^{-kt} \qquad [14.9]$$

where $S(t)$ is the survival fraction, $t$ ($\mu s^{-1}$) is the treatment time and $D$ is the decimal reduction time ($D = 2.303/\mathrm{k}$) corresponding to the reciprocal of the first-order rate constant.

Pruit and Kamau (1993) assumed there to be two populations of microorganisms which differ in their sensitivity to PEF and proposed a non-linear kinetic model:

$$S = S_c e^{-k_1 t} + (1 - S_c) e^{-k_2 t} \qquad [14.10]$$

where $S_c$ is the critical survivor ratio in population 1 (PEF-sensitive), $(1 - S_c)$ is the ratio of survivors in population 2 (PEF-resistant), $k_1$ is the specific death rate of subpopulation 1 ($\mu s^{-1}$) and $k_2$ is the specific death rate of subpopulation 2 ($\mu s^{-1}$).

Amiali *et al.* (2004) showed that Pruit and Kamau's model adequately predicts the inactivation of *E. coli* O157:H7 population suspended in dialysed liquid eggs. The inactivation kinetics followed the exponential decay equation with two population sensitivity to PEF treatment. The same authors also

© Woodhead Publishing Limited, 2012

reported that the inactivation rate of *E. coli* O157:H7 followed the exponential decay kinetic model with some tailing effect due to the resistance of a survival fraction to PEF treatment. They proposed the following model:

$$S = S_t + (1 - S_t)e^{-kt} \quad [14.11]$$

where $S_t$ is the tailing survival ration, $(1 - S_t)$ is the ratio of survivors in population and $k$ is the first specific death rate ($\mu s^{-1}$),

Hülsheger *et al.* (1981) proposed another model of the inactivation kinetics:

$$\ln(S) = -b_t \cdot \ln\left(\frac{t}{t_c}\right) \quad [14.12]$$

where $b_t$ is the regression coefficient, $t$ is the treatment time (μs) and $t_c$ is the extrapolated value of $t$ for 100% survival.

Finally, the same authors proposed the following empirical equation for the calculation of the survival fraction:

$$\log(S) = \frac{(E_c - E)}{k} \times \log\left(\frac{t}{t_c}\right) \quad [14.13]$$

where $t_c$ is the maximum treatment time (μs) that results in an $S$ value of 1 and $k$ is a first-order kinetics constant or microorganism constant (kV/cm).

### 14.3.3 Treatment temperature

Bazhal *et al.* (2006) showed that PEF combined with electric field treatment at moderate temperature (~50–60°C) exhibited a synergistic effect on the inactivation of microorganisms. If the electric field strength remains constant, the rate of inactivation increases with the temperature. However, it is important to note that the temperature should be kept lower than pasteurization temperature.

PEF processing does cause a slight increase in the temperature of the treated food; proper cooling, high flow rates, and sufficient time between pulses are therefore essential in order to maintain the desired food temperature.

Treatment of cultures of *L. monocytogenes* with PEF at 50°C resulted in more than 4 log CFU/ml reduction in whole milk, 2% and skim milk; however, when treatment was carried out at 25°C, only 1–3 log CFU/ml reduction was observed (Reina *et al.*, 1998). This difference may be a result of the greater thermal energy delivered to the cells at higher temperatures, or may be attributable to damage induced by thermal energy in *L. monocytogenes* cells which were more susceptible to PEF treatment (Fleischman *et al.*, 2004). Similar findings have also been reported for other bacteria, including *E. coli* O157:H7 and *Salmonella* Typhimurium DT104: in these cases, where the increased sensitivity at higher temperature was only related to the increased thermal energy (Jin *et al.*, 2009; Ravishankar *et al.*, 2002).

© Woodhead Publishing Limited, 2012

Bazhal *et al.* (2006) treated liquid whole egg inoculated with *E. coli* O157:H7 using thermal and pulsed electric field (PEF) batch treatments, alone and in combination with each other. Electric field intensities in the range from 9 to 15 kV/cm were used. The threshold temperature for thermal inactivation alone was 50°C. PEF enhanced the inactivation of *E. coli* O157:H7 when the sample temperature was higher than the thermal threshold temperature. The maximum inactivation of *E. coli* O157:H7 obtained using thermal treatment alone was 2 logs at 60°C. Nevertheless, combined heat and PEF treatments resulted in up to 4 log reductions.

The temperature of the medium in which bacteria cells are suspended plays a significant role in determining membrane fluidity (Ohshima *et al.*, 1997). The lipid bilayer of cell membranes can present different phases. At low temperatures, the well-ordered gel state is observed, and the fatty side chains assume an extended conformation, nestling together with a maximum Van der Waals contact. The phospholipids are assumed to be closely packed in a rigid 'gel structure'. However, at higher temperatures, the fatty acyl groups are less ordered and the thickness of the membrane decreases by about 15%. The phospholipids are less ordered and the membrane has a 'liquid crystalline' structure. Therefore, the phase transition of the phospholipids which occurs with shifts in temperature can affect the physical stability of the cell membrane (Russell *et al.*, 2000; Stanley, 1991; Jayaram, 2000).

### 14.3.4 Treatment energy

In the PEF treatment, the energy consumed and/or used to treat liquid food products can be expressed in a number of different ways. The most commonly used are listed below.

*Energy transfer*

When using PEF, the energy delivered to a product is determined by the product's resistivity/conductivity, temperature and the characteristics of the pulse (wave shape, width peak voltage and current). The energy delivered/input in each pulse is (Barbosa-Cánovas *et al.*, 1999):

$$W = \int_0^t P(t) \cdot dt \qquad [14.14]$$

where

$$P(t) = V_p(t) \cdot I(t) \qquad [14.15]$$

or

$$W = \int_0^t V_p(t) \cdot I(t) \cdot dt \qquad [14.16]$$

where $I$ is the current (A), $P$ is power (W), $V_p$ is the pulse voltage (V), $W$ is the energy (J) and $t$ is the treatment time (s).

© Woodhead Publishing Limited, 2012

*Energy stored in the capacitors*

The energy from the high voltage power supply is stored in capacitors and is discharged through the food material to generate the necessary electrical field in the food. The capacitance, $C_0$(F), of the energy storage capacitor is given by:

$$C_0 = \frac{\tau}{R} = \frac{\tau \cdot \sigma \cdot A}{d} \qquad [14.17]$$

where $A$ is the area of electrode surface (m$^2$), $R$ is the resistance ($\Omega$), $d$ is the gap between electrodes (m), $\sigma$ is the electrical conductivity of the food (S m$^{-1}$) and $\tau$ is the pulse duration or width (s), The energy stored in a capacitor [$Q$ (J)] is given by:

$$Q = \frac{1}{2} C_0 V_c^2 \qquad [14.18]$$

where $V_c$ is the charging voltage (V). This energy can be discharged instantaneously (< 1 ms) at a very high level of power.

*Energy input density*

Depending on the kind of pulse generator, the energy density can be expressed in different ways. For example, the energy density, $Q_s$ for a generator delivering a square pulse waveform is:

$$Q_s = \frac{V \cdot I \cdot t}{v} = \frac{V \cdot I \cdot \tau \cdot n}{v} = \frac{V^2 \cdot \tau \cdot n}{R \cdot v} = n \cdot \sigma \cdot \tau \cdot E^2 \qquad [14.19]$$

where $E$ is the electric field (V/m), $n$ is the pulse number and $v$ is the treatment volume (m$^3$).

Some researchers claim that the process in its present state is not cost-effective, as a relatively high energy input gives only about a 3–4 log reduction in the population of microorganisms. Van Heeschan *et al.* (2000) treated water inoculated with *Pseudomonas fluorescens* with a 66 kV/cm electric field, 150 ns pulse duration, and pulse frequency of 215 Hz. At a maximum energy input of 430 kJ/L they obtained a reduction of more than 3 log units. However, under the same treatment conditions, an energy input of 500 kJ/L was required to reduce *Bacillus cereus* spore viability by 1 log cycle.

Ohshima *et al.* (1997) reported a 3 log reduction in *E. coli* K12 suspended in distilled water using an applied voltage of 12 kV and energy of about 200 kJ/L. Similarly, Zhang *et al.* (1994) obtained a 3 log reduction in both *E. coli* and *S. cerevisiae* using an electric field intensity of 25 kV/cm, and energy inputs of 604 and 558 J, respectively. Heinz and Knorr (2000) obtained a 4 log reduction of *Bacillus subtilis* inoculated in MacIlvain buffer using 100 pulses of a 52.8 kV/cm electric field for an energy input of only 2.1 kJ/kg. The earlier study by Heinz *et al.* (1999) also showed that under such treatment conditions temperatures did not exceed 30°C. However, they reported a

© Woodhead Publishing Limited, 2012

maximum energy input of 310 kJ/kg after 200 applied pulses. Góngora-Nieto *et al.* (2003) treated liquid whole egg by PEF using a mean electric field of 30 kV/cm and energy input of 6331 kJ/L. The treatment time was 489 μs (266 pulses). The maximum shelf life of the PEF processed liquid whole egg under refrigeration at 4°C was 20 days.

Guerrero-Beltràn *et al.* (2010) used a combination of PEF and thermal treatment to inactivate *Listeria innocua*. The maximum electric field strength of 40 kV/cm was applied at a selected number of pulses (1–30) and a temperature of 72°C for less than 10 s. About 4.3 log cycle inactivation of *L. innocua* was observed after 20 pulses. In addition, in milk treated with 40 kV/cm of electric field, a small number of pulses and at a temperature close to 55°C, a higher inactivation rate of *L. innocua* was observed, along with lower energy consumption. An energy expenditure of 244 J/mL was achieved, which can be further reduced to 44 J/mL using a thermal regeneration system.

Zhang *et al.* (1994) reported a 9 log reduction of *E. coli* after treatment at 70 kV/cm and 80 pulses of 2 μs each. The energy required to achieve this inactivation was evaluated at only 97 kJ/L. According to Ho *et al.* (1995), PEF treatment could be more energy efficient than heat treatment. In thermal processing, such as high-temperature short-time (HTST) processing, the energy required to heat 49.5 ml of water from 20°C to 71.1°C was 12 kJ, while the energy required by PEF to treat the same amount of fluid was only 10.7 kJ.

*Lactobacillus plantarum* inoculated in an orange juice-milk based beverage was treated with PEF at 40 kV/cm electric field intensity. For any given quantity of energy applied, the highest degree of inactivation was achieved with high field strength and short treatment time. A maximum energy level of 1170 kJ/L was observed; up to 60% of this energy can be saved by raising the process temperature to 55°C (Sampedro *et al.*, 2007). Craven *et al.* (2008) reported a 5 log cycle reduction of *Pseudomonas* suspended in milk using 31 kV/cm electric field intensity, thus consuming energy of 139 kJ/L in combination with a temperature of 55°C.

Walkling-Ribeiro *et al.* (2011) pasteurized skim milk using three different processing technologies (thermal pasteurization (TP), PEF and microfiltration (MF)). Using PEF, indigenous microorganisms were inactivated by an electric field of 42 kV/cm for treatment time up to 2105 μs, accounting for energy densities between 407 and 815 kJ/L, while MF was applied with a transmembrane flux of 660 L/h $m^2$ and TP at 75°C for 24s. Combined processing with MF followed by PEF (MF/PEF) produced a 4.1 (at 407 and 632 kJ/L), 4.4 (at 668 kJ/L) and 4.8 (at 815 kJ/L) log cycle reduction in count of milk microorganisms, which was comparable to that of TP. Reversed processing (PEF/MF) achieved comparable reductions of 4.9, 5.3 and 5.7 log cycle (at 407, 632 and 668 kJ/L, respectively) and higher inactivation of 7.1 log cycle (at 815 kJ/mL) in milk than for TP. With higher field strength, shorter treatment time, larger energy density, and rising temperature, the efficacy

© Woodhead Publishing Limited, 2012

of PEF/MF increased while that of MF/PEF did not. The authors concluded that the combination of PEF and MF represented a potential alternative to the non-thermal pasteurization of milk with improved quality.

The impact of pulse energy dissipation must be taken into account, as the media temperature will increase. This energy may be removed by cooling; otherwise it could lead to a rise in the temperature of the medium, depending on the system. However, data concerning the energy input required for microbial decontamination are not available from all research groups, as electric field intensity and pulse number are the main parameters reported for treatment intensity.

### 14.3.5 Effects of properties of PEF treated food

*Effect of medium pH and ions on PEF*

The pH enhances PEF treatment by causing additional stress to microorganisms, resulting in an increase in microbial inactivation. PEF microbial inactivation increased with a decrease in the ionic strength of a food material (Barbosa-Cánovas *et al.*, 1999). However, the influence of pH on bacterial inactivation by PEF is unclear. Sale and Hamilton (1967) and later Hülsherger *et al.* (1981) reported that pH had no effect on the inactivation of *E. coli*. Vega-Mercado *et al.* (1996) found that the inactivation of *E. coli* was slightly higher in a simulated milk ultrafiltrate (SMUF) at low pH (5.69) than at a higher pH (6.82). This difference was attributed to the increased number of pulses and greater electric field intensity applied at the lower pH (pH has a hurdle effect, particularly for acidic products). The reduced inactivation rate in solutions of high ionic strength can be explained by the stability of the cell membrane when exposed to a medium with several ions. The PEF treatment and ionic strength were responsible for poration and compression of the membrane cell, while the pH of the medium affected the cytoplasm when the poration was completed. These factors disturbed the homeostasis of microorganisms, thereby leading to increased inactivation (Vega-Mercado *et al.*, 1996).

Jeantet *et al.* (1999) treated *S.* Enteritidis at pH levels of 7.0, 8.0 and 9.0, and found inactivation to be greater at pH 9.0 than at pH 7.0 or 8.0. The authors believed that a pH of 9.0 caused additional stress on the cell and consequently increased its susceptibility to PEF.

Wouter *et al.* (1999) studied the influence of the pH of the treatment medium on the inactivation of *L. innocua*, *S. cerevisiae* and *Lactobacillus plantarum* under continuous PEF treatment, observing that lower pH values resulted in increased inactivation. For example, a PEF treatment that inactivated 0.6 log cycles of *L. innocua* at pH 6 inactivated more than 6 log cycles at pH 4. However, Álvarez *et al.* (2000) found that *S. senftenberg* showed greater resistance to PEF at an acidic pH than at neutral pH, when studying the inactivation of this microorganism in different buffers with the same conductivity (2 μS/cm). Since the pH influences the microbial inactivation

© Woodhead Publishing Limited, 2012

by PEF, acid foods such as juices are good candidates for processing with this technology.

Ions dissolved in treatment media have also been found to produce different effects on inactivation patterns. Cations such as $Na^{2+}$, $K^+$, $Ca^{2+}$ and $Mg^{2+}$ reportedly play a role in the integrity of bacterial cell membranes. Hülsherger *et al.* (1981) reported that the presence of $Ca^{2+}$ and $Mg^{2+}$ induced a protective mechanism against PEF treatment in cell membranes.

*Conductivity and medium ionic strength*

The ionic strength of the food product plays an important role in microbial inactivation. The electric conductivity of a medium is defined as the ability to conduct electric current (Halden *et al.*, 1990; Palaniappan and Sastry, 1991). It is an important variable that determines the extent of biological changes, such as electropermeabilization, electrofusion, motility and microbial inactivation produced during PEF treatment (Barbosa-Cánovas *et al.*, 1999). The conductivity (S/m), $\sigma$, is given by:

$$\sigma = \frac{d}{R \times A} = \frac{1}{\rho} \qquad [14.20]$$

where $R$ is the resistance of the food ($\Omega$), $A$ is the surface area ($m^2$), $d$ is the gap between the electrodes (m) and $\rho$ is the resistivity ($\Omega$-m).

An increase in ionic strength results in an increase in the electron mobility through a solution and a decrease in the inactivation rate by PEF. The reduced inactivation rate in high ionic strength solutions can be explained by the stability of the cell membrane when exposed to a medium containing several ions (Tsong, 1990). In general, foods with high electrical conductivities are difficult to treat since they generate low peak electric fields across their treatment chambers due to the high current that is typically generated during PEF treatment of such products (Barbosa-Cánovas *et al.*, 1999).

It is therefore important to lower the electrical conductivity in order to obtain greater microbial inactivation for the same applied electric field. An increase in the difference between the conductivities of the medium and the microbial cytoplasm weakens the membrane structure due to an increased flow of ionic substances across the membrane (Barbosa-Cánovas *et al.*, 1999; Barsotti *et al.*, 1999a). The antimicrobial effect of PEF is thus inversely proportional to the ionic strength of the suspension material (Hülsherger *et al.*, 1981).

A product with high resistivity (low conductivity) can present a higher degree of inactivation. For example, the PEF inactivation of *L. brevis* cells in a suspension of phosphate buffer solutions of different conductivities (0.17–2.23 μS/cm) was investigated by Jayaram *et al.* (1992). The maximum reduction in the number of viable microorganisms ($N/N_0 \approx 10^{-7}$) was obtained in the liquid possessing the lowest conductivity (0.17 μS/cm). Vega-Mercado *et al.* (1996) treated *E. coli* in media with potassium chloride (KCl) concentrations varying from 22.8 mM to 168.0 mM. The inactivation

© Woodhead Publishing Limited, 2012

of *E. coli* treated with 0–8 pulses of 2 μs each, at an electric field intensity of 40 kV/cm, increased as the ionic strength of the KCl solution decreased. A difference of 2.5 logs was obtained between the 168 mM and 22.8 mM solutions. For the same applied electric field, Sensoy *et al.* (1997) reported an increase in the inactivation rate of *Salmonella dublin* with increasing conductivity of the medium. In contrast, Wouter *et al.* (1999), employing a field intensity of 26 kV/cm, showed that a higher electrical conductivity (7.9 μS/cm vs. 2.7 or 5.1 μS/cm) of the medium resulted in a decrease in the inactivation rate of *L. innocua*. At the outlet of the treatment chamber, with the medium temperature at 50°C at the outlet, and the with electrical conductivity ($\sigma$) at 7.9 μS/cm, a 2 log reduction was observed, whereas with an outlet temperature of 40°C and $\sigma$ of 2.7 μS/cm, the inactivation rate was higher (4.5 log cycles). However, Ho *et al.* (1995) stated that a variation in $\sigma$ between 0.65 μS/cm (0% NaCl w/v) and 10.2 μS/cm (0.5% NaCl w/v), had no significant effect on microbial inhibition. Álvarez *et al.* (2000) found contradictory results after suspending *S. senftenberg* in citrate-phosphate buffer (pH 7.0) diluted to different concentrations to vary the electrical conductivity. The authors obtained a lower applied electric field and inactivation rates when the electrical conductivity was high. The input voltage was increased to obtain the same applied electric field for a different electrical conductivity level; however, this did not affect the rate of microbial inactivation.

Depending on the specific PEF equipment and high voltage switch used, special care must be taken to remain within the operating limits specified by the manufacturer. When the electrical conductivity of a food sample is too high or too low with respect to the resistivity limits permitted by the electrical circuit, electrical components of the apparatus may become damaged. For example, the high intensity current that crosses a low resistivity food may overheat some switches unless a current limiting device is employed in the circuit. In contrast, a high resistivity food maintains a high voltage across the switch, and it may be necessary to place a shunt (a device used for bypassing a specific amount of current around a precision electronic instrument) between the electrodes in parallel with the food sample to protect the equipment (Barsotti *et al.*, 1999a,b).

### 14.3.6 Effect of microbial characteristics on PEF

*Size and shape of the cell*

Different types of microorganisms vary in their sensitivity to PEF due to their different membrane and cell wall constructions. The size and shape of a microorganism play an important role in its inactivation. Microbial cells with large diameters, such as yeast, are killed at lower electric field strengths than cells of smaller diameter (typical cells) (Grahl and Märkl, 1996). Zimmermann *et al.* (1974, 1976) reported a decrease in critical breakdown potential when the cell volume increases. This was also observed by Qin *et al.* (1998) who studied the inactivation of *E. coli*, *S. aureus* and *S. cerevisiae*.

© Woodhead Publishing Limited, 2012

Yeasts, whose cells are larger than those of bacteria, exhibit a lower critical external electric field for breakdown and were the most sensitive to PEF treatment. Conversely, *S. aureus* is smaller than *E. coli* and the induced voltage across the former bacteria membrane was smaller under similar PEF treatment conditions. Grahl and Märkl (1996) defined electric field values of 13.5 and 4.7 kV/cm for *E. coli* and *S. cerevisiae*, respectively. However, in some cases it appears that protozoan cells, which are larger than those of bacteria or yeasts, exhibited greater resistance to PEF, probably due to their ability to form cysts and thick-walled oocysts (Haas and Aturaliye, 1999).

*Microbial cell wall and cell membrane*

Gram-negative bacteria are, in general, more sensitive to PEF than Gram-positive bacteria, since the cell wall of the latter is much thicker and more rigid. When the membrane is punctured by the PEF process, the cell starts to leak small compounds and water begins to flush into the cell to equalize the osmotic imbalance across the membrane (Barbosa-Cánovas *et al.*, 1999). A more rigid cell wall might help to resist the osmotic forces thereby preventing cell lysis. Pothakamury *et al.* (1995) treated *E. coli* (Gram-negative) and *S. aureus* (Gram-positive) in their early stages of growth using 50 exponentially decaying pulses with durations of 200 and 300 μs at electric field strength of 16 kV/cm. While *E. coli* were inactivated by 4.5 log cycles, *S. aureus* were only inactivated by 2.5 log cycles. When the field intensity was low, the difference in inactivation rates between the two bacteria was more obvious than at higher applied electric field intensity. Other researchers have reported similar observations (Vega-Mercado *et al.*, 1996; Mazurek *et al.*, 1995).

Hülsherger *et al.* (1983) showed that bacteria and yeast do not exhibit similar membrane characteristics in terms of cell inactivation in pulsed electric fields. Assuming that the electric field induced membrane processes are responsible for cell inactivation, yeast cell membranes exhibit greater stability than bacterial cell membranes. These results contradict those of Sale and Hamilton (1967) who concluded that yeast cells were more sensitive to electric fields than bacterial cells. This might be attributable to the fact that the electric field strength applied by Sale and Hamilton (1967) (25 kV/cm) was greater than that applied by Hülsherger *et al.* (1983) (20 kV/cm). In addition, the pulse treatment of yeast resulted in greater inactivation when a larger number of pulses (30 pulses) or a greater electric field (25 kV/cm) was applied.

*Growth phase*

Microbial growth in the logarithmic phase is characterized by a higher number of cells undergoing division, during which the cell membrane is more susceptible to the applied electric field (Hülsheger *et al.*, 1983; Qin *et al.*, 1996; Barbosa-Cánovas *et al.*, 1999). Gásková *et al.* (1996) reported that the inactivation efficiency of PEF in the logarithmic phase was some 30% greater than in the stationary phase. This could be attributable to the

© Woodhead Publishing Limited, 2012

fact that the area between two dividing cells is narrower during growth and cell division, making them more susceptible to an applied electric field. This view is supported by the work of Pothakamury *et al.* (1996). Who observed the following effects of PEF treatment with 4 pulses, 2 μs, and 36 kV/cm: cells in the logarithmic phase were more sensitive and their viability reduced by more than 2 logs; cells in the stationary phase were reduced by only 1.4 logs; finally, cells in the lag phase were reduced by less than 0.5 logs. Álvarez *et al.* (2000) also studied the influence of cell age on resistance to PEF by treating *Salmonella senftenberg* at different growth stages with 200 pulses of 2 μs each at an electric field intensity of 19 kV/cm. They found that cells in the stationary phase and at the start of the logarithmic phase were more resistant to PEF than cells in the logarithmic phase. However, the influence of cell age was relatively minor, since the maximum difference in the number of survivors between the most and the least resistant cells was only a 1.5 log reduction.

Unfortunately, PEF has only been proven to be effective on vegetative cells (Amiali *et al.*, 2010; Ngadi *et al.*, 2009). Spores of sporulating bacteria exhibit a greater resistance to PEF treatment because of their dehydrated cytoplasm, which reduces their electrical conductivity. This means that it is difficult to achieve a sufficiently high voltage gradient to breach the surrounding membrane (Gould, 2000). Bacterial spores can only be damaged at the time of their germination and outgrowth. According to Grahl and Märkl (1996), the inactivation of *Clostridium tyrobutyrium* spores by PEF was negligible. No lethal effects on *Bacillus cereus* spores were detected after treatment with 30 exponentially decaying pulses at an electric field intensity of 22.4 kV/cm. Therefore, it is impossible to achieve commercial sterility of foods through PEF technology (Ngadi *et al.*, 2009). However, the authors reported that PEF did have an effect on *Bacillus* spores at field strengths exceeding 35 kV/cm. By using 30 exponential decaying pulses of 2 μs each, at an electric field intensity of 50 kV/cm, *B. cereus* was inactivated after up to 3.4 log cycles. Furthermore, *B. subtilis* activity was reduced by 5.0 logs after treatment with 50 pulses under similar treatment parameters, confirming the effect of higher field strengths. In addition, the applied number of pulses, temperature and treatment time would increase the efficiency of PEF treatment.

Yeast ascospores and bacterial spores are more resistant to adverse physical and chemical agents than vegetative cells. Grahl and Märkl (1996) found that *Byssochlamys nivea* ascospores were resistant to PEF treatment. However, ascospores are less resistant than bacterial spores, due to the absence of cortex (Raso *et al.*, 1998). These authors treated both vegetative cells and ascospores of *Zygosaccharomyces bailii*. Ascospores were more resistant than vegetative cells: with the former, the reduction was between 3.5 and 4 logs, while the latter were reduced by 4.5 to 5 logs. Furthermore, the yeast ascospore wall did not protect the ascospore population against the PEF treatment, as the bacterial spore cortex would do. Raso *et al.* (1998) reported that the inactivation of *Neosartoria fischeri* ascospores was

© Woodhead Publishing Limited, 2012

negligible, even after 40 pulses of 51 kV/cm electric field intensity. As reported above, cells of larger diameters were killed at lower electric field intensity than cells of smaller diameter. Owing to the comparatively greater diameter of ascospores (5.0–8.0 μm) one might expect their electrical conductivity to be much lower than that of the smaller bacteria, but this is not the case. The resistance of ascospores to PEF treatments is attributable to their structure, which includes an extremely thick intermediate space between the cell wall and the cytoplasmic membrane of the ascospores. Within this intermediate space another cell layer can be detected. This thick space in the cell wall of ascospores is considered a possible factor in their high heat resistance to PEF treatments.

*Effect of microbial population*

The efficiency of PEF on microbial inactivation is dependent on the microbial concentration in the food to be treated (Barbosa-Cánovas *et al.*, 1999). Different initial populations of *S. cerevisiae* treated with a 25 kV/cm electric field and a pulse duration of 25 μs indicated that the initial cell population was inversely correlated to the survival fraction under constant treatment conditions (Zhang *et al.*, 1994). This is in agreement with Jeantet *et al.* (1999), who reported that the greater the population, the lower the antimicrobial effect of PEF. However, the fact that high population decreased the lethal effect of PEF is not fully understood. Jayaram *et al.* (1992) proposed an explanation based on a higher transmembrane potential developed across clusters of cells rather than across an individual cell for the same electric field strength. However, for all preservation methods, the larger the number of microbial cells, the less effective the given treatment (Jay, 1992). Vega-Marcado *et al.* (1996) showed that a mixed population of microorganisms decreased the effectiveness of the PEF: 30 exponentially decaying pulses of 2 μs in duration at a field intensity of 30 kV/cm were applied to four different batches of pea soup. The first batch was inoculated with *E. coli* only, the second with *B. subtilis* only, the third with *E. coli* added to *B. subtilis*, and the last with *B. subtilis* added to *E. coli*. The treatment gave inactivations of 6.5, 5.0, 4.0 and 2.0 log units, for the first, second, third and fourth batches, respectively. However, the inactivation of *E. coli* in simulated milk ultrafiltrate (SMUF) after PEF treatment (70 kV/cm, 16 × 2 μs pulses) was not affected when the concentration of this bacterium was reduced by 5 log unit (Zhang *et al.*, 1995b). When the cell concentration was increased, a slightly lower inactivation of *S. cerevisiae* inoculated in apple juice occurred after a single 25 μs pulse under an applied electric field of 25 kV/cm (Qin *et al.*, 1995b). This was attributed to the formation of clusters of yeast cells and/or possibly concealed microorganisms in the regions of low electric field.

© Woodhead Publishing Limited, 2012

## 14.4 Mode of microbial inactivation in foods by pulsed electric fields (PEFs)

Although a complete understanding of the exact mechanism by which PEF inactivates microorganisms has not yet been achieved, many of the studies performed have drawn a similar conclusion, namely that the perturbation of the cell membrane and loss of membrane permeability play a key role in the inactivation process (Heinz *et al.*, 2001). It is also possible that changes in the transport of ions and in the structural arrangement of the microbial enzymes could be caused by the application of PEF to microbial cells (Góngora-Nieto *et al.*, 2002). It has been suggested that high-intensity pulsed electric fields may have additional effects, such as DNA damage and generation of toxic compounds; however, some later studies have rejected these hypotheses (Barbosa-Cánovas and Sepulveda, 2005). Although no definitive explanation of the antimicrobial mechanism(s) of PEF is currently available, it would appear that the integrity of the membrane is not the only relevant factor (Rajkovic *et al.*, 2010). However, while other explanations have been put forward, such as electromechanical compression and osmotic imbalance (Ngadi *et al.*, 2009), transmembrane potential remains the most frequently cited.

### 14.4.1 Transmembrane potential

The cell membrane acts as an insulator to the cytoplasm with electrical conductivity in the range of six to eight orders of magnitude greater than that of the membrane (Chen and Lee, 1994). Thus, the cell membrane can be regarded as a capacitor filled with a material with a low dielectric constant, i.e., $\varepsilon \approx 2$. When a certain electric field is applied to the cell, the ions inside the cell move along the field until the free charges are accumulated at both membrane surfaces. This accumulation of charge increases the electromechanical stress or transmembrane potential (VR), such that it becomes greater than the applied electric field (Zimmermann, 1986). When VR exceeds a critical value (~1 V), the resulting repulsion between charge-carrying molecules induces the formation of pores and subsequent weakening of the membrane structure, resulting in cell damage.

When cells are exposed to an electric field, the induced membrane potential, $V_R$, for spherical cells surrounded by a non-conducting membrane, is given by (Zimmermann, 1986; Hülsherger *et al.*, 1983):

$$V_R = f \cdot E_R \cdot r \cdot \cos(\theta) \quad [14.21]$$

where $f$ is a shape factor (1.5 for spheres), $r$ is the cell radius (μm), $E_R$ is the rupture electric field (V m$^{-1}$) and $\theta$ is the angle between the vector of the electric field and the vector of the cell radius $r$ at the relevant point on the membrane.

For given PEF treatment conditions, the induced voltage across the cell

© Woodhead Publishing Limited, 2012

membrane is proportional to the geometric size of the cell under investigation. When the cell is assumed to be a cylinder with two hemispheres at each end, then the shape factor *f* is given by (Zimmermann *et al.*, 1974):

$$f = \frac{L}{L - 0.33d} \qquad [14.22]$$

where *L* is the length of the cylindrical cell (μm) and *d* is the diameter of the cell (μm).

As expressed in Eqs [14.21] and [14.22], the size and shape of the cell affect its resistance to PEF treatment. This implies that the induced membrane potential depends upon cell size (Liu *et al.*, 1990).

## 14.5 Application of food treatment by pulsed electric fields (PEFs)

PEF has been proven to preserve the quality of different food products, and has been specifically used to improve the shelf life of milk, liquid egg products, and fruit juices.

### 14.5.1 Milk processing

Because milk is one of the most important foods in human nutrition and is susceptible to both spoilage and pathogenic microorganisms, pasteurization is mandatory. The inactivation of different pathogens in milk through the application of PEF treatment has been demonstrated, including *E. coli* O157:H7 (Martín *et al.*, 1997; Evrendilek and Zhang, 2005; Rivas *et al.*, 2006; Shin *et al.*, 2007), *Salmonella* dublin (Sensoy *et al.*, 1997), *Staphylococcus aureus* (Sobrino-López and Martín-Belloso, 2006), *Listeria innocua* (Fernandez-Molina *et al.*, 2006; Bermudez-Aguirre *et al.*, 2009; Guerrero-Beltràn *et al.*, 2010) and *Bacillus cereus* (Pina-Pérez *et al.*, 2009).

Evrendilek and Zhang (2005) studied the effect of pulse polarity and pulse delaying time on *E. coli* O157:H7 in skim milk ($\sigma$: 6.2 ± 3.4 μS/cm; pH 6.7 ± 0.65). Skim milk was processed at a pulse relaxation time of 20 μs, pulse frequency of 700 Hz, and outlet temperature of 30°C in a continuous flow system that discharged square wave pulses. They reported reductions of 1.27 and 1.88 log cycles in skim milk using monopolar and bipolar pulses, respectively. Generally, bipolar pulses are slightly more efficient than monopolar pulses at destroying microorganisms (Qin *et al.*, 1994; Ho *et al.*, 1995; Elez-Martínez, 2004, 2005).

Dutreux *et al.* (2000) applied PEF to skim milk ($\sigma$: 4.8 μS/cm; pH: 6.8) inoculated with *E. coli* ATCC 11775. The authors found a reduction of 4.0 log using 63 pulses of 2.5 μs duration and an electric field intensity of 41 kV/cm. The experiments were carried out using different media to test their

© Woodhead Publishing Limited, 2012

effects on microbial reduction when PEF was applied. The authors concluded that the composition of the medium did not seem to affect the inactivation of *E. coli* ATCC 11775 by PEF. However, Grahl and Märkl (1996), Martín *et al.* (1997) and Martín-Belloso *et al.* (1997) found that the presence of fats and proteins in UHT milk (1.5 and 3.5% fat) and skim milk limited the effectiveness of the PEF treatment. Other authors subjected UHT milk 1.5 or 3.5% fat inoculated with *E. coli* ATCC 11229 to PEF (Grahl and Märkl, 1996). They observed that UHT milk with 1.5% fat had a lower inactivation constant than UHT milk with 3.5% fat. They suggest that this indicates that the fat particles of milk seemed to protect bacteria against the induced electric field.

On the other hand, Martín-Belloso *et al.* (1997) and Martín *et al.* (1997) indicated that proteins decrease the lethal effect of PEF on microorganisms by absorbing free radicals and ions, which are active in the cell breakdown. Moreover, the inactivation of bacteria by PEF is a function of the resistivity of the solution, which is inversely proportional to ionic strength (Martín-Belloso *et al.*, 1997). Thus, microbial inactivation is more difficult in real foods than buffer solutions and model foods, due to the complex composition of food.

Whole milk product inoculated with *E. coli*, *Pseudomonas fluorescens* and *Bacillus stearothermophilus* were exposed to an electric field with strength of 30–60 kV/cm, with pulse duration of 1 μs pulse duration and treatment time of 210 μs in continuous PEF. A maximum of 8 log cycles reduction were obtained for *E. coli* and *P. fluorescens* with treatment time of 210 μs, pulse intensity of 60kV/cm and temperature of 50°C. However, only a 3 log cycles reduction was obtained for *B. stearothermophilus* (Shin *et al.*, 2007).

Reina *et al.* (1998) studied the effect of the PEF treatment of *L. monocytogenes* inoculated in milk samples with different fat contents (whole milk, 2% milk, and skim milk). They obtained 3.0 log reductions in all cases, when using 400 square wave pulses with a width of 1.5 μs pulse duration in a continuous flow treatment. No significant effects caused by the composition (fat content) of the medium were observed when the PEF treatment was applied. Picart *et al.* (2002) studied the influence of the fat content and pulse frequency on *L. innocua* in UHT sterilized whole milk (3.6% fat), skim milk (0% fat), and sterilized liquid dairy cream (20% fat) by PEF treatment. The authors reported a higher inactivation rate at 100 Hz than at 1.1 Hz for whole milk (up to 2.0 log cycles) and skim milk (up to 1.25 log cycles); however, they did not find a significant difference in the inactivation rate of microorganisms in dairy cream, since a 2.0 log reduction was achieved at both 100 Hz and 1.1 Hz.

Floury *et al.* (2006) inactivated *Salmonella* Enteritidis in skim milk by using a combined treatment of PEF and heat. They used an electric field of 47 kV/cm (500 ns/60 Hz) and a treatment temperature of 62°C for 19 s. The total inactivation was 2.3 ± 0.4 log cycles reduction. The authors suggested

© Woodhead Publishing Limited, 2012

that the combination of PEF and heating was more effective than each on its own and the lethality was additive rather than synergistic.

Rivas *et al.* (2006) treated a mixed beverage of milk and orange juice inoculated with *E. coli* CECT 516 (ATCC 8739). Bipolar square pulses with a width of 2.5, an electric field of 15–40 kV/cm and a treatment time of up to 700 μs were applied. A maximum of 3.83 log cycle reductions was achieved at only 15 kV/cm and 700 μs. The authors modelled the results using the Weibull distribution function.

A mixed beverage of whole egg and skim milk inoculated with *Bacillus cereus* was treated by PEF in combination with 12% antimicrobial cocoanOX powder (Pina-Pérez *et al.*, 2009). The maximum reduction rate was around 3 log cycles at 40kV/cm, 360 μs, 20°C. This indicates that the combination of PEF with an antimicrobial agent has a synergistic effect on the bacterial cell under study.

Reina *et al.* (1998) found that when the treatment temperature was raised from 25 to 50°C in whole milk, the inactivation of *L. monocytogenes* reached about 4.0 log reductions when an electric field of 30 kV/cm was applied for 600 μs. Fleischman *et al.* (2004) observed that increasing the temperature from 35 to 55°C in skim milk with gellan gum inoculated with *L. monocytogenes* resulted in a reduction of 1.0 to 4.5 log cycles respectively, using 10 pulses of 3.25 μs pulse duration. However, the authors demonstrated that temperatures above 50°C were sufficient to inactivate *L. monocytogenes* up to 4.0 log cycles, since it is sensitive at a temperature up to 45°C. Thus, this reduction was due to heat and not due to the PEF treatment. Finally, it was concluded that *L. monocytogenes* seemed not to be easily destroyed by PEF as a single treatment but that a combination of the treatment with heat would be adequate for the microbial inactivation.

Fernandez-Molina (2001) reported a reduction of 2.7 log cycles of *L. innocua* in raw skim milk when an electric field of 50 kV/cm was applied for 60 μs pulse duration at 4 Hz, and with an outlet temperature of 28°C, in a continuous flow system with exponential decay pulses of 2 μs width.

Pina-Pérez *et al.* (2007) obtained only a 1.2 log cycle reduction when *Enterobacter sakazakii* inoculated in infant formula milk (IFM) was exposed to PEF for 360 μs (2.5 μs pulse duration) at 40 kV/cm. The authors suggested that PEF has good potential for use in hospitals to achieve safe reconstituted infant formula before storage at refrigerated temperatures.

Yu (2009) treated whole milk (3.25%) inoculated with *E. coli* O157:H7 and *S.* Enteritidis using a biphasic instant reversal PEF waveform with a pulse duration of 2 μs pulse duration, applying a maximum of 120 pulses. The whole milk was inoculated with $10^7$ CFU/mL of either *E. coli* O157:H7 or *S.* Enteritidis, and was then treated in a continuous flow process at 20, 35, 45 and 50°C in combination with pulsed electric field intensities of 20 and 30 kV/cm. The maximum reduction of *E. coli* O157:H7 and *S.* Enteritidis was 4.1 and 5.2 logs at 30 kV/cm and 50°C, respectively (Figs 14.3 and 14.4). The inactivation rate constant increased from 0.016 to 0.104 $\mu s^{-1}$

© Woodhead Publishing Limited, 2012

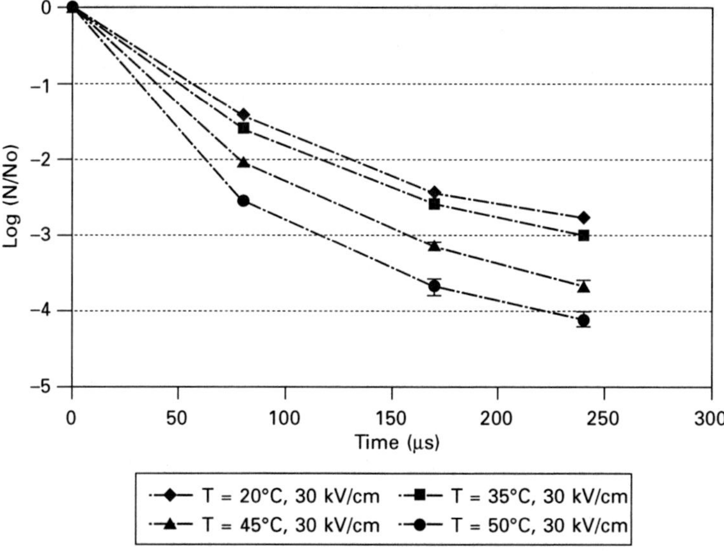


**Fig. 14.3** Survival fraction of *E. coli* O157:H7 in whole milk as a function of PEF treatment time and temperature at 30 kV/cm of electric field intensity.

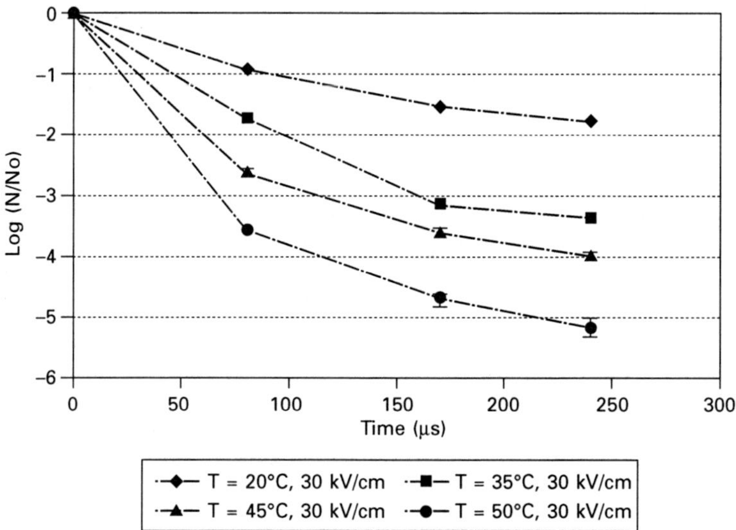


**Fig. 14.4** Survival fraction of *S.* Enteritidis O157:H7 in whole milk as a function of PEF treatment time and temperature at 30 kV/cm of electric field intensity.

for *S.* Enteritidis and from 0.035 to 0.075 $\mu s^{-1}$ for *E. coli* O157:H7 as the processing temperature increased from 20 to 50°C. The authors observed that *E. coli* O157:H7 seemed more resistant to the combination of PEF and heat than *S.* Enteritidis under the same treatment conditions.

© Woodhead Publishing Limited, 2012

PEF technology has demonstrated effectiveness in inactivating pathogenic bacteria and reducing the number of spoilage microorganisms in milk (Sobrino-López and Martín-Belloso, 2010; Bendicho *et al.*, 2002; Odriozola-Serrano *et al.*, 2006; Craven *et al.*, 2008, Yu, 2009). It may also be feasible to extend the shelf life of milk with PEF, retarding the growth of mesophilic bacteria by up to 80 days with only a slight impact on quality (Sepúlveda *et al.*, 2005).

Despite the potential of the new techniques for food preservation, most studies have demonstrated their application and effects on the maintenance of the chemical and nutritional characteristics of the final product, while studies focusing on stability and microbiological safety are still rare.

### 14.5.2 Liquid egg processing

Liquid egg products are widely used in a large number of food industries and establishments. Of the total number of eggs consumed, more than 30% are in the form of egg derivative products. Various studies have examined the PEF inactivation of different target microorganisms in eggs such as *S.* Enteritidis (Hermawan *et al.*, 2004; Jeantet *et al.*, 1999, 2004; Amiali *et al.*, 2004, 2006, 2007; Monfort *et al.*, 2011), *L. innocua* (Calderón-Miranda *et al.*, 1999), *E. coli* (Martín-Belloso *et al.*, 1997; Amiali *et al.*, 2004, 2005, 2006, 2007; Bazhal *et al.*, 2006) and *P. fluorescens* (Góngora-Nieto *et al.*, 1999).

The first work on the treatment of liquid egg product inoculated by *E. coli* ATCC 11229 was reported by Martín-Belloso *et al.* (1997). By maintaining the treatment temperature at 37°C, they exposed liquid whole egg (LWE) to an electric field intensity of 26 kV/cm with 2 and 4 μs pulse duration, 1.25 and 2.5 Hz pulse rate, 100 pulses/unit volume, and continuous and stepwise treatment conditions. Interesting results were obtained (up to 6 log cycle reductions) and no protein coagulation was observed. Calderon-Miranda *et al.* (1999) carried out a later study on the inactivation of *L. innocua* in LWE. The inactivation rate was up to 5.5 log cycle reductions using 50 kV/cm electrical field strength and a total of 32 pulses in combination with an antimicrobial agent (nisin: 100 IU/mL). Góngora-Nieto *et al.* (1999) inactivated *P. fluorescens* suspended in LWE using a combined treatment (PEF and citric acid) where they obtained 1 log cycle reductions. Hermawan *et al.* (2004) inactivated *S.* Enteritidis in liquid egg white using an electric field strength of 25 kV/cm, pulse frequency of 200 Hz, pulse duration of 2.12 μs, and total treatment time of 250 μs. However, the inactivation rate was only 1 log reduction.

Amiali *et al.* (2004) reported the inactivation of *E. coli* O157:H7 suspended in three different egg products (LWE, liquid egg white (LEW) and liquid egg yolk (LEY)). The total treatment for both products was 10000 μs (maximum of 500 pulses of 200 μs pulse duration = 10 000 μs total treatment), the electric field strength was 15 kV/cm and the pulse frequency was 1 Hz pulse

© Woodhead Publishing Limited, 2012

duration. The results were 3.5, 2.9 and 1 log cycle reductions for LWE, LEY and LEW, respectively. In continuous PEF treatment, the same authors inactivated *E. coli* O157:H7 and *S.* Enteritidis inoculated in LWE, LEY and LEW by using PEF in combination with moderate heat (maximum of 40°C). The treatment conditions were as follows: electric field of 30 and 40 kV/cm, biphasic instant reversal waveform with pulse duration of 2 μs, pulse frequency of 1 Hz and total treatment time of 105 μs. For *E. coli* O157:H7, the authors obtained up to 5, 3.6 and 2.9 log cycle reductions for LEY, LWE and LEW, respectively. For *S.* Enteritidis, the results were 5, 3.7 and 3.6 log cycle reductions for LEY, LWE and LEW, respectively. The authors observed that *E. coli* O157:H7 was more resistant than *S.* Enteritidis in LEW and LWE, whereas in LEY the opposite is true (Amiali *et al.*, 2006, 2007).

Bazhal *et al.* (2006) observed 4 log cycle reductions of *E. coli* O157:H7 using combined treatment (PEF and heat). They used a treatment temperature of 60°C and an electric field intensity of up to 15 kV/cm (Fig. 14.5). However, they used a static treatment chamber. Wesierska and Trziszka (2007) evaluated the inactivation of several microorganisms by PEF treatment including *Escherichia coli*, *Acinetobacter lwoffii*, *Citrobacter freundii*, *Serratia liquefaciens*, *Pseudomonas aeruginosa*, *Staphylococcus gallinarum*, *Staphylococcus xylosus* and *Oligella* sp., which were isolated from the surface of egg shells. PEF treatment was carried out using an electric field intensity of 25 kV/cm maximum and 200 pulses. At the end of the process the number of cells was reduced in population by more than 3–5 log cycles, depending on the bacteria. The authors concluded that the death of microorganisms was not due to the heat generated during the treatment (47°C) but due only to the effect of electric field intensity.

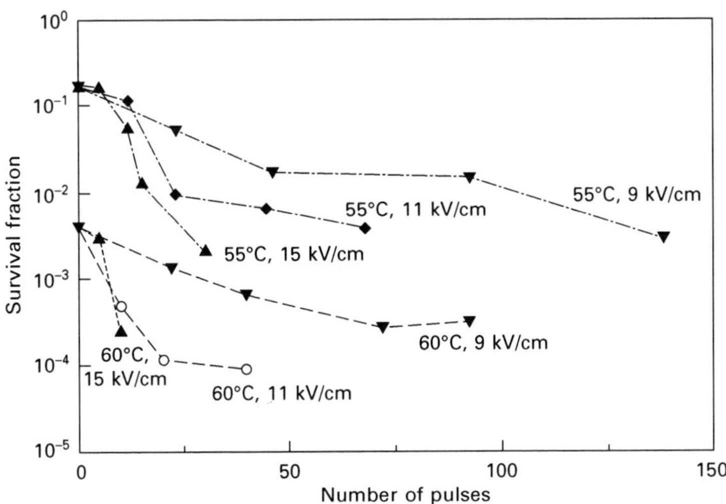


**Fig. 14.5** Viability of *E. coli* cells after combined thermal and PEF treatments at different temperatures and EF treatments. Thermal treatment was 4 minutes.

© Woodhead Publishing Limited, 2012

Jin *et al.* (2009) studied the effect of PEF in combination with pH and temperature on the inactivation of *S.* Typhymurium DT104 suspended in LWE. The pH was adjusted to 6.6, 7.2 or 8.2 at 15, 25, 30 and 40°C treatment temperatures. The PEF field strength, pulse duration and total treatment time were 25 kV/cm, 2.1 μs and 250 μs, respectively. The combination of PEF with a temperature of 55°C achieved 3 log cycle reductions and was comparable to heat treatment at 60° for 3.5 min.

Monfort *et al.* (2010a) pasteurized LWE suspended with *S.* Thyphimurum and *Staphylococcus aureus*. A maximum inactivation of 4 and 3 log cycle reductions for *S.* Thyphimurum and *S. aureus* were achieved with a treatment of 45 kV/cm, 30 μs and 419 kJ/kg and 40 kV/cm, 15 μs and 166 kJ/kg, respectively. The authors indicated that PEF treatment must be used in combination with other technologies in order to be a viable alternative to heat pasteurization of LWE. In addition, Monfort *et al.* (2010b) inactivated *S.* Enteritidis using an electric field ranging in intensity from 20 to 45 kV/cm. The level of inactivation was dependent only on the specific energy applied (106, 272 and 472 kJ/kg for 1.2 and 3 log reduction, respectively). These authors also noted that PEF technology alone cannot guarantee the safety of LWE based on US and European regulations. Furthermore, a recent study by Monfort *et al.* (2011) evaluated the effect of PEF in combination with both a temperature of 55°C and additives (10 mM EDTA or 2% triethyl citrate) on the inactivation of *S.* Enteritidis in LWE. A synergistic reduction in the strain was observed when LWE samples containing additives were treated with PEF (25 kV/cm, 100 and 200 kJ/kg), heat, or with PEF followed by heat. A PEF treatment of 25 kV/cm and 200 kJ/kg followed by heat treatment (55°C for 2 min), combined with 2% triethyl citrate, reduced the population of *S.* Enteritidis by more than 8 log cycles, with a minimal impact on its soluble protein content.

### 14.5.3 Fruit juice processing

The pasteurization of fruit juices by PEF has been widely reported in the literature (El-Hag *et al.*, 2006; Mosqueda-Melgar *et al.*, 2007; Noci *et al.*, 2008; Evrendilek *et al.*, 2008; García *et al.*, 2009; Altuntas *et al.*, 2010; Caminiti *et al.*, 2011). Most of the studies have focused on enzyme inactivation and shelf life of the product. In this section only studies on the decontamination of fruit juices are reported.

García *et al.* (2009) reported that PEF treatment with an electric field intensity of 35 kV/cm and 80 pulses resulted in an inactivation rate of about 5 log cycle reduction for *E. coli* O157:H7 inoculated in apple juice. However, this inactivation rate was obtained after 3 days of storage at 4°C which is due to the injury cell that died during storage. An equation based on the Weibullian-like distribution accurately described the kinetics of cell inactivation.

El Hag *et al.* (2006) investigated the effect of PEF on the inactivation of

© Woodhead Publishing Limited, 2012

natural microorganisms in orange juice. They applied 120 pulses/mL with an electric field intensity of 46 kV/cm to obtain 2 log cycle reductions. However, using the same PEF conditions, but with inoculated microorganisms, the result was only a 1 log cycle reduction. The authors suggested that other factors such as increased temperature and antimicrobial additives might be required to enhance the killing efficiency. The same authors also investigated the effect of PEF treatment on the inactivation of *E. coli* K12 strain (ATCC 10798) and naturally grown bacteria inoculated in apple juice (El Hag *et al.*, 2010). For an electric field strength higher than 40 kV/cm, and a temperature of about 41°C, a synergistic effect was observed. The PEF killing effect results in up to 6 and 2 log cycle reductions for the inoculated bacteria and naturally grown bacteria, respectively.

The hurdle approach is used to produce minimally processed food by applying several sub-lethal treatments to achieve microbial stability, rather than focusing solely on one lethal preservation method (McNamee *et al.*, 2010). Microbial stability is achieved by combining these hurdles to increase destruction of the microbial cytoplasmic membrane as well as preventing cell repair of survivors from PEF treatments, such as sub-lethally injured cells or bacterial endospores (Gálvez *et al.*, 2007; Leistner, 2000). Viedma *et al.* (2008) studied the effect of an antimicrobial agent (enterocin AS-48) in combination with PEF (35 kV/cm, 150 Hz, 4 μs and biopolar pulse) on the inactivation of *Salmonella enteric* serovar Choleraesuis CECT 915 (serotype 6,7:c:1,5) (ATCC 13312) inoculated in apple juice. The maximum inactivation of 4.5 log cycles was achieved with a PEF treatment of 1000 μs in combination with 60 μg/ml of AS-48 and a temperature of 40°C. The authors concluded that the combined treatment could improve the safety of freshly-made apple juice by helping to protect against *S. enterica* transmission.

McNamee *et al.* (2010) also reported the effect of combined PEF and bacteriocin treatment on the microbial inactivation of *Pichafermentans*, *E.coli* K12 or *Listeria innocua* in orange juice (OJ). They used PEF in combination with nisin (2.5 ppm), natamycin (10 ppm), benzoic acid (BA: 1000 ppm), or lactic acid (LA: 500 ppm). The PEF treatment conditions were 40 kV/cm, 100 μs and 56°C for electric field intensity, treatment time and temperature respectively. When no antimicrobial or PEF was applied (i.e. relying on the natural acidic effect), they obtained 1.5 and 0.7 log cycle reductions for *E. coli* K12 and *L. innocua* respectively, but no effect was observed for *P. fermentans*. However, with PEF alone, the inactivation rate was 4.8, 3.7 and 6.8 log cycle reductions for *P. fermentans*, *L. innocua* and *E. coli* K12, respectively. Nisin combined with PEF inactivated *L. innocua* and *E. coli* K12 in synergistic manner resulting in a total reduction of 5.6 and 7.9 log cycles, respectively. The same synergy was shown between LA and PEF in the inactivation of *L. innocua* and *P. fermentans* (6.1 and 7.8 log cycle reductions), but not for *E. coli* K12. The BA-PEF combination caused an additive inactivation of *P. fermentans*, whereas the natamycin-PEF combination against *P. fermentans* was not significantly different from the effect caused by PEF alone. The

© Woodhead Publishing Limited, 2012

authors demonstrated that appropriate combinations of nisin and PEF are capable of satisfying the FDA requirement of reductions above 5 log for *E. coli*, and are thus suitable for use as an alternative method for orange juice preservation, allowing the negative sensory effects of thermal treatments to be avoided. The PEF-biopreservative combination hurdles could therefore provide the beverage industry with effective non-thermal alternatives to prevent microbial spoilage, and improve the safety of fruit juice.

The physicochemical parameters of fruit juices (e.g. orange juice) vary significantly due to different pH levels during harvest season (3.4–3.8) and could have an important impact on the microbial rate, stability and safety of the juice (Sampedro *et al.*, 2011). Therefore, a mixture of different juices and milk, for example, which has a pH varying between 3.5 and 4.5, requires the addition of an adequate stabilizer, such as pectin, to maintain the physicochemical stability of the product. Sampedro *et al.* (2011) treated a mixed orange-milk beverage inoculated with *Salmonella enterica* serovar Typhymurium using PEF processing at different pH (3.5, 4, and 4.5), electric field intensity (15, 25, 35 and 40 kV/cm) and pectin concentration (0.1, 0.3, and 0.6%). After only 500 ms of treatment time, they obtained up to 2.5 log cycle reductions using an electric field strength of 40 kV/cm, a pH of 4.5 and 0.3% of pectin. A secondary model, based on the Weibull distribution function, was used together with Monte Carlo simulation to establish the most influential factors on the final number of *Salmonella* cells after PEF treatment. The result showed that Monte Carlo simulation can be useful in establishing the factors with the greatest influence on food safety.

Charles-Rodriguez *et al.* (2007) compared PEF and thermal processing (HTST) in the pasteurization of apple juice. The effect of process variables on the inactivation of *E. coli* ATCC 8739, such as electric field intensity (12, 24 and 36 kV/cm) and frequency (400, 600 and 800 Hz) for the PEF treatment, as well as temperature (73, 80 and 83°C) and time (27 s) for the HTST pasteurization were investigated. Both techniques achieved more than 5 log cycle reductions in microbial inactivation, which is the standard fruit juice pasteurization. However, PEF proved better able to preserve the pH than HTST, as the latter caused an increase in this property. Some variability was observed in terms of colour for all the treatments.

Walkling-Ribeiro *et al.* (2008) treated apple juice inoculated with *Staphylococcus aureus* using hurdle processes including ultraviolet irradiation (UV; 30 min, 20°C), pre-heating and PEF. Four different levels of pre-heating temperature (35–50°C), electric field strength (28–40 kV/cm) and total treatment (25–100 μs), were used in an orthogonal design, evaluating their impact on *S. aureus*. A higher reduction was achieved with a hurdle approach (UV; 46–58°C, 40 kV/cm and 100 μs) in comparison to conventional pasteurization (9.5 vs 8.2 log cycle reductions, respectively). The treatment showed little effect on measured quality attributes.

The same group (Noci *et al.*, 2008) reported the effect of UV and PEF on microbial inactivation, and on selected quality and antioxidant and enzymatic

© Woodhead Publishing Limited, 2012

activity of fresh apple juice. The two technologies were applied as stand-alone treatments (UV or PEF) or in combination (UV+PEF or PEF+UV). The UV treatment was a batch process whereas PEF was continuous, consisting of 100 square-wave pulses (1 μs, 15 Hz) and 40 kV/cm. The apple juice samples were processed by heat exchanger at 72°C or 94°C for 26 s as controls. The UV and PEF treatment resulted in 2.2 and 5.5 log cycle reductions, respectively, while the respective reductions for heat treatment at 72°C and 94°C were 6 and 6.7 log cycles. The combination of PEF+UV and UV+PEF treatment achieved a similar reduction as heat treatment at 94°C (6.2 and 7.1 log cycles, respectively) on an incubated sample (48 h at 37°C), with UV+PEF treatment producing a greater microbial reduction than PEF alone. This experiment showed the potential for combining UV irradiation and PEF to obtain satisfactory total microbial inactivation and improved product quality compared to heat pasteurization.

Azhuvalappil *et al.* (2010) achieved an inactivation rate of 6 log cycles of *E. coli* K12 inoculated in apple cider using three processing techniques (heat, PEF and UV). However, the result suggested that PEF-treated apple cider had a longer storage shelf life than UV-treated cider, and a better aroma and colour than the thermally processed sample.

Finally, Caminiti *et al.* (2011) studied the impact of a selected combination of non-thermal processing techniques on the quality of an apple and cranberry juice blend. The juices were processed by a combination of UV (5.3 J/cm$^2$) or high intensity light pulses (HILP) (3.3 J/cm$^2$) in combination with PEF (34 kV/cm, 18 Hz, 93 μs) or manothermosonication (MTS) (5 bars, 43°C, 750 W, 20 Hz). Selected physical and chemical attributes were evaluated pre- and post-processing and the sensory attributes of non-thermally treated samples were compared to conventional pasteurization (72°C for 26 s). The results showed that no significant changes were observed in terms of non-enzymatic browning, total phenolic content and antioxidant activity of the juices. The colour of the products was not affected by UV+PEF and HILP+PEF treatment. The HILP+PEF processing retained more monomeric anthocyanins than any other combined treatment. In addition, the combination of UV+PEF and HILP did not affect the odour and flavour of juice, while combinations that included MTS adversely affected those attributes.

## 14.6 Limitations and challenges to adoption of pulsed electric field (PEF) technology

Although a great deal of research in food engineering has investigated new preservation technologies, very few of these methods have as yet been implemented in the industry. This is because, alongside the multiple possibilities and advantages of PEF technology, there are also a number of limitations and drawbacks, among which the most significant are: difficulties in scaling up of

© Woodhead Publishing Limited, 2012

the system; bubble formation, leading to electrical breakdown of the treated product; suitability for use with particulate foods; availability of commercial units; and resistance of some microbial species, including bacterial spores (Barbosa-Cánovas and Altunakar, 2006; Singh and Kumar 2011).

A report by the US Food and Drug Administration (FDA, 2000) lists the following as the main drawbacks involved in the use of PEF technology:

(a) The availability of commercial units, which is limited to one by PurePulse Technologies, Inc., and one by Thomson-CSF. Many pulse-power suppliers are capable of designing and constructing reliable pulsers, but except for these 2 mentioned, the complete PEF systems must be assembled independently. The systems (including treatment chambers and power supply equipments) need to be scaled up to commercial systems.
(b) The presence of bubbles, which may lead to non-uniform treatment as well as operational and safety problems. When the applied electric field exceeds the dielectric strength of the gas bubbles, partial discharges take place inside the bubbles that can volatize the liquid and therefore increase the volume of the bubbles. The bubbles may become big enough to bridge the gap between the 2 electrodes and may produce a spark. Therefore, air bubbles in the food must be removed, particularly with batch systems. Vacuum degassing or pressurizing the treatment media during processing, using positive back pressure, can minimize the presence of gas. In general, however, the PEF method is not suitable for most of the solid food products containing air bubbles when placed in the treatment chamber.
(c) Limited application, which is restricted to food products that can withstand high electric fields. The dielectric property of a food is closely related to its physical structure and chemical composition. Homogeneous liquids with low electrical conductivity provide ideal conditions for continuous treatment with the PEF method. Food products without the addition of salt have conductivity in the range of 0.1 to 0.5 S/m. Products with high electrical conductivity reduce the resistance of the chamber and consequently require more energy to achieve a specific electrical field. Therefore, when processing high salt products, the salt should be added after processing.
(d) The particle size of the liquid food in both static and flow treatment modes. The maximum particle size in the liquid must be smaller than the gap of the treatment region in the chamber in order to maintain a proper processing operation.
(e) The lack of methods to accurately measure treatment delivery. The number and diversity in equipment limit the validity of conclusions that can be drawn about the effectiveness of particular process conditions. A method to measure treatment delivery would prevent inconsistent results due to variations in PEF systems. Such a method is not available yet.

© Woodhead Publishing Limited, 2012

To these we could also add:

- Only pumpable food products can be treated. Industrial PEF equipment is currently both rare and expensive, having a limited capacity of around 1800 L/h (Mittal *et al.*, 2000).
- PEF treatment can only inactivate vegetative bacteria, and is not effective on bacterial spores. It can therefore be used for acid products and those that are distributed in the refrigerated chain.
- PEF treatment does not inactivate enzymes. Products therefore remain at risk of enzymatic spoilage even after treatment.

However, a number of different attempts have been made to increase the availability of high capacity PEF equipment. Studies on the effect of PEF on dairy products have been conducted on skim milk, whole milk and yoghurt (Alvarez and Ji, 2003).

PEF can be increased by applying it in combination with other stressing factors, including antimicrobial compounds such as nisin and organic acids, increased water activity, pH and mild heat treatments. All of these have a synergistic effect on inactivation by PEF treatment, but more research is necessary to understand the mechanisms behind these synergisms, especially with reference to their effects on spores.

## 14.7 Food safety of pulsed electric field (PEF) processing

Mastwijk and Pol-Hofstad (2004) provide an excellent summary of the food safety aspects of PEF processing:

> Food safety policy is aimed at the reduction of the number of infections among certain risk groups, which has led the US Food and Drug Administration (FDA) to warn consumers about the risks from pathogenic organisms such as *Salmonella* and *E. coli* involved when consuming untreated fruit juice products. As a result, unpasteurized fruit and vegetable juices must be labeled (Federal Register, 1998). [...] The FDA's 1998 labeling action was followed by the development of a specific Hazard Analysis and Critical Control Points (HACCP) program for fruit and vegetable juices and juice products. In January 2001, the agency published a final rule designed to improve the safety of fruit and vegetable juice and juice products. Under the rule, juice processors must use HACCP principles for juice processing. It does not matter whether these companies produce unpasteurized or pasteurized juices. The food safety objective is aimed at the reduction in the number of contaminated products.
>
> The final FDA Juice HACCP rule requires that juice processors assess their manufacturing processes to identify any microbiological, chemical, or physical hazards that could contaminate their products. If a potential hazard is identified, processors are required to implement control measures

© Woodhead Publishing Limited, 2012

to prevent, reduce or eliminate those hazards. Juice manufacturers must use processes that achieve a 5-log reduction in the numbers of the most resistant pathogen in their finished products compared to levels that may be present in untreated juice. [...]

The problem is that thermal processing of fresh juices not only kills unwanted microbes to the required 5-log reduction, but heat treatment at certain levels also kills the product's taste. This is a problem for the juice processor, who must comply with the HACCP rule; achieve the 5-log reduction of target organisms for food safety, maintain a good product shelf-life and meet his customer's expectations of taste. [...]

To date, FDA has approved UV irradiation (21 CFR 179.39) and pulsed light (21 CFR 179.41) for use on fruit and vegetable juices and juice products to achieve the 5-log reduction of target microorganisms as part of HACCP rule compliance. Processors also may use chemical antimicrobial agents, such as certain sanitizers, on the surface of citrus fruit as long as FDA has approved the chemical agent or it is considered Generally Recognized As Safe (GRAS) by the agency. According to FDA, alternative treatment technologies 'that do not involve the use of a source of radiation or a chemical agent, e.g., high pressure processing, are not likely to require FDA approval.' Along these lines, pulsed electric field (PEF), is emerging as another commercially viable option for juice processors to use in reducing microbial hazards in fresh juices and juice products. [...]

*Microbial Safety*: Since PEF technology is used as a preservation step in the manufacture of beverages, Novel Food Regulation requires that its effectiveness is demonstrated. This includes assessment of target pathogens that are known to date, supplemented with data on a number of pathogens that have recently caused foodborne illness outbreaks in the US. Studies have shown that acidic juices (pH 4.6) can contain enteric bacterial pathogens such as *E. coli* O157:H7, various *Salmonella* species and the protozoan parasite *Cryptosporidium parvum*. Illness-causing organisms that are ubiquitous in nature, such as *Listeria monocytogenes*, also have been identified as possible contaminants in juice. The ability of some of these pathogens to survive in acidic foods requires adequate control during processing.

Inactivation by PEF is efficient for a number of microorganisms, including some well-known pathogens like *Listeria*, *Bacillus* and *E. coli*. In addition to inactivation data, it is important that industry and researchers provide shelf life data for specific products and include challenge tests.

*Toxicological Safety*: The most contestable issues related to PEF and substantial equivalence deals with chemical changes that may occur inside the products as a result of PEF treatment.

Because charged electrodes are in contact with the food, it is almost impossible to control the formation of electrolytic products and the release of electrode material into the product stream. It is therefore crucial to select

© Woodhead Publishing Limited, 2012

the correct design of electrodes and treatment devices, and the correct pulse shape, in order to minimize risk. (Mastwijk and Pol-Hofstad, 2004). The specific composition of the electrodes and the pulse geometry, along with the specific type of product, all have an effect on the amount of metal released into the product. Electrochemical action can also bring about changes in food components: PEF-treated tomato product has been shown to undergo minor chemical changes (Lelieveld *et al.*, 2001).

## 14.8 Conclusions and future trends

The development and introduction of new processing technologies should lead to enhanced product quality and improved equipment performance, while also lowering costs, in order to be acceptable to both the industry and the market (Mellbin, 1999). One strong argument in favour of PEF as an alternative method of food processing is that it is non-thermal, making it suitable for the reduction of microorganisms in food products containing heat sensitive components, which are normally difficult to pasteurize by conventional thermal processing. Research into PEF techniques needs to consider not only the inactivation of microorganisms, such as bacteria and yeast, but also the inactivation of spores and enzymes, the retention of vitamins and the effects of PEF treatments on other food components (Ngadi *et al.*, 2009; Amiali *et al.*, 2010).

Although many sectors of the food industry are considering the application of PEF technology, primarily to process liquid foods, it is not currently in use in the field. Detailed investigations are necessary to address regulatory and commercial concerns about this technology before it is used commercially.

Pulsed electric fields can be used as a single hurdle technology in combination with other hurdle methods, or as a complementary step with mild thermal processes, as the former offers the potential to produce and develop high quality products with properties resembling those of minimally processed (fresh-like) products, while maintaining the required safety aspects. Applying PEF to food material on an industrial scale requires an understanding of the electric field strength necessary to inactivate target microorganisms, as well as of the electric field enhancement, dielectric breakdown phenomena and dielectric and electrical properties of the food material. When all these parameters can be more accurately controlled, the commercial use of PEF products will become more viable.

## 14.9 Sources of further information and advice

- http://www.foodsafetymagazine.com/article.asp?id=1445&sub=sub1#13

© Woodhead Publishing Limited, 2012

- http://www.foodtech-international.com/papers/PulsedElectricField.htm
- http://www.fda.gov/food/scienceresearch/researchareas/safepracticesforfoodprocesses/ucm101662.htm
- http://www.fda.gov/Food/ScienceResearch/ResearchAreas/SafePracticesforFoodProcesses/ucm105791.htm

## 14.10 References


ALKHAFAJI, S.R. and FARID, M., 2007. An investigation on pulsed electric fields technology using new treatment chamber design. *Innovative Food Science and Emerging Technologies*, 8, 205–212.

ALTUNTAS, J., EVRENDILEK, G.A., SANGUM, M.K. and ZHANG, H.Q., 2010. Effects of pulsed electric field processing on the quality and microbial inactivation of sour cherry juice. *International Journal of Food Science and Technology*, 45, 899–905.

ÁLVAREZ, I., RASO, J., PALPO, A. and SALA, F.J., 2000. Influence of different factors on the inactivation of *Salmonella* senftenberg by pulsed electric fields. *International Journal of Food Microbiology*, 55, 143–146.

ALVAREZ, V.B. and JI, T., 2003. Emerging technologies and new processing procedures for milk and dairy products. In Gutierrez-Lopez, G.F. and Barbosa-Cánovas, G.V. (eds) *Food Science and Food Biotechnology: Facing the New Century*. Lancaster, PA: Technomic Publishing, pp. 313–327.

AMIALI, M., NGADI, M.O., SMITH, J.P. and RAGHAVAN, V.G.S., 2004. Inactivation of *Escherichia coli* O157:H7 in liquid dialyzed egg using pulsed electric fields. *Transaction of IChemE, Part C. Food and Bioproducts Processing*, 82(C2), 151–156.

AMIALI, M., NGADI, M.O., SMITH, J.P. and RAGHAVAN, G.S.V., 2005. Inactivation of *Escherichia coli* O157:H7 and *Salmonella* Enteritidis in liquid egg using continuous pulsed electric field system. *International Journal of Food Engineering*, 1(5), Art. 8.

AMIALI, M., NGADI, M.O., RAGHAVAN, G.S.V and SMITH, J.P., 2006. Inactivation of *Escherichia coli* O157:H7 and *Salmonella* Enteritidis in liquid egg white using pulsed electric field. *Journal of Food Science*, 71, 88–94.

AMIALI, M., NGADI, M.O., SMITH, J.P. and RAGHAVAN, G.S.V., 2007. Synergistic effect of temperature and pulsed electric field on inactivation of *Escherichia coli* O157:H7 and *Salmonella* Enteritidis in liquid egg yolk. *Journal of Food Engineering*, 79, 689–694.

AMIALI, M., NGADI, M.O., MUTHUKUMARAN, A. and RAGHAVAN, G.S.V., 2010. Physicochemical property changes and safety issues of foods during pulsed electric field processing. In Devahastion, S. (ed.) *Physicochemical Aspects of Food Engineering and Processing*, Boca Raton, FL: CRC Press, pp. 178–208.

AZHUVALAPPIL, Z., FAN, X., GEVEKE D.J. and ZHANG, H.Q., 2010. Thermal and nonthermal processing of apple cider: storage quality under equivalent process conditions. *Journal of Food Quality*, 33, 612–631.

BARBOSA-CÁNOVAS, G.V. and ALTUNAKAR, B., 2006. Pulsed electric fields processing of foods: an overview. In Raso, J. and Heinz, V. (eds) *Pulsed Electric Fields Technology for the Food Industry: Fundamentals and Applications*. New York: Springer Science + Business Media.

BARBOSA-CÁNOVAS, G.V. and SEPULVEDA, D., 2005. Present status and the future of PEF technology. In Barbosa-Cánovas, G.V., Tapia, M.S. and Cano, M.P. (eds), *Novel Food Processing Technologies*. Boca Raton, FL: CRC Press, pp. 1–44.

BARBOSA-CÁNOVAS, G.V., POTHAKAMURY, U.R., PALOU, E. and SWANSON, B.G., 1998. *Nonthermal Preservation of Foods*. New York: Marcel Dekker.


© Woodhead Publishing Limited, 2012

BARBOSA-CÁNOVAS, G.V., GÓNGORA-NIETO, M.M., POTHAKAMURY, U.R. and SWANSON, B.G., 1999. *Preservation of Foods with Pulsed Electric Fields*. San Diego, CA: Academic Press.

BARBOSA-CÁNOVAS, G.V., PIERSON, M.D., ZHANG, Q.H. and SCHAFFNER, D.W., 2000. Kinetics of microbial inactivation for alternative food processing technologies. *Journal of Food Science*, Supplement, 65–77.

BARSOTTI, L., MERLE, P. and CHEFTEL, J.C., 1999a. Food processing by pulsed electric fields. I. Physical aspects. *Food Review International*, 15(2), 163–180.

BARSOTTI, L., MERLE, P. and CHEFTEL, J.C., 1999b. Food processing by pulsed electric fields. II. Biological aspects. *Food Review International*, 15(2), 181–213.

BAZHAL, M.I., NGADI, M.O., SMITH, J.P. and RAGHAVAN, V.G.S., 2006. Inactivation of *Escherichia coli* in liquid whole egg using combined PEF and thermal treatments. *Lebensmittel-Wissenschaft und Technologie*, 39, 419–425.

BENDICHO, S., BARBOSA-CÁNOVAS, G.V. and MARTÍN, O., 2002. Milk processing by high intensity pulsed electric fields. *Trends in Food Science & Technology*, 13, 195–204.

BERMUDEZ-AGUIRRE, D., CORRADINI, M.G., MAWSON, R. and BARBOSA-CÁNOVAS, G.V., 2009. Modeling the inactivation of *Listeria innocua* in raw whole milk treated under thermo-sonication. *Innovative Food Science and Emerging Technologies*, 10, 172–178.

BIGELOW, W.D., 1921. The logarithmic nature of thermal death time curves. *Journal of Infectious Diseases*, 29(5), 528–536.

CALDERÓN-MIRANDA, M.L., BARBOSA-CÁNOVAS, G.V. and SWANSON, B.G. 1999. Transmission electron microscopy of *Listeria innocua* treated by pulsed electric fields and nisin skimmed milk. *International Journal of Food Microbiology*, 51, 31–38.

CAMINITI, I.M., NOCI, F., MUNOZ, A., WHYTE, P., MORGAN, D.J., CRONIN, D.A. and LYNG, J.G., 2011. Impact of selected combinations of non-thermal processing technologies on the quality of an apple and cranberry juice blend. *Food Chemistry*, 124, 1387–1392.

CASTRO, A., BARBOSA-CÁNOVAS, G.V. and SWANSON, B.G., 1993. Microbial inactivation of foods by pulsed electric fields. *Journal of Food Preservation*, 17, 47–73.

CHARLES-RODRIGUEZ, A.V., NEVÀREZ-MOORILLON, G.V., ZHANG, Q.H. and ORTEGA-RIVAS, E., 2007. Comparison of thermal processing and pulsed electric fields treatment in pasteurization of apple juice. *Transaction of IChemE, Part C, Food and Bioproducts Processing*, 85(C2), 93–97.

CHEN, W. and LEE, R., 1994. Altered ion channel conductance and ionic selectivity induced by large imposed membrane potential pulse. *Biophysics Journal*, 67, 603–612.

CRAVEN, H.M., SWIERGON, P., MIDGELY, S., NG, J., VERSTEEG, C., COVENTRY, M.J. and WAN J., 2008. Evaluation of pulsed electric field and minimal heat treatments for inactivation of *Pseudomonas* and enhancement of milk shelf-life. *Innovative Food Science & Emerging Technologies*, 9(2), 211–216.

CSERHALMI, ZS., VIDÁCS, I., BECZNER, J. and CZUKOR, B., 2002. Inactivation of *Saccharomyces cerevisiae* and *Bacillus cereus* by pulsed electric fields technology. *Innovative Food Science & Emerging Technologies*, 3, 41–45.

DUTREUX, N., NOTERMANS, S., WIJTZES, T., GÓNGORA-NIETO, M.M., BARBOSA-CÁNOVAS, G.V. and SWANSON, B.G., 2000. Pulsed electric fields inactivation of attached and free-living *Listeria innocua* under several conditions. *International Journal of Food Microbiology*, 54, 91–98.

ELEZ-MARTÍNEZ, P., ESCOLÀ-HERNÁNDEZ, J., SOLIVA-FORTUNY, R.C. and MARTÍN-BELLOSO, O., 2004. Inactivation of *Saccharomyces cerevisiae* suspended in orange juice using high-intensity pulsed electric fields. *Journal of Food Protection*, 67, 2596–2602.

ELEZ-MARTÍNEZ, P., ESCOLÀ-HERNÁNDEZ, J., SOLIVA-FORTUNY, R.C. and MARTÍN-BELLOSO, O., 2005. Inactivation of *Lactobacillus brevis* in orange juice by high-intensity pulsed electric fields. *Food Microbiology*, 22, 311–319.

EL-HAG, A.H., JAYARAM, S.H. and GRIFFITHS, M.W., 2006. Inactivation of natural grown microorganisms in orange juice using pulsed electric fields. *IEEE Transactions on Plasma Science*, 34(4), 1412–1415.

© Woodhead Publishing Limited, 2012

EL-HAG, A.H., DADARWAL, R., RODRIGUEZ-GONZALES, O., JAYARAM, S.H. and GRIFFITHS, M.W., 2010. Survivability of inoculated versus naturally grown bacteria in apple juice under pulsed electric fields. *IEEE Transactions on Industry Applications*, 46(1), 9–15.

EVRENDILEK, G.A. and ZHANG, Q.H., 2005. Effects of pulse polarity and pulse delaying time on pulsed electric fields-induced pasteurization on *Escherichia coli* O157:H7. *Journal of Food Engineering*, 68, 271–276.

EVRENDILEK, G.A., TOK, F.M., SOYLU, E.M. and SOYLU S., 2008. Inactivation of *Penicillum expansum* in sour cherry juice, peach and apricot nectars by pulsed electric fields. *Food Microbiology*, 25(5), 662–667.

FDA (Us Food and Drug Administration) (2000). *Kinetics of microbial inactivation for alternative food processing technologies*. Washington, DC: US FDA.

FEDERAL REGISTER. Docket No. 97N–0524. Vol. 63, No. 130. Wed., July 8, 1998.

FERNANDEZ-MOLINA, J.J., 2001. Inactivation of *Listeria innocua* and *Pseudomonas fluorescens* in skim milk treated with pulsed electric fields. In Barbosa-Cánovas, G.V. and Zhang, Q.H. (eds), *Pulsed Electric Fields in Food Processing: Fundamental aspects and applications*. Lancaster, PA: Technomic Publishing Company Inc. pp: 149–166.

FERNANDEZ-MOLINA, J., BERMUDEZ-AGUIRRE, D., ALTUNAKAR, B., SWANSON, B.G. and BARBOSA-CÁNOVAS, G.V., 2006. Inactivation of *Listeria innocua* and *Pseudomonas fluorescens* by pulsed electric field in skim milk: energy requirements. *Journal of Food Processing and Engineering*, 29(6), 561–573.

FLEISCHMAN, G.J., RAVISHANKAR, S. and BALASUBRAMANIAM, V.M., 2004. The inactivation of *Listeria monocytogenes* by pulsed electric field (PEF) treatment in static chamber. *Food Microbiology*, 21, 91–95.

FLOURY, J., GROSSET, N., LECONTE, N., PASCO, M., MADEC, M. and JEANTET, R., 2006. Continuous raw skim milk processing by pulsed electric field at non-lethal temperature: effect on microbial inactivation and functional properties. *Le Lait*, 86, 43–57.

GÁLVEZ, A., ABRIOUEL, H., LÓPEZ, L.R. and BEN OMAR, N., 2007. Bacteriocin-based strategies for food biopreservation. *International Journal of Food Microbiology*, 120(1–2), 51–70.

GARCÍA, D., SOMOLINOS, M., HASSANI, M., ALVARE, I. and PAGAN, R., 2009. Modeling the inactivation kinetics of *Escherichia coli* O157:H7 during the storage under refrigeration of apple juice treated by pulsed electric fields. *Journal of Food Safety*, 29, 546–563.

GÁSKOVÁ, D., SIGLER, K., JANDEROVA, B. and PLASEK, J. 1996. Effect of high-voltage electric pulses on yeast cells: factors influencing the killing efficiency. *Bioelectrochemistry Bioenergetics*, 39, 195–202.

GÓNGORA-NIETO, M.M., SEIGNOUR, L., RIQUET, P., DAVIDSON, P.M., BARBOSA-CÁNOVAS, G.V. and SWANSON, B.G., 1999. Hurdle approach for the inactivation of *Pseudomonas fluorescens* in liquid whole egg. *Food Engineering: Nonthermal Processing*. Chicago, IL: IFT annual meeting, 83A–2.

GÓNGORA-NIETO, M., SEPULVEDA, D., PEDROW, P., BARBOSA-CÁNOVAS, G.V. and SWANSON, B., 2002. Food processing by pulsed electric fields: treatment delivery, inactivation level, and regulatory aspects. *Lebensmittel-Wissenschaft und-Technologie*, 35(5), 375–388.

GÓNGORA-NIETO, M.M., PEDROW, P.D., SWANSON, B.G. and BARBOSA-CÁNOVAS, G.V., 2003. Energy analysis of liquid whole egg pasteurized by pulsed electric field. *Journal of Food Engineering*, 57, 209–216.

GOULD, G.W., 2000. Preservation: past, present and future. *British Medical Bulletin*, 56(1), 84–96.

GRAHL, T. and MÄRKL, H., 1996. Killing of microorganisms by pulsed electric fields. *Applied Microbiology & Biotechnology*, 45, 148–157.

GUERRERO-BELTRÀN, J.A., SEPULVEDA, D.R., GONGORA-NIETO, M.M., SWANSON, B. and BARBOSA-CÁNOVAS, G.V., 2010. Milk thermization by pulsed electric fields (PEF) and electrically induced heat. *Journal of Food Engineering*, 100, 56–60.

© Woodhead Publishing Limited, 2012

HAAS, C.N. and ATURALIYE, D.N., 1999. Kinetics of electroporation-assisted chlorination of *Giardia muris*. *Water Resource*, 33(8), 1761–1766.

HALDEN, K., DE ALWIS, A.A.P. and FRYER, P.J., 1990. Change in electric conductivity of foods during ohmic heating. *International Journal of Food Science & Technology*, 25, 9–25.

HEINZ, V. and KNORR, D., 2000. Effect of pH, ethanol addition and high hydrostatic pressure on the inactivation of *Bacillus subtilis* by pulsed electric fields. *Innovative Food Science & Emerging Technology*, 1, 151–159.

HEINZ, V., PHILLIPS, S.T., ZENKER, M. and KNORR, D., 1999. Inactivation of *Bacillus subtilis* by high intensity pulsed electric fields under close to isothermal conditions. *Food Biology*, 13(2), 155–168.

HEINZ, V., ALVAREZ, I., ANGERSBACH, A. and KNORR, D., 2001. Preservation of liquid foods by high intensity pulsed electric fields – basic concepts for process design. *Trends in Food Science and Technology*, 12, 103–111.

HERMAWAN, G.A., EVRENDILEK, W.R., ZHANG, Q.H. and RICHTER, E.R., 2004. Pulsed electric field treatment of liquid whole egg inoculated with *Salmonella* Enteritidis. *Journal of Food Safety*, 24(1), 71–85.

HO, S.Y., MITTAL, G.S., CROSS, J.D. and GRIFFITHS, M.W., 1995. Inactivation of *Pseudomonas fluorescens* by high voltage electric pulses. *Journal of Food Science*, 60(6), 1337–1340, 1343.

HUANG, K. and WANG, J., 2009. Designs of pulsed electric fields treatment chambers for liquid foods pasteurization process: a review. *Journal of Food Engineering*, 95(2), 227–239.

HÜLSHEGER, H. and NIEMANN, E.G., 1980. Lethal effects of high voltage pulses on *Escherichia coli* K12. *Radiation & Environmental Biophysics*, 18, 281–288.

HÜLSHEGER, H., POTEL, J. and NIEMANN, E.G., 1981. Killing of bacteria with electric pulses of high field strength. *Radiation & Environmental Biophysics*, 20, 53–65.

HÜLSHEGER, H., POTEL, J. and NIEMANN, E.G., 1983. Electric field effects on bacteria and yeast cells. *Radiation & Environmental Biophysics*, 22, 149–162.

JAY, J.M., 1992. *Modern Food Microbiology*, 4th edn. New York: Chapman & Hall.

JAYARAM, S.H., 2000. Sterilization of liquid foods by pulsed electric fields. *IEEE Electrical Insulation Magazine*, 16(6), 17–25.

JAYARAM, S., CASTLE, G.S.P. and MARGARITIS, A., 1992. Kinetics of sterilization of *Lactobacillus brevis* cells by the application of high voltage pulses. *Biotechnology Bioengineering*, 40(11), 1412–1420.

JEANTET, R., BARON, F., NAU, F., ROIGNANT, M. and BRULÉ, G., 1999. High intensity pulsed electric fields applied to egg white: effect on *Salmonella* Enteritidis inactivation and protein denaturation. *Journal of Food Protection*, 62(12), 1381–1386.

JEANTET, R., MC KEAG, J.R., FERNÁNDEZ, J.C., GOSSET, N., BARON, F. and KOROLCZUK, J., 2004. Pulsed electric field continuous treatment of egg products. *Sciences des Aliments*, 24, 137–158.

JEYAMKONDAN, S., JAYAS, D.S. and HOLLEY, R.A., 1999. Pulsed electric field processing of foods: a review. *Journal of Food Protection*, 62(9), 1088–1096.

JIN, T., ZHANG, H.Q., HERMAWAN, N. and DANTZER, W., 2009. Effect of pH and temperature on inactivation of *Salmonella typhimurium* DT104 in liquid whole egg by pulsed electric fields. *International Journal of Food Science and Technology*, 44, pp. 367–372.

KHADRE, M.A. and YOUSEF, A.E., 2002. Susceptibility of human rotavirus to ozone, high pressure, and pulsed electric field. *Journal of Food Protection*, 65, 1441–1446.

KNORR, D., GEULEN, M., GRAHL, T. and SITZMANN, W., 1994. Food application of high electric field pulses. *Trends in Food Science and Technology*, 5(3), 71–75.

LEISTNER, L., 2000. Basic aspects of food preservation by hurdle technology. *International Journal of Food Microbiology*, 55, 181–186.

LELIEVELD, H.L.M., WOUTERS, P.C. and LEON, A.E., 2001. *Pulsed Electrical Fields in Food Processing*. Lancaster, PA: Technomic Publishing.

© Woodhead Publishing Limited, 2012

LIU, D.-S., ASTUMIAN, R.D. and TSONG, T.Y., 1990. Activation of $Na^+$ and $K^+$ pumping modes of (Na,K)-ATPase by an oscillating electric field. *Journal of Biological Chemistry*, 265, 7260–7267.

LIU, X., YOUCEF, A.E. and CHISM, G.W., 1997. Inactivation of *Escherichia coli* O157:H7 by the combination of organic acids and pulsed electric field. *Journal of Food Safety*, 16, 287–299.

MARQUEZ, V.O., MITAL, G.S. and GRIFFITHS, M.W., 1997. Destruction and inhibition of bacterial spores by high voltage pulsed electric fields. *Journal of Food Science*, 62(2), 399–409.

MARTÍN, O., QIN, B.L., CHANG, F.J., BARBOSA-CÁNOVAS, G.V. and SWANSON, B.G., 1997. Inactivation of *Escherichia coli* in skim milk by high intensity pulsed electric fields. *Journal of Food Process Engineering*, 20, 317–336.

MARTÍN-BELLOSO, O., VEGA-MERCADO, H., QIN, B.L., CHANG, F.J., BARBOSA-CÁNOVAS, G.V. and SWANSON, B.G., 1997. Inactivation of *Escherichia coli* suspended in liquid egg using pulsed electric fields. *Journal of Food Processing and Preservation*, 21, 193–208.

MASTWIJK, H. and POL-HOFSTAD, I., 2004. Quality quaffs: Using PEF to assure safety of fresh juices. Available at: http://www.foodsafetymagazine.com/article.asp?id=1445&sub=sub1#13.

MAZUREK, B., LUBICKI, P. and STARONIEWICZ, Z., 1995. Effect of short HV pulses on bacteria and fungi. *IEEE Transactions on Dielectric Insulation*, 2, 418–425.

MCNAMEE, C., NOCI, F., CRONIN, D.A., LYNG, J.G., MORGAN, D.J. and SCANNELL, A.G.M., 2010. PEF based hurdle strategy to control *Pichia Fermantans*, *Listeria innocua* and *Escherichia coli* K12 in orange juice. *International Journal of Food Microbiology*, 138, 13–18.

MELLBIN, P., 1999. Business and cost considerations in fluid and juice applications. *Book of Abstracts of FoST Conference*, Tampere, Finland. November.

MIN, S., EVRENDILEK, G.A. and ZHANG, H.Q., 2007. Pulsed electric fields: processing system, microbial and enzyme inhibition, and shelf life extension of foods. *IEEE Transactions on Plasma Science*, 35(1), 59–73.

MITTAL, G.S., HO, S.Y.W., CROSS, J.D. and GRIFFITHS, M.W., 2000. Method and apparatus for electrically treating foodstuffs for preservation. US Patent 6,093,432.

MONFORT, S., GAYAN, E., SALDANA, G., PUERTOLAS, E., CONDON, S., RASO, J. and ALVAREZ, I., 2010a. Inactivation of *Salmonella typhimurium* and *Staphylococcus aureus* by pulsed electric fields in liquid whole egg. *Innovative Food Science & Emerging Technologies*, 11, 306–313.

MONFORT, S., GAYAN, E., CONDON, S., RASO, J. and ÁLVAREZ, I., 2010b. Evaluation of pulsed electric fields technology for liquid whole egg pasteurization. *Food Microbiology*, 27, 845–852.

MONFORT, S., GAYAN, E., CONDON, S., RASO, J. and ÁLVAREZ, I., 2011. Design of a combined process for the inactivation of *Samonella Enteritidis* in liquid whole egg at 55°C. *International Journal of Food Microbiology*, 145, 476–482.

MOSQUEDA-MELGAR, J., RAYBAUDI-MASSILIA, R.M. and MARTÍN-BELLOSO, O., 2007. Influence of treatment time and pulse frequency on *Salmonella* Enteritidis, *Escherichia coli* and *Listeria monocytogenes* populations inoculated in melon and watermelon juices treated by pulsed electric fields. *International Journal of Food Microbiology*, 117, 192–200.

NARSETTI, R., CURRY, R.D., MCDONALD, K.F., CLEVENGER, T.E. and NICHOLS, L.M., 2006. Microbial inactivation in water using pulsed electric fields and magnetic pulse compressor technology. *IEEE Transaction on Plasma Science*, 34(4), 1386–1393.

NGADI, M.O., YU, L., AMIALI, M. and ORTEGA-RIVAS, E., 2009. Food quality and safety issues during pulsed electric field processing. In Ortega-Rivas, E. (ed.) *Processing Effects on Safety and Quality of Foods*. Boca Raton, FL: CRC Press, pp. 446–472.

NOCI, F., RIENER, J., WALKLING-RIBEIRO, M., CRONIN, D.A., MORGAN, D.J. and LYNG, J.G., 2008. Ultraviolet irradiation and pulsed electric fields (PEF) in a hurdle

© Woodhead Publishing Limited, 2012

strategy for the preservation of fresh apple juice. *Journal of Food Engineering*, 85, pp. 141–146.

ODRIOZOLA-SERRANO, I., BENDICHO-PORTA, S. and MARTÍN-BELLOSO, O., 2006. Comparative study on shelf-life of whole milk processed by high intensity pulsed electric fields or heat treatment. *Journal of Dairy Science*, 89(3), 905–911.

OHSHIMA, T., SATO, K., TERAUCHI, M. and SATO, M., 1997. Physical and chemical modifications of high-voltage pulse sterilization. *Journal of Electrostatics*, 42, 159–166.

PALANIAPPAN, S. and SASTRY, S.K., 1991. Electrical conductivity of selected juices: influence of temperature, solids content, applied voltage, and particle size. *Journal of Food Process Engineering*, 14, 247–260.

PELEG, M., 1995. A model of microbial survival after exposure to pulsed electric fields. *Journal of Food Science and Agriculture*, 67, 93–99.

PERNI, S., CHALISE, P.R., SHAMA, G. and KONG, M.G., 2007. Bacterial cells exposed to nanosecond pulsed electric fields show lethal and sublethal effects. *International Journal of Food Microbiology*, 120, 311–314.

PICART, L., DUMAY, E. and CHEFTEL, J.C., 2002. Inactivation of *Listeria innocua* in dairy fluids by pulsed electric fields: influence of electric parameters and food composition. *Innovative Food Science & Emerging Technologies*, 3, 357–369.

PINA-PÉREZ, M.C., RODRIGO-ALIAGA, D., FERRER-BERNAT, C., RODRIGO-ENGUIDANOS, M. and MARTÍNEZ-LÓPEZ, A., 2007. Inactivation of *Enterobacter sakazakii* by pulsed electric field in buffered peptone water and infant formula milk. *International Dairy Journal*, 17, 1441–1449.

PINA-PÉREZ, M.C., SILVA-ANGULO, A.B., RODRIGO, D. and MARTINEZ-LÓPEZ, A., 2009. Synergistic effect of pulsed electric fields and cocoanOX 12% on the inactivation kinetics of *Bacillus cereus* in a mixed beverage of liquid whole egg and skim milk. *International Journal of Food Microbiology*, 130, 196–204.

POTHAKAMURY, U.R., MONSALVE-GONZÁLEZ, A., BARBOSA-CÁNOVAS, G.V. and SWANSON, B.G., 1995. Inactivation of *Escherichia coli* and *Staphylococcus aureus* in model foods by pulsed electric field technology. *Food Research International*, 28(2), 167–171.

PRASANNA, G.L. and PANDA, T., 1997. Electroporation: basic principles, practical considerations and applications in molecular biology. *Bioprocess Engineering*, 16, 261–264.

PRUITT, K.M. and KAMAU, D.N., 1993. Mathematical models of bacterial growth, inhibition and death under combined stress conditions. *Journal of Industrial Microbiology*, 12, 221–231.

QIN, B.L., ZHANG, Q., BARBOSA-CÁNOVAS, G.V., SWANSON, B.G. and PEDROW, P.D., 1994. Inactivation of microorganisms by pulsed electric fields of different voltage waveforms. *IEEE Transactions on Dielectrics and Electrical Insulation*, 1(6), 1047–1057.

QIN, B., ZHANG, Q.H., BARBOSA-CÁNOVAS, G.V., SWANSON, B.G. and PEDROW, P.D., 1995a. Pulsed electric field treatment chamber design for liquid food pasteurization using a finite element method. *ASAE Transaction*, 38(2), 557–565.

QIN, B.L., CHANG, F.J., BARBOSA-CÁNOVAS, G.V. and SWANSON, B.G., 1995b. Nonthermal inactivation of *Saccharomyces cerevisiae* in apple juice using pulsed electric fields. *Lebensmittel-Wissenschaft und-Technologie*, 28(6), 564–568.

QIN, B.L., POTHAKAMURY, U.R., BARBOSA-CÁNOVAS, G.V. and SWANSON, B.G., 1996. Nonthermal pasteurization of liquid foods using high-intensity pulsed electric field. *Critical Review in Food Science and Nutrition*, 36, 603–627.

QIN, B.L., BARBOSA-CÁNOVAS, G.V., SWANSON, B.G. and PEDROW, P.D., 1998. Inactivating microorganisms using a pulsed electric field continuous treatment system. *IEEE Transactions on Industry Applications*, 34(1), 43–49.

RAJKOVIC, A., SMIGIC, N. and DEVLIEGHERE, F., 2010. Contemporary strategies in combating microbial contamination in food chain. *International Journal of Food Microbiology*, 141, S29–S42.

© Woodhead Publishing Limited, 2012

RASO, J., CALDERÓN, M.L., GÓNGORA-NIETO, M.M., BARBOSA-CÁNOVAS, G.V. and SWANSON, B.G., 1998. Inactivation of *Zygosaccharomyces bailii* in fruit juices by heat, hydrostatic pressure and pulsed electric fields. *Journal of Food Science*, 63(6), 1042–1044.

RAVISHANKAR, S., FLEISCHMAN, G.J. and BALASUBRAMANIAM, V.M., 2002. The inactivation of *Escherichia coli* O157:H7 during pulsed electric field (PEF) treatment in a static chamber. *Food Microbiology*, 19, 351–361.

REINA, L.D., JIN, Z.T., YOUCEF, A.E. and ZHANG, Q.H., 1998. Inactivation of *Listeria monocytogenes* in milk by pulsed electric fields. *Journal of Food Protection*, 61(9), 1203–1206.

RIVAS, A., SAMPEDRO, F., RODRIGO, D. and MARTÍNEZ, A., 2006. Nature of the inactivation of *Escherichia coli* suspended in an orange juice and milk beverage. *European Food Research and Technology*, 223, 541–545.

RUSSELL, N.J., COLLEY, M., SIMPSON, R.K., TRIVETT, A.J. and EVANS, R.I., 2000. Mechanism of action of pulsed high electric field (PHEF) on the membranes of food-poisoning bacteria is an 'all-or-nothing' effect. *International Journal of Food Microbiology*, 55, 133–136.

SALE, A.J.H. and HAMILTON, W.A., 1967. Effects of high electric fields on microorganisms. I. Killing of bacteria and yeasts. *Biochemica & Biophysica Acta*, 143, 781–788.

SAMPEDRO, F., RIVAS, A., RODRIGO, D., MARTÍNEZ, A. and RODRIGO, M., 2007. Pulsed electric fields inactivation of *Lactobacillus plantarum* in an orange juice–milk based beverage: effect of process parameters. *Journal of Food Engineering*, 80, 931–938.

SAMPEDRO, F., RODRIGO, D. and MARTÍNEZ, A., 2011. Modeling the effect of pH and pectin concentration on the PEF inactivation of *Salmonella enterica* serovar Typhimurium by using the Monte Carlo simulation. *Food Control*, 22, 420–425.

SCHOENBACH, K.H., JOSHI, R.P., STARK, R.H., DOBBS, F.C. and BEEBE, S.J., 2000. Bacterial decontamination of liquids with pulsed electric fields. *IEEE Transaction on Dielectrics and Electrical Insulation*, 7(5), 637–645.

SENSOY, I., ZHANG, Q.H. and SASTRY, S.K., 1997. Inactivation kinetics of *Salmonella dublin* by pulsed electric field. *Journal of Food Process Engineering*, 20, 367–381.

SEPÚLVEDA, D.R., GÓNGORA-NIETO, M.M., GUERRERO-BELTRAN, J.A. and BARBOSA-CÁNOVAS, G.V., 2005. Production of extended shelf-life milk by processing pasteurized milk with pulsed electric fields. *Journal of Food Engineering*, 67, 81–86.

SHIN, J.K., JUNG, K.J., PYUN, Y.R. and CHUNG, M.S., 2007. Application of pulsed electric fields with square wave pulse to milk inoculated with *Escherichia coli*, *Pseudomonas fluorescens*, and *Bacillus stearothermophilus*. *Food Science and Biotechnology*, 16(6), 1082–1084.

SINGH, R. and KUMAR, A., 2011. Pulsed electric fields, processing and application in food industry. *European Journal of Food Research & Review*, 1(2), 71–93.

SOBRINO-LÓPEZ, A. and MARTÍN-BELLOSO, O., 2006. Enhancing inactivation of *Staphylococcus aureus* in skim milk by combining high-intensity pulsed electric fields and nisin. *Journal of Food Protection*, 69, 345–353.

SOBRINO-LÓPEZ, A. and MARTÍN-BELLOSO O., 2010. Potential of high-intensity pulsed electric field technology for milk processing. *Food Engineering Review*, 2, 17–27.

STANLEY, D.W., 1991. Biological membrane deterioration and associated quality losses in food tissues. In Clydesdale, F.M. (ed.) *Critical Reviews in Food Science and Nutrition*. New York: CRC Press, 235–245.

SU, Y., ZHANG, Q.H. and YIN, Y., 1996. Inactivation of *Bacillus subtilis* spores using high voltage pulsed electric fields. Poster # 26A-14, Book of Abstracts, Institute of Food Technology, p. 48.

TSONG, T.Y., 1990. Review: on electroporation of cell membranes and some related phenomena. *Bioelectrochemistry & Bioenergetic*, 24, 271–295.

VAN HEESCHAN, E.J.M., PEMEN, A.J.M., HUIJBRECHTS, A.H.J., VAN DER LAAN, P.C.T., KRZYSTOF, J.P., ZANSTRA, G.J. and DE JONG, P., 2000. A fast pulsed power source

© Woodhead Publishing Limited, 2012

applied to treatment of conducting liquid and air. *IEEE Transactions on Plasma Science*, 28(1), 137–143.

VEGA-MERCADO, H., POTHAKAMURY, U.R., CHANG, F.J., BARBOSA-CÁNOVAS, G.V. and SWANSON, B.G. 1996. Inactivation of *Escherichia coli* by combining pH, ionic strength and pulsed electric fields hurdles. *Food Research International*, 29(2), 117–121.

VIEDMA, P.M., LÓPEZ, A.S., BEN OMAR, N., ABRIOUEL, H., LOPEZ, R.L., VALDINIA, E., MARTIN BELLOSO, O. and GÁLVEZ, A., 2008. Enhanced bactericidal effect of enterocin AS-48 in combination with high-intensity pulsed-electric field treatment against *Salmonella enterica* in apple juice. *International Journal of Food Microbiology*, 128, 244–249.

WALKLING-RIBEIRO, M., NOCI, F., CRONIN, D.A., LYNG, J.G. and MORGAN, D.J., 2008. Inactivation of *Escherichia coli* in a tropical fruits smoothie by a combination of heat and pulsed electric fields. *Journal of Food Science*, 73(8), M395–M399.

WALKLING-RIBEIRO, M., RODRIGUEZ-GONZALEZ, O., JAYARAM, S. and GRIFFITHS, M.W., 2011. Microbial inactivation and shelf life comparison of 'cold' hurdle processing with pulsed electric fields and microfiltration and conventional thermal pasteurization in skim milk. *International Journal of Food Microbiology*, 144, 379–386.

WESIERSKA, E. and TRZISZKA, T., 2007. Evaluation of the use of pulsed electrical field as a factor with antimicrobial activity. *Journal of Food Engineering*, 78, 1320–1325.

WOUTER, P.C., DUTREUX, N., SMELT, J.P.M. and LELIEVELD, H.L.M., 1999. Effects of pulsed electric fields on inactivation kinetics of *Listeria innocua. Applied Environmental Microbiology*, 65, 5364–5371.

YU, L.-J., 2009. Application of pulsed electric field treated milk on cheese processing: coagulation properties and flavour development. McGill university Thesis.

ZHANG, Q.H., CHANG, F.J., BARBOSA-CÁNOVAS, G.V. and SWANSON, B.G., 1994. Inactivation of microorganisms in a semisolid model food using high voltage pulsed electric fields. *Lebensmittel-Wissenschaft und-Technologie*, 27, 538–543.

ZHANG, Q.H., CHANG, F.J., BARBOSA-CÁNOVAS, G.V. and SWANSON, B.G., 1995a. Engineering aspects of pulsed electric field pasteurization. *Journal of Food Engineering*, 25, 261–291.

ZHANG, Q.H., QIN, B.L., BARBOSA-CÁNOVAS, G.V. and SWANSON, B.G. 1995b. Inactivation of *Escherichia coli* for food pasteurization by high-strength pulsed electric fields. *Journal of Food Processing & Preservation*, 19, 103–118.

ZIMMERMANN, U., 1986. Electrical breakdown, electropermeabilization and electrofusion. *Reviews of Physiology Biochemistry & Pharmacology*, 105, 176–256.

ZIMMERMANN, U., PILWAT, G. and RIEMANN, F., 1974. Dielectric breakdown of cell membranes. *Biophysical Journal*, 14, 881–899.

ZIMMERMANN, U., PILWAT, G., BECKER, F. and RIEMANN, F., 1976. Effects of external electrical fields on cell membranes. *Bioelectrochemistry and Bioenergetics*, 3, 58–83.

© Woodhead Publishing Limited, 2012

# 15

# Microbial decontamination of food by infrared (IR) heating

**R. Ramaswamy, Heinz North America, USA, K. Krishnamurthy, Illinois Institute of Technology, USA and S. Jun, University of Hawaii, USA**

**Abstract**: Infrared (IR) heating is the application of electromagnetic radiation (in the wavelength range of 0.78–1000 μm) to generate heat in the exposed materials. This generated heat energy can be used to achieve many desirable effects, including decontamination in foods. Of the various novel thermal processing applications presently under consideration, IR heating is the most promising for adoption in both small- and large-scale food processing plants. The technology shows great potential in various applications because of its inherent advantages, such as controlled, rapid heating and precise targeted application. Apart from surface heating of foods and dehydration of agricultural products, IR radiation could conveniently be used for decontamination and disinfection of food and food-contact surfaces. This chapter provides a comprehensive review of IR radiation and its applications with an emphasis on food decontamination. The chapter also covers selective heating, the synergistic use of IR radiation with other technologies, the impact of IR radiation on food quality, and finally future trends in research and development in this area.

**Key words**: infrared, radiation, wavelength, emitters, selective heating, inactivation, disinfestation, food quality.

## 15.1 Introduction

Food safety has become one of the most challenging concerns for the food industry, government regulatory agencies, and consumers. With an ever-increasing choice of fresh produce and processed foods flooding the market, public concern on the safety of these foods has become the primary focus for food manufacturers. Consumers are unwilling to compromise on food quality in order to ensure the safety of foods. Frequent outbreaks of

© Woodhead Publishing Limited, 2012

foodborne illnesses have prompted regulatory authorities to introduce new food safety laws and regulations that food manufacturers must adhere to. The recent enactment of the US Food and Drug Administration (FDA) Food Safety Modernization Act (FSMA) by the US government is a significant step forward in entrusting the responsibility and accountability for food safety to all players in the food production system (FDA, 2011).

Food safety researchers are challenged as never before because of the complex demands from various sections of the food industry to ensure food safety without compromising on quality. This has prompted continuous improvements in existing technologies as well as exploration of novel processing technologies that target both food safety and quality. Most of these existing and novel processing technologies fall into two broad categories: thermal and non-thermal. Thermal processing involves the application of heat to prepare foods that are free from pathogenic and spoilage microorganisms. As with traditional heating methods, novel thermal technologies such as infrared heating, ohmic heating, and microwave heating result in an appreciable increase in the temperature of the foods, but the mode of heating in each case is different. However, some of the non-thermal technologies – namely high pressure processing (HPP), pulsed electric field (PEF), and pulsed ultraviolet (UV) light processing – also involve an associated temperature rise in the food product, depending on the process conditions. This temperature increase contributes to additive or synergistic effects on microbial inactivation, thereby increasing treatment efficiencies in most foods. Each of these technologies has their own advantages and disadvantages in applications for various foods. Readers interested in these technologies may refer to other chapters in this book, other books, review articles, and technical papers/bulletins available from various sources. The emphasis of this chapter will be on infrared (IR) heating and its potential use as a decontamination agent in various food applications. Although this technology, like many others, has been in use for over a century (mainly for drying applications), its potential for use in food safety applications has received increased attention and gained research momentum over the past two decades.

### 15.1.1 Basics of infrared heating

All matter above the temperature of absolute zero (–273°C) possesses electromagnetic energy and emits radiation, in a wide range of electromagnetic spectral frequencies (Pidwirny, 2006; Susek, 2010). These frequencies are produced by the oscillation of individual atoms or molecules with electric charges. The temperature of the emitting surface has a direct impact on these frequencies and the total amount of energy radiated. Since the maximum radiated power at room temperature occurs in the IR region (0.78–1000 μm) of the electromagnetic spectrum, utilizing this frequency of radiation holds special significance, especially in food applications. The broad IR range is classified into near infrared (NIR), mid infrared (MIR), and far infrared (FIR);

© Woodhead Publishing Limited, 2012

their spectral wavelengths are 0.78–1.4 μm for NIR, 1.4–3 μm for MIR, and 3–1000 μm for FIR (Sakai and Hanzawa, 1994). Overall, IR wavelengths fall between those of visible light (0.38–0.78 μm) and microwaves (1–1000 mm) (Decareau, 1985). As a result of its wide spectral wavelength range and its absorption by various food components, FIR is the most commonly used spectrum for food applications (Sakai and Hanzawa, 1994; Sandu, 1986). Most of the commercially available FIR heaters operate in the range 2.5–30 μm.

The wavelength of the electromagnetic spectrum plays a very important role in determining how the atoms or molecules behave under radiation (Halford, 1957). In the IR range of 2.5–100 μm, atoms or molecules are in vibration state, leading to the generation of heat. Atoms or molecules subjected to incident radiation wavelengths of <2.5 μm are in the electronic state, as in UV rays; those subjected to wavelengths >100 μm are in a rotational state, as is the case in microwaves. This explains the reason for the improved absorption efficiency of IR waves by various food components. In addition, the temperature of the surface emitting the radiation dictates the peak wavelength, i.e. the wavelength at which maximum radiation is emitted by the emitter. This is explained by Wien's displacement law, according to which the peak wavelength ($\lambda_{max}$) is indirectly proportional to the temperature ($T$) of the emitter (Dagerskog and Osterstrom, 1979):

$$\lambda_{max} = \frac{2898}{T} \qquad [15.1]$$

Furthermore, it is clear from Stefan–Boltzmann's law, which states that the higher the temperature of the emitter, the higher the energy radiated will be and the lower the peak wavelength will be, that the penetration capability of NIR radiation is superior compared with FIR radiation. Krishnamurthy *et al.* (2008) listed the basic laws (Planck's law, Wien's displacement law, Stefan–Boltzmann's law and modified Beer's law) that pertain to IR radiation and described the parameters that influence these laws (Table 15.1). Since most of the incident radiation energy in the FIR range is absorbed on the surface, FIR is most suited to the surface heating of foods. This has been verified experimentally (Hashimoto *et al.*, 1990; Sato *et al.*, 1992) and through mathematical models (Sakai *et al.*, 1993). For example, the application of NIR radiation to wheat flour batter during baking resulted in a wet crust due to distribution of heat to inner layers; in contrast, FIR radiation resulted in a dry crust with more surface color because of the higher surface temperatures (Sato and Shibukawa, 1989; Sato *et al.*, 1992).

Unlike conventional heating ovens, where food is heated by convective air on the surface followed by conduction inside, IR heating involves radiation heating of the surface followed by conductive heating inside (Trivittayasil *et al.*, 2011). This facilitates improved energy efficiency of the IR heating system since no energy is lost in heating the air around the food. Air is transparent to radiation and the process chamber remains unheated, facilitating better

© Woodhead Publishing Limited, 2012

**Table 15.1** Basic laws pertaining to infrared radiation

| Basic laws | Aspects addressed/explanation |
|---|---|
| Planck's law: $E_{b\lambda}(T,\lambda) = \dfrac{2\pi hc_0^2}{n^2\lambda^5[e^{hc_0/n\lambda kT} - 1]}$ | Gives spectral blackbody emissive power distribution $E_{b\lambda}(T, \lambda)$. |
| Wien's displacement law: $\lambda_{max} = \dfrac{2898}{T}$ | Gives the peak wavelength ($\lambda_{max}$), where spectral distribution of radiation emitted by a black body reaches a maximum emissive power. |
| Stefan–Boltzmann's law: $E_b(T) = n^2\sigma T^4$ | Gives the total power radiated ($E_b(T)$) at a specific temperature from an infrared source. |
| Modified Beer's law: $H_\lambda = H_{\lambda 0}\exp(-\sigma^*_\lambda u)$ | Gives the transmitted spectral irradiance ($H_\lambda$ W/m$^2\cdot$ µm) in non-homogeneous systems. |

$k$: Boltzmann's constant ($1.3806 \times 10^{-23}$ J/K), $n$: refractive index of the medium ($n$ for vacuum is 1 and for most gases, $n$ is very close to unity), $\lambda$: the wavelength (µm), $T$: source temperature (K), $c_0$: speed of light (km/sec), $h$: Planck's constant ($6.626 \times 10^{-34}$ J·s), $\sigma$: Stefan–Boltzmann constant $\left(5.670 \times 10^{-8}\dfrac{W}{m^2K^4}\right)$, $\lambda_{max}$: peak wavelength, $H_{\lambda 0}$: incident spectral irradiance (W/m$^2$ · µm), $u$: mass of absorbing medium per unit area (kg/m$^2$) and $\sigma^*_\lambda$: spectral extinction coefficient (m$^2$/kg).

control during continuous processing operations. IR radiation has also been reported to be beneficial in the penetrative heating of potatoes (Afzal and Abe, 1998) and meat patties/burgers (Sheridan and Shilton, 1999). IR heating has many advantages over convective heating of foods (Das and Das, 2010; Sakai and Hanzawa, 1994; Sheridan and Shilton, 1999). These advantages include:

- very efficient energy input;
- precise or targeted application;
- reduced cycle times;
- high levels of control (precise temperature over precise time);
- rapid heating;
- suitable for food-grade manufacturing and applications;
- easy to use on food and food-contact surfaces;
- effective in unfavorable conditions (dust particles, porous surfaces);
- reliable and practical (easy fitting and handling);
- can be easily programmed to prevent over heating;
- leaves no residues (organic or inorganic).

These advantages are being exploited in various applications including roasting, frying, broiling, cooking, baking, thawing, and drying; and these applications have been reviewed and reported by many researchers (Krishnamurthy *et al.*, 2008, 2010b; Sakai and Hanzawa, 1994). In contrast, applications for IR radiation heating in the decontamination of food and food-contact

© Woodhead Publishing Limited, 2012

surfaces have received very little attention to date (Krishnamurthy *et al.*, 2010b) and this area needs to be researched in detail taking into account the most recent developments.

## 15.2 Infrared heating equipment and design

Any object that is above absolute temperature emits radiation, i.e acts as an emitter. The temperature of the source/heater element, the color of light, power density and response times can all be controlled by careful selection of the emitter. Although there is no precise demarcation between wavelength ranges, emitters can be broadly classified into those contributing to short-, medium- and long-wavelength radiation. Some of the salient features of these emitters are presented in Table 15.2. Most of the IR radiations that fall into the short-, medium-, and long-wavelength ranges bring about changes in the vibration and rotational states of atoms and molecules and hugely affect water, proteins, starches and other organic materials, but not the surrounding air. This selective heating is exploited in drying, concentration and disinfection procedures. Das and Das (2010) also classified the emitters based on their source of heat, type of filament, embedded environment, and radiation reflectors. This classification is summarized in Table 15.3.

The aforementioned salient features need to be considered when selecting IR emitters for a specified application. Apart from these features, the distance between the emitter and the object, and the distance between the emitters when arranged in arrays, need to be carefully considered for effective use of any IR system. Lambert's cosine law of radiation states that the maximum

**Table 15.2** Emitter characteristics based on their wavelength range

| Wavelength | Short | Medium | Long |
|---|---|---|---|
| Range | 0.7–1.4 μm | 1.4–3.0 μm | > 3.0 μm |
| Color of radiation | Bright white | Bright orange | Dull orange |
| Radiator temperature | 1300–2600 K | 850–1200 K | 500–800 K |
| Time to maximum temperature or peak emissive power | Few seconds | One minute | Five minutes |
| Power density | 300 kWm$^{-2}$ | 90 kWm$^{-2}$ | 40 kWm$^{-2}$ |
| Common applications | Powder coating, adhesive bonding, metal castings, preheating | Drying and curing of food products | In processes requiring both convection and IR heating |
| Source of heating | Electric | Electric, gas | Electric, gas |
| Dominant mechanism of energy absorption | Vibration and rotation of atoms and molecules | Vibration and rotation of atoms and molecules | Vibration and rotation of atoms and molecules |

Source: Das and Das, 2010.

© Woodhead Publishing Limited, 2012

**Table 15.3** Classification of emitters based on their mode of fabrication and arrangement

| Source of heat | Types | Salient features | Application |
|---|---|---|---|
| Electric-fired (metal filament placed inside a vacuum or inert gas filled enclosure) | Reflector type incandescent lamps (vacuum, inert-gas filled, halogen lamps) | Precise application of heat by reflective shield; high emission efficiency; long shelf life | Most widely used in food service industry, curing, baking, dehydration of foods |
| | Quartz tube | High energy transmission efficiency; oxidation may limit the life | |
| | Ceramic | Suited for modular arrangement; best for controlled zonal heating; concave, flat and convex faces available | |
| | Tubular metal sheathed elements | More sturdy, durable and resistant to thermal shock; influenced by direct airflow | |
| | Radiant panels | High power intensities at lower source temperature; variety of panels (low and high temperature) available | |
| Gas-fired (perforated ceramic or steel tubes heated by natural gas or liquefied gas) | Direct flame radiator | Gas flames heat the radiating surface; ventilation essential; no electricity needed | Conveyor-type bread-baking ovens |
| | Ceramic burner | Combustion inside a perforated ceramic panel | Industrial applications like textile, paint, power heating, paper drying, coal coating |
| | Metal fiber burner | Combustion inside a thin steel wire mesh | |
| | High intensity porous burner | Heating in two stages; high radiant efficiency; resistant to shock | |
| | Catalytic gas-fired emitter | No visible flame; operates on exothermic oxidation-reduction in the presence of a catalyst (platinum); preheating the catalyst will improve the efficiency | Grain disinfection |

*(Continued)*

© Woodhead Publishing Limited, 2012

**Table 15.3** Continued

| Source of heat | Types | Salient features | Application |
|---|---|---|---|
| Carbon twin IR emitter (combination of carbon IR emitter and shortwave IR emitter in twin quartz glass tubes) | | Higher depth of penetration; low operating costs; high emitter efficiency | Products where both shortwave and longwave IR radiation are needed (simultaneous cooking and drying of foods) |

Source: Das and Das, 2010.

radiation can be observed on a surface located at right angles to the radiation source and any deviation of the angle of the object to the source (angle $\theta$) will directly affect the radiation intensity by the cosine of the angle $\theta$. Air velocity on the surface is another factor that influences the water removal flux (Datta and Ni, 2002). Figure 15.1 shows a schematic diagram of IR systems used for processing foods. In the batch-type system (Fig. 15.1(a)), the entire food is exposed to the same process conditions for the whole treatment time. The system can be customized to any size and different loading methods can be employed. The simplest form consists of an IR lamp and an aluminum sheet, layered reflector. The distance between the heat source and the product can be adjusted for optimum effect. Low- and high-temperature models with optional convective heating arrangements can be made to suit various applications. For large-scale treatment, conveyor ovens (Fig. 15.1(b)) are available in various styles such as flat belt, spindle, chain-on-edge, and tow-line. Again, these systems can be tailored to fit any floor plan. The control of temperatures in individual zones is achieved using reflectors, side reflectors, and burners in-series. In both the batch-type and the conveyor oven configurations, heat can be applied from the top, bottom, sides or any of these combinations based on the products being processed.

## 15.3 Mechanisms of microbial inactivation

IR heating is a thermal technology and therefore it is presumed to inactivate microorganisms by heat, and the effect of heat on the internal contents of microorganisms, e.g. thermal denaturation of proteins and nucleic acids (Sakai and Hanzawa, 1994). However, as the IR spectrum lies between that of microwaves and UV light, an overlapping effect involving induction heating and damage to DNA has also been reported to be responsible for microbial inactivation (Hamanaka *et al.*, 2000; Krishnamurthy, 2006). Krishnamurthy (2006) verified and demonstrated the effect of IR heating on *Staphylococcus aureus* cells in milk using transmission electron microscopic and infrared

© Woodhead Publishing Limited, 2012

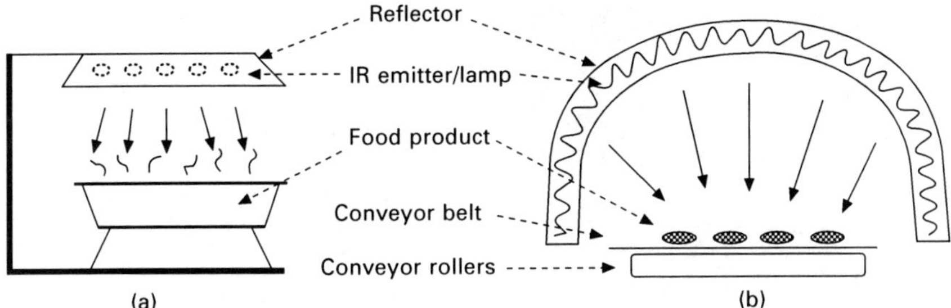


**Fig. 15.1** Schematic diagram of typical IR heating systems (a) batch system, (b) continuous system.

spectroscopic observations. It was reported that IR heating resulted in both external and internal damage to the cells with the images revealing cell wall damage, shrinkage of cytoplasmic membranes, disintegration of mesosomes, and leakage of internal cellular contents (Fig. 15.2). The rise in temperature effected by IR heating is expected to have effects similar to those of conventional heating methods, i.e. damage to cell walls and leakage of intracellular components. However, a more pronounced effect of IR radiation was observed by Hashimoto *et al.* (1991, 1992a,b) in their experiments with *Escherichia coli* and *S. aureus* in a phosphate buffer medium. FIR radiation was found to perform better than both conductive water-bath heating and convective hot-air drying. This may be due to the increased energy levels of FIR concentrated on the surface of the phosphate buffer and agar, resulting in more inactivation compared with the distributed energy in other modes of heating.

Even when they are not killed, microbial cells may be sub-lethally injured, making it difficult for them to survive in a non-supportive environment. In an experiment with *E. coli* in phosphate-buffered saline, IR heating was compared with conductive heating along with four inhibitory agents: penicillin, chloramphenicol, rifampicin, and nalidixic acid (Sawai *et al.*, 1995). While nalidixic acid was found to have no additive effect on the inactivation (1.8 $\log_{10}$ colony forming units (CFU)/ml with and without inhibitor), penicillin, chloramphenicol, and rifampicin were reported to have post-treatment effects resulting in additional log reductions of the bacteria. Although the results did not suggest any difference between IR and conductive heating in the inactivation levels, the cell components were found to be more susceptible to IR heating than they were to conductive heating. Protein denaturation was the main cause for the loss in cell function, followed by damage to RNA, cell walls, and DNA.

Further research in the same laboratory (Sawai *et al.*, 1997) revealed that IR heating inflicted more damage to the cell walls of cells that were in their exponential phase; for cells that were in their stationary phase, IR heating inflicted more damage to RNA. Damage to the cell walls was more effective in elimination of microbes: the authors reported a 3.9 $\log_{10}$CFU/ml reduction

© Woodhead Publishing Limited, 2012

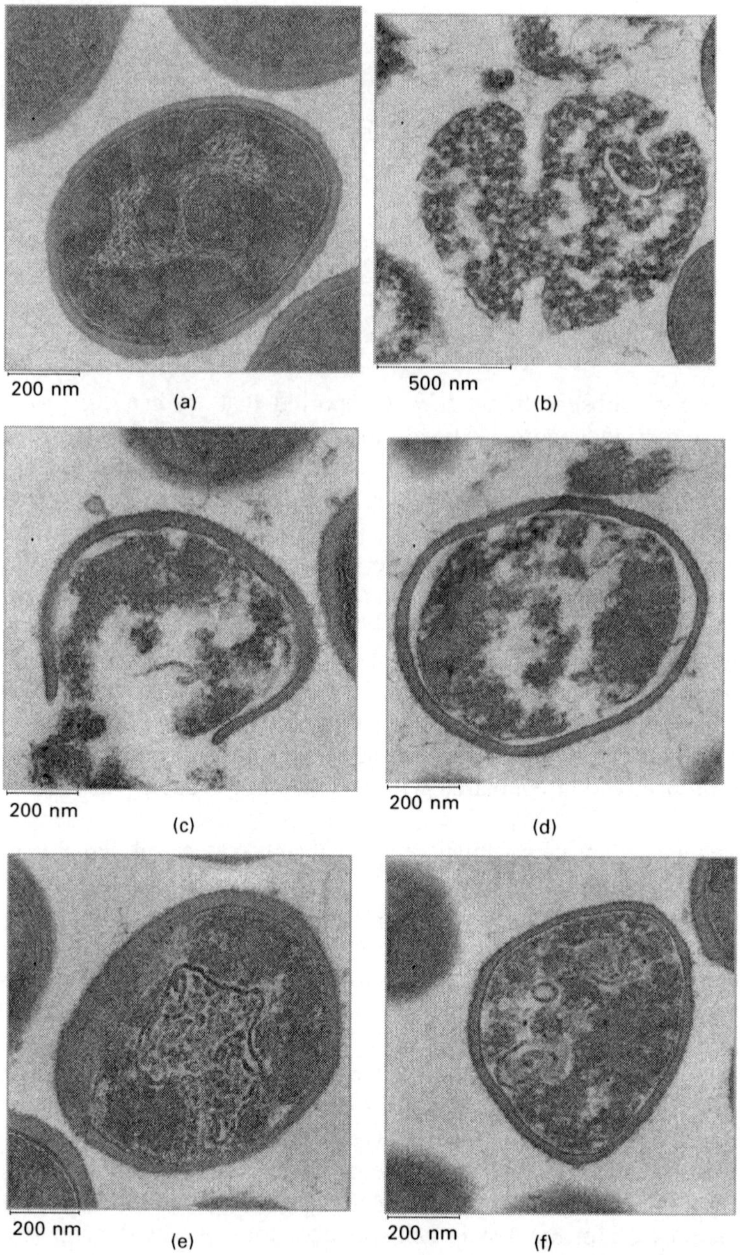

200 nm (a)

500 nm (b)

200 nm (c)

200 nm (d)

200 nm (e)

200 nm (f)

**Fig. 15.2** Microscopic images of infrared heat (700°C for 20 min) treated *S. aureus*: (a) control sample, (b) lack of cell wall, (c) cell wall breakage and cytoplasm content leakage, (d) cytoplasm shrinkage, (e) breakage in mesosomes, and (f) cytoplasm damage (Krishnamurthy *et al.*, 2006, 2010a) (with kind permission from Springer Science+Business Media: 'Microscopic and spectroscopic evaluation of inactivation of *Staphylococcus aureus* by pulsed UV light and infrared heating').

© Woodhead Publishing Limited, 2012

in exponential phase cells when compared to a 1.8 $\log_{10}$ CFU/ml reduction in stationary phase cells, when cells were treated with IR radiation at 3.2 $KW/m^2$ for 5 min.

## 15.4 Application of infrared (IR) in food decontamination

### 15.4.1 Inactivation of vegetative bacteria

In general, vegetative cells are easier to inactivate than their spore counterparts, although there is a difference in their tolerance to various processing conditions. The bacteria of major public health concern or the bacteria with the highest resistance in various foods need to be targeted to ensure the safety of these foods. When the bacteria of highest resistance are targeted, other harmful bacteria present in the food are also taken care of by the extremely severe process conditions required for the inactivation of the highly tolerant bacterial population. This will ensure the safety of the food while at the same time not subjecting the foods to over-processing.

A classical first-order kinetic model (eq. [15.2]) was used by Trivittayasil *et al.* (2011) to describe the surface inactivation kinetics of IR treatments, since IR mainly influences the surface because of its low penetration capacity:

$$\frac{dN}{dt} = -k(T) \cdot N \qquad [15.2]$$

As the inactivation rate is influenced directly by the applied temperature level, an Arrhenius equation (eq. [15.3]) was applied to include this temperature effect (eq. [15.4]).

$$k = Ae^{-\frac{Ea}{RT}} \qquad [15.3]$$

$$k(T) = k_{\mathrm{ref}} \exp\left\{\frac{E_a}{R}\left(\frac{1}{T_{\mathrm{ref}}} - \frac{1}{T}\right)\right\} \qquad [15.4]$$

where $k$ is the inactivation rate ($s^{-1}$) at temperature $T$ (K), $k_{\mathrm{ref}}$ is the inactivation rate at reference temperature ($T_{\mathrm{ref}}$), $E_{\mathrm{a}}$ is the Arrhenius activation energy (kJ/kmol) and $R$ is the universal gas constant (kJ/K/kmol). Yamada (2010) and Marquenie (2002) have determined the values of these parameters for *Cladosporium* spp. and *Penicillium* spp., and *Botrytis cinerea* and *Monilinia fructigena*, respectively. In all these studies, no inactivation effect was observed during the initial stages (shown by shoulders in the inactivation curve) because the temperature did not reach the reference level.

*Listeria monocytogenes*

*Listeria monocytogenes* is the pathogen of most public health concern in ready-to-eat (RTE) meat products. Because of its high mortality rate among

© Woodhead Publishing Limited, 2012

immune-compromised people, including pregnant women and infants, the USDA's Food Safety and Inspection Service (FSIS) has adopted a zero-tolerance policy for this organism in RTE meats (FDA/CFSAN 2003; Mead *et al.*, 1999). This organism is especially dangerous because of its ability to grow under refrigerated environments, but it can be easily eliminated by thermal processing. However, most of the reported outbreaks have been due to post-process contamination during the various handling stages after the cooking stage. This recontamination occurs mostly on the surface of RTE products. IR treatment could be very effective in achieving safety of RTE meats against *L. monocytogenes*.

Attempts were made by Gande and Murina (2003) and Huang (2004) to inactivate *L. monocytogenes* from the surface of turkey frankfurters. Although the control measures used were not as extreme as those used in a standard thermal processing operation (see below), the authors could achieve a 3.5–4.5 log reduction of the pathogenic bacteria from an initial population of $10^6$–$10^7$ CFU/cm$^2$, which is a relatively high population for RTE meats. The average reported level of *L. monocytogenes* contamination in RTE meats is in the range of $10^3$–$10^4$ CFU/g (Gombas *et al.*, 2003). The reduction described by Gande and Murina (2003) and Huang (2004) was achieved at a surface temperature of 70–80°C from a ceramic IR source maintained at 545±1°C. The frankfurters were not held at these target temperatures for any length of time to effect complete inactivation, as would be done in any effective thermal processing operation.

Huang and Sites (2008) developed an IR surface pasteurization process for hotdogs with provisions for automatic temperature control. With a 3 min holding time at 80°C and a 2 min holding time at 85°C, they could achieve a more than 6.4 log reduction from an initial high population of 7.32 $\log_{10}$ CFU/g of a cocktail mixture of four strains of *L. monocytogenes* (H7763, H7776, H7778, and 46877). The temperature used in this study was 70–85°C using a quartz IR emitter maintained at less than 330°C. The wavelength of the quartz emitter was reported to be in the range from FIR to NIR (5–350 μm). The authors have cautioned that a longer holding time should be used to eliminate the pathogens completely, because of the uneven surface of hotdogs. However, this may not be necessary considering the very high level of initial inoculums used in the experiments and the high level of inactivation that could be achieved with the process conditions tested. Proving this hypothesis with further research on various RTE meats could lead to successful commercial applications.

*Salmonella*

Salmonellosis is an illness inflicted by foodborne pathogens of the genus *Salmonella*, in particular *Salmonella* Typhimurium and *Salmonella* Enteritidis. *Salmonella* is a rod-shaped, Gram-negative, non-spore-forming enterobacteria responsible for many gastrointestinal outbreaks worldwide. Although poultry and egg products are the major sources of infection of this pathogen, fruit

© Woodhead Publishing Limited, 2012

and vegetables, and tree nuts (especially almonds) are susceptible to cross-contamination through land, water and processing practices (CDC, 2004).

According to the Almond Board of California (2003) almanac report infected raw almonds contain a maximum of 2 log CFU/g *Salmonella*. Several methods of decontamination have been found to be useful for complete inactivation of *Salmonella* on the surface of raw almonds. These include treatments with propylene oxide (Danyluk *et al.*, 2005), chlorine dioxide (Wihodo *et al.*, 2005), citric acid (Pao *et al.*, 2006), or steam (Lee *et al.*, 2006). Each of these methods, though effective in *Salmonella* inactivation, has their own disadvantages. Chemical treatments leave residues in the almonds and steam treatment increases the moisture content of almonds, thereby reducing the final quality. Brandl *et al.* (2008) investigated the use of IR heat in reducing the *Salmonella* population of raw almond kernels. While a 35 s steam treatment reduced the *Salmonella* population by 3.7–3.9 $\log_{10}$ CFU/g, dry IR heat treatment with a radiation density of 5458 W/m$^2$ for 35 s only resulted in a 1.03 log reduction. The higher inactivation efficiency of steam was due to its better penetration capability but results in the infusion of moisture into the almonds (Lee *et al.*, 2006). In their experiments with a double-sided catalytic IR heating system, Brandl *et al.* (2008) achieved a faster temperature increase in the surface of almonds with minimal difference between the top and bottom surfaces. Holding the temperature for 60 min at an initial temperature of more than 100°C resulted in a greater than 4.2 log CFU/g reduction of *Salmonella* Enteritidis. As this process was a dry treatment, it resulted in a moisture loss of less than 1.06%. Brandl *et al.* (2008) have also shown that similar inactivation levels could be achieved at a lower temperature by pre-treating (wetting) the raw almonds.

The almond industry sets a target of a 4 log reduction of *Pediococcus* spp. In their experiments with IR roasting, sequential infrared and hot air roasting (SIRHA), and hot air roasting, Yang *et al.* (2010) inactivated *Pediococcus* spp. NRRL B-2354 (a surrogate for *Salmonella* Enteriditis PT 30). Compared with traditional hot air roasting, SIRHA resulted in a 62% time saving and achieved a 6.96 log reduction at 150°C. Decimal reduction times at temperatures of 130, 140, and 150°C for SIRHA at a thermal resistance constant of 25.4°C were reported to be 8.68, 3.72, and 1.42 min, respectively. Yang *et al.* (2010) concluded that SIRHA could serve as a potential medium-roasting pasteurization method based on the fact that there was no significant quality change as a result of the treatment. Heating almonds to 120°C by IR treatment, followed by ambient temperature cooling to 90°C and holding at 90°C for 5 min, resulted in a more than 5 log reduction of *Pediococcus* spp. (Bingol *et al.*, 2011). Reducing the process temperature (80°C) required longer periods of holding (22 min) to obtain the targeted 4 log reduction. Further reduction in the process temperature did not result in the mandatory pasteurization requirement.

© Woodhead Publishing Limited, 2012

### 15.4.2 Inactivation of spores

The spores of various pathogenic microorganisms have proved more difficult to eliminate than vegetative cells. Their increased resistance to inactivation treatments including IR is attributed to their physiological adaptations when faced with stressors such as heat, chemicals, and radiation (Stumbo, 1973; Sawai *et al.*, 1997). The most effective method of inactivating spores is thermal, i.e. heating the foods to extremely high temperatures (>121°C).

*Bacillus cereus* spores are commonly found in dry powders, such as spices and herbs. They can produce toxins and have the potential to cause illness at >3 $\log_{10}$ CFU/g (Jaquette and Beuchat, 1998). IR radiation has been used to produce effective *B. cereus* decontamination of grain surfaces (Hamanaka *et al.*, 2006) and paprika powder (Staack *et al.*, 2008a). Staack *et al.* (2008b) used NIR radiation at high heat flux to increase the temperature of paprika powder to 95–100°C followed by low heat flux to maintain (for 6 min) the temperature in a closed sample holder to effect a 4.5 log reduction in *B. cereus* (water activity, $a_w = 0.88$). It was also reported that a significant reduction in the number of spores could be attained by increasing the temperature and $a_w$ values. However, changing the pH (from 4.5 to 4.0) did not lead to a significant reduction in spore counts.

Sawai *et al.* (2009) compared the effect of application of NIR (1 μm) and FIR (3–6 μm) radiation on the inactivation of *Bacillus subtilis* spores in phosphate-buffered saline suspensions. FIR radiation was found to be more effective because of the ease with which it was absorbed by the spore cell components; this resulted in heat activation and germination of spores into vegetative cells, which were then killed by either IR heat treatment or other methods. NIR treatment resulted in gradual reduction of spore colony counts but did not induce heat activation of spores. This study demonstrated the effects of IR spectral variation on the efficiency of spore inactivation.

### 15.4.3 IR radiation as a grain disinfestation tool

IR radiation is considered to be a safe, non-chemical and rapid treatment for grain disinfestation, and is seen as a potential alternative to conventional disinfection methods (mostly chemicals, e.g. methyl bromide) used in the grain industry. The application of IR irradiation to control grain insects was reported as early as the 1960s and 1970s, but serious efforts at a commercial level have only been reported in the last decade. Pan *et al.* (2006) used a catalytic IR emitter to eliminate adult insects, larvae, and eggs in rice. They heated single-layer, rough rice samples infested with the adults and eggs of lesser grain borers (*Rhizopertha dominica*) and Angoumois grain moths (*Sitotroga cerealella*) to 61°C for 1 min. They reported complete killing of these insects by IR heating and subsequent tempering treatment. Although IR heating reduced the moisture content by 1.7% and an additional 1.4% moisture content was removed by the tempering treatment, the process did not affect the milling quality of the grains. This is a result of the selective

© Woodhead Publishing Limited, 2012

heating of the higher moisture content present in insects compared with the low moisture level of grains (Erdogdu *et al.*, 2010).

Subramanyam (2004, 2005) reported the various factors influencing the susceptibility of insect species to IR radiation. The age of the insects, distance of the grain and insect samples from the heaters, duration of exposure to IR radiation, heat intensity (pressure), and grain moisture were all cited as factors that will affect the disinfestation. Early research in the 1970s by Kirkpatrick (1975) and his USDA team of scientists (Kirkpatrick *et al.*, 1972) observed that 99.7% of rice weevils and 99.3% of lesser grain borers were killed at a grain temperature of 48.6°C. However, the mortality rates for immature insects developing inside the kernels were lower (75 and 83%, respectively). The reason for this was attributed to the immediate cooling to 26°C after IR exposure. Delayed cooling (to 38°C in 48 h) led to 99.8 and 93% mortality in immature rice weevils and lesser grain borers, respectively. The effect was reported to be more pronounced when IR radiation was applied under vacuum (25 mmHg).

Subramanyam (2004, 2005) used flameless catalytic IR radiation to achieve similar results on rice weevils, red flour beetle, sawtoothed grain beetle, merchant grain beetle, and lesser grain borers. The author also cited grain quantity as one of the influencing factors in addition to the factors cited above. Shorter exposures of 60 s or less were reported to be sufficient to control stored-product insects developing internally or externally in stored grains. As with microorganisms, consideration of the life-cycle stages of the insects and the location of the insects within the grains were both reported to be important for effective disinfestation by IR radiation (Khamis *et al.*, 2010). Kirkpatrick *et al.* (1972) reported that the older life-cycle stages of the lesser grain borer and the younger life-cycle stages of the rice weevil were highly susceptible to IR radiation.

## 15.5 Effectiveness of infrared (IR) and ultraviolet (UV) irradiation on food safety and quality

### 15.5.1 Synergistic effect of IR and UV irradiation

IR radiation can be used in combination with UV irradiation for effective inactivation of microorganisms. The damage to DNA caused by UV irradiation makes the cells susceptible to IR heating. By combining UV with IR, sub-lethal temperatures could be used to inactivate microorganisms. Hamanaka *et al.* (2011) reported effective inactivation (>3 log CFU/cm$^2$) of *Rhodotorula mucilaginosa* yeast on the surfaces of figs during storage (3 days) by sequential treatment with UV and IR radiation. Fresh figs are susceptible to damage by yeasts and molds, especially during storage and distribution. Surface-decontaminated (3 log CFU/cm$^2$; 30 s IR followed by 30 s UV) fresh figs were stored for up to 3 days and the most frequently detected yeast *R.*

© Woodhead Publishing Limited, 2012

*mucilaginosa* was isolated and used ($10^7$ CFU/ml) for the fungal inactivation test. The cell suspension was diluted to 100 CFU/ml in potato dextrose agar (PDA) plates and subjected to IR (5.6 μW/cm$^2$/nm; 5, 10, 20, and 30 s) and UV (20 mW/cm$^2$; 1, 3, 5 s) treatments by setting the lamp at 130 mm from the surface of the PDA medium. While independent treatments (UV and IR for 30 s) were not very effective (only 1.5 log reduction) in inactivation of these fungal cells, the sequential treatment was found to be effective (>3 log) with minor changes in color, hardness, and respiration characteristics. IR pre-treatment was found to accelerate the cell-inactivating efficiency of UV irradiation. The longer the pre-treatment time with IR, the greater was the mortality rate reported for the same UV treatment (3 log reduction with 20 s IR followed by 5 s UV versus 4 log reduction with 30 s IR followed by 5 s UV). The inactivation efficiencies were also reported to be independent of the order of treatment.

### 15.5.2 Selective IR heating

Selective heating is the preferential heating of particular portions of the food to a specified degree to achieve commercial sterilization or decontamination while preventing quality degradation from overheating of internal tissues (Trivittayasil *et al.*, 2011). Since IR radiation is available over a wide range of often overlapping wavelengths and because many types of IR emitters are now available for use, researchers have been able to explore various options to achieve selective heating for specific applications. The research is further complicated by the varying IR radiation absorbing capacities of particular components of food products. Most of the work carried out and reported on has been related to quality control using selective heating (Lentz *et al.*, 1995; Shuman and Staley, 1950; Dagerskog and Osterstrom, 1979; Jun, 2002).

As mentioned above, the different components of foods absorb IR radiation in different wavelength ranges (Sandu, 1986; Rosenthal, 1992; Krishnamurthy *et al.*, 2008). The absorption wavelengths of various chemical groups and their associated food components are presented in Table 15.4. The wavelengths of the IR spectrum that are absorbed by various components of

**Table 15.4** Absorption wavelengths of various food components and their responsible chemical groups

| Food component | Chemical group responsible for absorption of IR | Absorption wavelength (μm) |
|---|---|---|
| Water, sugars | Hydroxyl group (O–H) | 2.7–3.3 |
| Lipids, sugars, proteins | Aliphatic carbon-hydrogen bond | 3.25–3.7 |
| Lipids | Carbonyl group (C=O) ester | 5.71–5.76 |
| Proteins | Carbonyl group (C=O) amide | 5.92 |
| Proteins | Nitrogen–hydrogen group (–NH–) | 2.83–3.33 |
| Unsaturated lipids | Carbon–carbon double bond (C=C) | 4.44–4.76 |

Source: Rosenthal, 1992.

© Woodhead Publishing Limited, 2012

foods overlap with one another, with water (hydroxyl group) standing apart and dominating the absorption. By eliminating the wavelength pertaining to the water component, it is possible to target specific food components in a more effective way. In this way, it is possible to increase the inactivation rate of microorganisms in specific food components.

Selective IR heating based on the use of optical bandpass filters to deliver energy selectively to targeted food components appears to be a promising technique. Jun (2002) demonstrated this by selectively heating soy protein by 6°C more than glucose in 5 min, when they were treated under controlled IR radiation. Jun and Irudayaraj (2003) developed simulation models that could support the experimental results of selective IR heating. It would also be feasible to heat the microorganisms present in food products selectively, without adversely increasing the temperature of sensitive food components. Jun and Irudayaraj (2003) utilized selective IR heating in the wavelength range of 5.88–6.66 μm using optical bandpass filters for the inactivation of *Aspergillus niger* and *Fusarium proliferatum* in cornmeal. The selected wavelength denatured the protein in the microorganisms, leading to a 40% increase in inactivation of *A. niger* and *F. proliferatum*. It was presumed that the absorption of energy by the fungal spores increased as a result of selective heating, leading to an increased inactivation. This is an area of application for IR heating that needs further attention. There is great potential, in particular for microbial inactivation applications, to develop combination systems that use selected bandwidths and effectively block other bandwidths.

### 15.5.3 Effect of IR heating on quality of foods

Maintaining food quality while achieving the desired microbial inactivation is always a challenge in thermal processing methods. This becomes much more important in products that are sensitive to temperature, for example fruits. This has prompted researchers to strive to optimize process parameters through various model-based approaches to prevent quality deterioration as a result of overheating. The various approaches that have been used include a Monte Carlo ray-tracing simulation technique (Tanaka *et al.*, 2006, 2007) and computation fluid dynamics (Trivittayasil *et al.*, 2011). Product sample temperatures were predicted using convection–diffusion airflow and heat transfer simulation. Surface evaporation due to surface heating by IR radiation was included to improve the accuracy of the prediction models.

The main parameters influencing the food quality and the inactivation efficiency are the distance of the IR sources from the product and the treatment time for the same power input. Maintaining the temperature just below the critical level is important to keep the heat injury to a minimum. Increasing the heating distance increases the treatment time as it takes longer for the product to reach the reference temperature for effective microbial inactivation. However, increasing the heating distance also lowers the product exposure temperature which is beneficial for achieving commercial quality

© Woodhead Publishing Limited, 2012

in heat-sensitive products. There is therefore the potential to optimize this parameter for systems treating heat-sensitive foods.

IR heating, when applied in optimized conditions was reported to maintain and in some cases improve the natural quality of food and agricultural materials when compared with conventional heating techniques (Tanaka *et al.*, 2007). This optimization includes quality parameters like color (Chua and Chou, 2005), shrinkage or appearance (Lin *et al.*, 2007), fracture in cell walls (Arntfield *et al.*, 2001; Galindo *et al.*, 2005), enzyme inactivation (Hebbar *et al.*, 2004), rancidity prevention (Kouzeh *et al.*, 1982), oxidative stability (Kumar *et al.*, 2009), reduced bitterness and protein solubility (McCurdy, 1992), increased chlorophyll content (Mongpreneet *et al.*, 2002), and taste (Burgheimer *et al.*, 1971; Khan and Vandermey, 1985). Development of brown color on the surface is considered to be an advantage of IR radiation processing (Gabel *et al.*, 2006) and recent efforts indicate that combining IR radiation with other radiation techniques, e.g. microwave, can exploit the advantages of both (Datta and Ni, 2002).

## 15.6 Conclusions and future trends

Novel technologies have been the subject of increased attention in recent years because of the constant demand from consumers for quality products that are safe to eat. Convenience of use and availability of resources give IR heating a head start in commercial adoption. Furthermore, the technique can easily be integrated into any in-line processing system.

IR applications and their various benefits for food processing are well known in the food industry. However, the use of IR irradiation for food decontamination applications has not been widely researched and reported; the literature available is limited to a very few microorganisms and some raw foods. This lack of published research reflects the inherent drawback of the use of IR, i.e. not being able to reach the interior of foods, and also the wide variation in the surface characteristics of foods. It is clear that the equipment (lamp, waveguide, power, etc.) and process parameters (time, power of exposure, distance of application, etc.) need to be optimized for specific applications.

It is difficult to compare published results because of the above-mentioned variability in surface characteristics and size and shape of the foods treated, and also because the research is often carried out using custom-made equipment. The use of a standard method of testing and reporting by the scientific community would make comparison of results much easier and would ensure faster adoption of these IR techniques. As this is mostly a thermal application, the process would need to satisfy the Federal thermal process requirements. Regulatory agencies should therefore come forward with their recommendations for adoption of this technology for specific foods. The situation becomes a little more complicated when IR is used in

© Woodhead Publishing Limited, 2012

conjunction with other technologies that are also novel (as discussed below). Coming up with a code of conduct of research in this field will push towards a concerted effort to satisfy the quality and safety demands of consumers and regulatory agencies.

Synergistic application with other conventional and novel technologies, as well as the selective targeting of specific components of the food system, remain a challenge and need further study. Standardizing the equipment and process parameters will encourage the adoption of the techniques. Integrating product development with IR radiation capabilities could lead to more new products with highly desirable characteristics in the marketplace. This requires a deeper understanding of the interaction of IR radiation with various food components and how this interaction may affect food flavor and color. Advanced theoretical and 3D radiation modeling to better understand and predict the chemical kinetics and microbial death kinetics will help to optimize IR applications for an increased number of food products.

Information on IR applications in other disciplines of science needs to be studied so that developments can be transferred effectively to the food industry. Coordinating research through consortiums and increasing industry–academic–government interactions would accelerate innovations in IR radiation for the overall benefit of food science.

## 15.7 References

AFZAL T M and ABE T (1998), 'Diffusion in potato during far infrared radiation drying', *J Food Eng*, 37, 353–365.

ALMOND BOARD OF CALIFORNIA. *Almond Almanac*. 2003. Modesto, CA: www.almondboard.com

ARNTFIELD S D, SCANLON M G, MALCOLMSON L J, WATTS B M, CENKOWSKI S, RYLAND D and SAVOIE V (2001), 'Reduction in lentil cooking time using micronization: comparison of 2 micronization temperatures', *J Food Sci*, 66, 500–505.

BINGOL G, YANG J, BRANDL M T, PAN Z, WANG H and MCHUGH T H (2011), 'Infrared pasteurization of raw almonds', *J Food Eng*, 104, 387–393.

BRANDL M T, PAN Z, HUYNH S, ZHU Y and MCHUGH T H (2008), 'Reduction of *Salmonella* Enteritidis population sizes on almond kernels with infrared heat', *J Food Prot*, 71, 897–902.

BURGHEIMER F, STEINBERG M P and NELSON A I (1971), 'Effect of near infrared energy on rate of freeze-drying of beef spectral distribution', *J Food Sci*, 36, 273–276.

CDC (2004), 'Outbreak of *Salmonella* serotype Enteritidis infections associated with raw almonds – United States and Canada 2003–2004', *Morbidity and Mortality Weekly Report*, 28, 97–99.

CHUA K J and CHOU S K (2005), 'A comparative study between intermittent microwave and infrared drying of bioproducts', *Int J Food Sci Technol*, 40, 23–39.

DAGERSKOG M and OSTERSTROM L (1979), 'Infrared radiation for food processing I: A study of the fundamental properties of infrared radiation', *Lebensm.-Wiss. Technol*, 12, 237–242.

DANYLUK M D, UESUGI A R and HARRIS L J (2005), 'Survival of *Salmonella* Enteritidis PT 30 on inoculated almonds after commercial fumigation with propylene oxide', *J Food Prot*, 68, 1613–1622.

© Woodhead Publishing Limited, 2012

DAS I and DAS S K (2010), 'Emitters and infrared heating system design', in Pan Z and Atungulu G, *Infrared heating for food and agricultural processing*, Boca Raton, FL: CRC Press.

DATTA A K and NI H (2002), 'Infrared and hot-air-assisted microwave heating of foods for control of surface moisture', *J Food Eng*, 51, 355–364.

DECAREAU R V (1985), *Microwaves in the food processing industry*, Orlando, FL: Academic Press.

ERDOGDU B S, EKIZ I H, ERDOGDU F, ATUNGULU G G and PAN Z (2010), 'Industrial applications of infrared radiation heating and economic benefits in food and agricultural processing', in Pan Z and Atungulu G, *Infrared heating for food and agricultural processing*, Boca Raton, FL: CRC Press.

FDA (2011), *Food Safety Modernization Act*, Public law 111–253, 111th Congress, Jan. 4.

FDA/CFSAN (2003), *Quantitative assessment of relative risk to public health from food borne Listeria monocytogenes among selected categories of ready-to-eat foods*, Washington DC: Food and Drug Administration.

GABEL M M, PAN Z, AMARATUNGA K S P, HARRIS L J and THOMPSON J F (2006), 'Catalytic infrared dehydration of onions', *J Food Sci*, 71, E351–357.

GALINDO F G, TOLEDO R T and SJOHOLM I (2005), 'Tissue damage in heated carrot slices. Comparing mild hot water blanching and infrared heating', *J Food Eng*, 67, 381–385.

GANDE N and MURINA P (2003), 'Prepackage surface pasteurization of ready-to-eat meats with a radiant heat oven for reduction of *Listeria monocytogenes*', *J Food Prot*, 66, 1623–1630.

GOMBAS D E, CHEN Y, CLAVERO R S and SCOTT V N (2003), 'Survey of *Listeria monocytogenes* in ready-to-eat foods', *J Food Prot*, 66, 559–569.

HALFORD R S (1957), 'The influence of molecular environment on infrared spectra', *Ann NY Acad Sci*, 69, 63–69.

HAMANAKA D, DOKAN S, YASUNAGA E, KUROKI S, UCHINO T and AKIMOTO K (2000), 'The sterilization effects on infrared ray of the agricultural products spoilage microorganisms (part 1)', in *ASABE* (American Society of Agricultural and Biological Engineering) Meeting, Milwakee, WI, 9–12 July 2000.

HAMANAKA D, UCHINO T, FURUSE N, HAN W and TANAKA S (2006), 'Effect of the wavelength of infrared heaters on the inactivation of bacterial spores at various water activities', *Int. J. Food Microbiol.*, 108, 281–285.

HAMANAKA D, NORIMURA N, BABA N, MANO K, KAKIUCHI M, TANAKA F and UCHINO T (2011), 'Surface decontamination of fig fruit by combination of infrared radiation heating with ultraviolet radiation', *Food Control*, 22, 375–380.

HASHIMOTO A, TAKAHASHI M, HONDA T, SHIMUZU M and WATANABE A (1990), 'Penetration of infrared radiation energy into sweet potato', *Nippon Shokuhin Kogyo Gakkaishi*, 37, 876–893.

HASHIMOTO A, SAWAI J, IGARASHI H and SHIMUZU M (1991), 'Effect of far-infrared radiation on pasteurization of bacteria suspended in phosphate-buffered saline', *Kagaku Kogaluc Ronbunshu*, 17, 627–633.

HASHIMOTO A, SAWAI J, IGARASHI H and SHIMIZU M (1992a), 'Effect of far-infrared irradiation on pasteurization of bacteria suspended in liquid medium below lethal temperature', *J Chem Eng Jpn*, 25, 275–281.

HASHIMOTO A, IGARASHI H and SHIMIZU M (1992b), 'Far-infrared irradiation effect on pasteurization of bacteria on or within wet-solid medium', *J Chem Eng Jpn*, 25, 666–671.

HEBBAR H U, VISHWANATHAN K H and RAMESH M N (2004), 'Development of combined infrared and hot air dryer for vegetables', *J Food Eng*, 65, 557–563.

HUANG L (2004), 'Infrared surface pasteurization of turkey frankfurters', *Inno Food Sci Emerg Technol*, 5, 345–351.


© Woodhead Publishing Limited, 2012

HUANG L and SITES J (2008), 'Elimination of *Listeria monocytogenes* on hotdogs by infrared surface treatment', *J Food Sci*, 73, M27–M31.

JAQUETTE C B and BEUCHAT L R (1998), 'Survival and growth of psychrotrophic *Bacillus cereus* in dry and reconstituted infant rice cereal', *J Food Prot*, 61, 1629–1635.

JUN S (2002), 'Selective Far Infrared Heating of Food Systems', PhD Dissertation, Pennsylvania, PA: Pennsylvania State University, Department of Agricultural and Biological Engineering.

JUN S and IRUDAYARAJ J (2003), 'A dynamic fungal inactivation approach using selective infrared heating', *Trans ASAE*, 46, 1407–1412.

KHAMIS M, SUBRAMANYAM B, DOGAN H, FLINN D W and GWIRTZ J A (2010), 'Effectiveness of flameless catalytic radiation against life stages of three stored-product insect species in stored wheat', *Julius-Kuhn-Archiv*, 425, 695–700.

KHAN M A and VANDERMEY P A (1985), 'Quality assessment of ground beef patties after infrared heat processing in a conveyorized tube broiler for foodservice use', *J Food Sci*, 50, 707–709.

KIRKPATRICK R L (1975), 'Infrared radiation for control of lesser grain borers and rice weevils in bulk wheat', *J Kansas Entomol Soc*, 48, 100–104.

KIRKPATRICK R L, BROWER J H and TILTON E W (1972), 'A comparison of microwave and infrared radiation to control rice weevils (Coleoptera: Curculionidae) in wheat', *J Kansas Entomol Soc*, 45, 434–438.

KOUZEH K M, VAN ZUILICHEM D J, ROOZEN J P and PILNIK W (1982), 'A modified procedure for low temperature infrared radiation of soybeans. II. Inactivation of lipoxygenase and keeping quality of full fat flour', *Lebensm. Wiss. Technol*, 15, 139–142.

KRISHNAMURTHY K (2006), 'Decontamination of Milk and Water by Pulsed UV Light and Infrared Heating', PhD Dissertation, Pennsylvania, PA: Pennsylvania State University, Department of Agricultural and Biological Engineering.

KRISHNAMURTHY K, KHURANA H K, JUN S, IRUDAYARAJ J and DEMIRCI A (2008), 'Infrared heating in food processing', *Comp Rev Food Sci Food Safety*, 7, 2–13.

KRISHNAMURTHY K, TEWARI J C, IRUDAYARAJ J and DEMIRCI A (2010a), 'Microscopic and spectroscopic evaluation of inactivation of *Staphylococcus aureus* by pulsed UV light and infrared heating', *Food Bioproc Technol*, 3, 93–104.

KRISHNAMURTHY K, JUN S C, IRUDAYARAJ J and DEMIRCI A (2010b), 'Infrared radiation heating for food safety improvement', in Pan Z and Atungulu G, *Infrared heating for food and agricultural processing*, Boca Raton, FL: CRC Press.

KUMAR C M, RAO A G A and SINGH S A (2009), 'Effect of infrared heating on the formation of sesamol and quality of defatted flours from *Sesamum indicum* L', *J Food Sci*, 74, H105–H111.

LEE S-Y, OH S-W, CHUNG H-J, REYES-DE-CORCUERA J I, POWERS J R and KANG D-H (2006), 'Reduction of *Salmonella* enteric serovar Enteritidis on the surface of raw shelled almonds by exposure to steam', *J Food Prot*, 69, 591–595.

LENTZ R R, PESHECK P S, ANDERSON G R, DEMARS J and PECK T R, (THE PILLSBURY CO.), (1995), *Method of processing food utilizing infra-red radiation.* US patent 5,382,441.

LIN Y P, LEE T Y, TSEN J H and KING AN-ERL V (2007), 'Dehydration of yam slices using FIR-assisted freeze-drying', *J Food Eng*, 79, 1295–1301.

MCCURDY S M (1992), 'Infrared processing of dry peas, canola, and canola screenings', *J Food Sci*, 57, 941–944.

MARQUENIE D (2002), 'Evaluation of Physical Techniques for Surface Disinfection of Strawberry and Sweet Cherry', PhD Thesis, Belgium: Katholieke Universiteit Leuven.

MEAD P S, SLUTKER L, DIETZ V, MCCRAIG L F, BRESEE S, SHAPIRO C, GRIFFIN P and TAUXE R V (1999), 'Food related illness and death in the United States', *Emerg Infect Dis*, 5, 607–625.

MONGPRENEET S, ABE T and TSURUSAKI T (2002), 'Accelerated drying of welsh onion by far infrared radiation under vacuum conditions', *J Food Eng*, 55, 147–156.

© Woodhead Publishing Limited, 2012

PAN Z, KHIR R, LEWIS R, GODFREY L, SALIM A and THOMPSON J F (2006), 'Simultanous rough rice drying and disinfestations using infrared radiation', in *ASABE* (American Society of Agricultural and Biological Engineering) Meeting, Portland, OR, 9–12 July.

PAO S, KALANTARI A and HUANG G (2006), 'Utilizing acidic sprays for eliminating *Salmonella* enteric on raw almonds', *J Food Sci*, 71, M14–M19.

PIDWIRNY M (2006), 'The Nature of Radiation', in *Fundamentals of Physical Geography*, 2nd edn. Available from: http://www.physicalgeography.net/fundamentals/6f.html (accessed 11 July 2010).

ROSENTHAL I (1992), *Electromagnetic radiations in food science*, Berlin, Heidelberg: Springer Verlag.

SAKAI N and HANZAWA T (1994), 'Applications and advances in far-infrared heating in Japan', *Trends Food Sci Technol*, 5, 357–362.

SAKAI N, FUJII A and HANZAWA T (1993), 'Heat transfer analysis in a food heated by far infrared radiation', *Nippon Shokuhin Kogyo Gakkaishi*, 40, 469–477.

SANDU C (1986), 'Infrared radiative drying in food engineering: a process analysis', *Biotech Prog*, 2, 109–119.

SATO H and SHIBUKAWA S (1989), 'Effects of radiant characteristics of heaters on food processing', *Nippon Kasei Gakkaishi*, 40, 987–994.

SATO H, HATAE K and SHIMADA A (1992), 'Effects of radiant characteristics on crust formation and coloring process of food surface. Studies of radiation heating condition of the food: Part 1', *Nippon Shokuhin Kogyo Gakkaishi*, 39, 784–789.

SAWAI J, SAGARA K, IGARASHI H, HASHIMOTO A, KOKUGAN T and SHIMIZU M (1995), 'Injury of *Escherichia coli* in physiological phosphate buffered saline induced by far-infrared irradiation', *J Chem Eng Jpn*, 28, 294–299.

SAWAI J, KOJIMA H, IGARASHI H, HASHIMOTO A, FUJISAWA M, KOKUGAN T and SHIMUZU M (1997), 'Pasteurization of bacterial spores in liquid-medium by infra-red irradiation', *J Chem Eng Jpn*, 30, 170–172.

SAWAI J, MATSUMOTO K, SAITO T-A, ISOMURA Y and WADA R (2009), 'Heat activation and germination-promotion of *Bacillus subtilis* spores by infrared radiation', *Int Biodeterior Biodegrad*, 63, 196–200.

SHERIDAN P and SHILTON N (1999), 'Application of infra-red radiation to cooking of meat products', *J Food Eng*, 41, 203–208.

SHUMAN A C and STALEY C H (1950), 'Drying by infra-red radiation', *Food Technol*, 4, 481–484.

STAACK N, AHRNE L, BORCH E and KNORR E (2008a), 'Effect of infrared heating on quality and microbial decontamination in paprika powder', *J Food Eng*, 86, 17–24.

STAACK N, AHRNE L, BORCH E and KNORR E (2008b), 'Effects of temperature, pH, and controlled water activity on inactivation of spores of *Bacillus cereus* in paprika powder by near-IR radiation', *J Food Eng*, 89, 319–324.

STUMBO C R (1973), *Thermobacteriology in food processing*, Orlando, FL: Academic Press.

SUBRAMANYAM B (2004), 'Hot technology for killing insects', *Milling Journal*, 48–50.

SUBRAMANYAM B (2005), 'Hot technology for killing insects [part II]', *Milling Journal*, 58–59.

SUSEK W (2010), 'Thermal microwave radiation for subsurface absolute temperature measurement', *Acta Physica Polonica A*, 118, 1246–1249.

TANAKA F, MORITA K, IWASAKI K, VERBOVEN P, SCHERLINCK N and NICOLAI B (2006), 'Monte Carlo simulation of far infrared radiation heat transfer: theoretical approach', *J Food Proc Eng*, 29, 349–361.

TANAKA F, VERBOVEN P, SCHEERLINCK N, MORITA K, IWASAKI K and NICOLAI B (2007), 'Investigation of far infrared radiation heating as an alternative technique for surface decontamination of strawberry', *J Food Eng*, 79, 445–452.

TRIVITTAYASIL V, TANAKA F and UCHINO T (2011), 'Investigation of deactivation of mold conidia by infrared heating in a model-based approach', *J Food Eng*, 104, 565–570.

© Woodhead Publishing Limited, 2012

WIHODO M, HAN Y, SELBY T L, LORCHEIM P, CZARNESKI M, HUANG G and LINTON R H (2005), 'Decontamination of raw almonds using chlorine dioxide gas', in *Institute of Food Technologists Annual Meeting*, New Orleans, LA, 15–20 July 2005.

YAMADA H (2010), 'Investigation of Factors Affecting Microbial Sterilization Effect by Electromagnetic Waves', MSc. Thesis, Japan: Kyushu University.

YANG J, BINGOL G, PAN Z, BRANDL M T, MCHUGH T H and WANG H (2010), 'Infrared heating for dry-roasting and pasteurization of almonds', *J Food Eng*, 101, 273–280.

© Woodhead Publishing Limited, 2012

# 16

# Microbial decontamination of food by non-thermal plasmas

**M. G. Kong, Loughborough University, UK**

**Abstract**: Gas plasmas contain an electrically modulated group of transient agents including reactive species, excited atoms and molecules, electrons and ions, and photons that carry various forms of energies (e.g. chemical, kinetic, optical and electromagnetic, and thermal). Both individually and collectively, they induce a range of biological effects on microbes, animal cells, and living animal and plant tissues. Of particular interest are their antimicrobial effect and its implication for food decontamination. This chapter provides an overview of the principles and mechanisms, the technological capabilities and limitations, and selected examples of applications of non-thermal gas plasmas generated in open air at atmospheric pressure, often known as cold atmospheric plasmas (CAPs). There is an extensive and growing body of evidence that cold atmospheric plasmas are very effective against a wide range of microbes (e.g., bacterial spores and biofilm-forming microorganisms) as well as biomolecules such as proteins and lipids. As a technology, CAPs can be adapted and scaled up to treat both contaminated fluids (both air and liquids) and solid surfaces of different shapes and composition (e.g., fresh vegetables and dry herbs). When designed and engineered appropriately, they are inexpensive and environmentally friendly. Gas plasma-based food decontamination represents a new and novel route to food safety control, but there remain significant gaps in scientific understanding and technological capability. Yet gas plasma techniques are already being used in the food industry and they have the potential to become an essential technology platform for food decontamination.

**Key words**: ionized gases, electrical discharges, microbial inactivation, protein denaturation, arrays of gas plasma jets.

© Woodhead Publishing Limited, 2012

## 16.1 Introduction

An atom or a molecule has an equal number of electrons and protons with the electrons orbiting around the proton-containing nucleus of the atom or the molecule (hereafter 'the atom' is used). The electrostatic force between the electrons and the protons binds the electrons to the atom, and normally the atom is electrically neutral. When sufficient energy is supplied externally to a gas atom, some of its orbiting electrons may gain adequate energy to escape the electrostatic confinement of protons and become freed as an independent identity. At this point, the gas atom becomes positively charged and hence an ion whereas the freed electrons are further accelerated to gain more and more kinetic energy by the external energy source. These electrons then pass their kinetic energy to atoms of the background gas through collisions between them, thus inducing chemical dissociation, excitation of atom-bound electrons to orbits of higher energy, and indeed release of more free electrons from collided gas atoms. This is an electron multiplication process. On the other hand, the free electrons may also be lost, for example, through recombination with ions and surface absorption on a nearby solid (e.g., an electrode). When the number of electrons generated can compensate the number of electrons lost, the gas is said to be ionized and the ionized gas is gas plasma (Kogelschatz, 2002).

The most common form of gas plasma is an electrical discharge. Gas plasma is the fourth state of matter after solid, liquid and gas, and it offers a novel form of non-equilibrium chemistry, not easily accessible in other states of matter (Walsh and Kong, 2007). In addition to charged particles (i.e. electrons and ions), products of electron collisions with gas atoms or molecules include free radicals, excited state atoms and molecules, other reactive species, UV photons, and transient electromagnetic fields, all capable of inducing significant biological effects (Kong *et al.*, 2009; Fridman *et al.*, 2008). Depending on its chemical composition and its dose, a gas plasma can inflict lethal damage to microbes and indeed infectious biomolecules (Kong *et al.*, 2009; von Keudell *et al.*, 2010). Therefore gas plasmas offer a new route to biological decontamination and form a generic technology capability to serve many industrial, medical and healthcare sectors including the food industry (Vleugels *et al.*, 2005; Deng *et al.*, 2007; Knorr *et al.*, 2011).

While many plasma agents are known to be bactericidal, their use for biological decontamination has in the past been limited by the need for a vacuum chamber to contain instabilities and overheating in gas plasma (Walsh and Kong, 2007). Many materials important in the food industry are not vacuum-compatible, such as packed meat, salads, fruits and vegetables, seeds, nuts and powders. A technology with a vacuum chamber as an essential item is also undesirable in terms of capital cost, the need for large floor space, and the additional time required to achieve the vacuum environment. Since the late 1980s, however, there has been a sequence of technological breakthroughs that collectively make it possible now to obtain near room-temperature gas

© Woodhead Publishing Limited, 2012

plasma under ambient conditions without a vacuum chamber (Kogelschatz, 2002). In particular, the use of nanosecond-pulsed electrical excitation allows for large quantities of bactericidal plasma agents to be produced with little heat generation (Walsh and Kong, 2007). These discharges are commonly known as cold atmospheric plasmas (CAPs). They have been shown to achieve 6 log reductions of many microorganisms including *Escherichia coli*, *Listeria*, Methicillin-resistant *Staphylococcus aureus* (MRSA), *Bacillus subtilis* and *Clostridium difficile* spores, as well as biofilm-forming bacteria such as *Pantoea agglomerans* and *Pseudomonas aeruginosa* (Kayes *et al.*, 2007; Kong *et al.*, 2009). Controlled heat generation and effective antimicrobial effects open the door possibly for CAPs to disinfect many heat-labile materials such as plastics, cotton fabrics, and indeed plant and animal tissues directly.

## 16.2 Plasma-based food decontamination: principles and mechanisms

Depending on the amount of energy dissipation in an electrical discharge, the main plasma effects may be dominated by macroscopic physical phenomena, a complex group of reaction chemistry, or a combination of the two. Macroscopic physical phenomena typically associated with gas plasmas include considerable heat generation, shock wave, strong gas flow and electromagnetic fields (Locke *et al.*, 2006). These plasma-mediated physical forces are useful for destruction and/or removal of solid matter, and they can be applied specifically to the local contaminated areas. In the context of contamination in food and food processing, some applications can benefit from such plasma-mediated physical forces. An example is plasma cleaning of metallic surfaces and components of food processing equipment, from which surface-borne contaminants may be very effectively removed by physical forces generated in gas plasmas. Electrical energy deposition in such plasmas is typically much above 1 Joule per electrical pulse (Locke *et al.*, 2006), and for the convenience of the application community they may be regarded as high-energy atmospheric plasmas. In comparison with low-energy atmospheric plasmas, particularly those with gas temperature much less than 100°C (see below), high-energy atmospheric plasmas tend to have greater intensity to their optical emission and presence of unstable and moving filaments.

For heat-labile materials such as plastics, cotton fabric, and indeed living plant and animal tissues, the gas temperature of the plasma should ideally be less than 60°C to avoid permanent material damage (De Boer *et al.*, 1998; Liu and Kong, 2011). Bactericidal effects of sub-60°C gas plasma predominantly rely on its reaction chemistry involving reactive oxygen species (ROS) such as singlet oxygen ($^1O_2$), atomic oxygen (O), superoxide ($O_2^-$), ozone ($O_3$), hydroxyl oxide (OH) and hydrogen peroxide ($H_2O_2$), reactive nitrogen species

© Woodhead Publishing Limited, 2012

(RNS) such as nitric oxide (NO), charged particles, and UV photons (Kong *et al.*, 2009). Compared to high-energy atmospheric plasmas, sub-60°C atmospheric plasmas and their applications to food decontamination are more recent after it has become possible to separate effective production of reactive plasma species from significant electrical energy deposition in the plasma. In addition to the use of nanosecond-pulsed electrical excitation (Walsh and Kong, 2007; Pai *et al.*, 2009), cold atmospheric plasmas may be configured as plasma plumes or an array of plasma plumes for which the region of the electrical energy input is separated physically from the downstream region of plasma processing. This physical separation makes the properties of the plasma less dependent on the properties of the sample.

An example of plasma plumes and plasma plume arrays (Kong *et al.*, 2011) is shown in Fig. 16.1. Such atmospheric plasma sources are particularly useful for treatment of three-dimensionally structured objects, and are scalable to suit industrial-scale applications. Their low gas temperature is directly associated with the level of the electrical energy deposition in the gas plasma, and the latter is typically less than 100 μJ/pulse (Liu and Kong, 2011). Unlike high-energy atmospheric plasmas, sub-60°C atmospheric plasmas have more efficient production of highly reactive plasma species from the deposited electrical energy and their biological effects stem largely from plasma chemistry.

Figure 16.2 shows two SEM images of *B. subtilis* spores on a filter membrane, one before (Fig. 16.2a) and the other after plasma treatment for 5 min (Fig. 16.2b) (Deng *et al.*, 2006). The plasma source used was a room-temperature plasma jet in atmospheric helium-oxygen flow at 3 slm (standard litre per minute), similar to that shown in Fig. 16.1(a). Oxygen flow rate was 2 sccm (standard cubic centimetres per minute), representing about 0.07% of the helium content. The plasma jet was sustained at 5 kV

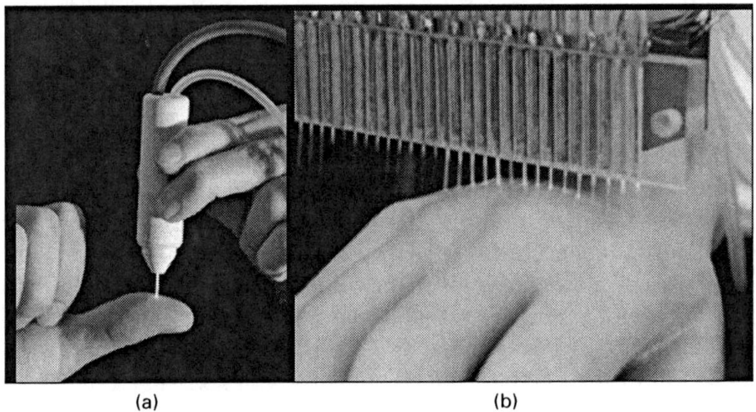


**Fig. 16.1** (a) A cold atmospheric plasma (CAP) jet treating a thumb and (b) a CAP jet array treating a hand (adapted from Kong *et al.*, 2011).

© Woodhead Publishing Limited, 2012

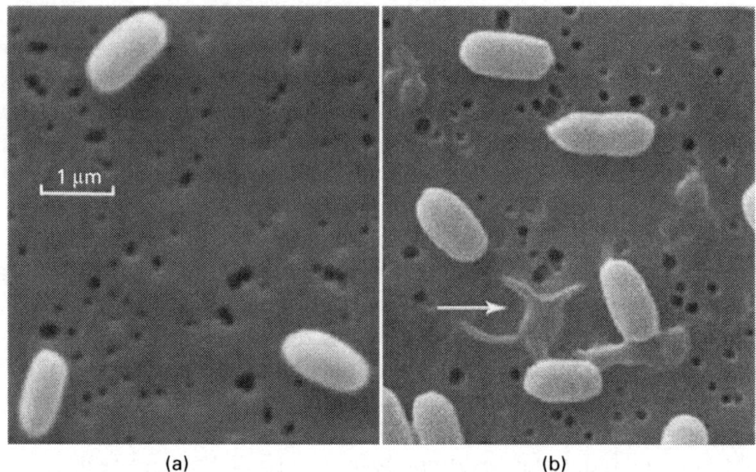


**Fig. 16.2** Scanning electron microscope (SEM) images of *B. Subtilis* spores (a) before plasma treatment and (b) after plasma treatment for 5 min. Arrow indicates ruptured spores (adapted from Deng *et al.*, 2006).

and 33 kHz, the electrode nozzle was placed 1 cm away from the spore-laden membrane filter, and the membrane was placed on top of a layer of technical no. 3 agar in a Petri dish. Prior to plasma treatment, the spore samples were washed four times to remove unwanted spore debris and then heat-shocked to eliminate any vegetative cells. It is clear from Fig. 16.2 that treatment of the helium-oxygen plasma jet caused a range of damage, including slight spore shrinkage and deformation, through leakage of cytoplasm contents, to spore fragmentation and lyses as well as rupture of spore membrane (marked with an arrow in Fig. 16.2b).

To see which cell components may have been damaged by plasma, the spore samples were stained with propidium iodide which allowed viable bacteria to fluoresce green and fatally injured bacteria to fluoresce red. Fluorescence images of stained *B. subtilis* samples before and after plasma treatment indicated severe spore damage (Deng *et al.*, 2006). The substantial spore damage demonstrated by the SEM in Fig. 16.2 was interesting as this was inflicted with only a small amount of oxygen mixed into the background helium flow (i.e. 0.07%). Figure 16.3 shows the inactivation kinetics of the spore samples (Deng *et al.*, 2005). It is shown that spore inactivation efficacy is significantly influenced by the temperature at which spores were formed and so physiological conditions of the microbial samples are important in how the microbe community reacts to the impact of the impinging plasma jet. Spores formed at lower temperatures tend to be less resistant to plasma treatment and this is similar to other stresses. At a sporulation temperature of 22°C, a 6 log reduction of *B. subtilis* spores was achieved over a plasma treatment period of 5 min. Since the plasma plume had a diameter much smaller than the size of the spore sample, it was necessary to scan the plasma

© Woodhead Publishing Limited, 2012

jet across the spore sample. The actual plasma contact time in this case was about one tenth of the plasma treatment time, or 60 s at most.

Inactivation kinetics curves shown in Fig. 16.3 have multiple phases and this is characteristic of plasma inactivation reported so far (Kayes *et al.*, 2007; Kong *et al.*, 2009). Current consensus on possible reasons is associated with properties of the microbial sample, both at the individual and community levels. At an individual cell level, it is possible that a sufficient amount of plasma-mediated damage, such as that inflicted on the cell membrane, needs to be accumulated before cell death is facilitated. The initial phase of accumulation of sub-lethal damage would manifest itself with a slow killing of microorganisms, before a phase of more rapid killing. Yet for many plasma inactivation experiments, the initial phase of slow killing may be too short to be observed, as shown in Fig. 16.3. To this end, multiple phases are also explained by the fact that surface-borne microbes are often formed in multiple layers. Microorganisms on the uppermost layers of the sample are directly exposed to an impinging plasma and therefore become inactivated rapidly. However, after their inactivation, debris from the microorganisms on the uppermost layers remains on the top of the microbial sample and represents a physical barrier to the infusion of incoming plasma species. This restricts the access of plasma species to microbes lying on lower layers within the biofilm and results in a slower inactivation rate (Moisan *et al.*, 2001; Deng *et al.*, 2005; Yu *et al.*, 2006). Multiple phases in inactivation kinetics are indicative of subpopulations of the microbe community to the treatment of gas plasma. In addition to cell clumping due to surface deposition of

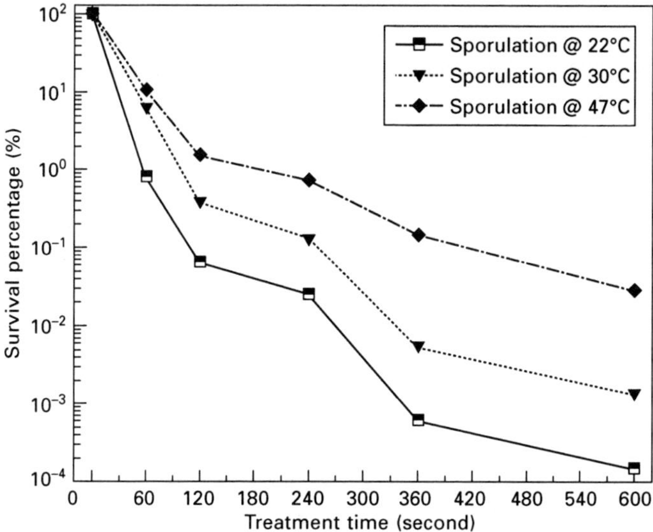


**Fig. 16.3** Survival curves of *Bacillus subtilis* spores at different sporulation temperatures and similar initial spore densities of around $5 \times 10^7$ per filter (adapted from Deng *et al.*, 2005).

© Woodhead Publishing Limited, 2012

microbes, there are a number of sources of heterogeneity in the original microbial sample. These include physical protection of microbes by pores and crevices on the substratum surface, physiological heterogeneity in the original microbial population, and possible change in cell phenotypes (e.g. switching from planktonic cells to sessile cells).

Many of the above heterogeneities are to be expected as they are also observed in inactivation studies of non-plasma techniques such as toxic gases and UV light. Much less obvious is the heterogeneity mediated by plasma interaction with the surface-borne microbial community. In a recent study of plasma jet inactivation of *Listeria innocua* by the author's group, the listerial cells were prepared in one single monolayer on a membrane filter by means of a vacuum filtration technique. Despite the initial listerial sample being homogeneously distributed on the substratum surface without cell clumping, the inactivation kinetics again exhibited a multi-phasic character (data not shown). Close examination of the plasma treated *Listeria* sample at different instances of plasma treatment revealed an intriguing sequence of events that appear to indicate some form of microbial self-organization stimulated by plasma action, not dissimilar to those observed for non-plasma stimulation (e.g. deprivation of nutrient supply) (Ben-Jacob *et al.*, 2000).

Specifically, the SEM image after 15 s of plasma treatment shows a circular pattern of cell aggregates mainly made of heavily damaged *Listeria* cells (Fig. 16.4a). Seemingly viable cells do exist (circled in Fig. 16.4a), and this is consistent with inactivation kinetics at this time point when there were still at least three orders of magnitude of viable *Listeria* cells. Formation of this circular pattern of cell aggregates is believed to be associated with physical forces caused by plasma-mediated sample drying. In other words, the microbial community developed heterogeneity in terms of the topology of their surface distribution under the influence of the plasma jet. This plasma-mediated heterogeneity is novel and indicative of the complexity of plasma interaction of surface-borne microorganisms. From 0 to 15 s, the inactivation kinetics were characterized with a phase of rapid reduction of viable cells by about 3 logs. This was then followed by a distinct second phase with a slower reduction rate of viable cells between 15 and 60 s. Figure 16.4(b) shows a typical SEM image in the second phase in which the nodes of the circular pattern became more compact with severely damaged listerial cells but each node had an opening at its top. Interestingly, viable cells (marked with a white circle) now appeared to reside underneath a lump of heavily damaged cells formed as part of the node rim. As a result, the nodes of damaged cells acted as a refuge for the viable cells and as such the node structure with a top opening is known as the cell refuge.

Severely damaged cells after 15 s of plasma treatment made up the majority of the cell aggregates and are likely to remain near their physical locations in Fig. 16.4(a), as their motility would have been substantially compromised. Compared to the regions devoid of cells in Fig. 16.4(a), these cell aggregates appeared more hydrated in a way that is reminiscent of a biofilm community

© Woodhead Publishing Limited, 2012

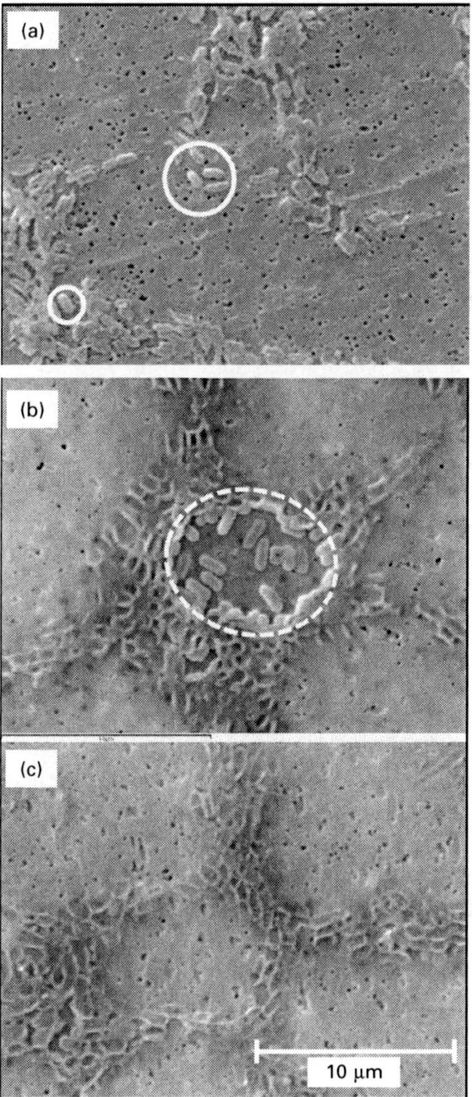


**Fig. 16.4** SEM images of initially homogeneous *Listeria* cells at $4 \times 10^7$ CFU after being treated by a pulsed radio-frequency atmospheric He-$O_2$ plasma jet at $V_{pp} = 6.88$ kV for (a) 15 s (formation of circular pattern of microbes); (b) 45 s (initiation of cell refuge formation); and (c) 90 s. Cells within the white circles in (a) and (b) are viable.

that contains water channels. Cells of *L. innocua* surviving 15 s of plasma treatment would retain many of their functionalities including motility, and by means of pilus-mediated twitching or gliding, they may have retained the ability to move to regions favouring their subsequent survival. The movement of viable cells may have been helped by the hydrophilic cell surface of

© Woodhead Publishing Limited, 2012

listerial cells (Tresse *et al.*, 2006) that makes them more likely to move to a more hydrated region than to adhere to the dried surface of the void regions. Consequently there is an anisotropic movement of the viable cells towards the much less motile filamentous aggregates of severely damaged cells. Therefore, the circular cell aggregation patterning may be regarded as isles of discrete craters created by the impinging plasma plume and the rims of these cell refuges are stationary due to loss of motility following plasma treatment. On the other hand, viable cells would have retained their motility to move to the more hydrated rim region of cell aggregates and in doing so they would attain refuge by means of the physical protection offered by cell aggregates against subsequent plasma treatment. Despite the formation of cell refuges, further plasma treatment to 90 s is seen in Fig. 16.4(c) to have inflicted severe damage to all *Listeria* cells including those which were protected by the cell refuge in Fig. 16.4(b). At this point, a 6 log reduction of viable *Listeria* cells was achieved.

It is of interest to identify the main plasma agents responsible for microbial inactivation by sub-60°C atmospheric plasmas. This depends critically on the chemical composition of the plasma-forming gas and the choices for the latter are in principle numerous. However, given the physical environment of and the cost consideration for food decontamination, plasma-forming gases are most likely to be a mixture containing oxygen, nitrogen, trace amount of water moisture, and possibly an inert gas (e.g. argon, low-grade helium or nitrogen) at atmospheric pressure. Sub-60°C gas plasmas thus generated contain a panel of ROS and RNS in addition to charged particles and UV photons. Figure 16.5 shows optical emission spectra of an atmospheric plasma jet, similar to that in Fig. 16.1a, in (a) a helium flow and (b) in a helium-oxygen flow ($O_2$/He = 0.1%) (Perni *et al.*, 2007). In both cases, there are emission lines associated with excited oxygen atoms, superoxide ($O_2(^1\Sigma_g^+)$ and $O_2(^1\Delta_g)$ metastables), hydroxyl radicals (OH), excited nitrogen atoms, excited nitrogen molecules, molecular nitrogen ions, nitric oxide (NO), helium metastables, and UV photons. A 0.1% admixture of oxygen in the helium flow alters the intensity profile of these emission lines suggesting a change in the concentration profile of relevant plasma species (Perni *et al.*, 2007). One important feature of gas plasmas is that it is in general very difficult to produce one plasma species without the presence of others because many of them are involved in the same set of chemical reactions and that different plasma species may work synergistically in their biological effects. This adds complexity to mechanistic studies of plasma inactivation. However, despite a large array of possible bactericidal plasma species as indicated in Fig. 16.5, not all plasma species reach their concentration thresholds above which their biological effects become significant. In other words, some plasma species may remain as bystanders in a given plasma inactivation experiment.

To link reactive species and UV photons generated in gas plasma to bacterial inactivation, bacterial mutants were considered to identify whether specific repair mechanisms may be triggered by a sub-60°C atmospheric plasma jet

© Woodhead Publishing Limited, 2012

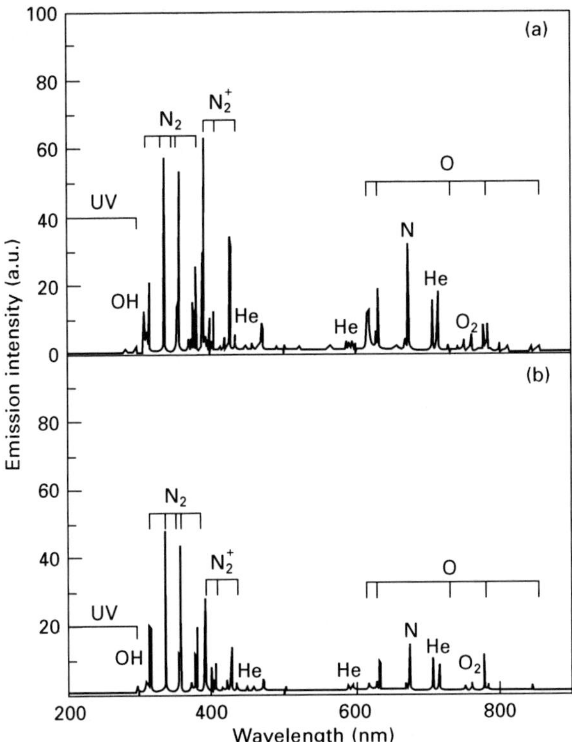


**Fig. 16.5** Optical emission spectra of a sub-60°C atmospheric plasma jet in (a) a helium flow and (b) in a helium-oxygen flow, both at the sample point (adapted from Perni *et al.*, 2007).

(Perni *et al.*, 2007). *E. coli* K12 MG1655 single gene mutants (Blattner *et al.*, 1997) were constructed using the method of Datsenko and Wanner (2000), where a kanamycin resistance gene cassette was used to replace the *rpoS* gene, and a chloramphenicol resistance was used to replace the *recA* and *soxS* genes. Wild type *E. coli* K12 strain MG1655 cells and Δ*recA*, Δ*rpoS*, Δ*soxS* mutants of this strain were stored and prepared as described elsewhere (Yu *et al.*, 2006), except that the Δ*recA* and Δ*soxS* cells were stored at 4°C on solid LB plates containing 100 μg/ml of chloramphenicol, whilst the Δ*rpo*S cells were stored on plates containing 50 μg/ml kanamycin to select against loss of the antibiotic resistance genes that had replaced the *recA*, *soxS* or *rpoS* genes.

Figure 16.6 shows the results of plasma inactivation experiments. The Δ*recA* mutants lack the *recA* gene which is associated with DNA repair (Cox, 1998), and in the case of plasma treatment this is most likely to be facilitated by UV photons. Both the wild-type *E. coli* and its Δ*recA* mutant are seen to undergo approximately a 2.5 log reduction over 3 min of plasma treatment. This suggests that little DNA damage was taking place, and therefore there

© Woodhead Publishing Limited, 2012

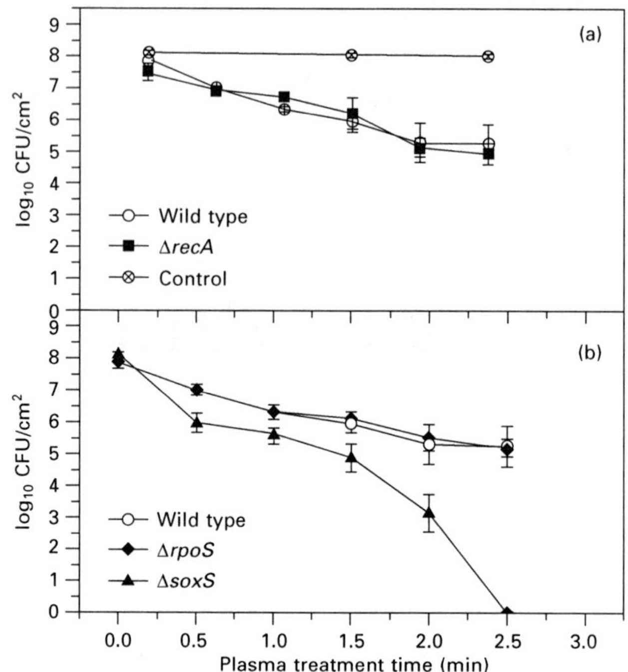


**Fig. 16.6** Inactivation kinetics of wild type *E. coli* K12 and its (a) Δ*recA* mutant; and (b) Δ*rpoS* and Δ*soxS* mutants. The atmospheric He-$O_2$ plasma jet used is the same as that used in Fig. 16.4. Control in (a) is the kinetics of the most vulnerable Δ*soxS* subject to a unionized He-$O_2$ gas flow (adapted from Perni *et al.*, 2007).

was an insignificant UV effect over the time course of the experiment. The Δ*rpoS* mutant was considered to understand the role of OH radicals, as it lacks the *rpoS* gene and has a reduced resistance to OH radicals (Visick and Clarke, 1997). Again inactivation kinetics of the Δ*rpoS* mutants are very similar to that of the wild type, suggesting that OH radicals were unlikely to be important in the inactivation seen in Fig. 16.6. As the *rpoS* gene also regulates aspects of *E. coli* resistance to environmental stresses such as heat, acid and salt (Cheville *et al.*, 1996), this result indicates that any thermal effects in this inactivation study were likely to be negligible.

Figure 16.6(b) shows that the Δ*soxS* mutants suffered an 8 log reduction after 3 min of plasma treatment, a clear contrast to the susceptibility of the wild type and the two other mutants to plasmas. The Δ*soxS* mutant lacks the *soxS* gene that regulates it, and is therefore required for *E. coli* resistance against reactive oxygen species, particularly nitric oxide and superoxide-generating agents (Li and Demple, 1994). The observation of a very significant susceptibility of the Δ*soxS* mutants to plasma suggests that NO and/or superoxide-generating agents play a dominant role in the observed *E. coli* inactivation.

© Woodhead Publishing Limited, 2012

## 16.3 Capabilities and limitations of non-thermal plasma

As shown in Section 16.2 and numerous reports in the literature, sub-60°C atmospheric plasmas possess the basic capability as a novel disinfection technology to compete with and complement the current decontamination strategies such as dry and moist heat, freezing, high gas pressure, aggressive and stable chemicals (in both gas and liquid forms), and enzyme-based treatments. There is already a sound scientific underpinning to plasma decontamination, in particular the role of non-equilibrium and electrically modulated reaction chemistry of highly transient reactive species (e.g. reactive oxygen species and reactive nitrogen species). In the context of biomolecule degradation and microbial inactivation, these transient and highly reactive plasma species represent a novel form of physicochemical energies that are not easily accessible under normal conditions, and as such they offer a new dimension in our defence against infection. Through diffusion and/or convection, these plasma species can be delivered in either gas or liquid form to contaminated surfaces, crevices, pores and surface cracks. Essentially, the current understanding and technological capability is already adequate in driving forward some applications and this has increased focus on engineering implementation.

For decontamination of food and food preparation equipment, one of the first engineering questions is whether the usually small atmospheric plasma could be sufficiently scaled up to be useful in an industrial setting. An example of practical challenges is decontamination of a conveyor belt, which is typically 1–2 meters wide. Figure 16.7 shows a one-metre wide atmospheric helium plasma formed in 8 mm gap (Walsh *et al.*, 2008), which can be adapted for disinfecting the surface of a conveyor belt or similar food processing surfaces. The one-metre wide plasma source is known as a plasma screen (Walsh *et al.*, 2008), and its use may be upscaled further with several plasma screens configured as an array to disinfect a moving conveyor belt simultaneously. The spatial homogeneity of each plasma screen seen in Fig. 16.7 is useful to ensure consistent disinfection efficacy across the full width of the conveyor belt. Electrical power consumption was found to be about 0.3–1.0 W/cm$^3$ under sinusoidal electrical excitation, or up to 32 W for one plasma screen. This is based on a conventional plasma source technology and can be further

**Fig. 16.7** Plasma screen: a radio-frequency atmospheric dielectric barrier discharge of 1 m wide. It was generated at 4.8 MHz and between two parallel electrodes separated by a helium gap of 8 mm. The dissipated power was 70 W, and the helium flow rate was 5 slm. The exposure time of the image was 50 ms (adapted from Walsh *et al.*, 2008).

© Woodhead Publishing Limited, 2012

enhanced with nanosecond pulsed excitation (Walsh and Kong, 2007; Pai *et al.*, 2009). Using the similar sinusoidal excitation but in a molecular gas such as air, electrical consumption tends to increase by a factor of 10, or up to 320 W. Therefore in its upscaled form of several plasma screens working in an array, such sub-60°C atmospheric plasma screen systems are likely to consume similar electrical power to that of a kitchen kettle. Given that the timescale for a 6 log reduction of spores is 60 s for helium-oxygen plasmas and 10–30 s for air plasmas, the electrical energy cost of such a scaled-up plasma system is in the order of 0.05–0.5 kWh per disinfection process. The short timescale of the plasma decontamination process also suggests a small cost of gas consumption.

For uneven surfaces or surfaces of three-dimensional structures, an array of plasma jets such as that shown in Fig. 16.1(b) is more effective and in principle such plasma jet arrays can be upscaled to span a length of more than 1 m (Cao *et al.*, 2010). An additional benefit of plasma jet arrays is that plasma generation usually takes place in the upper-stream electrode region and is distant from the location of the downstream sample, in contrast to the plasma screen where the object to be treated is often placed on one electrode and so the plasma generation region co-locates with the plasma processing region. A spatial separation of these two regions significantly reduces the influence of plasma-sample interactions on plasma generation. This is particularly useful when water-containing samples need to be reliably plasma-treated. In fact, sub-60°C plasma jets can be realized in a plasma-forming gas containing significant water vapour (e.g. 2000 ppm) (Liu and Kong, 2011). From a plasma chemistry standpoint, this is also beneficial as it introduces abundant $H_2O_2$ and OH radicals. It is worth mentioning that many bactericidal agents such as $H_2O_2$, OH radicals, superoxide, NO and ozone may be generated individually using non-plasma methods. However, the advantage of gas plasma is that these species are generated at the location of contamination and at the point of need. Clearly this removes the need to store these reactive agents in gas bottles and is in practice desirable. In addition, their production by means of gas plasma brings many other and more important advantages such as the participation of short-lived species (e.g. OH radicals in a moist environment), a location-specific application or even on-site generation of reactive species thus better controlling possible damage to uncontaminated materials, and synergistic effects among different reactive species. It has already been discussed in the example of *E. coli* mutants (Fig. 16.6) that bacterial resistance against OH radicals, superoxide and NO is associated with different genes, and as such simultaneous attack on different cellular machineries by these plasma species is likely to be synergistic (Kong *et al.*, 2009).

As already discussed, gas plasma is a surface treatment process and therefore their penetration into a porous material is restricted by the extent of the diffusion and convection of plasma species and their half-lives. While depending on specific properties of the porous material, microorganisms,

© Woodhead Publishing Limited, 2012

and relevant plasma species, it is accepted that plasma penetration depth is no more than 100 μm with today's technology. This limitation is evident in the case of biofilm disinfection, for which the effectiveness of plasma inactivation has been shown to be compromised by the presence of biofilm (Vleugels *et al.*, 2005; Abramzon *et al.*, 2006; Goree *et al.*, 2006). Figure 16.8 shows that when surface-borne *P. agglomerans* were allowed to grow for 24 h, plasma inactivation of viable cells was reduced by about 2 logs compared to samples grown for 12 h (Vleugels *et al.*, 2005) While this is characteristic of plasma treatment, it is possible that penetration depth of future plasma technologies may go beyond the current limit of 100 μm, particularly when in combination with other technologies. For example, the use of nanoparticles to trap reactive plasma species can in principle tap into the ability of nanoparticles to penetrate many centimetres into plant and animal tissues (Kong *et al.*, 2011). This possibility and others will need further studies and development.

Compared to many conventional decontamination technologies (e.g. heat and chemicals), gas plasmas are a comparatively new and novel technology for practical decontamination with a development history of only about 10 years or so. As the plasma decontamination technology is challenged by an increasing number of practical applications and by emerging scientific questions of plasma-microbe interactions, additional questions about its technological capability are expected to emerge. One such question is whether gas plasmas can be optimized so as to effectively disinfect contaminated surfaces with little damage to the substratum material including living plant and animal tissues. For plant tissues, early evidence is encouraging as indicated in two independent studies of bacterial contamination of the skins of bell peppers, melon and mango (Vleugels *et al.*, 2005; Perni *et al.*, 2008a). The skins of these plant tissues were found to be free of any gross damage after plasma

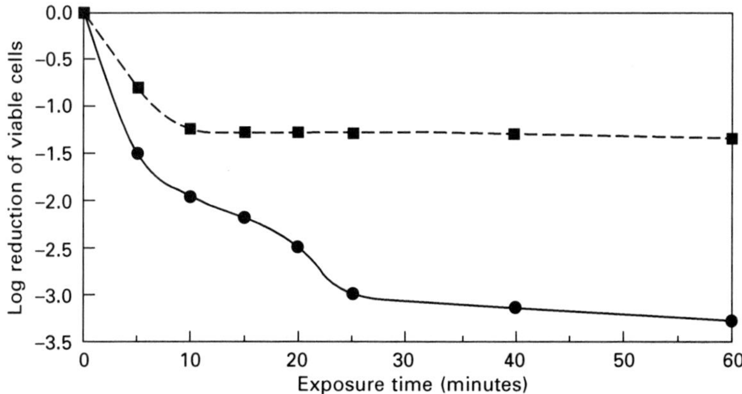


**Fig. 16.8** Inactivation kinetics of a sub-60°C atmospheric He-$O_2$ plasma jet for 12-h-old *P. agglomerans* samples (solid curve) and 24-h-old *P. agglomerans* samples (dashed curve) (adapted from Vleugels *et al.*, 2005).

© Woodhead Publishing Limited, 2012

treatment. The case of plant tissues is perhaps less complex than animal and human tissues, since plant tissues are in general used for human consumption shortly after their decontamination and as such possible mutation of plasma-treated plant cells, if any, is of little relevance. A similar argument could also be made for plasma disinfection of animals intended for human consumption. However, the issue becomes more important if gas plasmas are used as an infection-containment strategy used routinely through the entire farming cycle. The concern is not so much of possible plasma-mediated mutation, but rather a lack of certainty of whether or how it may be triggered in the first place.

The use of sub-60°C atmospheric plasmas in the food industry presents a much lower regulatory challenge than in the case of their use as a therapeutic solution in medicine, with the latter case needing to provide direct *in-vitro* and *in-vivo* evidence of possible plasma damage (or the lack of it). It is, however, important to effectively control plasma interactions with plant tissues and to mitigate any potential plasma damage. In addition, it is important to understand and control the effluent of gas plasmas as some may become an environmental hazard depending on their concentrations (e.g. ozone). These are examples of relevant technical topics that need to be addressed as part of regulatory considerations. One of the less obvious limitations of plasma technology is that some seemingly very similar plasma sources may behave differently and their inactivation efficacy could also differ significantly. While many reasons may contribute to such differences, one is related to a lack of control of the plasma-microbe interactions including plasma chemistry, plasma dynamics, and heterogeneity of microbe-surface and microbe-plasma interactions. Sub-60°C atmospheric plasmas have recently witnessed much change in terms of their technological capability, which may not be obvious to the user community who may have relied on an insufficiently satisfactory experience in the past as a static indicator of the technological capability of the fast evolving sub-60°C atmospheric plasmas. Therefore a key component of successful application of plasma technology is an appropriate engineering realization and a continued update and implementation of the current understanding of plasma-microbe interactions for any given application. This is best achieved through sustained partnerships between the plasma and user communities.

## 16.4 Selected applications and effect on food quality

Plasma decontamination of contaminated air and plasma de-odour are relatively well developed and have reached the point of practical use with commercial products (Plasmacat, 2011; Tri-air, 2011; plasma-clean, 2011). Treatment of contaminated gases presents lower requirements to plasma technologies, because gases can tolerate both heat and plasma instabilities much better than solid materials. For contaminated abiotic solid materials

© Woodhead Publishing Limited, 2012

such as food processing surfaces and food packaging materials, current sub-60°C atmospheric plasma sources such as those in Figs 16.1(b) and 16.7 should be directly amendable and there have been some industry-facing development projects (NovelQ, 2011) that may have been significantly advanced towards successful commercialization. With an appropriately focused R&D programme, this is one of the areas that may bring early success of the gas plasma technology to the food industry.

Direct plasma treatment of food products presents a greater challenge since this may also impact on the appearance, taste, texture, nutrient content, and other essential markers of food quality. Before these questions are addressed, one common general question from the user community is whether gas plasmas are a source of strong radiation at UV and shorter wavelengths such as γ-radiation or a source of energetic electron beams, both being used widely in food decontamination. As discussed above, sub-60°C atmospheric plasmas are low-energy discharges and their optical emission is typically weak at UV wavelengths of 200–300 nm (Fig. 16.5) and very little, if any at all, at γ-ray wavelengths (i.e. $< 10$ picometers). Radiation at wavelengths shorter than 100 nm is associated with energies above 12 electron volts (eV) and is unlikely to be supported by sub-60°C atmospheric plasmas whose mean electron energy is typically less than 2–5 eV. The low mean electron energy of the sub-60°C plasmas also means that it is very unlikely for them to form an energetic electron beam at an energy level similar to the electron beam technology for food decontamination (e.g. mega-electron volts). Therefore sub-60°C atmospheric plasmas are very different in their decontamination principle from the conventional decontamination technologies of electron beam and γ-radiation. They represent a much milder and less harsh technique.

It is not surprising that one of the early applications of direct plasma disinfection of foods is for nuts such as almonds (Deng *et al.*, 2007), given the protection of the nut shells. Using a pair of parallel-plate electrodes and in air, three types of almonds contaminated with microbes were bathed in the air plasma for up to 40 s and this led to a 5 log reduction of viable *E. coli*. This particular study did not report whether the plasma treatment may compromise the raw quality of almonds. A literature search has not yielded any indication of industrial uptake of this interesting research.

Plasma treatment of fresh fruits and vegetables is more challenging because they are less hardened than nuts and are more susceptible to plasma damage. Of such fresh produce, those with skins (e.g. whole oranges and apples) are clearly more protected. Effective bacterial inactivation by low-temperature atmospheric plasmas has been demonstrated for bell peppers inoculated with *P. agglomerans* (Vleugels *et al.*, 2005), melons and mangos with *E. coli* K12 and three spoilage microorganisms (*Saccharomyces cerevisae*, *Pantoea agglomerans*, and *Gluconacetobacter liquefaciens*) (Perni *et al.*, 2008a), apples, cantaloupe melons, and lettuce with *E. coli* O157:H7, *Salmonella* and *Listeria monocytogenes* (Crizer *et al.*, 2007), and apples with *E. coli* O157:H7 and *Salmonella* Stanley (Niemira and Sites, 2008). An interesting

© Woodhead Publishing Limited, 2012

and related study was plasma disinfection of cut surfaces of mangos and cantaloupe melons inoculated with *S. cerevisiae*, *G. liquefaciens* and *L. monocytogenes* Scott (Perni *et al.*, 2008b). It was shown that after inoculation the microorganisms started to migrate into the porous structure of the cut fruit tissues and became protected by the fruit tissues. This led to reduced efficacy of plasma inactivation from 6 log reduction to 2.5 log reduction (Perni *et al.*, 2008b). Unpublished data with several plasma-treated fruits from the author's group suggest that there was very little difference in taste and the changes in vitamin C and water content were also very small for a range of fruits treated with a sub-60°C atmospheric He-$O_2$ plasma jet. Taken together, the above results provide encouraging evidence of sub-60°C atmospheric plasmas for disinfection of fresh fruits and vegetables.

Low-temperature atmospheric plasmas have also found success in disinfection of dairy and meat products. Using a 13.56 MHz atmospheric plasma afterglow in helium mixed with air, sliced cheese and ham inoculated with a mixture of three different strains of *Listeria monocytogenes* were treated (Song *et al.*, 2009). In the case of sliced cheeses, a 5–6 log reduction was achieved and this was much better than the 1 log reduction achieved for sliced ham. The lower inactivation efficacy with sliced ham was speculated to be related to the rougher ham surface which may harbour *Listeria* cells in surface cracks and crevices. An independent study of plasma jet disinfection of chicken breast muscles and skins provided consistent data, with log reduction of viable listerial cells becoming progressively better from rough skin, through smooth skins, to chicken muscles without skins (Noriega *et al.*, 2011). These results confirm the findings of the biofilm study (Vleugels *et al.*, 2005; Abramzon *et al.*, 2006; Goree *et al.*, 2006) and the cut fruit tissue study (Perni *et al.*, 2008b) and suggest strongly that microbe-surface interactions are critically important in plasma disinfection. Data on plasma impact on quality of plasma-treated dairy and meat products are yet to be reported.

Most studies of direct plasma treatment of fresh fruits and vegetables focus on efficacy of plasma inactivation with little direct data on possible plasma effects on the quality of fresh produce. The study of bell peppers (Vleugels *et al.*, 2005) did measure directly plasma-mediated coloration and found that the amount of colour change was not distinguishable to the naked eye.

## 16.5 Conclusions and future trends

Notwithstanding successful and currently promising attempts to industrialize and commercialize the low-temperature atmospheric plasma technology, the vast majority of the experience and expertise is at present related to laboratory studies. As a novel decontamination technique, low-energy atmospheric plasmas are much less harsh than many of the current techniques and should in principle cause much less damage to foods themselves and

© Woodhead Publishing Limited, 2012

food processing surfaces. There is clear scope to optimize plasma sources and their reaction chemistry so as to minimize or even eliminate damage to the substratum material. Similarly, there are ways to control the production of some of the plasma effluents that are environmentally less desirable. It is expected that the next few years are likely to witness improvements in these technical aspects.

Greater challenges are found in other areas, in particular effective inactivation of microbes protected in surface structures (e.g. crevices, pores), biofilms, and plasma-mediated microbe self-organization on surfaces. While it is difficult to predict future innovations, it is expected that combination of the plasma technology with other techniques may be an important area of future investigation. Another practical challenge is the upscaling of the sub-60°C atmospheric plasmas. Although several forms of scaled-up plasma sources have already been successfully demonstrated, they need to be further developed taking into consideration the concurrent needs for plasma penetration, safety to food and food processing materials as well as to the environment, and electricity and gas consumption. Some of these requirements could become mutually exclusive, and as such it is important to explore plasma-microbe and plasma-tissue interactions over a wide parametric range. In addition, key plasma-induced chemical pathways in plasma-microbe interactions are likely to become increasingly important. While there is currently relatively little successful commercialization of plasma disinfection of foods, laboratory results so far are both encouraging and indicative of possible routes for future improvement.

In addition to technological innovation, there is a need for strong drivers from the standpoints of food safety, reliability of food processing, and regulatory concerns of alternative food decontamination technologies. The new *E. coli* strain that caused both bloody diarrhoea and neurological disorders in Germany in May 2011 (EHEC, 2011) serves as a good reminder that new decontamination technologies are not only desirable but also necessary in our war against antimicrobial resistance. Basic studies in this area and others will become more important.

## 16.6 Sources of further information and advice

Most current plasma disinfection studies are disseminated at major international conferences in plasma science, particularly the annual *IEEE International Conference on Plasma Science* which has a subject area devoted to plasma medicine, the biannual *International Symposium on Plasma Chemistry*, the recently established *International Conference for Plasma Medicine* that takes place every two years, and the biannual *Gaseous Electronics Conference*. In fact, most gas plasma conferences now have a special session on plasma medicine and plasma decontamination. There is also a growing trend for

© Woodhead Publishing Limited, 2012

plasma decontamination studies to be presented at microbiology, biology and medicine conferences.

There have been several comprehensive topical reviews of low-temperature atmospheric plasma decontamination, recent examples of which are discussed above and provided in the reference list (Kong *et al.*, 2009, 2011; Moisan *et al.*, 2001; Laroussi *et al.*, 2006; Fridman *et al.*, 2008). Although this chapter focuses on food decontamination using atmospheric pressure vacuum-less plasmas, it is important to note that low-pressure plasma inactivation is also a very active research area with similar general prospects for commercialization. Examples of relevant topical reviews are provided in the reference list (Moisan *et al.*, 2001; von Keudell *et al.*, 2010). Finally, there are about 100 and growing research groups working on the general interdisciplinary field involving plasma science and biological sciences, and the work of some of these groups is discussed and referenced here, and further reading is recommended.

## 16.7 References

ABRAMZON N, JOAQUIN J C, BRAY J and BRELLES-MARINO G (2006), 'Biofilm destruction by RF high-pressure cold plasma jet', *IEEE Trans. Plasma Sci.*, 34, 1304–1309.

BEN-JACOB E, COHEN I and LEVINE H (2000), 'Cooperative self-organization of microorganisms', *Advance Phys*, 49, 395–554.

BLATTNER F R *et al.* (1997), 'The complete genome sequence of *Escherichia coli* K-12', *Science*, 277, 1453–1462.

CAO Z, NIE Q, BAYLISS D L, WALSH J L, REN C S, WANG D Z and KONG M G (2010), 'Spatially extended atmospheric plasma arrays', *Plasma Sources Sci Technol*, 19(2), 025003.

CHEVILLE A M, ARNOLD K W, BUCHRIESER C, CHENG C M and KASPAR C W (1996), 'rpoS regulation of acid, heat, and salt tolerance in *Escherichia coli* O157:H7', *Appl Environ Microbiology*, 62(5), 1822–1824.

COX M M (1998), 'A broadening view of recombinational DNA repair in bacteria', *Genes to Cells*, 3, 65–78.

CRIZER F J, KELLY-WINTENBERG K, SOUTH S L and GOLDEN D A (2007), 'Atmospheric plasma inactivation of foodborne pathogens on fresh produce surfaces', *J Food Protection*, 70(10), 2290–2296.

DATSENKO K A and WANNER B L (2000), 'One-step inactivation of chromosomal genes in *Escherichia coli* K-12 using PCR products', *Proc Natl Acad Sci USA*, 97, 6640–6645.

DE BOER J F, SRINIVAS S, MALEKAFZALI A, CHEN Z-P and NELSON J (1998), 'Imaging thermally damaged tissue by polarization sensitive optical coherence tomography', *Optics Express*, 3, 212–218.

DENG X T, SHI J J, SHAMA G and KONG M G (2005), 'Effects of microbial loading and sporulation temperature on atmospheric plasma inactivation of *Bacillus subtilis* spores', *Appl Phys Lett*, 87, 153901.

DENG X T, SHI J J and KONG M G (2006), 'Physical mechanisms of inactivation of *Bacillus subtilis* spores using cold atmospheric plasmas', *IEEE Trans Plasma Sci*, 34, 1310–1316.

DENG S, RUAN R, MOK C K, HUANG G, LIN X and CHEN P (2007), 'Inactivation of *Escherichia coli* on almonds using nonthermal plasma', *J Food Sci*, 72(2), M62–M66.

© Woodhead Publishing Limited, 2012

EHEC (2011), *E. coli*: Germany says worst of illness is over. Available from: http://www.bbc.co.uk/news/world-europe-13691087 (accessed 11 June 2011).

FRIDMAN G, FRIEDMAN G, GUTSOL A, SHEKTER A B, VASILETS V N and FRIDMAN A (2008), 'Applied plasma medicine', *Plasma Process and Polymer*, 5, 503–533.

GOREE J, LIU B, DRAKE D and STOFFELS E (2006), 'Killing of *S. mutans* bacteria using a plasma needle at atmospheric pressure', *IEEE Trans Plasma Sci*, 34, 1317–1324.

KAYES M M, CRITZER F J, KELLY-WINTENBERG K, ROTH J R, MONTIE T C and GOLDEN D A (2007), 'Inactivation of foodborne pathogens using a one atmosphere uniform glow discharge plasma', *Foodborne Pathogens and Diseases*, 4, 50–59.

KNORR D, FROEHLING A, JAEGER H, REINEKE K, SCHLUETER O and SCHOESSLER K (2011), 'Emerging technologies in food processing', *Annu Rev Food Sci Technol*, 2, 203–235.

KOGELSCHATZ U (2002), 'Filamentary, patterned, and diffuse barrier discharges', *IEEE Trans Plasma Sci*, 30, 1400–1408.

KONG M G, KROESEN G, MORFILL G, NOSENKO T, SHIMIZU T, VAN DIJK J and ZIMMERMANN J L (2009), 'Plasma medicine: an introductory review', *New J Phys*, 11, 115012.

KONG M G, KEIDAR M and OSTRIKOV K (2011), 'Plasmas meet nanoparticles – where synergies can advance the frontier of medicine', *J Phys D: Appl Phys*, 44, 174018.

LAROUSSI M, TENDERO C, LU X, ALLA S and HYNES W L (2006), 'Inactivation of bacteria by the plasma pencil', *Plasma Process and Polymer*, 3, 470–473.

LI Z and DEMPLE B (1994), 'SoxS, an activator of superoxide stress genes in *Escherichia coli*. Purification and interaction with DNA', *J Biological Chem*, 269, 18371–18377.

LIU J J and KONG M G (2011), 'Sub-60°C atmospheric helium-water plasma jets: modes, electron heating and downstream reaction chemistry', *J Phys D: Appl Phys*, 44(34), 345203.

LOCKE B R, SATO M, SUNKA P, HOFFMANN M R and CHANG J-S (2006), 'Electrohydraulic discharge and nonthermal plasma for water treatment', *Industrial and Engineering Chemistry Research*, 45, 882–905.

MOISAN M, BARBEAU J, MOREAU S, PELLETIER J, TABRIZIAN M and YAHIA L H (2001), 'Low-temperature sterilization using gas plasmas: a review of the experiments and an analysis of the inactivation mechanisms', *Int J Pharm*, 226, 1–21.

NIEMIRA B A and SITES J (2008), 'Cold plasma inactivates *Salmonella* Stanley and *Escherichia coli* O157:H7 inoculated on golden delicious apples', *J Food Protection*, 71(7), 1357–1365.

NORIEGA E, SHAMA G, LACA A, DIAZ M and KONG M G (2011), 'Cold atmospheric gas plasma disinfection of chicken meat and chicken skin contaminated with *Listeria innocua*', *Food Microbiology*, 28(7), 1293–1300.

NOVELQ (2011), Novel processing methods for the production and distribution of high-quality and safe foods. Available from: http://www.novelq.org/Default.aspx (accessed 20 August 2011).

PAI D Z, STANCU G D, LACOSTE D A and LAUX C O (2009), 'Nanosecond repetitively pulsed discharges in air at atmospheric pressure – the glow regime', *Plasma Sources Sci Technol*, 18, 045030.

PERNI S, SHAMA G, HOBMAN J L, LUND P A, KERSHAW C J, HIDALGO-ARROYO G A, PENN C W, DENG X T, WALSH J L and KONG M G (2007), 'Probing bactericidal mechanisms induced by cold atmospheric plasmas with *Escherichia coli* mutants', *Appl Phys Lett*, 90, 073902.

PERNI S, SHAMA G and KONG M G (2008a), 'Cold atmospheric plasma decontamination of the pericarps of fruit', *J Food Protection*, 71(2), 302–308.

PERNI S, SHAMA G and KONG M G (2008b), 'Cold atmospheric plasma disinfection of cut fruit surfaces contaminated with migrating microorganisms', *J Food Protection*, 71(8), 1619–1625.

PLASMACAT (2011), Odour control and waste air treatment. Available from http://www.plasmacat.com/index.html (accessed 20 August 2011).

© Woodhead Publishing Limited, 2012

PLASMA-CLEAN (2011), Odour control and air purification system. Available from: http://www.plasma-clean.com/. (accessed 20 August 2011).

SONG H P, KIM B, CHOE J H, JUNG S, MOON S Y, CHOE W and JO C (2009), 'Evaluation of atmospheric pressure plasma to improve the safety of sliced cheese and ham inoculated by 3-strain cocktail *Listeria monocytogenes*', *Food Microbiology*, 26(4), 432–436.

TRESSE O, LEBRET V, BENEZECH T and FAILLE C (2006), 'Comparative evaluation of adhesion, surface properties, and surface protein composition of *Listeria monocytogenes* strains after cultivation at constant pH of 5 and 7', *J Appl Microbiol*, 101, 53–62.

TRI-AIR (2011), Fresh air technology. Available from: http://www.tri-airdevelopments.com/ (accessed 20 August 2011).

VISICK J E and CLARKE S (1997), 'RpoS- and OxyR-independent induction of HPI catalase at stationary phase in *Escherichia coli* and identification of rpoS mutations in common laboratory strains', *J. Bacteriology*, 179, 4158–4163.

VLEUGELS M, SHAMA G, DENG X T, GREENACRE E, BROCKLEHURST T and KONG M G (2005), 'Atmospheric plasma inactivation of biofilm-forming bacteria for food safety control', *IEEE Trans Plasma Sci*, 33, 824–828.

VON KEUDELL A *et al.* (2010), 'Inactivation of bacteria and biomolecules by low-pressure plasma discharges', *Plasma Process and Polymer*, 7(3-4), 327–352.

WALSH J W and KONG M G (2007), '10 ns pulsed atmospheric air plasma for uniform treatment of polymeric surfaces', *Appl Phys Lett*, 91, 251504.

WALSH J L, CAO Z and KONG M G (2008), 'Atmospheric dielectric-barrier discharges scalable from 1 mm to 1 m', *IEEE Trans Plasma Sci*, 34, 1314–1315.

YU H, PERNI S, SHI J J, WANG D Z, KONG M G and SHAMA G (2006), 'Effects of cell surface loading and phase of growth in cold atmospheric plasma inactivation of *Escherichia coli* K12', *J Appl Microbiol*, 101, 1323–1330.

© Woodhead Publishing Limited, 2012

# Part III

# Current and emerging chemical decontamination methods

© Woodhead Publishing Limited, 2012

# 17

# Microbial decontamination of food using ozone

**A. S. Chawla, D. R. Kasler, S. K. Sastry and A. E. Yousef, The Ohio State University, USA**

**Abstract**: Ozone has the potential to fill a substantial gap in today's technologies that are used to ensure food safety. If can be generated directly from water, air or pure oxygen by several methods; the most efficient being the corona discharge process. Food processors can use ozone in its gaseous or aqueous state. In all states, concentration of ozone during food treatment should be measured with reasonable accuracy. Inconsistent use of equipment, procedures and units to measure ozone make it difficult to compare decontamination results from different sources. Despite these technical hurdles, ozone has been proven effective at decontaminating various foods. The benefits of ozone, in many cases, outweigh the drawbacks and its applications in food are expected to increase in popularity in the coming years.

**Key words**: ozone, ozone generators, ozone measurements, food safety, ozone toxicity.

## 17.1 Introduction

The rising interest in novel food processing and preservation systems is driven by a number of factors including (a) consumer preference for minimally processed food free of chemical preservatives; (b) recent highly-publicized outbreaks of foodborne diseases caused by pathogens such as *Salmonella* sp., enterohemorrhagic *Escherichia coli*, and *Listeria monocytogenes*; and (c) the passage of new food safety legislations in the US and other countries (e.g., FDA, 2011). Consistent with these developments, food processors are searching for more potent antimicrobial agents than those currently in use, or re-evaluating existing antimicrobials that may have been overlooked.

© Woodhead Publishing Limited, 2012

Ozone has been applied effectively in drinking water since 1906. It is the strongest oxidizing sanitizer that can be used in food processing. In response to promising antimicrobial results, several small companies have already incorporated ozone into their processing lines. Ozone can be adapted to existing food processing operations with few modifications. However, there are concerns about the costs and risks associated with using ozone in food processing.

Application of aqueous sanitizers for decontamination of foods and the processing environment is a common practice in the food industry. Ozone has limited solubility in water as compared to these sanitizers. Aqueous applications of ozone would require efficient gas injection systems and closed treatment vessels. Problems related to the potential of off gassing, and monitoring of dissolution and residual ozone in processing water, add further complexities to the system. Being a strong oxidizer, ozone is not compatible with some commonly used equipment parts (e.g., vulcanized rubber), and may cause corrosion of existing equipment (e.g., pumps). These points need to be considered when upgrading from a traditional sanitizer to an ozone-based system. Interestingly, the fact that ozone has low solubility in water paves the way for the beneficial application of its gaseous state in food sanitization and decontamination.

## 17.2 Ozone properties, generation and decomposition

Ozone is an allotrope of oxygen comprising three oxygen atoms arranged at an obtuse angle. It has a characteristic and distinctive odor that is detectable by humans at concentrations as low as 0.02 ppm (Horváth *et al.*, 1985). Ozone is generated naturally in the earth's stratosphere by the action of short-wave ultraviolet (UV) light (<240 nm) on molecular oxygen (Fig. 17.1). Longer UV radiation (240–320 nm) decomposes ozone to oxygen. This decomposition occurs at a lower altitude, generating the ozone layer in the stratosphere. Gaseous ozone is heavier than air, and it is colorless at low concentrations, but with a bluish color at high concentrations (Rice *et al.*, 1982). Ozone has an oxidation potential of 2.07 V (Horváth *et al.*, 1985), which is higher than that of chlorine (1.36 V) and hydrogen peroxide (1.8 V); therefore, ozone is the strongest oxidant currently available for food applications.

Industrially, ozone is produced using commercial ozone generators. Since ozone is very reactive, it cannot be stored for significant periods of time; therefore, it must be generated at the point of application as needed. Reactivity of ozone causes its rapid decomposition. This instability is a problem when extended treatments are required, but it is beneficial during disposal of excess ozone.

© Woodhead Publishing Limited, 2012

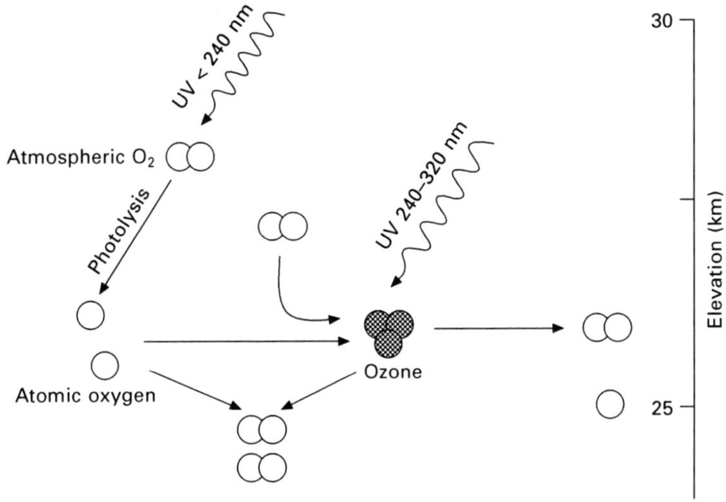


**Fig 17.1** Ozone formation and decomposition in the stratosphere according to Chapman mechanism (Velaco *et al.*, 2008).

### 17.2.1 Ozone generators

Ozone can be generated by several techniques using UV radiation, cold plasma, chemical or chemonuclear reactions, electrolysis, and corona discharge. Generators based on a few of these techniques are available commercially; these will be discussed in some detail.

*Ultraviolet radiation*

Experimental and commercial photochemical production of ozone is similar to the natural process that occurs in the stratosphere, which was described earlier. Briefly, oxygen is photo-dissociated by short wavelength UV radiation (< 240 nm) to produce oxygen atoms; these react with oxygen molecules to form ozone (Velaco *et al.*, 2008). The UV source could be a low-pressure mercury lamp that produces not only the 185 nm radiation responsible for the production of ozone, but also the 254 nm radiation that destroys ozone; therefore, only small concentrations of ozone are typically generated by this method (Dohan and Masschelein, 1987; DuRon, 1982). The quantities of ozone generated per 40 W UV bulb are low (0.5 g/h) and maximum concentrations is 0.25% mass. Medium pressure UV lamps generate a higher proportion of the 185 nm radiation and thus ozone is produced at a greater concentration than that produced by the low-pressure lamp. Jeong *et al.* (2005) used an ozone-producing UV lamp having a maximum emission at 254 nm and a minor emission (<5%) at 185 nm; the authors reported relatively high ozone levels (140 ppm, volume basis) from this source.

Vacuum UV (VUV) lamps that emit radiation at 172 nm during xenon excimer decay, have been found to produce 1–2% ozone, on a volume basis

© Woodhead Publishing Limited, 2012

(Eliasson and Kogelschatz, 1991). The VUV lamp consists of a coaxial quartz cylinder which has a high voltage electrode in the center, Xenon gas is filled in the annulus and an electric discharge is applied. During the electric discharges, which last a few microseconds, electrons with suitable energy excite xenon atoms, which react to form an exited dimer (excimer). These xenon excimers have a very short lifetime of 100 ns and decay rapidly releasing the absorbed energy as UV radiation at 172 nm. This short UV radiation can then be used for photo-dissociation of oxygen to produce ozone. It was suggested that high ozone concentrations (15%, mass basis) theoretically could be produced from these lamps using dry oxygen as feed gas at atmospheric pressure (Salvermoser *et al.*, 2008). Commercial VUV lamps based on xenon excimer are available (e.g., Osram, 2011); a schematic of such a lamp is shown in Fig. 17.2.

*Electrolysis*
Ozone can be generated efficiently using an electrochemical procedure (Andrews and Murphy, 2002). Ozone generation by this method involves electrolysis of water into hydrogen and oxygen atoms. Hydrogen gas is vented from the gas-water mixture, and oxygen atoms recombine to produce ozone in an oxygen mix. Electrochemically generated ozone is self-pressurized (≤20 psig), and ozone concentration up to 12–14% (mass basis, in oxygen) can be produced (Lynntech Inc., 1998).

*Corona discharge*
In this process, dry air or oxygen gas is fed through a gap between two electrodes, separated by a dielectric material. A strong electric discharge is then applied producing free electrons that cause the dissociation of molecular oxygen to the atomic form. When atomic oxygen collides with molecular oxygen, ozone is formed (Carlins and Clark, 1982). Concentration of ozone produced in this system depends on the voltage, current, frequency, dielectric material, discharge gap, absolute pressure within the discharge gap, and concentration of oxygen in the gas passing through the electrodes

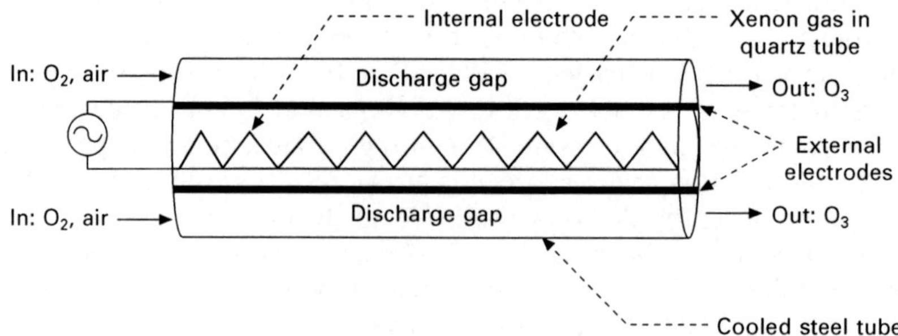


**Fig. 17.2** Xenon eximer-based vacuum ultraviolet lamp (modified from Osram, 2011).

© Woodhead Publishing Limited, 2012

(Horváth *et al.*, 1985). High efficiency corona-discharge ozone generators that can produce up to 14% (mass basis) ozone in oxygen gas mixture are currently available. Figure 17.3 presents a schematic of a high efficiency corona discharge system. The use of air as a feed gas for ozone generators is discouraged. Nitrogen and moisture in the air can combine with ozone, producing nitric acid which causes corrosion and damage to the generator (Horváth *et al.*, 1985). It is more economical to produce ozone from oxygen than air (Geering, 1999).

### 17.2.2 Ozone decomposition

Ozone is more stable in the gaseous than in the aqueous phase (Stumm, 1958). It has been calculated that the theoretical half-life of ozone gas (1.5% $O_3$ in $O_2$, mass basis) at 25, 100, and 250°C is 19.3 years, 5.2 h, and 0.1 s, respectively (Wojtowicz, 2004). In practice, the half-life of gaseous ozone is shorter than these theoretical values. For aqueous ozone, the half-life could vary from seconds to hours depending on water quality and temperature (Kim *et al.*, 2003; Weavers and Wickramanayake, 2001). It has been reported that the half-life of ozone in deionized and tap waters at 25°C was 12 min and 6 min, respectively (Kim, 1998). The stability of aqueous ozone is influenced

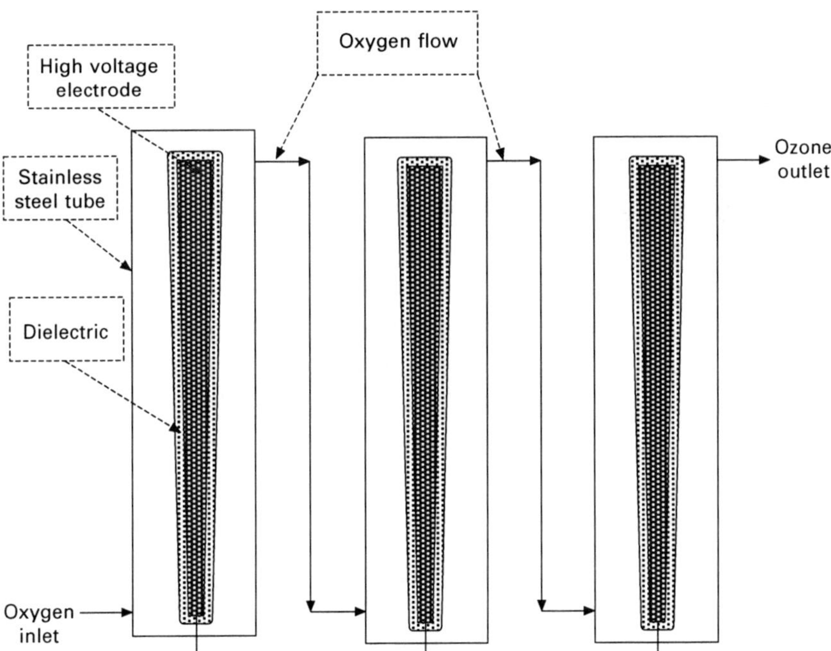


**Fig. 17.3** Multi-tube coronal discharge with intelligent gap system (modified from Degremont Technologies, 2011).

© Woodhead Publishing Limited, 2012

by the presence of ozone-demand material in the water, ozone concentration, temperature, pH, incidence of UV radiation, mechanical stirring, and the presence of metal ions and radical scavengers (Horváth *et al.*, 1985; Kim *et al.*, 2003; Weavers and Wickramanayake, 2001). Ozone in aqueous phase continuously decomposes in a step-wise mode with the generation of free radical species, which include hydroperoxyl ($HO_2^{\cdot}$), hydroxyl ($^{\cdot}OH$), and superoxide ($^{\cdot}O_2^-$) radicals (Bablon *et al.*, 1991; Grimes *et al.*, 1983; Hoigné and Bader, 1975). The generated free radicals have a strong oxidizing power, a half-life of microseconds, and are responsible for the high reactivity of ozone (Kim *et al.*, 2003).

Due to its highly reactive nature, ozone gas is harmful to humans, even at low concentrations; hence any excess gas must be destroyed before discharge to the local environment (Dhandapani and Oyama 1997). Ozone is readily destroyed at high temperature (> 350°C) or catalytically by passage of the gas through wet granular activated carbon (GAC) beds (Alvarez *et al.*, 2006; Horváth *et al.*, 1985; Subrahmanyam *et al.*, 2005). The latter method, wet GAC, is not recommended for large-scale ozone destruction, due to the large amount of heat liberated, which can build up sufficiently to cause the GAC medium to deflagrate (Subrahmanyam *et al.*, 2005).

The off-gas destruction unit must be designed to reduce ozone concentration to 0.1 ppm by volume, the current limit set by OSHA for worker exposure in an eight-hour shift (Dhandapani and Oyama, 1997). The system design ideally should include a blower for directing ozone gas toward the destruction unit; alternatively, the discharge side should be under slight vacuum to ensure that no ozone escapes conversion into oxygen.

## 17.3 Ozone measurement

Because ozone is extremely reactive and has a high redox potential, small amounts can be useful in food applications. Ozone's high reactivity makes it decompose rapidly during decontamination processes; hence, it is difficult to measure its concentration accurately during applications. The following is a summary of different methods available to measure ozone concentration in gaseous and aqueous phases.

### 17.3.1 Gaseous phase

Gaseous ozone concentration may need to be measured to monitor (a) the output of an ozone generator; (b) the level of ozone during gaseous treatment of food; (c) concentrations in the off-gas streams of treatment systems; and (d) the ambient concentration in the environment where ozone is used. A number of methods have been implemented; these rely on ultraviolet absorption, amperometry, calorimetry, iodometry and chemiluminescence (Khadre *et al.*, 2001; Rice, 1986; Weavers and Wickramanayake, 2001).

© Woodhead Publishing Limited, 2012

*UV absorption method*

Ozone molecules in an ozone-oxygen or ozone-air stream absorbs UV light with maximum absorbance at a wavelength of 253.7 nm and a molar absorption coefficient of 3000 ± 30 $M^{-1}cm^{-1}$ at 273°K and 1 atm (Gottschalk *et al.*, 2000; Langlais *et al.*, 1991; Stanley and Johnson, 1984). In a UV-based measuring device, a decrease in UV intensity at 254 nm is proportional to the concentration of ozone, based on Lambert–Beer's law of absorption:

$$I_l = I_o \times 10^{-\varepsilon c(M) l} \qquad [17.1]$$

where $c(M)$ is concentration in mol/liter, $I_l$ is the intensity of UV light passing through the absorption cell containing the sample, $I_o$ is the intensity of UV light passing through the absorption cell containing the reference, $l$ is the internal width of the absorption cell in cm, and $\varepsilon$ is the molar extinction coefficient in $M^{-1}$ $cm^{-1}$.

A laboratory UV-spectrophotometer, with single- or double-beam design, can be used to measure ozone concentration (Bollyky, 2003). Similar principles apply to UV-based ozone monitors. Special gas-flow cells are used in these devices. In the single-beam design, one UV absorption cell is used. The intensity of UV light passing through this sample cell is measured by a detector, as ozone and ozone-free (reference) gas streams flowing alternately through the cell every 5–20 s. The difference in UV intensity between the reference and ozone gas streams is used to determine the ozone concentration (Eq. [17.1]). In the double-beam design, there are two UV absorption cells that separately measure the reference and ozone gas streams. This design is simpler compared to the single-beam design as it does not require frequent switching of gas streams. Pressure and temperature compensation of the inlet gas stream is provided in both types of UV monitors. The UV method is not subject to interference by extraneous compounds since gases normally present in air do not absorb UV at a wavelength of 253.7 nm (Stanley and Johnson, 1984; Gottschalk *et al.*, 2000).

*Electrochemical methods*

Several electrochemical methods, including galvanic (Mancy *et al.*, 1962), amperometric (Stanley and Johnson, 1979), steady-state voltametric and pulsed voltametric (Smart *et al.*, 1979), have been investigated for the *in situ* measurement of ozone. Selected electrochemical methods are described below.

Membrane amperometric method

Amperometric electrodes (Fig. 17.4) are based on the principle of electrochemical reduction of ozone on a noble metal surface such as gold or platinum:

$$O_3 + 2H^+ + 2e \rightarrow O_2 + H_2O \qquad [17.2]$$

The measuring device consists of a cathode, which is made of gold or platinum,

© Woodhead Publishing Limited, 2012

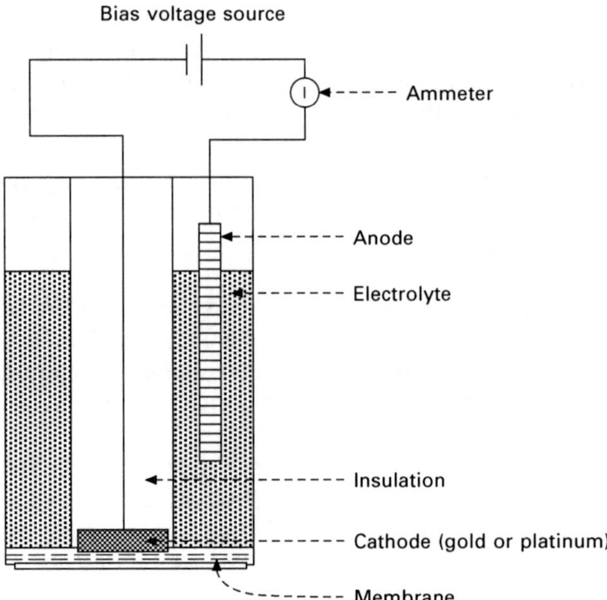


**Fig. 17.4** Schematic of an amperometric membrane ozone probe (modified from Lee and Tsao, 1979).

and a reference electrode. Two types of reference electrodes are commonly used: silver/silver chloride (Ag/AgCl) or mercury-containing electrode. The latter is referred to as standard calomel electrode, SCE (Bohn, 1971; Fiedler *et al.*, 2007). When the cathode is made electronegative by applying an external bias-voltage with respect to the reference electrode (anode), the dissolved ozone is reduced at the surface of the cathode. As the negative bias-voltage increases, the current increases initially, followed by a constant current period. Ozone reaction at the cathode is fast, and it is limited only by the rate of ozone diffusing to the cathode. The current produced by the reduction of ozone at the cathode is proportional to the concentration of ozone (Stanley and Johnson, 1984). In order to increase the sensitivity and selectivity of these electrodes, several electrode materials, electrolyte media, membrane materials and applied voltage potentials have been introduced (Stanley and Johnson, 1984). A linear and reproducible response to dissolved ozone was detected in the presence of other oxidizers such as chlorine, hydrogen peroxide, and chlorine dioxide. Because the electrochemical cell is isolated from the sample gas by a gas-permeable but liquid-impermeable membrane, this method permits the use of the same monitor for gas and liquid phase measurements (Bollyky, 2003).

Thin-film semiconductor based methods

Thin-film semiconductor-type monitors are small portable devices that are well suited for measuring ozone or detecting the leakage of the gas. Briefly, the

© Woodhead Publishing Limited, 2012

sensor design consists of a platinum electrode coated with a semiconductor material, and a platinum heating element (Fig. 17.5). The heater maintains the semiconductor at an elevated temperature. When this hot electrode comes into contact with ozone, the resistance of the semiconductor changes due to adsorption and decomposition of ozone. A voltage signal proportional to the change in resistance can then be generated (Bollyky, 2003). As just indicated, the use of metal oxide films as ozone sensors requires an active heating element and subsequent operations at high temperatures (above 573 K), demanding a high energy consumption for these types of sensor elements. To overcome this difficulty, metal oxide sensors were developed wherein sensor films are first exposed to UV radiation emitted by LED diodes, followed by oxidation with ozone. With this design, the conductivity of the metal oxide sensors changes by several orders of magnitude, enabling their use as ozone sensors at ambient temperature (Martins *et al.*, 2004). Several ozone sensing materials, such as ZnO, $SmFeO_3$, $CuAlO_2$, and $In_2O_3$ have been explored in view of their excellent ozone sensing properties at ambient temperature (Knake and Hauser, 2002). Novel sensors based on tungsten trioxide ($WO_3$) semiconductors have emerged as inexpensive alternatives for ozone monitoring (Utembe *et al.*, 2006).

Many workers in the field recommend using the UV spectrophotometric method to measure ozone in the gas phase (e.g., Gordon *et al.*, 1988). Hence, instruments that measure ozone concentration on the basis of UV absorption are widely used in various applications.

### 17.3.2 Aqueous phase

Methods for measuring ozone in the aqueous phase can be grouped into chemical, physical, or physicochemical-based methods (Horváth *et al.*, 1985; Kinman, 1975; Stanley and Johnson, 1984; Weavers and Wickramanayake, 2001). The following is a summary of the commonly used methods.

*Indigo trisulfonate method*

This method is based on the decolorization of indigo trisulfonate upon its reaction with ozone (Bader and Hoigne, 1981). Ozone disrupts the sole

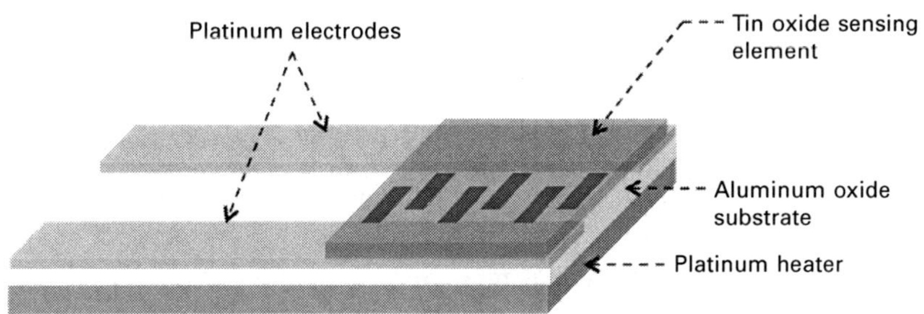


**Fig. 17.5** Schematic of thin film semiconductor based ozone sensor (modified from Barsan and Weimer, 2003).

© Woodhead Publishing Limited, 2012

carbon–carbon double bond of the reagent. Change in color is determined spectrophotometrically at 600 nm, and the measured absorbance is used to calculate ozone concentration with a minimum detection limit of 0.005 μg/ml. Compared to the iodometric procedures, which are described later, the indigo method is not subject to much interference from other oxidant compounds (Stanley and Johnson, 1984); thus, it is recommended as a standard to measure residual ozone (APHA, 1995; Gordon *et al.*, 1988). The indigo method is accurate within 2% (Grunwell *et al.*, 1983) and is often used to measure ozone in the aqueous phase.

*Iodometric method*

The iodometric method is based on the oxidation of potassium iodide (KI) by ozone and the liberation of free iodine (APHA, 1971). Ozone is reduced to oxygen and the iodide anion is oxidized to iodine:

$$O_3 + 2I^- + H_2O \rightarrow O_2 + I_2 + 2\ OH^- \qquad [17.3]$$

Therefore, one mole of iodine is produced for every mole of ozone reduced (Stanley and Johnson, 1984). The quantity of liberated iodine can be determined by one of the following methods:

- Colorimetrically, by measuring changes in absorbance at 352 nm.
- Iodometric titration: the liberated iodine is titrated with a reducing agent such as sodium thiosulfate or phenylarsine oxide using starch as an indicator. The iodine in the starch-iodine complex (dark blue) is reduced to iodide (pale yellow) indicating the endpoint of this reaction.

Ozone measurement by the iodometric method is subject to interference from other reducing or oxidizing agents present in solution, e.g., Cl, Br, $H_2O_2$, Mn, and organic peroxides (Gottschalk *et al.*, 2000; Stanley and Johnson, 1984).

*Diethyl phenylenediamine colorimetric method*

Ozone reacts directly with the *N,N*-diethyl-*p*-phenylenediamine (DPD) reagent to form a red colored product that can be detected, qualitatively, by titration with ferrous ammonium sulfate or by measuring absorbance at 525 nm. This method is influenced by the presence of free halogens and oxidized manganese (Stanley and Johnson, 1984).

*Oxidation reduction potential (ORP) method*

Oxidation reduction (redox) potential refers to the availability of electrons in a system and is a measure of electron transfer between oxidized and reduced species in the system (Gambrell and Patrick, 1980; Gambrell *et al.*, 1991). An oxidizer such as ozone loses electrons and a reduced substance gains electrons. Redox potential measurements require three components: a platinum or gold (test) electrode, a reference electrode and a voltmeter (Bohn, 1971). Platinum is routinely used as a test electrode due to its non-

© Woodhead Publishing Limited, 2012

selectivity (i.e., it accepts electrons from all redox reactions) and stability (i.e., it does not release chemical species that affect the redox reaction). The reference could be an Ag/AgCl electrode or a standard calomel electrode (Bohn, 1971; Fiedler *et al.*, 2007). The test and the reference electrode, that together constitute the OPR probe, are then connected to a voltmeter (Fig. 17.6). When an ORP probe is placed in a sample, it gives an OPR reading, in mV, on the voltmeter.

*UV absorption method*
Molecular ozone dissolved in water retains UV absorption characteristics similar to those of the gaseous ozone; the gas has a maximum absorbance at 254 nm (Adler and Hill, 1950). Based on Lambert–Beer's law (Eq. [17.1]), the decrease in the intensity of transmitted UV is proportional to the concentration of ozone. Therefore, concentration of ozone in a solution can be measured spectrophotometrically at 254 nm by converting the optical density reading into dissolved ozone concentration (Gottschalk *et al.*, 2000; Stanley and Johnson, 1984). This method is more selective but less sensitive than the iodometric method, and readings are susceptible to interference from dissolved organic and inorganic compounds (e.g. chlorine, bromine)

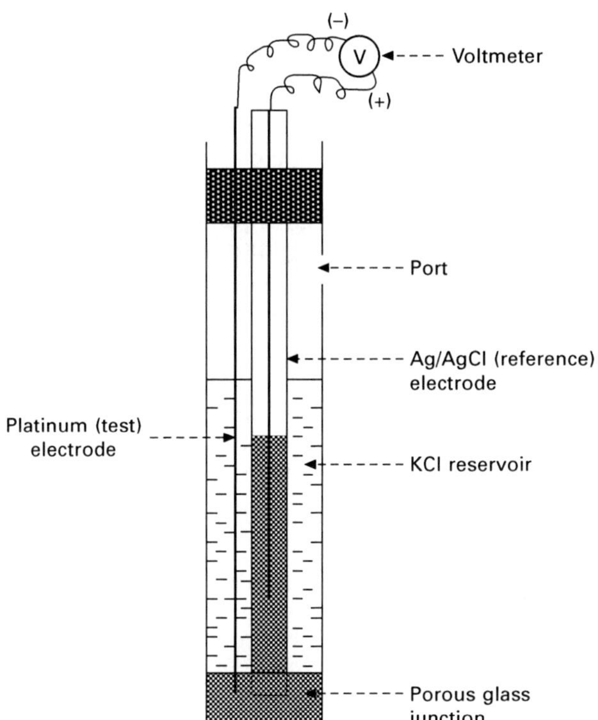


**Fig. 17.6** Schematic of oxidation reduction probe with a silver (Ag)/silver chloride (AgCl) reference and platinum (test) electrodes.

© Woodhead Publishing Limited, 2012

which absorb UV light at wavelengths between 200 and 300 nm (Stanley and Johnson, 1984).

## 17.4 Units for expressing ozone concentration

Ozone is always in a solution state, whether the medium that contains it is a gas or a liquid. Aqueous ozone is dissolved in water, whereas gaseous ozone is dissolved in a carrier gas or a gas mixture such as oxygen or air, respectively. Therefore, ozone quantity is often expressed as a ratio or a concentration. The most commonly used units to quantify ozone are $ppm_{mass}$ for aqueous ozone, $ppm_{volume}$ for low concentrations of gaseous ozone, and $g/m^3$ for the high gaseous concentrations. These are not the only units used to measure ozone, and are not always the best choices. Different measuring devices report ozone quantity as $g/Nm^3$, g/kg, % mass, % volume, $ppm_{mass}$, and others (IN USA Inc., 1996).

Diversity of methods used to measure ozone and units to express its concentration makes it difficult to replicate sanitization processes in a different facility. This problem can be attributed to incomplete data in many published reports and research papers, improper unit conversion of ozone concentration, and inadequate documentation provided with ozone monitors. These challenges make it difficult for new researchers in the field to recognize what the reported ozone measurements actually are. It is plausible to represent ozone as (i) mass ozone per total volume (e.g., $g_{ozone}/m^3_{total}$), (ii) mass ozone to mass of solvent (e.g., $g_{ozone}/kg_{solvent}$), typically denoted as a 'mass ratio,' or (iii) mass ozone to total mass [e.g., $(kg_{ozone}/kg_{total})*100$]; the latter is also known as 'percent mass.' Each of these unit expressions has its own advantages and disadvantages.

In a liquid or a gas solution, molecules move randomly thoughout the entire volume, and this would include the dissolved ozone molecules (Incropera and Dewitt, 2002). When a surface or a product is exposed to this solution, the targeted microorganisms are not exposed instantaneously to ozone. It takes time for ozone molecules to diffuse and randomly collide with the target to create a successful decontamination of product surface or bulk. Logically, the more ozone molecules are available, the more likely these collisions are to occur, and the quicker is the decontamination. However, the ratio of ozone to solvent molecules (or mass ratio) also is important for process efficacy. The random motion of particles, which is a form of mass transfer, is controlled by the molecular spacing, with smaller spacing yielding slower transfer (Incropera and Dewitt, 2002). For clarification, let us consider several systems containing the same mass of ozone but with different ozone-to-solvent mass ratios. Among these systems, the lower the ratio of ozone molecules to other molecules in the system, the greater the number of solvent molecules, and the smaller the space available for ozone molecules to accommodate their random motion that allows successful collisions with product to be

© Woodhead Publishing Limited, 2012

decontaminated. Therefore, both the mass of ozone and the mass ratio are important for process efficacy. Therefore, generators are evaluated not only on the basis of ozone output (e.g., g/hr) but also on how rich this gaseous output in ozone is (i.e., mass ratio).

### 17.4.1 Commonly used units

A mass-to-volume ratio is the simplest way to express an ozone concentration. Once this ratio is known, only the volume of the mixture is needed for calculating the ozone mass. Another advantage of a mass-to-volume ratio is that it is the unit that most analysts measure and report. UV spectrophotometers measure ozone mass per volumetric flow (Dosoretz *et al.*, 1998). Similarly, the amperometric method requires a volumetric flow of a solution past the membrane for proper diffusion of ozone to occur into the electrodes (ATI, 2009; Knake and Hauser, 2002). The indigo and iodometric measurements use a volume of ozone solution that must be defined to get the actual measurement, which is calculated as a mass-to-volume ratio. However, the mass-to-volume ratio gives no information about the mass of the solvent, which is also an important factor in regulating decontamination. The quantity of solvent in the system can be calculated using a mass-to-volume ratio but other parameters must be measured to complete the calculations; without knowing these other parameters, the solvent mass is not known.

The mass ratio is the most useful measurement unit for these reasons:

- it is the measurement that matters the most for ozone efficacy,
- output of ozone generators is often expressed in this unit, since equipment manufacturers usually provide this information based on specific operating conditions, and
- knowing the mass ratio makes it easy to convert to other measurement units.

However, there are limitations for using these units: (a) additional information should be provided before the mass of ozone in the system can be determined, and (b) measuring devices must use internal calculations to yield this unit, but it is not clear to the user how these calculations are done exactly.

'Percent mass' and 'mass ratio' are closely related and equally useful units. It should be cautioned that percent mass occasionally is reported just as a percentage ('%') which can be taken, mistakenly, as a '% volume' measurement. Therefore, it is advised to use the mass ratio whenever possible.

Ozone concentration is often expressed in ppm, but the unit should be clearly stated. The unit should be designated as $ppm_{mass}$, $ppm_{volume}$, or $ppm_{molar}$ to eliminate any confusion and misinterpretation. Figure 17.7 shows the variation in ozone mass calculated using these different assumptions about the units of ppm. Researchers sometimes present equal ozone ppm in aqueous and gaseous solutions as comparable concentrations which is not the case.

© Woodhead Publishing Limited, 2012

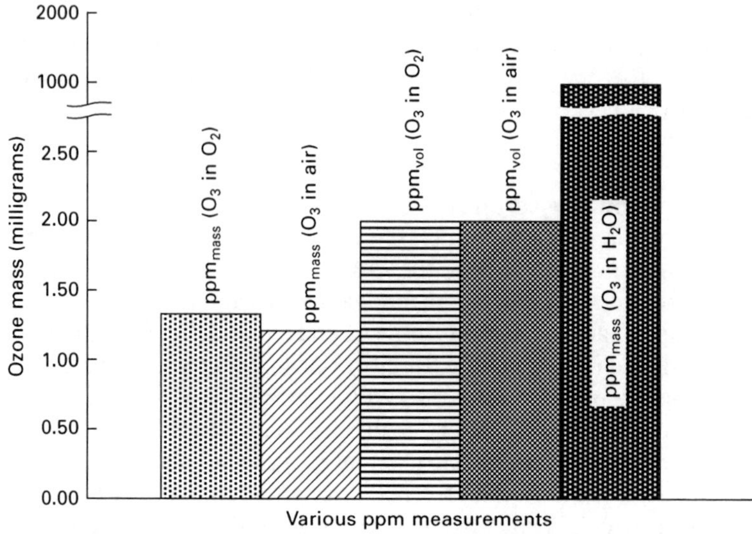


**Fig. 17.7** Mass of ozone calculated from 10 ppm measurements indicated by five hypothetical detectors, using the following assumptions. The ozone detectors are calibrated for two types of ppm measurements (mass and volume basis) in three different solvents (oxygen, air, and water). Volume of the vessel is 100 liters, pressure of sanitizer (ozone + solvent) is 14.7 psia, and effect of temperature on concentration is considered negligible. Note that ozone-water $ppm_{volume}$ (i.e., volume $O_3$/volume $H_2O$) is not included since the value may not be of practical use.

The '% volume' should be used with caution or avoided entirely. In a gaseous solution, for example, volume ratios mean very little as gases expand and contract to fill containers. Another questionable measurement is the mass-to-normal volume ratio (e.g., g/Nm$^2$). Normal or 'standard' volume-based measurement is of practical value to engineers as it eliminates the need for knowing other gas parameters while converting this to other units. However, normal volume is defined at normal atmospheric temperature and pressure, which often are considered to be 21°C and 14.7 psi, respectively, but these values vary between disciplines, industries and also when using different unit systems (i.e., metric vs. imperial). Conversion of normal volume ratio to other units requires a large number of steps, compared to most other conversion exercises.

### 17.4.2 Gaseous ozone

Similar to all gasses, ozone gas is compressible; therefore, pressure and temperature greatly change its density, i.e., the mass of gas held in a certain volume. The relationship between these variables is stated in the ideal gas law:

$$PV = nRT \quad [17.4]$$

© Woodhead Publishing Limited, 2012

where $P$ is the absolute pressure of a control volume, $V$ is the control volume such as an ozone treatment vessel, $n$ is the number of moles of gas in the control volume, $R$ is the gas constant, which equals 8314 J/(kg mol.K), and $T$ is the absolute temperature of the control volume.

According to this equation, the mass ratio or mass-to-volume ratio means little without taking into account the temperature, pressure, and volume of a system. For instance, if decontamination processes in two containments are run with the same ozone gas mass ratio, temperature and volume, an increase in pressure in one of the processes (i.e., by introducing more of the gas mixture) increases the actual mass of ozone that is available for the decontamination process (n), without changing the ozone-to-solvent gas mass ratio. The concept also is applicable to the 'mass-to-volume' ratio. Similarly, ignoring the temperature factor can lead to inaccurate comparisons and deductions about the two processes. However, it should be cautioned that applying the ideal gas equation requires that pressure and temperature are in absolute form and that all units are the same and they reduce to proper values. Using a mass ratio or mass-to-volume ratio is the most representative measurement of ozone, but chamber volume, temperature and pressure still need to be reported to properly compare different experiments. Figure 17.8 reports change in values of various measurement units when a 100-liter vessel is pressurized, while introducing a constant stream of ozone-oxygen mixture. Temperature is presumed constant during pressure increase. Although ozone mass and mass-to-volume ratio ($g/m^3$) increase linearly during the pressurization, measurements expressed in other units are non-linear (Fig. 17.8).

### 17.4.3 Aqueous ozone

The most common unit of measurement for ozone in aqueous solution is ppm by mass, which can effectively be considered a form of mass ratio, e.g., μg ozone/g solvent (Ingram and Barnes, 1954). It should be obvious that 1 $ppm_{mass}$ actually means 1 μg ozone/g solution. Considering that levels of ozone in aqueous solution rarely exceed 20 $ppm_{mass}$, the mass of dissolved ozone does not significantly affect the total system mass. Therefore, 1 μg ozone/g solution is considered similar to 1 μg ozone/g solvent.

Temperature and pressure cause only limited changes in water density, particularly in the ranges applicable to ozone applications in food. Most aqueous ozone treatments of food occur at 0–30°C, and atmospheric pressure. For these aqueous systems, a relationship like the ideal gas law is not needed. Assuming that water density is one gram per milliliter, this makes it easy to determine ozone and water on mass basis. Ozone mass in solution simply can be determined by multiplying the ozone mass ratio by the volume of the water.

© Woodhead Publishing Limited, 2012

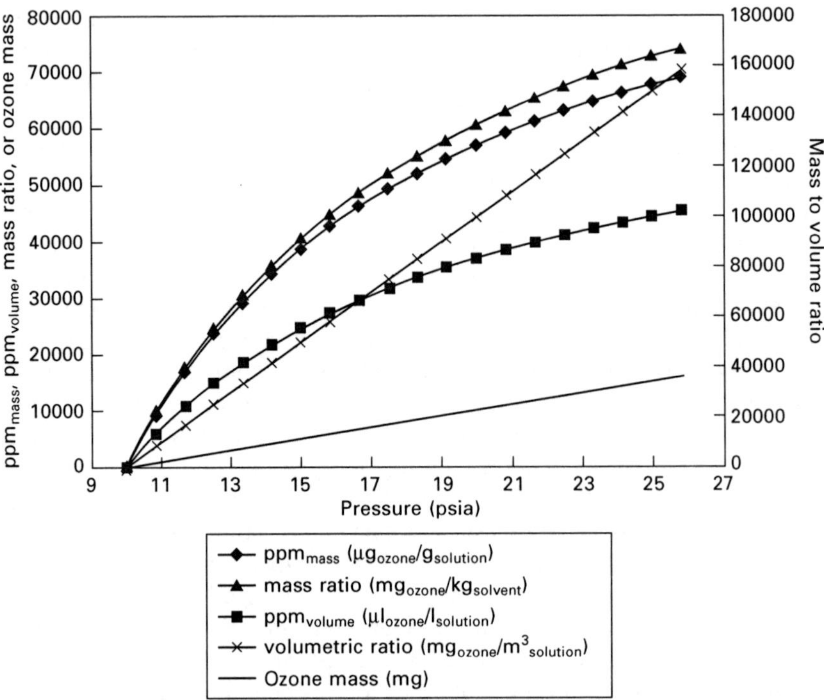


**Fig. 17.8** Predicated ozone concentration, expressed in different units, during pressurization of a 100-liter vessel with an ozone-oxygen mixture. The following conditions are assumed during vessel pressurization. Ozone is generated at 100 g/hour and a mass-ratio of $120g_{ozone}/kg_{oxygen}$. The vessel initially contains air at 10 psia (i.e., under mild vacuum). The solvent gas is air, initially, and air-oxygen mixture at the end, and the solution is a mixture of ozone, oxygen, and air.

## 17.5 Mode of microbial inactivation by ozone

Ozone has a broad antimicrobial spectrum. Its efficacy has been demonstrated against both Gram-positive and Gram-negative bacteria (Ingram and Haines, 1949; Guzel-Seydim *et al.*, 2004), bacterial spores (Ishizaki *et al.*, 1986; Khadre and Yousef, 2001b), fungi (Palou *et al.*, 2001; Allen *et al.*, 2003; Oztekin *et al.*, 2006), viruses (Kim *et al.*, 1980; Roy *et al.*, 1981; Ohmine, 2005), and protozoa (Khalifa *et al.*, 2001). Thus, most microorganisms are sensitive to ozone which suggests that cells possess several sites for ozone action, and reaction of ozone with these sites leads to cell lethality. One of the proposed sites of ozone interaction is the unsaturated lipids of the cell membrane (Victorin 1992; Thanomsub *et al.*, 2002). In studies performed by Scott and Lesher (1963), it was demonstrated that ozone reacts with unsaturated lipids of *E. coli* cell membrane, ultimately resulting in leakage of cell contents and death. Other researchers implicated ozone in DNA strand breakage and mutagenesis in bacteria (Ishizaki *et al.*, 1986). Treatment of

© Woodhead Publishing Limited, 2012

viruses with ozone has revealed that damage occurs via the destruction of the capsid protein, compromised integrity of the lipid envelopes as well as damage to RNA or DNA (Kim *et al.*, 1980; Roy *et al.*, 1981; Ohmine, 2005). Spore inactivation is thought to relate to damage to spore coatings, which may allow ozone to penetrate into the core of the spore (Khadre and Yousef, 2001b).

Reactive oxygen species produced as ozone decomposes, especially the hydroxyl radical, are linked to the antimicrobial action of the sanitizer (Block, 2001). This radical is the strongest known oxidizer and can cause damage to many components within viable cells. In a study conducted on *Bacillus subtilis* spores, hydroxyl radicals were found to be principally responsible for inactivation of the bacterium (Cho *et al.*, 2002). In contrast, Hunt and Marinas (1997) reported that addition of radical scavengers to the treatment medium did not significantly decrease inactivation of *E. coli*, indicating the involvement of molecular ozone in microbial inactivation.

## 17.6 Applications of ozone for food decontamination

### 17.6.1 Process design considerations

Exposure to ozone can lead to serious health hazards in humans. Material selection and equipment maintenance are particularly important to minimize this hazard. Ozone reacts with several commonly used materials, including rubber and plastic. Surfaces exposed to ozone should consist only of compatible materials (Kim *et al.*, 2003).

Ozone, with its high reactivity and relatively rapid decomposition to non-toxic oxygen, has promising applications as a aqueous or gaseous sanitizer. When used in the aqueous state, ozone sanitization is usually a stand-alone unit operation. Decontamination of food with aqueous ozone (e.g., fresh produce sanitization) is often applied after a pre-cleaning step. In contrast, gaseous ozone treatment can be incorporated into several processing operations such as refrigerated storage or transportation, allowing more flexibility to food processors. Gaseous diffusion is many orders of magnitude faster than diffusion by liquids (Cussler, 1997); thus, gaseous sanitizers can reach the targeted contaminants within a commodity, in a relatively short time. Additionally, gaseous sanitization is less likely to modify the composition of the water soluble compounds contained in the food matrix.

Process design depends greatly on the characteristics of the treated food. Treatment of smooth surfaces of a solid food (e.g., tomatoes) with ozone is a relatively easy process to design. The process complexity increases when the surface is rough in texture or interrupted with crevices (e.g., cantaloupe). Additional design complications arise when the treated food is not solid. The physical properties of the food matrix (e.g., viscosity, texture, and shape in case of solids) also influence the decontamination efficiency. Components in the food matrix may compete with contaminating microorganisms for the

© Woodhead Publishing Limited, 2012

sanitizer. The organic material in food, especially proteins and fats, greatly reduces the efficacy of ozone (Ingram and Haines 1949, Guzel-Seydim *et al.*, 2004).

Availability of water is critical for enhancing the efficacy of ozone treatments. Increasing the water activity improved the antifungal properties of ozone during the treatment of barley grains (Allen *et al.*, 2003). Similarly, Wu *et al.* (2006) treated wheat with ozone and observed greater inactivation of fungal spores when they increased water activity and treatment temperature (from 10 to 40°C). A positive correlation was also observed between relative humidity and antimicrobial efficacy of gaseous ozone during fresh produce sanitization (Vurma *et al.*, 2009). Increased relative humidity has been demonstrated to aid in bacterial spore inactivation (Ishizaki *et al.*, 1986). Contribution of humidity could be attributed to the fact that the gas molecules might have to pass from the gaseous to aqueous phase to be effective against targeted microorganisms. The pH of the medium also greatly influences the efficacy of aqueous ozone, as stability of the sanitizer decreases when medium pH increases. A low pH is desired if molecular ozone reactions are necessary for inactivation; however, higher pH has been demonstrated to encourage formation of hydroxyl radicals, contributing to spore inactivation (Cho *et al.*, 2002).

### 17.6.2 Selected food applications

Ozone is approved in the US for use in food (FDA, 2001). There are many reports on the potential applications of ozone as an antimicrobial agent in the food industry; these have been reviewed in the previously published literature (e.g., Kim *et al.*, 2003; Perry and Yousef, 2011; Yousef *et al.*, 2011). Ozone has been tested on many products including meat, fish, poultry, fruits, and vegetables. A few of these products are treated commercially with ozone. Ozone has also been used in other areas of food production such as removal of residual pesticides on fruits, and decontamination of food contact surfaces, food packaging material, and storage areas of fresh fruits and vegetables. A summary of selected food applications using aqueous and gaseous ozone is presented in Tables 17.1 and 17.2, respectively.

*Fresh produce*

Fresh fruits and vegetables are consumed raw, often minimally processed. These products are subject to many types of microbial contaminants, originating from multiple sources. Fresh produce is susceptible to contamination beginning from the pre-harvesting stage to retail distribution. The microbial contaminants include many pathogenic and spoilage bacteria, spoilage fungi, Norwalk-like and hepatitis A viruses, and parasites, most notably *Cyclospora* and *Cryptosporidium*. The lack of treatments that ensure microbial lethality creates a unique challenge for maintaining the safety and quality of fresh produce.

© Woodhead Publishing Limited, 2012

**Table 17.1** Food decontamination using aqueous ozone

| Product | Ozone measurement method | Ozone concentration and treatment time | Targeted microorganism | Result | Reference |
|---|---|---|---|---|---|
| Shredded lettuce | Amperometric | 5 ppm, 5 min | *Shigella sonnei* | 1.8 log reduction | Selma *et al.*, 2007 |
| Apples (surface) | UV spectrophotometer ($A_{258}$ nm) and indigo method | 22–25 mg $l^{-1}$, 3 min | *Escherichia coli* O157:H7 | 2.6 log reduction | Achen and Yousef, 2001 |
| Alfalfa sprouts | UV spectrophotometer ($A_{258}$ nm) | 5–23 μg $ml^{-1}$, 2 min | *Listeria monocytogenes* | < 1 log reduction<br>Reduced sensory quality | Wade *et al.*, 2003 |
| | | 21 ppm, 64 min | *E. coli* O157:H7 | 2 log reduction | Sharma *et al.*, 2003 |
| Lettuce | DPD kit (CHEMetrics, Vacu-vials, Ozone K-7403, Calverton, VA, US) | > 10 mg $ml^{-1}$, 10 min | Natural spoilage microbiota | 1.1 log reduction | Garcia *et al.*, 2003 |
| Beef | UV spectrophotometer ($A_{254}$ nm) | 95 mg $l^{-1}$, spray | *E. coli* O157:H7<br>*Salmonella* Typhimurium | Not more effective than spraying with plain water | Castillo *et al.*, 2003 |
| Mussels (shucked) | Indigo trisulfonate (Bader and Hoigne, 1981) | 1 mg $l^{-1}$, 60–90 min | Natural microbiota | 0.7–2.1 log reduction | Manousaridis *et al.*, 2005 |
| Green pepper (cut) | Amperometric (Orbisphere Analyzer 3660, Orbisphere Laboratories, UK) | 0.3–3.95 mg $l^{-1}$ 20 sec–30 min | Natural microbiota | Not more effective than plain water | Ketteringham *et al.*, 2006 |
| Potato | Amperometric (B&C Electronics Srl, Carnate, Milan, Italy) | 5 ppm, 1 min | *Yersinia enterocolitica*<br>*L. monocytogenes* | 1.6 log reduction<br>0.8 log reduction | Selma *et al.*, 2006 |
| Shrimp | Amperometric | 3 ppm, 1 min | Natural microbiota | 3 log reduction<br>No negative oxidation effects | Chawla *et al.*, 2007 |

© Woodhead Publishing Limited, 2012

**Table 17.2** Food decontamination using gaseous ozone

| Product | Ozone measurement method | Treatment conditions (concentration, time) | Targeted microorganism | Result | Reference |
|---|---|---|---|---|---|
| Baby carrots | UV spectrophotometric | 7.6 mg $l^{-1}$, 15 min | *Escherichia coli* O157:H7 | 2.6 log reduction | Singh *et al.*, 2002 |
| Green pepper | UV spectrophotometric | 8 mg $l^{-1}$, 25 min | *E. coli* O157:H7 | 7 log reduction | Han *et al.*, 2002 |
| Chicken skin | UV spectrophotometric | >2000 ppm, 30 min | *Salmonella* Infantis<br>*Pseudomonas aeruginosa* | < 1 log reduction (both organisms)<br>No negative sensory outcomes | Al-Haddad *et al.*, 2005 |
| Pistachios (in-shell) | Amperometric Eco Sensor (model OS-3, Santa Fe, NM, USA) | 1 ppm, 240 min | *E. coli*<br>*Bacillus cereus* | > 2.5 log reduction (both organisms), no increase in rancidity | Akbas and Ozdemir, 2006 |
| Figs (dried) | Portable ozone detector (Model OZO21ZX) | 5 or 10 ppm, 3 or 5 hours | Natural microbiota | 1.46 log reduction in coliform count | Oztekin *et al.*, 2006 |
| Figs (dried) | Amperometric Eco Sensor (model OS-3, Santa Fe, NM, USA) | 1 ppm, 360 min<br>1 ppm, 360 min<br>5–9 ppm, 360 min | *E. coli*<br>*B. cereus* (cells)<br>*B. cereus* (spores) | 3.5 log reduction<br>3.5 log reduction<br>2 log reduction | Akbas and Ozdemir, 2008a, 2008b |
| Cherry tomato | Iodometric titration (Rakness *et al.*, 1996) | 10 mg $l^{-1}$, 10–20 min | *Salmonella* Enteritidis | ~7 log reduction | Das *et al.*, 2006 |
| Tomato (sliced) | Amperometric (EcoSensor, Inc., A-21ZX model, USA) | 4 μl $l^{-1}$, 30 min every 3 hours up to 15 days | Natural microbiota | 1.1–1.2 log reduction (bacteria)<br>0.5 log reduction (fungi)<br>No off-flavors produced | Aguayo *et al.*, 2006 |

© Woodhead Publishing Limited, 2012

| | | | | | |
|---|---|---|---|---|---|
| Blueberries | UV spectrophotometric | 5% (wt/wt), 64 min (without pressurized ozone) | *E. coli* O157:H7 | 2.2 log reduction | Bialka and Demirci, 2007 |
| | | 5% (wt/wt), 64 min (with pressurized ozone) | *Salmonella* spp. | 3 log reduction<br>No detectable difference | |
| Cantaloupe (cut) | UV (model H1-SPT, IN USA Inc., Needham, MA, USA) | 10,000 ppm, 30 min | *Salmonella* spp. | 2.8 log reduction (12.6 $cm^2$) | Selma *et al.*, 2008b |
| Black pepper | Amperometric (model OS-3, Eco Sensors, Inc., Santa Fe, NM, USA) | 1 ppm, 120 min | *E. coli* | ~7 log reduction | Emer *et al.*, 2008 |
| Baby spinach | UV spectrophotometric | (1.5 g $O_3$ $kg^{-1}$ gas mixture, or 935 ppm v ozone/v gas mixture), 30 min | *E. coli* O157:H7 | 1.8 log reduction<br>No apparent change in visual quality | Vurma *et al.*, 2009 |

© Woodhead Publishing Limited, 2012

Ozone is one of several sanitizers that are being explored as alternatives to chlorine, which is commonly used in fresh produce cleaning and sanitization. Fresh produce sanitization is one of the most promising applications of ozone (Vurma, 2009). The natural mesophilic and psychrotrophic microbial populations in shredded lettuce were reduced by 3.9 and 4.6 logs, respectively, when treated with 1.3 mM aqueous ozone for 5 min (Kim *et al.*, 1999). *Shigella sonnei* was reduced by 1.8 logs when inoculated in lettuce and treated with 5 ppm aqueous ozone for 5 min (Selma *et al.*, 2007). A process involving generation of ozone within spinach packaging decreased inoculated *E. coli* O157:H7 by 3–5 logs, but negatively impacted product quality (Klockow and Keener, 2009). Vurma *et al.* (2009) investigated application of gaseous ozone for inactivation of *E. coli* O157:H7 in spinach during simulated vacuum cooling and transport operations. The researchers found that applying (935 ppm) ozone during vacuum cooling, combined with application of 5–10 ppm ozone during simulated transportation, resulted in ≥ 4-log reduction of this pathogen. When apple surfaces were inoculated with *E. coli* O157:H7 and dipped in 23–30 mg/l aqueous ozone, the treatment inactivated 2.6 log of pathogen's population. Inactivation of the pathogen increased to 3.7 log when ozone was bubbled into the wash water used to treat the inoculated apples (Achen and Yousef, 2001). However, only a small reduction (<1 log) was achieved when this pathogen was inoculated in the stem-calyx region of the apple, regardless of ozone delivery method. A gaseous ozone treatment (20 mg/l for 15 min) completely inactivated *Salmonella* Enteritidis that was spot-inoculated onto cherry tomatoes (Das *et al.*, 2006). Continuous exposure of blueberries to very high 5% (mass basis) gaseous ozone concentrations for 64 min resulted in a 2.2-log reduction of *E. coli* O157:H7; similar conditions under pressure inactivated 3 log of *Salmonella* population (Bialka and Demirci, 2007). When cantaloupe rind was inoculated with *Salmonella* and treated with gaseous ozone (10,000 ppm for 30 min) under vacuum, the treatment inactivated up to 4.2 log of the pathogen's population (Selma *et al.*, 2008b).

*Shell eggs*

Egg-borne salmonellosis is a disease of great public health concern and is transmitted by unpasteurized shell eggs (Perry, 2010). The use of ozone for decontamination of shell eggs has been investigated by several researchers (Koidis *et al.*, 2000; Perry *et al.*, 2008; Perry, 2010; Rodriguez-Romo and Yousef, 2005). When surfaces of shell eggs were inoculated with *Salmonella* Enteritidis and treated with 1.4 ppm aqueous ozone at 22°C for 90 s, the population of the pathogen decreased by 1 log. Greater reduction (2 log) in *Salmonella* population was achieved when inoculated eggs were treated with 3 ppm ozone at 4°C for 90 s (Koidis *et al.*, 2000). Synergy between ozone and UV was demonstrated against *Salmonella* on shell eggs (Rodriguez-Romo and Yousef, 2005). When shell eggs were surface-inoculated with *Salmonella* Enteritidis and treated with gaseous ozone (5%, mass

© Woodhead Publishing Limited, 2012

basis) and UV radiation for 2 min of total treatment time, the combination eliminated 4.6 log of the pathogen. In a subsequent study, *Salmonella* Enteritidis was inoculated in the egg yolk and treated with a combination of mild heat (55–58°C) and pressurized gaseous ozone (Perry, 2010). The combined process inactivated > 6 log of the pathogen without negatively impacting the egg quality.

*Meats and poultry*

Pathogenic microorganisms associated with meat and meat products (e.g. *Salmonella* with poultry, enterohemorrhagic *E. coli* with beef, and *L. monocytogenes* with uncooked or ready-to-eat sliced products) continue to pose health risks to consumers. Meat and meat products may come in contact with spoilage and pathogenic microorganisms during slaughtering, handling, processing and distribution (Zhao *et al.*, 2001). Application of ozone at low levels may not be sufficient to decontaminate meat surfaces, but high levels of ozone may produce undesirable quality changes in meat (Khadre *et al.*, 2001). The presence of many components on the meat surface that readily react with the sanitizer reduces its efficacy against the targeted pathogens. Some researchers, however, reported some success in application of ozone on poultry and meat (e.g., Dave, 1999).

*Seafood*

Soaking peeled shrimp in 3 ppm aqueous ozone for 1 min reduced viable bacterial load by 3 logs and extended the shelf life of shrimp stored on ice by 4 days (Chawla, 2006). Viable bacterial load decreased by 2–3 logs when the skin of gutted fresh jack mackerel (*Trachurus trachurus*) and shimaaji (*Caranx mertensi*) was treated with 0.6 ppm ozone dissolved in 3% NaCl solution for 30–60 min. A previous study showed that a similar treatment increased the shelf life of fish by 60% (Haraguchi *et al.*, 1969). A 3-day extension in the shelf life of shucked mussels was achieved by washing with 1 mg/l ozone for 90 min, as compared to untreated controls (Manousaridis *et al.*, 2005). Washing with 4.5 ppm aqueous ozone for 10 s significantly reduced natural microbiota on live catfish and processed catfish fillets (Sopher *et al.*, 2007).

*Dried foods*

Dried food products are often laden with bacterial and fungal spores. It has been suggested that the water activity, humidity in the environment, temperature of the product and applied ozone concentration influence the efficacy of gaseous ozone against microbial contaminants on dry food and food ingredients (Kim *et al.*, 2003). The surface properties of foods also influence ozone inactivation of microorganisms in dried foods. Higher ozone concentration and longer treatment time were needed for cereal flour and ground pepper to achieve comparable microbicidal effect for the whole cereal and pepper (Naitoh *et al.*, 1992; Zagon *et al.*, 1992). Similarly, higher detoxification of aflatoxins

© Woodhead Publishing Limited, 2012

was achieved in whole pistachio kernels as compared to ground pistachios upon treatment with gaseous ozone (Akbas and Ozdemir, 2006).

Ozone was effectively used as an anti-fungal fumigant to preserve quality of stored wheat and higher ozone efficacy was achieved as both temperature and water activity of wheat was increased (Wu *et al.*, 2006). A 3-log decrease in *Bacillus* spp. and *Micrococcus* spp. populations was observed on cereal grains, peas, beans, and spices, and the efficacy of treatment was found to depend on the temperature, relative humidity, and ozone concentration applied. Similarly, increased water activity and temperature during gaseous ozone application resulted in a greater than 1 log reduction of fungal spores in barley (Allen *et al.*, 2003).

There is increasing concern about insect tolerance to routinely used fumigants such as phosphine in stored grains; therefore, ozone has been investigated to remove insect infestations during grain storage. A greater than 1 log reduction of three common stored grain pests was observed upon 50 ppm gaseous ozone application for 3 days (Kells *et al.*, 2001).

*Fruit juices*

Fruit juices contain several bioactive compounds, including carotenoids, phenolic compounds, vitamins and anthocyanins (Abeysinghe *et al.*, 2007). Ozone can be added directly to food (FDA, 2001), hence some researchers suggested using the agent as an alternative to thermal pasteurization of fruit juices (Steenstrup and Floros, 2004). These authors observed a rapid inactivation of *E. coli* O157:H7 in apple cider when treated with 860 ppm (v/v) ozone. Greater inactivation of the pathogen in apple cider was achieved when ozone concentration exceeded 1000 ppm (v/v). In a different investigation, application of ozone (0.9 g/h at a flow rate of 2.4 l/min) resulted in less than 1 log reduction of *E. coli* O157:H7 inoculated in either apple cider or orange juice. A similar treatment inactivated 1 log of *Salmonella* population in apple cider and 1.8 log of the pathogen in orange juice (Williams *et al.*, 2005). Ozone has also been found to destroy the mycotoxin patulin in apple juice (Cataldo, 2008). According to the experience of the authors of this chapter, ozone concentrations that are high enough to inactive bacterial pathogens cause great damage to the quality of fruit juices.

### 17.6.3 Food processing environment and equipment

Processing environments are susceptible to biofilm formation on wet food contact surfaces, within microscopic crevices, or in joints and unions. Biofilm microorganisms are protected against sanitizers which cannot effectively penetrate the biofilm polymers (Dixon, 1998; Zottola and Sasahara, 1994). To ensure product safety, it is essential to clean and disinfect processing areas on a regular basis. Food processing equipment also needs to be effectively cleaned and sanitized in order to maintain its functionality and to prevent contamination of processed food (Urano and Fukuzaki, 2001).

© Woodhead Publishing Limited, 2012

However, there is an increasing concern about the efficacy, environmental impact, and costs associated with the use of current chemical sanitizers. Hence, alternative methods for decontamination of processing environment and equipment are sought. Ozone, in gaseous and aqueous states, has been explored for surface disinfection of packages and equipment (Pascual *et al.*, 2007). According to these authors, moderate doses of gaseous and aqueous ozone (0.5–3.5 ppm) were sufficient to cause significant microbial reductions. However, they noticed that longer exposure times were needed for gaseous than aqueous forms of ozone to achieve the microbial reduction. Efficacy of ozone against natural contaminants, *B. subtilis* spores, and biofilms of *Pseudomonas fluorescence*, was tested on surfaces of multilaminated food packaging material and stainless steel (Khadre and Yousef, 2001a). Application of low level (5.9 μg/ml) aqueous ozone for 1 min completely eliminated natural microbiota from the packaging material. The population of *P. fluorescence* on the multilaminated packaging material decreased by 8 log/12.5 $cm^2$ upon repeated treatment with 3.6 μg/ml ozone. Counts of *B. subtilis* spores ($10^8$/6.3 $cm^2$) on packaging material and stainless steel decreased below detection level upon washing them with 13 μg/ml ozone and 8 μg/ml aqueous ozone, respectively. The authors also found that ozone inactivated *P. fluorescence* biofilms more effectively on stainless steel than on the packaging material. Aqueous ozone may be effective against microbial contaminants on processing equipment, walls or floors. It has been reported that ozone decreased surface microbiota by 3 log, when tested in wineries for barrel cleaning, tank sanitation, and clean-in-place processes (Hampson, 2000). Ozone has been effectively used to disinfect oak barrels and to control *Brettanomyces* sp., a fungus that causes off-taste and other defects during wine ageing in Australia (Day, 2004).

### 17.6.4 Food storage

Use of ozone in food packing houses and storage rooms can minimize post-harvest diseases in fruit, control fruit ripening by decomposing ethylene (a growth hormone), and generally decrease quality loss caused by mold growth. Ozone can also be a viable substitute to sulfur dioxide as a fumigant. When used on fresh produce, sulfur dioxide leaves sulfite residues and may cause bleaching of the product (Smilanick *et al.*, 1990; Luvisi *et al.*, 1992). Some researchers reported beneficial effect when ozone is used in fruit and vegetable storage atmosphere, whereas others found no benefits from this application. A slight but significant decrease in the rate of decay of strawberries by *Botrytis cinerea* (Nadas *et al.*, 2003), lemons and oranges by *Penicillium digitatum* and *P. italicum* (Palou *et al.*, 2001), and peaches by *Monilinia fructicola* (Palou *et al.*, 2002) was observed when these fruits were stored under low levels of ozone (0.3–1.0 ppm). On the contrary, applying gaseous ozone (0.3–5 ppm) in a storage environment for extended periods did not control fungus infestation in wounds or under the fruit surface (field infections) for

© Woodhead Publishing Limited, 2012

products such as table grapes (Shimizu *et al.*, 1982), apples (Schomer and McColloch, 1948), pears (Spotts and Cervantes, 1992), citrus (Hopkins and Loucks, 1949; Smilanick *et al.*, 2002) and peaches (Palou *et al.*, 2002). It was suggested that fungal structures in fruit wounds are protected from the oxidative effects of sanitizers like ozone, chlorine and chlorine dioxide due to the limited penetration of the gas and consumption of the sanitizer in reactions with fruit components, thereby decreasing the effective concentration of the antimicrobial agent (Adaskaveg, 1995; Palou *et al.*, 2002; Spotts and Peters, 1980). Palou *et al.* (2002) found that applying ozone in cold storage facility prevented 'nesting' that occurs when rots due to growth of *Mucor piriformis*, *P. expansum*, and *B. cinerea* spread from decayed fruit to adjacent healthy ones. Additionally, ozone was found to reduce the rate of re-infection of stored produce, inhibit mold growth on the surface of walls, packages, and floors of rooms by decreasing the load of airborne fungal spores (Schomer and McColloch, 1948; Song *et al.*, 2000).

Ethylene is generated by fruits in storage and it accelerates their ripening. The rapid reaction of ethylene and ozone in air is a well-documented phenomenon (Dickson *et al.*, 1992). It is presumed that ozone can oxidize ethylene in storage facility, reducing fruit respiration and controlling spoilage microorganisms (Aguayo *et al.*, 2006). According to Skog and Chu (2001), ozone at 0.4 μl/l could prevent ethylene accumulation in apple and pear storage rooms. Mushrooms, apples, pears, broccoli, and cucumbers were not harmed by this low ozone concentration. Whole and fresh-cut cantaloupe melons maintained good visual quality, aroma, and firmness after exposure to gaseous ozone (Selma *et al.*, 2008a). The shelf life of strawberries stored under a gas mixture (10 ppm ozone and carbon dioxide) increased by 8 days and maintained better quality, as compared to untreated strawberries (Vurma, 2009).

### 17.6.5 Quality benefits and concerns for ozone-treated foods

Despite several reports on food quality improvements (Aguayo *et al.*, 2006; Vurma, 2009), or retention (e.g., Kamotani *et al.*, 2010; Selma *et al.*, 2008b; Perry *et al.*, 2011) by ozone treatments, the process may damage the quality of some foods. Ozone is not universally beneficial and in some cases may promote oxidative spoilage in foods (Rice *et al.*, 1982). Treatment with gaseous ozone was reported to degrade ascorbic acid content of orange juice (Tiwari *et al.*, 2008). Apple cider treated with ozone and stored for a period of 21 days was found to have increased sedimentation, decreased soluble solids and lower sucrose content, as compared to untreated control (Choi and Nielsen, 2005). Thus, treatment of fruit juices with ozone may not be advisable; however, Dock (1999) observed no detrimental changes in quality of ozone-treated apple cider.

There are no conclusive results on the effects of ozone treatment on the ascorbic acid content of fresh cut vegetables and fruits. Although levels of

© Woodhead Publishing Limited, 2012

ascorbic acid decreased upon treating broccoli florets with gaseous ozone (Lewis *et al.*, 1996), these were changes observed in fresh cut celery treated with aqueous ozone (Zhang *et al.*, 2005). On the contrary, the ascorbic acid level seems to increase in spinach leaves and strawberries after exposure to gaseous ozone (Liew and Prange, 1994; Perez *et al.*, 1999). Aqueous ozone treatment decreased slightly the ascorbic acid content of fresh cut iceberg lettuce, but the final phenolic content was not affected (Beltran *et al.*, 2005). Some of these changes in ascorbic acid content of plants may be attributed to an oxidative stress response upon exposure to ozone (Perez *et al.*, 1999).

Change in food color serves as an indicator of quality loss. Oxidation of carotenoids during ozone treatment produced color changes in orange juice (Tiwari *et al.*, 2008), blackberry juice (Tiwari *et al.*, 2009), broccoli, cucumbers, mushrooms (Skog and Chu, 2001), and carrots (Liew and Prange, 1994). Other researchers observed a slight, but not significant, change in the anthocyanin content of strawberries (Perez *et al.*, 1999) and blackberries (Barth *et al.*, 1995) upon treatment with ozone. An increase in the total phenolics content of seedless grapes was observed during cold-storage under ozone (Artes-Hernandez *et al.*, 2007).

Flavor and aroma are essential quality attributes of food; protection of these attributes is essential for successful ozonation treatment. Zhao and Cranston (1995) treated ground black peppers with ozone and reported an increase in the oxidation of volatile compounds in the product. Other researchers reported 40% reduction of volatile esters in strawberries in response to ozone treatment (Perez *et al.*, 1999). A trained sensory panel observed undesirable changes in the aroma of tomatoes that were ozone treated and stored for 15 days at 5°C (Aguayo *et al.*, 2006). High concentrations of ozone had a slightly negative effect on sensory scores for flavor, appearance, overall palatability for flaked red peppers, but no differences in sensory characteristics were seen with dried figs (Akbas and Ozdemir, 2008a, 2008b). In ground pistachios, there were no differences in fatty acid composition, except at higher ozone concentrations where a difference in peroxide values was observed (Akbas and Ozdemir, 2006). Storage of fresh whole scad (*Trachurus trachurus*) on ice, under gaseous ozone atmosphere, extended the shelf life by 2 days with no detectable signs of rancidity or fatty acid oxidation (Da Silva *et al.*, 1998). Shrimp treated with aqueous ozone had extended shelf life by a week and better consumer acceptability, compared to untreated product (Chawla, 2006).

The overall quality of several foods was not affected by the ozone treatment. Ozone used during commercial storage of onions produced no changes in the chemical composition or sensory quality of this commodity (Song *et al.*, 2000). Aqueous ozone treatment of fresh cut celery and strawberries caused no changes in their total sugar content (Perez *et al.*, 1999; Zhang *et al.*, 2005). When shell eggs were pasteurized with heat-ozone combination, the structure of egg proteins and activity of lysozyme were not affected by the treatment (Perry *et al.*, 2011). Similarly treated eggs were rated by

© Woodhead Publishing Limited, 2012

consumers to have a better visual appeal than the commercially pasteurized shell eggs (Kamotani *et al.*, 2010).

## 17.7 Ozone safety considerations and limitations

Ozone is an unstable molecule that has a short half-life. Therefore, the gas cannot be stored or transported, and it has to be generated at the site of application. These are some of the drawbacks to using ozone as a decontaminating agent in food applications. Although ozone generators are relatively inexpensive, successful application requires customized processing equipment. Careful optimization and control of process variables and proper disposal of excess ozone are critical factors to be observed. Factors affecting the stability of the gas such as temperature, relative humidity, and exposure to UV must be controlled to maintain effective concentration during treatment.

Ozone is a highly corrosive gas; therefore, processing equipment must be made of ozone-compatible materials such as 316 stainless steel and Teflon to maintain functionality of the process and safety of operators. As discussed earlier, accurate measurement of ozone is difficult due to its rapid decomposition. Measurement difficulties increase when ozone is mixed with the food or the presence of other oxidizing agents (e.g., chlorine) in the medium.

Ozone is toxic upon prolonged exposure even at very low levels. Human exposure to ozone above a low threshold (0.2–0.5 ppm) can lead to a number of negative health effects (Horváth *et al.*, 1985). Ozone is a respiratory irritant above these low levels and causes coughing, headaches, nausea and dizziness. Exposure above 6 ppm for a long period can cause pulmonary edema that leads to diminished breathing capacity (Horváth *et al.*, 1985). Permanent lung damage can occur upon repeated exposure to ozone (Scheel *et al.*, 1959). Other symptoms such as loss of vision have also been reported upon human exposure to ozone (Lagerwerff, 1963).

Standards set forth by the Occupational Safety and Health Administration of the United States specify that workers may not be exposed to concentrations exceeding 0.1 ppm for extended periods of time, or 0.2 ppm for short-term exposure (OSHA, 2004).

Ozone is used as an antimicrobial agent in both gaseous and aqueous forms. Low solubility of ozone in water makes it necessary to use specialized contact chambers and dispersion systems to prepare its aqueous phase. The discrepancy in reporting and monitoring concentrations of gaseous ozone is a challenge when comparing its efficacy in different systems. The decontamination efficacy of ozone is affected by several factors such as pH, temperature, humidity and ozone-demand of the treated medium. Accessibility of ozone to the target microorganism may also play a role in its efficacy; microorganisms are not always found on food surfaces and may be internalized or strongly attached to the food. Ozone stability increases in

© Woodhead Publishing Limited, 2012

a medium with low pH and decreases as the pH of the medium increases. Instability of ozone at high temperatures makes it an ineffective sanitizer at these temperatures. Ozone has high decontamination efficacy in low-ozone demand medium such as water or buffer systems. However, the presence of unsaturated fats and soluble proteins in foods may decrease its efficacy against target microorganisms. Foods rich in unsaturated fat such as meats may require higher ozone concentrations for effective decontamination as compared to low fat foods like fresh produce. The strong oxidizing power of ozone may cause sensory defects such as discoloration and undesirable odors and may alter nutritional components such as vitamins, enzymes, amino acids and essential fatty acids in some treated foods. Ozone may also cause physiological tissue damage to treated fruits and vegetables. These adverse effects of ozone depend on the applied dosage, treatment conditions and the composition of the food.

## 17.8 Conclusion and future trends

Reviewing the published literature on ozone clearly illustrates the great discrepancy in results reported by different researchers. Additionally, food processors who test ozone in their facility report different experimental methodologies. The authors of this chapter believe that deficiencies in ozone measuring devices, inconsistent reporting of measurement units, and general methodology flaws contribute to these discrepancies. Recommendations have been made throughout the chapter in an attempt to correct these shortcomings. Planning successful studies to decontaminate food by ozone requires (a) a good understanding of the physical and chemical properties of this unique compound, (b) availability of adequate ozone generation, containment, and destruction equipment, (c) an expertise in microbiology so that pathogen monitoring in food is performed correctly, and (d) well-trained analysts who can work safely with this hazardous chemical.

Despite the inconsistencies, there have been successful attempts to use ozone in decontaminating food, and the transfer of the technology from laboratory settings to production facility is under way. As the industry currently suffers from increased risk of disease transmission by fresh produce, use of ozone could eliminate pathogens in these foods, a goal that conventional sanitization sometimes fails to achieve. There is a robust growth in the organic food segment; these foods could be made safer with proper ozone application. Use of ozone, in sequence with physical treatments such as mild heat, could lead to safe products having high consumer acceptability. Elimination of *Salmonella* in shell eggs is an example of emerging applications that use heat-ozone combinations. It is foreseen that ozone replaces commonly used chemicals in cleaning-in-place (CIP) operations, with ozone serving as both a cleaning and sanitizing agent. If these efforts are successful, use of ozone in CIP can reduce cleaning and sanitization times, minimize consumption of

© Woodhead Publishing Limited, 2012

chemicals, save water, and decrease costs. Past attempts to decontaminate dry foods (e.g., nuts) and food ingredients (e.g., spices) by ozone generally have not been successful. However, recent investigations show that effective decontamination of certain tree nuts such as almonds and pistachios, without significantly affecting product quality, seems feasible. Ozone offers the industry another antimicrobial tool in the ongoing quest for safer foods. Correct implementation of ozone could improve the safety and quality of food when it is prepared in environmentally friendly processing establishments.

## 17.9 Sources of further information and advice

Supplementing the cited references, the authors suggest the following additional sources of information on ozone. The International Ozone Association (2011) publishes research on various aspects of ozone, including water and food applications. The association hosts the *Ozone: Science and Engineering* journal, which is a unique publication dedicated to ozone. Readers also can visit various websites of ozone equipment manufacturers and those of companies focusing on ozone integration into processing lines. Various ozone factsheets have been published, including those by Ramaswamy *et al.* (2007) and Rice (2011).

## 17.10 Acknowledgments

Compilation of this work was funded by the National Integrated Food Safety Initiative (NIFSI), United States Department of Agriculture (USDA), under Agreement No. 2009-51110-05902 titled 'Pathogen Inactivation in Fresh Produce by Incorporation of Sanitizers into Existing Operations within the Produce Chain', in cooperation with The Ohio State University, Iowa State University and New Mexico State University.

## 17.11 References

ABEYSINGHE D C, LI X, SUN C D, ZHANG W S, ZHON C H and CHEN K S (2007), 'Bioactive compounds and antioxidant capacities in different edible tissues of citrus fruit of four species', *Food Chem*, 104, 1338–1344.

ACHEN M and YOUSEF A E (2001), 'Efficacy of ozone against *Escherichia coli* O157:H7 on apples', *J Food Sci*, 66, 1380–1384.

ADASKAVEG J E (1995), 'Postharvest sanitation to reduce decay of perishable commodities', *Perishables Handling Newslett*, 82, 21–25.

ADLER M G and HILL G R (1950), 'The kinetics and mechanism of hydroxide ion catalyzed ozone decomposition in aqueous solution', *J Am Chem Soc*, 72, 1884–1887.

AGUAYO E, ESCALONA V H and ARTES F (2006), 'Effect of cyclic exposure to ozone gas

© Woodhead Publishing Limited, 2012

on physicochemical, sensorial and microbial quality of whole and sliced tomatoes', *Postharvest Biol Technol*, 39, 169–177.

AKBAS M Y and OZDEMIR M (2006), 'Effect of different ozone treatments on aflatoxin degradation and physicochemical properties of pistachios', *J Sci Food Agric*, 86, 2099–2104.

AKBAS M Y and OZDEMIR M (2008a), 'Application of gaseous ozone to control populations of *Escherichia coli*, *Bacillus cereus* and *Bacillus cereus* spores in dried figs', *Food Micro*, 25, 386–391.

AKBAS M Y and OZDEMIR M (2008b), 'Effect of gaseous ozone on microbial inactivation and sensory of flaked red peppers', *Int J Food Microbiol*, 43, 657–662.

AL-HADDAD K S H, AL-QASSEMI R A S and ROBINSON R K (2005), 'The use of gaseous ozone and gas packaging to control populations of *Salmonella infantis* and *Pseudomonas aeruginosa* on the skin of chicken portions', *Food Cont*, 16, 405–410.

ALLEN B, WU J and DOAN H (2003), 'Inactivation of fungi associated with barley grain by gaseous ozone', *J Environ Sci Health*, 35, 617–630.

ALVAREZ P M, GARCIA-ARAYA J F, BELTRAN F J, GIRALDEZ I, JARAMILLO J and GOMEZ-SERRANO V (2006), 'The influence of various factors on aqueous ozone decomposition by granular activated carbons and the development of a mechanistic approach', *Carbon*, 44, 3102–3112.

ANDREWS C C and MURPHY O J (2002), Generation and delivery device for ozone gas. US patent 6,461,487.

APHA (1971), *Standard Methods for the Examination of Water and Wastewater*, 13th edn, American Public Health Association, Washington, DC.

APHA (1995), *Standard Methods for the Examination of Water and Wastewater*, 19th edn, American Public Health Association, Washington, DC.

ARTES-HERNANDEZ F, AGUAYO E, ARTES F and TOMAS-BARBERAN F A (2007), 'Enriched ozone atmosphere enhances bioactive phenolics in seedless table grapes after prolonged shelf life', *J Sci Food Agric*, 87, 824–831.

ATI (2009), *Model Q45H/64 Dissolved ozone measuement system operation and maintenance manual*, Collegeville, PA, Analytical Technology Inc.

BABLON G, BELLAMY W D, BOURBIGOT M M, DANIEL F B, DORÉ M, ERB F, GORDON G, LANGLAIS B, LAPLANCHE A, LEGUBE B, MARTIN G, MASSCHELEIN W J, PACEY G, RECKHOW D A and VENTRESQUE C (1991), 'Fundamental aspects', in Langlais G, Reckhow DA and Brink D R, *Ozone in water treatment application and engineering*, Boca Raton, FL, Lewis Publishers, 11–132.

BADER H and HOIGNE J (1981), 'Determination of ozone in water by the indigo method', *Water Res*, 15, 449–456.

BARSAN N and WEIMAR U (2003), 'Understanding the fundamental principles of metal oxide based gas sensors, the example of CO sensing with $SnO_2$ sensors in the presence of humidity', *J Phys Condens Matter*, 15, R813–R839.

BARTH M M, ZHOU C, MERCIER J and PAYNE F A (1995), 'Ozone storage effects on anthocyanin content and fungal growth in blackberries', *J Food Sci*, 60, 1286–1288.

BELTRAN D, SELMA M V, MARIN A and GIL M I (2005), 'Ozonated water extends the shelf life of fresh-cut lettuce', *J Agric Food Chem*, 53, 5654–5663.

BIALKA K L and DEMIRCI A (2007), 'Decontamination of *Escherichia coli* O157:H7 and *Salmonella enterica* on blueberries using ozone and pulsed UV-light', *J Food Sci*, 72, M391–396.

BLOCK S S (2001), 'Peroxygen compounds', in Block S S, *Disinfection Sterilization and Preservation*, 5th edn, Philadelphia, PA, Williams & Wilkins, 185–204.

BOHN H L (1971), 'Redox potentials', *Soil Sci*, 112, 39–45.

BOLLYKY L J (2003), 'Ozone in gas', in Liptak BG, *Instrument engineers handbook, Process measurement and analysis*, 4th edn, Boca Raton, FL, CRC Press, 1130–1131.

CARLINS J J and CLARK R G (1982), 'Ozone generation by corona discharge', in Rice R G and Netzer A, *Handbook of ozone technology and applications, Vol 1*, Ann Arbor, MI, Ann Arbor Science, 41–76.

© Woodhead Publishing Limited, 2012

CASTILLO A, MCKENZIE K S, LUCIA L M and ACUFF G R (2003), 'Ozone treatment for reduction of *Escherichia coli* O157:H7 and *Salmonella* Typhimurium on beef carcass surfaces', *J Food Prot*, 66, 775–779.

CATALDO F (2008), 'Ozone decomposition of patulin, a mycotoxin and food contaminant', *Ozone Sci Eng*, 30, 197–201.

CHAWLA A S (2006), 'Application of ozonated water technology for improving quality and safety of peeled shrimp meat', MS thesis, Baton Rouge, LA, Louisiana State University.

CHAWLA A, BELL J W and JANES M E (2007), 'Optimization of ozonated water treatment of wild caught and mechanically peeled shrimp meat', *J Aquat Food Prod Tech*, 16(2), 41–56.

CHO M, CHUNG H and YOON J (2002), 'Effect of pH and importance of ozone initiated radical reactions in inactivating *Bacillus subtilis* spore', *Ozone Sci Eng*, 24, 145–150.

CHOI L H and NIELSEN S S (2005), 'The effects of thermal and nonthermal processing methods on apple cider quality and consumer acceptability', *J Food Qty*, 28, 3–29.

CUSSLER E L (1997), *Diffusion mass transfer in fluid systems*, Cambridge, Cambridge University Press.

DA SILVA M V, GIBBS P A and KIRBY R M (1998), 'Sensorial and microbial effects of gaseous ozone on fresh scad (*Trachurus trachurus*)', *J App Microbiol*, 84, 802–810.

DAS E, CANDAN G and ALEV B (2006), 'Effect of controlled atmosphere storage, modified atmosphere packaging and gaseous ozone treatment on the survival of *Salmonella* Enteritidis on cherry tomatoes', *Food Microbiol*, 23, 430–438.

DAVE S A (1999), *Effect of ozone against Salmonella enteritidis in aqueous suspensions and on poultry meat*, MSc Thesis, Ohio State University.

DAY C (2004), 'Developments in the management of Brettanomyces', *Wine Ind J*, 19, 18–19.

DEGREMONT TECHNOLOGIES (2011), IGS Technology, Leonia, Degremont Technologies. Available from: http://www.degremont-technologies.com/dgtech.php?article1080 (accessed 12 October 2011).

DHANDAPANI B and OYAMA S T (1997), 'Gas phase ozone decomposition catalysts', *App Catalysis B Env*, 11, 129–166.

DICKSON R G, LAW S E, KAYS S J and EITEMAN M A (1992), 'Abatement of ethylene by ozone treatment in controlled atmosphere storage of fruits and vegetables', in *Proc International Winter Meeting, Am Soc Agric Eng*, 1–9.

DIXON B (1998), 'Biofilms, cultural diversity in action', *ASM News*, 64(9), 484–485.

DOCK L L (1999), *Development of thermal and non-thermal preservation methods for producing microbially safe apple cider*, PhD Thesis, Lafayette, IN, Purdue University.

DOHAN J M and MASSCHELEIN W J (1987), 'Photochemical generation of ozone present state-of-the-art', *Ozone Sci Eng*, 9, 315–334.

DOSORETZ V, BEHR D, KELLER S (IN USA INC.) (1998), Gas Detection and Measurement System, US patent 5770156.

DURON B (1982), 'Ozone generation by corona discharge', in Rice R G and Netzer A, *Handbook of ozone technology and applications, volume 1*, Ann Arbor, MI, Ann Arbor Science, 41–76.

ELIASSON B and KOGELSCHATZ U (1991), 'Ozone generation with narrow-band UV radiation', *Ozone Sci Eng*, 13(3), 365–373.

EMER Z, AKBAS M Y and OZDEMIR M (2008), 'Bactericidal activity of ozone against *Escherichia coli* in whole and ground black peppers', *J Food Prot*, 71(5), 914–917.

FDA (2001), 'Secondary direct food additives permitted in food for human consumption', *Federal Register*, 21(3), 173368.

FDA (2011), 'FDA food safety modernization act'. Available from: http://www.fda.gov/Food/FoodSafety/FSMA/ucm247548.htm (accessed 20 October 2011).

FIEDLER S, VEPRASKAS M J and RICHARDSON J L (2007), 'Soil redox potential importance

© Woodhead Publishing Limited, 2012

field measurements and observations', in Sparks D L, *Advances in Agronomy, Vol. 94*, San Diego, CA, Academic Press, 1–54.

GAMBRELL R P and PATRICK W H JR (1980), 'Chemical and microbiological properties of anaerobic soils and sediments', in Hook D D, *Plant life in anaerobic soils and sediments*, Ann Arbor, MI, Ann Arbor Science, 375–413.

GAMBRELL R P, DELAUNE R D and PATRICK W H JR (1991), 'Plant life under oxygen deprivation', in Jackson M B, *Plant life under oxygen deprivation ecology physiology and bio-chemistry*, The Hague, SPB Academic.

GARIA A, MOUNT J R and DAVIDSON P M (2003), 'Ozone and chlorine treatment of minimally processed lettuce', *J Food Sci*, 68, 2747–2751.

GEERING F (1999), 'Ozone applications – the state-of-the-art in Switzerland', *Ozone Sci Eng*, 21(2), 187–200.

GORDON G, COOPER W J, RICE R G and PACEY G E (1988), 'Methods for measuring disinfectant residuals', *J Am Water Works Assoc*, 80, 94–108.

GOTTSCHALK C, LIBRA J A and SAUPE A (2000), *Ozonation of Water and Waste Water, A Practical Guide to Understanding Ozone and its Application*, Weinheim, Wiley-VCH, 68–73.

GRIMES H D, PERKINS K K and BOSS W F (1983), 'Ozone degrades into hydroxyl radical under physiological conditions: a spin trapping study', *Plant Physiology*, 72, 1016–1020.

GRUNWELL J, BENGA J, COHEN H and GORDON G (1983), 'A detailed comparison of analytical methods for residual ozone measurement', *Ozone Sci Eng*, 5, 203–223.

GUZEL-SEYDIM Z, BEVER P I JR and GREENE A K (2004), 'Efficacy of ozone to reduce bacterial populations in the presence of food components', *Food Microbiol*, 21, 475–479.

HAMPSON B C (2000), 'Use of ozone for winery and environmental sanitation', *Practical Winery and Vineyard*, Jan/Feb, 27–30.

HAN Y, FLOROS J D, LINTON R H, NIELSEN S S and NELSON P E (2002), 'Response surface modeling for the inactivation of *Escherichia coli* O157:H7 on green peppers (*Capsicum annuum*) by ozone gas treatment', *J Food Sci*, 67, 1188–1193.

HARAGUCHI T, SIMIDU U and AISO K (1969), 'Preserving effect of ozone on fish', *Bull Jap Soc Sci Fish (Nippon Suisan Gakkaishi)*, 35, 915–919.

HOIGNÉ J and BADER H (1975), 'Ozonation of water role of hydroxyl radicals as oxidizing intermediates', *Science*, 190, 782–784.

HOPKINS E F and LOUCKS K W (1949), 'Has ozone any value in the treatment of citrus fruit for decay?' *Citrus Ind*, 30 (5–7), 22.

HORVÁTH M, BILITZKY L and HUTTNER J (1985), *Ozone*, Budapest, Elsevier.

HUNT N K and MARINAS B J (1997), 'Kinetics of *Escherichia coli* inactivation with ozone', *Wat Res*, 31, 1355–1362.

INCROPERA F and DEWITT D (2002), *Fundamentals of Heat and Mass Transfer*, 5th edn, New York, John Wiley & Sons, 917–920.

INGRAM M and BARNES E (1954), 'Sterilization by means of ozone', *Symposium on Industrial Methods of Sterilization*, 264–271.

INGRAM M and HAINES R B (1949), 'Inhibition of bacterial growth by pure ozone in the presence of nutrients', *J Hyg*, 47, 146–158.

INTERNATIONAL OZONE ASSOCIATION (2011), Available from: http://www.io3a.org/ (accessed 12 October 2011).

IN USA INC. (1996), *Mini-HiCon High Concentration Ozone Monitor Operating and Maintenance Instructions*, Needham, MA, IN USA Inc.

ISHIZAKI K, SHINRIKI N and MATSUYAMA H (1986), 'Inactivation of *Bacillus* spores by gaseous ozone', *J Appl Bacteriol*, 60, 67–72.

JEONG J, SEKIGUCHI K, LEE W and SAKAMOTO K (2005), 'Photodegradation of gaseous volatile organic compounds (VOCs) using $TiO_2$ photoirradiated by an ozone-producing UV lamp: decompostion characteristics identification of by-products and water-soluble organic intermediates', *J Photochem Photobiol: A Chem*, 169, 279–287.

© Woodhead Publishing Limited, 2012

KAMOTANI S, HOOKER N, SMITH S and LEE K (2010), 'Consumer acceptance of ozone-treated whole shell eggs', *J Food Sci*, 75, S103–S107.

KELLS S A, MASON L J, MAIER D E and WOLOSHUK C P (2001), 'Efficacy and fumigation characteristics of ozone in stored maize', *J Stored Prod Res*, 37, 371–382.

KETTERINGHAM L, GAUSSERES R, JAMES S J and JAMES C (2006), 'Application of aqueous ozone for treating pre-cut green peppers (*Capsicum annuum* L)', *J Food Eng*, 76, 104–111.

KHADRE M A and YOUSEF A E (2001a), 'Sporicidal action of ozone and hydrogen peroxide, a comparative study', *Int J Food Microbiol*, 71, 131–138.

KHADRE M A and YOUSEF A E (2001b), 'Decontamination of multilaminated aseptic food packaging material and stainless steel by ozone', *J Food Saf*, 21, 1–13.

KHADRE M A, YOUSEF A E and KIM J G (2001), 'Microbiological aspects of ozone application in food: a review', *J Food Sci*, 66, 1–11.

KHALIFA A M, EL TEMSAHY M M, ABOU EL and NAGA I F (2001), 'Effect of ozone on the viability of some protozoa in drinking water', *J Egypt Soc Parasitol*, 31, 603–616.

KIM C K, GENTILE D M and SPROUL O J (1980), 'Mechanism of ozone inactivation of bacteriophage f2', *Appl Environ Microbiol*, 39, 210–218.

KIM J G (1998), '*Ozone as an antimicrobial agent in minimally processed food*', PhD thesis, Columbus, OH, Ohio State University.

KIM J G, YOUSEF A E and CHISM G W (1999), 'Use of ozone to inactivate microorganisms on lettuce', *J Food Saf*, 19, 17–34.

KIM J G, YOUSEF A E and KHADRE M A (2003), 'Ozone and its current and future application in the food industry', *Adv Food Nutr Res*, 45, 167–218.

KINMAN R N (1975), 'Analysis of ozone fundamental principles', in *Proceedings of the first international symposium on ozone for water and wastewater treatment*, International Ozone Association, 56–68.

KLOCKOW P A and KEENER K M (2009), 'Safety and quality assessment of packaged spinach treated with a novel ozone generation system', *LWT Food Sci and Tech*, 42, 1047–1053.

KNAKE R and HAUSER P (2002), 'Sensitive electrochemical detection of ozone', *Analytica Chimica Acta*, 459, 199–207.

KOIDIS P, BORI M and VARELTZIS K (2000), 'Efficacy of ozone treatment to eliminate *Salmonella* Enteritidis from eggshell surface', *Archiv für Lebensmittelhygiene*, 51(1), 4–6.

LAGERWERFF J M (1963), 'Prolonged ozone inhalation and its effects on visual parameters', *Aerosp Med*, 34, 479–486.

LANGLAIS B, RECKHOW D A and BRINK D R (1991), *Ozone in water treatment, application and engineering*, Boca Raton, FL, Lewis Publishers.

LEE Y H and TSAO G T (1979), 'Dissolved oxygen electrodes', *Adv Biochem Eng*, 13, 35–86.

LEWIS L, ZHUANG H, PAYNE F A and BARTH M M (1996), 'Beta-carotene content and color assessment in ozone treated broccoli florets during modified atmosphere packaging', in *Book of Abstracts*, Institute of Food Technologists Annual Meeting, 99.

LIEW C L and PRANGE R K (1994), 'Effect of ozone and storage temperature on postharvest diseases and physiology of carrots (*Daucus carota* L,)', *J Amer Soc Hortic Sci*, 119, 563–567.

LUVISI D A, SHOREY H H, SMILANIC J L, THOMPSON J F, GUMP B H and KNUTSON J (1992), 'Sulfur dioxide fumigation of table grapes', *Univ Cal Div Agric Nat Res*, *Bull*, 1932.

LYNNTECH INC. (1998), *LT 1 Electrochemical ozone generator instruction's manual*. College Station, TX, Lynntech Inc.

MANCY K H, OKUN D A and REILLEY C N (1962), 'A galvanic cell oxygen analyzer', *J Electroanal Chem*, 4, 65–92.

MANOUSARIDIS G, NERANTZAKI A, PALEOLOGOS E K, TSIOTSIAS A, SAVVAIDIS I N and

© Woodhead Publishing Limited, 2012

KONTOMINAS M G (2005), 'Effect of ozone on microbial, chemical and sensory attributes of shucked mussels', *Food Microbiol*, 22, 1–9.

MARTINS R, FORTUNATO E, NUNES P, FERREIRA I, MARQUES A, BENDER A, KATSARAKIS N, CIMALLA V and KIRIAKIDIS G (2004), 'Zinc oxide as an ozone sensor', *J App Physics*, 96(3), 1398–1408.

NADAS A, OLMO M and GARCIA J M (2003), 'Growth of *Botrytis cinerea* and strawberry quality in ozone-enriched atmospheres', *J Food Sci*, 68, 1798–1802.

NAITOH S, OKADA Y and SAKAI T (1992), 'Studies on utilization of ozone in food preservation. V. Changes in microflora of ozone-treated cereals, grain, peas, beans and spices during storage', *J Jap Soc Food Sci Technol*, 35, 69–77.

OHMINE S (2005), '*Investigation of the mechanisms of ozone-mediated viral inactivation*', MS thesis, Utah State University.

OSHA (2004), 'Chemical sampling information: ozone'. Available from: http://www.osha.gov/dts/chemicalsampling/data/CH_259300.html (accessed 10 September 2011).

OSRAM (2011), 'XERADEX VUV Xenon Excimer lamp'. Available from: http://www.osram.com/osram_com/Professionals/DisplayOptic_Lighting/Products/XERADEX_VUV_Xenon_Excimer_Lamps/index.html (accessed 1 November 2011).

OZTEKIN S, ZORLUGENC B and ZORLUGENC F K (2006), 'Effects of ozone treatment on microflora of dried figs', *J Food Eng*, 75, 396–399.

PALOU L, SMILANICK J L, CRISOTO C H and MANSOUR M (2001), 'Effect of gaseous ozone exposure on development of green and blue molds on cold stored citrus fruit', *Plant Dis*, 85, 632–638.

PALOU L, CRISOSTO C H, SMILANICK J L, ADASKAVEG J E and ZOFFOLI J P (2002), 'Effects of continuous 0.3 ppm ozone exposure on decay development and physiological responses of peaches and table grapes in cold storage', *Postharvest Biol Technol*, 24, 39–48.

PASCUAL A, LLORCA I and CANUT A (2007), 'Use of ozone in food industries for reducing the environmental impact of cleaning and disinfection activities', *Trends Food Sci Technol*, 18, S29–S35.

PEREZ A G, SANZ C, RIOS J J, OLIAS R and OLIAS J M (1999), 'Effects of ozone treatment on postharvest strawberry quality', *J Agric Food Chem*, 47, 1652–1656.

PERRY J J (2010), '*Ozone based treatments for inactivation of* Salmonella enterica *serovar Enteritidis in shell eggs*', PhD thesis, Columbus, OH, Ohio State University.

PERRY J J and YOUSEF A E (2011), 'Decontamination of raw foods using ozone-based sanitization techniques', *Annu Rev Food Sci Technol*, 2, 281–298.

PERRY J J, RODRIGUEZ-ROMO L A and YOUSEF A E (2008), 'Inactivation of *Salmonella enterica* serovar Enteritidis in shell eggs by sequential application of heat and ozone', *Lett Appl Microbiol*, 46, 620–625.

PERRY J J, RODRIGUEZ-SAONA L E and YOUSEF A E (2011), 'Quality of shell eggs pasteurized with heat or heat-ozone combination during extended storage', *J Food Sci*, 76, S437–S444.

RAKNESS K, GORDON G, LANGLAIS B, MASSCHELEIN W, MATSUMOTO N, RICHARD Y, ROBSON C M and SOMIYA I (1996), 'Guildeline for measurement of ozone concentration in the process gas from an ozone generator', *Ozone Sci Eng*, 18, 209–229.

RAMASWAMY R, RODRIGUEZ-ROMO L, VURMA M, BALASUBRAMANIAM V M and YOUSEF A E (2007), 'Ozone Technology: Fact Sheet for Food Processors'. Available from: http://ohioline.osu.edu/fse-fact/0005.html (accessed 12 October 2011).

RICE R G (1986), 'Applications of ozone in water and wastewater treatment', in Rice R G, Bollyky L J and Lacy W J, *Analytical aspects of ozone treatment of water and wastewater*, Boca Raton, FL, Lewis Publisher, 7–26.

RICE R G (2011), 'Ozone: Fact Sheet for Agri-Food Processors' Available from: http://www.worldfoodscience.org/cms/?pid=1006025 (accessed 12 October 2011).

RICE R G, FARQUHAR J W and BOLLYKY L J (1982), 'Review of the applications of ozone for increasing storage times of perishable foods', *Ozone Sci Eng*, 4, 147–163.

© Woodhead Publishing Limited, 2012

RODRIGUEZ-ROMO L A and YOUSEF A E (2005), 'Inactivation of *Salmonella enterica* serovar Enteritidis on shell eggs by ozone and UV radiation', *J Food Prot*, 68, 711–717.

ROY D, WONG P K Y, ENGELBRECHT R S and CHIAN E S K (1981), 'Mechanism of enteroviral inactivation by ozone', *Appl Environ Microbiol*, 41, 718–723.

SALVERMOSER M, MURNICK D E and KOGELSCHATZ U (2008), 'Influence of water vapor on photochemical ozone generation with efficient 172 nm xenon excimer lamps', *Ozone Sci Eng*, 30, 228–237.

SCHEEL L D, DOBROGORSKI O J, MOUNTAIN J T, SVIRBELY J L and STOKINGER H E (1959), 'Physiologic, biochemical, immunologic and pathologic changes following ozone exposure', *J Appl Physiol*, 14, 67–80.

SCHOMER H A and MCCOLLOCH L P (1948), 'Ozone in relation to storage of apples', *USDA Circular*, 765, 24.

SCOTT D B M and LESHER E C (1963), 'Effect of ozone on survival and permeability of *Escherichia coli*', *J Bacteriol*, 5, 67–76.

SELMA M V, BELTRAN D, CHACON-VERA E and GIL M I (2006), 'Effect of ozone on the inactivation of *Yersinia enterocolitica* and the reduction of natural flora on potatoes', *J Food Prot*, 69, 2357–2363.

SELMA M V, BELTRAN D, ALLENDE A, CHACON-VERA E and GIL M I (2007), 'Elimination by ozone of *Shigella sonnei* in shredded lettuce and water', *Food Microbiol*, 24, 492–499.

SELMA M V, IBANEZ A M, ALLENDE A, CANTWELL M and SUSLOW T (2008a), 'Effect of gaseous ozone and hot water on microbial and sensory quality of cantaloupe and potential transference of *Escherichia coli* O157:H7 during cutting', *Food Microbiol*, 25, 162–168.

SELMA M V, IBANEZ A M, CANTWELL M and SUSLOW T (2008b), 'Reduction by gaseous ozone of *Salmonella* and microbial flora associated with fresh-cut cantaloupe', *Food Microbiol*, 25, 558–565.

SHARMA R R, DEMIRCI A, BEUHAT L R and FETT W F (2003), 'Application of ozone for inactivation of *Escherichia coli* O157:H7 on inoculated alfalfa sprouts', *J Food Proc Pres Res*, 27, 52–64.

SHIMIZU Y, MAKINOTT J, SATO J and IWAMOTO S (1982), 'Preventing rot of Kyoho grapes in cold storage with ozone', *Res Bull Aichi Agric Res Ctr*, 14, 225–238.

SINGH N, BHUNIA A K and STROSHINE R L (2002), 'Efficacy of chlorine dioxide, ozone, and thyme essential oil or a sequential washing in killing *Escherichia coli* O157:H7 on lettuce and baby carrots', *Lebensm-Wiss u -Technol*, 35, 720–729.

SKOG L J and CHU C L (2001), 'Effect of ozone on qualities of fruits and vegetables in cold storage', *Can J Plant Sci*, 81, 773–778.

SMART R B, DORMOND-HERRERA R and MANCY K H (1979), '*In situ* voltammetric membrane ozone electrode', *Anal Chem*, 51, 2315–2319.

SMILANICK J L, HARVEY J M, HARTSELL P L, HENSEN D J, HARRIS C M, FOUSE D C and ASSEMI M (1990), 'Factors influencing sulfite residues in table grapes after sulfur dioxide fumigation', *Am J Enol Vitic*, 41(2), 131–136.

SMILANICK J L, MARGOSAN D M and GABLER F M (2002), 'Impact of ozonated water on the quality and shelf-life of fresh citrus fruit, stone fruit, and table grapes', *Ozone Sci Eng*, 24, 343–356.

SONG J, FAN L, HILDEBRAND P D and FORNEY C F (2000), 'Biological effects of corona discharge on onions in a commercial storage facility', *Hort Tec*, 10(3), 608–612.

SOPHER C C, BATTLES G T and KNUEVE E A (2007), 'Ozone applications in catfish processing', *Ozone Sci Eng*, 29, 221–228.

SPOTTS R A and CERVANTES L A (1992), 'Effect of ozonated water on postharvest pathogens of pear in laboratory and packinghouse tests', *Plant Dis*, 76, 256–259.

SPOTTS R A and PETERS B B (1980), 'Chlorine and chlorine dioxide for control of d'Anjou pear decay', *Plant Dis*, 64, 1095–1097.

© Woodhead Publishing Limited, 2012

STANLEY J H and JOHNSON J D (1979), 'Amperometric membrane electrode for measurement of ozone in water', *Anal Chem*, 51, 2144–2147.

STANLEY J H and JOHNSON J D (1984), 'Analysis of ozone in aqueous solutions', in Rice R G and Netzer A, *Handbook of ozone technology and applications, Vol 2, Ozone for drinking water treatment*, Boca Raton, FL, Lewis Butterworth Publishers, 255–276.

STEENSTRUP L D and FLOROS J D (2004), 'Inactivation of *E. coli* O157:H7 in apple cider by ozone at various temperatures and concentrations', *J Food Proc Preserv*, 28, 103–116.

STUMM W (1958), 'Ozone as a disinfectant for water and sewage', *J Boston Soc Civil Eng*, 46, 68.

SUBRAHMANYAM C, BULUSHEV D A and KIWI-MINSKER L (2005), 'Dynamic behaviour of activated carbon catalysts during ozone decomposition at room temperature', *Applied Catalysis: B Env*, 61, 98–106.

THANOMSUB B, ANUPUNPISIT V, CHANPHETCH S, WATCHARACHAIPONG T, POONKHUM R and SRISUKONTH C (2002), 'Effects of ozone treatment on cell growth and ultrastructural changes in bacteria', *J Gen Appl Microbiol*, 48, 193–199.

TIWARI B K, MUTHUKUMARAPPAN K, O'DONNELL C P and CULLEN P J (2008), 'Kinetics of freshly squeezed orange juice quality changes during ozone processing', *J Agric Food Chem*, 56, 6416–6422.

TIWARI B K, O'DONNELL C P, MUTHUKUMARAPPAN K and CULLEN P J (2009), 'Anthocyanin and color degradation in ozone treated blackberry juice', *Innov Food Sci Emerg Technol*, 10, 70–75.

URANO H and FUKUZAKI S (2001), 'Facilitation of cleaning of alumina surfaces fouled with heat-treated bovine serum albumin by ozone treatment', *J Food Prot*, 61, 108–112.

UTEMBE S R, HANSFORD G M, SANDERSON M G, FRESHWATER R A, PRATT K F E, WILLIAMS D E, COX R A and JONES R L (2006), 'An ozone monitoring instrument based on the tungsten trioxide ($WO_3$) semiconductor', *Sens Actuators B: Chem*, 114, 507–512.

VELACO R M, URIBE F J and PEREZ-CHAVELA (2008), 'Stratospheric ozone dynamics according to the Chapman mechanism', *J Math Chem*, 44, 529–539.

VICTORIN K (1992), 'Review of the genotoxicity of ozone', *Mutat Res*, 277, 221–238.

VURMA M (2009), '*Development of ozone based processes for decontamination of fresh produce to enhance safety and enhance shelf life*', PhD thesis, Columbus, OH, Ohio State University.

VURMA M, PANDIT R B, SASTRY S K and YOUSEF A E (2009), 'Inactivation of *Escherichia coli* O157:H7 and natural microbiota on spinach leaves using gaseous ozone during vacuum cooling and simulated transportation', *J Food Prot*, 72, 1538–1546.

WADE W N, SCOUTEN A J, MCWATTERS K H, WICK R L and DEMIRCI A (2003), 'Efficacy of ozone in killing *Listeria monocytogenes* on alfalfa seeds and sprouts and effects on sensory quality of sprouts', *J Food Prot*, 66, 44–51.

WEAVERS L K and WICKRAMANAYAKE G B (2001), 'Disinfection and sterilization using ozone', in Block S S, *Disinfection, sterilization and preservation*, Baltimore, MD, Lippincott Williams and Wilkins, 205–214.

WILLIAMS R C, SUMMER S S and GOLDEN D A (2005), 'Inactivation of *Escherichia coli* O157:H7 and *Salmonella* in apple cider and orange juice treated with combinations of ozone, dimethyl dicarbonate and hydrogen peroxide', *J Food Sci*, 70(4), M197–M201.

WOJTOWICZ J B A (2004), 'Ozone', *Kirk-Othmer encyclopedia of chemical technology*, New York, John Wiley & Sons.

WU J, DOAN H and CUENCA M A (2006), 'Investigation of gaseous ozone as an antifungal fumigant for stored wheat', *J Chem Technol Biotechnol*, 81, 1288–1293.

YOUSEF A E, VURMA M and RODRIGUEZ-ROMO L A (2011), 'Basics of ozone sanitization and food applications', in Zhang H Q, Barbosa-Canovas G V, Balasubramaniam V M, Dunne C P, Farkas D F and Yuan J T C, *Nonthermal processing technologies for food*, Ames, IA, Blackwell Publishing.

© Woodhead Publishing Limited, 2012

ZAGON J, DEHNE L I, WIRZ J, LINKE B and BOEGL K W (1992), 'Ozone treatment for removal of microorganisms from spices as an alternative to ethylene oxide fumigation or irradiation: results of a practical study', *Bundesgesundheitsblatt*, 35, 20–33.

ZHANG L, LU Z, YU Z and GAO X (2005), 'Preservation of fresh-cut celery by treatment of ozonated water', *Food Cont*, 16, 279–283.

ZHAO C, GE B, DE VILLENA J, SUDLER R, YEH E, ZHAO S, WHITE D G, WAGNER D and MENG J (2001), 'Prevalence of *Campylobacter* spp., *Escherichia coli*, and *Salmonella* serovars in retail chicken, turkey, pork, and beef from the Greater Washington, D.C. area', *Appl Environ Microbiol*, 67, 5431–5436.

ZHAO J and CRANSTON P M (1995), 'Microbial decontamination of black pepper by ozone and the effect of the treatment on volatile oil constitutes of the spice', *J Sci Food Agric*, 68, 11–18.

ZOTTOLA E A and SASAHARA K C (1994), 'Microbial biofilms in the food processing industry – should they be a concern?', *Intern J Food Microbiol*, 23, 125–148.

© Woodhead Publishing Limited, 2012

# 18

# Chlorine dioxide for microbial decontamination of food

**V. Trinetta and M. Morgan, Purdue University, USA and R. Linton, The Ohio State University, USA**

**Abstract**: The food industry is continually striving to develop new preservation and sanitation techniques. Chlorine dioxide ($ClO_2$) can be used to control pathogenic and spoilage microorganisms and increase product shelf life. This chapter offers an overview of the use of $ClO_2$ in food decontamination. A novel explanation of $ClO_2$ oxidation action is given, the regulatory status of $ClO_2$ is discussed, and current uses of $ClO_2$ in the food industry are reported. Furthermore, relevant research results are presented including investigations concerning $ClO_2$ applications for different environmental and food-contact surfaces and food matrices.

**Key words**: chlorine dioxide, microbial inactivation, food safety.

## 18.1 Introduction

Since Sir Humphrey Davy discovered chlorine dioxide gas in 1815, there has been recognition of its effective use as a bleaching agent and disinfectant (Simpson, 2005). The textile and paper industries have broadly applied $ClO_2$ as a bleaching agent, while the food industry has used this agent as an extraction solvent and powerful antimicrobial. This chapter offers an overview of the use of $ClO_2$ as an antimicrobial agent for food decontamination. A novel explanation of $ClO_2$ oxidation action is given, the regulatory status of $ClO_2$ has been updated, and current uses of $ClO_2$ in the food industry are reported. Furthermore, relevant research results, about $ClO_2$ applications on different environmental and food-contact surfaces and food matrices, are presented.

© Woodhead Publishing Limited, 2012

### 18.1.1 $ClO_2$: Chemical-physical properties and generation systems

$ClO_2$ is a yellow-green gas at ambient temperature; it smells similar to chlorine. The odor threshold is about 0.1–0.3 mg/l and it becomes irritating for eyes at concentrations above 0.5 mg/l. $ClO_2$ has a molecular weight of 67.45 (Knapp and Battisti, 2001). Below –59°C, $ClO_2$ exists as a solid, while the boiling point is 11°C. The gas is rapidly solubilized in water: the partition coefficient ($C_{ClO_2(H_2O)}/C_{ClO_2(air)}$) of $ClO_2$ is ~38 at 22°C and 101 KPa (Masschelein and Rice, 1979). $ClO_2$ solutions are stable for months and even for years if properly stored at refrigeration temperatures and away from light. At concentrations above 10% by volume in air, the gas is potentially explosive and unstable, making delivery at these concentrations impossible and dangerous.

Since $ClO_2$ gas must be produced at the site of use, several methods exist for generation. The choice of system depends on the form and the amount of $ClO_2$ required (Knapp and Battisti, 2001). In the oil, textile and paper industries, $ClO_2$ is generated in large quantities (> 1 ton per day); these large-scale generation systems are usually based on chlorate ion reduction to be most economical. For smaller-scale generators, sodium chlorite is the common solution used to produce chlorine dioxide (Simpson, 2005).

In general, aqueous $ClO_2$ solutions are generated by acidification of sodium chlorite solution (acid-chlorite solution), and reducing agents, such as sulfuric acid or hydrochloric acid, are used to activate sodium chlorite. Higher $ClO_2$ yields are obtained by the chlorine-chlorite solution method: chlorite ion reacts in aqueous solution with chlorine or HCl to form chlorine dioxide, but longer reaction times are required. Sodium chlorite solution can also be vaporized and react under vacuum with chlorine gas (gaseous chlorine-chlorite solution). This is a faster process with high chlorine dioxide production rates.

For practical reasons, generators for industrial applications principally use three chemical feedstocks, hydrochloric acid, sodium hypochlorite, and sodium chlorite, in order to avoid the restriction of storing liquid chlorine on site. $ClO_2$ gas is commonly generated by bubbling 4% $Cl_2$ in a nitrogen gas carrier, subsequently passed through solid sodium chlorite cartridges to produce high-purity $ClO_2$ gas. The activation of solid sodium chlorite by water or humid environment is also used to deliver controlled amounts of $ClO_2$ gas through sachets, tablets, or pouches over time. A more detailed explanation of $ClO_2$ generation methods can be found in *White's Handbook of Chlorination and Alternative Disinfectants* (White, 2010).

### 18.1.2 Mechanisms of microbial inactivation by $ClO_2$

$ClO_2$ is an oxidizing agent that reacts taking electrons from several cellular constituents, breaking molecular bonds and consequently causing the death of microorganisms. Different detailed explanations of $ClO_2$ action within a biological environment have been offered. Benarde *et al.* (1967) and Olivieri

© Woodhead Publishing Limited, 2012

(1968) stated that protein synthesis was inhibited by the disinfectant action. Roller *et al.* (1980) also demonstrated that compromised protein synthesis was one of the causes of cell death. Several other mechanisms of action have been proposed, such as the inhibition of amino acid synthesis, the inactivation of messenger RNA (and consequently the lack in coded information for protein translation), and the destruction of ribosome. Up to now, the damage of protein synthesis is the most common accepted theory of microbial inactivation by $ClO_2$, even if the specific mechanisms are not known.

Bakhmutova-Albert *et al.* (2008) investigated the oxidation of dihydronicotinamide adenine dinucleotide (NADH) by $ClO_2$. NADH is a key co-enzyme in many biological redox reactions, such as mitochondrial electron-transport chain and ATP synthesis. The study reports the rapid oxidation of NADH by $ClO_2$ and the evaluation of the stoichiometry, kinetics, and mechanism of the reaction. Through UV spectra, molar absorbance, and kinetic measurements, the authors were able to demonstrate that 2 mol of $ClO_2$ are needed to convert 1 mol of NADH and that $NAD^+$ is the only product of the oxidation (Bakhmutova-Albert *et al.*, 2008). The reaction is completed in 30 ms at room temperature and even when the temperature is decreased (~ 4°C) the reaction is still very rapid. An electron-transfer mechanism is involved in the oxidation of NADH, based on the results obtained by ion chromatography. The authors proposed a multiple step reaction, where $ClO_2$ accepts 1 $e^-$ from NADH to form $ClO_2^-$ and $NADH^{\bullet +}$, a series of very fast deprotonation occurs and a second electron transfer proceeds with the formation of another equivalent of $ClO_2^-$ and the final product $NAD^+$ (Fig. 18.1). Although this sequence can be very rapid and consequently very difficult to measure, the data obtained by Bakhmutova-Albert *et al.* (2008) strongly suggest that NADH is the primary oxidation target of $ClO_2$ in a biological environment (e.g. microbial cells).

### 18.1.3 Handling and safe use

The main safety concern for $ClO_2$ is inhalation by people when exposed to the gas. Inhalation and excessive exposure to $ClO_2$ can cause dangerous health effects, such as irritation of the eyes, nose, throat and consequently severe respiratory damage. Chemical exposure limits established by the American Conference of Governmental Industrial Hygienist (ACGIH), the Occupational Safety and Health Administration (OSHA) and the National Institute for

$$NADH + ClO_2^{\bullet} \rightleftharpoons NADH^{\bullet +} + ClO_2^-$$

$$NADH^{\bullet +} + H_2O \xrightarrow{\text{fast deprotonation}} NADH^{\bullet} + H_3O^+$$

$$NADH^{\bullet} + ClO_2^{\bullet} \longrightarrow NAD^+ + ClO_2^-$$

**Fig. 18.1** Proposed mechanism of NADH oxidation by $ClO_2$ (Bakhmutova-Albert *et al.*, 2008).

© Woodhead Publishing Limited, 2012

Occupational Safety and Health (NIOSH) have proposed exposure limits and several precautions to be considered when handling the chemical (Table 18.1). Off-gassing of $ClO_2$ is a significant exposure risk when these products are handled. Therefore, most manufacturers require the use of a chemical cartridge respirator when handling chlorine dioxide in confined spaces.

### 18.1.4 $ClO_2$ regulatory status

The use of $ClO_2$ in the food industry is currently regulated by both the US Food and Drug Administration (FDA) and the Environmental Protection Agency (EPA). The EPA regulates the use of pesticides and antimicrobial pesticides, under the authority of two federal statutes in the US: the Federal Insecticide Fungicide and Rodenticide Act (FIFRA) and the Federal Food, Drug, and Cosmetic Act (FFDCA). More detailed delineation of the regulatory authority of the FDA and EPA can be found in 'Guidance for Industry: Antimicrobial Food Additives' (FDA, 1999).

Several sections of the Code of Federal Regulations (CFR) include applications of $ClO_2$ in foods or on food-contact surfaces. Under FDA, Title 21, some of the applications include bleaching of flour (21 CFR 137.105), sanitizing food contact surfaces (21 CFR 178.1010), and as a secondary

**Table 18.1** Chlorine dioxide chemical exposure limits and precautions for handling and storage

| | Agency[a] | Value |
|---|---|---|
| **Exposure limits** | | |
| Time-weighted average threshold limit value for an 8 h workday (TLV-TWA) | ACGIH | 0.1 ppm |
| Time-weighted average permissible exposure limit for an 8 h workday (PEL-TWA) | OSHA | 0.1 ppm |
| Short term exposure limit, 15 min (STEL) | ACGIH | 0.3 ppm |
| Concentrations Immediately Dangerous to Life or Health (IDLH) | NIOSH | 5 ppm |
| **Precautions** | | |
| • Do not store the $ClO_2$ at temperature > 38°C; | | |
| • Do not expose $ClO_2$ to UV lights; | | |
| • Plastics, rubber and coatings will be corroded over time; | | |
| • Store the chemical in containers approved for its use; | | |
| • Continuously ventilate the spaces where $ClO_2$ is handled; | | |
| • Do not breathe chlorine dioxide vapors, always use respirator mask, protective glasses and gloves; | | |
| • After handling $ClO_2$ always wash hands; | | |
| • Never flush $ClO_2$ solutions to a sanitary sewer or outlet; | | |
| • All personnel handling $ClO_2$ have to be trained; | | |
| • In case of danger call 911. | | |

[a]ACGIH = American Conference of Governmental Industrial Hygienist.
OSHA = Occupational Safety and Health Administration.
NIOSH = National Institute for Occupational Safety and Health.

© Woodhead Publishing Limited, 2012

direct additive in food (21 CFR 173.300). Under EPA, Title 40, applications include the use of $ClO_2$ in primary drinking water (40 CFR 141), antimicrobial formulations for food-contact surface sanitizing solutions (40 CFR 180), and pulp and paper bleaching (40 CFR 430). Moreover, the final Disinfectants and Disinfection Byproducts Rule (DBPR) and National Primary Drinking Water Regulation (NPDWR) (Federal Register, 1998) established the maximum residual disinfectant level (MRDL) for $ClO_2$ at 0.8 mg/l and a maximum contaminant level (MCL) for chlorite ions at 1.0 mg/l in public drinking water. These regulatory limits were derived from the oral reference dose (RfD) for chlorite, but no limits were established for chlorate or chloride residuals (EPA, 2003).

In 1998, the FDA (through Regulation 21 CFR 173.300) approved the use of $ClO_2$ as an antimicrobial agent in water used to wash fruit and vegetables (not to exceed 3 mg/l residual), with the requirement of a potable water rinse, blanching, cooking, or canning step in order to assure that there are no residues of concern in or on the surfaces (FDA, 1998).

Currently, no direct applications of chlorine dioxide gas on fresh fruits and vegetables, either designated as raw agricultural commodities, or processed foods, can be found in the regulations, although some Food Contact Notifications recognized by the FDA may include such gaseous applications. One recently approved application of chlorine dioxide by the EPA includes the use of $ClO_2$ as a spray or fogging agent to control plant pathogens and various forms of rot or other tuber disease-causing organisms in potato storage areas and on potatoes in storage (NPIRS, 2011). It is assumed that a significant amount of $ClO_2$ gas will be generated by spraying or fogging an aqueous solution. The main limitation for more approvals of the direct application of $ClO_2$ gas to foods is the quantification of residues. More data are required by the EPA to demonstrate that residues of chlorite or chlorine dioxide are below safe levels which would require the establishment of a tolerance, or that levels are undetectable and could be approved with an exemption from tolerance.

## 18.2 Chlorine dioxide ($ClO_2$) as a food decontamination technology: research updates

### 18.2.1 Potable drinking water

The first use of $ClO_2$ in municipal drinking water dates back to 1944 when it was used to control taste and odor at a Niagara Falls water plant (Synan *et al.*, 1944). Several applications of $ClO_2$ to protect drinking water from disease-causing organisms and pathogens are listed in Table 18.2. Huang *et al.* (1997) investigated the disinfection rates of $ClO_2$ on *Bacillus subtilis*, *Escherichia coli* and *Staphylococcus aureus* in water. Each bacterium showed different resistance responses to the disinfectant, but after 20 min at 2, 3 or

© Woodhead Publishing Limited, 2012

**Table 18.2** Examples of studies on the effect of aqueous $ClO_2$ for drinking water treatments

| Microbial target | Generation method | Treatment conditions | Reduction observed | References |
|---|---|---|---|---|
| *B. subtilis* | $ClO_2$ gas + water | 2 mg/l, 20 min | 3 log | Huang *et al.*, 1997 |
| *E. coli* | $ClO_2$ gas + water | 3 mg/l, 20 min | 3 log | Huang *et al.*, 1997 |
| *E. coli* ATCC 11229 | $NaClO_2 + (CH_3CO)_2O$ | 1.4 mg/l, 30 sec | 5 log | Foschino *et al.*, 1998 |
| *Legionella* | 5% stabilized $ClO_2$ | 50–80 mg/l, 60 min | 3 log | Walker *et al.*, 1995 |
| *S. aureus* | $ClO_2$ gas + water | 2.5 mg/l, 20 min | 3 log | Huang *et al.*, 1997 |

© Woodhead Publishing Limited, 2012

2.5 mg/l $ClO_2$, a 3-log reduction or higher was observed. The disinfectant had a broad bactericidal spectrum, within a wide range of pH values. In another study, the bactericidal activity of $ClO_2$ was investigated against *E. coli* ATCC 11229 (Foschino *et al.*, 1998). The efficacy of the treatment was strongly influenced by the physiological state of the cells. A 5-log reduction was achieved after only 30 s at 1.4 mg/l $ClO_2$ in suspension test and a linear relationship between pathogen inactivation rate and disinfectant concentration was observed. This notable difference in exposure time and treatment efficacy is probably due to the generation methods used. Foschino *et al.* (1998) applied acidified sodium chlorite solutions to generate $ClO_2$, consequently the disinfectant solution will have a much lower pH compared to the aqueous solutions used by Huang *et al.* (1997).

Great bacterial inactivation by $ClO_2$ was also reported by Walker *et al.* (1995) and, again, the efficacy of the disinfectant depended on the culture conditions: *Legionella* biofilms were more resistant to the treatments than planktonic cells. A treatment at 50–80 mg/l $ClO_2$ was performed at all outlets of a hospital hot-water system for 1 h, followed by a constant concentration of 3–5 mg/l over an 8-hour period. After disinfection, *Legionella* population was significantly reduced, and less bacterial biofilm was detected.

The promising results obtained from the use of $ClO_2$ to control pathogens and spoilage microorganisms in water, together with some advantages, such as the formation of less by-product, the easy generation systems developed, and the effectiveness over a broad range of pH, have fostered the potential use of $ClO_2$ for food decontamination applications.

### 18.2.2 Decontamination of solid surfaces

$ClO_2$ can be used to decontaminate food contact surfaces in aqueous form or as a gas. Solutions have been applied in several food industry processing facilities, where the sanitizer showed effectiveness in controlling microbial populations. Also the use of gaseous $ClO_2$ has been reported for paper, metal, plastic, and glass disinfection (Table 18.3).

With increasing concerns about bioterrorism, food, containers and production, facilities are considered potential vehicles for intentional microbial contamination. Several researchers have investigated $ClO_2$ effectiveness against *Bacillus* strains and spores. Kreske *et al.* (2006) evaluated the efficacy of aqueous $ClO_2$ in killing spores of *B. cereus* and *B. thuringiensis* on stainless steel surfaces. During experiments, it was observed that the composition of spore suspensions affected the lethality of the sanitizer, in particular the presence of bovine serum decreased $ClO_2$ efficacy. Nevertheless, a greater than 5-log CFU/cm$^2$ reduction was observed in *Bacillus* spore suspensions after a treatment with 75 mg/l $ClO_2$ for 5 min.

Aqueous $ClO_2$ was also investigated to inactivate biofilms on stainless steel surfaces. Microbial biofilm formation represents an important concern in food processing facilities, as cells in biofilms often develop resistance

© Woodhead Publishing Limited, 2012

**Table 18.3** Examples of studies on the effect of $ClO_2$ (aqueous and gas form) to decontaminate food contact surfaces

| Surfaces | Microbial target | Generation methods[a] | Treatment conditions | Reduction observed[b] | References |
|---|---|---|---|---|---|
| $ClO_2$ aqueous solution applications | | | | | |
| Stainless steel | *B. cereus* biofilms | 80% $NaClO_2$ | 200 mg/l, 5 min | 2.8 log CFU/$cm^2$ | Ryu and Beuchat, 2005 |
| Stainless steel | *B. cereus* C1 spores | Electro generator | 75 mg/l, 5 min | >5.7 log CFU/$cm^2$ | Kreske *et al.*, 2006 |
| Stainless steel | *B. cereus* F3812/84 spores | Electro generator | 75 mg/l, 5 min | >6 log CFU/$cm^2$ | Kreske *et al.*, 2006 |
| Stainless steel | *B. cereus* F4616A/90 spores | Electro generator | 75 mg/l, 5 min | >5.6 log CFU/$cm^2$ | Kreske *et al.*, 2006 |
| Stainless steel | *B. cereus* F4810/72 spores | Electro generator | 75 mg/l, 5 min | 5 log CFU/$cm^2$ | Kreske *et al.*, 2006 |
| Stainless steel | *B. cereus* O38-2 spores | Electro generator | 75 mg/l, 5 min | >5.5 log CFU/$cm^2$ | Kreske *et al.*, 2006 |
| Stainless steel | *B. thuringiensis* spores | Electro generator | 75 mg/l, 5 min | >6.4 log CFU/$cm^2$ | Kreske *et al.*, 2006 |
| Stainless steel | *Listeria biofilms* | $ClO_2$ gas + $H_2O$ | 7 mg/l, 10 min | 3.7 log CFU/$cm^2$ | Vaid *et al.*, 2010 |
| $ClO_2$ gas applications | | | | | |
| Epoxy | *B. thuringensis* spores | 4% $Cl_2$ + 80% $NaClO_2$ | 15 mg/l, 12 h, 85–92% RH | >3.7 log CFU/$cm^2$ | Han *et al.*, 2003 |
| Paper | *B. thuringensis* spores | 4% $Cl_2$ + 80% $NaClO_2$ | 15 mg/l, 12 h, 85–92% RH | >4.8 log CFU/$cm^2$ | Han *et al.*, 2003 |
| Plastic | *B. thuringensis* spores | 4% $Cl_2$ + 80% $NaClO_2$ | 15 mg/l, 12 h, 85–92% RH | >3.5 log CFU/$cm^2$ | Han *et al.*, 2003 |
| Stainless steel | *L. buchneri* | 4% $Cl_2$ + 80% $NaClO_2$ | 10 mg/l, 30 min, 85–95% RH | >4.7 log CFU/$cm^2$ | Han *et al.*, 1999 |
| Stainless steel | *L. mesenteroides* | 4% $Cl_2$ + 80% $NaClO_2$ | 10 mg/l, 30 min, 85–95% RH | >4.7 log CFU/$cm^2$ | Han *et al.*, 1999 |
| Stainless steel | *Listeria* biofilms | 2% $Cl_2$ + 80% $NaClO_2$ | 0.3 mg/l,10 min, 75% RH | 3.2 log CFU/$cm^2$ | Vaid *et al.*, 2010 |
| Stainless steel | Yeast and molds | 4% $Cl_2$ + 80% $NaClO_2$ | 10 mg/l, 30 min, 85–95% RH | >3.9 log CFU/$cm^2$ | Han *et al.*, 1999 |
| Wood | *B. thuringensis* spores | 4% $Cl_2$ + 80% $NaClO_2$ | 15 mg/l, 12 h, 85–92% RH | >3.5 log CFU/$cm^2$ | Han *et al.*, 2003 |

[a]80% $NaClO_2$: acidification of sodium chlorite solution; Electro generator: electrolysis of a chlorite solution; 4% $Cl_2$ + 80% $NaClO_2$: solid sodium chlorite reaction with chlorine gas; 2% $Cl_2$ + 80% $NaClO_2$; solid sodium chlorite reaction with chlorine gas.
[b]Log reduction was reported as log CFU/$cm^2$, taking in account coupon dimension and/or dilution factors, as reported in the original papers.

© Woodhead Publishing Limited, 2012

to sanitizers routinely used in food industry. Biofilm cells may survive the treatment, detach, and contaminate processed food products. A 2.8-log CFU/cm$^2$ reduction was observed after a treatment with 200 mg/l $ClO_2$ for 5 min in *Bacillus cereus* biofilms (Ryu and Beuchat, 2005). Biofilms exposed to air were more resistant to $ClO_2$ compared to those immersed in broth, suggesting that the characterization of biofilm development and maturation is essential to identify favorable treatment conditions. Moreover, the authors suggested that a synergistic treatment of oxidizing disinfectants and wet heat may enhance the lethality rate in *B. cereus*, since cells treated with $ClO_2$ showed a decrease in heat resistance.

Vaid *et al.* (2010) compared microbial reduction obtained in *Listeria* biofilms after treatment with aqueous $ClO_2$ solutions and gas. A mixture of five *L. monocytogenes* strains was used and biofilms were developed at room temperature with 100% relative humidity (RH) in 4 days. Based on the assumption that $ClO_2$ is 23 times more concentrated in aqueous phase at 22°C compared to the gas phase (Taube and Dodgen, 1949), a solution of 7 mg/l and a gas concentration of 0.3 mg/l were used. A 3.7-log CFU/cm$^2$ reduction was observed after 10 min treatment with $ClO_2$ aqueous solutions, while a 3.2-log CFU/cm$^2$ reduction was reported following 10 min gas injection. At their equilibrium concentrations, aqueous solution and gas had statistically similar effectiveness in *Listeria* inactivation within biofilms.

Different $ClO_2$ gas concentrations (2–14 mg/l), relative humidity (70–93%), temperatures (9–29°C) and exposure times (10–30 min) were selected to sanitize tanks used for aseptic juice storage (Han *et al.*, 1999). Stainless steel surfaces were inoculated with *Lactobacilllus buchneri*, *Leuconostoc mesenteroides*, yeast and molds. Data suggested that temperature and relative humidity were both significant factors and microbial inactivation enhanced with higher values. A greater than 3.9-log CFU/cm$^2$ reduction for all the spoilage microorganisms was observed after a treatment at 10 mg/l $ClO_2$ for 30 min. Several other surfaces, i.e. paper, wood, and plastic materials were experimentally contaminated with *Bacillus* spores and treated with gaseous $ClO_2$ (Han *et al.*, 2003). A greater than 3.5-log CFU/cm$^2$ reduction was observed on each surface tested following a treatment with 15 mg/l $ClO_2$ for 12 h. Interestingly, spores on paper and wood seemed to be more resistant, as a 30 mg/l $ClO_2$ gas concentration was necessary to achieve complete inactivation, compared to 25 mg/l for epoxy surfaces and 20 mg/l for plastic. *Bacillus* spores survived better on paper than on plastic, and wood surfaces were not discolored or damaged by $ClO_2$ gas treatments.

Both $ClO_2$ aqueous and gas form demonstrated effectiveness as antimicrobial agents on environmental and food contact surfaces. The contact time required to achieve similar microbial reduction was in general lower for aqueous treatments than gaseous injections, and the concentration needed to inactivate spores was higher than the concentration necessary to kill bacteria.

© Woodhead Publishing Limited, 2012

## 18.3 Decontamination of fruits and vegetables

Numerous studies have reported successful applications of $ClO_2$ for fruit and vegetable decontamination with aqueous and gaseous forms (Tables 18.4–18.6). The main advantage of gas over aqueous solutions is that gaseous $ClO_2$ has more penetration ability and can reach microorganisms protected by surface irregularities (especially those that may be hydrophobic) (Han *et al.*, 2001), although handling of solutions is easier and existing processes do not need to be modified in order to adopt aqueous technology.

### 18.3.1 Apple, cantaloupes, and berries

$ClO_2$ solutions at 5 mg/l for 5 min were used to decontaminate apples, inoculated with *E. coli* O157:H7 and *L. monocytogenes* (Rodgers *et al.*, 2004). After treatment, the pathogen populations were under detectable levels and remained unchanged for 9 days at 4°C. Inherent mesophilic bacteria increased constantly during storage time. A 1.5-log CFU/g reduction was observed in the population of yeast and molds following $ClO_2$ treatment, but after 9 days the count increased by 2 log. No color changes were observed and treatment was judged acceptable by the consumers (untrained panel). In the same study, aqueous solutions were also evaluated for microbial inactivation on cantaloupe surfaces (Rodgers *et al.*, 2004). A reduction greater than 5-log CFU/g in *E. coli* O157:H7 and *L. monocytogenes* was reported after a treatment with 5 mg/l aqueous solution for 5 min, and a significant increase in produce shelf life compared to the controls was observed.

Effectiveness of gaseous $ClO_2$ to decontaminate fruits has also been studied. Treatment of apples inoculated with *L. monocytogenes* was evaluated (Du *et al.*, 2002). Bacteria attached to the apple skin surfaces were easier to inactivate: a 4.8-log CFU/g reduction was observed after a treatment with 4 mg/l $ClO_2$ gas for 10 min, while a lower reduction was observed on the calyx and stem cavities. Bacteria might be protected by the cavities, or the exposure time was not sufficient to reach and detach the bacteria in those sites.

Gaseous $ClO_2$ was also effectively used to reduce *E. coli* O157:H7 on apple surfaces (Du *et al.*, 2003). The authors reported that *E. coli* was more resistant than *L. monocytogenes*, and again, the bacteria spot-inoculated on calyx and stem cavity were not significantly inactivated. Several exposure times and concentrations were evaluated, and a treatment with 4.8 mg/l $ClO_2$ gas for 20 min was considered the ideal processing condition for *E. coli* reduction on apples.

$ClO_2$ gas efficacy was investigated against *Alicyclobacillus acidoterrestris* spores on apples (Lee *et al.*, 2006). Using sachets releasing 4.32 or 1.78 mg/l $ClO_2$ over 1 h, spores were reduced to undetectable levels, even if the overall quality of treated samples was compromised (e.g. small black spots developed). Conversely, a 4-log CFU/g reduction was observed with a lower

© Woodhead Publishing Limited, 2012

**Table 18.4** Examples of studies on the effect of $ClO_2$ (aqueous and gas form) to reduce microbial load on fruit

| | Microbial target | Generation method[a] | Treatment conditions | Reduction observed[b] | Shelf life[c], quality | References |
|---|---|---|---|---|---|---|
| $ClO_2$ aqueous solution applications | | | | | | |
| Apple | *E. coli* O157:H7 | 2% $NaClO_2 + H_3PO_4$ | 5 mg/l, 5 min | > 5 log CFU/g | Yes, no defects | Rodgers *et al.*, 2004 |
| Apple | *L. monocytogenes* | 2% $NaClO_2 + H_3PO_4$ | 5 mg/l, 5 min | > 5 log CFU/g | Yes, no defects | Rodgers *et al.*, 2004 |
| Cantaloupe | *E. coli* O157:H7 | 2% $NaClO_2 + H_3PO_4$ | 5 mg/l, 5 min | > 5 log CFU/g | Yes, no defects | Rodgers *et al.*, 2004 |
| Cantaloupe | *L. monocytogenes* | 2% $NaClO_2 + H_3PO_4$ | 5 mg/l, 5 min | > 5 log CFU/g | Yes, no defects | Rodgers *et al.*, 2004 |
| Blueberries | *L. monocytogenes* | Sachet | 15 mg/l, 30 min | 3.9 log CFU/g | Yes, no defects | Wu and Kim, 2007 |
| Blueberries | *P. aeruginosa* | Sachet | 15 mg/l, 30 min | 3.8 log CFU/g | Yes, no defects | Wu and Kim, 2007 |
| Blueberries | *S. aureus* | Sachet | 15 mg/l, 30 min | 4.6 log CFU/g | Yes, no defects | Wu and Kim, 2007 |
| Blueberries | *S. Typhimurium* | Sachet | 15 mg/l, 30 min | 3.2 log CFU/g | Yes, no defects | Wu and Kim, 2007 |
| Blueberries | Yeast and molds | Sachet | 15 mg/l, 30 min | 2.2 log CFU/g | Yes, no defects | Wu and Kim, 2007 |
| Blueberries | *Y. enterocolitica* | Sachet | 15 mg/l, 30 min | 3.3 log CFU/g | Yes, no defects | Wu and Kim, 2007 |
| Strawberries | *E. coli* O157:H7 | 2% $NaClO_2 + H_3PO_4$ | 5 mg/l, 5 min | 5.6 log CFU/g | Yes, no defects | Rodgers *et al.*, 2004 |
| Strawberries | *L. monocytogenes* | 2% $NaClO_2 + H_3PO_4$ | 5 mg/l, 5 min | 5.6 log CFU/g | Yes, no defects | Rodgers *et al.*, 2004 |
| $ClO_2$ gas applications | | | | | | |
| Apples | *A. acidoterrestris* | Sachet | 0.60 mg/l, 3 h | 4 log CFU/g | Yes, black spots | Lee *et al.*, 2006 |
| Apples[d] | *E. coli* O157:H7 | 4% $Cl_2$ + 80% $NaClO_2$ | 4.8 mg/l, 20 min, 90% RH | >4.5 log CFU/$cm^2$ | NR | Du *et al.*, 2003 |
| Apples[d] | *L. monocytogenes* | 4% $Cl_2$ + 80% $NaClO_2$ | 4 mg/l, 10 min, 90% RH | 4.8 log CFU/g | NR | Du *et al.*, 2002 |
| Cantaloupe | *L. monocytogenes* | 2% $Cl_2$ + 80% $NaClO_2$ | 5 mg/l, 10 min, 95% RH | 4.3 log CFU/$cm^2$ | Yes, no defects | Mahmoud *et al.*, 2008 |

*(Continued)*

© Woodhead Publishing Limited, 2012

**Table 18.4** Continued

| | Microbial target | Generation method[a] | Treatment conditions | Reduction | Shelf lifec, | References |
|---|---|---|---|---|---|---|
| Cantaloupe | *S.* Poona | 2% $Cl_2$ + 80% $NaClO_2$ | 5 mg/l, 10 min, 95% RH | 5 log $CFU/cm^2$ | Yes, no defects | Mahmoud *et al.*, 2008 |
| Blueberries | *E. coli* O157:H7 | Sachet | 4 mg/l, 12 h | 3.6 log CFU/g | Yes, no defects | Popa *et al.*, 2007 |
| Blueberries | *L. monocytogenes* | Sachet | 4 mg/l, 12 h | 3.9 log CFU/g | Yes, no defects | Popa *et al.*, 2007 |
| Blueberries | *Salmonella* spp. | Sachet | 4 mg/l, 12 h | 4.2 log CFU/g | Yes, no defects | Popa *et al.*, 2007 |
| Blueberries[d] | *S. enterica* | Sachet | 8 mg/l, 120 min | >2.4 log CFU/g | Yes, no defects | Sy *et al.*, 2005a |
| Blueberries[d] | Yeast and molds | Sachet | 8 mg/l, 120 min | 2.1 log CFU/g | Yes, no defects | Sy *et al.*, 2005a |
| Blueberries | Yeast and molds | Sachet | 4 mg/l, 12 h | 3 log CFU/g | Yes, no defects | Popa *et al.*, 2007 |
| Strawberries | *E. coli* O157:H7 | 2% $Cl_2$ + 80% $NaClO_2$ | 5 mg/l, 10 min | 3.3 log CFU/g | Yes, no defects | Mahmoud *et al.*, 2007 |
| Strawberries | *L. monocytogenes* | 2% $Cl_2$ + 80% $NaClO_2$ | 5 mg/l, 10 min | 3.2 log CFU/g | Yes, no defects | Mahmoud *et al.*, 2007 |
| Strawberries | *Salmonella* spp. | 2% $Cl_2$ + 80% $NaClO_2$ | 5 mg/l, 10 min | 2.9 log CFU/g | Yes, no defects | Mahmoud *et al.*, 2007 |
| Strawberries[d] | *Salmonella* | Sachet | 8 mg/l, 120 min | >3.8 log CFU/g | Yes, no defects | Sy *et al.*, 2005a |
| Strawberries[d] | Yeast and molds | Sachet | 8 mg/l, 120 min | >4 log CFU/g | Yes, no defects | Sy *et al.*, 2005a |

[a]2% $NaClO_2$ + $H_3PO_4$: acidification of sodium chlorite solution; Sachet: acid and solid sodium chlorite in a sachet with humidity; 4% $Cl_2$ + 80% $NaClO_2$: solid sodium chlorite reaction with chlorine gas; 2% $Cl_2$ + 80% $NaClO_2$: solid sodium chlorite reaction with chlorine gas.
[b]When possible, log reduction was reported as log $CFU/cm^2$ or log CFU/g, taking in account sample dimension, weight and/or dilution factors, as reported in the original papers.
[c]Significant improvement in microbial shelf life. NR: not reported.
[d]Inoculation sites were calyx, stem, and skin.

© Woodhead Publishing Limited, 2012

$ClO_2$ concentration and a longer exposure time, 0.60 mg/l for 3 h, without the development of cosmetic damage. Furthermore, Mahmoud *et al.* (2008) treated cantaloupe inoculated with *E. coli* O157:H7, *L. monocytogenes* and *S. Poona* with gaseous $ClO_2$. A 5-log CFU/cm$^2$ reduction was achieved in *Salmonella* population with a treatment at 5 mg/l for 10 min, while 4.6- and 4.3-log CFU/cm$^2$ reductions were observed for *E. coli* and *Listeria*, respectively. The effect of $ClO_2$ gas increased with increasing time and concentration, and inactivation kinetics were best determined using the Weibull model. Treatments did not affect sample color, and shelf life of treated samples was extended by up to 9 days, when stored at 22°C.

Other commodities of particular interest for $ClO_2$ applications are berries. These small fruits may be exposed to potential microbial contamination during harvesting and packing. Several studies have been conducted to evaluate the bactericidal effectiveness of the disinfectant on berries. Wu and Kim (2007) generated aqueous $ClO_2$ using a chemical pouch and evaluated the antimicrobial activity against *L. monocytogenes*, *Pseudomonas aeruginosa*, *Salmonella Typhimurium*, *Staphilococcus aureus*, *Yersinia enterecolitica*, and inherent yeast and molds. Solutions at 1, 3 and 5 mg/l reacted rapidly with fruit samples; therefore a decrease in $ClO_2$ concentration over time was noticed. However, a concentration of 15 mg/l for 30 min showed a significant microbial reduction (Wu and Kim, 2007), as shown in Table 18.4. The visual appearance of blueberries was not compromised by treatments and this simple aqueous $ClO_2$ method presented several advantages, especially for small berry producers. It was efficient, not expensive, relatively short in time, and effective for microbial decontamination.

Aqueous solutions have also been applied for strawberry decontamination. A greater than 5-log CFU/g reduction was observed for *E. coli* O157:H7 and *Listeria* populations after treatment with a solution of $ClO_2$ at 5 mg/l for 5 min (Rodgers *et al.*, 2004). During storage at 4°C for 9 days, the pathogen population remained under detectable levels. Mesophilic bacteria count increased gradually over time, but at the end of the study, the count was lower than the initial control levels. In contrast, the final yeast and mold population was significantly higher than the initial counts. No color changes were observed on treated strawberries and $ClO_2$ was rated positively by the panelists (trained judges).

The use of $ClO_2$ gas has showed promising results as a sanitizer to control pathogens and inherent microflora on berries. $ClO_2$ gas (8 mg/l for 120 min) effectively reduced *Salmonella* population on blueberry skin, calyx and stem scar: 3.67-, 2.44- and 3.24-log CFU/g reductions were observed, respectively (Sy *et al.*, 2005a). Cells attached to the skin were in general largely exposed to $ClO_2$ and consequently more sensitive to the treatment, compared to the cells protected by calyx and stem scar. Yeast and mold populations were also significantly reduced by gaseous $ClO_2$, and the sensory quality of treated fruit was not affected by treatments.

Similar observations were reported by Popa *et al.* (2007). *L. monocytogenes*,

© Woodhead Publishing Limited, 2012

*Salmonella*, *E. coli* O157:H7, yeast and molds were significantly reduced after a treatment with 4 mg/l of $ClO_2$ gas for 12 h on blueberries. Gaseous $ClO_2$ was used to inactivate *E. coli* O157:H7, *L. monocytogenes* and *S. enterica* experimentally inoculated on strawberries (Mahmoud *et al.*, 2007). Following a treatment with 5 mg/l $ClO_2$ for 10 min, >2.9-log CFU/g reductions were reported for all tested bacteria (Table 18.4). The overall quality appearance of treated strawberries was not compromised. Moreover, inherent microflora was significantly reduced by gas treatment and the shelf life of treated samples was extended up to 16 days, compared to 8 days of total shelf life for the controls, at refrigerated temperatures. Similar findings were reported by Sy *et al.* (2005a), where a greater than 3.8-log CFU/g reduction was observed in *Salmonella* population, after $ClO_2$ treatments. Yeast and mold populations on treated samples were significantly reduced by gaseous $ClO_2$, and appearance, color, and aroma were not significantly different between untreated and treated samples during storage at 4°C for 10 days. Gas effectiveness was not influenced by the inoculation site, as previously reported for blueberries, probably because strawberries' skin and stem scar surfaces presented similar porosity.

Overall, the use of $ClO_2$ as a disinfectant to decontaminate fruits and berries has shown promising results. In general, increased shelf life and no external damage were reported after gaseous treatment or immersion in disinfectant solutions.

### 18.3.2 Raw and minimally processed vegetables (MPV)`

Bactericidal activity of $ClO_2$ was investigated on inoculated green pepper surfaces. Han *et al.* (2001) compared the reduction obtained following aqueous and gas $ClO_2$ treatments at 3 mg/l for 10 min. The gaseous form was more effective in reducing the pathogen on both uninjured and injured sample surfaces, compared to aqueous solutions. $ClO_2$ gas was also effective against *E. coli* O157:H7 on green peppers (Han *et al.*, 2000), where a greater than 6-log CFU reduction was observed on sample surfaces, following a treatment with 1.24 mg/l $ClO_2$ for 30 min. The attachment of *E. coli* cells to peppers was influenced by the surface properties: small lesions, lenticels, and micro-cracks protected microorganisms from the sanitizer antimicrobial action.

Decontamination effects of $ClO_2$ on lettuce were investigated by Rodgers *et al.* (2004). A greater than 5 log CFU/g reduction was observed in *E. coli* O157:H7 and *L. monocytogenes* population after a treatment with 5 mg/l $ClO_2$ aqueous solutions for 5 min. Pathogen populations remained under detectable levels throughout the experiment duration (9 days at 4°C). Conversely, inherent microflora populations increased gradually over time, but in general the count was lower than untreated samples. No color changes were observed and treated lettuce was considered acceptable by a trained panel.

Negative quality impacts were instead reported after gaseous $ClO_2$

© Woodhead Publishing Limited, 2012

**Table 18.5** Examples of studies on the effect of $ClO_2$ (aqueous and gas form) to reduce microbial load on vegetables

| | Microbial target | Generation method[a] | Treatment conditions | Reduction observed[b] | Shelf life[c], quality | References |
|---|---|---|---|---|---|---|
| $ClO_2$ aqueous solution applications | | | | | | |
| Carrots – shredded | Aerobic bacteria | Stock solution | 20 mg/l, 20 min | 1.9 log CFU/g | Yes, no defects | Gomez-Lopez *et al.*, 2008b |
| Green pepper | *L. monocytogenes* | $ClO_2 + H_2O$ | 3 mg/l, 10 min | 3 log CFU/g | NR | Han *et al.*, 2001 |
| Lettuce | *E. coli* O157:H7 | 2% $NaClO_2 + H_3PO_4$ | 5 mg/l, 5 min | > 5 log CFU/g | Yes, no defects | Rodgers *et al.*, 2004 |
| Lettuce | *L. monocytogenes* | 2% $NaClO_2 + H_3PO_4$ | 5 mg/l, 5 min | > 5 log CFU/g | Yes, no defects | Rodgers *et al.*, 2004 |
| Lettuce – shredded | Aerobic bacteria | Stock solution | 20 mg/l, 20 min | No reduction | No, browning | Gomez-Lopez *et al.*, 2008b |
| Lettuce – shredded | *E. coli* O157:H7 | Sachet | 200 mg/l, 2 min | ~ 1 log CFU/g | NR | Keskinen *et al.*, 2009 |
| Tomatoes | *E. carotovaca* | Pouch | 20 mg/l, 1 min | 5 log $CFU/cm^2$ | NR | Pao *et al.*, 2007 |
| Tomatoes | *S. enterica* | Pouch | 20 mg/l, 1 min | 5 log $CFU/cm^2$ | NR | Pao *et al.*, 2007 |
| $ClO_2$ gas applications | | | | | | |
| Cabbage – shredded | Microflora | Stock solution + air | 1.29 mg/l, 10 min | > 0.3 log CFU/g | No, no defects | Gomez-Lopez *et al.*, 2008c |
| Cabbage – shredded | *L. monocytogenes* | Sachet | 4.1 mg/l, 29.3 min | 3.6 log CFU/g | NR, browning | Sy *et al.*, 2005b |
| Cabbage – shredded | *E. coli O157:H7* | Sachet | 3.65 mg/l, 20.5 min | 3.2 log CFU/g | NR, browning | Sy *et al.*, 2005b |
| Cabbage – shredded | *Salmonella* spp | Sachet | 4.1 mg/l, 30.8 min | 4.4 log CFU/g | NR, browning | Sy *et al.*, 2005b |
| Carrots – shredded | Microflora | Stock solution + air | 1.33 mg/l, 6 min | > 0.7 log CFU/g | Yes, no defects | Gomez-Lopez *et al.*, 2007 |
| Carrots – shredded | *L. monocytogenes* | Sachet | 4.1 mg/l, 29.3 min | 5.8 log CFU/g | NR, whitening | Sy *et al.*, 2005b |
| Carrots – shredded | *E. coli* O157:H7 | Sachet | 4.1 mg/l, 20.5 min | 5.6 log CFU/g | NR, whitening | Sy *et al.*, 2005b |
| Carrots – shredded | *Salmonella* spp | Sachet | 4.1 mg/l, 30.8 min | 5.1 log CFU/g | NR, whitening | Sy *et al.*, 2005b |
| Green pepper | *E. coli* O157:H7 | 4% $Cl_2$ + 80% $NaClO_2$ | 24 mg/l, 30 min | > 6 log CFU | NR | Han *et al.*, 2000 |
| Green pepper | *L. monocytogenes* | 4% $Cl_2$ + 80% $NaClO_2$ | 3 mg/l, 10 min, 95% RH | > 5.3 log CFU/g | NR | Han *et al.*, 2001 |

*(Continued)*

© Woodhead Publishing Limited, 2012

**Table 18.5** Continued

| | Microbial target | Generation method[a] | Treatment conditions | Reduction observed[b] | Shelf life[c], quality | References |
|---|---|---|---|---|---|---|
| Lettuce | *E. coli O157:H7* | 2% $Cl_2$ + 80% $NaClO_2$ | 5 mg/l, 10 min, 95% RH | 3.9 log CFU/$cm^2$ | Yes, whitening | Mahmoud and Linton, 2008 |
| Lettuce | *S. enterica* | 2% $Cl_2$ + 80% $NaClO_2$ | 5 mg/l, 10 min, 95% RH | 2.7 log CFU/$cm^2$ | Yes, whitening | Mahmoud and Linton, 2008 |
| Lettuce – shredded | Microflora | Stock solution + air | 1.74 mg/l, 10 min | > 0.6 log CFU/g | No, no defects | Gomez-Lopez *et al.*, 2008c |
| Lettuce – shredded | *L. monocytogenes* | Sachet | 4.1 mg/l, 29.3 min | 1.5 log CFU/g | NR, browning | Sy *et al.*, 2005b |
| Lettuce – shredded | *E. coli* O157:H7 | Sachet | 3.65 mg/l, 20.5 min | 1.6 log CFU/g | NR, browning | Sy *et al.*, 2005b |
| Lettuce – shredded | *Salmonella* spp | Sachet | 4.1 mg/l, 30.8 min | 1.6 log CFU/g | NR, browning | Sy *et al.*, 2005b |
| Tomatoes[d] | *E. carotovora* | Sachet | 0.001–2.3 mg, 2–24 h | >5.7 log CFU/$cm^2$ | Yes, bleaching | Mahovic *et al.*, 2007 |
| Tomatoes | *L. monocytogenes* | 2% $Cl_2$ + 80% $NaClO_2$ | 0.5 mg/l,12 min, 95% RH | >5 log CFU/$cm^2$ | Yes, no defects | Bhagat *et al.*, 2010 |
| Tomatoes | *Salmonella* | Sachet | 4.1 mg/l, 25 min | 2 log CFU/g | No, no defects | Sy *et al.*, 2005b |
| Tomatoes[d] | *Salmonella* | Sachet | ~2.4 mg/l, 60 min | >6 log CFU/$cm^2$ | NR | Yuk *et al.*, 2005 |
| Tomatoes – wound | *S. enterica* | Sachet | 0.5 mg, 2 h | 3.8 log CFU/$cm^2$ | NR | Mahovic *et al.*, 2009 |
| Tomatoes | *S. enterica* | 2% $Cl_2$ + 80% $NaClO_2$ | 0.5 mg/l,12 min, 95% RH | >5 log CFU/$cm^2$ | Yes, no defects | Bhagat *et al.*, 2010 |
| Tomatoes | *S. enterica* | 2% $Cl_2$ + 80% $NaClO_2$ | 10 mg/l, 180 sec, 95% RH | ~ 5 log CFU/$cm^2$ | Yes, no defects | Trinetta *et al.*, 2010 |

[a]Stock solution: sodium chlorite reaction with citric acid; 2% $NaClO_2$ + $H_3PO_4$: acidification of sodium chlorite solution; Sachet, Pouch: acid and solid sodium chlorite in a sachet with humidity; Stock solution + air: sodium chlorite reaction with citric acid to form $ClO_2$ that was subsequently stripped from the solution by air bubbling; 4% $Cl_2$ + 80% $NaClO_2$: solid sodium chlorite reaction with chlorine gas; 2% $Cl_2$ + 80% $NaClO_2$: solid sodium chlorite reaction with chlorine gas.
[b]When possible, log reduction was reported as log CFU/$cm^2$ or log CFU/g, taking in account sample dimension, weight and/or dilution factors, as reported in the original papers.
[c]Significant improvement in microbial shelf life; NR: not reported.
[d]Inoculation sites were wound and stem.

© Woodhead Publishing Limited, 2012

treatments on lettuce (Mahmoud and Linton, 2008). The green color of iceberg lettuce turned into white-brown following exposure to 5 mg/l for 10 min. The phenomenon observed was probably due to chlorophyll oxidation reaction, as reported by Singh *et al.* (2002). Even though the treatments selected led to undesirable quality attributes, a 3.9-log CFU/cm$^2$ reduction in *E. coli* O157:H7 population and a 2.7-log CFU/cm$^2$ reduction in *S. enterica* were achieved. $ClO_2$ gas inactivation effectiveness increased with increasing treatment time and concentration, as stated previously for other commodities. The gas significantly reduced inherent microflora on lettuce leaves, and even if a gradual growth was noticed during storage at 4°C for 7 days, treated lettuce maintained always a lower count than the controls.

Tomatoes were recently associated with foodborne illness outbreaks (CDC, 2010) and researchers concluded that contamination occurred at farm or packing level. Soft rot bacteria, such as *Erwinia carotovora*, represent the most important cause of decay on tomato growth and again contamination takes place during packing house dumping and/or hydrocooling operations (Pao *et al.*, 2007). Therefore, the use of $ClO_2$ has been largely investigated on this commodity, in order to inactivate spoilage bacteria and offer a suitable alternative for microbial decontamination. Pao *et al.* (2007) explored the sanitizing effects of $ClO_2$ solutions on *S. enterica* and *E. carotovora* on tomato surfaces freshly inoculated and after 24 h of drying. A treatment of 1 min at 20 mg/l $ClO_2$ was sufficient to achieve a 5-log CFU/cm$^2$ reduction of bacteria populations on freshly spot-inoculated tomatoes; but once the contaminants were dried, $ClO_2$ had no significant inactivation effects, compared to regular tap water washing. The cells' attachment phase seemed important for disinfectant effectiveness, as reported by Foschino *et al.* (1998). Bactericidal activity of $ClO_2$ solutions was compromised when microbial cells were strongly attached to hard surfaces.

Decontamination effectiveness of gaseous $ClO_2$ on tomatoes has been widely explored too. High efficacy of $ClO_2$ gas compared to aqueous solution, was observed for bacterial soft rot inactivation on fresh tomatoes (Mahovic *et al.*, 2007). Samples remained firm and dry with no evidence of *E. carotovora* activity after exposure to $ClO_2$ gas. Tomatoes treated with 88 mg for 24 h (~0.2 mg/l) and with 99 mg for 2 h (~2.3 mg/l) remained free of decay, but wounds became bleached and sunken, and stem scars cracked. Conversely, tomatoes treated with 0.75 or 7.5 mg $ClO_2$ for 2 or 24 h respectively (~ 0.01 mg/l), showed no quality damage after treatments, but bacteria activity on wounds was reported.

The efficacy of $ClO_2$ gas for tomato decontamination was also investigated by Bhagat *et al.* (2010). A greater than 5-log CFU/cm$^2$ reduction was achieved in *Listeria* and *Salmonella* spp. populations after treatments with 0.5 mg/l $ClO_2$ gas for 12 min. Treated tomatoes stored at 22°C for 28 days, did not present any color difference, compared to the controls. After 3 weeks of storage, untreated samples were completely infested with molds, while treated tomatoes did not show any mold growth. Successful results for *Salmonella*

© Woodhead Publishing Limited, 2012

inactivation using $ClO_2$ gas were also reported by Sy *et al.* (2005b). A 2-log CFU/g reduction in pathogen population was observed after treatments with 4.1 mg/l $ClO_2$ gas for 25 min (Table 18.5). However, inherent microflora was not significantly different on treated and untreated samples during the storage period. Gas penetration was affected by the presence of fungal propagules on tomato skins. Treated samples were rated significantly higher for appearance, color, aroma, and overall quality, compared to controls, but at the end of the study all produce was compromised by over-ripening and decay caused by molds.

In another study, $ClO_2$ gas was compared to individual or combined sanitizer treatments for *Salmonella* spp. inactivation on green tomatoes (Yuk *et al.*, 2005). Tomatoes were placed in a 22 quart vessel chamber with a $ClO_2$ sachet that produced 100 mg gas over a 1 h period (~2.4 mg/l). Pathogen population on smooth surfaces and stem scar was decreased to undetectable levels and, moreover, *Salmonella* growth was completely inhibited during all the 5 days of storage. Conversely, there was no difference at the puncture wound sites between treated and untreated tomatoes, showing that the gas was not effective in these locations.

Mahovic *et al.* (2009) reported instead a reduction in *Salmonella* population on tomato wounds following $ClO_2$ gas exposure. Effectiveness of the gas varied with concentrations used, and after a treatment with 0.5 mg/l $ClO_2$ for 2 h, the number of *Salmonella* viable cells was reduced by 3.8-log CFU/ $cm^2$, compared to the controls.

With the intent of applying $ClO_2$ gas technology to large-scale tomato-packing facilities and enhance safety and quality of produce, Trinetta *et al.* (2010) evaluated several high gas concentrations (2, 5, 8 and 10 mg/l) for relatively short exposure times (from 10 to 180 s). Treatments with 10 mg/l $ClO_2$ for 180 s achieved an almost 5-log CFU/$cm^2$ reduction in *Salmonella* spp. population. Treated tomatoes were subsequently stored at room temperature for 28 days, and $ClO_2$ significantly reduced inherent microflora. After 3 weeks, control tomatoes were visibly moldy, while molds were not observed on treated samples. No changes in fruit color were reported and $ClO_2$ by-product residuals were under detectable levels within 24 h of treatment, showing that the technology did not pose any additional chemical safety risk for consumers and moreover increased produce shelf life.

When fresh vegetables are physically modified from the original form, through trimming, peeling, washing, cutting and packaging, they are defined as minimally processed vegetables (MPV). Decontamination of MPV has been investigated with the aim of reducing pathogens and spoilage microflora, and moreover trying to keep fresh sensory attributes and nutritional quality (Gomez-Lopez *et al.*, 2008a, 2009). The efficacy of aqueous $ClO_2$ was evaluated for decontamination of minimally processed cabbage, carrots, and lettuce (Gomez-Lopez *et al.*, 2008b). Aerobic plate count was not significantly reduced by the treatment on lettuce and cabbage, but nearly 2-log CFU/g reduction was observed on carrots after immersion in 20 mg/l $ClO_2$ solutions

© Woodhead Publishing Limited, 2012

for 20 min. Lettuce was the only produce whose quality was compromised by $ClO_2$ washing; green leaves turned brown after treatments.

Conversely, no quality reduction was observed by Rodgers *et al.* (2004) on lettuce treated with aqueous $ClO_2$. No damage was also observed for treated cabbage and carrots (Gomez-Lopez *et al.*, 2008b). Keskinen *et al.* (2009) noticed only a 1-log CFU/g reduction in *E. coli* O157:H7 population on cut lettuce leafs, after 2 min washing with 200 mg/l $ClO_2$ solutions. Keskinen *et al.* (2009) investigated also the limiting causes to treatment efficacy, and following a confocal scanning laser microscopy analysis, areas like damaged tissue and stomata were identified as harborage sites, that were inaccessible to the sanitizer.

The efficacy of gaseous $ClO_2$ was investigated for MPV shelf life extension. Gomez-Lopez *et al.* (2008c) reported that a pre-treatment by immersion in a cysteine solution could completely inhibit bleaching and browning of lettuce and cabbage, but treatments failed to prolong produce shelf life. As reported in Table 18.5, only about 0.3-log CFU/g reduction on inherent microflora was observed after treatment. Promising results were instead obtained with $ClO_2$ gas treatments on carrots (Gomez-Lopez *et al.*, 2007). Respiration rate and sensory attributes were not compromised by the gas, inherent microflora was significantly reduced and treatment prolonged carrot shelf life by one more day, compared to control samples. High microbial inactivation effectiveness was also reported by Sy *et al.* (2005b), on shredded lettuce using $ClO_2$ gas, but also in this study produce quality was compromised (Table 18.5).

Overall, all these studies on vegetable decontamination demonstrate the potential of $ClO_2$, both in aqueous and gaseous forms, as effective microbial inactivation techniques. Results on shelf life extension and overall quality appearance were not always successful, especially for leaf-greens and minimally processed products, where chlorophyll was oxidized by $ClO_2$ and consequently leaves turned white-brown.

### 18.3.3 Seeds and sprouts

Microbial contamination may not occur only during post-harvest operations, but also pre-harvest, through contaminated manure, soil, poor water quality, and sick animals. Bacteria could enter in the plant root systems and arrive at the edible portion of the produce. Therefore several researchers have focused on the study of technologies to control and inactivate bacteria attached onto seeds and sprout surfaces, and examples of $ClO_2$ applications are listed in Table 18.6.

Efficacy of aqueous $ClO_2$ combined with air drying and dry heat was evaluated on radish seeds (Bang *et al.*, 2011). Experimentally inoculated seeds (110 g) were washed for 5 min with 500 mg/l $ClO_2$ solutions, air dried for 2 h and heated at 55°C for 6 h. More than 5-log CFU/g reduction was achieved in *E. coli* O157:H7 and inherent aerobic microflora after treatment; however, the pathogen was detected in 5-day-old sprouts. Germination rate

© Woodhead Publishing Limited, 2012

**Table 18.6** Examples of studies on the effect of $ClO_2$ (aqueous and gas form) to decontaminate seeds and sprouts

| | Microbial target | Generation method[a] | Treatment conditions | Reduction observed | Quality[b] | References |
|---|---|---|---|---|---|---|
| $ClO_2$ aqueous solution applications | | | | | | |
| Alfalfa seeds | *E. coli* O157:H7 | $HCl + NaClO_2$ | 25 mg/l, 5 min | No reduction | NA, NA | Singh *et al.*, 2003 |
| Radish seeds | *E. coli* O157:H7 | $ClO_2 + H_2O$ | 500 mg/l, 5 min | >5 log CFU/g | NR, NA | Bang *et al.*, 2011 |
| Radish seeds | Aerobic bacteria | $ClO_2 + H_2O$ | 500 mg/l, 5 min | >5 log CFU/g | NR, NA | Bang *et al.*, 2011 |
| Radish seeds | Yeast and molds | $ClO_2 + H_2O$ | 500 mg/l, 5 min | ~1 log CFU/g | NR, NA | Bang *et al.*, 2011 |
| $ClO_2$ gas applications | | | | | | |
| Alfalfa sprouts | *Salmonella* | 2% $Cl_2$ + 80% $NaClO_2$ | 5 mg/l, 20 min, 90% RH | >3 log CFU/g | A, A | Bhagat *et al.*, 2010 |
| Cantaloupe seeds | *S. Poona* | 2% $Cl_2$ + 80% $NaClO_2$ | 10 mg/l, 3 min, 75% RH | 2 log CFU/g | NA, A | Trinetta *et al.*, 2011a |
| Lettuce seeds | *E. coli* O157:H7 | 2% $Cl_2$ + 80% $NaClO_2$ | 10 mg/l, 3 min, 75% RH | 1.8 log CFU/g | NA, NA | Trinetta *et al.*, 2011a |
| Tomato seeds | *S. Poona* | 2% $Cl_2$ + 80% $NaClO_2$ | 10 mg/l, 3 min, 75% RH | >5 log CFU/g | NA, NA | Trinetta *et al.*, 2011a |

[a] $HCl + NaClO_2$: acidification of sodium chlorite solution; 2% $Cl_2$ + 80% $NaClO_2$: solid sodium chlorite reaction with chlorine gas.
[b] Sprout vigor and % germination; NA: not affected; A: affected; NR: not reported.

© Woodhead Publishing Limited, 2012

was not affected by $ClO_2$, but the treatment was ineffective for yeast and mold inactivation. Antimicrobial activity of aqueous $ClO_2$ was also investigated on alfalfa seeds, experimentally inoculated with *E. coli* O157:H7 before and during sprouting (Singh *et al.*, 2003). In this study as well, $ClO_2$ did not compromise germination of seeds. However, the treatment did not significantly reduce pathogen population that instead increased after sprouting.

The efficacy of $ClO_2$ gas was also evaluated for seed and sprout decontamination. Alfalfa sprouts were experimentally inoculated with *Salmonella* and treated with 5 mg/l for 20 min in order to achieve a > 3-log CFU/g reduction (Bhagat, 2010). The color of sprouts was negatively affected by the gas. As reported previously for lettuce, chlorophyll was oxidized by $ClO_2$, thus compromising the overall quality. In another study, the efficacy of $ClO_2$ gas was evaluated against *S. enterica* and *E. coli* O157:H7 on pre-inoculated cantaloupe, lettuce, and tomato seeds (Trinetta *et al.*, 2011a). The treatment was noticeably more effective against *Salmonella* on contaminated tomato seeds, compared to *Salmonella* inactivation on cantaloupe seeds or to *E. coli* on lettuce seeds (Table 18.6). In general, sprout vigor and germination percentage of seeds and sprouts were not affected by $ClO_2$. Moreover, high microbial log reductions were obtained on treated surfaces. Again, the quality of produce with high chlorophyll content was negatively affected by $ClO_2$ oxidation reactions.

### 18.3.4 Decontamination of meat and seafood products

Extensive research on the application of $ClO_2$ for meat and seafood products is less readily available (Table 18.7), even though the US Food and Drug Administration and the US Department of Agriculture have approved the use of $ClO_2$ in ice water in order to minimize microbiological cross-contamination of poultry carcasses (Federal Register, 1995), while sodium chlorite is approved for water and/or ice in contact with seafood (Federal Register, 1998).

The application of $ClO_2$ spray washes to reduce fecal contamination on pre-rigor beef carcasses was investigated by Cutter and Dorsa (1995). Surprisingly, 20 mg/l $ClO_2$ tank concentration was no more effective than spray washing with water for reducing fecal bacteria population: a 2-log CFU/cm$^2$ reduction was observed after 60 s contact time.

Decontamination of chicken breasts was studied using $ClO_2$ gas combined with modified atmosphere packaging (Ellis *et al.*, 2006). Samples were experimentally inoculated with *Salmonella* Typhimurium and microbial count and quality of treated and untreated samples was monitored for 15 days at refrigerated temperature. Two different conditions were used: slow-release rate sachets, able to produce ~2.25 mg $ClO_2$ for 22 days and fast-release rate sachets at ~6.6 mg $ClO_2$ for 26 h. The microbial growth on the surface of chicken breast was reduced. Within the first week of shelf life study, approximately 1-log CFU/chicken breast reduction was observed in *Salmonella* counts for treated samples with both $ClO_2$ rate release, compared

© Woodhead Publishing Limited, 2012

**Table 18.7** Examples of studies on the effect of $ClO_2$ (aqueous and gas form) to reduce microbial load on meat and seafood products

| | Microbial target | Generation method[a] | Treatment conditions | Reduction observed[b] | Shelf life[c], quality | References |
|---|---|---|---|---|---|---|
| $ClO_2$ aqueous solution applicaltions | | | | | | |
| Beef carcasses | Fecal bacteria | $HCl + NaClO_2$ | 20 mg/l, 1 min | 2 log CFU/$cm^2$ | No, NR | Cutter and Dorsa 1995 |
| Mangrove snapper | *L. monocytogenes* | Commercial solution | 200 mg/l, 5 min | ~1.5 log CFU/g | NR, browning | Lin *et al.*, 1996 |
| Red grouper | Aerobic bacteria | 2% $NaClO_2 + H_3PO_4$ | 200 mg/l, 5 min | >4.9 log CFU/g | Yes, discoloration | Kim *et al.*, 1999 |
| Salmon | Aerobic bacteria | 2% $NaClO_2 + H_3PO_4$ | 200 mg/l, 5 min | >4.9 log CFU/g | Yes, discoloration | Kim *et al.*, 1999 |
| Scallops | Aerobic bacteria | 2% $NaClO_2 + H_3PO_4$ | 200 mg/l, 5 min | >4.9 log CFU/g | Yes, discoloration | Kim *et al.*, 1999 |
| Shrimp | Aerobic bacteria | 2% $NaClO_2 + H_3PO_4$ | 200 mg/l, 5 min | >4.9 log CFU/g | Yes, discoloration | Kim *et al.*, 1999 |
| $ClO_2$ gas applications | | | | | | |
| Chicken breast | Aerobic bacteria | Sachet | 2.25 mg, 22 days | 1 log CFU/chicken breast | Yes, discoloration | Ellis *et al.*, 2006 |
| Chicken breast | Aerobic bacteria | Sachet | 6.6 mg, 26 h | 1 log CFU/chicken breast | Yes, discoloration | Ellis *et al.*, 2006 |
| Chicken breast | *S.* Typhimurium | Sachet | 2.25 mg, 22 days | 1 log CFU/chicken breast | Yes, discoloration | Ellis *et al.*, 2006 |
| Chicken breast | *S.* Typhimurium | Sachet | 6.6 mg, 26 h | 1.5 log CFU/ chicken breast | Yes, discoloration | Ellis *et al.*, 2006 |

[a] $HCl+NaClO_2$: acidification of sodium chlorite solution; Commercial solution: sodium chlorite reaction with citric acid; 2% $NaClO_2 + C_6H_8O_7$: acidification of sodium chlorite solution; 2% $NaClO_2 + H_3PO_4$: acidification of sodium chlorite solution; Sachet: acid and solid sodium chlorite in a sachet with humidity.
[b] When possible, log reduction was reported as log CFU/$cm^2$ or log CFU/g, taking in account sample dimension, weight and/or dilution factors, as reported in the original papers.
[c] Significant improvement in microbial shelf life: NR: not reported.

© Woodhead Publishing Limited, 2012

to the control, and after two weeks no differences were noticed in pathogen population. The inherent microbial load was also reduced by 1-log CFU/chicken breast within the first week by the fast-release sachets, while in the second week no difference was observed compared to the samples without sachets. On the contrary, samples treated with slow-release $ClO_2$ sachets had inherent bacteria count still lower than untreated samples after 15 days of storage. $ClO_2$ adversely affected the color of treated samples. Green-brown color was observed in the areas close to $ClO_2$ sachets, while the surrounding areas were yellow. No off-odors were detected by the sensory panel.

A fish model system (Mangrove Snapper cubes) was used to evaluate the bactericidal activity of aqueous $ClO_2$ against *Listeria monocytogenes* for potential application in seafood processing operations (Lin *et al.*, 1996). A linear trend in bactericidal activity was observed with increasing disinfectant concentrations and a ~1.5 log CFU/g reduction in pathogen population was achieved treating the samples with a solution at 200 mg/l $ClO_2$ for 5 min. A light brown color in fish cubes was noticed with $ClO_2$ solutions at 400 mg/l, while no change in pH and no detectable chlorine residues were detected. The same treatment was applied to Red Grouper, Atlantic salmon, shrimps and scallops in order to reduce bacterial loads (Kim *et al.*, 1999). All samples after treatments with $ClO_2$ solution at 200 mg/l for 5 min contained no viable bacteria and again a dose-related $ClO_2$ bactericidal effect was observed.

In general, treatment of seafood at low $ClO_2$ solution concentrations resulted in no appearance or discoloration defects. Instead at high levels, discoloration due to organic reactions was reported. Red Grouper fillets developed a rusty color when treated with 200 mg/l $ClO_2$, and the same phenomenon was observed for treated salmon fillets. Odor formation and discoloration occurred also in treated scallops, but the overall quality after 7 days was better than the untreated ones (Kim *et al.*, 1999). Shrimp had slight discoloration after $ClO_2$ treatments and chlorine smell. The skin of whole fishes was bleached, and moreover $ClO_2$ interacted with fish blood, developing a light chocolate color in the gills (Kim *et al.*, 1999).

Some defects were noticed after $ClO_2$ treatments both on meat and seafood products, even though the disinfectant was still very effective in reducing the microbial load and extending shelf life. These observations suggest that $ClO_2$ presents potential for use as a decontamination technique, but more research has to be conducted in order to determine the appropriate levels that do not compromise product quality.

### 18.3.5 Possible reactions with food components

$ClO_2$ can react with organic compounds found in food, and antimicrobial activity might be reduced, since carbohydrates, lipids, and proteins can interfere with this strong oxidizer (Vandekinderen *et al.*, 2009). Studies have demonstrated that $ClO_2$ was less effective as a decontamination agent when used with products rich in proteins and lipids such as meat and seafood,

© Woodhead Publishing Limited, 2012

compared to vegetables, fruits, and bakery goods, characterized by higher carbohydrate contents (Vandekinderen *et al.*, 2009).

The predominant reaction of $ClO_2$ with carbohydrates is oxidation and consequently formation of carboxylic acid (Fukayama *et al.*, 1986). Also lipids react immediately with $ClO_2$, through oxidation with unsaturated fatty acids at their double bonds. Furthermore protein, peptides and amino acids are readily oxidized by $ClO_2$ and aromatic amino acids are in general more vulnerable to oxidation action, compared to aliphatic ones. Little information is available regarding the toxicity of these by-products for humans (Fukayama *et al.*, 1986). It is clear that some food components compromise the antimicrobial activity of $ClO_2$ by acting as an 'organic demand', lowering the concentration available to inactivate microorganisms. When $ClO_2$ reacts with these components, residues such as chlorite can be produced as a by-product. Since chlorite is known to have toxic effects when fed to animals, its levels must be minimized for any food application. Better knowledge about chemical reactions and the levels of residues still needs to be gained in order to assess $ClO_2$ food treatment risks for consumers.

### 18.3.6 Decontamination using $ClO_2$: effects on food quality

When choosing a technology for food decontamination, some aspects have to be considered, such as the ability to:

- reduce the risk of contamination of both pathogenic and spoilage organisms;
- increase product shelf life;
- keep fresh food quality attributes;
- not compromise consumers' health.

The use of $ClO_2$ has been shown to be a very effective antimicrobial agent against several pathogens, while showing promising results in prolonging the shelf life of fruits and vegetables by reducing the load of inherent microflora (Du *et al.*, 2003; Han *et al.*, 2000; Mahmoud *et al.*, 2007). Gomez-Lopez *et al.* (2007) report a reduction of psychrotrophic microorganisms, and consequently an extension of 1 day in shelf life of grated carrots treated with $ClO_2$; Mahmoud *et al.* (2008) were able to prolong cantaloupe shelf life up to 9 days compared to untreated fruit during a storage study at room temperature.

However, several negative observations of discoloration, browning, and bleaching have been reported, which is likely due to chlorophyll and/or phenol oxidation reactions (Singh *et al.*, 2002). Lettuce leaves turned from green to white (Mahmoud and Linton, 2008), and bleaching spots were observed on blueberry and strawberries (Popa *et al.*, 2007; Sy *et al.*, 2005a), whereas the sensory and quality characteristics of raspberries, tomatoes, and onions were not compromised by $ClO_2$ (Sy *et al.*, 2005a, 2005b). These diverse observations related to exposure to $ClO_2$ make the selection of $ClO_2$

© Woodhead Publishing Limited, 2012

concentration and time very important. Since $ClO_2$ is a strong bleaching agent, the quality of any food product will be degraded when exposed to high enough concentrations or long exposure times.

The major benefit of using $ClO_2$ for food decontamination is its higher antimicrobial effectiveness, when compared to chlorine, in the presence of organics. $ClO_2$ also forms less disinfection by-products than chlorine-based solutions (Gomez-Lopez *et al.*, 2009). Since the use of $ClO_2$ is approved for drinking water disinfection, many studies about toxicity have been conducted and no adverse effects have been reported for consumers (Gomez-Lopez *et al.*, 2009). $ClO_2$ reacts mainly by oxidation and rapidly dissociates into chlorate ($ClO_3^-$) and chlorite ($ClO_2^-$) ions and eventually to chloride ($Cl^-$).

Few studies have reported the measurement of $ClO_2$ by-products after treatment of food products. No chlorite was detected in potatoes skin (Tsai *et al.*, 2001) and similar results were observed in sea scallops, mahi-mahi, and shrimp (Kim *et al.*, 1999). Conversely, Trinetta *et al.* (2011b) evaluated residues of $ClO_2$, chlorite, chlorate, and chloride on selected produce by rinsing the surfaces and observed detectable residual levels on lettuce and sprouts. Very low residuals, compared to EPA acceptable levels for drinking water, were detected in tomatoes, oranges, apples, strawberries, and cantaloupe.

## 18.4 Limitations and challenges to adoption of chlorine dioxide ($ClO_2$) technology

Chlorine dioxide, like all sanitizers, has its limits and challenges for adoption. As an equipment sanitizer, all surfaces still must be properly cleaned prior to using $ClO_2$ as a sanitizer. The levels of $ClO_2$ recommended (50–200 ppm) in solution, will quickly off-gas dangerous levels of gas if used in enclosed spaces. Due to the dangers of off-gassing, some early adopters have abandoned $ClO_2$ for use in their processing plants.

Within a food manufacturing facility, strict operating procedures must be in place to prevent and mitigate any leaks in the gassing system. Room monitors are typically installed to warn of dangerous levels of $ClO_2$, and all unnecessary personnel are normally excluded from potential exposure areas when gas is being used. For all of these reasons, the safety of $ClO_2$ for employees in a facility is the most significant challenge for adoption of either aqueous or gaseous $ClO_2$ in the food industry.

## 18.5 Conclusion and future trends

Overall $ClO_2$ research has shown promising results for controlling pathogen and spoilage microorganisms on different food matrices, environmental and food-contact surfaces. The technology seems to be more suitable for

© Woodhead Publishing Limited, 2012

hard surfaces and fruits, compared to vegetables and meat products. Most vegetable and meat applications resulted in quality loss after treatment. Several aqueous solutions are available on the market, while regulatory approvals for gaseous antimicrobial products (sachets and direct gas generators) are still limited. $ClO_2$ gas might represent the future technology, since more penetration ability on product surfaces against pathogens has been shown, compared to the aqueous form.

Even if governmental agencies have recognized the potential of the disinfectant, currently no approved applications of aqueous or gaseous $ClO_2$ claim to reduce human pathogens on the produce surface. The EPA still does not have an established protocol for assessing the effectiveness of such a treatment against etiological agents. Up to now, all approved products are legally limited to claims for the control of only spoilage organisms. Technology manufacturers are continuing to pursue regulatory approval for $ClO_2$ applications on fresh fruits and vegetables. Therefore, in order to support the use of $ClO_2$ as a food disinfectant, extensive research on by-product chemistry and their potential safety needs to be conducted.

## 18.6 Sources of further information and advice

Suggested sources for further information are: EPA (1999); Gates (1998); Linton *et al.* (2006); Masschelein and Rice (1979).

## 18.7 References and further reading


APHA (1995), *Standard Methods for the Examination of Water and Wastewater*, Washington DC, American Public Health Association.

BANG J, KIM H, KIM H, BEUCHAT LR and RYU JH (2011), 'Combined effects of chlorine dioxide, drying, and dry heat treatments in inactivating microorganisms on radish seeds', *Food Microbiol*, 28, 114–118.

BAKHMUTOVA-ALBERT EV, MARGERUM DW, AUER JG and APPLEGATE BM (2008), 'Chlorine dioxide oxidation of dihydronicotinamide adenine dinucleotide (NADH)', *Inorganic Chem*, 47, 2205–2211.

BENARDE MA, SNOW BW, OLIVIERI VP and DAVIDSON B (1967), 'Kinetics and mechanism of bacterial disinfection by chlorine dioxide', *J Appl Microbiol*, 15, 257–265.

BHAGAT A (2010), *Modeling critical factors to optimize the treatment of selected fruit and vegetables with chlorine dioxide gas using a miniaturized industrial-size tunnel system*, PhD Thesis, West Lafayette, IN, Purdue University.

BHAGAT A, MAHMOUD B and LINTON R (2010), 'Inactivation of *Salmonella enterica* and *Listeria monocytogenes* inoculated on hydroponic tomatoes using chlorine dioxide gas', *Foodborne Pathog Dis*, 7, 677–685.

CDC (2010), Outbreak surveillance data. Available from: http://www.cdc.gov/outbreaknet/surveillance_data.html (accessed 16 November 2010).

CUTTER CN and DORSA WJ (1995), 'Chlorine dioxide spray washes for reducing fecal contamination on beef', *J Food Prot*, 58, 294–296.


© Woodhead Publishing Limited, 2012

DU J, HAN Y and LINTON RH (2002), 'Inactivation by chlorine dioxide gas ($ClO_2$) of *Listeria monocytogenes* spotted onto different apple surfaces', *Food Microbiol*, 19, 481–490.

DU J, HAN Y and LINTON RH (2003), 'Efficacy of chlorine dioxide gas in reducing *Escherichia coli* O157:H7 on apple surfaces', *Food Microbiol*, 20, 583–591.

ELLIS M, COOKSEY K, DAWSON P, HAN I and VERGANO P (2006), 'Quality of fresh chicken breasts using a combination of modified atmosphere packaging and chlorine dioxide sachets', *J Food Prot*, 69, 1991–1996.

EPA (1999), 'Chlorine dioxide'. Available from: http://www.epa.gov/ogwdw/mdbp/pdf/alter/chapt_4.pdf (accessed 9 April 2012).

EPA (2003), 'Information on chlorine dioxide'. Available from: http://www.epa.gov/iris/subst/0496.htm (accessed 9 April 2012).

FDA (1998), 'Secondary Direct Food Additives Permitted in Food for Human Consumption: final rule', 21 CFR 173.300. Available from: http://www.fda.gov/ohrms/dockets/98fr/081399a.txt (accessed 9 April 2012).

FDA (1999), 'Guidance for Industry: Antimicrobial Food Additives'. Available from: http://www.fda.gov/Food/GuidanceComplianceRegulatoryInformation/GuidanceDocuments/FoodIngredientsandPackaging/ucm077256.htm (accessed 17 April 2010).

FEDERAL REGISTER (1995), 'Secondary Direct Food Additives Permitted in Food for Human Consumption: final rule', 21 CFR Part 173. 60 (42), 1189921–11901.

FEDERAL REGISTER (1998), 'National Primary Drinking Water Regulations: Disinfectants and Disinfection Byproducts: final rule', 40 CFR Parts 9, 141, and 142. 63(241): 69390–69476. Available from: www.gpoaccess.gov (accessed 8 December 2010).

FOSCHINO R, NERVEGNA I, MOTTA A and GALLI A (1998), 'Bactericidal activity of chlorine dioxide against *Escherichia coli* in water and on hard surfaces', *J Food Prot*, 61, 668–672.

FUKAYAMA MY, TAN H, WHEELER WB and WEI CI (1986), 'Reaction of aqueous chlorine and chlorine dioxide with model food compounds', *Environ Health Perspec*, 69, 267–274.

GATES DJ (1998), *The Chlorine Dioxide Handbook*, Denver, CO, American Water Works Association.

GOMEZ-LOPEZ VM, DEVLIEGHERE VM, RAGAERT F and DEBEVERE J (2007), 'Shelf-life extension of minimally processed carrots by gaseous chlorine dioxide', *Int J Food Microbiol*, 116, 221–227.

GOMEZ-LOPEZ VM, RAGAERT P, DEBEVERE J and DEVLIEGHERE F (2008a), 'Decontamination methods to prolong the shelf-life of minimally processed vegetables, state-of-the-art', *Crit Rev Food Sci Nutr*, 48, 487–495.

GOMEZ-LOPEZ VM, DEVLIEGHERE F, RAGAERT P, CHEN L, RYCKEBOER J and DEBEVERE J (2008b), 'Reduction of microbial load and sensory evaluation of minimally processed vegetables treated with chlorine dioxide and electrolysed water', *Ital J Food Sci*, 20, 321–333.

GOMEZ-LOPEZ VM, RAGAERT P, JEYACHCHANDRAN V, DEBEVERE J and DEVLIEGHERE F (2008c), 'Shelf-life of minimally processed lettuce and cabbage treated with gaseous chlorine dioxide and cysteine', *Int J Food Microbiol*, 121, 74–83.

GOMEZ-LOPEZ VM, RAJKOVIC A, RAGAERT P, SMIGIC N and DEVLIEGHERE F (2009), 'Chlorine dioxide for minimally processed produce preservation: a review', *Trends Food Sci Technol*, 20, 17–26.

HAN Y, GUENTERT AM, SMITH RS, LINTON RH and NELSON PE (1999), 'Efficacy of chlorine dioxide gas as a sanitizer for tanks used for aseptic juice storage', *Food Microbiol*, 16, 53–61.

HAN Y, LINTON RH, NIELSEN SS and NELSON PE (2000), 'Inactivation of *Escherichia coli* O157:H7 on surface-uninjured and -injured green pepper (*Capsicum annuum* L.) by chlorine dioxide gas as demonstrated by confocal laser scanning microbiology', *Food Microbiol*, 17, 643–655.

© Woodhead Publishing Limited, 2012

HAN Y, LINTON RH, NIELSEN SS and NELSON PE (2001), 'Reduction of *Listeria monocytogenes* on green peppers (*Capsicum annuum* L.) by gaseous and aqueous chlorine dioxide and water washing and its growth at 7°C', *J Food Prot*, 64, 1730–1738.

HAN Y, APPLEGATE B, LINTON RH and NELSON PE (2003), 'Decontamination of *Bacillus thuringiensis* spores on selected surfaces by chlorine dioxide gas', *J Environ Health*, 66, 16–21.

HUANG J-L, WANG L, REN N and MA FJ (1997), 'Disinfection effect of chlorine dioxide on bacteria in water', *Water Res*, 31, 607–613.

KESKINEN LA, BURKE A and ANNOUS BA (2009), 'Efficacy of chlorine, acid electrolyzed water and aqueous chlorine dioxide solutions to decontaminate *Escherichia coli* O157:H7 from lettuce leaves', *Int J Food Microbiol*, 132, 134–140.

KIM JM, HUANG TS, MARSHALL MR and WEI CI (1999), 'Chlorine dioxide treatment of seafood to reduce bacterial loads', *J Food Sci*, 64, 1098–1093.

KNAPP J and BATTISTI D (2001), 'Chlorine dioxide', in SS Block, *Disinfection, Sterilization, Preservation*, Philadelphia, PA, Lippincott Williams & Wilkens, 215–227.

KRESKE AC, RYU JH and BEUCHAT LR (2006), 'Evaluation of chlorine, chlorine dioxide, and a peroxyacetic acid-based sanitizer for effectiveness in killing *Bacillus cereus* and *Bacillus thuringiensis* spores in suspensions, on the surface of stainless steel, and on apples', *J Food Prot*, 69, 1892–1903.

LEE SY, DANCER GI, CHANG S, RHEE MS and KANG DH (2006), 'Efficacy of chlorine dioxide gas against *Alicyclobacillus acidoterrestris* spores on apple surfaces', *Int J Food Microbiol*, 108, 364–368.

LIN WF, HUANG TS, CORNELL JA, LIN CM and CHENG W (1996), 'Bactericidal activity of aqueous chlorine and chlorine dioxide solutions in a fish model system', *J Food Sci*, 61, 1030–1034.

LINTON RH, HAN YC, SELBY TL and NELSON PE (2006), 'Gas-/vapor-phase sanitation (decontamination) treatments', in GM Sapers, JR Gorny and AE Yousef, *Microbiology of fruits and vegetables*, Boca Raton, FL, CRC Press, 401–435.

MAHMOUD BSM and LINTON RH (2008), 'Inactivation kinetics of inoculated *Escherichia coli* O157:H7 and *Salmonella enterica* on lettuce by chlorine dioxide gas', *Food Microbiol*, 25, 244–252.

MAHMOUD BSM, BHAGAT AR and LINTON RH (2007), 'Inactivation kinetics of inoculated *Escherichia coli* O157:H7, *Listeria monocytogenes* and *Salmonella enterica* on strawberries by chlorine dioxide gas', *Food Microbiol*, 24, 736–744.

MAHMOUD BSM, VAIDYA NA, CORVALAN CM and LINTON RH (2008), 'Inactivation kinetics of inoculated *Escherichia coli* O157:H7, *Listeria monocytogenes* and *Salmonella Poona* on whole cantaloupe by chlorine dioxide gas', *Food Microbiol*, 25, 857–865.

MAHOVIC MJ, TENNEY JD and BARTZ JA (2007), 'Application of chlorine dioxide gas for control of bacterial soft rot in tomatoes', *Plant Dis*, 91, 1316–1320.

MAHOVIC M, BARTZ JA, SCHNEIDER KR and TENNEY JD (2009), 'Chlorine dioxide gas from an aqueous solution: reduction of *Salmonella* in wounds on tomato fruit and movement to sinks in a treatment chamber', *J Food Prot*, 72, 952–958.

MASSCHELEIN WJ and RICE RG (1979), *Chlorine dioxide chemistry and environmental impact of oxychlorine compounds*, Ann Arbor, MI, Ann Arbor Science Publishers, 111–145.

NPIRS (2011), 'The National Pesticide Information Retrieval System'. Available from: http://ppis.ceris.purdue.edu/npublic.htm (accessed 18 June 2011).

OLIVIERI V P (1968), *Chlorine dioxide and protein synthesis*, Masters Thesis, Morgantown, University of West Virginia.

PAO S, KELSEY DF, KHALID MF and ETTINGER MR (2007), 'Using aqueous chlorine dioxide to prevent contamination of tomatoes with *Salmonella enterica* and *Erwinia carotovora* during fruit washing', *J Food Prot*, 70, 629–634.

POPA I, HANSON EJ, TODD ECD, SCHILDER AC and RYSER ET (2007), 'Efficacy of chlorine

© Woodhead Publishing Limited, 2012

dioxide gas sachets for enhancing the microbiological quality and safety of blueberries', *J Food Prot*, 70, 2084–2088.

RODGERS SL, CASH JN, SIDDIQ M and RYSER ET (2004), 'A comparison of different chemical sanitizers for inactivating *Escherichia coli* O157:H7 and *Listeria monocytogenes* in solution and on apples, lettuce, strawberries, and cantaloupe', *J Food Prot*, 67, 721–731.

ROLLER SD, OLIVIERI VP and KAWATA K (1980), 'Mode of bacterial inactivation by chlorine dioxide', *Water Res*, 14, 635–641.

RYU JH and BEUCHAT LR (2005), 'Biofilm formation and sporulation by *Bacillus cereus* on a stainless steel surface and subsequent resistance of vegetative cells and spores to chlorine, chlorine dioxide, and a peroxyacetic acid-based sanitizer', *J Food Prot*, 68, 2614–2622.

SIMPSON GD (2005), *Practical Chlorine Dioxide*, Colleyville, TX, Simpson GD & Associates.

SINGH N, SINGH RK, BHUNIA AK and STROSHINE RL (2002), 'Effect of inoculation and washing methods on the efficacy of different sanitizer against *Escherichia coli* O157:H7 on lettuce and baby carrots', *Lebensm-Wiss Technol*, 35, 720–729.

SINGH N, SINGH RK and BHUNIA AK (2003), 'Sequential disinfection of *Escherichia coli* O157:H7 inoculated alfalfa seeds before and during sprouting using aqueous chlorine dioxide, ozonated water, and thyme essential oil', *Lebens Wiss Technol*, 36, 235–243.

SY KV, MCWATTERS KH and BEUCHAT LR (2005a), 'Efficacy of gaseous chlorine dioxide as a sanitizer for killing *Salmonella*, yeasts, and molds on blueberries, strawberries, and raspberries', *J Food Prot*, 68, 1165–1175.

SY KV, MURRAY MB, HARRISON MD and BEUCHAT LR (2005b), 'Evaluation of gaseous chlorine dioxide as a sanitizer for killing *Salmonella*, *Escherichia coli* O157:H7, *Listeria monocytogenes*, and yeasts and molds on fresh and fresh-cut produce', *J Food Prot*, 68, 1176–1187.

SYNAN JF, MACMAHON JD and VINCENT GP (1944), 'Chlorine dioxide – development in the treatment of potable water'. *Water Works Sewer*, 91, 423–426.

TAUBE H and DODGEN H (1949), 'Applications of radioactive chlorine to the study of the mechanisms of reactions involving changes in the oxidation state of chlorine', *J Am Chem Soc*, 71, 3330–3336.

TRINETTA V, MORGAN M and LINTON R (2010), 'Use of high-concentration-short-time chlorine dioxide gas treatments for the inactivation of *Salmonella enterica* spp. inoculated onto Roma tomatoes', *Food Microbiol*, 27, 1009–1015.

TRINETTA V, VAIDYA N, LINTON R and MORGAN M (2011a), 'A comparative study on the effectiveness of chlorine dioxide gas, ozone gas and e-beam irradiation treatments for inactivation of pathogens inoculated onto tomato, cantaloupe and lettuce seeds', *Int J Food Microbiol*, 146, 203–206.

TRINETTA V, VAIDYA N, LINTON R and MORGAN M (2011b), 'Evaluation of chlorine dioxide gas residues on selected food produce', *J Food Sci*, 76, T11–T15.

TSAI LS, HUXSOLL CC and ROBERTSON G (2001), 'Prevention of potato spoilage during storage by chlorine dioxide', *J Food Sci*, 66, 472–477.

VAID R, LINTON RH and MORGAN MT (2010), 'Comparison of inactivation of *Listeria monocytogenes* within a biofilm matrix using chlorine dioxide gas, aqueous chlorine dioxide and sodium hypochlorite treatments', *Food Microbiol*, 27, 979–984.

VANDEKINDEREN I, DEVLIEGHERE F, VAN CAMP J, KERKAERT B, CUCU T, RAGAERT P, DE BRUYNE J and DE MEULENAER B (2009), 'Effects of food composition on the inactivation of foodborne microorganisms by chlorine dioxide', *Int J Food Microbiol*, 131, 138–144.

YUK HG, BARTZ JA and SCHNEIDER KR (2005), 'Effectiveness of individual or combined sanitizer treatments for inactivating *Salmonella* spp. on smooth surface, steam scar, and wounds of tomatoes', *J Food Sci*, 80, M409–M414.

© Woodhead Publishing Limited, 2012

WALKER JT, MACKERNESS CW, MALLON D, MAKIN T, WILLIETS T and KEEVIN CW (1995), 'Control of *Legionella pneumophila* in a hospital water system by chlorine dioxide', *J Ind Microbiol*, 15, 384–390.

WU VC and KIM B (2007), 'Effect of simple chlorine dioxide method for controlling five foodborne pathogens, yeast and molds on blueberries', *Food Microbiol*, 24, 794–800.

WHITE GC (2010), 'Chlorine dioxide', *White's Handbook of Chlorination and Alternative Disinfectants*, 5th edn, Hoboken, NJ, John Wiley & Sons, 700–766.

© Woodhead Publishing Limited, 2012

# 19

# Electrolyzed oxidizing water for microbial decontamination of food

**K.-C. Cheng, National Taiwan University, Taiwan and S. R. S. Dev, K. L. Bialka and A. Demirci, The Pennsylvania State University, USA**

**Abstract**: The growing demand for safe, minimally processed food and the disadvantages of traditional chemical- and heat-based methods of microbial decontamination have meant that new technologies are emerging for high quality, effective decontamination of food. Electrolyzed oxidizing water (EOW) is electrolyzed soft tap water with sodium chloride added. The user- and environmentally-friendly status of this method, coupled with its low cost, makes it an effective and suitable method for microbial decontamination. This chapter provides an overview of the production, properties and applications of EOW, as well as a section on potential future trends.

**Key words**: electrolyzed oxidizing water (EOW), food decontamination, food industry, novel sanitation technology.

## 19.1 Introduction

Industrial food decontamination methods, such as pasteurization, thermal sterilization, irradiation, washing and addition of antimicrobials play a significant role in preventing food-related outbreaks of disease. Chemicals (such as oxidizing and reducing agents, salts and organic acids) have recently been used as antimicrobials and disinfecting agents in order to maintain both food quality and safety, due to the adverse effects that traditional thermal processes may have on the quality of minimally processed foods such as fresh produce, raw meat, seafood, etc. Meanwhile, these chemicals must have certain properties including effectiveness against a broad spectrum of microorganisms. They must also be non-toxic and non-irritating to both food

© Woodhead Publishing Limited, 2012

products and consumers, and cost-effective. However, some of the methods currently in use have been found to be unacceptable due to factors such as chemical residues, limited effectiveness, discoloration of food, and high cost. Consequently, both food providers and consumers increasingly demand novel sanitation technologies that are suitable for minimally processed foods, i.e. which provide effective microbial decontamination while maintaining the freshness and wholesomeness of the product.

Electrolyzed oxidizing water (EO water or EOW) is an emerging decontamination technology that has attracted much attention in recent years. EO water is produced by electrolyzing soft tap water containing sodium chloride in an electrolysis chamber. It is not only an effective solution for the inactivation of microorganisms, but is also user- and environmentally-friendly; the on-site generation process needs only electricity and table salt, which eliminates the potential hazards and costs of transportation and storage. This chapter provides an overview of the production, properties, and current applications of EO water, as well as potential future trends.

There are several academic researchers and professional and industry consortia currently interested in EOW for diverse applications. Industrial-scale electrolysis units sold for institutional disinfection and municipal water treatment are commonly known as bleach generators. The use of these chlorine generators avoids the need to ship and store chlorine gas, as well as the weight penalty of shipping prepared bleach solutions.

The field of electrochemical activation (ECA) technology has existed for more than a century. In 1907, the York & Pennsylvania Co. used an electrolytic bleach generator for its paper mill, which generated enough bleach for its paper industry. This electrolytic bleach is now referred to as the anolyte. But the equipment producing such high concentrations of anolyte were not consistent and the technological expertise took a very long time to develop.

Only recently, the US Environmental Protection Agency (EPA) has approved a few companies for producing anolyte solutions for industrial use, as their equipment provides an anolyte of the consistent quality needed to meet and pass all the various EPA product registration tests. The EPA has reviewed the data related to EcaFlo Anolyte, a brand of electrolyzed water developed by Integrated Environmental Technologies, Ltd ('IET'), Little River, SC, USA. After several independent laboratory tests, EcaFlo Anolyte is the first anolyte solution that is registered by the US EPA as a broad spectrum disinfectant. A broad spectrum disinfectant is a product generally approved for use in hospitals, universities, public school systems, medical and veterinary schools, cleaning services, athletic departments, veterinary clinics, medical research labs, state, county, city and federal government offices, and professional sports teams (US EPA, 2010).

© Woodhead Publishing Limited, 2012

## 19.2 Electrolyzed oxidizing water (EOW): principles and technology

### 19.2.1 Electrolysis and EOW production

The concept of EOW was originally conceived in Russia and used for pre-surgical preparation of surgeons' hands (Nikitin and Vinnik, 1965). In 1992, the EOW system was developed in Japan, a country that consumes lots of raw food, and has been applied for food safety due to its strong bactericidal effects on most pathogenic bacteria (Shimizu and Hurusawa, 1992). EOW is produced through electrodialysis, which is a membrane separation process driven by an electrical current. More specifically, electrodialysis involves the dissociation of an ionic solution using a direct current (DC) to attract negatively charged ions to the anode, and positively charged ions to the cathode (American Water Works Association, 1995). During production, a relatively weak sodium chloride solution is passed through an electrolytic chamber, where the anode and cathode are separated by a membrane (Fig. 19.1). When a current is passed through this chamber, the solution is allowed to dissociate into two streams; acidic and alkaline EOW. Negatively charged ions of the salt and water, $Cl^-$ and $OH^-$ are attracted to the anode where they bind with dissociated water molecules ($H^+$ and $O_2$) to form HCl, HOCl, $Cl_2$, $OCl^-$, and $O_2$ (Hsu, 2005). Thus, the acidic portion of EOW (acidic electrolyzed oxidizing water; acidic EOW) comprises hypochlorous acid and is produced at the anode. Similarly, the alkaline solution (alkaline electrolyzed oxidizing water; alkaline EOW) is produced at the cathode, where water dissociates

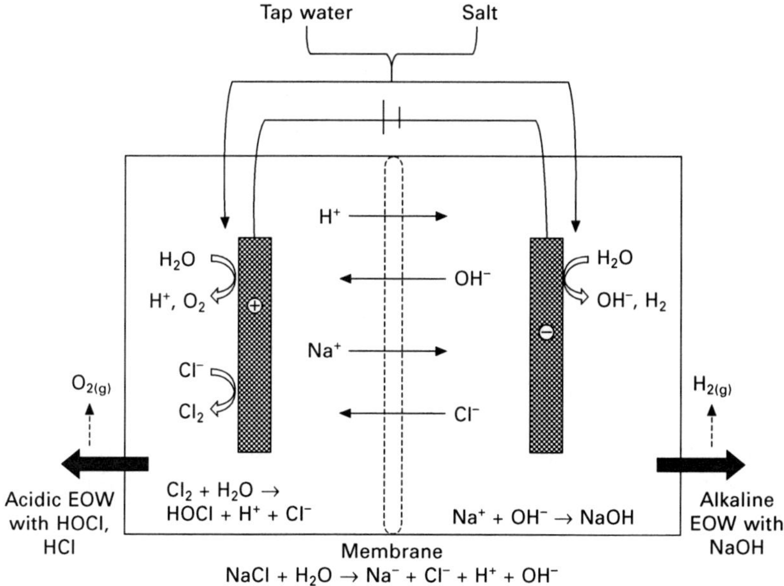


**Fig. 19.1** Schematic of electrolyzed oxidizing water (EOW) generation system.

© Woodhead Publishing Limited, 2012

into hydroxyl ions which later bind with sodium to form sodium hydroxide and hydrogen gas.

The production conditions of EOW affect the properties of the solutions. Factors such as flow rate, water temperature, and salt concentration have been investigated by Hsu (2003, 2005). The effects of these factors on the oxidation reduction potential or ORP, total residual chlorine, dissolved oxygen, and electrical conductivity of acidic EOW were studied and will be covered in this chapter.

### 19.2.2 EOW properties

EOW consists of two solutions with different uses, properties, and modes of action. Alkaline EOW has a pH of greater than 11 and a strong reducing capacity which makes it an ideal solution for cleaning (Hsu, 2005; Walker *et al.*, 2005a). On the other hand, acidic EOW is an ideal biocide with a high oxidation-reduction potential, low pH, and moderate levels of free chlorine (Kim *et al.*, 2000a; Park *et al.*, 2002a).

*Alkaline EOW*

The alkaline portion of EOW is generated from the cathode and consists of sodium hydroxide ions giving it a basic pH (10.0–11.5) and a low ORP (–800 to –900 mV). Due to the presence of sodium hydroxide and negative ORP, alkaline EOW can be used as detergent replacement (Fabrizio *et al.*, 2002). The strong reducing ORP also make alkaline EOW a candidate for the reduction of free radicals (Al-Haq *et al.*, 2005).

*Acidic EOW*

The acidic EOW had a low pH (2.5–3.5), high ORP (1000–1200 mV), high dissolved oxygen and free chlorine (30–90 ppm) depending on the amperage used for its production. Higher amperage results in lower pH and higher ORP and free chlorine concentration. For example, 14 amp amperage yields acidic EOW with a pH of 2.6, ORP of 1150 mV, and about 90 ppm free chlorine (Kim *et al.*, 2001; Sharma and Demirci, 2003).

*Neutralized EOW*

Neutralized EOW is produced at the anode and mixed with hydroxide ions or by using a single-cell chamber (Al-Haq *et al.*, 2005). The pH of neutralized EOW is approximately 7 with an ORP of 700 mV (Yang *et al.*, 2003; Deza *et al.*, 2007). Although neutralized EOW has different pH and ORP levels, the biocidal elements of neutralized EOW are similar to that of acidic EOW, but also contain $HO_2$ and $O_2$. It is believed to be less corrosive and has longer shelf life compared with acidic EOW (Rahman *et al.*, 2010).

© Woodhead Publishing Limited, 2012

### 19.2.3 EOW generators

Many companies manufacture EOW generators. Figure 19.2 shows the typical EOW generator of Hoshizaki Electric Co. (Japan). The production capacity varies depending on the manufacturer and model. Commercial systems are

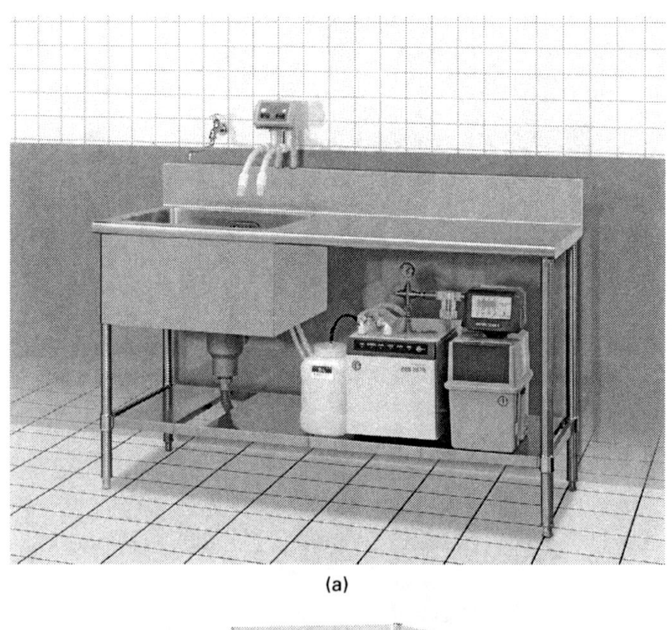

(a)

(b)

**Fig. 19.2** EOW generators (Hoshizaki Electric Co. Ltd, Japan).

© Woodhead Publishing Limited, 2012

also available for the production of other specific types of EOW. For example, the EOW generators made by the MIOX® Corporation (New Mexico, USA) are currently available for the production of neutralized EOW (Fig. 19.3).

(a)

(b)

**Fig. 19.3** Neutralized EOW generators (MIOX Corporation, Albuquerque, NM, USA).

© Woodhead Publishing Limited, 2012

Commercial EOW generators can be divided into three major types based on their automatic control systems (Hsu, 2003). The first type of EOW generators, made by the Hoshizaki Electric Co., allows the user to select the amperage and voltage, while the equipment adjusts the brine flow rate accordingly. The second type of EOW generators, made by the ARV® and the Amano® companies, allows the user to select the brine flow rate while the machine adjusts the voltage and amperage automatically. The third type of EOW generator, made by Nippon Intek® and Toyo® companies (Japan), allows the user to select a desired chlorine concentration level of EOW from a display panel, and the machine changes brine flow rate, amperage and voltage automatically (Hsu, 2003; Huang *et al.*, 2008). Some models can be controlled by Ethernet connection. The AQUAOX® Company (Florida, USA), for example, allows customers to set up desired pH and ORP, while they can remotely adjust the amperage, voltage, and brine flow rate (Fig. 19.4).

*Storage effect*

The changes in physical properties of EOW during storage have also been of great interest. Len *et al.* (2002) studied the effects of pH and chlorine on the storage of acidic EOW. They found that the loss of chlorine followed first rate kinetics, and decreased approximately 5-fold after 30 h of agitation. Acidic EOW lost 100% of chlorine due to evaporation after 100 h of open storage. Degradation of chlorine was not obvious when EOW was stored in a closed vessel. It took approximately 1400 h for the chlorine concentration to decrease by 60% in a closed container. Hsu and Kao (2004) also investigated the effects of storage on EOW properties. After 12 days of semi-open storage

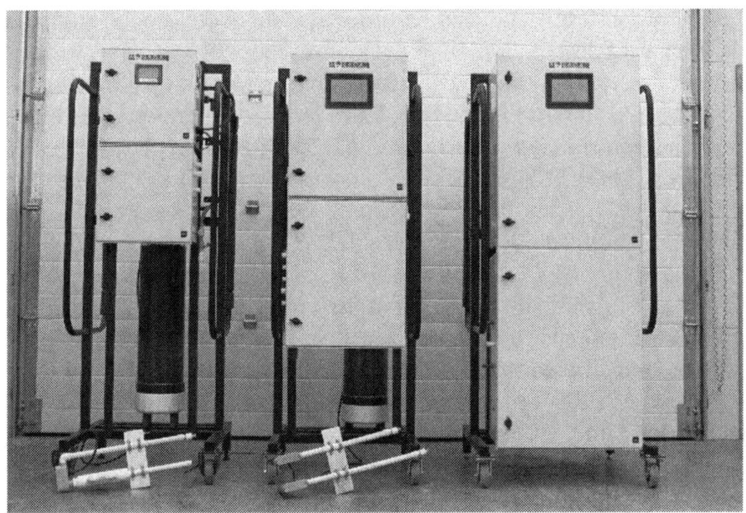

**Fig. 19.4** EOW generators (AQUAOX Corporation, Coconut Creek, FL, USA).

© Woodhead Publishing Limited, 2012

(i.e. glass bottles with four periodical openings of the screw caps), they found that the pH remained stable but the ORP dropped from 1150 mV to 1124 mV, and the concentration of chlorine decreased by 81%. They also noted that the length of storage time is not as important as whether EOW has been exposed to the atmosphere. Fabrizio and Cutter (2003) also reported that EOW stored at 4°C was more stable than that stored at 25°C.

### 19.2.4 Advantages of EOW

The main advantage of EOW lies in its safety. Although EOW is a strong acid, it is not corrosive to the skin, animal tissues, or mucous membranes. The cost of using EOW is much less (~1 US cent/L) compared with glutaraldehyde (1 USD/L) (Sakurai *et al.*, 2003). In addition, EOW is environmentally friendly, because it is only generated by the electrolysis of water and salt. The EOW becomes ordinary water when it contacts organic matter or is diluted by tap water or reverse osmosis (RO) water. Therefore, there is no need for special handling, storage or transportation for EOW (Al-Haq *et al.*, 2005). The limitations and challenges of the use of EOW will be discussed in Section 19.5.

## 19.3 Mechanisms of microbial inactivation in foods using electrolyzed oxidizing water (EOW)

### 19.3.1 Antimicrobial activity of acidic EOW

Acidic EOW works similarly to chlorine when used as an antimicrobial. This includes: the inhibition of glucose oxidation (mainly aldolase enzyme), disruption of protein synthesis, reactions with nucleic acids, unbalanced metabolism after destruction of key enzymes, induction of DNA lesions, and inhibition of oxygen uptake and oxidative phosphorylation (Marriott, 1999). The efficacy of acidic EOW has been linked to a combination of its high ORP, low pH, and chlorine content (Len *et al.*, 2000; Park *et al.*, 2002a; Sharma and Demirci, 2003). It has also been reported that acidic EOW may contain reactive oxygen species like $O_3$, ·O, and ·OH, which are known as strong oxidants and therefore good antimicrobials (Stan and Daeschel, 2003). ORP refers to the ability of a solution to oxidize or reduce and has, by some, been used as a measure of disinfection capability. Microorganisms require specific ranges of ORP in order to survive, usually between +200 mV to +800 mV (Jay, 2000), which clearly falls outside of the ranges of ORP produced by EOW. Kim *et al.* (2000a) reported that the ORP of a treatment solution is the primary factor, which affects microbial inactivation, and can be a much better indicator of a sanitizer's efficacy than free chlorine concentration. However, Len *et al.* (2000) found a high correlation ($r = 0.95$) between hypochlorous acid (HOCl) content of EOW and its antimicrobial activity. They suggested

© Woodhead Publishing Limited, 2012

that HOCl is the primary bactericidal agent due to its neutral charge, which would allow it to better penetrate the cell wall of a microorganism.

Other researchers have attributed the disinfection properties to a combination of the properties possessed by EOW. The bactericidal properties of acidic EOW were explained by Venkitanarayanan *et al.* (1999a) as a combination of high ORP, hypochlorous acid, and low pH. They hypothesized that the low pH of EOW sensitizes the outer membrane of the microorganism allowing for the more efficient transfer of hypochlorous acid into the cell. Kim *et al.* (2000b) found that the interaction of residual chlorine, ORP, and pH was necessary to achieve inactivation of foodborne pathogens. Liao *et al.* (2007) reported that high ORP could damage the outer and inner membrane of *E. coli* O157:H7 and leads to the inactivation mechanism. They proposed that high ORP could affect and damage the redox state of the glutathione disulfate-glutathione couple (GSSG/2GSH) first, and then penetrated the outer and inner membrane of *E. coli*, resulting in the necrosis of the cells.

### 19.3.2 Inactivation of suspended cells

The efficacy of EOW has been widely documented for the inactivation of microorganisms. Both bacteria and fungi have been found to be effectively inactivated by acidic EOW primarily due to its free chlorine concentration, ORP, temperature, and pH.

Venkitanarayanan *et al.* (1999a) evaluated the effects of acidic EOW at four different temperatures, 4, 23, 35, and 45°C on suspensions of *E. coli* O157:H7, *S.* Enteritidis, and *L. monocytogenes*. At a temperature of 4°C, complete inactivation of *E. coli* O157:H7 was observed after 10 min at a pH of 2.4, an ORP of 1150 mV, and a free chlorine concentration of 86 ppm. At a pH of 2.5 and free chlorine level of 83 ppm, complete inactivation was achieved after 10 min at 4°C. For *L. monocytogenes* complete inactivation at 4°C was achieved after 10 min at a lower level of free chlorine. Only 43 ppm of free chlorine was needed for complete inactivation at a pH of 2.6 and an ORP of 1160 mV. When the temperature was increased to 23°C, there was no observable difference in the amount of time needed for complete inactivation of the microorganism. However, when the temperature was increased to either 35 or 45°C, there was a noticeable difference in the time required for complete inactivation. At 35°C and free chlorine concentration of 84 ppm, only 2 min of exposure was needed to completely inactivate *E. coli* O157:H7. Four-minute exposure at a pH of 2.4 and free chlorine concentration of 79 ppm was required to completely inactivate *S.* Enteritidis, whereas only 2 min exposure was enough to inactivate *L. monocytogenes* at a pH of 2.5 and a free chlorine concentration of 73 ppm. At 45°C a 1 min exposure to acidic EOW at a pH of 2.4 and a free chlorine concentration of 85 ppm was required to inactivate *E. coli* O157:H7. A 3 min exposure to EOW at a pH of 2.4 and a free chlorine concentration of 79 ppm, was required for complete inactivation of *S.* Enteritidis as well as *L. monocytogenes*, but at

© Woodhead Publishing Limited, 2012

a free chlorine concentration of 73 ppm. Stan and Daeschel (2003) found a 6.6 $\log_{10}$ CFU/ml reduction of *Salmonella enterica* in an aqueous system. *Campylobacter jejuni* was successfully inactivated in solution by Park *et al.* (2002b) after 10 s of exposure to acidic EOW.

Park *et al.* (2004) evaluated the effects of acidic EOW in terms of both residual chlorine concentration and pH on the inactivation of *E. coli* O157:H7 and *L. monocytogenes*. Residual chlorine concentrations between 0.1 and 5.0 ppm and pHs of 3.0, 5.0, and 7.0 were assessed. They found that the level of residual chlorine of EOW maintained at 1.0 ppm or higher was needed for complete inactivation of these microorganisms. The residual chlorine was the more important characteristic in a pH range between 2.6 and 7.0 for inactivation. Sufficient residual chlorine (greater than 2.0 ppm) was needed to achieve complete inactivation of *E. coli* O157:H7 and *L. monocytogenes*. The individual properties of EOW have also been evaluated with respect to their impact on microorganisms. Kim *et al.* (2000a) assessed ORP, pH, and residual chlorine levels on *E. coli* O157:H7, *L. monocytogenes*, and *B. cereus*. In order to reduce the ORP readings of acidic EOW, they added iron and found that very little inactivation of the microorganisms was possible. The same results were observed when the chlorine of acidic EOW was reduced by the addition of neutralizing buffer. They found that the interaction of residual chlorine, ORP, and pH was necessary to achieve inactivation of foodborne pathogens. They also found that spore-forming bacteria were much more resistant to treatment with EOW than non-spore-forming bacteria. Complete inactivation of *E. coli* O157:H7 and *L. monocytogenes* was achieved after 10 s of exposure, while *B. cereus* spores were never completely inactivated. After 120 s of treatment with 56 ppm residual chlorine EOW, a reduction of 3.52 $\log_{10}$ CFU/ml of spores was achieved.

Fabrizio and Cutter (2003) found that acidic EOW was very effective in inactivating the food pathogens *L. monocytogenes* and *S.* Typhimurium. They compared acidic EOW, alkaline EOW, one-day stored acidic EOW, and chlorinated water (20 ppm). They found a reduction of 5.07 log CFU/ml of *S.* Typhimurium and a reduction of 6.76 log CFU/ml of *L. monocytogenes* by both acidic EOW and one-day stored acidic EOW treatments. One-day stored acidic EOW did not lose any biocidal activity. This was significantly higher than the reductions achieved using chlorinated water, which resulted in reductions of 0.03 and 0.17 log CFU/ml of *S.* Typhimurium and *L. monocytogenes*, respectively, after 15 min of treatment. Meanwhile, they found that alkaline EOW was not effective at reducing either microorganism.

The effectiveness of EOW has also been studied for fungi. Okull and LaBorde (2004) evaluated the effect of acidic EOW on *Penicillium expansum* in suspension. They found a 4.85 log CFU/ml reduction of *P. expansum* spores after 30 s of exposure to acidic EOW compared to a reduction of 3.59 and 3.88 log CFU/ml of spores using 100 and 200 ppm chlorine solutions, respectively. Buck *et al.* (2002) evaluated the effectiveness of acidic EOW on a wide variety of fungal species including *Botrytis*, *Monilinia*, *Curvularia*,

© Woodhead Publishing Limited, 2012

and *Helminthosporium*. All tested species of fungus either failed to germinate or had reduced levels of germination. They found that thin-walled species, like *Botrytis* and *Monilinia* were killed after 30 s exposure and that thicker-walled species, like *Curvularia*, required 2 min exposure to kill. They also found that the addition of surfactants eliminated the germicidal activity of EOW.

### 19.3.3 Inactivation of biofilms

A biofilm is defined as naturally occurring microbial cell layers that are irreversibly attached to solid surfaces. These microbial cells are embedded in an exopolysaccharide matrix and exhibit different growth and bioactivity compared to suspended cells. The exopolysaccharide matrix may also delay the uptake of antimicrobial agents by the embedded bacteria (Stewart *et al.*, 2001).

Relatively less data were reported on the efficiency of EOW for inactivating bacteria in biofilms. Kim *et al.* (2001) reported that acidic EOW reduced *L. monocytogenes* in biofilms on stainless steel surface by 9 log CFU/stainless steel coupon (2 by 5 cm) after 5 min treatment. The highest inactivation rate was reported within the first seconds of treatment. Thus, acidic EOW needed longer exposure times to reach the cells inside the biofilm. Ayebah and Hung (2005) reported that acidic EOW can reduce *L. monocytogenes* population by 4.3 to 5.2 log CFU/stainless steel coupon (2 by 5 cm), depending on the treatment time. A combined treatment of alkaline EOW followed by acidic EOW contributed an additional 0.3–1.2 log reduction. The reductions of *L. monocytogenes* population in biofilms obtained in the presence of organic matter have also been studied. Ayebah *et al.* (2006) reported that the bactericidal activity of acidic EOW decreased with the increased chicken serum covered on the biofilms, because free available chlorine in these solutions disappeared quickly. Chicken serum decreased the ORP and chlorine concentration but not the pH of acidic EOW.

## 19.4 Applications of electrolyzed oxidizing water (EOW) in the food industry

### 19.4.1 Surfaces and utensils

There has been great interest in using EOW as a sanitizer for food contact surfaces and utensils since they are important sources of direct and indirect contamination of food products with spoilage and pathogenic microorganisms. EOW has been applied for the treatment of many utensils, such as processing gloves, cutting boards, stainless steel, tiles and china (Hricova *et al.*, 2008). Ayebah and Hung (2005) attributed this to its 'high antimicrobial activity, low cost and ease of production and use'. Acidic EOW with antimicrobial

© Woodhead Publishing Limited, 2012

activity is therefore adopted, and alkaline EOW is recommended as sequential treatment to remove food residues and make the adherent bacteria more susceptible to acidic EOW (Ayebah *et al.*, 2006).

*E. coli* O157:H7 and *L. monocytogenes* were successfully inactivated on kitchen cutting boards with acidic EOW (Venkitanarayanan *et al.*, 1999b). They were able to completely inactivate both *E. coli* O157:H7 and *L. monocytogenes* at all treatment times (5–20 min) and temperatures (23–55°C).

The effectiveness of inactivating *Staphylococcus aureus* and *Enterobacter aerogenes* on a variety of surfaces has been evaluated by Park *et al.* (2002a). After examining glass, stainless steel, glazed ceramic tiles, unglazed ceramic tiles, and vitreous china, they found more than a 2 log CFU/cm$^2$ reduction of *E. aerogenes* on all surfaces and no surviving bacteria in the wash solutions after 5 min of acidic EOW treatment. Reductions of *S. aureus* were slightly lower, averaging about 1.73 log CFU/cm$^2$ on all surfaces, and again there were no surviving bacteria in the wash water after 5 min. However, when the materials were treated with agitated acidic EOW for 5 min at 50 rpm agitation, there were no survivors of both *E. aerogenes* and *S. aureus* on any of the surfaces. The increased efficacy of EOW with agitation was attributed to the efficient removal of cells from the surface and subsequent death in solution, increased penetration of the EOW, and more efficient interaction of chlorine and cells under agitated conditions.

Liu *et al.* (2006) investigated acidic EOW's effectiveness against *L. monocytogenes* on seafood processing equipment surfaces. Stainless steel, ceramic tile, and floor tile were contaminated with crabmeat residue and inoculated with *L. monocytogenes*. Log reductions of 2.33, 2.33, and 1.52 CFU/chip were achieved on stainless steel, ceramic tile, and floor tile, respectively. They also found that the use of acidic EOW can contribute more than 4.46 log CFU/cm$^2$ reductions of *L. monocytogenes* on seafood processing gloves (Liu and Su, 2006).

The use of neutralized EOW has also been evaluated for its efficacy at inactivating foodborne microorganisms on surfaces. Deza *et al.* (2005) evaluated it for the purpose of inactivating *E. coli*, *L. monocytogenes*, *Psuedomonas aeruginosa*, and *S. aureus* on stainless steel and glass. More than 7 log CFU/ml reductions of all microorganisms tested were found with neutralized EOW, which proved to be just as effective as treating these surfaces with sodium chlorite solution. However, neutralized EOW, as well as acidic EOW, has the advantages of being safe to handle and being produced on site. Guentzel *et al.* (2008) reported a spray treatment on food service surfaces with neutralized EOW containing 278–310 ppm of total residue chlorine (pH 6.38), which resulted in a 79–100% reduction in microbial growth of mixed cultures of *E. coli*, *L. monocytogenes*, *S. typhinurium*, *E. faecalis* and *S. aureus*. Handojo *et al.* (2009) investigated residual bacteria and different food types left on tableware items such as plates, forks, spoons, and knives after various washing and sanitization protocols. The results demonstrated that neutralized EOW is as effective (>6.9 log CFU/tableware reduction for

© Woodhead Publishing Limited, 2012

mechanical dishwashing and >5.4 log CFU/tableware reduction for manual dishwashing) as the other chemical sanitizers for food contact surface sanitization in both manual and mechanical ware washing.

### 19.4.2 Treatment of animal-based products

*Raw meat*

The Food Safety Inspection Services (FSIS) of the USDA has established a zero tolerance policy that does not allow any fecal contaminants on the surfaces of poultry carcasses (USDA, 1994). The efficacy of EOW in reducing pathogens on poultry has been investigated in recent years. Park *et al.* (2002a) reported that for chicken wings (50 ± 5 g) inoculated with *Campylobacter jejuni*, soaking in EOW (pH of 2.57, ORP of 1082 mV and free chlorine of 50 mg/L) with 100 rpm agitation for 30 min has achieved 3 log CFU/g reductions. Meanwhile, Fabrizio *et al.* (2002) reported that spray-washing with alkaline EOW followed by immersion in acidic EOW had a better effectiveness than spraying with acetic acid and trisodium phosphate followed by immersion in chlorine solution at the end of a 7-day refrigerated storage. Later, Kim *et al.* (2005) recommended spray-washing chicken carcasses with alkaline EOW to remove feces before defeathering and evisceration to reduce the potential cross-contamination. However, combining immersion with spray-washing did not significantly improve the bactericidal effect of EOW compared to the immersion-only treatment.

Fabrizio and Cutter (2004) examined the spray-washing with EOW for 15 s to disinfect pork bellies inoculated with feces containing *L. monocytogenes*, *S.* Typhimurium and *Campylobacter coli*. This study demonstrated that a 15 s spraying with EOW (pH of 2.4, ORP of 1160 mV, and free chlorine of 50 mg/L) had the ability to reduce the populations of *L. monocytogenes*, *S.* Typhimurium and *C. coli* (1.23, 1.67 and 1.81 CFU/mL, respectively) on the pork surfaces and inferred that longer contact times might strengthen the disinfection effectiveness. The hides of cattle are the primary source of pathogens such as *E. coli* O157:H7 to the beef processing industry. Bosilevac *et al.* (2005) reported that sequentially applied alkaline EOW and acidic EOW containing 70 mg/L of free chlorine at 60°C for 10 s can reduce aerobic bacteria counts by 3.5 log CFU/100 $cm^2$ and reduce *Enterobacteriaceae* counts by 4.3 log CFU/ 100 $cm^2$. Fabrizio and Cutter (2005) dipped or sprayed frankfurters and ham inoculated with *L. monocytogenes* with acidic EOW containing 45 mg/L of free chlorine (pH 2.3 and ORP 1150 mV) and/or alkaline EOW for 30 min while no significant difference ($p < 0.05$) between treatments on Hunter $L^*$, $a^*$, $b^*$ values for frankfurters and ham after 7-day storage at 4°C was found. The results indicated that EOW has no detrimental bleaching effects on the surface of tested read-to-eat meats.

Economic evaluation of reducing *Campylobacter* within the poultry meat chain (farm, processing plant, and consumer) has been investigated by Gellynck *et al.* (2008). They found that the decontamination of carcasses

© Woodhead Publishing Limited, 2012

with acidic EOW in the processing plant was the most efficient (cost–benefit ratio: 17.66) when compared with other evaluated measures (lactic acid (4.06), crust freezing (2.54), or irradiation (0.31)).

*Chicken egg*

Egg-borne salmonellosis, which is related to the consumption of raw or undercooked eggs, is a significant public health problem in the United States and European countries (CDCP, 2010). *Salmonella* Enteritidis is the main causative pathogen of salmonellosis related to eggs. The majority of *S.* Enteritidis outbreaks have been related to the consumption of raw or undercooked eggs or egg-containing foods. Other pathogenic microorganisms associated with eggshell surfaces are *Yersinia enterocolitica*, *E. coli* O157:H7, and *Campylobacter jejuni* (Erdogrul, 2004).

Several attempts have been made to apply EOW to fresh chicken eggs. The results demonstrated that the populations of *E. coli*, *S. aureus*, and *S.* Typhimurium on the surface of shell eggs can be reduced by 3–6 log CFU/egg with acidic EOW spray (Russell, 2003). Bialka *et al.* (2004) performed a pilot-scale study. EOW treatment and detergent-sanitizer treatment were compared by using *E. coil* K12. Log reductions of ≥2.98 and ≥2.91 CFU/g of shell were found, respectively. The results indicate that EO water has the potential to be used as a sanitizing agent for the egg washing process. Immersion of eggs in acidic EOW for 5 min (100 rpm of agitation) reduced *L. monocytogenes* and *S.* Enteritidis by 3.7 to 2.3 log CFU/egg, respectively (Park *et al.*, 2005). The effect of slightly acidic electrolyzed water (SA EOW) for inactivation of *S.* Enteritidis on the surface of shell eggs was also evaluated (Cao *et al.*, 2009). The results demonstrated that SA EOW was effective for inactivating *S.* Enteritidis. A reduction of 6.5 log CFU/g was observed for SA EOW containing 15 ppm of chlorine for 3 min. Another study also reported that the spraying of acidic EOW on the shell eggs did not negatively affect hatchability or chick quality (Fasenko *et al.*, 2009).

### 19.4.3 Fruits and vegetables

Fresh fruit and vegetables are essential for people around the world. These products, including pre-washed and pre-cut salads, are generally eaten without processing. The sanitation process is therefore of considerable importance. Much research has been concentrated on the use of EOW to inactivate pathogens on fresh produce. Alfalfa seeds and sprouts have received a lot of attention due to their notoriety in *E. coli* O157:H7 outbreaks and difficulty in decontamination (NACMCF, 1999). Kim *et al.* (2003) evaluated acidic EOW and two other sources of chlorine, chlorine water (84 ppm of active chlorine) and calcium chlorite (90 and 20,000 ppm active chlorine), on their ability to disinfect alfalfa seeds and sprouts inoculated with *Salmonella*. After 10 min of treatment with each solution, a 1.09 log CFU/g reduction was achieved using EOW which was not significantly different from the reductions with

© Woodhead Publishing Limited, 2012

the two chlorine treatments. Stan and Daeschel (2003) also evaluated the effect of acidic EOW on *Salmonella enterica*. They found 2.04 and 1.96 log CFU/g reductions for seeds treated for 15 and 60 min, respectively. It was also noted that no *Salmonella* were able to be recovered in the wash solution after acidic EOW treatment. Sharma and Demirci (2003) evaluated the use of acidic EOW to inactivate *E. coli* O157:H7 on alfalfa seeds and sprouts in conjunction with mechanical agitation. They achieved a maximum reduction of 96.9% (1.56 log CFU/g) on alfalfa seeds after treatment for 64 min and 99.8% (2.72 log CFU/g) on alfalfa sprouts after 64 min using a solution with 65 ppm free chlorine.

Bari *et al.* (2003a) evaluated the effect of EOW in conjunction with dry heat and sonication on *E. coli* O157:H7 on a variety of seeds, alfalfa, radish and mung bean. Acidic EOW was compared to distilled water, a chlorine solution (200 ppm), and a sanitizer called califresh-S. For all treatments acidic EOW resulted in the highest log reductions. When combined with dry heat and a solution temperature of 50°C, a 3.42 CFU/g reduction was reached; when this was combined with sonication the reduction of *E. coli* O157:H7 was increased to 4.56 log CFU/g on mung bean seeds. Reductions of *E. coli* O157:H7 on radish seeds were less than those on mung bean seeds. A maximum 1.94 log CFU/g reduction was achieved when the seeds were treated with dry heat, sonication, and hot EOW. The results for alfalfa seeds are comparable to those for mung bean seeds, with a maximum reduction of 4.29 log CFU/g resulting from a treatment that consisted of dry heat, sonication, and hot EOW.

Koseki *et al.* (2004a) assessed the efficacy of acidic EOW and alkaline EOW followed by acidic EOW for the purpose of decontaminating cucumbers and strawberries. When washing with acidic EOW alone for 10 min, reductions of 1.5 log CFU/g, 1.8 log CFU/g, and 2.1 log CFU/g were obtained for aerobic mesophilic bacteria, coliform bacteria, and fungi, respectively on cucumbers. When acidic EOW was combined with alkaline EOW each at 5 min treatment, the reduction in aerobic mesophilic bacteria was significantly increased up to 2 log CFU/g. However, the reductions in coliform bacteria and fungi were not significantly increased. After 10 min of acidic EOW treatment on strawberries, reductions of 1.6, 2.4, and 1.7 log CFU/g were obtained for aerobic mesophilic bacteria, coliform bacteria, and fungi, respectively. When combined with alkaline, EOW reductions were not significantly increased. Aerobic mesophilic bacteria were reduced by 1.1 log CFU/g and fungi by 0.8 log CFU/g. There was no change in coliform bacteria reduction since both treatments resulted in below detectable levels.

The efficacy of EOW has also been extensively studied for its application on lettuce. Park *et al.* (2001) evaluated the effects of acidic EOW on lettuce. When treated for 3 min, reductions of 2.41 and 2.65 log CFU/leaf were achieved for *E. coli* O157:H7 and *L. monocytogenes*, respectively. Koseki *et al.* (2004b) washed lettuce with both alkaline and acidic EOW, which resulted in a 2 CFU/g log reduction of aerobic bacteria. Koseki *et al.* (2004b)

© Woodhead Publishing Limited, 2012

also used acidic EOW in the form of ice to inactivate *L. monocytogenes* and *E. coli* O157:H7. Acidic EOW was frozen and allowed to emit chlorine gas at concentrations of 30, 70, 150, and 240 ppm. A maximum reduction of 2.0 log CFU/g for *E. coli* O157:H7 was achieved at a chlorine concentration of 150 ppm without any adverse quality effects. *L. monocytogenes* was reduced by 1.5 log CFU/g at 240 ppm of chlorine. Neutralized EOW was also used for reducing bacterial contamination on lettuce (Guentzel *et al.*, 2008; Abadias *et al.*, 2008; Vandekinderen *et al.*, 2009). The bactericidal activity of diluted neutralized EOW against tested pathogens is similar to that of chlorinated water (1–2 log reduction), and the use of neutralized EOW did not change the nutritional and sensory qualities of lettuce.

Tomatoes washed with acidic EOW showed significantly greater reductions in pathogenic microorganisms than 200 ppm chlorine water. Bari *et al.* (2003b) reported a 7.85 log CFU/tomato reductions of *E. coli* O157:H7, 7.46 log CFU/tomato reductions of *Salmonella*, and 7.54 log CFU/g reductions of *L. monocytogenes*. However, using chlorine water only resulted in a reduction of approximately 4.7 log CFU/tomato. Furthermore, no adverse quality effects were observed on tomatoes treated with acidic EOW.

Abbasi and Lazarovits (2006) also studied the germicidal effect of acidic EOW on bacterial and fungal pathogens. The results demonstrated that the viability of propagules of *Xanthomonas campestris* pv. *vesicatoria* (bacterial spot pathogen), *Streptomyces scabies* (potato scab pathogen), and *Fusarium oxysporum* f.sp. *lycopersici* (root rot pathogen) was significantly reduced 4–8 log CFU/g within 2 min of exposure to acidic EOW. E J Park *et al.* (2009) further studied the bactericidal effect of acidic EOW with the presence of organic materials. They concluded that the residual chlorine concentration and bactericidal activity of acidic EOW decreased proportional to the addition of bovine serum while the bacteria population of *E. coli* O157:H7, *S.* Typhimunium, and *L. monocytogenes* was still below the detection limit (0.7 log CFU/g) within 3 min. Deza *et al.* (2003) used neutralized EOW to decontaminate tomatoes inoculated with *E. coli* O157:H7, *S.* Enteritidis, and *L. monocytogenes*. After treatment for 60 s, reductions of 4.92, 4.30, and 4.74 log CFU/cm$^2$ were achieved on the surfaces for *E. coli* O157:H7, *Salmonella* Enteritidis, and *L. monocytogenes*, respectively.

Acidic EOW was also used to extend the shelf life of pears inoculated with *Botryosphaeria berengeriana* (Al-Haq *et al.*, 2002). They found that after immersion in acidic EOW and storage, disease incidence and severity were significantly decreased. Okull and LaBorde (2004) demonstrated substantial reduction in cross-contamination of spores of *Penicillium expansum* via wounded apples from decayed fruit or by direct addition of spores to a simulated dump tank. Full-strength (pH of 3.1, ORP of 1133 mV and free chlorine of 59.6 ppm) and 50% EOW (pH of 6, ORP of 895 mV and free chlorine of 33.5 ppm) decreased viable spore populations by greater than 4 and 2 log CFU/mL, respectively. Udompijitkul *et al.* (2007) also reported that acidic EOW can effectively reduce *E. coli* O157:H7 and *L. monocytogenes*

© Woodhead Publishing Limited, 2012

population on fresh strawberries by at least 2 log CFU/g after 10–15 min treatment. Other applications of EOW on sanitizing fresh produce, such as spinach, carrots, and almonds are summarized in Table 19.1.

### 19.4.4 Fish and seafood

Seafood is consumed all over the world and provides the world's prime source of protein (Huang *et al.*, 2006a). Food processing safety is therefore important, especially in the countries that consume lots of raw seafood materials as cuisine (i.e. Japan and Taiwan). Mahmoud *et al.* (2004) evaluated the efficiency of acidic EOW at pH of 2.2, ORP of 1.137 mV and free chlorine of 41 ppm on both carp skin and fillets. They reported reductions of 2.8 log CFU/cm$^2$ and 2.0 log CFU/g for the skin and fillets after 15 min treatment. Huang *et al.* (2006a) reported that acidic EOW can effectively reduce the population of *E. coli* O157:H7 and *V. parahaemolyticus* on tilapia. Around 0.8 and 2.6 log CFU/cm$^2$ reductions were obtained after 10 min treatment. Ozer and Demirci (2006) also reported that acidic EOW can reduce *E. coli* O157:H7 and *Vibrio vulnificus* on salmon fillets from about 0.4 to 1.1 log CFU/g depending on exposure time and temperature. Ren and Su (2006) evaluated the efficiency of acidic EOW on fresh oysters inoculated with *V. parahaemolyticus* and *V. vulnificus*. Their results showed that both *V. parahaemolyticus* and *V. vulnificus* were reduced about 1 log CFU/g after 4 h of exposure. Further treatment did not increase the reduction. Kim *et al.* (2006) introduced electrolyzed water ice (EW-ice; derived from acidic EOW) for preserving fresh pacific saury. Chemical analysis demonstrated that EW-ice retarded the formation of volatile basic nitrogen and thiobarbituric acid-reactive (TBAR) substances, and reduced the accumulation of alkaline compounds in the fresh fish in comparison with tap water ice. Sensory analysis also confirmed that EW-ice preserved more fresh saury, which had a 4–5 days longer shelf life. Abou-Taleb and Kawai (2008) reported that two commonly used essential oils can increase the shelf life if coated on semi-fried tuna slices after treatment with acidic EOW. Treatment of 1% eugenol and linalool significantly suppressed lipid oxidation on the semi-fried tuna. The shelf life of semi-fried tuna was extended to 15 days compared with 5 days for control samples when stored at 5°C.

The ice (PRO-SAN®) prepared by neutralized EOW and freezing solution, was used for tilapia whole fish and fillet preservation (Feliciano *et al.*, 2010). The results showed more than 2 and 3 log CFU/g reductions of *E. coli* K12, *Listeria innocua*, and *Pseudomonas putida* were obtained, respectively, which were significantly ($p < 0.05$) higher when compared with conventional ice.

### 19.4.5 Clean-in-place (CIP) cleaning of food processing equipments

CIP can be defined as the 'cleaning of complete items of plant or pipeline circuits without dismantling or opening of the equipment and with little or

© Woodhead Publishing Limited, 2012

**Table 19.1** Applications of EOW on vegetables

| Food Product | EOW Type | Exposure time (min.) | Indicator(s) | Reduction (log CFU) | EOW property | | | | Ref. |
|---|---|---|---|---|---|---|---|---|---|
| | | | | | pH | ORP (mV) | Temp. (°C) | Free chloride (ppm) | |
| Carrot | Neutralized EOW | 4 | Aerobic bacteria | 1.0/g | 6.8 | NA | 23 | 20 | Izumi (1999) |
| Bell pepper | Neutralized EOW | 4 | Aerobic bacteria | 1.0/g | 6.8 | NA | 23 | 20 | Izumi (1999) |
| Lemon | Acidic EOW | 15 | *E. coli* O157:H7 | 4.7/lemon | 3.0 | 1176 | 4 | 42–52 | Pangloli *et al.* (2009) |
| Cabbage | Alkaline EOW with 1% citric acid | 5 | Bacteria, yeast, and mold | 3.98 CFU(total)/g; 3.45 CFU (yeast and mold)/g | 11.0 | –830 | 50 | – | Rahman *et al.* (2010) |
| Cereal grain | Acidic EOW with 1% citric acid | 3 h | *B. cereus* | ~5.4 CFU (vegetative cell)/g; ~3.0 CFU (spore)/g | 2.5 | 1160 | 50 | 70 | E J Park *et al.* (2009) |
| | Alkaline EOW with 1% citric acid | 3 h | *B. cereus* | ~4.7 CFU (vegetative cell)/g; 2.8 CFU (spore)/g | NA | NA | 50 | – | E J Park *et al.* (2009) |
| Broccoli | Neutralized EOW | 3 | Coliform counts | No colifrom observed | 7.2 | NA | 23 | 100 | Das and Kim (2010) |
| Apple | Acidic EOW | 5 | *E. coli* O157:H7 | 2.0 CFU/g | 2.7 | 1150 | 4 | 70 | Wang *et al.* (2007) |
| Orange | Acidic EOW | 5 | *E. coli* O157:H7 | 4.5 CFU/g | 2.7 | 1150 | 4 | 74 | Wang *et al.* (2009) |
| Avocadoes | Acidic EOW | 5 | *E. coli* O157:H7 | 4.3 CFU/g | 2.7 | 1150 | 4 | 74 | Wang *et al.* (2009) |
| Brown rice | Acidic EOW | Wash | Aerobic bacteria | 4.82 CFU/g | 3.0 | 1079 | NA | 20 mM | Lu *et al.* (2010) |
| Almond | Acidic EOW with IR heat | 3 | *Salmonella* | 3.0 CFU/g | NA | NA | 23 | NA | Bari *et al.* (2009) |

© Woodhead Publishing Limited, 2012

no manual involvement on the part of the operator' (Bremer *et al.*, 2006). Under high turbulence and high velocity, jets are sprayed on the surfaces or cleaning solutions are circulated through the plant (Romney, 1990). Effective periodical CIP prevents the formation of biofilms on equipment surfaces. A large number of factors influence the effectiveness of cleaning which includes the nature and age of the biofilm, composition and concentration of the cleaning agent, time and temperature of cleaning, degree of turbulence of the cleaning solution; and the characteristics of the surface being cleaned (Stewart and Seiberling, 1996; Changani *et al.*, 1997; Lelievre *et al.*, 2002; Boulange-Petermann *et al.*, 2004). Sodium hydroxide (NaOH) at concentrations in the range of 1–5%, is the most common alkaline cleaner used in CIP (Flint *et al.*, 1997). Proteins and carbohydrates are effectively removed by the caustic alkali (Chisti, 1999). The CIP process usually involves a caustic wash, which is followed by an acid wash and then a sanitizer.

Stainless steel has been the most commonly used material for food contact surfaces in the food industry. Nitric acid with other acids or surfactants is the most common acid used in the industry. Acid wash removes traces of alkali products on the surfaces and adroitly removes the mineral scales left on the surfaces after due caustic chemical exposure while enhancing the cleaning (Bremer *et al.*, 2006). The bacteriostatic conditions provided by the acid wash delay the growth of any remaining microorganisms (White and Rabe, 1970; Stewart and Seiberling, 1996). The third most important part of CIP is the sanitizer. Some of the common sanitizers include ammonium compounds, anionic acids, iodophores, chlorine-based compounds, and peroxyacetic acid/hydrogen peroxide (Bremer *et al.*, 2006).

EOW is less dangerous, more cost-effective, and easily adaptable into CIP cleaning. EOW has been used as a disinfectant for food processing equipment. A sequential treatment with basic EOW and acidic EOW is generally recommended to replicate the use of chemical in CIP. The principle behind this recommendation is that basic EOW would effectively remove food residues and make the adherent bacteria more susceptible to acidic EOW, which in turn has a strong bactericidal effect. Moreover, acidic EOW is effective in preventing cross-contamination (Huang *et al.*, 2006a, 2006b; Kim *et al.*, 2003; Koseki *et al.*, 2004c; Park *et al.*, 2002a). Koseki *et al.* (2001, 2004c) found even higher efficiencies of this sequential treatment. Basic EOW helps destabilize and/or dissolve the glycocalyx formed by the bacterial biofilms and thus facilitates the penetration of the active acidic EOW components into the bacterial cells. Thus the acidic EOW replaces the need for the sanitizer in the third step. Instead, just water can be used for rinsing the pipelines and surfaces.

Using EOW to clean milking systems eliminates many of the dangers associated with storing and using the cleaning chemicals. Since EOW is produced as both an alkaline solution and an acidic solution, it should fit easily into the accepted three-step washing process. The effect of EOW in reducing bacteria in the pipelines of the milking system has been investigated

© Woodhead Publishing Limited, 2012

by Walker *et al.* (2005a, 2005b). Initially they evaluated the efficacy of EOW for cleaning surfaces made of five different materials, namely stainless steel sanitary pipe, PVC milk hose, rubber liners, rubber gasket material, and polysulfone plastic, that are associated with milking systems (Walker *et al.*, 2005a). They developed a response surface model to predict the optimal treatment time within the range of 5–20 min and optimal treatment temperatures within the range of 25–60°C, for the effective use of EO water for the CIP of milking systems. The temperatures investigated by them are lower than the temperatures currently used (70–75°C) for conventional dairy cleaners (NDPC, 1993), which could prove the CIP using EO water to be a more energy efficient process as well. After applying a linear regression model to the data from the five materials tested, the stainless steel specimens resulted in the most useful model. Using the stainless steel data, a regression model of the response surface design showed that the alkaline EO water treatment, the acid EOW treatment, and the treatment temperature were all significant parameters in the CIP of milking systems using EO water.

In a further study, a pilot-scale milking system with all the major components of a typical pipeline milking system was constructed (Walker *et al.*, 2005b). The milking unit was connected via a PVC milk hose and washing manifold to eight 10 ft (3.05 m) segments of 1.5 in. (3.94 cm) diameter sanitary pipe. Before the experiments, the system was adjusted to provide acceptable flow dynamics for CIP cleaning. A two-channel vacuum recorder (Reinemann, 1995) was used to monitor slug velocity for the initial adjustment of the system. The timing and admission rate of the air injector was adjusted to achieve slug flow with a slug velocity of approximately 30 ft/s (9.1 m/s) and a delay of approximately 6 s between each slug. A cocktail of *Pseudomonas fluorescens* B2, *Micrococcus luteus* (ATCC 10240), *Enterococcus faecalis* (ATCC 51299), and *Escherichia coli* (ATCC 25922) was used as the inoculum in this study to increase microbial population of raw milk. The efficacy of the EO treatment was evaluated for both short term (single cycle of milking and cleaning) and long term (10 consecutive cycles of milking and cleaning). To evaluate the surviving microorganisms, a sterile calcium alginate swab was used to swab an area of approximately 30 $cm^2$. For each method, eight locations on the milking system were swabbed: one on the inside of the plastic claw, one on the neck of the glass receiver, two on the inside of the rubber liners, two on the inside of the stainless steel elbows, and two on the inside of the stainless steel pipeline. All detectable bacteria from the non-porous milk contact surfaces were successfully removed by a 10 min wash with basic EOW at 60°C followed by a 10 min wash with acidic EOW at 60°C. Also ATP residue tests were negative. Thus EOW has shown to have the potential to be used as a cleaning and sanitizing agent for CIP of on-farm milking systems.

Turbulence in EO water used for CIP is induced by injecting air into the flow regime thereby creating a slug flow (Walker *et al.*, 2005b). Turbulent EO water (pH of 2.53, ORP of 1178 mV and chlorine of 53 mg/L) on stainless

© Woodhead Publishing Limited, 2012

steel for 5 min had reduced *Enterobacter aerogenes* and *Staphylococcus aureus* to <1 CFU/cm$^2$ (Park *et al.*, 2002a). Listeriosis, a potentially life-threatening foodborne illness, is caused by the pathogen *Listeria monocytogenes*. Major outbreaks of listeriosis have been associated with processed foods and the formation of *L. monocytogenes* biofilms in the processing environment (Carpentier and Chassaing, 2004). Appropriate CIP techniques could prevent this from happening.

*L. monocytogenes* biofilms on stainless steel surfaces were successfully inactivated with a combination of alkaline EOW and acidic EOW. The combination produced a 4.3–5.2 log CFU/cm$^2$ reduction. It is reported that the acidic EOW (pH of 2.42, ORP of 1077 mV and free chlorine of 50 mg/L) and modified EO water (pH of 6.12, ORP of 774 mV and free chlorine of 50 mg/L) did not have any adverse effect on stainless steel for a period of 8 days (Ayebah *et al.*, 2005).

EOW produced more than 5.0 log CFU/cm$^2$ reduction in the population of *Vibrio parahaemolyticus* within only 0.5 min (Chiu *et al.*, 2006). Moreover the use of neutral EOW for 1 min reduced the population of *E. coli* O157:H7, *L. monocytogenes*, *P. aeruginosa*, and *S. aureus* on stainless steel and glass by nearly 6 orders of magnitude. High reductions were also obtained for these pathogens on glass (Deza *et al.*, 2005). Kim *et al.* (2001) found that AEW reduced *L. monocytogenes* in biofilms on stainless steel to non-detectable levels within 5 min. In the presence of organic matter, *L. monocytogenes* was reduced by 1.5 to 2.3 orders of magnitude. Some studies suggest that there is a threshold concentration of chlorine beyond which increase in concentration does not enhance its effectiveness (Lee and Frank, 1991; Rossoni and Gaylarde, 2000). Therefore application parameters such as time duration, velocity of flow, and temperature should be studied before implementing the technology in industry to achieve the required level of cleaning depending on the specific application. Easy handling, low hazard risk, and low cost are additional advantages of using electrolyzed water in industrial CIP applications. Thus pending further research and approval of the appropriate authorities, CIP using EOW can be an economical and environmentally friendly alternative for use of harmful and dangerous chemicals.

### 19.4.6 Other applications

There are many other applications of EOW in industries including the pharmaceutical and biotechnological industries. For example, alkaline EOW (pH at 11.0) was reported to be effective against the $H_2O_2$-induced oxidative damage on DNA, RNA, and protein (Lee *et al.*, 2006). Hydrogen-enriched neutralized EOW (dissolved hydrogen: 0.9–1.1 ppm; ORP: –150–80 mV) also demonstrated its ability to alleviate the oxidative stress. There were no reverse mutations, chromosome aberrations, and subchronic toxicity effects observed during the analysis (Saitoh *et al.*, 2010). The no-observable-adverse-effect level (NOAEL) for hydrogen-enriched neutralized EOW was greater than

© Woodhead Publishing Limited, 2012

20 mL/kg/day. Acidic EOW (pH of 2.7 and ORP was 1000–1100 mV) was also applied on peritoneal irrigation of experimental perforated peritonitis (Kubota *et al.*, 2009). There were no adverse effects of acidic EOW observed in the experimental group. The 5-day survival rates were 25% and 86% in group S (irrigated with saline) and group E (irrigated with strong acidic EOW), respectively.

## 19.5 Limitations and challenges to adoption of electrolyzed oxidizing water (EOW) technology

Since $Cl_2$ and $H_2$ gases are generated as by-products, good ventilation is required at the production site. Moreover, due to the low pH and chlorine content, it might cause pitting and minor corrosions of the container if the quality of stainless steel is not good (Huang *et al.*, 2008). EOW loses its ORP fairly quickly when exposed to the atmosphere; therefore, it should be stored in a closed container (Hsu and Kao, 2004; Huang *et al.*, 2008). The electrolysis process also needs to be monitored frequently to obtain the desired ORP if the auto-adjustment control is not applicable.

Electrolysis machines can be expensive since this technique has not been widely adopted. Some manufacturers selling water ionizers for home use claim the alkaline water provides health benefits (Callaway *et al.*, 2005), but these claims are not supported by scientific research and contradict basic aspects of chemistry and physiology (*Wounds magazine*, 2006).

## 19.6 Conclusions and future trends

EOW demonstrates a great potential to be used as a cleaning and sanitizing agent in the food industry. The generators produce EOWs that possess different properties and benefits. Acidic EOW works as a sanitizing agent due to its low pH, high ORP, and presence of chlorine in the form of hypochlorous acid, while alkaline EOW, with high pH and negative ORP, makes itself an attractive cleaning agent. Neutralized EOW also demonstrated similar biocidal function as acidic EOW. EOW has been successfully shown to inactivate a variety of pathogenic and spoilage microorganisms in solution and has been effectively used to decontaminate both food contact surfaces as well as a variety of foods. Its ease of production and handling, as well as its proven biocidal properties, make it an attractive product and a suitable alternative to hazardous sanitizers and cleaning compounds.

Therefore, a greater understanding of EOW deserves further research. First, the roles and interaction of ORP, pH, and free chlorine are not fully understood; second, the exact biocidal component of EOW is not completely studied and needs more investigation; third, the role of the reactive oxygen

© Woodhead Publishing Limited, 2012

species present in acidic EOW needs to be investigated; fourth, the effects and value of alkaline EOW have been overlooked by many researchers for both contact surface and food decontamination and should be further investigated; fifth, the long-term effect of EOW on processing equipment needs to be evaluated. Finally, the environmental effect of used EOW needs to be documented. EOW has a great potential to become an alternative tool in agricultural, food, and even pharmaceutical industries, as long as it complies with good manufacturing practice (GMP) and good hygiene practice (GHP).

## 19.7 Sources of further information and advice

When it comes to applications in the food industry, EPA approval is not enough. An approval from the Food and Drug Administration (FDA) is mandatory. The FDA is an operating division of the US Department of Health and Human Services (DHHS). While the responsibility of DHHS is to protect the overall health of Americans, the FDA has a more specialized role in the oversight of food, drugs, and related products. Also, though not mandatory, the approval of the state's USDA (United States Department of Agriculture) Directorate of Food Safety is recommended and its recommendation to the FDA can expedite the process.

Many EOW generator suppliers have been working on the approval for use in the food industry by the US Department of Agriculture (USDA), the Food and Drug Administration (FDA), and the US Environmental Protection Agency (EPA) (Park *et al.*, 2002a) while several advantages of the use of EOW and its applications have been addressed. Nonetheless for specific industrial use, there are several EOW generators in the market place. To name a few, Hoshizaki Inc. (Shimane, Japan), i-Chlor Inc. (Brainard, NE, USA), Aquaox LLC (West Palm Beach, FL, USA) are worth mentioning which, while still awaiting the EPA and FDA approval for industrial application, have received the US FDA permission for use in food services and restaurants and other research investigations. Internationally, equipment manufacturers like Envirolyte (Northland, New Zealand), Aquaeca (Saint Petersburg, Russia), SAFIC (Johannesburg, South Africa) supply equipment globally.

A complete list of Food Safety and Sanitation Resources is available at the National Agricultural Library website at http://www.nal.usda.gov/fnic/service/foodsafety.pdf. A detailed explanation of all the food sanitation rules is available at the Department of Human Services website at http://www.oregon.gov/DHS/ph/foodsafety/docs/foodsanitationrulesweb.pdf.

© Woodhead Publishing Limited, 2012

## 19.8 References

ABADIAS M, USALL J, OLIVEIRA M, ALEGRE I and VIÑAS I (2008), 'Efficacy of neutral electrolyzed water (NEW) for reducing microbial contamination on minimally processed vegetables', *Int J Food Microbiol*, 123, 151–158.

ABBASI P A and LAZAROVITS G (2006), 'Effect of acidic electrolyzed water on the viability of bacterial and fungal plant pathogens and on bacterial spot disease of tomato', *Canad J Microbiol*, 52, 915–923.

ABOU-TALEB M and KAWAI Y (2008), 'Shelf life of semifried tuna slices coated with essential oil compounds after treatment with anodic electrolyzed NaCl solution', *J Food Prot*, 71, 770–774.

AL-HAQ M I, SEO Y, OSHITA S and KAWAGOE Y (2002), 'Disinfection effects of electrolyzed oxidizing water on suppressing fruit rot of pear caused by *Botryosphaeria berengeriana*', *Food Res Int*, 35, 657–664.

AL-HAQ M I, SUGIYAMA J and ISOBE S (2005), 'Applications of electrolyzed water in agricultural and food industries', *Food Sci Technol Res*, 11, 135–140.

AMERICAN WATER WORKS ASSOCIATION (1995), Eletrodialysis and electrodialysis reversal. American Water Works Association, Denver, CO.

AYEBAH B and HUNG Y (2005), 'Electrolyzed water and its corrosiveness on various surface materials commonly found in food processing facilities', *J Food Proc Eng*, 28, 247–264.

AYEBAH B, HUNG Y C and FRANK J F (2005), 'Enhancing the bactericidal effect of electrolyzed oxidizing water on *Listeria monocytogenes* biofilms formed on stainless steel', *J Food Prot*, 68, 1375–1380.

AYEBAH B, HUNG Y C, KIM C and FRANK J F (2006), 'Efficacy of electrolyzed water in the inactivation of planktonic and biofilm *Listeria monocytogenes* in the presence of organic matter', *J Food Prot*, 69, 2143–2150.

BARI M L, NAZUKA E, SABINA Y, TODORIKI S and ISSHIKI K (2003a), 'Chemical and irradiation treatments for killing *Escherichia coli* O157:H7 on alfalfa, radish, and mung bean seeds', *J Food Prot*, 66, 767–774.

BARI M L, SABINA Y, ISOBE S, UEMRUA T and ISSHIKI K (2003b), 'Effectiveness of electrolyzed acidic water in killing *Escherichia coli* O157:H7, *Salmonella* Enteritidis, and *Listeria monocytogenes* on the surfaces of tomatoes', *J Food Prot*, 66, 542–548.

BARI M L, NEI D, SOTOME I, NISHINA I, ISOBE S and KAWAMOTO S (2009), 'Effectiveness of sanitizers, dry heat, hot water, and gas catalytic infrared heat treatments to inactivate *Salmonella* on almonds', *Foodborne Pathog Dis*, 6, 953–958.

BIALKA K L, DEMIRCI A, KNABEL S J, PATTERSON P H and PURI V M (2004), 'Efficacy of electrolyzed oxidizing water for the microbial safety and quality of eggs', *J Poult Sci*, 83, 2071–2078.

BOSILEVAC J M, SHACKELFOR S D, BRICHTA D M and KOOHMARAIE M (2005), 'Efficacy of ozonated and electrolyzed oxidizing waters to decontaminate hides of cattle before slaughter', *J Food Prot*, 68, 1393–1398.

BOULANGE-PETERMANN L, JULLIEN C, DUBOIS P E, BENEZECH T and FAILLE C (2004), 'Influence of surface chemistry on the hygienic status of industrial stainless steel', *Biofouling*, 20, 25–33.

BREMER P J, FILLERY S and MCQUILLAN J (2006), 'Laboratory scale clean-in-place (CIP) studies on the effectiveness of different caustic and acid wash steps on the removal of dairy biofilms', *Intl J Food Microbiol*, 106, 254–262.

BUCK J W, VAN IERSEL M W, OETTING R D and HUNG Y C (2002), '*In vitro* fungicidal acitivity of acidic electrolyzed oxidizing water', *Plant Dis*, 86, 278–281.

CALLAWAY C S, WU C, YEH J Y and SAALIA F K (2005), 'The evaluation of electrolysed water as an agent for reducing micro-organisms on vegetables', *Int J Food Sci Tech*, 40, 495–500.

CAO W, ZHU Z W, SHI Z X, WANG C Y and LI B M (2009), 'Efficiency of slightly acidic

© Woodhead Publishing Limited, 2012

electrolyzed water for inactivation of *Salmonella enteritidis* and its contaminated shell eggs', *Int J Food Microbiol*, 130, 88–93.

CARPENTIER B and CHASSAING D (2004), 'Interactions in biofilms between *Listeria monocytogenes* and resident microorganisms from food industry premises', *Int J Food Microbiol*, 97, 111–122.

CENTERS FOR DISEASE CONTROL and PREVENTION (2010), *Salmonella* Enteritidis. National Center for National Center for Zoonotic, Vector-Borne, and Enteric Diseases. Available at: http://www.cdc.gov/nczved/divisions/dfbmd/diseases/salmonella_enteritidis/(accessed 11 February 2011).

CHANGANI S D, BELMAR-BEINY M T and FRYER P J (1997), 'Engineering and chemical factors associated with fouling and cleaning in milk processing', *Exp Therm Fluid Sci*, 14, 392–406.

CHISTI Y (1999), Modern systems of plant cleaning, In *Encyclopedia of Food Microbiology*. London: Academic Press.

CHIU T H, DUAN J, LIU C and SU Y C (2006), 'Efficacy of electrolyzed oxidizing water in inactivating *Vibrio parahaemolyticus* on kitchen cutting boards and food contact surfaces', *Lett Appl Microbiol*, 43, 666–672.

DAS B K and KIM J G (2010), 'Microbial quality and safety of fresh-cut broccoli with different sanitizers and contact times', *J Microbiol Biotechnol*, 20, 363–369.

DEZA M A, ARAUJO M and GARRIDO M J (2003), 'Inactivation of *Escherichia coli* O157:H7, *Salmonella enteritidis*, and *Listeria monocytogenes* on the surface of tomoatoes by neutral electrolyzed water', *Lett Appl Microbiol*, 37, 482–487.

DEZA M A, ARAUJO M and GARRIDO M J (2005), 'Inactivation of *Escherichia coli*, *Listeria monocytogenes*, *Pseudomonas aeruinosa* and *Staphylococcus aureus* on stainless steel and glass surfaces by neutral electrolysed oxidizing water', *Lett Appl Microbiol*, 40, 341–346.

DEZA M A, ARAUJO M and GARRIDO M J (2007), 'Efficacy of neutral electrolyzed water to inactivate *Escherichia coli*, *Listeria monocytogenes*, *Pseudomonas aeruginos*, and *Straphylococcus aureus* on plastic and wooden kitchen cutting boards', *J Food Prot*, 70, 102–108.

ERDOGRUL O (2004), '*Listeria monocytogenes*, *Yersinia enterocolitica* and *Salmonella* Enteritidis in quail eggs', *Turk J Vet Anim Sci*, 28, 597–601.

FABRIZIO K A and CUTTER C N (2003), 'Stability of electrolyzed oxidizing water and its efficacy against cell suspensions of *Salmonella* Typhimurium and *Listeria monocytongenes*', *J Food Prot*, 66, 1379–1384.

FABRIZIO K A and CUTTER C N (2004), 'Comparison of electrolyzed oxidizing water with other antimicrobial interventions to reduce pathogens on fresh pork', *Meat Sci*, 68, 463–468.

FABRIZIO K A and CUTTER C N (2005), 'Application of electrolyzed oxidizing water to reduce *Listeria monocytogenes* on ready-to-eat meats', *Meat Sci*, 71, 327–333.

FABRIZIO K A, SHARMA R R, DEMIRCI A and CUTTER C N (2002), 'Comparison of electrolyzed oxidizing water with various antimicrobial interventions to reduce *Salmonella* on poultry', *Poult Sci*, 81, 1598–1605.

FASENKO G M, O'DEA CHRISTOPHER E E and MCMULLEN L M (2009), 'Spraying hatching eggs with electrolyzed oxidizing water reduces eggshell microbial load without compromising broiler production parameters', *Poult Sci*, 88, 1121–1127.

FELICIANO L, LEE J, LOPES J A and PASCALL M A (2010), 'Efficacy of sanitized ice in reducing bacterial load on fish fillet and in the water collected from the melted ice', *J Food Sci*, 75, 231–238.

FLINT S H, BREMER P J and BROOKS J D (1997), 'Biofilms in dairy manufacturing plant – description, current concerns and methods of control', *Biofouling*, 11, 81–97.

GELLYNCK X, MESSENS W, HALET D, GRIJSPEERDT K, HARTNETT E and VIAENE J (2008), 'Economics of reducing *Campylobacter* at different levels within the Belgian poultry meat chain', *J Food Prot*, 71, 479–485.

© Woodhead Publishing Limited, 2012

GUENTZEL J L, LAM K L, CALLAN M A, EMMONS S A and DUNHAM V L (2008), 'Inactivation of bacteria on spinach and lettuce using neutral electrolyzed oxidizing water', *Food Microbiol*, 25, 36–41.

HANDOJO A, ZHAI Y, FRANKEL G and PASCALL M A (2009), 'Measurement of adhesion strengths between various milk products on glass surfaces using contact angle measurement and atomic force microscopy', *J Food Eng*, 92, 305–312.

HRICOVA D, STEPHAN R and ZWEIFEL C (2008), 'Electrolyzed water and its application in the food industry', *J Food Prot*, 71, 1934–1947.

HSU S Y (2003), 'Effects of water flow rate, salt concentration and water temperature on efficiency of an electrolyzed oxidizing water generator', *J Food Eng*, 60, 469–473.

HSU S (2005), 'Effects of flow rate, temperature and salt concentration on chemical and physical properties of electrolyzed oxidizing water', *J Food Eng*, 66, 171–176.

HSU S and KAO H (2004), 'Effects of storage conditions on chemical and physical properties of electrolyzed oxidizing water', *J Food Eng*, 65, 465–471.

HUANG Y R, HSIEH H S, LIN S Y, LIN S J, HUNG Y C and HWANG D F (2006a), 'Application of electrolyzed water on the reduction of bacterial contamination for seafood', *Food Control*, 17, 987–993.

HUANG Y R, SHIAU C Y, HUNG Y C and HWANG D F (2006b), 'Change of hygienic quality and freshness in tuna treated with electrolyzed water and carbon monoxide gas during refrigerated and frozen storage', *J Food Sci*, 71, M127–M133.

HUANG Y R, HUNG Y C, HSU S Y, HUANG Y W and HWANG D F (2008), 'Application of electrolyzed water in the food industry, *Food Control*, 19, 329–345.

IZUMI H (1999), 'Electrolyzed water as a disinfectant for fresh-cut vegetables', *J Food Sci*, 64, 536–539.

JAY J M (2000), *Modern Food Microbiology*, 6th edn, Gaithersburg, MD: Aspen Publications.

KIM C, HUNG C Y and BRACKETT R E (2000a), 'Roles of oxidation-reduction potential in electrolyzed oxidizing and chemically modified water for the inactivation of food-related pathogens', *J Food Prot*, 63, 19–24.

KIM C, HUNG C Y and BRACKETT R E (2000b), 'Efficacy of electrolyzed oxidizing and chemically modified water on different types of foodborne pathogens', *Int J Food Microbiol*, 61, 199–207.

KIM C, HUNG Y C, BRACKETT R E and FRANK J F (2001), 'Inactivation of *Listeria monocytogenes* biofilms by electrolyzed oxidizing water', *J Food Process Pres*, 25, 91–100.

KIM C, HUNG C Y and BRACKETT R E (2003), 'Efficacy of electrolyzed oxidizing water in inactivating *Salmonella* on alfalfa seeds and sprouts', *J Food Prot*, 66, 208–214.

KIM C, HUNG Y C and RUSSELL S M (2005), 'Efficacy of electrolyzed water in the prevention and removal of fecal material attachment and its microbicidal effectiveness during simulated industrial poultry processing', *Poultry Sci*, 84, 1778–1784.

KIM W T, LIM Y S, HIN I S, PARK H, CHUNG D and SUZUKI T (2006), 'Use of electrolyzed water ice for preserving freshness of pacific saury (*Cololabis saira*)', *J Food Prot*, 69, 2199–2204.

KOSEKI S, YOSHIDA K, ISOBE S and ITOH K (2001), 'Decontamination of lettuce using acidic electrolyzed water', *J Food Prot*, 64, 652–658.

KOSEKI S, YOSHIDA K, ISOBE S and ITOH K (2004a), 'Efficacy of acidic electrolyzed water for microbial decontamination of cucumbers and strawberries', *J Food Prot*, 67, 1247–1251.

KOSEKI S, ISOBE S and ITOH K (2004b), 'Efficacy of acidic electrolyzed water ice for pathogen control on lettuce', *J Food Prot*, 67, 2544–2549.

KOSEKI S, YOSHIDA Y, KAMITANI K, ISOBE S and ITOH K (2004c), 'Effect of mild heat pre-treatment with alkaline electrolyzed water on the efficacy of acidic electrolyzed water against *Escherichia coli* O157:H7 and *Salmonella* on lettuce', *Food Microbiol*, 21, 559–566.

© Woodhead Publishing Limited, 2012

KUBOTA A, NOSE K, YONEKURA T, KOSUMI T, YAMAUCHI K and OYANAGI H (2009), 'Effect of electrolyzed strong acid water on peritoneal irrigation of experimental perforated peritonitis', *Surg Today*, 39, 514–517.

LEE M Y, KIM Y K, RYOO K K, LEE Y B and PARK E J (2006), 'Electrolyzed-reduced water protects against oxidative damage to DNA, RNA, and protein', *Appl Biochem Biotechnol*, 135, 133–144.

LEE S H and FRANK J F (1991), 'Inactivation of surface adherent *Listeria monocytogenes* by hypochlorite and heat', *J Food Prot*, 54, 4–6.

LELIEVRE C, LEGENTILHOMME P, GAUCHER C, LEGRAND J, FAILLE C and BENEZECH T (2002), 'Cleaning in place: effect of local wall shear stress variation on bacterial removal from stainless steel equipment', *Chem Eng Sci*, 57, 1287–1297.

LEN S, HUNG Y, ERICKSON M and KIM C (2000), 'Ultraviolet spectrophotometric characterization and bactericidal properties of electrolyzed oxidizing water as influenced by amperage and pH', *J Food Prot*, 63, 1534–1537.

LEN S, HUNG Y, CHUNG D, ANDERSON J L, ERICKSON M C and MORITA K (2002), 'Effects of storage conditions and pH on chlorine loss in electrolyzed oxidizing water', *J Agric Food Chem*, 50, 209–212.

LIAO L B, CHEN W M and XIAO X M (2007), 'The generation and inactivation mechanism of oxidation-reduction potential of electrolyzed oxidizing water', *J Food Eng*, 78, 1326–1332.

LIU C and SU Y C (2006), 'Efficiency of electrolyzed oxidizing water on reducing *Listeria monocytogenes* contamination on seafood processing surface', *Int J Food Microbiol*, 110, 149–154.

LIU C, DUAN J and SU Y (2006), 'Effects of electrolyzed oxidizing water on reducing *Listeria monocytogenes* contamination on seafood processing surfaces', *Int J Food Microbiol*, 106, 248–253.

LU Z H, ZHANG Y, LI L T, CURTIS B R, KONG X L, FULCHER G R, ZHANG G and CAO W (2010), 'Inhibition of microbial growth and enrichment of c-aminobutyric acid during germination of brown rice by electrolyzed oxidizing water', *J Food Prot*, 73, 483–487.

MAHMOUD B S M, YAMAZAKI K, MIYASHITA K, SHIN S and SUZUKI T (2004), 'Bacterial microflora of carp (*Cyprinus carpio*) and its shelf-life extension by essential oil compounds', *Food Microbiol*, 21, 657–666.

MARRIOTT N G (1999), *Principles of food sanitation*, 4th edn, Gaithersburg, MD: Aspen Publishers.

NACMCF (NATIONAL ADVISORY COMMITTEE ON MICROBIOLOGICAL CRITERIA FOR FOODS) (1999), Microbiological safety evaluations and recommendations on sprouted seeds. *Int J Food Microbiol*, 52, 123–153.

NIKITIN B A and VINNIK L A (1965), 'Pre-surgical preparation of surgeon's hands with the products of electrolysis of a 3% solution of sodium chloride', *Khirurgiia*, 41, 104–105.

NORTHEAST DAIRY PRACTICES COUNCIL (NDPC) (1993), *Guidelines for Effective Installation, Cleaning, and Sanitizing of Milking Systems*. Syracuse, NY: NDPC.

OKULL D O and LABORDE LF (2004), 'Activity of electrolyzed oxidizing water against *Penicillium expansum* in suspension and on wounded apples', *J Food Sci*, 69, M23–M27.

OZER N P and DEMIRCI A (2006), 'Electrolyzed oxidizing water treatment for decontamination of raw salmon inoculated with *Escherichia coli* O157:H7 and *Listeria monocytogenes* Scott A and response surface modeling', *J Food Eng*, 72, 234–241.

PANGLOLI P, HUNG Y C, BEUCHAT L R, KING C H and ZHAO Z H (2009), 'Reduction of *Escherichia coli* O157:H7 on produce by use of electrolyzed water under simulated food service operation conditions', *J Food Prot*, 72, 1854–1861.

PARK E J, ALEXANDER E, TAYLOR G A, COSTA R and KANG D H (2009), 'The decontaminative effects of acidic electrolyzed water for *Escherichia coli* O157:H7, *Salmonella*

© Woodhead Publishing Limited, 2012

Typhimurium, and *Listeria monocytogenes* on green onions and tomatoes with differing organic demands', *Food Microbiol*, 26, 386–390.

PARK H, HUNG Y C, DOYLE M P, EZEIKE G O I and KIM C (2001), 'Pathogen reduction and quality of lettuce treated with electrolyzed oxidizing and acidified chlorinated water', *J Food Sci*, 66, 1368–1372.

PARK H, HUNG Y C and KIM C (2002a), 'Effectiveness of electrolyzed water as a sanitizer for treating different surfaces', *J Food Prot*, 65, 1276–1280.

PARK H, HUNG Y C and BRACKETT R E (2002b), 'Antimicrobial effect of electrolyzed water for inactivating *Campylobacter jejuni* during poultry washing', *Int J Food Microbiol*, 72, 77–83.

PARK H, HUNG Y C and CHUNG D (2004), 'Effects of chlorine and pH on efficacy of electrolyzed water for inactivating *Escherichia coli* O157:H7 and *Listeria monocytogenes*', *Int J Food Microbiol*, 91, 13–18.

PARK H, HUNG Y C, LIN C and BRACKETT R E (2005), 'Efficacy of electrolyzed water in inactivating *Salmonella* Enteritidis and *Listeria monocytogenes* on shell eggs', *J Food Prot*, 68, 986–990.

RAHMAN S M E, JIN Y G and OH D H (2010), 'Combined effects of alkaline electrolyzed water and citric acid with mild heat to control microorganisms on cabbage', *J Food Sci*, 75, M111–115.

REINEMANN D J (1995), System design and performance testing for cleaning milking systems. In *Designing a Modern Milking Center: Parlors, Milking Systems, Management, and Economics*, 207–224. NRAES-73. Ithaca, NY: Natural Resource, Agriculture, and Engineering Service.

REN T and SU Y C (2006), 'Effects of electrolyzed oxidizing water on reducing *Vibrio parahaemolyticus* and *Vibrio vulnificus* in raw oysters', *J Food Prot*, 69, 1829–1834.

ROMNEY A (1990), *CIP: Cleaning in Place*. Cambridge: The Society for Dairy Technology.

ROSSONI E M and GAYLARDE C C (2000), 'Comparison of sodium hypochlorite and peracetic acid as sanitizing agents for stainless steel food processing surfaces using epifluorescence microscopy', *Int J Food Microbiol*, 61, 81–85.

RUSSELL S M (2003), 'The effect of electrolyzed oxidative water applied using electrostatic spraying on pathogenic and indicator bacteria on the surface of eggs', *Poult Sci*, 82, 158–162.

SAITOH Y, HARATA Y, MIZUHASHI F, NAKAJIMA M and MIWA N (2010), 'Biological safety of neutral-pH hydrogen-enriched electrolyzed water upon mutagenicity, genotoxicity and subchronic oral toxicity', *Toxicol Ind Health*, 26, 203–216.

SAKURAI Y, NAKATSU M, SATO Y and SATO K (2003), 'Endoscope contamination from HBV- and HCV-positive patients and evaluation of a cleaning/disinfecting method using strong acidic electrolyzed water', *Digest Endosc*, 15, 19–24.

SHARMA R R and DEMIRCI A (2003), 'Treatment of *Escherichia coli* O157:H7 inoculated alfalfa seeds and sprouts with electrolyzed oxidizing water', *Int J Food Microbiol*, 86, 231–237.

SHIMIZU Y and HURUSAWA T (1992), 'Antiviral, antibacterial, and antifungal actions of electrolyzed oxidizing water through electrolysis', *Dental J*, 37, 1055–1062.

STAN S D and DAESCHEL M A (2003), 'Reduction of *Salmonella* Enterica on alfalfa seeds with acidic electrolyzed oxidizing water and enhanced uptake of acidic electrolyzed oxidizing water into seeds by gas exchange', *J Food Prot*, 66, 2017–2022.

STEWART J C and SEIBERLING D A (1996), 'The secrets out: clean in place', *Chem Eng*, 103, 72–79.

STEWART P S, RAYNER J, ROE F and REES W M (2001), 'Biofilm penetration and disinfection efficacy of alkaline hypochlorite and chlorosulfamates', *J Appl Microbiol*, 91, 525–532.

UDOMPIJITKUL P, DAESCHEL M A and ZHAO Y (2007), 'Antimicrobial effect of electrolyzed

© Woodhead Publishing Limited, 2012

oxidizing water against *Escherichia coli* O157:H7 and *Listeria monocytogenes* on fresh strawberries (*Fragaria x ananassa*)', *J Food Sci*, 72, M397–M406.

UNITED STATES DEPARTMENT OF AGRICULTURE (1994), Enhanced poultry inspection. Proposed rule. Fed. Regist. 59: 35659.

US ENVIRONMENTAL PROTECTION AGENCY (US EPA) (2010), Office of Pesticide Programs, All Actively Registered Hospital Disinfectants Products (including ATP-tested products and those in need of testing), Published March 2010. http://www.epa.gov/oppad001/active-hospital-disinf.pdf

VANDEKINDEREN I, DEVLIEGHERE F, DE MEULENAER B, RAGAERT P and VAN CAMP J (2009), 'Optimization and evaluation of a decontamination step with peroxyacetic acid for fresh-cut produce', *Food Microbiol*, 8, 882–888.

VENKITANARAYANAN K S, EZIEKE G O, HUNG Y and DOYLE M P (1999a), 'Efficacy of electrolyzed oxidizing water for inactivating *Escherichia coli* O157:H7, *Salmonella* Enteritidis, and *Listeria monocytogenes*', *Appl Environ Microbiol*, 65, 4276–4279.

VENKITANARAYANAN K S, EZIEKE G O, HUNG Y and DOYLE M P (1999b), 'Inactivation of *Escherichia coli* O157:H7 and *Listeria monocytogenes* on plastic kitchen cutting boards by electrolyzed oxidizing water', *J Food Prot*, 62, 857–860.

WALKER S P, DEMIRCI A, GRAVES R E, SPENCER S B and ROBERTS R F (2005a), 'Response surface modeling for cleaning and disinfecting materials used in milking systems with electrolyzed oxidizing water', *Int J Dairy Technol*, 58, 65–73.

WALKER S P, DEMIRCI A, GRAVES R E, SPENCER S B and ROBERTS R F (2005b), 'Cleaning of a pipeline milking system using electrolyzed oxidizing water', *Trans ASAE*, 48, 1827–1833.

WANG H, FENG H and LUO Y (2007), 'Control of browning and microbial growth on fresh-cut apples by sequential treatment of sanitizers and calcium ascorbate', *J Food Sci*, 72, M1–M7.

WANG H, FENG H, LIANG W, LUO Y and MALYARCHUK V (2009), 'Effect of surface roughness on retention and removal of *Escherichia coli* O157:H7 on surfaces of selected fruits', *J Food Sci*, 74, E8–E15.

WHITE J C and RABE G O (1970), 'Evaluating the use of nitric acid as a detergent in dairy cleaned-in-place systems', *J Milk Food Technol*, 33, 25–28.

*Wound magazine* (2006), available at: http://www.oculusis.com/us/news/articles/WoundMagazine012706.pdf

YANG H, SWEM B L and LI Y (2003), 'The effect of pH on inactivation of pathogenic bacteria of fresh-cut lettuce by dipping treatment with electrolyzed water', *J Food Sci*, 68, 1013–1017.

© Woodhead Publishing Limited, 2012

# 20

# Organic acids and other chemical treatments for microbial decontamination of food

**A. Lianou and K. P. Koutsoumanis Aristotle University of Thessaloniki, Greece and J. N. Sofos, Colorado State University, USA**

**Abstract**: Decontamination of food products using organic acids and other chemical treatments has been and continues to be one of the most important interventions for controlling their microbiological safety and quality. This chapter first covers aspects pertinent to the principles and technology of decontamination with chemical agents, and then reviews food decontamination applications of chemical treatments, with a particular emphasis on organic acids, as well as information regarding their mode of action and effectiveness against spoilage and/or pathogenic bacteria. Additional topics discussed in the chapter include potential effects of chemical decontamination on food quality, concerns and risks other than food quality associated with this type of intervention, and regulatory aspects of its implementation.

**Key words**: chemical agents, food decontamination, microbiological safety and quality, organic acids.

## 20.1 Introduction

Aiming at enhancing food safety and reducing the incidence of foodborne diseases, numerous technologies have been developed for the control of foodborne pathogenic bacteria, utilizing both physical and chemical methods (Rajkovic *et al.*, 2010). Decontamination of food products using chemical treatments has been and continues to be one of the most important interventions for controlling their microbiological safety and quality. Organic acids constitute an inexpensive and effective means of reducing both the population levels and the prevalence of pathogenic bacteria, and as such they are frequently used in decontamination applications in various food commodities (Loretz

© Woodhead Publishing Limited, 2012

*et al.*, 2010, 2011a, 2011b; Ölmez and Kretzschmar, 2009; Rajkovic *et al.*, 2010; Smulders and Greer, 1998).

The present chapter first covers aspects pertinent to the principles and technology of chemical decontamination. It then reviews food decontamination applications of chemical treatments, with a particular emphasis on organic acids, as well as information on their effectiveness against spoilage and/or pathogenic bacteria, as evaluated in research studies and documented in the scientific literature. Information on the mode of action (i.e., mechanisms of microbial inactivation) of different chemical compounds utilized for the purpose of food decontamination, as well as on applications of chemical treatments for specific food commodities (i.e., meat and poultry, seafood and produce) is also provided. Additional subjects discussed in the chapter include potential effects of chemical decontamination on food quality as well as other concerns and risks associated with this type of intervention. Finally, information on regulatory aspects of the implementation of organic acids and other chemical agents is presented, and future trends regarding chemical decontamination of foods are discussed.

## 20.2 Chemical decontamination of food

### 20.2.1 Principles

The insufficiency of the classical defence lines, relying exclusively on high-level hygiene in food production, harvesting and processing, to control bacterial agents of enteric infections was pointed out as early as four decades ago. The first decontamination studies were carried out on meat in the early 1970s, while the concept of 'intervention' was introduced by Mossel (1984) who proposed carcass decontamination using organic acids as an additional line of defence against meatborne pathogens. Aiming at reducing the prevalence and populations of bacterial contaminants, decontamination of foods results in an improvement in their microbiological status (Hugas and Tsigarida, 2008). Various decontamination technologies have been developed and found to be effective against microbial contaminants of foods; such interventions can be applied at different points of food processing (i.e., on raw materials, during processing or on processed final products), and include physical (e.g., hot water, steam vacuuming), chemical (e.g., organic acids, chlorine) and biological (e.g., bacteriophages, bacteriocins) interventions, or combinations of these intervention types (Bolder, 1997; Hugas and Tsigarida, 2008; Koutsoumanis *et al.*, 2004).

Chemical decontamination treatments involve the application of chemical compounds exhibiting antimicrobial activity at pre-determined points of food processing, and can be incorporated in hazard analysis and critical control points (HACCP) systems as hazard-reducing critical control points (CCP) (Bolton *et al.*, 2001; Hugas and Tsigarida, 2008). Food safety management systems integrating chemical decontamination interventions are expected

© Woodhead Publishing Limited, 2012

to be advantageous over non-intervention systems in achieving consistent reductions in bacterial contamination, as well as in relying less on human effort (Bolton *et al.*, 2001). Nevertheless, in some cases, it may be better for chemical decontamination treatments to constitute part of the prerequisite programs of HACCP and be implemented as good manufacturing practices (GMP) rather than in the form of CCP (Bolton *et al.*, 2001). In any case, and similarly to what is the case for any type of intervention, in order for chemical decontamination to be regarded as a positive measure for improving the microbiological safety and quality of food, it should be part of an integrated approach for the control of pathogenic and spoilage microorganisms, since it cannot, and should not be applied with the objective to, compensate for poor process hygiene or inappropriate manufacturing practices (Bolder, 1997; Huffman, 2002; Hugas and Tsigarida, 2008; Loretz *et al.*, 2010; Sofos, 2008). Moreover, chemical decontamination treatments should not exert adverse toxicological or other health effects on food workers during their application or on consumers, they should not negatively affect the organoleptic properties of foods, while environmental issues (e.g., quantities of generated waste water that need to be recycled) should also be taken into account (Bolder, 1997; Bolton *et al.*, 2001; Hugas and Tsigarida, 2008; Loretz *et al.*, 2011a; Sofos *et al.*, 1999). Finally, additional criteria that should be considered when selecting a decontamination treatment are its applicability in the production process and the setting up and operational costs (Bolder, 1997; Bolton *et al.*, 2001; Loretz *et al.*, 2011a).

Chemical decontamination interventions, aiming at improving the microbiological safety and quality of food, are usually implemented at the post-harvest level and at different points of the food processing chain. The efficacy of decontamination treatments, using organic acids or other chemical agents, will depend on the type of food tissue/product being treated, its initial microbial load and generally its microbial ecology, the type of bacterial contaminants to be inactivated, as well as on the ability of the latter to attach to the treated food and form biofilms (Chaiyakosa *et al.*, 2007; Hugas and Tsigarida, 2008; Kim and Marshall, 2000a, 2001, 2002; Rajkovic *et al.*, 2010). The decontamination efficacy of applied chemical treatments may also depend on the contamination conditions, since the latter can affect decisively the strength of bacterial attachment. It has been demonstrated that the longer the time interval between contamination and decontamination and the higher the temperature encountered during this interval, the more difficult it is to decontaminate fresh produce commodities due to firmer attachment of bacterial cells (Buchanan *et al.*, 1999; Ells and Hansen, 2006; Kondo *et al.*, 2006; Ölmez, 2010; Sapers, 2001; Singh *et al.*, 2002; Ukuku and Sapers, 2001; Ukuku *et al.*, 2001). Similarly, Cabedo *et al.* (1996) reported that the reductions of *Escherichia coli* achieved by chemical decontamination treatments (i.e., acetic acid, trisodium phosphate, and hydrogen peroxide) on beef carcass tissue samples decreased as the time lapse between exposure to contamination and application of the treatments increased.

© Woodhead Publishing Limited, 2012

There are factors associated with the chemical agents *per se* that can also have a significant impact on their decontamination efficacy. Such factors include the type of the applied chemical compound, its stability in solution and its concentration (Loretz *et al.*, 2011a, 2011b; Rajkovic *et al.*, 2010). In general, the bactericidal activity of organic acids has been shown to increase with increasing concentrations of their solutions (Anderson *et al.*, 1992a; Bautista *et al.*, 1997; Cutter and Siragusa, 1994; Greer and Dilts, 1992; Kotula and Thelappurate, 1994; Tamblyn and Conner, 1997). However, this is not always the case, and the relationship between the decontamination efficacy of organic acid solutions and their concentrations may depend on the technological parameters (e.g., the solution's temperature) of the applied intervention. For instance, according to the findings of Harris *et al.* (2006), the higher tested concentrations (i.e., 4%) of acetic and lactic acids were not associated with considerably higher pathogen reductions than those caused by the lower concentrations (i.e., 2%). Furthermore, Anderson and Marshall (1990b) reported that, overall, the concentration of lactic acid applied on beef tissues was an insignificant variable with regard to its decontamination efficacy at high temperatures, but became significant as the temperature of the solutions was lowered.

### 20.2.2 Technology

In addition to the basic principles of decontamination using chemical agents and the abovementioned factors upon which their bactericidal activity may depend, there are certain operational parameters pertaining to the technology of chemical food decontamination that need to be considered. Operational factors of chemical decontamination treatments that may affect their efficacy and should, therefore, be taken into account when designing and implementing such interventions include the stage of application during food processing, the method of application, the equipment design, the temperature and duration of application, the pH of the applied chemical solution, and the coverage and contact time (setting time) of tissue/product surface with the solution (Castillo *et al.*, 2001; Cutter and Siragusa, 1994; Escudero *et al.*, 1999; Hardin *et al.*, 1995; Kondo *et al.*, 2006; Kotula and Thelappurate, 1994; Sofos and Smith, 1998; Tamblyn *et al.*, 1997; Theron and Lues, 2007; Ukuku and Fett, 2004). The most important operational/process parameters of chemical decontamination that may affect considerably its efficacy are discussed below.

*Stage of application*

The stage of application of chemical decontamination treatments during food processing is a technological aspect that warrants special consideration. For instance, the efficacy of organic acids against *Salmonella enterica* serotype Typhimurium attached to broiler skin has been shown to be different for different application stages including simulated chiller, post-process dip and

© Woodhead Publishing Limited, 2012

scalder applications (Tamblyn and Conner, 1997; Tamblyn *et al.*, 1997). The stage during food processing that decontamination treatments are implemented reflects the time of exposure to bacterial contamination preceding the application of such treatments, with the latter time, as already mentioned previously, affecting the attachment of bacterial cells and, thus, the efficacy of the applied decontamination interventions (Sofos and Smith, 1998). Moreover, decontamination treatments applied at early stages of food processing may reduce microbial adherence at subsequent processing stages due to changes in the surface physical characteristics of the treated tissue/product. For example, such an observation was made by Dickson (1995) with regard to carcass decontamination, and for washing treatments applied at early stages (i.e., prior to evisceration) of the slaughtering process. Nevertheless, the sole application of decontamination treatments at early stages of processing may not be sufficient to assure food safety, because, even if the applied treatments are effective against pathogenic bacteria, re-contamination of food products at later stages of processing may take place. Thus, combined application of chemical interventions both at early and later stages of food processing may be needed. Furthermore, the stage of application of decontamination treatments should be decided in conjunction with the chemical compound applied as well as the microorganisms to be inactivated. For instance, in a recent review of the literature on the antibacterial activity of different decontamination treatments for poultry carcasses, it was noted that, with the exception of bacteria belonging to the family Enterobacteriaceae, the highest bacterial reductions were frequently observed after immersion chilling and using acetic acid (Loretz *et al.*, 2010).

*Method of application*

Decontamination technologies assessed for their efficacy, and with some of them being practiced commercially, include application of chemical solutions via immersion or spraying, depending on the food commodity being treated. However, research findings regarding the method of application of decontamination treatments have not been conclusive; the decontamination efficacy of organic acids and other chemicals applied by spraying has been demonstrated as superior, equivalent or inferior to that of the same agents applied as dipping solutions, depending on the chemical agent and the type of microbial contamination (Okolocha and Ellerbroek, 2005; Sakhare *et al.*, 1999; Sinhamahapatra *et al.*, 2004).

In addition to conventional methods of application of chemical treatments (i.e., immersion and spraying), research findings demonstrate that alternative decontamination technologies also hold promise for future utilization by the food industry. Such promising decontamination technologies are vacuum infiltration, vapor-phase application, and surface pasteurization with decontamination solutions (Delaquis *et al.*, 1997; Deumier, 2004; Sapers, 2001). Vacuum infiltration has been shown to be an effective means of decontamination of various food commodities including produce and meat/

© Woodhead Publishing Limited, 2012

poultry products. This technology is believed to improve contact between the applied chemical agent and bacterial cells attached in inaccessible sites by removing gas or liquid barriers that block penetration of the agent (Sapers, 2001). Application of vacuum infiltration for the treatment of fresh produce items with chemical compounds (e.g., hydrogen peroxide, chlorine) has been shown to be effective in reducing bacterial populations without affecting the sensorial quality of these commodities (Sapers, 2001).

Furthermore, immersion of poultry skin in organic acid solutions under pulsed vacuum has been shown to improve their decontaminating effect, most likely by facilitating the infiltration of these solutions into the follicles and pores of this type of tissue (Deumier, 2004). The latter process could be implemented under industrial conditions by adapting its application with that of a vacuum tumbler, which is commonly used by meat/poultry product manufacturers (Deumier, 2004). Indeed, the decontamination effectiveness of vacuum tumbling of meat and poultry products has been specifically assessed during the last decade (Deumier, 2006; Pohlman *et al.*, 2002; Stivarius *et al.*, 2002b). Research data indicate that the use of vacuum tumbled (in organic acid solution) poultry in the manufacture of further processed poultry products could enhance the microbiological safety of these products without adversely affecting their sensory quality, and the tumbling speed has been identified as the most significant process parameter (Deumier, 2006).

Vapor-phase application of volatile chemical agents, such as hydrogen peroxide, has also been investigated and shown to be promising as an alternative means of application of decontamination agents in fresh fruits; however, the lengthy application of such treatments and the potential injury caused to some commodities constitute important limitations of this technology (Aharoni *et al.*, 1994; Sapers, 2001).

*Temperature and duration of application*

With reference to the attributes of a decontamination technique, temperature is one of the first parameters that needs to be decided. Cutter *et al.* (1997) concluded that temperature had no measurable effect on the decontamination efficacy of an acetic acid solution applied by spray washing on beef carcasses. However, a positive, up to a certain extent, correlation between temperature of the applied chemical solutions and their decontamination effectiveness has been shown to exist according to the findings of other researchers (Anderson and Marshall, 1990b; Anderson *et al.*, 1992a; Greer and Dilts, 1992; Xiong *et al.*, 1998b). For instance, microbial reductions on lean beef muscle samples dipped in solutions of organic acids (individual or mixtures) have been shown to increase with temperature of applied solutions (Anderson *et al.*, 1992a).

Application of organic acids at elevated temperatures (55°C) has been generally associated with significant microbial reductions and identified as an effective method for decontamination of beef carcass surfaces (Hardin *et al.*, 1995). More specifically, based on the findings of Hardin *et al.* (1995), lactic acid (2%, 55°C) reduced significantly the levels of *E. coli* O157:H7 inoculated

© Woodhead Publishing Limited, 2012

on beef carcass surfaces, demonstrating a higher effectiveness against this pathogen than did acetic acid of the same concentration and temperature. It seems that the most effective decontamination temperature is specific to the chemical agent applied, while potential negative effects of elevated temperatures on the overall decontamination procedure or the organoleptic attributes of treated products also need to be taken into account. According to the findings of a study assessing the decontamination efficacy of pre-chill chemical spraying against *S.* Typhimurium attached to chicken breast skins, the use of 0.1% cetylpyridinium chloride sprays above 55°C led to excessive foam formation, while spraying with 10% trisodium phosphate or 2% lactic acid at 70°C resulted in tissue swelling and irreversible discoloration (Xiong *et al.*, 1998b).

The relationship between the duration of chemical treatments and their efficacy against food microbial contaminants has been shown to be variable, with some chemical agents demonstrating a higher effectiveness when applied for longer periods, while for other chemical solutions the exposure time does not appear to affect significantly their performance (Kotula and Thelappurate, 1994; Riedel *et al.*, 2009; Xiong *et al.*, 1998b). In any case, the time of exposure to chemical treatments depends on the existing processing chain speeds and the application equipment used (Sofos and Smith, 1998).

*Spraying attributes*

When the application method of choice is spraying, then, in addition to temperature and time, spraying pressure also needs to be evaluated. A potential concern with increasing pressure, which anyhow has not been consistently associated with significantly enhanced decontamination effectiveness of chemical solutions (Bautista *et al.*, 1997; Li *et al.*, 1997; Xiong *et al.*, 1998b), is bacterial penetration into treated tissues (Anderson *et al.*, 1992b). When no appreciable effect of increasing spraying pressure on the efficacy of chemical treatments is observed, then its selection should be oriented towards minimization of chemical consumption (Xiong *et al.*, 1998b). Furthermore, the equipment used needs to meet some requirements in order for events of bacterial redistribution and spread, potentially occurring during spraying of chemical solutions, to be avoided. Although such events are not as likely in the case of decontamination treatments, whose target is not only the removal (as in spray-washing treatments applying solely water) but also the inactivation of microorganisms, the likelihood of their occurrence can be minimized through proper design of the spraying equipment used. The equipment's characteristics that may determine the extent of contamination redistribution during spraying include the type, number, distribution and positions of spraying nozzles, the spraying angle, as well as the overall design of the spraying chamber and the spraying system (Sofos and Smith, 1998).

© Woodhead Publishing Limited, 2012

*Wash water*

With particular reference to produce decontamination, the physicochemical properties of process wash water may also affect considerably its efficacy and should, therefore, be taken into account when designing its technological aspects; such properties are the pH, temperature, turbidity and presence of organic matter (Gil *et al.*, 2009). For instance, in the context of a standard procedure for washing lettuce leaves in tap water including 100 ppm of available free chlorine described by Adams *et al.* (1989), adjustment of the solution's pH from 9.0 to 4.5–5.0 using inorganic and organic acids resulted in an increase in the microbicidal effect of the hypochlorite solution used. Moreover, although the decontamination efficacy of some chemical solutions may not be significantly affected by temperature (Escudero *et al.*, 1999; Sapers *et al.*, 1999), inappropriate temperature control of the process water used for produce washing in dump tanks, as commonly practiced in the fresh produce industry, may promote pathogen internalization (Buchanan *et al.*, 1999). Research findings suggest that the turbidity created by the presence of organic and inorganic material in wash water may impair the efficacy of applied disinfectants (Chaidez *et al.*, 2003). Indeed, some chemical agents, such as chlorine, have been shown to be neutralized in the presence of organic matter before manifesting their decontamination activity, and this is particularly true in the case of fresh-cut produce items where the juice released from the cut produce commodities contains very high levels of organic matter (Beuchat and Ryu, 1997; Beuchat *et al.*, 2004; Gonzalez *et al.*, 2004; Sapers, 2001; Zhang *et al.*, 2009). Thus, this problem should be taken into consideration when selecting a chemical agent for use in food decontamination, and if chlorine is the decontaminant of choice, then methods for maintaining its efficacy in the presence of organic matter should be in place. For example, monitoring the oxidation-reduction potential and the pH of process water and using their values to control hypochlorite addition and pH adjustment, have been proposed as means of improving the efficacy of chlorine when the latter is utilized in fresh produce decontamination (Sapers, 2001).

## 20.3 Types of chemical treatments

### 20.3.1 Organic acids

Organic acids are either naturally present in foods or chemically synthesized and added, directly or indirectly, to food products, with some of them constituting typical products of microbial fermentations (i.e., formed during fermentation of carbohydrates present in foods) (Stratford and Eklund, 2003; Theron and Lues, 2011). Organic acids have been utilized for a long time as food additives and preservatives aiming at preventing food deterioration and extending the shelf life of perishable food ingredients (Ricke, 2003). Some organic acids such as acetic, benzoic, citric, formic, lactic, and propionic acid, are well-established chemical preservative agents exhibiting a broad-

© Woodhead Publishing Limited, 2012

spectrum antimicrobial activity (Brul and Coote, 1999; Helander *et al.*, 1997; Siragusa, 1995; Theron and Lues, 2007). Acetic and lactic acids, in particular, are commonly used as inexpensive, environmentally friendly and effective interventions to reduce the levels and the prevalence of bacterial pathogens in food products (Rajkovic *et al.*, 2010; Siragusa, 1995). Due to their bactericidal properties, organic acids have constituted the basic active ingredients of antimicrobial products developed and evaluated for the reduction of foodborne pathogenic bacteria on food surfaces (Gonzalez *et al.*, 2004; Laury *et al.*, 2009; López-Gálvez *et al.*, 2009; Okolocha and Ellerbroek, 2005).

Organic acids are organic compounds with acidic properties and constitute a non-homologous series of chemical compounds, as they differ in the number of the carboxy groups, hydroxy groups and carbon–carbon double bonds in their molecules (Theron and Lues, 2011). Organic acids can be classified based on the following criteria: (i) the type of carbon chain (aliphatic, alicyclic, aromatic, and heterocyclic), (ii) being saturated or unsaturated, (iii) being substituted or non-substituted, and (iv) the number of functional groups (mono-, di-, tricarboxylic, etc.) (Theron and Lues, 2011). This group of chemical compounds primarily includes the saturated straight-chain monocarboxylic acids and their respective derivatives (i.e., unsaturated, hydroxylic, phenolic, and multicarboxylic versions) (Ricke, 2003). The acidity of an organic acid is determined mainly by the relative stability of the conjugate base of the molecule (Theron and Lues, 2011). The most common organic acids are the carboxylic acids, which are generally weak acids and whose acidity is associated with their carboxylic group –COOH (Ricke, 2003; Theron and Lues, 2011).

Organic acids exist in two basic forms, pure acids (lactic acid, propionic acid, acetic acid, citric acid, and benzoic acid) or buffered acids (calcium and sodium salts of propionic, acetic, citric, and benzoic acids), with the latter being safer to handle and less caustic to machinery (Theron and Lues, 2011). In contrast to what is the case for strong mineral acids, organic acids do not dissociate completely in water. Furthermore, although the lower molecular weight acids (e.g., formic and acetic acids) are water soluble, the high molecular weight compounds (e.g., benzoic acid) are insoluble in the molecular (neutral) form. Most organic acids, however, are very soluble in organic solvents (Theron and Lues, 2011). The lowest ($C_1$–$C_4$) monocarboxylic aliphatic acids are rather volatile liquids with a distinct pungency, whereas those acids containing more carbon atoms are of relatively oily substance and slightly water soluble. In comparison, dicarboxylic acids are colorless solids with melting points at approximately 100°C (Theron and Lues, 2011). The organic acids most commonly used in foods along with their basic physical and chemical properties, including their dissociation constants ($pK_a$), are presented in Table 20.1.

The decontamination efficacy of organic acids has been assessed in numerous research studies involving various foods and microorganisms,

© Woodhead Publishing Limited, 2012

**Table 20.1** Organic acids commonly used in foods

| Acid | Molecular formula[a] | Molecular weight $(gmol^{-1})$[a] | Appearance[a] | Melting point (°C)[a] | Boiling point (°C)[a] | p$K_a$ at 25°C[b] |
|---|---|---|---|---|---|---|
| Acetic | $C_2H_4O_2$ | 60.1 | Colorless liquid | 16.6 | 118.2 | 4.75 |
| Benzoic | $C_7H_6O_2$ | 122.1 | Colorless crystalline solid | 122.4 | 250.0 | 4.19 |
| Citric | $C_6H_8O_7$ | 192.1 (anhydrous)<br>210.1 (monohydrate) | White crystalline solid | 153 (anhydrous) | 175 (decomposes) | 3.14 (Step 1)<br>4.77 (Step 2)<br>6.39 (Step 3) |
| Formic | $CH_2O_2$ | 46.0 | Colorless liquid | 8.4 | 100.8 | 3.75 |
| Lactic | $C_3H_6O_3$ | 90.1 | L: hygroscopic crystalline<br>D: hygroscopic plates<br>DL: hygroscopic crystalline or syrupy | L: 53<br>D: 53<br>DL: 16.8 | 122 | 3.08 |
| Propionic | $C_3H_6O_2$ | 74.1 | Colorless liquid | –21 | 141 | 4.87 |

[a]As derived from Dawson *et al.* (1986).
[b]p$K_a$, dissociation constant; as derived from Lindsay (2008).

© Woodhead Publishing Limited, 2012

both spoilage and pathogenic. Information regarding the effectiveness (i.e., reductions in bacterial populations) of organic acid decontamination applications in meat, poultry, seafood, and produce is summarized in Tables 20.2–20.5. The quantitative data included in these tables are data documented in research studies either as explicit references in the text of the studies or as table contents; in order for arbitrary approximations to be avoided, studies presenting pertinent data solely in the form of graphs were not included. The bacterial reductions presented in the tables are immediate reductions (i.e., at most 60 min post-treatment), and refer to differences in bacterial populations between treated (with organic acids) and control samples, with the latter being, depending on the study, either untreated or water-treated samples.

As demonstrated by the data summarized in Tables 20.2–20.5, various organic acids (i.e., acetic, citric, fumaric, lactic, malic, mandelic, polylactic, propionic, and tartaric) have been investigated for their decontamination efficacy when applied in miscellaneous food products, and different quantitative and qualitative (i.e., general trends) data have been reported by different investigators. The diversity and, often, discrepancy of research findings can be attributed to differences pertinent to the decontamination technology applied, as well as to other parameters that may also affect the decontamination efficacy of organic acids, including the type of acid, the type of treated tissue (e.g., adipose or lean meat tissue, poultry skin or meat, etc.) and its pH and buffering capacity, the bacterial target and the initial bacterial population (Bell *et al.*, 1997; Cutter and Siragusa, 1994; Dickson, 1992; Greer and Dilts, 1992, 1995; Hardin *et al.*, 1995; Rajkovic *et al.*, 2010; Riedel *et al.*, 2009).

It seems that the effects of the type of organic acid being used and that of the tissue being treated on the decontamination efficacy of these chemical agents cannot be evaluated separately. This happens because, given the weak acid nature of most of these compounds, pH is regarded as a primary determinant of their effectiveness (Ricke, 2003). The equilibrium between the undissociated and dissociated states of a given organic acid and, thus, its antimicrobial activity as will be discussed in more detail later in the chapter, are pH-dependent. With the p$K_a$ of an acid representing the pH at which equal proportions of dissociated and undissociated molecules are present in solution, the antimicrobial activity of organic acids increases as the environmental pH approaches the p$K_a$ (Stratford and Eklund, 2003). For instance, given the above and that lactic acid is a stronger acid (i.e., with lower p$K_a$) than acetic acid (Table 20.1), the latter compound would be expected to be a more potent antimicrobial than the former in foods with a moderately low pH (i.e., pH 4.0–6.0). Nevertheless, this is partially counterbalanced in the case of lactic acid, since equimolar concentrations of this stronger acid produce a greater pH reduction and, thus, an increased fraction of its undissociated state (Adams and Hall, 1988). Indeed, there are research data on beef decontamination demonstrating a higher or similar

© Woodhead Publishing Limited, 2012

**Table 20.2** Applications of meat decontamination using organic acids

| Product | Agent (concentration) | Application[a] | Microorganism | Reduction (log cfu)[b] | Reference |
|---|---|---|---|---|---|
| **Fresh meat** | | | | | |
| Beef carcass | Acetic acid (1%) | SP, 25°C, 0.5 min | *Escherichia coli* | 3.0 $cm^{-2}$ | Bell *et al.*, 1997 |
| | | | *Listeria innocua* | 2.4 $cm^{-2}$ | |
| | | | *Salmonella* Wentworth | 3.2 $cm^{-2}$ | |
| | Acetic acid (1.5–3.0%) | SP, 32°C, 0.3 min | Aerobic bacteria | 1.3–2.0 $cm^{-2}$ | Dorsa *et al.*, 1997 |
| | | | *Escherichia coli* O157:H7 | >2.7 $cm^{-2}$ | |
| | Acetic acid (2%); after water washing | SP, 55°C, 0.2 min | *Escherichia coli* O157:H7 | 2.4–3.7 $cm^{-2}$ | Hardin *et al.*, 1995 |
| | | | *Salmonella* Typhimurium | 3.2–5.1 $cm^{-2}$ | |
| | Lactic acid (2%) | SP, ~42°C | Aerobic bacteria | 1.6/100 $cm^2$ | Bosilevac *et al.*, 2006 |
| | | | Enterobacteriaceae | 1.0/100 $cm^2$ | |
| | Lactic acid (1.5–3.0%) | SP, 32°C, 0.3 min | Aerobic bacteria | 1.3–2.0 $cm^{-2}$ | Dorsa *et al.*, 1997 |
| | | | *Escherichia coli* O157:H7 | >2.7 $cm^{-2}$ | |
| | | | *Listeria innocua* | 2.8–4.0 $cm^{-2}$ | |
| | Lactic acid (4%) | SP | Aerobic bacteria | ≥2.0 $cm^{-2}$ (distal) | Gill and Badoni, 2004 |
| | | | | ≤2.0 $cm^{-2}$ (medial) | |
| | Lactic acid (2%); after water washing | SP, 55°C, 0.2 min | *Escherichia coli* O157:H7 | 3.0–4.9 $cm^{-2}$ | Hardin *et al.*, 1995 |
| | | | *Salmonella* Typhimurium | 3.4–5.0 $cm^{-2}$ | |
| Beef meat pieces | Acetic acid (0.5–1.5%) | SP, 20 and 55°C | *Escherichia coli* O157:H7 | <0.1 $g^{-1}$ | Brackett *et al.*, 1994 |
| | Acetic acid (2%); after water washing | SP, 21°C, 0.2 min | *Escherichia coli* | 3.7 $cm^{-2}$ | Cabedo *et al.*, 1996 |
| | Acetic acid (0.6–1.2%) | IM, 1–2°C, 0.3–2 min | Aerobic bacteria | 0.5–0.8 | Kotula and Thelappurate, 1994 |

*(Continued)*

© Woodhead Publishing Limited, 2012

**Table 20.2** Continued

| Product | Agent (concentration) | Application[a] | Microorganism | Reduction (log cfu)[b] | Reference |
|---|---|---|---|---|---|
| | | | *Escherichia coli* | 0.5–0.7 | |
| | Acetic acid (1%) | IM, 55°C, 0.1 min | *Escherichia coli* O157:H7 | 0.8 $cm^{-2}$ | Podolak *et al.*, 1996 |
| | | | *Listeria monocytogenes* | 0.4 $cm^{-2}$ | |
| | Citric acid (0.5–1.5%) | SP, 20 and 55°C | *Escherichia coli* O157:H7 | <0.2 $g^{-1}$ | Brackett *et al.*, 1994 |
| | Fumaric acid (0.5–2.5%) | IM, 55°C, 0.1 min | *Escherichia coli* O157:H7 | 0.6–1.1 $cm^{-2}$ | Podolak *et al.*, 1996 |
| | | | *Listeria monocytogenes* | 0.5–0.8 $cm^{-2}$ | |
| | Lactic acid (2%) | IM, 23°C, 5 min | *Listeria monocytogenes* | 1.6 $cm^{-2}$ | Ariyapitipun *et al.*, 2000 |
| | Lactic acid (0.5–1.5%) | SP, 20 and 55°C | *Escherichia coli* O157:H7 | <0.3 $g^{-1}$ | Brackett *et al.*, 1994 |
| | Lactic acid (2–4%) | SP | Aerobic bacteria | ≥1.0 $cm^{-2}$ (2%)<br>≥1.5 $cm^{-2}$ (4%) | Gill and Badoni, 2004 |
| | Lactic acid (0.6–1.2%) | IM, 1–2°C, 0.3–2 min | Aerobic bacteria | 0.4 | Kotula and Thelappurate, 1994 |
| | | | *Escherichia coli* | 0.2–0.4 | |
| | Lactic acid (2%) | IM, 55°C, 0.5 min | *Listeria monocytogenes* | 1.4 $cm^{-2}$ | Koutsoumanis *et al.*, 2004 |
| | Lactic acid (1–2%) | IM, 24–25°C, 0.3 min | *Listeria monocytogenes* | 0.7–1.1 $g^{-1}$ | Özdemir *et al.*, 2006a |
| | | | *Salmonella* Typhimurium | 0.05–0.70 $g^{-1}$ | |
| | Lactic acid (2%) | IM, 35°C, 0.3 min | *Listeria monocytogenes* | 0.8 $g^{-1}$ | Özdemir *et al.*, 2006b |
| | | | *Staphylococcus aureus* | 1.2 $g^{-1}$ | |
| | Lactic acid (1%) | IM, 55°C, 0.1 min | *Escherichia coli* O157:H7 | 0.8 $cm^{-2}$ | Podolak *et al.*, 1996 |
| | | | *Listeria monocytogenes* | 0.5 $cm^{-2}$ | |
| | Polylactic acid (low-molecular-weight; 2%) | IM, 23°C, 5 min | *Listeria monocytogenes* | 1.2 $cm^{-2}$ | Ariyapitipun *et al.*, 2000 |

© Woodhead Publishing Limited, 2012

| | | | | | |
|---|---|---|---|---|---|
| Beef trim | Acetic acid (2–4%) | SP, ambient temp. | *Escherichia coli* O157:H7 | 1.5–2.0 $g^{-1}$ | Harris *et al*., 2006 |
| | | | *Salmonella* Typhimurium | 1.5–2.0 $g^{-1}$ | |
| | Lactic acid (2–4%) | SP, ambient temp. | *Escherichia coli* O157:H7 | 1.5–2.0 $g^{-1}$ | |
| | | | *Salmonella* Typhimurium | 1.5–2.0 $g^{-1}$ | |
| | Lactic acid (2%) | SP, 15°C | Coliforms (fecal) | 0.7–1.1 $cm^{-2}$ | Kang *et al*., 2001 |
| Beef variety meats | Acetic acid (2%) | IM, 50°C, 0.2 min | Aerobic bacteria | 0.5–2.6 $g^{-1}$ | Delmore Jr *et al*., 2000 |
| | | | Coliforms | 0.5–2.4 $g^{-1}$ | |
| | | | *Escherichia coli* | –0.1–2.3 $g^{-1}$ | |
| | | SP, 40–50°C, 0.2 min | Aerobic bacteria | 0.3–1.6 $g^{-1}$ | |
| | | | Coliforms | 0.5–2.5 $g^{-1}$ | |
| | | | *Escherichia coli* | –0.1–2.1 $g^{-1}$ | |
| | Lactic acid (2%) | IM, 50°C, 0.2 min | Aerobic bacteria | 0.7–1.6 $g^{-1}$ | |
| | | | Coliforms | 0.5–2.2 $g^{-1}$ | |
| | | | *Escherichia coli* | –0.5–2.0 $g^{-1}$ | |
| | | SP, 40–50°C, 0.2 min | Aerobic bacteria | 0.4–1.7 $g^{-1}$ | |
| | | | Coliforms | –0.1–2.5 $g^{-1}$ | |
| | | | *Escherichia coli* | 0.0–1.8 $g^{-1}$ | |
| Pork carcass | Acetic acid (1.8%) | SP | Aerobic bacteria | 1.0–1.5 $cm^{-2}$ | Eggenberger-Solorzano *et al*., 2002 |
| | Lactic acid (1–5%) | SP, 55°C, 1.5 min | Aerobic bacteria | 0.3–1.3 $cm^{-2}$ (mesophilic) 0.1–0.9 $cm^{-2}$ (psychrotrophic) | van Netten *et al*., 1997a |
| | | | Enterobacteriaceae | 0.2–>1.1 $cm^{-2}$ | |

(Continued)

© Woodhead Publishing Limited, 2012

**Table 20.2** Continued

| Product | Agent (concentration) | Application[a] | Microorganism | Reduction (log cfu)[b] | Reference |
|---|---|---|---|---|---|
| Pork meat pieces | Acetic acid (1–3%) | IM, 4°C, 1 min | *Escherichia coli* | 0.5–0.8 $cm^{-2}$ | Choi *et al.*, 2009 |
| | | | *Escherichia coli* O157:H7 | 0.5–0.9 $cm^{-2}$ | |
| | | | *Listeria monocytogenes* | 0.9–1.1 $cm^{-2}$ | |
| | | | *Salmonella* Typhimurium | 0.5–0.8 $cm^{-2}$ | |
| | Lactic acid (1–3%) | IM, 4°C, 1 min | *Escherichia coli* | 0.5–0.8 $cm^{-2}$ | Choi *et al.*, 2009 |
| | | | *Escherichia coli* O157:H7 | 0.6–0.9 $cm^{-2}$ | |
| | | | *Listeria monocytogenes* | 0.9–1.2 $cm^{-2}$ | |
| | | | *Salmonella* Typhimurium | 0.6–0.9 $cm^{-2}$ | |
| | Lactic acid (3%) | IM, 55°C, 0.3 min | *Aeromonas hydrophila* | >4.0 $cm^{-2}$ (lean/fat) | Greer and Dilts, 1995 |
| | | | *Brochothrix thermosphacta* | 1.0 $cm^{-2}$ (lean) | |
| | | | | >4.0 $cm^{-2}$ (fat) | |
| | | | *Listeria monocytogenes* | 1.0 $cm^{-2}$ (lean) | |
| | | | | 2.0–3.0 $cm^{-2}$ (fat) | |
| | | | *Pseudomonas fragi* | 1.0 $cm^{-2}$ (lean) | |
| | | | | >4.0 $cm^{-2}$ (fat) | |
| | | | *Yersinia enterocolitica* | 1.0 $cm^{-2}$ (lean) | |
| | | | | 2.0–3.0 $cm^{-2}$ (fat) | |

© Woodhead Publishing Limited, 2012

| | | | | | |
|---|---|---|---|---|---|
| Pork trim | Lactic acid (2%) | SP, 15°C, 0.3–2 min | Coliforms | 1.0–2.0 $cm^{-2}$ | Castelo *et al.*, 2001 |
| Sheep carcass | Lactic acid (1–2%) | SP, 0.5 min | Aerobic bacteria | 1.6–1.8 $cm^{-2}$ | Beyaz and Tayar, 2010 |
| | | | Coliforms | 2.7–3.0 $cm^{-2}$ | |
| | | | *Escherichia coli* | 2.1–2.2 $cm^{-2}$ | |
| Sheep/goat carcass | Lactic acid (2%) | SP, 2–4 min | Aerobic bacteria | 0.5 $g^{-1}$ | Dubal *et al.*, 2004 |
| | | | *Escherichia coli* | 0.4 $g^{-1}$ | |
| **Processed meat products** | | | | | |
| Frankfurters | Acetic acid (2.5%) | IM, 23°C, 2 min | *Listeria monocytogenes* | 0.7–1.7 $cm^{-2}$ | Barmpalia *et al.*, 2004 |
| | Lactic acid (2.5%) | IM, 23°C, 2 min | *Listeria monocytogenes* | 0.7–2.1 $cm^{-2}$ | |
| | Lactic acid (5%) | SP, 23°C, 0.2 min | *Listeria monocytogenes* | 1.8 $cm^{-2}$ | Byelashov *et al.*, 2008 |
| | Lactic acid (3.4%) | IM, 20°C, 0.5 min | *Listeria monocytogenes* | 2.0 $frank^{-1}$ | Nuñez de Gonzalez *et al.*, 2004 |

[a] IM, immersion; SP, spraying.
[b] Bacterial reductions achieved by the applied agent, as compared to untreated or water-treated samples, expressed per $cm^2$ or g of sample, or per sample.

© Woodhead Publishing Limited, 2012

**Table 20.3** Applications of poultry decontamination using organic acids

| Product | Agent (concentration) | Application[a] | Microorganism | Reduction (log cfu)[b] | Reference |
|---|---|---|---|---|---|
| Chicken breast | Acetic acid (0.1–0.3%) | SP, 20–55°C, 0.1–0.3 min | *Escherichia coli* | 0.5–1.2/10 $cm^2$ | Jiménez *et al.*, 2005 |
| | Acetic acid (1.0–2.5%) | SP, 25°C, 0.2–0.4 min | *Salmonella* Hadar | 1.2–1.8/10 $cm^2$ | Jiménez *et al.*, 2007 |
| | Acetic acid (2.5%) | SP, 55°C, 0.5 min | *Salmonella* Hadar | 1.8–2.0/10 $cm^2$ | |
| | Acetic acid (0.5–6.0%) | IM, 0°C, 60 min | *Salmonella* Typhimurium | 0.0–0.7 $skin^{-1}$ | Tamblyn and Conner, 1997 |
| | | IM, 23°C, 0.3 min | *Salmonella* Typhimurium | 0.0–0.3 $skin^{-1}$ | |
| | | IM, 50°C, 2 min | *Salmonella* Typhimurium | 0.7–2.4 $skin^{-1}$ | |
| | Acetic acid (5%) | IM, 0°C, 60 min | *Salmonella* Typhimurium | 0.0–2.5 $skin^{-1}$ | Tamblyn *et al.*, 1997 |
| | | IM, 23°C, 0.3 min | *Salmonella* Typhimurium | 0.0–0.4 $skin^{-1}$ | |
| | | IM, 50°C, 2 min | *Salmonella* Typhimurium | 1.7–2.0 $skin^{-1}$ | |
| | Citric acid (0.5–6.0%) | IM, 0°C, 60 min | *Salmonella* Typhimurium | 0.0–1.9 $skin^{-1}$ | Tamblyn and Conner, 1997 |
| | | IM, 23°C, 0.3 min | *Salmonella* Typhimurium | 0.0–0.1 $skin^{-1}$ | |
| | | IM, 50°C, 2 min | *Salmonella* Typhimurium | 0.3–1.9 $skin^{-1}$ | |
| | Lactic acid (0.5–2.0%) | IM, 25°C, 10–30 min | *Escherichia coli* O157:H7 | 0.5–2.6 $g^{-1}$ | Anang *et al.*, 2007 |
| | | | *Listeria monocytogenes* | 1.0–2.0 $g^{-1}$ | |
| | | | *Salmonella* Enteritidis | 0.8–1.7 $g^{-1}$ | |
| | Lactic acid (4%) | IM, 55°C, 15 min | *Listeria monocytogenes*[c] | 2.4 $g^{-1}$ | Gonçalves *et al.*, 2005 |
| | Lactic acid (1%) | IM, 25°C, 30 min | *Listeria monocytogenes* | 2.3 $cm^{-2}$ | Hwang and Beuchat, 1995 |
| | | | *Salmonella* spp. | 2.0 $cm^{-2}$ | |
| | Lactic acid (0.5–6.0%) | IM, 0°C, 60 min | *Salmonella* Typhimurium | 0.2–2.3 $skin^{-1}$ | Tamblyn and Conner, 1997 |
| | | IM, 23°C, 0.3 min | *Salmonella* Typhimurium | 0.0–1.2 $skin^{-1}$ | |
| | | IM, 50°C, 2 min | *Salmonella* Typhimurium | 0.9–3.1 $skin^{-1}$ | |
| | Lactic acid (1–2%) | SP, 20°C, 0.5 min | Aerobic bacteria | 2.2–2.3 $ml^{-1}$ | Xiong *et al.*, 1998a |
| | | | *Salmonella* Typhimurium | 2.2 $ml^{-1}$ | |
| | Lactic acid (2%) | SP, 25–70°C, 0.5–3 min | *Salmonella* Typhimurium | 0.9–1.7 $skin^{-1}$ | Xiong *et al.*, 1998b |
| | Malic acid (0.5–6.0%) | IM, 0°C, 60 min | *Salmonella* Typhimurium | 0.4–2.7 $skin^{-1}$ | Tamblyn and Conner, 1997 |

© Woodhead Publishing Limited, 2012

| | | | | | |
|---|---|---|---|---|---|
| | | IM, 23°C, 0.3 min | *Salmonella* Typhimurium | 0.0–0.3 skin$^{-1}$ | |
| | | IM, 50°C, 2 min | *Salmonella* Typhimurium | 0.7–1.8 skin$^{-1}$ | |
| | Mandelic acid (0.5–6.0%) | IM, 0°C, 60 min | *Salmonella* Typhimurium | 0.2–2.2 skin$^{-1}$ | Tamblyn and Conner, 1997 |
| | | IM, 23°C, 0.3 min | *Salmonella* Typhimurium | 0.0–1.8 skin$^{-1}$ | |
| | | IM, 50°C, 2 min | *Salmonella* Typhimurium | 0.4–2.1 skin$^{-1}$ | |
| | Propionic acid (0.5–6.0%) | IM, 0°C, 60 min | *Salmonella* Typhimurium | 0.5–2.2 skin$^{-1}$ | Tamblyn and Conner, 1997 |
| | | IM, 23°C, 0.3 min | *Salmonella* Typhimurium | 0.0–1.7 skin$^{-1}$ | |
| | | IM, 50°C, 2 min | *Salmonella* Typhimurium | 0.4–1.4 skin$^{-1}$ | |
| | Tartaric acid (0.5–6.0%) | IM, 0°C, 60 min | *Salmonella* Typhimurium | 0.0–1.7 skin$^{-1}$ | Tamblyn and Conner, 1997 |
| | | IM, 23°C, 0.3 min | *Salmonella* Typhimurium | 0.0–0.2 skin$^{-1}$ | |
| | | IM, 50°C, 2 min | *Salmonella* Typhimurium | 0.4–2.0 skin$^{-1}$ | |
| Chicken carcass | Acetic acid (0.5–1.5%) | IM, 0.3 min | *Escherichia coli* | 0.7–1.4 cm$^{-2}$ | Bin Jasass, 2008 |
| | Acetic acid (2%) | IM, 4°C | Aerobic bacteria | 2.0 ml$^{-1}$ | Fabrizio *et al.*, 2002 |
| | | | Coliforms | 3.0 ml$^{-1}$ | |
| | | | *Escherichia coli* | 2.8 ml$^{-1}$ | |
| | | | *Salmonella* Typhimurium | 1.4 ml$^{-1}$ | |
| | | SP, 0.3 min | Aerobic bacteria | 0.4 ml$^{-1}$ | |
| | | | Coliforms | 0.2 ml$^{-1}$ | |
| | | | *Salmonella* Typhimurium | 0.8 ml$^{-1}$ | |
| | Acetic acid (0.5%) | IM | Aerobic bacteria | 0.2–0.7 cm$^{-2}$ | Sakhare *et al.*, 1999 |
| | | | Coliforms | 0.9–2.2 cm$^{-2}$ | |
| | | | *Staphylococcus aureus* | 0.5–0.7 cm$^{-2}$ | |
| | | SP | Aerobic bacteria | 0.5–0.8 cm$^{-2}$ | |
| | | | Coliforms | 0.1–1.0 cm$^{-2}$ | |
| | | | *Staphylococcus aureus* | 1.3–1.8 cm$^{-2}$ | |
| | Citric acid (2–10%) | IM, 1 min | Aerobic bacteria | 0.2–1.1 g$^{-1}$ | Doležalová *et al.*, 2010 |
| | | | Coliforms | 0.4–2.4 g$^{-1}$ | |
| | Formic acid (2%) | IM, 1 min | *Campylobacter jejuni* | 1.6 ml$^{-1}$ | Riedel *et al.*, 2009 |
| | Lactic acid (1–3%) | IM, 0.3 min | *Escherichia coli* | 0.5–2.1 cm$^{-2}$ | Bin Jasass, 2008 |

*(Continued)*

© Woodhead Publishing Limited, 2012

**Table 20.3** Continued

| Product | Agent (concentration) | Application[a] | Microorganism | Reduction (log cfu)[b] | Reference |
|---|---|---|---|---|---|
| | Lactic acid (1–2%) | IM, 1 min | Aerobic bacteria | 1.1–1.8 $g^{-1}$ | Doležalová *et al.*, 2010 |
| | | | Coliforms | 1.1–3.4 $g^{-1}$ | |
| | | | Psychrotrophs | 1.1–1.8 $g^{-1}$ | |
| | Lactic acid (10%) | IM, 10°C | *Campylobacter jejuni* | 0.3 $g^{-1}$ | Ellerbroek *et al.*, 2007 |
| | | SP, 10°C | *Campylobacter jejuni* | 0.2 $g^{-1}$ | |
| | Lactic acid (15%) | IM, 30°C | *Campylobacter jejuni* | 1.5 $g^{-1}$ | |
| | | SP, 30°C | *Campylobacter jejuni* | 0.9 $g^{-1}$ | |
| | Lactic acid (2%) | IM, ambient temp., 3 min | Aerobic bacteria | >2.0 $carcass^{-1}$ | Killinger *et al.*, 2010 |
| | | | Coliforms | >2.0 $carcass^{-1}$ | |
| | Lactic acid (1%) | SP, 22°C, 0.5 min | *Salmonella* Typhimurium | 0.6–0.7 $carcass^{-1}$ | Li *et al.*, 1997 |
| | Lactic acid (1%) | SP, 0.2 min | Aerobic bacteria | 0.4–1.3 $ml^{-1}$ | Okolocha and Ellerbroek, 2005 |
| | | | Enterobacteriaceae | 0.9–1.0 $ml^{-1}$ | |
| | | | Lactobacilli | –0.2–1.0 $ml^{-1}$ | |
| | | | Pseudomonads | 0.3–0.4 $ml^{-1}$ | |
| | | IM | Aerobic bacteria | 0.6 $ml^{-1}$ | |
| | | | Enterobacteriaceae | 1.1 $ml^{-1}$ | |
| | | | Lactobacilli | 0.4 $ml^{-1}$ | |
| | | | Pseudomonads | 0.4 $ml^{-1}$ | |
| | Lactic acid (2.5%) | IM, 1 min | *Campylobacter jejuni* | 1.7 $ml^{-1}$ | Riedel *et al.*, 2009 |
| | Lactic acid (0.25%) | IM | Aerobic bacteria | 0.8–0.9 $cm^{-2}$ | Sakhare *et al.*, 1999 |
| | | | Coliforms | 3.0 $cm^{-2}$ | |
| | | | *Staphylococcus aureus* | 0.3–0.6 $cm^{-2}$ | |
| | | SP | Aerobic bacteria | 1.0–1.1 $cm^{-2}$ | |
| | | | Coliforms | 0.2–2.0 $cm^{-2}$ | |

© Woodhead Publishing Limited, 2012

| | | | | | |
|---|---|---|---|---|---|
| | | | *Staphylococcus aureus* | 1.3–1.9 $cm^{-2}$ | |
| | Lactic acid (2%) | IM, 20°C, 0.5 min | Aerobic bacteria | 1.4 $cm^{-2}$ | Sinhamahapatra *et al.*, 2004 |
| | | | Coliforms | 1.0 $cm^{-2}$ | |
| | | SP, 20°C, 0.5 min | Aerobic bacteria | 1.1 $cm^{-2}$ | |
| | | | Coliforms | 1.1 $cm^{-2}$ | |
| | Lactic acid (2%) | SP, 35°C, 0.3 min | Aerobic bacteria | 1.0 $carcass^{-1}$ | Yang *et al.*, 1998 |
| | | | *Salmonella* Typhimurium | 1.8 $carcass^{-1}$ | |
| | Tartaric acid (2%) | IM, 1 min | *Campylobacter jejuni* | 0.9 $ml^{-1}$ | Riedel *et al.*, 2009 |
| Chicken legs | Citric acid (2%) | IM, 18°C, 15 min | Aerobic bacteria | 1.2 $g^{-1}$ | del Río *et al.*, 2007a |
| | | | *Brochothrix thermosphacta* | 1.1 $g^{-1}$ | |
| | | | Coliforms | 1.3 $g^{-1}$ | |
| | | | Enterobacteriaceae | 1.5 $g^{-1}$ | |
| | | | Enterococci | 1.2 $g^{-1}$ | |
| | | | Lactic acid bacteria | 0.2 $g^{-1}$ | |
| | | | Micrococcaceae | 0.9 $g^{-1}$ | |
| | | | Psychrotrophs | 1.1 $g^{-1}$ | |
| | Citric acid (2%) | IM, 18°C, 15 min | *Bacillus cereus* | 1.6 $g^{-1}$ | del Río *et al.*, 2007b |
| | | | *Escherichia coli* | 0.9 $g^{-1}$ | |
| | | | *Listeria monocytogenes* | 1.3 $g^{-1}$ | |
| | | | *Salmonella* Enteritidis | 0.2 $g^{-1}$ | |
| | | | *Staphylococcus aureus* | 0.9 $g^{-1}$ | |
| | | | *Yersinia enterocolitica* | 1.3 $g^{-1}$ | |
| | Lactic acid (0.11–0.55 mol $l^{-1}$) | IM, 5 min | Aerobic bacteria | 0.3–1.3 $g^{-1}$ | González-Fandos and Dominguez, 2006 |
| | | | *Listeria monocytogenes* | 0.5–1.1 $g^{-1}$ | |
| | Lactic acid (5%) | IM, 20°C, 1min | *Listeria innocua* | 0.7 $cm^{-2}$ | Lecompte *et al.*, 2008 |
| | Lactic acid (10%) | IM, 20°C, 1 min | *Listeria innocua* | 1.5 $cm^{-2}$ | |

*(Continued)*

© Woodhead Publishing Limited, 2012

**Table 20.3** Continued

| Product | Agent (concentration) | Application[a] | Microorganism | Reduction (log cfu)[b] | Reference |
|---|---|---|---|---|---|
| | Lactic acid (10%) | IM, 20°C, 30 min | *Listeria innocua* | 2.5 $cm^{-2}$ | |
| Chicken wings | Acetic acid (2%) | IM, 4°C, 0.3–0.8 min | *Campylobacter jejuni* | 1.2–1.4 $cm^{-2}$ | Zhao and Doyle, 2006 |
| | Lactic acid (2–8%) | IM, 21°C, 1 min | Aerobic bacteria | 0.8–1.5 $g^{-1}$ | Ismail *et al.*, 2001 |
| | Lactic acid (2%) | IM, ambient temp., 3 min | *Salmonella enterica* | >5.0 $wing^{-1}$ | Killinger *et al.*, 2010 |
| Turkey carcass | Lactic acid (1.24–8.50%) | SP, 22°C, 0.2 min | Aerobic bacteria | 1.3–3.4 $ml^{-1}$ | Bautista *et al.*, 1997 |
| | | | Coliforms | 2.0–6.0 $ml^{-1}$ | |

[a]IM, immersion ; SP, spraying.
[b]Bacterial reductions achieved by the applied agent, as compared to untreated or water-treated samples, expressed per $cm^2$ or g of sample, per sample (i.e., skin, wing or carcass), or per ml of sample rinse.
[c]*L. monocytogenes* populations were determined using the three-tube most probable number (MPN) technique.

© Woodhead Publishing Limited, 2012

**Table 20.4** Applications of seafood decontamination using organic acids

| Product | Agent (concentration) | Application[a] | Microorganism | Reduction (log cfu)[b] | Reference |
|---|---|---|---|---|---|
| Catfish fillets | Acetic acid (2%) | IM, 4°C, 10 min | Coliforms | 2.2 $g^{-1}$ | Bal'a and Marshall, 1998 |
| | Lactic acid (2%) | IM, 4°C, 10 min | Coliforms | 1.2 $g^{-1}$ | |
| | | | *Listeria monocytogenes* | 0.8 $g^{-1}$ | |
| | Lactic acid (1–2%) | SP, 15°C | Aerobic bacteria | ~1.0 $g^{-1}$ | Fernandes *et al.*, 1998 |
| | Lactic acid (1.70–2.55%) | IM, 7°C, 10 min | Aerobic bacteria | 0.9–1.9 $g^{-1}$ | Ingham, 1989 |
| | Lactic acid (0.85–2.55%) | IM, ambient temp., 30 min | *Listeria monocytogenes* | 0.9–1.4 $g^{-1}$ | Verhaegh *et al.*, 1996 |
| | Malic acid (2%) | IM, 4°C, 10 min | Aerobic bacteria | 1.2 $g^{-1}$ | Bal'a and Marshall, 1998 |
| | Propionic acid (1–2%) | SP, 15°C | Aerobic bacteria | ~1.0 $g^{-1}$ | Fernandes *et al.*, 1998 |
| | Tartaric acid (2%) | IM, 4°C, 10 min | Aerobic bacteria | 1.2 $g^{-1}$ | |
| | | | Coliforms | 1.2 $g^{-1}$ | |
| Catfish skin | Lactic acid (2%) | IM, 21°C, 5 min | *Salmonella* Typhimurium | >3.9 $skin^{-1}$ | Kim and Marshall, 2000a |
| | Lactic acid (2%) | IM, 21°C, 1 min | *Salmonella* Typhimurium | >2.0 $skin^{-1}$ | |
| | Lactic acid (0.5–2.0%) | IM, 21°C, 1–10 min | *Edwardsiella tarda* | 0.3–3.5 $skin^{-1}$ | Kim and Marshall, 2001 |
| | | | *Listeria monocytogenes* | 0.4–3.9 $skin^{-1}$ | |
| Mussels | Citric acid (5.88%); lemon juice | ambient temp., 0–15 min[c] | *Salmonella* Typhimurium | 0.1–0.6 $g^{-1}$ | Kişla, 2007 |
| | Lactic acid (0.5–2.0%) | IM, 5–60 min | *Vibrio parahaemolyticus* | 1.9–>3.4 $g^{-1}$ | Terzi and Gucukoglu, 2010 |
| Shrimp | Lactic acid (1.5–3.0%) | IM, 25°C, 10–30 min | *Escherichia coli* | 2.0–4.7 $g^{-1}$ | Shirazinejad *et al.*, 2010 |
| | | | *Salmonella* Enteritidis | 3.3–4.3 $g^{-1}$ | |
| | | | *Vibrio cholerae* | 2.5–3.8 $g^{-1}$ | |
| | | | *Vibrio parahaemolyticus* | 2.6–3.9 $g^{-1}$ | |

[a]IM, immersion; SP, spraying.
[b]Bacterial reductions achieved by the applied agent, as compared to untreated or water-treated samples, expressed per g of sample or per sample.
[c]Not specified whether or not mussels were completely immersed in the treatment solution (10 ml of treatment solution were added to jars containing 50 g portions of the product and mixed well); treatment time of 0 min refers to application of treatment for only a few seconds.

© Woodhead Publishing Limited, 2012

**Table 20.5** Applications of produce decontamination using organic acids

| Product | Agent (concentration) | Application[a] | Microorganism | Reduction (log cfu)[b] | Reference |
|---|---|---|---|---|---|
| Alfalfa seeds | Acetic acid (2–5%) | IM, 23°C, 10 min | *Salmonella enterica* | 0.4–1.7 $g^{-1}$ | Weissinger and Beuchat, 2000 |
| | Citric acid (2–5%) | IM, 23°C, 10 min | *Salmonella enterica* | 1.4–3.0 $g^{-1}$ | |
| | Lactic acid (2–5%) | IM, 23°C, 10 min | *Salmonella enterica* | 1.2–3.0 $g^{-1}$ | |
| Bell peppers | Lactic acid (2%) | IM, 55–60°C, 0.3 min | *Escherichia coli* O157:H7 | 2.4 $cm^{-2}$ | Alvarado-Casillas *et al.*, 2007 |
| | | | *Salmonella* Typhimurium | 2.9 $cm^{-2}$ | |
| | | SP, 55–60°C, 0.3 min | *Escherichia coli* O157:H7 | 3.6 $cm^{-2}$ | |
| | | | *Salmonella* Typhimurium | >2.9 $cm^{-2}$ | |
| Cabbage | Acetic acid (0.50 ml/100 g cabbage); gaseous | F, ambient temp., 10 min | Aerobic bacteria | 1.1 $g^{-1}$ | Delaquis *et al.*, 1997 |
| | | | Lactic acid bacteria | >2.0 $g^{-1}$ | |
| Carrots | Acetic acid (4.03%); grape vinegar | IM, ambient temp., 0–60 min | *Salmonella* Typhimurium | 1.6–3.6 $g^{-1}$ | Sengum and Karapinar, 2004 |
| | Citric acid (4.46%); lemon juice | IM, ambient temp., 0–60 min | *Salmonella* Typhimurium | 0.8–4.0 $g^{-1}$ | |
| | Lactic acid (2%) | IM, 4°C, 1 min | *Aeromonas hydrophila* | 2.7 $ml^{-1}$ | Uyttendaele *et al.*, 2004 |
| | | | Psychrotrophs | 1.8 $ml^{-1}$ | |
| Lettuce | Acetic acid (0.5–1.0%) | IM, 20°C, 2–5 min | *Escherichia coli* | 2.1–2.4 $g^{-1}$ | Akbas and Ölmez, 2007a |
| | | | *Listeria monocytogenes* | 1.2–1.4 $g^{-1}$ | |
| | Acetic acid (0.05–0.50%); rice vinegar | IM, 25°C, 5 min | *Escherichia coli* O157:H7 | <1 $g^{-1}$ | Chang and Fang, 2007 |
| | Acetic acid (5%); rice vinegar | IM, 25°C, 5 min | *Escherichia coli* O157:H7 | ≥3 $g^{-1}$ | |
| | Acetic acid (0.5%) | IM, 22°C, 10 min | *Yersinia enterocolitica*[c] | 0.7–2.7 $g^{-1}$ | Escudero *et al.*, 1999 |
| | Acetic acid (0.5–1.0%) | IM, 20°C, 1.5 min | *Listeria monocytogenes* | <1.0 $cm^{-2}$ | Samara and Koutsoumanis, 2009 |

© Woodhead Publishing Limited, 2012

| | | | | | |
|---|---|---|---|---|---|
| | Acetic acid (0.3%); apple | IM, 4°C, 10 min | Aerobic bacteria | 1.2 $g^{-1}$ | Vijayakumar and Wolf–Hall, 2002 |
| | cider vinegar | IM, 21°C, 10 min | *Escherichia coli* | 2.7 $g^{-1}$ | |
| | Acetic acid (1.9%); | IM, 21°C, 10 min | Aerobic bacteria | 2.3 $g^{-1}$ | |
| | white vinegar | IM, 21°C, 5–10 min | *Escherichia coli* | 5.4 $g^{-1}$ | |
| | Acetic acid (0.5–1.0%) | IM, 22°C, 10 min | *Listeria monocytogenes* | 0.0–0.2 $g^{-1}$ | Zhang and Farber, 1996 |
| | Ascorbic acid (0.5–1.0%) | IM, 20°C, 2–5 min | *Escherichia coli* | 1.7–2.1 $g^{-1}$ | Akbas and Ölmez, 2007a |
| | | | *Listeria monocytogenes* | 1.1–1.3 $g^{-1}$ | |
| | Citric acid (0.5–1.0%) | IM, 20°C, 2–5 min | *Escherichia coli* | 2.7–3.1 $g^{-1}$ | Akbas and Ölmez, 2007a |
| | | | *Listeria monocytogenes* | 1.4–1.8 $g^{-1}$ | |
| | Citric acid (0.5%) | IM, 20°C, 2 min | Aerobic bacteria | ~1.7 $g^{-1}$ | Akbas and Ölmez, 2007b |
| | | | Enterobacteriaceae | 0.7 $g^{-1}$ | |
| | | | Psychrotrophs | ~1.5 $g^{-1}$ | |
| | Citric acid (0.5–1.0%) | IM, 20°C, 1.5 min | *Listeria monocytogenes* | <1.0 $cm^{-2}$ | Samara and Koutsoumanis, 2009 |
| | Citric acid (0.6%); lemon juice | IM, 21°C, 10 min | Aerobic bacteria | 1.8 $g^{-1}$ | Vijayakumar and Wolf–Hall, 2002 |
| | | IM, 4°C, 10 min | *Escherichia coli* | 2.1 $g^{-1}$ | |
| | Fumaric acid (50 mM) | IM, ambient temp., 10 min | Aerobic bacteria | 1.4 $g^{-1}$ | Kondo *et al.*, 2006 |
| | Lactic acid (0.5–1.0%) | IM, 20°C, 2–5 min | *Escherichia coli* | 2.7–3.0 $g^{-1}$ | Akbas and Ölmez, 2007a |
| | | | *Listeria monocytogenes* | 2.0–2.2 $g^{-1}$ | |
| | Lactic acid (0.5%) | IM, 20°C, 2 min | Aerobic bacteria | ~1.7 $g^{-1}$ | Akbas and Ölmez, 2007b |
| | | | Enterobacteriaceae | 2.2 $g^{-1}$ | |
| | | | Psychrotrophs | ~1.5 $g^{-1}$ | |
| | Lactic acid (0.5%) | IM, 22°C, 5 min | *Yersinia enterocolitica*[c] | 2.3–3.2 $g^{-1}$ | Escudero *et al.*, 1999 |
| | Lactic acid (0.5–1.0%) | IM, 20°C, 1.5 min | *Listeria monocytogenes* | <1.0 $cm^{-2}$ | Samara and Koutsoumanis, 2009 |
| | Lactic acid (0.5–1.0%) | IM, 22°C, 10 min | *Listeria monocytogenes* | 0.1–0.5 $g^{-1}$ | Zhang and Farber, 1996 |
| | Propionic acid (0.5–1.0%) | IM, 20°C, 1.5 min | *Listeria monocytogenes* | <1.0 $cm^{-2}$ | Samara and Koutsoumanis, 2009 |
| Melons[d] | Citric acid (0.5%) | IM, 19°C, 5 min | *Listeria monocytogenes* | ~0.8 $cm^{-2}$ | Ukuku *et al.*, 2005 |

*(Continued)*

© Woodhead Publishing Limited, 2012

**Table 20.5** Continued

| Product | Agent (concentration) | Application[a] | Microorganism | Reduction (log cfu)[b] | Reference |
|---|---|---|---|---|---|
| | Lactic acid (2%) | IM, 55–60°C, 0.3 min | *Escherichia coli* O157:H7 | 1.6 $cm^{-2}$ | Alvarado-Casillas *et al.*, 2007 |
| | | | *Salmonella* Typhimurium | 2.9 $cm^{-2}$ | |
| | | SP, 55–60°C, 0.3 min | *Escherichia coli* O157:H7 | 2.0 $cm^{-2}$ | |
| | | | *Salmonella* Typhimurium | 3.0 $cm^{-2}$ | |
| | Lactic acid (1.5%) | IM, 25–35°C, 1–10 min | *Escherichia coli* O157:H7 | 6.2–7.3 $cm^{-2}$ | Materon, 2003 |
| Mung bean seed | Acetic acid (121 μl/l of air); gaseous | F, 22°C, 12 h | Aerobic bacteria | 2.0 $g^{-1}$ | Delaquis *et al.*, 1999 |
| | Acetic acid (242 μl/l of air); gaseous | F, 22–45°C, 12–24 h | Aerobic bacteria | >2.0 $g^{-1}$ | |
| Parsley | Acetic acid (2–5%) | IM, 15–30 min | *Yersinia enterocolitica* | >7.0 $g^{-1}$ | Karapinar and Gönül, 1992 |
| | Acetic acid (2.0–2.5%); vinegar | IM, 15–30 min | *Yersinia enterocolitica* | >7.0 $g^{-1}$ | |
| | Acetic acid (0.5–2.6%); white vinegar | IM, 21°C, 5 min | *Shigella sonnei* | 1.0–3.3 $g^{-1}$ | Wu *et al.*, 2000 |
| | Acetic acid (5.2%); white vinegar | IM, 21°C, 5 min | *Shigella sonnei* | >6.0 $g^{-1}$ | |

[a] F, fumigation; IM, immersion; SP, spraying.
[b] Bacterial reductions achieved by the applied agent, as compared to untreated or water-treated samples, expressed per $cm^2$ or g of sample, or per ml of sample rinse.
[c] Maximum population reductions corresponding to two *Yersinia enterocolitica* strains.
[d] Cantaloupe and/or honeydew melons.

© Woodhead Publishing Limited, 2012

bactericidal efficacy of lactic acid compared to that of acetic acid (Anderson *et al.*, 1992a; Greer and Dilts, 1992; Podolak *et al.*, 1996).

The balance between the abovementioned events and, thus, the antimicrobial activity of a certain organic acid relative to that exerted by another one, depend on the buffering capacity of the food matrix and the quantity of the acidulant used (Adams and Hall, 1988). With particular reference to the impact of tissue type on the effectiveness of meat decontamination using organic acids, research findings have not been consistent, with some studies reporting higher bacterial reductions on adipose while others on lean beef tissue (Bell *et al.*, 1997; Cutter and Siragusa, 1994; Cutter *et al.*, 1997; Dickson, 1992; Greer and Dilts, 1995). The dependency of the bactericidal efficacy of organic acids on the type of bacterial contaminant has been well recognized (Greer and Dilts, 1992, 1995; Hardin *et al.*, 1995), and so has the considerable intra-species (i.e., strain) variability with regard to the susceptibility of foodborne bacteria to these chemical agents (Birk *et al.*, 2010; Cutter and Siragusa, 1994). However, regarding the impact of the initial numbers of bacterial contaminants on the antimicrobial activity of organic acids, research findings have not been conclusive. Dickson (1992) reported that the effectiveness of acetic acid decontamination of beef tissue was independent of the level of initial contamination, while according to the findings of other investigators (Cutter *et al.*, 1997; Greer and Dilts, 1992), the bacterial levels initially present on beef significantly affected the bactericidal efficacy of lactic and acetic acid. Greer and Dilts (1992), in the context of a study undertaken on lean beef, reported that the effect of initial bacterial numbers on the decontamination efficacy of organic acids appeared to depend on the test organism and was often influenced through interaction with acid temperature and concentration. Nonetheless, as reported by the above researchers, meatborne pathogenic bacteria were, overall, more susceptible to organic acids at initial levels of $\geq 10^4$ cfu/cm$^2$ (Greer and Dilts, 1992).

In addition to food decontamination using solutions of organic acids individually, the application of mixtures of organic acids (i.e., combined application of two or more compounds) has also been described in the scientific literature, and it has been evaluated as an effective decontamination intervention in various foods including meat and fresh produce (Anderson and Marshall, 1990a; Anderson *et al.*, 1992a; Dubal *et al.*, 2004; Ölmez, 2010; Podolak *et al.*, 1996). When organic acids are combined, a synergistic interaction with regard to their antimicrobial activity can be observed. The synergistic effects that have been observed when mixtures of lactic and acetic acids are used on weakly buffered media, for instance, have been ascribed to the potentiation of acetic acid at the lower pH environment created by lactic acid (Adams and Hall, 1988; Helander *et al.*, 1997).

The natural presence of acetic acid and citric acid in vinegar and lemon juice, respectively, offers the opportunity for their application as food decontamination agents throughout the food chain, even at food-service establishments by

© Woodhead Publishing Limited, 2012

kitchen staff or at domestic settings by consumers. Solutions of 'household sanitizers' such as vinegar (e.g., apple cider, grape, and rice vinegar) and lemon juice hold promise as decontamination treatments for produce commodities including carrots, lettuce and parsley (Chang and Fang, 2007; Karapinar and Gönül, 1992; Sengum and Karapinar, 2004; Vijayakumar and Wolf-Hall, 2002; Wu *et al.*, 2000). Various levels of decontamination effectiveness of the above compounds (i.e., vinegar and lemon juice) when applied on produce items have been documented in research studies, depending on the organic acid concentration and the duration of the treatment application. According to Sengum and Karapinar (2004), treatment of carrots with lemon juice was more effective against *S.* Typhimurium than treatment with grape vinegar, while the maximum antimicrobial effect was obtained by dipping carrots in lemon juice-vinegar mixture (1:1) for at least 30 min. Finally, both organic acids and organic acid-containing natural compounds can be used as marinade ingredients for meat and poultry, aiming at enhancing their microbiological safety and quality without, however, adversely or negatively affecting the taste of the prepared products (Birk *et al.*, 2010; Kargiotou *et al.*, 2011).

Apart from exhibiting an immediate bactericidal effect, which characterizes their decontamination effectiveness, organic acids can also demonstrate a residual efficacy against foodborne bacteria and, thus, a noticeable preservation potential for various food products. As indicated by the data of numerous research studies, organic acids (e.g., acetic, citric, fumaric, lactic, malic, oxalic, low-molecular-weight polylactic, propionic and tartaric acids) are capable of suppressing bacterial proliferation during refrigerated storage of various food products, allowing in this way for a considerable extension of their shelf life. Such a growth inhibition potential of organic acid treatments has been observed against both spoilage and pathogenic foodborne bacteria, and with reference to various foods including beef (Ariyapitipun *et al.*, 2000; Dorsa *et al.*, 1997, 1998a; Kotula and Thelappurate, 1994; Mustapha *et al.*, 2002; Özdemir *et al.*, 2006a, 2006b; Podolak *et al.*, 1996), pork (Greer and Dilts, 1995), sheep/goat meat (Dubal *et al.*, 2004), poultry (Anang *et al.*, 2006, 2007; del Río *et al.*, 2007a, 2007b; Doležalová *et al.*, 2010; González-Fandos and Dominguez, 2006; González-Fandos *et al.*, 2009; Okolocha and Ellerbroek, 2005), seafood such as catfish and mussels (Bal'a and Marshall, 1998; Ingham, 1989; Marshall and Kim, 1996; Terzi and Gucukoglu, 2010), and produce items such as cantaloupes, bell peppers, carrots and lettuce (Akbas and Ölmez, 2007b; Alvarado-Casillas *et al.*, 2007; Samara and Koutsoumanis, 2009; Uyttendaele *et al.*, 2004).

The application of mixtures of organic acids also has been associated with prolonged bacteriostasis during storage of food products resulting in significant shelf life extension (Dubal *et al.*, 2004; Goddard *et al.*, 1996; Marshall and Kim, 1996; Palumbo and Williams, 1994). The bacteriostatic effect exerted by organic acids and their mixtures is even more important in the case of processed meat products, due to the long shelf life as well as the ready-to-eat (RTE) status of many of these products. With regard

© Woodhead Publishing Limited, 2012

to processed meat products, there are relatively few research data on the decontamination effectiveness of organic acid treatments, while a limited immediate bactericidal effect of such treatments has been reported in some cases (Geornaras *et al.*, 2005, 2006a, 2006b). Nevertheless, as demonstrated by several research findings, the application of organic acids, either individually or in mixtures, is capable of considerably restricting the growth of pathogenic bacteria during refrigerated storage of RTE meat products (Byelashov *et al.*, 2008; Nuñez de Gonzalez *et al.*, 2004; Palumbo and Williams, 1994; Samelis *et al.*, 2001a). Finally, the application of organic acid treatments on raw materials may enhance the microbiological safety and quality of further processed final products. Such examples include the reduction and growth suppression of pathogenic and spoilage bacteria in ground beef produced by organic acid-treated carcass parts or beef trim (Castillo *et al.*, 2001; Dorsa *et al.*, 1998b, 1998c; Harris *et al.*, 2006; Stivarius *et al.*, 2002a, 2002b), in sausages prepared from decontaminated (with organic acids) pork meat (Wan *et al.*, 2007), and in coleslaw made from acetic acid-treated cabbage (Delaquis *et al.*, 1997).

Salts of organic acids, known mainly for their antimicrobial activity when incorporated in the formulation of processed meat products (Barmpalia *et al.*, 2005; Mbandi and Shelef, 2002), have also been evaluated with regard to their decontamination efficacy in meat, poultry, and seafood products (Degnan *et al.*, 1994; Gonçalves *et al.*, 2005; Lin and Chuang, 2001; Stivarius *et al.*, 2002a). For instance, dipping of pork loin chops in 10% sodium lactate solution appeared to be an effective treatment in terms of decontamination as well as shelf life extension of this product (Lin and Chuang, 2001). However, in the context of decontamination interventions, salts of organic acids including sodium benzoate, sodium lactate and potassium sorbate have been evaluated primarily in combination with lactic acid (Doležalová *et al.*, 2010; Ismail *et al.*, 2001; Özdemir *et al.*, 2006b; Zeitoun and Debevere, 1990, 1991). Sodium lactate has also been utilized in buffered lactic acid systems which, by maintaining the pH of treated samples at low levels, appear to be more effective decontamination and preservation (i.e., shelf life extension) agents than unbuffered lactic acid solutions, particularly in the case of foods with high buffering capacity such as poultry meat (Zeitoun and Debevere, 1990, 1991).

### 20.3.2 Other chemical treatments

*Trisodium phosphate*

Phosphates have been used as additives to enhance the functional properties of meats (i.e., increase water binding and meat binding, enhance emulsification, retard oxidative rancidity and color deterioration, and enhance cured-color development) (Lee *et al.*, 1994). Moreover, the bactericidal effect of phosphate-based compounds and particularly of trisodium phosphate ($Na_3PO_4$; TSP) is well documented in the scientific literature and confirmed in several industrial

© Woodhead Publishing Limited, 2012

trials. Trisodium phosphate, naturally present in tissues and bones (Capita *et al.*, 2002c), is a white, free-flow crystalline material that complies with the specifications of the Food Chemicals Codex (Keener *et al.*, 2004).

Most of the studies on the decontamination effectiveness of TSP have been undertaken on poultry (Capita *et al.*, 2002c; Keener *et al.*, 2004; Loretz *et al.*, 2010). Treatment of poultry carcasses and parts with TSP has been carried out at various concentrations (mainly 8–12%), and has yielded reductions of aerobic bacteria, coliforms, Enterobacteriaceae, and *E. coli* ranging from 0.5 to 2.7, 0.8 to 4.1, 0.7 to 3.8, and 0.5 to 4.0 orders of magnitude, respectively (Bashor *et al.*, 2004; Bin Jasass, 2008; Capita *et al.*, 2000a; Castillo *et al.*, 2005; del Río *et al.*, 2007a, 2007b; Kim *et al.*, 1994b; Loretz *et al.*, 2010; Whyte *et al.*, 2001). With reference to pathogenic bacteria, TSP treatments of poultry have been shown to reduce *Campylobacter* spp. and *Listeria monocytogenes* by 0.2–1.7 and 1.0–4.5 logs, respectively (Arritt *et al.*, 2002; Capita *et al.*, 2001; del Río *et al.*, 2007a; Gonçalves *et al.*, 2005; Loretz *et al.*, 2010), while reductions of approximately 1.0–1.5 log have been reported for *Staphylococcus aureus*, *Bacillus cereus*, and *Yersinia enterocolitica* (del Río *et al.*, 2007a). Poultry decontamination with TSP solutions has been associated with *Salmonella* reductions between 0.6 and 2.3 log (del Río *et al.*, 2007a; Lillard, 1994; Loretz *et al.*, 2010). However, Bautista *et al.* (1997), who evaluated the potential bactericidal effect of TSP on fecally contaminated turkey carcasses, reported that, irrespective of concentration and spraying pressure, it was not effective against *Salmonella* spp., while a potential interference of the high pH values in test solutions resulting from TSP treatments with the efficient recovery of this pathogen has also been pointed out (Lillard, 1994). Moreover, the research findings of Capita *et al.* (2002a) suggested that poultry carcass skin sampling site is an important factor that needs to be considered when TSP decontamination protocols are developed since the former may affect the efficacy of this chemical treatment.

Despite the fact that the majority of research studies assessing the antimicrobial activity of TSP have been carried out on poultry, other foods have also been tested. Such foods include principally beef, with the decontamination treatment being applied either on carcasses or on further processed meat products (Cabedo *et al.*, 1996; Delmore Jr. *et al.*, 2000; Dickson *et al.*, 1994; Dorsa *et al.*, 1997, 1998a, 1998b, 1998c; Fratamico *et al.*, 1996; Gorman *et al.*, 1995; Hwang and Beuchat, 1995; Xiong *et al.*, 1998a, 1998b), and various produce items including alfalfa seeds, apples, oranges, green peppers, cabbage, lettuce, and tomatoes (Annous *et al.*, 2001; Liao and Cooke, 2001; Liao and Sapers, 2000; Pao and Davis, 1999; Taormina and Beuchat, 1999b; Weissinger and Beuchat, 2000; Zhang and Farber, 1996; Zhuang and Beuchat, 1996). Trisodium phosphate has been shown to hold promise as a decontamination agent for the above foods, resulting frequently in significant reductions in the populations of bacterial pathogens. Research data on the bactericidal activity of this compound on seafood are relatively

© Woodhead Publishing Limited, 2012

few, with its decontamination potential being evaluated on products such as catfish, rainbow trout, shrimp and blue crab meat (Degnan *et al.*, 1994; Kim and Marshall, 2002; Marshall and Jindal, 1997; Mu *et al.*, 1997).

Factors that need to be considered when phosphate-based chemicals, such as TSP, are selected for food decontamination purposes include their potentially costly application (Keener *et al.*, 2004), as well as the environmental implications of their utilization associated with a potential increase in the phosphate content of the generated wastewater (Oyarzabal, 2005). In addition to TSP, other phosphate-based compounds that have been tested, exhibiting, however, relatively limited decontamination effectiveness, include sodium acid pyrophosphate, monosodium phosphate, sodium hexametaphosphate, and sodium tripolyphosphate (Hwang and Beuchat, 1995; Loretz *et al.*, 2010; Skandamis *et al.*, 2010).

*Chlorine*

Chlorine is active against a wide range of microorganisms, and its efficacy in killing bacterial pathogens has been studied extensively. Due to its low cost and ample availability, chlorine (primarily as aqueous solutions of sodium hypochlorite (NaClO)) constitutes the most widely used chemical in the food industry for sanitation of equipment, utensils and water supplies (Beuchat and Ryu, 1997; Francis *et al.*, 1999; Oyarzabal, 2005; Sofos and Smith, 1998). When sodium hypochlorite is added to water, hypochlorous acid (HOCl) is formed, with the latter being responsible for chlorine's antimicrobial activity (Keener *et al.*, 2004). At pH 6.0, chlorine hydrolyzes completely to hypochlorous acid, and as the pH increases, the amount of hypochlorous acid and, thus, the bactericidal effectiveness of chlorine decrease (Keener *et al.*, 2004). Nevertheless, organic impurities contained in water are expected to react with the initial amount of added chlorine reducing the amount of available chlorine to form hypochlorous acid. The amount of chlorine required to completely oxidize such organic impurities is known as the chlorine demand of the water, and any amount of chlorine in excess of this amount exists as combined residual or free residual chlorine (Keener *et al.*, 2004). The antimicrobial activity of chlorine depends on the concentration of free residual chlorine (as hypochlorous acid) in water that comes into contact with microbial cells (Beuchat and Ryu, 1997; Keener *et al.*, 2004).

Chlorine, used in various forms, is the most traditional chemical decontamination treatment for beef and poultry carcasses, with its main application being its addition in the water used for final carcass washing or chilling (Hugas and Tsigarida, 2008; Skandamis *et al.*, 2010). Furthermore, chlorine has been routinely added to wash, spray and flume waters used in the fresh fruit and vegetable industry (Beuchat and Ryu, 1997; Gil *et al.*, 2009; Ölmez and Kretzschmar, 2009; Sapers, 2001). However, given that the antimicrobial activity of chlorine is counteracted by organic material (due to oxidizing-reducing reactions) and at the same time remarkably affected by the temperature and pH of the solution, this chemical agent frequently

© Woodhead Publishing Limited, 2012

exhibits a limited decontamination effectiveness when applied on meat, poultry or produce (Oyarzabal, 2005; Sapers, 2001; Skandamis *et al.*, 2010). Furthermore, upon reaction with organic matter, chlorine can result in the formation of potentially mutagenic or carcinogenic halogenated by-products (e.g., trihalomethanes) (Gil *et al.*, 2009; Komulainen, 2004; Ölmez and Kretzschmar, 2009), while its use has been associated with the production of high amounts of wastewater with very high levels of biological oxygen demand (Ölmez and Kretzschmar, 2009). Another limitation with regard to chlorine's use is its corrosive properties (Ölmez and Kretzschmar, 2009). Hence, during the last decade, the food industry has been seeking alternatives to chlorine, and chlorine dioxide, acidified sodium chlorite, hydrogen peroxide and peroxyacetic acid are the chemical agents that mainly have been evaluated for this purpose.

*Chlorine dioxide*

Chlorine dioxide ($ClO_2$) is a powerful oxidizing and sanitizing agent, exhibiting a broad biocidal activity, and known for its antimicrobial activity since the early 1900s (Keener *et al.*, 2004; Rajkovic *et al.*, 2010). It is a synthetic yellow-greenish gas with chlorine odor (Keener *et al.*, 2004), and it can be produced via two different reactions: an acid reacting with sodium chlorite, or sodium chlorite reacting with chlorine gas (Ölmez and Kretzschmar, 2009). Chlorine dioxide is unstable as a gas, undergoing decomposition into chlorine gas, oxygen gas and heat (Keener *et al.*, 2004). However, it is highly soluble in water and stable in aqueous solution, without, unlike chlorine, reacting with water or forming hypochlorous acid (Keener *et al.*, 2004; Rajkovic *et al.*, 2010). Given the above and the fact that chlorine dioxide is less affected by pH and organic matter, it demonstrates a higher oxidizing power and it can, therefore, be used in a more cost-effective way (i.e., at lower doses) than chlorine (Lillard, 1979; Ölmez and Kretzschmar, 2009; Oyarzabal, 2005).

Various levels of decontamination effectiveness of chlorine dioxide have been documented in the scientific literature, with its efficacy, when utilized as decontamination agent for meat and poultry, not being consistently or considerably higher than that exerted by chlorine or water only (Byelashov and Sofos, 2009; Loretz *et al.*, 2010; Skandamis *et al.*, 2010). Nevertheless, this chemical appears to be a promising chlorine alternative for use in fresh produce decontamination (Gil *et al.*, 2009; Pao and Davis, 1999; Rodgers *et al.*, 2004; Zhang and Farber, 1996). Chlorine dioxide produces fewer organochlorine compounds and is less corrosive than chlorine (Ölmez and Kretzschmar, 2009; Rajkovic *et al.*, 2010). Nevertheless, in the presence of iodide in the water source, chlorine dioxide forms more iodinated by-products than chlorine (Hua and Reckhow, 2007). Moreover, due to its explosive nature, this chemical needs to be generated on-site (Ölmez and Kretzschmar, 2009).

© Woodhead Publishing Limited, 2012

*Acidified sodium chlorite*

This is another broad-spectrum oxidative antimicrobial, with its chemistry being related to that of chlorine dioxide (Keener *et al.*, 2004). Acidified sodium chlorite (ASC) solutions contain sodium chlorite at specified concentrations, depending on the intended application, and an acidulant at concentrations sufficient to achieve a pH of 2.3–3.2, with citric acid being commonly used for this purpose (Keener *et al.*, 2004; Oyarzabal, 2005). When sodium chlorite and a weak organic acid are combined in solution, chlorous acid is instantaneously formed, with the proportion of chlorite that dissociates to chlorous acid increasing as the pH of the mixed solution decreases (Keener *et al.*, 2004).

This chemical agent has demonstrated an important decontamination efficacy when tested on meat and poultry products (Bashor *et al.*, 2004; del Río *et al.*, 2007a, 2007b; Echeverry *et al.*, 2010; Gill and Badoni, 2004; Harris *et al.*, 2006; Oyarzabal *et al.*, 2004) as well as in fresh produce commodities (Gonzalez *et al.*, 2004; Stopforth *et al.*, 2008). With particular reference to poultry carcasses, ASC is used either as a spray or dip solution before the carcass immersion in the pre-chill or chill tank, or as an additive in the pre-chill or chill tank (Oyarzabal, 2005). Furthermore, research data indicate that ASC can also be effectively applied for decontamination purposes as a post-chill dip application (Oyarzabal *et al.*, 2004). In addition to ASC, acidified calcium chlorite has also been evaluated with regard to its decontamination efficacy and shown to significantly reduce *Salmonella* and native microflora populations on whole and cut cantaloupes (Fan *et al.*, 2009).

*Hydrogen peroxide*

Hydrogen peroxide ($H_2O_2$) is an easy-to-use and low-cost chemical whose antimicrobial properties have been recognized for years, and various applications have been proposed and developed (Juven and Pierson, 1996; Ölmez and Kretzschmar, 2009). Hydrogen peroxide is probably generated in small amounts by almost all bacteria growing aerobically, but it is usually detected in aerobic cultures of catalase-negative bacteria (Juven and Pierson, 1996). It is a clear, colorless liquid which is completely miscible in water and soluble in a range of organic solvents (Jones, 1999). This chemical compound may exert either bacteriostatic or bactericidal effects, depending mainly on its concentration.

With reference to food decontamination, hydrogen peroxide has been primarily investigated as a produce disinfecting agent (Ölmez and Kretzschmar, 2009). Dilute hydrogen peroxide solutions as well as commercial disinfectants containing this chemical agent have been shown to be effective in controlling post-harvest decay and extending the shelf life of fruits and vegetables (Aharoni *et al.*, 1994; Sapers, 2001). Hydrogen peroxide produces no residues since it is rapidly decomposed into water and oxygen by the enzyme catalase which is naturally abundant in plants (Ölmez and Kretzschmar, 2009; Sapers, 2001), and its decontamination effectiveness has been assessed and noted in various

© Woodhead Publishing Limited, 2012

produce commodities including apples, cantaloupe and honeydew melons, lettuce as well as rice and alfalfa seeds (Annous *et al.*, 2001; Beuchat, 1997; Holliday *et al.*, 2001; Lin *et al.*, 2002; Piernas and Guiraud, 1997; Sapers *et al.*, 1999; Ukuku, 2004, 2006; Ukuku *et al.*, 2001; Weissinger and Beuchat, 2000). Sapers *et al.* (2001) demonstrated that a hydrogen peroxide wash, applied to cantaloupe melons prior to cutting, shows promise in improving the microbiological quality and shelf life of fresh-cut fruit. On the other hand, the antibacterial activity of hydrogen peroxide has been only occasionally evaluated in beef (Bell *et al.*, 1997; Cabedo *et al.*, 1996; Gorman *et al.*, 1995) and poultry (Loretz *et al.*, 2010) carcasses and parts, with the decontamination efficacy of the agent appearing to be generally limited.

*Peroxyacetic acid*

Peroxyacetic acid (PAA), also referred to as peracetic acid, is the peroxide of acetic acid. It has an oxidation potential larger than that of chlorine or chlorine dioxide, and belongs to the class of man-made chemicals known as organic peroxides (Kitis, 2004). Peroxyacetic acid is commercially available in the form of an aqueous quaternary equilibrium mixture of acetic acid and hydrogen peroxide (Kitis, 2004; Ölmez and Kretzschmar, 2009). Although considerably less stable than hydrogen peroxide, PAA is a more potent antimicrobial agent than the latter, being active at low concentrations against a wide spectrum of microorganisms (Kitis, 2004). This chemical is a clear, colorless liquid, soluble in water and in polar organic solvents, with a strong pungent acetic acid odor and an acidic pH of less than 2.0. Peroxyacetic acid has a $pK_a$ of 8.2, and its bactericidal activity is greater at lower pH values. Nevertheless, it can be active over a wide range of temperatures, and it produces little, if any, toxic or mutagenic by-products upon reaction with organic material (Kitis, 2004).

Peroxyacetic acid is well known for its considerable bactericidal activity when used in produce process water, with its efficacy not being affected by the organic load of water or by temperature changes, while only harmless by-products (i.e., acetic acid, water, and oxygen) are formed from its spontaneous decomposition (Ölmez and Kretzschmar, 2009; Sapers, 2001). Peroxyacetic acid and a PAA-based sanitizer have been evaluated in numerous studies, and have been shown to be particularly efficacious when used as water disinfection agents, preventing cross-contamination of fruits and vegetables with pathogenic bacteria during the washing process (Baert *et al.*, 2009; Beuchat *et al.*, 2004; do Socorro Rocha Bastos *et al.*, 2005; Fan *et al.*, 2009; Gonzalez *et al.*, 2004; López-Galvez *et al.*, 2009; Pao and Davis, 1999). Despite the fact that a commercial PAA-based solution for application in carcasses has been available in the United States, findings regarding the effectiveness of PAA or peroxyacid preparations as decontamination treatments of meat and poultry have been conflicting (del Río *et al.*, 2007a, 2007b; Skandamis *et al.*, 2010).

© Woodhead Publishing Limited, 2012

*Other chemical agents*
Chemical agents, in addition to the abovementioned, that have been evaluated for their efficacy and approved or proposed for use in food decontamination are: cetylpyridinium chloride (Li *et al.*, 1997; Özdemir *et al.*, 2006b; Pohlman *et al.*, 2002; Xiong *et al.*, 1998b), sulfate-based compounds (e.g., acidified calcium sulfate, sodium bisulfate, sodium thiosulfate) (Fan *et al.*, 2009; Li *et al.*, 1997; Loretz *et al.*, 2010; Zhang and Farber, 1996), sodium and calcium hydroxide (Alvarado-Casillas *et al.*, 2007; Capita *et al.*, 2002b; Holliday *et al.*, 2001; Loretz *et al.*, 2010), sodium bicarbonate (Bell *et al.*, 1997; Loretz *et al.*, 2011a), ozone (Gorman *et al.*, 1995; Ölmez, 2010; Rodgers *et al.*, 2004), electrolyzed oxidizing water (Fabrizio *et al.*, 2002; Phuvasate and Su, 2010), glucose monohydrate (Gögüs *et al.*, 2007), gluconic acid (Stivarius *et al.*, 2002a), saponin (Loretz *et al.*, 2011a), and various commercially available preparations.

### 20.3.3 Multiple hurdle approach

Chemical decontamination treatments can be applied simultaneously with or sequentially to other interventions, either chemical or physical, aiming at the metabolic exhaustion of bacterial cells through their exposure to several stress factors. In the context of this approach, known as 'multiple hurdle' approach, two or more hurdles, intelligently applied at suboptimal levels, can have a synergistic effect on the prevalence or population levels of microbial contaminants in foods, and be more effective than each hurdle applied individually at optimal level (Leistner, 2000). Synergistic or additive effects of various factors pertinent to the application of decontamination technologies have frequently been described in the literature, provided that initial contamination levels were high enough to allow for such effects to be measured (Samelis *et al.*, 2001b; Sofos and Smith, 1998).

One of the most important factors that need to be considered when multiple hurdle technology interventions are developed, and refer to the sequential application of treatments, is the order in which treatments are applied (Koutsoumanis *et al.*, 2004). Numerous applications embracing the basic principles of the multiple hurdle approach and combining chemical and physical meat decontamination interventions have been described (Geornaras and Sofos, 2005). The simplest example of such an application is the enhanced decontaminating effect of acid solutions exhibited under conditions of increased temperature (Cutter *et al.*, 1997). Similarly, the application of organic salt solutions at 55°C produced a significant synergistic reduction of *S.* Typhimurium on poultry carcasses (Milillo and Ricke, 2010). Lecompte *et al.* (2008) reported that the sequential application of steam and lactic acid treatments in poultry constitutes an intervention of considerable decontamination and at the same time preservation potential. Another example of a multiple hurdle approach, combining chemical and physical treatments, is the utilization of organic acids and other chemical compounds

© Woodhead Publishing Limited, 2012

as components of pre-drying marinades, with the latter being applied during beef jerky processing as a means of enhancing the effectiveness of drying treatments in inactivating foodborne pathogens (Calicioglu *et al.*, 2002b, 2002c, 2003; Yoon *et al.*, 2006).

With regard to the combined application (either simultaneous or sequential) of chemical treatments, in addition to the use of organic acids in mixtures which has been discussed earlier in this chapter, there are several reports in the scientific literature referring to various food products (Bell *et al.*, 1997; Calicioglu *et al.*, 2002a; Fabrizio *et al.*, 2002; Sapers *et al.*, 2001; Ukuku *et al.*, 2005; Venkitanarayanan *et al.*, 2002; Zhao *et al.*, 2009). For instance, as demonstrated by the results of a study undertaken by Zhao *et al.* (2009), the combination of levulinic acid with sodium dodecylsulfate increased remarkably the activity of these two chemicals against *E. coli* O157:H7 and *Salmonella* on lettuce, chicken wings and skin, as well as in water contaminated with chicken feces or feathers.

## 20.4 Mechanisms of microbial inactivation

The bactericidal activity of chemical compounds resides, in general, in their ability to cause disruption of the cell wall, the cytoplasmic membrane and other cellular components, as well as in the metabolic perturbation resulting from their interference with specific metabolic functions (e.g., replication, protein synthesis) and physiological responses (Denyer and Stewart, 1998). Basic aspects of the mechanisms of microbial inactivation by organic acids and other chemical agents applied in the context of food decontamination are presented in the following sections.

### 20.4.1 Organic acids

Most organic acids are capable of freely moving throughout bacterial cells, due to their simple structure and small molecular size or mass (Theron and Lues, 2007), and exhibit bacteriostatic or bactericidal properties depending on the physiological status of the target organism and the physicochemical characteristics of the external environment (Ricke, 2003). The antimicrobial action of weak organic acids, particularly lactic acid and acetic acid, has been studied extensively, and has been proposed to depend on (i) the pH-lowering effect solely; (ii) the extent of dissociation of the acid; and (iii) a specific effect associated with the acid molecule (Smulders *et al.*, 1986). The relative significance of each one of the above contributions has been a point of discussion and frequent debate among researchers.

Although the exact antibacterial mechanism(s) have not been completely elucidated, bacterial inactivation by weak organic acids has been traditionally attributed to the ability of the lipophilic, undissociated acid molecules to

© Woodhead Publishing Limited, 2012

penetrate the bacterial cytoplasmic membrane and to dissociate inside the cell (Adams and Hall, 1988; Booth, 1985). Undissociated organic acids possess high antimicrobial activity, much stronger than that exerted by their dissociated forms, and the extent of dissociation (i.e., the concentration of undissociated acid) and, thus, the antimicrobial effectiveness of an organic acid in solution is determined by its $pK_a$ and the pH of the external medium (Adams and Hall, 1988; Birk *et al.*, 2010; Helander *et al.*, 1997; Ricke, 2003; Stratford and Eklund, 2003). Low pH levels favor the uncharged, undissociated state of the compound with the latter being capable of passively crossing the cell membrane and entering the cell (Birk *et al.*, 2010; Booth, 1985; Brul and Coote, 1999). Upon entering the bacterial cell and encountering the higher pH environment of the cytoplasm, the acid molecule dissociates and charged ions (i.e., protons and anions) are released and accumulate inside the cell. The latter procedure, which takes place until equilibrium in accordance with the pH gradient across the membrane is reached, results in reduced intracellular pH and disruption of the membrane proton-motive force (Adams and Hall, 1988; Booth, 1985; Brul and Coote, 1999). Hence, actions proposed to be responsible for the antimicrobial activity of organic acids include membrane disruption, stress on intracellular pH homeostasis resulting in energy depletion, and inhibition of essential metabolic reactions (e.g., ATP synthesis, RNA and protein synthesis, DNA replication) (Booth, 1985; Brul and Coote, 1999; Ricke, 2003). Furthermore, intracellular accumulation of toxic anions has also been proposed as being associated with the antibacterial activity of organic acids (Brul and Coote, 1999; Russell, 1992).

Due to the complex nature of the interaction between energy dissipation and disruption of ATP generation capabilities of bacteria, the exact mechanisms characterizing the mode of action of organic acids have been difficult to determine (Ricke, 2003). Nevertheless, various research findings have been presented towards this direction. According to the results of early studies referring to formic and propionic acids, sublethal concentrations of these acids may have an impact on overall cell physiology and result in various responses, such as enlargement of bacterial cells, most likely as a result of inhibition of the synthesis of macromolecules (e.g., DNA, RNA, lipids, peptidoglycan and proteins) (Cherrington *et al.*, 1990; Thompson and Hinton, 1996). Sublethal concentrations of these acids were not associated with cell membrane damage, with the phospholipid and fatty acid composition of the latter, however, being possibly altered (Thompson and Hinton, 1996). Moreover, the DNA synthesis inhibition caused by the above organic acids did not appear to involve a physical damage of the DNA molecule, nor an induction of an SOS response (Cherrington *et al.*, 1991a). Bacterial cell death caused by formic acid or propionic acid was not associated with a reduction in culture turbidity or a loss of membrane integrity, indicating that cell death most likely results from the irreversible denaturation of acid-labile molecules (e.g., proteins and DNA) rather than from cell lysis (Cherrington *et al.*, 1991b).

© Woodhead Publishing Limited, 2012

Lactic acid is well known for its pH-lowering activity, and low-molecular-weight polylactic acid (a lactic acid polymer) is capable of continuously releasing free lactic acid and, therefore, maintaining low pH levels for a longer period of time (Ariyapitipun *et al.*, 2000; Lim and Mustapha, 2003). Lactic acid also functions as a permeabilizer of the outer membrane of Gram-negative bacteria and potentially as a potentiator of the effect of other antimicrobial substances, as evidenced by its ability to cause lipopolysaccharide release and bacterial sensitization to detergents or lysozyme, respectively (Alakomi *et al.*, 2000). In addition to intracellular acidification, cell membrane damage and inhibition of synthesis of functional macromolecules, lactic acid-mediated death of *E. coli* has also been associated with decreased intracellular potassium levels and limited oxygen uptake (Lim and Mustapha, 2003; Wang, 2002).

Citric acid does not fit under the description of classic weak organic acids (i.e., lipophilic, undissociated acids), meaning that it acts more as a chelator, exerting its antibacterial activity by sequestering metal ions ($Ca^{2+}$, $Mg^{2+}$, $Fe^{3+}$) from the external medium required for bacterial homeostasis (Brul and Coote, 1999; Stratford and Eklund, 2003; Theron and Lues, 2011). Similarly, however, to lactic acid, citric acid also may act as a permeabilizing agent of the outer membrane of Gram-negative bacteria, as well as a potentiator of the effect of other antibacterial agents (Brul and Coote, 1999).

### 20.4.2 Other chemical antimicrobial compounds

The exact mechanisms underlying TSP bactericidal activity have not been completely elucidated, and various possible modes of action have been proposed. The high alkalinity of TSP solutions (pH 10–12) and its potential effect on cell membrane components have been identified as the most important factors defining the antimicrobial activity of this chemical compound (Capita *et al.*, 2002c; Mendonca *et al.*, 1994). Research findings on the activity of TSP against *Salmonella* on chicken carcasses suggest that one of the major mechanisms of bacterial reduction is the detachment of contaminants from the carcass surface, facilitating bacterial removal by subsequent washing (Kim *et al.*, 1994a). Although the latter mode of action has not been explained in detail yet, there are research data suggesting that detachment of bacterial cells may be associated with sequestration of metal ions (Capita *et al.*, 2002c; Lee *et al.*, 1994), or with the ability of this compound to remove fat from food surfaces known as 'detergent effect' (Capita *et al.*, 2002b, 2002c; Keener *et al.*, 2004; Kim *et al.*, 1994a). In general, Gram-negative bacteria are much more sensitive to the effect of alkaline compounds, such as TSP, than Gram-positive bacteria, and such a difference has been attributed to an outer membrane-specific activity exerted by these agents (Capita *et al.*, 2002c; Keener *et al.*, 2004). In addition to the above, as evidenced by electron microscopic examination, TSP is a potent membrane-acting agent, with its action occurring at both functional and structural levels (Sampathkumar *et*

© Woodhead Publishing Limited, 2012

*al.*, 2003). It appears that it is primarily the high pH during TSP treatments and to a lesser extent sequestration of metal ions that lead to disruption of fatty molecules in the cell membrane, resulting in membrane damage, release of intracellular contents and cell death (Sampathkumar *et al.*, 2003). Most likely, the bactericidal activity exerted by TSP results from a combination of the aforementioned factors. The antimicrobial effects of phosphates, including TSP, can be altered substantially by the metal-ion content of the environment; thus, conditions such as hard water, metal containers, chemical additives containing divalent cations and ingredients with high iron contents should be avoided or minimized, if a maximum bactericidal efficacy of these compounds is to be achieved (Lee *et al.*, 1994).

Hypochlorous acid, the compound that is responsible for chlorine's bactericidal activity, is a highly destructive, non-selective oxidant, reacting with various subcellular constituents and affecting multiple metabolic processes (Dukan and Touati, 1996). The bactericidal action of this compound has primarily been associated with alterations in the cytoplasmic membrane's permeability, loss of energy-linked respiration due to destruction of cellular electron transport chains, and reaction with nucleotides and DNA damage (Albrich *et al.*, 1981; Dukan and Touati, 1996). In contrast to chlorine, which reacts both via oxidation and electrophilic substitution, chlorine dioxide reacts only by oxidation (Rajkovic *et al.*, 2010). Although the exact mechanism of microbial inactivation by chlorine dioxide is not completely understood (Rajkovic *et al.*, 2010), it has been assumed that this chemical compound exerts its bactericidal activity via direct action in the cellular membrane and oxidation of cellular constituents (Oyarzabal, 2005). Early research findings indicated that the mode of action of chlorine dioxide is directly related to protein synthesis, while loss of permeability control due to non-specific oxidative damage of the outer membrane, leading to destruction of the trans-membrane ionic gradient, was later identified as the lethal effect at the physiological level of this chemical on cells of Gram-negative bacteria (Rajkovic *et al.*, 2010). Another event that has also been associated with the bactericidal activity of chlorine dioxide is the denaturation of constituent proteins critical to cellular integrity and function, through the covalent oxidative modification of tryptophan and tyrosine residues (Ogata, 2007). With regard to ASC, this compound provides a non-specific attack on the amino acid component of the cell membrane (Keener *et al.*, 2004). The oxychlorous antimicrobial compounds that are generated upon combination of ASC with organic matter are the main active compounds against microbial cells by oxidizing sulfide and disulfide bonds in proteins on cell membrane surfaces (Keener *et al.*, 2004; Kemp *et al.*, 2000).

The antimicrobial activity of hydrogen peroxide is attributed mainly to its strong oxidizing power. The oxidative killing of metabolically active cells by this compound generally involves formation of radicals, with the latter being capable of damaging fundamental cellular components such as nucleic acids, proteins, and lipids (Juven and Pierson, 1996). With reference

© Woodhead Publishing Limited, 2012

to PAA, although the available research data regarding its mode of action as an antimicrobial agent are limited, it may be speculated that it functions similarly to other peroxides and oxidizing agents, with its bactericidal activity being based on the release of active oxygen (Kitis, 2004). Events that are likely to be associated with this compound's antimicrobial activity include:

- reaction with double bonds and oxidation of sensitive sulfhydryl and sulfur bonds in proteins, enzymes, and other metabolites;
- disruption of the chemiostatic function of the lipoprotein cytoplasmic membrane (or outer membrane in Gram-negative bacteria) and transport through dislocation or rupture of cell walls;
- oxidation of intracellular essential enzymes and impairment of vital biochemical pathways, active transport across membranes, and intracellular solute levels; and
- action on the basis of the DNA molecule (Kitis, 2004).

Furthermore, PAA has the advantage of being capable of inactivating catalase, an enzyme known to detoxify free hydroxyl radicals (Kitis, 2004).

## 20.5 Applications of chemical treatments for specific food products

### 20.5.1 Meat and poultry

The majority of meat and poultry decontamination technologies that have been developed and evaluated over the last three decades aim at reducing microbial contamination at the carcass level, and as such they have been carried out in research abattoirs using meat model systems such as artificially inoculated carcasses, meat cuts, and meat tissue samples (Loretz *et al.*, 2010, 2011a, 2011b; Smulders and Greer, 1998; Theron and Lues, 2007). The application of chemical decontamination treatments at early stages of the slaughtering sequence, when bacterial attachment and organic matter load are still minimal, is of great importance for the advancement of meat microbiological safety and quality (Cabedo *et al.*, 1996; Dickson, 1995; Sofos and Smith, 1998). Carcass decontamination technologies assessed for their efficacy, and with some of them being practiced commercially, include application of chemical solutions via immersion or spraying (Loretz *et al.*, 2010, 2011a, 2011b; Sofos, 2008). Application of chemical solutions, including organic acids, may take place either prior to carcass evisceration after hide removal, or after evisceration prior to chilling, during chilling as well as after chilling and before fabrication (Byelashov and Sofos, 2009; Edwards and Fung, 2006; Loretz *et al.*, 2010, 2011a, 2011b; Skandamis *et al.*, 2010). Organic acids, mainly acetic and lactic acid, are widely used in the United States and Canada for beef and pig carcass decontamination,

© Woodhead Publishing Limited, 2012

and are frequently applied using spraying cabinets (Loretz *et al.*, 2011a, 2011b).

Despite the fact that there is a considerable amount of research data regarding chemical decontamination of meat/poultry carcasses, the potential for utilization of such decontamination treatments in further processed products has been assessed in relatively few studies. There are trials describing the application, primarily in the form of dipping solutions, of organic acids and other chemicals in various raw meat/poultry products including poultry parts (breasts, legs or wings) and beef cuts of various carcass regions (Anang *et al.*, 2007; Castillo *et al.*, 2005; Echeverry *et al.*, 2010; Ismail *et al.*, 2001; Kim and Marshall, 1999; Kim *et al.*, 1994a; Kotula and Thelappurate, 1994; Koutsoumanis *et al.*, 2004). While the above studies assess the decontamination efficacy of chemical treatments on samples collected either from processing plants after fabrication or from retail outlets, their findings indicate that such treatments may enhance the microbiological safety and quality of meat and poultry even when applied at late stages of the processing sequence (e.g., after fabrication and prior to packaging). In addition, as supported by recent research data, chemical decontamination interventions, applied by purveyors prior to mechanical tenderization or brine-enhancement of beef steaks, hold promise for increasing the microbiological safety of these types of products (Echeverry *et al.*, 2010). Finally, decontamination treatments utilizing solutions of organic acids or other chemicals may be applied in RTE meat products as a means of inactivating post-processing bacterial contamination. Such treatments have been evaluated in laboratory trials for both their decontamination and preservation efficacies when applied in RTE meat products, and have been shown to constitute promising pathogen control interventions for these products (Barmpalia *et al.*, 2004; Byelashov *et al.*, 2008; Geornaras *et al.*, 2005, 2006a, 2006b; Palumbo and Williams, 1994; Singh *et al.*, 2005).

### 20.5.2 Seafood

With seafood constituting a highly perishable food commodity, the development of effective decontamination interventions, aiming at the inactivation of both seafood specific spoilage organisms (e.g., *Pseudomonas* spp., *Shewanella* spp. and *Photobacterium* spp.) and bacterial pathogens of concern (e.g., *Vibrio* spp., *Aeromonas* spp., and *L. monocytogenes*), has attracted a significant research interest in the last two decades (Cortesi *et al.*, 2009; Huss *et al.*, 2000). Chemical decontamination of shrimp during handling and processing has been regarded as one of the most useful techniques employed in the control of pathogenic and/or spoilage contaminants of this product (Shirazinejad *et al.*, 2010). Although much research has been conducted since the 1990s on how to prevent spoilage of fishery products using organic acids (Cortesi *et al.*, 2009), relatively few studies, compared to other food commodities, have specifically assessed the decontamination effectiveness of these and

© Woodhead Publishing Limited, 2012

other chemical agents against pathogenic or spoilage microorganisms. In addition to chlorine, which exhibits an important decontamination efficacy against pathogenic bacteria surface-inoculated on seafood and constitutes the most widely used decontaminating agent in the seafood industry (Bremer and Osborne, 1998; Chaiyakosa *et al.*, 2007), organic acids also hold promise in this direction, with the decontamination treatments in all cases being applied as dipping solutions.

With particular reference to fish decontamination, it is important that chemical treatments are also evaluated for efficacy on fish skin, since the latter can be an important source of bacterial contamination during processing with its microbial load being potentially transferred to processing surfaces or skinned products such as fish fillets (Kim and Marshall, 2000a, 2001, 2002). Furthermore, fish skin mucus has been associated with enhanced bacterial accumulation (Krovacek *et al.*, 1987) and with decreased antimicrobial effects of chemical solutions including lactic acid and TSP (Kim and Marshall, 2000a, 2001, 2002). Based on research findings regarding catfish decontamination, it has been suggested that removal of skin mucus prior to application of chemical agents may improve their bactericidal efficacy (Kim and Marshall, 2000a, 2001, 2002).

An additional indirect effect that chemical treatments may have on the microbiological quality of seafood products is that exerted by the application of such treatments in ice used for chilling of these products. Seafood products are usually chilled with ice, aiming at delay of autolysis and quality retention and retardation of bacterial growth during storage, with melting ice, however, constituting a potential source of cross-contamination if not discarded properly (Feliciano *et al.*, 2010). Research findings indicate that the above cross-contamination prospect could be substantially reduced by the use of chemically treated ice (Feliciano *et al.*, 2010; Phuvasate and Su, 2010), suggesting an alternative decontamination technology of potential value for the seafood industry. More specifically, Phuvasate and Su (2010) demonstrated that a treatment of electrolyzed oxidizing ice (100 ppm chlorine) for 24 h resulted in considerable reductions of histamine-producing bacteria on tuna skin and that, consequently, electrolyzed oxidizing ice may be regarded as a promising post-harvest decontamination treatment for fish. According to the results of another study, although sanitized ice appeared not to significantly reduce the bacterial load on tilapia fish fillets, addition of sanitizers (i.e., a neutral electrolyzed water sanitizer or an organic acid formulation) to ice used to store seafood holds promise for reducing the bacterial load of melting water (Feliciano *et al.*, 2010).

### 20.5.3 Produce

Fruits and vegetables are agricultural commodities that are commonly consumed raw, and frequently undergo minimal processing, such as alterations in form (e.g., by trimming, peeling, slicing, chopping, coning or shredding), washing

© Woodhead Publishing Limited, 2012

and/or decontamination (Francis *et al.*, 1999; USFDA, 2008). Decontamination has been acknowledged as one of the most important processing steps in the fresh-cut fruit and vegetable industry, affecting the quality, safety, and shelf life of the end product (Gil *et al.*, 2009; Ölmez and Kretzschmar, 2009). Although conventional washing technology was originally developed primarily to remove soil from produce and not microorganisms (Sapers, 2001), evaluation of the decontamination efficacy of various chemical treatments, either as part of or following water washing, soon constituted the objective of numerous research studies. Processors of fruits and vegetables usually rely on mechanical washing in the presence of sanitizers followed by rinsing with potable water to reduce the microbial load of fresh produce (Gil *et al.*, 2009; Heaton and Jones, 2008). Fresh produce destined for the manufacture of RTE products, after removal of outer layers or surface dirt, is thoroughly washed, with the washing agent being either water alone or an aqueous solution of a decontaminant, and chlorine has been routinely used for this purpose (Beuchat and Ryu, 1997; Francis *et al.*, 1999; Ölmez and Kretzschmar, 2009). It is assumed that in the absence of a decontamination agent, large quantities of water would be required in order to achieve the same level of microbial reduction (Gil *et al.*, 2009). Chemical agents, including organic acids, that have been evaluated for their decontamination efficacy on fresh produce or commercially utilized in the fresh fruit and vegetable industry, have been applied mainly in the form of dipping treatments.

Bacterial cells may not reside in exposed sites of produce surfaces, and this is particularly true in the case of leafy vegetables where microbial cells may locally invade the leaf interior avoiding exposure to external stresses (Lindow and Brandl, 2003). Bacterial attachment tends to be localized in pores, stomata, indentations and other natural irregularities of intact produce surfaces, as well as on cut surfaces or in surface punctures, crevices and cracks, escaping in this way contact with washing and sanitizing treatments (Adams *et al.*, 1989; Annous *et al.*, 2001; Ells and Hansen, 2006; Gil *et al.*, 2009; Liao and Cooke, 2001; Liao and Sapers, 2000; Sapers, 2001; Sapers *et al.*, 1999). Furthermore, the hydrophobic nature of the waxy cuticle covering leaves and other produce surfaces, such as cantaloupe melons, may not only affect bacterial attachment, but also protect surface contaminants from exposure to decontamination treatments when the latter are incapable of penetrating or dissolving these waxes (Adams *et al.*, 1989; Beuchat and Ryu, 1997; do Socorro Rocha Bastos *et al.*, 2005; Ölmez and Kretzschmar, 2009; Zhang and Farber, 1996). In addition to inaccessibility of attachment sites, factors impeding or even precluding disinfection of contaminated fresh produce commodities using aqueous solutions of chemical agents, include internalization of microorganisms within growing plant tissues and biofilm formation (Annous *et al.*, 2005; Fett, 2000; Gil *et al.*, 2009; Koseki *et al.*, 2001; Ölmez and Kretzschmar, 2009; Ölmez and Temur, 2010; Sapers, 2001; Ukuku and Sapers, 2001).

Hence, decontamination technologies applied in fresh fruits and vegetables

© Woodhead Publishing Limited, 2012

should be able to, at least to some extent, overcome the above limitations. Surface-active agents (surfactants) should help in this direction by reducing the hydrophobicity on fruit and vegetable skins as well as on the surface of edible leaves, stems and flowers, provided, however, that they do not deteriorate the sensory attributes of fresh produce (Adams *et al.*, 1989; Beuchat and Ryu, 1997; Escudero *et al.*, 1999; Materon, 2003; Ölmez and Kretzschmar, 2009; Ölmez and Temur, 2010; Zhang and Farber, 1996). Moreover, novel decontamination technologies mentioned previously in the present chapter, such as vapor-phase application and vacuum infiltration of antimicrobial agents, hold promise with regard to reaching microbial contamination attached in inaccessible sites of produce commodities (Sapers, 2001).

Due to the above limitations of produce decontamination using chemical agents, it has been suggested that controlling points of potential contamination throughout the food chain by good agricultural practices, GMP, and good hygiene practices (GHP) should be preferred over decontamination technologies for the control of foodborne pathogens associated with raw produce (Beuchat and Ryu, 1997; Sapers, 2001; USFDA, 1998). Nevertheless, despite their limited ability to inactivate microorganisms attached to produce surfaces, chemical agents have demonstrated a high efficacy against microorganisms suspended in water and, thus, a great potential to be used as water disinfection agents (Gil *et al.*, 2009; Sapers, 2001; Tauxe *et al.*, 1997; USFDA, 1998). The quality of water used for produce washing and chilling after harvesting has been demonstrated by epidemiological data to be critical for the microbiological safety of this food commodity (Gil *et al.*, 2009). Water can play a dual contradictive role in the fresh produce industry serving both as a means of reducing and potentially transferring pathogenic microorganisms to fruits and vegetables. Given that water in the fresh produce industry is mostly re-used, process water may constitute a source of cross-contamination of successive loads of fruits and vegetables with microorganisms, including pathogens (Gil *et al.*, 2009). Since such events of cross-contamination may take place even when large quantities of water are used or even in the presence of sanitizers (Francis *et al.*, 1999; López-Gálvez *et al.*, 2009; Nguyen-the and Prunier, 1989), it is really important that a chemical decontaminating agent is used in process water to inactivate microorganisms before they attach or become internalized in produce (Chaidez *et al.*, 2003; Gil *et al.*, 2009; Pao *et al.*, 2007), and that both the type and concentration of the applied chemical compound are properly selected. Hence, in addition to their direct decontamination efficacy on fresh produce commodities, chemical agents should also be evaluated with regard to their suitability to maintain the microbiological quality of process water, preventing cross-contamination between contaminated and non-contaminated products (Baert *et al.*, 2009; Gil *et al.*, 2009; Gonzalez *et al.*, 2004; López-Gálvez *et al.*, 2009). Chemical treatments demonstrating a satisfactory water disinfection activity can be applied to hydrocooling, flume and wash water as a means of reducing microbial populations in recirculating water systems (Sapers, 2001), providing, however, that disinfectants used

© Woodhead Publishing Limited, 2012

for water treatment are applied at their minimum effective doses (Gil *et al.*, 2009).

Chemical decontamination has been recognized as a critical pathogen control intervention with regard to sprout production too, with the human risk for disease from the consumption of these commodities being strongly associated with seed production (Gill *et al.*, 2003). In the latter case, chemical disinfection treatments are recommended to be applied at the seed level prior to germination, and should ensure complete inactivation of the pathogens while at the same time maintaining the seeds' viability and vigor (Beuchat, 1997; Kumar *et al.*, 2006; NACMCF, 1999). If pathogenic bacteria are present on or in seeds, either in the absence of decontamination treatments or as residual contamination surviving the applied decontamination interventions, sprouting conditions may favor their proliferation and high bacterial populations may develop and survive on mature sprouts at the commercial production and marketing levels (Holliday *et al.*, 2001; Jaquette *et al.*, 1996; NACMCF, 1999; Taormina and Beuchat, 1999b).

Various chemical decontamination treatments applied as dips have been specifically described for sprouting vegetable seeds including sodium hypochlorite, hydrogen peroxide, calcium hydroxide, ethanol and calcium hypochlorite, with the latter constituting a seed disinfection intervention recommended for implementation by the industry (Beuchat, 1997; Beuchat *et al.*, 2001; Holliday *et al.*, 2001; Jaquette *et al.*, 1996; NACMCF, 1999; Piernas and Guiraud, 1997; Taormina and Beuchat, 1999a, 1999b; Weissinger and Beuchat, 2000). Factors that may affect the adhesion of bacterial cells and consequently the efficacy of applied decontamination treatments include seed cultivar, degree of hardness, age, seed coat damage, as well as the type and amount of organic material surrounding the target cells (Beuchat *et al.*, 2001). The potential impact of seed scarification (i.e., treatment aiming at enhancing water uptake and, thus, germination uniformity during sprout production) and/or polishing on the efficacy of chemical treatments against pathogenic bacteria remains to be clarified (Holliday *et al.*, 2001). Some chemical treatments, such as ethanol, have been associated with inhibition of seed germination (Piernas and Guiraud, 1997), while others, such as organic acids applied as dipping solutions, appear to be incapable of eliminating or even significantly reducing populations of pathogenic bacteria when used at concentrations that do not reduce considerably seed germination (Taormina and Beuchat, 1999a; Weissinger and Beuchat, 2000). Furthermore, the inaccessibility of bacterial cells (e.g., in crevices or between the cotyledon and testa of seeds) to lethal concentrations of chemical agents often impairs their decontamination efficacy (Beuchat, 1997). Hence, there is a need for more effective disinfection procedures for seeds if the safety of sprouted vegetables is to be enhanced. Fumigation may constitute an appealing alternative to dipping treatments for sprouting seeds, and gaseous acetic acid appears to be an effective decontamination treatment when applied with this technology. According to the findings of Delaquis *et al.* (1999), fumigation with gaseous

© Woodhead Publishing Limited, 2012

acetic acid was lethal to pathogenic bacteria inoculated on mung bean seeds as well as to the indigenous microflora of this commodity, without affecting the surface microstructure of the seeds. However, the observed slight loss in seed germination rate and the need to design appropriate fumigation equipment are factors that need to considered and weighed against the benefits of such a treatment (Delaquis *et al.*, 1999).

## 20.6 Effects of chemical decontamination on food quality

In order for a chemical intervention, whose decontamination efficacy has been assessed and validated, to be actually applicable at a commercial level, it needs to also be assured that it does not affect detrimentally the quality characteristics of food products. Potentially undesirable effects of chemical decontamination may be associated with alterations in the color, odor, flavor or overall appearance of treated foods, and such effects may be observed either immediately after decontamination or during storage/display of decontaminated products.

Decontamination with organic acids has frequently been associated with color and odor/flavor changes of treated foods. Organic acid decontamination treatments of raw meat, applied individually or as part of combination treatments, and primarily acetic and lactic acids, have been shown to result in considerable discoloration, with treated meat samples appearing to be paler and less red compared to untreated samples (Castelo *et al.*, 2001; Kotula and Thelappurate, 1994; Lin and Chuang, 2001; Pipek *et al.*, 2005; Stivarius *et al.*, 2002a, 2002b; van Netten *et al.*, 1995). Color changes may also constitute a concern with regard to processed meat products subjected to post-processing decontamination treatments with organic acids or organic acid salts (Geornaras *et al.*, 2005; Lu *et al.*, 2005). Moreover, color deterioration has been observed in acid-treated poultry (Anang *et al.*, 2006; Bautista *et al.*, 1997; Bilgili *et al.*, 1998; Deumier, 2004; González-Fandos and Dominguez, 2006; Kim and Marshall, 2000b; Mendonca *et al.*, 1989) and seafood (Bal'a and Marshall, 1998; Ingham, 1989; Kim and Marshall, 2001; Marshall and Kim, 1996; Shirazinejad *et al.*, 2010). The discrete decoloration of poultry observed after treatment with lactic acid solutions could be attributed to oxidation reactions (Mendonca *et al.*, 1989), or to protein denaturation caused by the quick decrease of pH (González-Fandos and Dominguez, 2006), which would also justify the absence of a similar observation when treating poultry with lactic acid buffer (Zeitoun and Debevere, 1990).

Organic acid, and particularly lactic acid, decontamination of fish has been associated with noticeable whitening of flesh surface and skin mucus (Bal'a and Marshall, 1998; Ingham, 1989; Kim and Marshall, 2001; Marshall and Kim, 1996). According to Bal'a and Marshall (1998), Hunter color analysis revealed that acid-treated catfish fillets were lighter and more yellow than untreated fillets. However, such sensory changes are not believed to have a

© Woodhead Publishing Limited, 2012

detrimental impact on the consumer acceptance of catfish fillets since their surface is usually coated with batter and breading prior to cooking (Ingham, 1989). With reference to produce, acetic and propionic acid treatments of lettuce have led to a significant negative effect on the color and overall appearance of this commodity (Samara and Koutsoumanis, 2009), while decontamination with fumaric acid has been shown to promote browning of lettuce even when applied at low concentrations (Kondo *et al.*, 2006). Furthermore, parsley treated with vinegar containing ≥2.6% acetic acid was noticeably discolored (Wu *et al.*, 2000).

Negative effects of organic acid treatments on the odor/flavor of treated foods have been documented at several instances, and include an acidic or vinegar-like odor and/or a sour flavor of treated poultry, seafood, or produce (Chang and Fang, 2007; Kim and Marshall, 2000b; Marshall and Kim, 1996; Vijayakumar and Wolf-Hall, 2002; Wu *et al.*, 2000), a less beef-like odor and more off odor during refrigerated display of ground beef originating from treated beef trimmings compared to untreated samples (Stivarius *et al.*, 2002a), as well as lower odor and flavor scores of treated processed meat products compared to untreated (Geornaras *et al.*, 2005).

Negative effects of chemical decontamination on food quality have also been documented for other agents including TSP and hydrogen peroxide. As reported by Lin and Chuang (2001), dipping of pork loin chops in 10% TSP resulted in the development of a dark and dry surface. According to the findings of Capita *et al.* (2000b), although dipping of poultry in 8 and 10% solutions of TSP did not adversely affect the sensory quality of the product, the color, flavor and overall acceptability of poultry dipped in 12% TSP were rated significantly lower than the control sample. Similarly, treatment of chicken legs with TSP resulted in darker, less red and less yellow color compared to untreated legs, as well as in detectable chemical odor when the chemical agent was applied at concentrations higher than 10% (Kim and Marshall, 1999). With reference to fresh produce, TSP may compromise its quality even at low concentrations, as supported by research data on fresh-cut lettuce, indicating that it may not be a suitable decontamination agent for this food commodity (Zhang and Farber, 1996). Due to its inherent phytotoxicity, hydrogen peroxide may also deteriorate remarkably the quality of some produce items. At the concentrations required in order to exert a noticeable antibacterial activity (i.e., 4–5%), hydrogen peroxide is likely to interfere with the overall quality of fruits and vegetables (Ölmez and Kretzschmar, 2009). It has been shown to cause bleaching of anthocyanins in mechanically damaged berries, as well as to induce extensive browning on apple skin, lettuce, and mushrooms when applied in the absence of an anti-browning agent (McWatters *et al.*, 2002b; Ölmez and Kretzschmar, 2009; Sapers, 2001). The combined application of hydrogen peroxide and lactic acid in decontaminating produce commodities has given different results with regard to quality effects for different products; treatment of lettuce resulted in severe browning of this commodity (Lin *et al.*, 2002; McWatters *et al.*,

© Woodhead Publishing Limited, 2012

2002b), while the sensory and quality characteristics of apples treated with such a chemical solution did not appear to be adversely affected (McWatters *et al.*, 2002a; Venkitanarayanan *et al.*, 2002).

In some cases, the negative effects on food quality caused by chemical decontamination are only transitory and can be easily overcome, while ways for avoiding such effects have also been proposed. The addition of color stabilizers such as nicotinic and ascorbic acid in organic acid solutions may reduce the discoloration caused in meat products (van Netten *et al.*, 1995). Furthermore, the issue of discoloration of meat/poultry products can be minimized with shorter contact times, immediate rinsing with water post-treatment and subsequent water chilling (Bautista *et al.*, 1997). With respect to fresh produce items, washing with water after the application of the chemical decontamination treatment may be beneficial for their quality attributes. Indeed, a simple wash of vinegar-treated parsley with tap water for 1 min was able to restore its color and diminish the undesirable odor given to the product by the decontamination treatment (Wu *et al.*, 2000). Similarly, washing with tap water of lettuce, previously treated with rice vinegar, improved its initially unacceptable sour flavor (Chang and Fang, 2007). Moreover, the combined application of hydrogen peroxide and mild heat followed by a cold water rinse has been shown to be capable of maintaining, and potentially improving, the sensory quality of fresh-cut lettuce (Lin *et al.*, 2002). Finally, with regard to fish fillets processing, detrimental sensory changes can be avoided by applying chemical decontamination treatments with skin on rather than treating skinless fillets (Bal'a and Marshall, 1998).

There are several research studies describing the application of chemical decontamination treatments that did not affect the quality and organoleptic attributes of treated products (del Río *et al.*, 2007b; McWatters *et al.*, 2002a; Samara and Koutsoumanis, 2009; Singh *et al.*, 2005; Sinhamahapatra *et al.*, 2004; Venkitanarayanan *et al.*, 2002). Sinhamahapatra *et al.* (2004) investigated the effects of different decontaminants applied on the surface of dressed broilers (2% lactic acid for 30 s, 1200 ppm ASC for 5 s, and 50 ppm chlorine for 5 min), and concluded that none of the tested treatments affected muscle pH, water holding capacity, extract release volume, appearance, smell, tenderness, or overall acceptability of broilers significantly. In addition, as demonstrated by the findings of other investigations, chemical decontamination treatments may even improve the quality of food products, outperforming control samples (i.e., untreated samples) for sensory/quality characteristics such as oxymyoglobin and color stability, as well as overall acceptability (del Río *et al.*, 2007b; Doležalová *et al.*, 2010; Gögüs *et al.*, 2007; Pohlman *et al.*, 2002; Quilo *et al.*, 2010). Given the above, it becomes evident that the development of chemical treatments exerting a significant decontamination efficacy without compromising food quality is attainable. Parameters pertinent to decontamination that may have a considerable impact on food quality attributes and should, therefore, be carefully selected include the chemical agents' concentrations, the exposure time (i.e., duration of

© Woodhead Publishing Limited, 2012

the applied chemical treatment), as well the application conditions (Bilgili *et al.*, 1998; Capita *et al.*, 2000b; Pipek *et al.*, 2005; Shirazinejad *et al.*, 2010). The product-specific optimization of these parameters is expected to facilitate the minimization or even the avoidance of negative effects of chemical decontamination on food quality.

## 20.7 Potential concerns and risks associated with chemical decontamination

In order for chemical decontamination treatments to be acceptable for use in the food industry, certain criteria, in addition to food quality retention, need to be met. The absence of adverse toxicological or other health effects on food workers and consumers needs to be assured, while the potential impact of their use on the environment is another parameter of major significance. For instance, as already mentioned previously, chlorine and chlorine-based disinfectants can react with organic matter and result in the formation of potentially mutagenic or carcinogenic halogenated by-products such as trihalomethanes, haloacetic acids, haloketones and chloropicrin (Gil *et al.*, 2009; Komulainen, 2004; Ölmez and Kretzschmar, 2009). Furthermore, chemical agents, including organic acids, may cause respiratory and skin/eye irritation of operators, particularly when used at high concentrations (Bolton *et al.*, 2001). The use of chlorine, in addition to the potential health effects mentioned above, has also been associated with the production of high amounts of wastewater with very high levels of biological oxygen demand (Ölmez and Kretzschmar, 2009). It is due to these health and environmental risks posed by the use of chlorine that its use is not universally accepted, with the regulatory authorities of many countries questioning its safety, and with the research community seeking alternative decontamination treatments.

Nevertheless, with regard to environmental pollution, it is not only the use of chlorine that warrants attention. The use of organic acids for disinfection purposes in the fresh-cut produce industry, for example, is also expected to have an impact on the wastewater quality (Ölmez and Kretzschmar, 2009). Hence, technological parameters of decontamination treatments, such as concentration of the applied chemical agent and extent of use, need to be decided on with the potential health and environmental impacts always being kept in mind. For example, as mentioned previously, the potential increase in the phosphate content of the generated wastewater has been acknowledged as one of the major limitations for the widespread use of TSP in food decontamination (Oyarzabal, 2005). In general, for the purpose of recycling wastewater, high concentration applications of chemical treatments should be avoided, irrespective of the chemical agent used (Li *et al.*, 1997).

Another factor that should be taken into account when carrying out studies on the efficiency of chemical decontamination treatments is the potential impact

© Woodhead Publishing Limited, 2012

of the latter on the microbial ecology of treated products. Such an impact may be indirect or expressed as a direct effect on the microbial composition of the treated product. An example of a potential indirect effect would be an increase in the humidity on the surface of decontaminated products, as is likely to occur with certain chemical treatments, affecting consequently microbial growth and, therefore, the products' shelf life (Dorsa *et al.*, 1998a; Heller *et al.*, 2007). Direct effects of decontamination treatments on the microbiology of treated products may be associated with bacterial injury and potential subsequent repair. Spoilage or pathogenic bacteria, injured by the application of a chemical treatment, may recover during subsequent processing and storage of food products, and the significance and implications of such phenomena in microbiological food safety and quality need to be assessed (Anderson and Marshall, 1990b; Dickson, 1992; Dickson and Siragusa, 1994; Gill and Badoni, 2004; van Netten *et al.*, 1997a). Moreover, a concern that has been raised with regard to decontamination interventions in general, including chemical decontamination, is whether changes in the natural microflora of food products caused by these interventions will result in augmented growth of bacterial pathogens due to possible reduction of microbial competition. With this in mind, it is evident that assessment of the spoilage microbial profile of chemically treated foods is also important.

Koutsoumanis *et al.* (2004) reported that the predominant spoilage association of lactic acid-treated fresh beef was shifted during aerobic storage from Gram-negative to Gram-positive bacteria and yeasts, and hypothesized that this change could be beneficial from a food safety perspective, since *L. monocytogenes* growth in treated samples was limited compared to untreated samples. A similar lactic acid-induced shift in the bacterial community of fresh meat was documented by van Netten *et al.* (1994), who, however, did not observe an antagonistic reduction against meatborne pathogens. As frequently demonstrated by *in vitro* experimental data, the microbial changes caused by chemical decontamination treatments may compromise the antagonistic control of foodborne pathogens surviving decontamination or contaminating food products post-treatment (del Río *et al.*, 2006, 2008; Nissen *et al.*, 2001; Ukuku, 2006). Indeed, according to the results of a study carried out by Nissen *et al.* (2001), *E. coli* O157:H7 growth was 2–3 logs greater on beef treated with a combination of steam vacuum and lactic acid (0.2 M) than on untreated samples stored at 10°C in air and vacuum packages.

A competitive advantage given to bacterial pathogens by decontamination treatments such as the above is of particular importance for vacuum-packaged products, since the extended shelf life of the latter may allow for prolific pathogen growth, deteriorating in this way their safety, provided that they are stored at growth allowing temperatures. The concentration of a chemical decontaminant appears to be a fundamental technological aspect with regard to the aforementioned issue too. When the growth kinetic behavior of pathogenic and spoilage bacteria was evaluated at 28°C in culture broths containing

© Woodhead Publishing Limited, 2012

different chemical agents, it was suggested that low TSP and high citric acid concentrations could favour the outgrowth of pathogenic bacteria, such as *Salmonella* and *L. monocytogenes*, relative to spoilage bacteria, rendering these treatments potentially dangerous for the consumer (del Río *et al.*, 2008). Given the above, research studies need to determine the likelihood of occurrence of this scenario, as well as its implications for food safety, for different products and decontamination and storage conditions, in order to make sure that extensive growth of foodborne pathogens will not occur within the shelf life of a given product. The addition of harmless microflora, such as a protective culture of lactic acid bacteria, as a means of partially restoring the background microflora of chemically treated products has been proposed (Jay, 1995; Nissen *et al.*, 2001). In any case, proper handling of decontaminated products after treatment and before packaging in the context of a strict HACCP approach is expected to be of vital importance for controlling the risk discussed above (Nissen *et al.*, 2001).

An important potential risk associated with chemical decontamination is the development of enhanced microbial resistance to stressful environmental conditions. Such a risk is particularly pertinent to organic acid decontamination in the context of which acid adaptation phenomena may be induced (Davidson and Harrison, 2003). The potential induction of acid resistance mechanisms in the surviving population under sublethal acid stress conditions, resulting in alterations in its virulence characteristics, constitutes a significant concern in the scientific community, as well as among food processors and legislators (Hill *et al.*, 1995; Theron and Lues, 2007). The term 'acid adaptation', also known as 'acid habituation' (Goodson and Rowbury, 1989) or 'acid tolerance response' (Foster and Hall, 1990), refers to the increased microbial resistance to extreme pH conditions after adaptation to sublethal acidic environments. Bacterial strains capable of adapting to sublethal pH environments encountered during the application of organic acid decontamination treatments, may recontaminate treated food products, exhibit an enhanced survival and growth in such foods, provided that the stressing pressure is still present, or even resist subsequently applied decontamination treatments.

The findings documented in the literature with regard to the likelihood of occurrence of the above scenarios, and particularly with regard to the extent to which acid adaptation may contribute to pathogen survival in decontaminated foods, have not been consistent (Berry and Cutter, 2000; Beuchat and Scouten, 2004; Ikeda *et al.*, 2003; Stopforth *et al.*, 2004; van Netten *et al.*, 1997b, 1998). For example, Berry and Cutter (2000) reported that acid-adapted *E. coli* O157:H7 exhibited an enhanced survival in vacuum-packaged beef treated with acetic acid and stored at 4°C compared to non-adapted cultures. In contrast to these findings, acid adaptation of *L. monocytogenes* did not appear to affect the survival and growth of the pathogen in beef treated with hot water and/or lactic acid and stored at 10°C (Ikeda *et al.*, 2003). It seems that parameters such as the inherent acid resistance of different bacterial species or strains, the applied product storage conditions, as well as the

© Woodhead Publishing Limited, 2012

method employed for the induction of acid adaptation (e.g., pH adjustment with organic acids or addition of glucose to the culture medium) may account for conflicting results such as the above (Samelis, 2005).

In addition to the above, acid-adapted bacterial pathogens may survive in the sublethal pH environments created in acidic decontamination runoff fluids, and, if contained in microbial niches established in the processing environment that favor their survival, growth, and acid tolerance responses, they may also serve as cross-contamination sources (Samelis *et al.*, 2001b, 2001c, 2002, 2004, 2005). Under certain conditions, and depending on the natural microflora also present in acidic decontamination washings, acid-stressed pathogen survivors may resuscitate, with such resuscitation potentially resulting in enhanced pathogen survival and prevalence in biofilms (Stopforth *et al.*, 2003). Acid adaptation, induced by organic acid decontamination treatments, may result in enhanced pathogen survival during transit through the stomach, increasing the likelihood of intestinal colonization and, therefore, their virulence potential. Decontamination of lettuce with organic acids did not seem to increase the acid tolerance of *L. monocytogenes* during subsequent exposure to simulated gastric fluid; on the contrary, lactic acid treatment appeared to sensitize the pathogen to acidic simulated stomach conditions (Samara and Koutsoumanis, 2009).

Additional concerns that have been expressed with regard to chemical decontamination, and which require further investigation in order for a conclusive scientific opinion to be formed, include the development of microbial resistance to the chemical agents used as well as of cross-tolerance (or cross-protection) to other types of environmental stress (e.g., heat, osmotic) (Skandamis *et al.*, 2010; Theron and Lues, 2007).

Despite the above, when chemical agents are selected and applied properly and at appropriate concentrations, chemical decontamination treatments are expected to constitute valuable pathogen control interventions with the food safety risks associated with stress-adaptation phenomena being minimized. Given the availability of many different chemical agents that can be utilized in decontamination systems, rotation of the use of differing agents over time within a certain food processing facility has been proposed as a means of preventing selection for bacterial resistance (Samelis *et al.*, 2001b). Stress responses of foodborne pathogens and potential strategies for control of stress-adapted bacteria with the objective of enhancing food safety are reviewed and discussed by Samelis and Sofos (2003a). It should be noted that resistance associated with stress adaptation is lost upon exposure of the stressed cells to environments without the stressor, as they regain their original sensitivity because the resistance is acquired through adaptation without mutation. Thus, proper cleaning and sanitation on a daily basis should eliminate niches or potential harborage sites of stress-adapted cells.

© Woodhead Publishing Limited, 2012

## 20.8 Legislative aspects of chemical decontamination

Organic acids, by constituting inexpensive and safe chemical agents, were very soon acknowledged as having considerable potential for acceptance and use by the food industry (Siragusa, 1995). Indeed, organic acids have been assigned a 'generally recognized as safe' (GRAS) status by the United States Food and Drug Administration (US FDA) and have been approved for use, with American abattoirs being, most likely, the first to incorporate organic acid sprays as part of the beef carcass dressing process (Smulders and Greer, 1998). The United States Department of Agriculture Food Safety and Inspection Service (USDA-FSIS) has approved organic acids (acetic, citric and lactic acid) as acceptable interventions for the reduction of microbial pathogens on meat carcasses, cuts, and trimmings (USDA-FSIS, 2010). More specifically, organic acids may be applied at concentrations of up to 5% (aqueous solutions) and at temperatures of up to 55°C in the form of sprays/rinses or dips (USDA-FSIS, 2010), and the application of such treatments (primarily lactic and acetic acid) is widely practiced in the United States for the purpose of beef carcass decontamination (Smulders and Greer, 1998). Importation of American beef from acid-treated carcasses was approved by Canadian regulatory authorities much earlier than the actual approval for use of organic acids in Canadian abattoirs (Smulders and Greer, 1998). The utilization of organic acid sprays on red meat has been included as a microbial control agent in directives of the Canadian Food Inspection Agency, with their use being identified as a 'processing aid' and approved as an adjunct to GMP during the carcass dressing process (Smulders and Greer, 1998; Theron and Lues, 2007).

In addition to organic acids, several other chemical agents have been approved for use in decontamination systems in the United States, such as TSP, chlorine dioxide, ASC and PAA. Trisodium phosphate has a GRAS status as a food processing aid (i.e., no label declaration needed), and the USDA-FSIS has approved its use for decontamination of beef and broiler carcasses (Capita *et al.*, 2002c). Furthermore, treatment with 10% TSP (AvGard™, Rhône, Poulenc, France) is patented and officially accepted and implemented in the United States as part of poultry slaughter process (Bolder, 1997). With reference to chloride dioxide's application in the United States, this chemical is only allowed to be used in whole produce or poultry, with a maximum concentration of 3 ppm being allowable in water that is in contact with fresh fruits and vegetables or in chilled poultry water, and with rinsing of treated products with potable water being necessary (Ölmez and Kretzschmar, 2009; Rajkovic *et al.*, 2010).

In the United States, wash water disinfectants used for fresh-cut produce are regulated by the FDA as secondary direct food additive, unless they are regarded as GRAS. If the product to be treated is a raw agricultural commodity and is washed in a food processing facility, then both the Environmental Protection Agency (EPA) and the FDA have regulatory jurisdiction, and

© Woodhead Publishing Limited, 2012

the disinfectant products are required to be registered as pesticides with the EPA (Gil *et al.*, 2009). A list of wash water disinfectants and sanitizing solutions approved by the FDA is reported in the Code of Federal Regulations 21 CFR Sections 173.315 and 178.1010 (CFR, 2011a, 2011b). The FDA, EPA, and USDA have approved ASC as an antimicrobial compound for use on fruits and vegetables, seafood, and red meat and poultry (Keener *et al.*, 2004). Regarding PAA and its use in the decontamination of fruits and vegetables, it is allowed up to a concentration of 80 ppm in wash water (Ölmez and Kretzschmar, 2009). Despite the fact that it has been assigned a GRAS status, hydrogen peroxide's use in the food industry is limited only to certain products (milk, dried egg, starch, tea, and wine) as an antimicrobial or bleaching agent, with the United States CFR necessitating that residual hydrogen peroxide is removed by appropriate physical and chemical methods during food processing (Ölmez and Kretzschmar, 2009). Agents approved by the USDA-FSIS for use in meat and poultry are listed in Directive 7120.1, Revision 2, 4/12/10 (USDA-FSIS, 2010).

The European Union (EU) has traditionally been reluctant to embrace the use of chemical agents for the removal of microbial contamination of foods. Meat hygiene regulations within the EU do not allow any method of decontamination of red meat/poultry carcasses, parts or viscera, other than washing with potable water (Capita *et al.*, 2002c; Theron and Lues, 2007). The unwillingness of European legislators to grant permission for adoption of chemical decontamination interventions stems from the fear that the latter may conceal or compensate for poor hygienic practices, with existing legislation highlighting, on the other hand, the significance of applying GMP and/or food safety management systems throughout the production line (Hugas and Tsigarida, 2008; Capita *et al.*, 2002c; Theron and Lues, 2007). Nevertheless, the EU has provided a legal basis for the use of substances other than potable water to decontaminate foods of animal origin (OJEU, 2004), and a draft Regulation proposal setting the conditions for such decontamination treatments is under discussion with the Member States and stakeholders (Hugas and Tsigarida, 2008).

Upon official request from the European Commission, the European Food Safety Authority (EFSA), being the risk assessment body in food safety, is responsible for evaluating the safety and efficacy of substances intended to be used for decontamination purposes (Hugas and Tsigarida, 2008). Several scientific opinions have been issued so far by EFSA regarding the efficacy of chemical agents for use in decontamination of foods of animal origin, such as lactic acid and peroxyacids (EFSA, 2005, 2006a, 2006b; Hugas and Tsigarida, 2008), with their use, however, still not being authorized in the EU (Rajkovic *et al.*, 2010). Recently, EFSA issued a scientific opinion on the assessment of the safety and efficacy of lactic acid when used to reduce microbial surface contamination on beef hides, carcasses, cuts, and trimmings (EFSA, 2011); it was concluded that, with reference to human toxicological effects, such treatments would be of no safety concern provided that the substance

© Woodhead Publishing Limited, 2012

used complies with the EU specifications for food additives (OJEU, 2008). Furthermore, it was pointed out that, although variable, microbial reductions achieved by lactic acid treatment of beef are generally significant compared to untreated or water treated controls. As should be normal practice, it is recommended that, according to HACCP principles, during use, business operators verify lactic acid concentration, temperature of application and other factors affecting its efficacy as a decontaminating agent, and validate its efficacy under their specific processing conditions (EFSA, 2011).

With regard to surface decontamination of fresh produce, although the application of chemical agents is not legalized in the EU (Rajkovic *et al.*, 2010), there are opportunities to use substances, such as chlorine and chlorine dioxide, for fruit and vegetable washing, provided that they function as processing aids (Gil *et al.*, 2009). It is really important that wash water chemicals are used as processing aids and not as additives, since the former are more likely to receive consumer acceptance when it comes to raw agricultural food commodities. Nevertheless, legislation on processing aids is not yet harmonized at EU level, and developing more detailed regulations governing their use is one of the Commission's goals for the near future (Gil *et al.*, 2009).

## 20.9 Future trends

The development and implementation of chemical decontamination treatments capable of assuring the safety and quality of foods without, however, posing unacceptable risks to the consumers and the environment, still constitute a challenge for researchers and food processors. Future research on decontamination treatments should focus on low-cost and safe applications that do not result in residues in treated products, thereby facilitating consumer acceptance (Dinçer and Baysal, 2004), with issues associated with environmental pollution and sustainable use of resources constituting major targets in the context of food safety improvement (Ölmez and Kretzschmar, 2009; Sofos, 2008). Maximization of the antimicrobial activity of decontamination interventions may be achieved through proper selection of processes/treatments, and optimization of their intensities, combinations, sequences, and timing of application (Sofos, 2008). Nevertheless, given that extensive use of chemical agents such as organic acids resulting in bacterial adaptation also represents an important challenge for future research (Theron and Lues, 2007), the objective is for all the above to be done without leading to pathogen stress adaptation, resistance selection, or cross-protection phenomena (Sofos, 2008). Further qualification and quantification of the physiological and molecular responses to stresses by which foodborne bacterial pathogens adapt, acquire resistance and modify their virulence characteristics, are expected to be useful in this direction;

© Woodhead Publishing Limited, 2012

a better understanding of the general and specific mechanisms underlying the adaptive responses of foodborne pathogens is likely to lead to improved strategies for their control (Rajkovic *et al.*, 2010; Ricke, 2003).

Molecular methods, which are anyhow rapidly replacing the more traditional culture-based methods for pathogen detection when assessing the efficacy of decontamination procedures (Gil *et al.*, 2009), are expected to also be very useful in evaluating bacterial responses under various environmental conditions (Ricke, 2003; Ricke *et al.*, 2005). Utilization of molecular approaches allowing for quantification of gene responses at the mRNA synthesis level is expected to provide more detailed information regarding pathogen responses, favoring, in this way, the optimization of hurdle technologies (Ricke *et al.*, 2005). Furthermore, and with particular reference to produce decontamination, treatments capable of eliminating or decomposing biofilms should be further investigated, and a better understanding of the mechanisms involved in bacterial attachment and biofilm formation (e.g., cell physiology and morphology, cell–cell and cell–food surface interactions, adhesion kinetics) is expected to contribute to the improvement of the existing technologies and to the development of new interventions (Ölmez and Kretzschmar, 2009).

When assessing the antibacterial efficacy of chemical agents, it is important that, in addition to its demonstration under laboratory conditions, it is also validated under commercial conditions, and this is another issue that future research should address. Future food safety regulations and research activities should be based on the findings of proper risk assessment, with the contribution of quantitative microbiology (utilizing mathematical models that predict microbial behavior) being of vital importance in this direction (Sofos, 2008). Moreover, methods, treatments, processes, interventions, and hurdles applied for the purpose of foodborne pathogen control should be managed properly, and in order for this to be feasible, collaboration, cooperation and coordination among the various sectors involved (i.e., industry, scientific community, regulatory authorities, and public health agencies) need to be in place (Sofos, 2008). Finally, harmonization of food safety related regulatory activities and requirements, both at national and international levels, is expected to become more important in the near future; global harmonization of activities pertinent to pathogen control should contribute to optimization of the use of resources, facilitate trade, and enhance food safety (Gil *et al.*, 2009; Sofos, 2008).

## 20.10 Sources of further information and advice

Additional information with regard to the use of organic acids and other chemical treatments for the purpose of food decontamination can be obtained from the books/book chapters, and the food safety authorities and trade associations suggested below.

© Woodhead Publishing Limited, 2012

### 20.10.1 Books and book chapters

- Byelashov and Sofos, 2009
- Davidson and Harrison, 2003
- Samelis and Sofos, 2003a
- Samelis and Sofos, 2003b
- Skandamis *et al.*, 2010
- Sofos, 2005
- Stratford and Eklund, 2003
- Theron and Lues, 2011

### 20.10.2 Food safety authorities

- European Food Safety Authority (http://www.efsa.europa.eu)
- United States Department of Agriculture Food Safety and Inspection Service (http://www.fsis.usda.gov)
- United States Food and Drug Administration (http://www.fda.gov)

### 20.10.3 Food trade associations

- American Association of Meat Processors (http://www.aamp.com)
- American Meat Institute (http://www.meatami.org)
- American Meat Science Association (http://www.meatscience.org)
- Canadian Institute of Food Science and Technology (http://www.cifst.ca)
- Food Processors of Canada (http://foodnet.fic.ca)
- Grocery Manufacturers of America (http://www.gmaonline.org)
- Institute of Food Technologists (http://www.ift.org)
- International Association for Food Protection (http://www.foodprotection.org)
- International Food Information Council Foundation (http://ific.org)
- National Cattlemen's Beef Association (http://www.beef.org)
- National Meat Association (http://www.nmaonline.org)
- National Pork Board (http://www.pork.org)
- North American Meat Processors Association (http://www.namp.com)
- Produce Marketing Association (http://www.pma.com)
- US Poultry and Egg Association (http://www.poultryegg.org)
- United Fresh Fruit and Vegetable Association (http://www.unitedfresh.org)

## 20.11 References

ADAMS MR and HALL CJ (1988), 'Growth inhibition of food-borne pathogens by lactic and acetic acids and their mixtures', *Int J Food Sci Technol*, 23, 287–292.

© Woodhead Publishing Limited, 2012

ADAMS MR, HARTLEY AD and COX LJ (1989), 'Factors affecting the efficacy of washing procedures used in the production of prepared salads', *Food Microbiol*, 6, 69–77.

AHARONI Y, COPEL A and FALLIK E (1994), 'The use of hydrogen peroxide to control postharvest decay on "Galia" melons', *Ann Appl Biol*, 125, 189–193.

AKBAS MY and ÖLMEZ H (2007a), 'Inactivation of *Escherichia coli* and *Listeria monocytogenes* on iceberg lettuce by dip wash treatments with organic acids', *Lett Appl Microbiol*, 44, 619–624.

AKBAS MY and ÖLMEZ H (2007b), 'Effectiveness of organic acid, ozonated water and chlorine dippings on microbial reduction and storage quality of fresh-cut iceberg lettuce', *J Sci Food Agric*, 87, 2609–2616.

ALAKOMI H-L, SKYTTÄ E, SAARELA M, MATTILA-SANDHOLM T, LATVA-KALA K and HELANDER IM (2000), 'Lactic acid permeabilizes Gram-negative bacteria by disrupting the outer membrane', *Appl Environ Microbiol*, 66, 2001–2005.

ALBRICH JM, MCCARTHY CA and HURST JK (1981), 'Biological reactivity of hypochlorous acid: implications for microbicidal mechanisms of leukocyte myeloperoxidase', *Proc Natl Acad Sci*, 78, 210–214.

ALVARADO-CASILLAS S, IBARRA-SÁNCHEZ S, RODRÍGUEZ-GARCÍA O, MARTÍNEZ-GONZÁLES N and CASTILLO A (2007), 'Comparison of rinsing and sanitizing procedures for reducing bacterial pathogens on fresh cantaloupes and bell peppers', *J Food Prot*, 70, 655–660.

ANANG DM, RUSUL G, RADU S, BAKAR J and BEUCHAT LR (2006), 'Inhibitory effect of oxalic acid on bacterial spoilage of raw chilled chicken', *J Food Prot*, 69, 1913–1919.

ANANG DM, RUSUL G, BAKAR J and LING FH (2007), 'Effects of lactic acid and lauricidin on the survival of *Listeria monocytogenes*, *Salmonella enteritidis* and *Escherichia coli* O157:H7 in chicken breast stored at 4°C', *Food Control*, 18, 961–969.

ANDERSON ME and MARSHALL RT (1990a), 'Reducing microbial populations on beef tissues: concentration and temperature of an acid mixture', *J Food Sci*, 55, 903–905.

ANDERSON ME and MARSHALL RT (1990b), 'Reducing microbial populations on beef tissues: concentration and temperature of lactic acid', *J Food Saf*, 10, 181–190.

ANDERSON ME, MARSHALL RT and DICKSON JS (1992a), 'Efficacies of acetic, lactic and two mixed acids in reducing numbers of bacteria on surfaces of lean meat', *J Food Saf*, 12, 139–147.

ANDERSON ME, MARSHALL RT and DICKSON JS (1992b), 'Estimating depths of bacterial penetration into post-rigor carcass tissue during washing', *J Food Saf*, 12, 191–198.

ANNOUS BA, SAPERS GM, MATTRAZZO AM and RIORDAN DCR (2001), 'Efficacy of washing with a commercial flatbed brush washer, using conventional and experimental washing agents, in reducing populations of *Escherichia coli* on artificially inoculated apples', *J Food Prot*, 64, 159–163.

ANNOUS BA, SOLOMON EB, COOKE PH and BURKE A (2005), 'Biofilm formation by *Salmonella* spp. on cantaloupe melons', *J Food Saf*, 25, 276–287.

ARIYAPITIPUN T, MUSTAPHA A and CLARKE AD (2000), 'Survival of *Listeria monocytogenes* Scott A on vacuum-packaged raw beef treated with polylactic acid, lactic acid, and nisin', *J Food Prot*, 63, 131–136.

ARRITT FM, EIFERT JD, PIERSON MD and SUMNER SS (2002), 'Efficacy of antimicrobials against *Campylobacter jejuni* on chicken breast skin', *J Appl Poult Res*, 11, 358–366.

BAERT L, VANDEKINDEREN I, DEVLIEGHERE F, VAN COILLIE E, DEBEVERE J and UYTTENDAELE M (2009), 'Efficacy of sodium hypochlorite peroxyacetic acid to reduce murine norovirus 1, B40-8, *Listeria monocytogenes*, and *Escherichia coli* O157:H7 on shredded iceberg lettuce and in residual wash water', *J Food Prot*, 72, 1047–1054.

BAL'A MFA and MARSHALL DL (1998), 'Organic acid dipping of catfish fillets: effect on color, microbial load, and *Listeria monocytogenes*', *J Food Prot*, 61, 1470–1474.

BARMPALIA IM, GEORNARAS I, BELK KE, SCANGA JA, KENDALL PA, SMITH GC and SOFOS JN (2004), 'Control of *Listeria monocytogenes* on frankfurters with antimicrobials in the formulation and by dipping in organic acid solutions', *J Food Prot*, 67, 2456–2464.

© Woodhead Publishing Limited, 2012

BARMPALIA IM, KOUTSOUMANIS KP, GEORNARAS I, BELK KE, SCANGA JA, KENDALL PA, SMITH GC and SOFOS JN (2005), 'Effect of antimicrobials as ingredients of pork bologna for *Listeria monocytogenes* control during storage at 4 or 10°C', *Food Microbiol*, 22, 205–211.

BASHOR MP, CURTIS PA, KEENER KM, SHELDON BW, KATHARIOU S and OSBORNE JA (2004), 'Effects of carcass washers on *Campylobacter* contamination in large broiler processing plants', *Poult Sci*, 83, 1232–1239.

BAUTISTA DA, SYLVESTER N, BARBUT S and GRIFFITHS MW (1997), 'The determination of efficacy of antimicrobial rinses on turkey carcasses using response surface designs', *Int J Food Microbiol*, 34, 279–292.

BELL KY, CUTTER CN and SUMNER SS (1997), 'Reduction of foodborne micro-organisms on beef carcass tissue using acetic acid, sodium bicarbonate, and hydrogen peroxide spray washes', *Food Microbiol*, 14, 439–448.

BERRY ED and CUTTER CN (2000), 'Effects of acid adaptation of *Escherichia coli* O157:H7 on efficacy of acetic acid spray washes to decontaminate beef carcass tissue', *Appl Environ Microbiol*, 66, 1493–1498.

BEUCHAT LR (1997), 'Comparison of chemical treatments to kill *Salmonella* on alfalfa seeds destined for sprout production', *Int J Food Microbiol*, 34, 329–333.

BEUCHAT LR and RYU J-H (1997), 'Produce handling and processing practices', *Emerg Infect Dis*, 3, 459–465.

BEUCHAT LR and SCOUTEN AJ (2004), 'Viability of acid-adapted *Escherichia coli* O157:H7 in ground beef treated with acidic calcium sulfate', *J Food Prot*, 67, 591–595.

BEUCHAT LR, WARD TE and PETTIGREW CA (2001), 'Comparison of chlorine and a prototype produce wash product for effectiveness in killing *Salmonella* and *Escherichia coli* O157:H7 on alfalfa seeds', *J Food Prot*, 64, 152–158.

BEUCHAT LR, ADLER BB and LANG MM (2004), 'Efficacy of chlorine and a peroxyacetic acid sanitizer in killing *Listeria monocytogenes* on iceberg and romaine lettuce using simulated commercial processing conditions', *J Food Prot*, 67, 1238–1242.

BEYAZ D and TAYAR M (2010), 'The effect of lactic acid spray application on the microbiological quality of sheep carcasses', *J Anim Vet Adv*, 9, 1858–1863.

BILGILI SF, CONNER DE, PINION JL and TAMBLYN KC (1998), 'Broiler skin color as affected by organic acids: influence of concentration and method of application', *Poult Sci*, 77, 751–757.

BIN JASASS FM (2008), 'Effectiveness of trisodium phosphate, lactic acid, and acetic acid in reduction of *E. coli* and microbial load on chicken surfaces', *Afr J Microbiol Res*, 2, 50–55.

BIRK T, GRØNLUND AC, CHRISTENSEN BB, KNØCHEL S, LOHSE K and ROSENQUIST H (2010), 'Effect of organic acids and marination ingredients on the survival of *Campylobacter jejuni* on meat', *J Food Prot*, 73, 258–265.

BOLDER NM (1997), 'Decontamination of meat and poultry carcasses', *Trends Food Sci Technol*, 8, 221–227.

BOLTON DJ, DOHERTY AM and SHERIDAN JJ (2001), 'Beef HACCP: intervention and non-intervention systems', *Int J Food Microbiol*, 66, 119–129.

BOOTH IR (1985), 'Regulation of cytoplasmic pH in bacteria', *Microbiol Rev*, 49, 359–378.

BOSILEVAC JM, NOU X, BARKOCY-GALLAGHER GA, ARTHUR TM and KOOHMARAIE M (2006), 'Treatments using hot water instead of lactic acid reduce levels of aerobic bacteria and *Enterobacteriaceae* and reduce the prevalence of *Escherichia coli* O157:H7 on preevisceration beef carcasses', *J Food Prot*, 69, 1808–1813.

BRACKETT RE, HAO Y-Y and DOYLE MP (1994), 'Ineffectiveness of hot acid sprays to decontaminate *Escherichia coli* O157:H7 on beef', *J Food Prot*, 57, 198–203.

BREMER PJ and OSBORNE CM (1998), 'Reducing total aerobic counts and *Listeria monocytogenes* on the surface of king salmon (*Oncorhynchus tshawytscha*)', *J Food Prot*, 61, 849–854.

© Woodhead Publishing Limited, 2012

BRUL S and COOTE P (1999), 'Preservative agents in foods. Mode of action and microbial resistance mechanisms', *Int J Food Microbiol*, 50, 1–17.

BUCHANAN RL, EDELSON SG, MILLER RL and SAPERS GM (1999), 'Contamination of intact apples after immersion in an aqueous environment containing *Escherichia coli* O157:H7', *J Food Prot*, 62, 444–450.

BYELASHOV OA and SOFOS JN (2009), 'Strategies for on-line decontamination of carcasses', in Toldrá F, *Safety of meat and processed meat*, New York, Springer, 149–182.

BYELASHOV OA, KENDALL PA, BELK KE, SCANGA JA and SOFOS JN (2008), 'Control of *Listeria monocytogenes* on vacuum-packaged frankfurters sprayed with lactic acid alone or in combination with sodium lauryl sulfate', *J Food Prot*, 71, 728–734.

CABEDO L, SOFOS JN and SMITH GC (1996), 'Removal of bacteria from beef tissue by spray washing after different times of exposure to fecal material', *J Food Prot*, 59, 1284–1287.

CALICIOGLU M, KASPAR CW, BUEGE DR and LUCHANSKY JB (2002a), 'Effectiveness of spraying with Tween 20 and lactic acid in decontaminating inoculated *Escherichia coli* O157:H7 and indigenous *Escherichia coli* biotype I on beef', *J Food Prot*, 65, 26–32.

CALICIOGLU M, SOFOS JN, SAMELIS J, KENDALL PA and SMITH GC (2002b), 'Inactivation of acid-adapted and nonadapted *Escherichia coli* O157:H7 during drying and storage of beef jerky treated with different marinades', *J Food Prot*, 65, 1394–1405.

CALICIOGLU M, SOFOS JN, SAMELIS J, KENDALL PA and SMITH GC (2002c), 'Destruction of acid- and non-adapted *Listeria monocytogenes* during drying and storage of beef jerky', *Food Microbiol*, 19, 545–559.

CALICIOGLU M, SOFOS JN, SAMELIS J, KENDALL PA and SMITH GC (2003), 'Effect of acid adaptation on inactivation of *Salmonella* during drying and storage of beef jerky treated with marinades', *Int J Food Microbiol*, 89, 51–65.

CAPITA R, ALONSO-CALLEJA C, GARCÍA ARIAS MT, MORENO B and GARCÍA-FERNÁNDEZ MC (2000a), 'Effect of trisodium phosphate on mesophilic and psychrotrophic bacterial flora attached to the skin of chicken carcasses during refrigerated storage', *Food Sci Tech Int*, 6, 345–350.

CAPITA R, ALONSO-CALLEJA C, SIERRA M, MORENO B and DEL CAMINO GARCÍA-FERNÁNDEZ M (2000b), 'Effect of trisodium phosphate solutions washing on the sensory evaluation of poultry meat', *Meat Sci*, 55, 471–474.

CAPITA R, ALONSO-CALLEJA C, GARCÍA-FERNÁNDEZ C and MORENO B (2001), 'Efficacy of trisodium phosphate solutions in reducing *Listeria monocytogenes* populations on chicken skin during refrigerated storage', *J Food Prot*, 64, 1627–1630.

CAPITA R, ALONSO-CALLEJA C, RODRÍGUEZ-PÉREZ R, MORENO B and DEL CAMINO GARCÍA-FERNÁNDEZ M (2002a), 'Influence of poultry carcass skin sample site on the effectiveness of trisodium phosphate against *Listeria monocytogenes*', *J Food Prot*, 65, 853–856.

CAPITA R, ALONSO-CALLEJA C, DEL CAMINO GARCÍA-FERNÁNDEZ M and MORENO B (2002b), 'Activity of trisodium phosphate with sodium hydroxide wash solutions against *Listeria monocytogenes* attached to chicken skin during refrigerated storage', *Food Microbiol*, 19, 57–63.

CAPITA R, ALONSO-CALLEJA C, GARCÍA-FERNÁNDEZ MC and MORENO B (2002c), 'Review: trisodium phosphate (TSP) treatment for decontamination of poultry', *Food Sci Tech Int*, 8, 11–24.

CASTELO MM, KANG D-H, SIRAGUSA GR, KOOHMARAIE M and BERRY ED (2001), 'Evaluation of combination treatment processes for the microbial decontamination of pork trim', *J Food Prot*, 64, 335–342.

CASTILLO A, LUCIA LM, ROBERSON DB, STEVENSON TH, MERCADO I and ACUFF GR (2001), 'Lactic acid sprays reduce bacterial pathogens on cold beef carcass surfaces and in subsequently produced ground beef', *J Food Prot*, 64, 58–62.

CASTILLO LA, MÉSZÁROS L and KISS IF (2005), 'Effect of trisodium phosphate on the microbial contamination of chicken meat', *Acta Aliment Hung*, 34, 5–11.

© Woodhead Publishing Limited, 2012

CFR (2011a), 'Chemicals used in washing or to assist in the peeling of fruits and vegetables', Code of Federal Regulations 21 CFR Part 173, Section 173.315. Available from: http://www.accessdata.fda.gov/scripts/cdrh/cfdocs/cfcfr/CFRSearch.cfm?fr=173.315 (accessed 30 August 2011).

CFR (2011b), 'Sanitizing solutions', Code of Federal Regulations 21 CFR Part 178, Section 178.1010. Available from: http://www.accessdata.fda.gov/scripts/cdrh/cfdocs/cfcfr/CFRSearch.cfm?fr=178.1010 (accessed 30 August 2011).

CHAIDEZ C, MORENO M, RUBIO W, ANGULO M and VALDEZ B (2003), 'Comparison of the disinfection efficacy of chlorine-based products for inactivation of viral indicators and pathogenic bacteria in produce wash water', *Int J Environ Health Res*, 13, 295–302.

CHAIYAKOSA S, CHARERNJIRATRAGUL W, UMSAKUL K and VUDDHAKUL V (2007), 'Comparing the efficiency of chitosan with chlorine for reducing *Vibrio parahaemolyticus* in shrimp', *Food Control*, 18, 1031–1035.

CHANG J-M and FANG TJ (2007), 'Survival of *Escherichia coli* O157:H7 and *Salmonella enterica* serovars Typhimurium in iceberg lettuce and the antimicrobial effect of rice vinegar against *E. coli* O157:H7', *Food Microbiol*, 24, 745–751.

CHERRINGTON CA, HINTON M and CHOPRA I (1990), 'Effect of short-chain organic acids on macromolecular synthesis in *Escherichia coli*', *J Bacteriol*, 68, 69–74.

CHERRINGTON CA, HINTON M, PEARSON GR and CHOPRA I (1991a), 'Inhibition of *Escherichia coli* K12 by short-chain organic acids: lack of evidence for induction of the SOS response', *J Appl Bacteriol*, 70, 156–160.

CHERRINGTON CA, HINTON M, PEARSON GR and CHOPRA I (1991b), 'Short-chain organic acids at pH 5.0 kill *Escherichia coli* and *Salmonella* spp. without causing membrane perturbation', *J Appl Bacteriol*, 70, 161–165.

CHOI YM, KIM OY, KIM KH, KIM BC and RHEE MS (2009), 'Combined effect of organic acids and supercritical carbon dioxide treatments against nonpathogenic *Escherichia coli*, *Listeria monocytogenes*, *Salmonella typhimurium* and *E. coli* O157:H7 in fresh pork', *Lett Appl Microbiol*, 49, 510–515.

CORTESI ML, PANEBIANCO A, GIUFFRIDA A and ANASTASIO A (2009), 'Innovations in seafood preservation and storage', *Vet Res Commun*, 33, S15–S23.

CUTTER CN and SIRAGUSA GR (1994), 'Efficacy of organic acids against *Escherichia coli* O157:H7 attached to beef carcass tissue using a pilot scale model carcass washer', *J Food Prot*, 57, 97–103.

CUTTER CN, DORSA WJ and SIRAGUSA GR (1997), 'Parameters affecting the efficacy of spray washes against *Escherichia coli* O157:H7 and fecal contamination on beef', *J Food Prot*, 60, 614–618.

DAVIDSON PM and HARRISON MA (2003), 'Microbial adaptation to stresses by food preservatives', in Yousef AE and Juneja VK, *Microbial stress adaptation and food safety*, Boca Raton, FL, CRC Press, 55–73.

DAWSON RMC, ELLIOT DC, ELLIOT WH and JONES KM (1986), 'Carboxylic acids, alcohols, aldehydes, and ketones', in *Data for biochemical research*, 3rd edn, New York, Oxford University Press, 33–53.

DEGNAN AJ, KASPAR CW, OTWELL WS, TAMPLIN ML and LUCHANSKY JB (1994), 'Evaluation of lactic acid bacterium fermentation products and food-grade chemicals to control *Listeria monocytogenes* in blue crab (*Callinectes sapidus*) meat', *Appl Environ Microbiol*, 60, 3198–3203.

DELAQUIS PJ, GRAHAM HS and HOCKING R (1997), 'Shelf-life of coleslaw made from cabbage treated with gaseous acetic acid', *J Food Process Preserv*, 21, 129–140.

DELAQUIS PJ, SHOLBERG PL and STANICH K (1999), 'Disinfection of mung bean seed with gaseous acetic acid', *J Food Prot*, 62, 953–957.

DELMORE JR RJ, SOFOS JN, SCHMIDT GR, BELK KE, LLOYD WR and SMITH GC (2000), 'Interventions to reduce microbiological contamination of beef variety meats', *J Food Prot*, 63, 44–50.

© Woodhead Publishing Limited, 2012

DEL RÍO E, CAPITA R, PRIETO M and ALONSO-CALLEJA C (2006), 'Comparison of pathogenic and spoilage bacterial levels on refrigerated poultry parts following treatment with trisodium phosphate', *Food Microbiol*, 23, 195–198.

DEL RÍO E, MURIENTE R, PRIETO M, ALONSO-CALLEJA C and CAPITA R (2007a), 'Effectiveness of trisodium phosphate, acidified sodium chlorite, citric acid, and peroxyacids against pathogenic bacteria on poultry during refrigerated storage', *J Food Prot*, 70, 2063–2071.

DEL RÍO E, PANIZO-MORÁN M, PRIETO M, ALONSO-CALLEJA C and CAPITA R (2007b), 'Effect of various chemical decontamination treatments on natural microflora and sensory characteristics of poultry', *Int J Food Microbiol*, 115, 268–280.

DEL RÍO E, DE CASO BG, PRIETO M, ALONSO-CALLEJA C and CAPITA R (2008), 'Effect of poultry decontaminants concentration on growth kinetics for pathogenic and spoilage bacteria', *Food Microbiol*, 25, 888–894.

DENYER SP and STEWART GSAB (1998), 'Mechanisms of action of disinfectants', *Int Biodeter Biodegr*, 41, 261–268.

DEUMIER F (2004), 'Pulsed-vacuum immersion of chicken meat and skin in acid solutions. Effects on mass transfers, colour and microbial quality', *Int J Food Sci Technol*, 39, 277–286.

DEUMIER F (2006), 'Decontamination of deboned chicken legs by vacuum-tumbling in lactic acid solution', *Int J Food Sci Technol*, 41, 23–32.

DICKSON JS (1992), 'Acetic acid action on beef tissue surfaces contaminated with *Salmonella typhimurium*', *J Food Sci*, 57, 297–301.

DICKSON JS (1995), 'Susceptibility of preevisceration washed beef carcasses to contamination by *Escherichia coli* O157:H7 and Salmonellae', *J Food Prot*, 58, 1065–1068.

DICKSON JS and SIRAGUSA GR (1994), 'Survival of *Salmonella typhimurium*, *Escherichia coli* O157:H7 and *Listeria monocytogenes* during storage on beef sanitized with organic acids', *J Food Saf*, 14, 313–327.

DICKSON JS, CUTTER CN and SIRAGUSA GR (1994), 'Antimicrobial effects of trisodium phosphate against bacteria attached to beef tissue', *J Food Prot*, 57, 952–955.

DINÇER AH and BAYSAL T (2004), 'Decontamination techniques of pathogen bacteria in meat and poultry', *Crit Rev Microbiol*, 30, 197–204.

DOLEŽALOVÁ M, MOLATOVÁ Z, BUŇKA F, BŘEZINA P and MAROUNEK M (2010), 'Effect of organic acids on growth of chilled chicken skin microflora', *J Food Saf*, 30, 353–365.

DORSA WJ, CUTTER CN and SIRAGUSA GR (1997), 'Effects of acetic acid, lactic acid and trisodium phosphate on the microflora of refrigerated beef carcass surface tissue inoculated with *Escherichia coli* O157:H7, *Listeria innocua*, and *Clostridium sporogenes*', *J Food Prot*, 60, 619–624.

DORSA WJ, CUTTER CN and SIRAGUSA GR (1998a), 'Long-term effect of alkaline, organic acid, or hot water washes on the microbial profile of refrigerated beef contaminated with bacterial pathogens after washing', *J Food Prot*, 61, 300–306.

DORSA WJ, CUTTER CN and SIRAGUSA GR (1998b), 'Bacterial profile of ground beef made from carcass tissue experimentally contaminated with pathogenic and spoilage bacteria before being washed with hot water, alkaline solution, or organic acid and then stored at 4 or 12°C', *J Food Prot*, 61, 1109–1118.

DORSA WJ, CUTTER CN and SIRAGUSA GR (1998c), 'Long-term bacterial profile of refrigerated ground beef made from carcass tissue, experimentally contaminated with pathogens and spoilage bacteria after hot water, alkaline, or organic acid washes', *J Food Prot*, 61, 1615–1622.

DO SOCORRO ROCHA BASTOS M, DE FÁTIMA FERREIRA SOARES N, DE ANDRADE NJ, ARRUDA AC and ALVES RE (2005), 'The effect of the association of sanitizers and surfactant in the microbiota of the Cantaloupe (*Cucumis melo* L.) melon surface', *Food Control*, 16, 369–373.

DUBAL ZB, PATURKAR AM, WASKAR VS, ZENDE RJ, LATHA C, RAWOOL DB and KADAM MM (2004),

© Woodhead Publishing Limited, 2012

'Effect of food grade organic acids on inoculated *S. aureus*, *L. monocytogenes*, *E. coli* and *S.* Typhimurium in sheep/goat meat stored at refrigeration temperature', *Meat Sci*, 66, 817–821.

DUKAN S and TOUATI D (1996), 'Hypochlorous acid stress in *Escherichia coli*: resistance, DNA damage, and comparison with hydrogen peroxide stress', *J Bacteriol*, 178, 6145–6150.

ECHEVERRY A, CHANCE BROOKS J, MILLER MF, COLLINS JA, LONERAGAN GH and BRASHEARS MM (2010), 'Validation of lactic acid bacteria, lactic acid, and acidified sodium chlorite as decontaminating interventions to control *Escherichia coli* O157:H7 and *Salmonella* Typhimurium DT 104 in mechanically tenderized and brine-enhanced (nonintact) beef at the purveyor', *J Food Prot*, 73, 2169–2179.

EDWARDS JR and FUNG DYC (2006), 'Prevention and decontamination of *Escherichia coli* O157:H7 on raw beef carcasses in commercial beef abattoirs', *J Rapid Methods Autom Microbiol*, 14, 1–95.

EFSA (2005), 'Opinion of the Scientific Panel on Biological Hazards on the evaluation of the efficacy of peroxyacids for use as an antimicrobial substance applied on poultry carcasses', *The EFSA Journal*, 306, 1–10.

EFSA (2006a), 'Opinion from the Scientific Panel on Biological Hazards on the evaluation of the efficacy of L (+) Lactic acid for carcass decontamination', *The EFSA Journal*, 342, 1–6.

EFSA (2006b), 'Opinion of the Scientific Panel on Biological Hazards on the request from the European Commission related to the evaluation of the efficacy of SAN-PEL® for use as an antimicrobial substance applied on carcasses of chickens, turkeys, quails, pigs, beef, sheep, goats and game and in washing the shells of eggs', *The EFSA Journal*, 352, 1–6.

EFSA (2011), 'Scientific Opinion on the evaluation of the safety and efficacy of lactic acid for the removal of microbial surface contamination of beef carcasses, cuts and trimmings', *The EFSA Journal*, 9, 2317–2351.

EGGENBERGER-SOLORZANO L, NIEBUHR SE, ACUFF GR and DICKSON JS (2002), 'Hot water and organic acid interventions to control microbiological contamination on hog carcasses during processing', *J Food Prot*, 65, 1248–1252.

ELLERBROEK L, LIENAU J-A, ALTER T and SCHLICHTING D (2007), 'Effectiveness of different chemical decontamination methods on the *Campylobacter* load of poultry carcasses', *Fleischwirtschaft*, 4, 224–227.

ELLS TC and HANSEN LT (2006), 'Strain and growth temperature influence *Listeria* spp. attachment to intact and cut cabbage', *Int J Food Microbiol*, 111, 34–42.

ESCUDERO ME, VELÁZQUEZ L, DI GENARO MS and DE GUZMÁN AMS (1999), 'Effectiveness of various disinfectants in the elimination of *Yersinia enterocolitica* on fresh lettuce', *J Food Prot*, 62, 665–669.

FABRIZIO KA, SHARMA RR, DEMIRCI A and CUTTER CN (2002), 'Comparison of electrolyzed oxidizing water with various antimicrobial interventions to reduce *Salmonella* species on poultry', *Poult Sci*, 81, 1598–1605.

FAN X, ANNOUS BA, KESKINEN LA and MATTHEIS JP (2009), 'Use of chemical sanitizers to reduce microbial populations and maintain quality of whole and fresh-cut cantaloupe', *J Food Prot*, 72, 2453–2460.

FELICIANO L, LEE J, LOPES JA and PASCALL MA (2010), 'Efficacy of sanitized ice in reducing bacterial load on fish fillet and in the water collected from the melted ice', *J Food Sci*, 75, M231–M238.

FERNANDES CF, FLICK GJ, COHEN J and THOMAS TB (1998), 'Role of organic acids during processing to improve quality of channel fish fillets', *J Food Prot*, 61, 495–498.

FETT WF (2000), 'Naturally occurring biofilms on alfalfa and other types of sprouts', *J Food Prot*, 63, 625–632.

FOSTER JW and HALL HK (1990), 'Adaptive acidification tolerance response of *Salmonella typhimurium*', *J Bacteriol*, 172, 771–778.

© Woodhead Publishing Limited, 2012

FRANCIS GA, THOMAS C and O'BEIRNE D (1999), 'The microbiological safety of minimally processed vegetables', *Int J Food Sci Technol*, 34, 1–22.

FRATAMICO PM, SCHULTZ FJ, BENEDICT RC, BUCHANAN RL and COOKE PH (1996), 'Factors influencing attachment of *Escherichia coli* O157:H7 to beef tissues and removal using selected sanitizing rinses', *J Food Prot*, 59, 453–459.

GEORNARAS I and SOFOS JN (2005), 'Combining physical and chemical decontamination interventions for meat', in Sofos JN, *Improving the safety of fresh meat*, Cambridge, Woodhead Publishing, 433–460.

GEORNARAS I, BELK KE, SCANGA JA, KENDALL PA, SMITH GC and SOFOS JN (2005), 'Postprocessing antimicrobial treatments to control *Listeria monocytogenes* in commercial vacuum-packaged bologna and ham stored at 10°C', *J Food Prot*, 68, 991–998.

GEORNARAS I, SKANDAMIS PN, BELK KE, SCANGA JA, KENDALL PA, SMITH GC and SOFOS JN (2006a), 'Postprocess control of *Listeria monocytogenes* on commercial frankfurters formulated with and without antimicrobials and stored at 10°C', *J Food Prot*, 69, 53–61.

GEORNARAS I, SKANDAMIS PN, BELK KE, SCANGA JA, KENDALL PA, SMITH GC and SOFOS JN (2006b), 'Post-processing application of chemical solutions for control of *Listeria monocytogenes*, cultured under different conditions, on commercial smoked sausage formulated with and without potassium lactate-sodium diacetate', *Food Microbiol*, 23, 762–771.

GILL CJ, KEENE WE, MOHLE-BOETANI JC, FARRAR JA, WALLER PL, HAHN CG and CIESLAK PR (2003), 'Alfalfa seed decontamination in a *Salmonella* outbreak', *Emerg Infect Dis*, 9, 474–479.

GILL CO and BADONI M (2004), 'Effects of peroxyacetic acid, acidified sodium chlorite or lactic acid solutions on the microflora of chilled beef carcasses', *Int J Food Microbiol*, 91, 43–50.

GIL MI, SELMA MV, LÓPEZ-GÁLVEZ F and ALLENDE A (2009), 'Fresh-cut product sanitation and wash water disinfection: problems and solutions', *Int J Food Microbiol*, 134, 37–45.

GODDARD BL, MIKEL WB, CONNER DE and JONES WR (1996), 'Use of organic acids to improve the chemical, physical, and microbial attributes of beef strip loins stored at –1°C for 112 days', *J Food Prot*, 59, 849–853.

GÖGÜS U, BOZOGLU F and ALPAS H (2007), 'A comparative study on the effects of glucose monohydrate, hot water, and sodium pyrophosphate on quality parameters and microbial flora of deboned and matured brisket', *J Food Sci*, 72, M215–M221.

GONÇALVES AC, ALMEIDA RCC, ALVES MAO and ALMEIDA PF (2005), 'Quantitative investigation on the effects of chemical treatments in reducing *Listeria monocytogenes* populations on chicken breast meat', *Food Control*, 16, 617–622.

GONZALEZ RJ, LUO Y, RUIZ-CRUZ S and MCEVOY JL (2004), 'Efficacy of sanitizers to inactivate *Escherichia coli* O157:H7 on fresh-cut carrot shreds under simulated process water conditions', *J Food Prot*, 67, 2375–2380.

GONZÁLEZ-FANDOS E and DOMINGUEZ JL (2006), 'Efficacy of lactic acid against *Listeria monocytogenes* attached to poultry skin during refrigerated storage', *J Appl Microbiol*, 101, 1331–1339.

GONZÁLEZ-FANDOS E, HERRERA B and MAYA N (2009), 'Efficacy of citric acid against *Listeria monocytogenes* attached to poultry skin during refrigerated storage', *Int J Food Sci Technol*, 44, 262–268.

GOODSON M and ROWBURY RJ (1989), 'Habituation to normally lethal acidity by prior growth of *Escherichia coli* at a sub-lethal acid pH value', *Lett Appl Microbiol*, 8, 77–79.

GORMAN BM, SOFOS JN, MORGAN JB, SCHMIDT GR and SMITH GC (1995), 'Evaluation of hand-trimming, various sanitizing agents, and hot water spray-washing as decontamination interventions for beef brisket adipose tissue', *J Food Prot*, 58, 899–907.

© Woodhead Publishing Limited, 2012

GREER GG and DILTS BD (1992), 'Factors affecting the susceptibility of meatborne pathogens and spoilage bacteria to organic acids', *Food Res Int*, 25, 355–364.

GREER GG and DILTS BD (1995), 'Lactic acid inhibition of the growth of spoilage bacteria and cold tolerant pathogens on pork', *Int J Food Microbiol*, 25, 141–151.

HARDIN MD, ACUFF GR, LUCIA LM, OMAN JS and SAVELL JW (1995), 'Comparison of methods for decontamination from beef carcass surfaces', *J Food Prot*, 58, 368–374.

HARRIS K, MILLER MF, LONERAGAN GH and BRASHEARS MM (2006), 'Validation of the use of organic acids and acidified sodium chlorite to reduce *Escherichia coli* O157 and *Salmonella* Typhimurium in beef trim and ground beef in a simulated processing environment', *J Food Prot*, 69, 1802–1807.

HEATON JC and JONES K (2008), 'Microbial contamination of fruit and vegetables and the behaviour of enteropathogens in the phyllosphere: a review', *J Appl Microbiol*, 104, 613–626.

HELANDER IM, VON WRIGHT A and MATTILA-SANDHOLM T-M (1997), 'Potential of lactic acid bacteria and novel antimicrobials against Gram-negative bacteria', *Trends Food Sci Technol*, 8, 146–150.

HELLER CE, SCANGA JA, SOFOS JN, BELK KE, WARREN-SERNA W, BELLINGER GR, BACON RT, ROSSMAN ML and SMITH GC (2007), 'Decontamination of beef subprimal cuts intended for blade tenderization or moisture enhancement', *J Food Prot*, 70, 1174–1180.

HILL C, O'DRISCOLL B and BOOTH I (1995), 'Acid adaptation and food poisoning microorganisms', *Int J Food Microbiol*, 28, 245–254.

HOLLIDAY SL, SCOUTEN AJ and BEUCHAT LR (2001), 'Efficacy of chemical treatments in eliminating *Salmonella* and *Escherichia coli* O157:H7 on scarified and polished alfalfa seeds', *J Food Prot*, 64, 1489–1495.

HUA G and RECKHOW DA (2007), 'Comparison of disinfection byproduct formation from chlorine and alternative disinfectants', *Water Res*, 41, 1667–1678.

HUFFMAN RD (2002), 'Current and future technologies for the decontamination of carcasses and fresh meat', *Meat Sci*, 62, 285–294.

HUGAS M and TSIGARIDA E (2008), 'Pros and cons of carcass decontamination: the role of the European Food Safety Authority', *Meat Sci*, 78, 43–52.

HUSS HH, JØRGENSEN LV and VOGEL BF (2000), 'Control options for *Listeria monocytogenes* in seafoods', *Int J Food Microbiol*, 62, 267–274.

HWANG, C-A and BEUCHAT LR (1995), 'Efficacy of selected chemicals for killing pathogenic and spoilage microorganisms on chicken skin', *J Food Prot*, 58, 19–23.

IKEDA JS, SAMELIS J, KENDALL PA, SMITH GC and SOFOS JN (2003), 'Acid adaptation does not promote survival or growth of *Listeria monocytogenes* on fresh beef following acid and nonacid decontamination treatments', *J Food Prot*, 66, 985–992.

INGHAM SC (1989), 'Lactic acid dipping for inhibiting microbial spoilage of refrigerated catfish fillet pieces', *J Food Qual*, 12, 433–443.

ISMAIL SAS, DEAK T, ABD EL-RAHMAN HA, YASSIEN MAM and BEUCHAT LR (2001), 'Effectiveness of immersion treatments with acids, trisodium phosphate, and herb decoctions in reducing populations of *Yarrowia lipolytica* and naturally occurring aerobic microorganisms on raw chicken', *Int J Food Microbiol*, 64, 13–19.

JAQUETTE CB, BEUCHAT LR and MAHON BE (1996), 'Efficacy of chlorine and heat treatment in killing *Salmonella stanley* inoculated onto alfalfa seeds and growth and survival of the pathogen during sprouting and storage', *Appl Environ Microbiol*, 62, 2212–2215.

JAY JM (1995), 'Foods with low numbers of microorganisms may not be the safest foods, or why did human listeriosis and hemorrhagic colitis become foodborne diseases?', *Dairy Food Environ Sanit*, 15, 674–677.

JIMÉNEZ SM, DESTEFANIS P, SALSI MS, TIBURZI MC and PIROVANI ME (2005), 'Predictive model for reduction of *Escherichia coli* during acetic acid decontamination of chicken skin', *J Appl Microbiol*, 99, 829–835.

JIMÉNEZ SM, CALIUSCO MF, TIBURZI MC, SALSI MS and PIROVANI ME (2007), 'Predictive models for reduction of *Salmonella* Hadar on chicken skin during single and double sequential spraying treatments with acetic acid', *J Appl Microbiol*, 103, 528–535.

© Woodhead Publishing Limited, 2012

JONES CW (1999), 'Introduction to the preparation and properties of hydrogen peroxide', in Clark JH, *Applications of hydrogen peroxide and derivatives*, Cambridge, Royal Society of Chemistry, 1–35.

JUVEN BJ and PIERSON MD (1996), 'Antibacterial effects of hydrogen peroxide and methods for its detection and quantitation', *J Food Prot*, 59, 1233–1241.

KANG D-H, KOOHMARAIE M, DORSA WJ and SIRAGUSA GR (2001), 'Development of a multiple-step process for the microbial decontamination of beef trim', *J Food Prot*, 64, 63–71.

KARAPINAR M and GÖNÜL SA (1992), 'Removal of *Yersinia enterocolitica* from fresh parsley by washing with acetic acid or vinegar', *Int J Food Microbiol*, 16, 261–264.

KARGIOTOU C, KATSANIDIS E, RHOADES J, KONTOMINAS M and KOUTSOUMANIS K (2011), 'Efficacies of soy sauce and wine based marinades for controlling spoilage of raw beef', *Food Microbiol*, 28, 158–163.

KEENER KM, BASHOR MP, CURTIS PA, SHELDON BW and KATHARIOU S (2004), 'Comprehensive review of *Campylobacter* and poultry processing', *Compr Rev Food Sci F*, 3, 105–116.

KEMP GK, ALDRICH ML and WALDROUP AL (2000), 'Acidified sodium chlorite antimicrobial treatment of broiler carcasses', *J Food Prot*, 63, 1087–1092.

KILLINGER KM, KANNAN A, BARY AI and COGGER CG (2010), 'Validation of a 2 percent lactic acid antimicrobial rinse for mobile poultry slaughter operations', *J Food Prot*, 73, 2079–2083.

KIM CR and MARSHALL DL (1999), 'Microbiological, colour and sensory changes of refrigerated chicken legs treated with selected phosphates', *Food Res Int*, 32, 209–215.

KIM J and MARSHALL DL (2000a), 'Lactic acid inactivation of *Salmonella typhimurium* attached to catfish skin', *J Food Saf*, 20, 53–64.

KIM CR and MARSHALL DL (2000b), 'Quality evaluation of refrigerated chicken wings treated with organic acids', *J Food Qual*, 23, 327–335.

KIM J and MARSHALL DL (2001), 'Effect of lactic acid on *Listeria monocytogenes* and *Edwardsiella tarda* attached to catfish skin', *Food Microbiol*, 18, 589–596.

KIM J and MARSHALL DL (2002), 'Influence of catfish skin mucus on trisodium phosphate inactivation of attached *Salmonella* Typhimurium, *Edwardsiella tarda*, and *Listeria monocytogenes*', *J Food Prot*, 65, 1146–1151.

KIM J-W, SLAVIK MF and BENDER FG (1994a), 'Removal of *Salmonella typhimurium* attached to chicken skin by rinsing with trisodium phosphate solution: scanning electron microscopic examination', *J Food Saf*, 14, 77–84.

KIM J-W, SLAVIK MF, PHARR MD, RABEN DP, LOBSINGER CM and TSAI S (1994b), 'Reduction of *Salmonella* on post-chill chicken carcasses by trisodium phosphate ($Na_3PO_4$) treatment', *J Food Saf*, 14, 9–17.

KIŞLA D (2007), 'Effectiveness of lemon juice in the elimination of *Salmonella* Typhimurium in stuffed mussels', *J Food Prot*, 70, 2847–2850.

KITIS M (2004), 'Disinfection of wastewater with peracetic acid: a review', *Environ Int*, 30, 47–55.

KOMULAINEN H (2004), 'Experimental cancer studies of chlorinated by-products', *Toxicol*, 198, 239–248.

KONDO N, MURATA M and ISSHIKI K (2006), 'Efficiency of sodium hypochlorite, fumaric acid, and mild heat in killing native microflora and *Escherichia coli* O157:H7, *Salmonella* Typhimurium DT 104, and *Staphylococcus aureus* attached to fresh-cut lettuce', *J Food Prot*, 69, 323–329.

KOSEKI S, YOSHIDA K, ISOBE S and ITOH K (2001), 'Decontamination of lettuce using acidic electrolyzed water', *J Food Prot*, 64, 652–658.

KOTULA KL and THELAPPURATE R (1994), 'Microbiological and sensory attributes of retail cuts of beef treated with acetic and lactic acid solutions', *J Food Prot*, 57, 665–670.

© Woodhead Publishing Limited, 2012

KOUTSOUMANIS KP, ASHTON LV, GEORNARAS I, BELK KE, SCANGA JA, KENDALL PA, SMITH GC and SOFOS JN (2004), 'Effect of single or sequential hot water and lactic acid decontamination treatments on the survival and growth of *Listeria monocytogenes* and spoilage microflora during aerobic storage of fresh beef at 4, 10, and 25°C', *J Food Prot*, 67, 2703–2711.

KROVACEK K, FARIS A, AHNE W and MASSON I (1987), 'Adhesion of *Aeromonas hydrophila* and *Vibrio anguillarum* to mucus-coated glass slides', *FEMS Microbiol Lett*, 42, 85–89.

KUMAR M, HORA R, KOSTRZYNSKA M, WAITES WM and WARRINER K (2006), 'Inactivation of *Escherichia coli* O157:H7 and *Salmonella* on mung beans, alfalfa, and other seed types destined for sprout production by using an oxychloro-based sanitizer', *J Food Prot*, 69, 1571–1578.

LAURY AM, ALVARADO MV, NACE G, ALVARADO CZ, BROOKS JC, ECHEVERRY A and BRASHEARS MM (2009), 'Validation of a lactic acid- and citric acid-based antimicrobial product for the reduction of *Escherichia coli* O157:H7 and *Salmonella* on beef tips and whole chicken carcasses', *J Food Prot*, 72, 2208–2211.

LECOMPTE J-Y, KONDJOYAN A, SARTER S, PORTANGUEN S and COLLIGNAN A (2008), 'Effects of steam and lactic acid treatments on inactivation of *Listeria innocua* surface-inoculated on chicken skins', *Int J Food Microbiol*, 127, 155–161.

LEE RM, HARTMAN PA, OLSON DG and WILLIAMS FD (1994), 'Metal ions reverse the inhibitory effects of selected food-grade phosphates in *Staphylococcus aureus*', *J Food Prot*, 57, 284–288, 300.

LEISTNER L (2000), 'Basic aspects of food preservation by hurdle technology', *Int J Food Microbiol*, 55, 181–186.

LI Y, SLAVIK MF, WALKER JT and XIONG H (1997), 'Pre-chill spray of chicken carcasses to reduce *Salmonella typhimurium*', *J Food Sci*, 62, 605–607.

LIAO C-H and COOKE PH (2001), 'Response to trisodium phosphate treatment of *Salmonella* Chester attached to fresh-cut green pepper slices', *Can J Microbiol*, 47, 25–32.

LIAO C-H and SAPERS GM (2000), 'Attachment and growth of *Salmonella* Chester on apple fruits and *in vivo* response of attached bacteria to sanitizer treatments', *J Food Prot*, 63, 876–883.

LILLARD HS (1979), 'Levels of chlorine and chlorine dioxide of equivalent bactericidal effect in poultry processing water', *J Food Sci*, 44, 1594–1597.

LILLARD HS (1994), 'Effect of trisodium phosphate on salmonellae attached to chicken skin', *J Food Prot*, 57, 465–469.

LIM K and MUSTAPHA A (2003), 'Reduction of *Escherichia coli* O157:H7 and *Lactobacillus plantarum* numbers on fresh beef by polylactic acid and vacuum packaging', *J Food Sci*, 68, 1422–1427.

LIN C-M, MOON SS, DOYLE MP and MCWATTERS KH (2002), 'Inactivation of *Escherichia coli* O157:H7, *Salmonella enterica* serotype Enteritidis, and *Listeria monocytogenes* on lettuce by hydrogen peroxide and lactic acid and by hydrogen peroxide with mild heat', *J Food Prot*, 65, 1215–1220.

LIN KW and CHUANG CH (2001), 'Effectiveness of dipping with phosphate, lactate and acetic acid solutions on the quality and shelf-life of pork loin chop', *J Food Sci*, 66, 494–499.

LINDOW SE and BRANDL MT (2003), 'Microbiology of the phyllosphere', *Appl Environ Microbiol*, 69, 1875–1883.

LINDSAY RC (2008), 'Food additives', in Damodaran S, Parkin KL and Fennema OR, *Fennema's food chemistry*, 4th edn, Boca Raton, FL, CRC Press, 689–749.

LÓPEZ-GÁLVEZ F, ALLENDE A, SELMA MV and GIL MI (2009), 'Prevention of *Escherichia coli* cross-contamination by different commercial sanitizers during washing of fresh-cut lettuce', *Int J Food Microbiol*, 133, 167–171.

LORETZ M, STEPHAN R and ZWEIFEL C (2010), 'Antimicrobial activity of decontamination treatments for poultry carcasses: a literature survey', *Food Control*, 21, 791–804.

© Woodhead Publishing Limited, 2012

LORETZ M, STEPHAN R and ZWEIFEL C (2011a), 'Antibacterial activity of decontamination treatments for cattle hides and beef carcasses', *Food Control*, 22, 347–359.

LORETZ M, STEPHAN R and ZWEIFEL C (2011b), 'Antibacterial activity of decontamination treatments for pig carcasses', *Food Control*, 22, 1121–1125.

LU Z, SEBRANEK JG, DICKSON JS, MENDONCA AF and BAILEY TB (2005), 'Effects of organic acid salt solutions on sensory and other quality characteristics of frankfurters', *J Food Sci*, 70, S123–S127.

MARSHALL DL and JINDAL V (1997), 'Microbiological quality of catfish frames treated with selected phosphates', *J Food Prot*, 60, 1081–1083.

MARSHALL DL and KIM CR (1996), 'Microbiological and sensory analyses of refrigerated catfish fillets treated with acetic and lactic acids', *J Food Qual*, 19, 317–329.

MATERON LA (2003), 'Survival of *Escherichia coli* O157:H7 applied to cantaloupes and the effectiveness of chlorinated water and lactic acid as disinfectants', *World J Microbiol Biotechnol*, 19, 867–873.

MBANDI E and SHELEF LA (2002), 'Enhanced antimicrobial effects of combination of lactate and diacetate on *Listeria monocytogenes* and *Salmonella* spp. in beef bologna', *Int J Food Microbiol*, 76, 191–198.

MCWATTERS KH, DOYLE MP, WALKER SL, RIMAL AP and VENKITANARAYANAN K (2002a), 'Consumer acceptance of raw apples treated with an antibacterial solution designed for home use', *J Food Prot*, 65, 106–110.

MCWATTERS KH, HASHIM IB, WALKER SL, DOYLE MP and RIMAL AP (2002b), 'Acceptability of lettuce treated with a lactic acid and hydrogen peroxide antibacterial solution', *J Food Qual*, 25, 223–242.

MENDONCA AF, MOLINS RA, KRAFT AA and WALKER HW (1989), 'Microbiological, chemical, and physical changes in fresh, vacuum-packaged pork treated with organic acids and salts', *J Food Sci*, 54, 18–21.

MENDONCA AF, AMOROSO TL and KNABEL SJ (1994), 'Destruction of Gram-negative food-borne pathogens by high pH involves disruption of the cytoplasmic membrane', *Appl Environ Microbiol*, 60, 4009–4014.

MILILLO SR and RICKE SC (2010), 'Synergistic reduction of *Salmonella* in a model raw chicken media using a combined thermal and acidified organic acid salt intervention treatment', *J Food Sci*, 75, M121–M125.

MOSSEL DAA (1984), 'Intervention as the rational approach to control diseases of microbial etiology transmitted by foods', *J Food Saf*, 6, 89–104.

MU D, HUANG Y-W, GATES KW and WU W-H (1997), 'Effect of trisodium phosphate on *Listeria monocytogenes* attached to rainbow trout (*Oncorhynchus mykiss*) and shrimp (*Penaeus* spp.) during refrigerated storage', *J Food Saf*, 17, 37–46.

MUSTAPHA A., ARIYAPITIPUN T and CLARKE AD (2002), 'Survival of *Escherichia coli* O157:H7 on vacuum-packaged raw beef treated with polylactic acid, lactic acid, and nisin', *J Food Sci*, 67, 262–267.

NACMCF (NATIONAL ADVISORY COMMITTEE ON MICROBIOLOGICAL CRITERIA FOR FOODS) (1999), 'Microbiological safety evaluations and recommendations on sprouted seeds', *Int J Food Microbiol*, 52, 123–153.

NGUYEN-THE C and PRUNIER JP (1989), 'Involvement of pseudomonads in deterioration of "ready-to-use" salads', *Int J Food Sci Technol*, 24, 47–58.

NISSEN H, MAUGESTEN T and LEA P (2001), 'Survival and growth of *Escherichia coli* O157:H7, *Yersinia enterocolitica* and *Salmonella enteritidis* on decontaminated and untreated meat', *Meat Sci*, 57, 291–298.

NUÑEZ DE GONZALEZ MT, KEETON JT, ACUFF GR, RINGER LJ and LUCIA LM (2004), 'Effectiveness of acidic calcium sulphate with propionic and lactic acid and lactates as postprocessing dipping solutions to control *Listeria monocytogenes* on frankfurters with or without potassium lactate and stored vacuum packaged at 4.5°C', *J Food Prot*, 67, 915–921.

© Woodhead Publishing Limited, 2012

OGATA N (2007), 'Denaturation of protein by chlorine dioxide: oxidative modification of tryptophan and tyrosine residues', *Biochemistry-US*, 46, 4898–4911.

OJEU (2004), 'Regulation (EC) No 853/2004 of the European Parliament and of the Council of 29 April 2004 laying down specific hygiene rules for food of animal origin', *Official Journal of the European Union* L226 (25/06/2004), 22–82.

OJEU (2008), 'Regulation (EC) No 1333/2008 of the European Parliament and of the Council of 16 December 2008 on food additives', *Official Journal of the European Union* L354 (31/12/2008), 16–33.

OKOLOCHA EC and ELLERBROEK L (2005), 'The influence of acid and alkaline treatments on pathogens and the shelf life of poultry meat', *Food Control*, 16, 217–225.

ÖLMEZ H (2010), 'Effect of different sanitizing methods and incubation time and temperature on inactivation of *Escherichia coli* on lettuce', *J Food Saf*, 30, 288–299.

ÖLMEZ H and KRETZSCHMAR U (2009), 'Potential alternative disinfection methods for organic fresh-cut industry for minimizing water consumption and environmental impact', *LWT-Food Sci Technol*, 42, 686–693.

ÖLMEZ H and TEMUR SD (2010), 'Effects of different sanitizing treatments on biofilms and attachment of *Escherichia coli* and *Listeria monocytogenes* on green leaf lettuce', *LWT-Food Sci Technol*, 43, 964–970.

OYARZABAL OA (2005), 'Reduction of *Campylobacter* spp. by commercial antimicrobials applied during the processing of broiler chickens: a review from the United States perspective', *J Food Prot*, 68, 1752–1760.

OYARZABAL OA, HAWK C, BILGILI SF, WARF CC and KEMP GK (2004), 'Effects of postchill application of acidified sodium chlorite to control *Campylobacter* spp. and *Escherichia coli* on commercial broiler carcasses', *J Food Prot*, 67, 2288–2291.

ÖZDEMIR H, YILDIRIM Y, KÜPLÜLÜ Ö, KOLUMAN A, GÖNCÜOĞLU M and İNAT G (2006a), 'Effects of lactic acid and hot water treatments on *Salmonella* Typhimurium and *Listeria monocytogenes* on beef', *Food Control*, 17, 299–303.

ÖZDEMIR H, GÜCÜKOĞLU A and PAMUK Ş (2006b), 'Effects of cetylpyridinium chloride, lactic acid and sodium benzoate on populations of *Listeria monocytogenes* and *Staphylococcus aureus* on beef', *J Food Saf*, 26, 41–48.

PALUMBO SA and WILLIAMS AC (1994), 'Control of *Listeria monocytogenes* on the surface of frankfurters by acid treatments', *Food Microbiol*, 11, 293–300.

PAO S and DAVIS CL (1999), 'Enhancing microbiological safety of fresh orange juice by fruit immersion in hot water and chemical sanitizers', *J Food Prot*, 62, 756–760.

PAO S, KELSEY DF, KHALID MF and ETTINGER MR (2007), 'Using aqueous chlorine dioxide to prevent contamination of tomatoes with *Salmonella enterica* and *Erwinia carotovora* during fruit washing', *J Food Prot*, 70, 629–634.

PHUVASATE S and SU Y-C (2010), 'Effects of electrolyzed oxidizing water and ice treatments on reducing histamine-producing bacteria on fish skin and food contact surface', *Food Control*, 21, 286–291.

PIERNAS V and GUIRAUD JP (1997), 'Disinfection of rice seeds prior to sprouting', *J Food Sci*, 62, 611–615.

PIPEK P, ŠIKULOVÁ M, JELENÍKOVÁ J and IZUMIMOTO M (2005), 'Colour changes after carcasses decontamination by steam and lactic acid', *Meat Sci*, 69, 673–680.

PODOLAK RK, ZAYAS JF, KASTNER CL and FUNG DYC (1996), 'Inhibition of *Listeria monocytogenes* and *Escherichia coli* O157:H7 on beef by application of organic acids', *J Food Prot*, 59, 370–373.

POHLMAN FW, STIVARIUS MR, MCELYEA KS and WALDROUP AL (2002), 'Reduction of *E. coli*, *Salmonella typhimurium*, coliforms, aerobic bacteria, and improvement of ground beef color using trisodium phosphate or cetylpyridinium chloride before grinding', *Meat Sci*, 60, 349–356.

QUILO SA, POHLMAN FW, DIAS-MORSE PN, BROWN JR AH, CRANDALL PG and STORY RP (2010), 'Microbial, instrumental color and sensory characteristics of inoculated ground beef produced using potassium lactate, sodium metasilicate or peroxyacetic acid as multiple antimicrobial interventions', *Meat Sci*, 84, 470–476.

© Woodhead Publishing Limited, 2012

RAJKOVIC A, SMIGIC N and DEVLIEGHERE F (2010), 'Contemporary strategies in combating microbial contamination in food chain', *Int J Food Microbiol*, 141, S29–S42.

RICKE SC (2003), 'Perspectives of the use of organic acids and short chain fatty acids as antimicrobials', *Poult Sci*, 82, 632–639.

RICKE SC, KUNDINGER MM, MILLER DR and KEETON JT (2005), 'Alternatives to antibiotics: chemical and physical antimicrobial interventions and foodborne pathogen response', *Poult Sci*, 84, 667–675.

RIEDEL CT, BRØNDSTED L, ROSENQUIST H, HAXGART SN and CHRISTENSEN BB (2009), 'Chemical decontamination of *Campylobacter jejuni* on chicken skin and meat', *J Food Prot*, 72, 1173–1180.

RODGERS SL, CASH JN, SIDDIQ M and RYSER ET (2004), 'A comparison of different chemical sanitizers for inactivating *Escherichia coli* O157:H7 and *Listeria monocytogenes* in solution and on apples, lettuce, strawberries, and cantaloupe', *J Food Prot*, 67, 721–731.

RUSSELL JB (1992), 'Another explanation for the toxicity of fermentation acids at low pH: anion accumulation versus uncoupling', *J Appl Bacteriol*, 73, 363–370.

SAKHARE PZ, SACHINDRA NM, YASHODA KP and NARASIMHA RAO D (1999), 'Efficacy of intermittent decontamination treatments during processing in reducing the microbial load on broiler chicken carcass', *Food Control*, 10, 189–194.

SAMARA A and KOUTSOUMANIS KP (2009), 'Effect of treating lettuce surfaces with acidulants on the behaviour of *Listeria monocytogenes* during storage at 5 and 20°C and subsequent exposure to simulated gastric fluid', *Int J Food Microbiol*, 129, 1–7.

SAMELIS J (2005), 'Meat decontamination and pathogen stress adaptation', in Sofos JN, *Improving the safety of fresh meat*, Cambridge, Woodhead Publishing, 562–591.

SAMELIS J and SOFOS JN (2003a), 'Strategies to control stress-adapted pathogens', in Yousef AE and Juneja VK, *Microbial stress adaptation and food safety*, Boca Raton, FL, CRC Press, 303–351.

SAMELIS J and SOFOS JN (2003b), 'Organic acids', in Roller S, *Natural antimicrobials for the minimal processing of foods*, Cambridge, Woodhead Publishing, 98–132.

SAMELIS J, SOFOS JN, KAIN ML, SCANGA JA, BELK KE and SMITH GC (2001a), 'Organic acids and their salts as dipping solutions to control *Listeria monocytogenes* inoculated following processing of sliced pork bologna stored at 4°C in vacuum packages', *J Food Prot*, 64, 1722–1729.

SAMELIS J, SOFOS JN, KENDALL PA and SMITH GC (2001b), 'Fate of *Escherichia coli* O157:H7, *Salmonella* Typhimurium DT 104, and *Listeria monocytogenes* in fresh meat decontamination fluids at 4 and 10°C', *J Food Prot*, 64, 950–957.

SAMELIS J, SOFOS JN, KENDALL PA and SMITH GC (2001c), 'Influence of the natural microbial flora on the acid tolerance response *of Listeria monocytogenes* in a model system of fresh meat decontamination fluids', *Appl Environ Microbiol*, 67, 2410–2420.

SAMELIS J, SOFOS JN, KENDALL PA and SMITH GC (2002), 'Effect of acid adaptation on survival of *Escherichia coli* O157:H7 in meat decontamination washing fluids and potential effects of organic acid interventions on the microbial ecology of the meat plant environment', *J Food Prot*, 65, 33–40.

SAMELIS J, KENDALL P, SMITH GC and SOFOS JN (2004), 'Acid tolerance of acid-adapted and nonadapted *Escherichia* coli O157:H7 following habituation (10°C) in fresh beef decontamination runoff fluids of different pH values', *J Food Prot*, 67, 638–645.

SAMELIS J, SOFOS JN, KENDALL PA and SMITH GC (2005), 'Survival or growth of *Escherichia coli* O157:H7 in a model system of fresh meat decontamination runoff waste fluids and its resistance to subsequent lactic acid stress', *Appl Environ Microbiol*, 71, 6228–6234.

SAMPATHKUMAR B, KHACHATOURIANS GG and KORBER DR (2003), 'High pH during trisodium phosphate treatment causes membrane damage and destruction of *Salmonella enterica* serovar Enteritidis', *Appl Environ Microbiol*, 69, 122–129.

© Woodhead Publishing Limited, 2012

SAPERS GM (2001), 'Efficacy of washing and sanitizing methods for disinfection of fresh fruit and vegetable products', *Food Technol Biotechnol*, 39, 305–311.

SAPERS GM, MILLER RL and MATTRAZZO AM (1999), 'Effectiveness of sanitizing agents in inactivating *Escherichia coli* in Golden Delicious apples', *J Food Sci*, 64, 734–737.

SAPERS GM, MILLER RL, PILIZOTA V and MATTRAZZO AM (2001), 'Antimicrobial treatments for minimally processed cantaloupe melon', *J Food Sci*, 66, 345–349.

SENGUM IY and KARAPINAR M (2004), 'Effectiveness of lemon juice, vinegar and their mixture in the elimination of *Salmonella typhimurium* on carrots (*Daucus carota* L.)', *Int J Food Microbiol*, 96, 301–305.

SHIRAZINEJAD A, ISMAIL N and BHAT R (2010), 'Lactic acid as a potential decontaminant of selected foodborne pathogenic bacteria in shrimp (*Penaeus merguiensis* de Man)', *Foodborne Pathog Dis*, 7, 1531–1536.

SINGH M, GILL VS, THIPPAREDDI H, PHEBUS RK, MARSDEN JL, HERALD TJ and NUTSCH AL (2005), 'Antimicrobial activity of cetylpyridinium chloride against *Listeria monocytogenes* on frankfurters and subsequent effect on quality attributes', *J Food Prot*, 68, 1823–1830.

SINGH N, SINGH RK, BHUNIA AK and STROSHINE RL (2002), 'Effect of inoculation and washing methods on the efficacy of different sanitizers against *Escherichia coli* O157:H7 on lettuce', *Food Microbiol*, 19, 183–193.

SINHAMAHAPATRA M, BISWAS S, DAS AK and BHATTACHARYYA D (2004), 'Comparative study of different surface decontaminants on chicken quality', *Br Poult Sci*, 45, 624–630.

SIRAGUSA GR (1995), 'The effectiveness of carcass decontamination systems for controlling the presence of pathogens on the surfaces of meat animal carcasses', *J Food Saf*, 15, 229–238.

SKANDAMIS PN, NYCHAS G-JE and SOFOS JN (2010), 'Meat decontamination', in Toldrá F, *Handbook of meat processing*, Ames, IA, Blackwell Publishing, 43–85.

SMULDERS FJM and GREER GG (1998), 'Integrating microbial decontamination with organic acids in HACCP programmes for muscle foods: prospects and controversies', *Int J Food Microbiol*, 44, 149–169.

SMULDERS FJM, BARENDSEN P, VAN LOGTESTIJN JG, MOSSEL DAA and VAN DER MAREL GM (1986), 'Review: Lactic acid: considerations in favour of its acceptance as a meat decontaminant', *J Food Technol*, 21, 419–436.

SOFOS JN (ed.) (2005), *Improving the safety of fresh meat*, Cambridge, Woodhead Publishing.

SOFOS JN (2008), 'Challenges to meat safety in the 21st century', *Meat Sci*, 78, 3–13.

SOFOS JN and SMITH GC (1998), 'Nonacid meat decontamination technologies: model studies and commercial applications', *Int J Food Microbiol*, 44, 171–188.

SOFOS JN, BELK KE and SMITH GC (1999), 'Processes to reduce contamination with pathogenic microorganisms in meat', *Proc 45th Intl Congress of Meat Sci and Technol, August 1–6, Yokohama, Japan*, pp. 596–605.

STIVARIUS MR, POHLMAN FW, MCELYEA KS and APPLE JK (2002a), 'The effects of acetic acid, gluconic acid and trisodium citrate treatment of beef trimmings on microbial, color and odor characteristics of ground beef through simulated retail display', *Meat Sci*, 60, 245–252.

STIVARIUS MR, POHLMAN FW, MCELYEA KS and WALDROUP AL (2002b), 'Effects of hot water and lactic acid treatment of beef trimmings prior to grinding on microbial, instrumental color and sensory properties of ground beef during display', *Meat Sci*, 60, 327–334.

STOPFORTH JD, SAMELIS J, SOFOS JN, KENDALL PA and SMITH GC (2003), 'Influence of organic acid concentration on survival of *Listeria monocytogenes* and *Escherichia coli* O157:H7 in beef carcass wash water and on model equipment surfaces', *Food Microbiol*, 20, 651–660.

STOPFORTH JD, YOON Y, BELK KE, SCANGA JA, KENDALL PA, SMITH GC and SOFOS JN (2004), 'Effect of simulated spray chilling with chemical solutions on acid-habituated and

© Woodhead Publishing Limited, 2012

non-acid-habituated *Escherichia coli* O157:H7 cells attached to beef carcass tissue', *J Food Prot*, 67, 2099–2106.

STOPFORTH JD, MAI T, KOTTAPALLI B and SAMADPOUR M (2008), 'Effect of acidified sodium chlorite, chlorine, and acidic electrolyzed water on *Escherichia coli* O157:H7, *Salmonella* and *Listeria monocytogenes* inoculated into leafy greens', *J Food Prot*, 71, 625–628.

STRATFORD M and EKLUND T (2003), 'Organic acids and esters', in Russell NJ and Gould GW, *Food preservatives*, 2nd edn, New York, Kluwer Academic/Plenum Publishers, 48–84.

TAMBLYN KC and CONNER DE (1997), 'Bactericidal activity of organic acids against *Salmonella typhimurium* attached to broiler chicken skin', *J Food Prot*, 60, 629–633.

TAMBLYN KC, CONNER DE and BILGILI SF (1997), 'Utilization of the skin attachment model to determine the antibacterial efficacy of potential carcass treatments', *Poultry Sci*, 76, 1318–1323.

TAORMINA PJ and BEUCHAT LR (1999a), 'Comparison of chemical treatments to eliminate enterohemorrhagic *Escherichia coli* O157:H7 on alfalfa seeds', *J Food Prot*, 62, 318–324.

TAORMINA PJ and BEUCHAT LR (1999b), 'Behavior of enterohemorrhagic *Escherichia coli* O157:H7 on alfalfa sprouts during the sprouting process as influenced by treatments with various chemicals', *J Food Prot*, 62, 850–856.

TAUXE R, KRUSE H, HEDBERG C, POTTER M, MADDEN J and WACHSMUTH K (1997), 'Microbial hazards and emerging issues associated with produce. A preliminary report to the National Advisory Committee on Microbiological Criteria for Foods', *J Food Prot*, 60, 1400–1408.

TERZI G and GUCUKOGLU A (2010), 'Effects of lactic acid and chitosan on the survival of *V. parahaemolyticus* in mussel samples', *J Anim Vet Adv*, 9, 990–994.

THERON MM and LUES JFR (2007), 'Organic acids and meat preservation: a review', *Food Rev Int*, 23, 141–158.

THERON MM and LUES JFR (2011), *Organic acids and food preservation*, Boca Raton, FL, CRC Press.

THOMPSON JL and HINTON M (1996), 'Effect of short-chain fatty acids on the size of enteric bacteria', *Lett Appl Microbiol*, 22, 408–412.

UKUKU DO (2004), 'Effect of hydrogen peroxide treatment on microbial quality and appearance of whole and fresh-cut melons contaminated with *Salmonella* spp.', *Int J Food Microbiol*, 95, 137–146.

UKUKU DO (2006), 'Effect of sanitizing treatments on removal of bacteria from cantaloupe surface, and re-contamination with *Salmonella*', *Food Microbiol*, 23, 289–293.

UKUKU DO and FETT WF (2004), 'Method of applying sanitizers and sample preparation affects recovery of native microflora and *Salmonella* on whole cantaloupe surfaces', *J Food Prot*, 67, 999–1004.

UKUKU DO and SAPERS GM (2001), 'Effect of sanitizer treatments on *Salmonella* Stanley attached to the surface of cantaloupe and cell transfer to fresh-cut tissues during cutting practices', *J Food Prot*, 64, 1286–1291.

UKUKU DO, PILIZOTA V and SAPERS GM (2001), 'Influence of washing treatments on native microflora and *Escherichia coli* population of inoculated cantaloupes', *J Food Saf*, 21, 31–47.

UKUKU DO, BARI ML, KAWAMOTO S and ISSHIKI K (2005), 'Use of hydrogen peroxide in combination with nisin, sodium lactate and citric acid for reducing transfer of bacterial pathogens from whole melon surfaces to fresh-cut pieces', *Int J Food Microbiol*, 104, 225–233.

USDA-FSIS (2010), 'Safe and suitable ingredients used in the production of meat, poultry, and egg products', US Department of Agriculture Food Safety and Inspection Service,

© Woodhead Publishing Limited, 2012

Directive 7120.1, Revision 2, 4/12/10. Available from: http://www.fsis.usda.gov/OPPDE/rdad/FSISDirectives/7120.1Rev2.pdf (accessed 20 September 2011).

USFDA (1998), 'Guidance for industry: Guide to minimize microbial food safety hazards for fresh fruits and vegetables', US Food and Drug Administration. Available from: http://www.fda.gov/Food/GuidanceComplianceRegulatoryInformation/GuidanceDocuments/ProduceandPlanProducts/ucm064574.htm (accessed 30 August 2011).

USFDA (2008), 'Guidance for industry: Guide to minimize microbial food safety hazards of fresh-cut fruits and vegetables', US Food and Drug Administration. Available from: http://www.fda.gov/Food/GuidanceComplianceRegulatoryInformation/GuidanceDocuments/ProduceandPlanProducts/ucm064458.htm (accessed 30 August 2011).

UYTTENDAELE M, NEYTS K, VANDERSWALMEN H, NOTEBAERT E and DEBEVERE J (2004), 'Control of *Aeromonas* on minimally processed vegetables by decontamination with lactic acid, chlorinated water, or thyme essential oil solution', *Int J Food Microbiol*, 90, 263–271.

VAN NETTEN P, HUIS IN 'T VELD JH and MOSSEL DAA (1994), 'The effect of lactic acid decontamination on the microflora on meat', *J Food Saf*, 14, 243–257.

VAN NETTEN P, MOSSEL DAA and HUIS IN 'T VELD J (1995), 'Lactic acid decontamination of fresh pork carcasses: a pilot plant study', *Int J Food Microbiol*, 25, 1–9.

VAN NETTEN P, MOSSEL DAA and HUIS IN 'T VELD JHJ (1997a), 'Microbial changes on freshly slaughtered pork carcasses due to "hot" lactic acid decontamination', *J Food Saf*, 17, 89–111.

VAN NETTEN P, VALENTIJN A, MOSSEL DAA and HUIS IN 'T VELD JHJ (1997b), 'Fate of low temperature and acid-adapted *Yersinia enterocolitica* and *Listeria monocytogenes* that contaminate lactic acid decontaminated meat during chill storage', *J Appl Microbiol*, 82, 769–779.

VAN NETTEN P, VALENTIJN A, MOSSEL DAA and HUIS IN 'T VELD JHJ (1998), 'The survival and growth of acid-adapted mesophilic pathogens that contaminate meat after lactic acid decontamination', *J Appl Microbiol*, 1998, 559–567.

VENKITANARAYANAN KS, LIN C-M, BAILEY H and DOYLE MP (2002), 'Inactivation of *Escherichia coli* O157:H7, *Salmonella* Enteritidis, and *Listeria monocytogenes* on apples, oranges, and tomatoes by lactic acid with hydrogen peroxide', *J Food Prot*, 65, 100–105.

VERHAEGH EGA, MARSHALL DL and OH D-H (1996), 'Effect of monolaurin and lactic acid on *Listeria monocytogenes* attached to catfish fillets', *Int J Food Microbiol*, 29, 403–410.

VIJAYAKUMAR C and WOLF-HALL CE (2002), 'Evaluation of household sanitizers for reducing levels of *Escherichia coli* on iceberg lettuce', *J Food Prot*, 65, 1646–1650.

WAN T-C, LIN L-C and SAKATA R (2007), 'Effect of organic acids on the microbial quality of Taiwanese-style sausages', *Anim Sci J*, 78, 407–412.

WANG J (2002), 'Death of *Escherichia coli* upon exposure to organic acid', Doctoral dissertation, Columbia, MO, University of Missouri-Columbia. Available from: WorldCat Online Computer Library Center Inc., Dublin, OH, accession nr 51808626.

WEISSINGER WR and BEUCHAT LR (2000), 'Comparison of aqueous chemical treatments to eliminate *Salmonella* on alfalfa seeds', *J Food Prot*, 63, 1475–1482.

WHYTE P, COLLINS JD, MCGILL K, MONAHAN C and O'MAHONY H (2001), 'Quantitative investigation of the effects of chemical decontamination procedures on the microbiological status of broiler carcasses during processing', *J Food Prot*, 64, 179–183.

WU FM, DOYLE MP, BEUCHAT LR, WELLS JG, MINTZ ED and SWAMINATHAN B (2000), 'Fate of *Shigella sonnei* on parsley and methods of disinfection', *J Food Prot*, 63, 568–572.

XIONG H, LI Y, SLAVIK MF and WALKER JT (1998a), 'Spraying chicken skin with selected chemicals to reduce attached *Salmonella typhimurium*', *J Food Prot*, 61, 272–275.

XIONG H, LI Y, SLAVIK M and WALKER J (1998b), 'Chemical spray conditions for reducing bacteria on chicken skins', *J Food Sci*, 63, 699–701.

© Woodhead Publishing Limited, 2012

YANG Z, LI Y and SLAVIK M (1998), 'Use of antimicrobial spray applied with an inside-outside birdwasher to reduce bacterial contamination on prechilled chicken carcasses', *J Food Prot*, 61, 829–832.

YOON Y, SKANDAMIS PN, KENDALL PA, SMITH GC and SOFOS JN (2006), 'A predictive model for the effect of temperature and predrying treatments in reducing *Listeria monocytogenes* populations during drying of beef jerky', *J Food Prot*, 69, 62–70.

ZEITOUN AAM and DEBEVERE JM (1990), 'The effect of treatment with buffered lactic acid on microbial decontamination and on shelf life of poultry', *Int J Food Microbiol*, 11, 305–312.

ZEITOUN AAM and DEBEVERE JM (1991), 'Inhibition, survival and growth of *Listeria monocytogenes* on poultry as influenced by buffered lactic acid treatment and modified atmosphere packaging', *Int J Food Microbiol*, 14, 161–170.

ZHANG S and FARBER JM (1996), 'The effects of various disinfectants against *Listeria monocytogenes* on fresh-cut vegetables', *Food Microbiol*, 13, 311–321.

ZHANG G, MA L, PHELAN VH and DOYLE MP (2009), 'Efficacy of antimicrobial agents in lettuce leaf processing water for control of *Escherichia coli* O157:H7', *J Food Prot*, 72, 1392–1397.

ZHAO T and DOYLE MP (2006), 'Reduction of *Campylobacter jejuni* on chicken wings by chemical treatments', *J Food Prot*, 69, 762–767.

ZHAO T, ZHAO P and DOYLE MP (2009), 'Inactivation of *Salmonella* and *Escherichia coli* O157:H7 on lettuce and poultry skin by combinations of levulinic acid and sodium dodecyl sulfate', *J Food Prot*, 72, 928–936.

ZHUANG R-Y and BEUCHAT LR (1996), 'Effectiveness of trisodium phosphate for killing *Salmonella montevideo* on tomatoes', *Lett Appl Microbiol*, 22, 97–100.

© Woodhead Publishing Limited, 2012

21

# Dense phase $CO_2$ (DPCD) for microbial decontamination of food

**M. O. Balaban, University of Auckland, New Zealand and G. Ferrentino and S. Spilimbergo, University of Trento, Italy**

**Abstract**: This chapter discusses the application of dense phase $CO_2$ (DPCD) technology as an alternative to thermal food preservation. This technology has been applied mostly to liquid foods; however, applications for solid foods exist, and research in this area is accelerating. This chapter first reviews the principles of the DPCD process, its advantages, its modes of application, and the theories about its mechanisms of microbial inactivation. Then, the efficacy of DPCD treatment in preservation of various liquid and solid foods and its effects on their quality are discussed.

**Key words**: dense phase carbon dioxide, microbial safety, quality attributes, microbial inactivation mechanisms.

## 21.1 Introduction

Thermal treatment of foods to assure their safety has been well known for centuries. Heating kills pests, inactivates microorganisms (bacteria in vegetative and spore forms, yeasts and their spores, molds and their spores, and some viruses), and inactivates enzymes that can cause undesirable effects in foods. Its mode of action is known, and the kinetics of microbial and enzyme inactivation have been defined. However, heating also destroys some nutrients, and changes physical, chemical, and sensory attributes of foods. Since consumer attitudes are moving towards 'less processed', 'more natural', and 'more healthy' foods, other processing methods such as ultra-high pressure, pulsed electric, pulsed magnetic, high intensity ultrasound, UV lights, etc., have been advanced as alternatives to thermal processing.

Dense phase carbon dioxide (DPCD) is one of these alternative non-thermal

© Woodhead Publishing Limited, 2012

technologies to thermal methods of food preservation. Carbon dioxide is a gas that exists naturally in foods (bread, beer, wine, etc.). It is a non-polar molecule that has been used as an effective solvent of non-polar materials such as lipids in supercritical extraction. Carbon dioxide is also an inert molecule and does not react with food components. It is one of the least expensive industrial gases. The technology to extract $CO_2$ from flue gases exists and does not add to the 'carbon footprint' of the operation regarding the source of the gas. Carbon dioxide is non-flammable and not explosive. This makes it easy to transport and store. These properties make it an attractive gas to use on an industrial scale. DPCD is mostly applied to liquid foods as a pasteurization method. However, applications for solid foods also exist. Effects on spores, viruses, and other pests have been described in the literature. The application of this technology to sterilize medical equipment is outside the scope of this chapter. Instead, a brief overview of the principles of this technology, its modes of application, the theories about its mechanisms of microbial inactivation, existing reports on its efficacy in the preservation of various liquid and solid foods and its effects on their quality, as well as future trends and recommendations involving this technology will be described.

## 21.2 Food decontamination using dense phase $CO_2$ (DPCD): principles and technology

The $CO_2$ used in the DPCD process is not only a powerful solvent for a wide range of compounds of interest in food processing, but is inert, non-toxic, non-flammable, recyclable, and readily available in high purity, leaving no residue when removed after the process. It has been considered to be a GRAS (Generally Recognized as Safe) material since 1979 (FDA, 1979), which means it can be used in food products. Because of the need for a preservation method that is safe, inexpensive, and non-destructive to heat sensitive compounds, the use of dense phase $CO_2$ has been tested as a food preservation method on microbial cells in cultures or broths.

DPCD technology is based on the contact between the food to be decontaminated and the pressurized $CO_2$. The efficiency of the process depends on the way this contact is achieved: since $CO_2$ is a non-polar molecule, its dispersion into polar compounds is difficult. Water is a special case, since $CO_2$ interacts with water, first dissolving in it, and then chemically reacting with it to form carbonic acid.

There are three main steps that occur during the DPCD process:

1. the compression of $CO_2$;
2. the pasteurization of the product in contact with pressurized $CO_2$;
3. the decompression of the system with the release of $CO_2$ and the recovery of the product.

© Woodhead Publishing Limited, 2012

Like any technology involving the pressurization of gases, safety aspects of handling pressurized vessels are important. Special precaution is required for the heating of the gases while pressurizing, and especially their significant cooling during depressurization. The depressurizing valve needs to be heated since expanding $CO_2$ cools the valve to such an extent that the formation of 'dry ice' may plug it. In addition, $CO_2$ requires more attention than, for example, pressurized air, since it can suffocate, and it can accumulate in closed spaces since it is heavier than air. Finally, since $CO_2$ is a 'greenhouse' gas, its release into the atmosphere is of concern, and further regulations (such as cap-and-trade rules) may be implemented in the future regarding its capture and re-use during the DPCD process.

### 21.2.1 States of $CO_2$

Depending on temperature and pressure applied, $CO_2$ exists in the solid, liquid, gas, or supercritical fluid states (Fig. 21.1). The 'dense phase' region includes mostly the supercritical phase, but also some of the liquid and gaseous regions.

The most important property of $CO_2$ is the density, which varies with temperature and pressure. By changing $CO_2$ density, it is possible to change its solvent (or extraction) power. Also, its solubility and thus its capability of coming in contact with the liquid food containing microorganisms depend on temperature, pressure, and the composition of liquid. For this reason it is necessary to choose the proper pressure and temperature conditions for the DPCD process. Physical properties such as density, diffusivity, solubility

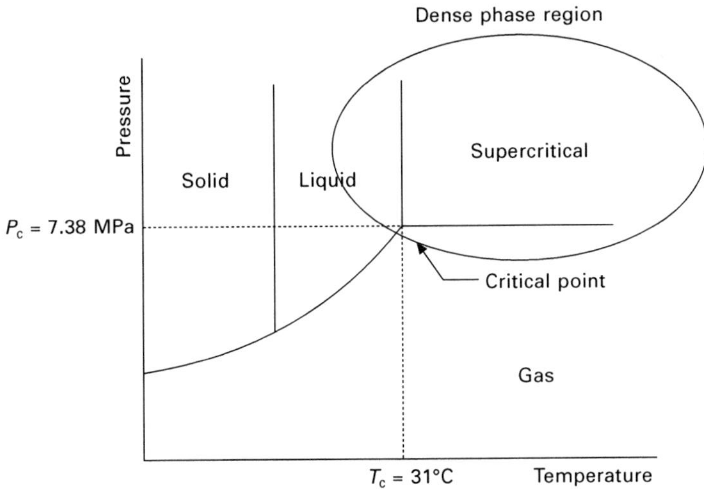


**Fig. 21.1** Pressure–temperature phase diagram of carbon dioxide. The coordinates of the critical point, as well as the conceptual location of the dense phase region are as shown.

© Woodhead Publishing Limited, 2012

in aqueous solution, and extraction power vary dramatically around the critical point. $CO_2$ has a low critical pressure (7.3 MPa) and a low critical temperature (31.7°C). The supercritical state is characterized by gas-like diffusivity and liquid-like density. The gas-like diffusivity allows supercritical $CO_2$ to quickly diffuse through complex matrices; the liquid-like density confers high extraction power. Relevant physical data of $CO_2$ are compiled in Table 21.1.

### 21.2.2 Solubility of $CO_2$

Dodds *et al.* (1956) stated that solubility of $CO_2$ in water depends on its pressure and temperature in equilibrium with the solution. Pressure has a direct effect on $CO_2$ solubility: an increase in pressure increases $CO_2$ solubility. On the other hand, with increasing temperature $CO_2$ solubility decreases like any other gas. The presence of other substances in the liquid food could have a positive or negative effect on solubility (Descoins *et al.*, 2006; Meyssami *et al.*, 1992; Ferrentino *et al.*, 2010b). There is a wealth of information in the literature regarding the solubility of $CO_2$ in aqueous systems, with experimental measurement and thermodynamic prediction of solubility under various conditions (Ferentino *et al.*, 2010a). As an example, the solubility of $CO_2$ in pure water and grapefruit juice at 40°C and different pressures is shown in Fig. 21.2 (Ferrentino *et al.*, 2009b).

### 21.2.3 Process parameters

The three process parameters most considered in the literature are pressure, temperature, and treatment time. In particular, pressure and temperature affect the efficiency of the treatment to the biggest extent due to their dominating effect on $CO_2$ mass transfer, solubility and on biological activities of the microbial cells. In particular, as far as temperature is concerned, it has been demonstrated that the pasteurization/sterilization effect induced by $CO_2$ is more pronounced as temperature is increased (Hong *et al.*, 1999). This is probably due to the fact that a higher temperature increases the fluidity of a microorganism's cell membrane, making it easier to penetrate, and, additionally, increases the diffusivity of $CO_2$. On the other hand, higher temperatures may

**Table 21.1** Comparison of physical properties of gases, liquids and supercritical fluids

| Property | Gas | Supercritical fluid | Liquid |
|---|---|---|---|
| Density ($g/cm^3$) | $10^{-3}$ | 0.3 | 1 |
| Viscosity (Pa·s) | $10^{-5}$ | $10^{-4}$ | $10^{-3}$ |
| Diffusivity ($cm^2/s$) | 0.1 | $10^{-3}$ | $5 \times 10^{-6}$ |
| Surface tension $H_2O/CO_2$ (mN/m) | < 1 | < 1 | 72/1 |

Source: Balaban and Meireles, 1999.

© Woodhead Publishing Limited, 2012

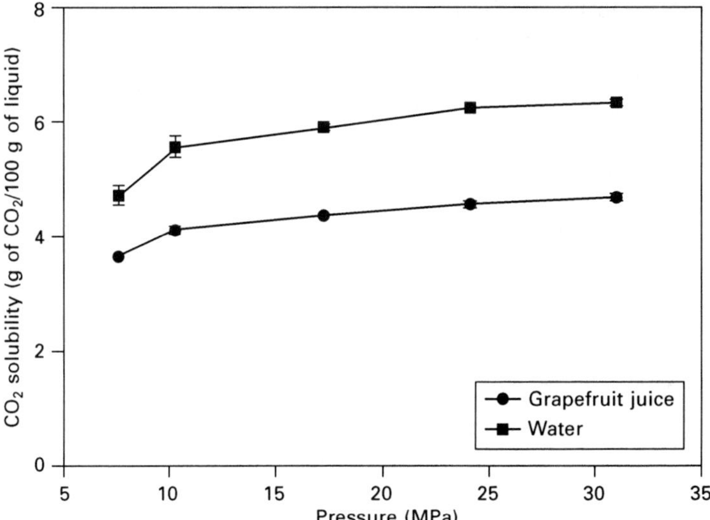


**Fig. 21.2** Experimental solubility of $CO_2$ at 40°C in pure water and grapefruit juice.

reduce the ability of $CO_2$ to extract low volatility materials and decrease $CO_2$ solubility in aqueous media. Hong and Pyun (1999) reported that the inactivation of *Lactobacillus*, *plantarum* at 30°C and 7 MPa was higher than that at 40°C and 7 MPa.

The efficiency of the DPCD process is also a strong function of the operating $CO_2$ pressure. At higher pressures, a shorter exposure time is needed to induce the same pasteurizing effect (Lin *et al.*, 1993, 1994; Hong *et al.*, 1997; Hong and Pyun, 1999). The effect of $CO_2$ pressure, however, does not go on indefinitely and is limited by the saturation solubility of $CO_2$ in the suspending medium (Sims and Estigarribia, 2003). Spilimbergo *et al.* (2002) demonstrated that above 10 MPa the solubility of $CO_2$ was a weak function of pressure. An increase in pressure from 10 to 30 MPa at 55–60°C did not influence appreciably the solubility of $CO_2$ in water. Published microbial inactivation kinetics data demonstrate that the treatment time is an additional parameter which controls the process. Dense phase $CO_2$ process is more effective as treatment time increases, although a long process may be difficult to implement in industrial settings.

In a batch system the mixing speed is considered as a key process parameter. Some studies reported that an increase in the mixing speed can enhance the solubilization of $CO_2$ and consequently its contact with bacterial cells, making cellular penetration easier and generally improving the microbial inactivation (Lin *et al.*, 1992; Hong *et al.*, 1997).

The effect of the depressurization rate is another process parameter which needs to be controlled. One of the first papers addressing the dense phase $CO_2$ process suggested that the cells were mechanically ruptured like a 'popped

© Woodhead Publishing Limited, 2012

balloon' by the fast expansion of $CO_2$ within the cells during the flash discharge of pressure (Fraser, 1951). In subsequent years other researchers (Lin *et al.*, 1991; Nakamura *et al.*, 1994; Castor and Hong, 1992) have claimed that rapid decompression is a parameter that enhances the disruption of bacterial cells. However, although mechanical cell rupture may be a cause of the cell death, secondary effects such as intensive localized cooling due to the Joule–Thomson effect (for gases, heating due to pressurization, and cooling due to depressurization) when pressurized $CO_2$ is expanded, may also play some role in cell lysis – and hence in bacterial inactivation – during a fast decompression rate.

### 21.2.4 Types of DPCD systems

For batch systems, both $CO_2$ and the product are stationary in the reactor throughout the treatment time. A typical batch system has a pressure vessel where the sample is placed and $CO_2$ is injected from a gas cylinder to reach the desired pressure. The pressure vessel is equipped with a water bath or a heater to keep the temperature inside the system at the desired experimental value. The sample is left in the vessel for a fixed treatment time after which the $CO_2$ release valve is opened to release the gas. Finally, the sample is removed from the treatment vessel. Some setups contain a stirring system to increase the mixing between the fluid and the sample, thus to decrease the time to saturate the sample with $CO_2$ (Hong and Pyun, 1999). The conceptual operation of the batch system is shown in Fig. 21.3.

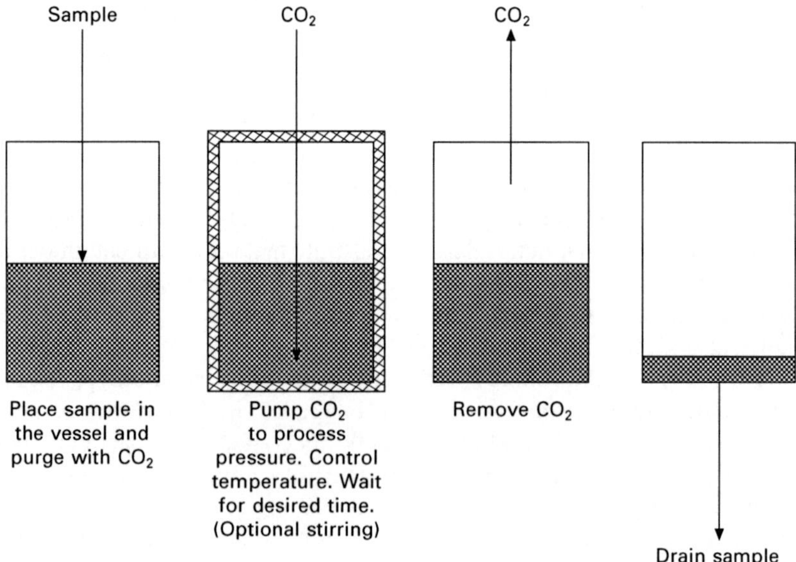


**Fig. 21.3** Conceptual steps for the operation of a batch system.

© Woodhead Publishing Limited, 2012

An improved design of batch system has been carried out by Spilimbergo and Mantoan (2006), who developed a multi-batch apparatus, specifically constructed to monitor the inactivation kinetics of different microorganisms under the same operating conditions. The multi-batch apparatus consists of identical reactors connected in parallel, so that each experimental run provides a set of experimental data taken in identical process conditions but different treatment times. Each reactor is connected to an on-off valve that can be used to depressurize it independently of the others. The ten reactors are submerged in a single temperature-controlled water bath so that the temperature is uniform in all of them. The bath is provided with ten magnetic stirrers, each one serving one reactor. These stirrers enable enough mixing to ensure fast mass transport of $CO_2$ from the fluid to the liquid phase. The $CO_2$ reservoir is heated by a water circuit in order to precisely control the desired process temperature without overheating the samples during pressurization (Fig. 21.4).

The semi-continuous apparatus represents an improvement over the batch design (Spilimbergo *et al.*, 2003). The system consists of a high pressure vessel in which the sample is loaded and where the liquid $CO_2$ is continuously fed by a high pressure pump. A 5 μm porous metallic filter, placed at the bottom of the cylinder, allows the atomization of $CO_2$ flow

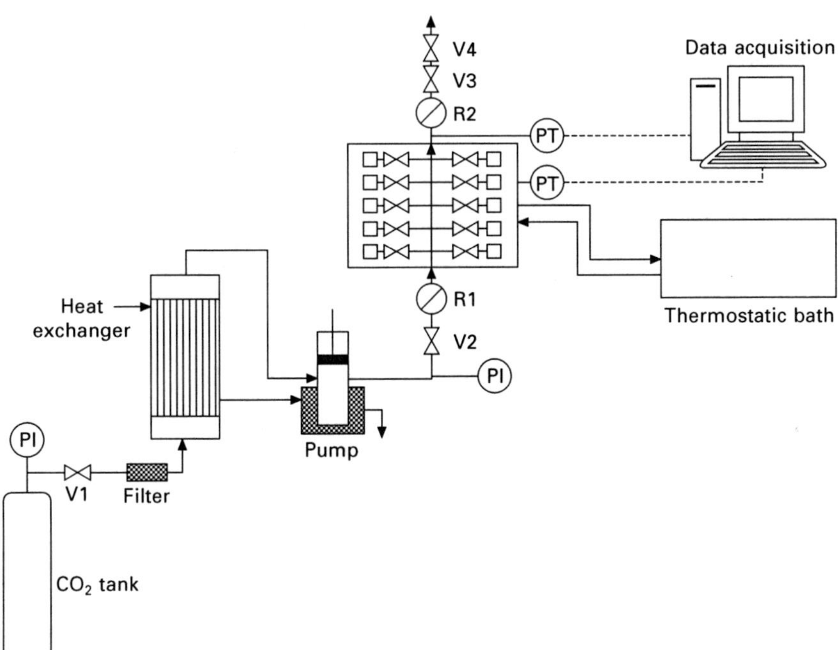


**Fig. 21.4** A multi-batch operation system, showing the multiple chambers that can be run in parallel. V1–V4: valves; R1, R2: electrical resistance; PT: pressure transducer; PI: pressure manometer.

© Woodhead Publishing Limited, 2012

into micro-bubbles; the vessel is thermally insulated and equipped with a resistance temperature probe located inside, a temperature control and an electric resistor. The depressurization of the system is carried out through the opening of an outlet valve, which regulates the $CO_2$ pressure in the autoclave and is heated by another resistor to prevent freezing during the $CO_2$ expansion. The conceptual operation of a semi-continuous operation is shown in Fig. 21.5.

The subsequent step in the development and optimization of the $CO_2$ technology is represented by the continuous plug flow system designed by Shimoda *et al.* (2001). The liquid $CO_2$ and the sample are simultaneously pumped in the $CO_2$ dissolving vessel. Liquid $CO_2$ is heated to the gaseous or supercritical state while passing through an evaporator, and is then dispersed into the liquid by a stainless steel mesh filter with 10 μm pore size attached to the bottom of the dissolving vessel. The micro-bubbles of pressurized $CO_2$ migrate upwards while being dissolved in the suspension. The presence of the stainless steel mesh filter assures optimized contact between the $CO_2$ and the liquid food. A simplified diagram of the operation is shown in Fig. 21.6.

An interesting apparatus was designed by Porocrit LLC (Sims, 2001) which is based on the same principle as the one described above, but the unique feature of their system is the use of a microporous polypropylene membrane contactor to carbonate the stream rapidly to saturation. This system appears particularly efficient as $CO_2$ is not mixed with the liquid but instantaneously diffuses into it at saturation levels in the membrane contactor. In addition

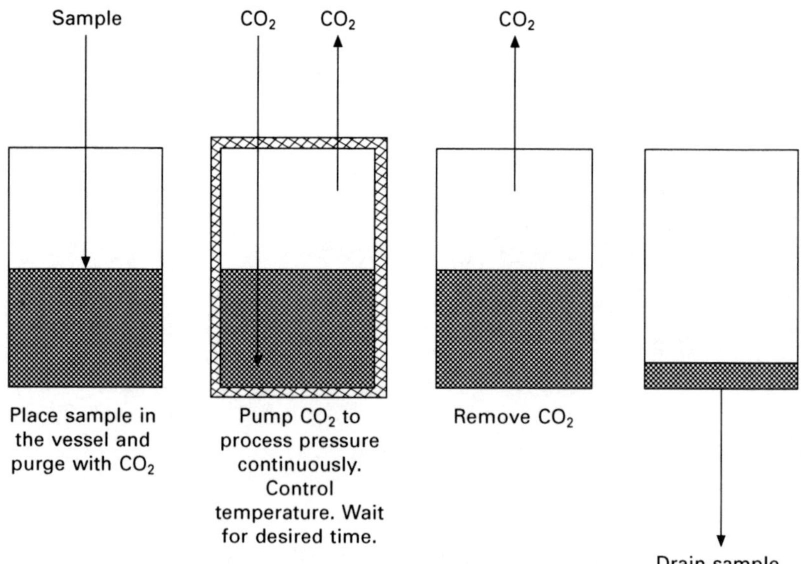


**Fig. 21.5** Conceptual steps for the operation of a semi-continuous operation.

© Woodhead Publishing Limited, 2012

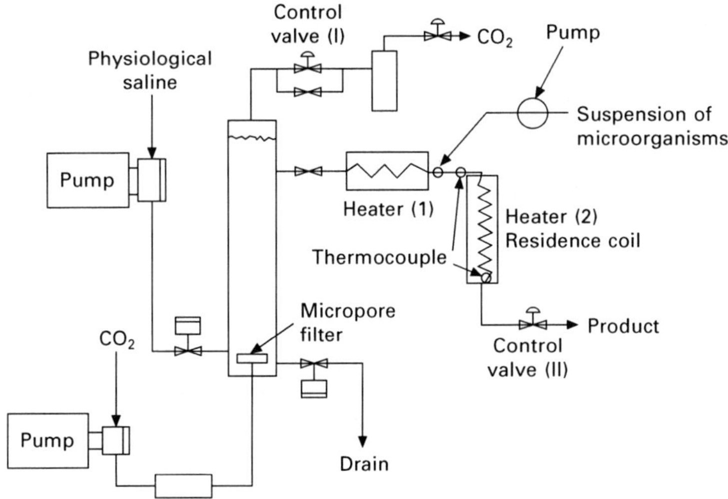


**Fig. 21.6** Conceptual diagram of a continuous micro-bubble system.

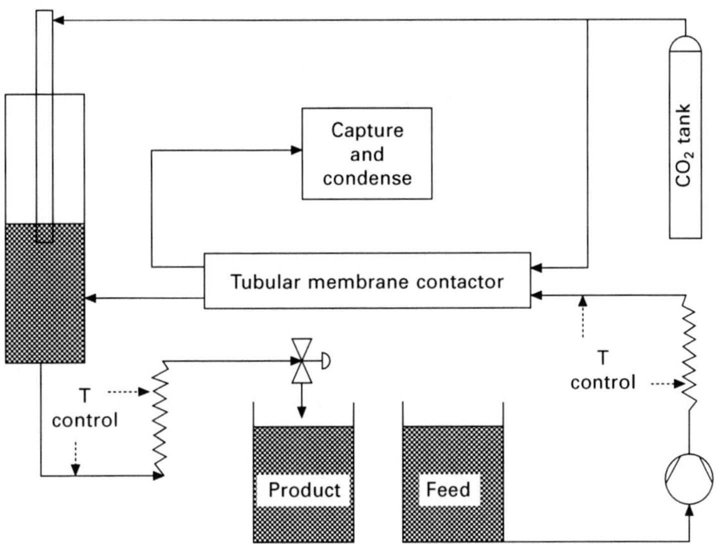


**Fig. 21.7** Simplified diagram of a continuous membrane-contactor system.

dense $CO_2$ is continuously recirculated without depressurization. In this way, the fastest killing rate of a large range of bacteria and yeasts could be obtained at a particular pressure and temperature that never exceeded 45°C (Sims and Estigarribia, 2003). A simplified conceptual diagram is shown in Fig. 21.7.

© Woodhead Publishing Limited, 2012

The design of a continuous dense phase $CO_2$ apparatus to be used on an industrial scale was realized for the first time by Praxair, Inc. (Chicago, IL) in 1999. The system allows $CO_2$ and the product to be pumped through the system and mixed before passing through the high pressure pump, which increases the mixture pressure to the process levels. Product temperature is controlled in holding coils. Residence time is adjusted by setting the flow rate of the product through the coils. If the solubility of $CO_2$ in the liquid under process conditions is unknown, then an excess amount of $CO_2$ is used to assure saturation. The amount of $CO_2$ used per unit weight of liquid can be controlled. This is called the $CO_2$ to liquid ratio. At the end of the process, an expansion valve is used to release $CO_2$ from the mixture; the residual $CO_2$ in the food can be pulled out through a vacuum tank. This system has been shown to be very effective in killing pathogens and spoilage bacteria in a short time, in the order of 5 min (Damar and Balaban, 2006). The conceptual diagram of the continuous operation is shown in Fig. 21.8.

## 21.3 Mechanisms of microbial inactivation

The bacteriostatic action and inhibitory effect of $CO_2$ on growth and metabolism of microorganisms has been reported since 1951 (Fraser, 1951). For instance, *Pseudomonas* was found to be very sensitive while other types of microorganisms, such as *Lactobacillus* and *Clostridium* were less sensitive. However, it was with the work published by Kamihira *et al.* (1987) that the inhibitory effect of $CO_2$ started to be addressed systematically. These authors tested the inactivation effect of $CO_2$ at supercritical, liquid and gaseous states on wet and dry *Escherichia coli*, *Staphilococcus aureus* and conidia

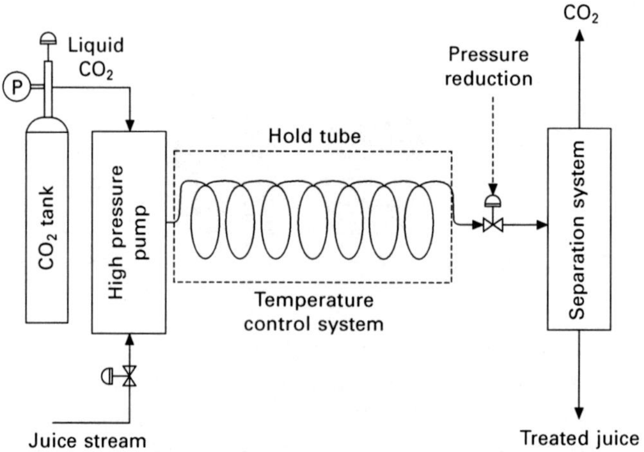


**Fig. 21.8** Conceptual diagram of a continuous system.

© Woodhead Publishing Limited, 2012

of *Aspergillus niger* by using a supercritical fluid extraction apparatus. At 20 MPa, 35°C and 2 h treatment with supercritical $CO_2$, wet cells of Baker's yeast were reduced by 7 logs, *E. coli* by 6 logs, *S. aureus* by 5 logs, and conidia of *A. niger* by 5 logs, while their dry counterparts were all reduced by less than 1 log.

Since then, many studies have investigated the effects of $CO_2$ on pathogenic and spoilage microorganisms, vegetative cells and spores. Some viruses have also been studied. A number of studies investigated the application of the treatment to vegetative bacteria, both Gram-positive and Gram-negative. These two groups of cells have different responses to the treatment based on the differences in their cell wall structures. Gram-positive bacteria show a higher resistance to inactivation than Gram-negative bacteria. In a semi-continuous apparatus, Gram-positive *Bacillus subtilis* suspended in phosphate buffered saline was completely inactivated after 2.5 min exposure under 7.4 MPa $CO_2$ at 38°C, while the same concentration of Gram-negative *Serratia marcescens* was inactivated at 0 min (not considering the time needed to pressurize the system to the desired pressure) under the same conditions (Spilimbergo *et al.*, 2003).

Spores are highly resistant to DPCD treatment. Vegetative *Geobacillus stearothermophilus* cells were reduced by more than 6 log cycles after 1.5 h exposure to $CO_2$ at 2.75 MPa and 25°C. However, even with 2 h exposure to $CO_2$ at 20 MPa and 35°C, 80% of *Geobacillus stearothermophilus* spores remained viable. Even with the addition of ethanol or acetic acid to increase $CO_2$ solubility, less than 60% of the *Geobacillus stearothermophilus* spores were inactivated (Kamihira *et al.*, 1987).

For more inactivation, increasing treatment time, raising temperature, using pressure cycling or pretreatment have been employed. Enomoto *et al.* (1997) achieved approximately 7 log reduction of *Bacillus megaterium* spores with a 50 h treatment at 7.8 MPa and 60°C. However a 50 h treatment would be problematic for a practical application of the process and the high temperatures used (55–90°C) could damage heat-sensitive materials.

The exact $CO_2$ microbial inactivation mechanism is still unknown. However, some mechanisms have been suggested by Garcia-Gonzalez *et al.* (2007) such as lowering of extracellular pH, disruption of the cell membrane, lowering of the intracellular pH, and extraction of key cellular components. These are described as follows.

### 21.3.1 Lowering the extracellular pH

Pressurized $CO_2$ dissolves in the water of the food matrix forming $H^+$ and $HCO_3^-$ ions. In this way it lowers the extracellular pH, inhibiting the microbial growth and diminishing microbial resistance to inactivation because of the increased energy consumption to maintain pH homeostasis by the proton motive force exerted by the protonic pump of the cells. This lower external pH contributes to an increase in cell permeability, which facilitates the penetration

© Woodhead Publishing Limited, 2012

of $CO_2$ into microbial cells and its accumulation in the cytoplasmic interior of the cells (Lin *et al.*, 1993, 1994; Spilimbergo *et al.*, 2002).

### 21.3.2 Disruption of the cell membrane

Due to its hydrophobic and liposoluble nature, during the process pressurized $CO_2$ can diffuse and accumulate in the cellular membrane. The presence of $CO_2$ increases the disordering of the hydrocarbon lipid chains and the fluidity of the membrane, leading to the rupture of the lipid–protein interactions according to the so-called 'anesthetic effect' (Isenschmid *et al.*, 1995).

### 21.3.3 Lowering of the intracellular pH

Once inside the cells, $CO_2$ is able to lower the internal pH causing inhibition and/or inactivation of key enzymes that are essential for metabolic and regulatory processes, such as glycolysis, amino acid and peptide transport, active transport of ions, and proton translocation (Hutkins and Nannen, 1993). In addition, $CO_2$ accumulated in the cytoplasmic interior of the bacterial cells may convert $HCO_3^-$ to $CO_3^{2-}$, which could precipitate intracellular inorganic electrolytes (such as $Ca^{2+}$, $Mg^{2+}$ and similar ions) from cells and cell membranes (Lin *et al.*, 1993). Since these inorganic electrolytes help in maintaining the osmotic relationships between cells and their surrounding media, this could have deleterious effects on the volume of the cells.

### 21.3.4 Extraction of key cellular components

Several authors (Kamihira *et al.*, 1987; Lin *et al.*, 1992, 1993) suggested that accumulated $CO_2$ could, due to its relatively high solvating power, 'extract' vital constituents from the cells or cell membranes. In this mechanism, pressurized $CO_2$ first penetrates into the cells to build up density to a critical level within the cells after which it removes intracellular constituents, such as phospholipids and hydrophobic compounds, in order to disturb or alter the structure of the bio-membrane and the balance of the biological system, promoting inactivation (Lin *et al.*, 1992, 1993). This removal process appeared to be stimulated by a sudden release of the applied pressure, leading to a rapid transfer of intracellular materials out of the biological system into the extracellular environment (Lin *et al.*, 1992, 1993).

Oulé *et al.* (2006) carried out a study reporting the effect of the physical state of $CO_2$ on the cells by plating and by observation using transmission and scanning electron microscopy. According to these authors, the $CO_2$ physical state influenced its effects on microorganism's vitality inducing a bacteriostatic or bactericidal effect. $CO_2$ in vapor phase generated a bacteriostatic effect, while $CO_2$ in liquid or supercritical states provided a bactericidal effect. Scanning electron microscopy images showed that during vapor $CO_2$ treatment, the cells seemed to be sustaining a stress translated by

© Woodhead Publishing Limited, 2012

a slight depression of the cellular envelope. The accumulated $CO_2$ in the lipid phase of the membrane can induce a decrease in membrane viscosity because the phospholipids are rendered soluble. The alterations in the cell membrane were visible as surface modifications generated by the cell wall collapse and caused stress to the cells. It is believed that this effect is reversible and as soon as the contact between cells and $CO_2$ ceases, proteins to repair damage are synthesized by the cells to continue cellular division and growth.

When $CO_2$ is in liquid or supercritical states, its effect is not limited to the cellular membrane since it can penetrate the cell to cause irreversible damage and its effect becomes bactericidal. In the liquid state, the mechanism of cell inactivation of $CO_2$ involves two stages (Ballestra *et al.*, 1996). The first stage consists of stressing the cells, making them more sensitive to the treatment, blocking the biological functions, and altering the membrane. Then, $CO_2$ diffusion across the biological membrane causes a collapse and formation of depressions in the cell wall with a retraction of the cytoplasm and precipitation of the cytoplasmic contents. In the inactivation kinetics, this first stage represents the lag phase corresponding to the time during which the number of viable cells remains constant before inactivation. After the first stage, the inactivation phase starts. The cells are flaccid and emptied of their cytoplasmic content with holes on the surface of their cellular envelope, completely sunken and wrinkled.

In supercritical conditions the effect of $CO_2$ is drastic and the cells do not pass through a stressed step. $CO_2$ diffuses rapidly inside the cells to exercise its bactericidal effects which involve the supercritical extraction of intracellular substances and the rupture of the cytoplasmic membrane, causing loss of vital cellular constituents and the fragmentation of cell envelope. The cells are completely destroyed and the wall fragments are found in the suspending medium.

## 21.4 Decontamination of liquid and solid foods

### 21.4.1 Liquid foods

Dense phase $CO_2$ technology has been applied mainly to juices and a few beverages. A number of publications demonstrate that the treatment is sufficient for microbial safety. Table 21.2 presents a compilation of some experimental results that can be found in the literature with an indication of the type of liquid foods, the target microorganisms, the process conditions, and the corresponding microbial inactivation. A great number of studies reported the application of the treatment to apple juice. Spilimbergo *et al.* (2007) showed that *Saccharomyces cerevisiae* inoculated in an apple juice was reduced by 4.5 log cycles when subjected to supercritical $CO_2$ at 36°C, 20 MPa and 50 min.

Ferrentino *et al.* (2009a) treated apple juice prepared from an 'Annurca' apple purée. Microbial inactivation kinetics showed that 5-log reduction

© Woodhead Publishing Limited, 2012

**Table 21.2** Summary of the studies on microbial inactivation in liquid foods by dense phase $CO_2$

| Food system | Target microorganism | Process conditions | System | Reduction | Reference |
|---|---|---|---|---|---|
| Milk | *L. monocytogenes* | 7 MPa, 45°C, 1 h | Batch | 3 log | Lin *et al.* (1994) |
| Whole-skim milk | *S. aureus* | 9 MPa, 25°C, 5 h | Batch | 7 log | Erkmen (1997) |
| Fruit juice | *E. faecalis* | 6 MPa, 45°C, 3 h | Batch | 5 log | Erkmen (1999) |
| Milk | | 6 MPa, 45°C, 24 h | | | |
| Orange juice | *L. monocytogenes* | 6 MPa, 45°C, 8 h | Batch | 6 log | Erkmen (2000) |
| Peach juice | | 6 MPa, 45°C, 4 h | | | |
| Carrot juice | | 6 MPa, 45°C, 12 h | | | |
| Whole milk | *E. coli* | 10 MPa, 30°C, 6 h | Batch | 6.4 log | Erkmen (2001) |
| Skim milk | | | | 7.2 log | |
| Natural orange juice | Molds and yeasts | 30 MPa, 28°C, 15 min | Semi-continuous process | Total inactivation | Spilimbergo *et al.* (2002) |
| Watermelon juice | Natural micro flora | 34.4 MPa, 40°C, 5 min | Continuous plant | 6 log | Lecky and Balaban (2004) |
| Orange juice | *E. coli* | 10.7 MPa, 25°C, 10 min | Continuous plant | 5 log | Kincal *et al.* (2005) |
| | *S. typhimurium* | 21 MPa, 25°C, 10 min | | 6 log | |
| | *L. monocytogenes* | 38 MPa, 25°C, 10 min | | 6 log | |
| Grape juice | *S. cerevisiae* | 49 MPa, 25°C, 170 g $CO_2$/kg juice | Continuous plant | 5.5 log | Gunes *et al.* (2005) |
| | *C. stellata* | 49 MPa, 35°C, 170 g $CO_2$/kg juice | | 4.5 log | |
| Beer | Yeasts | 26.5 MPa, 21°C, 4.77 min | Continuous plant | 7.3 log | Dagan and Balaban (2006) |
| Mandarin juice | Total aerobic count | 41.1 MPa, 35°C, 9 min | Continuous plant | 3.47 log | Lim *et al.* (2006) |
| Muscadine grape juice | Molds and yeasts | 40 MPa, 30°C, 6.5 min | Continuous plant | 7 log | Del Pozo-Insfran *et al.* (2006) |
| | Total aerobic count | | | 5.7 log | |
| Apple juice | *S. cerevisiae* | 20 MPa, 36°C, 50 min | Multi-batch system | 4.52 log | Spilimbergo *et al.* (2007) |
| Apple juice | *A. acidoterrestris* | 10.0 MPa, 65°C, 40 min | Batch | Total inactivation | Bae *et al.* (2009) |

© Woodhead Publishing Limited, 2012

| | | | | | |
|---|---|---|---|---|---|
| Coconut water | Total aerobic count | 34.5 MPa, 25°C, 6 min | Continuous plant | 5 log | Damar *et al.* (2009) |
| Melon juice | Total aerobic count | 35 MPa, 55°C, 60 min | Batch | Total inactivation | Chen *et al.* (2010) |
| Liquid whole egg | Natural microflora | 13 MPa, 45°C, 10 min | Batch | Total inactivation | Garcia-Gonzalez *et al.* (2009) |
| Apple juice | Natural microflora | 15.0 MPa, 35°C, 15 min | Continuous plant | Total inactivation | Da Porto *et al.* (2010) |
| Grapefruit juice | Molds and yeasts<br>Total aerobic count | 34.5 MPa, 40°C, 7 min | Continuous plant | 5 log | Ferrentino *et al.* (2009b) |
| Guava purée | Molds and yeasts<br>Total aerobic count | 34.5 MPa, 35°C, 6.5 min | Continuous plant | >3.2 log | Plaza (2010) |
| Kiwi juice<br>Peach juice | Total aerobic count | 10.0 MPa, 35°C, 15 min | Batch | Total inactivation | Spilimbergo and Ciola (2010) |
| Jamaica beverage | Molds and yeasts | 13.8 MPa, 35°C, 5 min | Continuous plant | 5 log | Rodrigues (2010) |
| Apple juice | Molds and yeasts | 20 MPa, 57°C, 30 min | Batch | Total inactivation | Liao *et al.* (2010) |
| Blood orange juice | Total aerobic count<br>Molds and yeasts<br>Spoilage microflora | 23 MPa, 36°C, 15 min | Continuous plant | Total inactivation | Fabroni *et al.* (2010) |

© Woodhead Publishing Limited, 2012

of natural flora in the apple juice was achieved at 16.0 MPa, 60°C and 40 min. Temperature was an important variable for the process efficiency, with inactivation significantly enhanced when it increased from 35 to 60°C. The effect of pressure was less significant.

*Alicyclobacillus acidoterrestris* is of special interest in the fruit juice industry. Since common pasteurization techniques do not target spores, they do not deactivate the spores of *A. acidoterrestris*. The organism is resistant to acidic conditions and high temperatures and causes spoilage of juices. Bae *et al.* (2009) investigated the lethal effect of the process at 65°C and 70°C and 8–12 MPa, after 10–40 min of treatment on *Alicyclobacillus acidoterrestris* spores ($10^6$–$10^7$ spores/ml) suspended in apple juice. The study reported that *A. acidoterrestris* spores were completely inactivated to undetectable levels above 65°C, 10 MPa after 40 min and 70°C, 8 MPa after 30 min. Da Porto *et al.* (2010) treated apple juice with a continuous system demonstrating that microbial safety of the juice was achieved at 15 MPa, 35°C and after 15 min of treatment.

Several fruit juices and beverages (orange, apple, tangerine, grapefruit, watermelon, grape, guava, cane sugar juices and coconut water, kava, beer, and hibiscus drink) have been treated at the University of Florida using a continuous dense phase $CO_2$ apparatus designed by Praxair (Chicago, IL) (Balaban *et al.*, 1995; Balaban, 2004). This will be called the UF system. Watermelon juice was treated by DPCD with pressure from 10.3 to 34.4 MPa, the $CO_2$ to juice ratio from 5 to 15%; the treatment time from 4 to 6 min and the temperature from room temperature to 40°C. It was shown that for fresh watermelon juice a treatment at 34.4 MPa, 40°C, 10% $CO_2$, and 5 min treatment time resulted in 6 log cycle reduction of native aerobic microorganisms (Lecky and Balaban, 2004).

Kincal *et al.* (2005) reported 5 log reduction of selected pathogenic bacteria (*Escherichia coli* O157:H7, *Salmonella typhimurium*, and *Listeria monocytogenes*) in orange juice. They obtained at least 5 log reductions, enough to satisfy juice HACCP requirements, using the prototype Praxair system. Natural flora and yeasts were also studied for grape juice (Del Pozo-Insfran *et al.*, 2006; Gunes *et al.*, 2005), and coconut water (Damar *et al.*, 2009). Decontamination of mandarin juice processed with the UF system achieved a maximum log reduction of 3.47 for total aerobic count at 35°C, 41.1 MPa, 9 min of residence time, and 7% $CO_2$ (Lim *et al.*, 2006).

With the UF system, Ferrentino *et al.* (2009b) carried out experiments to detect the efficiency to inactivate yeasts and molds and total aerobic microorganisms in a fresh squeezed red blush grapefruit juice. A central composite design was used with pressure (13.8, 24.1, and 34.5 MPa) and treatment time (5, 7, and 9 min) as variables at constant temperature (40°C), and $CO_2$ level (5.7%) after experimentally measuring $CO_2$ solubility in the juice. Five log reductions for yeasts and molds and total aerobic microorganisms occurred at 34.5 MPa and after 7 min of treatment. A storage study of 4 weeks

© Woodhead Publishing Limited, 2012

at 4°C was performed showing no growth of total aerobic microorganisms and yeasts and molds over the entire storage period.

Chen *et al.* (2010) studied the application of thermal and dense phase $CO_2$ pasteurization to Hami melon juice. They found $CO_2$ efficiently induced microbial inactivation in terms of total aerobic count. Spilimbergo and Ciola (2010) applied the supercritical pasteurization to peach and kiwi juices. Total inactivation of both naturally occurring microorganisms and inoculated *Saccharomyces cerevisiae* strain was obtained after 15 min of treatment at 10 MPa and 35°C, for both juices.

Fabroni *et al.* (2010) tested the effectiveness of using supercritical $CO_2$ treatment to stabilize freshly squeezed blood orange juice with a continuous pilot system. After a process at 13 MPa, 36 ± 1°C, 5.08 L/h juice flow rate, 1.96 L/h $CO_2$ flow rate, corresponding to a 0.385 $gCO_2$/g juice ratio (time 0), there were no culturable organisms present in the juice. As storage continued, counts began to increase. Vitamin C, total anthocyanins, and antioxidant levels were very similar to untreated juice. Total flavanones and total phenolics decreased less than those in thermally treated juice. The $L^*$, $a^*$, and $b^*$ values decreased after $CO_2$ treatment. However, these changes were not drastic enough to significantly alter the characteristic color of the juice. The authors concluded that DPCD was a new mild technology for producing a stabilized blood orange juice with a shelf life of 20 days.

DPCD can also be used as a decontamination process for different liquid foods such as beer (Dagan and Balaban, 2006), guava purée (Plaza, 2010), and Jamaica beverage (Rodrigues, 2010). Beer was treated with dense $CO_2$ since the freshness is the top priority for brewers and the use of a non-thermal method in order to inhibit the growth of spoilage microorganisms has been considered of great importance for the quality of this product. Dagan and Balaban (2006) demonstrated that the continuous dense $CO_2$ system could pasteurize beer at 27.6 MPa, 21°C, 5% $CO_2$, and 5 min. Dense $CO_2$ was also efficient in the pasteurization of Jamaica beverage obtained from the extraction of *Hibiscus sabdariffa* red calyces. The optimal process conditions to inactivate yeasts and molds were 13.8 MPa, 35°C, and 5 min.

Guava purée obtained from *Psidium guajava* L. fruit was processed with dense $CO_2$ (Plaza, 2010) to investigate if this non-thermal method would minimize or prevent undesirable changes in phytochemical composition compared to traditional heat pasteurization. Microbial reduction was quantified as a function of pressure and residence time using 8% $CO_2$ and a temperature of 35°C. Optimum treatment conditions for microbial inactivation were 34.5 MPa for 6.9 min and 8% $CO_2$ at 35°C.

Dense phase $CO_2$ process has also been applied to milk as an alternative process to thermal pasteurization. Erkmen (1997, 2001) studied the effect of the treatment on the inactivation of *E. coli* and *S. aureus* inoculated to whole and skim milk. The treatment was performed in a batch system and the objective was to define the best process conditions in terms of pressure, temperature and treatment time to induce microbial inactivation. The results

© Woodhead Publishing Limited, 2012

showed that a treatment at 10 MPa, 30°C and 6 h caused a decrease of 6.42 and 7.24 log cycles on *E. coli* inoculated in whole and skim milk, respectively. It was observed that microorganisms were inactivated more easily when they were suspended in skim milk than in whole milk, probably due to the protective effect of the fat globules on the cells, retarding the penetration of $CO_2$. The pasteurization effects of $CO_2$ on *S. aureus* and aerobic bacteria were observed at 14.6 MPa for 5 h and 9 MPa for 2 h at 25°C in whole and skim milk, respectively. Studies to investigate the effect of dense phase $CO_2$ on lipase and lipolytic enzymes in raw whole milk were also carried out by Tisi (2004). A prototype continuous high pressure $CO_2$ equipment designed by Praxair Inc. (Chicago, IL) was used for the treatment of milk with liquid $CO_2$ at pressures between 7 and 62 MPa and temperatures between 15 and 40°C. At higher temperatures (40°C) and $CO_2$ concentrations, the process was shown to significantly reduce the proteolytic and lipolytic end-products. However, these conditions also changed the structure of the casein protein. After processing raw whole milk at 15°C, no inactivation of the lipolytic enzymes was detected.

Garcia-Gonzalez *et al.* (2009) investigated the effect of $CO_2$ at high pressure on the inactivation of naturally occurring microorganisms in liquid whole eggs; they concluded that $CO_2$ processing extended the shelf life of the sample up to 5 weeks at 4°C, which is the current shelf life of heat pasteurized liquid whole eggs.

### 21.4.2 Solid foods

Compared to liquid foods, fewer reports have been published regarding the application of dense phase $CO_2$ to solid foods. Table 21.3 reports the experimental results found in the literature. If also shows the types of food, microorganisms, and the conditions of $CO_2$ treatment with the corresponding microbial inactivation.

Typically, the solid food is placed in the treatment vessel, and dense phase $CO_2$ (gaseous or supercritical) is introduced into the vessel. The authors are not aware of any continuous system. Treatments have been performed under batch or semi-continuous modes.

Haas *et al.* (1989) reported the treatment of flour, fresh herbs (thyme, mint, chives, and oregano), fresh strawberries, honeydew melon, and cucumber in order to delay surface molding. The treatment reduced molds by 2 logs in strawberries, at 6 MPa and 22°C for 2 h. It also caused a complete microbial inactivation in fresh spices of thyme, mint, chives, and oregano at 50°C, 6.2 MPa for 2 h.

The treatment of fresh vegetables with DPCD has been studied for the reduction of microbial loads (Kühne and Knorr, 1990; Hong and Park, 1999; Zhong *et al.*, 2008). Kühne and Knorr (1990) carried out experiments on fresh celery and leafstalks showing a substantial inactivation effect. Their data indicated a reduction in the total plate count by about $10^4$ cfu $g^{-1}$ with

© Woodhead Publishing Limited, 2012

**Table 21.3** Summary of the studies on microbial inactivation in solid foods by dense phase $CO_2$

| Food System | Target microorganism | Process conditions | Reduction | Reference |
|---|---|---|---|---|
| Flour | Mold | 6.2 MPa, 23°C, 2 h | 99.8% | Haas *et al.* (1989) |
| | Bacteria | | 99.6% | |
| Strawberries | | 6.2 MPa, –22°C, 2 h | 99% | |
| Mozzarella cheese | | 6.2 MPa, 23°C, 16 h | 87% | |
| Parmesan cheese | | 1.4 MPa, 23°C, 168 h | 50% | |
| Romano cheese | | 1.4 MPa, 23°C, 168 h | 99% | |
| Onions | | 5.5 MPa, 23°C, 2 h | 90% | |
| Dry peppers (30% moisture added) | Bacteria | 5.5 MPa, 23°C, 2 h | 90% | |
| Chives | | 5.5 MPa, 45°C, 2 h | Total inactivation | |
| Thyme | | | | |
| Oregano | | | | |
| Parsley | | | | |
| Mint | | | | |
| Fresh celery leaves and leafstalks | Natural microorganisms | 6.9, 31.4 and 62.8 MPa, 40 or 60°C, 30 or 60 min | 4 log (cfu/g) | Kühne and Knorr (1990) |
| Chicken meat | *Salmonella typhimurium* | 13.7 MPa, 35°C, 2 h | 94–98% | Wei *et al.* (1991) |
| | *Listeria monocytogenes* ATCC15313 | | 79–84% | |
| Shrimp | *Listeria monocytogenes* ATCC15313 | | 99% | |
| Ground beef systems | *Escherichia Coli* | 31.03 MPa, 42.5°C, 180 min | 1 log (cfu/g) | Sirisee *et al.* (1998) |
| | *Staphylococcus aureus* | 31.03 MPa, 42.5°C, 120 min | 3 log (cfu/g) | |
| Kimchi vegetables | Lactic acid bacteria | 6.9 MPa, 10°C, 24 h | 4 log (cfu/ml) | Hong and Park (1999) |
| Skinned beef meat | *Brochothrix thermosphacta* | 6.1 MPa, 45°C, 150 min | 5 log | Erkmen (2000) |
| Minced beef meat | | | 1 log | |
| Alfalfa seeds | *Escherichia coli* K12 | 27.6 MPa, 50°C, 60 min | 92.8% | Mazzoni *et al.* (2001) |
| | Total aerobic bacteria | | 85.6% | |

*(Continued)*

© Woodhead Publishing Limited, 2012

**Table 21.3** Continued

| Food System | Target microorganism | Process conditions | Reduction | Reference |
|---|---|---|---|---|
| Beef trimmings | Total plate count | 10.3 MPa, 36°C, 15 min | 0.83 log | Meurehg (2006) |
| | *Escherichia coli* O157:H7 | | 0.93 log | |
| | *Escherichia coli* | | 1.00 log | |
| | *Salmonella* spp. | | 1.06 log | |
| Ground beef | Total plate count | 10.3 MPa, 36°C, 15 min | 0.78 log | |
| | *Escherichia coli* O157:H7 | | 0.94 log | |
| | *Escherichia coli* | | 0.94 log | |
| | *Salmonella* spp. | | 1.23 log | |
| Cocoa powder | Aerobic mesophilic spores | 30.0 MPa, 65°C, 40 min | Total inactivation | Calvo *et al.* (2007) |
| | Aerobic thermophilic spores | | | |
| | Mesophilic thermoresistant spores | | | |
| | Thermophilic thermoresistant spores | | | |
| | Total plate count | | | |
| Fresh spinach leaves | *Escherichia coli* K12 | 10 MPa, 40°C, 10 min | 5 log (cfu per leaf) | Zhong *et al.* (2008) |
| Ginseng powder | Total aerobic microbial count | 10 MPa, 60°C, 15 h | 2.67 log (cfu/g) | Dehghani *et al.* (2008) |
| Alfalfa seeds | *Escherichia coli* O157:H7 | 15 MPa, 35°C, 10 min | 3.51 log (cfu/g) | Jung *et al.* (2009) |
| | *Listeria monocytogenes* | 10 MPa, 45°C, 5 min | 2.65 log(cfu/g) | |
| | *Salmonella typhimurium* | | 2.48 log(cfu/g) | |
| Pears | *S. cerevisiae* | 10 MPa, 50°C, 10 min | 4 log (cfu/g) | Valverde *et al.* (2010) |
| Oyster | Aerobic plate count | 17.2 MPa, 60°C, 60 min | 3 log (cfu/g) | Meujo *et al.* (2010) |
| Paprika powder | Mesophilic aerobic microorganisms | 30.0 MPa, 90°C, 45 min | 5.5 log (cfu/g) | Calvo and Torres (2010) |

© Woodhead Publishing Limited, 2012

a treatment of 30 min at 40°C and 62.8 MPa. They also reported that the process cannot be applied to inactivate spores on the same substrates and under the same experimental conditions.

Promising results were obtained by Hong and Park (1999) treating *Baechu kimchi*, a Chinese cabbage, a traditional fermented vegetable food in Korea. DPCD treatment at 7 MPa for 24 h was able to maintain less than 40% of the bacterial population existing in the untreated *kimchi* for 6 days. However, it was insufficient for the reduction of lactic acid bacteria. The same research group isolated *Lactobacillus* sp. from the late stages of *kimchi* fermentation showing that the treatment induced inactivation if microbial cells were suspended in the liquid culture broth (Hong *et al.*, 1997). The conclusion of the study evidenced the different level of microbial inactivation by changing the substrate from solid to liquid. The different result was attributed to the lowered diffusivity and the limited mass transfer of $CO_2$ due to the compact structure of the salted Chinese cabbage and to the lack of *kimchi* juice in the solid sample which could act as an easy penetration medium.

A recent application of the treatment to vegetables was reported by Zhong *et al.* (2008) who evaluated the potential of the treatment on the inactivation of inoculated *Escherichia coli* K12 and background microflora, primarily Gram-positive rod-shaped microorganisms, on fresh spinach leaves. The study demonstrated that the microbial reduction obtained in supercritical $CO_2$ conditions (7.5 and 10 MPa at 40°C and 40 min) was significantly higher than that in subcritical state (5 MPa at 40°C and 40 min). At 5 MPa and 10 min, the microbial reduction of *E. coli* and background microflora was about 2 and 1 log cycles, respectively, while in supercritical conditions a reduction to undetectable level (~ 5 log cycles) was reached. Valverde *et al.* (2010) reported the effect of the DCPD on yeasts in pear and the final pear quality. At 55°C, in an order of minutes of treatment, and less than 6 MPa, total inactivation (5 log cycles) of *Saccharomyces cerevisiae* was obtained. However, the pears lost their texture and became darker due to enzymatic browning.

There is interest in the application of the $CO_2$ technology in supercritical state to meats, considering the ability of $CO_2$ to extract and fractionate fats from ground beef into lower-temperature melting components, as well as the removal of cholesterol (Chao *et al.*, 1991; Meurehg, 2006). The efficacy of supercritical $CO_2$ for the inactivation of bacterial strains in meats has been investigated. Sirisee *et al.* (1998) tested the effects on *E. coli* and *S. aureus* in ground beef highlighting that longer treatment times were needed to inactivate both target organisms, compared to the same treatment carried out on the same microorganisms but in a liquid phosphate buffer solution. The difference was attributed to the presence of fats and proteins, which could play an important role in protecting microorganisms from high pressure $CO_2$ bactericidal action, and to the lower moisture content (72%), which reduced the amount of $CO_2$ that can dissolve in the matrix. The protective effect of carbohydrate and other organic compounds in foods was also reported by

© Woodhead Publishing Limited, 2012

Erkmen (2000) who inoculated *Brocothrix thermosphacta* microbial cells on minced and skinned beef and carried out the treatment in a batch device at 6.1 MPa, 45°C for 150 min. Also in this study, the results demonstrated that the treatment was not as effective as in a liquid substrate. In a further study, DPCD treatment was applied to chicken meat strips (breast meat with no skin) to inactivate *Salmonella* and *Listeria* culture into which the chicken samples were dipped (Wei *et al.*, 1991). The treatment was carried out at 13.7 MPa and 35°C for 2 h and the samples spiked with *Salmonella* were shown to reduce the bacterial numbers by 94–98% while those spiked with *Listeria* were reduced by only 79–84%.

Calvo *et al.* (2007) tested the efficacy of the treatment on *Aspergillus niger* and *Aspergillus ochraceus* spores in a non-fermented high polyphenol cocoa powder. Results demonstrated that an increase in pressure from 13 to 30 MPa at 65°C for 40 min did not show any effect on the microbial inactivation. No microbial inactivation was also observed after the treatment at 30 MPa and increasing the temperature to 40, 65 and 80°C. The same results were also obtained applying 12 cycles of compression/decompression up to 30 MPa to the cocoa sample heated at 80°C. Previously published works reported that this procedure was very effective in the inactivation of spores of different bacteria mainly suspended in liquid media. It was demonstrated that the compression/decompression action induced spores' germination by an abrupt change in the surrounding conditions (Spilimbergo *et al.*, 2002; Dillow *et al.*, 1999).

More recently the applicability of the treatment has been tested on seafood. Meujo *et al.* (2010) proposed an innovative approach to post-harvest processing of oysters focusing on the effects of supercritical carbon dioxide on bacterial contaminants trapped in the digestive system of oysters. Experiments were performed *in vitro* on bacterial culture of a non-pathogenic strain of *Vibrio* used as a model for *Vibrio* spp. and several bacterial isolates from an oyster homogenate. The authors reported that the level of total bacterial inactivation achieved with the DPCD treatment (10 MPa for 30 min at 37°C or 17.2 MPa for 60 min at 60°C), was comparable to that achieved with several FDA-approved post-harvest processing for oysters, namely, high hydrostatic pressure and quick freezing (Prapaiwong *et al.*, 2009).

Calvo and Torres (2010) studied the effects of DPCD on the inactivation of microorganisms in paprika. Dehghani *et al.* (2008) reported on the treatment of ginseng powder contaminated with fungi and $5 \times 10^7$ bacteria/g using dense phase $CO_2$. A 2.67-log reduction of bacteria in the sample was achieved after long treatment time of 15 h at 60°C and 10 MPa, when using $CO_2$ alone. The addition of a small quantity of water/ethanol/$H_2O_2$ mixture, as low as 0.02 ml of each additive/g ginseng powder, was sufficient for complete inactivation of fungi within 6 h at 60°C and 10 MPa. Under these conditions the bacterial count was decreased from $5 \times 10^7$ to $2.0 \times 10^3$ CFU/g.

© Woodhead Publishing Limited, 2012

## 21.5 Effects on food quality

### 21.5.1 Liquid foods

The effects of the DPCD treatment on the quality attributes of liquid foods have been investigated. Lack of oxygen and lower temperature are two attributes that contribute to better preservation of nutrients and quality compared to thermal pasteurization.

Arreola *et al.* (1991) evaluated quality attributes (pH, °Brix, cloud stability, total acidity, color, ascorbic acid content and sensory attributes) of orange juice treated with dense $CO_2$ at 7–34 MPa, 35–60°C and 15–180 min in a batch system. They showed that there was no significant difference ($p < 0.01$) in pH or °Brix of the original juice and treated juice. Ascorbic acid retention was significantly higher in the treated samples (71–95% of original) than the temperature controls (62–83% of original). The higher ascorbic acid retention was explained by the higher stability of ascorbic acid under low pH provided during the process and also under the $O_2$-excluded environment.

Cloud was enhanced in the treated samples from 1.27 to 4.1 times, and was stable even in the presence of residual pectinesterase (PE). Cloud stability of DPCD samples (29 MPa, 50°C, 4 h) was retained after 66 days of refrigerated storage, whereas temperature control (50°C, 4 h) and room temperature control (25°C, 4 h) lost the cloud completely.

The DPCD-treated orange juice showed a higher brightness compared to the fresh untreated juice. Sensory evaluations of untreated, treated, and commercial but unpasteurized samples by 30 untrained panelists showed that overall acceptability, flavor, and aroma of fresh and treated orange juice were not significantly different. Gui *et al.* (2006) reported a significant reduction of the browning degree in dense phase $CO_2$-treated cloudy apple juice processed at 55°C, 30 MPa for 60 min, and stored at 4°C. The study by Gasperi *et al.* (2009) focused on possible sensory modifications caused by the treatment in apple juice. Difference from control and ranking tests were performed on fresh untreated control and treated juices. The results showed no significant differences between the samples.

Del Pozo-Insfran *et al.* (2006) observed no significant changes in total anthocyanins, total soluble phenolics and antioxidant capacity of dense phase $CO_2$-treated muscadine grape juice whereas heat treatment of samples caused a decrease of 16, 26 and 10%, respectively. Moreover, treated juices retained higher total anthocyanins, total soluble phenolics and antioxidant capacity content than thermally pasteurized juices after 10 weeks of storage at 4°C.

A storage study was performed on a fresh grapefruit juice treated by dense phase $CO_2$ (34.5 MPa, 40°C and 7 min). °Brix, pH, titratable acidity, pectinesterase inactivation, cloud, color, hue tint and color density, total phenolics, antioxidant capacity, and ascorbic acid were measured after the treatment and during 6 weeks' storage at 4°C. The treated juice showed an increase in the cloud value (91%), and a partial inactivation of pectinesterase (69.17%). No significant differences were detected between treated and

© Woodhead Publishing Limited, 2012

untreated juices for °Brix, pH, and titratable acidity. Treated juice had higher lightness and redness and lower yellowness. The study showed that the treatment and the storage did not affect the total phenolic content of the juice. Slight differences were detected for the ascorbic acid content and the antioxidant capacity. Overall, the experimental results highlighted that the treatment can maintain the physical and quality attributes of the juice, extending its shelf life and safety (Ferrentino *et al.*, 2009b).

Damar *et al.* (2009) investigated consumer likeability and flavor profile of a dense phase $CO_2$ (34.5 MPa, 25°C, 13%$CO_2$, 6 min) treated coconut water beverage in comparison to that of fresh, untreated, and heat processed (74°C, 15 sec) samples. Sensory panels that were conducted throughout 9 weeks of refrigerated storage (4°C) showed that dense phase $CO_2$-treated and fresh coconut water beverages were liked similarly, whereas heat-treated coconut water beverage was liked significantly less. Gas chromatography-olfactory analysis of flavor compounds in dense phase $CO_2$ and heat processed coconut water beverages showed that there were differences in the aroma profiles of dense phase $CO_2$ and heat-treated samples. Heat-treated samples had more aroma compounds described as green, fruity, nutty, rancid, unpleasant, fatty, and burnt aromas. These could have possibly been developed by decomposition of aroma compounds during heating.

The study carried out by Spilimbergo and Ciola (2010) reported no significant changes in chemical, physical (pH, sugar content, titratable acidity, absorbance at 420 nm and turbidity), and sensory attributes between untreated and dense phase $CO_2$-treated peach and kiwi juices. The results demonstrated the feasibility and the potential of the treatment as an alternative low temperature pasteurization process for peach and kiwi juices.

Niu *et al.* (2010) treated orange juice under 40 MPa at 55°C between 10 to 60 min. A control was thermally treated at 90°C for 60 s. Particle size, consistency coefficient and $a^*$ (redness) value for orange juice tended to become smaller as the treatment time was extended. The proportion of the small particles increased after the treatment, suggesting that the increase in cloud and its stability in orange juice is primarily due to homogenization during the process.

### 21.5.2 Solid foods

Solid foods may require more severe conditions (higher pressure, temperature, and longer treatment time) for microbial inactivation compared to liquid foods. Therefore, it is possible that the sensory properties of the foods could be adversely affected compared to liquid samples. In almost all the studies that are described below, a visual observation of the samples after the treatment was made; few systematic or quantitative analyses were performed to investigate the chemical and physical effects of the treatment for solid foods.

The study by Haas *et al.* (1989) reported that the treatment caused gross tissue destruction of strawberries, honeydew melon, and cucumber. The

© Woodhead Publishing Limited, 2012

quality analyses performed on fresh herbs exposed to 5.5 MPa at 45°C for 2 h demonstrated that thyme, mint, chives, and oregano had enhanced aromas after the treatment. In addition some herbs showed a different taste after the treatment; in particular, parsley tasted similar to the untreated, but developed a slight off-aroma, while untreated thyme and mint tasted better than the treated sample.

The same effect has been found in the treatment of fresh celery and leafstalks (Kühne and Knorr, 1990) and of fresh spinach leaves (Zhong *et al.*, 2008). The fresh celery and leafstalks underwent a color change turning whitish similar to a color resulting from being cooked or soaked in acid. The same change happened to the spinach leaves which resulted in greater leaf discoloration and decrease of leaf firmness. The discoloration phenomenon was attributed to the dissolution of the $CO_2$ into leaf tissues, which acidified the leaves and degraded the chlorophylls (Mingotaud *et al.*, 1996).

The discoloration phenomenon was also observed when the treatment was applied to ground beef (Sirisee *et al.*, 1998). The color after the treatment looked like that of cooked ground beef due to the high concentrations of $CO_2$ which caused darkening in tissues by combining with myoglobin to form metmyoglobin. Also the chicken samples treated by Wei *et al.* (1991) turned whitish and seemed to be cooked or soaked in acid. In addition, the treated samples showed a liquid loss from the tissue which was absent in the untreated samples.

The treatment carried out on other food matrices was demonstrated to be less disruptive, preserving the quality attributes of the foods. Calvo *et al.* (2007) reported that the water content of treated cocoa powder decreases, which is considered an important aspect for the storage stability, shelf life, and physical aspect of the cocoa powder, since the humidity should not exceed 9%, as specified for this type of product.

Mazzoni *et al.* (2001) showed that dense $CO_2$ did not have any detrimental effect on the viability of alfalfa seeds. The percent of germination of the treated samples was over 90% and no significant differences in the germination rate were detected between the treated and untreated seeds, indicating that the process could be effective for the treatment of alfalfa seeds at commercial level without compromising the germination quality. Alfalfa sprouted seeds were also studied by Jung *et al.* (2009). Without impairing the seed germination capability, the maximum reduction level of *E. coli* O157:H7 was 3.51 CFU/g with supercritical $CO_2$ treatment at 15 MPa and 35°C for 10 min. Maximum reductions of *L. monocytogenes* and *S. typhimurium* were 2.65 and 2.48 log CFU/g, respectively, with treatment at 10 MPa and 45°C for 5 min.

The study published by Meujo *et al.* (2010) reported on oysters treated by supercritical $CO_2$ and subjected to a sensory analysis assessed by a panel of 13 people to judge the physical appearance, smell, and texture of the samples after the treatment. The results revealed that oysters remained acceptable regarding their physical appearance, texture, and smell after exposure to the $CO_2$ process (10 and 20 MPa for 20 and 50 min at 37°C).

© Woodhead Publishing Limited, 2012

### 21.5.3 Effects on enzymes

Studies on enzyme inactivation by dense phase $CO_2$ indicate its good potential, especially in fruit and vegetable juice processing where these enzymes cause quality deterioration if not inactivated. The inactivation of enzymes affecting food quality has been reported by several researchers (Balaban *et al.*, 1991; Chen *et al.*, 1992, 1993; Park *et al.*, 2002).

Dense phase $CO_2$ can inactivate certain enzymes at temperatures where thermal inactivation is not effective (Balaban *et al.*, 1991). Among these enzymes, pectinesterase (PE) causes cloud loss in some fruit juices; polyphenol oxidase (PPO) causes undesirable browning in fruits, vegetables, juices, and some seafood; lipoxygenase (LOX) causes chlorophyll destruction and off-flavor development in frozen vegetables; peroxidase (POD) has an important role in the discoloration of foods and is used as an index of heat treatment efficiency in fruit and vegetable processing.

Balaban *et al.* (1991) studied PE inactivation in orange juice. Without $CO_2$, the pH of orange juice must be lowered to 2.4 for substantial PE inactivation. $CO_2$ lowered pH only to 3.1. Therefore, the pH lowering effect alone was not sufficient to explain enzyme inactivation. The results of Chen *et al.* (1992) supported this conclusion. The extent of enzyme inactivation by dense $CO_2$ is affected by the type and source of the enzyme, treatment conditions such as pressure, temperature, time, and treatment medium properties. Balaban *et al.* (1991) observed that higher temperatures and pressures of $CO_2$ treatment result in a higher percentage PE inactivation.

It has also been demonstrated that an enzyme isolated from different sources has different resistance to the treatment, as it has also been shown for heat inactivation. For example, potato PPO was more resistant to inactivation by $CO_2$ compared with spiny lobster and shrimp PPOs (Chen *et al.*, 1992). In addition, the presence of other soluble compounds in the treatment medium may have a protective effect against this treatment. Tedjo *et al.* (2000) showed that residual percentage LOX and percentage POD activity increased by increasing the sucrose concentration up to 40%. This could be explained by the decrease of $CO_2$ solubility as sucrose concentration increases (Ferrentino *et al.*, 2010a).

Zhou *et al.* (2009a) studied the effects of dense phase $CO_2$ on the activity and structure of pectin methylesterase (PME) extracted from Valencia orange peel. The extract was dissolved in phosphate buffer and NaCl was added before treatment with dense $CO_2$ (8–30 MPa, 55°C and 10 min) and heat (55°C for 10 min). Dense phase $CO_2$ caused significant inactivation of PME, with the lowest residual activity about 9.3% at 30 MPa. The SDS-PAGE electrophoretic behavior of treated PME was not altered; however, changes in the secondary and tertiary structures were detected. The $\beta$-structure fraction in the secondary structure decreased as the process pressures increased. After 7-day storage at 4°C, no alteration of PME activity and no reversion of its $\beta$-structure fraction were observed.

Zhou *et al.* (2009b) also studied the inactivation of PME from carrot and

© Woodhead Publishing Limited, 2012

peach in buffer after treatment with dense $CO_2$ at 55°C. Both PMEs were inactivated during the process, and their residual activity decreased with increasing pressure. Niu *et al.* (2010) reported a study on the efficiency of the process on apple PPO demonstrating the complete inactivation of the enzyme after a treatment carried out in the range of temperatures from 25 to 65°C at 20 MPa and 20 min treatment time. Chen *et al.* (2010) investigated the effects of the treatment on the enzymes of Hami melon juice. The results demonstrated that minimum residual activities of PPO, POD and LOX were 25.26%, 38.46 and 0.02%, respectively after treatment at 35 MPa, 55°C for 60 min.

## 21.6 Future trends and recommendations

There are two major issues to resolve for this technology to be implemented in industry. The first issue is 'economics'. This involves several points. The cost of equipment is currently not optimized since there is no mass production of this type of system. High pressure gas handling requires that safety concerns are addressed. With increasing restrictions in venting $CO_2$ to the atmosphere, additional components must be added to the system to capture and re-use the $CO_2$. Optimization of the use of energy for pressurization (e.g. use of the de-pressurization step to recover some of the pressurization energy) has not been studied in this technology. The economics also depend on the size of the operation; with increasing size the economies of scale becoming more beneficial. However, the cost of processing compared to simple thermal pasteurization may never be competitive. Instead, other 'advantages' must be demonstrated and publicized, such as better quality, greater retention of nutrients, better 'healthy' products, etc., to justify the added cost. The comparison with competing 'non-thermal' technologies such as ultra-high pressure, and pulsed electric treatments must be conducted both in terms of cost, and in terms of safety, quality, and nutrient retention.

The second issue is regulatory. Safety regulations dictate that the process must 'demonstrate' a certain level of pathogen reduction. So far, most of the research in this area has been on a case-by-case basis. In other words, for a certain product, the effects of pressure, temperature, residence time, $CO_2$ levels, etc., have been studied. When the product is different, the process starts over. This is because for different products, the effect of the treatment on microorganisms is different. Recently, there have been studies to try to predict the effect of DPCD on microorganisms from the solubility of $CO_2$ in the substrate (Ferrentino *et al.*, 2010a). This offers advantages since the solubility of $CO_2$ in a given substrate at different process conditions can be thermodynamically predicted with good accuracy. This approach may reduce the number of confirmatory experiments to be performed to comply with the regulations.

© Woodhead Publishing Limited, 2012

Ultimately, the adoption of the process by the food industry will depend on the identification of 'niche' applications where no other process would assure safety while at the same time preserving quality and nutrients, and is therefore sought by the consumer who will accept a certain price for the perceived benefits.

## 21.7 Sources of further information and advice

The US Food and Drug Administration (FDA) issued a 'Guidance for Industry: Juice HACCP Hazards and Controls Guidance' in March 2004 (FDA, 2004). The section that involves DPCD is under section V, Process Validation, subsection 5.34 Dense Phase $CO_2$ Processing Systems. The section mentions:

> Dense phase carbon dioxide processing, a technology in which carbon dioxide under moderate pressure (1200–1500 psig, 8.25 to 10.3 MPa) is the principal anti-microbial agent, has been shown to be effective in reducing vegetative pathogens. In the gas industry, supercritical and liquid carbon dioxide ($CO_2$) are known collectively as dense phase $CO_2$. Continuous processes have been developed using this technology. Pathogen challenge tests showed that microbial inactivation increases as $CO_2$ concentration increases. It appears that pressure and residence time may be used to optimize the bactericidal effects of $CO_2$. For these processes, $CO_2$ concentration is critical to the process. The process is performed under ambient conditions, and temperature is not monitored as a critical factor.

Praxair Inc. demonstrated greater than 5-log reduction of juice pathogens to meet the Juice HACCP guidelines set by the US FDA. As a consequence, 'Better Than Fresh' (BTF) technology is currently being considered by the FDA as an alternative to thermal pasteurization (Connery *et al.*, 2005).

In European Union (EU) countries, the national regulations for new products have been replaced by the 'Novel Food Regulation (NFR)', a Community regulation for novel foods and ingredients (EC No. 258/97), which has been in force since 1997. The objective of the Regulation is to protect public health by ensuring food safety. The NFR defines novel foods as 'food ingredients that were not used for human consumption to a significant degree within the EU before 15 May 1997'. The NFR legislation addresses food safety concerns in the context of (i) foods and food ingredients with a new molecular structure, (ii) those consisting of or isolated from microorganisms, plants or animals, or (iii) those derived from novel production processes. However, products extracted and/or treated by supercritical $CO_2$ are already on the market. For instance the Advisory Committee on Novel Foods and Processes (ACNFP) considered an application made by Algatechnologies (1998) Ltd.,

© Woodhead Publishing Limited, 2012

for an opinion on the equivalence of their astaxanthin-rich extract compared with an existing astaxanthin-rich extract from the same source marketed by Valensa (formerly known as US Nutra). Both products were obtained from *Haematococcus pluvialis* algae using supercritical carbon dioxide extraction technology. Dense phase $CO_2$ products probably could be regarded as novel foods since (i) they have no history of human consumption in the EU so far, and (ii) they have been produced by a new manufacturing process. To this extent before dense phase $CO_2$ processed foods could be sold in the EU market, applications must be submitted in accordance with Commission Recommendation 97/618/EC concerning the scientific information and the safety assessment report required.

## 21.8 References

ARREOLA A G, BALABAN M O, MARSHALL M R, PEPLOW A J, WEI C I and CORNELL J A (1991), 'Supercritical $CO_2$ effects on some quality attributes of single strength orange juice', *J Food Sci*, 56, 1030–1033.

BAE Y Y, LEE H J, KIM S A and RHEE M S (2009), 'Inactivation of *Alicyclobacillus acidoterrestris* spores in apple juice by supercritical carbon dioxide', *Int J Food Microbiol*, 136, 95–100.

BALABAN M O (2004), Method and apparatus for continuous flow reduction of microbial and/or enzymatic activity in a liquid product using carbon dioxide. US Patent 6,723,365.

BALABAN M O and MEIRELES M A A (1999), Supercritical fluid technology applications for the food industry. In Francis F J *Wiley Encyclopedia of Food Science and Technology*, 2nd edn. New York, John Wiley & Sons, pp. 2220–2226.

BALABAN M O, ARREOLA A G, MARSHALL M R, PEPLOW A, WEI C I and CORNELL J (1991), 'Inactivation of pectinesterase in orange juice by supercritical carbon dioxide', *J Food Sci*, 6, 743–746.

BALABAN M O, MARSHALL M R and WICKER L (1995), Inactivation of enzymes in foods with pressurized $CO_2$. US Patent 5,393,547.

BALLESTRA P, DA SILVA A A and CUQ J L (1996), 'Inactivation of *Escherichia coli* by carbon dioxide under pressure', *J Food Sci*, 61, 829–831.

CALVO L and TORRES E (2010), 'Microbial inactivation of paprika using high-pressure $CO_2$', *J Supercritical Fluids*, 52, 134–141.

CALVO L, MUGUERZA B and CIENFUEGOS-JOVELLANOS E (2007), 'Microbial inactivation and butter extraction in a cocoa derivative using high pressure $CO_2$', *J Supercritical Fluids*, 42, 80–87.

CASTOR T P and HONG G T (1992), Supercritical fluid disruption of and extraction from microbial cells. US Patent 5,380,826.

CHAO R R, MLVANEY S J, BAILEY M E and FERNANDO L N (1991), 'Supercritical $CO_2$ conditions affecting extraction of lipid and cholesterol from ground beef', *J Food Sci*, 56, 183–187.

CHEN J S, BALABAN M O, WEI C I, MARSHALL M R and HSU W Y (1992), 'Inactivation of polyphenol oxidase by high pressure $CO_2$', *J Agric Food Chem*, 40, 2345–2349.

CHEN J S, BALABAN M O, WEI C I, GLEESON R A and MARSHALL M R (1993), 'Effect of $CO_2$ on the inactivation of Florida spiny lobster polyphenol oxidase', *J Sci Food Agric*, 61, 253–259.

CHEN J L, ZHANG J, SONG L, JIANG Y, WU J and HU X S (2010), 'Changes in microorganism, enzyme, aroma of hami melon (*Cucumis melo* L.) juice treated with dense phase

© Woodhead Publishing Limited, 2012

carbon dioxide and stored at 4°C', *Innovative Food Sci and Emerg Technol*, 11, 623–629.

CONNERY K A, SHAH P, COLEMAN L and HUNEK B (2005), 'Commercialization of Better Than Fresh™ dense phase carbon dioxide processing for liquid food', ISSF 2005, Orlando, FL.

DA PORTO C, DECORTI D and TUBARO F (2010), 'Effects of continuous dense-phase $CO_2$ system on antioxidant capacity and volatile compounds of apple juice', *Int J Food Sci Technol*, 45, 1821–1827.

DAGAN G F and BALABAN M O (2006), 'Pasteurization of beer by continuous dense phase $CO_2$ system', *J Food Sci*, 71, E164–E169.

DAMAR S and BALABAN M O (2006), 'Review of dense phase $CO_2$ technology: microbial and enzyme inactivation, and effects on food quality', *J Food Sci*, 71, R1–R11.

DAMAR S, BALABAN M O and SIMS C A (2009), 'Dense phase $CO_2$ processing of coconut water', *Int J Food Sci Technol*, 44, 666–673.

DEHGHANI F, ANNABI N, TITUS M, VALTCHEV P and TUMILAR A (2008), 'Sterilization of ginseng using a high pressure $CO_2$ at moderate temperatures', *Biotech and Bioengineering*, 102, 569–576.

DEL POZO-INSFRAN D, BALABAN M O, and TALCOTT S T (2006), 'Microbial stability, phytochemical retention and organoleptic attributes of dense phase $CO_2$ processed Muscadine grape juice', *J Agric Food Chem*, 54, 5468–5473.

DESCOINS C, MATHLOUTHI M, MOUAL M and HENNEQUIN J (2006), 'Carbonation monitoring of beverage in a laboratory scale unit with on-line measurement of dissolved $CO_2$', *J Food Chem*, 95, 541–553.

DILLOW A K, DEHGHANI F, HRKACH J S, FOSTER N R and LANGER R (1999), 'Bacterial inactivation by using near- and supercritical $CO_2$', *Proc Natl Acad Sci USA*, 96, 10344–10348.

DODDS W S, STUTZMAN L F and SOLLAMI B J (1956), '$CO_2$ solubility in water', *Ind Eng Chem*, 1, 92–95.

ENOMOTO A, NAKAMURA K, HAKODA M and AMAYA N (1997), 'Lethal effect of high-pressure carbon dioxide on a bacterial spore', *J Ferment Bioeng*, 83, 305–307.

ERKMEN O (1997), 'Antimicrobial effect of pressurized carbon dioxide on *Staphylococcus aureus* in broth and milk', *Lebensm Wiss Technol*, 30, 826–829.

ERKMEN O (1999), 'Antimicrobial effect of pressurized carbon dioxide on *Enterococcus faecalis* in physiological saline and foods', *J Sci Food Agr*, 80, 465–470.

ERKMEN O (2000), 'Effect of carbon dioxide on *Listeria monocytogenes* in physiological saline and foods', *Food Microbiol*, 17, 589–596.

ERKMEN O (2001), 'Mathematical modelling of *Escherichia coli* inactivation under high pressure carbon dioxide', *J Biosci Bioeng*, 92, 39–43.

FABRONI S, AMENTA M, TIMPANARO N and RAPISARDA P (2010), 'Supercritical carbon dioxide-treated blood orange juice as a new product in the fresh fruit juice market', *Innov Food Sci Emerg Technol*, 11, 477–484.

FERRENTINO G, BRUNO M C, FERRARI G, POLETTO M and BALABAN M O (2009a), 'Microbial inactivation and shelf life of apple juice treated with high pressure carbon dioxide', *J Biol Eng*, 3, doi: 10.1186/1754-1611-3-3.

FERRENTINO G, PLAZA, M L, RAMIREZ M, FERRARI M G and BALABAN M O (2009b), 'Effects of pasteurization of red grapefruit juice with dense phase carbon dioxide on physical and quality attributes', *J Food Sci*, 74, E333–E341.

FERRENTINO G, BALABAN M O, FERRARI G and POLETTO M (2010a), 'Food treatment with high pressure carbon dioxide: *S. cerevisiae* inactivation kinetics expressed as a function of $CO_2$ solubility', *J Supercritical Fluids*, 52, 151–160.

FERRENTINO G, BARLETTA D, DONSI F, FERRARI G and POLETTO M (2010b), 'Experimental measurements and thermodynamic modeling of $CO_2$ solubility at high pressure in model apple juices', *Ind Eng Chem Res*, 49, 2992–3000.

FOOD AND DRUG ADMINISTRATION (FDA) (1979), Database of Select Committee on GRAS Substances Reviews, Carbon Dioxide. Available from: http://www.accessdata.

© Woodhead Publishing Limited, 2012

fda.gov/scripts/fcn/fcnDetailNavigation.cfm?rpt=scogsListing&id=69 (accessed June 2011).

FOOD AND DRUG ADMINISTRATION (FDA) (2004), Guidance for Industry: Juice HACCP Hazards and Controls guidance, first editions; final guidance. Available from: http://www.fda.gov/Food/GuidanceComplianceRegulatoryInformation/GuidanceDocuments/Juice/ucm072557.htm (accessed February 2011).

FRASER D (1951), 'Bursting bacteria by release of gas pressure', *Nature*, 167, 33–34.

GARCIA-GONZALEZ L, GEERAERD A H, SPILIMBERGO S, ELST K, VAN GINNEKEN L, DEBEVERE J, VAN IMPE J F and DEVLIEGHERE F (2007), 'High pressure carbon dioxide inactivation of microorganisms in foods: the past, the present and the future', *Int J Food Microbiol*, 117, 1–28.

GARCIA-GONZALEZ L, GEERAERD A H, ELST K, VAN GINNEKEN L, VAN IMPE J F and DEBEVERE L (2009), 'Inactivation of naturally occurring microorganisms in liquid whole egg using high pressure carbon dioxide processing as an alternative to heat pasteurization', *J Supercritical Fluids*, 51, 74–82.

GASPERI F, APREA E, BIASIOLI F, CARLIN S, ENDRIZZI I, PIRRETTI G and SPILIMBERGO S (2009), 'Effects of supercritical $CO_2$ and $N_2O$ pasteurization on the quality of fresh apple juice', *Food Chem*, 115, 19–136.

GUI F Q, WU J H, CHEN F, LIAO X J, HU X S, ZHANG Z H and WANG Z F (2006), 'Change of polyphenol oxidase activity, color, and browning degree during storage of cloudy apple juice treated by supercritical carbon dioxide', *Eur Food Res Technol*, 223, 427–432.

GUNES G, BLUM L K and HOTCHKISS J H (2005), 'Inactivation of yeasts in grape juice using a continuous dense phase carbon dioxide processing system', *J Sci Food Agric*, 85, 2362–2368.

HAAS G J, PRESCOTT H E, DUDLEY E, DIK R, HINTLIAN C and KEANE L (1989), 'Inactivation of microorganisms by carbon dioxide under pressure', *J Food Safety*, 9, 253–265.

HONG S I and PARK W S (1999), 'High pressure carbon dioxide effect on kimchi fermentation', *Biosci Biotechnol Biochem*, 63, 1119–1121.

HONG S I and PYUN Y R (1999), 'Inactivation kinetic of *Lactobacillus plantarum* by high pressure carbon dioxide', *J Food Sci*, 64, 728–733.

HONG S I, PARK W S and PYUN Y R (1997), 'Inactivation of *Lactobacillus* sp. from Kimchi by high pressure $CO_2$', *Lebensm Wiss Technol*, 30, 681–685.

HONG S I, PARK W S and PYUN Y R (1999), 'Non-thermal inactivation of *Lactobacillus plantarum* as influenced by pressure and temperature of pressurized carbon dioxide', *Int J Food Sci Technol*, 34, 125–130.

HUTKINS R W and NANNEN N L (1993), 'pH homeostasis in lactic-acid bacteria', *J Dairy Sci*, 76, 2354–2365.

ISENSCHMID A, MARISON I W and VON STOCKAR U (1995), 'The influence of pressure and temperature of compressed $CO_2$ on the survival of yeast cells', *J Biotechnol*, 39, 229–237.

JUNG W Y, CHOI Y M and RHEE M S (2009), 'Potential use of supercritical carbon dioxide to decontaminate *Escherichia coli* O157:H7, *Listeria monocytogenes*, and *Salmonella typhimurium* in alfalfa sprouted seeds', *Int J Food Microbiol*, 136, 66–70.

KAMIHIRA M, TANIGUCHI M and KOBAYASHI T (1987), 'Sterilization of microorganisms with supercritical $CO_2$', *Agric Biol Chem*, 51, 407–412.

KINCAL D, HILL W S, BALABAN M O, PORTIER K M, WEI C I and MARSHALL M R (2005), 'A continuous high pressure carbon dioxide system for microbial reduction in orange juice', *J Food Sci*, 70, M249–M254.

KÜHNE K and KNORR D (1990), 'Effects of high pressure carbon dioxide on the reduction of microorganisms in fresh celery', *Int Z Lebensm, Market, Verpack, Analy*, 41, 55–57.

LECKY M and BALABAN M O (2004), 'Continuous high pressure carbon dioxide processing of watermelon juice', Abstract 49H-3, p. 133. *Institute of Food Technologists Annual Meeting*, Las Vegas, Nevada.

© Woodhead Publishing Limited, 2012

LIAO H, ZHANG L, HU X and LIAO X (2010), 'Effect of high pressure $CO_2$ and mild heat processing on natural microorganisms in apple juice', *Int J Food Microbiol*, 137, 81–87.

LIM S, YAGIZ Y and BALABAN M O (2006), 'Continuous high pressure carbon dioxide processing of mandarin juice', *Food Sci Biotechnol*, 15, 13–18.

LIN H M, CHAN E C, CHEN C and CHEN L F (1991), 'Disintegration of yeast cells by pressurized $CO_2$', *Biotechnol Prog*, 7, 201–204.

LIN H M, YANG Z Y and CHEN L F (1992), 'An improved method for disruption of microbial cells with pressurized carbon dioxide', *Biotechnol Prog*, 8, 165–166.

LIN H M, YANG Z Y and CHEN L F (1993), 'Inactivation of *Leuconostoc dextranicum* with carbon dioxide under pressure', *Chem Eng J*, 52, B29–B34.

LIN H M, CAO N and CHEN L F (1994), 'Antimicrobial effect of pressurized carbon dioxide on *Listeria monocytogenes*', *J Food Sci*, 59, 657–659.

MAZZONI A M, SHARMA R R, DEMERCI A and ZIEGLER G R (2001), 'Supercritical carbon dioxide treatment to inactivate aerobic microorganisms on alfalfa seeds', *J Food Safety*, 21, 215–223.

MEUJO D A F, KEVIN D A, PENG J, BOWLING J J, LIU J and HAMANN M T (2010), 'Reducing oyster-associated bacteria levels using supercritical fluid $CO_2$ as an agent of warm pasteurization', *Int J Food Microbiol*, 138, 63–70.

MEUREHG T C A (2006), 'Control of *Escherichia coli* O157:H7, generic *Escherichia coli*, and *Salmonella* spp. on beef trimmings prior to grinding using a controlled phase carbon dioxide ($_{cp}CO_2$) system', PhD Thesis, Kansas State University, Manhattan, KS.

MEYSSAMI B, BALABAN M O and TEIXEIRA A A (1992), 'Prediction of pH in model systems pressurized with carbon dioxide', *Biotechnol Progr*, 8, 149–154.

MINGOTAUD C, CHAUVET J P and PATTERSON L K (1996), 'Stability of chlorophyll A at the gas–water interface in pure and mixed monolayers. An evaluation of interfacial pH', *J Phys Chem*, 100, 18554–18561.

NAKAMURA K, ENOMOTO A, FUKUSHIMA H, NAGAI K and HAKODA M (1994), 'Disruption of microbial cells by the flash discharge of high-pressure carbon dioxide', *Biosc, Biotechnol Biochem*, 58, 1297–1301.

NIU L, HU X, WU J, LIAO X, CHEN F, ZHAO G and WANG Z (2010), 'Effect of dense phase carbon dioxide process on physicochemical properties and flavor compounds of orange juice', *J Food Process Preserv*, 34, 530–548.

OULÉ M K, TANO K, BERNIER A M and ARUL J (2006), '*Escherichia coli* inactivation mechanism by pressurized $CO_2$', *Can J Microbiol*, 52, 1208–1217.

PARK S J, LEE J I and PARK J (2002), 'Effects of a combined process of high-pressure carbon dioxide and high hydrostatic pressure on the quality of carrot juice', *J Food Sci*, 67, 1827–1834.

PLAZA M L (2010), 'Quality improvement of guava purée by dense phase carbon dioxide treatment', PhD dissertation, University of Florida, Gainesville, FL.

PRAPAIWONG N, WALLACE R K and ARIAS C R (2009), 'Bacterial loads and microbial composition in high pressure treated oysters during storage', *Int J Food Microbiol*, 131, 145–150.

RODRIGUES M M R (2010), 'Processing of a hibiscus beverage using dense phase carbon dioxide', PhD. dissertation, University of Florida, Gainesville, FL.

SHIMODA M, COCUNUBO-CASTELLANOS J, KAGO H, MIYAKE M, OSAJIMA Y and HAYAKAWA I (2001), 'The influence of dissolved $CO_2$ concentration on the death kinetics of *Saccharomyces cerevisiae*', *J Appl Microbiol*, 91, 306–311.

SIMS M (2001), (Porocrit, LLC, USA), *Method and membrane system for sterilizing and preserving liquids using carbon dioxide*. US Patent 6331272.

SIMS M and ESTIGARRIBIA E (2003), 'Membrane carbon dioxide sterilization of liquid foods: scale up of a commercial continuous process', *Proc. of the 6th International Symposium on Supercritical Fluids, Versailles, France*, pp. 1457–1460.

SIRISEE U, HSIEH F and HUFF H E (1998), 'Microbial safety of supercritical carbon dioxide processes', *J Food Process Preserv*, 22, 387–403.

© Woodhead Publishing Limited, 2012

SPILIMBERGO S and CIOLA L (2010), 'Supercritical $CO_2$ and $N_2O$ pasteurisation of peach and kiwi juice', *Int J Food Sci Technol*, 45, 1619–1625.

SPILIMBERGO S and MANTOAN D (2006), 'Kinetic analysis of microorganisms inactivation in apple juice by high pressure carbon dioxide', *Int J Food Eng*, 2, 1–9.

SPILIMBERGO S, ELVASSORE N and BERTUCCO A (2002), 'Microbial inactivation by high pressure', *J Supercritical Fluids*, 22, 55–63.

SPILIMBERGO S, ELVASSORE N and BERTUCCO A (2003), 'Inactivation of microorganisms by supercritical $CO_2$ in a semi-continuous process', *Ital J Food Sci*, 1, 115–124.

SPILIMBERGO S, MANTOAN D and DALSER A (2007), 'Supercritical gases pasteurization of apple juice', *J Supercritical Fluids*, 40, 485–489.

TEDJO W, ESHTIAGHI M N and KNORR D (2000), 'Impact of supercritical $CO_2$ and high pressure on lipoxygenase and peroxidase activity', *J Food Sci*, 65, 1284–1287.

TISI D A (2004), 'Effect of dense phase carbon dioxide on enzyme activity and casein proteins in raw milk', PhD dissertation, Cornell University, Ithaca, NY.

VALVERDE M T, MARÍN-INIESTA F and CALVO L (2010), 'Inactivation of *Saccharomyces cerevisiae* in conference pear with high pressure carbon dioxide and effects on pear quality', *J Food Eng*, 98, 421–428.

WEI C I, BALABAN M O, FERNANDO S Y and PEPLOW A J (1991), 'Bacterial effect of high pressure $CO_2$ treatment on foods spiked with *Listeria* or *Salmonella*', *J Food Prot*, 54, 189–193.

ZHONG Q, BLACK D G, DAVIDSON P M and GOLDEN D A (2008), 'Nonthermal inactivation of *Escherichia coli* K-12 on spinach leaves, using dense phase $CO_2$', *J Food Prot*, 71, 1015–1017.

ZHOU L, WU J, HU X, ZHI X and LIAO X (2009a), 'Alterations in the activity and structure of pectin methylesterase treated by high pressure carbon dioxide', *J Agric Food Chem*, 57, 1890–1895.

ZHOU L, ZHANG Y, HU X, LIAO X and HE J (2009b), 'Comparison of the inactivation kinetics of pectin methylesterases from carrot and peach by high-pressure carbon dioxide', *Food Chem*, 115, 449–455.

© Woodhead Publishing Limited, 2012

# Part IV

# Current and emerging packaging technologies and post-packaging decontamination

© Woodhead Publishing Limited, 2012

# 22

# Packaging technologies and their role in food safety

**M. Lalpuria, R. Anantheswaran and J. Floros, The Pennsylvania State University, USA**

**Abstract**: The importance of food packaging for food safety and quality is discussed. A review of mechanical, barrier, thermal, and optical properties of packaging materials is presented. Edible and bio-based materials for food packaging applications are also reviewed. Formation of rigid, semi-rigid, and flexible packages is described. Package integrity and food-package interactions are discussed with emphasis on food safety related examples. Packaging needs of various thermal, non-thermal, and other food processing technologies are reviewed. Current and future trends in food packaging such as active and intelligent packaging, nanotechnology, and traceability, are addressed.

**Key words**: food packaging, safety, thermal, non-thermal, integrity.

## 22.1 Introduction

Most foods are perishable and susceptible to spoilage. Improper handling, chemical reactions, and microbial growth are the leading causes of food deterioration and spoilage (Driscoll & Paterson, 1999). Bruising of fruits and vegetables during harvest and post-harvest handling leads to rotting. Storing of high moisture foods such as leafy vegetables at low humidity causes wilting. Condensation of moisture on food surface promotes microbial growth, thus contaminating the food. It is essential to maintain food safety for our well-being and to enjoy the food in its highest quality.

Food packaging can be defined as 'a complex and dynamic system aiming to safely prepare foods for transportation, distribution, storage, retailing, handling and end-use, and safely deliver these foods to the consumer in a

© Woodhead Publishing Limited, 2012

sound condition (maximum quality) at a minimum cost' (Floros, 1993). The principal objective of packaging is protection of its contents. In addition it must also inform the consumers about the product, be cost effective, provide convenience, have appealing graphics, and be compatible with the product and the environment. An enclosure or container that performs one or more of these functions is a package.

The food packaging industry has witnessed enormous growth in the past 50 years, largely due to advances in food science and technology, material science and engineering, and improved methods of food preservation. In today's society, the demand for food is constantly changing in response to changing consumer lifestyles, technological innovations, and improved logistics. The fast pace of life, increases in single-person households, and multiple-income families have led to changes in food preparation and consumption habits. Consequently, there have been rapid advances in food processing technology to make food preparation and consumption more convenient, as well as ensure food safety. Examples include microwaveable, canned or frozen meals, aseptically processed foods that can be stored at room temperature for several months, and several others.

In spite of the advances in food processing technology, contamination of food by natural means, accidental incorporation or malpractice, does occur. In order to preserve the quality of processed foods and avoid recontamination or cross-contamination, it is important to package it properly. After processing and packaging, the products are distributed to reach the final consumers and most likely, stored for some time (in warehouses, supermarkets or homes) before consumption. During distribution and storage, the packaged product interacts with its environment as shown in Fig. 22.1. Environmental factors such as gases, humidity, odors, light, temperature, and mechanical handling can trigger adverse reactions causing food deterioration (Floros, 1993; Rahman, 1999; Robertson, 2010). The deteriorative effect of major environmental factors on food quality is presented in Table 22.1. In order to ensure food safety throughout distribution and storage, it is important to understand the role of packaging in achieving safety and quality, and to design a packaging system accordingly.

The goal of this chapter is to review various food packaging technologies

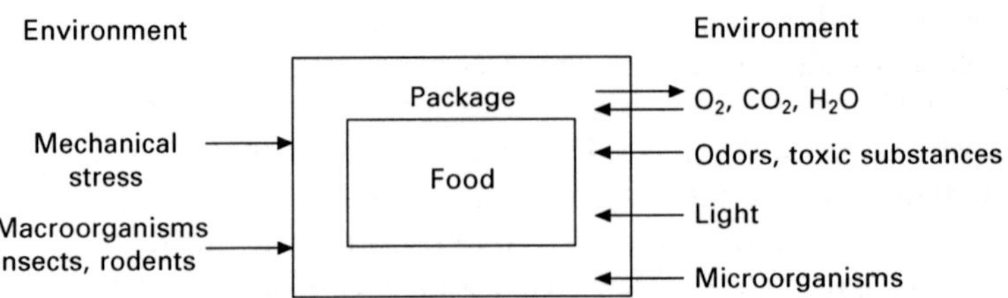


**Fig. 22.1** Interaction between packaged food and its environment during distribution and storage.

© Woodhead Publishing Limited, 2012

**Table 22.1** Deteriorative effect of various environmental factors on food quality

| Environmental factor | Deteriorative effect on food |
|---|---|
| Light | Color, flavor and nutrient degradation, rancidity |
| Oxygen | Growth of microorganisms, lipid oxidation (rancidity, loss of vitamin activity), Mailliard browning, protein loss |
| Water vapor/moisture | |
| Environment to food | Stickiness, swelling, microbial growth |
| Food to environment | Dehydration, texture loss |
| Microorganisms | Food spoilage, potential health hazard |
| Odors, toxic chemicals | Off-flavor formation, chemical changes, toxic hazards |
| Macroorganisms (insects, rodents, etc.) | Infestation of food, spread of disease (fruitfly, rats) |
| Mechanical stress (drop, compression, abrasion, improper handling) | Deformation, crushing or breaking the product, breaking of seal, leakage of contents |

Sources: Floros, 1993; Rahman, 1999; Roberson, 2010.

and their impact on food safety. To begin with, it is essential to understand the different levels of packaging and the various functions of packaging for food applications. A detailed review is presented of advantages and disadvantages of various packaging materials, package formation and issues such as package integrity and food package interactions (migration and sorption). Packaging needs of various thermal, non-thermal, and other food processing technologies are then discussed. Finally, the environmental impact of packaging materials and future trends in food packaging are addressed.

## 22.2 Levels of packaging

There are four levels of packaging depending on proximity to the product and end use: primary, secondary, tertiary and quaternary.

- Primary packaging is in direct contact with the food product and provides major protection against adverse environmental factors. Examples of primary packaging include bags, boxes, pouches, cans and bottles.
- Secondary packaging contains one or more primary packages and provides protection to the food and the primary package. It is sometimes designed to display the primary package in retail outlets. Examples include paperboard or corrugated cases.
- Tertiary packaging is composed of a number of secondary packages and is usually made of corrugated material.

© Woodhead Publishing Limited, 2012

- Quaternary packaging is frequently used for distribution of tertiary packages. An example of quaternary packaging is a palletized box.

## 22.3 Role of packaging

The most important objective of packaging is to provide optimum protective properties to maintain the product in good condition throughout its anticipated shelf life (Paine & Paine, 1983). The package should be of the right shape and size, it should have appropriate and sufficient information, attractive graphics and convenience features that meet consumer expectations. Therefore, fundamental functions of packaging for food applications include containment, protection, preservation, convenience and communication.

In addition to performing the above functions, good packaging should also be cost-effective, have minimal adverse environmental impact, and should not impart any undesirable contaminants to the food (Robertson, 2010). The role of packaging will continue to evolve in response to technological innovations and changing consumer demands. The significance of each function of packaging depends on the nature of the food, level of packaging (primary/secondary/tertiary/quaternary) and its end-use. If the package fails to perform adequately, a good deal of the energy and expenditure spent at the production and processing stage will be wasted (Mauer & Ozen, 2004).

## 22.4 Packaging materials

The right selection of packaging materials plays a key role in maintaining product quality and freshness during distribution and storage. The shape and form of a package are important from a marketing standpoint. The US Food and Drug Administration (FDA) regulates packaging materials under section 409 of the Food, Drug, and Cosmetic Act. Once known as indirect food additives, the FDA now refers to these materials as food contact substances (FCS).

Due to the large variety of food products, a wide range of packaging materials and package types exist. Materials commonly used for food packaging applications include one or more of the following: paper and paperboard, glass, metals, and plastics. Table 22.2 presents a comparison between the various properties of these materials. Edible and bio-based materials have generated tremendous interest in recent years as they are derived from renewable sources.

Each of these materials has unique properties that cater to different product needs. Today's packaging technology often combines two or more of these materials to obtain a composite with the desired functional and

© Woodhead Publishing Limited, 2012

**Table 22.2** Comparison of various performance properties of commonly used packaging materials

| Property | Paper and paperboard | Glass | Metal | Plastic | Bio-based materials[a] |
|---|---|---|---|---|---|
| Barrier | Poor barrier unless coated, laminated or wrapped | Perfect barrier to gases, vapor, micro- and macroorganisms | Excellent barrier to gases, vapor, micro- and macro-organisms | Wide range of barrier properties | Polysaccharides, proteins and PLA are poor water vapor barrier; lipid-based film is good water vapor barrier; PHAs are good barriers for water vapor, oxygen and aroma |
| Optical | Opaque | Transparent | Opaque | Wide range of optical properties | Polysaccharide, protein, PLA and PHA films range from clear to translucent; lipid films are generally opaque |
| Thermal | Not heat sealable; can be heated to high temperatures in a microwave | High thermal resistance | High thermal resistance | Wide range of thermal properties | Poor thermal resistance |
| Mechanical | Good stiffness, tears easily | Rigid, brittle | Rigid, good tensile strength | Usually low stiffness, can be rigid, semi-rigid and flexible, tensile and tear strength variable | Polysaccharide, protein PLA and PHA films have good mechanical properties |
| Weight | Light weight | Heavy weight | Heavy weight | Light weight | Light weight |
| Fragility | Not fragile | Fragile | Sturdy | Not fragile | Lipid films are fragile |
| Recyclability | Recyclable; recycled paper generally not used for food contact application | Recyclable; recycled glass can be used for food contact application | Recyclable; recycled metal can be used for food contact application | Recyclable; concerns regarding use of recycled plastics for food contact application | Recyclable |

[a]PLA: Polylactic acid; PHA: Polyhydroxyalkanoates.
Sources: Kirwan *et al.*, 2003; Paine & Paine, 1983; Petersen, *et al.*, 1999; Robertson, 2006; Rudnik, 2008.

© Woodhead Publishing Limited, 2012

aesthetic properties. ASTM methods for quantifying various properties of the packaging materials can be found in Section 22.10.

### 22.4.1 Paper and paperboard

Paper and paperboard are sheet materials made from an interlaced network of cellulose fibers, derived from wood. Paperboard is thicker than paper with a grammage (area density) above 250 $g/m^2$. They are the only widely used packaging material made from renewable sources. They can be made into rigid or flexible packages, depending on the type and amount of fibre used and how these fibres are processed during manufacturing. They are used for making different types of packages such as tea bags, fast food and frozen food boxes, milk and juice cartons, overwraps for sandwiches and delis, rigid corrugated, and solid boxes.

Paper and paperboard are advantageous for food packaging because of their low cost and light weight. They can be used over a wide range of temperatures, from storage of frozen foods to heating at high temperatures in microwave ovens. They are good barriers to light transmittance, but are permeable to water vapor, oxygen and other volatiles. Hence, plain paper is not used to protect foods for long periods of time. They are commonly used for packaging inexpensive and low moisture foods like sugar, flour, pasta or rapidly consumed goods such as bread and confectionery. Paperboard cartons are popular forms of secondary or tertiary packaging materials as they are economical, collapsible, printable and have excellent mechanical strength (Floros & Matsos, 2003).

Paper and paperboard are hygroscopic and hygroexpansive in nature (Mauer & Ozen, 2004). They absorb and lose moisture with varying environmental conditions. They expand in hot and humid conditions, which can adversely affect the integrity of packages, particularly of laminated structures, or distort package graphics. They cannot be used for packaging high moisture or fatty foods unless coated with wax or laminated with plastics or aluminum foil. Also, they do not offer protection from pests and may absorb aromas from the environment. Owing to these limitations, they are mostly used as laminations or secondary and tertiary packages for food applications.

### 22.4.2 Glass

Glass has been used as a food packaging material for centuries. Glass is produced by heating a mixture of silica, sodium carbonate, limestone and alumina to high temperatures until the materials melt into a thick liquid mass and is then poured into molds. Cullet, which is recycled broken glass, is also added to this mixture. The two main types of glass containers used in food packaging are bottles with narrow necks and jars and pots with wide openings.

Glass has several advantages for food packaging applications. It is odorless,

© Woodhead Publishing Limited, 2012

chemically inert to most food products, impermeable to gases, vapors and liquids, thermally stable for use in hot filling and in-bottle pasteurization and sterilization, withstands high pressure, high aesthetic appeal and quality perception, resealability, and microwaveability. The transparency of glass allows consumers to see the product, yet colored glass can protect light-sensitive contents such as in beer and wines.

The disadvantages of using glass for food packaging are its heavy weight and fragility. The heavy weight of glass contributes to increased transportation cost. The fragility of glass poses a safety risk due to the possible presence of broken or chipped glass in foods. Hence glass packaging has been replaced by plastic packaging for many products such as milk, cooking oil, carbonated beverages, and salad dressings (Girling, 2003; Mauer & Ozen, 2004; Floros & Matsos, 2003).

### 22.4.3 Metals

The four metals predominantly used in food packaging are: aluminum, steel, tin and chromium. All of them have excellent barrier properties for moisture, air, light, odor and microorganisms, high mechanical strength, rigidity, and thermal stability.

Aluminum is resistant to most forms of corrosion. It is generally used for light weight packaging needed for soft-drink cans, pet foods, seafood, etc. Aluminum cannot be soldered due to its ductility; hence it is used in making two-piece cans, foils, collapsible tubes, and laminated paper or plastic packaging.

Tin and steel are used as composites to form tinplate, made by coating low carbon steel with thin layers of tin on both sides. Even though tin coating protects the steel from rust and corrosion, tinplate containers are often lacquered to provide an additional inert barrier between the metal and the food product. Tinplate is very ductile and formable, has good strength, solderability and weldability. Hence, it can be used for making containers of many different shapes. It is widely used for making three-piece cans, which consist of two ends and one body. Tinplate is used for packaging beverages, processed foods, powdered foods, and confections.

Tin-free steel, also referred to as electrolytic chromium/chrome oxide coated steel, is a composite of chromium and steel. It has good formability and strength and is used for making can ends, trays, bottle caps and closures, and drums for bulk storage of ingredients or finished goods (Marsh & Bugusu, 2007). It is more heat resistant and cheaper than tinplate.

Metal containers commonly called 'tin cans' are widely used for packaging commercially sterilized or canned food products. They provide excellent protection from micro- and macroorganisms, are stackable, tamperproof, hermetically sealed and they can be thermally processed (Floros & Matsos, 2003). However, they are heavy and there may be safety issues related to can opening (cuts and bruises).

© Woodhead Publishing Limited, 2012

### 22.4.4 Plastics

Plastics are widely used in food packaging for making rigid containers like bottles, cups, trays and jars, and flexible films in the form of bags, pouches, and sachets. The use of plastics for food packaging has several advantages: light weight, good moisture and gas barrier, good puncture resistance, heat sealability, microwaveability, recyclability, non-fragility, the ability to form numerous shapes and sizes, and consumer acceptability and preference.

The most commonly used plastics for food packaging are: low density polyethylene (LDPE), high density polyethylene (HDPE), polypropylene (PP), polyethylene terephthalate (PET), polyvinyl chloride (PVC), and polystyrene (PS). Their properties are presented in Table 22.3.

The use of plastic food packaging is widespread due to its low cost and various functional advantages over traditional materials such as glass and metals. However, there are issues related to migration of additives from plastic packages into the food, and absorption of flavor compounds by the package in direct contact with the food. For example, bisphenol A, used in production of polycarbonate plastic containers and epoxy resins (protective lining inside metal cans), has been reported to leach out from these containers into the food (Biles *et al.*, 1997, 1999; Carwile *et al.*, 2009). The FDA has also expressed concern about its potential adverse effects on the brain, behavior, and prostate gland in fetuses, infants, and young children. On the other hand, absorption of the aroma compound d-limonene from orange juice packaged in plastic bottles has been shown to cause loss of organoleptic quality in the juice during storage (Kutty *et al.*, 1994; Mannheim *et al.*, 1987). Even though most plastics are recyclable, there are safety concerns regarding use of recycled plastics for primary packaging due to possible contamination.

### 22.4.5 Edible coatings and other bio-based materials

An edible coating is a thin layer of edible material applied on the surface of a food or placed between two food components to retard deterioration and extend shelf life. These coatings can be applied to the food by dipping the product in a coating solution or by spray-coating the food followed by drying. Also, dry films can be cast and placed directly onto the food.

Edible coatings are usually categorized into three types: polysaccharides, proteins, and lipids. Films composed primarily of polysaccharides or proteins have good mechanical and optical properties, but they are highly sensitive to moisture and have poor water vapor barrier properties. In contrast, lipid-based films have good water vapor barrier properties, but they are opaque, fragile, and prone to rancidity (Gontard & Guilbert, 1994). These materials can be combined into a composite to obtain a film with desired structural integrity and characteristic functionality. Edible coatings in direct contact with foods can also act as vehicles for carrying certain functional agents such as antimicrobials, flavors and nutrients that enhance the overall quality of the product.

© Woodhead Publishing Limited, 2012

**Table 22.3** Properties of commonly used plastics for food packaging

| Property | Low density polyethylene (LDPE) | High density polyethylene (HDPE) | Polypropylene (PP) | Polyethylene terephthalate (PET) | Polyvinyl chloride (PVC) | Polystyrene (PS) |
|---|---|---|---|---|---|---|
| Mechanical | Excellent elasticity; good resistance to mechanical abuse | Excellent tensile strength, stiffness and resistance to mechanical abuse | Excellent tensile strength, stiffness and flex cracking resistance | Excellent tensile strength, stiffness and elasticity | High rigidity and stiffness (unplasticized) | Good stiffness, but brittle |
| Moisture barrier[a] | Good, water vapor permeability in range 1.0–1.5 | Good, water vapor permeability ~ 0.3 | Good, water vapor permeability in range 0.25–0.7 | Excellent, water vapor permeability in range 0.18–1.80 | Good, water vapor permeability in range 0.9–5.0 (unplasticized) | Poor, water vapor permeability in range 7–10 |
| Oxygen barrier[b] | Poor, oxygen permeability ~ 500 | Good aroma barrier, oxygen permeability ~ 185 | Good, oxygen permeability in range 100–240 | Excellent, oxygen permeability in range 0.2–26.7 | Good, oxygen permeability in range 4–30 (unplasticized) | Poor, oxygen permeability in range 250–350 |
| Thermal | Low thermal resistance; excellent heat sealing strength | Low thermal resistance; good heat sealing strength | High thermal resistance; heat sealable | Excellent thermal resistance; poor heat sealing strength | Low thermal resistance; good heat sealing strength | Good thermal resistance |
| Chemical | Excellent chemical resistance (except to oils and organic solvents) | Excellent chemical resistance | Excellent grease resistance | Moderate chemical resistance | Excellent oil and grease resistance | Excellent resistance |
| Optical | Good clarity | Good clarity | Good clarity, glossy | Transparent | Transparent, glossy | Transparent |

*(Continued)*

© Woodhead Publishing Limited, 2012

**Table 22.3** Continued

| Property | Low density polyethylene (LDPE) | High density polyethylene (HDPE) | Polypropylene (PP) | Polyethylene terephthalate (PET) | Polyvinyl chloride (PVC) | Polystyrene (PS) |
|---|---|---|---|---|---|---|
| Applications | Shrink and stretch wrap, aseptic juice box | Milk containers, grocery bags, tubs for butter, yogurt, ice cream, twist wrapping sugar confectionery | Bread, hot-fill bottles, ice-cream containers | Carbonated soda and beer bottles, metalized films for snack foods, 'ovenable' trays for prepared/frozen meals | Films for meat packaging, containers for juices, edible oils, trays for chocolate assortments | Trays for fresh produce like meat and fish, containers for yogurt, ice cream, fresh cheese |

[a] Water vapor permeability values are given in g mil/100 $in^2$ day.
[b] Oxygen permeability values are given in cc mil/100 $in^2$ day atm.
Sources: Floros, 1993; Kirwan *et al.*, 2003; Paine & Paine, 1983; Piringer & Baner, 2000; Robertson, 2006.

© Woodhead Publishing Limited, 2012

In addition to protecting the foods from various deteriorative environmental factors, edible coatings also provide additional benefits:

- reduce waste and environmental pollution as they are made from natural substances;
- allow packaging of individual food pieces (fruits, vegetables, nuts); and
- enhance the overall quality of food.

However, they cannot ever replace synthetic packaging for long shelf life storage of foods. They show great potential as a complementary packaging for improving the overall quality of food and extending its shelf life (Kester & Fennema, 1986).

Bio-based materials have been defined as materials derived from renewable sources (Haugaard & Mortensen, 2003). They can be divided into three types based on their historical development. The first generation consists of synthetic polymers that contain 5–20% starch. These materials do not biodegrade fully, but break down into smaller molecules (biofragments). The second generation consists of synthetic polymers with 40–75% starch. Complete degradation of an entire film of this type takes a minimum of 2–3 years. The third generation materials are fully bio-based and biodegradable and can be further divided into three categories based on their origin and production (Robertson, 2006).

- Category 1 – Polymers directly extracted from agricultural and marine sources such as polysaccharides, proteins (plant and animal) or lipids. Some examples include cellulose, starch, whey and alginate. Paper and paperboard are the most widely used packaging material in this category.
- Category 2 – Polymers produced by chemical synthesis of renewable bio-based monomers. Examples include polylactic acid (PLA) and other polyesters. PLA, synthesized from lactic acid monomers, is one of the most promising biopolymers. It is versatile, recyclable, has high transparency and good processability. However, its mechanical behavior is highly sensitive to temperature and humidity. Since the degradation product of this polymer is lactic acid, which occurs naturally in foods and the body, they can be used in food contact applications (Vlieger, 2003). Currently it is used in food service applications, mainly for short shelf life products such as salads and beverages stored at low temperatures (Robertson, 2006; Siracusa *et al.*, 2008).
- Category 3 – Polymers produced by natural or genetically modified microorganisms. Examples include polyhydroxyalkanoates (PHA) and bacterial cellulose. PHAs are produced in nature by bacterial fermentation of sugar and lipids. A common type is polyhydroxybutyrate (PHB). Its mechanical and barrier properties are comparable to that of synthetic polymers. Despite this, they are not being used for commercial applications as yet, due to their high cost.

© Woodhead Publishing Limited, 2012

Research and development on food bio-packaging has intensified during the last decade. Nevertheless, the use of bio-based materials for food packaging applications is limited. They are mostly used as edible films and coatings on minimally processed fruits and vegetables, meat, seafoods, and dairy products. PLA is being used for making beverage cups, trays, and films.

### 22.4.6 Composites

Two or more packaging materials can be combined to produce a single composite structure. This can be achieved in two ways: lamination and co-extrusion. Laminates are formed by bonding two or more films using water, solvent, or solid-based adhesives. Co-extrusion involves combining two or more layers of molten plastic to obtain composite films. This process is accomplished in one step as compared to the multiple steps needed for lamination, but requires materials that have thermal characteristics suitable for co-extrusion. Combining different materials to get a single packaging structure can provide improved performance characteristics such as barrier properties, mechanical strength, heat sealability and printability at relatively low cost. Composites can then be used to make a variety of packages such as cups, trays, bags, bottles and other flexible packages.

Paper and plastic films are often laminated with an aluminum sheet (thickness in μm range) to improve their barrier properties. The aluminum sheet provides a virtually total barrier. However, this laminate is expensive and it is typically used to package high value foods such as dried soups, herbs, and spices (Marsh & Bugusu, 2007). If the foil layer is very thin, there is risk of flex-cracking or pin-hole development. These issues adversely affect the barrier properties of laminates, jeopardizing food safety and quality.

Metalized films are cheaper alternatives to laminated films and are less prone to flex-cracking and pin-holing (Floros & Matsos, 2003). They are made by depositing a thin layer (nanometer range) of aluminum on plastic films under vacuum. They provide good barrier properties and flexibility at relatively low cost, and are commonly used for packaging coffee and snacks such as potato chips and pretzels.

Aluminum and metalized laminates are opaque and consumers perceive them to be environmentally unfriendly as a high amount of energy goes into their production (Lange & Wyser, 2003). This led to the development of thin, glass-like $SiO_x$ films formed by depositing gaseous organosilane and oxygen on PET, PP, and polyamides. These films are transparent, retortable, and microwaveable, and their barrier properties are comparable to metalized films. However, limited flex and crack-resistance, and relatively high production costs have restricted their extensive commercial application.

The presence of different materials in a composite makes recycling complicated and economically not feasible. However, combining different materials to obtain the desired functional property can prolong the product's shelf life. This often reduces the total amount of packaging materials used

© Woodhead Publishing Limited, 2012

at the source. It could be argued that using composites results in waste minimization and the failure to recycle is only a small compromise.

## 22.5 Formation of packages

Package design is an important aspect of packaging technology. Developments in package design have improved package integrity, appearance, and convenience, and reduced the amount of packaging material required (Krochta, 2007). Selecting the right size and shape of a package depends on various factors: (a) nature of the product (physical state, chemical nature, amount, factors causing deterioration); (b) environmental, transportation, and distribution hazards; (c) market, and (d) packaging material, machinery, and technology (Floros, 1993; Paine & Paine, 1983).

The form of a package defines the degree of package rigidity and packages can be broadly classified as rigid, semi-rigid, and flexible (Krochta, 2007). Rigid packages are stiff and do not change their shape upon filling. They cannot be deformed without causing damage. Examples include glass bottles and cans. Semi-rigid packages can be deformed to some extent without damage, but they return to their original shape on removing the applied force. Examples of semi-rigid packages are cartons and plastic bottles. Flexible packages assume a shape only when the product is filled and lose it when the product is dispensed. Examples are bags and pouches. Below we describe the formation of some of these package types.

### 22.5.1 Forming paper packages

Paper and paperboard are the most abundantly used material for flexible packaging. They are used for making flexible sheets, semi-rigid boards, and rigid containers. Flexible single and multi-wall bags are made by cutting paper into desired dimensions, folding and gluing the seams and bases. They are commonly used as grocery bags and packaging dry foods like sugar and flour. Sachets and pouches are paper-based flexible packages comprising aluminum foil or metalized PET. They are usually associated with form, fill, seal machines.

Semi-rigid structures like trays and folding cartons are made from paperboard. Paperboard trays are made by heat sealing, locking tabs and slots or by gluing with hot melt adhesive depending on the end-use. Folding cartons are made by creasing and cutting an outline, folding and gluing, and printing on the surface. These semi-rigid containers can be used for a wide range of products such as cereals, ice-cream, coffee, tea, confectionery, dried foods, and frozen and chilled foods.

Rigid cartons or boxes are mainly used for luxury food products such as chocolate confectionery, expensive wines, and spirits. For making rigid

© Woodhead Publishing Limited, 2012

cartons and boxes, a baseboard is cut, and a thin film of adhesive coated paper is placed on it. This paper is wrapped around the corner flaps to secure them. For the corners or sides of the box to stay up, a heat activated tape (stay tape) is applied before covering the paper.

Various other treatments are given to paper and paperboard to improve their physical performance and to form different packages. For example, paper is impregnated with resins such as ethyleneamines to impart wet-strength. For making flexible boxes and cartons with improved water and oxygen barrier properties, paper can be laminated with aluminum foil and extrusion coated with polyethylene. Clay coatings are applied to paper and paperboard to improve their appearance and printability.

### 22.5.2 Forming glass containers

Glass can be molded into various shapes: bottles, jars, jugs, tumblers, carboys, vials, and ampoules. These containers are made by a press-and-blow process (wide-mouth containers) or by a blow-and-blow process (narrow neck containers). Gobs (solid cylindrical mass) of molten glass is fed into a blank mold, where it is pressed with a metal plunger or blown with compressed air into thick, hollow, semi-molten containers called parisons. A plunger in the base of the mold forms the neck ring and threaded region (for closure) on the parisons. This roughly shaped glass is then transferred to blow molds, where a jet of compressed air forces it into the final container shape. This two-step process allows better control of glass thickness throughout the container. After formation, the glass container is cooled quickly to avoid crystallization and maintain its transparency. After cooling, it is reheated in an annealing oven and gradually cooled to remove any internal stresses. Coatings may be applied to the glass to strengthen it or for lubrication.

Although glass is impermeable to oxygen and water vapor, the contained food may still deteriorate if the bottle closure is faulty. Closure for glass packaging is usually metal or plastic, although wooden cork is still commonly used for wines and spirits. A wide range of closures is available and the decision on which one to use depends on the end use. Normal seals, made from composites of plastic/foil, are used for non-pressure filled products like milk and coffee. Pressure seals, made from metal or plastic with a composite liner, are used for carbonated beverages. Lastly, vacuum seals made from metal with a composite lining are used for non-carbonated beverages and sauces.

### 22.5.3 Forming metal cans

*Three-piece cans*

A three-piece can contains two ends and one body with a welded side seam. One of the ends is applied during can manufacturing and the other, after filling the product. They are made from tinplate or tin-free steel. For

© Woodhead Publishing Limited, 2012

making the can body, the material is slit into metal strips of appropriate size and rolled into a cylinder with slight overlap. The side seam is formed by welding the ends of the can body at the overlap. A flange is formed on both ends of the can body for fitting the can ends. It is then usually beaded (series of ridges on the sidewall) to prevent can implosion or collapse due to pressure differences between the inside and outside of the can encountered during thermal processing. The ends are fitted onto the can body by double seaming (Fig. 22.2) for hermetic sealing.

A double seam is formed by rolling the curled edge of the cover and the can body together to form a hermetic seal. The double seam is the most vulnerable part of the can, and it is critical to the overall safety of the food. It is important to ensure that every component of the double seam, particularly the overlap, has precise dimensions and conforms to strict guidelines making a tight seal (Floros & Matsos, 2003).

*Two-piece cans*

A two-piece can consists of a can body and one end, applied after filling the product. Two-piece cans are rapidly replacing three-piece cans due to their low cost and visual appeal. They are made from tin-free steel or aluminum sheets by two processes: draw and redraw (DRD) & draw and iron (D&I). In DRD, a shallow cylindrical can is stamped or drawn from a disc of metal by passing through a die. To increase the height of this can without decreasing the thickness, it is then passed through additional draws (second or third). The cans are flanged, beaded (for thermal processing), filled and the end is double seamed. The first step of the D&I process is similar to the DRD process. After drawing a shallow can, it is redrawn to achieve the desired can diameter and more material is forced to the can side. This can is then ironed to reduce the sidewall thickness. The sidewall of these cans is thinner than those obtained by the DRD process. Two-piece cans cannot be used for

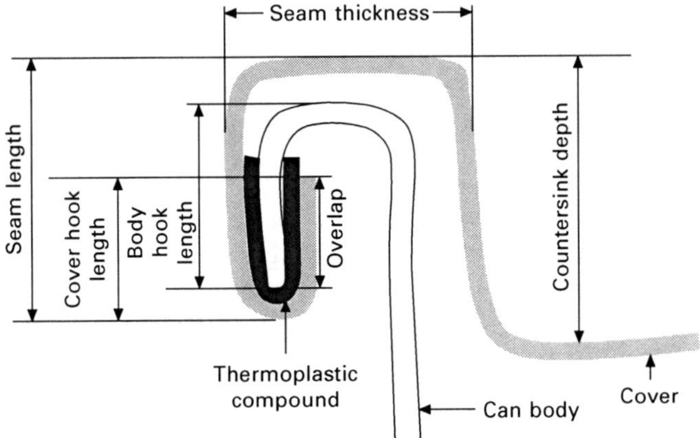


**Fig. 22.2** Double seam of a metal can and its components.

© Woodhead Publishing Limited, 2012

thermal processing and are most commonly used for carbonated beverages because high internal pressure prevents the sidewalls from collapsing/denting. For non-carbonated beverages, nitrogen is injected to pressurize the can.

### 22.5.4 Forming plastic packages

*Extrusion*

Plastics can be made into films, sheets, or tubes by extrusion. The plastic resin granules are fed into an extruder, compressed, heated, and passed through a die to form a sheet or a tube. Thin films of plastic can be made by a tubular or blown film process. These thin films can be used as liners, wraps, overwraps or used for making bags, envelopes, or pouches.

*Injection molding*

In this process, a definite amount of molten plastic is injected into a closed mold. After cooling, the container is ejected out of the die. It is used for making caps, closures, snap-on lids, wide-mouth tubs, and parisons for the injection blow molding process (Mauer & Ozen, 2004).

*Blow molding*

In this process, a hot plastic tube (parison) is placed into a mold and air is blown to force it into the shape of the mold, followed by cooling. It is mostly used for making bottles with narrow necks. There are three different types of blow molding: extrusion blow molding, injection blow molding and injection stretch blow molding.

*Thermoforming*

In this process, a previously extruded plastic sheet is heated to soften it and then molded into the desired shape mechanically or by vacuum/air pressure. It is used for making tubs, trays, cartons, cups, retortable and dual ovenable containers (Krochta, 2007).

## 22.6 Packaging for various processed foods

### 22.6.1 Thermally processed foods

Thermal processing is a term widely used in the food industry to describe the process of heating, holding, and cooling to produce microbiologically safe products of acceptable quality (Holdsworth & Simpson, 2007). Thermal processes employ a time–temperature schedule. The process is described in terms of the temperature to which the product is exposed and the time of exposure (holding time). In this section we will be confined to thermal processing of packaged foods. Various package factors can affect the efficiency of thermal processing (Chung *et al.*, 2006):

© Woodhead Publishing Limited, 2012

(1) *thermal conductivity of the packaging material* determines the rate of heat penetration. For example, heat penetrates glass more slowly than tinned iron walls of a can (Paine & Paine, 1983);
(2) *shape and dimension of the package* influences heat penetration into the package. Heat penetration is greater for packages with larger surface area and smaller thickness;
(3) *amount of headspace* is important for agitation and rotation, which can enhance internal heat transfer and also provide the appropriate vacuum; and
(4) *position of the package* inside a retort and the stacking can affect heat transfer to individual packages.

To preserve the benefits of thermal processing and ensure food safety throughout its anticipated shelf life, it is very important to select the right packaging material and package type. Processing should not alter the functional properties of the packaging material. Also, the package should be able to withstand the rigors of thermal processing and maintain its integrity. Below we review the packaging materials and package types used with different thermal processes.

*Pasteurization*
Pasteurization can be applied to packaged products either after packaging (in-container pasteurization) or before packaging, in which case the product is first heated to pasteurization temperature and then filled hot into the container (hot-fill). *In-container pasteurization* is generally applied to products like acidified fruit and vegetables, beers or other products without preservatives (Driscoll & Paterson, 1999). For glass bottles, water is used as the heating medium to prevent any damage due to thermal shock. After processing, the container is cooled to about 40°C. The residual heat facilitates surface water evaporation and minimizes corrosion of metal containers or closures. This rapid cooling also aids in the setting of adhesives used in labels (Driscoll & Rahman, 2007). *Hot-filling* is a cheaper alternative to aseptic packaging. In this technique, the food product is continuously pasteurized at high temperature (85–100°C), the hot product is then filled into the container, the container is sealed and held at this high temperature for a short time, and then cooled to less than 50°C. The residual heat of the product is usually sufficient to kill any pathogens or spoilage microorganisms residing in the container or its closure, but sometimes (for very small containers) an additional step of supplemental heat treatment may be necessary. Hot-filling is generally used for production of shelf-stable, high acid foods and beverages such as tomato products, ready-to-drink teas, fruit juices, sports and energy drinks, and many others (Beswick & Dunn, 2002).

Hot-filling bottles have to withstand high temperature and resist paneling on cooling (Hannay, 2002). When using glass, care must be taken during heating and cooling to avoid the risk of breakage due to thermal shock. Due

© Woodhead Publishing Limited, 2012

to its heavy weight and design constraints, glass is rapidly being replaced by plastic containers for hot-filling operations. For plastic packages, the temperature of filling is extremely important. If a plastic bottle is hot-filled above the glass transition temperature of the plastic material, the bottle and finish could distort (Mauer & Ozen, 2004). In addition, hot-fill plastic bottles must have thick sidewalls and they require expansion panels to allow for package contraction during cooling (Beswick & Dunn, 2002). PET bottles are most commonly used for hot-filling applications. Some other plastics used for hot-filling include special grades of PEN, PP, PS, and PVC (Rosato *et al.*, 2001).

*Retorting*

Retorting is heating of low acid foods prone to microbial spoilage in hermetically sealed containers to extend their shelf life. The goal of retort processing is to obtain commercial sterilization by application of heat. Foods can be sterilized in rigid containers like glass, metal cans and plastic or flexible retort pouches. Metal cans and glass jars are widely used for retort processes, because of their high mechanical strength, thermal stability, resistance to pressure, and excellent barrier properties. The advantage of using glass is product visibility; however, careful handling is required to prevent breakage (Holdsworth & Simpson, 2007). Three-piece cans made from tinplate or lacquered tin-free steel are commonly used for retort packaging. Two-piece cans made from aluminum and other flexible containers are slowly replacing them. Some advantages of aluminum over tinplate cans are lower shipping costs, no rusting and easier puncture opening. Plastic containers are sometimes used for retorting of foods, especially ready-to-eat military rations (Chung *et al.*, 2006). A typical retortable plastic container is made of an oxygen barrier layer such as ethylene vinyl chloride (EVOH) or polyvinylidene chloride (PVdC) sandwiched between two polypropylene layers. Flexible retortable pouches are laminated structures that must provide excellent barrier properties for long shelf life, seal integrity, puncture resistance and must withstand rigorous thermal processing (Driscoll & Paterson 1999). A typical retort pouch contains an Al foil layer sandwiched between two polypropylene layers.

*Aseptic processing*

Aseptic processing is the processing of commercially sterile and cooled food products being filled into commercially sterile containers under aseptic conditions. The package is hermetically sealed to produce a shelf-stable product that can be stored at ambient conditions. As the food product and package are sterilized separately, a wide variety of package designs and materials can be used for aseptic processing as compared to pasteurization or retorting. In addition to rigid glass and metal containers, flexible polymeric materials and semi-rigid paper/Al/plastic composites that offer cost benefits and convenience can be used without sacrificing product quality (Floros,

© Woodhead Publishing Limited, 2012

1993). However, flexible pouches lack rigidity, and package integrity may get compromised during handling and distribution. Aseptic filling systems have also been designed for HPDE and PET bottles (Ammann, 2001). Bottles made by blow molding are sterile and can be used directly for filling the product. For aseptic bulk packaging and storage, containers like tanks, totes, pouches, and bag-in-box systems are used. Factors like seasonality, cost, convenience, quality, and global competition have made aseptic bulk packaging popular, particularly for processed fruit and vegetable products (Floros *et al.*, 1998).

Packages for aseptic processing can be sterilized by heat treatment (saturated steam, superheated steam, hot air, mixture of hot air and steam), mechanical processes (water/rinsing/flushing, air blasting, brushing, ultrasound), irradiation (ionizing rays, ultraviolet, infrared) or chemical treatments (hydrogen peroxide, ozone, chlorine, peracetic acid, etc). These treatments can be used alone or in combination (Floros, 1993).

*Microwaveable foods*

Microwaveable foods are generally treated in packages during microwave heating. Hence, this method requires full compatibility between the food product and the package. Materials that are transparent to microwaves and do not appreciably react with the microwave field or modify the power distribution in microwave ovens are called passive packages (Bohrer & Brown, 2001). Paper and paperboard, glass and all plastics currently used for food packaging are microwave transparent.

Paper and paperboard contain some water and mobile ions in their structure and, therefore, they will heat in a microwave oven, but the rate is generally slow. Even though glass is transparent to microwaves, has excellent barrier properties and product visibility, it is not a popular choice for microwaveable foods, because of its heavy weight, cost, and fragility. Plastics commonly used for microwave heating applications are polyethylene (PE), polypropylene (PP), polystyrene (PS) and polyethylene terephthalate (PET). Flexible and semi-rigid composite structures made of paperboard and plastics are used for popping corn and boil-in-bag frozen vegetables, vegetables or meats with sauce, rice and other side dishes. In general, the problems encountered with microwave heating of foods packed in transparent materials are:

- sogginess or lack of crispness development;
- lack of browning or proper color development;
- localized hardening of food due to non-uniform moisture evaporation;
- non-uniform temperature distribution;
- boil-out or run-off of viscous products like sauces and purées; and
- inappropriate heating rates in multi-component food products (Bohrer & Brown, 2001).

The following factors should be considered when designing microwaveable packages:

© Woodhead Publishing Limited, 2012

(1) avoid arcing or rough edges as heat can accumulate in such spaces and may lead to overheating or burning;
(2) cover foods during heating to protect against physical hazards such as erupting due to boiling;
(3) handle heated foods safely;
(4) take into account the heat resistance of the packaging material as it will also get heated along with the food; and
(5) understand migration issues (Robertson, 2006).

To achieve localized effects such as browning and crisping, thin films of packaging material referred to as susceptors are used. They can absorb large amounts of microwave energy and re-emit this energy as heat. These films are made by depositing a thin layer of a metal like aluminum on a heat resistant polymer such as PS or PET which is laminated to paperboard. An ideal susceptor would heat up rapidly to a predetermined temperature, and then remain at that temperature throughout the heating process. They can be used as freestanding devices or incorporated into other packing forms such as trays, pouches, cartoons, and pads.

To overcome temperature variability and achieve appropriate heating rates for multi-component foods, thick metal films referred to as shields are used. These films are thick enough so as not to heat themselves, but rather reflect microwaves in the same way as oven walls. Examples of shield materials are foil, foil laminations or aluminum sheets converted into pans and trays. A package can also be designed to modify the electromagnetic field and achieve uniform heating and browning. This electromagnetic field modification can be achieved by creating large or small-scale patterns in thick metals depending on the heating objectives. A typical example is aluminum pan with specially constructed lids containing these special patterns (Bohrer & Brown, 2001). They are used for microwave heating of products like pot pies, chicken dinners, and macaroni cheese.

### 22.6.2 Non-thermally processed foods

*Irradiation*

Irradiation (IR) is the process of exposing a material to ionizing radiation and can be achieved using electron beams, gamma rays, or X-rays. Depending on the dose of radiation energy applied, foods may be pasteurized to eliminate vegetative cells or sterilized to eliminate all microorganisms. Radiation processing is widely used for medical product sterilization and food irradiation. The use of irradiation is becoming a common treatment to sterilize packages in aseptic processing of foods. IR is very effective in reducing microbial population, inhibiting sprouting and controlling insect infestation. Foods are generally prepackaged for IR treatments to avoid recontamination. This leads to safety concern regarding the effect of IR on packaging material properties and formation of radiolysis by-products.

© Woodhead Publishing Limited, 2012

IR can cause changes in some chemical and physical properties of polymeric packaging materials (Table 22.4). The change depends on both polymer composition and radiation conditions. Irradiation can lead to cross-linking and/or chain scission of polymers. Cross-linking is the predominant reaction during irradiation in most plastics used for food packaging such as PP, PE, and PS. Cross-linking can decrease elongation, crystallinity, and solubility, and may increase the mechanical strength of polymers. Chain scission, on the other hand, can decrease the chain length of polymers and produce hydrogen, methane, and hydrogen chloride for chlorine-containing polymers under vacuum (Ozen & Floros, 2001). In the presence of oxygen, additional chain scissions would form peroxides, alcohols, and various low molecular weight oxygen-containing compounds (Buchalla *et al.*, 1993). These low molecular weight by-products can migrate into the food and affect its quality and safety. A list of packaging materials and related maximum dose approved by the FDA for irradiation of prepackaged foods can be found in 21 CFR 179.45.

*Ultraviolet (UV) light*

Recent advances in research on UV light irradiation has demonstrated that UV treatment shows considerable potential in food processing as an alternative to traditional thermal processing. It can be used for inactivation of microorganisms in liquid products like fresh juices, soft drinks and beverages, ready-to-eat meats and for shelf life extension of fruits and vegetables (Koutchma, 2009). It can also be used for sterilization of packaging materials such as wrappers or bottle caps, by placing UV lamps over conveyors (Bintsis *et al.*, 2000). UV radiation in the range of 250–280 nm is lethal to most microorganisms. It inactivates microorganisms by DNA mutations induced by absorption of UV light by the DNA (Butz & Tauscher, 2002).

UV sterilization has been applied to foil caps of HDPE bottles (Nicolas, 1995), and cartons for liquid products (Kuse, 1982) during the manufacture of aseptically filled UHT dairy products. Treating packaging materials with UV light can cause modification of surface properties such as surface tension, hydrophilicity, adhesion and barrier properties (Ouyang *et al.*, 2000; Ozen & Floros, 2001). It can also alter some functional properties of packaging materials as shown in Table 22.4. UV-absorbing polymers such as LDPE and PET can undergo oxidation, and may produce oxidation products that can potentially migrate into the food (Ozen & Floros, 2001). PET, which was highly stable under ionizing treatment, was found to be more sensitive than LDPE under UV light treatment. An accumulation of oxidation products was found in UV-treated PET. This high sensitivity of PET was attributed to its polar nature that could enhance oxidation of film surfaces (Berends, 1996).

*Ozone*

Gaseous or aqueous ozone is used as a disinfectant for treating foods and

© Woodhead Publishing Limited, 2012

**Table 22.4** Effects of various non-thermal processes on properties of packaging materials

| Process | Material[a] | Effect | Reference |
|---|---|---|---|
| Ionizing radiation | Paper | Cellulose undergoes chain scission on irradiation, resulting in loss of mechanical properties. With increasing dose, paper becomes increasingly brittle | Ozen & Floros, 2001; Diehl, 1995 |
| | Tinplate cans | No effect on coatings, seals | Killoran, 1983 |
| | Glass | Glass acquires a discoloration (brown tint); evolution of oxygen due to disruption of glass structure | Diehl, 1995 |
| | PET/PVdC/PE | Decrease in oxygen permeability | Kim-Kang & Gilbert, 1991 |
| | EVA, HDPE, PS, LDPE, BOPP | No significant changes in $O_2$, $CO_2$ and water vapor permeabilities. No significant change in mechanical properties at radiation dose of 5 and 10 kGy | Goulas *et al.*, 2002 |
| | PET | Formation of volatiles | Morehouse & Komolprasert, 2004 |
| | LDPE, polyamide | Changes in optical properties | Moura *et al.*, 2004 |
| Ultraviolet light | PLA | Enhanced rate of degradation | Ho & Pometto, 1999 |
| | Nylon | Surface treatment imparted antimicrobial activity | Paik *et al.*, 1998 |
| Ozone | Virgin silicon membrane | Increase in oxygen permeability | Shanbhag & Sirkar, 1998 |
| | Polyamide | Increase in tensile strength; decrease in oxygen permeability | Ozen *et al.*, 2003 |
| | PE | Change in mechanical properties, decrease in oxygen permeability | Ozen *et al.*, 2003 |
| High pressure processing | PET/SiOx/LDPE, PET/$Al_2O_3$/LDPE, PET/PVdC/nylon/HDPE/PE, PE/nylon/EVOH/PE, PE/nylon/PE, metalized PET/EVA/LLDPE, PP/nylon/PP, PET/EVA/PET | No significant change in mechanical properties | Caner *et al.*, 2004 |
| | Metalized PET/EVA/LLDPE | Delamination, significant increase in $O_2$, $CO_2$ and water vapor permeability | Caner *et al.*, 2004 |

© Woodhead Publishing Limited, 2012

| | | |
|---|---|---|
| PA/PE,PA/PE, PET/PVDC/PE, PA/PP/PE | No significant change in $O_2$ and water vapor permeability | Lambert *et al.*, 2000 |
| LDPE, EVA | Decrease in sorption of D-limonene | Masuda *et al.*, 1992 |
| PA/PE | Delamination, detachment, cracks and folds | Götz & Weisser, 2002 |

[a]PE: polyethylene; LDPE: low density polyethylene; LLDPE: linear low density polyethylene; HDPE: high density polyethylene; PP: polypropylene; PET: polyethylene terephthalate; PVC: polyvinyl chloride; PS: polystyrene; PVdC: polyvinylidene chloride; EVA: ethylene vinyl acetate; BOPP: biaxially oriented polypropylene; PLA: polylactic acid; SiOx: silicon oxide; $Al_2O_3$: aluminum oxide; EVOH: ethylene vinyl alcohol.

© Woodhead Publishing Limited, 2012

packaging materials. Ozone is effective against a broad range of microorganisms due to its strong oxidant behavior. Due to its powerful oxidizing nature, it can also interact with polymers used as packaging material and alter its properties (Table 22.4). Ozone reacts with polymer surfaces and causes modification of surface properties due to formation of oxygen-containing functional groups and degradation of polymer chains. Ozone treatment significantly increased the surface tension and hydrophilicity of polymers such as PP, PE, and PS, and improved their adhesion properties (Ozen & Floros, 2001). The effect of ozone on the mechanical properties of plastic films depends on the chemical structure of polymers and on treatment conditions such as concentration of ozone and temperature. The tensile strength of oriented PP films decreased up to 75% after ozone treatment, while that of biaxially oriented nylon increased by 30% (Ozen *et al.*, 2003). Treatment of PS powder with ozone caused a yellow discoloration of the powder. Films made from this powder were opaque and brittle (Razumovskii & Zaikov, 1982).

There is concern regarding use of ozone-treated plastic packaging films as ozone might increase the migration of additives or monomers from these films. Steiner (1991) investigated the effects of ozone treatment on LDPE films and commonly used antioxidants such as butylated hydroxyanisole (BHA) and butylated hydroxyl toluene (BHT). Low molecular weight substances such as phthalic esters, alkanes, alkenes, ketones, and peroxides formed from the degradation of LDPE and LDPE additives were detected.

*High pressure processing*

High pressure processing (HPP) is a promising method for food processing as it inactivates spoilage and pathogenic microorganisms and deactivates enzymes with minimal loss of quality, sensory, and nutritional attributes of foods. HPP applies high pressure to the food by mechanical action at a relatively low temperature. The food is pressurized to 100–800 MPa for a short period, ranging from a few seconds to several minutes (Han, 2007). It is important to study the effect of high pressure on packaging materials as foods are prepackaged for HPP treatment. Packaging materials used for HPP applications must be able to withstand the compressive forces encountered during processing, while maintaining overall package integrity. Hence, rigid materials like metal, glass, and paperboard are not well suited for HPP applications (Caner *et al.*, 2004).

Flexible packages made of PP, polyester, PE, and nylon pouches or tubes are currently being used for HPP (Han, 2007). The effects of HPP treatment on the functional properties of various packaging films and laminates are presented in Table 22.4. Water vapor and oxygen permeabilities of several laminated films such as PP/EVOH/PP, OPP/PVOH/PE, KOP/CPP and PET/Al/CPP were not affected by HPP treatments between 400 and 600 MPa (Masuda *et al.*, 1992; Ochiai & Nakagawa, 1992). However, the presence of less elastic materials like Al (ruptures during compression) in metalized plastic films negatively affected its barrier properties (Caner *et al.*, 2004).

© Woodhead Publishing Limited, 2012

Studies evaluating the effect of HPP on the mechanical properties of polymeric packaging materials did not report any significant negative effects of HPP treatments (Han, 2007).

Sorption of some aroma compounds (p-cymene and acetophenone) was lower by films exposed to 500 MPa pressure as compared to non-pressurized films (Kübel *et al.*, 1996). Transition of plastic films to glassy state at higher pressures was speculated as the reason for decreased sorption of aroma compounds. During HPP treatment, packaging materials undergo compression and lose their capacity to absorb flavor compounds from foods (Caner *et al.*, 2004). When the pressure is released, these materials should ideally regain their original dimensions. However, some materials fail to regain their original free volume, and sorption of flavor compounds into these materials is reduced (Caner *et al.*, 2004).

It is important to test the overall integrity of HPP-treated packages in order to ensure food safety. Lambert *et al.* (2000) reported that the heat-seal strength of multilayer plastic packages (PA/PE, PET/PVDC/PE, PA/PE surlyn, PA/PP/PE) was unaffected by HPP treatment at 200–500 MPa for 30 min. However, delamination between polypropylene and aluminum layers was observed in ready-to-eat meal pouches (Schauwecker *et al.*, 2003). Al has lower compressibility than PP, and this could have resulted in delamination.

### 22.6.3 Modified atmosphere packaging

Modified atmosphere packaging (MAP) involves modifying or altering the atmosphere inside a package to provide optimum conditions for maintaining the quality of the food and increasing its shelf life. MAP retards deteriorative chemical and biochemical reactions, and slows down the growth of microorganisms. However, MAP cannot improve the shelf life of a poor quality food product. High quality foods, good hygiene practices, and strict temperature control throughout the distribution and storage chain are essential for maximizing the benefits of MAP (Mullan & McDowell, 2003). MAP has been used to extend the shelf life of food products such as fresh or minimally processed fruits and vegetables, meat, seafood, cheese, bakery products, milk and milk powders, pasta, coffee, nuts and snacks, and many others.

Selection of appropriate packaging materials is essential to maintain the quality and safety of MAP foods. The main factors to be considered when selecting a packaging material for MAP applications are gas and vapor barrier properties, optical, antifogging and mechanical properties, and heat sealability (Mullan & McDowell, 2003; Robertson, 2006). Packaging materials for MAP must have appropriate gas and vapor barrier properties to maintain optimum quality of a particular food product. They should have high gloss and transparency for bags, pouches, and top webs to enable the consumer to have a clear view of the product. Condensation of water droplets inside the package leads to poor product visibility and affects

© Woodhead Publishing Limited, 2012

consumer appeal. Hence, antifogging agents are applied to some packaging materials. Resistance to tear and puncture are important for maintaining package integrity. Effective heat sealing is essential for maintaining gas composition within the pack.

Glass and metals are rigid, virtually absolute barriers to water vapor and other gases, and opaque, and hence cannot be used for MAP applications. Plastics have variable barrier, mechanical and optical properties and are most suited for MAP applications. Plastics used for MAP are polyethylene (PE), polypropylene (PP), polystyrene (PS), polyethylene terephthalate (PET), polyvinyl chloride (PVC), polyvinylidene chloride (PVdC) and ethylenevinyl alcohol (EVOH). Because a single plastic material generally does not meet all the packaging requirements, multilayer composite structures are commonly used for MAP applications. Examples of laminated structures include PA/PP and PVdC coated PP/PE for lidding films, PA/EVOH/PE for pre-formed base tray, and PET/EVOH/PE and PS/EVOH/PE for thermoformed base trays (Mullan & McDowell, 2003). Flexible plastic films are used for bags, pouches, pillow packs, and top webs. Rigid and semi-rigid plastic containers are used for base trays. Preformed plastic-coated paperboard trays can also be used. For overall product safety, it is essential that the lidding film is adequately sealed onto the tray.

For non-respiring foods, MAP is designed without oxygen to minimize deteriorative oxidative reactions (off-flavor in poultry, rancidity in fish) or reduce microbial growth (molds in cheese). Hence high barrier films or laminates are used for prohibiting exchange of gases, especially oxygen, through the package (Driscoll & Paterson, 1999).

MAP packaging systems must allow for selective exchange of gases to maintain the desired gas composition within a package. For respiring products, MAP has to maintain >2% $O_2$ concentration within the package, to prevent growth of undesirable microorganisms. Generally, reduced levels of oxygen and increased levels of carbon dioxide in the atmosphere around the fresh produce are recommended (Floros & Matsos, 2005). The other challenge is that the respiration of the product varies with the gas composition within the package. Due to significant respiratory activity of the product, the gas composition inside the package keeps changing during storage and a good understanding of the dynamics of changes is required to tailor package design for an individual product. Mathematical models are available to predict the dynamic changes within a package under these dynamic conditions.

Package integrity is critical for the success of MAP foods. If the package leaks, the optimized atmosphere around the food produce will be disturbed as the headspace gas will interact with the normal atmosphere surrounding the package. Eventually, the beneficial effects of modified atmosphere packaging will be lost. It is therefore essential to monitor package integrity throughout the distribution chain, from the manufacturer to the consumer. The most effective way of detecting a leakage non-destructively is to use a visual leak indicator that is permanently attached to the package (Smolander

© Woodhead Publishing Limited, 2012

*et al.*, 1997). These indicators can detect the alteration of package atmosphere through a stimulus, such as an irreversible color change.

## 22.7 Package integrity

To ensure food safety throughout handling, distribution and storage, it is not only important to select the right packaging material and package type, but also to monitor package integrity. In the last few decades, the food packaging industry has witnessed tremendous increase in the use of flexible and semi-rigid packages. Plastics and composites are more prone to mechanical damage (punctures, cracks, etc.) than rigid materials like metals and glass (Gnanasekharan & Floros, 1994). Hence package integrity issues have become even more crucial from a food safety viewpoint.

Maintaining package integrity is achieved by ensuring the integrity of the package seals and the absence of any defects or leaks in the packaging material. Seal integrity refers to complete fusion of the packaging material around the seal without any discontinuities. Defects in the packaging material such as cracks or pinholes are the leading cause of post-process contamination of packaged food (Gnanasekharan & Floros, 1995). The degree of contamination is affected by variables such as pressure differential, microorganism type and concentration, shape and depth of the defect, and temperature and viscosity of the food product. A package with adequate seal strength could still have a channel microleak, making it susceptible to microbial contamination. Hence both testing schemes, seal integrity and pinhole detection, are important for the overall safety of packaged food products. In the food industry, Hazard Analysis and Critical Control Point (HACCP) generally includes container integrity as a critical control point (Harper *et al.*, 1995).

Package and seal integrity tests are classified as destructive and non-destructive (Table 22.5). Traditionally, destructive tests have been used to assess package and seal integrity. These tests simulate package performance in a 'real world' situation and provide information about conditions that may induce package failure. However, they are time consuming and expensive as the packages for testing cannot be reused. Hence, there is increased interest in developing non-destructive tests for evaluation of package integrity. Unlike destructive tests, non-destructive methods do not damage the package or its contents. They can be performed in a laboratory or in-line, as they do not interfere with the production line. Most non-destructive tests are based on a stimuli-response technique. The stimuli could be pressure, trace gas (helium, $CO_2$) or ultrasound and the corresponding measured response can be pressure differential, trace gas detection or sound attenuation.

There should be definite guidelines to accept or reject a package after performing these tests. Certain defects like corner dents, abrasions, and delamination may affect only the aesthetic value of the package and not pose any risk to food safety (Harper *et al.*, 1995). The size and location

© Woodhead Publishing Limited, 2012

**Table 22.5** Tests for package integrity evaluation

| Destructive tests | Non-destructive tests |
|---|---|
| 1. Biotest/Microbial challenge test | 1. Acoustic |
| 2. Bubble test | 1.1 Tap test |
| 3. Electrolytic test | 1.2 Ultrasound |
| 4. Dye penetration test | 2. Optical test |
| 5. Seal strength tests | 2.1 Infrared imaging |
| 5.1 Burst | 2.2 Machine vision imaging |
| 5.2 Tear/tensile | 2.3 X-ray |
| 6. Storage and distribution tests | 3. Pressure difference tests |
| 6.1 Vibration | 3.1 Pressure |
| 6.2 Compression | 3.2 Vacuum |
| 6.3 Impact | 4. Trace gas detection |
| | 5. Other tests |
| | 5.1 Thermography |
| | 5.2 Resistance to mechanical pressure |
| | 5.3 Capacitance |
| | 5.4 Eddy current probe |
| | 5.5 Spectrophotometer |

Source: Gnanasekharan & Floros, 1994; Harper *et al.*, 1995.

of the leak must be established to evaluate its consequences on the safety of food products. Leak size critical for maintaining package sterility has been found to be different based on the type of package and food, and the pressure imposed on the package during handling. Packages under vacuum, such as cans, may permit ingress of microorganisms due to the pressure differential. Similar size holes in flexible and semi-rigid packages, whose interior pressure is similar to ambient conditions, may not permit microbial ingress (Gnanasekharan & Floros, 1995). The reported values of minimum defect size for bacterial penetration vary between 0.2 to 80 μm (Harper *et al.*, 1995).

An automated, reliable and 100% in-line system for non-destructive testing of every package is of great interest to food companies. Non-destructive package testing serves as an immediate process control tool and ensures a stringent control on food safety. It eliminates waste of packages used for destructive testing, and reduces the overall cost of package integrity testing.

## 22.8 Migration and sorption

Consumer rejection of food products is most commonly associated with the presence of unacceptable flavor (Reineccius, 1991). In addition to protecting the food and providing adequate shelf life, the packaging material must also participate in the overall flavor management of the packaged food.

© Woodhead Publishing Limited, 2012

Packaging materials are not inert and may interact with the food. They can alter the flavor profile of foods by chemically reacting with food compounds to produce off-flavors, releasing components that produce off-flavors, and/or absorbing certain aroma compounds from the food (Gnanasekharan & Floros, 1997). Some of these migrating compounds may be toxic to humans posing a serious problem. Migration is the transfer of low molecular weight components from the packaging material to the contained food. Sorption or scalping is the absorption of flavor compounds by the packaging material. Both of these interactions, migration and sorption, are generally undesirable. Figure 22.3 represents various food-package interactions and their effect on the overall quality and safety of a food product.

Sorption can affect the food adversely in two ways: it can alter the organoleptic profile of the food; and it can lead to loss of package integrity. Absorption of food components by the packaging material could lead to loss of nutrients like vitamins, color changes, loss of aroma and/or taste, therefore resulting in overall quality deterioration and shelf life reduction. The permeation of absorbed components into the packaging material can sometimes alter the barrier and/or mechanical properties of the packaging material. Sorption of fats or organic acids by a polymer can cause delamination of the packaging structure. Most aroma compounds are volatile and can transfer from the headspace in to the package (Sajilata *et al.*, 2007). Hence, flavor sorption does not always require food to be in direct contact with the package.

Migration can affect the food adversely by imparting off-flavor and by adding toxic substances. Often, substances migrating from the packaging

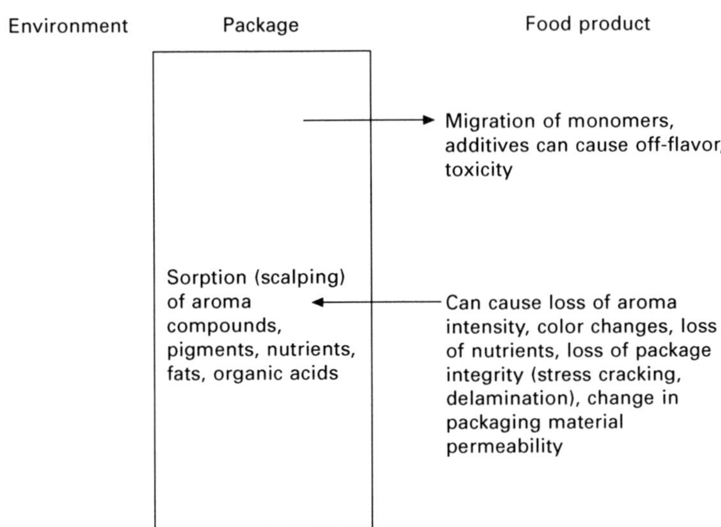


**Fig. 22.3** Food-package interactions (sources: Ackermann *et al.*, 1995; Nielsen & Jagerstad, 1994; Gnanasekharan & Floros, 1997).

© Woodhead Publishing Limited, 2012

material into food are odor-active and can impart off-flavors to the food. For example, migration of styrene monomers from polystyrene (PS) cups or films can impart a 'plastic-like' off-flavor to the food. Toxicological implications of component migration from packaging materials into foods are another serious problem. For example, there is still concern regarding the migration of vinyl chloride monomer into foods packaged in polyvinyl chloride (PVC). Thus migration can have both quality and safety implications.

Migration of potentially toxic substances has become a regulatory concern. Pre-market approval by the Food and Drug Administration (FDA) is currently required for food packaging materials that are not GRAS (generally recognized as safe). These components or food additives must be shown, through the food additive petition process, to be safe for their intended use. In the Federal Register of July 17, 1995, the FDA established a 'threshold of regulation' process wherein the likelihood of a substance posing a health risk depends on its dietary concentration and its toxic potency. European Union (EU) Directives on indirect food additives that could transfer to food serve as a template for member states to formulate their own legislation. Several Directives have been established related to migration of packaging materials into foods, and more are under revision (Arvanitoyannis & Bosnea, 2004; Knight & Creighton, 2004).

Even though migration and sorption are related to food-package interaction issues pertaining to all packaging materials, most of the research in this area is focused on plastics and composites. In general, most packaging materials are stable. The primary sources of migrants are the additives incorporated in the packaging materials to improve their functionality. Food package compatibility must be taken into consideration when selecting packaging materials and package types. Below we review migration and sorption issues related to each packaging material.

### 22.8.1 Paper and paperboard

Migrants from paper and paperboard packaging are mainly components transferred from solvents and adhesives used in paper manufacture and printing inks. Solvents used in wax coating paper and board were reported to impart a hydrocarbon-like off-odor in donuts packaged with wax-coated paperboard dividers (Reineccius, 1991). Migration of solvent residues from printing paper imparted a 'chemical taste' to packaged yogurt (Reineccius, 1991). Pentachlorophenol, used as a biocide in certain adhesives, causes a moldy off-flavor in food products. Moldy off-flavor due to migration of 2,4,6-trichloroanisole has been reported in cocoa powder packaged in paperboard containers (Kim-Kang, 1990).

Resins used in bonding of paper layers can also be a culprit for causing off-flavors in packaged foods. Quaker Oats Co. discovered a piney/spruce off-flavor in packaged ready-to-eat cereals that gave an uncharacteristic flavor note to the cereal. The inner glassine liner of the paperboard box (used as

© Woodhead Publishing Limited, 2012

secondary package) was found to be the source of this piney odor (Sajilata *et al.*, 2007). Aurela *et al.* (1999) reported exceptionally high migration of phthalates from paper and board packaging materials intended for packaging of sugar. These phthalates have toxicological significance and were found to originate from adhesives used in packaging joints.

Paper or paperboard manufacturing processes could also result in formation of potential migrants such as chlorophenols and nitrosamines. Chlorophenols can be formed during bleaching of wood pulp, and may migrate into foods packaged in containers made from such pulp (Kim-Kang, 1990; Kim-Kang *et al.*, 1990). Two nitrosamines (morpholine and *N*-nitrosomorpholine) were found in various paperboard packages. The source of these carcinogens was found to be a corrosion inhibitor used with boiler feedwater (Hotchkiss & Vecchio, 1983).

### 22.8.2 Glass

Even though glass is chemically inert, there have been incidences of off-flavor in food packaged in glass containers. The source of this off-flavor has been identified as the plastic or metal closures used for sealing glass bottles or the cleaning agents in reused glass containers. Titanium oxide, a whitener present in caps of carbonated beverage bottles, has been shown to catalyze reactions that may result in turpentine-like odor in the packaged product (Reineccius, 1991). A grease lubricant used in the washing lines of glass bottles has been shown to cause mineral oil-like off-flavors in mineral water packaged in these bottles (Ewender *et al.*, 1995).

### 22.8.3 Metal

The interior surfaces of most metal cans are coated with organic or metallic coatings to prevent corrosion. These coatings can migrate into the food and interact with the food components or cause taints. Bisphenol-A diglycidyl ether (BADGE) is used as an additive or starting agent in coatings for cans. BADGE and its chlorohydroxy compounds (BADGE.HCl, BADGE.2HCl and BADGE.HCl.$H_2O$) were detected in high amounts in canned fish (Hammarling *et al.*, 2000; Simoneau *et al.*, 1999). *In vitro* genotoxicity studies have shown that BADGE is mutagenic in many test systems. Mesityl oxide, a solvent used in side seam coating of three-piece cans, imparts a 'catty' taint to packaged pork products (Reineccius, 1991).

Commercial tin-lead alloys used for soldering the seams of tin cans contain about 98% lead. There is a possibility of lead migration into canned foods. Canned infant foods are therefore soldered with pure tin to avoid lead contamination and welding has been replaced by soldering for the same reason (Mannheim *et al.*, 1987). Lubricants are commonly used with tinplate to prevent abrasion and facilitate handling during can manufacture. Typical lubricants contain fatty acids and esters that are prone to oxidation, and

© Woodhead Publishing Limited, 2012

some oxidation products may migrate into packaged food (Gnanasekharan & Floros, 1997). Lubricants can impart stale, rancid, woody, and cardboard-like off-flavors in canned beer (Kim-Kang, 1990).

### 22.8.4 Plastic

Plastics commonly used in food packaging are polyethylene (PE), polypropylene (PP), polyethylene terephthalate (PET), polyvinyl chloride (PVC), aluminum sheet (Surlyn) and polystyrene (PS). Among these, PEs and PPs are mostly used for food contact applications, both in monolayer and composite structures, because of their good chemical resistance, barrier properties to oxygen and water, and thermosealability. Their polyolefinic nature makes them highly lipophilic, enabling retention of non-polar molecules such as most aroma compounds (Sajilata *et al.*, 2007). Potential migrants from plastic packaging are residual monomers, oligomers and additives such as plasticizers, stabilizers, slip additives, antioxidants, residual solvents, antifogging agents, pigments, etc. (Arvanitoyannis & Bosnea, 2004; Gnanasekharan & Floros, 1997). Table 22.6 lists some of the case studies pertaining to migration and sorption from plastic packaging, and their effects on food quality and safety.

## 22.9 Current and future trends

A continued trend in food packaging technology will be development of new high barrier materials, nanotechnology, improved convenience features related to production, distribution, sales, marketing, consumption and waste disposal, increased food safety, environmentally friendly and smart or intelligent packaging (Han, 2005). Use of high barrier materials will reduce the cost of material handling, transportation and distribution. Nanotechnology will be an important tool in improving barrier and structural/mechanical properties of packaging materials and development of sensing technologies (Brody *et al.*, 2008). Partial or complete substitution of petroleum-based materials with environmentally friendly materials, such as bio-based materials, will increase the use of recyclable and reusable materials (Han, 2005). Active and intelligent packaging technologies such as gas scavengers and regulators, freshness indicators, biosensors and traceability devices such as RFIDs will play an important role in improving food safety.

However, for these technologies to be adopted in packaging, they need to be inexpensive relative to the value of the product, reliable, accurate, reproducible in their range of operation, environmentally benign and food contact safe (Butler, 2001). Below we have listed some current and future trends in food packaging.

© Woodhead Publishing Limited, 2012

**Table 22.6** Migration and sorption from plastic packaging materials

| Plastic material | Food | Migrant | Effect on food | Reference |
|---|---|---|---|---|
| Polystyrene (PS) | Instant foods | Styrene dimer and trimer | Toxic and carcinogenic | Kawamura *et al.*, 1998 |
| Polystyrene (PS) | Coffee creamer | Styrene monomer | Styrene taint in the product | Piringer & Baner, 2000 |
| Polyamide/ionomer laminate | Cooked ham | From printing ink | Cat urine odor | Piringer & Baner, 2000 |
| High densitypolyethylene (HDPE) | Corn chips | 8-nonenal | 'Plastic' odor | Sanders *et al.*, 2005 |
| Polyvinyl chloride (PVC) | Olive oil, bread, cheese, meat | DEHA | Toxic | Petersen *et al.*, 1997 |
| Polyethylene (PE) | Milk | Antioxidant | Wooden off flavor | Nijssen, 1991 |
| Polyethylene terephthalate (PET) | Bottled water | Aldehydes formed due to thermal degradation of polymer during bottle production | Undesirable taste/flavor | Dabrowska *et al.*, 2003 |
| High density polyethylene (HDPE) | Rapeseed oil | Oil | Increased oxygen transmission rate | Johansson & Leufven, 1994 |
| Low density polyethylene/ paper/aluminum foil laminate | Fatty acid solution | Fatty acids | Delamination | Olafsson *et al.*, 1995 |
| Polyethylene terephthalate (PET) | Apulia table wines | Sorption of flavor compounds | Decline of aroma, taste and flavor | Mentana *et al.*, 2009 |
| High density polyethylene (HDPE) | Artificially flavored yogurt drink | Sorption of flavor compounds | Flavor balance disturbed | Linssen & Roozen, 1994 |

© Woodhead Publishing Limited, 2012

### 22.9.1 Active packaging

Active packaging implies an 'active' interaction of the package with its internal atmosphere or with the food. With this interaction, the packaging material changes the condition of the food it contains to provide extended shelf life, improved safety or better sensory properties, while maintaining other food quality parameters as high as possible (Ahvenainen, 2003). Active packaging is the application of specific packaging properties to specific situations. For instance, addition of ethylene releasing sachets in high moisture bakery products to suppress mold growth when low oxygen barrier packaging is used (Rooney, 1995); or use of moisture regulator packs for packaging Portuguese cheese, saloio (Pantaleao *et al.*, 2007). There are opportunities for reducing packaging cost by complementing active packaging with cheaper traditional passive packaging. Also, consumers may prefer foods with active packaging systems because they may contain fewer additives, and have better quality and improved safety (Han & Floros, 2007).

Most promising active packaging technologies are: antimicrobial packaging, oxygen scavengers, moisture regulators for fresh produce and dried foods, and ethylene absorbing systems for fruits and vegetables.

*Antimicrobial packaging*

Antimicrobials can be used to control the growth of undesirable microorganisms in food. They can be added to or immobilized (by cross-linking) in the packaging material. Antimicrobial activity can be achieved by release of antimicrobial from the packaging material on the food surface or package headspace; or inhibition of microbial growth by the immobilized antimicrobial at food-package contact surface. Active substances that can be used as antimicrobials include ethanol and other alcohol, organic acids such as sorbate, benzoate and propionate, bacteriocins such as nisin and pediocin, enzymes such as lysozyme, metals, plant/spice extracts and fungicides. Antimicrobial packaging systems have characteristic antimicrobial activity and spectrum to different microorganisms. The following factors must be considered for designing the most efficient system:

- antimicrobial agent,
- target microorganisms,
- internal package atmosphere,
- packaging material and
- food (Han & Floros, 2007).

*Oxygen scavenging systems*

An oxygen scavenger is a substance that reacts with oxygen chemically or enzymatically, thus protecting the packaged food against oxidative deterioration and quality changes due to oxygen (Vermeiren *et al.*, 2003). Most commercially available oxygen scavenging systems are based on the oxidation of iron powder or enzymes such as glucose oxidase, by oxygen present in the package's headspace. These reactions absorb most of the available oxygen in the package, minimize

© Woodhead Publishing Limited, 2012

the interaction of oxygen with food, and thus prevent deterioration. Most often, oxygen scavengers are packed in a sachet highly permeable to oxygen and placed inside the package. These type of sachets can be used for fresh pasta, meat products, baked goods such as bread, pizza crust, cookies and sponge cake, coffee, nuts, and snack foods like potato chips. Some safety concerns regarding the use of sachet-type oxygen scavenging systems in packaged foods are accidental ingestion of the sachet by consumers and potential spill/leak of the sachet contents into food (Lopez-Rubio *et al.*, 2004). Careful research and development must be undertaken in selecting the appropriate sachet material and its seal integrity before commercial application, and appropriate warnings must be declared on the package.

*Moisture regulators*

Control of moisture in foods is important to prevent moisture loss from fresh produce such as meat, poultry, fish, fruits and vegetables, moisture absorption by dried and semi-dried foods, and condensation of moisture inside packages. The undesirable loss and absorption of moisture by foods can be prevented by using packaging materials with high water vapor barrier. Excess water inside a food package usually occurs due to respiration of fresh produce, temperature fluctuations in high equilibrium relative humidity food products, or drip of tissue fluids from cut meat and poultry (Rooney, 1995). This build-up of excess moisture in the package can promote bacterial and mold growth leading to quality loss and shelf life reduction. Accumulation of excess water in the food package can be controlled by using desiccants such as silicates, clay, molecular sieves or humectant salts such as sodium chloride, magnesium chloride and calcium sulfate. A pad or sachet of the desiccant is wrapped around the food or placed in the package.

*Ethylene scavenging system*

Ethylene gas is a plant hormone that accelerates respiration in fresh fruits and vegetables leading to maturity, senescence and softening of tissues. Ethylene accumulation can also cause yellowing of green vegetables. Ethylene scavengers can be used to absorb ethylene from package headspace to prolong the shelf life and maintain an acceptable visual quality of respiring fruits and vegetables. The most commonly used ethylene scavenging system consists of potassium permanganate embedded in silica. The silica absorbs ethylene and potassium permanganate oxidizes it to acetate and ethanol. Silica can be filled in a sachet and placed in the package or incorporated in the packaging material. Other ethylene scavengers include clay, zeolite, ceramic powder, and mineral oxide powder.

### 22.9.2 Intelligent packaging

Intelligent packaging implies that the package has an indicator that monitors the condition of packaged food and gives information about its quality during

© Woodhead Publishing Limited, 2012

transportation and storage (Ahvenainen, 2003). It may sense the condition of packaged foods specifically or respond to the changes in food condition (Rodrigues & Han, 2003). There are two basic types of intelligent packaging devices:

1. data carriers such as barcode labels and radio frequency identification (RFID) tags, which record and transmit data and
2. package indicators such as time-temperature, gas composition and biosensors, which monitor package headspace/external environment and, whenever appropriate, issue warnings (Yam *et al.*, 2005).

*Data carriers*

Barcodes printed on packaging material or labels are the most inexpensive and popular form of data carriers. Depending on its storage capacity, it can contain information regarding manufacturer, item number, packed date, batch number and package weight. An advanced form of data carrier is the RFID tag, which uses radio frequencies to read information on a small device known as a tag. It contains information for automatic product identification and traceability. It has several advantages over a barcode such as higher storage capacity, can be used for real-time updates as tagged items move through the supply chain and multiple tags can be read simultaneously and rapidly.

*Package indicators*

Temperature abuse is common during storage, transportation and handling of food products (Ozdemir & Floros, 2004). It causes food wastage, quality and nutrition loss and can also lead to food poisoning. A time–temperature indicator (TTI) is a simple device attached to individual consumer packages or shipping containers and gives information about the temperature history at selected control points. They usually express an irreversible mechanical, chemical, electrochemical, enzymatic, or microbiological change as a visible response in the form of a mechanical deformation or color change (Taoukis & Labuza, 2003). They can also be used as a 'freshness indicator' for estimating the remaining shelf life of perishable products.

*Gas indicators*

The package label or prints on packaging films can monitor changes in gas composition inside a package, thereby providing means for assessing quality and safety of food products. Oxygen indicators are most common for food applications as large number of foods are prone to oxidative deterioration. Most oxygen indicators are designed to show color changes due to leaky or tampered packages (Yam *et al.*, 2005). For example, oxygen indicators can be used to detect improper sealing and quality deterioration of modified atmosphere packages containing foods such as pizza or cooked chicken patties (Smiddy *et al.*, 2002; Smolander *et al.*, 1997). Indicators

© Woodhead Publishing Limited, 2012

for monitoring carbon dioxide, water vapor, ethanol and other gases will also be useful.

*Biosensors*
Biosensors are compact analytical devices that detect, record and transmit information pertaining to biochemical reactions. Their incorporation in food packages will help monitor food contamination and alert the producer, retailer or consumer. In the case of contamination, the device will display a clear visual signal, such as development or change of color. Presently, biosensors for intelligent packaging of food products are not commercially available.

### 22.9.3 Nanotechnology

Nanotechnology can be used to produce foods with desired functional properties. By incorporating appropriate nanomaterials into food packaging, it will be possible to produce packages with improved mechanical, barrier, and thermal properties. Edible nanocomposite coatings have applications for fresh produce, bakery products and confectionery, where they can protect the food from moisture, gases, odors, and off-flavors (Sozer & Kokini, 2009). Polymeric packaging materials containing silicate or clay nanoparticles have improved mechanical and barrier properties (Neethirajan & Jayas, 2010). Synthetic or natural biopolymers (polysaccharides, proteins, lipids) of nanoscale size can be used for encapsulation and controlled delivery of bioactives such as vitamins, prebiotics, probiotics, and other nutraceuticals. They can also be used in active or intelligent packaging for developing nanosensors for detecting changes in package atmosphere, microorganisms, and other contaminants.

The impact of this highly promising technology on human health, environment, and general public perception needs to be taken into account. Currently, the potential risks of nanomaterials to human health and environment are unknown (Dowling, 2004). The use of nanoparticles for food applications must be employed only after rigorous safety testing and special attention should be given to consumer attitudes towards food nanotechnology.

### 22.9.4 Traceability

The increasing consumer demand for healthy safe foods has set exacting requirements for a well-structured traceability system. Such a system can make consumers feel safer by providing detailed information about where a product comes from, what its components are, and its processing and handling history (consumers can get this information by entering the tracking code on the company's website). An efficient tracing system will also be beneficial to the producers as it makes it possible to promptly trace all the packages in the case of a product recall and it also provides information to help identify the cause of the problem.

© Woodhead Publishing Limited, 2012

In practice, different technical solutions can be used in a traceability system. RFID is a promising traceability technology. However, the cost of RFIDs is a major hurdle, especially in the food sector where the value of most products is normally very low, and so the solutions adopted for the tracking system must also be very cheap. Nevertheless, with some big retail chain such as Wal-Mart pursuing traceability systems to get real-time visibility in their supply chains, RFIDs might become a common practice in the near future.

### 22.9.5 Recycling packaging materials

Recycling involves recovery of packaging materials from the waste stream and reprocessing them into new products. A typical recycling program entails collection, sorting and processing, manufacturing, and sale of recycled materials and products (Marsh & Bugusu, 2007). Increased environmental concerns have created a need for recycling packaging materials. Recycling reduces the volume of packaging materials entering the waste stream and land filling. It also saves materials and energy as long as the energy to ship and reprocess these materials does not exceed that of processing virgin materials (Marsh, 1991).

Almost all packaging materials (glass, metal, thermoplastic, paper and paperboard) are recyclable. They can be recycled by mechanical (glass, metal, paper and paperboard, and some plastics), chemical (plastics), or biological (renewable and biodegradable polymers) means. There is concern regarding use of recycled materials for food contact use due to potential contamination that might jeopardize food quality and safety. Generally, recycled glass and metals are considered safe for food contact use, since the heat used to melt and reform these materials is sufficient to kill all microorganisms and pyrolyze organic contaminants (Marsh & Bugusu, 2007). Glass can be crushed, melted and reformed an infinite number of times without any loss of structure and properties (Girling, 2003).

Most types of paper and paperboard are recycled; however, recycled paper is generally not suitable for use in food contact applications as it might not be completely free of contaminants. The outer linerboard used for packaging frozen foods and dried foods such as cereals are examples of recycled paper as secondary packages. For plastics, the recycling process utilizes sufficient heat to destroy microorganisms, but it is not sufficient to pyrolyze all organic contaminants. There is also a possibility that the plastic containers were used for some other purpose (cleaning agents, motor oil, etc.) before entering the waste stream. Hence, post consumer recycled plastics are not generally used in food contact applications. Some recycled plastics such as monolayer PET and HDPE bottles have been approved for direct or limited food contact applications by the US FDA and some countries in Europe (Franz & Welle, 2003).

© Woodhead Publishing Limited, 2012

## 22.10 Sources of further information and advice

### 22.10.1 Publications

AHVENAINEN R (2003), 'Active and intelligent packaging: an introduction', in R Ahvenainen (ed.), *Novel food packaging techniques*, Woodhead Publishing Limited/CRC Press, Cambridge, pp. 5–19.

AMERICAN SOCIETY FOR TESTING MATERIALS (ASTM) (2003), *Selected ASTM standards on packaging*, Institute of Packaging.

AMERICAN SOCIETY FOR TESTING MATERIALS (2002), *Consumer and healthcare packaging standards*, ASTM International.

BROWN WE (1992), *Plastics in food packaging: properties, design, and fabrication*, Marcel Dekker, New York.

COLES R & KIRWAN M (2011), *Food and beverage packaging technology*, Blackwell Publishing, Oxford.

EMBUSCADO ME & HUBER KC (2009), *Edible Films and Coatings for Food Applications*, Springer, New York.

GRAY JI HARTE BR & MILTZ J (1987), *Food product-package compatibility: proceedings of a seminar held at the School of Packaging, Michigan State University, East Lansing,* July 1986, Technomic Publ. Co., Lancaster, PA.

HAN JH (2005), *Innovations in food packaging*, Elsevier Academic, Maryland Heights, MO.

KATAN LL (1996), *Migration from food contact materials*, Blackie Academic & Professional, London.

KERRY J & BUTLER P (2008), *Smart packaging technologies for fast moving consumer goods*, John Wiley & Sons Ltd., Chichester.

LEE DS, YAM KL & PIERGIOVANNI L (2008), *Food packaging science and technology*, CRC Press, Boca Raton, FL.

PIRINGER OG & BANER AL (2000), *Plastic packaging materials for food: barrier function, mass transport, quality assurance, and legislation*, Wiley-VCH, Weinheim.

RJIK R & VERAART R (2010), *Global legislation for food packaging materials*, Wiley-VCH, Weinheim.

ROBERTSON GL (2009), *Food packaging and shelf life: a practical guide*, CRC Press, Boca Raton, FL.

ROONEY ML (1995), *Active food packaging*, Chapman & Hall, London.

### 22.10.2 Websites

Department of Food Science, Rutgers School of Environmental and Biological Sciences: http://foodsci.rutgers.edu/aboutUs/faculty.html

Department of Agricultural and Biological Engineering, University of Florida: http://www.abe.ufl.edu/

Department of Food Science and Technology, The Ohio State University: http://www-fst.ag.ohio-state.edu/

© Woodhead Publishing Limited, 2012

Department of Food, Nutrition, and Packaging Sciences, Clemson University: http://www.clemson.edu/cafls/departments/fnps/

Department of Food Science, The Pennsylvania State University: http://foodscience.psu.edu/

European Food Safety Authority: http://www.efsa.europa.eu/

Food ingredients and packaging, US Food and Drug Administration: http://www.fda.gov/Food/FoodIngredientsPackaging/default.htm

Food Packaging Division, Institute of Food Technologists: http://www.ift.org/divisions/food_pack/

Institute of Packaging Professionals: http://www.iopp.org/i4a/pages/index.cfm?pageid=1

School of Packaging, The Michigan State University: http://packaging.msu.edu/

Society of Plastic Engineers: http://www.4spe.org/

## 22.11 References

ACKERMANN P, JAGERSTAD M & OHLSSON T (1995), *Food and packaging materials – chemical interactions*, The Royal Society of Chemistry, Cambridge.

AHVENAINEN R (2003), 'Active and intelligent packaging: an introduction', in R Ahvenainen (ed.), *Novel Food Packaging Techniques*, Woodhead Publishing Limited/CRC Press, Cambridge, pp. 5–19.

AMMANN R (2001), 'Aseptic filling of HDPE and PET bottles', *Fruit Processing*, vol. 11, pp. 449–51.

ARVANITOYANNIS I & BOSNEA L (2004), 'Migration of substances from food packaging materials to foods', *Critical Reviews in Food Science and Nutrition*, vol. 44, no. 2, pp. 63–76.

AURELA B, KULMALA H & SODERHJELM L (1999), 'Phthalates in paper and board packaging and their migration into Tenax and sugar', *Food Additives and Contaminants*, vol. 16, no. 12, pp. 571–7.

BERENDS CL (1996), 'Stability of aseptically packaged food as a function of oxidation initiated by a polymer contact surface', PhD thesis, Virginia Polytechnic Institute and State University.

BESWICK RD & DUNN DJ (2002), *Plastics in Packaging*, Rapra Technology Limited, Shropshire.

BILES J, MCNEAL T, BEGLEY T & HOLLIFIELD H (1997), 'Determination of bisphenol-A in reusable polycarbonate food-contact plastics and migration to food-simulating liquids', *Journal of Agricultural and Food Chemistry*, vol. 45, no. 9, pp. 3541–4.

BILES J, WHITE K & MCNEAL T (1999), 'Determination of the diglycidyl ether of bisphenol A and its derivatives in canned foods', *Journal of Agricultural and Food Chemistry*, vol. 47, no. 5, pp. 1965–9.

BINTSIS T, LITOPOULOU-TZANETAKI E & ROBINSON R (2000), 'Existing and potential applications of ultraviolet light in the food industry – a critical review', *Journal of the Science of Food and Agriculture*, vol. 80, no. 6, pp. 637–45.

BOHRER TH & BROWN RK (2001), 'Packaging techniques for microwaveable foods', in AK Datta & RC Anantheswaran (eds), *Handbook of microwave technology for food applications*, Marcel Dekker, New York, pp. 397–467.

BRODY AL, BUGUSU B, HAN JH, SAND CK & MCHUGH TH (2008), 'Innovative food packaging solutions' *Journal of Food Science*, vol. 73, no. 8, pp. R107–R116.

© Woodhead Publishing Limited, 2012

BUCHALLA R, SCHUTTLER C & BOGL K (1993), 'Effects of ionizing radiation on plastic food packaging materials – a review. Part 1', *Journal of Food Protection*, vol. 56, no. 11, pp. 991–7.

BUTLER P (2001), 'Smart packaging goes back to nature', *Materials World*, vol. 9, no. 3, pp. 11–13.

BUTZ P & TAUSCHER B (2002), 'Emerging technologies: chemical aspects', *Food Research International*, vol. 35, no. 2–3, pp. 279–84.

CANER C, HERNANDEZ R & HARTE B (2004), 'High-pressure processing effects on the mechanical, barrier and mass transfer properties of food packaging flexible structures: a critical review', *Packaging Technology and Science*, vol. 17, no. 1, pp. 23–9.

CARWILE J, LUU H, BASSETT L, DRISCOLL D, YUAN C & CHANG J (2009), 'Polycarbonate bottle use and urinary bisphenol A concentrations', *Environmental Health Perspectives*, vol. 117, no. 9, pp. 1368–72.

CHUNG D, PAPADAKIS SE & YAM K (2006), 'Thermal processing of packaged foods', in YH Hui (ed.), *Handbook of food science, technology and engineering*, vol. 3, CRC Press, Boca Raton, FL.

DABROWSKA A, BORCZ A & NAWROCKI J (2003), 'Aldehyde contamination of mineral water stored in PET bottles', *Food Additives & Contaminants*, vol. 20, no. 12, pp. 1170–7.

DIEHL JF (1995), *Safety of irradiated foods*, Marcel Dekker, New York.

DOWLING AP (2004), 'Development of nanotechnologies', *Materials Today*, vol. 7, no. 12, pp. 30–5.

DRISCOLL RH & PATERSON JL (1999), 'Packaging and food preservation', in MS Rahman (ed.), *Handbook of food preservation*, Marcel Dekker, New York, pp. 687–734.

DRISCOLL RH & RAHMAN MS (2007), 'Types of packaging materials used for foods', in MS Rahman (ed.), *Handbook of food preservation*, 2nd edn, CRC Press, Boca Raton, FL.

EWENDER J, LINDNER–STEINERT A, RUTER M & PRINGER O (1995), 'Sensory problems caused by food and packaging interactions: overview and treatment of recent case studies', in P Ackermann, M Jagerstad & T Ohlsson (eds), *Food and packaging materials-chemical interactions*, The Royal Society of Chemistry, Cambridge, pp. 33–44.

FLOROS JD (1993), 'Aseptic packaging technology', in JV Chambers & PE Nelson (eds), *Principles of aseptic processing and packaging*, 2nd edn, The Food Processors Institute, Washington, DC, pp. 115–48.

FLOROS JD & MATSOS KI (2003), *Packaging and Canning, Modern*, Scribner's Sons, New York.

FLOROS JD & MATSOS KI (2005), 'Introduction to modified atmosphere packaging', in JH Han (ed.), *Innovations in food packaging*, Elsevier Academic Press, London, pp. 159–72.

FLOROS JD, OZDEMIR M & NELSON PE (1998), 'Trends in aseptic bulk storage and packaging', *Food Cosmetics and Drug Packaging*, vol. 21, pp. 236–9.

FRANZ R & WELLE F (2003), 'Recycling packaging materials', in R Ahvenainen (ed.), *Novel food packaging techniques*, Woodhead Publishing, Cambridge, pp. 497–518.

GIRLING PJ (2003), 'Packaging of food in glass containers', in R Coles, D McDowell & MJ Kirwan (eds), *Food packaging technology*, Blackwell Publishing, Oxford, pp. 152–73.

GNANASEKHARAN V & FLOROS JD (1994), 'Package integrity evaluation: criteria for selecting a method', *Packaging Technology & Engineering*, vol. 3, no. 6, pp. 44–8.

GNANASEKHARAN V & FLOROS JD (1995), 'A theoretical perspective on the minimum leak size for package integrity evaluation', in BA Blackistone & CL Harper (eds), *Plastic package integrity testing – Assuring seal quality*, Institute of Packaging Professionals, Herndon, VA, pp. 55–65.

GNANASEKHARAN V & FLOROS J (1997), 'Migration and sorption phenomena in packaged foods', *Critical Reviews in Food Science and Nutrition*, vol. 37, no. 6, pp. 519–59.

© Woodhead Publishing Limited, 2012

GONTARD N & GUILBERT S (1994), 'Biopackaging: technology and properties of edible and/or biodegradable material of agricultural origin', in M Mathlouthi (ed.), *Food packaging and preservation*, Chapman & Hall, New York, pp. 159–81.

GÖTZ J & WEISSER H (2002), 'Permeation of aroma compounds through plastic films under high pressure: *in-situ* measuring method', *Innovative Food Science & Emerging Technologies*, vol. 3, no. 1, pp. 25–31.

GOULAS A, RIGANAKOS K, BADEKA A & KONTOMINAS M (2002), 'Effect of ionizing radiation on the physicochemical and mechanical properties of commercial monolayer flexible plastics packaging materials', *Food Additives & Contaminants*, vol. 19, no. 12, pp. 1190–9.

HAMMARLING L, GUSTAVSSON H, SVENSSON K & OSKARSSON A (2000), 'Migration of bisphenol-A diglycidyl ether (BADGE) and its reaction products in canned foods', *Food Additives and Contaminants*, vol. 17, no. 11, pp. 937–43.

HAN JH (2005), 'New technologies in food packaging: overview', in JH Han (ed.), *Innovations in food packaging*, Elsevier Academic, London, pp. 3–10.

HAN JH (2007), 'Packaging for nonthermally processed foods', in JH Han (ed.), *Processing for nonthermal processing of food*, 1st edn, Blackwell Publishing, Oxford, pp. 3–16.

HAN JH & FLOROS JD (2007), 'Active packaging: a non-thermal process', in G Tewari & VK Juneja (eds), *Advances in thermal and non-thermal food preservation*, Blackwell Publishing, Ames, IA.

HANNAY F (2002), *Rigid Plastics Packaging – Materials, Processes and Applications*, Rapra review reports, Rapra Technology Limited, Shropshire.

HARPER CL, BLAKISTONE BA, LITCHFIELD J B & MORRIS SA (1995), 'Developments in food packaging integrity testing', *Trends in Food Science & Technology*, vol. 6, no. 10, pp. 336–40.

HAUGAARD VK & MORTENSEN G (2003), 'Biobased food packaging', in B Mattsson & U Sonesson (eds), *Environmentally friendly food processing*, Woodhead Publishing, Cambridge, pp. 180–204.

HO K & POMETTO A (1999), 'Effects of electron-beam irradiation and ultraviolet light (365 nm) on polylactic acid plastic films', *Journal of Environmental Polymer Degradation*, vol. 7, no. 2, pp. 93–100.

HOLDSWORTH D & SIMPSON R (2007), *Thermal processing of packaged foods*, Springer Science+Business Media, New York.

HOTCHKISS J & VECCHIO A (1983), 'Analysis of direct contact paper and paperboard food packaging for n-nitrosomorpholine and morpholine', *Journal of Food Science*, vol. 48, no. 1, pp. 240–2.

JOHANSSON F & LEUFVEN A (1994), 'Influence of sorbed vegetable oil and relative humidity on the oxygen transmission rate through various polymer packaging films', *Packaging Technology and Science*, vol. 7, no. 6, pp. 275–81.

KAWAMURA Y, KAWAMURA M, TAKEDA Y & YAMADA T (1998), 'Determination of styrene dimers and trimers in food contact polystyrene', *Journal of Food Hygienic Society of Japan*, vol. 39, no. 3, pp. 199–205.

KESTER J & FENNEMA O (1986), 'Edible films and coatings – a review', *Food Technology*, vol. 40, no. 12, pp. 47–59.

KILLORAN J (1983), 'Packaging irradiated foods', in JE Peterson (ed.), *Preservation of food by ionizing radiation*, vol. 2, CRC Press, Boca Raton, FL, pp. 317–26.

KIM-KANG H (1990), 'Volatiles in packaging materials', *CRC Criticial Reviews in Food Science and Nutrition*, vol. 29, no. 4, pp. 255–71.

KIM-KANG H & GILBERT SG (1991), 'Permeation characteristics of and extractables from gamma-irradiated and non-irradiated plastic laminates for a unit dosage injection device', *Packaging Technology and Science*, vol. 4, no. 1, pp. 35–48.

KIM-KANG H, GILBERT S, MALICK A & JOHNSON J (1990), 'Methods for predicting migration to packaged pharmaceuticals', *Journal of Pharmaceutical Sciences*, vol. 79, no. 2, pp. 120–3.

© Woodhead Publishing Limited, 2012

KIRWAN M, MCDOWELL D & COLES R (2003), *Food packaging technology*, Blackwell Publishing, Oxford.

KNIGHT DJ & CREIGHTON LA (2004), *Regulation of food packaging in Europe and the USA*, Rapra review reports, Rapra Technology Limited, Shropshire.

KOUTCHMA T (2009), 'Advances in ultraviolet light technology for non-thermal processing of liquid foods', *Food and Bioprocess Technology*, vol. 2, no. 2, pp. 138–55.

KROCHTA JM (2007), 'Food packaging', in DR Heldman & DB Lund (eds), *Handbook of food engineering*, 2nd edn, CRC Press, Boca Raton, FL, pp. 847–928.

KÜBEL J, LUDWIG H, MARX H & TAUSCHER B (1996), 'Diffusion of aroma compounds into packaging Films under High Pressure', *Packaging Technology and Science*, vol. 9, no. 3, pp. 143–52.

KUSE D (1982), 'UV-C sterilization of packaging materials in the dairy industry', *D Milchwirtschaft*, vol. 33, pp. 1134–7.

KUTTY V, BRADDOCK R & SADLER G (1994), 'Oxidation of d-limonene in presence of low density polyethylene', *Journal of Food Science*, vol. 59, no. 2, pp. 402–5.

LAMBERT Y, DEMAZEAU G, LARGETEAU A, BOUVIER J, LABORDE-CROUBIT S & CABANNES M (2000), 'Packaging for high-pressure treatments in the food industry', *Packaging Technology and Science*, vol. 13, no. 2, pp. 63–71.

LANGE J & WYSER Y (2003), 'Recent innovations in barrier technologies for plastic packaging – a review', *Packaging Technology and Science*, vol. 16, no. 4, pp. 149–58.

LINSSEN JPH & ROOZEN JP (1994), 'Food flavor and packaging interactions', in M Mathlouthi (ed.), *Food packaging and preservation*, Blackie Academic and Professional, Glasgow.

LOPEZ-RUBIO A, ALMENAR E, HERNANDEZ-MUNOZ P, LAGARON J, CATALA R & GAVARA R (2004), 'Overview of active polymer-based packaging technologies for food applications', *Food Reviews International*, vol. 20, no. 4, pp. 357–87.

MANNHEIM C MILTZ J & LETZTER A (1987), 'Interaction between polyethylene laminated cartons and aseptically packed citrus juices', *Journal of Food Science*, vol. 52, no. 3, pp. 737–40.

MARSH KS (1991), 'Effective management of food packaging: from production to disposal', *Food Technology*, vol. 45, pp. 225–34.

MARSH K & BUGUSU B (2007), 'Food packaging – roles, materials, and environmental issues', *Journal of Food Science*, vol. 72, no. 3, pp. R39–55.

MASUDA M, SAITO Y, IWANAMI T & HIRAI Y (1992), 'Effect of hydrostatic pressure on packaging materials for food', in C Balny, R Hayashi, K Heremans & P Masson (eds), *High pressure and biotechnology*, Colloque INSERM/John Libbery, London, pp. 545–7.

MAUER LJ & OZEN BF (2004), 'Food packaging', in JS Smith & YH Hui (eds), *Food processing principles and applications*, Blackwell Publishing, Oxford, pp. 101–32.

MENTANA A, PATI S, LA NOTTE E & DEL NOBILE M (2009), 'Chemical changes in Apulia table wines as affected by plastic packages', *Food Science & Technology*, vol. 42, no. 8, pp. 1360–6.

MOREHOUSE KM & KOMOLPRASERT V (2004), 'Irradiation of food and packaging: an overview', in KM Morehouse & V Komolprasert (eds), *Irradiation of food and packaging: recent developments* American Chemical Society, Boston, MA, pp. 1–11.

MOURA E, ORTIZ A, WIEBECK H, PAULA A, SILVA A & SILVA L (2004), 'Effects of gamma radiation on commercial food packaging films – study of changes in UV/VIS spectra', *Radiation Physics and Chemistry*, vol. 71, no. 1–2, pp. 201–4.

MULLAN M & MCDOWELL D (2003), 'Modified atmosphere packaging', in R Coles, D McDowell & MJ Kirwan (eds), *Food packaging technology*, Blackwell Publishing, Oxford, pp. 303–31.

NEETHIRAJAN S & JAYAS D (2010), 'Nanotechnology for the food and bioprocessing industries', *Food and Bioprocess Technology*, vol. 4, no. 1, pp. 39–47.

NICOLAS R (1995), 'Aseptic filling of UHT dairy products in HDPE bottles', *Food Technology Europe*, vol. 2, pp. 52–8.

© Woodhead Publishing Limited, 2012

NIELSEN T & JAGERSTAD M (1994), 'Flavor scalping by food packaging', *Trends in Food Science & Technology*, vol. 5, no. 11, pp. 353–6.

NIJSSEN B (1991), 'Off-flavors', in H Maarse (ed.), *Volatile compounds in foods and beverages*, Marcel Dekker, New York.

OCHIAI S & NAKAGAWA Y (1992), 'Packaging for high pressure food processing', in C Balny, R Hayashi, K Heremans & P Masson (eds), *High pressure and biotechnology*, Colloque INSERM/John Libbery, London, pp. 515–19.

OLAFSSON G, HILDINGSSON I & BERGENSTAHL B (1995), 'Transport of oleic and acetic acids from emulsions into low-density polyethylene – effects on adhesion with aluminum foil in laminated packaging', *Journal of Food Science*, vol. 60, no. 2, pp. 420–5.

OUYANG M, KLEMCHUK PP & KOBERSTEIN JT (2000), 'Exploring the effectiveness of SiOx coatings in protecting polymers against photo-oxidation', *Polymer Degradation and Stability*, vol. 70, no. 2, pp. 217–28.

OZDEMIR M & FLOROS J (2004), 'Active food packaging technologies', *Critical Reviews in Food Science and Nutrition*, vol. 44, no. 3, pp. 185–93.

OZEN B & FLOROS J (2001), 'Effects of emerging food processing techniques on the packaging materials', *Trends in Food Science & Technology*, vol. 12, no. 2, pp. 60–7.

OZEN B, MAUER L & FLOROS J (2003), 'Effects of ozone exposure on the structural, mechanical and barrier properties of select plastic packaging films', *Packaging Technology and Science*, vol. 15, no. 6, pp. 301–11.

PAIK JS, DHANASEKHARAN M & KELLY MJ (1998), 'Antimicrobial activity of UV-irradiated nylon film for packaging applications', *Packaging Technology and Science*, vol. 11, no. 4, pp. 179–87.

PAINE FA & PAINE HY (1983), *A handbook of food packaging*, Chapman & Hill, London.

PANTALEAO I, PINTADO M & POCAS M (2007), 'Evaluation of two packaging systems for regional cheese', *Food Chemistry*, vol. 102, no. 2, pp. 481–7.

PETERSEN J, LILLEMARK L & LUND L (1997), 'Migration from PVC cling films compared with their field of application', *Food Additives and Contaminants*, vol. 14, no. 4, pp. 345–53.

PETERSEN K, VÆGGEMOSE NIELSEN P, BERTELSEN G, LAWTHER M, OLSEN MB, NILSSON NH & MORTENSEN G (1999), 'Potential of biobased materials for food packaging', *Trends in Food Science & Technology*, vol. 10, no. 2, pp. 52–68.

PIRINGER OG & BANER AL (2000), *Plastic packaging materials for food*, Wiley-VCH, Weinheim.

RAHMAN MS (1999), 'Food preservation: overview', in MS Rahman (ed.), *Handbook of food preservation*, CRC Press, Boca Raton, FL.

RAZUMOVSKII, SD & ZAIKOV, GY (1982), 'Effect of ozone on saturated polymers: a review', *Polymer Science USSR*, vol. 24, no. 10, pp. 2305–25.

REINECCIUS G (1991), 'Off-flavors in foods', *Critical Reviews in Food Science and Nutrition*, vol. 29, no. 6, pp. 381–402.

ROBERTSON GL (2006), *Food packaging: principles and practice*, 2nd edn, CRC Press, Boca Raton, FL.

ROBERTSON GL (2010), *Food packaging and shelf-life: a practical guide*, CRC Press, Boca Raton, FL.

RODRIGUES ET & HAN JH (2003), *Intelligent packaging*, Taylor and Francis Informa Ltd, London.

ROONEY ML (1995), *Active food packaging*, Chapman & Hall, London.

ROSATO DV, ROSATO DV, ROSATO MG & SCHOTT NR (2001), *Plastics engineering, manufacturing and data handbook*, vol. 2, Kluwer Academic Publishers, Norwell, MA.

RUDNIK E (2008), *Compostable polymer materials*, Elsevier, London.

SAJILATA M, SAVITHA K, SINGHAL R & KANETKAR V (2007), 'Scalping of flavors in packaged foods', *Comprehensive Reviews in Food Science and Food Safety*, vol. 6, no. 1, pp. 17–35.

SANDERS R, ZYZAK D, MORSCH T, ZIMMERMAN S, SEARLES P & STROTHERS M (2005), 'Identification

© Woodhead Publishing Limited, 2012

of 8-nonenal as an important contributor to "plastic" off-odor in polyethylene packaging', *Journal of Agricultural and Food Chemistry*, vol. 53, no. 5, pp. 1713–16.

SCHAUWECKER A, BALASUBRAMANIAM V, SADLER G, PASCALL M & ADHIKARI C (2003), 'Influence of high-pressure processing on selected polymeric materials and on the migration of a pressure-transmitting fluid', *Packaging Technology and Science*, vol. 15, no. 5, pp. 255–62.

SHANBHAG P & SIRKAR K (1998), 'Ozone and oxygen permeation behavior of silicon capillary membranes employed in membrane ozonators', *Journal of Applied Polymer Science*, vol. 69, no. 7, pp. 1263–73.

SIMONEAU C, THEOBALD A, HANNAERT P, RONCARI P, RONCARI A & RUDOLPH T (1999), 'Monitoring of bisphenol-A-diglycidyl-ether (BADGE) in canned fish in oil', *Food Additives and Contaminants*, vol. 16, no. 5, pp. 189–95.

SIRACUSA V, ROCCULI P, ROMANI S & ROSA MD (2008), 'Biodegradable polymers for food packaging: a review', *Trends in Food Science and Technology*, vol. 19, no. 12, pp. 634–43.

SMIDDY M, PAPKOVSKAIA N, PAPKOVSKY D & KERRY J (2002), 'Use of oxygen sensors for the non-destructive measurement of the oxygen content in modified atmosphere and vacuum packs of cooked chicken patties: impact of oxygen content on lipid oxidation', *Food Research International*, vol. 35, no. 6, pp. 577–84.

SMOLANDER, M, HURME, E & AHVENAINEN, R (1997), 'Leak indicators for modified-atmosphere packages', *Trends in Food Science and Technology*, vol. 8, no. 4, pp. 101–6.

SOZER N & KOKINI J (2009), 'Nanotechnology and its applications in the food sector', *Trends in Biotechnology*, vol. 27, no. 2, pp. 82–9.

STEINER I (1991), 'Changes of a polyethylene foil for food packing after sterilization with ozone', *Deutsche Lebensmittel-Rundschau*, vol. 87, no. 4, pp. 107–12.

TAOUKIS PS & LABUZA TP (2003), 'Time-temperature indicators', in R Ahvenainen (ed.), *Novel food packaging techniques*, Woodhead Publishing, Cambridge, pp. 103–26.

VERMEIREN L, HEIRLINGS L, DEVLIEGHERE F & DEBEVERE J (2003), 'Oxygen, ethylene and other scavengers', in R Ahvenainen (ed.), *Novel food packaging techniques*, Woodhead Publishing, Cambridge, pp. 22–49.

VLIEGER JJ (2003), 'Green plastics for food packaging', in R Ahvenainen (ed.), *Novel food packaging techniques*, Woodhead Publishing, Cambridge, pp. 519–33.

YAM K, TAKHISTOV P & MILTZ J (2005), 'Intelligent packaging: Concepts & applications', *Journal of Food Science*, vol. 70, no. 1, pp. R1–R10.

© Woodhead Publishing Limited, 2012

# 23

# Emerging methods for post-packaging microbial decontamination of food

**H. Neetoo, H. Chen and D. G. Hoover, University of Delaware, USA**

**Abstract**: The increased demand by consumers for fresh-like, safe, nutritious, convenient, and flavorful packaged foods have paved the way for the continuous emergence of novel food processing technologies. Today, the food industry is more diverse, competitive, and efficient than ever. In addition, a number of novel thermal methods including microwave, radiofrequency and infrared heating, and non-thermal methods including high hydrostatic pressure, irradiation and pulsed light technology for processing of packaged foods, are in active research and development in academia, industry and government institutions. This chapter reviews the common approaches used for decontamination of packaged foods with current discussion relating to their mechanisms of microbial control, applications to food, possible process/food/package interactions, and limitations associated with the implementation of these technologies.

**Key words**: post-packaging decontamination, thermal processing, non-thermal processing, active packaging, packaging interactions.

## 23.1 Introduction

Traditionally, food has been processed in bulk followed by packaging to mostly act as a barrier to prevent access to spoilage and pathogenic microorganisms. Packaging materials have customarily been chosen to contain, protect, and preserve food for a certain period of time. However, with the changes in lifestyle and burgeoning consumer demands for safe, minimally processed, healthful, and tasty packaged foods, the primary functions of packaging have shifted. Greater emphasis is nowadays placed on the use of appropriate materials to package foods prior to processing with the view to avoiding

© Woodhead Publishing Limited, 2012

post-process recontamination of the product. The suitability and performance of each packaging material is thus a function of (i) the properties of the commodity, (ii) the nature of the in-package process, (iii) characteristics of the packaging material itself, as well as, (iv) the dynamic interplay among the various aforementioned factors. Research and development in the area of in-package or post-package processing methods therefore not only has to achieve a satisfactory level of decontamination but also be compatible with the food product of interest.

Technological advances in processing and packaging machinery have been quite considerable thanks to progress in food engineering technology and packaging material science, delivering higher standards of hygiene, safety and quality assurance. With the increased demand for products of higher nutritional and sensorial quality, novel thermal processing methods such as sous-vide cooking, microwave and radiofrequency heating as well as non-thermal methods such as high pressure processing, irradiation, and UV-light have received considerable attention.

A concomitant of in-package processing is that the packaging material deemed suitable for a particular product is exposed to processing conditions which may alter its structure and consequently its mechanical and mass transfer (barrier and migration) properties (Guillard *et al.*, 2010; Ozen and Floros, 2001). In this respect, post-package processing of foods may require extra functions of packaging in addition to the traditional functions of containment, protection, and preservation. Packaging materials would be expected to have strong physical and mechanical resistance to the process mechanisms. For example, in the case of high pressure processing, the packaging material must be able to withstand the operating pressures, have good sealing properties, and the ability to prevent quality deterioration during the application of pressure (Min and Zhang, 2007). The packaging materials for irradiation should be chemically stable and resistant to polymer degradation and prevent leaching of undesirable compounds into the adjacent food (Morehouse and Komolprasert, 2004). Critical protective barrier properties of packaging materials must be maintained to avoid deterioration of processed foods by microbiological, chemical, and physical factors. Therefore, as these decontamination technologies are adapted for in-package or post-package processing, the need for novel food packaging compatible with the processing methods cannot be overemphasized. This chapter reviews the various thermal and non-thermal methods that have been used for post-package processing of foods and discusses the mechanisms of microbial control and food products affecting these technologies as well as offering insight into the relevant packaging parameters to optimize the efficacy of these processes. Tables 23.1 and 23.2 provide examples of studies evaluating the efficacy of thermal and nonthermal interventions used to enhance the safety of packaged foods as covered in this chapter.

© Woodhead Publishing Limited, 2012

**Table 23.1** Studies investigating the application of thermal post-packaging decontamination methods

| Technology | Food matrix | Target microorganism(s) | Packaging material | Outcome | Reference |
|---|---|---|---|---|---|
| Steam/hot water pasteurization | Cooked chicken breast fillets | *L. monocytogenes* | Thick gas/moisture barrier bags | > 7 LR | Murphy *et al.* (2005) |
| Steam pasteurization | Raw frankfurters | *E. coli* O157:H7 | Thick gas/moisture barrier bags | > 7 LR | Murphy *et al.* (2004) |
| Hot water pasteurization | RTE beef sticks | *L. monocytogenes* | Natural casing | > 2 LR | Ingham *et al.* (2003) |
| Ambient/pressurized steam | Cooked bologna | *L. monocytogenes* | Thick packaging films | > 2 LR | Murphy *et al.* (2005) |
| Submerged water pasteurization | RTE deli meats | *L. monocytogenes* | Shrink-wrap vacuum-packaging bags | ≤ 4 LR | Muriana *et al.* (2002) |
| Microwave heating | Beef frankfurters | *L. monocytogenes* | Gas/moisture barrier bags | > 7 LR | Huang and Sites (2007) |
| Microwave heating | Frankfurters | *L. monocytogenes* | Zip-top bags (nylon/EVA copolymer) | 1.5–5.9 LR | Rodriguez-Marval *et al.* (2004) |
| Radiofrequency heating | Mashed potatoes | *C. sporogenes* (PA 3679) | Polymeric tray with Al foil lid | > 7 LR | Luechapattanaporn *et al.* (2004) |
| Infrared heating | Almonds | *S.* Enteritidis | Four-layer Al foil pouch | > 7.5 LR | Brandl *et al.* (2008) |
| Infrared heating + submerged pasteurization | RTE meat products | *L. monocytogenes* | Shrink-wrap vacuum-packaging bags | < 3.5 LR | Gande and Muriana (2003) |
| Sous-vide/cook-chill | Fish cakes | TVC | Cryovac 3-ply laminates (PE-Nylon-PE copolymer) | 8-fold increase in shelf life | Shakila *et al.* (2009) |
| Sous-vide cooking | Turkey | *C. perfringens* | Nylon-polyethylene film | No outgrowth | Juneja and Marmer (1994) |
| Sous-vide cooking | Salmon slices | *S. aureus*, *B. cereus*, *L. monocytogenes* | Polyethylene-polyamide pouch | Undetectable counts | Gonźalez-Fandos *et al.* (2005) |

Key: RTE = ready-to-eat, EVA = ethylene vinyl acetic acid, Al = aluminum, PE = polyethylene, *L. monocytogenes* = *Listeria monocytogenes*, *E. coli* O157:H7 = *Escherichia coli* O157:H7, *C. sporogenes* PA 3679 = *Clostridium sporogenes* PA 3679, *S.* Enteritidis = *Salmonella* Enteritidis, TVC = total viable count, *C. perfringens* = *Clostridium perfringens*, *S. aureus* = *Staphylococcus aureus*, *B. cereus* = *Bacillus cereus*, LR = log reduction.

© Woodhead Publishing Limited, 2012

**Table 23.2** Studies investigating the application of non-thermal post-packaging decontamination methods

| Technology | Food matrix | Target microorganism(s) | Packaging material | Outcome | Reference |
|---|---|---|---|---|---|
| HHP | Sausage | *L. monocytogenes* | Sterile PE bag | 6–7 LR | Chung *et al.* (2005) |
| HHP | Turkey meat | *L. monocytogenes* | Sterile LDPE pouches | 3.8 LR | Chen (2007) |
| HHP | Strained chicken baby food | *S. Senftenberg* 775W | PS bag | < 2 LR | Metrick *et al.* (1989) |
| Irradiation | Minced meat | Enterobacteriacea | Vacuum package | 4 LR | Farkas and Andrassy (1993) |
| Irradiation | Fresh catfish fillets | Background microflora | PE bag | 4-fold increase in shelf life | Przybylski *et al.* (1989) |
| Irradiation | Fresh pork | Background microflora | Moisture/gas barrier bags | > 2 LR | Lambert *et al.* (2000) |
| Irradiation | RTE ham and cheese sandwich | *S. aureus* | PET | 6.2 LR | Lamb *et al.* (2002) |
| Irradiation/ antimicrobials | Beef product | *C. botulinum* | Thick multilayer bag | > 4.4 LR | Suarez Rebollo *et al.* (1997) |
| Pulsed UV light | Raw poultry | *C. jejuni* | Leak-proof packaging | < 4.6 LR | Haughton *et al.* (2010) |
| Pulsed UV light | Chicken breast | *S.* Typhimurium | Vacuum pouch | < 2.4 LR | Keklik *et al.* (2010) |
| Antimicrobial packaging | Frankfurters | *L. monocytogenes* | Cellulose casings | 1.2 log lower than control | Luchansky and Call (2004) |
| Antimicrobial packaging | Hamsteaks | *L. monocytogenes* | Surlyn® films | 6.3 log lower than control | Ye *et al.*, (2008b) |

Key: HHP = high hydrostatic pressure, RTE = ready-to-eat, PE = polyethylene, LDPE = low density polyethylene, PS = polyester, PET = polyethylene terephthalate, *L. monocytogenes* = *Listeria monocytogenes*, *S.* Senftenberg 775W = *Salmonella* Senftenberg 775W, *S. aureus* = *Staphylococcus aureus*, *C. botulinum* = *Clostridium botulinum*, *C. jejuni* = *Campylobacter* jejuni, *S.* Typhimurium = *Salmonella* Typhimurium, LR = log reduction.

© Woodhead Publishing Limited, 2012

## 23.2 Conventional thermal processing (CTP)

Depending on the intensity, thermal preservation processes can be classified into two categories:

1. pasteurization, where heat processing occurs at temperatures ranging from 70 to 100°C to target the destruction of vegetative cells, and
2. sterilization, where heat processing occurs above 100°C with the aim of destroying all forms of microorganisms including spores.

Heat preservation methods can be conveniently divided into 'in-pack' (batch retort) and 'in-line' systems (Tucker and Featherstone, 2011). This section focuses particularly on 'in-pack' systems. Batch retorts can operate with a variety of heating media including condensing steam, mixtures of steam and air, water immersion, or water droplets either sprayed or rained onto the packs (Britt, 2008; May, 2000). Typical steps during a batch or 'in-pack' process consist of (i) filling the packs, (ii) hermetically sealing them, and (iii) thermal processing in a retort. During the heating step, caution has to be exercised to ensure that the coldest point reaches the target temperature to achieve process homogeneity. Consideration also needs to be paid to the shape and type of package. From the manufacturers' perspective, a metal can is the package of choice since it offers advantages such as high throughput, flexibility in package dimensions, and high compression strength to withstand mechanical abuse during processing and distribution. However, development of modern retorts has enabled more delicate packages such as plastic packs, pouches, trays, and glass jars to be processed (Tucker and Featherstone, 2011). These are addressed in greater detail in Section 23.2.2.

### 23.2.1 Microbial control using CTP

The heat resistance of microorganisms is one of the main factors that affect the kinetics of thermal microbial inactivation and must be known or determined for a specified process. Thermal processing makes use of two factors that dictate the time/temperature process: the $D$ value which is defined as the time required at any given temperature to destroy 90% of cells and the $z$ value which refers to the increase in temperature required for a 1-log reduction of $D$ values. A number of intrinsic factors (product-related factors) influence the heat resistance of microorganisms such as water activity, pH, and the composition and consistency of the food. The heat resistance of microorganisms in dry products or foods having a high lipid content is markedly higher than in foods with a higher moisture content and/or lower fat composition. The pH of the product can also have a substantial effect on microbial inactivation. Foods with a pH $< 4.5$ can be stabilized by a mild heat treatment whereas harsher conditions are needed for low-acid products with a pH $> 4.5$. Other factors unique to the product such as the presence of ions, oxygen content, and presence of endogenous antimicrobial substances can all significantly affect

© Woodhead Publishing Limited, 2012

heat requirements (Fellows, 2000). In addition, several factors can alter the rate of heat penetration from the heating medium to the actual foods. These can essentially be categorized as (i) process-related factors, (ii) product-related factors, and (iii) packaging-related factors. Process-related factors include the treatment temperature, the process time, the nature of the heating medium, and container agitation (Holdsworth and Simpson, 2008). Product-related factors include the product viscosity, initial temperature, thermal properties, and chemical or biochemical characteristics. The packaging-related factors are discussed in greater detail in the next section.

### 23.2.2 Packaging considerations for CTP

Materials for packaging should purport to contain, protect, preserve, portion, inform, promote, and render portable. However, containers used for post-package thermal processing should possess additional functionalities to render them suitable. The packaging material should possess several desirable attributes, including the ability to maintain a hermetic seal, to withstand the physical stresses during processing and not react adversely with the food. The packaging options for thermally processed foods include metal, glass, and specific plastics (laminates and composites) (Tucker and Featherstone, 2011).

*Metal containers*

Metal containers or cans are the most common form of packaging for thermally processed products. The physical strength of metal makes it a popular material for food packaging. Metal is also an excellent light and gas barrier, is quick and easy to seal, and can withstand the harsh processing conditions (Hutton, 2003). Cans are made of tinplate, tin-free steel or aluminum. The size of headspace in cans is critical and must be controlled. An inadequate headspace could prevent efficient mixing of the product, thereby reducing the rate of heating. At the same time, an excessive headspace could result in too much air in the container. Oxygen present in excessive levels could accelerate oxidation of the product resulting in discoloration and nutrient degradation (Tucker and Featherstone, 2011).

*Glass containers*

Glass is almost completely chemically inert and therefore is suitable for packing many products and all foods. The principles of processing in glass are basically the same as for cans, although it calls for certain modifications that are necessary to improve the thermal properties of glass to make it less prone to damage at very high temperatures.

The use of glass for post-package treatment offers several advantages, including its chemical stability, its reliable seal, and its physical strength. The main disadvantage for using glass is that it is brittle and therefore can be broken into pieces that can represent a physical hazard (Fellows, 2000). In

© Woodhead Publishing Limited, 2012

addition, clear glass is not a barrier to light so deterioration of light-sensitive nutrients such as vitamin B3 (niacin) and pigment compounds (bleaching) can occur, although this can be minimized using tinted glass (Bansal and Doremus, 1986). Moreover, glass containers are heavy and cumbersome and not suitable for products intended for freezing.

*Plastics containers*

Plastic polymers have also been used for packaging foods prior to thermal processing. Plastic-based packaging can take the form of (i) a flexible container, (ii) a semi-rigid container, and (iii) a rigid container. Some of the main advantages for using plastics include:

- the ease and low cost of transportation and storage of plastic packages,
- the thin profile of pouches which ensures rapid heat penetration and therefore faster sterilization, resulting in products of higher quality and saving of energy,
- the ease of opening plastic pouches, and
- its resistance to corrosion (Tucker and Featherstone, 2011).

Examples of synthetic polymers or plastics used include polypropylene (PP), polyethylene terephthalate (PET), ethylene vinylalcohol (EVOH), polyvinylidene chloride (PVDC), polyamide (PA) as well as glass (silicon dioxide)-coated barrier films (May, 2000). The different polymers have different mechanical and barrier properties, process adaptability and cost, and are used either singly or in multilayer applications. Depending on the final use of the package (i.e., pasteurization or sterilization), different films or laminates are used. Regardless of the final intent, the integrity of the hermetic seal must be maintained. Retort pouches used for packaging certain foods can take the form of pillow pouches or stand-up pouches (SUP). Rigid containers also known as plastic pots are also available and they are usually sealed with a flexible lid. Other developments in this area include bottles made of laminated polypropylene (PP)/ ethylene vinyl alcohol (EVOH) with foil laminated caps, polyvinylidene chloride (PVDC)/polypropylene (PP) containers, and polyethylene terephthalate (PET) bottles, which can be hot-filled or pasteurized (May, 2004). With rigid containers, care has to be taken to avoid the build-up of pressure during processing at high temperature.

### 23.2.3 Post-package decontamination using CTP

Conventional processing has been shown to be effective in reducing levels of human pathogens on various packaged animal and plant-derived food products (Annous and Kozempel, 2006). Huang (2007) previously demonstrated that pasteurization of frankfurters in single-layer packages by hot water immersion achieved satisfactory inactivation of *Listeria monocytogenes*. Samelis *et al.* (2001) also investigated the efficacy of hot water immersion to reduce

© Woodhead Publishing Limited, 2012

microbial contamination in post-package meat products. Selby *et al.* (2006) previously demonstrated the efficacy of hot water immersion as an effective post-packaging pasteurization technology to control *L. monocytogenes* on ready-to-eat (RTE) meat products (bologna) and the importance of meat formulation in optimizing product quality retention. Murphy *et al.* (2004) demonstrated the effectiveness of hot water and a steam cooker set at 90°C to achieve *ca.* 7 $\log_{10}$ CFU/g reduction of *L. monocytogenes* in vacuum-packaged fully cooked chicken breast fillets and strips. Murphy *et al.* also compared the thermal death time of *Escherichia coli* O157:H7 in pre-packaged raw and fully cooked franks during steam pasteurization (Murphy *et al.*, 2005). In addition, the use of ambient and pressurized steam pasteurization to reduce the population of *L. monocytogenes* on RTE deli meats was compared with reported higher efficacy with pressurized steam (Murphy *et al.*, 2005).

Recent progress in thermal processing of beef products has made possible the commercial availability of meat products cooked, stored and retailed in flexible packages. Different types of 'cook-in' packages are already available on the market for different purposes: the 'cook-in-ship,' where the same package is used throughout processing and distribution, and the 'cook-in-strip,' where the packaging used in processing is removed after cooking, and the product is repackaged before distribution. Hence, selection of the appropriate packaging material is key to minimizing microbiological and other deteriorative processes (Terlizzi, 1984).

### 23.2.4 Challenges and limitations associated with the use of CTP

There are important critical factors to consider during CTP. The type of retort used (e.g. still, horizontal, vertical, agitating) can significantly affect the rate of heat transfer into cans (Holdsworth and Simpson, 2008). Heat transfer through a packaged product is typically slow especially for bulky foods. As a result, products can be thermally over-processed leading to loss of nutritional and sensorial quality. Moreover, inconsistency in the operation of the processing systems can introduce batch-to-batch variation. Hence, the type of processing system and the way it operates should be understood and carried out reproducibly. Furthermore, it is vital to operate the heating unit at the correct processing temperature. Small deviations from the product or processing specification can translate into large differences in the process lethality. Hence, validations of the adequacy of the process should be carried out for any changes or deviations (Tucker and Featherstone, 2011).

There are also certain drawbacks with the use of flexible films or pouches for heating. The type and design of the retort needed for plastic-packaged food are more complex and expensive. Moreover, since pouches are not rigid, some products are bound to deform. It has also been reported that the shelf life of products packaged in pouches is shorter than their canned counterparts. Glass containers are also problematic as they can shatter in transit causing injury (May, 2004).

© Woodhead Publishing Limited, 2012

## 23.3 Sous-vide processing (SVP)

Sous-vide is a French term literally meaning 'under vacuum'. This technology allows food to be thermally processed using vaccum-packaging in heat-stable, high barrier or air-impermeable multi-laminate plastics. This form of processing is especially amenable for food consisting of partially cooked ingredients alone or combined with raw foods, requiring low temperature storage until the packaged food is thoroughly heated immediately prior to serving (Ghazala, 2004). In short, sous-vide is an 'assemble-package-pasteurize-cool-store' process. Figure 23.1 provides a simplified flow diagram that outlines the basic steps in sous-vide processing (SVP). On an industrial scale, pre-packed solid foods are pasteurized in a jacketed tank with a large-capacity and subjected to specific time–temperature regimes and then cooled in a chill tank (3–4°C) to achieve inactivation of microbes and enzymes with minimal chemical changes and alterations in the original organoleptic characteristics (color, flavor, and appearance). In addition to requiring minimal preparation and serving steps, it ensures consistency in the quality and presentation of the products (Ghazala, 2004).

### 23.3.1 Packaging considerations for SVP

Generally speaking, packaging can be considered a hurdle since this step or process prevents microorganisms outside of the hermetic seal from reaching

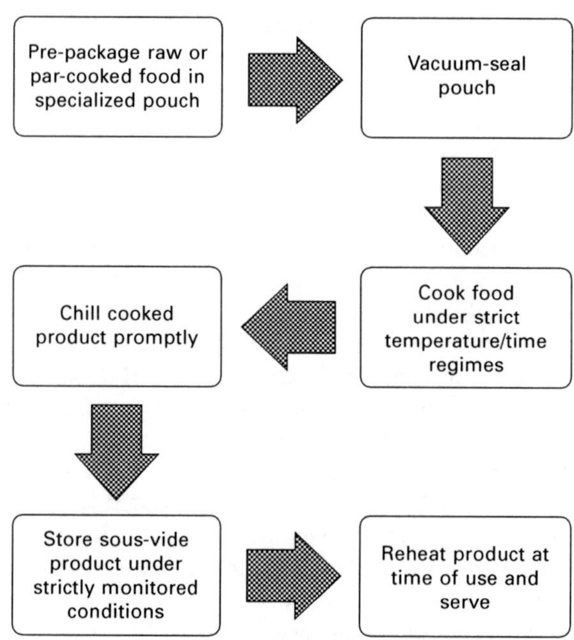


**Fig. 23.1** A simplified flow diagram that outlines the basic steps in sous-vide processing.

© Woodhead Publishing Limited, 2012

the food. Not only should the package be esthetically pleasing, the dimensions of the package should be such that the portions of food are enough to meet the required serving size.

After vacuum-packing, the plastic film acts as a skin on the surface of the food. Heat transfer through the packaging material is governed by its thermophysical properties (Martens, 1995). In addition, the choice of plastics is dependent on the highest temperatures reached during the heating and reheating stages in the life cycle of the product. For example, LDPE can only be used for temperatures lower than 85°C while other polymers such as medium density polyethylene (MDPE), polypropylene (PP), polyamide (PA) or polyethylene terephthalate (PET) may be used for heating applications requiring temperatures above 85°C, while high density polyethylene (HDPE) can be used for temperatures of up to 100°C. Packages made of polyethylene terephthalate (PET) are often used for SVP and consist of a pouch formed from a flexible film or a tray formed out of a rigid film covered by a flexible film (Dodds, 1995). Besides the chemical composition, the thickness, thermal resistance, and orientation of the material are also important physical considerations in the selection of the package.

Another requirement is that packages need to be impermeable to gases such as oxygen, carbon dioxide and water vapor. Oxygen transmission rate (OTR) is an important packaging factor helping to control the proliferation of anaerobic and aerobic bacteria. Low OTR or total imperviousness of the package promotes anaerobic growth while a high OTR promotes aerobic growth at the expense of anaerobic growth. Examples of gas-impervious plastics include PVDC, EVOH, and PA. Unlike PVDC, which has the ability to maintain its imperviousness to gases even at high moisture, EVOH barriers are usually compromised in moist conditions (Brown, 1992).

In addition to acting as a barrier to moisture and gases, the package also has to control the migration of low molecular weight compounds originating from the packaging into the product. These include monomeric compounds, plasticizers, additives or processing aids (Martens, 1995). The package must also be clean, pathogen-free, of high quality and free of package leaks before and after processing, as this could constitute an opportunity for recontamination or post-process spoilage (Ghazala, 2004). Thus, the plastic film needs to have a high tensile strength and resistance to puncture, tear and burst. In particular, the seam should have a high enough mechanical strength to resist rupture during all the steps of the entire life cycle of the packaged product including heating, cooling, transportation, etc. Indeed, microbial re-contamination of the product could occur especially if a leakage pathway is present and microorganisms are present in high numbers. The presence of leaks in the package could be attributed to too high sealing temperatures or the presence of dirt on the sealing equipment that can compromise the integrity of the seal in hermetically sealed packages. Hence, the package integrity must be ensured through the use of appropriate sealing and packaging systems (Ghazala, 2004).

© Woodhead Publishing Limited, 2012

Most of the films used in SVP are multilayer complexes of different plastics, produced by extrusion or lamination (Martens, 1991). While some plastics exhibit excellent moisture barrier properties, they tend to act as poor gas barriers. Since plastics can rarely act as effective water and gas barriers simultaneously, typical films comprise at least two polymers (Martens, 1995). For example, some films have an intermediate layer of EVOH to improve gas barrier properties. Moreover, barrier films such as PA (nylon) are often sandwiched between layers of heat-resistant plastics compatible with foods to impart imperviousness to gases such as $O_2$ and mechanical strength to the multi-laminate. Pouches consisting of an outer layer of PA and an inner layer of PE are also common. While PA imparts mechanical strength and oxygen impermeability, the PE layer confers water vapor imperviousness (Martens, 1995). Plascon Packaging, a Plascon Group company, specializes in an extensive line of flexible food packaging including 'Cook-Chill Bags' used in the sous-vide method of food preparation. These bags are made from Cryovac film, a multi-layer film that provides an oxygen barrier and excellent seal strength to prevent breaks or leakage (The Plascon Group, 2011).

Taken together, the pre-packaging material should typically be made of specialized plastic bags or pouches which (i) are non-transmissible to oxygen, (ii) can be hermetically sealed, (iii) can be layered or multi-laminated for improved barrier properties, (iv) can control moisture permeability, and (v) are flexible, tough and resistant to puncture during handling and transport (Brown, 1992).

### 23.3.2 Post-package decontamination using SVP

The application of sous-vide has been used widely in the processing of various types of vacuum-packaged raw or par-cooked meat, poultry, fish and even vegetable-based products for the purpose of enhanced sensorial and organoleptic characteristics, and enhanced microbiological safety and quality of these foods with extended shelf-lives. The pie chart in Fig. 23.2 shows the relative number of published studies on the application of sous-vide for processing foods of animal and plant origin. Nyati (2000) compared the microbiological status of various sous-vide animal-derived products and showed that foodborne pathogens, such as *L. monocytogenes*, *Clostridium perfringens*, *Bacillus cereus*, *Salmonella*, and other Enterobacteriaceae, were rendered undetectable after processing. González-Fandos *et al.* (2004) demonstrated the capacity of sous-vide cooking to reduce the counts of *Staphylococcus aureus*, *B. cereus*, *C. perfringens* and *L. monocytogenes* on rainbow trout and salmon and extend shelf life to >45 days during storage at 2°C. Similarly, Shakila *et al.* (2009) showed an improvement in the microbiological quality of sous-vide fish cakes during chilled storage (3°C) with an eight-fold increase in its shelf life compared to their conventionally cooked counterparts.

Significant research has in particular focused on spore-forming microorganisms

© Woodhead Publishing Limited, 2012

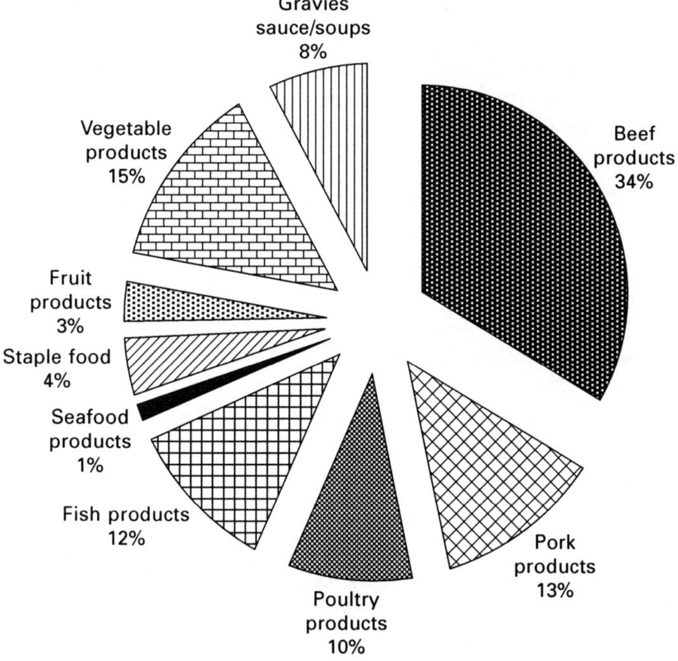


**Fig. 23.2** The relative number of published studies on the application of sous-vide for processing foods of animal and plant origin.

that constitute a safety risk in sous-vide products, such as *Clostridium botulinum*, *C. perfringens* and *Bacillus* spp. Several studies have demonstrated outgrowth of spores in sous-vide spaghetti and meat-sauce products (Simpson, 1995), carrot, cod and chicken homogenates (Brown and Gaze, 1990) as well as products containing mixtures of beef, pork, vegetables, rice and seafood. Researchers in recent years have also characterized the behavior of *C. perfringens* in sous-vide cooked products. Juneja and Marmer (1996) investigated the growth potential of *C. perfringens* in sous-vide cooked turkey products formulated with 0–3% salt and stored at temperatures of 4–28°C. Overall, storage at 4°C and a salt level of 3% proved to be most effective in controlling spore outgrowth. Outgrowth of *Bacillus* spores in sous-vide products is also a concern and has been extensively studied. Overall, mixed results have been found with respect to outgrowth of *Bacillus* spores. While Knochel *et al.* (1997) and Chavez-Lopez *et al.* (1997) found detectable populations of *Bacillus* spores in vacuum-cooked green beans and pilaf rice, respectively, Aran (2001) demonstrated that addition of calcium lactate (1.5%) and sodium lactate (3%) completely inhibited *B. cereus* outgrowth in beef goulash samples.

© Woodhead Publishing Limited, 2012

### 23.3.3 Challenges and limitations associated with the use of SVP

Although foods subjected to SVP generally have enhanced nutritional and sensorial properties compared to their more thoroughly heat-treated counterparts, they also raise several microbiological safety concerns, especially if process controls are not in place. Generally, short shelf life products (<10–14 days) subjected to sous-vide cooking undergo a 6-*D* reduction in the numbers of vegetative pathogens (process lethality equivalent to 12 min at 70°C) (Holdsworth and Simpson, 2008). This process is considered adequate for targeting *L. monocytogenes*, a psychrotrophic vegetative pathogen (Holdsworth and Simpson, 2008); however, it is unable to achieve a satisfactory inactivation of psychrophilic nonproteolytic *C. botulinum* type B and E spores, which can potentially outgrow during refrigeration and produce toxins (Holdsworth and Simpson, 2008). In addition, the existing preservation hurdles can also inhibit the spoilage microbiota and the absence of apparent signs of spoilage in terms of product color, odor and taste may mislead consumers into eating products that are unfit for consumption due to compromised safety. Moreover, improper chilling in the post-processing stage provides an environment that is conducive for the growth of psychrotrophic and mesophilic pathogens. Non-sporeforming facultative anaerobic psychrotrophs considered prime hazards in sous-vide processed products include *L. monocytogenes*, *Yersinia enterocolitica*, and *Aeromonas hydrophila* (Juneja, 2003). Non-sporeforming facultative anaerobic mesophilic pathogens such as *Salmonella*, *S. aureus*, or enteropathogenic strains of *E. coli* may constitute a risk if foods are subjected to temperature abuse (Juneja, 2003). Post-process recontamination can also occur. Even products such as meats and cheese that have been processed to kill *L. monocytogenes*, can still be recontaminated with the pathogen when opened, sliced or repackaged at retail if stringent hygienic practices are not observed (Ghazala, 2004). Hence, it is recommended to vacuum-package the food together with another hurdle such as the addition of preservatives, a low pH, or low water activity to reduce the risk of foodborne illness (Juneja, 2003).

The sous-vide process is quite a gentle cooking method for delicate foods. This process relies on the application of an extended mild heat treatment at a precisely controlled temperature, followed by chilled storage at temperatures of 3°C to ensure longevity of the product. Since these foods are not sterile, they are only expected to have a shelf life of less than 42 days (Ghazala, 2004), however, the quality and shelf life of the product also rely on the use of high-quality raw materials and clean equipment to package every item. It is thus necessary to exercise care and good hygienic practices at every step in the manufacturing, distribution, storage, and reuse stages of this type of product.

Moreover, accidental recontamination of the products during the on-going processing phase is also a potential concern. Packaged products can be cooked in different types of industrial equipment using different heating media. Water is a common heating and cooling medium. However, water also

© Woodhead Publishing Limited, 2012

acts as a vehicle for microbes. During heating of vacuum-packed products, fluid can enter through package leaks and cause recontamination. One leaky pouch could cause the heating and cooling water as well as other pouches in the entire batch to become contaminated (Martens, 1995).

## 23.4 Microwave (MW) heating

Microwaves are a form of electromagnetic radiation characterized by the wavelength and frequency of the waves used in food processing (915–2450 Hz). Microwave heating is generated by the conversion of electromagnetic energy to thermal energy. The technology of microwave heating of foods has garnered scientific and consumer interest due to its volumetric origin, rapid increase in temperature and relative ease of cleaning (Ahmed and Ramaswamy, 2007). Unlike more traditional forms of thermal processing, such as pasteurization and retorting which are characterized by a slow thermal diffusion process, the volumetric nature of heat generated by microwaves can substantially reduce the total heating time, thereby minimizing the overall severity of the process and leading to a greater retention of the desirable quality attributes of the product (Sumnu and Sahin, 2005). According to Tewari (2007), the time required to come to target process temperature is attained within one-quarter of the time typically reached by conventional heating processes. Microwave technology is also amenable to batch processing and therefore can be used to pasteurize pre-packaged products as well as offer the flexibility of being easily turned on or off.

### 23.4.1 Packaging considerations for microwave heating

Considerations given to the choice of packaging material is of paramount importance in the development of microwave-friendly packages. Packages should possess suitable thermal and electrical properties, be compatible with MW heating, and provide desired post-process barrier mechanical properties (Ramaswamy and Tang, 2008). Hence, positive interaction between package, MW and foods, good barrier and mechanical performances of packages, controlled migration of packaging components into foods, and costs of the packaging are equally important factors (Ramaswamy and Tang, 2008).

Suitable materials for MW are classified as transparent, absorbers, reflectors, or susceptors (Lefeuvre and Audhuy-Peaudecerf, 1994). Glass, paper, ceramics, composites, or plastic containers are suitable materials for microwave cooking because they do not constitute a barrier for microwaves. Examples of plastics commonly used include PET/PP laminates with a barrier layer of EVOH, PVDC, PE or susceptor materials. Tight-fit metallic containers, on the contrary, are inappropriate because metals reflect microwaves. For sterilization purposes, PET, high-density polypropylene (HDPP) and various

© Woodhead Publishing Limited, 2012

other polyester-based materials are particularly suitable materials, especially for high-quality trays, pouches, and bags (Ahmed and Ramaswamy, 2007).

Since foods are generally treated 'in-package' during microwave processing, this method also requires the complete harmony of the food and the package with no adverse food-package interaction (Ozen and Floros, 2001). Packages used in microwave processing have gained special attention in migration studies due to high temperatures reached during cooking. Castle (1989) reported that total levels of migration of oligomers into food from PET containers, such as roasting bags and susceptor pads, were ten times greater than that observed for the same type of food heated in a conventional oven in a PET tray. Hence, factors such as the temperature attained during cooking, the time of exposure, the extent of contact with the food, and the nature of the food surface, can substantially affect the amount of migrated compounds.

The cost and design of the package also influence the choice of the material. HDPP is a low-cost solution for microwave processes compared to other packaging polymers (Ahmed and Ramaswamy, 2007). Recently, food processors have started manufacturing 'new-generation' microwaveable convenience meals packaged in trays of crystallized polyethylene terephthalate (CPET) and olefin thermoplastics. Design flexibility is another driving factor for the use of thermoplastics to ensure optimal heat transfer. An example of optimal design for a microwaveable container is an oval-shaped tray with a central well for heating of food in the main body and sauce in the central well (Anon, 1989). Golden Valley Foods, Inc. previously developed microwaveable entrées packaged in a doughnut-shaped 'ThermaCore' container to eliminate hot and cold spots in the food product (Morris, 1979). Entrées can also be tray-packed with a slight modified atmosphere headspace to avoid the deformation of product surface appearance resulting from vacuum packaging (Morris, 1979).

### 23.4.2 Post-package decontamination using microwaves

Microwave heating has been the most widely utilized form of electric heating. Considerable research has focused on the use of microwave heating for pasteurization and sterilization applications. Microwave sterilization operates in the temperature range 110–130°C while pasteurization is a more gentle heat treatment occurring between 60 and 82°C (Orsat and Raghavan, 2005). Since microwave energy can heat foods effectively and rapidly, its use in food decontamination by pasteurization and sterilization has been the subject of intense study. The advantages of using microwaves for food decontamination include the rapid and homogeneous heating in certain food products. In other words, microwave heating can offer high temperature and short time processing (HTST), thereby ensuring a product of enhanced quality. Examples of packaged foods treated by microwaves include yogurt as well as pouch-packed meals (Decareau, 1985). Lau and Tang (2002) showed that microwaves, applied at a frequency of 915 MHz, were able to

© Woodhead Publishing Limited, 2012

pasteurize asparagus pickled in jars. The process was demonstrated to achieve uniform heating for a shorter holding time, with minimal quality deterioration compared to conventional pasteurization. Canumir *et al.* (2002) previously demonstrated the suitability of microwave pasteurization to inactivate *E. coli* in pre-packaged apple juice.

Modernization of microwave technologies has resulted in equipment with a longer operating life at lower costs and therefore a greater 'penetration' of microwaveable foods on the market. There are several lines of new packaged microwaveable products commercially available including the Eggology's On-the-go 100% egg whites, and Marks and Spencer's Steam Cuisine. The latter is a ready-to-cook product that takes ~6 min to cook from the raw stage using an 'intelligent double pressure cooking technology' (Bertrand, 2005). Although the microwave oven is a common household appliance, microwave heating has also found increasing applications in the food industry in various processing operations. Microwave processing has been successfully applied on a commercial scale for pasteurizing prepared meals, fresh pasta, bread, granola and milk (Giese, 1992). Microwave heating has also been used on an industrial scale to produce pre-packaged shelf-stable products, accompanied with strict temperature control to ensure predictability and reproducibility of the thermal process (Mullin and Bows, 1993). An example of a successful microwave sterilizing plant is located in Belgium and produces a line of 'Ready To Heat And Eat' dishes in their 'Top Cuisine' line (Bengtsson, 1998).

### 23.4.3 Challenges and limitations associated with the use of microwave heating

Microwave heating faces both process-related and package-related challenges. A major hurdle with microwave processing is attaining heating uniformity (Ohlsson, 1991). Prediction of temperature profiles is particularly difficult with foods of different dielectric properties, size, and geometry. Temperature mapping at different locations throughout the product is thus critical in order to control the temperature at the coldest point of the product where microbial destruction is likely to be least efficient (Ahmed and Ramaswamy, 2007). This non-uniform heat distribution has made comparison of the microbial inactivation kinetics of this technology to other conventional forms of heating rather challenging due to inherent differences in the monitoring of temperature (Ahmed and Ramaswamy, 2007). As a result, the lack of reproducibility and predictability of the process lethality can lead to survival of pathogens, such as *Salmonella* and *L. monocytogenes*, in 'cold spots' of the food (Schnepf and Barbeau, 1989). According to Ohlsson (1991), one way to compensate for the limitation of microwave heating is to use it in combination with conventional heating, using rapid volumetric heating for the final burst of 10–13°C to achieve HTST conditions.

Certain packaging materials such as glass and several types of plastics

© Woodhead Publishing Limited, 2012

can be problematic for microwave heating. Although glass is an acceptable microwaveable material, it has a relatively higher cost, high shipping weight, breakability and has the tendency to gain thermal energy from the surrounding food. Their tall cylindrical geometry can sometimes lead to focused interior heating (Schiffmann and Sacharow, 1992). Moreover, glass may also break during microwaving especially if there is a sharp temperature differential between the internal and external temperature. In addition, tall glass jars can also cause overflow or eruption of the product, especially for particulate foods (Schiffmann, 1988). Plastics are also prominent packaging materials for microwaveable foods. However, a caveat with the use of plastic pouches for packaging fatty foods is that fat separation can occur, causing the lipid layer to stick to the pouches (Schiffmann, 2001). During microwaving, this interfacial fat can rapidly gain heat and can cause the pouch to melt. Thus, product formulation to avoid fat separation as well as design of stand-up pouches to allow venting should be ensured to maintain package integrity and to avoid pressure build-up.

In addition, the use of microwave heating does not always necessarily guarantee a better retention of quality of the food products. The rate of deterioration and thermal degradation of the sensorial and nutritional attributes depend on several intrinsic (product-related) factors, as well as processing factors related to the design and dielectric properties of the microwave process. Sumnu and Sahin (2005) mentioned that microwave heating can lead to the development of quality problems in microwave-processed foods due to the occurrence of hot spots at the corner and edges of the product (Sumnu and Sahin, 2005). In addition, quality could also be compromised due to insufficient time for some biochemical reactions to occur. However, modernization in the design of microwave equipment to include features such as phase control heating, variable frequency or the combination of microwave energy with other thermal methods may improve the performance of MW food processing (Sumnu and Sahin, 2005). Advances in process development and controls also need to be accompanied with innovations in the area of packaging research and design.

## 23.5 Infrared (IR) and radiofrequency (RF) heating

Another processing method involving dielectric heating is RF heating which has the potential for the rapid heating of solid and semi-solid foods. RF heating refers to the heating of dielectric materials with electromagnetic energy at frequencies between 1 and 300 MHz (Orfeuil, 1987). Research and application of RF sterilization of packaged foods has been fairly limited. Previously, researchers at Washington State University, in conjunction with laboratories at Strayfield UK, the US Army Natick Soldier Center and the US Army Combat Ration Network have designed and created a prototype 27

© Woodhead Publishing Limited, 2012

MHz RF sterilization system to process foods pre-packaged in a polymeric tray or a multi-tray system to mimic large-scale industrial production systems (Ramaswamy and Tang, 2008). The research group demonstrated the efficacy of RF energy to inactivate heat-resistant spores to produce shelf-stable pre-packed foods such as RF-processed eggs, pasta and meats. In addition, research work undertaken by the group also involved the development of computer models to predict RF energy generation in packaged foods as well as improvement in the design of RF sterilization systems (Ramaswamy and Tang, 2008).

IR uses electromagnetic radiation generated from a hot source (quartz lamp, quartz tube, or metal rod) resulting from the vibrational and rotational energy of molecules. Thermal energy is thus generated following the absorption of radiating energy. Although IR is an emerging technology for the food industry, research on its application to process food has been conducted on an experimental or pilot scale. IR provides instant heating, thereby cutting down on the need for heat build-up. In addition, compared to conventional heating equipment, the operating and maintenance costs are lower and it is a safer and cleaner process.

### 23.5.1 Post-package decontamination using IR and RF and packaging considerations

Short (1 μm) and intermediate (5 μm) wave IR heating have gained wide acceptance and have been applied for the rapid baking, drying and cooking of foods of even geometry and modest thickness (Tewari, 2007). Short-wave IR has a penetration depth of several mm in many foods. In addition, it is not absorbed by transparent plastic packaging and therefore has been successfully applied for the surface pasteurization of packaged bakery products (Tewari, 2007). Long-wave IR (30 μm) has been in use for industrial cooking and drying applications, achieving shorter processing times when compared to convective heating.

Significant research on the applications of RF has also been carried out. Various materials have been used to package foods for processing by RF including waxed paper (Cathcart *et al.*, 1947), cellophane (Cathcart *et al.*, 1947), Cryovac tubings (Bengtsson and Green, 1970), or even tubes made of glass (Houben *et al.*, 1991). Advances in packaging research and design for RF-heated foods, such as the development of high-barrier plastic films through the application of EVOH or silicon oxide (SiOx) surface coatings, have also been reported (Tewari, 2007). Future developments should take into account (i) the compatibility and possible interaction among the packaging material, electromagnetic waves, and foods, (ii) the barrier and mechanical performances of the package, (iii) the possible migration of packaging components into foods, and (iv) the associated costs.

© Woodhead Publishing Limited, 2012

## 23.6 High hydrostatic pressure (HHP)

Among the array of non-thermal processing technologies, HHP has garnered the most attention since the early 1990s. For the last 20 years, HHP has been explored quite intensively by food research institutions as well as the food industry with the goal of enhancing the safety, quality, nutritional, and functional properties of a wide variety of packaged foods with minimal deleterious effects on their nutritional and organoleptic characteristics (Welti-Chanes *et al.*, 2005). During HHP, pre-packaged foods are exposed to pressures of the order of 200–600 MPa for a few minutes. Process temperatures during treatment can vary from subzero temperatures to above the boiling point of water (100°C) (Caner *et al.*, 2004a). Because HHP is performed on packaged foods, cost-intensive aseptic package sterilization can be avoided (Rastogi *et al.*, 2007).

### 23.6.1 Mechanisms of microbial control of HHP

The mechanism of microbial control and inactivation lies in a combination of processes including the breakdown of non-covalent bonds in large macromolecules, as well as the disruption and permeabilization of the cell membrane (Rastogi *et al.*, 2007; Welti-Chanes *et al.*, 2005). Food microorganisms, such as vegetative bacteria, human infectious viruses, fungi, protozoa, and parasites, can be reduced significantly when subjected to high pressure. The mechanisms of inactivation by HHP are discussed in greater detail in Chapter 13.

### 23.6.2 Packaging considerations for HHP

The nature, characteristics, and integrity of the packaging material are critical determining factors. Ensuring that the food package fits into the chamber, is made of a material that is compatible with the product, and encloses minimal headspace are important considerations as well (Han and Floros, 2007). Indeed, air is much more compressible than water and contains reactive oxygen, therefore the headspace within the package should be kept to a minimum (Lambert *et al.*, 2000). Packaging materials also need to have strong physical and mechanical properties to be able to withstand the high pressures. Ideally, the packaging material should be elastic and flexible, i.e., undergo reversible deformation with no permanent change to the shape and geometry of the material (Caner *et al.*, 2004a, b). Hence, rigid containers made of glass or metal are generally not suitable for HHP. Typical flexible polymer-based containment systems commonly used include plastic stomacher bags, polyester tubes, PE pouches, nylon cast PP pouches, and various other flexible pouch systems (Tewari, 2007). In addition, the material needs to be chemically inert and have good moisture and gas barrier properties (Han, 2007). Package and seal integrity should also not be compromised after pressure treatment in order to avoid post-process recontamination and ensure

© Woodhead Publishing Limited, 2012

food safety. To that end, several researchers have evaluated the suitability of packages for high pressure processing of food products with respect to barrier and mechanical properties (Caner *et al.*, 2004a; Lambert *et al.*, 2000; Masuda *et al.*, 1992), migration of compounds (Caner *et al.*, 2004b; Kubel *et al.*, 1996), delamination (Lambert *et al.*, 2000; Schauwecker *et al.*, 2002) and sealing integrity (Caner *et al.*, 2004a; Dobias *et al.*, 2004).

### 23.6.3 Post-package decontamination using HHP

HHP is applied today at a wide range of temperatures in order to pasteurize foods in their packaging while preserving their organoleptic and nutritional qualities. The food and the package are treated together so that post-processing contamination can be prevented. Several researchers have investigated the efficacy of HHP to inactivate foodborne pathogens *L. monocytogenes*, *S. aureus* and *Salmonella* on vacuum-packaged RTE meat and poultry products to enhance their microbiological safety (Garriga and Aymerich, 2009). Similarly, the commercial use of Safe Pac™ high pressure pasteurization to inactivate microorganisms and extend the shelf life of pre-packaged RTE products including meats, soups, wet salads, sauces, fruit smoothies, shellfish, and seafood by 200–300% has also been reported (Safe Pac LLC, 2010). Jofré *et al.* (2008a,b) and Marcos *et al.* (2008) demonstrated the synergistic effect of HHP and antimicrobial packaging incorporating natural antimicrobials to control *Salmonella* on sliced cooked ham. Jofré *et al.* (2008b) also demonstrated that antimicrobial packaging, HHP, and refrigerated storage acted as an effective triple combination of hurdles to inhibit *Salmonella* sp., *L. monocytogenes* and *S. aureus* in cooked ham. Another innovative example of 'hurdle technology' involves the use of HHP at 300–800 MPa to treat foods in contact with an antimicrobial packaging material to ensure a delayed release of the antibacterial compound, allyl isothiocyanate (AIT). AIT is encapsulated within cyclodextrins inside a polylactic acid (PLA) matrix to ensure a controlled release of the agent during ambient temperature storage (INRA, 2010).

Recent progress and development in the area of pressure-assisted thermal sterilization (PATS) have enabled the use of very high pressures (up to 900 MPa) combined with temperatures above 100°C to achieve a degree of microbiological inactivation akin to thermal sterilization but with a more refined temperature control (Heinz and Knorr, 2002). The overriding advantages of PATS lie in the better preservation of organoleptic and nutritional properties of foods as well as the safety and convenience of processing foods already in their packaging to minimize post-processing recontamination.

### 23.6.4 Challenges and limitations associated with the use of HHP

Although various flexible packaging systems have been used for HHP products, other forms of packaging including metal cans, glass bottles,

© Woodhead Publishing Limited, 2012

and paperboard-based packages are inappropriate (Min and Zhang, 2007). Moreover, the presence of metals such as aluminum in multilayered packaging was also affected during HHP processing (Caner *et al.*, 2004a). The researcher attributed it to the differential elasticity of the different components of the package, resulting in rupture of the aluminum layer given its rigidity. Other authors have reported that HHP also affected the barrier, mechanical and mass transfer properties of the package. In some instances, the general integrity of the material was compromised after pressure treatment. Delamination of the package due to the type or quality of the materials and adhesive has also been reported (Schauwecker *et al.*, 2002).

Since bacterial spores are quite baro-resistant and difficult to inactivate using pressure, HHP has greater commercial relevance for pasteurization rather than sterilization applications. Pressure-pasteurized foods need low temperature storage, and distribution to retain their sensory and nutritional qualities. Hence, the potential for temperature abuse to occur during refrigerated storage and distribution has to be carefully monitored and minimized. In addition, because of the high capital cost and costs associated with operations and service of the high-pressure equipment, the use of HHP in food processing is limited to certain niche products. HHP also has limited application for foods with a large number of air spaces such as plant-derived foods (Michel and Autio, 2001). Air within the package of the food can be detrimental to quality. Therefore, headspace must be limited to maximize efficient utilization of the chamber space and the package space (Rastogi *et al.*, 2007). Removal of air from the package also considerably reduces the time to reach target pressure and thus cuts down costs (Hogan *et al.*, 2005). With regard to the regulatory aspects, establishment of microbiological criteria for HHP-processed foods (Rastogi *et al.*, 2007), and adherence to Good Manufacturing Practices (GMP) and Hazard Analysis Critical Control Point (HACCP) are also critical.

The comparison of experimental results produced in different laboratories is also challenging due to insufficient description of methodologies used and differences in sample preparation and packaging procedures, as well as differences in the strains of inocula used. Furthermore, the HHP equipment used has different designs and configurations (Rastogi *et al.*, 2007). It is thus necessary to provide an adequate description of the HHP equipment such as chamber size, material of construction, wall thickness, pressure-transmitting medium, packaging material, heating and cooling systems, power specification, data acquisition system, and any other relevant information (Rastogi *et al.*, 2007). HHP treatments are also dependent on the dual effect of pressure and temperature. Documentation of the temperature of the pressure vessel and pressure-transmitting fluid, and the temperature history of the packaged product under pressure is essential for the optimization and design of industrial processes (Denys *et al.*, 1997).

© Woodhead Publishing Limited, 2012

## 23.7 Irradiation

Food irradiation is a process involving the exposure of food to ionizing radiations, such as gamma rays emitted from the radioisotopes cobalt-60 and cesium-137, and high energy electrons and X-rays produced by machine sources (Ohlsson, 2002). Ionizing radiation can be applied to decontaminate pre-packaged food. Since food and packaging materials are irradiated concurrently, the radiation stability of the packaging material is of paramount importance if the technology is to be implemented successfully.

### 23.7.1 Mechanisms of microbial control using irradiation

During irradiation, molecules absorb energy to form highly labile and reactive ions or free radicals that result in breakage of a small percentage of chemical bonds (Ohlsson, 2002). High-energy electrons fracture molecules (typically water molecules) into highly reactive species such as radicals, which can disrupt other cellular macromolecules such as DNA, proteins, or structures such as cell membranes (Fellows, 2000). The mechanisms of inactivation by irradiation are discussed in greater detail in Chapter 11.

### 23.7.2 Packaging considerations for irradiation

Foods to be irradiated are typically packaged in the final form prior to being processed, thus reducing the likelihood that new pathogens will be introduced after the irradiation step. The radiation must therefore have the ability to penetrate the packaging materials of interest (Han, 2007). In addition, the seal integrity and barrier properties of the package must be uncompromised since this may nullify the benefits of post-packaging decontamination (Morehouse and Komolprasert, 2004). Careful selection of packages and packaging atmospheres should also aim to minimize nutrient degradation during irradiation (Morehouse and Komolprasert, 2004). Packaging foods in the absence of oxygen, with subsequent exposure to radiation at low temperatures, will reduce vitamin loss, which can be further minimized by storing sealed food packages at low temperatures (Sommers, 2006).

Retention of the chemical and structural integrity of packaging materials during and post-treatment is also critical to ensure the microbial and toxicological safety of the food (Han, 2007). The majority of food packaging materials are composed of polymers that may be susceptible to chemical changes induced by ionizing radiation. Two important chemical reactions are cross-linking (polymerization) and chain scission (degradation) (Guillard *et al.*, 2010). Cross-linking of polymers occurs primarily under vacuum or an inert atmosphere, while chain scission is predominant in the presence of oxygen. Hence, the oxygen content within the package influences the reaction pathways affecting the packaging materials. Radiolytic products formed upon irradiation and present at high enough levels can potentially

© Woodhead Publishing Limited, 2012

migrate into food and affect the odor, taste, or safety of the irradiated food. Hence, careful choice of suitable packaging materials is necessary to avoid contamination of food with products of radiolysis (Ozen and Floros, 2001). The formation of radiolytic products is also dependent on the conditions of irradiation, including the absorbed dose, dose rate, the amount of oxygen in the atmosphere, the temperature and storage duration after irradiation. Research has shown that single-layer plastics are not able to meet the requirements of food packaging subject to irradiation while multilayered structures produced by co-extrusion, lamination, or co-injection molding are suitable candidates (Chuaqui-Offermanns, 1989). For any newly introduced food packaging material to be used with irradiation treatment, the safety of these materials must be assessed and then approved by the Food and Drug Administration (FDA) before the materials can be legally used in direct contact with food (Ozen and Floros, 2001). Besides the nature of the material, the geometrical shape of the container is also important (Rahman, 2007). Graham (1980) mentioned that a cubical form is the geometrical shape of choice since it allows optimum dose distribution during irradiation.

### 23.7.3 Post-package decontamination using irradiation

Various studies have demonstrated the ability of irradiation to destroy foodborne pathogens and spoilage microbiota on various pre-packaged meats and plant-derived products. Fan and Sokorai (2005) investigated the effects of various doses of irradiation on the quality of fresh-cut iceberg lettuce packaged in film bags and determined that 1–2 kGy was the optimum dose to ensure satisfactory inactivation of spoilage bacteria while minimizing sensorial changes. Lamb *et al.* (2002) showed that low-dose gamma irradiation effectively reduced *S. aureus* in ready-to-eat ham-and-cheese sandwiches and proved to be more effective than refrigeration alone. The integrity of the post-irradiated packaging material was not significantly changed. Other authors also showed the feasibility of irradiation at 2–2.5 kGy to reduce the population of aerobic bacteria (Niemand *et al.*, 1983) and bacterial members of the Enterobacteriaceae (Farkas *et al.*, 1997) on vacuum-packaged chilled meat products for shelf life extension.

In addition to the application of irradiation alone, combined application of ionizing radiation and modified atmosphere packaging (MAP) has also been investigated. Fan *et al.* (2003) demonstrated enhanced safety and improved quality of fresh-cut lettuce irradiated in MAP at doses of 1–2 kGy. Lambert *et al.* (2000) showed that the combination of irradiation (1 kGy) and MAP (10–20% $O_2$, balance $N_2$) significantly reduced the microbial populations of fresh pork, although to the detriment of organoleptic quality. Similarly, Przybylski *et al.* (1989) showed that application of low-dose irradiation and MAP brought about a four- to five-fold extension in the shelf life of fresh catfish fillets.

© Woodhead Publishing Limited, 2012

### 23.7.4 Challenges and limitations associated with the use of irradiation

Compared to other processing technologies, irradiation has relatively low operating costs due to its low energy requirements; however, the high investment costs, the requirement for a critical minimum capacity and product volume for economically feasible operation (Rahman, 2007), and the complexity of regulatory actions have been obstacles to the successful implementation of this technology (Arvanitoyannis and Tserkezou, 2010). Irradiation of food can also result in nutritional degradation and undesirable organoleptic changes with concomitant off-flavor development (Arvanitoyannis and Tserkezou, 2010). Factors affecting the kinetics of nutrient degradation that need to be carefully monitored include the radiation dose, treatment temperature, oxygen availability, and product composition. Irradiated foods should also be properly handled, stored and distributed to prevent chemical and biochemical deterioration, spoilage or further loss of nutritive value (Sommers, 2006; Thayer, 1994). Moreover, although irradiated foods have been declared as safe and wholesome, consumer attitudes toward this technology still remain one of the major factors hindering wide application of this technology. Griffith (1992) attributed the poor growth and acceptance of food irradiation to lack of consumer education and consultation with customers.

## 23.8 Pulsed light (PL) technology

PL technology is an innovative method for the decontamination and sterilization of foods using very high power and very short duration pulses of light emitted by inert gas flashlamps (Palmieri and Cacace, 2005). The high power pulses of radiation can be in the spectra of UV, visible (VL) and IR light. In addition to being used for the sterilization of packaging materials and online sterilization of transparent fluids, PL has also successfully been used for the surface decontamination of foods in plastic packaging. This process has shorter treatment times with concomitantly higher throughput, rendering the process very efficient.

### 23.8.1 Mechanisms of microbial control of PL

UV light exhibits germicidal properties in the wavelength range of 100–280 nm (UV-C region). UV light energy is thought to have a photochemical, photothermal, and/or photophysical impact on microorganisms. Wekhof (2000) summarized that the lethal effect of pulsed UV light is a combination of the germicidal nature of UV light (photochemical) and cellular disruption caused by local heating (photothermal). It is thought that the constant disturbance caused by the high-energy pulses also induces morphological damage to bacterial cells, ultimately leading to cell wall damage, cytoplasmic membrane shrinkage, cell content leakage, mesosome rupture, and genetic material

© Woodhead Publishing Limited, 2012

leakage (Krishnamurthy *et al.*, 2010). The mechanisms of inactivation by pulsed light technology are discussed in greater detail in Chapter 12.

### 23.8.2 Packaging considerations for PL

One of the packaging considerations for PL treatment of foods is the degree of transparency of the containment material. Polymers suitable for PL treatment of packaged food include PE, PP, polybutylene, ethylene vinyl acetic acid (EVA), nylon, Aclar®, and EVOH, which have the ability to transmit light and thus meet the requirements for PL processing (Anon, 2000). Most plastic packages are able to effectively transmit broadband light, except for PET, polycarbonate, polystyrene, and polyvinyl chloride (PVC). In addition to the transparency of the outer package, it is also essential that the surface topography is smooth and clear (Elmnasser *et al.*, 2007). Presence of pores and grooves on the surface could shadow microbial cells from light, causing incomplete light diffusion and reducing process effectiveness (Elmnasser *et al.*, 2007).

### 23.8.3 Post-package decontamination using PL

Dunn *et al.* (1995) reported that microbial log reductions of pathogens achieved by PL on relatively simple surfaces are considerably higher than in foods with more complex surface aspects. Indeed, efficient light absorption depends on the distance through which light is passing, the thickness of the product, and the thickness of the package. Hillegas and Demirci (2003) demonstrated the effect of product thickness on PL-induced microbial inactivation and showed a significant increase in reduction of *C. sporogenes* spores in clover honey in samples of 2 mm thickness compared to samples that were four-fold thicker. Similarly, Haughton *et al.* (2011) demonstrated the efficacy of PL to inactivate *Campylobacter* on packaged chicken and also showed an inverse correlation between the microbial reduction and film thickness. Keklik *et al.* (2010) demonstrated the optimal efficacy of pulsed UV light to decontaminate vacuum-packaged boneless chicken breasts for 15 s when placed at a distance of 5 cm from the quartz window in the pulsed UV light chamber. The authors reported that the treatment resulted in approximately 2 log reduction in the population of *Salmonella* Typhimurium, with minimal impact on the quality and color of the sample tested. The same authors demonstrated that PL was effective in reducing the population of *L. monocytogenes* on vacuum-packaged chicken frankfurters by 1.9 log, although the treatment significantly affected the color parameters of the samples (Keklik *et al.*, 2009). Moreover, the applicability of PL to pasteurize bottled beer has also been demonstrated previously (Palmieri and Cacace, 2005).

The use of PL to enhance the microbiological quality and extend the shelf life of foods has also been examined. Shuwaish *et al.* (2000) showed that the application of PL to high-density polyethylene (HDPE) pre-packaged catfish

© Woodhead Publishing Limited, 2012

fillets led to a decrease in coliform and psychrotrophic bacterial counts. Dunn *et al.* (1991) previously showed that the application of PL to freshly baked cakes packaged in clear plastic containers resulted in non-detectable molds when stored at ambient temperature for 11 days while their untreated counterparts produced visible molds. Rice (1994) similarly showed that PL can be used to inactivate mold spores and consequently extend the shelf life of foods, such as packaged sliced white bread and cakes, from a few days to over two weeks. The shelf life of cakes, pizza and bagels wrapped in clear plastic films was also extended to 11 days during storage at ambient temperature following PL treatment (Ohlsson, 2002). Djenane *et al.* (2003) reported that UV irradiated beef steak packaged in PE pouches with modified atmosphere followed by chilled storage at 1°C had an extended shelf life of 16 days.

### 23.8.4 Challenges and limitations associated with the use of PL

The efficiency of PL depends on the degree of exposure and susceptibility of microbes to the treatment. The sterilization of foods is possible as long as the packaging and packaged content are UV transparent; however, because of the opacity and irregular surface topography of most foods, PL is mostly effective for surface decontamination. Microorganisms penetrating into the deep crevices or surface irregularities of food are thus protected from the effects of PL due to the 'shadow effect' (Elmnasser *et al.*, 2007). Haughton *et al.* (2010) reported that this 'shadowing effect' can be exacerbated with the use of certain packaging materials such as polyethylene-polypropylene (PET-PP) and polystyrene (PS). The authors mentioned that the porous nature of the polymers can enable bacteria to hide in the package 'sub-surface', thereby shielding from UV light. In addition, the efficacy of PL is also limited by the thickness of the film (Haughton *et al.*, 2011). Furthermore, PL also has diminished efficacy on heavily contaminated foods, i.e., products with a high initial load of the target microorganisms due to the shadow effect on the underlying layer of cells (Wuytack *et al.*, 2003).

In addition, researchers have also demonstrated that UV light can interact with the packaging materials yielding undesirable products. UV can cause the package constituents to undergo certain chemical reactions. Materials made of PET are strong UV absorbers and undergo extensive oxidation following UV exposure (Ozen and Floros, 2001). Examples of surface oxidation products formed after treatment of certain plastics include carbonyls, carboxylic acids and hydroperoxides (Peeling and Clark, 1983). Researchers showed that a food solution containing d-limonene had a higher detectable level of oxidation products following contact with a UV-treated PET film. UV-treated LDPE also resulted in detectable amounts of oxidation products in the tested food solution albeit to a lesser extent (Berends, 1996). The higher abundance of oxidation products observed in samples contacted with PET may be attributed to the polar nature of PET, thought to accelerate the oxidation of

© Woodhead Publishing Limited, 2012

the film surface during UV treatment (Berends, 1996). In addition, exposure of a model food solution containing linoleic acid to UV light-treated PET resulted in significant accumulation of a major oxidation product, hexanal, over time. UV-treated LDPE produced lower amounts of oxidation products compared to PET; although oxidation was still significantly higher than of the samples contacted with untreated LDPE (Peeling and Clark, 1983). These observations are therefore a testament to the need to choose an optimal packaging material to assure the safety and lack of toxicity of the 'in-package' UV-treated food. In addition to the effect of UV on the migration properties of plastic films, the mechanical properties of films were also affected after UV treatment. Berends (1996) observed extensive crystallinity in PET film after UV exposure compared to LDPE, which presented limited structural modifications. Polypropylene is a fairly UV light transmissible polymer; however, it has been reported to be quite sensitive to ultraviolet degradation (Guillard *et al.*, 2010). UV aging caused embrittlement of polypropylene and also negatively affected the mechanical properties of the polymer, including its yield strength and percent elongation.

Another limitation is that local sample heating could be a concern for extended PL treatments lasting longer than 5s (Krishnamurthy *et al.*, 2010). Heat can originate from absorption of light by the food or by the lamp itself. For example, when studying the inactivation of *A. niger* spores on corn meal under certain settings, sample temperatures peaked to as high as 120°C, potentially burning the products (Jun *et al.*, 2003).

In addition, since UV treatment can interact with packaging materials and result in the formation of certain oxidation products, foods rich in proteins and lipids are not suitable for decontamination by PL (Elmnasser *et al.*, 2007). Discoloration of the product (Djenane *et al.*, 2003) and/or the packaging material could also occur and should be minimized.

## 23.9 Active packaging

Conventional mainstream packaging acts as a straightforward food container to protect, transport, and sell the products. Active packaging is more dynamic and is able to sense changes in the environments in the interior and exterior of a package by adjusting properties of the package (Han, 2007).

### 23.9.1 Mechanisms of microbial control of active packaging

Active packaging serves as an effective intervention or hurdle to reduce or prevent microbial growth during food storage. This is achieved by:

- actively or constantly changing permeation properties or the gas composition in the package headspace during storage, or
- actively adding antimicrobial agents incorporated into the packaging

© Woodhead Publishing Limited, 2012

materials to be released in minute concentrations over time (Hurme *et al.*, 2003).

The functionality of active-packaging systems can be categorized as: absorbers, releasers, or carriers of active agents (Han, 2003). The following section will deal with the various forms of active packaging that have found wide applications in food protection and preservation, notably oxygen control (oxygen scavengers), carbon dioxide control, antimicrobial release and modified and controlled atmosphere packaging.

### 23.9.2 Types of active packaging and uses in food decontamination

*Oxygen absorbers/scavengers*

Oxygen is a major stimulator of microbial deterioration in food products. As a result, removal or exclusion of oxygen using a high gas barrier is a key factor in controlling microbial growth. As the name implies, oxygen scavengers are sometimes called absorbers as they remove excess oxygen from the internal package environment. Scavengers may be incorporated in sachets (small pouches) or labels either affixed to the package or incorporated into the package material or resin (Han, 2007). Commercial oxygen scavengers can be inorganic (e.g., ferrous iron) or organic (e.g., unsaturated hydrocarbons, nylons or benzyl acrylate polymers) in nature. Most common food-related applications of oxygen scavengers relate to ready-made foods such as hamburgers, fresh pastas, noodles, sliced ham, sausages, cakes and shelf-stable bread, confections, nuts, coffee, herbs and spices.

*Ethanol emitters*

Ethanol has been used as an antifungal preservative by virtue of its ability to denature proteins in yeasts and molds. Ethanol can be administered into the package prior to sealing the food inside to achieve a prolonged desired antimicrobial effect during storage or through the use of ethanol-generating sachets. A commercially available ethanol emitter is Ethicarp®, consisting of an ethanol/water mixture adsorbed onto a substrate of silicon dioxide powder, which is held inside a sachet made of laminated paper-EVA copolymer. Ethanol vapor generators have been used in Japan to inhibit mold growth on high moisture bakery goods and fish products (Hurme *et al.*, 2003). Cakes with generally high moisture content did not become moldy when packaged with an ethanol-vapor generating sachet; however, the main drawback with using ethanol vapor for preservation purposes is the unpleasant odor and off-flavor formation. Moreover, the absorption of ethanol into the food could result in significant levels of ethanol in the products, which could raise regulatory concerns.

*Antimicrobial packaging*

Antimicrobial agents can be incorporated into packaging materials, which can then slowly leach into the food. Antimicrobial packaging is especially

© Woodhead Publishing Limited, 2012

suitable for vacuum-packaged foods by virtue of the intimate contact between the packaging material and the food surface. Examples of antimicrobial substances are bacteriocins (e.g., nisin), organic acids and esters, as well as inorganic substances such as silver, synthetic zeolite and aluminum silicate (Abe, 1990). These antimicrobial substances can be introduced into the packages by insertion of a separate porous sachet, immobilization of the antimicrobial onto the surface of the package by covalent bonds (Haynie *et al.*, 1995), impregnation onto the package substrate (Grower *et al.*, 2006) or incorporation into the material resin.

Smith and Rollin (1954) demonstrated the antimycotic effectiveness of a sorbic acid wrapper made of moisture-proof cellophane applied to natural and processed cheeses. Vojdani and Torres (1990) created cellulose-based edible films with antifungal functionality by the incorporation of various food-grade agents such as sorbic, propionic and benzoic acids. The application of bacteriocins, such as nisin (Neetoo *et al.*, 2008) and enterocins and sakacin (Jofré *et al.*, 2008a) to packaging films to inhibit the growth of Gram-positive bacteria, such as *L. monocytogenes*, on cold-smoked salmon has also been demonstrated. Figure 23.3 compares the development of *L. monocytogenes* on cold-smoked salmon samples packaged in antimicrobial films and stored at refrigeration and abusive temperatures. Similarly, Natrajan and Sheldon (2000) showed that nisin-coated PVC films delayed the growth of *Salmonella* Typhimurium on fresh broiler chicken skin and drumsticks. Chitosan is another antimicrobial agent that has been extensively studied for its antimicrobial functionality in food packaging. The incorporation of natural antimicrobials such as salts of organic acids (e.g., sodium lactate) in chitosan-coated plastic films has been reported to hinder outgrowth of *L. monocytogenes* and extend the shelf life of RTE products such as ham steaks (Ye *et al.*, 2008a) and cold-smoked salmon (Ye *et al.*, 2008b) with a high degree of efficacy throughout their refrigerated shelf life of 12 and 8 weeks, respectively.

*Edible films and coatings*

The use of edible films and coatings is an application of active food packaging, since the antimicrobial activity, edibility and biodegradability of the films and coatings are auxiliary functions that are not conferred by the conventional packaging systems (Han, 2007). Edible films and coatings are useful materials originating from edible biopolymers, including proteins, polysaccharides (carbohydrates and gums), and lipids (Gennadios *et al.*, 1994). Films and coatings not only provide an inherent form of protection but can also allow the slow release of incorporated additives, such as antimicrobial compounds, to impede microbial growth on the surface of foods. Films made of chitosan (Jiang *et al.*, 2010) and coatings made of alginate (Neetoo *et al.*, 2010), chitosan (Jiang *et al.*, 2010) and pectin (Juck *et al.*, 2010) were shown to act as effective carriers of antimicrobial agents, including nisin and salts of organic acids, to delay the growth of *L. monocytogenes* on various types of

© Woodhead Publishing Limited, 2012

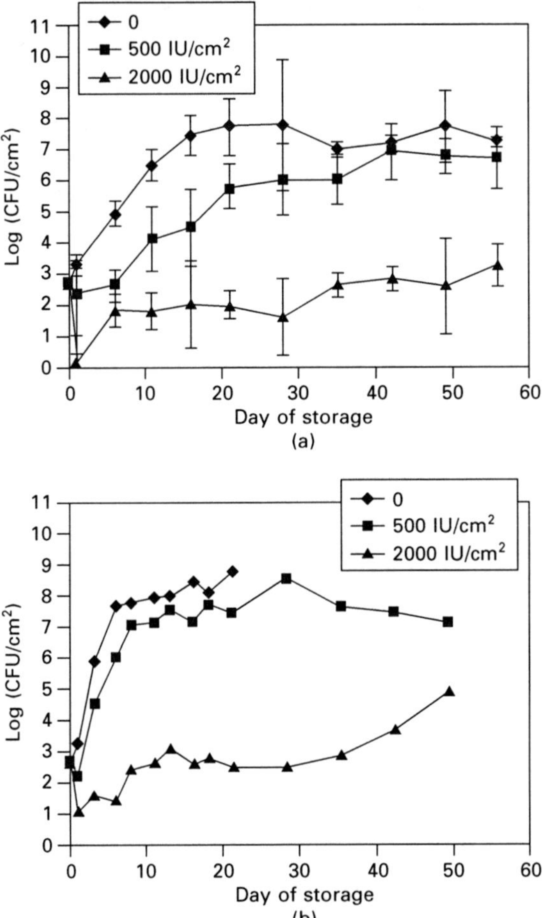


**Fig. 23.3** Effect of nisin concentration on LDPE film, inoculation level, and storage temperature on the growth of *L. monocytogenes* on cold-smoked salmon. Samples were inoculated with low ($5 \times 10^2$ CFU/cm$^2$) or high ($5 \times 10^5$ CFU/cm$^2$) levels of *L. monocytogenes* and stored at 4 or 10°C. (a) Low inoculation level and 4°C; (b) low inoculation level and 10°C; (c) high inoculation level and 4°C; and (d) high inoculation level and 10°C. Error bars represent ± 1 standard deviation. To make the figures easy to read, error bars are only shown for (a).

RTE foods during refrigerated storage. The evaluated foods were sliced and filleted cold-smoked salmon (Jiang *et al.*, 2010; Neetoo *et al.*, 2010) and roasted and poached turkey (Juck *et al.*, 2010).

*Modified atmosphere packaging*

Modified atmosphere packaging (MAP) relies on the replacement of air in a food package with a specified gas mixture. The most commonly used gases

© Woodhead Publishing Limited, 2012

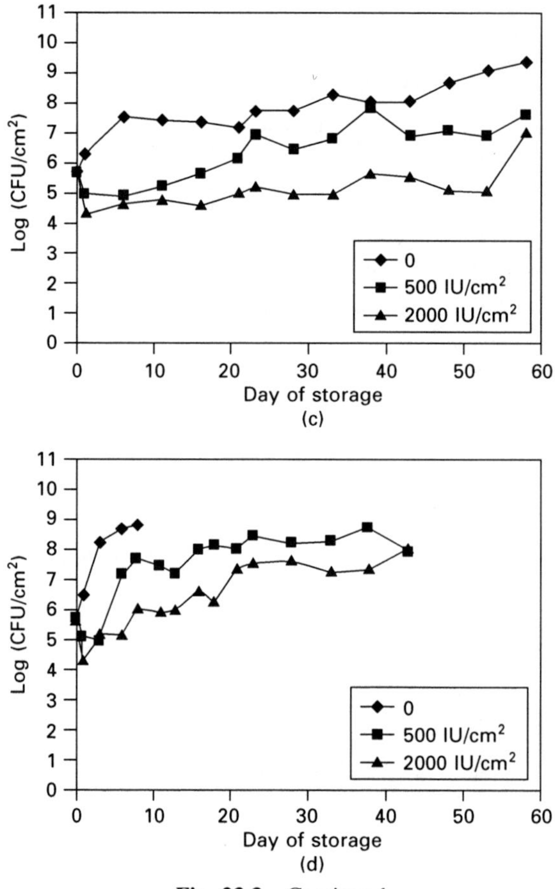


**Fig. 23.3** Continued

in MAP are nitrogen, oxygen, carbon dioxide and to a lesser extent noble gases such as argon. Depending on the final product under consideration, the composition of the gaseous atmosphere will vary. In addition to enhancing the quality and presentation of foods, MAP has also been used in food packaging to retard the growth rates of spoilage and pathogenic microorganisms. Carbon dioxide is the most important gas in MAP foods by virtue of its bacteriostatic and fungistatic properties. According to Farber (1991) and Dixon and Kell (1989), the mechanism of bacterial inactivation by carbon dioxide relies on:

- alteration of cell membrane functions, including effects on nutrient uptake and absorption,
- direct inhibition or reduction in the rate of enzymatic reactions,
- alteration of intracellular pH, and
- modification of the physicochemical properties of proteins.

© Woodhead Publishing Limited, 2012

The bacteriostatic ability of elevated levels of carbon dioxide (20–50%) has been shown to significantly control the growth of psychrotrophic pathogens relevant to MA-packaged produce (Bennik *et al.*, 1995).

### 23.9.3 Challenges and limitations associated with the use of active packaging

Addition of agents to packaging materials may negatively alter the chemical and physical properties of polymers. Critical properties of polymers, such as permeability, mechanical strength, color, appearance, and gloss, can deteriorate (Han *et al.*, 2004). The incorporation of certain components in the polymer film can also lead to their undesirable migration into food (Lopez-Rubio *et al.*, 2004). Moreover, the use of antimicrobials or other active agents should follow the guidelines of regulatory agencies. Active agents and packaging materials are regarded as food contact substances or food additives and require clear classification prior to commercialization. In addition, consumers may have health-related concerns about active packaging due to the possible interaction between the packaging and the food (Meroni, 2000). Han and Floros (2007) also mentioned that the manufacturing of new packaging systems requires processing procedures different from traditional packaging, which in turn may increase installation costs for establishing new systems, as well as add manufacturing costs to these novel materials.

One of the main drawbacks of MAP is that the resulting shelf life extension and microbiota suppression may provide ample opportunity for pathogens to grow (Phillips, 1996). Careful monitoring of parameters such as the food matrix and formulation, the pathogen load, and the packaging material itself, as well as storage parameters such as gas composition and storage temperatures, is therefore vital. Moreover, to maximize the efficacy of MAP, it is important to use high-quality raw materials in conjunction with processing, packaging, and distribution under good sanitary and hygienic practices to minimize cross-contamination. A quality assurance system through the implementation of HACCP throughout the production line needs to be established to ensure safety of MAP foods (Sivertsvik *et al.*, 2002).

## 23.10 Conclusion and future trends

Recalls, largely caused by the transference of pathogens such as *L. monocytogenes* into food products between the final processing step and packaging of perishable foods, have spurred interest in post-packaging decontamination. Improvements to existing designs and development of new technologies for thermal post-packaging applications have included hot water bath immersion, steam pasteurization, sous-vide processing, microwaving, and RF heating. Non-thermal post-packaging decontamination systems can include high hydrostatic pressure, irradiation, pulsed light

© Woodhead Publishing Limited, 2012

technology and active packaging. One can expect continuing research into these technologies in tandem with advances in food packaging materials, given increasing consumer awareness of product safety linked to human health and the potential of economic losses associated with contaminated products. Since non-thermal methods do not incorporate additional exposure of packaged products to elevated temperatures which can further degrade sensory quality and nutrient content, one can easily envision post-packaging decontamination using a non-thermal method, such as HHP, becoming more commonplace in the food industry in order to assure enhanced food safety at a moderate cost.

## 23.11 Sources of further information and advice

### 23.11.1 Books

BROWN W E (1992), *Plastics in food packaging: properties, design and fabrication*, New York, Marcel Dekker.

HAN J H (2007), *Packaging for nonthermal processing of foods*, Ames, IA, Blackwell.

HOLDSWORTH D and SIMPSON R (2008), *Thermal processing of packaged foods*, New York, Springer Science+Business Media.

MATHLOUTHI M (1994), *Food Packaging and Preservation*, London, Blackie Academic and Professional.

### 23.11.2 Journal articles

GALIC K, SCETAR M and KUREK M (2011), 'The benefits of processing and packaging', *Trends Food Sci Technol*, 22, 127–137.

GUILLARD V, MAURICIO-IGLESIAS M and GONTARD N (2010), 'Effect of novel food processing methods on packaging: structure, composition and migration properties', *Crit Rev Food Sci Nut*, 50, 969–988.

OZEN B F and FLOROS J D (2001), 'Effects of emerging food processing techniques on the packaging materials', *Trends Food Sci Technol*, 12, 60–67.

### 23.11.3 Media

Food processing safety and related microbiology (Video), Food Safe Series (www.agcmedia.com).

### 23.11.4 Guidelines

Campden and Chorleywood Food Research Association (1992), Guidelines for the good manufacture and handling of modified atmosphere packed food products, Technical Manual No. 34.

US Food and Drug Administration (2009), 'Kinetics of microbial inactivation

© Woodhead Publishing Limited, 2012

for alternative food processing technologies'. Available from: http://www.fda.gov/Food/ScienceResearch/ResearchAreas/SafePracticesforFoodProcesses/ucm100158.htm (accessed 29 July 2011).

### 23.11.5 Professional organizations

Institute of Food Technologists (IFT) Non Thermal Division
Institute of Food Science and Technology

### 23.11.6 Federal agencies/Regulatory bodies

US Department of Agriculture
US Food and Drug Administration
US Environmental Protection Agency
Center for Disease Control and Prevention (CDC)
IFT Science Advisory Board

### 23.11.7 Research centers

National Institute of Health
National Science Foundation
Illinois Institute of Technology
Campden and Chorleywood Food Research Association
Food Science Australia
European High Pressure Research Group
UK High Pressure Club for Food Processing
Japanese Research Group of High Pressure

### 23.11.8 Consortium

National Center for Food Safety and Technology

## 23.12 References

ABE Y (1990), 'Active packaging: a Japanese perspective', *International conference on modified atmosphere packaging*, The Campden Food and Drink Research Association, England, 15–17 October 1991.

AHMED J and RAMASWAMY H S (2007), 'Microwave pasteurization and sterilization of foods', in Rahman M S, *Handbook of food preservation*, Boca Raton, FL, CRC Press, 691–712.

ANNOUS B and KOZEMPEL M F (2006), 'Surface pasteurization with hot water and steam', in Sapers G M, Gorny J R and Yousef A E, *Microbiology of fruits and vegetables*, Boca Raton, FL, CRC Press, 479–497.

ANON (1989), 'Omelette tray is well positioned', *Packaging Week*, 3, 1–3.

ANON (2000), 'PureBright® Sterilization Systems', *PurePulse Technologies (Maxwell Technologies)*, San Diego, USA.

© Woodhead Publishing Limited, 2012

ARAN N (2001), 'The effect of calcium and sodium lactates on growth from spores of *Bacillus cereus* and *Clostridium perfringens* in a sous-vide beef goulash under temperature abuse', *Int J Food Microbiol*, 63, 117–123.

ARVANITOYANNIS I S and TSERKEZOU P (2010), 'Legislation on food irradiation: European Union, United States, Canada and Australia', in Arvanitoyannis I S, *Irradiation of food commodities: techniques, applications, detection, legislation, safety and consumer opinion*', London, Elsevier, 3–21.

BANSAL N P and DOREMUS R H (1986), *Handbook of glass properties*, New York, Academic Press.

BENGTSSON N (1998), 'New developments and trends in food processing and packaging in Europe', *7th annual conference of the Society of Packaging Science and Technology*, Tokyo, Japan, 7, 277–292.

BENGTSSON N and GREEN W (1970), 'Radio-frequency pasteurization of cured hams', *J Food Sci*, 35, 681–687.

BENNIK M H J, SMID E J, ROMBOUTS F M and GORRIS L G M (1995), 'Growth of psychrotrophic foodborne pathogens in a solid surface model system under the influence of carbon dioxide and oxygen', *Food Microbiol*, 12, 509–517.

BERENDS C L (1996), 'Stability of aseptically packaged food as a function of oxidation initiated by a polymer contact surface', PhD thesis, Virginia Polytech. Inst., Blacksburg, VA.

BERTRAND K (2005), 'Microwavable foods satisfy need for speed and palatability', *Food Technol*, 59, 30–34.

BRANDL M T, ZHU Y, MCHUGH T H, PAN Z and HUYNH S (2008), 'Reduction of *Salmonella* Enteritidis population sizes on almond kernels with infrared heat', *J Food Prot*, 71, 897–902.

BRITT I J (2008), 'Thermal processing', in Tucker G, *Food biodeterioration and preservation*, Oxford, Blackwell, 55–73.

BROWN G D and GAZE J E (1990), 'Determination of the growth potential of *Clostridium botulinum* types E and non-proteolytic B in sous-vide products at low temperatures', *Campden Food and Drink Research Association Technical Memorandum No. 593*, Chipping Campden.

BROWN W E (1992), *Plastics in food packaging: properties, design and fabrication*, New York, Marcel Dekker.

CANER C, HERNANDEZ R J and HARTE B R (2004a), 'High-pressure processing effects on the mechanical, barrier and mass transfer properties of food packaging flexible structures: a critical review', *Packag Technol Sci*, 17, 23–29.

CANER C, HERNANDEZ R J, PASCALL M, BALASUBRAMANIAM V M and HARTE B R (2004b), 'The effect of high pressure food processing on the sorption behavior of selected packaging materials', *Packag Technol Sci*, 17, 139–153.

CANUMIR J A, CELIS J E, DE BRUIJN J and VIDAL L V (2002), 'Pasteurisation of apple juice by using microwaves', *Lebensm Wiss Technol*, 35, 389–392.

CASTLE L (1989), 'Migration of polyethylene terephthalate (PET) oligomers from PET plastics into foods during microwave and conventional cooking and into bottled beverages', *J Food Prot*, 2, 337–342.

CATHCART W H, PARKER J J and BEATTIE H G (1947), 'The treatment of packaged bread with high frequency heat', *Food Technol*, 1, 174–177.

CHAVEZ-LOPEZ C, GIANOTTI A, TORRIANI S and GUERZONI M E (1997), 'Evaluation of the safety and prediction of the shelf-life of vacuum cooked foods', *Ital J Food Sci*, 9, 99–110.

CHEN H (2007), 'Temperature-assisted pressure inactivation of *Listeria monocytogenes* in turkey breast meat', *Int J Food Microbiol*, 117, 55–60.

CHUAQUI-OFFERMANNS N (1989), 'Food packaging materials and radiation processing of food: a brief review', *Int J Radiat Applic Instrum Part C*, 34, 1005–1007.

CHUNG Y K, CHISM G W, YOUSEF A E, VURMA M and TUREK E J (2005), 'Inactivation of barotolerant *Listeria monocytogenes* in sausage by combination of high-pressure processing and food-grade additives', *J Food Prot*, 68, 744–750.

© Woodhead Publishing Limited, 2012

DECAREAU RV (1985), *Microwaves in the food processing industry*, Orlando, FL, Academic Press.

DENYS S, VAN LOEY A M, HENDRICKX M E and TOBBACK P P (1997), 'Modeling heat transfer during high pressure freezing and thawing', *Biotechnology Progress*, 13, 416–423.

DIXON N M and KELL D B (1989), 'The inhibition of $CO_2$ on the growth and metabolism of microorganisms', *J Appl Bact*, 67, 109–136.

DJENANE D, SANCHEZ A, BELTRAN J and RONCALES P (2003), 'Extension of the shelf life of beef steaks packaged in a MAP by treatment with rosemary and display under UV-free lighting', *Meat Sci*, 64, 417–426

DOBIAS J, VOLDRICH M, MAREK M and CHUDACKOVA K (2004), 'Changes of properties of polymer packaging films during high pressure treatment', *J Food Eng*, 61, 545–549.

DODDS K L (1995), 'Introduction', in Farber J M and Dodds K L, *Principles of modified atmosphere and sous-vide product packaging*, Technomic Publishing, Lancaster, PA, 1–11.

DUNN J E, CLARK W R, ASMUS J F, PEARLMAN J S, BOYER K, PAIRCHAUD F and HOFFMAN G (1991), Methods for preservation of foodstuffs. US Patent 5 034 235.

DUNN J, CLARK W and OTT T (1995), 'Pulsed light treatment of food and packaging', *Food Technol*, 49, 95–98.

ELMNASSER N, GUILLOU S, LEROI F, ORANGE N, BAKHROUF A and FEDERIGHI M (2007), 'Pulsed-light system as a novel food decontamination technology: a review', *Can J Microbiol*, 53, 813–821.

FAN X and SOKORAI K J B (2005), 'Assessment of radiation sensitivity of fresh-cut vegetables using electrolyte leakage measurement', *Postharvest Biol Tech*, 36, 191–197.

FAN X, TOIVONEN P M A, RAJKOWSKI K T and SOKORAI K J B (2003), 'Warm water treatment in combination with modified atmosphere packaging reduced undesirable effects of irradiation on the quality of fresh-cut iceberg lettuce', *J Agric Food Chem*, 50, 1231–1236.

FARBER J M (1991), 'Microbiological aspects of modified atmosphere packaging technology – a review', *J Food Prot*, 54, 58–70.

FARKAS J and ANDRASSY E (1993), 'Interaction of ionising radiation and acidulants on the growth of the microflora of a vacuum-packaged chilled meat product', *Int J Food Microbiol*, 19, 145–152.

FARKAS J, SARAY T, MOHACSI-FARKAS C, HORTI C and ANDRASSY E (1997), 'Effects of low-dose gamma irradiation on shelf life and microbiological safety of precut/prepared vegetables', *Adv Food Sci*, 19, 111–119.

FELLOWS P J (2000), *Food processing technology: principles and practice*, Cambridge, Woodhead Publishing, 196–208.

GANDE N and MURIANA P (2003), 'Prepackage surface pasteurization of ready-to-eat meats with a radiant heat oven for reduction of *Listeria monocytogenes*', *J Food Prot*, 66, 1623–1630.

GARRIGA M and AYMERICH T (2009), 'Advanced decontamination technologies: high hydrostatic pressure on meat products', in Toldra F, *Safety of meat and processed meat: food microbiology and food safety*, New York, Springer Science + Business Media, 183–208.

GENNADIOS A, MCHUGH T H, WELTER C L and KROCHTA J M (1994), 'Edible coatings and films based on proteins', in Krochta J M, Baldwin E A and Nisperos-Carriedo M O, *Edible coatings and films to improve food quality*, Lancaster, Technomic Publishing, 201–207.

GHAZALA S (2004), 'Developments in cook-chill and sous-vide processing', in Richardson P, *Improving the thermal processing of foods*, Cambridge, Woodhead Publishing, 152–173.

GIESE J (1992), 'Advances in microwave food processing', *Food Tech*, 46, 118–123.

GONZÁLEZ-FANDOS E, GARCÍA-LINARES M C, VILLARINO-RODRÍGUEZ A, GARCÍA-ARIAS M T and GARCÍA-FERNÁNDEZ M C (2004), 'Evaluation of the microbiological safety and sensory

© Woodhead Publishing Limited, 2012

quality of rainbow trout (*Oncorhynchus mykiss*) processed by the sous-vide method', *Food Microbiol*, 21, 193–201.

GONZÁLEZ-FANDOS E, VILLARINO-RODRÍGUEZ A, GARCÍA-LINARES M C, GARCÍA-ARIAS M T and GARCÍA-FERNÁNDEZ M C (2005), 'Microbiological safety and sensory characteristics of salmon slices processed by the sous-vide method', *Food Control*, 16, 77–85.

GRAHAM H D (1980), 'Safety and wholesomeness of irradiated foods', in Graham H D, *The safety of foods*, Westport, CT, Avi Publishing, 546.

GRIFFITH G (1992), 'Irradiated foods: new technology, old debate', *Trends Food Sci Technol*, 3, 251.

GROWER J L, COOKSEY K and GETTY K (2006), 'Release of nisin from methylcellulose-hydroxypropyl methylcellulose film formed on low density polyethylene film', *J Food Sci*, 69, 107–111.

GUILLARD V, MAURICIO-IGLESIAS M and GONTARD N (2010), 'Effect of novel food processing methods on packaging: structure, composition and migration properties', *Crit Rev Food Sci Nut*, 50, 969–988.

HAN J H (2003), 'Antimicrobial packaging materials and films', in Ahvenainen R, *Novel food packaging techniques*, Cambridge, Woodhead Publishing, 50–70.

HAN J H (2007), 'Packaging for non-thermally processed foods', in Han J H, *Packaging for non-thermal processing of food*, Ames, IA, Blackwell, 3–17.

HAN J H and FLOROS J D (2007), 'Active packaging: a non-thermal process', in Tewari G and Juneja V K, *Advances in thermal and non–thermal food preservation*, Ames, IA, Blackwell, 167–184.

HAN J H, GOMES-FEITOSA C L, CASTELL-PEREZ M E, MOREIRA R G and DA SILVA P C F (2004), 'Quality of packaged romaine lettuce hearts exposed to low-dose electron bean irradiation', *Lebensm Wiss Technol*, 37, 705–715.

HAUGHTON P N, LYNG J G, CRONIN D A, MORGAN D J, FANNING S and WHYTE P (2010), 'Efficacy of UV light treatment for the microbiological decontamination of chicken associated packaging, and contact surfaces', *J Food Prot*, 74, 565–572.

HAUGHTON P N, LYNG J G, MORGAN D J, CRONIN D A, FANNING S and WHYTE P (2011), 'Efficacy of high-intensity pulsed light for the microbiological decontamination of chicken, associated packaging, and contact surfaces', *Foodborne Pathog Dis*, 8, 109–117.

HAYNIE S L, CRUM G A and DOELE B A (1995), 'Antimicrobial activities of amphiphilic peptides covalently bonded to a water-insoluble resin', *Antimicrob Agents Chemother*, 39, 301–307.

HEINZ V and KNORR D (2002), 'Effects of high pressure on spores', in Hendrickx M E G and Knorr D, *Ultra high pressure treatments of foods*, New York, Kluwer/Plenum, 77–114.

HILLEGAS S L and DEMIRCI A (2003), 'Inactivation of *Clostridium sporogenes* in clover honey by pulsed-UV light treatment', *Agricultural Engineering International: the CIGR Journal of Scientific Research and Development*, 5, December 1–7.

HOGAN E, KELLY A L and SUN D-W (2005), 'High pressure processing of foods: an overview', in Sun D-W, *Emerging technologies for food processing*, London, Elsevier Academic Press, 3–31.

HOLDSWORTH D and SIMPSON R (2008), *Thermal processing of packaged foods*, New York, Springer, 123–138.

HOUBEN J H, SCHOENMAKERS L, VAN PUTTEN E, VAN ROON P and KROL B (1991), 'Radio-frequency pasteurization of sausage emulsions as a continuous process', *J Microw Power Electromagn Energy*, 26, 202–205.

HUANG L (2007), 'Computer simulation of heat transfer during in-package pasteurization of beef frankfurters by hot water immersion', *J Food Eng*, 80, 839–849.

HUANG L and SITES J (2007), 'Automatic control of a microwave heating process for in-package pasteurization of beef frankfurters', *J Food Eng*, 80, 226–233.

HURME E, SIPILAINEN-MALM T and AHVENAINEN R (2003), 'Active and intelligent packaging',

© Woodhead Publishing Limited, 2012

in Novak J S, Sapers G M and Juneja V K, *Microbial safety of minimally processed foods*, Boca Raton, FL, CRC Press, 87–123.

HUTTON T (2003), 'Food packaging: an introduction', *Key Topics in Food Science and Technology*, Chipping Campden, Campden BRI.

INGHAM S C, FANSLAU M A, BUEGE D R, DEVITA M D and WADHERA R K (2003), 'Evaluation of small-scale hot-water postpackaging pasteurization treatments for destruction of *Listeria monocytogenes* on ready-to-eat beef snack sticks and natural-casing wieners', *J Food Prot*, 68, 2059–2067.

INRA (2010), Live from the labs, France, INRA/DPE. Available from: http://www.international.inra.fr/partnerships/with_the_private_sector/live_from_the_labs/high_pressure_processing_systems (accessed 27 February 2011).

JIANG Z, NEETOO H and CHEN H (2010), 'Control of *Listeria monocytogenes* on ready to eat foods using chitosan-based antimicrobial coatings and films', *J Food Sci*, 76, 22–26.

JOFRÉ A, AYMERICH T, MONFORT J M and GARRIGA M (2008a), 'Application of enterocins A and B, sakacin K and nisin to extend the safe shelf-life of pressurized ready-to-eat meat products', *Eur Food Res Technol*, 228, 159–162.

JOFRÉ A, GARRIGA M and AYMERICH T (2008b), 'Inhibition of *Salmonella* sp., *Listeria monocytogenes* and *Staphylococcus aureus* in cooked ham by combining antimicrobials, high hydrostatic pressure and refrigeration', *Meat Sci*, 78, 53–59.

JUCK G, NEETOO H and CHEN H (2010), 'Application of an active alginate coating to control the growth of *Listeria monocytogenes* on poached and deli turkey products', *Int J Food Microbiol*, 142, 302–308.

JUN S, IRUDAYARAJ J, DEMIRCI A and GEISER D (2003), 'Pulsed UV-light treatment of corn meal for inactivation of *Aspergillus niger* spores', *Int J Food Sci Technol*, 38, 883–888.

JUNEJA V K (2003), 'Sous-vide processed foods: safety hazards and control of microbial risks', in Novak J S, Sapers G M and Juneja V K, *Microbial safety of minimally processed foods*, Boca Raton, FL, CRC Press, 97–124.

JUNEJA V K and MARMER B S (1996), 'Growth of *Clostridium perfringens* from spore inocula in sous-vide turkey products', *Int J Food Microbiol*, 32, 115–123.

KEKLIK N M, DEMIRCI A and PURI V M (2009), 'Inactivation of *Listeria monocytogenes* on unpackaged and vacuum-packaged chicken frankfurters using pulsed UV-light', *J Food Sci*, 74, M431–M439.

KEKLIK N M, DEMIRCI A and PURI V M (2010), 'Decontamination of unpackaged and vaccum-packaged boneless chicken breast with pulsed ultraviolet light', *Poultry Sci*, 89, 570–581.

KNOCHEL S, VANGSGAARD R and SOHOLM JOHANSEN L (1997), 'Quality changes during storage of sous-vide cooked green beans (*Phaseolus vulgaris*)', *Z Lebensm Unters Forsch*, A205, 370–374.

KRISHNAMURTHY K, TEWARI J, IRUDAYARAJ J and DEMIRCI A (2010), 'Microscopic and spectroscopic evaluation of inactivation of *Staphylococcus aureus* by pulsed UV light and infrared heating', *Food Bioproc Technol*, 3, 93–104.

KUBEL J, LUDWIG H, MARX H and TAUSCHER B (1996), 'Diffusion of aroma compounds into packaging films under high pressure', *Packag Technol Sci*, 9, 143–152.

LAMB J L, GOGLEY J M, THOMPSON M J, SOLIS D R and SEN S (2002), 'Effect of low-dose gamma irradiation on *Staphylococcus aureus* and product packaging in ready-to-eat ham and cheese sandwiches', *J Food Prot*, 65, 1800–1805.

LAMBERT Y, DEMAZEAU G, LARGETEAU A, BOUVIER J M, LABORDE-CROUBIT S and CABANNES M (2000), 'Packaging for high pressure treatments in the food industry', *Packag Technol Sci*, 13, 63–71.

LAU M H and TANG J (2002), 'Pasteurization of pickled asparagus using 915 MHz microwaves', *Journal of Food Eng*, 51, 283–290.

LEFEUVRE S A E and AUDHUY-PEAUDECERF M B M (1994), 'Microwavability of packaged foods', in Mathlouthi M, *Food packaging and preservation*, London, Blackie Academic and Professional, 62–87.

© Woodhead Publishing Limited, 2012

LOPEZ-RUBIO A, ALMENAR E, HERNANDEZ-MUNOZ P, LAGARON J M, CATALA R and GAVARA R (2004), 'Overview of active polymer-based packaging technologies for food applications', *Food Rev Int*, 20, 357–387.

LUCHANSKY J B and CALL J E (2004), 'Evaluation of nisin-coated cellulose casings for the control of *Listeria monocytogenes* inoculated onto the surface of commercially prepared frankfurters', *J Food Prot*, 67, 1017–1021.

LUECHAPATTANAPORN K, KANG D H, TANG J, HALLBERG L M, WANG Y, WANG J and AL-HOLY M (2004), 'Microbial safety in radio-frequency processing of packaged foods', *J Food Sci*, 69, 201–206.

MARCOS B, JOFRÉ A, AYMERICH T, MONFORT J M and GARRIGA M (2008), 'Combined effect of natural antimicrobials and high pressure processing to prevent *Listeria monocytogenes* growth after a cold chain break during storage of cooked ham', *Food Control*, 19, 76–81.

MARTENS T (1991), 'Sous-vide, an opportunity for small and medium sized enterprise', FNK Symposium Proceedings, Zurich, Switzerland.

MARTENS T (1995), 'Introduction', in Farber J M and Dodds K L, *Current status of sous-vide in Europe*, Lancaster PA, Technomic Publishing, 1–11.

MASUDA M, SAITO Y, IWANAMI T and HIRAI Y (1992), 'Effects of hydrostatic pressure on packaging materials for food', in Balny C, Hayashi K, Heremans K and Massons P, *High pressure and biotechnology*, London, John Libbey Eurotext, 545–547.

MAY N S (2000), 'Retort technology', in Richardson P, *Thermal technologies in food processing*, Cambridge, Woodhead Publishing, 7–28.

MAY N (2004), 'Development in packaging formats for retort processing', in Richardson P, *Improving the thermal processing of foods*, Cambridge, Woodhead Publishing, 138–151.

MERONI A (2000), 'Active packaging as an opportunity to create package design that reflects the communicational, functional and logistical requirements of food products', *Packag Technol Sci*, 13, 243–248.

METRICK C, FARKAS D F and HOOVER D G (1989), 'Effects of high hydrostatic pressure on heat-resistant and heat-sensitive strains of *Salmonella*', *J Food Sci*, 54, 1547–1549.

MICHEL M and AUTIO K (2001), 'Effects of high pressure on protein- and polysaccharide-based structures', in Hendricks M E G and Knorr D, *Ultra high pressure treatments of foods*, New York, Kluwer Academic/Plenum Publishers, 189–210.

MIN S and ZHANG H Q (2007), 'Packaging for high-pressure processing, irradiation, and pulsed electric field processing', in Han J H, *Packaging for non-thermal processing of food*, Ames, IA, Blackwell, 67–86.

MOREHOUSE K M and KOMOLPRASERT V (2004), 'Irradiation of food and packaging: an overview', in Komolprasert V and Morehouse K M, *Irradiation of food and packaging*, Washington, DC, American Chemical Society.

MORRIS C E (1979), 'Developing microwave foods ... some surprises', *Chiltons Food Eng*, 51, 67–74.

MULLIN J and BOWS J (1993), 'Temperature measurement during microwave cooking', *Food Addit Contam*, 10, 663–672.

MURIANA P M, GROOMS J, DAVIDSON C A and QUIMBY W (2002), 'Postpackage pasteurization of ready-to-eat deli meats by submersion heating for reduction of *Listeria monocytogenes*', *J Food Prot*, 65, 963–969.

MURPHY R Y, MARCY J A and DAVIDSON M A (2004), 'Process lethality prediction for *Escherichia coli* O157:H7 in raw franks during cooking and fully cooked franks during post-cook pasteurization', *J Food Sci*, 69, 112–116.

MURPHY R Y, SCOTT L L, DUNCAN L K, JOHNSON N R, HANSON R E and FEZE N (2005), 'Eradicating *Listeria monocytogenes* from fully cooked franks by using an integrated pasteurization-packaging system', *J Food Prot*, 68, 507–511.

NATRAJAN N and SHELDON B W (2000), 'Efficacy of nisin-coated polymer films to inactivate *Salmonella* Typhimurium on fresh broiler skin', *J Food Prot*, 63, 1189–1196.

© Woodhead Publishing Limited, 2012

NEETOO H, YE M, JOERGER R, HICKS D T, HOOVER D G and CHEN H (2008), 'Use of nisin-coated plastic films to control *Listeria monocytogenes* and spoilage microflora on vacuum-packaged cold-smoked salmon', *Int J Food Microbiol*, 122, 8–15.

NEETOO H, YE M and CHEN H (2010), 'Bioactive alginate coatings to control *Listeria monocytogenes* on cold-smoked salmon slices and fillets', *Int J Food Microbiol*, 136, 326–331.

NIEMAND J G, VAN DER LINDE H J and HOLZAPFEL W H (1983), 'Shelf-life extension of minced beef through combined treatments involving radurization', *J Food Prot*, 46, 791–796.

NYATI H (2000), 'Survival characteristics and the applicability of predictive modeling to *Listeria monocytogenes* growth in sous-vide products', *Int J Food Microbiol*, 56, 123–132.

OHLSSON T (1991), 'Microwave processing in the food industry', *European Food and Drink Review*, 7, 9–11.

OHLSSON T (2002), 'Minimal processing of foods with non-thermal methods', in Ohlsson T and Bengtsson N, *Minimal processing technologies in the food industry*, Cambridge, Woodhead Publishing, 34–60.

ORFEUIL M (1987), *Electric process heating*, Columbus, OH, *Battelle Press.*

ORSAT V and RAGHAVAN V (2005), 'Microwave technology for food processing: an overview,' in Schubert H and Regier M, *The microwave processing of foods*, Cambridge, Woodhead Publishing, 105–115.

OZEN B F and FLOROS J D (2001), 'Effects of emerging food processing techniques on the packaging materials', *Trends Food Sci Technol*, 12, 60–67.

PALMIERI L and CACACE D (2005), 'High intensity pulsed light technology', in Sun D-W, *Emerging technologies for food processing*, London, Elsevier Academic Press, 279–306.

PEELING J and CLARK D T (1983), 'Surface ozonation and photooxidation of polyethylene film', *J Polym Sci Polym Chem Ed*, 21, 2047–2055.

PHILLIPS C A (1996), 'Review: modified atmosphere packaging and its effects on the microbiological quality and safety of produce', *Int J Food Sci Technol*, 31, 463–479.

PRZYBYLSKI L A, FINERTY M W, GRODNER R M and GERDES D L (1989), 'Extension of shelf-life of iced fresh channel catfish fillets using modified atmosphere packaging and low dose irradiation', *J Food Sci*, 54, 269–273.

RAHMAN M S (2007), 'Irradiation preservation of foods', in Rahman M S, *Handbook of food preservation*, Boca Raton, FL, CRC Press, 762–777.

RAMASWAMY H and TANG J (2008), 'Microwave and radiofrequency heating', *Food Sci Technol Int*, 14, 423–430.

RASTOGI N K, RAGHAVARAO K S M S, BALASUBRAMANIAM V M, NIRANJAN K and KNORR D (2007), 'Opportunities and challenges in high pressure processing of foods', *Crit Rev Food Sci Nut*, 47, 69–112.

RICE J (1994), 'Sterilizing with light and electrical impulses', *Food Proc*, July, 66.

RODRIGUEZ-MARVAL M, KENDALL P, BELK K E and SOFOS J N (2009), 'Inactivation of *Listeria monocytogenes* during reheating of frankfurters with hot water before consumption', *Food Prot Trends*, 30, 16–24.

SAFE PAC PASTEURIZATION LLC™ (2010), Safe Pac™ expands the food safety universe. Philadelphia, PA. Available from: http://www.safepac.biz (accessed 27 February 2011).

SAMELIS J, SOFOS J N, KENDALL P A and SMITH G C (2001), 'Fate of *Escherichia coli* O157:H7, *Salmonella* Typhimurium DT104, and *Listeria monocytogenes* in fresh meat decontamination fluids at 4 and 10°C', *J Food Prot*, 64, 950–957.

SCHAUWECKER A, BALASUBRAMANIAM V M, SADLER G, PASCALL M A and ADHIKARI C (2002), 'Influence of high pressure processing on selected polymeric materials and on the migration of a pressure-transmitting fluid', *Packag Technol Sci*, 15, 255–262.

SCHIFFMANN R F (1988), 'Hazards of microwave heating', *Microwave Foods* '88, Milltown, NJ, The Packaging Group.

© Woodhead Publishing Limited, 2012

SCHIFFMANN R F (2001), 'Microwave processes for the food industry', in Datta A K and Anantheswaran R C, *Handbook of microwave technology for food applications*, New York, Marcel Dekker, 299–337.

SCHIFFMANN R F and SACHAROW S (1992), *Microwave packaging*, Leatherhead, Pira International.

SCHNEPF M and BARBEAU W E (1989), 'Survival of *Salmonella* Typhimurium in roasting chickens cooked in a microwave, convention microwave and conventional electric oven', *J Food Saf*, 9, 245–252

SELBY T L, BERZINS A, GERRARD D E, CORVALAN C M, GRANT A L and LINTON R H (2006), 'Microbial heat resistance of *Listeria monocytogenes* and the impact on ready-to-eat meat quality after post-package pasteurization', *Meat Sci*, 425–434.

SHAKILA J R, JEYASEKARAN G, VIJAYAKUMAR A and SUKUMAR D (2009), 'Microbiological quality of sous-vide cook-chill fish cakes during chilled storage (3°C)', *Int J Food Sci Technol*, 44, 2120–2126.

SHUWAISH A, FIGUEROA J E and SILVA J L (2000), 'Pulsed-light treated pre-packaged catfish fillets', *IFT Annual Meeting*, Dallas, TX.

SIMPSON M V (1995), 'Challenges studies with *Clostridium botulinum* in a sous-vide spaghetti and meat-sauce product', *J Food Prot*, 58, 229–234.

SIVERTSVIK M, ROSNES J T and BERGSLIEN H (2002), 'Modified atmosphere packaging', in Ohlsson T and Bengtsson N, *Minimal processing technologies in the food industry*, Cambridge, Woodhead Publishing, 61–86.

SMITH D P and ROLLIN N J (1954), 'Sorbic acid as a fungistatic agent for foods: effectiveness of sorbic acid in protecting cheese', *Food Res*, 19, 59–65.

SOMMERS C (2006), 'Recent advances in food irradiation: mutagenicity testing of 2-dodecylcyclobutanone', in Juneja V K, Cherry J P and Tunick M T, *Advances in microbial food safety*, Washington, DC, American Chemical Society (ACS), 109–120.

SUAREZ REBOLLO M P, LASTA J A, MASANA M O and RODRIGUEZ H R (1997), 'Sodium propionate to control *Clostridium botulinum* toxigenesis in a shelf-stable beef product prepared by using combined processes including irradiation', *J Food Prot*, 60, 771–776.

SUMNU G and SAHIN S (2005), 'Recent developments in microwave heating', in Sun D-W, *Emerging technologies for food processing*, London, Elsevier Academic Press, 419–444.

TERLIZZI F M (1984), 'Processing and distributing cooked meats in flexible films', *Food Technol*, 38, 67–71, 73.

TEWARI G (2007), 'Microwave and radio-frequency heating', in Tewari G and Juneja V K, *Advances in thermal and non-thermal food preservation*, Ames, IA, Blackwell, 91–98.

THAYER D W (1994), 'Wholesomeness of irradiated foods', *Food Technol*, 48, 132–135.

THE PLASCON GROUP (2011), Traverse City, MI. Available from: http://www.plascongroup.com (accessed 26 July 2011).

TUCKER G and FEATHERSTONE S (2011), *Essentials of thermal processing*, Oxford, Blackwell.

VOJDANI F and TORRES J A (1990), 'Potassium sorbate permeability of methylcellulose and hydroxylpropyl methylcellulose coatings: effect of fatty acids', *J Food Sci*, 55, 841–846.

WEKHOF A (2000), 'Disinfection with flash lamps', *PDA J Pharm Sci Technol*, 54, 264–276.

WELTI-CHANES J, LÓPEZ-MALO A, PALOU E, BERMÚDEZ D, GUERRERO-BELTRÁN J A and BARBOSA-CÁNOVAS G V (2005), 'Fundamentals and applications of high pressure processing to foods', in Barbosa-Cánovas G V, Tapia S and Cano M P, *Novel food processing technologies*, New York, Marcel Dekker.

WUYTACK E Y, PHUONG L D T, AERTSEN A, REYNS K M, MARQUENIE D, DE KETELAERE B, MASSCHALCK B, VAN OPSTAL I, DIELS A M and MICHIELS C W (2003), 'Comparison of sublethal injury

© Woodhead Publishing Limited, 2012

induced in *Salmonella enterica* serovar Typhimurium by heat and by different non-thermal treatments', *J Food Prot*, 66, 31–37.

YE M, NEETOO H and CHEN H (2008a), 'Effectiveness of chitosan-coated plastic films incorporating antimicrobials in inhibition of *Listeria monocytogenes* on cold-smoked salmon', *Int J Food Microbiol*, 127, 235–240.

YE M, NEETOO H and CHEN H (2008b), 'Control of *Listeria monocytogenes* on ham steaks by antimicrobials incorporated into chitosan-coated plastic films', *Food Microbiol*, 25, 260–268.

© Woodhead Publishing Limited, 2012

# Index



© Woodhead Publishing Limited, 2012

© Woodhead Publishing Limited, 2012

© Woodhead Publishing Limited, 2012

© Woodhead Publishing Limited, 2012

© Woodhead Publishing Limited, 2012

© Woodhead Publishing Limited, 2012

© Woodhead Publishing Limited, 2012

© Woodhead Publishing Limited, 2012

© Woodhead Publishing Limited, 2012

© Woodhead Publishing Limited, 2012

© Woodhead Publishing Limited, 2012

© Woodhead Publishing Limited, 2012

© Woodhead Publishing Limited, 2012

© Woodhead Publishing Limited, 2012

© Woodhead Publishing Limited, 2012

© Woodhead Publishing Limited, 2012

© Woodhead Publishing Limited, 2012